Fish Physiology

The 50th Anniversary Issue of Fish Physiology: Physiological Applications

Volume 40B

This is Volume 40B in the

FISH PHYSIOLOGY series
Edited by Anthony P. Farrell, Colin J. Brauner and Erika J. Eliason
Honorary Editors: William S. Hoar and David J. Randall

A complete list of books in this series appears at the end of the volume

Fish Physiology

The 50th Anniversary Issue of Fish Physiology: Physiological Applications

Volume 40B

Edited by

David J. Randall
Department of Zoology, The University of British Columbia, Vancouver, British Columbia, Canada

Anthony P. Farrell
Department of Zoology, and Faculty of Land and Food Systems, The University of British Columbia, Vancouver, British Columbia, Canada

Colin J. Brauner
Department of Zoology, The University of British Columbia, Vancouver, British Columbia, Canada

Erika J. Eliason
Department of Ecology, Evolution, and Marine Biology, University of California - Santa Barbara, Santa Barbara, CA, United States

Academic Press is an imprint of Elsevier
125 London Wall, London EC2Y 5AS, United Kingdom
525 B Street, Suite 1650, San Diego, CA 92101, United States
50 Hampshire Street, 5th Floor, Cambridge, MA 02139, United States

First edition 2024

ISBN: 978-0-443-13731-0
ISSN: 1546-5098

For information on all Academic Press publications
visit our website at https://www.elsevier.com/books-and-journals

Publisher: Zoe Kruze
Acquisitions Editor: Mariana Kuhl
Editorial Project Manager: Palash Sharma
Production Project Manager: A. Maria Shalini
Cover Designer: Gopalakrishnan Venkatraman

Typeset by STRAIVE, India

Contents

Contributors

Céline Audet (765), Institut des sciences de la mer de Rimouski (ISMER), Université du Québec à Rimouski, Rimouski, QC, Canada

F.W.H. Beamish (299)

Quentin Bone (229)

J.R. Brett (603,691)

F.E.J. Fry (501)

T.D.D. Groves (603)

W.S. Hoar (9), Department of Zoology, University of British Columbia, Vancouver, British Columbia, Canada V6T 2A9

David R. Jones (415)

Jim Kieffer (287), Department of Biological Sciences, University of New Brunswick, Saint John, NB, Canada

Shaun S. Killen (591), School of Biodiversity, One Health, and Veterinary Medicine, University of Glasgow, Glasgow, United Kingdom

C.C. Lindsey (125)

Stephen D. McCormick (1,85), Department of Biology, University of Massachusetts, Amherst, MA, United States

David J. McKenzie (677), UMR Marbec, CNRS, IRD, Ifremer, INRAE, Université de Montpellier, Montpellier, France

Tommy Norin (591), DTU Aqua: National Institute of Aquatic Resources, Technical University of Denmark, Lyngby, Denmark

David J. Randall (415)

W.E. Ricker (771)

Jodie L. Rummer (399), College of Science and Engineering, James Cook University, Townsville, QLD, Australia

Patricia M. Schulte (487), Department of Zoology and Biodiversity Research Centre, The University of British Columbia, Vancouver, BC, Canada

Robert E. Shadwick (213), Department of Zoology, University of British Columbia, Vancouver, BC, United States

J. Mark Shrimpton (71), Ecosystem Science & Management Department, University of Northern British Columbia, Prince George, BC, Canada

Peter V. Skov (677), Technical University of Denmark, Hirtshals, Denmark

Emily M. Standen (115), University of Ottawa, Ottawa, ON, Canada

Foreword

Sadly, David (Dave) J. Randall passed away peacefully and unexpectedly at his home on April 4, 2024, close to completion of this volume. He will be dearly missed by his family, friends, colleagues, and the fish physiology community as a whole.

Dave obtained a B.Sc. in 1960 from the University of Southampton in classical Zoology with a parallel course in comparative animal physiology, taught partly by Dr. Gerald Kerkut, founder of the journal *Comparative Biochemistry and Physiology*. Dave went on to be the first postgraduate student of Professor Graham Shelton and obtained a Ph.D. in 1963, entitled *The Regulation of Breathing and Heart Rate in Teleost Fish*. Shortly after graduation, he accepted an assistant professor position in the Department of Zoology at the University of British Columbia (UBC), Canada, where he remained until his retirement and became Professor Emeritus. During his 40 years at UBC, Dave was a highly inspirational faculty member and a key member of a dynamic group within that faculty that helped establish the field of comparative physiology. He also served as Associate Dean, Graduate Studies, and as an elected member of the University Senate. Dave served two terms as Chair and Professor in Biology and Chemistry at the City University in Hong Kong (2000–2005, 2009–2012) and was an honorary professor in the Department of Anatomy, Chinese University of Hong Kong (2005–2008), while in Hong Kong, he served as a trustee of the World Wildlife Fund.

Dave was an elected fellow of both the Royal Society of Canada (1981) and the Brazilian Academy of Sciences (2013). He received the Fry Medal, the highest honor bestowed by the Canadian Society of Zoology (CSZ; 1993), an Award of Excellence from the American Fisheries Society (1994), and delivered the Bidder lecture for the Society of Experimental Biology (2000). He was a Guggenheim Fellow (1968-69), a Killam Fellow (1981–1982), and a NATO Scholar (1991) and received the Murray Newman Award from the Vancouver Aquarium (2008). Dave was President of the CSZ (1984–1985) and of the Western Canadian Universities Marine Biological Society (1998–2000).

As described in the preface of this volume, Dave conceived and began coediting the *Fish Physiology* series in the late 1960s with Dr. Bill Hoar, his lifelong colleague at UBC. This monograph series has been hugely successful, and its influence greatly contributed to establishing fish physiology as a mainstream field of research.

Dave's own research interests and expertise in the field of fish physiology were broad. He and his first graduate students pioneered techniques for the placement of chronic indwelling catheters in blood vessels and the buccal and opercular cavities of fish to visualize what was actually happening inside the fish in real time. They measured blood gases, heart rate, and blood pressures during exercise and hypoxia for the first time in fish. Dave then spent much of his career investigating the mechanisms of oxygen, carbon dioxide, and ammonia transport and exchange in fish. As he got more interested in mechanisms of ionoregulation, he came up with the novel idea that the design of fish gills reflects a trade-off between the needs for high permeability for respiratory gas exchange and low permeability for ion conservation. Although Dave did not use the exact term, we now know this concept as the "osmorespiratory compromise," which remains a very active research area today.

Dave moved easily between physiology and toxicology, driven by his belief that ideas were paramount and could help him cross disciplinary boundaries. For example, his idea that ammonia was a respiratory gas helped him to understand ammonia toxicity and develop environmental water quality criteria for ammonia used by the US EPA that are widely used today. His understanding of the osmorespiratory compromise enabled him to develop models predicting the uptake of organic pollutants across the gills as a function of the O_2 uptake rate in fish.

Dave participated in and organized many research expeditions. These included aboard the Research Vessel (R/V) Alpha Helix to Palau to study land crabs; more on land crabs in Moorea; R/V Alpha Helix to study air-breathing fishes of the Amazon, spawning salmon in Bella Coola, Canada; blood flow regulation in skates in Port Aransas, Texas; mudskippers and weather loaches in Singapore; baby wallabies in Australia; sturgeons in Italy; and the unique physiology of a fish that thrives in very alkaline waters in Lake Magadi, Kenya, among others.

Great people make the world a better place, and Dave Randall did just that. He did much to shape and promote the field of fish physiology. His interests were broad, his knowledge base was huge, and he was extremely generous with his ideas and traits that influenced all of those who knew him. We are happy that we were able to coedit this 50th anniversary issue with Dave, despite a sudden, bittersweet ending. Dave will be missed more than he knew.

Tony Farrell, Colin Brauner, and Erika Eliason
Vancouver, Canada
August 2024

Preface

In the mid-1960s, Bill Hoar and Dave Randall prepared a six-volume treatise entitled "Fish Physiology" to be published with the hope that it would serve biologists of the 1970s as Margaret Brown's "The Physiology of Fishes" had served readers earlier. They had no intention of continuing the series beyond six volumes, but the response was so positive that additional volumes were developed covering in-depth analysis of specific areas of fish biology. Advice from the scientific community when developing volumes was an important factor. The interval between volumes depended on developments in the field and how ready they were to spend time on the editing process. Bill Hoar eventually retired, and Dave Randall had had enough! At an international meeting, he discussed this with several colleagues and was persuaded to continue with guest editors to spread the load. Tony Farrell agreed to join Dave Randall as a coeditor in this enterprise. Dave Randall retired, and Colin Brauner assumed the challenge. Tony Farrell retired recently and was replaced by Erika Eliason. The series "Fish Physiology" recently celebrated its 50th anniversary.

The middle of the 20th century saw the rapid growth of comparative physiology. For example, the huge success of studies of the squid giant axon in unraveling the nature and transmission of the action potential and ideas of model systems in biological research engendered interest in comparative physiology. The Fisheries Research Board (now Department of Fisheries and Oceans) was established in Canada. Fred Fry at the University of Toronto had established a framework for the environmental study of fish. He was also reputed to have an electric suite, which he could plug in and survive the cold Canadian winters. The increased research activity in fish physiology created a market for the series. What then followed was a highly successful in series, which has produced a total of 47 books (several volumes have 2 books) that contain almost 500 chapters since the inaugural volume published in 1969 (i.e., averaging almost 1 book per year for nearly half a century!). Contributing authors were from everywhere, and the books were sold and used worldwide.

The treatise "Fish Physiology" is the culmination of many contributors from the international consortium of fish physiologists and other interested persons. This broad collection of like-minded people involved many subgroups but no overarching structure or organization, yet they all wished to bring some new level of synthesis and integration to a timely aspect of fish physiology. Academic Press, the original publishers, imposed length and layout and decided the books would have a green cover. Content was entirely left to the editors. Each volume represents the views of individual specialists on the subject under discussion. However, while discussing functional processes, the authors have referred to a wealth of comparative material so that the treatise has become more than an account of the physiology of fishes; it contains many fundamental concepts and principles important in the broad field of comparative animal physiology. The treatise has survived where others (e.g., Physiology of Reptiles) have long disappeared. We think the longevity is because the volumes serve and belong to the community of "fish physiologists." During the life of "Fish Physiology," publishers have always been supportive, despite changes in the business with the advent of the Internet. In addition, the world has become a much more bureaucratic and conservative place, with sound bites, fake news, and less regard for the rigor of scientific study. Careful analysis of data and ideas remained central to the presentation of "Fish Physiology," avoiding celebrity status, promotion, and politics; yet every chapter continues to be peer reviewed before publication. The volumes have maintained their initial focus to serve a broad range of fish physiologists.

The content of the "Fish Physiology" volumes has evolved over time. The initial volumes were devoted to understanding the basic mechanisms and principles of fish physiology, with a focus on a few model species and some application to natural environmental conditions. Then, as the field better understood mechanisms, the approach was broadened to not only delve deeper into system physiology (e.g., chapters in early volumes were expanded to become books), but interspecific differences in physiology were explored, permitting a more evolutionary framework. Finally, as interspecific physiological mechanisms were further resolved, it became possible to discuss physiology in light of a changing world. Thus, physiology can now inform on conservation, sustainability, and management, as exemplified with the most recent volumes.

Narrowing down the 50th anniversary issue to less than 20 chapters was a huge challenge. In general, the majority of chapters written throughout the series' history are very strong, expertly written, and well cited; many chapters could be candidates. Therefore, we limited ourselves to the first 14 volumes in the series to highlight the very important early work in the field that was published in the series. In particular, we wished to (re)introduce new researchers (such as graduate students) to this research that has stood the test of time and that has shaped the field. We want students and early career researchers to gain an appreciation for the history of the field and also see how earlier research remains applicable. Too often, research conducted in an earlier era is forgotten or not thought relevant simply due to the date of publication. However, this is far from the case, and this anniversary issue aims to address this for the field in general, and the series in particular. Second, and within this time window, we carefully reviewed selected chapters that were particularly novel or impactful on the field of fish physiology.

The standout chapters that were our final selection in no way should undermine the quality and impact of the many chapters that were not selected. In the first 25 years of the series, there were 155 chapters published. To appropriately highlight this tremendous body of work, we required two volumes. Each volume includes 9–10 republished chapters. We recruited acknowledged experts in the field of each republished chapter to write a short review, which precedes each of the republished chapters in their original form. Volume A focuses on physiological systems and development, while Volume B focuses on physiological applications to altered environments and growth. Thus, Volumes A and B reflect the evolution of the series as described above.

In closing, Bill Hoar and Dave Randall wrote in the preface to Volume 1 their hope that "Fish Physiology" would prove as valuable in fisheries research laboratories as in university reference libraries and that it would be a rich source of detailed information for the comparative physiologist and the zoologist as well as the specialist in fish physiology. It has more than fulfilled this hope because of the efforts of many people and the enduring nature of the series. Therefore, we thank all of the contributors over the past 50 years for their hard work and continued support; we feel fortunate to be part of such a welcoming and supportive community.

David J. Randall, Anthony P. Farrell, Colin J. Brauner and Erika J. Eliason
Vancouver, Canada
August 2024

Chapter 1

How salmon prepare for life in the ocean: An introduction to William Hoar's "The physiology of smolting salmonids"

Stephen D. McCormick*
Department of Biology, University of Massachusetts, Amherst, MA, United States
**Corresponding author: e-mail: mccormick@umext.umass.edu*

In this paper Steve McCormick discusses the impact of William (Bill) S. Hoar's chapter "The Physiology of Smolting Salmonids" in Fish Physiology, Volume XIB, published in 1988.

As part of their anadromous life cycle most anadromous salmon undergo a parr-smolt transformation that prepares juveniles for downstream migration and seawater entry. This preparation is critical for survival in a new environment and has fascinated biologists for decades, teaching us about life history strategies, migration, osmoregulation, imprinting and navigation. Understanding these changes also has important applications to salmon conservation, the use of hatcheries for restoring and supplementing populations of salmon and the salmon aquaculture industry.

In the introduction to his review Professor Hoar provides a comparative overview of anadromy and smolting. The "degree of anadromy" or how much a species relies on the marine environment (Rounsefell, 1958) differs among salmonids with the most recently evolved *Oncorhynchus* species showing the greatest marine tilt. Later in the review it becomes clear that the timing of the development of salinity tolerance differs among and within species and is related to the degree of anadromy: "There are many species of salmonid fishes, with smolting as variable as their morphology, physiology and behavior.", and provides us a warning that "The hazard of general comments is recognized." Thus, he provides a comparative perspective that is present throughout the manuscript and primes the reader to expect that there will be both commonalities and differences among species.

Fish Physiology, Vol. 40B. https://doi.org/10.1016/bs.fp.2024.05.003

FIG. 1 Difference in morphology between Atlantic salmon parr and smolt. In smolts, note the loss of parr marks (vertical dark bands on the fishes flank), intense silvering and black fin margins. *Photo credit: Stephen D. McCormick.*

In the second and largest section of the paper (The Physiology of the Salmon Smolt) Hoar first describes the morphological differences in parr and smolt that can be seen in Fig. 1. These differences in appearance were so large that for much of the 1700s the reigning wisdom was that these were two distinct species (Flowerdew, 1871). We now understand that the silvering, dark dorsal surface and fin margins characteristic of smolts "are adaptive to the survival of postsmolts in the marine habitat" through their facilitation of schooling behavior and predator avoidance in large rivers, estuaries, and the ocean. Size and growth are identified as critical to the decision to undergo smolting and the relative importance of the two appears to be species-specific. During smolting "metabolism and body composition are altered in many ways: the rate of oxygen consumption increases with heightened catabolism of carbohydrate, fat, and proteins." These changes could be traced to specific alterations in metabolic pathways that were under hormonal control. New forms of hemoglobin appear that are likely to support a more constantly swimming lifestyle necessary in a pelagic environment.

Since parr have a very limited ability to tolerate seawater, the development of salinity tolerance is a critical and well-studied aspect of smolting. In one of the most synthetic and insightful parts of the paper, Hoar identifies nine "important points" from the many papers examining osmoregulation in smolts. All salmonid species can be gradually acclimated to seawater and returned to freshwater at any time during their life cycle. Within a species there appears to be a size-dependence of seawater tolerance which is independent of the size-dependent development of smolting. Only fish in the smolt stage can tolerate direct transfer to seawater, but this capacity is reversible, the timing of which is dependent on time, temperature, and species. The capacity for high growth in seawater that characterizes smolts appears to be closely tied to osmoregulatory development. Changes in the major osmoregulatory organs (gill, gut, and kidney/urinary bladder) are described, emphasizing the preparatory nature of physiological changes and their connection to increased salinity tolerance.

Hormones have an important role in the parr-smolt transformation. Bill Hoar was the first to demonstrate that the thyroid is activated during smolting (Hoar, 1939), and it was thought that thyroid hormones may have a central, coordinating role in stimulating smolting, similar to what is seen in amphibian and flounder metamorphosis. But, by 1988, it was "now clear that thyroid hormones do not trigger the parr-smolt transformation." Rather, Hoar recognized that "a number of endocrine systems are stimulated during smolting" (what Howard Bern termed a "pan-hyperendocrine process") (Bern, 1978), and that different hormones are responsible for different aspects of smolt physiology. His fig. 3 summarizes increases in thyroxine, cortisol, insulin, and catecholamines, and he also discusses changes in growth hormone and prolactin (see Fig. 2 of this paper). Thyroid hormones stimulate some of the morphological and metabolic changes that occur during smolting and interact with other endocrine systems. Hoar correctly concluded that "growth hormone is the important factor in hydromineral regulation in the sea" though we now know that cortisol is also critical to the development of salinity tolerance and that the GH/IGF-I and interrenal axes have important interactions (McCormick, 2013).

Smolting and sexual maturation are both critical life history events for most salmon, with the former normally preceding the latter by 1 or 2 years. In some species and populations, however, there is an alternative strategy wherein maturation of males can precede smolting. Most of the work in this area has been done with Atlantic salmon, and Hoar reviews several concepts introduced by John Thorpe and colleagues (Thorpe, 1987). Prior maturation can influence subsequent growth rates which can in turn affect whether individual fish delay smolting. Sex steroids can interfere with smolt development and salinity tolerance, though in most cases fall maturation and spring smolt migration are sufficiently separated in time that this is not an issue.

With the possible exception of pink and chum salmon that smolt very early in development, the environment is a critical modulator of the timing of smolt development and downstream migration. Hoar provides evidence that photoperiod controls the timing of smolting (as it does for many other physiological processes in salmon), with some of the earliest work on Atlantic salmon coming from Richard (Dick) Saunders lab and in Pacific salmon from Craig Clarke's group. Temperature appears to control the rate at which fish respond to increased daylength, but also affects how quickly smolts lose their salinity tolerance and migratory tendency. The potential role of lunar rhythms in thyroid hormones and downstream migration is also summarized.

In his last section of the paper, Hoar tackles some of the "Practical Problems" relating to smolt physiology. Commercial salmon aquaculture (primarily Atlantic salmon) was in a strong growth phase in 1988 and continues to expand today. Then as now commercial operations usually produce smolts in freshwater hatcheries and then transfer fish to ocean net pens. Provincial, state, federal, tribal, and private

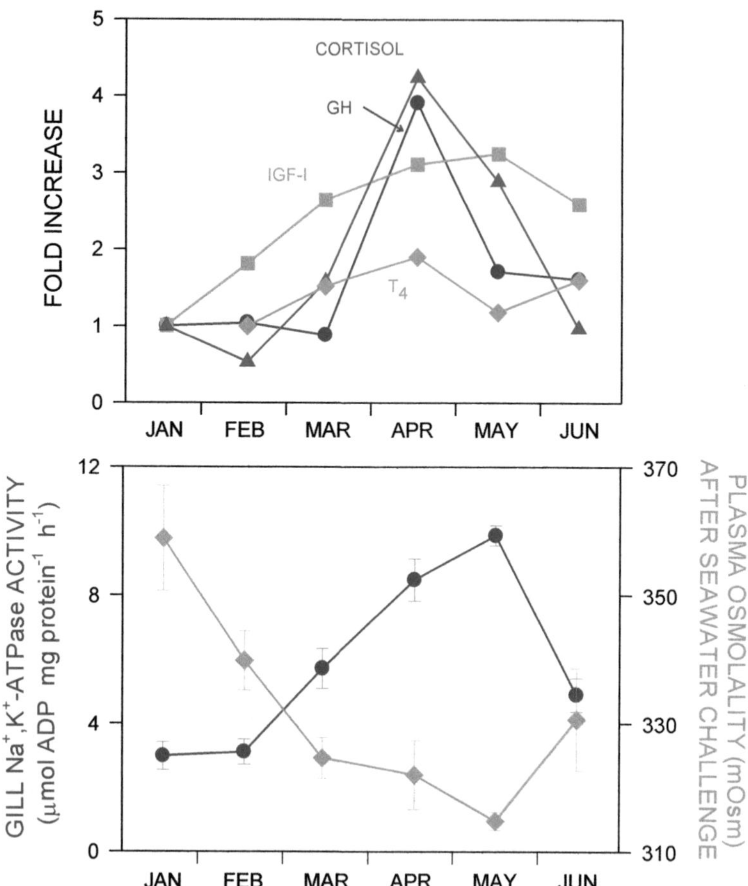

FIG. 2 Plasma hormone levels (upper panel), gill Na^+, K^+-ATPase activity and plasma osmolality 24h after exposure to seawater (lower panel) in juvenile Atlantic salmon during the parr–smolt transformation. Note that most endocrine changes peak in April, preceding the peaks in gill Na^+, K^+-ATPase activity and osmoregulatory capacity in May. *Abbreviations*: GH = growth hormone, IGF-I = insulin-like growth factor I, T_4 = thyroxine. Increase in hormone levels were expressed by setting the mean value of the January sample (February in the case of thyroxine) to 1 and expressing all subsequent changes relative to that value. *Modified with permission from McCormick, S.D., Lerner, D.T., Monette, M.Y., Nieves-Puigdoller, K., Kelly, J.T., Björnsson, B.T., 2009. Taking it with you when you go: how perturbations to the freshwater environment, including temperature, dams, and contaminants, affect marine survival of salmon. Am. Fish. Soc. Symp. 69, 195–214.*

hatcheries release juvenile salmon as fry, parr or smolts to conserve or restore damaged populations or to mitigate the impacts of dams and habitat loss. Hoar recognized that both commercial and restoration/mitigation hatcheries needed smolts that could survive transfer to and grow well in SW, but that fish released into the wild must also have the attributes of wild smolts to survive; the capacity for imprinting, navigation, high levels of swimming performance (for predator avoidance and prey capture) and immunocompetence among others. Commercial salmon hatcheries had highly variable success in rearing high quality smolts, with poor survival and growth after SW transfer a rampant problem. While the return rates of wild smolts vary by species, river system, and year, the return rates from restoration/mitigation hatcheries were almost always lower (often up to 10-fold lower) than their wild counterparts. Hoar identified two areas that would be the focus of advances in commercial and mitigation/restoration culture of salmon. The first was minimizing the juvenile freshwater phase, which was particularly important to save costs in commercial culture. The second was successful transfer to the marine environment, which could be achieved by producing high quality smolts with knowledge of the optimal time for transfer to SW. Hoar advocated that basic and applied research into the nutritional, physiological, and genetic control of growth and smolt development could be used to achieve these goals.

In the intervening years the use of improved food formulations, optimal growth temperatures, exposure to continuous light and other innovations have reduced the rearing time of smolts to 8 months or less. Photoperiod manipulation of the timing of smolt development (by moving from continuous light to 12h of light for at least 6 weeks and back to continuous light) allows for the nearly year-round production of smolts. Yet the problem of poor survival and growth after transfer to SW persists, perhaps because these artificial conditions have unintended consequences for smolt quality. And while more natural rearing conditions such as normal photoperiod and temperature cycles, more complex habitats and predator conditioning at restoration/mitigation hatcheries have been shown to improve return rates, they still lag behind those of wild fish.

We have learned a tremendous amount about smolting in the intervening 35 years since this review was published. Our knowledge of smolt biology has also helped us address the many aspects of conserving wild species (McCormick et al., 2009). For example, smolts are particularly sensitive to acid rain, which can rob them of their salinity tolerance, such that survival in the marine environment is compromised (Thorstad et al., 2013). Knowledge of the effects that sex steroids have on smolt development was critical to understanding the negative impact of nonylphenol (an estrogenic mimic used in pesticide sprays) on salmon populations (Fairchild et al., 1999). Knowledge of the behavior of smolts has allowed for the development of downstream passage for smolts around dams. The advent of molecular approaches such as qPCR, RNA seq, and whole genome sequencing has allowed us to identify the gill ion transporters that are involved in the development of salinity tolerance. Similarly, our increased capacity to measure the abundance of hormones, binding proteins and receptors at both the transcriptional and protein levels has led to a more accurate and nuanced view of the endocrine regulation of the physiological changes that occur during smolting.

But much remains to be learned. The molecular and cellular pathways of gill ionocyte development, critical for development of salinity tolerance, have yet to be elucidated. Knowledge of smolt-related changes in the kidney and intestine lag behind our understanding of changes in the gill. The complexity of endocrine changes during smolting that include tissue-specific changes in receptors, binding proteins, and interaction among hormones are still active areas of research. Even more challenging has been how these hormones control the many aspects of smolt development; while salinity tolerance and metabolic changes are highly tractable, changes in behavior, imprinting, navigation, and immunocompetence are more difficult to address. And what of the return migration of adult salmon to freshwater? Is there a "reverse smolting" where the physiologies necessary for freshwater entry and upstream migration occur in advance of riverine entry, or are they simply induced by exposure to freshwater? While there is some evidence for the former (Flores et al., 2012), much remains to be done.

Why has this review been so widely read and cited? It came at a time of intense interest in understanding the basic biology of smolting and applying that knowledge to commercial and restoration aquaculture. By 1988 there had already been two International Smolt Workshops (1981 in La Jolla California USA and 1984 in Stirling, Scotland), and these continue to be held every 4–5 years. Sufficient research on most salmonid species had been conducted by the mid-1980s to result in a large, complex and sometimes contradictory literature. Some of these apparent contradictions dissolved when placed in the context of evolution shaping important differences in timing, which were second nature to Hoar's understanding of smolting. Bill Hoar showed us that a great deal could be learned from using reductionist approaches inherent in physiology, but we should expect to be challenged (and fascinated) by the diversity and complexity of salmon life history.

This was a comprehensive, exhaustive review, one that was full of robust conclusions that have stood the test of time. Bill Hoar had a compelling writing style that managed to be highly accessible and authoritative. Basic and practical aspects were addressed that were important and useful to the developing aquaculture industry and conservation hatcheries. By presenting a clear-eyed synthesis of our knowledge and its gaps, this review also paved the way for a generation of researchers (and perhaps two or three) to continue examining the fascinating life history of salmon and the physiological challenges that it presents.

A Brief Biography of William Hoar

Professor William Stewart Hoar was born in Moncton, New Brunswick, Canada on August 31, 1913. He received a B.A. in Biology and Geology from the University of New Brunswick in 1934. After obtaining a Master's degree in Zoology from the University of Western Ontario and a Doctorate in Medical Sciences from Boston University, Professor Hoar held academic positions at the University of New Brunswick (1939–42 and 1943–46) and the University of Toronto (1942–43). During this time he also worked for several summers for the Fisheries Research Board of Canada. In 1945 Professor Hoar was appointed Professor of Zoology and Fisheries at the University of British Columbia. Here he became well known as an excellent teacher and scientist and played a major role in shaping the Department of Zoology, particularly from 1964 to 71 when he was Head of Department.

Continued

A Brief Biography of William Hoar

In addition to well over 100 scientific publications, Professor Hoar was the editor of the Canadian Journal of Zoology and coeditor of the multivolume series on Fish Physiology. In 1966 he wrote an influential textbook of General and Comparative Physiology which had three editions and was translated into a number of foreign languages. Professor Hoar has received no less than seven Honourary Degrees from Canadian Universities. In 1955 he was appointed a Fellow of the Royal Society of Canada (Academy III), received its Flavelle Medal in 1965 and was its president from 1971 to 1973. In 1974 he received the Fry Medal from the Canadian Society of Zoologists and in the same year became an Officer of the Order of Canada. Bill Hoar passed away in Vancouver on June 13, 2006.

Information obtained from the UBC Hoar Memorial Award website: https://zoology.ubc.ca/events/special-seminars-and-events/william-s-hoar-memorial-lecture.

References

Bern, H.A., 1978. Endocrinological studies on normal and abnormal salmon smoltification. In: Gaillard, P.J., Boer, H.H. (Eds.), Comparative Endocrinology. Elsevier/North Holland Biomedical Press, Amsterdam, pp. 77–100.

Fairchild, W.L., Swansburg, E.O., Arsenault, J.T., Brown, S.B., 1999. Does an association between pesticide use and subsequent declines in catch of Atlantic salmon (*Salmo salar*) represent a case of endocrine disruption? Environ. Health Perspect. 107, 349–357.

Flores, A.M., Shrimpton, J.M., Patterson, D.A., Hills, J.A., Cooke, S.J., Yada, T., Moriyama, S., Hinch, S.G., Farrell, A.P., 2012. Physiological and molecular endocrine changes in maturing wild sockeye salmon, *Oncorhynchus nerka*, during ocean and river migration. J. Comp. Physiol. B: Biochem. Syst. Environ. Physiol. 182, 77–90.

Flowerdew, H. 1871. The Parr and Salmon Controversy, With Authentic Reports of the Legal Judgements and Judge's Notes in the Various Lawsuits on the Parr Question, and Also a Brief Sketch of Some Incidents Connected With the Dessemination of the Modern Parr Theory.

Hoar, W.S., 1939. The thyroid gland of the Atlantic salmon. J. Morphol. 65, 257–295.

McCormick, S.D., 2013. Smolt physiology and endocrinology. In: McCormick, S.D., Farrell, A.P., Brauner, C.J. (Eds.), Euryhaline Fishes. Academic Press, Amsterdam, pp. 199–251.

McCormick, S.D., Lerner, D.T., Monette, M.Y., Nieves-Puigdoller, K., Kelly, J.T., Björnsson, B.T., 2009. Taking it with you when you go: how perturbations to the freshwater environment, including temperature, dams, and contaminants, affect marine survival of salmon. Am. Fish. Soc. Symp. 69, 195–214.

Rounsefell, G.A., 1958. Anadromy in north American salmonidae. Fish. Bull. 58, 171–185.

Thorpe, J.E., 1987. Smolting versus residency: Developmental conflict in salmonids. Amer. Fish. Soc. Symp. 1, 244–252.

Thorstad, E.B., Uglem, I., Finstad, B., Kroglund, F., Einarsdottir, I.E., Kristensen, T., Diserud, O., Archavala-Lopez, P., Mayer, I., Moore, A., Nilsen, R., Björnsson, B.T., Okland, F., 2013. Reduced marine survival of hatchery Atlantic salmon post-smolts exposed to aluminum and moderate acidification in freshwater. Estuar. Coast. Shelf Sci. 123, 1–10.

Chapter 2

THE PHYSIOLOGY OF SMOLTING SALMONIDS☆

W.S. HOAR
Department of Zoology, University of British Columbia, Vancouver, British Columbia, Canada V6T 2A9

Chapter Outline

I. INTRODUCTION

Many salmonids, fish of the genera *Oncorhynchus*, *Salmo*, and *Salvelinus*, are anadromous and undergo a distinct transformation prior to seaward migration. Typically, the cryptically colored, stream-dwelling juvenile (usually called a *parr*) changes into a more stream-lined, silvery and active pelagic individual referred to as a *smolt*, physiologically adapted for life in ocean waters. The prototype to which the term smolt was first applied is the Atlantic salmon *Salmo salar*. In this species, the gay markings of the stream-dwelling parr,

☆This is a reproduction of a previously published chapter in the Fish Physiology series, "1988 (Vol. 11B)/The Physiology of Smolting Salmonids: ISBN: 978-0-12-350434-0; ISSN: 1546-5098".

Fish Physiology, Vol. 40B. https://doi.org/10.1016/bs.fp.2024.06.001

aged 1, 2, or several years, are covered with a silvery layer of purines (guanine and hypoxanthine), while the body form becomes slender in relation to that of the parr with a decline in the weight per unit length (condition factor); in addition, the fins-particularly the pectorals and caudal-develop distinctly black margins (Wedemeyer *et al.*, 1980; Gorbman *et al.*, 1982). Details of the smolt transformation vary in different salmonid species; indeed, the salmonids (salmon, trout, char) form a spectrum extending from pink salmon *O. gorbuscha*, which are already silvery as emerging fry and able to enter saltwater when they come out of the gravel, to some species of *Salvelinus* (*alpinus, fontinalis, malma)* that migrate only short distances into the sea for a few months in the summer and the lake char *Salvelinus* namaycush, which is not known to smolt or to enter the sea at all; most species of *Oncorhynchus* and *Salmo* are intermediate between these extremes and spend 1–3 or more years in fresh water before smolting.

Many years ago, Rounsefell (1958) arranged the North American salmonid species in a decreasing order of anadromy (Fig. 1). His histogram is still instructive and appropriate for coastal regions, even though it is now recognized that there are no obligatory ocean migrants and that all species can complete their life cycles

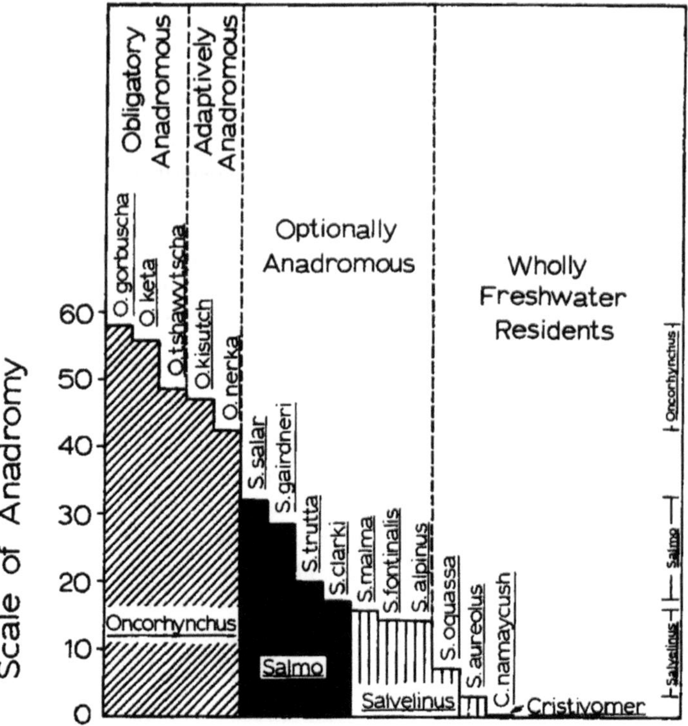

Fig. 1. Degree of anadromy of North American Salmonidae based on six different criteria. [Based on Rounsefell (1958).]

in lakes and streams and may become well adapted to a totally freshwater environment (Andrews, 1963; Berg, 1979; Collins, 1975; Peden and Edwards, 1976). The smolting changes associated with anadromy are characteristic of most species of *Oncorhynchus* and *Salmo*, but in the anadromous brook trout *Salvelinus fontinalis*, several of the most typical of them (elevated plasma T4, elevated gill ATPase, and increased osmoregulatory ability in sea water) are absent and smolting in this anadromous species has been questioned (McCormick *et al.*, 1985). Further, it is of interest that some populations of nonanadromous Atlantic salmon do not appear to smolt (Birt and Green, 1986).

The spectrum of anadromous salmonids has been discussed elsewhere with speculation concerning the evolution of the anadromous habit and the likely phylogenetic relationships of the different species (Hoar, 1976). This chapter focuses on the physiology of the typical smolt transformation, its control and modulation by the environment, and the implications of smolting physiology in salmon culture—an important and growing area of aquaculture, where high priority must be given to the production of smolts at as early an age as possible. A related topic considered here is the tendency of male salmonids to mature at an earlier age than the females. This interesting feature of salmonid development has important implications in salmon culture and, in some spieces, seems to have relevance to the age of smolting. Again the prototype is the Atlantic salmon in its early freshwater stages. Sexually mature and reproductively active male parr (aged 1+ years or older) were carefully investigated by Jones (1959) and others (King *et al.*, 1939) a quarter of a century ago. In several recent papers, Lundqvist (1983) has discussed the physiology of sexual maturation of male parr in relation to the smolt transformation.

The literature on the physiology of smolting is now extensive. There are several excellent reviews and key references may be readily located in their bibliographies (see Folmar and Dickhoff, 1980; Wedemeyer *et al.*, 1980) and in three recent symposia (Bern and Mahnken, 1982; Thorpe *et al.*, 1985; McCormick and Saunders, 1987).

II. THE PHYSIOLOGY OF THE SALMON SMOLT

A. Body Form and Coloration

Several morphological as well as physiological changes occur during smolting (Gorbman *et al.*, 1982; Winans, 1984; Winans and Nishioka, 1987). Superficially, however, both scientist and layman recognize a salmon smolt by its silvery appearance and its relatively slim, streamlined body. Smolt contrast sharply with parr or presmolts, which weigh more per unit length and display brightly colored pigment spots and bars formed by a variety of chromatophores in which the distribution of pigment granules can change cryptically. During smolting, body lipids not only decrease quantitatively but also change qualitatively (Section II,C).

Silvering is due to the synthesis of two purines, guanine and hypoxanthine. The needle-like crystals of these substances are deposited in two distinct skin layers: one directly beneath the scales and the other deep in the dermis adjacent to the muscles. Both the scale layer and the skin layer of purines are present in parr but become thicker in smolts; further, both layers contain guanine with lesser amounts of hypoxanthine, and there is a sharp increase in the ratio of guanine to hypoxanthine during the course of smolting (Markert and Vanstone, 1966; Vanstone and Markert, 1968; Eales, 1969; Hayashi, 1970). The formation of these reflecting layers occurs more rapidly in larger fish and at higher temperatures, but neither temperature nor alteration in photoperiod is required to induce purine deposition; this radical change in appearance seems to be controlled endogenously—a part of the smolt transformation (Johnston and Eales, 1970).

Clearly, the smolt transformation involves changes in purine nitrogen metabolism with an end result (a silvery fish) that is adaptive to the survival of the pelagic postsmolts in the marine habitat. However, the physiological basis and biochemical significance of the phenomenon are not at all clear. Silvering in juvenile salmon has most often been discussed in relation to the increased secretion of thyroid hormones, known to occur in many smolting salmonids (Section II,E). Experimentally, the feeding of thyroid gland tissue or the injection of thyroid stimulating hormone (TSH) or treatment with thyroid hormones will induce purine deposition in salmonids as well as in several other species of fish (Chua and Eales, 1971; Primdas and Eales, 1976). Thyroid treatment can also alter melanophore density, resulting in a lighter-colored fish. Eales (1979) reviews the experimental work on which these conclusions are based. Although a silvery body is advantageous in a pelagic life in deep waters, the physiological basis of purine deposition and the biochemical significance of the altered metabolism remain open to debate.

Guanine and hypoxanthine are insoluble by-products of nitrogen metabolism. They are disposed of in many different ways in various animal groups; their excretion has most often been examined in relation to the quantity of water available for their removal. In some animals with limited supplies of water, these purines and others such as uric acid are permanently stored in the tissues; in other animals they are metabolized to more soluble products. The smolt transformation prepares the juvenile salmon for life in a hyperosmotic environment where excretion of water (and hence substances dissolved in it) must be sharply reduced. These matters are discussed in textbooks of comparative physiology (for example, Hoar, 1983). The suggestion may be made that, in the first instance, purine deposition in salmon skin is metabolically more economic than the several enzymatic oxidations that would be required to turn the purines into more soluble substances such as allantoic acid or urea; in the end, the deposition of this reflecting material may have served as a preadaptation to a successful life in the sea. A number of tantalizing biological questions remain.

B. Growth and Size Relations

Several species of juvenile salmon, trout, and char have been shown to become progressively more tolerant of saltwater as they grow older and larger (Houston, 1961; Conte and Wagner, 1965; Parry, 1966; Wagner *et al.*, 1969; McCormick and Naiman, 1984). However, in salmonids that undergo a distinct smolt transformation, size alone is not the determining factor for a successful life in ocean waters. The transition from a freshwater to a marine life requires a smolt transformation—the coordinated physiological, biochemical and behavioral processes that occur during a very limited time span (Wedemeyer *et al.*, 1980). Nevertheless, the importance of a minimum size, age, or state of growth is emphasized since this appears necessary before smolting can occur.

In an oft-quoted paper, Elson (1957) argued that juvenile Atlantic salmon must attain a minimum length before smolting will occur. After examining data from both sides of the Atlantic, he concluded that parr that reach or exceed 10 cm in length at the end of the growing season are likely to become smolts the next season of smolt descent; otherwise, they remain an additional year or longer in the parr state. Although this rule of minimum length is useful, it has been shown that age is also important; Evans *et al.* (1984) reported that older fish tend to smolt at a relatively smaller size. Further, rate of growth, rather than length or age, appears to be a significant factor in some species of salmonids (Wagner *et al.*, 1969; Thorpe, 1977, 1986).

Thorpe (1977) and his associates described a bimodality in the growth of Atlantic salmon from Scottish rivers. In populations of hatchery-reared fish, individuals grow at similar rates until the late summer or autumn, when a distinct bimodality develops, which becomes progressively more marked as the season advances (Thorpe *et al.*, 1980). The more rapidly growing fish become 1+year smolts, while the more slowly growing ones remain in fresh water to become 2+year or older smolts. This bimodality is not related to sex or precocious male maturity (Thorpe *et al.*, 1982; Evans *et al.*, 1985; Villarreal and Thorpe, 1985). Genetic factors appear to be responsible for much of the variation in growth rate of Atlantic salmon (Gunnes and Gjedrem, 1978; Refstie and Steine, 1978; Thorpe and Morgan, 1978; Thorpe *et al.*, 1980). Growth bimodality is also the normal pattern in laboratory stocks of Atlantic salmon in Eastern Canada (Bailey *et al.*, 1980; Kristinsson *et al.*, 1985), where the growing conditions are similar to those in Scotland. However, it has not been observed in some more slowly developing laboratory populations of *Salmo salar* in the Baltic region (Eriksson *et al.*, 1979). Bimodality is difficult to demonstrate in nature, probably because of overlapping year classes and local environmental effects.

Several metabolic differences have been associated with the two modal groups of salmon. In the upper-mode fish, Higgins (1985) recorded faster rates of growth and metabolism characteristic of smolts, while Kristinsson *et al.* (1985) noted that, in December, the faster-growing, upper-mode fish had plasma thyroxine levels that were three times greater than those in the lower mode fish, indicating basic changes in the physiology at this stage. Effects of thyroid hormones on growth will be noted in a later section (Section II,E). Kristinsson *et al.* (1985) argue that entrance into the faster-growing phase marks a new developmental stage, which young salmon must reach in the autumn in order to smolt the following spring. This implies that the "decision" to smolt is made during the autumn under a decreasing photoperiod; thus, the short autumn period of rapid growth would be the first of a series of events that culminate in a smolt.

C. Metabolic and Biochemical Changes

Biochemically, the fully transformed smolt is quite a different fish from the parr that gave rise to it. Metabolism and body composition are altered in many ways: the rate of oxygen consumption increases with heightened catabolism of carbohydrate, fat, and protein. There are qualitative as well as quantitative changes in the lipids and blood proteins; gill enzyme systems concerned with ion regulation adjust adaptively. A markedly altered pattern of endocrinology accompanies these many biochemical changes (Section II,E).

Baraduc and Fontaine (1955) reported that the oxidative metabolism of the Atlantic salmon smolt was about 30% higher than that of the parr, even though smolts were in general larger than parr. These authors speculated concerning the possible involvement of thyroid hormone in the altered metabolism. This elevation in metabolism during smolting appears to be characteristic of typical *Salmo* smolts (Power, 1959; Malikova, 1957; Withey and Saunders, 1973; Higgins, 1985). Associated with the changes in oxidative metabolism, well-documented alterations in body composition indicate an elevated catabolism of the body reserves. There are predictable changes in plasma glucose, amino acid nitrogen, and free fatty acids; glycogen and lipid reserves are depleted and there is an elevation in moisture content. Atlantic salmon, steelhead trout (*Salmo gairdneri*), and coho salmon (*Oncorhynchus kisutch*) have been investigated by a number of workers; the literature can be readily traced through several papers and the reviews already cited (Fontaine and Hatey, 1950; Wendt and Saunders, 1973; Farmer *et al.*, 1978; Saunders and Henderson, 1978; Woo *et al.*, 1978; Sheridan *et al.*, 1985a; Sweeting *et al.*, 1985; Sheridan, 1986). Sheridan *et al.* (1985b) showed that the decline in tissue glycogen and fat is due not only to their increased breakdown but also to a greatly reduced synthesis. In spite of the marked loss of fat reserves, Atlantic salmon smolts are more buoyant than parr—probably an important factor in their rapid migration to sea (Saunders, 1965).

Changes in fatty acid composition of smolting Atlantic salmon were noted by Lovern (1934) more than half a century ago. Fatty acids as well as other lipid components have been investigated by many others since that time. Most recently, Sheridan *et al.* (1983, 1985a) have carefully studied the lipids of smolting steelhead trout and the effects of several hormones on lipid metabolism related to smolting in coho salmon (Sheridan, 1986). These studies confirm earlier reports of smolting salmon developing relatively high amounts of long-chain, polyunsaturated fatty acids and relatively low amounts of linoleic acid—values characteristic of typical marine fish, in contrast to freshwater species. Sheridan (1986) emphasizes that the lipid changes of smolting salmon anticipate life in the ocean—a preadaptation for the change in environment—and that growth hormone, prolactin, thyroid hormones, and cortisol are involved (Section II,E). Presumably, these changes in tissue lipids are significant in adaptation to the marine habitat, There are several physiological possibilities: degree of fatty acid saturation, cholesterol/phospholipid ratios, and fluidity of fats have been found important in the control of cell permeability and compensation for temperature change; mechanisms concerned with both permeability and temperature compensation are involved in the marine and adult life of the salmon. The physiology of lipids in relation to cell permeability and temperature compensation are discussed in many places (see, for example, Hoar, 1983; Isaia, 1984).

Several workers have investigated the blood proteins of smolting salmon—especially the multiple hemoglobins. In general, there is an increase in the complexity of the hemoglobin system with additional components added prior to migration (Vanstone *et al.*, 1964; Wilkins, 1968; Giles and Vanstone, 1976). Giles and Vanstone (1976) emphasize that in coho salmon the increased complexity occurs during smolting, with the appearance of two new anodic and four new cathodic components; this new pattern is retained throughout the remainder of life and cannot be induced in parr by manipulation of environmental oxygen, salinity, or temperature. Sullivan *et al.* (1985) confirm the increase in complexity of the hemoglobins at the time of coho smolting and show that triiodothyronine (T3) treatment accelerates the change at certain environmental salinities while thiourea has the opposite effect. A change in the hemoglobin system also occurs in smolting Atlantic salmon; Koch (1982) likewise argues that the thyroid hormones are probably involved in its expression.

Giles and Randall (1980) investigated the oxygen equilibria of the polymorphic hemoglobins of coho fry and adults, comparing oxygen affinity, Bohr shift, heat of oxygenation, and the influence of adenosine triphosphate. Adult hemoglobins preadapt emigrating smolt to the ocean environment, where several factors create lower oxygen tensions than those encountered by fry and parr (see also Vanstone *et al.*, 1964); in later life, the adult types of hemoglobin may be important in the spawning migration when salmon are liable to experience rapid changes in temperature and pH and may require sudden bursts of swimming activity (Giles and Randall, 1980). The functional significance of the polymorphs found in coho fry may relate to the unloading of oxygen in the tissues of a relatively small animal living in a well-oxygenated environment.

Electrophoretic comparisons of the hemoglobins of several species of salmon in the freshwater and marine stages show that there are consistent changes that probably preadapt the fish for changes in habitat (Vanstone *et al.*, 1964; Hashimoto and Matsuura, 1960; Bradley and Rourke, 1984). In chum salmon (*Oncorhynchus gorbuscha*), a species that migrates to saltwater during the fry stage without a typical parr–smolt transformation, there is a considerable increase in the proportion of hemoglobin with high oxygen affinity as the fish grow longer and heavier (Hashimoto and Matsuura, 1960). Again, the adaptive nature of these changes in gas transport is indicated. The interesting point emerging from the work on hemoglobins is that smolting salmon of both *Salmo* and *Oncorhynchus* species experience an adaptive change in their gas transport proteins at the time of the parr–smolt transformation and, further, that these changes usually occur well in advance of the actual change in habitat (Bradley and Rourke, 1984).

A third important group of biochemical changes relate to the gill enzyme systems and problems of ionic balance. In early life, salmon live in a hypoosmotic environment and are subject to osmotic flooding with water that leaches out essential ions during its removal by the kidneys. In later life, the marine fish lives in a hyperosmotic environment, is subject to loss of water by osmosis, and must make good the deficit by drinking seawater and excreting the salts. In the first case, gill cells actively absorb salt from the fresh waters; in the second instance, gill cells excrete ions, particularly the monovalent ones, sodium and chloride. These matters are considered more fully in the next section; here, the discussion focuses on the enzyme systems responsible for ion balance.

Zaugg and McLain (1970) were the first to demonstrate an increase in gill Na^+,K^+-ATPase activity during the parr–smolt transformation of coho salmon. Subsequently, comparable changes have been reported in Atlantic salmon, steelhead trout, and chinook salmon (Zaugg and McLain, 1972; Zaugg and Wagner, 1973; Giles and Vanstone, 1976; McCartney, 1976; Johnson *et al.*, 1977; Saunders and Henderson, 1978; Ewing and Birks, 1982; Boeuf *et al.*, 1985, and reviews cited). The smolting change in gill ATPases begins well in advance of migration both in *Salmo* and *Oncorhynchus*; it peaks during the migratory phase and entry into seawater. In seawater, it rises somewhat after 4–5 days and stabilizes at the higher level, but if smolts remain in fresh water beyond the normal time of migration, the gill enzyme activity declines to the freshwater level (Section IV,A). Nonsmolting and nonmigratory strains of brown trout (*S. trutta*) and rainbow trout (*S. gairneri*) do not show a seasonal increase in the gill Na^+,K^+-ATPases (Boeuf and Harache, 1982); moreover, neither the anadromous nor the nonanadromous forms of the brook trout (*Salvelinus fontinalis*) shows a seasonal change in gill ATPase (McCormick *et al.*, 1985). The contrasting picture of ATPase activity in freshwater parr and migrating salmon smolts is similar to that seen in comparisons of typical freshwater and marine teleosts. It is another of the major adaptations found in the typical parr–smolt transformation.

The physiology of ion transport and the role of the gill ATPases will not be detailed here. These topics were reviewed in Volume X,B of this series (Hoar and Randall, 1984); see in particular the chapters by Isaia (1984) and de Renzis and Bornancin (1984). Hormones involved are also considered in Volume X,B (Rankin and Bolis, 1984) and will be noted in Section II,E of this chapter.

The oxidative metabolism of gill tissues is high and may amount to as much as 7% of the fish's total oxygen consumption (Mommsen, 1984). Thus, several enzyme systems other than the Na^+,K^+-ATPases may be expected to reflect the changes in metabolic demands when freshwater salmon migrate into the ocean. The higher levels of succinic dehydrogenase (SDH) and cytochrome *c* oxidase found in smolting juvenile salmon reflect these altered metabolic demands (Chernitsky, 1980, 1986; Blake *et al.*, 1984; Langdon and Thorpe, 1984, 1985).

Carbonic anhydrase is another enzyme of importance in gill physiology. Its role in CO_2, H^+, HCO_3^-/Cl^- exchange mechanisms, and ammonia movement across fish gills is reviewed by Randall and Daxboeck (1984). The higher values of this ion reported in smolts adapted to seawater are probably related to the problems of gas exchange, ion, and acid–base regulation in waters of higher salinity (Milne and Randall, 1976; Dimberg *et al.*, 1981; Zbanyszek and Smith, 1984).

D. Osmotic and Ionic Regulation

Although the early developmental stages of all salmonids take place in fresh water, most species spend longer or shorter periods of their actively growing life in the sea. In the pink (*Oncorhynchus gorbuscha*) and chum salmon (*O. keta*), the capacity to hypoosmoregulate (excrete salts in the hyperosmotic marine environment and maintain the plasma electrolytes at about one-third seawater concentration) develops in the alevin stages (Weisbart, 1968; Kashiwagi and Sato, 1969); in some populations of chinook (*O. tshawytscha*) it occurs in fry or fingerlings (Clarke and Shelbourne, 1985). In other species of *Oncorhynchus* and in the *Salmo* and *Salvelinus* species, full hypoosmoregulatory ability is attained after a variable period of freshwater residency (usually 1 year or longer), and in *Salmo* and *Oncorhynchus* it requires the parr-smolt transformation. Studies of salinity relations of juvenile freshwater and downstream migrant salmon are now voluminous (Folmar and Dickhoff, 1980; Wedemeyer *et al.*, 1980; McCormick and Saunders, 1987).

The general physiological problems of an anadromous fish were stated in the previous section. The hypoosmotic freshwater habitat requires the excretion of large amounts of water and the acquisition of salts; the hyperosmotic marine environment demands the rigid conservation of water, the drinking of seawater, and the excretion of salt to provide fresh water for the tissues. Gills, opercular epithelia, kidneys, urinary bladder, and intestinal epithelia play active roles in this regulation. The glomerular filtration rate is altered during smolting [see Section II,D,2 and compare Holmes and Stainer (1966) with Eddy and Talbot (1985)]; the rectal and hindgut fluids of the marine salmon are strongly hypertonic and are responsible for the removal of the divalent ions acquired through body surfaces and by drinking sea water; transport of monovalent ions across the gut wall is increased to effect a concentration gradient that will move water from the gut into the body tissues; the gills excrete the excess monovalent ions (sodium and chloride) acquired from the gut.

Plasma electrolytes remain relatively constant throughout the freshwater life of juvenile salmonids. Folmar and Dickhoff (1980) review the literature and note values of 133–155 meq/l for Na^+, 3–6 meq/l for K^+, and 111–135 meq/l for Cl^-. Although some workers have found a decline in the plasma and tissue Cl^- during the presmolt stages (Fontaine, 1951; Kubo, 1955; Houston and Threadgold, 1963), more recent studies emphasize a relative

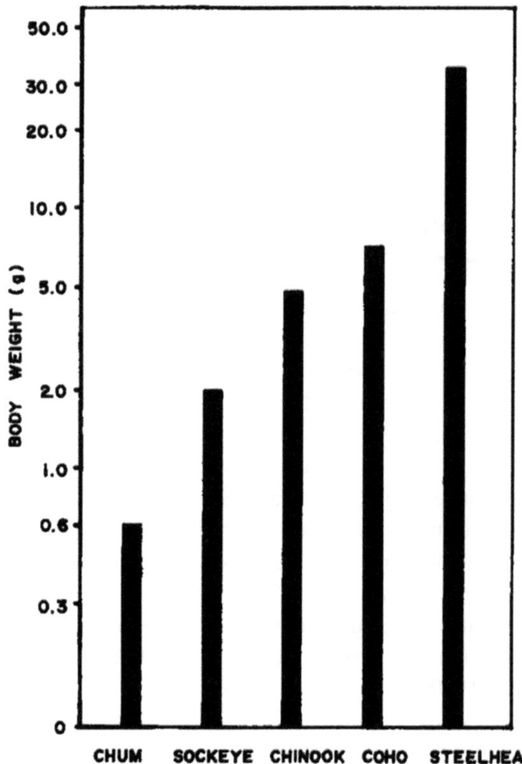

Fig. 2. Mean body weight at which samples of five salmonid species have demonstrated optimal hypoosmoregulatory capacity typical of smolts. [From Clarke (1982).]

constancy throughout the entire juvenile freshwater life (Parry, 1966; Conte *et al.*, 1966; Miles and Smith, 1968; Saunders and Henderson, 1970).

Even though plasma and tissue electrolytes remain constant throughout the parr stages, experiments have shown that salmonids differ in their capacities to deal with electrolytes when transferred to saline waters (Fig. 2). Although all species can tolerate mild changes in ambient salinity, the Oncorhynchids are in general more resistant to saltwater than species of *Salmo*, while the genus *Salvelinus* is the least resistant of the three genera. This is in line with Rounsefell's degrees of anadromy (Fig. 1), but the order of the species shown by Rounsefell (1958) is not in line with present-day studies (Hoar, 1976; Boeuf and Harache, 1982, 1984; Clarke, 1982). The important points to emerge from the many studies of salinity resistance of juvenile freshwater salmonids are:

1. All species can tolerate mild changes in ambient salinity.
2. Larger individuals have a greater tolerance than smaller members of a population (Houston, 1961; Conte and Wagner, 1965; Clarke and Blackburn, 1978; Jackson, 1981).
3. Salinity tolerance changes seasonally and is greater during the spring and early summer than in the autumn and winter (Conte and Wagner, 1965; Boeuf *et al.*, 1978; Lasserre *et al.*, 1978).
4. When given a choice between fresh water and saltwater, juvenile salmonids may show a distinct preference for one or the other. Saltwater preference is strongest at the peak of the smolt stage when the juveniles migrate, but like salinity tolerance, preference varies with the species, season, and status of smolt development (Baggerman, 1960a,b; McInerney, 1964; Otto and McInerney, 1970; Iwata *et al.*, 1985, 1986).
5. Transfer of young salmon to more saline waters, whether in the parr or smolt stage, is followed by transient changes in tissue electrolytes, after which the plasma values stabilize at or near the previous levels (*Oncorhychus* and *Salmo*), although the distribution of ions in the tissues may be different in the two environments (Eddy and Bath, 1979); in *Salvelinus alpinus* the stable values in sea water are about 25% higher than in fresh water (Gordon, 1957). In chum salmon, Black (1951) found that the adjustment period lasted 12 h; Miles and Smith (1968) reported 36 h for coho salmon; Houston (1959) gave 8–170 h for the adjustment phase in steelhead trout; Prunet and Boeuf (1985) found that Atlantic salmon smolts adjust to seawater in 4 days, while nonsmolting rainbow trout require 2 weeks; and Leray *et al.* (1981) and Jackson (1981) found a crisis period of 30–40 h in rainbow trout, with a new steady state established in 4–5 days. It is evident that this adjustment period varies with species, size, stage of smolting, temperature, and photoperiod. Literature reviewed in later sections shows that the adjustment period is marked by changes in the functional morphology of the osmoregulatory organs and in secretion

of hormones. Folmar and Dickhoff (1980) and Wedemeyer *et al.* (1980) summarize the extensive literature and provide bibliographies.

6. A capacity to acclimate is present in all species and stages of freshwater salmonids, and a gradual transfer to more saline waters is more successful than a sudden exposure even in species such as the chum salmon that are able to tolerate seawater as fry (Iwata and Komatsu, 1984).
7. Further, a smolt transformation is critical to successful growth in full seawater and, even though parr of some species (Section V) can be acclimated to seawater during favorable periods, they fail to grow normally and become "stunted" or die (Otto, 1971; Clarke and Nagahama, 1977; Bern, 1978; Folmar *et al.*, 1982a).
8. The smolt stage itself is brief and smolts that do not enter seawater during this narrow "window" of time (Boeuf and Harache, 1982) revert to the parr condition (Houston, 1961; Koch, 1982; Conte and Wagner, 1965, and reviews cited). Several workers have noted that this "narrow window" corresponds to the period of maximum gill enzyme activity.
9. Environmental factors, particularly temperature and photoperiod, modulate the time of smolting, desmolting, and the salinity preference and tolerance changes (Section IV).

The studies of the physiology of osmoregulation in marine fishes focus on the oral ingestion of seawater and the ionic extrusion mechanisms. Research on euryhaline species such as *Fundulus*, on the catadromous eels *Anguilla*, and on the anadromous salmonids have shown that transfer from fresh water to seawater activates these two processes. In rainbow trout, there is a rapid response of the drinking reflex, which reaches a peak within a few hours and then declines to a lower constant level, while the salt extrusion mechanisms are activated more slowly over a period of several days (Shehadeh and Gordon, 1969; Potts *et al.*, 1970; Bath and Eddy, 1979; Fig. 4 in Evans, 1984). The organs and tissues involved in these adjustments are the gills, opercular epithelia, kidneys, urinary bladder, intestine, and possibly the integument (Parry, 1966; Shehadeh and Gordon, 1969; Loretz *et al.*, 1982).

Studies of the euryhaline teleosts have shown the nature of the physiological changes required during the parr–smolt transformation. However, the research is spotty in respect to the temporal development of the changes during smolting. Present findings suggest that these changes occur well in advance of the smolt migration and preadapt the young salmon for life in a hyperosmotic habitat. In summarizing the physiology, the desirability of more studies of the actual development of the hyposmoregulatory mechanisms during smolting is noted.

1. *Gills*

The water permeability of gills has been measured in several euryhaline fishes (Platichthys, Fundulus, Anguilla, Salmo). In general, diffusional permeability of water is sharply reduced in the marine habitat (Isaia *et al.*, 1979; Evans,

1984; Isaia, 1984), but there do not seem to be permeability measurements that might demonstrate gradual changes during the parr–smolt transformation and preadaptation to the marine habitat.

The chloride cells of the gills and the opercular epithelia have been of major interest to salmonid physiologists. After many years of controversy, biologists now recognize the central role of these mitochondria-rich cells in processes of salt extrusion (Conte, 1969; Foskett and Scheffey, 1982; Zadunaisky, 1984). In general, these cells are more numerous and larger in the marine environment (Evans, 1984). In smolting salmonids, their proliferation commences well in advance of entrance into seawater. In studies of Atlantic salmon, Langdon and Thorpe (1984) describe seasonal changes in abundance and size of the gill chloride cells coinciding with variations in gill enzyme activity; a springtime peak occurred in both parr and smolts but was much more marked in the smolts. Loretz *et al.* (1982) studied samples of opercular epithelia of coho salmon at 2-week intervals from February to May. Only scattered chloride cells were found until late May when a three-fold increase in cells was noted. These studies suggest chloride-cell proliferation in the opercular epithelium during smolting. Several other studies record increased density and size of chloride cells in smolts (Threadgold and Houston, 1961; Chernitsky, 1980; Burton and Idler, 1984), but studies of the time course of their development from parr to smolt stages are few (Richman *et al.*, 1987).

2. *Kidney*

There are few studies of kidney function during the parr–smolt transformation. The changes that must occur are largely inferred from comparison of freshwater and marine teleosts. In general, the freshwater teleosts produce a copious, dilute (hypotonic) urine, while the marine teleosts excrete a scant volume of isotonic or slightly hypotonic urine (see Tables VII, IX, and X in Hickman and Trump, 1969). Comparative studies of euryhaline species in fresh water and saltwater show that they make the expected adjustments in physiological mechanisms when moved from one habitat to another (see Table XII in Hickman and Trump, 1969). It must be noted, however, that urine output in euryhaline teleosts is a resultant of adjustments in two basic renal mechanisms: glomerular filtration and tubular reabsorption (Fig. 47b in Hickman and Trump, 1969). In fresh water, glomerular filtration is high and tubular reabsorption is low, while in seawater the filtration rate is greatly reduced and the reabsorption accelerated.

Although salmon in fresh water and saltwater behave like other euryhaline fishes, the two basic mechanisms have not been carefully assessed during the parr–smolt transformation. In one of the few relevant studies, Holmes and Stainer (1966) found a reduction of almost 50% in the urine flow of *S. gairdneri* smolts in comparison with pre- and postsmolts (all stages studied in fresh water). Values for presmolts and postsmolts were essentially the same (near 4.5 ml kg^{-1} h^{-1}). At the same time, inulin clearance techniques gave a glomerular filtration rate (GFR) that was about 50% lower in the smolts than in the pre- and postsmolts. These results suggest a basic change in kidney function (GFR) during the parr–smolt transformation before migration; this could be preadaptive for a successful life in saltwater, where urine output is sharply reduced (Potts *et al.*, 1970).

In another relevant paper, Eddy and Talbot (1985) obtained quite different results with *S. salar*. These investigators reported a sharp *increase* in urine production coinciding with silvering indicative of the smolt stage; values for parr and presmolts were about 50% lower than those of smolts; again, all values were measured in fresh water (1–1.5 mg kg^{-1} h^{-1} in presmolts and 2.5–3 ml kg^{-1} h^{-1} in silvery "smolts"). Eddy and Talbot (1985) note that their Atlantic salmon smolts were much smaller (25–60 g) than the trout used by Holmes and Stainer (1966), which weighed 150–200 g and that the smaller fish were more difficult to catheterize. They also suggest that "handling diuresis" may have affected the steelhead data and note temperature differences between the two sets of experiments and the problems of accurately assessing the smolt stage. Clearly, more studies will be required to provide an accurate picture of changes in renal physiology during smolting. Structural as well as physiological differences have been described in comparisons of *Salmo gairdneri* adapted to fresh water and to seawater (Henderson *et al.*, 1978; Colville *et* d., 1983), and further investigations of renal physiology during smolting are likely to prove interesting.

3. *Urinary Bladder*

In many teleosts, the urinary bladder is an organ of significance in electrolyte balance (Lahlou and Fossat, 1971; Hirano *et al.*, 1973). In some species the organ is concerned with both water and salt transfers, but in the salmonids it may play only a minor role. In *S. gairdneri*, Hirano *et al.* (1973) found that the urinary bladder was osmotically impermeable in both fresh water and saltwater, although there seemed to be an active uptake of Na and Cl ions. In a study of fresh- and saltwater-adapted yearling coho salmon, Loretz *et al.* (1982) reported reduced electrolyte absorption in seawater. It is suggested that Na and Cl absorption may be necessary in freshwater salmonids to balance ion losses, but in seawater reabsorption is not required and may even be detrimental. Although it appears that the urinary bladder plays a relatively minor role in the water and ion balance of salmonids, its importance has not really been assessed in the smolt transformation.

4. *Intestine*

The intestine of the euryhaline fish is also an organ of water and ion regulation, greatly increasing the fluid absorption in the osmotically desiccating marine habitat (Shehadeh and Gordon, 1969; Lahlou *et al.*, 1975; Morley *et al.*,

1981). Collie and Bern (1982) used in vitro intestinal sac preparations to study fluid absorption. Fluid absorption was increased in smolting coho salmon prior to entrance into seawater; the higher level prevailed for several months before returning to the lower level. Thus, increased fluid absorption from the intestine seems to preadapt the young salmon for life in the ocean and occurs in concert with alterations in renal and branchial osmoregulatory mechanisms. The timing suggests a relationship with the thyroxin surge (Section II,E; Collie and Bern, 1982; Loretz *et al.*, 1982). The nutrient-absorbing role of the intestine also changes during smolting. An increased proline influx has been recorded and suggests that the higher nutritional demands of rapid growth initiated during smolting are met in part by an increased absorption efficiency (Collie, 1985). Experimentally, cortisol and growth hormone have been shown to increase intestinal proline absorption in coho and may regulate the process during smolting (Collie, 1985; Collie and Stevens, 1985).

E. Hormones and Smolting

The salmonids, like many other temperate-zone animals, have a seasonally changing physiology that is manifest in cycles of growth, precisely timed migrations, and seaons of reproduction geared to the most favorable seasons for the birth and development of the young. A working hypothesis developed in an earlier paper stated (Hoar, 1965):

> *... That several species of salmonids undergo physiological and behavioural cycles which, each springtime, preadapt them for life in the ocean; if they do not reach the ocean, the cyle is reversed and the physiology appropriate to life in fresh water again appears. Under natural conditions, changing photoperiods trigger the cycle at the appropriate season, but the cycle is endogenous and does not disappear under constant conditions of illumination. The theory is that this is a general phenomenon in the salmonids, and some evidence for it will probably be found in all species at all stages in their development. ... This cyclical physiology of the salmon ... has been very susceptible to modificaton through genetical processes. The smolt transformation, which is an obligatory part of the life cycle of Atlantic salmon, steelhead trout, and coho, is suppressed or lost in species such as the pinks and chums.*

The hypothesis is still relevant. In most salmonids, the smolt transformation is tightly timed by photoperiod, and by lunar cycles (*Oncorhynchus*) or flooding streams (*Salmo*), to occur in the springtime, but the amago salmon (*O. rhodurus*) and some populations of chinook salmon (*O. tshawytscha*) undergo a typical smolt transformation in the autumn (Ewing *et al.*, 1979; Nagahama *et al.*, 1982; Nagahama, 1985).

Seasonal changes in vertebrate physiology are timed and regulated by a neuroendocrine system. Environmental cues act through the peripheral sense organs and brain to trigger secretory activity in the hypothalamus;

hypothalamic releasing hormones regulate secretion of the hormones of the anterior pituitary gland, which in turn controls endocrine organs such as the thyroid and the interrenal glands through its trophic hormones or secretes hormones that act directly on target tissues (prolactin, growth hormone).

The parr–smolt transformation occurs in association with a general surge in endocrine activity that can be detected in most, if not all, of the endocrine organs. Hormones most thoroughly studied and considered most likely to be involved in the transformation are the thyroid hormones, prolactin, growth hormone, and the corticosteroids, but changes have also been studied in other endocrine factors, particularly the gonadal steroids (Hunt and Eales, 1979; Sower *et al.*, 1984; Patiño and Schreck, 1986; Ikuta *et al.*, 1987), and the secretions of the Stannius corpuscles and the urophysis (Bern, 1978; Aida *et al.*, 1980; Nishioka *et al.*, 1982) (Fig. 3).

1. *Thyroid Hormones*

A dramatic increase in thyroid activity is generally recognized in smolting salmon. This increase was first reported half a century ago in Atlantic salmon (Hoar, 1939). The histophysiological evidence presented at that time was later confirmed for *S. salar* (reviews by Fontaine, 1954, 1975) and extended to other smolting salmonids by Robertson (1948), Hoar and Bell (1950) and others.

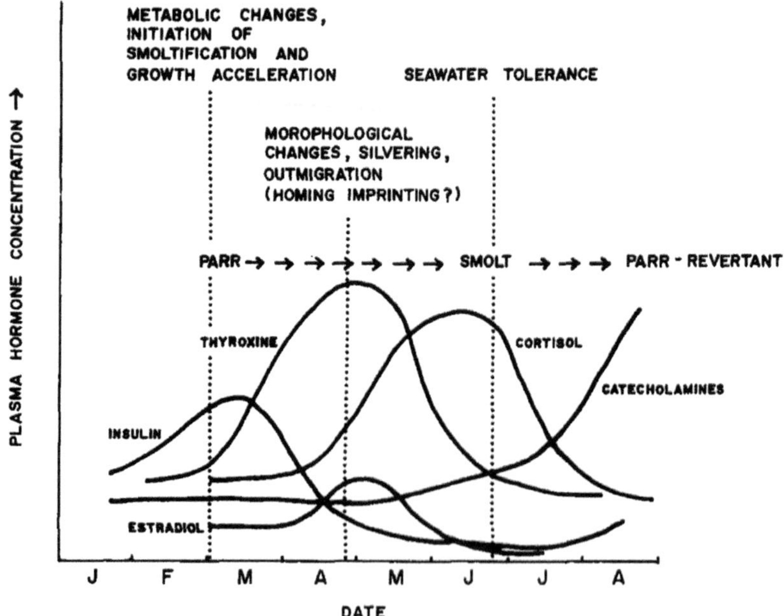

Fig. 3. Summary of typical patterns of hormone changes during smolting of coho salmon. (Graph presented at Tenth International Symposium of Comparative Endocrinology, Copper Mountain, Colorado, 1985. Courtesy of W. W. Dickhoff.)

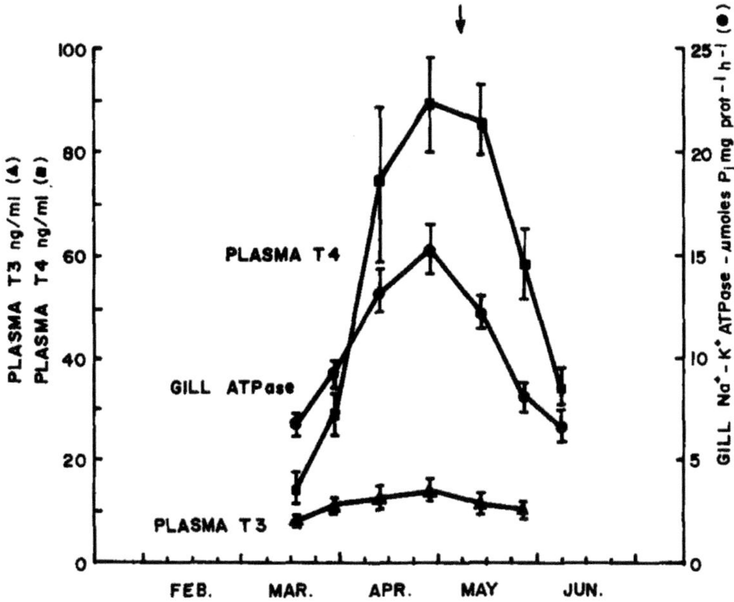

Fig. 4. Relationships between mean gill (Na+K)-ATPase activities and plasma T4 and T3 concentrations of coho salmon in fresh water at the Sandy Hatchery. Bars indicate ± SEM, $n = 10$ samples of three fish; arrow indicates hatchery release data. [From Folmar and Dickhoff (1981).]

Radioiodine techniques gave confirmatory results (Eales, 1963, 1965). The advent of radioimmunoassay (RIA) techniques has permitted measurements of the time course of the smolting surge in plasma thyroxine (T4) and correlated these changes with gill enzyme activity in both *Salmo* and *Oncorhynchus* (Folmar and Dickhoff, 1980, 1981; Boeuf and Prunet, 1985; Dickhoff *et al.*, 1985) (Fig. 4); with lunar cycles in *Oncorhynchus* and *S. gairdneri* (Grau *et al.*, 1981) (Fig. 5 and Section IVD); with stream flow-rate in *Salmo* (Youngson and Simpson, 1984; Lin *et al.*, 1985a; Youngson *et al.*, 1986); and with the migration time and subsequent yield (Fig. 6). Although T4 levels change dramatically, the triiodothyronine (T3) levels are quite stable (Fig. 4). The literature is now extensive for the different species of smolting salmonids, and only selected references are noted here (Nishikawa *et al.*, 1979; Ealcs, 1979; Folmar and Dickhoff, 1980; Dickhoff *et al.*, 1982; Specker and Richman, 1984; Specker *et al.*, 1984; Grau *et al.*, 1985c; Lin *et al.*, 1985a,b; Dickhoff and Sullivan, 1987).

Thus, there is ample evidence that thyroid activity is usually elevated at the time of the typical parr–smolt transformation. However, the evidence for a triggering role of the thyroid hormones in the onset of smolting is absent. Rather, it appears that thyroid hormones intensify several physiological and behavioral changes of smolting but that these changes are triggered by other hormones or

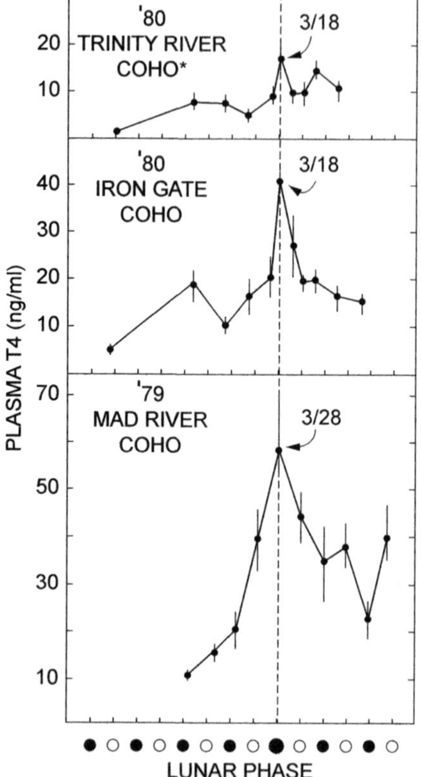

Fig. 5. Patterns of plasma T4 in three California coho salmon stocks plotted on a lunar calendar: ●, new moon; ○, full moon; asterisk, Trinity stock raised at Mad River Hatchery. [From Grau *et al.* (1981). Copyright 1981 by Am. Assoc. Adv. Sci.]

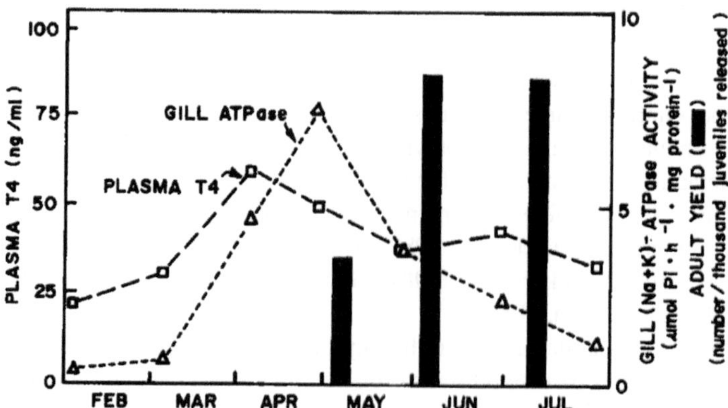

Fig. 6. Changes in gill ATPase activity and plasma T4 concentration of 1979-brood coho salmon reared at the Corvallis Fish Research Laboratory in relation to yield (bars) from groups of juveniles released from Big Creek Hatchery in May, June, and July. [From Ewing *et al.* (1985).]

occur endogenously (Hoar, 1976). Two groups of studies are relevant: those concerned with the general sensitivity of the thyroid to many different stimuli (what may be termed its "lability"), and the experimental attempts to alter the time of the parr–smolt transformation with thyroid hormones, thyroid-stimulating hormone (TSH), or antithyroid compounds such as thiourea.

First then, many studies have related changes in thyroid activity to altered environmental conditions: season, temperature, and light (Swift, 1959; Eales, 1965, 1979; White and Henderson, 1977; Osborne *et* al., 1978; Donaldson *et al.*, 1979; Eales *et al.*, 1982; Grau *et al.*, 1982; Leatherland, 1982; Specker *et al.*, 1984). Thyroid activity is also affected by the level of iodine in the ambient water (Black and Simpson, 1974; Sonstegard and Leatherland, 1976), by toxic substances (Moccia *et al.*, 1977; Leatherland and Sonstegard, 1978, 1981; Leatherland, 1982), and by salinity (Specker and Richman, 1984; Nishioka *et al.*, 1985a; Specker and Kobuke, 1987). Under certain conditions, changes in water flow (Youngson and Simpson, 1984; Youngson *et al.*, 1986) and the transfer from one freshwater environment to another freshwater environment may alter thyroid activity (Grau *et al.*, 1985b; Lin *et al.*, 1985b; Nishioka *et al.*, 1985a; Virtanen and Soivio, 1985). A relationship between nutritional status and thyroid activity has been noted (Leatherland *et al.*, 1977; Donaldson *et al.*, 1979; Eales, 1979; Eales and Shostak, 1985; Milne *et al.*, 1979) as well as changes in relation to the reproductive conditions of the fish (Sage, 1973; White and Henderson, 1977; Eales, 1979; Hunt and Eales, 1979; Sower *et al.*, 1984) and rearing density (Leatherland and Cho, 1985). Obviously, caution is required in the evaluation of thyroid activity in relation to the physiology of smolting.

Second, experiments designed to test the hypothesis of a causal relation between thyroid activity and the initiation of the parr–smolt transformation have provided scant evidence for this theory. Although body levels of thyroid hormone can be altered by injections of TSH, by feeding, injecting, or immersing fish in thyroid hormones, or by the use of some of the goitrogens such as thiourea (Chan and Eales, 1976; Milne and Leatherland, 1978, 1980; Eales, 1979, 1981; Eales and Shostak, 1985; Miwa and Inui, 1983; Specker and Schreck, 1984; Lin *et al.*, 1985a; Nishioka *et al.*, 1987), the development of typical smolts at seasons when smolting does not normally occur has not been achieved. Nonetheless, some components of the behavior of a typical smolt and some of the physiological changes characteristic of smolting can be induced. The most relevant of the studies seem to be those associated with salinity preference, general activity, growth, silvering, and some of the effects on metabolism.

Baggerman (1960a,b, 1963) induced changes in salinity preferences of juvenile Pacific salmon by the use of TSH, thyroid hormones, and thiourea. These changes were consistent with the theory that thyroid hormone is concerned with the change from a freshwater preference in the parr to a saltwater preference in a smolt. Increased swimming activity and other behavioral changes associated with downstream migration have been recorded in juvenile salmonids following thyroid treatment (Godin *et al.*, 1974), but it should be noted that these changes are not specific to the salmonids and have been similarly induced in some other teleosts—the stickleback *Gasterosteus aculeatus*, for example (Baggerman, 1962; Woodhead, 1970; Katz and Katz, 1978). Not all of the findings are consistent with Baggerman's hypothesis; Birks *et al.* (1985) injected T4 and/or thiourea into steelhead trout near the time of downstream migration and concluded that thyroid stimulation *reduced* the migration tendency. These workers argue that high thyroid activity antagonizes seaward migration and note that downstream migration of juvenile salmonids occurs at the time of falling thyroid levels.

Donaldson *et al.* (1979) review many experiments showing that dietary supplements of thyroid or thyroid hormones may stimulate body growth. Supplements of T3 are particularly active; these effects of thyroid treatment are much more marked when combined with pituitary growth hormone (Barrington *et al.*, 1961; Massey and Smith, 1968; Donaldson *et al.*, 1979; Eales, 1979; McBride *et al.*, 1982; Miwa and Inui, 1985). Donaldson *et al.* (1979), in their critical review, conclude that although thyroid hormone is probably necessary for normal growth, this hormone alone does not seem to stimulate growth but exerts its effect in combination with growth hormone or favorable nutrition.

Increased purine deposition with body silvering frequently follows thyroid treatment. This feature is also more marked when the thyroid hormone is combined with growth hormone; it too is not specific to nonsmolting salmonids. Moreover, the dark pigmentation of the dorsal and caudal (*Salmo*) or pelvic (*Oncorhynchus*) fins seen in typical smolts has not been induced in parr by thyroid treatment (Robertson, 1949; Chua and Eales, 1971; Ikuta *et al.*, 1985; Miwa and Inui, 1983, 1985). Fin darkening is probably caused by a pituitary factor. Komourdjian *et al.* (1976a) report darkening of fin margins and yellowing of the operculae and fin surfaces of Atlantic salmon parr injected with porcine growth hormone (GH); Langdon *et al.* (1984) noted fin darkening in juvenile Atlantic salmon treated with adrenocorticotropic hormone (ACTH). However, it should be remembered that the pituitary preparations used may have contained small amounts of different pituitary hormones. Moreover, ACTH and the melanophore-stimulating hormone (MSH) are biochemically related, having a common core of seven amino acids. Obviously, further studies of the endocrinology of fin pigmentation of smolts is required.

Thyroid hormone is involved in several aspects of intermediary metabolism. Eales (1979) reviews the literature concerning its role in fish metabolism; Donaldson *et al.* (1979) discuss its relationship to growth; Folmar and Dickhoff (1980) comment on the possible regulatory action of the thyroid hormones in smolting. Effects of growth (protein nitrogen metabolism) and purine nitrogen metabolism have been noted in previous sections of this chapter. Carbohydrate metabolism is also altered during the parr–smolt transformation (Section II,C). Lower glycogen and blood glucose appear to be characteristic of smolts, but parr treated with thyroid preparations have failed to show consistent changes in this regard. In a recent paper, Miwa and Inui (1983) treated amago salmon (*O. rhodurus*) with T4 and/or thiourea; the changes in carbohydrate metabolism characteristic of smolting salmon did not occur. The earlier literature is summarized in the reviews already cited.

Changes in the lipid metabolism of juvenile Atlantic salmon during the smolt transformation were reported by Lovern (1934) more than half a century ago. Subsequent research has confirmed Lovern's findings and described several other changes in the lipid metabolism of both smolting *Salmo* and *Oncorhynchus* (Sheridan *et al.*, 1983, 1985a,b; Sheridan, 1985). Treatment with thyroid hormone has been shown to decrease stored lipid in *Salvelinus* as well as *Salmo* and *Oncorhynchus* (Narayansingh and Eales, 1975; Sheridan, 1986), but evidently not in all teleost species (Eales, 1979). In a recent paper, Sheridan (1986) reports that treatment of coho salmon parr with T4 mobilized lipids in both liver and mesenteric fat with an accompanying increase in lipase activity; the plasma nonesterified fatty acids increased; lipase activity also increased in dark muscle. Smolts treated in a similar manner with T4 did not show these changes.

Finally, a very critical question concerns the possible role of thyroid hormones in salinity tolerance. Do thyroid hormones improve the resistance of salmon parr to saline waters and enable them to thrive in the sea? Most experiments have given negative answers, although there is suggestive evidence of a greater requirement for thyroid hormones in saltwater. Eales (1979) reviewed the experimental work and found negative or conflicting evidence of a role for the thyroid in osmotic and ionic regulation; experiments are difficult to interpret because of different levels of iodine in the ambient waters. More recent research has not altered these conclusions (Folmar and Dickhoff, 1979, 1980; Miwa and Inui, 1983; Specker and Richman, 1984; Specker and Kobuke, 1987). Miwa and Inui (1985) report that T4 alone does not improve sea water tolerance, but growth hormone alone or in combination with T4 increases the tolerance and induces a significant elevation in the gill Na^+, K^+-ATPase in amago salmon. Specker and Richman (1984) found that bovine TSH induced a thyroidal response during smolting (as measured by plasma T4); this response was greater during the early period of smolting, while the increase in T4 titers appeared sooner and was of greater duration in fish transferred to sea water. These results suggest an effect of salinity on the kinetics of T4 entry into and exiting from the bloodstream.

In summary, it is now clear that thyroid hormones do not trigger the parr-smolt transformation. At best, treatment of salmon parr with thyroid preparations creates "pseudo-smolts" (Eales, 1979). The thyroid seems to play an important role in enhancing smolting characteristics that are regulated endogenously or by other hormonal factors.

2. Growth Hormone and Prolactin

Growth hormone (GH), also called somatotropin (STH), is growth-promoting in teleost fishes, while prolactin (PRL), which is closely related to GH biochemically, serves several different functions. In the present context, the most relevant functions of prolactin are the conservation of sodium and the decrease of gill water permeability of euryhaline fishes in fresh water (Clarke and Bern, 1980; Loretz and Bern, 1982; Hirano, 1986). These recognized functions of GH and PRL suggest that they may both be important in the physiology of smolting. That the salmonids differ somewhat from euryhaline fish such as *Fundulus heteroclitus* was suggested many years ago by Smith (1956), who reported increased salinity tolerance in *Salmo trutta fario* injected with mammalian anterior pituitary powder and attributed this to growth hormone. A few years later, J. E. McInerney found that injections of purified STH (Nordic) induced a salinity preference change in young coho salmon (review by Hoar, 1966). Since that time, it has been established that in several salmonids, GH not only promotes growth but *increases survival in seawater*. There are several lines of evidence for the actions of these two closely related pituitary factors in smolting salmon: the effects of hypophysectomy, studies of the cytology and ultrastructure of the pituitary, hormone assays, and injections of GH or PRL.

Most hypophysectomized euryhaline fishes die if retained in fresh water. This is not true of several salmonids tested, although there is a decline in the sodium levels in fresh water (Komourdjian and Idler, 1977; Nishioka *et al.*, 1985b).

There have been several important cytological studies of the pituitary gland in relation to smolting. From these it is evident that both the growth-hormone-secreting and the prolactin-secreting cells are activated during the parr–smolt transformation. Among the more recent studies, Leatherland *et al.* (1974) reported greater activity of both GH and PRL pituitary cells in kokanee salmon smolts (*O.keta*). Likewise, Nishioka *et al.* (1982), in an ultrastructural study of various endocrine glands, noted that the PRL cells were active in coho salmon during the parr and smolt stages but substantially more active in freshwater smolts than seawater smolts; this argues for an involvement of PRL in the salmon smolt in fresh water. Nishioka *et al.* (1982) also found that GH cells were active in both parr and smolts but more active in the smolts *both in fresh water and in seawater*. Further cytological evidence (light and electron microscopy) of GH function in the marine environment comes from Nagahama *et al.* (1977), who noted greater GH activity in yearling coho parr when transferred to seawater. In contrast, prolactin cells of the parr were markedly more active in fresh water than in the seawater fish; when parr were transferred from saltwater to fresh water, however, a stimulation of the PRL cells was indicated—findings that again argue for a role of PRL in the freshwater coho salmon. Another paper by these workers provides confirmatory evidence (Clarke and Nagahama, 1977). Nagahama (1985) also finds that prolactin is involved in the adaptation of the amago salmon to fresh water.

Data on plasma PRL levels (RIA analyses) support the cytological findings. Prunet and Boeuf (1985) compared nonsmolting rainbow trout with smolting Atlantic salmon. Plasma prolactin levels declined significantly after transfer of the trout from fresh water to seawater, although the plasma osmotic pressure increased. In contrast, smolted Atlantic salmon adapted quickly to seawater and had similar prolactin levels in both environments. Prunet *et al.* (1985) measured plasma and pituitary prolactin levels in rainbow trout adapted to different salinities. Transfer from fresh water to sea water decreased plasma PRL; although a transient rise in pituitary PRL followed the transfer, the values after three weeks were lower than in the freshwater controls. The reverse transfer (SW to FW) induced, within one day, a rise in the plasma PRL. These findings indicate an important role for PRL in the freshwater adaptation of sedentary rainbow trout. Hirano's (1986) data for chum salmon are confirmatory.

Finally, there are many reports of the effects of GH and PRL injections into salmonids. Growth stimulation has been reported to follow GH injections into both intact (Higgs *et al.*, 1976; Komourdjian *et al.*, 1976a; Markert *et al.*, 1977) and hypophysectomized salmonids (Komourdjian *et al.*, 1978). Increased seawater tolerance was also reported in Komourdjian's experiments, which were carried out on rainbow trout and Atlantic salmon. Clarke *et al.* (1977) found that GH from either teleost or mammalian sources lowered plasma sodium in under-yearling sockeye salmon in both fresh water and seawater; survival was high in both environments. Miwa and Inui (1985) reported that ovine growth hormone increased significantly the survival of amago salmon parr transferred to 27% seawater. Bolton *et al.* (1987a) studied the effects of ovine or chum salmon GH on the plasma sodium, calcium, and magnesium levels of *S. gairdneri* transferred for 24 h to 80% sea water (3 i.p. injections of homone at 3 day intervals). They conclude that the seawater-adapting actions of GH are specific to the hormone and are not consequent to an increase in size.

In summary, the evidence from several lines of study is consistent in showing increased activity of both GH and PRL pituitary cells at the time of smolting. Further, prolactin plays a special role in the osmoregulatory physiology of the smolt in fresh water, while GH is the important factor in hydromineral regulation in the sea (review by Hirano, 1986).

3. *The Interrenal Gland*

Many years ago, M. Fontaine and his co-workers presented histophysiological evidence of increased interrenal activity in smolting Atlantic salmon (Fontaine and Olivereau, 1957, 1959; Olivereau, 1962, 1975). Their findings were subsequently confirmed in several species of *Salmo* and in *Oncorhynchus* by further light microscopy (review by Specker, 1982), by ultrastructure

(Nishioka *et al.*, 1982), and by biochemical studies of the plasma corticosteroids (Fontaine and Hatey, 1954; Specker and Schreck, 1982). The adrenocorticotropic hormone (ACTH) producing cells of the pituitary are also more active in the smolt that in the parr (Olivereau, 1975; Nishioka *et al.*, 1982).

Corticosteriods, especially cortisol, are essential to the life of a teleost fish. They function both in water-electrolyte homeostasis and in intermediary metabolism (Chester-Jones *et al.*, 1969; Chester-Jones and Henderson, 1980). Their importance in the salmonids has been amply demonstrated, but the details of their actions throughout the life cycle are still not clear (review by Specker, 1982). The action of the corticosteroids appears to differ somewhat among various species of euryhaline teleosts and may not be consistent among all the salmonids (see Nichols *et al.*, 1985). Investigations of interrenal physiology are complicated by "stress handling," which is known to alter corticosteriod dynamics (Wedemeyer, 1972; Donaldson, 1981; Redding *et al.*, 1984; Leatherland and Cho, 1985).

Elevation of the corticosteroids during the smolting episode of the coho salmon is shown in Fig. 7. The rise is correlated with the increase in gill ATPase activity noted in Section II, C (see also Patiño *et al.*, 1985), but a cause-and-effect relationship has not been established (compare Langhorne and Simpson, 1986 and Richman and Zaugg, 1987). However, the data are suggestive of an important function of the interrenal in improving the salinity

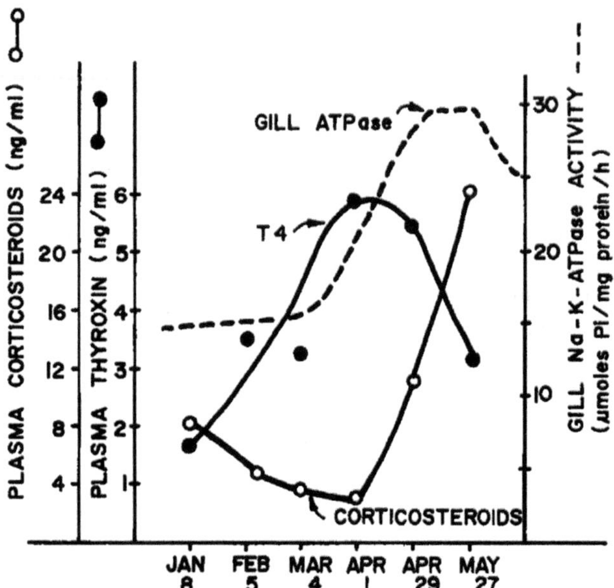

Fig. 7. Plasma concentrations of thyroxine and corticosteroids (from Specker and Schreck, 1982) and gill (Na+K)-ATPase activity (from Zaugg and McLain, 1970, 1976) in laboratory-reared, constant 10°C, coho salmon during smolting. [From Specker (1982).]

tolerance of the smolt by stimulating gill Na^+, K^+-ATPase. Young (1986), using an in vitro system for the incubation of interrenal tissue of coho salmon, found a marked increase in the sensitivity of the tissues to ACTH during April, and this was correlated with peak plasma T4 titers and enhanced osmoregulatory capacities. Langdon *et al.* (1984) reported that ACTH, but not cortisol, increases the gill ATPase activity in juvenile Atlantic salmon, while both ACTH and cortisol increase the succinic dehydrogenase in suspensions of gill cell homogenates; neither treatment affects size or abundance of chloride cells in intact gills. Studies of several salmonids transferred from fresh to salt water support the view that corticosteroids improve salinity tolerance (Nichols and Weisbart, 1985; Nichols *et al.*, 1985), but substantially more research will be required to clarify the details of interrenal physiology in the preadaptation of smolting salmon to marine life.

Sheridan (1986) provided recent evidence of an action of cortisol in regulating smolt-associated changes in intermediary metabolism. Cortisol implants into coho parr caused a significant reduction in total lipid and triacylglycerol content of liver and dark muscle—changes characteristic of the parr-smolt transformation. These changes were accompanied by increased lipase activity. The changes did not occur in coho smolt similarly treated. The data suggest that the rise in cortisol is responsible for some of the changes in lipid metabolism observed at the time of the smolt transformation. Cortisol may also play a special part in stress-related physiology and the functioning of the immune system during this phase of the salmon life cycle (Specker, 1982; Barton *et al.*, 1985), but substantially more research on the interrenal hormones is required.

III. SEXUAL MATURATION: AN ALTERNATE STRATEGY IN DEVELOPING MALE PARR

Variable numbers of male parr may become sexually mature. This interesting phenomenon has been examined most thoroughly in the Atlantic salmon, where it was noted more than a century ago (see Orton *et al.*, 1938). It is a true sexual maturation, resulting in functional males that participate in spawning with large adult females and produce viable offspring (Orton *et al.*, 1938; King *et al.*, 1939; Jones, 1959). The percentage of males that mature as parr varies in different populations. It may be as low as 5–10% (Evropeizeva, 1959; Bailey *et al.*, 1980) but is often in excess of 50% (Evropeizeva, 1959, 1960; Österdahl, 1969; Dodd *et al.*, 1978; Lundqvist, 1983; Myers, 1984). In an extreme situation, 100% of male Atlantic salmon in the Black Sea population are said to mature as parr, and only the females are anadromous (Leyzerovich, 1973). It would be interesting to have more information on the biology of these Black Sea populations; Österdahl (1969) suggested that even though parr may contribute substantially to the genetics of a population, large adult males are necessary in the spawning behavior, where they interact with the adult females in nesting, spawning reactions, and defense of the redd.

However, Myers and Hutchings (1985) have more recently shown that large males are not necessary for natural spawning; mature male parr mated successfully with Atlantic salmon grilse.

Precocious maturation is not confined to Atlantic salmon but has been described also in *S. gairdneri* (Schmidt and House, 1979; Skarphedinsson *et al.*, 1985), in *S. trutta* (Jonsson and Sandlund, 1979), in several species of *Oncorhynchus* (Robertson, 1957; Gebhards, 1960; MacKinnon and Donaldson, 1976; Hard *et al.*, 1985), and in *Salvelinus* (Leyzerovich, 1973). It appears to be a viable and biologically important alternate tactic in the life history of many salmonids (Saunders and Schom, 1985).

The age of precocious maturation is variable. In nature, male Atlantic salmon parr may first mature at 1+ years of age or later. They may mature again in subsequent years if they remain in fresh water, or they may become smolts but remain in fresh water, revert to the parr condition, and mature; further, postsmolts (i.e., smolts during their first year in sea water) may occasionally mature (Saunders and Henderson, 1965; Sutterlin *et al.*, 1978; Lundqvist and Fridberg, 1982); a year later they may return as grilse. In hatcheries with very favorable growing conditions, both Atlantic and coho salmon parr may mature at 0+ years of age (Saunders *et al.*, 1982); maturation of parr at 0+ has also been recorded in natural habitats but only in the lower reaches of the River Scorff (France), where growing conditions are very favorable (Baglinière and Maisse, 1985). Again, as in other salmonid characteristics, there is a spectrum in the biology of early maturation among different species and stocks of salmonids. The present discussion focuses on the relationship of early maturation to the smolting process.

There are several implications of early maturation in fish culture. The precocious males are likely to suffer increased mortality and/or delay in reaching the rapidly growing marine phase. Sex ratios of smolts and adults are affected (Leyzerovich, 1973; Mitans, 1973; Dalley *et al.*, 1983; Myers, 1984); further, there appear to be genetic factors determining precocious maturity, and this too has implications in salmon breeding programs (Saunders and Sreedharan, 1977; Glebe *et al.*, 1978; Naevdal *et* al., 1978; J. E. Thorpe, 1987; Thorpe *et al.*, 1983). It is therefore important to understand, and if possible regulate, the reproductive physiology of the male parr. Research interest has focused on the endocrinology, the genetics (Section V), the relative rates of growth of precocious parr and smolting juveniles, and the relationships between precocious sexual maturation and the parr–smolt transformation.

Studies of the reproductive physiology of maturing male parr have shown that changes in the pituitary gonadotrops (Lindahl, 1980) and in the pituitary and plasma gonadotropins parallel those of maturing adult fish (Crim and Evans, 1978). The levels of testosterone and 11-ketotestosterone rise in a comparable manner in the precocious parr and the maturing adults, although the actual values are lower in the parr (Stuart-Kregor *et al.*, 1981). In Atlantic salmon, at any rate, the endocrinological regulation of precocious maturation is evidently the same as that of the maturing adult males (Dodd *et al.*, 1978).

Is rate of growth a determining factor in the onset of precocious sexual maturation? Eriksson *et al.* (1979) summarized the conflicting literature and concluded that genetic factors rather than rates of growth were determinants of early maturity. In a study of hatchery-reared Baltic *Salmo salar*, all individuals were found to grow at comparable rates until the end of the second summer, when the fish were 1+ years old (unimodal population as discussed in Section 11). At that time, about 50% of the males matured. In short, maturation of 50% of the males took place at a time when immature males, maturing males, and all females were similar in size. Rates of growth did not seem to be the determining factor in the onset of maturity. The following season, when the fish were 2+ years old, another group of males matured—again at a time when the immatures, the maturing parr, and the females formed a unimodal group. Following maturation (at either 1+ or 2+ years), growth rate declines while the gonads are ripe, but the interesting point is that growth rate does not appear to determine the onset of maturation in these populations. In contrast, data on size dependence have been presented by several workers who claim that the larger, more rapidly growing individuals mature (Evropeizeva, 1960; Dalley *et al.*, 1983; Myers *et al.*, 1986 and citations in Eriksson *et al.*, 1979), while Sutterlin *et al.* (1978), in a study of postsmolts in sea water, found that it was the smallest individuals that matured. In support of Eriksson *et al.* (1979), Naevdal *et al.* (1978), in a study of a hatchery-reared population of Atlantic salmon, found indirect evidence supporting the concept that sexual maturation is randomly distributed and not related to size or rate of growth. These workers conclude that precocious maturation retards the growth of the male parr but that growth before maturation is uniform, thus supporting the idea that genetic factors are involved.

Thorpe and Morgan (1980) discuss a "two-threshold hypothesis" for critical sizes for smolting and precocious maturation. In a population of Scottish salmon where parr may smolt and emigrate at 1+ year (if they have attained the critical size), it is the more slowly developing individuals that remain in fresh water and become sexually mature at 1+ years. These workers discuss the "two-threshold hypothesis" and suggest that if the critical size for maturation is greater than the critical size for smolting, then populations with a high proportion of *small* 1-year smolts will not tend to mature in their first autumn; populations with a high proportion of *large* 1-year smolts may be expected to show a high incidence of precocious parr in the first autumn.

What is the effect of precocious maturation on the subsequent tendency to smolt and migrate? Thorpe, (1986, 1987) has examined the interrelationships among these three relevant factors: growth rate, sexual maturation, and time of smolting. Although he finds a strong positive correlation between growth rate and maturation rate, the "decision" to smolt is also critical to the onset of sexual maturity. In sibling populations of Atlantic salmon, relatively large individuals that smolted at 1 year did not mature until at least 2.5 years, while in the same population some smaller fish that did not smolt at 1 year matured 6 months later without emigrating to sea. Moreover, at age 2 years, the fish that matured at 1.5 years failed to undergo smolting as completely as their immature siblings (conclusion based on gill enzyme studies). Thorpe (1986, 1987) argues that the processes of smolting and reproduction are mutually incompatible. Smolting is a commitment to life in saline waters; reproduction demands a freshwater habitat. This concept of incompatibility is supported by several investigators. Nagahama (1985) compared plasma levels of testosterone and 11-ketotestosterone in amago salmon at three stages: in precociously maturing males, in smolting and in desmolting fish. The values were highest in early maturing male fish and lowest in the smolts; they increased during desmolting. Findings of Miwa and Inui (1986) also support this concept of incompatibility. These workers fed sex steriods to sterilized amago salmon and noted reduced gill Na^+, K^+-ATPase activity, reduced numbers of chloride cells, and a thicker skin and gill epithelia than in the smolting controls; the steroids prevented epidermal silvering and an increase in seawater tolerance (see also Aida *et al.*, 1984; Ikuta *et al.*, 1985, 1987; McCormick and Naiman, 1985).

Several other investigations are also relevant to the effects of parr maturation on the subsequent tendency of salmon to smolt and migrate. In the first place, Thorpe and Morgan (1980) and Saunders *et al.* (1982) found that male parr that matured in autumn may migrate the following spring while their body fats are depleted and their energy reserves at a low level as a consequence of spawning (Thorpe and Morgan, 1980; Saunders *et al.*, 1982). This situation seems to srgue against some theories that suggest a role for changing energy reserves in triggering the smolt transformation (see Saunders *et al.*, 1982), but further investigation on this point is indicated. In the second place, some workers have noted a strong tendency of precocious parr to remain in fresh water and to mature the following autumn (Leyzerovich, 1973; Eriksson *et al.*, 1979). This situation plus the increased mortality associated with sexual maturation reduce the proportion of males with respect to females in the migrating smolt population (Mitans, 1973; Dalley *et al.*, 1983; Myers, 1984). The increased mortality of the males may be related to their spawning activity, the harsh winter conditions that postspawners face while energy reserves are depleted, or predation; it occurs both in nature and under laboratory conditions (Clarke *et al.*, 1985).

In summary, the age of sexual maturation is very flexible in male salmon. Although maturation in the parr stages in best known in *S. salar*, it has also been recorded in other salmonids. There are several consequences of parr maturation: (a) sex ratios of migrant smolts and returning adults are biased toward the female, since sexual maturation reduces the relative numbers of males—at least in natural populations; (b) maturation seems to reduce the tendency to smolt and migrate the following spring (Österdahl, 1969)—in this connection, it may be relevant that precocious male parr have relatively low capacities to adapt to seawater (Aida *et al.*, 1984; Lundqvist *et al.*, 1986); (c) male smolts may adopt the alternative tactic of remaining in fresh water and spawning as parr that have readapted to fresh water (Lundqvist and Fridberg, 1982; Thorpe, 1986, 1987); and (d) variability in the life history of salmon may be an important safeguard against loss of small stocks through several successive years of reproductive failure (Saunders and Schom, 1985). Thus, several factors related to parr maturation may affect the smolt populations of Atlantic salmon and the productivity of a river system.

IV. ENVIRONMENTAL MODULATION OF THE SMOLT TRANSFORMATION

The season of the parr–smolt transformation is highly predictable. Like the dates of migration, reproduction, and the emergence of small fry from the gravel, parr become smolts in relation to the seasonal changes in their surroundings—especially the cyclical variations in day length and temperature. The environmental regulation of smolting has now been critically studied for more than two decades (see reviews by Hoar, 1965, 1976). The early experiments demonstrated a strong effect of photoperiod but indicated an underlying endogenous rhythmicity; such seasonal changes as skin silvering, growth, salinity preference, and tolerance could be advanced by accelerated photoperiods or delayed by retarded photoperiods, but the changes could not be entirely suppressed. An underlying circannual rhythm was suspected.

A circannual rhythm has been defined as one that persists under constant environmental conditions but deviates by a fixed amount from the annual cycle of 365.26 days. Circannual rhythms were first charted in the golden hamster (*Citellus lateralis*) by Pengelley and Asmundson (1969), but are now recognized in many animals. The pioneer work on the teleost fishes is summarized by Baggerman (1980), Eriksson and Lundqvist (1982), and Lam (1983).

A. The Circannual Rhythm of Smolting

Eriksson and Lundqvist (1982) kept individually marked Baltic *S. salar* under constant conditions of light [light-dark (LD) 12: 12] and water temperature (11.0 ± 0.5°C) for 14 months. The fish were given a surplus of food, and their condition factor (weight per unit length) and skin coloration were recorded at regular intervals. Under these constant conditions, a period of high condition was followed by a period of lower condition factor, and this in turn by another period of high condition. Likewise, parr-like appearance was followed by silver smolt, and this in turn by reverted parr. The mean time between two smoltings was about 10 months. In the absence of environmental cues, the cycle of growth and skin coloration runs at its own frequency, and this differs substantially fiom 1 year. The endogenous rhythmicity noted by earlier workers is truly circannual, as defined by workers in this field. Factors such as photoperiod and temperature synchronize or entrain the rhythm to the annual cycle; they serve as *zeitgebers* in the biological clock terminology. Eriksson and Lundqvist's (1982) findings are in line with Brown's (1945) experimental work on the growth of brown trout held under constant conditions (11.5°C and LD 12: 12); her fish showed seasonal changes in growth even though the conditions in the experimental tanks remained constant. Likewise, Wagner (1974a), in a study of juvenile steelhead trout raised at constant temperature in total darkness, found that some fish that reached a certain critical size developed smolt characteristics indicating an endogenous rhythm of smolting. Atlantic salmon, coho and sockeye salmon, and also steelhead trout in many investigations of photoperiod effects show evidence of endogenous rhythms of physiology that are synchronized by daylength (Section IV, B).

B. Modulation of the Rhythm by Photoperiod

Among salmonids, photoperiod is the most usual synchronizer of seasonally changing physiological processes of sexual maturation, spawning, growth, smolting, and migration (reviews by Poston, 1978; Wedemeyer *et al.*, 1980; Lundqvist, 1983). Many biologists have now studied the role of photoperiod (PP) in salmonids—especially the Atlantic salmon, steelhead trout, and sockeye and coho salmon. The most critical of the studies on *S. salar* were done by R. L. Saunders and his associates in Eastern Canada; at about the same time, H. H. Wagner carried out the first comprehensive analyses of smolting steelhead trout, and these studies have been followed by several noteworthy investigations of *Oncorhynchus*—particularly those of W. C. Clarke and associates on coho and sockeye salmon. Poston (1978) tabulates several photoperiod effects noted in other species of salmonids.

Saunders and Henderson (1970) compared Atlantic salmon held under (a) constant day lengths of 13 h light; (b) simulated natural PP (increasing daylength in springtime); and (c) the reciprocal of natural PP (decreasing daylength from early March and increasing daylength from late June). Growth rates, degree of silvering, thyroid activity (as shown by histology), excitability, and plasma osmotic and chloride levels were examined. The young salmon under constant and natural PP showed no particular departure from the normal conditions of smolting; however, reciprocal PP affected both growth and the excitability of the fish, indicating a disturbed endocrine physiology. In contrast to the natural sequence in smolting, the fish under reciprocal PP developed a high condition factor (weight in relation to length) in fresh water, while in the sea they grew more slowly, ate less, and had lower efficiencies of food conversion. The reciprocal-PP fish were also less sensitive to external stimuli than those under natural or constant PP. No differences between natural and reciprocal PPs were noted in the degree of silvering, thyroid histology, and plasma osmotic and chloride levels. Only growth processes and excitability were altered in these tests.

In a subsequent study with reciprocal photoperiods, the experimental fish were found to have lower metabolic rates (standard rate of O_2 consumption) compared with natural-PP fish when the measurements were made in total darkness in seawater (Withey and Saunders, 1973). In a third study, Saunders and Henderson (1978) compared gill ATPase activity, body lipids, and moisture in fish held under different photoperiods. Under both natural and reciprocal PPs, gill ATPases increased markedly during the late winter and spring while the levels of total body lipids declined and moisture increased. The point of interest, however, is that these changes occurred earlier and were more marked in fish held under reciprocal PP, indicating that the long nights of winter trigger preadaptations of Atlantic salmon for the smolt phase. Salinity resistance (measured at 40% salinity) increased in a comparable manner under all photoperiod regimes. The lipid and moisture data are in line with the findings of Komourdjian *et al.* (1976b), who attributed the changes to the stimulation of the pituitary secretion of growth hormone and ACTH by the longer day lengths. Saunders's investigations of Atlantic salmon in Eastern Canada have been followed by studies of photoperiod effects on *S. salar* of the Baltic region (Eriksson and Lundqvist, 1982; Clarke *et al.*, 1985); findings are similar with the two different populations of salmon.

Wagner's (1974a,b) investigation of the effects of daylength on steelhead trout differed from those of Saunders in that accelerated and decelerated as well as reversed or reciprocal PPs were tested. In the accelerated or decelerated regimes, the simulated seasonal change was advanced or retarded by 6 min per week. In this way, the anticipated season of smolting was advanced or retarded by 6 to 8 weeks. The coefficient of condition and migratory behavior were assessed in the various groups by periodic weight/length measurements and by releases of fish into a natural stream with subsequent trapping and enumeration downstream. Wagner's (1974a) experiments demonstrated the importance of both photoperiod and the rate of change of PP in regulating the time of smolting. Increasing day length, rather than day length or total exposure to light, was considered the prime stimulator of smolting. Reports by Zaugg and Wagner (1973) and Zaugg (1981) confirm photoperiod as the synchronizer of an endogenous rhythm of smolting and show that advanced PP will accelerate gill Na^+, K^+-ATPase changes characteristic of smolting in the steelhead by 1 month, although seawater survival does not seem to be closely associated with PP and smolting (Wagner, 1974b).

Clarke *et al.* (1978, 1981) studied osmoregulatory performance (24-h seawater challenge) and body lipid and liver glycogen levels in coho and sockeye salmon subject to constant or increasing or decreasing daylengths at different seasons and at different temperatures. Like the two species of *Salmo* considered above, these two oncorhynchids are physiologically responsive to photoperiod and temperature; advanced photoperiods accelerate growth and improve osmoregulatory ability, while temperature controls the rate of the response. Sensitivity of fry to photoperiod varies seasonally.

More recent studies have emphasized that the outcome of daylength manipulations may depend on the age/size of the fish *and* the period of exposure to short days before the initiation of the advanced PP regime. In studies of juvenile coho salmon, Brauer (1982) fixed daylength at 12.27 h for 1 month before starting to increase the day in late March; the delay (phase adjustment) resulted in improved growth, better food conversion, and seawater adaptability. Clarke and Shelbourn (1986) commenced their experiment at an earlier stage with free-swimming coho fry in February. Daylength was fixed at 9.75 h for 1 or for 2 months before starting to increase the light period at the natural rate of increase. Thus, three groups of coho were compared: fish under natural daylength, fish with 1 month delay and fish with 2 months delay. Delayed PPs produced fish of more uniform size (absence of bimodality seen under natural PP), with greater capacities to hypoosmoregulate following the 24-h seawater challenge, and improved growth in seawater. The sustained exposure to short days synchronized smolting in these coho.

Finally, photoperiod studies of chinook juveniles emphasize the species variations in responsiveness to day-length manipulation. Chinook salmon, unlike the four species discussed above, migrate to sea over an extended period and develop an early tolerance to seawater. There are two varieties of chinook salmon (Healey, 1983): the spring chinook, which spend a year or more in fresh water as juveniles and tend to produce yearling smolts (stream type), and the fall chinook, which produce underyearling migrants (ocean type). Ewing *et al.* (1979) studied a population of spring chinooks that showed peaks in gill ATPase activity in October of their first year, and the following May and October of their second year. Fish were reared under controlled photoperiods for 2 years; only the October peak of the first year was modified and found to be suppressed when photoperiods were advanced by 3 months. Clarke *et al.* (1981), in their study of fall chinooks, found no evidence of photoperiod regulation of growth or osmoregulation; these chinooks seemed insensitive to photoperiod manipulation; growth/size was considered the important factor in determining their entrance into the sea.

In summary, there is a circannual rhythm of physiological changes associated with the parr–smolt transformation. After juvenile salmon reach a certain critical size they become smolts with the capacity to hypoosmoregulate and grow in seawater; behavioral changes result in a downstream migration. Many of the smolting changes are reversible, and fish that cannot reach the sea revert to the parr condition. These cycles of physiology occur at intervals of about 1 year under constant light conditions; under natural light conditions they occur at predictable seasons according to the geographic location. If the seasonal changes in day length are advanced (simulating early spring conditions), the smolt transformation can be accelerated, while the prolongation of winter light conditions (short daylength) will delay the changes. The most important components of light in the manipulation of daylength (photoperiod) are *direction* and *rate of change*. Further, the period of darkness (short days of winter) that precedes the lengthening days is important in synchronizing the changes; prolongation of the season of short days accelerates the changes that occur in the subsequent period of lengthening photoperiod. Of the several changes associated with smolting, growth, hypoosmoregulatory capacity, and migratory tendency seem most sensitive to photoperiod manipulation, while body silvering seems independent. However, generalizations must be made cautiously, since the response varies not only with the photoperiod regime but also with the age/size of the fish, the season when light control is initiated, and with the temperature (Section IV,C).

C. Temperature Effects

The smolt migration was correlated with rising springtime temperatures by some of the very early fisheries scientists. Foerster (1937) correlated the commencement of sockeye smolt migration from Cultus Lake, British Columbia, with the vernal rise in lake temperature; the winter minimal water temperatures approximated 2.5°C and the threshold migration temperature seemed to be about 4.4°C. Cessation of migration appeared to be related to the warming of the lake surface waters, which formed a "temperature blanket." White (1939) reported that peaks of Atlantic salmon migration from Forest Glen Brook, Cape Breton Island, occurred at low light intensities (night) when the water temperature rose sharply above 10–12°C. In this section of the review, experimental work on the effects of temperature on the physiological changes of smolting is considered—particularly in relation to the photoperiod regulation of the parr–smolt transformation.

Temperature effects were studied in many of the photoperiod investigations reviewed in the previous section. Some experiments were performed at a single constant temperature (Saunders and Henderson, 1970, 1978; Komourdjian *et al.*, 1976b; Ewing *et al.*, 1980a; Brauer, 1982; Clarke *et al.*, 1985); in others, the ambient temperature was used (Lundqvist, 1980); in still others, comparisons were made at a lower and at a higher temperature (Zaugg *et al.*, 1972; Ewing *et al.*, 1979; Pereira and Adelman, 1985), and some experiments have been designed to test temperature specifically with three or more temperatures maintained at a constant level (Knutsson and Grav, 1976; Clarke *et al.*, 1978, 1981; Clarke and Shelbourn, 1980, 1985, 1986). Wagner's (1974a) studies of steelhead differed in the use of changing temperatures rather than constant temperatures; he compared fish subjected to a simulated normal stream temperature with fish subjected to a seasonally advanced temperature, an accelerated temperature, and a decelerated temperature regime.

In general, it is concluded that temperature controls the rate of the physiological response to photoperiod. Smolting occurs sooner at higher temperatures, and further, changing temperatures seem to be more stimulating than a constant temperature (Wagner, 1974a; Jonsson and Rudd-Hansen, 1985). However, these generalizations cannot be applied directly to the manipulation of the time of smolting without due regard to (a) species variations, (b) the inhibitory effects of high as well as low temperatures, and (c) the fact that high temperatures accelerate the reversal of the smolt to parr condition as well as the parr-smolt transformation (review by Wedemeyer *et al.*, 1980).

Although Atlantic salmon (*S. salar*) smolt at temperatures as high as 15°C (Saunders and Henderson, 1970; Komourdjian *et al.*, 1976b; Johnston and Saunders, 1981), the normal smolting increase in gill ATPase is suppressed or declines in steelhead trout (*S. gairdneri*) at temperatures in excess of 13°C (Zaugg *et al.*, 1972; Zaugg and Wagner, 1973; Zaugg, 1981). It should be noted, however, that the smolt runs of Atlantic salmon peak at about 10–12°C and are normally over before water temperatures of 15°C are reached (White, 1939). Further, it seems relevant that Knutsson and Grav (1976), in studies of *S. salar*, reported optimal growth under long photoperiods at 15°C but more pronounced effects of photoperiod on seawater adaptation at 11°C.

The experiments of Zaugg and associates (see also Adams *et al.*, 1973, 1975) suggest an optimum temperature for the parr–smolt transformation. Higher temperatures accelerate smolting up to about 10–12°C, but above these temperatures at least some of the smolting changes (gill enzyme activity and migratory behavior) are inhibited or occur only briefly (Zaugg and McLain, 1976) (Fig. 8).

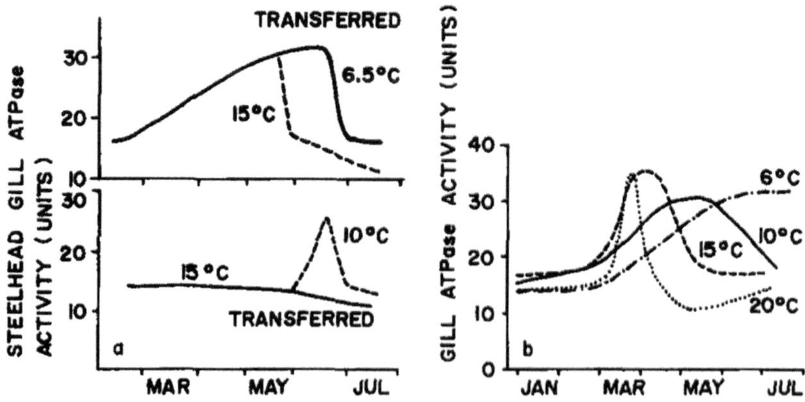

Fig. 8. Effects of temperature on gill Na^+, K^+–ATPase activity. (a) Yearling steelhead trout held at 6.5°C and transferred to 15°C (upper panel); held at 15°C and transferred to 10°C (lower panel). [From Wedemeyer *et al.* (1980) after Zaugg *et al.* (1972).] (b) Summary of changes during development of coho salmon. [From Wedemeyer *et al.* (1980) after Zaugg and McLain (1976). Courtesy *Marine Fisheries Review.*]

Finally, the significance of temperature in the smolt-to-parr reversal is emphasized. Zaugg and McLain (1976) noted that the time span of elevated gill ATPase activity was much longer in steelhead maintained at 6–10°C than at temperatures in excess of 13°C; only a very brief period of enzyme activity occurred at 20°C (Fig. 8). Higher temperatures accelerate the smolting changes, but unless the young salmon can enter saltwater within a very short time span, the physiology will revert to that of the parr and the young fish will be trapped in fresh water for another year (review by Wedemeyer *et al.*, 1980).

D. Other Modulating Environmental Factors

Although photoperiod and temperature are the most reliable cues for synchronizing the smolting season, three other environmental variables may be important in timing the transformation: the lunar cycle, the runoff associated with flooding rivers in springtime, and the nature of the dissolved solids, which may change naturally with stream flow or because of industrial activities.

1. *Lunar Cycle*

In juvenile coho salmon, chinook salmon, and steelhead trout, Grau (1982) and Grau *et al.* (1981, 1982) observed a strong correlation between the new moon and the peak surge in plasma thyroxine. Figure 5 shows a consistent peaking of plasma T4 at the time of the new moon when values for juvenile coho are plotted on a lunar calendar. In a subsequent study of coho, raised in Hawaii, where photoperiod has no significant effect on the timing or magnitude of changes in thyroid hormone, Grau *et al.* (1985a) found three peaks in plasma thyroid hormones; two of these occurred at the time of the new moon, while the third was associated with the full moon—suggesting that the periodicity is semilunar.

Juvenile chinook salmon (*O. tshawytscha*) were also examined in some detail (Grau, 1982; Grau *et al.*, 1982). Chinook juveniles do not show a single mass migration but move seaward in groups at several different times throughout the spring and summer. The hatchery fish examined showed four peaks in plasma T4, and these peaks occurred in the samples taken closest to the new moon, again supporting the theory that the lunar-phased peak in thyroid hormone is concerned with the initiation of migration. To test the practical implications of these findings, the return of coho to the Trinity River Hatchery in California was studied in relation to the time of juvenile release; recoveries from groups released on the new moon closest to the expected peak of plasma T4 were approximately twofold greater than the previous releases that were not lunar-based (Nishioka *et al.*, 1983). These arguments find some support in field studies of downstream migrating juvenile coho salmon—particularly Mason's (1975) study where the downstream movements of coho fry peaked with the new moon while seaward migration of smolts peaked with the full moon. Movements of fry and smolt showed no obvious relation to either water temperature or stream flow; lunar rhythmicity was the important factor in timing movements of coho in Lyman Creek, Vancouver Island. Mason (1975) suggests that in streams subject to greater variations in flow and temperature, these factors, rather than lunar periodicity, might trigger migration.

Yamauchi *et al.* (1984, 1985) examined the concept of lunar phasing of T4 surges and migration in masu salmon. Their fish were best able to osmoregulate in seawater at the peak of smolting (as judged by external characters). T4 was high during smolting and peaked at the time of the new moon in April, when the onset of migration occurred. The largest migrations occurred immediately after a rainfall around the time of the new moon, thus showing the importance of rainfall as well as the lunar cycle.

Although studies of juvenile oncorhynchids have usually supported a concept of lunar periodicity in thyroid function and downstream migration, data are less consistent in studies of *Salmo*. Boeuf and Prunet (1985) found suggestive evidence that T4 peaked at the new moon, but Youngson *et al.* (1983, 1985, 1986) emphasized the effects of stream flow on downstream migration; they considered stream flow rather than lunar phasing important in Atlantic salmon migrations and noted that elevated stream flow occurs around the time of the new moon and thus facilitates migration. In a study of *S. salar* smolts tagged internally with ultrasonic telemetry transmitters, movement through the estuary of the Penobscot River, Maine, was dependent on water currents and seemingly on no other environmental variable, although the rising springtime water temperature (above 5°C) initiated the migration (Fried *et al.*, 1978). The first reports of elevated plasma T4 in steelhead trout at the time of the new moon (Grau *et al.*, 1981, 1982) were not confirmed in a later investigation (Lin *et al.*, 1985a).

In summary, present evidence of a lunar or semilunar synchronization of some physiological changes during smolting (especailly thyroid secretion) is reasonably persuasive for several species of *Oncorhynchus* but questionable for *Salmo*. However, more research is required to confirm this suggested distinction between the two salmonid genera, since exceptions have also been found in coho salmon. The importance of further research is apparent. If the lunar cycle provides a *zeitgeber* in addition to photoperiod, then hatchery operations are likely to be improved by relating juvenile releases to the lunar calendar.

2. *Dissolved Solids and pH*

Composition of dissolved solids may change during seasons of heavy runoff with seepage from the land. It is conceivable that particular substances present during the spring runoff may affect the physiology of juvenile salmon. Although naturally occurring dissolved solids are considered significant in regulating reproduction of some tropical and subtropical fish (Lam, 1983), salmon smolting is not known to be affected by them. There is, however, ample evidence that industrial contaminants and pesticides affect the physiology of young salmon. Effects of such substances on thyroid activity were noted in Section II,E,l. There is also evidence that salinity tolerance and migratory tendencies are altered by contaminants. Lorz and McPherson (1976) found that chronic copper exposure during smolting partially or completely inactivated the gill ATPase system of coho salmon and that the normal migratory behavior of the fish was suppressed. Damage from copper was not apparent until the fish were moved into saltwater, when there was a high mortality. Davis and Shand (1978) obtained similar results with young sockeye salmon. Several other heavy metals and a number of organic pollutants have been tested and the damaging effects evaluated (Lorz and McPherson, 1976; review by Wedemeyer *et al.*, 1980). The damage varies, but the fact remains that smolting physiology is not normal; this stage of salmon development as well as the earlier stages is adversely affected by heavy metals and pesticides (Folmar *et al.*, 1982b; Nichols *et al.*, 1984).

Another topic of current concern is the acidification of fresh waters by acid precipitation ("acid rain") and its effect on fish populations. Saunders *et al.* (1983) reared Atlantic salmon in waters at about pH 6.5 and about pH 4.5. Smolting, as judged by salinity tolerance and gill ATPase, was impaired at the lower pH. These workers conclude that smolting does not proceed normally in *S. salar* subjected to low pH. This is not surprising, but the knowledge is important in culturing salmon and in predicting the effects of industrial activity on the future of the wild stocks.

V. SOME PRACTICAL PROBLEMS IN SMOLT PRODUCTION

The history of salmonid culture can be traced to a European monk, Dom Pinchon, in the fifteenth century. His simple methods of hatching eggs and culturing fish were the basis of more sophisticated techniques developed in the eighteenth century, leading in France to the first attempts (1842) to exploit these ideas practically for the enhancement of salmonid stocks (notes from Day, 1887). By the latter half of the nineteenth century, salmonid hatcheries were widely established in Asia, Europe, and North America (Bowen, 1970; Thorpe, 1980). The objective of these early efforts was to release large numbers of juvenile salmon and trout into natural environments; the assumption was that the larger the number of fish released, the greater would be the return to the fishermen. Hatcheries usually produced fry and fingerlings, but in some cases (sockeye salmon, for example) the fish were cultured to the smolt stage (Foerster, 1968). Realization that returns of mature fish may not be directly related to the numbers of eggs hatched and juveniles released has resulted in many changes in objectives and orientation. In the past quarter century, the emphasis has been to enhance salmon stocks through stream improvement, the production and release of juveniles under more natural conditions, and the protection of growing fish in ponds and sea pens (Thorpe, 1980).

Whether salmon and trout are cultured to improve commercial and sport fishing or to grow fish to marketable size in sea pens, an understanding of the physiology of smolting is basic to success. It is obvious that, in both cases, *early smolting* is economically advantageous in minimizing culture costs and losses through predation and disease in fresh water. To obtain maximum advantage from the more favorable growing conditions of the marine environment, the fish should migrate or be transferred to the sea at the earliest age consistent with normal growth and ability to survive in seawater.

Much of the recent research on the physiology of smolting has been directed to practical problems associated with salmonid ranching (Thorpe, 1980) and the culture of salmon in sea pens. There have been many workshops and discussions and several important symposia devoted to these topics. Two recent symposia have been particularly valuable: one held at La Jolla, California, in 1981 focused on Pacific salmon, especially the coho (*Aquaculture*, Vol. 28, pp. 1–270, 1982); the other at the University of Stirling in 1984 was directed primarily to the Atlantic salmon (*Aquaculture*, Vol. 45, pp. 1–404, 1985). The organizers of these symposia identified several problems as central to the production of quality smolts for release in streams or culture in sea pens.

In very broad terms, there are three major thrusts toward production of smolts of good quality: (a) achievement of rapid growth to smolting size; (b) successful osmotic and ionic regulation in the marine environment; and (c) a high growth rate in the sea with a successful return to fresh water. The recent research has been concentrated in three main areas: (1) the environmental regulation of growth, early smolting, and seawater adaptability; (2) the endocrinological basis of smolting with the objective of manipulating hormones in the regulation of smolting physiology; and (3) improvement of the stock through genetics. Space permits only brief reference to the recent literature with an emphasis on the importance of further studies. The hazard of general comments is recognized. There are many species of salmonid fishes, with smolting biology as variable as their morphology, physiology and behavior. Further, the smolt transformation is not a single event but involves many changes that occur at different rates over an extended time period. The generalizations that follow are made with some reservations.

A. Minimizing the Juvenile Freshwater Phase

For practical reasons, the abbreviation of the parr stage remains a major objective of salmon farmers. In nature the environment is dominant in regulating the age of smolting. For example, Atlantic salmon smolts in the southern part of their range may migrate seaward after one summer in the rivers (1+ year smolts), but at Ungava Bay, the most northern extent of their distribution, the average age of smolt migration is 5+ years and some juveniles are 8+ years at migration (Power, 1969). Manipulation of temperature and photoperiod is the most economical approach to the production of 1+ year smolts (Wedemeyer *et al.*, 1980) or even 0+ year smolts that have been produced under hatchery conditions (Brannon *et al.*, 1982; Ísaksson, 1985; Zaugg *et al.*, 1986.

Current research has shown that manipulation of temperature and day length must be based on a careful study of the species involved. There is an optimum temperature, which may vary with the species. Low temperatures delay growth and smolting; high temperatures increase growth rate (Donaldson and Brannon, 1975; Brannon *et al.*, 1982; Piggins and Mills, 1985) but may suppress some important changes of smolting (Fig. 8) and also accelerate the reverse change of smolt to parr (Wedemeyer *et al.*, 1980). Photoperiod control must also be based on studies of the particular species. Day length per se is less important than the direction and rate of change in day length; short days preceding long days are involved, and exposure to continuous light or days of constant length may result in satisfactory growth but a failure to smolt (Section IV, B). Saunders *et al.* (1985a) report that juvenile Atlantic salmon under continuous light grow faster than salmon under simulated natural photoperiods but fail to smolt and do not grow like natural smolts when transferred to sea cages.

Several rearing conditions, in addition to temperature and day length, are crucial to growth and early smolting. Adequate diet, in terms of energy-providing and growth-promoting foods, is the most obvious of these (Cowey, 1982). In addition, experiments with supplements of inorganic salts have shown their potential for elevating gill ATPase and increasing survival in seawater; tests have been carried out with Atlantic salmon parr (Basulto, 1976) and with coho and chinook salmon (Zaugg, 1982; Nishioka *et al.*, 1985a). Dietary supplements of thyroid hormones and anabolic steroids have been carefully tested for growth-promoting and smolt-inducing effects. Thyroid hormones, especially T3, increase growth in Atlantic salmon parr but do not induce early smolting (Saunders *et al.*, 1985b). Improved seawater tolerance as well as growth enhancement have been shown in several tests with T3 supplements to the diets of coho salmon, chinook salmon, and steelhead trout (Higgs *et al.*, 1979,1982; Fagerlund *et al.*, 1980; McBride *et al.*, 1982). The anabolic steroids (androgens) also have demonstrable growth-promoting effects when added to the diet of young Pacific salmon and steelhead trout, but there is no evidence of improved seawater tolerance (Fagerlund *et al.*, 1980; Higgs *et al.*, 1982; McBride *et al.*, 1982). Related to these dietary tests are several demonstrations of the growth-promoting effects of injected growth hormone, which also improves saltwater tolerance (Komourdjian *et al.*, 1976a,b; Higgs *et al.*, 1977). However, from an economic angle, injection procedures have considerably less potential for use in mass production of young salmon.

Rearing density is also important in smolt production. High density in rearing ponds depresses growth rate through competition for food and increases the hazards of infection, with resulting delays in smolting and reduced survival of juveniles (Refstie, 1977; Schreck *et al.*, 1985). Crowding also causes stress responses in the adrenocortical system, with elevated plasma corticosteroids and detectable histological changes in the interrenal gland (Noakes and Leatherland, 1977; Specker and Schreck, 1980; Schreck, 1982).

Finally, the importance of genetic factors is now widely recognized in attempts to improve growth and achieve early smolting (Wilkins and Gosling, 1983). Several different studies of Atlantic salmon have demonstrated growth variation and differences in the time of smolting of juveniles cultured from gametes obtained in different geographic regions (Refstie *et al.*, 1977; Refstie and Steine, 1978; Thorpe and Morgan, 1978; Riddell and Leggett, 1981; Riddell *et al.*, 1981). The potential of breeding programs in the production of early age smolts is recognized for both *Salmo* (Refstie and Steine, 1978; Bailey *et al.*, 1980) and *Oncorhynchus* (Saxton *et al.*, 1984). The hazards of reduced genetic variability as a result of selection programs has also been noted (Allendorf and Utter, 1979).

B. Successful Transfer to the Marine Habitat

Smolting does not commit a salmon or a trout to life in saltwater. In sea-going populations, reversion to the parr condition in fresh water is a viable option. The mechanisms triggering or regulating smolt–parr reversion have not been well investigated; studies of seawater adaptability should be investigated throughout the year in relation to temperature and photoperiod. However, this option of smolt–parr reversion in salmonids means that the time of smolt release or transfer from hatchery to sea pens is critical to successful salmon management. It is now felt that poor adult returns from cultured smolts are sometimes related to inappropriate times of release or transfer; losses of as many as 70% from premature transfer of smolts have been noted (Folmar and Dickhoff, 1981; see also Wedemeyer *et al.*, 1980; Bilton *et al.*, 1982). For both Atlantic and Pacific salmon, the physiological condition of the smolt-like freshwater fish is recognized as important to growth and survival in the marine habitat. The unsatisfactory saltwater growth of Atlantic salmon "smolts" cultured under continuous or unchanging light conditions has been noted (Saunders *et al.*, 1985a). Another example for Atlantic salmon has been reported by Gudjónsson (1972), who reared *S. salar* in constant light and obtained good growth and healthy juveniles using heated (geothermal) waters in Iceland; however, there were no returns of sea-run fish—juveniles of smolt size did not migrate when released but remained in the streams. In contrast, salmon raised in similar heated waters under outdoor conditions of changing day length gave good adult returns (see also Wedemeyer *et al.*, 1980).

Marine survival of young coho salmon has been intensively studied in relation to time of transfer from fresh water to seawater (Folmar and Dickhoff, 1980, 1981). This species seems to be extremely sensitive to time

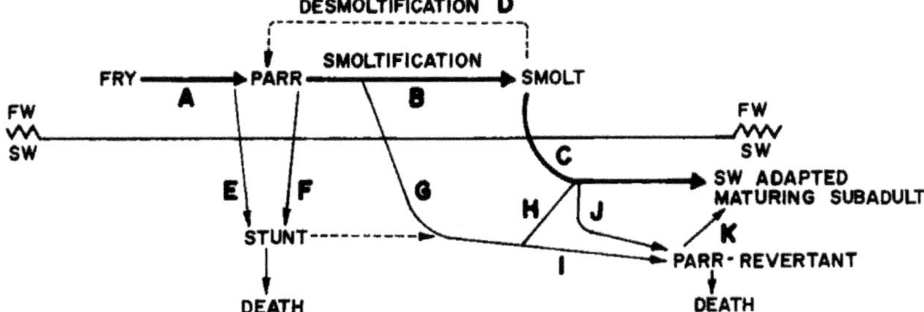

Fig. 9. Diagram showing possible sources of "stunts" and "parr-revertants" during the smolting and postsmolting stages in the life of the coho salmon. (A, B, C) Normal sequence in anadromous salmon. (D) Smolts that do not enter seawater revert to freshwater parr condition. (E, F) Premature transfer to seawater during the parr stage, resulting in "stunting." (G, I) Transfer to seawater before the completion of smolting or (H, J) failure of smolts to make adequate growth may result in death of undersize fish. Further details in original papers. [From Folmar *et al.* (1982a).]

of transfer, and fish moved to sea pens either too early or too late fail to realize their full potential and may become "stunted" or revert to the parr condition (Fig. 9). Morphological and physiological characteristics of "stunts" and "parr-revertants" have been tabulated and the literature summarized in some detail by Folmar *et al.* (1982a). Many of the studies have focused on the endocrine system, which appears to be hypofunctional (Clarke and Nagahama, 1977; Bern, 1978; Fryer and Bern, 1979); "stunts" are said to be "hypoendocrine" (Nishioka *et al.*, 1982; Grau *et al.*, 1985a). However, recent studies of plasma GH levels in normal and stunted yearling coho salmon show that the plasma levels of GH are consistently higher in the stunts, suggesting that deficiencies in the receptor and/or the mediating system may be the cause of stunting (Bolton *et al.*, 1987b). Several organs other than the endocrine organs are also different in the stunts from the normal fish; liver, muscle, and the organs concerned with osmotic and ion regulation show describable differences (see Folmar *et al.*, 1982a).

Various smolting characteristics have been considered in predicting the best time for smolt release or transfer from hatchery to sea pens. A minimum size is essential (Section II, B), and, if the fish are in the smolt stage, the larger the smolts, the better seem to be the returns (Mahnken *et al.*, 1982). Superficial features such as silvering and low condition factor are not reliable predictive indices, since these features may persist for a time after the readiness to migrate has passed and while the smolt is reverting to a parr (Wedemeyer *et al.*, 1980).

Physiological testing of osmoregulatory competence (seawater challenge test) can be used to follow smolting changes (Clarke, 1982; Saunders *et al.*, 1982; Hogstrand and Haux, 1985), but this, or the assessment of chloride-cell density and ion transport, may not always be practical in large-scale production programs. Also among the more technically demanding prediction indices are the well-marked changes in the endocrine system (especially the thyroid and interrenal glands) and in the gill ATPases. A surge in thyroid hormone occurs reliably in smolting salmonids (Grau *et al.*, 1985a). Tests indicate that the best time for coho release follows the T4 surge when the cortisol peaks. Of many smolting characteristics examined, Folmar and Dickhoff (1981) reported that only one was statistically related to the percent survival after 6 months in seawater; this statistic was the percent of the area beneath the T4 curve prior to seawater transfer. Comparisons of plasma T4 in smolts leaving fishways voluntarily with levels in fish that remain behind show that the former are consistently higher (Grau *et al.*, 1985a). Thus, the thyroxine surge seems to be a reliable index for coho, and this may be used to advantage with a lunar calendar. Loretz *et al.* (1982) suggested that the appropriate time for seawater entry of hatchery-reared coho salmon may be several weeks after the new moon–related thyroxine peak.

R. D. Ewing and associates have assessed gill ATPases in relation to the most favorable time for release of young chinooks (Ewing *et al.*, 1980a,b; Ewing and Birks, 1983). For this salmonid, gill ATPase activity seems to be a reliable index of smolting but one that is not always a necessary prerequisite to migration. Ewing *et al.* (1984) also found that the T4 surge is a poor index for predicting release time of chinooks.

The recognition of a *migratory readiness* as distinct from *physiological development* has been emphasized (Solomon, 1978). Physiological changes occur over an extended time span and are environmentally regulated (particularly by photoperiod and temperature); when the fish are in a proper state of migratory readiness, a proximal stimulus (lunar phase, stream flooding) initiates migration. Thus, behavioral and environmental factors as well as physiological changes must often be considered. Marked species variations are recognized, and, at this stage in fisheries science, there appears to be no simple and precise index of smolting climax and the most favorable time for release or transfer. The search for more reliable tests in predicting release times is likely to remain an important objective of salmonid biologists for some time.

ACKNOWLEDGMENTS

The manuscript was read critically by H. A. Bern, W. Craig Clarke, W. W. Dickhoff, Hans Lundqvist, and R. L. Saunders. I am grateful for their many helpful suggestions.

REFERENCES

Adams, B. L., Zaugg, W. S., and McLain, L. R. (1973). Temperature effect on parr–smolt transformation in steelhead trout (*Salmo gairdneri*) as measured by gill sodium–potassium stimulated adenosine triphosphatase. *Comp. Biochem. Physiol.* **44,** 1333–1339.

Adams, B. L., Zaugg, W. S., and McLain, L. R. (1975). Inhibition of saltwater survival and Na^+-K^+-ATPase elevation in steelhead trout (*Salmo gairdneri*) by moderate water temperatures. *Trans. Am. Fish. Soc.* **104,** 766–769.

Aida, K., Nishioka, R. S., and Bern, H. A. (1980). Changes in the corpuscles of Stannius of coho salmon (*Oncorhynchus kisutch*) during smoltification and seawater adaptation. *Gen. Comp. Endocrinol.* **41,** 296–304.

Aida, K., Kato, T., and Awaji, M. (1984). Effects of castration on the smoltification of precocious male masu salmon (*Oncorhynchus masou*). *Bull. Jpn. Soc. Sci. Fish.* **50,** 565–571.

Allendorf, F. W., and Utter, F. M. (1979). Population genetics. *In* "Fish Physiology" (W. S. Hoar, D. J. Randall, and J. R. Brett, eds.), Vol. 8, pp. 407–454. Academic Press, New York.

Andrews, J. (1963). Healthy salmon at U.B.C. show expert he's wrong. *Vancouver Sun*, Vancouver, British Columbia, Jan. 29.

Baggerman, B. (1960a). Salinity preference, thyroid activity and the seaward migration of four species of Pacific salmon (*Oncorhynchus*). *J. Fish. Res. Board Can.* **17,** 295–322.

Baggerman, B. (1960b). Factors in the diadromous migrations of fish. *Symp. Zool. Soc. London* **1,** 33–60.

Baggerman, B. (1962). Some endocrine aspects of fish migration. *Gen. Comp. Endocrinol., Suppl.* **1,** 188–205.

Baggerman, B. (1963). The effect of TSH and antithyroid substances on salinity preference and thyroid activity in juvenile Pacific salmon. *Can. J. Zool.* **41,** 307–319.

Baggerman, B. (1980). Photoperiodic and endogenous control of the annual reproductive cycle in teleost fishes. *In* "Environmental Physiology of Fishes" (M. A. Ali, ed.), pp. 533–567. Plenum, New York.

Baglinière, J. L., and Maisse, G. (1985). Precocious maturation and smoltification in wild Atlantic salmon in the Armorican Massif, France. *Aquaculture* **45,** 249–263.

Bailey, J. K., Saunders, R. L., and Buzeta, M. I. (1980). Influence of parental smolt age and sea age on growth and smoking of hatchery-reared Atlantic salmon (*Salmo salar*). *Can. J. Fish. Aquat. Sci.* **37,** 1379–1386.

Baraduc, M. M., and Fontaine, M. (1955). Étude comparée du métabolismé respiratoire du jeune Saumon sédentaire (parr) et migrateur (smolt). *C. R. Seances Soc. Biol. Ses. Fil.* **149,** 1327–1329.

Barrington, E. J. W., Barron, N., and Figgins, D. J. (1961). The influence of thyroid powder and thyroxine upon the growth of rainbow trout (*Salmo gairdneri*). *Gen. Comp. Endocrinol.* **1,** 170–178.

Barton, B. A., Schreck, C. B., Ewing, R. D., Hemmingsen, A. R., and Patino, R. (1985). Changes in plasma cortisol during stress and smoltification of coho salmon, *Oncorhynchus kisutch. Gen. Comp. Endocrinol.* **59,** 468–471.

Basulto, S. (1976). Induced saltwater tolerance in connection with inorganic salts in the feeding of Atlantic salmon (*Salmo salar* L.). *Aquaculture* **8,** 45–55.

Bath, R. N., and Eddy, F. B. (1979). Salt and water balance in rainbow trout (*Salmo gairdneri*) rapidly transferred from fresh water to sea water. *J. Exp. Biol.* **83,** 193–202.

Berg, R. E. (1979). External morphology of the pink salmon, *Oncorhynchus gorbuscha,* introduced into Lake Superior. *J. Fish. Res. Board Can.* **36,** 1283–1287.

Bern, H. A. (1978). Endocrinological studies on normal and abnormal salmon smoltification. *In* "Comparative Endocrinology" (P. J. Gaillard and H. H. Boer, eds.), pp. 97–100. Elsevier/ North-Holland, Amsterdam.

Bern, H. A., and Mahnken, C. V. W., eds. (1982). Salmonid smoltification. Proceedings of a symposium sponsored by the Pacific sea grant advisory program and the California Sea Grant College Program. *Aquaculture* **28,** 1–270.

Bilton, H. T., Alderdice, D. F., and Schnute, J. T. (1982). Influence of time and size at release of juvenile coho salmon (*Oncorhynchus kisutch*) on returns at maturity. *Can. J. Fish. Aquat. Sci.* **39,** 426–447.

Birks, E. K., Ewing, R. D., and Hemmingsen, A. R. (1985). Migration tendency in juvenile steelhead trout, *Salmo gairdneri* Richardson, injected with thyroxine and thiourea. *J. Fish Biol.* **26,** 291–300.

Birt, T. P., and Green, J. M. (1986). Parr–smolt transformation in sexually mature male anadromous and nonanadromous Atlantic salmon, *Salmo salar. Can. J. Fish. Aquat. Sci.* **43,** 680–686.

Black, J. J., and Simpson, C. L. (1974). Thyroid enlargement in Lake Erie coho salmon. *J. Natl. Cancer Inst.* (*U.S.*) **53,** 725–730.

Black, V. S. (1951). Changes in body chloride, density, and water content of chum (*Oncorhynchus keta)* and coho (O. *kisutch*) salmon fry when transferred from fresh water to sea water. *J. Fish. Res. Board Can.* **8,** 164–177.

Blake, R. L., Roberts, F. L., and Saunders, R. L. (1984). Parr–smolt transformation of Atlantic salmon (*Salmo salar*): Activities of two respiratory enzymes and concentrations of mitochondria in the liver. *Can. J. Fish. Aquat. Sci.* **41,** 199–203.

Boeuf, G., and Harache, Y. (1982). Criteria for adaptation of salmonids to high salinity seawater in France. *Aquaculture* **28,** 163–176.

Boeuf, G., and Harache, Y. (1984). Adaptation osmotique à l'eau de mer de differentes espèces (*Salmo trutta, Salmo gairdneri, Salvelinus fontinalis*) et hydride (*Salmo trutta* ♀ × *Salvelinus fontinalis* ♂) de salmonides. *Aquaculture* **40,** 343–358.

Boeuf, G., and Prunet, P, (1985). Measurements of gill (Na^+-K^+)-ATPase activity and plasma thyroid hormones during smoltification in Atlantic salmon (*Salmo salar* L.). *Aquaculture* **45,** 111–119.

Boeuf, G., Lasserre, P., and Harache, Y. (1978). Osmotic adaptation of *Oncorhynchus kisutch* Walbaum. II. Plasma osmotic and ionic variations, and gill (Na^+-K^+)-AT-Pase activity of yearling coho salmon transferred to sea water. *Aquaculture* **15,** 35–52.

Boeuf, G., Le Roux, A., Gaignon, J. L., and Harache, Y. (1985). Gill (Na^+-K^+)-ATPase activity and smolting in Atlantic salmon (*Salmo salar* L.) in France. *Aquaculture* 45, 73–81.

Bolton, J. P., Collie, N. L., Kawauchi, H., and Hirano, T. (1987a). Osmoregulatory actions of growth hormone in rainbow trout (*Salmo gairdneri*). *J. Endocrinol.* **112,** 63–68.

Bolton, J. P., Young, G., Nishioka, R. S., Hirano, T., and Bern, H. A. (1987b). Plasma growth hormone levels in normal and stunted yearling coho salmon (*Oncorhynchus kisutch*). *J. Exp. Zool.* **242,** 379–382.

Bowen, J. T. (1970). A history of fish culture as related to the development of fishery programs. *In* "A Century of Fisheries in North America" (N. G. Benson, ed.), Spec. Publ. No. 7. Am. Fish. Soc., Washington, D.C.

Bradley, T. M., and Rourke, A. W. (1984). An electrophoretic analysis of plasma proteins from juvenile *Oncorhynchus tshawytscha* (Walbaum). *J. Fish Biol.* **24,** 703–709.

Brannon, E., Feldmann, C., and Donaldson, L. (1982). University of Washington zero-age coho smolt production. *Aquaculture* **28,** 195–200.

Brauer, E. P. (1982). The photoperiod control of coho salmon smoltification. *Aquaculture* **28,** 105–111.

Brown, M. E. (1945). The growth of brown trout (*Salmo trutta* L.). II. The growth of two-year-old trout at a constant temperature of 11.5°C. *J. Exp. Biol.* **22,** 130–144.

Burton, M. P., and Idler, D. R. (1984). Can Newfoundland landlocked salmon, *Salmo solar* L., adapt to sea water? *J. Fish Biol.* **24,** 59–64.

Chan, H. H., and Eales, J. G. (1976). Influence of bovine TSH on plasma thyroxine levels and thyroid function in brook trout, *Salvelinus fontinalis* (Mitchill). *Gen. Comp. Endocrinol.* **28,** 461–472.

Chernitsky, A. G. (1980). Functional state of chloride cells of Baltic salmon (*Salmo solar* L.) at different stages of its life cycle. *Comp. Biochem. Physiol. A* **67A,** 519–522.

Chernitsky, A. G. (1986). Quantitative evaluation of the degree of parr-smolt transformation in wild smolts and hatchery juveniles of Atlantic salmon (*Salmo salar* L.) by SDH activity of chloride cells. *Aquaculture* **59,** 287–297.

Chester-Jones, I., and Henderson, I. W. (1980). "General, Comparative and Clinical Endocrinology," Vol. 3, pp. 395–523. Academic Press, London.

Chester-Jones, I., Chan, D. K. O., Henderson, I. W., and Ball, J. N. (1969). The adrenocortical steroids, adrenocorticotropin and the Corpuscles of Stannius. *In* "Fish Physiology" (W. S. Hoar and D. J. Randall, eds.), Vol. 2, pp. 321–376. Academic Press, New York.

Chua, D., and Eales, J. G. (1971). Thyroid function and dermal purines in brook trout, *Salvelinus fontinalis* (Mitchill). *Can. J. Zool.* **49,** 1557–1561.

Clarke, W. C. (1982). Evaluation of the seawater challenge test as an index of marine survival. *Aquacultures* **28,** 177–183.

Clarke, W. C., and Bern, H. A. (1980). Comparative endocrinology of prolactin. *In* "Hormonal Proteins and Peptides" (C. H. Li, ed.), Vol. 8, pp. 105–197. Academic Press, New York.

Clarke, W. C., and Blackburn, J. (1978). Seawater challenge test performed on hatchery stocks of chinook and coho salmon in 1977. *Tech. Rep.—Fish. Mar. Serv. (Can.)* **761,** 1–19.

Clarke, W. C., and Nagahama, Y. (1977). Effect of premature transfer to sea water on growth and morphology of the pituitary, thyroid, pancreas, and interrenal in juvenile coho salmon (*Oncorhynchus kisutch*). *Can. J. Zool.* **55,** 1620–1630.

Clarke, W. C., and Shelboum, J. E. (1980). Growth and smolting of under-yearling coho salmon in relation to photoperiod and temperature. *Proc. N. Pac. Aquacult. Symp., 1980,* pp. 209–216.

Clarke, W. C., and Shelboum, J. E. (1985). Growth and development of seawater adaptability by juvenile fall chinook salmon (*Oncorhynchus tshawytscha)* in relation to temperature. *Aquaculture* **45,** 21–31.

Clarke, W. C., and Shelboum, J. E. (1986). Delayed photoperiod produces more uniform growth and greater seawater adaptability in underyearling coho salmon (*Oncorhynchus kisutch*). *Aquaculture* **56,** 287–299.

Clarke, W. C., Farmer, S. W., and Hartwell, K. M. (1977). Effect of teleost pituitary growth hormone on growth of *Tilapia mossambica* and on growth and seawater adaptation of sockeye salmon *Oncorhynchus nerka*. *Gen. Comp. Endocrinol.* **33,** 174–178.

Clarke, W. C., Shelboum, J. E., and Brett, J. R. (1978). Growth and adaptation to sea water in 'underyearling' sockeye (*Oncorhynchus nerka*) and coho (O. *kisutch*) salmon subjected to regimes of constant or changing temperature and day length. *Can. J. Zool.* **56,** 2413–2421.

Clarke, W. C., Shelboum, J. E., and Brett, J. R. (1981). Effect of artificial photoperiod cycles, temperature, and salinity on growth and smolting in underyearling coho (*Oncorhynchus kisutch*), chinook (O. *tshawytscha)* and sockeye (O. *nerka*) salmon. *Aquaculture* **22,** 105–116.

Clarke, W. C., Lundqvist, H., and Eriksson, L.-O. (1985). Accelerated photoperiod advances seasonal cycle of seawater adaptation in juvenile Baltic salmon, *Salmo salar* L. *J. Fish Biol.* **26,** 29–35.

Collie, N. L. (1985). Intestinal nutrient transport in coho salmon (*Oncorhynchus kisutch*) and the effects of development, starvation, and seawater adaptation. *J. Comp. Physiol.* **B156,** 163–174.

Collie, N. L., and Bern, H. A. (1982). Changes in intestinal fluid transport associated with smoltification and seawater adaptation in coho salmon, *Oncorhynchus kisutch* (Walbaum). *J. Fish Biol.* **21,** 337–348.

Collie, N. L., and Stevens, J. J. (1985). Hormonal effects on L-proline transport in coho salmon (*Oncorhynchus kisutch*) intestine. *Gen. Comp. Endocrinol.* **59,** 399–409.

Collins, J. J. (1975). Occurrence of pink salmon (*Oncorhynchus gorbuscha*) in Lake Huron. *J. Fish. Res. Board Can.* **32,** 402–404.

Colville, T. P., Richards, R. H., and Dobbie, J. W. (1983). Variations in renal corpuscular morphology with adaptation to sea water in the rainbow trout, *Salmo gairdneri* Richardson. *J. Fish Biol.* **23,** 451–456.

Conte, F. P. (1969). Salt secretion. *In* "Fish Physiology" (W. S. Hoar and D. J. Randall, eds.), Vol. 1, pp. 241–292. Academic Press, New York.

Conte, F. P., and Wagner, H. H. (1965). Development of osmotic and ionic regulation in juvenile steelhead trout *Salmo gairdneri. Comp. Biochem. Physiol.* **14,** 603–620.

Conte, F. P., Wagner, H. H., Fessler, J., and Gnose, C. (1966). Development of osmotic and ionic regulation in juvenile coho salmon *Oncorhynchus kisutch. Comp. Biochem. Physiol.* **18,** 1–15.

Cowey, C. B. (1982). Special issue on fish biochemistry. *Comp. Biochem. Physiol. B* **73B,** 1–180.

Crim, L. W., and Evans, D. M. (1978). Seasonal levels of pituitary and plasma gonadotropin in male and female Atlantic salmon parr. *Can. J. Zool.* **56,** 1550–1555.

Dailey, E. L., Andrews, C. W., and Green, J. M. (1983). Precocious male Atlantic salmon parr (*Salmo salar*) in insular Newfoundland. *Can. J. Fish. Aquat. Sci.* **40,** 647–652.

Davis, J. C., and Shand, I. G. (1978). Acute and sublethal copper sensitivity, growth and saltwater survival in young Babine Lake sockeye salmon. *Tech. Rep.—Fish. Mar. Serv. (Can.)* **847,** 1–55.

Day, F. (1887). "British and Irish Salmonidae." Williams & Norgate, London.

de Renzis, G., and Bornancin, M. (1984). Ion transport and gill ATPases. *In* "Fish Physiology" (W. S. Hoar and D. J. Randall, eds.), Vol. 10B, pp. 65–104. Academic Press, New York.

Dickhoff, W. W., and Sullivan, C. V. (1987). The thyroid gland in smoltification. *Am. Fish* Soc. *Symp. 1,* 197–210.

Dickhoff, W. W., Darling, D. S., and Gorbman, A. (1982). Thyroid function during smoltification of salmonid fishes. *Gunma Symp. Endocrinol.* **19,** 45–61.

Dickhoff, W. W., Sullivan, C. V., and Mahnken, C. V. W. (1985). Thyroid hormones and gill ATPase during smoltification of Atlantic salmon (*Salmo salar*). *Aquaculture* **45,** 376.

Dimberg, K., Höglund, L. B., Knutsson, P. G., and Ridderstråle, Y. (1981). Histochemical localization of carbonic anhydrase in gill lamellae from young salmon (*Salmo salar* L.) adapted to fresh water and salt water. *Acta Physiol. Scand.* **112,** 218–220.

Dodd, J. M., Stuart-Kregor, P. A. C., Sumpter, J. P., Crim, L. W., and Peter, R. E. (1978). Premature sexual maturation in the Atlantic salmon (*Salmo salar* L.). *In* "Comparative Endocrinology" (P. J. Gaillard and H. H. Boer, eds.), pp. 101–104. Elsevier/North-Holland, Amsterdam.

Donaldson, E. M. (1981). The pituitary–interrenal axis as an indicator of stress in fish. *In* "Stress and Fish" (A. D. Pickering, ed.), pp. 11–48. Academic Press, New York.

Donaldson, E. M., Fagerlund, U. H. M., Higgs, D. A., and McBride, J. R. (1979). Hormonal enhancement of growth. *In* "Fish Physiology" (W. S. Hoar, D. J. Randall, and J. R. Brett, eds.), Vol. 8, pp. 455–597. Academic Press, New York.

Donaldson, L. R., and Brannon, E. L. (1975). The use of warmed water to accelerate the production of coho salmon. *Fisheries (Am. Fish. Soc., Bethesda)* **1**(4), 12–15.

Eales, J. G. (1963). A comparative study of thyroid function in migrant juvenile salmon. *Can. J. Zool.* **41,** 811–824.

Eales, J. G. (1965). Factors influencing seasonal changes in thyroid activity in juvenile steelhead trout, *Salmo gairdneri. Can. J. Zool.* **43,** 719–729.

Eales, J. G. (1969). A comparative study of purines responsible for silvering in several freshwater fishes. *J. Fish. Res. Board Can.* **26,** 1927–1931.

Eales, J. G. (1979). Thyroid in cyclostomes and fishes. *In* "Hormones and Evolution" (E. J. W. Barrington, ed.), Vol. 1, pp. 341–436. Academic Press, London.

Eales, J. G. (1981). Extrathyroidal effects of low concentrations of thiourea on rainbow trout, *Salmo gairdneri. Can. J. Fish. Aquat. Sci.* **38,** 1283–1285.

Eales, J. G., and Shostak, S. (1985). Correlation between food ration, somatic growth parameters and thyroid function in Arctic charr, *Salvelinus alpinus* L. *Comp Biochem. Physiol. A* **80A,** 553–558.

Eales, J. G., Chang, J. P., Van der Kraak, G., Omeljaniuk, R. J., and Uin, L. (1982). Effects of temperature on plasma thyroxine and iodine kinetics in rainbow trout, *Salmo gairdneri. Gen. Comp. Endocrinol.* **47,** 295–307.

Eddy, F. B., and Bath, R. N. (1979). Ionic regulation in rainbow trout (*Salmo gairdneri*) adapted to fresh water and dilute sea water. *J. Exp. Biol.* **83,** 181–192.

Eddy, F. B., and Talbot, C. (1985). Urine production in smolting Atlantic salmon, *Salmo salar* L. *Aquaculture* **45,** 67–72.

Elson, P. F. (1957). The importance of size in the change from parr to smolt in Atlantic salmon. *Can. Fish. Cult.* **21,** 1–6.

Eriksson, L.-O., and Lundqvist, H. (1982). Circannual rhythms and photoperiod regulation of growth and smolting in Baltic salmon (*Salmo salar* L.). *Aquaculture* **28,** 113–121.

Eriksson, L.-O., Lundqvist, H., and Johansson, H. (1979). On the normality of size distribution and the precocious maturation in a Baltic salmon, *Salmo solar* L., parr population. *Aquilo, Ser. Zool.* 19, 81–86.

Evans, D. H. (1984). The roles of gill permeability and transport mechanisms in euryhalinity. *In* "Fish Physiology" (W. S. Hoar and D. J. Randall, eds,), Vol. 10B, pp. 239–283. Academic Press, New York.

Evans, G. T., Rice, J. C., and Chadwick, E. M. P. (1984). Patterns of growth and smolting of Atlantic salmon (*Salmo salar*) parr. *Can. J. Fish. Aquat. Sci.* **41,** 783–797.

Evans, G. T., Rice, J. *C*., and Chadwick, E. M. P. (1985). Patterns of growth and smolting of Atlantic salmon (*Salmo salar*) parr in a southwestern Newfoundland river. *Can. J. Fish. Aquat. Sci.* **42,** 539–543.

Evropeizeva, N. V. (1959). Experimental analysis of the young salmon (*Salmo salar* L.) in the stage of transition to the sea. *Rapp P.-V. Réun, Cons. Int. Explor. Mer.* **148,** 29–39.

Evropeizeva, N. V. (1960). Correlation between the process of early gonad ripening, and transformation to the seaward-migrating stage, among male Baltic salmon (*Salmo salar* L.) held in ponds (in Russian). *Zool. Zh.* **39,** 777–779; *Fish. Res. Board Can., Transl. Ser.* **430.**

Ewing, R. D., and Birks, E. K. (1982). Criteria for parr–smolt transformation in juvenile chinook salmon (*Oncorhynchus tshawytscha). Aquaculture* **28,** 185–194.

Ewing, R. D., Johnson, S. L., Pribble, H. J., and Lichatowich, J. A. (1979). Temperature and photoperiod effects on gill (Na+K)-ATPase activity in chinook salmon (*Oncorhynchus tshawytscha). J. Fish. Res. Board Can.* **36,** 1347–1353.

Ewing, R. D., Pribble, H. J., Johnson, S. L., Fustish, C. A., Diamond, J., and Lichatowich, J. A. (1980a). Influence of size, growth rate, and photoperiod on cyclic changes in gill (Na

+K)-ATPase activity in chinook salmon (*Oncorhynchus tshawytscha*). *Can. J. Fish. Aquat. Sci.* **37,** 600–605.

Ewing, R. D., Fustish, C. A., Johnson, S. L., and Pribble, H. J. (1980b). Seaward migration of juvenile chinook salmon without elevated gill (Na+K)-ATPase activities. *Trans. Am. Fish. Soc.* **109,** 349–356.

Ewing, R. D., Evenson, M. D., Birks, E. K., and Hemmingsen, A. R. (1984). Indices of parr–smolt transformation in juvenile steelhead trout (*Salmo gairdneri*) undergoing volitional release at Cole River Hatchery, Oregon. *Aquaculture* **40,** 209–221.

Ewing, R. D., Hemmingsen, A. R., Evenson, M. D., and Lindsay, R. L. (1985). Gill (Na +K)-ATPase activity and plasma thyroxine concentrations do not predict time of release of hatchery coho (*Oncorhynchus kisutch*) and chinook salmon (*Oncorhynchus tshawytscha*) for maximum adult returns. *Aquaculture* **45,** 359–373.

Fagerlund, U. H. M., Higgs, D. A., McBride, J. R., Plotnikoff, M. D., and Dosanjh, B. S. (1980). The potential for using the anabolic hormones 17α-methyltestosterone and (or) 3,5,3′-triiodo-L-thyronine in fresh water rearing of coho salmon (*Oncorhynchus kisutch*) and the effects on subsequent seawater performance. *Can. J. Zool.* **58,** 1424–1432.

Farmer, G. J., Ritter, J. A., and Ashfleld, D. (1978). Seawater adaptation and parr–smolt transformation of juvenile Atlantic salmon, *Salmo salar*. *J. Fish. Res. Board Can.* **35,** 93–100.

Foerster, R. E. (1937). The relation of temperature to the seaward migration of young sockeye salmon (*Oncorhynchus nerka*). *J. Biol. Board Can.* **3,** 421–438.

Foerster, R. E. (1968). The sockeye salmon, *Oncorhynchus nerka*. *Bull. Fish. Res. Board Can.* **162,** 1–422.

Folmar, L. C., and Dickhoff, W. W. (1979). Plasma thyroxine and gill Na^+-K^+ ATPase changes during seawater acclimation of coho salmon, *Oncorhynchus kisutch*. *Comp. Biochem. Physiol. A* **63A,** 329–332.

Folmar, L. C., and Dickhoff, W. W. (1980). The parr–smolt transformation (smoltification) and seawater adaptation in salmonids: A review of selected literature. *Aquaculture* **21,** 1–37.

Folmar, L. C., and Dickhoff, W. W. (1981). Evaluation of some physiological parameters as predictive indices of smoltification. *Aquaculture* **23,** 309–324.

Folmar, L. C., Dickhoff, W. W., Mahnken, C. V. W., and Waknitz, F. W. (1982a). Stunting and parr-reversion during smoltification of coho salmon (*Oncorhynchus kisutch*). *Aquaculture* **28,** 91–104.

Folmar, L. C., Dickhoff, W. W., Zaugg, W. S., and Hodgins, H. O. (1982b). The effects of aroclor 1254 and No. 2 fuel oil on smoltification and seawater adaptation of coho salmon (*Oncorhynchus kisutch*). *Aquat. Toxicol.* **2,** 291–299.

Fontaine, M. (1951). Sur la diminution de la teneur en chlore du muscle des jeunes saumons (smolts) lors de la migration d'avalaison. *C. R. Hebd. Seances Acad. Sci.* **232,** 2477–2479.

Fontaine, M. (1954). Du déterminisme physiologique des migrations. *Biol. Rev. Cambridge Philos. Soc.* **29,** 390–418.

Fontaine, M. (1975). Physiological mechanisms in the migration of marine and amphihaline fish. *Adv. Mar. Biol.* **13,** 241–355.

Fontaine, M., and Hatey, J. (1950). Variations de la teneur du foie en glycogène chez le jeune saumon (*Salmo salar* L.) au cours de la smoltification. *C. R. Seances Soc. Biol. Ses Fil.* **144,** 953–955.

Fontaine, M., and Hatey, J. (1954). Sur la teneur en 17-hydroxycorticostéroides du plasma de saumon (*Salmo salar* L.). *C. R. Hebd. Seances Acad. Sci.* **239,** 319–321.

Fontaine, M., and Olivereau, M. (1957). Interrénal antérieur et smoltification chez *Salmo salar* (L,). *J. Physiol. (Paris)* **49,** 174–176.

Fontaine, M., and Olivereau, M. (1959). Interrénal antérieur et smoltification chez *Salmo salar* L. *Bull. Soc. Zool. Fr.* **84,** 161–162.

Foskett, J. K., and Scheffey, C. (1982). The chloride cell: Definitive identification as the salt-secreting cell in teleosts. *Science* **215,** 164–166.

Fried, S. M., McCleave, J. D., and LaBar, G. W. (1978). Seaward migration of hatchery-reared Atlantic salmon, *Salmo salar,* smolts in the Penobscot River estuary, Maine: Riverine movements. *J. Fish Res. Board Can.* **35,** 76–87.

Fryer, J. N., and Bern, H. A. (1979). Growth hormone binding to tissues of normal and stunted juvenile coho salmon, *Oncorhynchus kisutch. J. Fish Biol.* **15,** 527–533.

Gebhards, S. V. (1960). Biological notes on precocious male chinook salmon parr in the Salmon River drainage, Idaho. *Prog. Fish-Cult.* **22**(3), 121–123.

Giles, M. A., and Randall, D. J. (1980). Oxygenation characteristics of the polymorphic hemoglobins of coho salmon (*Oncorhynchus kisutch*) at different developmental stages. *Comp. Biochem. Physiol. A* **65A,** 265–271.

Giles, M. A., and Vanstone, W. E. (1976). Ontogenetic variation in the multiple hemoglobins of coho salmon (*Oncorhynchus kisutch*) and the effect of environmental factors on their expression. *J. Fish. Res. Board Can.* **33,** 1144–1149.

Glebe, B. D., Saunders, R. L., and Sreedharan, A. (1978). Genetic and environmental influence in expression of precocious sexual maturity of hatchery-reared Atlantic salmon (*Salmo salar*) parr. *Can. J. Genet. Cytol.* **20,** 444.

Godin, J.-G., Dill, P. A., and Drury, D. E. (1974). Effects of thyroid hormones on behaviour of yearling Atlantic salmon (*Salmo salar*). *J. Fish. Res. Board Can.* **31,** 1787–1790.

Gorbman, A., Dickhoff, W. W., Mighell, J. L., Prentice, E. F., and Waknitz, F. W. (1982). Morphological indices of developmental progress in the parr–smolt coho salmon, *Oncorhynchus kisutch. Aquaculture* **28,** 1–19.

Gordon, M. S. (1957). Observations on osmoregulation in the Arctic char (*Salvelinus alpinus* L.). *Biol. Bull.* (*Woods Hole, Mass.*) **112,** 28–32.

Grau, E. G. (1982). Is the lunar cycle a factor timing the onset of salmon migration? *In* "Salmon and Trout Migratory Behavior Symposium" (E. L. Brannon and E. O. Salo, eds.), pp. 184–189. Univ. of Washington Press, Seattle.

Grau, E. G., Dickhoff, W. W., Nishioka, R. S., Bern, H. A., and Folmar, L. C. (1981). Lunar phasing of the thyroxine surge preparatory to seaward migration of salmonid fish. *Science* **221,** 607–609.

Grau, E. G., Specker, J. L., Nishioka, R. S., and Bern, H. A. (1982). Factors determining the occurrence of the surge in thyroid activity in salmon during smoltification. *Aquaculture* **28,** 49–57.

Grau, E. G., Fast, A. W., Nishioka, R. S., Bern, H. A., Barclay, D. K., and Katase, S. A. (1985a). Variations in thyroid hormone levels and in performance in the seawater challenge test accompanying development in coho salmon raised in Hawaii. *Aquaculture* **45,** 121–132.

Grau, E. G., Fast, A. W., Nishioka, R. S., and Bern, H. A. (1985b). The effect of transfer to novel conditions on blood thyroxine in coho salmon. *Aquaculture* **45,** 377.

Grau, E. G., Nishioka, R. S., Specker, J. L., and Bern, H. A. (1985c). Endocrine involvement in the smoltification of salmon with special reference to the role of the thyroid gland. *In* "Current Trends in Comparative Endocrinology" (B. Lofts and W. N. Holmes, eds.), pp. 491–493. Hong Kong Univ. Press, Hong Kong.

Gudjónsson, T. (1972). Smolt rearing techniques, stocking and tagged adult salmon recaptures in Iceland. *Int. Atl. Salmon Found., Spec. Publ.* **4,** 227–235.

Gunnes, K., and Gjedrem, T. (1978). Selection experiments with salmon. IV. Growth of Atlantic salmon during two years in the sea. *Aquaculture* **15,** 19–33.

Hard, J. J., Wertheimer, A. C., Heard, W. R., and Martin, R. M. (1985). Early male maturity in two stocks of chinook salmon (*Oncorhynchus tshawytscha*) transplanted to an experimental hatchery in southeastern Alaska. *Aquaculture* 48, 351–359.

Hashimoto, K., and Matsuura, F. (1960). Comparative studies on two hemoglobins of salmon. V. Change in proportion of two hemoglobins with growth. *Bull. Jpn. Soc. Sci. Fish.* **26,** 931–937.

Hayashi, S. (1970). Biochemical studies on the skin of fish. II. Seasonal changes of purine content of masu salmon from parr to smolt. *Bull. Jpn. Soc. Sci. Fish.* **37,** 508–512.

Healey, M. C. (1983). Coastwide distribution and ocean migration patterns of stream- and ocean-type chinook salmon, *Oncorhynchus tshawytscha. Can. Field Nat.* **97,** 427–433.

Henderson, I. W., Brown, J. A., Oliver, J. A., and Haywood, G. P. (1978). Hormones and single nephron function in fishes. *In* "Comparative Endocrinology" (P. J. Gaillard and H. H. Boer, eds.), pp. 217–222. Elsevier/North-Holland, Amsterdam.

Hickman, C. P., Jr., and Trump, B. F. (1969). The kidney. *In* "Fish Physiology" (W. S. Hoar and D. J. Randall, eds.), Vol. 1, pp. 91–239. Academic Press, New York.

Higgins, P. J. (1985). Metabolic differences between Atlantic salmon (*Salmo salar*) parr and smolts. *Aquaculture* **45,** 33–53.

Higgs, D. A., Donaldson, E. M., Dye, H. M., and McBride, J. R. (1976). Influence of bovine growth hormone and L-thyroxine on growth, muscle composition, and histological structure of the gonads, thyroid, pacreas, and pituitary of coho salmon (*Oncorhynchus kisutch*). *J. Fish. Res. Board Can.* **33,** 1585–1603.

Higgs, D. A., Fagerlund, U. H. M., McBride, J. R., Dye, H. M., and Donaldson, S. M. (1977). Influence of combinations of bovine growth hormone, 17 α-methyltestosterone, and L-thyroxine on growth of yearling coho salmon (*Oncorhynchus kisutch*). *Can. J. Zool.* **55,** 1048–1056.

Higgs, D. A., Fagerlund, U. H. M., McBride, J. R., and Eales, J. G. (1979). Influence of orally administered L-thyroxine or 3,5,3′-triiodo-L-thyronine on growth, food consumption, and food conversion of underyearling coho salmon (*Oncorhynchus kisutch*). *Can. J. Zool.* **57,** 1974–1979.

Higgs, D. A., Fagerlund, U. H. M., Eales, J. G., and McBride, J. R. (1982). Application of thyroid and steriod hormones as anabolic agents in fish culture. *Comp. Biochem. Physiol. B* **73B,** 143–176.

Hirano, T. (1986). The spectrum of prolactin action in teleosts. *Prog. Clin. Biol. Res.* **205,** 53–74.

Hirano, T., Johnson, D. W., Bern, H. A., and Utida, S. (1973). Studies on water and ion movements in the isolated urinary bladder of selected freshwater, marine and euryhaline teleosts. *Comp. Biochem. Physiol. A* **45A,** 529–540.

Hoar, W. S. (1939). The thyroid gland of the Atlantic salmon. *J. Morphol.* **65,** 257–295.

Hoar, W. S. (1965). The endocrine system as a chemical link between the organism and its environment. *Tram. R. Soc. Can., Ser. IV,* **3,** 175–200.

Hoar, W. S. (1966). Hormonal activities of the pars distalis in cyclostomes, fish and amphibia. *In* "The Pituitary Gland" (G. W. Harris and B. T. Donovan, eds.), Vol. 1, pp. 242–294. Butterworth, London.

Hoar, W. S. (1976). Smolt transformation: Evolution, behavior, and physiology. *J. Fish. Res. Board Can.* **33,** 1233–1252.

Hoar, W. S. (1983). "General and Comparative Physiology," 3rd ed. Prentice-Hall, Englewood Cliffs, New Jersey.

Hoar, W. S., and Bell, G. M. (1950). The thyroid gland in relation to the seaward migration of Pacific salmon. *Can. J. Res., Sect. D* **28,** 126–136.

Hoar, W. S., and Randall, D. J., eds. (1984). "Fish Physiology," Vols. 10A and 10B. Academic Press, New York.

Hogstrand, C., and Haux, C. (1985). Evaluation of the sea-water challenge test on sea trout, *Salmo trutta. Comp. Biochem. Physiol., A* **82A,** 261–266.

Holmes, W. N., and Stainer, I. M. (1966). Studies on the renal excretion of electrolytes by the trout (*Salmo gairdneri*). *J. Exp. Biol.* **44,** 33–46.

Houston, A. H, (1959). Osmoregulatory adaptation of steelhead trout (*Salmo gairdneri* Richardson) to sea water. *Can. J. Zool.* **37,** 729–748.

Houston, A. H. (1961). Influence of size upon the adaptation of steelhead trout (*Salmo gairdneri*) and chum salmon (*Oncorhynchus keta*) to sea water. *J. Fish. Res. Board Can.* **18,** 401–415.

Houston, A. H., and Threadgold, L. T. (1963). Body fluid regulation in smolting Atlantic salmon. *J. Fish. Res. Board Can.* **20,** 1355–1369.

Hunt, D. W. C., and Eales, J. G. (1979). The influence of testosterone propionate on thyroid function of immature rainbow trout, *Salmo gairdneri* Richardson. *Gen. Comp. Endocrinol.* **37,** 115–121.

Ikuta, K., Aida, K., Okumoto, N., and Hanyu, I. (1985). Effects of thyroxine and methyltestosterone on smoltification of masu salmon (*Oncorhynchus masou*). *Aquaculture* **45,** 289–303.

Ikuta, K., Aida, K., Okumoto, N., and Hanyu, I. (1987). Effects of sex steriods on smoltification of masu salmon, *Oncorhynchus masou. Gen. Comp. Endocrinol.* **65,** 99–110.

Isaia, J. (1984). Water and nonelectrolyte permeation. *In* "Fish Physiology" (W. S. Hoar and D. J. Randall, eds.), Vol. 10B, pp. 1–38. Academic Press, New York.

Isaia, J., Payan, P., and Girard, J.-P. (1979). A study of the water permeability of the gills of freshwater- and seawater-adapted trout (*Saltno gairdneri*): Mode of action of epinephrine. *Physiol. Zool.* **52,** 269–279.

Ísaksson, A. (1985). The production of one-year smolts and prospects of producing zero-smolts of Atlantic salmon in Iceland using geothermal resources. *Aquaculture* **45,** 305–319.

Iwata, M., and Komatsu, S. (1984). Importance of estuarine residence for adaptation of chum salmon (*Oncorhynchus keta*) fry to seawater. *Can. J. Fish. Aquat.* **Sci. 41,** 744–749.

Iwata, M., Ogura, H., Komatsu, S., Suzuki, K., Nishioka, R. S., and Bern, H. A. (1985). Changes in salinity preference of chum and coho salmon during development. *Aquaculture* **45,** 380–381.

Iwata, M., Ogura, H., Komatsu, S., and Suzuki, K. (1986). Loss of seawater preference in chum salmon (*Oncorhynchus keta*) fry retained in fresh water after migration. *J. Exp. Zool.* **240,** 369–376.

Jackson, A. J. (1981). Osmotic regulation in rainbow trout (*Salmo gairdneri*) following transfer to sea water. *Aquaculture* **24,** 143–151.

Johnson, S. L., Ewing, R. D., and Lichatowick, J. A. (1977). Characterization of gill (Na +K)-ATPase activated adenosine triphosphatase from chinook salmon, *Oncorhynchus tshawytscha. J. Exp. Zool.* **199,** 345–354.

Johnston, C. E., and Eales, J. G. (1970). Influence of body size on silvering of Atlantic salmon (*Salmo salar*) at parr-smolt transformation. *J. Fish. Res. Board Can.* **27,** 983–987.

Johnston, C. E., and Saunders, R. L. (1981). Parr–smolt transformation of yearling Atlantic salmon (*Salmo salar*) at several rearing temperatures. *Can. J. Fish. Aquat. Sci.* **38,** 1189–1198.

Jones, J.W. (1959). "The Salmon," Chapter 7, pp. 116–129. Collins, London.

Jonsson, B., and Ruud-Hansen, J. (1985). Water temperature as primary influence on timing of seaward migration of Atlantic salmon (*Salmo salar*) smolts. *Can. J. Fish. Aquat. Sci.* **42,** 593–595.

Jonsson, B., and Sandlund, O. T. (1979). Environmental factors and life histories of isolated river stocks of brown trout (*Salmo trutta m. fario*) in Søre river system, Norway. *Environ. Biol. Fishes* **4,** 43–54.

Kashiwagi, M., and Sato, R. (1969). Studies on the osmoregulation of chum salmon, *Oncorhynchus keta* (Walbaum). 1. The tolerance of eyed period eggs, alevins and fry of chum salmon to sea water. *Tohoku J. Agric. Res.* **20,** 41–47.

Katz, A. H., and Katz, H. M. (1978). Effects of DL-thyroxine on swimming speed in the pearl danio, *Brachydanio albolineatus* (Blyth). *J. Fish Biol.* **12,** 527–530.

King, G. M., Jones, J. W., and Orton, J. H. (1939). Behaviour of mature male salmon parr, *Salmo salar* juv. L. *Nature (London)* **143,** 162–163.

Knutsson, S., and Grav, T. (1976). Seawater adaptation in Atlantic salmon (*Salmo salar* L.) at different experimental temperatures and photoperiods. *Aquaculture* **8,** 169–187.

Koch, H. J. A. (1982). Hemoglobin changes with size in the Atlantic salmon (*Salmo salar* L.). *Aquaculture* **28,** 231–240.

Komourdjian, M. P., and Idler, D. R. (1977). Hypophysectomy of rainbow trout, *Salmo gairdneri,* and its effect on plasmatic sodium regulation. *Gen. Comp. Endocrinol.* **32,** 536–542.

Komourdjian, M. P., Saunders, R. L., and Fenwick, J. C. (1976a). The effect of porcine somatotropin on growth, and survivial in seawater of Atlantic salmon (*Salmo salar*) parr. *Can. J. Zool.* **54,** 531–535.

Komourdjian, M. P., Saunders, R. L., and Fenwick, J. C. (1976b). Evidence for the role of growth hormone as a part of a 'light-pituitary axis' in growth and smoltification of Atlantic salmon (*Salmo salar*). *Can. J. Zool.* **54,** 544–551.

Komourdjian, M. P., Burton, M. P., and Idler, D. R. (1978). Growth of rainbow trout, *Salmo gairdneri,* after hypophysectomy and somatotropin therapy. *Gen. Comp. Endocrinol.* **34,** 158–162.

Kristinsson, J. B., Saunders, R. L., and Wiggs, A. J. (1985). Growth dynamics during the development of bimodal length–frequency distribution in juvenile Atlantic salmon (*Salmo salar* L.). *Aquaculture* **45,** 1–20.

Kubo, T. (1955). Changes in some characteristics of blood of smolts of *Oncorhynchus masou* during seaward migration. *Bull. Fac. Fish., Hokkaido Univ.* **6,** 201–207.

Lahlou, B., and Fossat, B. (1971). Méchanisme du transport de l'eau et du sel à travers le vessie urinaire d'un poisson téléostéen en eau douce, la truite arc-en-ciel. *C. R. Hebd. Seances Acad. Sci.* **273,** 2108–2110.

Lahlou, B., Crenesse, D., Bensahla-Talet, A., and Porthe-Nibelle, J. (1975). Adaptation de la truite d'élevage à l'eau de mer. Effets sur les concentrations plasmatiques, les échanges branchiaux et le transport intestinal du sodium. *J. Physiol. (Paris)* **70,** 593–603.

Lam, T. J. (1983). Environmental influences on gonadal activity in fish. *In* "Fish Physiology" (W. S. Hoar, D. J. Randall, and E. M. Donaldson, eds.), Vol. 9B, pp. 65–116. Academic Press, New York.

Langdon, J. S., and Thorpe, J. E. (1984). Responses of gill Na^+-K^+ ATPase activity, succinic dehydrogenase activity and chloride cells to seawater adaptation in Atlantic salmon, *Salmo salar* L., parr and smolt. *J. Fish Biol.* **24,** 323–331.

Langdon, J. S., and Thorpe, J. E. (1985). The ontogeny of smoltification: Developmental patterns of gill Na^+/K^+-ATPase, SDH, and chloride cells in juvenile Atlantic salmon, *Salmo salar* L. *Aquaculture* **45,** 83–95.

Langdon, J. S., Thorpe, J. E., and Roberts, R. J. (1984). Effects of cortisol and ACTH on gill Na^+/K^+-ATPase, SDH and chloride cells in juvenile Atlantic salmon *Salmo salar* L. *Comp. Biochem. Physiol. A* **77A,** 9–12.

Langhome, P., and Simpson, T. H. (1986). The interrelationship of cortisol, gill (Na+K) ATPase, and homeostasis during the parr-smolt transformation of Atlantic salmon. (*Salmo salar* L.). *Gen. Comp. Endocrinol.* **61,** 203–213.

Lasserre, P., Boeuf, G., and Harache, Y. (1978). Osmotic adaptation of *Oncorhynchus kisutch* Walbaum. I. Seasonal variation of gill Na^+-K^+ ATPase activity in coho salmon, 0^+-age and yearling, reared in fresh water. *Aquaculture* **14,** 365–382.

Leatherland, J. F. (1982). Environmental physiology of the teleostean thyroid gland: A reivew. *Environ. Biol. Fishes* **7,** 83–110.

Leatherland, J. F., and Cho, C. Y. (1985). Effect of rearing density on thyroid and interrenal gland activity and plasma and hepatic metabolite levels in rainbow trout, *Salmo gairdneri* Richardson. *J. Fish Biol.* **27,** 583–592.

Leatherland, J. F., and Sonstegard, R. A. (1978). Lowering of serum thyroxine and triiodothyronine levels in yearling coho salmon, *Oncorhynchus kisutch,* by dietary mirex and PCBs. *J. Fish. Res. Board Can.* **35,** 1285–1289.

Leatherland, J. F., and Sonstegard, R. A. (1981). Thyroid funciton, pituitary structure and serum lipids in Great Lakes coho salmon, *Oncorhynchus kisutch* Walbaum, 'jacks' compared with sexually immature spring salmon. *J. Fish Biol.* **18,** 643–653.

Leatherland, J. F., McKeown, B. A., and John, T. M. (1974). Circadian rhythm of plasma prolactin, growth hormone, glucose and free fatty acid in juvenile kokanee salmon, *Oncorhynchus nerka. Comp. Biochem. Physiol. A* **47A,** 821–828.

Leatherland, J. F., Cho, C. Y., and Slinger, S. J. (1977). Effects of diet, ambient temperature, and holding conditions on plasma thyroxine levels in rainbow trout (*Salmo gairdneri*). *J. Fish Res. Board Can.* **34,** 677–682.

Leray, C., Colin, D. A., and Florentz, A. (1981). Time course of osmotic adaptation and gill energetics of rainbow trout (*Salmo gairdneri*) following abrupt changes in external salinity. *J. Comp. Physiol.* **144,** 175–181.

Leyzerovich, K. A. (1973). Dwarf males in hatchery propagation of the Atlantic salmon [*Salmo solar* (L.)]. *J. Ichthyol.* (*Engl. Transl.*) **13,** 382–391.

Lin, R. J., Rivas, R. J., Nishioka, R. S., Grau, E. G., and Bern, H. A. (1985a). Effects of feeding triiodothyronine (T_3) on thyroxin (T_4) levels in steelhead trout, *Salmo gairdneri. Aquaculture* **45,** 133–142.

Lin, R. J., Rivas, R. J., Grau, E. G., Nishioka, R. S., and Bern, H. A. (1985b). Changes in plasma thyroxin following transfer of young coho salmon (*Oncorhynchus kisutch*) from fresh water to fresh water. *Aquaculture* **45,** 381–382.

Lindahl, K. (1980). The gonadotropic cell in parr, precocious parr male and smolt of Atlantic salmon, *Salmo salar*. An immunological, light- and electron microscopical study. *Acta Zool.* (*Stockholm*) **61,** 117–125.

Loretz, C. A., and Bern, H. A. (1982). Prolactin and osmoregulation in vertebrates. An update. *Neuroendocrinology* **35,** 292–304.

Loretz, C. A., Collie, N. L., Richman, N. H., III, and Bern, H. A. (1982). Osmoregulatory changes accompanying smoltification in coho salmon. *Aquaculture* **28,** 67–74.

Lorz, H. W., and McPherson, B. P. (1976). Effects of copper or zinc in fresh water on adaptation to sea water and ATPase activity, and the effects of copper on migratory disposition of coho salmon (*Oncorhynchus kisutch*). *J. Fish. Res. Board Can.* **33,** 2023–2030.

Lovern, J. A. (1934). Fat metabolism in fishes. V. The fat of the salmon in its young freshwater stages. *Biochem. J.* **28,** 1961–1963.

Lundqvist, H. (1980). Influence of photoperiod on growth in Baltic salmon parr (*Salmo salar* L.) with special reference to the effect of precocious sexual maturation. *Can. J. Zool.* **58,** 940–944.

Lundqvist, H. (1983). Precocious sexual maturation and smolting in Baltic salmon (*Salmo salar* L.): Photoperiodic synchronization and adaptive significance of annual biological cycles. Thesis, Dept. of Ecological Zoology, University of Umeå, Umeå, Sweden.

Lundqvist, H., and Fridberg, G. (1982). Sexual maturation versus immaturity: different tactics with adaptive values in Baltic salmon (*Salmo salar* L.) male smolts. *Can. J. Zool.* **60,** 1822–1827.

Lundqvist, H., Clarke, W. C., Eriksson, L.-O., Funegård, P., and Engström, B. (1986). Seawater adaptability in three different river stocks of Baltic salmon (*Salmo salar* L.) during smolting. *Aquaculture* **52,** 219–229.

McBride, J. R., Higgs, D. A., Fagerlund, U. H. M., and Buckley, J. T. (1982). Thyroid and steriod hormones: Potential for control of growth and smoltification of salmonids. *Aquaculture* **28,** 201–209.

McCartney, T. H. (1976). Sodium-potassium dependent adenosine triphosphatase activity in gills and kidneys of Atlantic salmon (*Salmo salar*). *Comp. Biochem. Physiol. A* **53A,** 351–353.

McCormick, S. D., and Naiman, R. J. (1984). Osmoregulation in brook trout, *Salvelinus fontinalis*. II. Effects of size, age and photoperiod on seawater survival and ionic regulation. *Comp. Biochem. Physiol. A* **79A,** 17–28.

McCormick, S. D., and Naiman, R. (1985). Hypoosmoregulation in an anadromous teleost: Influence of sex and maturation. *J. Exp. Zool.* **234,** 193–198.

McCormick, S. D., and Saunders, R. L. (1987). Physiological adaptations to marine life. *Amer Fish Soc. Symp. 1,* 211–229.

McCormick, S. D., Naiman, R. J., and Montgomery, E. T. (1985). Physiological smolt characteristics of anadromous and non-anadromous brook trout (*Salvelinus fontinalis*) and Atlantic salmon (*Salmo salar*). *Can. J. Fish. Aquat. Sci.* **42,** 529–538.

McInerney, J. E. (1964). Salinity preference: An orientation mechanism in salmon migration. *J. Fish. Res. Board Can.* **21,** 995–1018.

MacKinnon, C. N., and Donaldson, E. M. (1976). Environmentally induced precocious sexual development of the male pink salmon (*Oncorhynchus gorbuscha*). *J. Fish. Res. Board Can.* **33,** 2602–2605.

Mahnken, C. V. W., Prentice, E., Waknitz, W., Monan, G., Sims, C., and Williams, J. (1982). The application of recent smoltification research to public hatchery releases: An assessment of size/time requirements for Columbia River Hatchery coho salmon (*Oncorhynchus kisutch*). *Aquaculture* **28,** 251–268.

Malikova, E. M. (1957). Biochemical analysis of young salmon at the time of their transformation to a condition close to the smolt stage, and during retention of smolts in freshwater, (in Russian). *Tr. Latv. Otdel. VNIRO* **2,** 241–255; *Fish. Res. Board Can., Transl. Ser.* **232,** 1–19 (1959).

Markert, J. R., and Vanstone, W. E. (1966). Pigments in the belly skin of coho salmon (*Oncorhynchus kisutch*). *J. Fish. Res. Board Can.* **23,** 1095–1098.

Markert, J. R., Higgs, D. A., Dye, H. M., and MacQuarrie, D. W. (1977). Influence of bovine growth hormone on growth rate, appetite, and food conversion of yearling coho salmon (*Oncorhynchus kisutch*) fed two diets of different composition. *Can. J. Zool.* **55,** 74–83.

Mason, J. C. (1975). Seaward movements of juvenile fishes, including lunar periodicity in the movement of coho salmon (*Oncorhynchus kisutch*) fry. *J. Fish. Res. Board Can.* **32,** 2542–2547.

Massey, B. D., and Smith, C. L. (1968). The action of thyroxine on mitochondrial respiration and phosphorylation in the trout (*Salmo trutta fario* L.). *Comp. Biochem. Physiol.* **25,** 241–255.

Miles, H. M., and Smith, L. S. (1968). Ionic regulation in migrating juvenile coho salmon. *Oncorhynchus kisutch. Comp. Biochem. Physiol.* **26,** 381–398.

Milne, R. S., and Leatherland, J. F. (1978). Effect of ovine TSH, thiourea, ovine prolactin and bovine growth hormone of plasma thyroxine and triiodothyronine levels in rainbow trout, *Salmo gairdneri. J. Comp. Physiol.* **124,** 105–110.

Milne, R. S., and Leatherland, J. F. (1980). Changes in plasma thyroid hormone following administration of exogenous pituitary hormones and steroid hormones to rainbow trout (*Salmo gairdneri*). *Comp. Biochem. Physiol. A* **66A,** 679–686.

Milne, R. S., and Randall, D. J. (1976). Regulation of arterial pH during fresh water to sea water transfer in the rainbow trout *Salmo gairdneri. Comp. Biochem. Physiol. A.* **53A,** 157–160.

Milne, R. S., Leatherland, J. F., and Holub, B. J. (1979). Changes in plasma thyroxine, triiodothyronine and cortisol associated with starvation in rainbow trout (*Salmo gairdneri*). *Environ. Biol. Fishes* **4,** 185–190.

Mitans, A. R. (1973). Dwarf males and the sex structure of a Baltic salmon [Salmo *salar* (L.).] population. *J. Ichthyol (Engl. Transl.)* **13,** 192–197.

Miwa, S., and Inui, Y. (1983). Effects of thyroxine and thiourea on the parr-smolt transformation of amago salmon (*Oncorhynchus rhodurus*). *Bull. Natl. Res. Inst. Aquacult. (Jpn.)* **4,** 41–52.

Miwa, S., and Inui, Y. (1985). Effects of L-thyroxine and ovine growth hormone on smoltification of Amago salmon (*Oncorhynchus rhodurus*). *Gen. Comp. Endocrinol.* **58,** 436–442.

Miwa, S., and Inui, Y. (1986). Inhibitory effects of 17α-methyltestosterone and estradiol-17β on smoltification of sterilized amago salmon (*Oncorhynchus rhodurus*). *Aquaculture* **53,** 21–39.

Moccia, R. D., Leatherland, J. L., and Sonstegard, R. A. (1977). Increasing frequency of thyroid goitres in coho salmon (*Oncorhynchus kisutch*) in the Great Lakes. *Science* **198,** 425–426.

Mommsen, T. P. (1984). Metabolism of the fish gill. *In* "Fish Physiology‘ (W. S. Hoar and D. J. Randall, eds.), Vol. 10B, pp. 203–238. Academic Press, New York.

Morley, M., Chadwick, A., and El Tounsey, E. M. (1981). The effect of prolactin on the water absorption by the intestine of the trout (*Salmo gairdneri*). *Gen. Comp. Endocrinol.* **44,** 64–68.

Myers, R. A. (1984). Demographic consequences of precocious maturation of Atlantic salmon (*Salmo salar*). *Can. J. Fish. Aquat.* Sci. **41,** 1349–1353.

Myers, R. A., and Hutchings, J. A. (1985). Mating of anadromous Atlantic salmon, *Salmo salar* L., with mature male parr. *Int. Counc. Explor. Sea, Counc. Meet.* **M:8,** 1–6.

Myers, R. A., Hutchings, J. A., and Gibson, R. J. (1986). Variation in male parr maturation within and among populations of Atlantic salmon, *Salmo salar. Can. J. Fish. Aquat. Sci.* **43,** 1242–1248.

Naevdal, G., Holm, M., Ingebritsen, O., and Møller, D. (1978). Variation in age at first spawning in Atlantic salmon (*Salmo salar*). *J. Fish. Res. Board Can.* **35,** 145–147.

Nagahama, Y. (1985). Involvement of endocrine systems in the amago salmon, *Oncorhynchus rhodurus. Aquaculture* **45,** 383–384.

Nagahama, Y., Clarke, W. C., and Hoar, W. S. (1977). Influence of salinity on ultrastructure of the secretory cells of the adenohypophyseal pars distalis in yearling coho salmon *Oncorhynchus kisutch. Can. J. Zool.* **55,** 183–198.

Nagahama, Y., Adachi, S., Tashiro, F., and Grau, E. G. (1982). Some endocrine factors affecting the development of seawater tolerance during the parr–smolt transformation of the amago salmon, *Oncorhynchus rhodurus. Aquaculture* **28,** 81–90.

Narayansingh, T., and Eales, J. G. (1975). The influence of physiological doses of thyroxine on the lipid reserve of starved and fed brook trout, *Salvelinus fontinalis* (Mitchill). *Comp. Biochem. Physiol. B* **52B,** 407–412.

Nichols, D. J., and Weisbart, M. (1985). Cortisol dynamics during seawater adaptation of Adantic salmon *Salmo salar*. *Am. J. Physiol.* **248,** R651–R659.

Nichols, D. J., and Weisbart, M., and Quinn, J. (1985). Cortisol kinetics and fluid distribution in brook trout (*Salvelinus fontinalis*). *J. Endocrinol.* **107,** 57–69.

Nichols, J. W., Wedemeyer, G. A., Mayer, F. L., Dickhoff, W. W., Gregory, S. V., Yasutake, W. T., and Smith, S. D. (1984). Effects of freshwater exposure to arsenic trioxide on the parr–smolt transformation of coho salmon (*Oncorhynchus kisutch*). *Environ. Toxicol. Chem.* **3,** 143–149.

Nishikawa, K., Hirashima, T., Suzuki, S., and Suzuki, M. (1979). Changes in circulating L-thyroxine and L-triiodothyronine of the masu salmon, *Oncorhynchus masou,* accompanying the smoltification, measured by radioimmunoassay. *Endocrinol. Jpn.* **26,** 731–735.

Nishioka, R. S., Bern, H. A., Lai, K. V., Nagahama, Y., and Grau, E. G. (1982). Changes in the endocrine organs of coho salmon during normal and abnormal smoltification—An electron-microscope study. *Aquaculture* **28,** 21–38.

Nishioka, R. S., Young, G., Grau, E. G., and Bern, H. A. (1983). Environmental, behavioral and endocrine bases for lunar-phased hatchery releases of salmon. *Proc. N. Pac. Aquacult. Symp., 2nd, 1983,* pp. 161–172.

Nishioka, R. S., Young, G., Bern, H. A., Jochimsen, W., and Hiser, C. (1985a). Attempts to intensify the thyroxin surge in coho and king salmon by chemical stimulation. *Aquaculture* **45,** 215–225.

Nishioka, R. S., Richman, N. H., Young, G., and Bern, H. A. (1985b). Preliminary studies on the effects of hypophysectomy on coho and king salmon. *Aquaculture* **45,** 385–386.

Nishioka, R. S., Grau, E. G., Lai, K. V., and Bern, H. A. (1987). Effect of thyroid-stimulating hormone on the physiology and morphology of the thyroid gland in coho salmon, *Oncorhynchus kisutch*. *Fish. Physiol. Biochem.* **3,** 63–71.

Noakes, D. L. G., and Leatherland, J. F. (1977). Social dominance and interrenal cell activity in rainbow trout, *Salmo gairdneri* (Pisces, Salmonidae). *Environ. Biol. Fishes* **2,** 131–136.

Olivereau, M. (1962). Modifications de l'interrénal du smolt (*Salmo solar* L.) au cours du passage d'eau douce en eau de mer. *Gen. Comp. Endocrinol.* **2,** 565–573.

Olivereau, M. (1975). Histophysiologie de l'axe hypophysocorticosurrenalien chez le saumon de l'Atlantique (cycle en eau douce, vie thalassique et reproduction). *Gen. Comp. Endocrinol.* **27,** 9–27.

Orton, J. H., Jones, J. W., and King, G. M. (1938). The male sexual stage in salmon parr (*Salmo salar* L. juv.). *Proc. R. Soc. London, Ser. B* **125,** 103–114.

Osborne, R. H., Simpson, T. H., and Youngson, A. F. (1978). Seasonal and diurnal rhythms of thyroidal status in the rainbow trout, *Salmo gairdneri* Richardson. *J. Fish Biol.* **12,** 531–540.

Osterdahl, L. (1969). The smolt run of a small Swedish river. *In* "Symposium on Salmon and Trout in Streams" (T. G. Northcote, ed.), p. 205–215. Inst. Fish., Univ. of British Columbia, Vancouver, Canada.

Otto, R. G. (1971). Effects of salinity on the survival and growth of pre-smolt coho salmon (*Oncorhynchus kisutch*). *J Fish. Res. Board Can.* **28,** 343–349.

Otto, R. G., and Mclnerney, J. D. (1970). Development of salinity preference in pre-smolt coho salmon, *Oncorhynchus kisutch*. *J. Fish. Res. Board Can.* **27,** 793–800.

Parry, G. (1966). Osmotic adaptation in fishes. *Biol. Rev. Cambridge Philos. Soc.* **41,** 392–444.

Patiño, R., and Schreck, C. B. (1986). Sexual dimorphism of plasma sex steroid levels in juvenile coho salmon, *Oncorhynchus kisutch,* during smoltification. *Gen. Comp. Endocrinol.* **61,** 127–133.

Patiño, R, Schreck, C. B., and Redding, J. M. (1985). Clearance of plasma corticosteroids during smoltification of coho salmon, *Oncorhynchus kisutch. Comp. Biochem. Physiol. A* **82A,** 531–535.

Peden, A. E., and Edwards, J. C. (1976). Permanent residence in fresh water of a large chum salmon (*Oncorhynchus keta*). *Syesis* **9,** 363.

Pengelley, E. T., and Asmundson, S. J. (1969). Free-running periods of endogenous circannual rhythms in the golden-mantled ground squirrel, *Citellus lateralis. Comp. Biochem. Physiol.* **30,** 177–183.

Pereira, D. L., and Adelman, I. R. (1985). Interactions of temperature, size and photoperiod on growth and smoltification of chinook salmon (*Oncorhynchus tshawytscha*). *Aquaculture* **46,** 185–192.

Piggins, D. J., and Mills, C. P. R. (1985). Comparative aspects of the biology of naturally produced and hatchery-reared Atlantic salmon smolts (*Salmo salar* L.). *Aquaculture* **45,** 321–333.

Poston, H. A. (1978). Neuroendocrine mediation of photoperiod and other environmental influences on physiological responses in salmonids: A review. *Tech. Pap. U.S. Fish. Wild. Serv.* **96,** 1–14.

Potts, W. T. W., Foster, M. A., and Stather, J. W. (1970). Salt and water balance in salmon smolts. *J. Exp. Biol.* **52,** 553–564.

Power, G. (1959). Field measurements of the basal oxygen consumption of Atlantic salmon parr and smolts. *Arctic* **12,** 195–202.

Power, G. (1969). The salmon of Ungava Bay. *Tech. Paper—Arct. Inst. North Am.* **22,** 1–72.

Primdas, F. H., and Eales, J. G. (1976). The influence of TSH and ACTH on purine and pteridine deposition in the skin of rainbow trout (*Salmo gairdneri*). *Can. J. Zool.* **54,** 576–581.

Prunet, P., and Boeuf, G. (1985). Plasma prolactin level during transfer of rainbow trout (*Salmo gairdneri*) and Atlantic salmon (*Salmo solar*) from fresh water to sea water. *Aquaculture* **45,** 167–176.

Prunet, P., Boeuf, G., and Houdebine, L. M. (1985). Plasma and pituitary prolactin levels in rainbow trout during adaptation to different salinities. *J. Exp. Zool.* **235,** 187–196.

Randall, D. J., and Daxboeck, C. (1984). Oxygen and carbon dioxide transfer across fish gills. *In* "Fish Physiology" (W. S. Hoar and D. J. Randall, eds.), Vol. 10A, pp. 263–314. Academic Press, New York.

Rankin, J. C., and Bolis, L. (1984). Hormonal control of water movement across the gills. *In* "Fish Physiology" (W. S. Hoar and D. J. Randall, eds.), Vol. 10B, pp. 177–201. Academic Press, New York.

Redding, J. M., Patiño, R., and Schreck, C. B. (1984). Clearance of corticosteroids in yearling coho salmon, *Oncohynchus kisutch*, in fresh water and seawater and after stress. *Gen. Comp. Endocrinol.* **54,** 433–443.

Refstie, T. (1977). Effect of density on growth and survival of rainbow trout. *Aquaculture* **11,** 329–334.

Refstie, T., and Steine, T. A. (1978). Selection experiments with salmon. III. Genetic and environmental sources of variation in length and weight in the freshwater phase. *Aquaculture* **14,** 221–234.

Refstie, T., Steine, T. A., and Gjedrem, T. (1977). Selection experiments with salmon. II. Proportion of Atlantic salmon smoltifying at 1 year of age. *Aquaculture* **10,** 231–242.

Richman, N. H., III, and Zaugg, W. S. (1987). Effects of cortisol and growth hormone on osmoregulation in pre- and desmoltified coho salmon (*Oncorhynchus kisutch*). *Gen. Comp. Endocrinol.* **65,** 189–198.

Richman, N. H., Tai De Diaz, S. T., Nishioka, R. S., Prunet, P., and Bern, H. A. (1987). Osmoregulatory and endocrine relationships with chloride cell morphology and density during smoltification in coho salmon (*Oncorhynchus kisutch*). *Aquaculture* **60,** 265–285.

Riddell, B. E., and Leggett, W. C. (1981). Evidence of an adaptive basis for geographic variation in body morphology and time of downstream migration in juvenile Atlantic salmon (*Salmo salar*). *Can. J. Fish. Aquat. Sci.* **38,** 308–320.

Riddell, B. E., Leggett, W. C., and Saunders, R. L. (1981). Evidence of adaptive polygenic variation between two populations of Atlantic salmon (*Salmo salar*) native to tributaries of the S. W. Miramichi River, N. B. *Can. J. Fish. Aquat. Sci.* **38,** 321–333.

Robertson, O. H. (1948). The occurrence of increased activity of the thyroid gland in rainbow trout at the time of transformation from parr to silvery smolt. *Physiol. Zool.* **21,** 282–295.

Robertson, O. H. (1949). Production of silvery smolt stage in rainbow trout by intramuscular injection of mammalian thyroid extract and thyrotropic hormone. *J. Exp. Zool.* **110,** 337–355.

Robertson, O. H. (1957). Survival of precociously mature king salmon male parr (*Oncorhynchus tshawytscha* juv.) after spawning. *Calif. Fish Game* **43,** 119–130.

Rounsefell, G. A. (1958). Anadromy in North American Salmonidae. *U.S. Fish. Wildl. Serv. Fish. Bull.* **58,** 171–185.

Sage, M. (1973). The evolution of thyroidal function in fishes. *Am. Zool.* **13,** 899–905.

Saunders, R. L. (1965). Adjustment of buoyancy in young Atlantic salmon and brook trout by changes in swimbladder volume. *J. Fish. Res. Board Can.* **22,** 335–352.

Saunders, R. L., and Henderson, E. B. (1965). Precocious sexual development in male post-smolt Atlantic salmon reared in the laboratory. *J. Fish. Res. Board Can.* **22,** 1567–1570.

Saunders, R. L., and Henderson, E. B. (1970). Influence of photoperiod on smolt development and growth of Atlantic salmon (*Salmo salar*). *J. Fish. Res. Board Can.* **27,** 1295–1311.

Saunders, R. L., and Henderson, E. B. (1978). Changes in gill ATPase activity and smolt status of Atlantic salmon (*Salmo salar*). *J. Fish. Res. Board Can.* **35,** 1542–1546.

Saunders, R. L., and Schom, C. B. (1985). Importance of the variation in life history parameters of Atlantic salmon (*Salmo salar*). *Can. J. Fish. Aquat. Sci.* **42,** 615–618.

Saunders, R. L., and Sreedharan, A. (1977). The incidence and genetic implications of sexual maturity in male Atlantic salmon parr. *Int. Counc. Explor. Sea, Counc. Meet.* **M:21,** 1–8.

Saunders, R. L., Henderson, E. B., and Glebe, B. D. (1982). Precocious sexual maturation and smoltification in male Atlantic salmon (*Salmo salar*). *Aquaculture* **28,** 211–229.

Saunders, R. L., Henderson, E. B., Harmon, P. R., Johnston, C. E., and Eales, J. G. (1983). Effects of low environmental pH on smolting of Atlantic salmon (*Salmo salar*). *Can. J. Fish. Aquat. Sci.* **40,** 1203–1211.

Saunders, R. L., Henderson, E. B., and Harmon, P. R. (1985a). Effects of photoperiod on juvenile growth and smolting of Atlantic salmon and subsequent survival and growth in sea cages. *Aquaculture* **45,** 55–66.

Saunders, R. L., McCormick, S. D., Henderson, E. B., Eales, J. G., and Johnston, C. E. (1985b). The effect of orally administered 3,5,3′-triiodo-L-thyronine on growth and salinity tolerance of Atlantic salmon (*Salmo salar* L.). *Aquaculture* **45,** 143–156.

Saxton, A. M., Hershberger, W. K., and Iwamoto, R. N. (1984). Smoltification in the net-pen culture of coho salmon: Quantitative genetic analysis. *Trans. Am. Fish. Soc.* **113,** 339–347.

Schmidt, S. P., and House, E. W. (1979). Precocious sexual development in hatchery-reared and laboratory-maintained steelhead trout (*Salmo gairdneri*). *J. Fish. Res. Board Can.* **36,** 90–93.

Schreck, C. B. (1982). Stress and rearing of salmonids. *Aquaculture* **28,** 241–249.

Schreck, C. B., Patiño, R., Pring, C. K., Winton, J. R., and Holway, J. E. (1985). Effect of rearing density on indices of smoltification and performance of coho salmon, *Oncorhynchus kisutch*. *Aquaculture* **45,** 345–358.

Shehadeh, Z. H., and Gordon, M. S. (1969). The role of the intestine in salinity adaptation of the rainbow trout, *Salmo gairdneri*. *Comp. Biochem. Physiol.* **30,** 397–418.

Sheridan, M. A. (1985). Changes in the lipid composition of juvenile salmonids associated with smoltification and premature transfer to seawater. *Aquaculture* **45,** 387–388.

Sheridan, M. A. (1986). Effects of thyroxin, cortisol, growth hormone, and prolactin on lipid metabolism of coho salmon, *Oncorhynchus kisutch,* during smoltification. *Gen. Comp. Endocrinol.* **64,** 220–238.

Sheridan, M. A., Allen, W. V., and Kerstetter, T. H. (1983). Seasonal variations in the lipid composition of the steelhead trout, *Salmo gairdneri* Richardson, associated with the parr–smolt transformation. *J. Fish Biol.* **23,** 125–134.

Sheridan, M. A., Allen, W. V., and Kerstetter, T. H. (1985a). Changes in the fatty acid composition of steelhead trout, *Salmo gairdneri* Richardson, associated with parr-smolt transformation. *Comp. Biochem. Physiol. B* **80B,** 671–676.

Sheridan, M. A., Woo, N. Y. S., and Bern, H. A. (1985b). Changes in the rates of glycogenesis, glycogenolysis, lipogenesis, and lipolysis in selected tissues of the coho salmon associated with parr–smolt transformation. *J. Exp. Zool.* **236,** 35–44.

Skarphedinsson, O., Bye, V. J,, and Scott, A. P. (1985). The influence of photoperiod on sexual development in underyearling rainbow trout, *Salmo gairdneri* Richardson. *J. Fish Biol.* **27,** 319–326.

Smith, D. C. W. (1956). The role of the endocrine organs in the salinity tolerance of trout. *Mem. Soc. Endocrinol.* **5,** 83–101.

Solomon, D. J. (1978). Migration of smolts of Atlantic salmon (***Salmo*** *salar* L.) and sea trout (S. *trutta* L.) in a chalkstream. *Environ. Biol. Fishes* **3,** 223–229,

Sonstegard, R. A., and Leatherland, *J. F.* (1976). The epizootiology and pathogenesis of thyroid hyperplasia in coho salmon (*Oncorhynchus kisutch*) in Lake Ontario. *Cancer Res.* **36,** 4467–4475.

Sower, S. A., Sullivan, C. V., and Gorbman, A. (1984). Changes in plasma estradiol and effects of triiodothyronine on plasma estradiol during smoltification of coho salmon (*Oncorhynchus kisutch*). *Gen. Comp. Endocrinol.* **54,** 486–492.

Specker, J. L. (1982). Interrenal function and smoltification. *Aquaculture* **28,** 59–66.

Specker, J. L., and Kobuke, L. (1987). Seawater acclimation and thyroidal response to thyrotropin in juvenile coho salmon (*Oncorhynchus kisutch*). *J. Exp. Zool.* **241,** 327–332.

Specker, J. L., and Richman, N. H., III (1984). Environmental salinity and the thyroidal response to thyrotropin in juvenile coho salmon (*Oncorhynchus kisutch*). *J. Exp. Zool.* **230,** 329–333.

Specker, J. L., and Schreck, C. B. (1980). Stress responses to transportation and fitness for marine survival in coho salmon (*Oncorhynchus kisutch*) smolts. *Can. J. Fish. Aquat. Sci.* **37,** 765–769.

Specker, J. L., and Schreck, C. B. (1982). Changes in plasma corticosteroids during smoltification of coho salmon, *Oncorhynchus kisutch. Gen. Comp. Endocrinol.* **46,** 53-58.

Specker, *J.* L., and Schreck, C. B. (1984). Thyroidal response to mammalian thyrotropin during smoltification of coho salmon (*Oncorhynchus kisutch*). *Comp. Biochem. Physiol. A* **78A,** 441–444.

Specker, J. L., DiStefano, J. J., III, Grau, E. G., Nishioka, R. S., and Bern, H. A. (1984). Development-associated changes in thyroxine kinetics in juvenile salmon. *Endocrinology* **115,** 399–406.

Stuart-Kregor, P. A. C., Sumpter, J. P., and Dodd, J. M. (1981). The involvement of gonadotrophin and sex steroids in the control of reproduction in the parr and adults of Atlantic salmon, *Salmo salar* L. *J. Fish Biol.* **18,** 59–72.

Sullivan, C. V., Dickhoff, W. W., Mahnken, C. V. W., and Hershberger, W. K. (1985). Changes in the hemoglobin system of the coho salmon *Oncorhynchus kisutch* during smoltification and triiodothyronine and propylthiouracil treatment. *Comp. Biochem. Physiol. A* **81A,** 807–813.

Sutterlin, A. M., Harmon, P., and Young, B. (1978). Precocious sexual maturation in Atlantic salmon (*Salmo salar*) postsmolts reared in a seawater impoundment. *J. Fish Res. Board Can.* **35,** 1269–1271.

Sweeting, R. M., Wagner, G. F., and McKeown, B. A. (1985). Changes in plasma glucose, amino acid nitrogen and growth hormone during smoltification and seawater adaptation in coho salmon, *Oncorhynchus kisutch. Aquaculture* **45,** 185–197.

Swift, D. R. (1959). Seasonal variation in the activity of the thyroid gland of yearling brown trout *Salmo trutta* Linn. *J. Exp. Biol.* **36,** 120–125.

Thorpe, J. E. (1977). Bimodal distribution of length of juvenile Atlantic salmon (*Salmo salar* L.) under artificial rearing conditions. *J. Fish Biol.* **11,** 175–184.

Thorpe, J. E., ed. (1980). "Salmon Ranching." Academic Press, London.

Thorpe, J. E. (1986). Age at first maturity in Atlantic salmon, *Salmo salar:* Freshwater period influences and conflicts for smolting. *Can. Spec. Publ. Fish Aquat. Sci.* **89,** 7–14.

Thorpe, J. E. (1987). Smolting versus residency: Developmental conflict in salmonids. *Am. Fish Soc. Symp. 1,* 244–252.

Thorpe, J. E., and Morgan, R. I. G. (1978). Paternal influence on growth rate, smolting rate and survival in hatchery reared juvenile Atlantic salmon, *Salmo salar. J. Fish Biol.* **13,** 549–556.

Thorpe, J. E., and Morgan, R. I. G. (1980). Growth-rate and smolting-rate of progeny of male Atlantic salmon parr, *Salmo salar* L. *J. Fish Biol.* **17,** 451–459.

Thorpe, J. E., Morgan, R. I. G., Ottaway, E. M., and Miles, M. S. (1980). Time of divergence of growth groups between potential 1 + and 2+ smolts among sibling Atlantic salmon. *J. Fish Biol.* **17,** 13–21.

Thorpe, J. E., Talbot, C., and Villarreal, C. (1982). Bimodality of growth and smolting in Atlantic salmon, *Salmo salar* L. *Aquaculture* **28,** 123–132.

Thorpe, J. E., Morgan, R. I. G., Talbot, C., and Miles, M. S. (1983). Inheritance of developmental rates in Atlantic salmon, *Salmo salar* L. *Aquaculture* **33,** 119–128.

Thorpe, J. E., Bern, H. A., Saunders, R. L., and Soivio, A. (1985). Salmon Smoltification II. Proceedings of a workshop sponsored by the Commission of the European Communities. Univ. Stirling, Scotland 3-6 July, 1984. *Aquaculture* **45,** 1–404.

Threadgold, L. T., and Houston, A. H. (1961). An electron microscope study of the 'chloride-secretory cell' of *Salmo salar* L., with reference to plasma-electrolyte regulation. *Nature (London)* **190,** 612–614.

Vanstone, W. E., and Markert, J. R. (1968). Some morphological and biochemical changes in coho salmon, *Oncorhynchus kisutch,* during the parr–smolt transformation. *J. Fish. Res. Board Can.* **25,** 2403–2418.

Vanstone, W. E., Roberts, E., and Tsuyuki, H. (1964). Changes in the multiple hemoglobin patterns of some Pacific salmon, genus *Oncorhynchus,* during the parr–smolt transformation. *Can. J. Physiol. Pharmacol.* **42,** 697–703.

Villarreal, C. A., and Thorpe, J. E. (1985). Gonadal growth and bimodality of length frequency distribution in juvenile Atlantic salmon (*Salmo salar*). *Aquaculture* **45,** 265–288.

Virtanen, E., and Soivio, A. (1985). The patterns of T_3, T_4, cortisol and Na^+-K^+-ATPase during smoltification of hatchery-reared *Salmo salar* and comparison with wild smolts. *Aquaculture* **45,** 97–109.

Wagner, H. H. (1974a). Photoperiod and temperature regulation of smolting in steel-head trout (*Salmo gairdneri*). *Can. J. Zool.* **52,** 219–234.

Wagner, H. H. (1974b). Seawater adaptation independent of photoperiod in steelhead trout (*Salmo gairdneri*). *Can. J. Zool.* **52,** 805–812.

Wagner, H. H., Conte, F. P., and Fessler, J. L. (1969). Development of osmotic and ionic regulation in two races of chinook salmon, *Oncorhynchus tshawytscha. Comp. Biochem. Physiol.* **29,** 325–341.

Wedemeyer, G. (1972). Some physiological consequences of handling stress in the juvenile coho salmon (*Oncorhynchus kisutch*) and steelhead trout (*Salmo gairdneri*). *J. Fish. Res. Board Can.* **29,** 1780–1783.

Wedemeyer, G. A., Saunders, R. L., and Clarke, W. C. (1980). Environmental factors affecting smoltification and early marine survival of anadromous salmonids. *Mar. Fish. Rev.* **42**(6), 1–14.

Weisbart, M. (1968). Osmotic and ionic regulation in embryos, alevins, and fry of the five species of Pacific salmon. *Can. J. Zool.* **46,** 385–397.

Wendt, C. A. G., and Saunders, R. L. (1973). Changes in carbohydrate metabolism in young Atlantic salmon in response to various forms of stress. *Int. Atl. Salmon Found., Spec. Publ. Ser.* **4,**(1), 55–82.

White, B. A., and Henderson, N. E. (1977). Annual variations in the circulating levels of thyroid hormones in the brook trout, *Salvelinus fontinalis,* as measured by radioimmunoassay. *Can. J. Zool.* **55,** 475–481.

White, H. C. (1939). Factors influencing descent of Atlantic salmon smolts. *J. Fish. Res. Board Can.* **4,** 323–326.

Wilkins, N. P. (1968). Multiple hemoglobins of the Atlantic salmon (*Salmo salar*). *J. Fish. Res. Board Can.* **25,** 2651–2663.

Wilkins, N. P., and Gosling, E. M., eds. (1983). Genetics in aquaculture. Proceedings of the International Symposium, Galway, Ireland, 1982. *Aquaculture* **33,** 1–426.

Winans, G. A. (1984). Multivariate morphometric variability in Pacific salmon: Technical demonstration. *Can. J. Fish. Aquat. Sci.* **41,** 1150–1159.

Winans, G. A., and Nishioka, R. S. (1987). A multivariate description of changes in body shape of coho salmon (*Oncorhynchus kisutch*) during smoltification. *Aquaculture* **66** (in press).

Withey, K. G., and Saunders, R. L. (1973). Effect of a reciprocal photoperiod regime on standard rate of oxygen consumption of postsmolt Atlantic salmon (*Salmo salar*). *J. Fish. Res. Board Can.* **30,** 1898–1900.

Woo, N. Y. S., Bern, H. A., and Nishioka, R. S. (1978). Changes in body composition associated with smoltification and premature transfer to seawater in coho salmon (*Oncorhynchus kisutch*) and king salmon (*O. tshawytscha*). *J. Fish Biol.* **13,** 421–428.

Woodhead, P. M. J. (1970). The effect of thyroxine upon the swimming of cod. *J. Fish. Res. Board Can.* **27,** 2337–2338.

Yamauchi, K., Koide, K., Adachi, S., and Nagahama, Y. (1984). Changes in seawater adaptability and blood thyroxine concentrations during smoltification of the masu salmon, *Oncorhynchus masou,* and the Amago salmon, *Oncorhynchus rhodurus. Aquaculture* **42,** 247–256.

Yamauchi, K., Ban, M., Kasahara, N., Izumi, T., Kojima, H., and Harako, T. (1985). Physiological and behavioral changes occurring during smoltification in the masu salmon, *Oncorhynchus masou. Aquaculture* **45,** 227–235.

Young, G. (1986). Cortisol secretion *in vitro* by the interrenal of coho salmon (*Oncorhynchus kisutch*) during smoltification: Relationship with plasma thyroxine and plasma cortisol. *Gen. Comp. Endocrinol.* **63,** 191–200.

Youngson, A. F., and Simpson, T. H. (1984). Changes in serum thyroxine levels during smolting in captive and wild Atlantic salmon, *Salmo salar* L. *J. Fish Biol.* **24,** 29–39.

Youngson, A. F., Buck, R. J. G., Simpson, T. H., and Hay, D. W. (1983). The autumn and spring emigrations of juvenile Atlantic salmon, *Salmo salar* L., from the Girnock Bum, Aberdeenshire, Scotland: Environmental release of migration. *J. Fish Biol.* **23,** 625–639.

Youngson, A. F., Scott, D. C. B., Johnstone, R., and Pretswell, D. (1985). The thyroid system's role in the downstream migration of Atlantic salmon (*Salmo salar* L.) smolts. *Aquaculture* **45,** 392–393.

Youngson, A. F., McLay, H. A., and Olsen, T. C. (1986). The responsiveness of the thyroid system of Atlantic salmon (*Salmo salar* L.) smolts to increased water velocity. *Aquaculture* **56,** 243–255.

Zadunaisky, J. A. (1984). The chloride cell: The active transport of chloride and the paracellular pathways. *In* "Fish Physiology" (W. S. Hoar and D. J. Randall, eds.), Vol. 10B, pp. 129–176. Academic Press, New York.

Zaugg, W. S. (1981). Advanced photoperiod and water temperature effects on gill Na^+-K^+ adenosine triphosphatase activity and migration of juvenile steelhead (*Salmo gairdneri*). *Can. J. Fish. Aquat. Sci.* **38,** 758–764.

Zaugg, W. S. (1982). Some changes in smoltification and seawater adaptability of salmonids resulting from environmental and other factors. *Aquaculture* **28,** 143–151.

Zaugg, W. S., and McLain, L. R. (1970). Adenosine triphosphatase activity in gills of salmonids: Seasonal variations and salt water influence in coho salmon, *Oncorhynchus kisutch. Comp. Biochem. Physiol.* **35,** 587–596.

Zaugg, W. S., and McLain, L. R. (1972). Changes in gill adenosinetriphosphatase activity associated with parr–smolt transformation in steelhead trout, coho, and spring chinook salmon. *J. Fish. Res. Board Can.* **29,** 167–171.

Zaugg, W. S., and McLain, L. R. (1976). Influence of water temperature on gill sodium, potassium-stimulated ATPase activity in juvenile coho salmon (*Oncorhynchus kisutch*). *Comp. Biochem. Physiol. A* **54A,** 419–421.

Zaugg, W. S., and Wagner, H. H. (1973). Gill ATPase activity related to parr–smolt transformation and migration in steelhead trout (*Salmo gairdneri*): Influence of photoperiod and temperature. *Comp. Biochem. Physiol. B* **45B,** 955–965.

Zaugg, W. S., Adams, B. L., and McLain, L. R. (1972). Steelhead migration: Potential temperature effects as indicated by gill adenosine triphosphatase activities. *Science* **176,** 415–416.

Zaugg, W. S., Bodle, J. E., Manning, J. E., and Wold, E. (1986). Smolt transformation and seaward migration of 0-age progeny of adult spring chinook salmon (*Oncorhynchus tshawytscha*) matured early with photoperiod control. *Can. J. Fish. Aquat. Sci.* **43,** 885–888.

Zbanyszek, R., and Smith, L. S. (1984). Changes in carbonic anhydrase activity in coho salmon smolts resulting from physical training and transfer into seawater. *Comp. Biochem. Physiol.* A **79A,** 229–233.

Chapter 3

Have we figured out how hormones control gill Na^+,K^+-ATPase and chloride cell function?

J. Mark Shrimpton*

Ecosystem Science & Management Department, University of Northern British Columbia, Prince George, BC, Canada

**Corresponding author: e-mail: mark.shrimpton@unbc.ca*

Mark Shrimpton discusses the impact of Stephen McCormick's chapter "Hormonal Control of Gill Na^+,K^+-ATPase and Chloride Cell Function" in *Fish Physiology*, Volume 14 published in 1995.

Internal homeostasis of fish is maintained by biochemical and cellular mechanisms that oppose the large differences between the environment and the internal ionic and osmotic content. In fresh water, fish are hyperosmotic to their environment and marine fish are approximately one-third the osmotic concentration of seawater. Kidney, intestine, and gills are the organs involved in osmoregulation. Much research has focused on the gill and in particular the large mitochondrial rich cells known as chloride cells that have high concentrations of the enzyme Na^+,K^+-ATPase. The distribution and abundance of chloride cells is altered in response to changes in salinity as are the subunits and specific activity of Na^+,K^+-ATPase. These changes are under hormonal control with growth hormone functioning for saltwater adaptation and prolactin functioning for freshwater adaptation. Cortisol serves a dual role for fish to survive in both fresh and seawater. The use of molecular and immunohistochemical approaches has provided new insights into the hormonal control of ion balance and work on basal vertebrates has increased our understanding of the evolution of euryhalinity in fishes. This chapter describes some of the advances that have occurred since the publication in 1995 of Volume 14 in the *Fish Physiology* series and how chloride cells and Na^+,K^+-ATPase in the gill maintain homeostasis of fishes and the endocrine factors that regulate their function.

The ability of animals to survive in a variety of external osmotic environments required the evolution of mechanisms to maintain a stable internal environment. Organs important for osmotic and ionic balance in bony fishes are

Fish Physiology, Vol. 40B. https://doi.org/10.1016/bs.fp.2024.05.004

kidney, intestine, and gills, but it is the gills that has captivated our interest. Studies on biochemical and cellular mechanisms for maintaining homeostasis have shown the importance of the gill and particularly a type of large cells with abundant mitochondria. These cells are found in the gills of both seawater and freshwater fishes and are referred to as chloride cells (see review by Edwards and Marshall, 2013). Chloride cells in the gill of bony fishes have high concentrations of the enzyme Na^+,K^+-ATPase (NKA) which requires energy to translocate ions. This enzyme or more specifically different isoforms of NKA and the location of chloride cells in the gill drive the secretion of excess sodium and chloride in salt water, but also provide the mechanism for uptake of sodium and chloride in fresh water. Although research has examined how stenohaline species regulate ions, much of our research focus has examined the processes that diadromous or euryhaline species use to regulate ions and maintain stable internal osmotic pressures. Diadromous species include anadromous species that spawn in freshwater habitats and migrate to the ocean for part of their life cycle and catadromous species that spawn in the ocean and migrate to rivers for part of their life cycle. The economic importance of these species has fuelled much of the research on fishes that have developed these strategies.

Anadromous species such as Pacific and Atlantic salmon are remarkable in the extensive migrations they undertake from freshwater natal locations to productive marine environments which maximize their reproductive potential for spawning when they return to freshwater locations that have appropriate habitat for larval survival. Throughout the life cycle of salmonids, the highest rates of mortality occur during the incubation period (Quinn, 2018) but transitioning between marine and freshwater environments also affect physiological performance (Brauner et al., 1992) and ultimately survival (Thorstad et al., 2012). Work on lamprey, a group of jawless fishes that diverged from the rest of the vertebrate lineage more than 500 million years ago, has shown similarities in molecular and cellular mechanisms within the gill, kidney, and gut for osmoregulation in anadromous forms (Ferreira-Martins et al., 2021). The similarities and differences in ultrastructure and distribution of ion transport cells in the gills of lamprey suggest that the capacity to exploit both the marine and freshwater environments was acquired early in the evolution of fishes. While much research examining movement between freshwater and seawater environments has been conducted on anadromous species, catadromous species also undergo extensive migrations where juveniles migrate from the ocean to fresh water to feed, grow, and then return to the ocean when sexually mature; a pattern that has been most extensively studied in eels. There are vast numbers of bony fishes (estimated to be more than 25,000 species), but only a small proportion, approximately 250 species, successfully move between freshwater and seawater environments (McDowall, 2001).

For anadromous species, our emphasis on salmon has been due to their commercial, cultural, and recreational importance. The importance has also extended to the resources that have been developed to work on anadromous species. Steve McCormick's lab is located in the Silvio O. Conte Anadromous Fish Research Center in Turner's Falls, MA—a facility named after the western Massachusetts member of the United States House of Representatives from 1959 to 1991 who was influential in funding the lab and also an avid angler. Similarly, the considerable amount of work examining eel physiology is due to the species "gastronomic popularity," particularly in Europe and Asia (Tesch, 2003).

We know much about the factors that control the timing of movements between fresh and marine waters. These factors have been extensively relied on to establish fisheries for glass eels when they migrate into rivers and for salmon when they migrate along the coast before river entry. How diadromous species manage to successfully transition between waters that differ so much in salinity has long been of interest to physiologists. Almost a century ago it was known that fish in fresh water must actively take up ions to replenish those lost though urine and by diffusion across the epidermis and this is done by the gill (Krogh, 1937). In seawater, bony fish drink to replace water lost across the gill but must excrete the excess ions. The site of chloride secretion was found to be specialized cells in the gill termed chloride cells (Keys and Willmer, 1932) or "mitochondrion-rich cells" due to the large number of mitochondria present in the cells to power ion transport or more recently "ionocytes" because this cell type does more than just chloride secretion in seawater (Hiroi and McCormick, 2012). Thus, our fascination with the gill in fish has had a long history as not only an organ important for gas exchange, acid-base balance and ammonia excretion, but also for ionoregulatory balance. In fact, the gill is more important for ion regulation in early development than for oxygen uptake (Fu et al., 2010).

The chapter in *Fish Physiology* by Steve McCormick provides a comprehensive overview of the role of the Na^+,K^+-ATPase enzyme in ion regulation. The NKA enzyme plays a key role in extrusion of salts across the gill in seawater fishes and also uptake of ions in freshwater fishes. Much work has been done examining the remodeling of the gill when fish transition between marine and freshwater environments. The role of NKA, the structure of the enzyme, and stoichiometry are well explained in the chapter. Different isoforms of the subunits for NKA were known in mammals at the time of publication, but it was not until almost a decade later that multiple isoforms of the different subunits were described for teleosts. Four NKA mRNA α-isoforms were identified and found to be expressed in the gill of rainbow trout (*Oncorhynchus mykiss*): α1a, α1b, α1c and α3 (Richards et al., 2003). Two mRNA isoforms, NKA α1c and α3 were found in very low levels in freshwater trout gill and abundance did not change following transfer to higher salinity water. The differential expression in rainbow trout of the α subunits following transfer to 40% and 80% seawater was the first indication that there might be isoforms of NKA that have different roles in maintaining ion balance depending on the salinity where fish are found (Richards et al., 2003).

Further work has confirmed these findings as McCormick et al. (2009) showed the abundance of proteins for NKA α1a decreased and NKA α1b increased in gill tissue when Atlantic salmon (*Salmo salar*) parr were acclimated to seawater. They used western blots and isoform-specific antibodies to Atlantic salmon α1a and α1b. Additionally, Flores and Shrimpton (2012) transferred rainbow trout from freshwater to ion poor water or salt water and found differences in abundance of mRNA for the α1a and α1b subunits consistent with the earlier work; an upregulation of α1a when transferred to ion poor water and downregulation of the α1a when transferred to higher salinity water. The α1b isoform increased dramatically with transfer to salt water, but no change for fish transferred to ion poor water (Fig. 1).

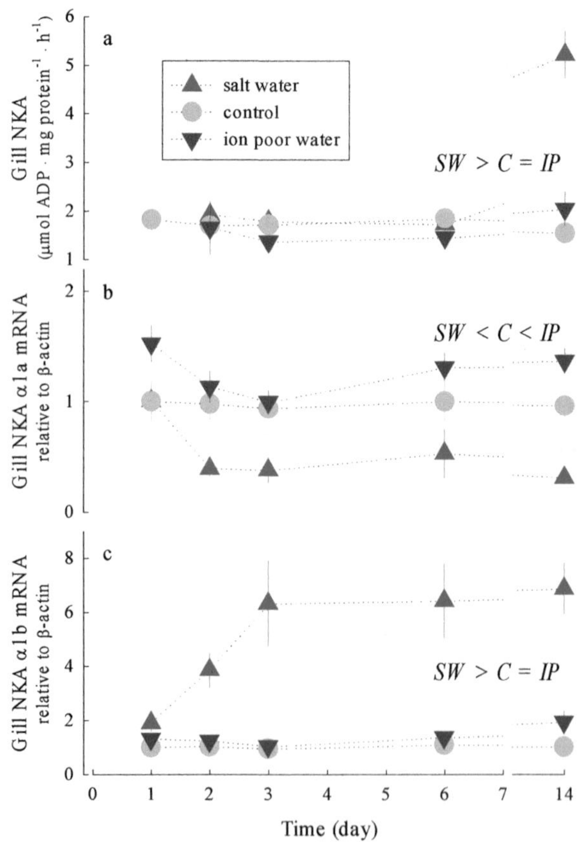

FIG. 1 (A) Gill Na^+, K^+-ATPase (NKA) activity (μmol ADP mg protein^{-1} h^{-1}), (B) Gill Na^+, K^+-ATPase (NKA) α1a expression (relative to β-actin) and (C) Gill NKA α1b expression (relative to β-actin) for fish in 3 experimental treatment groups: control (C), ion-poor water (IP) and salt water (SW). Overall effects of treatment determined by a Bonferroni test and the direction of change are shown ($P < 0.05$). Values are means ± se. *Adapted from Flores, A.-M., Shrimpton, J.M., 2012. Differential physiological and endocrine responses of rainbow trout,* Oncorhynchus mykiss, *transferred from fresh water to ion-poor or salt water. Gen. Comp. Endocrinol. 175, 246–247, with permission from Elsevier.*

Although changes in the α1-isoforms reveal the importance of these two subunits in experiments involving transfer to waters of different salinity, the seasonal changes that occur with development emphasize the importance of these changes. An increase in α1b mRNA and a decrease in α1a mRNA have been shown to occur during the spring in Atlantic salmon associated with development and smolting (Nilsen et al., 2007). An increase was found in the α1a isoform in gill tissue of maturing sockeye salmon (*O. nerka*) naturally migrating from the ocean and back to spawning grounds, but no change in α1b (Flores et al., 2012). Taken together, these findings indicate two subtypes of the α1 subunit for NKA in the gill: a saltwater subtype, α1b, and a freshwater subtype, α1a.

Although it was known that the site of salt extrusion for marine teleosts was the chloride cell, chloride cells were also considered to be the site of salt uptake in fresh water. How the same cell might function in this dual role was not clear, but the location of the chloride cells in the gill appeared to be important. Uchida et al. (1996) used immunocytochemical staining with an antiserum specific for the NKA α-subunit and detected chloride cells in filament and lamellar epithelia in freshwater chum salmon (*O. keta*) fry. The size of chloride cells on the filaments did not change, but the intensity of immunoreaction increased with transfer to seawater while chloride cells on the lamellae decreased in size. Much research has been conducted since then to define the difference in the two forms of the chloride cell. The use of immunofluorescent probes has provided clear evidence that chloride cells on the lamellae differ from those on the filaments. Immunolocalization of NKA with α1-isoform specific antibodies indicates that they are present in distinct chloride cells that differ between fresh water and seawater; α1a present in lamellar chloride cells in fresh water and α1b present in chloride cells on the filaments of the gill in seawater (Fig. 2). Chloride cells on the filaments were also more abundant in seawater, but a small number were also found in the gills of freshwater fish (McCormick et al., 2009). The combined patterns in abundance and immunolocalization of the two isoforms of the α-subunit explains the salinity-related changes in NKA and chloride cells that occur with transition between freshwater and marine environments. The use of fluorescent antibodies to other transport proteins, $Na^+/K^+/2Cl^-$ cotransporter (NKCC), cystic fibrosis transmembrane conductance regulator (CFTR), Na^+/Cl^- cotransporter (NCC), and Na^+/H^+ exchanger 3 (NHE3), has provided further validation of our models for ion secretion in saltwater fishes (McCormick et al., 2003) and presented new models for ion uptake in freshwater fishes (Edwards and Marshall, 2013; Hiroi et al., 2008).

Hormonal regulation of NKA activity and chloride cell morphology and development has been investigated for a long time and the principal endocrine factors are well described in the chapter by McCormick; cortisol, growth hormone (GH), insulin-like growth factor I (IGF-1), prolactin (PRL), thyroid hormones, and sex steroids. Of these endocrine factors, cortisol, GH, IGF-1, and PRL have a direct role in ion regulation. The role of thyroid hormones

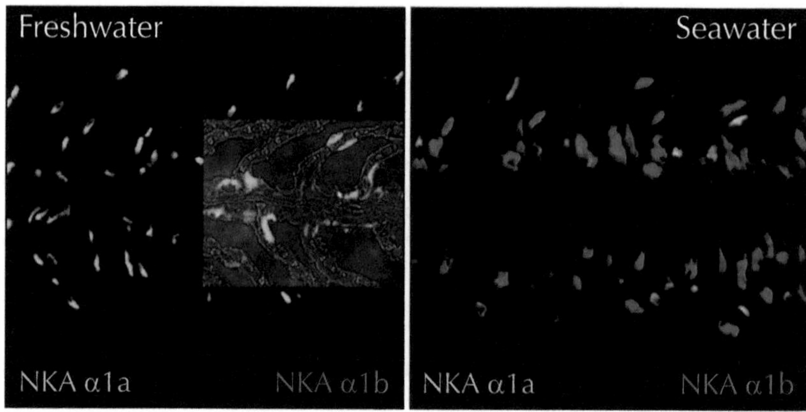

FIG. 2 Co-localization of NKA α1a (green) and NKA α1b (red) in gill tissue of Atlantic salmon parr in freshwater (left) and gradually acclimated to 30 ppt seawater (right). The α1 isoforms of NKA were only detected in chloride cells. *Reprinted from Hiroi, J., McCormick, S.D., 2012. New insights into gill ionocyte and ion transporter function in euryhaline and diadromous fish. Resp. Physiol. Neurobiol. 184, 261, with permission from Elsevier.*

and sex steroids appear to be more permissive and enhance or inhibit the role of cortisol, GH or PRL. For GH and PRL, the importance of these hormones is clear, and our knowledge of their role has not really changed—other than our approach for investigation. I recall listening to a talk by Thrandur Björnsson at a meeting hosted by the Physiology Section of the American Fisheries Society where he stated that "growth hormone" really should be called the "saltwater-adapting hormone" (Björnsson, 1997). Given the extensive research that has shown the importance of GH for enhancing saltwater tolerance, his suggestion has merit. I think an argument could be made that PRL should be called the "freshwater-adapting hormone" as it has long been known to be essential for survival of fishes in freshwater (Pickford and Phillips, 1959).

Our appreciation for the role these two hormones play in osmoregulation has a long history in the discipline of fish physiology, but PRL and GH have a much longer evolutionary history. Work on phylogenetically ancient vertebrates, sea lamprey (*Petromyzon marinus*) and Arctic lamprey (*Lethenteron camtschaticum*), found two distinct genes for GH and PRL receptors (Gong et al., 2020). The divergence of PRL and GH genes, therefore, occurred prior to emergence of an ancestral vertebrate to lamprey and suggests that a differential role for these two hormones arose early in the evolution of vertebrates. Mediation of GH as a "saltwater-adapting hormone" through IGF also appears to have a long evolutionary history. Recently, seasonal increases in mRNA for GH found in the pituitary of sea lamprey during metamorphosis were accompanied by elevated mRNA for IGF-1 in the gill and the development of saltwater tolerance (Ferreira-Martins et al., 2023). Regulation of ionic balance by corticosteroids is also a basal trait among vertebrates and evolved at least 500 million years ago. Cortisol, however, is absent in lamprey and it is the hormone 11-deoxycortisol that has a functional role and will increase gill NKA activity (Close et al., 2010) and improve saltwater tolerance (Shaughnessy et al., 2020). Endogenous increases in 11-deoxycortisol also coincide with metamorphosis and higher seasonal saltwater tolerance (Shaughnessy et al., 2020).

Although cortisol was once viewed as *the* seawater adapting hormone, the function of cortisol is now known to be much more complex. Cortisol is involved in stimulating ionoregulatory ability in both freshwater and seawater fishes. While working as a postdoctoral fellow in Steve McCormick's lab at the Conte Anadromous Fish Research Center in Turner's Falls MA, I still clearly remember him coming into my office and saying to me: "Here's a question for you. How does cortisol act to stimulate both hypo-osmoregulatory function and hyper-osmoregulatory function?" He did not need to add which taxonomic group he was referring to as it was clear the importance of this hormone on survival of fishes in seawater, fresh water, and particularly for the transition between the two environments. An area of research that we investigated while I was in Massachusetts and has continued in our individual labs and many other labs around the world for almost three decades.

Determining the dual role of cortisol in hypo-osmoregulation and hyper-osmoregulation is of much interest to fish physiologists. Cortisol has been shown to enhance ionoregulation in salt water by stimulating cellular recruitment of chloride cells (McCormick, 1990) and increase NKA activity (Madsen, 1990) in gill tissue of juvenile salmonids. Cortisol treatment also enhanced whole body Na^+ and Cl^- influxes through recruitment and proliferation of chloride cells for rainbow trout in ion poor water (Laurent and Perry, 1990). More recently, cortisol treatment has been shown to stimulate both the NKA α1a and NKA α1b subunits in Atlantic salmon (McCormick et al., 2008) and rainbow trout (Kiilerich et al., 2011). The interaction of cortisol with both GH and PRL, however, is particularly intriguing and provides insight into the dual role for cortisol. The effect of cortisol on ionoregulation is amplified when administered with GH or PRL. Treatment of brown trout (*S. trutta*) with GH had an additive effect on increasing NKA activity, chloride cell numbers and ionoregulatory ability in seawater when administered with cortisol (Madsen, 1990). Similarly, the effects of PRL were enhanced by treatment with cortisol on the Mozambique tilapia (*Oreochromis mossambicus*) in freshwater (Watanabe et al., 2016).

Understanding the dual role has been perplexing, but examination of the receptors for cortisol appeared to provide some clues. In sea lamprey, the corticosteroid 11-deoxycortisol acts through a single, ancestral corticosteroid receptor (CR) (Shaughnessy et al., 2020). It was long thought that the function of cortisol in teleosts was similar to the ancestral form and was through a single class of receptor, the glucocorticoid receptor (GR). Adding to the complexity, two distinct GR have been found in teleosts, GR1 and GR2 (Bury et al., 2003), and a second class of receptor for cortisol, the mineralocorticoid receptor (MR), has also been identified in fish (Colombe et al., 2000). The identification of multiple corticosteroid receptors in teleosts suggested a more complicated corticosteroid pattern of signaling in fish than previously recognized. Potentially, through these two classes of steroid receptors, cortisol may function in both fresh- and saltwater ionoregulation in concert with PRL or GH. Experimental evidence suggests that GR is important for the transition from fresh water to seawater. For example, an increase in GR abundance was detected during seawater adaptation for the European eel, *Anguilla anguilla* (Marsigliante et al., 2000), rainbow trout (Singer et al., 2007), and Atlantic salmon (Kiilerich et al., 2007). Conversely, it was suggested that cortisol may act through an MR in freshwater acclimation (Sloman et al., 2001).

The evidence for a differential role for GR and MR, however, is limited. MR mRNA levels did not change during smolting in Atlantic salmon, whereas the abundance of GR1 mRNA increased (Kiilerich et al., 2007; Nilsen et al., 2008). Interestingly, mRNA for both GR and MR significantly increased at the onset of de-smoltification (Kiilerich et al., 2007). Most studies, however, have shown little difference between the abundance of GR and MR mRNAs. Transfer from fresh water to seawater resulted in an increase in mRNA for GR and MR in steelhead (Yada et al., 2008) and Atlantic salmon smolts (Nilsen et al., 2008), but slight decreases for both classes of receptors in rainbow trout (Kiilerich et al., 2011). Transfer of rainbow trout from seawater to fresh water, resulted in significant increases in the abundance of GR and MR mRNAs (Kiilerich et al., 2011), but declines in mRNA for GR and MR were found after migrating adult sockeye salmon entered fresh water (Flores et al., 2012). Flores and Shrimpton (2012) found no significant change in GR or MR following transfer of rainbow trout from fresh water to salt water or ion poor water despite significant changes in receptor mRNA for GH and PRL and mRNA for the α1a and α1b subunits. Transcriptional patterns of GR1, GR2, and MR in the gill did not differ after cortisol implants were administered to rainbow trout with an initial decline in mRNA for all corticosteroid receptors and then an increase (Teles et al., 2013) suggesting autoregulation for all three receptors by cortisol (Sathiyaa and Vijayan, 2003).

The presence of both GR and MR could suggest that there may be two different ligands: one activating GR and one activating MR similar to the mammalian model. In this model high circulating levels of cortisol would activate a GR, but do not activate a MR due to the co-expression of MR with 11β-hydroxysteroid dehydrogenase type 2 which breaks down cortisol and allows mineralocorticoid specificity (Funder et al., 1988). Fish cannot synthesize aldosterone, the ligand for MR in higher vertebrates. They do produce

11-deoxycorticosterone (DOC) which has been detected in the plasma of fish, although at much lower concentrations than cortisol (Kiilerich et al., 2011), similar to the difference between cortisol and aldosterone in mammals. Additionally, DOC has been shown to be a potent agonist of the MR in rainbow trout with approximately a 10-fold greater affinity for the MR than cortisol (Sturm et al., 2005). DOC concentrations in plasma of rainbow trout, however, did not change in response to transfer from fresh water to seawater or the reciprocal transfers suggesting DOC may not be functional (Kiilerich et al., 2011). Other work also suggests that cortisol, not DOC, functions as the mineralocorticoid. Cortisol treatment of Atlantic salmon parr increased gill NKA activity and improved salinity tolerance, whereas DOC and aldosterone had no effect. NKA α1a and α1b mRNA levels were both upregulated by cortisol, whereas DOC and aldosterone had no effect (Fig. 3A). Cortisol, therefore, appears to be the only ligand for both GR and MR in euryhaline teleosts.

Further work suggests that functional roles for GR and MR may differ. The putative GR blocker RU486 was shown to inhibit cortisol-induced increases in NKA α1a and α1b mRNA (Kiilerich et al., 2011; McCormick et al., 2008). In contrast the putative MR blocker spironolactone did not inhibit the increase in NKA α1a or α1b mRNA levels from cortisol treatment (Fig. 3B). Such findings argue that for cortisol regulation of the NKA α1 subunits in the gill, only GR has a function. Further evidence that cortisol performs the role of both mineralocorticoid and glucocorticoid through a GR was reported by Cruz et al. (2013) using transcriptional knockdown in zebrafish (*Danio rerio*). Inhibition of GR protein synthesis in knockdown morphants resulted in a significant decrease in ionocytes; cells that were unaffected in MR knockdown morphants. Additionally, Sakamoto et al. (2016) created a MR-knockout mutant of medaka (*Oryzias latipes*) which survived in both fresh water and seawater, suggesting a limited role for MR in ionoregulation. Using transcriptional knockdown Kolosov and Kelly (2019), however, showed that the role of cortisol in freshwater ion regulation appears to be through an MR which resulted in reduced paracellular permeability due to a lack of formation of tight junctions. Both GR and MR appear to be important for ionoregulation, although, how genetic compensation may mediate the observed responses is unclear.

The sensitivity of gill tissue to cortisol is related to the number of intracellular receptors (Shrimpton and McCormick, 1999) and cortisol receptor dynamics take several hours to change in response to cortisol treatment (Sathiyaa and Vijayan, 2003). Consequently, the tissue response is not as rapid as changes in circulating cortisol levels normally. Non-genomic cortisol signaling has also been identified, providing evidence that cortisol activates multiple pathways that include rapid effects (Das et al., 2018). One pathway is through membrane bound GR and MR which have been identified on the surface of cells in the gill of rainbow trout using immunofluorescence (Aedo et al., 2023) and may be a pathway for rapid cortisol action during salinity acclimation.

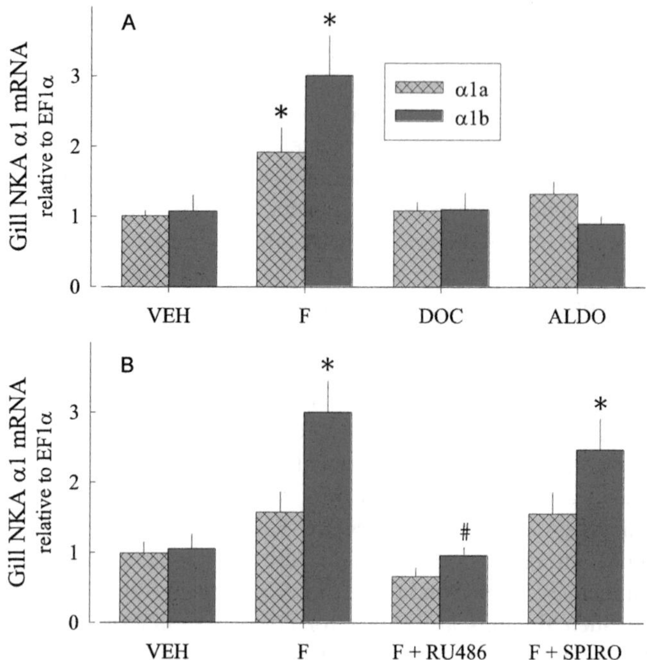

FIG. 3 (A) Gill NKA α1a and α1b mRNA levels relative to elongation factor 1α in Atlantic salmon parr treated with cortisol (F; 50 μg g^{-1}), 11-deoxycorticosterone (DOC; 45 μg g^{-1}) and aldosterone (ALDO; 50 μg g^{-1}) for 6 days. * indicates significant difference between the treated and vehicle group ($P < 0.05$, Student-Neuman-Keuls test). Values are means ± se ($n = 6$–8 fish per group). (B) Effect of RU 486 and spironolactone on the ability of cortisol to affect gill NKA α1a and α1b mRNA levels in Atlantic salmon parr. Cortisol was administered at 50 μg g^{-1}; RU 486 and spironolactone were administered at 400 μg g^{-1}. Values are mean ± standard error ($n = 8$–10 fish per group). * and # indicates significant difference between the vehicle and cortisol-treated groups, respectively ($P < 0.05$, Student-Neuman-Keuls test). *Adapted from McCormick, S.D., Regish, A., O'Dea, M.F., Shrimpton, J.M., 2008. Are we missing a mineralocorticoid in teleost fish? Effects of cortisol, deoxycorticosterone and aldosterone on osmoregulation, gill Na^+, K^+-ATPase activity and isoform mRNA levels in Atlantic salmon. Gen. Comp. Endocrinol. 157, 37–38, with permission from Elsevier.*

Despite all of the work examining the role of cortisol in ion regulation for both freshwater and seawater teleosts, the mechanism for the dual role of this hormone remains unclear. Although new molecular methods have furthered our understanding of the proteins and cells that are involved in ion regulation in marine and freshwaters, they focus on the transcription of the gene through the production of mRNA, a process that is often referred to as gene expression. Measures of mRNA using quantitative Real Time-Polymerase Chain Reaction (qRT-PCR), however, do not measure transcription rates, but are a "snapshot" at a particular moment in time and under physiological conditions that reflect both synthesis of RNA as well as degradation of RNA. Nevertheless, the extraction of RNA from tissues and conversion to complementary

DNA (cDNA) has increased our capacity to look at multiple genes simultaneously. Quantification of transcription at multiple selected gene loci allows us to detect changes and interactions that occur among multiple endocrine factors. Such techniques can quantify transcriptional responses that are preparatory to movement between freshwater and seawater or following experimental manipulations.

The expression of the gene, however, also requires translation of the mRNA into the protein molecule and post-translational modifications that are essential for functionality of hormones and receptors (Duma et al., 2006; Sadeghi et al., 2023). Translational regulation will alter synthesis of the protein and the mechanisms of modulation are not well understood in fishes. There is some evidence that translation modulation occurs in Atlantic salmon. Christensen et al. (2018) found the abundance of ion-transporter mRNA and protein do not always correlate well. Approaches that use qRT-PCR assays and incorporate them into high-throughput platforms can examine mRNA for enzymes that synthesize the hormone or mRNA for the receptor of the hormone. These techniques will measure changes in response to a hormone, rather than the hormone specifically. There are real advantages to understanding the actual and relative changes in the hormone. This is perhaps the greatest strength of the chapter in *Fish Physiology* examining hormonal regulation of NKA and chloride cells by Steve McCormick—he provided a summary of the actual endocrine signal and tissue responsiveness. For these reasons, the chapter is a classic and should be read by all interested in the mechanisms used by euryhaline fishes to survive in both hypoosmotic and hyperosmotic environments.

Acknowledgments

I am thankful for the privilege of working with so many talented individuals who have shared ideas to support my investigations on anadromous fishes, particularly Dave Randall and Steve McCormick as mentors and whose examples I have emulated with my own students. I also thank Mark Sheridan, Colin Brauner and Erika Eliason for their thoughtful comments and suggestions that greatly improved this chapter.

References

Aedo, J., Aravena, D., Zuloaga, R., Alegría, D., Valdés, J.A., Molina, A., 2023. Early regulation of corticosteroid receptor expression in rainbow trout (*Oncorhynchus mykiss*) gills is mediated by membrane-initiated cortisol signaling. Comp. Biochem. Physiol. A 281, 111423. https://doi.org/10.1016/j.cbpa.2023.111423.

Björnsson, B.T., 1997. The biology of salmon growth hormone: from daylight to dominance. Fish Physiol. Biochem. 17, 9–24.

Brauner, C.J., Shrimpton, J.M., Randall, D.J., 1992. The effect of short duration seawater exposure on plasma ion concentrations and swimming performance in coho salmon (*Oncorhynchus kisutch*). Can. J. Fish. Aquat. Sci. 49, 2399–2405.

Bury, N.R., Sturm, A., Le Rouzic, P., Lethimonier, C., Ducouret, B., Guiguen, Y., Robinson-Rechavi, M., Laudet, V., Rafestin-Oblin, M.E., Prunet, P., 2003. Evidence for two distinct functional glucocorticoid receptors in teleost fish. J. Mol. Endocrinol. 31, 141–156.

Christensen, A.K., Regish, A.M., McCormick, S.D., 2018. Shifts in the relationship between mRNA and protein abundance of gill iontransporters during smolt development and seawater acclimation in Atlantic salmon (*Salmo salar*). Comp. Biochem. Physiol. A 221, 63–73.

Close, D.A., Yun, S.-S., McCormick, S.D., Wildbill, A.J., Li, W., 2010. 11-Deoxycortisol is a corticosteroid hormone in the lamprey. Proc. Natl. Acad. Sci. U. S. A. 107, 13942–13947.

Colombe, L., Fostier, A., Bury, N., Pakdel, F., Guiguen, Y., 2000. A mineralocorticoid like receptor in the rainbow trout, *Oncorhynchus mykiss*: cloning and characterization of its steroid binding domain. Steroids 65, 319–328.

Cruz, S.A., Lin, C.-H., Chao, P.-L., Hwang, P.-P., 2013. Glucocorticoid receptor, but not mineralocorticoid receptor, mediates cortisol regulation of epidermal ionocyte development and ion transport in Zebrafish (*Danio rerio*). PLoS One 8, e77997. https://doi.org/10.1371/journal.pone.0077997.

Das, C., Thraya, M., Vijayan, M.M., 2018. Nongenomic cortisol signaling in fish. Gen. Comp. Endocrinol. 265, 121–127. https://doi.org/10.1016/j.ygcen.2018.04.019.

Duma, D., Jewell, C.M., Cidlowski, J.A., 2006. Multiple glucocorticoid receptor isoforms and mechanisms of post-translational modification. J. Steroid Biochem. Mol. Biol. 102, 11–21.

Edwards, S.L., Marshall, W.S., 2013. Principles and patterns of osmoregulation and euryhalinity in fishes. In: McCormick, S.D., Farrell, A.P., Brauner, C.J. (Eds.), *Fish Physiology*, Vol. 32, *Euryhaline Fishes*. Elsevier, New York, pp. 1–44.

Ferreira-Martins, D., Wilson, J.M., Kelly, S.P., Kolosov, D., McCormick, S.D., 2021. A review of osmoregulation in lamprey. J. Great Lakes Res. 47, S59–S71. https://doi.org/10.1016/j.jglr.2021.05.003.

Ferreira-Martins, D., Walton, E., Karlstrom, R.O., Sheridan, M.A., McCormick, S.D., 2023. The GH/IGF axis in the sea lamprey during metamorphosis and seawater acclimation. Mol. Cell. Endocrinol. 571, 111937. https://doi.org/10.1016/j.mce.2023.111937.

Flores, A.-M., Shrimpton, J.M., 2012. Differential physiological and endocrine responses of rainbow trout, *Oncorhynchus mykiss*, transferred from fresh water to ion-poor or salt water. Gen. Comp. Endocrinol. 175, 244–250. https://doi.org/10.1016/j.ygcen.2011.11.002.

Flores, A.-M., Shrimpton, J.M., Patterson, D.A., Hills, J.A., Cooke, S.J., Yada, T., Moriyama, S., Hinch, S.G., Farrell, A.P., 2012. Physiological and molecular endocrine changes in maturing wild sockeye salmon, *Oncorhynchus nerka*, during ocean and river migration. J. Comp. Physiol. B 182, 77–90. https://doi.org/10.1007/s00360-011-0600-4.

Fu, C., Wilson, J.M., Rombough, P.J., Brauner, C.J., 2010. Ions first: Na^+ uptake shifts from the skin to the gills before O_2 uptake in developing rainbow trout, *Oncorhynchus mykiss*. Proc. R. Soc. B 277, 1553–1560. https://doi.org/10.1098/rspb.2009.1545.

Funder, J.W., Pearce, P.T., Smith, R., Smith, A.I., 1988. Mineralocorticoid action: target tissue specificity is enzyme, not receptor, mediated. Science 242, 583–585.

Gong, N., Ferreira-Martins, D., McCormick, S.D., Sheridan, M.A., 2020. Divergent genes encoding the putative receptors for growth hormone and prolactin in sea lamprey display distinct patterns of expression. Sci. Rep. 10, 1674. https://doi.org/10.1038/s41598-020-58344-5.

Hiroi, J., McCormick, S.D., 2012. New insights into gill ionocyte and ion transporter function in euryhaline and diadromous fish. Respir. Physiol. Neurobiol. 184, 257–268.

Hiroi, J., Yasumasu, S., McCormick, S.D., Hwang, P.-P., Kaneko, T., 2008. Evidence for an apical Na–Cl cotransporter involved in ion uptake in a teleost fish. J. Exp. Biol. 211, 2584–2599.

Keys, A.B., Willmer, E.N., 1932. "Chloride-secreting cells" in the gills of fish with special reference to the common eel. J. Physiol. 76, 368–378.

Kiilerich, P., Kristiansen, K., Madsen, S.S., 2007. Hormone receptors in gill of smelting Atlantic salmon, Salmo salar: expression of growth hormone, prolactin, mineralocorticoid and glucocorticoid receptors and 11β-hydroxysteroid dehydrogenase type 2. Gen. Comp. Endocrinol. 152, 295–303.

Kiilerich, P., Milla, S., Sturm, A., Valotaire, C., Chevolleau, S., Giton, F., Terrien, X., Fiet, J., Fostier, A., Debrauwer, L., Prunet, P., 2011. Implication of the mineralocorticoid axis in rainbow trout osmoregulation during salinity acclimation. J. Endocrinol. 209, 221–235.

Kolosov, D., Kelly, S.P., 2019. The mineralocorticoid receptor contributes to barrier function of a model fish gill epithelium. J. Exp. Biol. 222, jeb.192096. https://doi.org/10.1242/jeb.192096.

Krogh, A., 1937. Osmotic regulation in fresh water fishes by active absorption of chloride ions. Z. Vergl. Physiol. 24, 656–666.

Laurent, P., Perry, S.F., 1990. Effects of cortisol on gill chloride cell morphology and ionic uptake in the freshwater troutv. Cell Tissue Res. 259, 429–442.

Madsen, S.S., 1990. The role of cortisol and growth hormone in seawater adaptation and development of hypoosmoregulatroy mechanisms in sea trout parr (*Salmo trutta trutta*). Gen. Comp. Endocrinol. 79, 1–11.

Marsigliante, S., Barker, S., Jimenez, E., Storelli, C., 2000. Glucocorticoid receptors in the euryhaline teleost *Anguilla anguilla*. Mol. Cell. Endocrinol. 162, 193–201.

McCormick, S.D., 1990. Cortisol directly stimulates differentiation of chloride cells in tilapia opercular membrane. Am. J. Physiol. 259, R857–R863.

McCormick, S.D., Sundell, K., Björnsson, B.T., Brown, C.L., Hiroi, J., 2003. Influence of salinity on the localization of Na^+/K^+-ATPase, $Na^+/K^+/2Cl^-$ cotransporter (NKCC) and CFTR anion channel in chloride cells of the Hawaiian goby (*Stenogobius hawaiiensis*). J. Exp. Biol. 206, 4575–4583.

McCormick, S.D., Regish, A., O'Dea, M.F., Shrimpton, J.M., 2008. Are we missing a mineralocorticoid in teleost fish? Effects of cortisol, deoxycorticosterone and aldosterone on osmoregulation, gill Na^+, K^+-ATPase activity and isoform mRNA levels in Atlantic salmon. Gen. Comp. Endocrinol. 157, 35–40.

McCormick, S.D., Regish, A.M., Christensen, A.K., 2009. Distinct freshwater and seawater isoforms of Na^+/K^+-ATPase in gill chloride cells of Atlantic salmon. J. Exp. Biol. 212, 3994–4001. https://doi.org/10.1242/jeb.037275.

McDowall, R.M., 2001. Diadromy, diversity and divergence: implications for speciation processes in fish. Fish Fish. 2, 278–285.

Nilsen, T.O., Ebbesson, L.O.E., Madsen, S.S., MCormick, S.D., Anderson, E., Björnsson, B.T., Prunet, P., Stefansson, S.O., 2007. Differential expression of gill Na^+,K^+-ATPase α- and β-subunits, $Na^+,K^+,2Cl^-$ cotransporter and CFTR anion channel in juvenile anadromous and landlocked Atlantic salmon *Salmo salar*. J. Exp. Biol. 210, 2885–2896.

Nilsen, T.O., Ebbesson, L.O.E., Kiilerich, P., Björnsson, B.T., Madsen, S.S., McCormick, S.D., Stefansson, S.O., 2008. Endocrine systems in juvenile anadromous and landlocked Atlantic salmon (*Salmo salar*): seasonal development and seawater acclimation. Gen. Comp. Endocrinol. 155, 762–772.

Pickford, G.E., Phillips, J.G., 1959. Prolactin, a factor in promoting survival of hypophysectomized killifish in fresh water. Science 130, 454–455.

Quinn, T.P., 2018. The Behavior and Ecology of Pacific Salmon and Trout, second ed. UBC Press, Vancouver, BC.

Richards, J.G., Semple, J.W., Bystriansky, J.S., Schulte, P.M., 2003. Na^{+}/K^{+}-ATPase α-isoform switching in gills of rainbow trout (*Oncorhynchus mykiss*) during salinity transfer. J. Exp. Biol. 206, 4475–4486.

Sadeghi, J., Chaganti, S.R., Heath, D.D., 2023. Regulation of host gene expression by gastrointestinal tract microbiota in Chinook Salmon (*Oncorhynchus tshawytscha*). Mol. Ecol. 32, 4427–4446.

Sakamoto, T., Yoshiki, M., Takahashi, H., Yoshida, M., Ogino, Y., Ikeuchi, T., Nakamachi, T., Konno, N., Matsuda, K., Sakamoto, H., 2016. Principal function of mineralocorticoid signaling suggested by constitutive knockout of the mineralocorticoid receptor in medaka fish. Sci. Rep. 6, 37991. https://doi.org/10.1038/srep37991.

Sathiyaa, R., Vijayan, M.M., 2003. Autoregulation of glucocorticoid receptor by cortisol in rainbow trout hepatocytes. Am. J. Physiol. 284, C1508–C1515.

Shaughnessy, C.A., Barany, A., McCormick, S.D., 2020. 11-Deoxycortisol controls hydromineral balance in the most basal osmoregulating vertebrate, sea lamprey (*Petromyzon marinus*). Sci. Rep. 10, 12148. https://doi.org/10.1038/s41598-020-69061-4.

Shrimpton, J.M., McCormick, S.D., 1999. Responsiveness of gill $Na^{+}K^{+}$ATPase to cortisol is related to gill corticosteroid receptor concentration in juvenile rainbow trout. J. Exp. Biol. 202, 987–995.

Singer, T.D., Raptis, S., Sathiyaa, R., Nichols, J.W., Playle, R.C., Vijayan, M.M., 2007. Tissue-specific modulation of glucocorticoid receptor expression in response to salinity acclimation in rainbow trout. Comp. Biochem. Physiol. B 146, 271–278.

Sloman, K.A., Desforges, P.R., Gilmour, K.M., 2001. Evidence for a mineralocorticoidlike receptor linked to branchial chloride cell proliferation in freshwater rainbow trout. J. Exp. Biol. 204, 3953–3961.

Sturm, A., Bury, N., Dengreville, L., Fagart, J., Flouriot, G., Rafestion-Oblin, M.E., Prunet, P., 2005. 11-Deoxycorticosterone is a potent agonist of the rainbow trout (*Onchrhynchus mykiss*) mineralocorticoid receptor. Endocrinology 146, 47–55.

Teles, M., Tridico, R., Callo, A., Fierro-Castro, C., Tort, L., 2013. Differential expression of the corticosteroid receptors GR1, GR2 and MR in rainbow trout organs with slow release cortisol implants. Comp. Biochem. Physiol. A 164, 506–511.

Tesch, F.-W., 2003. The Eel. Blackwell Science, London.

Thorstad, E.B., Whoriskey, F., Uglem, I., Moore, A., Rikardsen, A.H., Finstad, B., 2012. A critical life stage of the Atlantic salmon Salmo salar: behaviour and survival during the smolt and initial post-smolt migration. J. Fish Biol. 81, 500–542. https://doi.org/10.1111/j.1095-8649.2012.03370.x.

Uchida, K., Kaneko, T., Yamauchi, K., Hirano, T., 1996. Morphometrical analysis of chloride cell activity in the gill filaments and lamellae and changes in Na^{+},K^{+}-ATPase activity during seawater adaptation in chum salmon fry. J. Exp. Zool. 276, 193–200.

Watanabe, S., Itoh, K., Kaneko, T., 2016. Prolactin and cortisol mediate the maintenance of hyperosmoregulatory ionocytes in gills of Mozambique tilapia: exploring with an improved gill incubation system. Gen. Comp. Endocrinol. 232, 151–159.

Yada, T., Hyodo, S., Schreck, C.B., 2008. Effects of seawater acclimation on mRNA levels of corticosteroid receptor genes in osmoregulatory and immune systems in trout. Gen. Comp. Endocrinol. 156, 622–627.

Chapter 4

HORMONAL CONTROL OF GILL Na^+,K^+-ATPase AND CHLORIDE CELL FUNCTION☆

STEPHEN D. MCCORMICK

Chapter Outline

I. INTRODUCTION

The secretion of excess sodium and chloride by teleosts in seawater is carried out by gill chloride cells. These highly specialized cells are characterized by a large, columnar appearance, numerous mitochondria, an extensive tubular system, an apical crypt, and mucosal-serosal exposure (see reviews by Evans *et al.*, 1982; Zadunaisky, 1984; Pisam and Rombourg, 1991). The tubular

☆This is a reproduction of a previously published chapter in the Fish Physiology series, "1995 (Vol. 14)/Hormonal Control of Gill Na^+, K^+-ATPase and Chloride Cell Function: ISBN: 978-0-12-350438-8; ISSN: 1546–5098".

Fish Physiology, Vol. 40B. https://doi.org/10.1016/bs.fp.2024.07.001

system is continuous with the basolateral membrane, resulting in a large surface area for the placement of transport proteins. Perhaps the most important of these proteins is Na^+, K^+-ATPase. Also known as the sodium pump, Na^+, K^+-ATPase plays a central role in the salt-secretory function of chloride cells. Present in all animal cells, this energy-dependent, ion-translocating enzyme occurs in high concentrations in most transport epithelia; up to 10^8 sodium pumps may be present in a single chloride cell (Karnaky, 1986)! Na^+, K^+-ATPase participates in ion transport either directly through movement of sodium and potassium across the plasma membrane or indirectly through generation of ionic and electrical gradients.

Regulation of chloride cells and Na^+, K^+-ATPase is critical during movement of fish between fresh water and seawater and within estuaries, and is also important in stenohaline fish under several variable environmental conditions. Since the neuroendocrine system is the primary link between a changing environment and physiological adaptation, the hormonal control of chloride cells and Na^+, K^+-ATPase is critical to euryhalinity. Reviews in this general area have appeared previously (Foskett *et al.*, 1983; Mayer-Gostan *et al.*, 1987) though not recently, and this review will focus primarily on new approaches and recent evidence on the hormonal control of Na^+, K^+-ATPase and chloride cells in teleosts.

II. NA^+, K^+-ATPase AND CHLORIDE CELL FUNCTION

A. Seawater

In seawater the osmoregulatory function of the gill is to secrete excess monovalent ions. Although there are several models for ion movement across the gill (Payan *et al.*, 1984), the model proposed by Silva *et al.* (1977) has the most experimental support (Evans *et al.*, 1982). Na^+, K^+-ATPase, located on the basolateral membrane, creates low intracellular Na^+ and a highly negative charge within the cell. The Na^+ gradient is then used to transport Cl^- into the cell through a $Na^+/K^+/Cl^-$ cotransporter, then Cl^- leaves the cell "downhill" on an electrical gradient through an apical Cl^- channel. Na^+ is transported through a paracellular pathway down its electrical gradient (seawater being more negative than plasma).

In this model, Na^+, K^+-ATPase generates ionic and electrical gradients for use in excretion of Na^+ and Cl^-. Epstein *et al.* (1967) were the first to suggest a role for Na^+, K^+-ATPase in the salt excretory function of the gill, based on their observation of higher levels of gill Na^+, K^+-ATPase activity in seawater-adapted killifish (*Fundulus heteroclitus*). Ouabain, a specific inhibitor of Na^+, K^+-ATPase that binds to the potassium-binding site, strongly inhibits Cl^- excretion and short-circuit current (Silva *et al.*, 1977; Karnaky *et al.*, 1977). The basal location of Na^+, K^+-ATPase was inferred from the greater efficacy of injected ouabain compared to external ouabain exposure (Silva *et al.*, 1977). Cell fractions of dispersed gill tissue that are rich in chloride cells

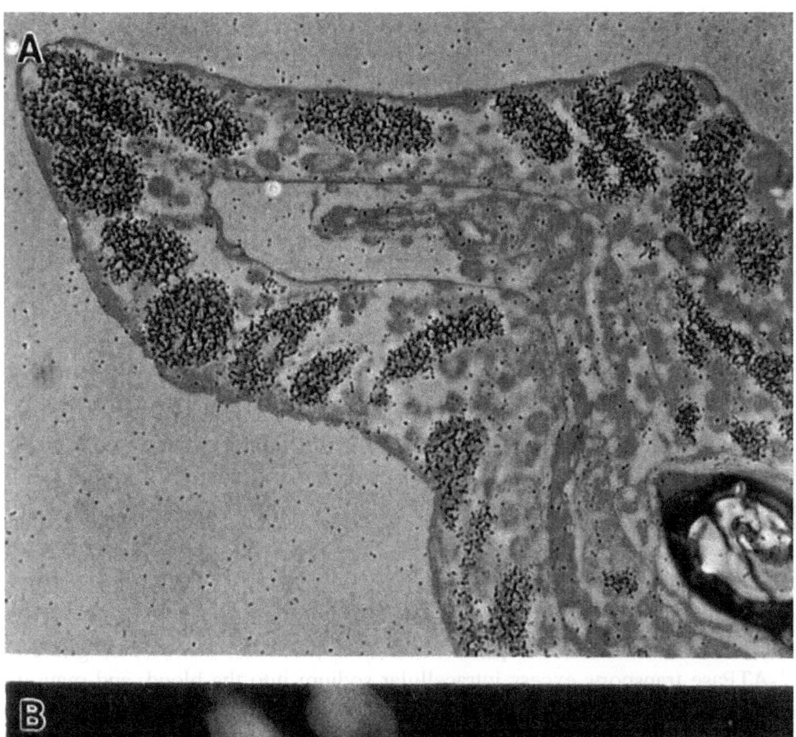

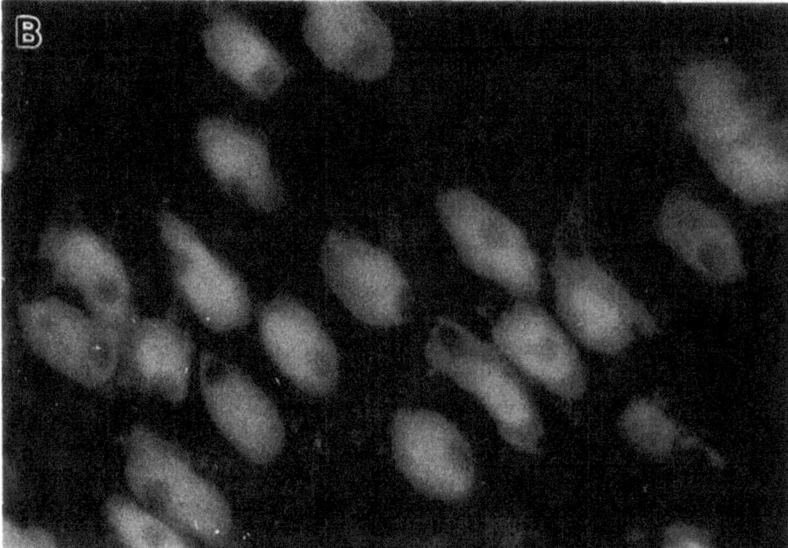

Fig. 1. (A) [^{3}H]Ouabain binding in chloride cells of seawater-adapted killifish. [Reproduced from the *Journal of Cell Biology,* Karnaky *et al.* (1976) **70,** pp. 157–177, by copyright permission of the Rockefeller University Press.] (B) Fluorescent staining of chloride cells in the jawskin of the long-jawed mudsucker following exposure to anthroylouabain. [Reproduced from McCormick (1990a) with permission from Springer-Verlag.] In both preparations, note the high concentration of ouabain binding throughout the cells (but not in the nucleus), indicative of its placement on the extensive tubular system of chloride cells.

have much higher Na^+, K^+-ATPase activity than other cell fractions (Kamiya, 1972; Sargent *et al.*, 1975). Karnaky *et al.* (1976) were able to visualize high concentration of Na^+, K^+-ATPase in chloride cells through the use of [^{3}H]ouabain autoradiography (Fig. 1). These results indicated that the sodium pump was not located on the apical surface, but rather throughout the cell on the extensive tubular system that is continuous with the basolateral membrane.

The model of Silva *et al.* (1977) offers several predictions of the biochemical properties of chloride cells that have yet to be verified. For example, a bumetanide-sensitive $Na^+/K^+/Cl^-$cotransporter should be located on the basolateral membrane of chloride cells, and an apical Cl^- channel should be present on the apical membrane of chloride cells. Molecular methods, along with more traditional immunological and pharmacological approaches, should allow exploration of these predictions.

B. Fresh Water

Our current understanding of chloride uptake by the gills in fresh water emphasize $Cl^- - HCO_3^-$ exchange at the apical surface (Avella and Bornancin, 1990). Sodium uptake is accomplished either by apical $Na^+ - H^+/NH_4^+$ exchange or by an apical Na^+ channel that is supported by an electrogenic, apical H^+-ATPase (see Lin and Randall, Chapter 9, this volume). In this model, gill Na^+, K^+-ATPase transports excess intracellular sodium into the blood, and may also provide an electrical gradient for apical transport of sodium and basolateral transport of chloride. However, the relative importance of Na^+, K^+-ATPase in sodium and chloride uptake is unclear. Ouabain has been reported to decrease ion uptake (Epstein *et al.*, 1967) or to have no effect (Kerstetter and Keeler, 1976) in rainbow trout (*Oncorhynchus gairdneri*). Increases in Na^+, K^+- ATPase in a few teleosts upon transfer from seawater to fresh water (discussed in the following) support a role for Na^+, K^+-ATPase in ion uptake. The role of the sodium pump in ion uptake is less certain than its role in ion excretion, and further studies that take account of potential differences among species are needed.

The site of gill ion uptake in teleosts is also uncertain. Foskett and Scheffey (1982) provided direct evidence for chloride cells as the site of chloride excretion in seawater teleosts, but similar direct evidence for chloride cells as the site of ion uptake is lacking. Proliferation in response to low ion content in fresh water is evidence for the role of chloride cells in ion uptake (Laurent and Dunel, 1980; Laurent *et al.*, 1985; Avella *et al.*, 1987; Perry and Laurent, 1989), but there is also evidence for involvement of other cells in the gill (see Goss *et al.*, Chapter 10, this volume). Although variation occurs among species, the appearance of chloride cells in fresh water is distinct from that of the same species in seawater (Pisam *et al.*, 1987; Laurent and Perry, 1991). In fresh water, chloride cells are generally smaller and less columnar, have a less pronounced tubular system, and often do not have apical and serosal contacts. Accessory cells, which interdigitate with seawater chloride cells to form "leaky" junctions for passive paracellular sodium efflux, are absent in fresh water (Laurent and Perry, 1991).

III. PROPERTIES OF NA^+,K^+-ATPase

A great deal is known about the structure and biochemistry of Na^+, K^+-ATPase, which is present in all animal cells (see reviews by De Renzis and Bornancin, 1984; Rossier *et al.* 1989; Skou and Esmann, 1992). The protein complex consists of a catalytic α-subunit and a β-subunit; the latter stabilizes folding of the α-subunit, confers K^+ affinities, and is involved in cell–cell interactions (Schmalzing and Gloor, 1994). Both subunits of Na^+, K^+-ATPase have been sequenced from one fish species to date, the Pacific electric ray (*Torpedo californica*), which contains an electric organ rich in Na^+, K^+-ATPase. The α-subunit consists of 1022 amino acids and has a calculated molecular mass of 112,000 and six to eight membrane-spanning domains (Kawakami *et al.*, 1985). The β-subunit is a 35-kDa glycoprotein of 300 amino acids with a single membrane-spanning domain (Noguchi *et al.*, 1986). Its amino acid sequence is less highly conserved among vertebrates (60%) than the α-subunit (Rossier *et al.*, 1989). The α-subunit has also been fully sequenced in one teleost, the white sucker (*Catostomus commersoni*), and consists of 1027 amino acids and a predicted eight membrane-spanning domains (Schörock *et al.*, 1991). There is substantial sequence similarity with other vertebrate Na^+, K^+-ATPase α-subunits (74–98%). A partial sequence (450 base pairs) of the gene encoding the rainbow trout α-subunit also shows high sequence similarity (Kisen *et al.*, 1994). Three isoforms of the α-subunit and four isoforms of the β-subunit have been isolated in mammals; these isoforms are present in varying proportions in different organs (Skou and Esmann, 1992).

The stoichiometry of Na^+, K^+-ATPase appears to be similar in all animals investigated (Skou and Esmann, 1992). Three internal sodium ions are transported outward in exchange for two potassium ions. The sodium pump is therefore electrogenic, creating a potential difference across the cell membrane. Each translocation of ions requires the hydrolysis of ATP and is accompanied by a conformational change. Binding sites for ions, ATP, and inhibitors are all contained on the α-subunit.

IV. METHODS

A. Quantitation and Localization of Na^+,K^+-ATPase

Measurement of Na^+, K^+-ATPase is most frequently accomplished by measurement of V_{max} in a microsomal preparation or crude homogenate and expressed as activity per unit wet weight or protein content. The breakdown of ATP is detected (most often by measuring inorganic phosphate or ADP) in the presence and absence of ouabain. Such biochemical measurements are, by their nature, method dependent. Efforts should be made to optimize the conditions for expression of enzyme activity, including tissue preparation, ion concentrations, pH, ouabain, detergent, and temperature. Several methods specific for measurement of gill Na^+, K^+-ATPase activity of teleosts have been presented (Johnson *et al.*, 1977; Zaugg, 1982; Mayer-Gostan and Lemaire, 1991), including a "microassay" suitable for nonlethal biopsies and organ cultured tissues (McCormick, 1993). These biochemical assays are an approximation of the total number of sodium pumps and rapid activation or deactivation (e.g., phosphorylation) may not be detected by these methods.

[^{3}H] Ouabain binding to gill homogenates and whole tissue has also been used to measure the number of sodium pumps. In all cases examined to date, gill ouabain binding and Na^+, K^+-ATPase activity are strongly correlated (Sargent and Thomson, 1974; Stagg and Shuttleworth, 1982). B_{max} values for teleost gills range from 3 to 20 pmol/mg dry weight in the three species examined (Hossler *et al.*, 1979; Stagg and Shuttleworth, 1982; McCormick and Bern, 1989). Since chloride cells usually represent less than 10% of the gill, the sodium pump content of individual chloride cells is much higher (Karnaky, 1986). Hootman and Ernst (1988) outline a method for using [^{3}H] ouabain to measure turnover (*in situ* activity) and sodium pump numbers in intact cells.

Autoradiographic methods using [^{3}H]ouabain have localized Na^+, K^+-ATPase to chloride cells and provided information on their intracellular location (Fig. 1). A fluorescent ouabain analog, anthroylouabain, has been recently used to examine Na^+, K^+-ATPase-rich chloride cells in the opercular membrane (Fig. 1). This probe has the advantage of ouabain specificity and increased fluorescence upon binding (Fortes, 1977). A major disadvantage is the high autofluorescence in the wavelengths necessary to view anthracene. Nonetheless, in tilapia this stain distinguishes the freshwater from the seawater form of the chloride cell. Histochemical methods have also been used to localize Na^+, K^+-ATPase in fish gills and are reviewed in De Renzis and Bornancin (1984).

Molecular methods have recently been applied to the measurement of Na^+, K^+-ATPase gene expression. Using a 450-base pair fragment of the α-subunit gene, Northern blot analysis was used by Kisen *et al.* (1994) to detect a prominent mRNA of 3.7 kilobases in rainbow trout gill. This probe was found to be useful for measuring gill Na^+, K^+-ATPase gene expression in several other teleosts. Northern blots can also be used to measure gill Na^+, K^+-ATPase mRNA in brown trout (*Salmo trutta*) using cDNA of the *Xenopus* α-subunit (S. Madsen, personal communication). These studies revealed that gill Na^+, K^+-ATPase α-subunit mRNA levels are higher in fish in seawater than in fish in fresh water; increases occur within 1 day of transfer to seawater in brown trout.

B. Morphology and Function of Chloride Cells

The distinguishing characteristics of chloride cells, especially numerous mitochondria and an extensive tubular system, permit their identification from other cells in the gill and opercular membrane at the light microscope level. The vital mitochondrial stain DASPEI (2-*p*-dimethylaminostyryl-ethylpyridiniumiodide) has been used widely on opercular membranes (Karnaky *et al.*, 1984). Such vital stains are rarely used on the gill because of tissue thickness and complex structure, although gill dispersions are possible (Perry and Walsh, 1989). Champy-Maillet's fixative (a 0.2% osmic acid with saturated zinc powder and 25% mg/ml iodine) stains plasma membranes and the extensive tubular system, rendering the entire chloride cell black (e.g., Avella *et al.*, 1987). Great care must be taken in the examination of chloride cells using postfixation staining methods such as hematoxylin and eosin, as positive identification of chloride cells can be problematic. Surprisingly, immunohistochemistry has not been widely used in studies of chloride cells. Currently available methods for protein isolation should permit production of antibodies specific for teleost Na^+, K^+-ATPase, transport proteins, or mitochondrial enzymes.

A variety of methods can be used to probe chloride cell function. Methods for direct and indirect measurement of ion movements have been widely used and are reviewed in detail elsewhere in this volume (see Marshall, Chapter 1). Methods for examining chloride cell function in skin and opercular membranes, particularly the use of Ussing chambers to examine ion transport and electrophysiology, have been reviewed (McCormick, 1994). Although the vibrating probe has been used successfully to localize chloride secretion to chloride cells (Foskett and Scheffey, 1982), it has not yet been used to examine regulation of chloride cell function. Perhaps the most powerful new tools for exploring chloride cell physiology are ion-sensitive fluorescent dyes. These have already proven useful in basic studies of ion transport (see Chapters 1 and 8, this volume), and should be equally useful in regulatory (endocrine) studies of ion uptake and secretion. To date, these methods have been used only with chloride cells in the opercular membrane, though confocal laser microscopy may permit their use in gill tissue.

C. Organ and Cell Culture

In vitro methods are useful for determining the direct action of hormones and other agents on function of cells and tissues. Recent advances in gill respiratory cell culture (see Pärt and Bergstrom, Chapter 8, this volume) and rectal gland cell culture for elasmobranchs (see Valentich *et al.*, Chapter 7, this volume) illustrate the contributions that such methods can make to our understanding of transport physiology. Unfortunately, a method for long-term culture of chloride cells is not yet available. Isolated chloride cells have relatively poor viability in culture using methods suitable for other gill cells

(Peter Pärt, personal communication). It is unclear whether chloride cell preparations would possess the transport capabilities of intact tissue, particularly if accessory cells are required for ion secretion. Nonetheless, chloride cell cultures would be very useful, if only to examine regulation of transport proteins such as Na^+, K^+-ATPase.

Long-term organ cultures of primary gill filaments and opercular membranes have been used in several studies of the hormonal regulation of Na^+, K^+-ATPase (McCormick and Bern, 1989; McCormick, 1990b; McCormick *et al.*, 1991a; Madsen and Bern, 1993; see McCormick, 1994, for detailed methods). Though more difficult to prepare, the opercular membrane culture has the advantage of being more accessible than gills for morphological and transport studies (e.g., Ussing chamber or ion-sensitive dye) following hormone exposure. In preparing culture media, physiological levels of pH, ions, and gases appropriate for the species being studied should be used. In many cases this means only a slight modification of existing commercial culture media. Minimal Essential Medium (MEM) has been widely used and apparently supports a variety of teleost cell types equally or better than media with more additives (Fernandez *et al.*, 1993). Gassing for physiological ranges of pH (7.7–8.0), pCO_2 (1-4 mm Hg), and HCO_3^- (3–10 m*M*) (Heisler, 1993) can theoretically be achieved by using 0.3–0.5% CO_2 gas. One percent CO_2 is commercially available and has been used successfully in gill organ culture (S. D. McCormick, unpublished). Saturation with 5% CO_2, widely used in fish cell and organ culture, results in supraphysiological levels (for teleosts) of CO_2 and HCO_3^-, though this is apparently not detrimental to growth of cells or maintenance of tissue in culture (McCormick, 1990b; Fernandez *et al.*, 1993). The health and responsiveness of cells in culture are the most important criteria, and these can be judged through vital stains, histological appearance, and biochemical analyses.

V. ENVIRONMENTAL AND DEVELOPMENTAL REGULATION

An excellent review has been written on the environmental influences on chloride cells (Laurent and Perry, 1991), though a similar overview on gill Na^+, K^+-ATPase is lacking. Since space limitations preclude an exhaustive review, only heuristic and recent research will be presented here.

The most widely recognized and investigated environmental determinant of gill Na^+, K^+-ATPase and chloride cell development is salinity. For almost all teleosts examined to date, which includes several dozen species, increasing salinity results in higher levels of gill or opercular Na^+, K^+-ATPase activity (Kirschner, 1980; De Renzis and Bornancin, 1984). This is true both for comparisons within species and, generally, when comparing freshwater and marine forms. Despite this generalization, it is enlightening to examine the exceptions. No difference in gill Na^+, K^+-ATPase activity was found between freshwater- and seawater-adapted flounder (*Platichthys flesus;* Stagg and

Shuttleworth, 1982) and striped bass (*Morone saxatilis;* Madsen *et al*., 1994). Gill Na^+, K^+-ATPase activity has been found to be higher in fresh water in sea bass (*Dicentrarchus labrax*), thick-lipped mullet (*Chelon labrosus*), and Australian bass (*Macquaria novemaculeata;* Lasserre, 1971; Langdon, 1987). These fish are generally considered to be of marine origin. One possible explanation, which lacks experimental support, is that these fish maintain higher gill permeability and ion fluxes than other fish in fresh water and therefore require higher levels of gill Na^+, K^+-ATPase.

Exposure of fish to low ion content in fresh water results in a dramatic increase in chloride cell density, particularly on the secondary lamellae, which is presumably involved in ion uptake. Surprisingly, we know little of the impact of ion-poor water on gill Na^+, K^+-ATPase. In Atlantic salmon (*Salmo salar*), gill Na^+, K^+-ATPase activity does not increase in ion-poor water (S. D. McCormick, unpublished).

Gill Na^+, K^+-ATPase activity generally increases following acclimation to low temperature, but there is contradictory evidence (even in the same species) that makes interpretations difficult (see De Renzis and Bornancin, 1984). In addition to methodological considerations, interpretation of temperature effects is made even more difficult by potential interaction of physiological demands on the gills, such as gas exchange and ion transport.

Although moderate changes in pressure have no effect on gill Na^+, K^+-ATPase activity *in vitro*, changes in pressure that might be experienced in the deep ocean result in rapid deactivation of gill Na^+, K^+-ATPase (Gibbs and Somero, 1989). This effect of pressure, which is presumably the result of changes in membrane fluidity and the structure of the enzyme, is less pronounced in fish living at greater depths. Deep-living species generally have lower gill Na^+, K^+-ATPase activity, which may reflect the lower activity and metabolic rate of these species (Gibbs and Somero, 1990).

Diet and nutrition may also play a role in regulation of chloride cell function. Food restriction in tilapia reduces chloride cell number and gill Na^+, K^+-ATPase activity, but this apparently has no effect on salinity tolerance (Kultz and Jurss, 1991). Diets high in NaCl can increase gill Na^+, K^+-ATPase activity and chloride cell numbers in rainbow trout, but this is not a universal finding among teleosts (Salman and Eddy, 1987).

Size has a potential but largely unexamined influence on chloride cell development. Gibbs and Somero (1990) found that total gill Na^+, K^+-ATPase (activity per fish) was related to size; this result was largely related to allometric growth of the gill rather than to changes within the gill. There is great variation of the scaling coefficient among species, indicating that gill Na^+, K^+-ATPase may increase or decrease with increasing size. No size relationship between protein-specific gill Na^+, K^+-ATPase activity and size was found in brook trout (*Salvelinus fontinalis*) weighing from 2 to 380 g (McCormick and Naiman, 1984a). Chloride cells on skin and yolk sac of small embryonic and larval fishes, present prior to gill development, may be important sites for ion exchange in fresh water and seawater (Ayson *et al.*, 1994).

Developmental events play a crucial role in regulating chloride cells in some diadromous fishes. In both downstream-migrating juvenile salmonids (known as smolts) and maturing "silver" European eels (*Anguilla anguilla*), chloride cell numbers and gill Na^+, K^+-ATPase activity increase prior to exposure to seawater (Zaugg and McLain, 1970; Thomson and Sargent, 1977; see reviews by McCormick and Saunders, 1987; Hoar, 1988). These preparatory adaptations result in increased salinity tolerance of the migrants. In anadromous salmonids, this developmental event is size dependent and cued by increasing daylength (Saunders and Henderson, 1970; Duston and Saunders, 1990). Some of the endocrine factors discussed in Section VI are likely to be responsive not only to salinity but also to environmental and developmental influences.

VI. HORMONAL REGULATION

A. Cortisol

Cortisol was the first hormone shown to stimulate gill Na^+, K^+-ATPase activity (Pickford *et al.*, 1970b); following this work with killifish, American eel (*Anguilla rostrata*), tilapia (*Oreochromis mossambicus*), and several salmonids have been shown to respond similarly (Epstein *et al.*, 1971; Richman and Zaugg, 1987; Björnsson *et al.*, 1987; Dange, 1986; Madsen, 1990c; Bisbal and Specker, 1991). Stenohaline fish have not been sufficiently examined, and it would be of some interest to determine if stenohalinity is in part due to an inability to respond to cortisol (or other hormones). Hypophysectomy reduces gill Na^+, K^+-ATPase activity in American eel, killifish, and coho salmon (*Oncorhynchus kisutch*), which can be at least partially restored by cortisol treatment (Pickford *et al.*, 1970b; Butler and Carmichael, 1972; Björnsson *et al.*, 1987; Richman and Zaugg, 1987). This effect is presumably through the loss of pituitary ACTH (a cortisol secretagogue), though other pituitary hormones are involved in regulating circulating cortisol and gill Na^+, K^+-ATPase (see the following).

Cortisol treatment also affects the morphology and development of chloride cells (Fig. 2). Exogenous cortisol causes increased chloride cell numbers in American eel, tilapia, brown trout (*Salmo trutta*), rainbow trout, and coho salmon (Foskett *et al.*, 1981; Madsen, 1990a,c; Doyle and Epstein, 1993; Richman and Zaugg, 1987). Chloride cell size is less often measured, but was found to increase following cortisol treatment in brown trout (Madsen, 1990c). Doyle and Epstein (1972) reported increased development of the tubular system in chloride cells of the American eel after cortisol injection. As might be expected from these effects on gill Na^+, K^+-ATPase activity and chloride cells, cortisol treatment also increases salinity tolerance in eel,

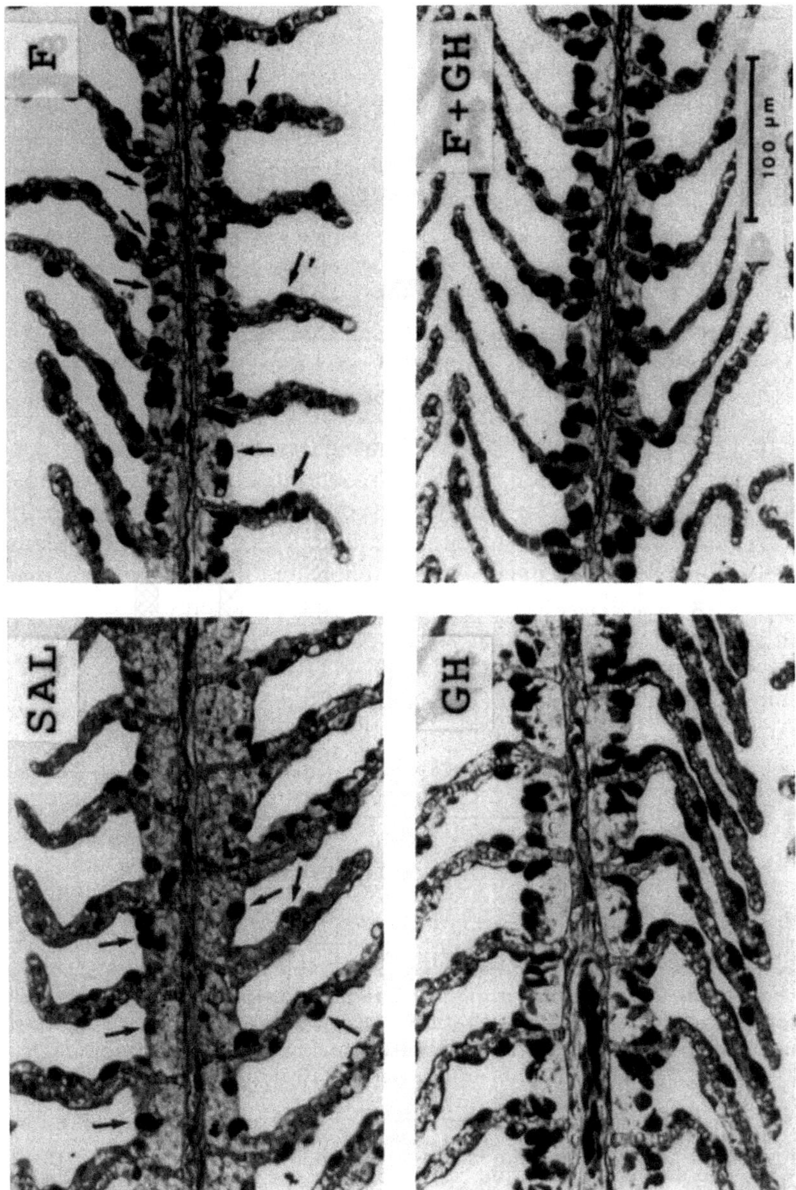

Fig. 2. The effect of cortisol (F) and growth hormone (GH) treatments on gill chloride cells of brown trout (SAL = saline-injected). Tissue was fixed in Champy-Maillet's fixative, which stains the extensive tubular system of chloride cells. Note that cortisol and growth hormone each increase cell number and cell size. When injected together, cell size is increased slightly (but not significantly) over treatment with either hormone alone. In the same experiment, cortisol and growth hormone acted additively to increase gill Na^+, K^+-ATPase activity. [Reproduced from Madsen (1990c) with permission from Academic Press.]

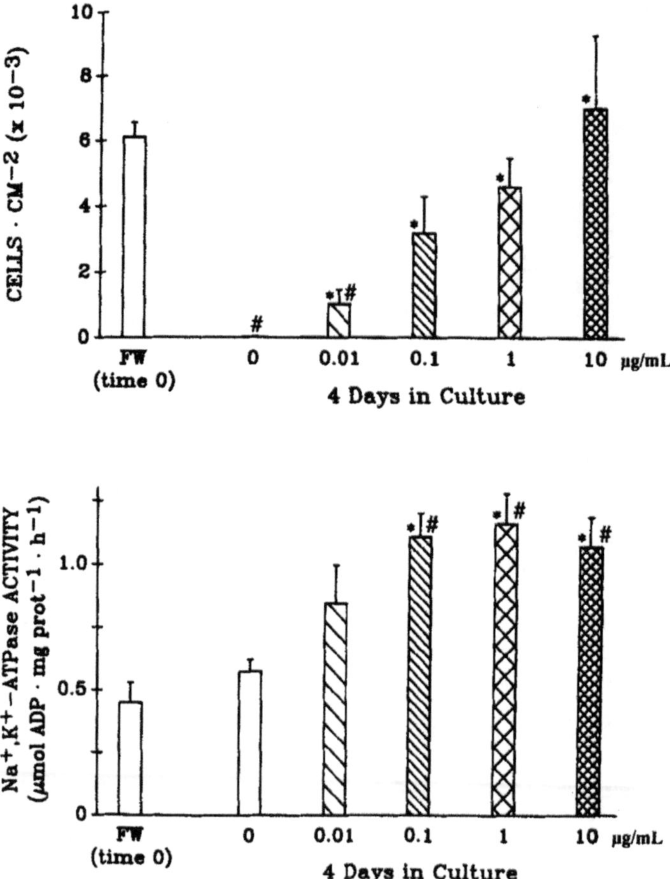

Fig. 3. Effect of cortisol *in vitro* on chloride cell number and Na^+, K^+-ATPase activity in opercular membrane of freshwater-adapted tilapia. Upper: In the absence of cortisol, chloride cell numbers decrease dramatically, and are increased in a dose-dependent manner by cortisol. Lower: Cortisol increases Na^+, K^+-ATPase activity to levels higher than seen initially. [*, indicates significant difference from control; #, indicates significant difference from initial (Time 0).] Cell height and the number of cells spanning the opercular membrane are also increased. [Reproduced from McCormick (1990b) with permission from The American Physiological Society.]

brown trout, and Atlantic salmon (Epstein *et al.*, 1971; Madsen, 1990c; Bisbal and Specker, 1991).

In vitro stimulation of gill and opercular membrane Na^+, K^+-ATPase by cortisol has been demonstrated in coho salmon and tilapia, respectively (Fig. 3; McCormick and Bern, 1989; McCormick, 1990b), indicating its direct effect on these tissues. Using anthroylouabain, increased gill Na^+, K^+-ATPase was shown to be primarily a response of chloride cells. Cortisol also increased cell height and the proportion of chloride cells spanning the opercular membrane, but chloride cell numbers did not increase. From this evidence it was concluded that cortisol directly causes differentiation of the seawater form of the chloride cell. These *in vitro* findings contrast with the *in vivo* effect of cortisol in tilapia, where chloride cell density increases, but chloride cell size does not (Foskett *et al.*, 1981). These differences suggest that other endocrine factors, stimulated by or acting with cortisol *in vivo*, are involved in chloride cell proliferation in tilapia. Electrophysiological properties of the opercular membrane are also not affected by cortisol *in vivo* (Foskett *et al.*, 1981), indicating that other hormones (or salinity itself) are required for activating sodium and chloride secretion.

Developmental differences in the responsiveness of gill Na^+, K^+-ATPase activity to cortisol *in vivo* and *in vitro* have been found in coho and Atlantic salmon (McCormick *et al.*, 1991a). Maximum responsiveness occurs when normal increases in gill Na^+, K^+-ATPase activity begin in early spring. As in the parr–smolt transformation itself, cortisol responsiveness can be altered by photoperiod. Changes in responsiveness may be due in large part to changes in cortisol receptors, which have been found in gill tissue of several euryhaline species (Sandor *et al.*, 1984; Chakraborti *et al.*, 1987; Maule and Schreck, 1990). In coho salmon and anadromous rainbow trout, gill cytosolic cortisol receptors are most numerous from January to March (Shrimpton *et al.*, 1994; McLeese *et al.*, 1994), the time of maximum *in vitro* responsiveness. Cortisol receptors may also be one avenue by which growth hormone interacts with cortisol (see the following). Weisbart *et al.* (1987) found that cytosolic cortisol receptors decrease and nuclear receptors increase following exposure of brook trout to seawater. This response is similar to that seen following injection of cortisol, suggesting that translocation of the cortisol receptor to the nucleus is involved in seawater acclimation. However, the location of cortisol receptors in gill tissue and their importance in chloride cell differentiation and/or proliferation have yet to be established. Corticoid receptor antagonists such as RU-486 (Baulieu, 1989) may prove to be useful in this regard.

Reported inabilities of cortisol treatment to stimulate gill Na^+, K^+-ATPase activity in salmonids and tilapia (Langdon *et al.*, 1984; Hegab and Hanke, 1984; Redding *et al.*, 1984) may in part relate to developmental differences in responsiveness to cortisol. The time course and method of administration of cortisol may also alter its effectiveness. Specker *et al.* (1994) have developed a vegetable oil-based implant that results in prolonged elevation of physiological levels of cortisol.

Often referred to as a "seawater-adapting hormone," it is interesting to find that cortisol may be more than permissively involved in ion uptake. Laurent and Perry (1990) have shown that cortisol treatment of freshwater rainbow trout results in increased influx of Na^+ and Cl^-, and increased number and apical surface area of gill chloride cells. Although it has yet to be demonstrated that cortisol increases net influx of Na^+ and Cl^-, this combination of ion transport and morphological evidence is particularly powerful, and similar results were obtained for European eel, bullhead catfish (*Ictalurus nebulosus*), and tilapia (Perry *et al.*, 1992). Cortisol also increases gill H^+-ATPase, which has been implicated in Na^+ uptake in coho salmon in fresh water (Lin and Randall, 1993). This potential dual effect of cortisol raises several fundamental questions in our understanding of chloride cell function. Does cortisol simultaneously increase both freshwater and seawater forms of the chloride cell? Or are chloride cells "bipolar," with the same cells able to switch rapidly from ion uptake to ion secretion? Answers to these questions must await the development of methods that will permit the *functional* differentiation of chloride cells.

A possible rapid action (minutes to hours) of cortisol on ion transport should not be discounted. Cortisol increases quickly following exposure to seawater (and other stressors) (Schreck, 1981). Mineralocorticoids in amphibians and mammals rapidly increase sodium retention (Minuth *et al.*, 1987). Forrest *et al.* (1973) found increased sodium efflux in American eels treated for 2 days with cortisol and then transferred to seawater. This effect occurred prior to an increase in gill Na^+, K^+-ATPase activity, which occurred after 14 days of cortisol treatment.

B. Growth Hormone

Sakamoto *et al.* (1993) present evidence for the role of growth hormone (GH) in seawater acclimation of salmonids, including the effects of exogenous GH, changes in circulating GH, and metabolic clearance of GH and GH receptors. Exogenous GH increases salinity tolerance in Atlantic salmon, sockeye salmon (*Oncorhynchus nerka*), and rainbow and brown trout; this effect was found to be independent of the hormone's influence on body size (Komourdjian *et al.*, 1976; Clarke *et al.*, 1977; Bolton *et al.*, 1987; Madsen 1990c). Treatment with GH for more than 1 week results in increased gill Na^+, K^+-ATPase activity and chloride cells in all of these species (Boeuf *et al.*, 1990; Madsen, 1990b,c). Growth hormone partially restores the decrease in gill Na^+, K^+-ATPase activity and salinity tolerance following hypophysectomy of coho salmon (Björnsson *et al.*, 1987; Richman *et al.*, 1987). As with cortisol in eels, growth hormone treatment for 2 days can be shown to improve ion regulatory performance of rainbow trout and Atlantic salmon in seawater, prior to a detectable increase in gill Na^+, K^+-ATPase activity (Collie *et al.*, 1989; McCormick *et al.*, 1991b). The short-term and long-term effects of cortisol and growth hormone may involve different mechanisms (such as protein activation, protein synthesis, mitogenesis, and hemodynamics), though much remains to be investigated in this regard.

Almost all the published work on the role of growth hormone in teleost hypoosmoregulation has been conducted with salmonids. However, Flik *et al.* (1993) found that GH treatment doubled chloride cells density in tilapia opercular membrane. GH also increases the salinity tolerance of hypophysectomized tilapia (T. Sakamoto and E. G. Grau, personal communication), suggesting that a significant role of GH in salt secretion may prove to be common among euryhaline fishes.

There is a strong interaction between growth hormone and cortisol in the regulation of salt secretion. GH and cortisol act in synergy to increase gill Na^+, K^+-ATPase activity and salinity tolerance in brown trout, rainbow trout, and Atlantic salmon (Madsen, 1990b,c). Findings of an increase in gill cortisol receptors following growth hormone treatment in coho salmon (Shrimpton *et al.*, 1995) suggest at least one pathway for this synergy. Although growth hormone receptors have been found in gill tissue of several salmonids (Gray *et al.*, 1990; Yao *et al.*, 1991; Sakamoto and Hirano, 1991), there is currently no evidence for a direct action of GH on Na^+, K^+-ATPase activity (McCormick *et al.*, 1991a). GH may also act through the interrenal to affect cortisol release. Young (1988) found that both *in vivo* and *in vitro* GH treatment increased responsiveness of the coho salmon interrenal to ACTH. However, an effect of GH on circulating levels of cortisol has yet to be demonstrated. In addition to its effect on cortisol receptors and cortisol secretion, at least one other avenue for the action of growth hormone is through its major influence on insulin-like growth factor I (IGF-I).

C. Insulin-like Growth Factor I

IGF-I is a 70-amino acid polypeptide that is produced primarily in the liver but also in several other tissues. Growth hormone is the most important secretogogue, and the physiological activity of IGF-I is controlled by several binding proteins, which like IGF-I itself are under complex endocrine and nutritional control. The cDNA of IGF-I has been isolated from five salmonids (Duan *et al.*, 1994), and the deduced amino acid sequence has an 80% similarity with mammalian IGF-I. Other aspects of the structure and actions of IGF-I in ectothermic vertebrates can be found in Bern *et al.* (1991).

IGF-I mRNA increases in gill and kidney of rainbow trout following seawater exposure (Sakamoto and Hirano, 1993). As with growth hormone, increases in salinity tolerance have been observed within 48 hr of injection of IGF-I in rainbow trout and Atlantic salmon (McCormick *et al.*, 1991b, S. D. McCormick, unpublished). It is not known whether this short-term effect involves the well-known mitogenic activity of IGF-I or some other mechanism of action. In other vertebrates, IGF-I has varied osmoregulatory actions, including rapid effects on glomerular filtration, direct stimulation of sodium transport in isolated epithelia, and mitogenesis of renal cells (Hammerman *et al.*, 1993).

IGF-I may also have a role in the long-term action of growth hormone and cortisol in stimulating gill Na^+, K^+-ATPase activity and chloride cells. Treatment of brown trout with cortisol, GH, or IGF-I increases gill Na^+, K^+-ATPase α-subunit mRNA (S. Madsen, personal communication). IGF-I treatment also increases responsiveness of gill Na^+, K^+-ATPase to cortisol in

Atlantic salmon (S. D. McCormick, unpublished). IGF-I increases Na^+, K^+-ATPase in gill organ culture of coho salmon if the fish are pretreated with GH (Madsen and Bern, 1993). It appears, however, that IGF-I cannot carry out all the long-term effects of GH. Treatment of Atlantic salmon with GH for 2 weeks results in two-fold increases in gill Na^+, K^+-ATPase activity, but the same period of treatment with IGF-I has no effect (S. D. McCormick, unpublished). Although the mitogenic actions commonly associated with IGF-I suggest that this hormone may play a role in GH-mediated increases in numbers (or differentiation) of chloride cells, this remains to be established. Some of the action of IGF-I may be through paracrine or autocrine pathways (local hormone production); in rainbow trout, IGF-I mRNA levels in the gill are highest in chloride cells (T. Sakamoto and S. Hyodo, personal communication).

D. Prolactin

Since the seminal findings of Pickford and Phillips (1959) on the role of prolactin in maintaining ion homeostasis in freshwater killifish, prolactin has been shown to have a role in ion uptake in most, if not all, teleosts (see Hirano, 1986, for a review of prolactin in fish). In euryhaline fish, plasma prolactin levels decrease rapidly following transfer from fresh water to seawater, and increase upon "reverse" transfer. Surprisingly few investigations have examined the biochemical and morphological effects of prolactin on fish osmoregulatory organs. Prolactin decreases gill Na^+, K^+-ATPase activity and increases kidney Na^+, K^+-ATPase activity in hypophysectomized, freshwater killifish (Pickford *et al.*, 1970a). In seawater tilapia there was no effect of prolactin on gill Na^+, K^+-ATPase activity in spite of a profound effect on plasma ions and chloride cells (Herndon *et al.*, 1991). Madsen and Bern (1992) reported that prolactin injections reduce gill Na^+, K^+-ATPase activity in freshwater rainbow trout and antagonize the action of GH in increasing salinity tolerance and gill Na^+, K^+-ATPase activity. In contrast, Atlantic salmon implanted with prolactin show increased gill Na^+, K^+-ATPase activity, with no antagonism of the actions of GH (Boeuf *et al.*, 1994). This confusing picture of the effects of prolactin may stem from the use of heterologous hormones, species differences or developmental differences.

Although limited, current information indicates that prolactin strongly affects chloride cell morphology. Based on reductions in opercular membrane conductance and short-circuit current following prolactin treatment of seawater-adapted tilapia, Foskett *et al.* (1982) postulated that prolactin reduces chloride cell numbers and active transport in the remaining chloride cells. Herndon *et al.* (1991) found that prolactin treatment of seawater-adapted tilapia resulted in a dramatic reduction in chloride cell size without changing chloride cell density. Chloride cell height and proportion of cells spanning the opercular membrane were reduced, suggesting that these cells were

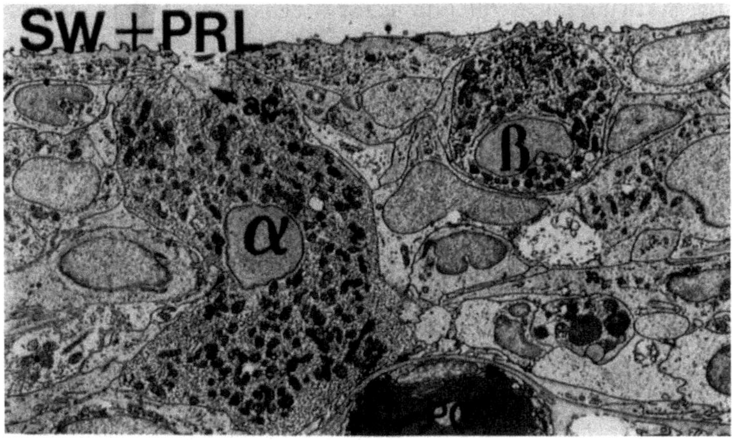

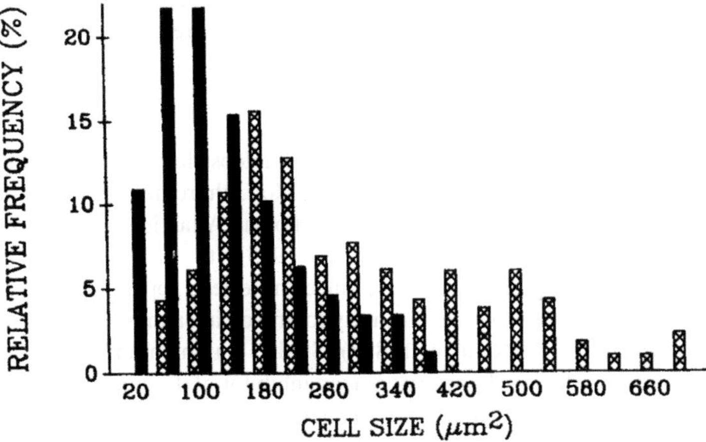

Fig. 4. The effects of prolactin (PRL) on chloride cell size and morphology in tilapia. Upper: α- and β-cells (the latter being the seawater form of the chloride cell) of the Nile tilapia (ac = apical crypt). Prolactin injection in seawater-adapted Nile tilapia caused the appearance of α-cells and decreased size of β-cells. [Reproduced from Pisam, M., Auperin, B., Prunet, P., Rentierdelrue, F., Martial, J., and Rambourg, A. (1993). Effects of prolactin on alpha and beta chloride cells in the gill epithelium of the saltwater adapted tilapia *Oreochromis niloticus*. *Anat. Rec.* **235,** 275–284. Copyright © (1993) by Wiley–Liss, Inc., a division of John Wiley & Sons, Inc. Reprinted by permission of John Wiley & Sons, Inc.] Lower: In Mozambique tilapia, prolactin (solid bars) treatment causes a dramatic shift in chloride cell size compared to saline (hatched bars) treatment. The vital mitochondrial stain DASPEI was used to identify chloride cells in this study. [Reproduced from Herndon *et al.* (1991) with permission from Academic Press.] The size distribution of chloride cells in these studies underscores the need to view them as a heterogeneous population of cells with varying degrees of development.

effectively "removed" as chloride secretory cells (Fig. 4). Working with the closely related but stenohaline Nile tilapia (*Oreochromis niloticus*), Pisam *et al.* (1993) found that prolactin injection caused the appearance of the smaller "β-chloride cells" that were previously absent from these fish in seawater. The remaining "α-chloride cells" were also reduced in size and had shallower apical crypts and reduced tubular systems (Fig. 4).

These effects of prolactin on chloride cells raise several interesting questions. Is decreased cell size the result of prolactin acting on existing (secretory) chloride cells, or are undeveloped cells affected? How rapid do morphological changes occur and are they sufficient to explain the effects of prolactin on ion regulation? The possible mode(s) of action of prolactin have not been examined. Prolactin receptors have been found in gill tissue (Dauder *et al.*, 1990; Prunet and Auperin, 1994), but as yet there is no evidence for direct action of prolactin on the gills or opercular membrane and regulation of prolactin receptors has not been investigated. Investigations into other possible endocrine factors involved in the prolactin response and interaction of prolactin with other endocrine systems are required.

E. Thyroid Hormones

The reported effects of thyroid hormones in teleost hypoosmoregulation are contradictory. Thyroxine (T_4) or 3,5′,3′-triiodo-L-thyronine (T_3) incorporated in food can improve salinity tolerance of coho and Atlantic salmon (Fagerlund *et al.*, 1980; Refstie, 1982; Saunders *et al.*, 1985). It is unclear whether this is an effect on osmoregulation per se, since the effect apparently depends on prior increases in body size, and salinity tolerance in salmonids is size dependent (McCormick and Naiman, 1984b). Studies on chum and amago salmon have shown no effect of T_4 or thiourea, an inhibitor of T_4 production, on salinity tolerance (Miwa and Inui, 1983; Iwata *et al.*, 1987) despite the fact that these treatments were effective in altering circulating levels of thyroid hormones and body silvering (a morphological change that is part of the parr–smolt transformation). The possible influence of thyroid hormones on salinity tolerance in nonsalmonids has not been examined.

Thyroxine has no apparent positive effect by itself on gill Na^+, K^+-ATPase activity in tilapia or amago, coho, and Atlantic salmon (Miwa and Inui, 1985; Saunders *et al.*, 1985; Björnsson *et al.*, 1987; Dange, 1986). In rainbow trout, T_4 injections had no effect on gill Na^+, K^+-ATPase activity (Madsen, 1990a), whereas T_3 immersion decreased activity (Omelanjiuk and Eales, 1986). In contrast to these studies, Madsen and Korsgaard (1989) found that multiple injections of thyroxine could advance increases in gill Na^+, K^+-ATPase activity and chloride cell numbers that occur during the parr–smolt transformation of Atlantic salmon.

Greater agreement in the literature is found regarding the interaction of thyroid hormones with other endocrine axes. Thyroxine increases the capacity of cortisol to increase gill Na^+, K^+-ATPase activity in tilapia (Dange, 1986). In amago salmon, T_4 and GH in combination were capable of elevating gill Na^+, K^+-ATPase, whereas each hormone alone had no effect. The most convincing evidence for a role of thyroid hormones in seawater acclimation utilized an inhibitor of 5′-monodeiodination of T_4 (Lebel and Leloup, 1992; Leloup and Lebel, 1993). In these studies, rainbow and brown trout have been found to require conversion of T_4 to T_3 for normal and GH-stimulated acclimation to seawater. From these results the authors suggest that the action of growth hormone on hypoosmoregulatory mechanisms is through its increase of T_4 to T_3 conversion (de Luze *et al.*, 1989). However, this seems to be an incomplete explanation given the equivocal effect of T_3 on salinity tolerance and its general inability to increase gill Na^+, K^+-ATPase activity. An alternative explanation is that T_3 is necessary for the peripheral action of GH (or IGF-I), possibly through regulation of receptors or binding proteins. Other indirect pathways for thyroid hormone action include their ability to increase pituitary GH production (Moav and McKeown, 1992) and interrenal sensitivity to ACTH (Young and Lin, 1988). Thyroid hormone receptors have been found in liver, kidney, and gill of rainbow trout, brown trout, and European eel (Bres and Eales, 1988; Lebel and Leloup, 1989), but their endocrine regulation and response to salinity have not been examined.

F. Sex Steroids

Normal sexual maturation and treatment with exogenous sex steroids have been shown to have a negative effect on the ability of several salmonids to adapt to seawater (McCormick and Naiman, 1985; Ikuta *et al.*, 1987; Lundqvist *et al.*, 1989; Schmitz and Mayer, 1993). Repeated injections of 17β-estradiol result in decreased gill chloride cell density and Na^+, K^+-ATPase activity in Atlantic salmon (Madsen and Korsgaard, 1989). The mode of action of sex steroids, particularly whether they are acting directly on the gill or through other endocrine systems, is currently unknown. It is also unclear whether the action of sex steroids on hypoosmoregulation is peculiar to the anadromous life history of salmonids in which sexual maturation occurs only in fresh water; research on other fishes is clearly warranted.

G. Rapid Activation

Little work has been done on the rapid activation of Na^+, K^+-ATPase. Although it can be assumed that any of the hormones involved in rapid changes in ion efflux will involve stimulation of Na^+, K^+-ATPase, it is not clear whether these will be direct (stimulation of the pump itself) or indirect (through increased substrate availability provided by stimulation or inhibition of other transporters or channels). Important advances in this area will come from methods that permit rapid, simultaneous measurement of ion transport and morphological or biochemical changes in chloride cells. Ion-sensitive dyes and fluorescent probes may be particularly valuable in this area.

VII. SUMMARY AND PROSPECTUS

Any summary of the endocrine control of physiological function in fishes must confront the diversity and evolutionary history of this large group and the sometimes contradictory results in the literature. Nevertheless, a hypothetical model of endocrine regulation of Na^+, K^+-ATPase and chloride secretory cells will apply to some species (Fig. 5). Evidence to date indicates that cortisol (F), growth hormone (GH), insulin-like growth factor I (IGF-I), and thyroid hormones (T_4 and T_3) increase Na^+, K^+-ATPase and/or promote the differentiation of chloride secretory cells. Cortisol has been shown to have direct action on Na^+, K^+-ATPase and chloride cells, and cortisol receptors have been found in gill tissue. Growth hormone acts in several ways: by

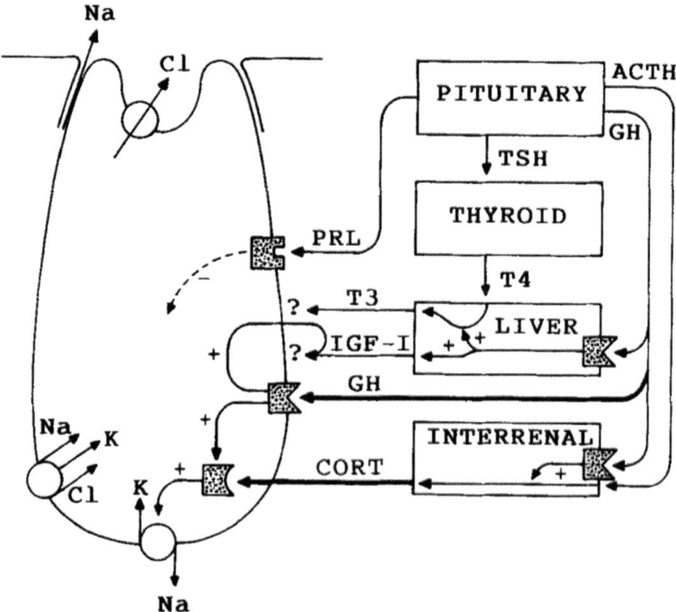

Fig. 5 Model of endocrine regulation of Na^+, K^+-ATPase and chloride secretory cells. Evidence to date indicates that cortisol (CORT), growth hormone (GH), insulin-like growth factor I(IGF-I), and thyroid hormones (T_4 and T_3) increase Na^+, K^+-ATPase and/or promote the differentiation of chloride secretory cells, whereas prolactin (PRL) is inhibitory. Adrenocorticotropic hormone (ACTH) and thyroid-stimulating hormone (TSH) are pituitary hormones involved in stimulating secretion of cortisol and thyroid hormones, respectively. See text for details.

increasing gill cortisol receptors, by increasing the sensitivity of the interrenal to adrenocorticotropic hormone (ACTH), and by production of IGF-I. Growth hormone receptors have been found in both gill and kidney. Although IGF-I is known to increase salinity tolerance and Na^+, K^+-ATPase activity, its mode of action is unknown and receptors in the gill have not been demonstrated. The degree to which the GH/IGF-I axis is important in nonsalmonids requires clarification. Thyroid hormones promote chloride cell development, possibly through their interaction with the GH/IGF-I axis, though this has yet to be fully explored. Prolactin inhibits the development of the chloride secretory cell and promotes the development of the freshwater form of the chloride cell. Prolactin's effects on gill Na^+, K^+-ATPase are equivocal, and although prolactin receptors have been found in the gill, its mode of action is unclear. The universality of this model is suspect, and particularly may not apply to those species in which increased Na^+, K^+-ATPase is associated with exposure to fresh water rather than seawater.

Recent and upcoming advances promote optimism about the solution of the questions posed in this review. The advance of molecular methods will allow for the greater use of homologous hormones and radioimmunoas-says and the detection of changes in receptor gene expression. Isolation of the cDNA for other ATPases and cotransport proteins of chloride cells (in fresh water and seawater) will be powerful tools for determining the functional attributes of chloride cells and their endocrine control. Continued development of methods for organ and cell culture will advance our understanding of direct hormone actions. Combining the new methods with classical physiology, histology, and endocrinology will be especially useful to our understanding of chloride cell regulation.

ACKNOWLEDGMENTS

I thank Mark Shrimpton and Joe Zydlewski for their helpful comments and discussion. Howard Bern, Tetsuye Hirano, Trevor Shuttleworth, and an anonymous reviewer made many helpful comments in review. I am grateful to Victoria McCormick for her rendition of Fig. 5.

REFERENCES

Avella, M., and Bornancin, M. (1990). Ion fluxes in the gills of freshwater and seawater salmonid fish. *In* "Animal Nutrition and Transport Processes. 2. Transport, Respiration and Excretion: Comparative and Environmental Aspects" (J. P. Truchet and B. Lahlou, eds.), Comp. Physiol. Vol. 6, pp. 1–13. Karger, Basel.

Avella, M., Masoni, A., Bornancin, M., and Mayer-Gostan, N. (1987). Gill morphology and sodium influx in the rainbow trout (*Salmo gairdneri*) acclimated to artificial freshwater environments. *J. Exp. Zool.* **241,** 159–169.

Ayson, F. G., Kaneko, T., Hasegawa, S., and Hirano, T. (1994). Development of mitochondrion-rich cells in the yolk-sac membrane of embryos and larvae of tilapia, *Oreochromis mossambicus*, in fresh water and seawater. *J. Exp. Zool.* **270,** 129–135.

Baulieu, E. E. (1989). Contragestion and other clinical applications of RU486, an antiprogesterone at the receptor. *Science* **245,** 1351–1357.

Bern, H. A., McCormick, S. D., Kelley, K. M., Gray, E. S., Nishioka, R. S., Madsen, S. S., and Tsai, P. I. (1991). Insulin-like growth factors "under water": Role in growth and function of fish and other poikilothermic vertebrates. *In* "Modern Concepts of Insulin-like Growth Factors" (E. M. Spencer, ed.), pp. 85–96. Elsevier, New York.

Bisbal, G. A., and Specker, J. L. (1991). Cortisol stimulates hypo-osmoregulatory ability in Atlantic salmon, *Salmo salar* L. *J. Fish Biol.* **39,** 421–432.

Björnsson, B. T., Yamauchi, K., Nishioka, R. S., Deftos, L. J., and Bern, H. A. (1987). Effects of hypophysectomy and subsequent hormonal replacement therapy on hormonal and osmoregulatory status of coho salmon, *Oncorhynchus kisutch. Gen. Comp. Endocrinol.* **68,** 421–430.

Boeuf, G., Prunet, P., and Le Bail, P.-Y. (1990). Is growth hormone treatment able to stimulate the smoltification in the Atlantic salmon? *Seances Acad.* Sci. **310,** 75–80.

Boeuf, G., Marc, A. M., Prunet, P., Le Bail, P.-Y., and Smal, J. (1994). Stimulation of the parr-smolt transformation by hormonal treatment in Atlantic salmon (*Salmo salar* L.). *Aquaculture* **121,** 195–208.

Bolton, J. P., Collie, N. L., Kawauchi, H., and Hirano, T. (1987). Osmoregulatory actions of growth hormone in rainbow trout (*Salmo gairdneri*). *J. Endocrinol.* **112,** 63–68.

Bres, O., and Eales, J. G. (1988). High-affinity, limited-capacity triiodothyronine-binding sites in nuclei from various tissues of the rainbow trout (*Salmo gairdneri*). *Gen. Comp. Endocrinol.* **69,** 71–79.

Butler, D. G., and Carmichael, F. J. (1972). Na^+-K^+-ATPase activity in eel (*Anguilla rostrata*) gill in relation to change in environmental salinity: Role of adrenocortical steroids. *Gen. Comp. Endocrinol.* **19,** 421–427.

Chakraborti, P. K., Weisbart, M., and Chakraborti, A. (1987). The presence of corticosteroid receptor activity in the gills of the brook trout, *Salvelinus fontinalis. Gen. Comp. Endocrinol.* **66,** 323–332.

Clarke, W. C., Farmer, S. W., and Hartwell, K. M. (1977). Effects of teleost pituitary growth hormone on growth of *Tilapia mossambica* and on growth and seawater adaptation of sockeye salmon (*Oncorhynchus nerka*). *Gen. Comp. Endocrinol.* **33,** 174–178.

Collie, N. L., Bolton, J. P., Kawauchi, H., and Hirano, T. (1989). Survival of salmonids in seawater and the time-frame of growth hormone action. *Fish Physiol. Biochem.* **7,** 315–321.

Dange, A. D. (1986). Branchial Na^+, K^+-ATPase activity in freshwater or saltwater acclimated tilapia *Oreochromis* (*Sarotherodon*) *mossambicus*: Effects of cortisol and thyroxine. *Gen. Comp. Endocrinol.* **62,** 341–343.

Dauder, S., Young, G., Hass, L., and Bern, H. A. (1990). Prolactin receptors in liver, kidney, and gill of the tilapia (*Oreochromis mossambicus*): Characterization and effect of salinity on specific binding of iodinated ovine prolactin. *Gen. Comp. Endocrinol.* **77,** 368–377.

de Luze, A., Leloup, J., Papkoff, H., Kikuyama, S., and Kawauchi, H. (1989). Effects of vertebrate prolactins and growth hormones on thyroxine 5'-monodeiodination in the eel (*Anguilla anguilla*): A potential bioassay for growth hormone. *Gen. Comp. Endocrinol.* **73,** 186–193.

De Renzis, G., and Bornancin, M. (1984). Ion transport and gill ATPases. *In* "Fish Physiology" (W. S. Hoar and D. J. Randall, eds.), Vol. 10B, pp. 65–104.

Doyle, W. L., and Epstein, F. H. (1972). Effects of cortisol treatment and osmotic adaptation on the chloride cells in the eel, *Anguilla rostrata. Cytobiologie* **6,** 58–73.

Duan, C., Duguay, S. J., Swanson, P., Dickhoff, W. W., and Plisetskaya, E. (1994). Tissue-specific expression of insulin-like growth factor I messenger ribonucleic acids in salmonids: Developmental, hormonal, and nutritional regulation. *In* "Perspectives in Comparative

Endocrinology" (K. G. Davey, R. E. Peter, and S. S. Tobe, eds.), pp. 365–372. National Research Council of Canada, Ottawa.

Duston, J., and Saunders, R. L. (1990). The entrainment role of photoperiod on hypoosmoregulatory and growth-related aspects of smolting in Atlantic salmon (*Salmo salar*). *Can. J. Zool.* **68,** 707–715.

Epstein, F. H., Katz, A. I., and Pickford, G. E. (1967). Sodium- and potassium-activated adenosine triphosphatase of gills: Role in adaptation of teleosts to salt water. *Science* **156,** 1245–1247.

Epstein, F. H., Cynamon, M., and McKay, W. (1971). Endocrine control of Na-K-ATPase and seawater adaptation in *Anguilla rostrata. Gen. Comp. Endocrinol.* **16,** 323–328.

Evans, D. H., Claiborne, J. B., Farmer, L., Mallery, C., and Krasny, E. J. (1982). Fish gill ionic transport methods and models. *Biol. Bull.* (*Woods Hole, Mass.*) **163,** 108–130.

Fagerlund, U. H. M., Higgs, D. A., McBride, J. R., Plotnikoff, M. D., and Dosanjh, B. S. (1980). The potential for using anabolic hormones 17a-methyltestosterone and (or) 3,5,3′-triiodo-L--thyronine in the fresh water rearing of coho salmon (*Oncorhynchus kisutch*) and the effects on subsequent seawater performance. *Can. J. Zool.* **58,** 1424–1432.

Fernandez, R. D., Yoshimizu, M., Ezura, Y., and Kimura, T. (1993). Comparative growth response of fish cell lines in different media, temperatures, and sodium chloride concentrations. *Gyobyo Kenkyu* **28,** 27–34.

Flik, G., Atsma, W., Fenwick, J. C., Rentierdelrue, F., Smal, J., and Bonga, S. E. W. (1993). Homologous recombinant growth hormone and calcium metabolism in the tilapia, *Oreochromis mossambicus*, adapted to fresh water. *J. Exp. Biol.* **185,** 107–119.

Forrest, J. N., Cohen, A. D., Schon, D. A., and Epstein, F. H. (1973). Na^+ transport and Na-K-ATPase in gills during adaptation to seawater: Effects of cortisol. *Am. J. Physiol.* **224,** 709–713.

Fortes, P. A. G. (1977). Anthroylouabain: A specific fluorescent probe for the cardiac glycoside receptor. *Biochemistry* **16,** 531–540.

Foskett, J. K., and Scheffey, C. (1982). The chloride cell: Definitive identification as the salt-secretory cell in teleosts. *Science* **215,** 164–166.

Foskett, J. K., Logsdon, C. D., Turner, T., Machen, T. E., and Bern, H. A. (1981). Differentiation of the chloride extrusion mechanism during seawater adaptation of a teleost fish, the cichlid *Sarotherodon mossambicus. J. Exp. Biol.* **93,** 209–224.

Foskett, J. K., Machen, T. E., and Bern, H. A. (1982). Chloride secretion and conductance of teleost opercular membrane: Effects of prolactin. *Am. J. Physiol.* **242,** R380–R389.

Foskett, J. K., Bern, H. A., Machen, T. E., and Conner, M. (1983). Chloride cells and the hormonal control of teleost fish osmoregulation. *J. Exp. Biol.* **106,** 255–281.

Gibbs, A., and Somero, G. N. (1989). Pressure adaptation of Na^+, K^+-ATPase in gills of marine teleosts. *J. Exp. Biol.* **143,** 475–492.

Gibbs, A., and Somero, G. N. (1990). Na^+-K^+-Adenosine triphosphatase activities in gills of marine teleost fishes: Changes with depth, size and locomotory activity level. *Mar. Biol.* (*Berlin*) **106,** 315–321.

Gray, E. S., Young, G., and Bern, H. A. (1990). Radioreceptor assay for growth hormone in coho salmon (*Oncorhynchus kisutch*) and its application to the study of stunting. *J. Exp. Biol.* **256,** 290–296.

Hammerman, M. R., Oshea, M., and Miller, S. B. (1993). Role of growth factors in regulation of renal growth. *Annu. Rev. Physiol.* **55,** 305–321.

Hegab, S. A., and Hanke, W. (1984). The significance of cortisol for osmoregulation in carp (*Cyprinus carpio*) and tilapia (*Sarotherodon mossambicus*). *Gen. Comp. Endocrinol.* **54,** 409–417.

Heisler, N. (1993). Acid-base regulation. *In* "The Physiology of Fishes" (D. H. Evans, ed.), pp. 343–378. CRC Press, Boca Raton, FL.

Herndon, T. M., McCormick, S. D., and Bern, H. A. (1991). Effects of prolactin on chloride cells in opercular membrane of seawater-adapted tilapia. *Gen. Comp. Endocrinol.* **83,** 283–289.

Hirano, T. (1986). The spectrum of prolactin action in teleosts. *In* "Comparative Endocrinology: Developments and Directions" (C. L. Ralph, ed.), pp. 53–74. A. R. Liss, New York.

Hoar, W. S. (1988). The physiology of smolting salmonids. *In* "Fish Physiology" (W. S. Hoar and D. Randall, eds.), Vol. 11B, pp. 275–343. Academic Press, Orlando, FL.

Hootman, S. W., and Ernst, S. A. (1988). Estimation of Na^+, K^+-pump numbers and turnover in intact cells with [^{3}H]ouabain. *In* "Methods in Enzymology" (S. Fleischer and B. Fleischer, eds.), Vol. 156. pp. 213–228. Academic Press, Orlando, FL.

Hossler, F. E., Ruby, J. R., and McIlwain, T. D. (1979). The gill arch of the mullet, *Mugil cephalus.* II. Modification in surface ultrastructure and Na,K-ATPase content during adaptation to various salinities. *J. Exp. Zool* **208,** 399–405.

Ikuta, K., Aida, K., Okumoto, N., and Hanyu, I. (1987). Effects of sex steroids on the smoltification of masu salmon, *Oncorhynchus masou. Gen. Comp. Endocrinol.* **65,** 99–110.

Iwata, M., Komatsu, S., Hasegawa, S., Ogasawara, T., and Hirano, T. (1987). Inconsistent effect of thyroid hormone alteration on seawater adaptability of fry of chum salmon *Oncorhynchus keta. Nippon Suisan Gakkaishi* **53,** 1969–1973.

Johnson, S. L., Ewing, R. D., and Lichatowich, J. A. (1977). Characterization of gill (Na + K)-activated adensine triphosphatase from chinook salmon, *Oncorhynchus tshawytscha. J. Exp. Zool.* **199,** 345–354.

Kamiya, M. (1972). Sodium-potassium-activated adenosine triphosphatase in isolated chloride cells from eel gills. *Comp. Biochem. Physiol., B* **43,** 611–617.

Karnaky, K. J. (1986). Structure and function of the chloride cell of *Fundulus heteroclitus* and other teleosts. *Am. Zool.* **26,** 209–224.

Karnaky, K. J., Kinter, L. B., Kinter, W. B., and Stirling, C. E. (1976). Teleost chloride cell. II. Autoradiographic localization of gill Na,K-ATPase in killifish *Fundulus heteroclitus* adapted to low and high salinity environments. *J. Cell Biol.* **70,** 157–177.

Karnaky, K. J., Degnan, K. J., and Zadunaisky, J. A. (1977). Chloride transport across isolated opercular epithelium of killifish: A membrane rich in chloride cells. *Science* **195,** 203–205.

Karnaky, K. J., Degnan, K. J., Garretson, L. T., and Zadunaisky, J. A. (1984). Identification and quantification of mitochondria-rich cells in transporting cells in transporting epithelia. *Am. J. Physiol.* **246,** R770–R775.

Kawakami, K., Noguchi, S., Noda, M., Takahashi, H., Toshiko, O., Kawamura, M., Nojima, H., Nagano, K., Hirose, T., Inayama, S., Hayashida, H., Miyata, T., and Numa, S. (1985). Primary structure of the a-subunit of *Torpedo californica* (Na^++K^+)-ATPase deduced from cDNA sequence. *Nature* (*London*) **316,** 733–736.

Kerstetter, T. H., and Keeler, M. (1976). On the interaction of $NH_4{}^+$ and Na^+ fluxes in the isolated trout gill. *J. Exp. Biol.* **64,** 517–527.

Kirschner, L. B. (1980). Comparison of vertebrate salt-excreting organs. *Am. J. Physiol.* **7,** R219–R223.

Kisen, G., Gallais, C., Auperin, B., Klungland, H., Sandra, O., Prunet, P., and Andersen, O. (1994). Northern blot analysis of the Na^+, K^+-ATPase α-subunit in salmonids. *Comp. Biochem. Physiol., B* **107,** 255–259.

Komourdjian, M. P., Saunders, R. L., and Fenwick, J. C. (1976). The effect of porcine somatotropin on growth and survival in seawater of Atlantic salmon (*Salmo salar*) parr. *Can. J. Zool.* **54,** 531–535.

Kultz, D., and Jurss, K. (1991). Acclimation of chloride cells and Na/K-ATPase to energy deficiency in tilapia (*Oreochromis mossambicus*). *Zool. Jahrb., Abt. Allg. Zool. Physiol.* **95,** 39–50.
Langdon, J. S. (1987). Active osmoregulation in the Australian bass, *Macquaria novemaculeata* (Steindachner), and the golden perch, *Macquaria ambigua* (Richardson) (Percichthyidae). *Aust. J. Mar. Freshwater Res.* **38,** 771–776.
Langdon, J. S., Thorpe, J. E., and Roberts, R. J. (1984). Effects of cortisol and ACTH on gill Na^+/K^+-ATPase, SDH and chloride cells in juvenile Atlantic salmon, *Salmo salar*. *Comp. Biochem. Physiol., A* **77,** 9–12.
Lasserre, P. (1971). Increase of (Na^++K^+)-dependent ATPase activity in gills and kidneys of two marine teleosts, *Crenimugil labrosus* (Risso, 1826) and *Dicentrarchus labrax* (Linnaeus, 1758), during adaptation to fresh water. *Life Sci.* **10,** 113–119.
Laurent, P., and Dunel, S. (1980). Morphology of gill epithelia in fish. *Am. J. Physiol.* **7,** R147–R159.
Laurent, P., and Perry, S. F. (1990). Effects of cortisol on gill chloride cell morphology and ionic uptake in the freshwater trout, *Salmo gairdneri*. *Cell Tissue Res.* **259,** 429–442.
Laurent, P., and Perry, S. F. (1991). Environmental effects on fish gill morphology. *Physiol. Zool.* **64,** 4–25.
Laurent, P., Hobe, H., and Dunel-Erb, S. (1985). The role of environmental sodium chloride relative to calcium in gill morphology of freshwater salmonid fish. *Cell Tissue Res.* **240,** 675–692.
Lebel, J. M., and Leloup, J. (1989). Triiodothyronine binding to putative solubilized nuclear thyroid hormone receptors in liver and gill of the brown trout (*Salmo trutta*) and the European eel (*Anguilla anguilla*). *Gen. Comp. Endocrinol.* **75,** 301–309.
Lebel, J. M., and Leloup, J. (1992). Triiodothyronine is required for the acclimation to seawater of the brown trout (*Salmo trutta*) and rainbow trout (*Oncorhynchus mykiss*). *C. R. Seances Acad. Sci.,* Ser. 3 **314,** 461–468.
Leloup, J., and Lebel, J. M. (1993). Triiodothyronine is necessary for the action of growth hormone in acclimation to seawater of brown (*Salmo trutta*) and rainbow trout (*Oncorhynchus mykiss*). *Fish Physiol. Biochem.* **11,** 165–173.
Lin, H., and Randall, D. J. (1993). H^+-ATPase activity in crude homogenates of fish gill tissue: Inhibitor sensitivity and environmental and hormonal regulation. *J. Exp. Biol.* **180,** 163–174.
Lundqvist, H., Borg, B., and Berglund, I. (1989). Androgens impair seawater adaptability in smolting Baltic salmon (*Salmo salar*). *Can. J. Zool.* **67,** 1733–1736.
Madsen, S. S. (1990a). Effect of repetitive cortisol and thyroxine injections on chloride cell number and Na^+/K^+-ATPase activity in gill of freshwater acclimated rainbow trout, *Salmo gairdneri*. *Comp. Biochem. Physiol., A* **95,** 171–176.
Madsen, S. S. (1990b). Enhanced hypoosmoregulatory response to growth hormone after cortisol treatment in immature rainbow trout, *Salmo gairdneri*. *Fish Physiol. Biochem.***8,** 271–279.
Madsen, S. S. (1990c). The role of cortisol and growth hormone in seawater adaptation and development of hypoosmoregulatory mechanisms in sea trout parr (*Salmo trutta trutta*). *Gen. Comp. Endocrinol.* **79,** 1–11.
Madsen, S. S., and Bern, H. A. (1992). Antagonism of prolactin and growth hormone: Impact on seawater adaptation in two salmonids, *Salmo trutta and Oncorhynchus mykiss*. *Zool. Sci.* **9,** 775–784.
Madsen, S. S., and Bern, H. A. (1993). *In vitro* effects of insulin-like growth factor-I on gill Na^+, K^+-ATPase in coho salmon, *Oncorhynchus kisutch*. *J. Endocrinol.* **138,** 23–30.
Madsen, S. S., and Korsgaard, B. (1989). Time-course effects of repetitive oestradiol-17β and thyroxine injections on the natural spring smolting of Atlantic salmon, *Salmo salar* L. *J. Fish Biol.* **35,** 119–128.

Madsen, S. S., McCormick, S. D., Young, G., Endersen, J. S., Nishioka, R. S., and Bern, H. A. (1994). Physiology of seawater acclimation in the striped bass, *Morone saxatilis* (Walbaum). *Fish Physiol. Biochem.* **13,** 1–11.

Maule, A. G., and Schreck, C. B. (1990). Glucocorticoid receptors in leukocytes and gill of juvenile coho salmon (*Oncorhynchus kisutch*). *Gen. Comp. Endocrinol.* **77,** 448–455.

Mayer-Gostan, N., and Lemaire, S. (1991). Measurements of fish gill ATPases using microplates. *Comp. Biochem. Physiol., B* **98,** 323–326.

Mayer-Gostan, N., Wendelaar Bonga, S. E., and Balm, P. H. M. (1987). Mechanisms of hormone actions on gill transport. In "Vertebrate Endocrinology: Fundamentals and Biomedical Implications" (P. K. T. Pang and M. P. Schreibman, eds.), Vol. 2, pp. 211–238. Academic Press, Boston.

McCormick, S. D. (1990a). Fluorescent labelling of Na^+, K^+-ATPase in intact cells by use of a fluorescent derivative of ouabain: Salinity and teleost chloride cells. *Cell Tissue Res.* **260,** 529–533.

McCormick, S. D. (1990b). Cortisol directly stimulates differentiation of chloride cells in tilapia opercular membrane. *Am. J. Physiol.* **259,** R857–R863.

McCormick, S. D. (1993). Methods for non-lethal gill biopsy and measurement of Na^+, K^+-ATPase activity. *Can. J. Fish. Aquat. Sci.* **50,** 656–658.

McCormick, S. D. (1994). Opercular membranes and skin. *In* "Biochemistry and Molecular Biology of Fishes" (P. W. Hochachka and T. P. Mommsen, eds.), Vol. 3, pp. 231–238. Elsevier, Amsterdam.

McCormick, S. D., and Bern, H. A. (1989). *In vitro* stimulation of Na^+, K^+-ATPase activity and ouabain binding by cortisol in coho salmon gill. *Am. J. Physiol.* **256,** R707–R715.

McCormick, S. D., and Naiman, R. J. (1984a). Osmoregulation in the brook trout, *Salvelinus fontinalis.* I. Diel, photoperiod and growth related physiological changes in freshwater. *Comp. Biochem. Physiol., A* **79,** 7–16.

McCormick, S. D., and Naiman, R. J. (1984b). Osmoregulation in the brook trout, *Salvelinus fontinalis.* II. Effects of size, age and photoperiod on seawater survival and ionic regulation. *Comp. Biochem. Physiol., A* **79,** 17–28.

McCormick, S. D., and Naiman, R. J. (1985). Hypoosmoregulation in an anadromous teleost: Influence of sex and maturation. *J. Exp. Zool.* **234,** 193–198.

McCormick, S. D., and Saunders, R. L. (1987). Preparatory physiological adaptations for marine life in salmonids: Osmoregulation, growth and metabolism. *Am. Fish. Soc. Symp.* **1,** 211–229.

McCormick, S. D., Dickhoff, W. W., Duston, J., Nishioka, R. S., and Bern, H. A. (1991a). Developmental differences in the responsiveness of gill Na^+, K^+-ATPase to cortisol in salmonids. *Gen. Comp. Endocrinol.* **84,** 308–317.

McCormick, S. D., Sakamoto, T., Hasegawa, S., and Hirano, T. (1991b). Osmoregulatory action of insulin-like growth factor I in rainbow trout (*Oncorhynchus kisutch*). *J. Endocrinol.* **130,** 87–92.

McLeese, J. M., Johnsson, J., Huntley, F. M., Clarke, W. C., and Weisbart, M. (1994). Seasonal changes in osmoregulation, cortisol, and cortisol receptor activity in the gills of parr/smolt of steelhead trout and steelhead rainbow trout hybrids, *Oncorhynchus mykiss. Gen. Comp. Endocrinol.* **93,** 103–113.

Minuth, W. W., Steckelings, U., and Gross, P. (1987). Complex physiological and biochemical action of aldosterone in toad urinary bladder and mammalian renal collecting duct cells. *Renal Physiol.* **10,** 297–310.

Miwa, S., and Inui, Y. (1983). Effects of thyroxine ant thiourea on the parr-smolt transformation of Amago salmon (*Oncorhynchus kisutch*). *Bull. Natl. Res. Inst. Aquacult.* (*Jpn.*) **4,** 41–52.

Miwa, S., and Inui, Y. (1985). Effects of L-thyroxine and ovine growth hormone on smoltification of amago salmon (*Oncorhynchus kisutch*). *Gen. Comp. Endocrinol.* **58,** 436–442.

Moav, B., and McKeown, B. A. (1992). Thyroid hormone increases transcription of growth hormone mRNA in rainbow trout pituitary. *Horm. Metab. Res.* **24,** 10–14.

Noguchi, S., Noda, M., Takahashi, H., Kasakami, K., Ohta, T., Nagano, K., Hirose, T., Inayama, S., Kawamura, M., and Numa, S. (1986). Primary structure of the B-subunit of *Torpedo californica* (Na^{+}+K^{+})-ATPase deduced from the cDNA sequence. *FEBS Lett.* **196,** 315–320.

Omelanjiuk, R. J., and Eales, J. G. (1986). The effect of 3,5,3′-triiodo-L-thyronine on gill Na^{+}/K^{+}-ATPase of rainbow trout (*Salmo gairdneri*), in fresh water. *Comp. Biochem. Physiol., A* **84,** 427–429.

Payan, P., Girard, J. P., and Mayer-Gostan, N. (1984). Branchial ion movements in teleosts: The roles of respiratory and chloride cells. *In* "Fish Physiology" (W. S. Hoar and D. J. Randall, eds.), Vol. 10B, pp. 39–63. Academic Press, New York.

Perry, S. F., and Laurent, P. (1989). Adaptational responses of rainbow trout to lowered external NaCl concentration: Contribution of the branchial chloride cell. *J. Exp. Biol.* **147,** 147–168.

Perry, S. F., and Walsh, P. J. (1989). Metabolism of isolated fish gill cells: Contribution of epithelial chloride cells. *J. Exp. Biol.* **144,** 507–520.

Perry, S. F., Goss, G. G., and Laurent, P. (1992). The interrelationships between gill chloride cell morphology and ionic uptake in four freshwater teleosts. *Can. J. Zool.* **70,** 1775–1786.

Pickford, G. E., and Phillips, J. G. (1959). Prolactin, a factor promoting survival of hypophysectomized killifish in freshwater. *Science* **130,** 454–455.

Pickford, G. E., Griffith, R. W., Torretti, J., Hendlez, E., and Epstein, F. H. (1970a). Branchial reduction and renal stimulation of Na^{+}, K^{+}-ATPase by prolactin in hypophysectomized killifish in freshwater. *Nature (London)* **228,** 378–379.

Pickford, G. E., Pang, P. K. T., Weinstein, E., Torreti, J., Hendlez, E., and Epstein, F. H. (1970b). The response of hypophysectomized cyprinodont, *Fundulus heteroclitus*, to replacement therapy with cortisol: *Effects on blood serum and sodium–potassium* activated adenosine triphosphatase in the gills, kidney and intestinal mucosa. *Gen. Comp. Endocrinol.* **14,** 524–534.

Pisam, M., and Rombourg, A. (1991). Mitochondria-rich cells in the gill epithelium of teleost fishes: An ultrastructural approach. *Int. Rev. Cytol.* **130,** 191–232.

Pisam, M., Caroff, A., and Rambourg, A. (1987). Two types of chloride cells in the gill epithelium of a freshwater-adapted euryhaline fish: *Lebistes reticulatus*; Their modifications during adaptation to saltwater. *Am. J. Anat.* **179,** 40–50.

Pisam, M., Auperin, B., Prunet, P., Rentierdelrue, F., Martial, J., and Rambourg, A. (1993). Effects of prolactin on alpha and beta chloride cells in the gill epithelium of the saltwater adapted tilapia *Oreochromis niloticus. Anat. Rec.* **235,** 275–284.

Prunet, P., and Auperin B. (1994). Prolactin receptors. *In* "Fish Physiology" (W. S. Hoar and D. Randall, eds.), Vol. 13, pp. 367–391. Academic Press, New York.

Redding, J. M., Schreck, C. B., Birks, E. K., and Ewing, R. D. (1984). Cortisol and its effects on plasma thyroid hormone and electrolyte concentrations in fresh water and during seawater acclimation in yearling coho salmon, *Oncorhynchus kisutch. Gen. Comp. Endocrinol.* **56,** 146–155.

Refstie, T. (1982). The effect of feeding thyroid hormones on saltwater tolerance and growth rates of Atlantic salmon. *Can. J. Zool.* **60,** 2706–2712.

Richman, N. H., and Zaugg, W. S. (1987). Effects on cortisol and growth hormone on osmoregulation in pre- and desmoltified coho salmon (*Oncorhynchus kisutch*). *Gen. Comp. Endocrinol.* **65,** 189–198.

Richman, N. H., Nishioka, R. S., Young, G., and Bern, H. A. (1987). Effects of cortisol and growth hormone replacement on osmoregulation in hypophysectomized coho salmon *Oncorhynchus kisutch*. *Gen. Comp. Endocrinol.* **67,** 194–201.

Rossier, B. C., Geering, K., and Kraehenbuhl, J. P. (1989). Regulation of the sodium pump: How and why? *Trends Biochem. Sci.* **12,** 483–487.

Sakamoto, T., and Hirano, T. (1991). Growth hormone receptors in the liver and osmoregulatory organs of rainbow trout: Characterization and dynamics during adaptation to seawater. *J. Endocrinol.* **130,** 425–433.

Sakamoto, T., and Hirano, T. (1993). Expression of insulin-like growth factor-I gene in osmoregulatory organs during seawater adaptation of the salmonid fish: Possible mode of osmoregulatory action of growth hormone. *Proc. Natl. Acad. Sci. U.S.A.* **90,** 1912–1916.

Sakamoto, T., McCormick, S. D., and Hirano, T. (1993). Osmoregulatory actions of growth hormone and its mode of action in salmonids: A review. *Fish Physiol. Biochem.* **11,**155–164.

Salman, N. A., and Eddy, F. B. (1987). Response of chloride cell numbers and gill Na^+/K^+-ATPase activity of freshwater rainbow trout (*Salmo gairdneri* Richardson) to salt feeding. *Aquaculture* **61,** 41–48.

Sandor, T., DiBattista, J. A., and Mehdi, A. Z. (1984). Glucocorticoid receptors in the gill tissue of fish. *Gen. Comp. Endocrinol.* **53,** 353–364.

Sargent, J. R., and Thomson, A. J. (1974). The nature and properties of the inducible sodium-plus-potassium ion-dependent adenosine triphosphatase in the gills of eels (*Anguilla anguilla*) adapted to fresh water and sea water. *Biochem. J.* **144,** 69–75.

Sargent, J. R., Thomson, A. J., and Bornancin, M. (1975). Activities and localization of succinic dehydrogenase and Na^+/K^+-activated adenosine triphosphatase in the gills of fresh water and sea water eels (*Anguilla anguilla*). *Comp. Biochem. Physiol., B* **51,**75–79.

Saunders, R. L., and Henderson, E. B. (1970). Influence of photoperiod on smolt development and growth of Atlantic salmon (*Salmo salar*). *J. Fish. Res. Board Can.* **27,** 1295–1311.

Saunders, R. L., McCormick, S. D., Henderson, E. B., Eales, J. G., and Johnston, C. E. (1985). The effect of orally administered 3,5,3′-triiodo-L-thyronine on growth and salinity tolerance of Atlantic salmon (*Salmo salar*). *Aquaculture* **45,** 143–156.

Schmalzing, G., and Gloor, S. (1994). Na^+/K^+-pump beta-subunits: Structure and functions. *Cell. Physiol. Biochem.* **4,** 96–114.

Schmitz, M., and Mayer, I. (1993). Effect of androgens on seawater adaptation in Arctic char, *Salvelinus alpinus*. *Fish Physiol. Biochem.* **12,** 11–20.

Schonrock, C., Morley, S. D., Okawara, Y., Lederis, K., and Richter, D. (1991). Sodium and potassium ATPase of the teleost fish *Catostomus commersoni*—sequence, protein structure and evolutionary conservation of the alpha-subunit. *Biol. Chem. Hoppeseyler.* **372,** 279–286.

Schreck, C. B. (1981). Stress and compensation in teleostean fishes: Response to social and physical factors. *In* "Stress and Fish" (A. D. Pickering, ed.), pp. 295–321. Academic Press, New York.

Shrimpton, J. M., Bernier, N. J., and Randall, D. J. (1994). Changes in cortisol dynamics in wild and hatchery reared juvenile coho salmon. *Can. J. Fish Aquat. Sci.* **51,** 2179–2187.

Shrimpton, J. M., Devlin, R. H., Donaldson, E. M., Mclean, E., and Randall, D. J. (1995). Increases in gill corticosteroid receptor abundance and saltwater tolerance in juvenile coho salmon (*Oncorhynchus kisutch*) treated with growth hormone and placental lactogen. *Gen. Comp. Endocrinol.* (in press).

Silva, P., Solomon, R., Spokes, K., and Epstein, F. H. (1977). Ouabain inhibition of gill Na-K-ATPase: Relationship to active chloride transport. *J. Exp. Zool.* **199,** 419–426.

Skou, J. C., and Esmann, M. (1992). The Na,K-ATPase. *J. Bioenerg. Biomembr.* **24,** 249–261.

Specker, J. L., Portesi, D. M., Cornell, S. C., and Veillette, P. A. (1994). Methodology for implanting cortisol in Atlantic salmon and effects of chronically elevated cortisol on osmoregulatory physiology. *Aquaculture* **121,** 181–193.

Stagg, R. M., and Shuttleworth, T. J. (1982). Na^+, K^+-ATPase, ouabain binding and ouabain-sensitive oxygen consumption in gill from *Platichthys flesus* adapted to seawater and freshwater. *J. Cell. Physiol.* **147,** 93–99.

Thomson, A. J., and Sargent, J. R. (1977). Changes in the levels of chloride cells and (Na^++K^+)-dependent ATPase in the gills of yellow and silver eels adapting to seawater. *J. Exp. Zool.* **200,** 33–40.

Weisbart, M., Chakraborti, P. K., Gallivan, G., and Eales, J. G. (1987). Dynamics of cortisol receptor activity in the gills of the brook trout, *Salvelinus fontinalis*, during seawater adaptation. *Gen. Comp. Endocrinol.* **68,** 440–448.

Yao, K., Niu, P.-D., Le Gac, F., and Le Bail, P.-Y. (1991). Presence of specific growth hormone binding sites in rainbow trout (*Oncorhynchus mykiss*) tissue: Characterization of the hepatic receptor. *Gen. Comp. Endocrinol.* **81,** 72–82.

Young, G. (1988). Enhanced response of the interrenal of coho salmon (*Oncorhynchus kisutch*) to ACTH after growth hormone treatment *in vivo* and *in vitro*. *Gen. Comp. Endocrinol.* **71,** 85–92.

Young, G., and Lin, R. (1988). Response of the interrenal to adrenocorticotropic hormone after short-term thyroxine treatment of coho salmon (*Oncorhynchus kisutch*). *J. Exp. Zool.* **245,** 53–58.

Zadunaisky, J. A. (1984). The chloride cell: The active transport of chloride and the paracellular pathway. *In* "Fish Physiology" (W. S. Hoar and D. Randall, eds.), Vol. 11B, pp. 275–343. Academic Press, New York.

Zaugg, W. S. (1982). A simplified preparation for adenosine triphosphatase determination in gill tissue. *Can. J. Fish. Aquat. Sci.* **39,** 215–217.

Zaugg, W. S., and McLain, L. R. (1970). Adenosine triphosphatase activity in gills of salmonids: Seasonal variations and salt water influences in coho salmon, *Oncorhynchus kisutch*. *Comp. Biochem. Physiol.* **35,** 587–596.

Chapter 5

Studying the locomotory habits in fish reveals six tenets of effective science

Emily M. Standen*
University of Ottawa, Ottawa, ON, Canada
**Corresponding author: e-mail: estanden@uottawa.ca*

Chapter Outline

Emily M. Standen discusses how she has been impacted by C.C. Lindsey's chapter on "Form, Function and Locomotory Habits in Fish," in Fish Physiology, Volume 7 published in 1978.

Fish locomotion has been a widely studied field. Physicists, mathematicians, engineers and biologists have added to our understanding of how fish move through water, through different substrates and over land. Since the 1978 publication of C.C. Lindsey's chapter on ***Form, function and locomotory habits in fish***, there have been remarkable advancements in the technology and tools we use to measure, quantify, calculate and model fish motion. This article argues that although new technologies are adding much higher resolution data on different aspects of fish swimming, they support the hypotheses and understanding that was already laid out by early experimentalists using simple and elegant techniques. As a celebration of the enduring impact C.C. Lindsey's 1978 chapter has had on the field of fish swimming, this article reflects upon how the history of fish swimming, and Lindsey's remarkable breadth and clarity in reviewing it, clearly shows the importance of how we approach our science, how we reflect back on the body of literature that make up whatever field we work in, and what we can do to ensure we move knowledge forward in the most beneficial way possible. Despite huge

Fish Physiology, Vol. 40B. https://doi.org/10.1016/bs.fp.2024.05.005

leaps forward in the technologies used to study and compute fish locomotor habits, Lindsey's chapter remains a thorough and impressive summary of the work of so many early scientists, that still guide and influence the questions that surround fish locomotion today.

I have always loved fish. I started loving them because I feared them; large fiercely taxidermized bodies on the walls of dimly lit Canadian hunting lodges. I continue loving them because of their diversity, their complexity and their incredible artistic beauty in motion. As an undergrad, I immersed myself in the scientific literature produced by the men that most likely caught those taxidermized fish at the turn of the last century. I identified with them in their love of fish and their curiosity about how animals in the aquatic world functioned. I was fascinated by the ingenuity, elegance, and creative approaches that they used to understand fluid motion in fishes. For me, Cas Lindsey's chapter opened a portal into the world of fish swimming, bringing together a tapestry of ideas, notions and approaches that stimulated my imagination and desire to pull the fish off the wall, put them back into the water, and understand them on a deeper scientific level.

What I find most remarkable about the overview provided by Lindsey in this chapter is the impressive summary of information across disciplines, species and techniques that has been reviewed. After hooking his reader with an introduction of the incredible diversity of fish morphology, behavior and physiology, Lindsey opens with an artful homage to the ancient greats. He discusses the fascination thinkers like Aristotle, Galileo, Borelli and other even more ancient Hindu texts held for swimming fish. What is beautiful about this work is how Lindsey weaves the ancient questions with the more modern (turn of the last century) development of hypotheses and paradoxes that surround the swimming fish. This reminder of the enduring human fascination with fish (History of Science) along with his clear description of the physical environment fish live in (physics and chemistry), the hydrodynamics fish use to produce forces (mathematics and modeling) and the diversity of fish functional morphology (biology) Lindsey reinforces the importance of interdisciplinarity in truly understanding something scientifically. In this chapter, Lindsey encapsulates the state of knowledge on fish swimming prior to 1978. For me one of the most impactful aspects of Lindsey's chapter comes from realizing the remarkable insight and critical thinking early scientists had without the bells and whistles provided by the ever expanding and advancing technological tools available to science. In my opinion, the true value of Lindsey's chapter is in how effectively it demonstrates the fundamental tenets of exceptional science.

1 Tenet #1: Respect the power of observation and reflection

As humans we observe phenomena. As scientists we reflect on these observations, formulate questions about them, re-observe, formulate hypotheses about how the phenomena work and then make predictions that test our hypotheses.

Science can happen at many levels. Great science comes from asking the most pertinent of questions. As is the way of Science, advancement in technology allows us to build on those that come before us. Critically, new technology gives us new data at higher temporal and spatial resolutions. While looking at our new data we must reference our predecessors, spend time reflecting and critically thinking about past observations and ensure that our new observations are filling in the appropriate gaps. It is easy to get caught into the excitement of data volume, slick presentation and overwhelming "wow-factor." Reading through the elegant and simple techniques used in the early days of fish swimming has taught me to momentarily put down my color palates and fancy statistical approaches and focus on biological relevance. I return to Lindsey's chapter from time to time to remind myself that in the rush of data collection I must not forget to afford enough time for thorough reflection and re-observation so that the questions, hypotheses, and predictions I build moving forward are appropriate and valuable.

Achieving the goal (informed hypotheses): *Give yourself time to think.* This isn't a luxury, it is a necessity for ensuring careful and thoughtful science. *Don't let the sizzle of technology blind you.* When reviewing or reading science be critical of the question despite the coolness of the approach. More tech is not necessarily better.

Benefits to Science: Time to think allows scientists to lay the foundation for their next hypothesis, ensuring biological relevance and objective interpretation of data. This also supports the movement towards accessible science; thinking, not technology is the power of great science.

2 Tenet #2: Apply the 80/20 rule

Most certainly Lindsey's chapter allows us to reflect on how advantageous certain technological advancements have been for fine tuning our understanding of fish swimming. Both fluid dynamics and telemetry technologies have rocketed forward since 1978 allowing us to fill gaps in our knowledge across scales from fish migration to fluid forces (most recently the elusive American Eel, Stahl et al., 2023). High speed and light sensitive video cameras combined with advanced computational power, track particles in flow to visualize the 2D and 3D flow structures produced by fishes (Guo et al., 2023) and computational fluid dynamics allow us to predict flow patterns and processes without the need for cooperative animals (Bravo-Córdoba et al., 2021; Doi et al., 2021). When diving into the papers summarized by Lindsey it becomes clear that, although modern technology is remarkable and allows us to evaluate at much finer resolutions, prior to 1978 most of the major hypotheses and predictions surrounding fish swimming were already laid out. For example, mathematical models of fish swimming proposed by Lighthill are still being discussed and incorporated into modern models in the context of how drag, thrust and recoil are components of efficient swimming (Lighthill, 1960, 1970; Paniccia et al., 2021).

The complexity of vortex formation during fish swimming and between schooling fishes were well defined hypotheses with experimental support based on echo location data of schooling fish (Olst and Hunter, 1970), counting tail-beat frequency as a metric of effort (Zuyev and Belyayev, 1970), or visualizing flow around fishes with dye (Aleev and Ovcharov, 1971, 1973). I am impressed by the amount of information that was gained using relatively low tech solutions. The time early researchers had, free of learning the intensive technical performance of modern technologies, perhaps left them free to do more critical thinking about their observations (see Section 1). The value of the Lindsey chapter for me, is in reminding me that sometimes the infinite resolution in millions of frames per second, and millimeter by millimeter movement patterns are not required to understand the system (Fig. 1A and B). In this way, I try to think critically about what is essential and apply the 80/20 rule from engineering (80% of the data can be gathered/analyzed in 20% of the

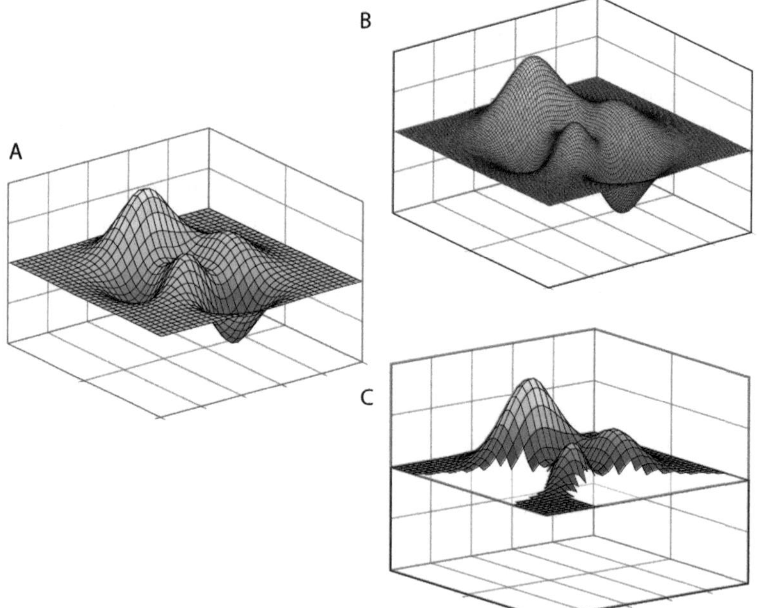

FIG. 1 The knowledge landscape. Both spatial and temporal resolution and negative data are important for understanding the complete knowledge landscape. Panel (B) is a sampling scale that is three times that of panel (A) but does not add substantially more information about the system as a whole. The same general landscape is observed in both panels however the amount of data collection and analysis effort required for panel (B) is much larger. Careful thought must be given to the importance of data resolution for maximizing the outcome per unit effort in science. Panel (C) Assumes anything below the plane of interest is considered negative data and represents the sections of the knowledge landscape that are missing when we do not report negative data. Axes are unlabeled and are meant to represent our knowledge (and subsequent ability to understand) a given scientific question.

time). Sometimes that last 20% of data is critical and worth every ounce of technology we throw at it, but sometimes it may just add to the noise, and we may want to consider its value before we spend too much time, and money, trying to sort out its meaning.

Achieve the goal (maximize data value to research effort): Before each research step consider the 80/20 rule and determine what the value of the increased data resolution is relative to the time and resources it will take to analyze.

Benefit for Science: Not everyone needs the latest gear. Science becomes more accessible to more researchers. We gain perspective more quickly with a broader view to help us focus on where and when to apply the tools that give us finer resolution.

3 Tenet #3: Be aware of the lens that technology and experience impose on you

Let the question choose the technology not the other way around. Even when we have chosen our tools correctly, we must constantly pull back from our scopes, turn off our computers and take a look at the larger picture, or at the very least, make an effort to see our biases. In his section on the early history of fish locomotion studies Lindsey reflects on the changing understanding of tail function in fish swimming. Aristotle (384–322 BCE) first postulated that the tail of fishes was used for steerage while Borelli (1680–1681), argued it was for thrust. Humans reflect on the world from their own experiences. Perhaps this is an ancient example of unconscious bias, the familiar oared Triremes of the Greek era that steered with a rudder may have swayed Aristotle's conclusion on fish tail function, while the Italian Borelli understood the possibility of a stern-drive propulsion based off of the Venetian gondola. Regardless, our life experience limits the possible questions we will form when we observe the world. Those questions will be further constrained when we ask ourselves, what technology do we have to help answer this? Technology can be pushed and developed to address our questions. In this way technology can provide us with a "new" lens with which to view the world. No matter how powerful, technology necessarily selects for us a particular set of data, a single lens view. Remaining aware of our data's limitations and our own perspectives will keep us aware of the limitation of the questions we are able to answer.

Achieve the goal (know your lens): *Let your questions choose your technology*. And be sure to explicitly state the limitations of that technology when presenting the data. *Work in collaboration with people that have different scientific and life experiences than you do*. Diversity leads to discovery by expanding the lens used to formulate and answer your questions.

Benefits to Science: Development of informed scientific questions that are aware of their biases and limitations. Effective and efficient use of technology that focuses questions and energy where progress needs to be made.

4 Tenet #4: Keep an eye on the invisible

Sometimes the invisible is simply what we choose not to see. Aspects of other fields (i.e. physics, chemistry and physiology) are often "invisible" because we are insulated from them as we work in our own realm. Exploration and collaboration between disciplines (see Section 3) can reveal huge landscapes of understanding that exist but have never been applied to one's area of research. Opening your eyes to what is known in other disciplines, particularly if it contradicts your observations of the world, can be extremely valuable. A classic example of the power of interdisciplinary exchange is reported by Lindsey when he discusses Gray's Paradox (Gray, 1936). The paradox is simple, dolphins swim much faster than would be expected because their muscle mass is too small to overcome their hydrodynamic drag. In 1936 measuring fluid forces on dolphins was not possible and so Gray borrowed tools from engineering to predict the drag forces experienced by a static dolphin shape. He then borrowed knowledge of mammalian muscle physiology to predict the mass of muscle required to produce enough force to overcome this drag and concluded, dolphin muscle mass was not enough to reach the swimming speeds seen in the wild. This contradiction between what is seen in the world and what is predicted to be reality based on current tools and understanding, allows scientists to see limitations in the current way of approaching scientific questions. In the case of the dolphin, it inspired a range of interesting questions about the drag reducing hydrodynamic properties of skin (Gray, 1936, and others), the benefits and limitations of static or quasi-static fluid force estimation (Dickinson and Götz, 1993; Ellington, 1984), traditional fluid mechanics (Lighthill, 1971), computational fluid mechanics (Fish, 1998; Long Jr et al., 1997; Moin and Kim, 1997) and many other aspects of swimming (for a review see Fish, 2006).

Achieve the goal (cross disciplines, explain inconsistencies): *When data contradicts reality, dig deeper*. Explore your world, keep refining your question, and question your techniques.

Benefits to science: Breaking down of silos and clarification of technical limitations. Work done in other disciplines is applied more broadly and can be refined through new applications. Efficiency, effectiveness and knowledge transfer means larger scientific steps forward with less effort.

5 Tenet #5: Embrace the positive and publish the negative

If we are to follow the previous three tenets it seems appropriate to quote the Nobel Laureate in Literature, Samuel Beckett, who famously wrote "Ever tried. Ever failed. No matter. Try again. Fail again. Fail better." (Worstward Ho, Samuel Beckett, 1983). No words better sum up experimental science which, in my experience, is roughly 95% failure if you are any good at all. If one succeeds more than that one may be asking the wrong questions.

Asking pertinent questions matters because science is limited by available funding. Often negative results are met with less enthusiasm than positive results. Publishing negative results is important because it ensures that the same experiment is not conducted again, freeing up grant dollars for other experiments. A negative result is valuable because it shows the outcome of a set of conditions on a given system, it can also be a message to the scientific community to say "Here is the outcome under these conditions, choose wisely how to proceed from here." The details of how a study failed to show a relationship are as important as the details of how a study succeeded in finding a relationship. Science is conditional. Results are repeatable under controlled conditions, the wider the conditions the more robust the conclusion; the change in condition that results in changed results (or negative results) is valuable knowledge about the system as a whole. Both positive and negative results allow us to define the biological landscape (Fig. 1C). Not publishing negative data is akin to leaving a portion of the landscape invisible to other researchers. At best this limits our capacity to understand the system as a whole, and at its worst, it may lead us down false paths. Although far from a systematic review, Lindsey's chapter outlines the landscape of fish swimming knowledge overtime. It provides an important and reflective history of our understanding of fish swimming and continues to guide the direction of fish research.

Achieve the goal (appreciate ALL results): Publish your negative results and review negative results mindful of their value.

Benefits to science: Grant dollars are not wasted on repeated experiments. A positive and realistic recognition of scientific hurdles is more transparent to the scientific community and the general public.

6 Tenet #6: Keep your science accessible

Accessible science means that when you speak about your work, people understand you the way you intend. The world of fish swimming, as clearly demonstrated by Lindsey's review, involves many disciplines requiring engineers, biomechanists, ecologists, physiologists and behavioralists to communicate effectively. Each field can use the same words but in slightly different ways which leads to confusion and misrepresentation (even within the field of biology the confusion between adaptation, acclamation and acclimatization is an excellent example, these words are often interchanged which obscures their specific meaning).

There are over 35,000 different species of fish in the world with a diversity of form and function that is staggering. Despite this remarkable variation, discussing fish swimming in a comparative context requires a common system of classification. Lindsey takes the time to recognize the limitations of classification systems and returns to the early paper by Breder (1926). By using Breeder's categories, even if blunt, Lindsey has established a baseline of terms, within which fish biologists can effectively discuss and argue about the discreet or continuous nature of the carangiform, anguilliform or other forms of fish locomotion. These conversations are important, they represent the progress of science and as such the clear definitions of terms used in the discussion are critical in ensuring understanding between biologists and more importantly between fields of research.

Scientists have a responsibility to simplify language used to describe scientific discovery. This will ensure the accessibility of science and the validity of the debate and discussion over complex biological function. Consistency of a system is important as well. Reinventing a classification system, even if it is flawed, can be confusing and is valuable only if a significant improvement will be made.

Achieve the goal (communicate simply): Use existing terms when you can, define them for interdisciplinary audiences, be aware of alternate interpretations and only introduce new systems if the improvement is significant.

Benefit for Science: Development of a clear, consistent and agreed upon system of communication will reduce interpretation error and facilitate knowledge transfer.

7 Conclusion

The 1978 Chapter on *Form, function and locomotory habits in fish* by Lindsey is a wonderful walk through the variable and fantastic world of fish swimming. It represents snapshots of the diversity and wealth of evolutionary invention that exists in the long history of fishes. Lindsey's summary, and all of the papers that are gathered within it, are a wonderful "starting" point for any fish biologist. It provides the historical ground work for fish locomotion and, if read with a philosophical air, can allow one to be inspired by the past and focus on the critical aspects of what makes good science so effective.

References

Aleev, Y.G., Ovcharov, O.P., 1971. The role of vortex formation in locomotion of fish and the influence of the boundary between two media upon the flowing pattern. Zool. Zhurnal. 50, 228–234.

Aleev, Y.G., Ovcharov, O.P., 1973. The three-dimensional pattern of flow round a moving fish. Ichthyol. (USSR) 13, 933–936.

Beckett, S., 1983. Worstward Ho. John Calder.

Borelli, G.A., 1680–1681. De motu animalium. Rome. Translated by Paul Maquet, 1989. On the Movement of Animals. Springler-Verlag, Berlin.

Bravo-Córdoba, F.J., Fuentes-Pérez, J.F., Valbuena-Castro, J., Martínez de Azagra-Paredes, A., Sanz-Ronda, F.J., 2021. Turning pools in stepped fishways: biological assessment via fish response and CFD models. Water 13 (9), 1186. https://doi.org/10.3390/w13091186.

Breder, C.M., 1926. The locomotion of fishes. Fortschr. Zool. 4, 159–297.

Dickinson, M.H., Götz, K.G., 1993. Unsteady aerodynamic performance of model wings at low Reynolds numbers. J. Exp. Biol. 174 (1), 45–64. https://doi.org/10.1242/jeb.174.1.45.

Doi, K., Takagi, T., Mitsunaga, Y., Torisawa, S., 2021. Hydrodynamical effect of parallelly swimming fish using computational fluid dynamics method. PLoS ONE 16 (5), e0250837. https://doi.org/10.1371/journal.pone.0250837.

Ellington, C.P., 1984. The aerodynamics of hovering insect flight. I. The quasi-steady analysis. Philos. Trans. R. Soc. Lond. B Biol. Sci. 305 (1122), 1–15. https://doi.org/10.1098/rstb.1984.0049.

Fish, F.E., 1998. Comparative kinematics and hydrodynamics of odontocete cetaceans: morphological and ecological correlates with swimming performance. J. Exp. Biol. 201 (20), 2867–2877. https://doi.org/10.1242/jeb.201.20.2867.

Fish, F.E., 2006. The myth and reality of Gray's paradox: implication of dolphin drag reduction for technology. Bioinspir. Biomim. 1 (2), R17. https://doi.org/10.1088/1748-3182/1/2/R01.

Gray, J., 1936. Studies in animal locomotion: VI. The propulsive powers of the dolphin. J. Exp. Biol. 13 (2), 192–199. https://doi.org/10.1242/jeb.13.2.192.

Guo, J., Han, P., Zhang, W., Wang, J., Lauder, G.V., Di Santo, V., Dong, H., 2023. Vortex dynamics and fin-fin interactions resulting in performance enhancement in fish-like propulsion. Phys. Rev. Fluids 8 (7), 073101. https://doi.org/10.1103/PhysRevFluids.8.073101.

Lighthill, M.J., 1960. Note on the swimming of slender fish. J. Fluid Mech. 9 (2), 305–317.

Lighthill, M.J., 1970. Aquatic animal propulsion of high hydromechanical efficiency. J. Fluid Mech. 44 (2), 265–301.

Lighthill, M.J., 1971. Large-amplitude elongated-body theory of fish locomotion. Proc. Roy. Soc. Lond. B Biol. Sci. 179 (1055), 125–138. https://doi.org/10.1098/rspb.1971.0085.

Long Jr., J.H., Pabst, D.A., Shepherd, W.R., Mclellan, W.A., 1997. Locomotor design of dolphin vertebral columns: bending mechanics and morphology of *Delphinus delphis*. J. Exp. Biol. 200 (1), 65–81. https://doi.org/10.1242/jeb.200.1.65.

Moin, P., Kim, J., 1997. Tackling turbulence with supercomputers. Sci. Am. 276 (1), 62–68. http://www.jstor.org/stable/24993565.

Olst, J.C.V., Hunter, J.R., 1970. Some aspects of the organization of fish schools. J. Fish. Res. Board Can. 27 (7), 1225–1238.

Paniccia, D., Graziani, G., Lugni, C., Piva, R., 2021. The relevance of recoil and free swimming in aquatic locomotion. J. Fluids Struct. 103, 103290.

Stahl, A., Larocque, S.M., Gardner-Costa, J., et al., 2023. Spatial ecology of translocated American eel (*Anguilla rostrata*) in a large freshwater lake. Anim. Biotelem. 11, 2. https://doi.org/10.1186/s40317-022-00308-9.

Zuyev, G.V., Belyayev, V.V., 1970. An experimental study of the swimming of fish in groups as exemplified by the horsemackerel (*Trachurus mediterraneus ponticus* Aleev). Zchthyol. (USSR) 10, 545–549.

Chapter 6

FORM, FUNCTION, AND LOCOMOTORY HABITS IN FISH☆

C.C. LINDSEY

Chapter Outline

I. INTRODUCTION

A fish moving through water is constrained by physical forces quite different from those affecting an animal moving on land or through the air. Some of the problems facing fish may be appreciated by examining the favorable and unfavorable features of water, in contrast to those of land or air, as a medium for locomotion.

☆This is a reproduction of a previously published chapter in the Fish Physiology series, "1978 (Vol. 7)/Form, Function, and Locomotory Habits in Fish: ISBN: 978-0-12-350407-4; ISSN: 1546-5098".

Fish Physiology, Vol. 40B. https://doi.org/10.1016/bs.fp.2024.07.002

Most fish swim by pushing back against the water with undulations of their body or their fins. Water is unfavorable in that it presents a yielding medium against which to push, and much energy may be wasted in making profitless eddies. Water is favorable in that it offers little drag from friction, but drag of another sort, due to inertial forces ("pressure drag"), is high, because water is so dense. At the same time, the density of water makes it a very favorable medium in which to live because it buoys up the body of an organism. Fish do not require strong internal structures to carry their weight, in contrast to land animals which are severely limited by their need for structural support. Also because of the buoyancy of water, work to keep from falling is minimal. Compared with terrestrial life, fish need expend little energy to move vertically. Because liquids are almost incompressible, pressure is not usually a problem to fish, except that it constrains the rapid vertical movements of any species carrying a chamber of compressible gas.

In contrast to animals underwater, those inhabiting the air live in an insubstantial medium even more yielding than water. Hence most terrestrial animals perform their locomotion by pushing against the earth rather than against the air. The ground is a medium with almost no "give," so that little energy is lost by imparting waste motion to the soil, and all goes into forward momentum of the animal. But, because frictional drag is very great, few animals can glide rapidly over (or within) the earth's surface. (Despite the give of water, fish can swim faster in water than snakes can crawl over the ground.) Most terrestrial animals have abandoned fishlike techniques of propulsion; instead they "walk," taking advantage of the frictional forces and resistance of the ground to provide thrust for the limbs, and of the negligible drag of the air to allow forward progression of the body and the individual limbs. Walking is an inappropriate technique for attaining speed underwater, as is apparent to a human who tries to run while partly immersed. Many aquatic invertebrates which are heavier than water do walk slowly on the bottom, but to attain speed they must leave the bottom and swim.

Few species of fish walk on the bottom underwater with their paired limbs. On the other hand, those fish which make excursions out onto land usually have to resort to walking or skipping in some fashion, since the air provides so insubstantial a medium against which their usual body undulations can act. Proof of this is the observation that a live fish on a slippery deck flaps futilely without achieving forward progression, even though the same muscular undulations performed underwater would have propelled it swiftly forward. Eels can progress overland through grass or on rough ground, but here the body undulations are thrusting against the ground, not the air; the eel is not swimming through the air, but is crawling like a snake.

Just as the fastest aquatic animals abandon the bottom and swim through the water, so the fastest terrestrial animals leave the ground and fly through the air. Roughly half the living species of animals can fly. Most are small animals (insects), in which the surface area for lift is great relative to the weight. For them the viscosity of the medium is important relative to its inertia (i.e., they operate at low Reynolds numbers, as explained in Chapter 3). Their physical constraints are therefore in some ways comparable to those of larval fish and other small animals underwater. One feature of locomotion under such conditions is that stopping and starting are no problem; when viscosity is dominant a flying insect or swimming fish larva stops as soon as it ceases to propel itself, and begins to move at full speed with negligible time-lag as soon as the propulsive movements begin (Lighthill, 1969). The larger flying species (birds and bats) on the other hand, operate with speeds and dimensions where the viscous forces are less significant; to them, as to larger fish, the inertial forces of the medium are dominant (although the absolute values are very different in water and in air). To these animals, extra energy is required to accelerate the body from rest; conversely a moving bird or large fish, if well streamlined, can glide a long way after it ceases propulsive movements.

Gravity, which is of minor concern underwater, is a major factor in air. To birds supporting themselves in the air, a severe constraint is weight. Extreme structural economies (such as hollow bones and feathers) are needed to minimize the effects of gravity in a large flying animal. Underwater, some of the midwater fishes which lack gas chambers display comparable weight economies (such as reduced skeleton, and substitution of light fatty tissues wherever possible). Fish can thereby achieve neutral buoyancy in water. But in air the density of the medium is so slight that even the most lightly built bird still weighs much more than the air it displaces. No animal has attained neutral buoyancy in air, although this should be theoretically possible through development of a large bladder of hydrogen or other light gas. In water, on the other hand, "weightlessness" is relatively easy to attain, by inclusion of a small gas chamber which displaces an equivalent volume (and 800 times the weight) of water. The majority of bony fish (but not sharks) have such a gas chamber; to them, structural weight economies are not necessary.

Animals flying through air are like winged aircraft; they must divert part of their locomotory effort into lift to overcome gravity. Neutrally buoyant fish in water are like dirigibles; they can concentrate all their effort into forward thrust. Only in sharks, rays, tunas, and other fish which may lack a gas chamber does lift become important, and so their locomotion through water has some features in common with the flight of birds or aircraft.

Most of the few fish which are capable of brief aerial locomotion progress through the air by passive gliding rather than by flying (and swim underwater by conventional body undulations). They are comparable to those restricted groups of mammals, amphibians, and reptiles which can glide but which are not primarily adapted to this mode of locomotion.

Only very few fish (including the Gasteropelecidae, and *Pantodon*) can fly in air by beating their wings, and this they do ineptly. The formidable problems of simultaneous adaptations for locomotion both in midair and underwater have been overcome, among the vertebrates, only by diving birds such as loons. Among the invertebrates a few types of insects (including water beetles and some bugs) can swim underwater and fly in air in the same life history stage. Most which can move effectively in both media do so at two radically different life history stages (e.g., dragonfly nymph and adult).

Another set of locomotory constraints arising from properties of water are those factors determining the oxygen available to the propulsive system. Water contains only about one-thirtieth as much oxygen as does an equal volume of air. Water is also much heavier and more viscous to move. To meet these problems, fish have a flow-through respiratory system which extracts a high proportion of the available oxygen. This is achieved without undue expansion of the respiratory surface, which in tunas is about the same as that in terrestrial vertebrates of equivalent weight, and in other fish is smaller. A very large gill area would present too great a drag, apart from allowing excessive ion exchange (a nonexistent problem in terrestrial lungs). But even with its highly efficient countercurrent design and moderate gill area, the energy expenditure for gill ventilation in an active fish may amount to about one-tenth of its total metabolic output. In contrast, an active man probably uses no more than 3% of his total oxygen consumption for breathing.

Some fish species move the water over the gills with their branchial muscles. Others simply swim with their mouths open, in which case their drag is increased and the energy for gill ventilation must come from additional work by the propulsive body muscles. The proportion of the total energy output used for ventilation is probably less in small than in large fish. This is just one of many ways in which physical constraints of the environment operate differently on small and on large fish.

An elegant survey of aquatic animal propulsion (including many major groups omitted above) is given by Lighthill (1969), whose view is that of an aerodynamicist. Other insights into the locomotory problems in different media are to be found throughout the works of Alexander (1967, 1968, 1971), Gray (1968 and earlier), Schmidt-Nielsen (1972a,b), and Tucker (1975).

II. EARLY HISTORY OF STUDIES ON FISH LOCOMOTION

Few of the features of fish locomotion are evident to the naked eye. The usual method of locomotion in fish is now thought to depend on passing alternating waves of contraction backward along the body muscles. Thrust against the water either is generated by the sides of the body pushing obliquely backward (as in a swimming eel), or else has become progressively more concentrated in the tail fin (as in fast swimmers such as the tuna). The paired fins seem to contribute little in conventional forward swimming, and are reserved for maneuvering. Not only are these points difficult to discern by direct observation; they also do not follow by logical extension from the locomotory habits of terrestrial animals. Consequently, the historical development of opinions on fish locomotion has been marked by contradictions and controversies.

In the fourth century B.C., Aristotle referred to fish locomotion in various passages in his three works "Parts of Animals," "Movement of Animals," and

"Progression of Animals." He wrote that fish with very long bodies and no paired fins, such as the moray, move along by an undulating motion of the body; "that is, they use the water just as serpents use the ground." He made the acute observation that eels move in the same way in water and on land but with fewer bends in the former medium (a hydrodynamic explanation for which might only now be attempted). Aristotle anticipated Newton by 20 centuries in his statement that there must always be something "immovable" outside an animal "supported upon which that which moves moves. For if that which supports the animal is to be always giving way ... there will be no progress, that is, no walking unless the ground were to remain still, and flying or swimming unless the air or sea were to offer resistance." Aristotle fell short of an understanding of undulatory locomotion in that he did not perceive that the crests of the body waves pass backward, and he tried to draw an analogy between a snake's motion and the walking of quadrupeds. He correctly described a ray as swimming by means of the edges of its flattened disc. He was apparently wrong with respect to the locomotion of most fishes, as he thought that the two pairs of fins were their principal means of propulsion. He suggested that the caudal fin was primarily for steering.

The ancient Hindu medical work *Susruta-samhita* probably reached its final form in the early years A.D., but it may contain components which predate Aristotle. According to the free translation by Hora (1935), it suggests a correlation between body form, habitat, and locomotion in some freshwater fishes; river fish are said to be bulky in the middle because they move with their head and tail; pool fish, having little space to move about, are deep-bodied; torrent fish are flattened because they crawl with their chests on the bottom.

The first attempts at mechanical analysis of fish locomotion were by Borelli, a disciple of Galileo, who in 1680 published a diagram of a fish swimming by sweeping its caudal fin and peduncle side-to-side in an arc (reproduced in Gray, 1968). Borelli thought of the tail fin as operating like an oar sculled behind a boat. He dispelled the notion of Aristotle that the paired fins when present are the main locomotory organs, stating they are held at the sides during swimming; Borelli pointed out also that the muscles of the body, which provide power for the tail strokes, are large in proportion to those of the paired fins. He outlined the role of the gas bladder in controlling the specific gravity and hence the position of the fish in the water. He departed from observation, however, when he described the tail as operating like a frog's foot, contracted during a "preparatory" outward sweep and expanded during a powerful inward propulsive sweep. Two fundamental points he made, which display the influence of Galileo, were: (1) the fish cannot move its tail without also moving its body, and (2) the force moving the fish forward is due to resistance of the water against the surface of the moving tail.

Pettigrew (1874) challenged Borelli's view that forward motion could result simply by lashing the tail from side to side. He argued that moving the tail from the midline outward during the "preparatory" sweep would produce a backward movement of the whole. Breder (1926) comments "His logic would seem to be correct, but that he was in error has been positively demonstrated by the construction of a model." (Breder built a model boat, which *did* swim forward simply by means of an oscillating rigid tail vane.) What Pettigrew overlooked was that the outward sweep of the tail, on a fish or on a boat, does not move the whole backward; instead, it swings the main body (ahead of the joint) slightly in a sidewise arc opposite to that of the tail. The subsequent inward stroke moves the whole forward as well as swinging the body back into line. Pettigrew invoked complex rolling motions of the fish to overcome his supposed difficulty. Breder (1926) attacked Pettigrew enthusiastically for these and other "absurdities," and complained that uncritical paraphrasing of Pettigrew's views by Bridge (1904) in the Cambridge Natural History has given them undeserved circulation. However, Pettigrew did observe that fish throw their bodies into a double or sigmoid curve rather than into the simple arc described by Borelli (although he believed this must occur in all fish, which it does not). Pettigrew correctly observed that the tail tip of a slowly moving sturgeon described a figure- 8 when viewed from above. He implied that this was an adaptation for efficient propulsion; Gray (1933a) showed it to be the inevitable result of the propagation of a wave of curvature along any inextensible body.

The possibility of precise observation on locomotion dates from Marey's (1895) use of cinematography. He produced sequential photographs of swimming fish, a technique since widely emulated. His pictures demonstrated that waves of curvature pass along the length of the body. Unfortunately, his Fig. 199 of a swimming conger eel shows one in which the body undulations are either stationary or are slowly moving *forward* relative to the water, so that the fish must actually have been slowing down when it was photographed. Since Marey did not attempt to analyze the forces involved in swimming, he missed the fact that this was an unhappily chosen figure to illustrate swimming. Although Gray (1933b) drew attention in a footnote to the aberration, Marey's figure has been widely copied. It still appears in the third edition of Norman and Greenwood's (1975) otherwise admirable text. [Nikolsky's (1963) text does not make this mistake, but it does use as an illustration of locomotion (his Fig. 32) a sequence from Gray (1933a) which actually shows an eel swimming backward.] Notwithstanding, Marey provided a strong incentive for studies on fish locomotion, and inspired later experimental studies in France by Houssay (1912), who tried to measure the thrust and drag of fish, and by Magnan (1929, 1930).

Dean (1895) reproduced one of Marey's illustrations showing the wave form in a swimming eel. Dean explicitly stated "It is the pressure of the fish's body against the water enclosed in these incurved places which causes the forward movement."

In an outstanding synthesis called "The Locomotion of Fishes," Breder (1926) gave an extensive treatment of the mechanical principles, as well as a systematic description of locomotion in different fish groups. Breder concluded that "All the movements of fishes when swimming (except exhalation) are fundamentally of an undulatory muscular nature even though obscured by various specializations, and are induced by the serial action of metameral muscles." He categorized the types of movement of the body and of the fins, and coined many of the terms (e.g., Anguilliform, Carangiform) now in use.

Th dominant figure in marrying precise measurement of moving animals with mathematical analysis has been Sir James Gray. The wide-ranging curiosity of Gray and his colleagues, notably H. W. Lissmann, has been brought to bear on the locomotion of large vertebrates, and of sperm, and of most groups in between. In a series of papers starting in 1933 Gray analyzed photographs to show how undulatory swimming movements generate thrust. Gray (1936b) compared the calculated drag of a swimming dolphin with the calculated power output of its muscles, and concluded that dolphins (and some fish) are observed to swim at speeds which according to theory are impossibly fast. This famous "Gray's Paradox" has stimulated much research and controversy ever since. In trying to balance the equation, some have sought improved measures of the muscular power output. Some have recorded much more precise observations on swimming fish [notably Bainbridge (1958 and later) using an ingenious "fish wheel"]. Some have looked for drag-reducing mechanisms. Most important, some have questioned the assumption that drag of a swimming fish could be equated to that of a rigid body or model. Some tried to measure the drag on actual fish, but their measurements seemed only to add to the growing confusion. Sir Geoffrey Taylor (1952), and Sir James Lighthill (1960), proposed mathematical models which might allow calculation of the drag of a swimming fish. Newer "bulk momentum" models concentrate attention on the kinematics of the trailing edge of the tail throughout one propulsive cycle, from which they attempt to calculate the thrust and power which must have been generated. Their significance is described in Webb's (1975) publication, which is itself an important bridge across the communication gap between biologists and physical scientists. Recently DuBois *et al.* (1974) implanted pressure sensors at various points on the surface of live and of dead fish in a water tunnel. Webb (1975; see also Chapter 3) brings up to date the developments in fish hydrodynamics and energetics. One might summarize that the gap between the swimming fish and the scientists is closing, but the fish is still well ahead.

III. MODES OF SWIMMING

A. Nomenclature of Modes

The different types, or modes, of propulsive movements of fish were classified by Breder (1926), whose nomenclature, somewhat expanded, is followed here. Examples of fish displaying these modes are shown in Fig. 1. As Breder

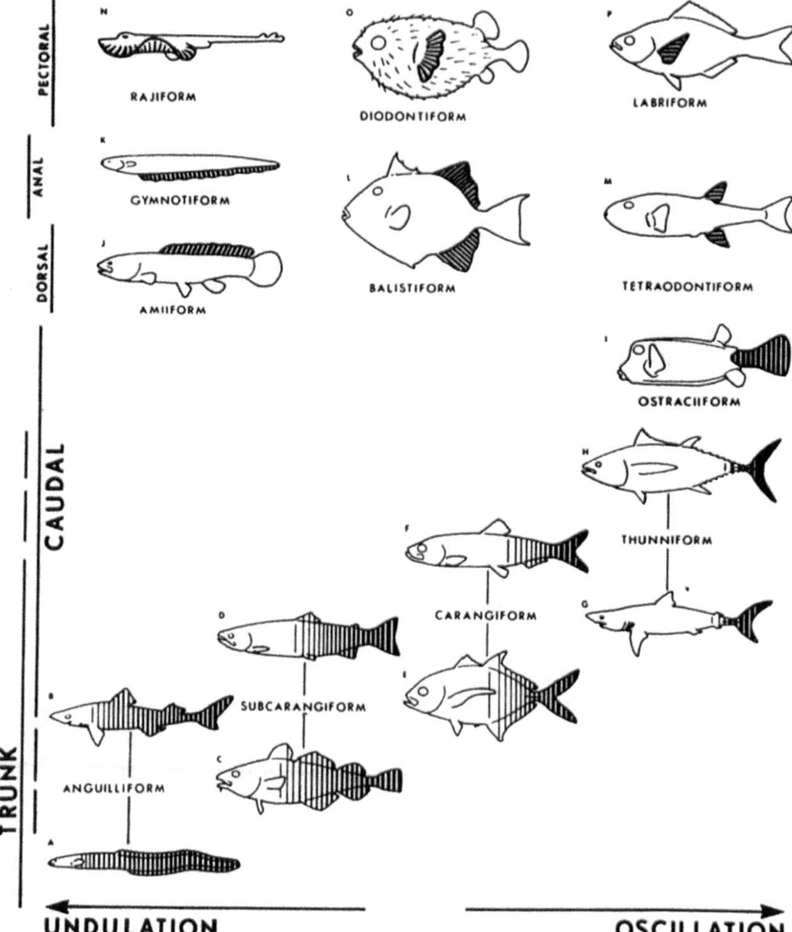

Fig. 1. Modes of forward swimming in fish, arranged along the vertical axis according to the propulsive contributions of body and fins (indicated by density of shading), and along the horizontal axis according to a scale running from serpentine undulation (more than one wavelength present) to oscillation (a rigid wigwag or fanlike motion). Species illustrated are: (A) *Anguilla anguilla*, (B) *Squalus acanthias*, (C) *Gadus morhua*, (D) *Salmo gairdneri*, (E) *Caranx hippos*, (F) *Clupea harengus*, (G) *Isurus glaucus*, (H) *Thunnus albacares*, (I) *Ostracion tuberculatum*, (J) *Amia calva*, (K) *Gymnotus carapo*, (L) *Balistes capriscus*, (M) *Lagocephalus laevigatus*, (N) *Raja undulata*, (O) *Diodon holocanthus*, and (P) *Cymatogaster aggregata*.

stated, the suffix "-form" (e.g., in anguilliform) refers to the types of movement and not to the body forms, and is therefore not strictly parallel to words such as "fusiform." Indeed, one fish may show more than one mode of swimming, as in the surfperch *Cymatogaster aggregata*, which usually swims with its pectoral fins (labriform mode) but switches to caudal fin locomotion (carangiform mode) at high speeds (Webb, 1973b). Breder (1926), Bainbridge (1963), and Webb (1975) have stressed that these classifications refer to average types within an essentially continuous range of swimming modes, and should not be applied too rigorously. Use of Breder's nomenclature in the following discussion and in Fig. 1 is simply for convenience.

The arrangement of swimming modes implies no evolutionary or taxonomic affinities; clearly there has often been functional convergence on one swimming mode by taxonomically remote groups (e.g., locomotion is anguilliform both in lampreys and in blennies). The rationale for a classification according to propulsive mode is that similar hydrodynamic analysis may be applicable to animals which swim in the same way, regardless of diverse phyletic origins.

In a swimming eel, most of the body is bent into backward-moving waves, whose amplitude is quite wide over the whole body length (Fig. 2A). In progressively shorter, thicker fish, propulsive waves tend to be increasingly concentrated in the tail region, so that only about half a wavelength is visible, and its amplitude rises rapidly in the region of the tail base (Fig. 2B, C). This propulsive mode reaches its climax in the very swift mackerels and tunas that

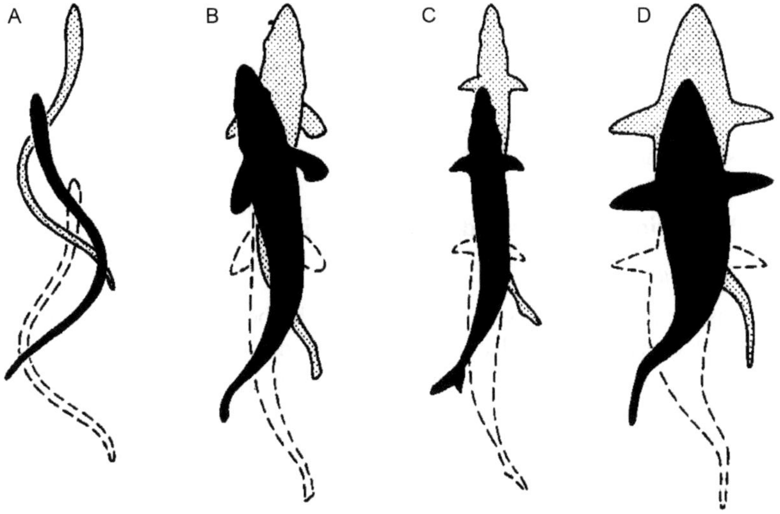

Fig. 2. Gradation of swimming modes from (A) anguilliform, through (B) subcarangiform, and (C) carangiform, to (D) thunniform. The black silhouette (dorsal view) is superimposed on successive positions one-half tail beat earlier (broken outline) and one-half tail beat later (stippled). (A) *Anguilla anguilla*, 7 cm long, about 1.5beats/sec (B) *Gadus merlangus*, 24 cm long, about 1.7 beats/sec. (C) *Scomber scombrus*, 40 cm long, about 2.4 beats/sec. (D) *Euthynnus affinis*, length unknown, perhaps about 40 cm, about 2.4 beats/sec. A, B, and C based on Gray (1933a, 1968); D based on Fierstine and Walters (1968).

appear to the naked eye to swim by moving only the caudal fin (Fig. 2D). A few species like the boxfish *Ostracion* have totally inflexible bodies which cannot be thrown into a wave; they swim by oscillating the caudal fin back and forth like a fan or pendulum.

This gradation from multi-waved undulation, through progressive and eventually exclusive concentration of propulsive movement in a pendulumlike oscillation, can be seen not only in the body but also in the median and paired fins. The nomenclature "anguilliform–carangiform–ostraciiform" was used by Breder (1926) as a scale to refer to flexures in the dorsal, anal, and pectoral fins as well as in the whole body, although he coined additional names for the swimming modes which involve the fins. A comparable "undulation–oscillation" scale is used in Fig. 1.

The modes of swimming dealt with in this section refer to straightforward horizontal motion in still water while free from contact with any solid. They therefore omit accelerating, turning, rising, stopping, and other maneuvers which may be vitally important components of locomotion. Nonswimming movements such as burrowing, creeping, jumping, and flying are dealt with in Section IV, as is jet propulsion.

A somewhat different analysis of locomotion types was presented by Kramer (1960). He grouped fish into ten categories using combinations of the swimming characteristics: use of trunk versus median or paired fin muscles; straight versus curving path; adaptation for high sustained speed versus rapid acceleration; inclination of fin axes on the body. Fish were also categorized according to ecological types: fast swimmers, roamers, swimmers between obstacles, slow and precise maneuverers, and bottom resters. Kramer proposed no nomenclature for his swimming modes.

B. Propulsion by Body and/or Caudal Fin

1. Classification

The classification of modes of propulsion by the body or tail has since been somewhat expanded from that designed by Breder (1926), but nomenclature has not been uniform. Table I shows the apparent equivalence between terms used by different authors. The two extremes of the spectrum "anguilliform" and "ostraciiform" have been used consistently. Breder's intermediate term "carangiform" covers a wide range of swimming patterns, which are now suspected to require more than one hydrodynamic model (Lighthill, 1969). Hence three terms, "subcarangiform," "carangiform," and "thunniform," will be used, the latter as a more convenient substitute for "carangiform with large lunate tail" (Lighthill, 1969) or "carangiform mode with semilunate tail" (Webb, 1975).

Some characteristics of each of these propulsive modes are given in Table II, and examples of taxonomic groups in which some (but not necessarily all) members use that mode.

Table I Classification by Various Authors of Methods of Fish Propulsion Involving the Body and/or Caudal Fin

Breder (1926)	Bainbridge (1961)	Marshall (1971)[a]	Webb (1975)	Present
Anguilliform	Anguilliform	Anguilliform	Anguilliform	Anguilliform
Carangiform	More anguilliform carangiform	Subanguilliform	Subcarangiform	Subcarangiform
Carangiform	Carangiform	Fusiform	Carangiform	Carangiform
Carangiform	More ostraciiform carangiform	Thunniform	Carangiform with lunate tail	Thunniform
Ostraciiform	Ostraciiform		Ostraciform	Ostraciiform

[a]*Marshall did not propose a formal classification, but used these terms in his text.*

Table II Comparison of Swimming Modes Involving the Body and/or the Caudal Fin

Character	Anguilliform mode	Subcarangiform mode	Carangiform mode	Thunniform mode	Ostraciiform mode
Wave length / Body length	Short, always < 1.0	Uusually <1.0, but can be 1.02 in *Oncorhynchus nerka*[a]	Usually > 1.0, but 0.93 in *scomber scombrus*[b]	1–2	Pendulum motion
Wave lengths visible on body	Always > 0.5, usually > 1.0	> 0.5 usually not more than 1.0	Up to 0.5	0.5–10	Pendulum motion
Amplitude / Body length	Large along whole body. *Anguilla* max. 0.36, *Clupea* larva max. 0.46	Undulations wide only in posterior ½ or ⅓ of body, max. about 0.2	Undulations confined to posterior ⅓, max. about 0.3	Undulations confined to peduncle and tail, max. > 0.3[c]	Tail pivots on caudal peduncle *Torpedo nobiliana max.* about 0.25[d]
Body shape	Long thin. Anterior cylindrical, posterior compressed	Fusiform Peduncle fairly deep	Mass concentrated anteriorly. Peduncle quite narrow	Massive rounded anterior. Surface streamlined. Extreme narrow-necking of peduncle, with laternal keels	Expanded or depressed, laterally inflexible, often armored, poorly streamlined
Span of body and median fins	Taper extreme: Chimaeridae, Saccopharyn-goidei, Notacanthiformes, Regalecidae, Cepolidae, Trichiuridae Taper moderate: Petroymzontidae, Anguillidae, *Chlamydoselache* Span expanded posteriorly by tail: *Squalus*, Siluriformes Expanded dorsal and anal fins: Osteoglossidae, Trachypteridae, Pleuronectiformes	2–3 dorsal, 1–3 anal fins, gaps filled by vortex sheets (Gadiformes). Or dorsal or anal may be long to reduce yawing (Cyprinidae). Or short dorsal followed by adipose (Salmonidae)	May have stiff median fins resisting yawing: Carangidae	High first dorsal fin fixed (*Lamna, Isurus, Carcharodon*) or collapsible (*Thunnus, Euthynnus, Acanthocybium*). Finlets (5–11), on peduncle (Scombridae), or small second dorsal and anal fins (fast sharks)	Span variable; median fins often small, not placed to reduce yawing

Caudal fin	Aspect-ratio low (Siluriformes) or moderate (*Squalus*). Often small, rounded or absent. Span alterable	Rather low aspect-ratio. Posterior margin almost straight or slightly scooped. Span alterable[e,f]	Rather high aspect-ratio. Posterior margin scooped or notched. Span alterable moderately (*Clupea*) or not at all (*Caranx*, *Scomber*). Stiff tips lead during beats, center follows	Very high aspect-ratio. Lunate margin. Span almost fixed. Center leads during beats, tips follow	Low aspect-ratio. Often rounded or square. Quite rigid, pivots on peduncle
Other examples (not necessarily including *all* species in group indicated)	Myxinidae, Blennioidei, Ophidioidei, Ammodytidae, Synbranchiformes, Trichiuridae,[g] Pleuronectiformes (which undulate on sides) Young of most fish even if adult mode differs	*Triakis*, Esocoidei, Poeciliidae, *Mugil*	Clupeidae Characidae, Mormyridae, *Pomatomus*, *Sarda*, *Sardinops*	Some Scombroidei, some Lamnoidei, whales and dolphins, ichthyosaurs, (Thunniform mode question-able in *Rhineodon*, Istiophoridae, Xiphiidae, Luvaridae, Stromateidae, Bramidae, Coryphaenidae)	Loricariidae, Ostraciidae, Tetraodontidae, Diodontidae, Lophiiformes, Lophotidae,[f] Trichiuridae[g]

[a]Webb (1973a).
[b]Gray (1933a).
[c]Fierstine and Walters (1968).
[d]Roberts (1969c).
[e]Bainbridge (1963).
[f]Webb (1975).
[g]Bone (1971).

2. Anguilliform Mode

Anguilliform is a purely undulatory mode of swimming, in which most or all of the length of the body participates. The side-to-side amplitude of the wave is relatively large along the whole body, and it increases toward the tail. The body is long and thin; in eels it may be nearly cylindrical anteriorly, and somewhat laterally compressed toward the posterior. The caudal fin is often small, or even absent.

Figure 2A shows three successive positions of a swimming eel *Anguilla*, after which the mode is named. It must not be supposed from these outlines that there are fixed pivots of nodes around which the body sections oscillate. The manner in which the waves of contraction move smoothly backward is better illustrated in the successive outlines of a swimming herring larva in Fig. 3. Each wave is generated by contractions of the body muscles in a few anterior segments on one side of the vertebral column, while those on the opposite side are relaxed and are slightly stretched. The resultant bending of the body toward the contracted side passes backward as the wave of muscle contractions moves toward the posterior. Meanwhile the anterior muscles on the side which had been contracted relax, their partners on the opposite side contract, and a bend in the reverse direction is initiated, and passed backward in turn. In Fig. 3 the crests of the propulsive waves can be seen to move backward with respect to the fixed background, producing a thrust which drives the whole fish forward. Quantification of the thrust generated by this type of swimming is discussed in Chapter 3.

Anguilliform swimming is widespread among fishes. The young of most forms probably swim in the anguilliform mode even though their adult locomotion (and body form) may be very different. [In herring *Clupea harengus* the swimming of the larva is anguilliform (Fig. 3) but in the adult is carangiform (Webb, 1975; see also Fig. 1F).] Several phylogenetically remote groups of fish with long flexible bodies swim as adults in this way. Performance in

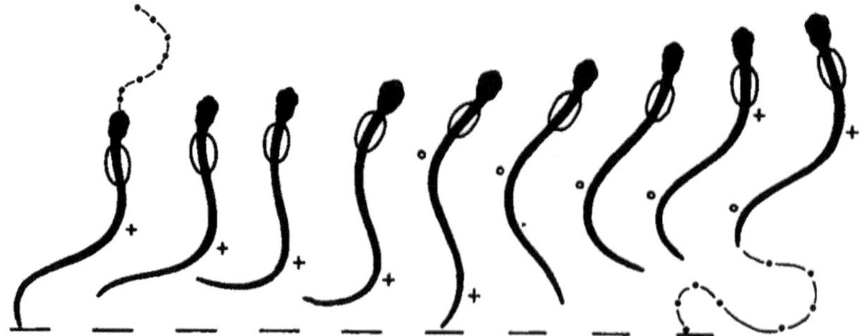

Fig. 3. Anguilliform swimming by herring larva *Clupea harengus*, about 6.5 mm long, with yolk sac. Successive cinephotos (at 0.021sec intervals) are displaced to the right. Lower broken line represents a fixed position on background. Movements of snout and tail tip indicated by dots. Position of wave crests shown by crosses and circles. (Based on Rosenthal, 1968.)

the anguilliform mode is evidently improved by a compressed cross section; this may be achieved by laterally compressing the body itself, or by augmenting the thrust with expanded dorsal or anal fins to produce a wide "span" (Fig. 1A). Highly flexible ribbon-shaped bodies with wide span are found in many species among the groups named in Table II.

High speed is not characteristic of the pure anguilliform mode. Many anguilliform swimmers live close to the bottom. Flatfish (Pleuronectiformes) usually swim in the anguilliform mode; they swim, however, on their side, so that the undulations are vertical rather than horizontal. A very wide body span in flatfish is achieved both by a high compressed body and by elongate dorsal and anal fins. Some flatfish augment the propulsive thrust of the body by also passing undulations of wider amplitude along the dorsal and anal fin rays (Aleev, 1963). The Greenland halibut *Reinhardtius hippoglossoides*, which is less flattened than other flatfish and has less complete asymmetry, swims like other flatfish when it is close to the bottom, but in midwater it adopts a vertical position. "It is still an open question whether the Greenland halibut is a round flatfish or a flat roundfish" (deGroot, 1970). A few other fish swim pelagically in the anguilliform mode, although perhaps not very fast; these include some cusk-eels—Ophidiidae, and presumably the eel *Anguilla* during its ocean migrations.

Anguilliform swimming is evidently least efficient (in the sense described in Chapter 3) when associated with bodies whose span tapers toward the posterior. Table II lists some groups which taper markedly, and some with an almost constant span along the whole body (partly by virtue of the median fins).

Toward the other extreme, in many catfishes, Siluriformes, and more particularly in many of the smaller sharks, the body moves in a distinctly anguilliform mode but the span is expanded posteriorly by a moderate sized although flexible caudal fin. The shape of the caudal may vary considerably as measured by its "aspect-ratio." [Aspect-ratio is the square of the maximum fin height (or span) divided by the fin area.] The aspect-ratio of the tail may be quite low in catfish, but is higher in some sharks. The dogfish shark *Squalus* in Fig. 1B has the body undulations more pronounced toward the posterior than has the eel, and.it has a heterocercal tail with moderate aspect-ratio. Here, the anguilliform mode merges imperceptibly into the subcarangiform mode.

3. *Subcarangiform Mode*

Body movements in subcarangiform swimming (e.g., trout) are essentially similar to those in anguilliform, the major difference being that the side-to-side amplitude of the undulations is slight at the anterior, and expands significantly only in the posterior half or one-third of the body. The tip of the snout does not move in the straight mean path of progression, but oscillates about this with a moderate amplitude (Fig. 2B). There is no fixed node; no point on the fish's body seems to move forward on, or parallel to, the mean path of progression. The whole body in fact generally executes a sinuous path through the water (Bainbridge, 1963).

The amplitude of subcarangiform undulations does not change with swimming speed except at speeds lower than 1 or 2 body lengths (ℓ)/sec. Amplitude is constant at moderate and high steady swimming speeds even in fish of different lengths. At low speeds, below 1 or 2, ℓ/sec, Bainbridge (1958) believed he detected a correlation between amplitude and tail beat frequency, but Hunter and Zweifel (1971) regarded his data as inconclusive on this point. The speed is altered by varying the velocity at which the waves are passed backward, and hence also the tail beat frequency. The maximum frequency attainable decreases with increasing size of the fish. Wavelength relative to body length is generally assumed to remain constant within a species, but there is some reason to suspect that longer fish may use slightly shorter specific wavelengths.

The body tends to be heavier and more rounded anteriorly compared to anguilliform swimmers. The caudal peduncle is fairly deep; the caudal fin which it bears has a rather low aspect-ratio, with its posterior margin almost straight or only moderately indented ("scooped out") in the center (Fig. 1C, D).

The caudal fin in a subcarangiform swimmer tends to be flexible, and is provided with intrinsic muscles (Fig. 7D) which can slightly open or shut the "fan" of caudal fin rays so as to alter the fin area by as much as 10% at different phases of one beat (Bainbridge, 1963). In rainbow trout, the depth of the whole tail is also increased at higher swimming speeds (Webb, 1975). The movements of the caudal fin during swimming are obviously very complex; they probably involve rapid adjustments which control the thrust, but they are not understood. Surprisingly, amputation of the caudal fin has little effect on straight forward swimming in the subcarangiform mode; the welldeveloped caudal in these fish has probably evolved primarily in response to requirements for high acceleration, fast turning, and high-speed maneuverability (Webb, 1973a).

Cods and their relatives (Gadiformes) have two or three dorsal and one or two anal fins which are separated from each other and from the caudal fin by narrow gaps (Fig. 1C). When the fish is moving forward, these gaps become filled by "vortex sheets" which behave hydrodynamically almost like a solid surface (see Chapter 3). Such fish therefore have functionally a wide span, and their subcarangiform swimming may approach the anguilliform. In other fish which use the subcarangiform mode the median fins are less continuous; the dorsal fin may be short and followed by a small adipose fin (Fig. 1D), or the dorsal or the anal fins may be relatively long and serve to reduce the side-to-side yawing of the body in response to the lateral tail beats.

The most complete descriptions of subcarangiform swimming are by Bainbridge (1958, 1963), who studied rainbow trout *Salmo gairdneri*, bream *Abramis brama*, dace *Leuciscus leuciscus*, and goldfish *Carassius auratus*, and by Webb (1973a), who studied sockeye salmon *Oncorhynchus nerka*.

4. Carangiform Mode

In carangiform swimmers only the posterior portion of the body is capable of wide flexure. Undulations are largely confined to the last third of the length, and the thrust is delivered by the rather stiff tail (Fig. 2C). Carangiform swimming is faster and probably more efficient than anguilliform swimming, since the same thrust can be delivered, provided the amplitude has risen to a large value in the region immediately ahead of the trailing edge, with less energy lost in displacing water laterally and forming vortices (Lighthill, 1969).

Several morphological adaptations are necessary for efficient carangiform swimming. Because there is never a complete wavelength on the body, and because the lateral flexures are concentrated at the posterior, the side forces produced by the flexures do not cancel out their net effect as they do in the anguilliform mode. There is thus a tendency for the body to recoil in response to the tail movement by sideslipping and yawing (Fig. 2C).

Recoil movements, which waste energy, are minimized in two ways. Wave amplitude increases very rapidly as it approaches the caudal peduncle. In order to avoid generating a large sideways thrust in this region, the depth of the peduncle is greatly reduced (Fig. 1E). This local reduction of the span is called narrow necking. Narrow necking occurs to a moderate degree in some of the subcarangiform swimmers, is well developed in the carangiform, and reaches its extreme in the thunniform (Fig. 1H). Lateral oscillations grow in amplitude mainly in the region of this slim peduncle, and reach a large approximately uniform value over the caudal fin.

A second adaptation which reduces recoil in carangiform fish is the concentration of mass and of body depth toward the anterior (Fig. 1E, F). Stiff median fins may further increase the overall span and help resist sideways movement of the body.

The caudal fin, which delivers most of the thrust, is stiff so as to allow little dorsoventral bending. The upper and lower edge are "swept back," and the center of the fin is scooped out to various degrees, often as a V-shaped notch (Fig. 1E, F). As the fin moves, the scooped out center area fills with a vortex sheet which works as effectively as the rest of the tail (Lighthill, 1969). The high span and reduction of area due to scooping of the center produce a high aspect-ratio compared with subcarangiform swimmers. The span is under less control by intrinsic muscles than in subcarangiform swimmers; the fan of caudal rays can be expanded only moderately (*Clupea*) or scarcely at all (*Caranx, Scomber*).

In the carangiform mode, the caudal fin is not simply wagged back and forth like a stiff blade on a hinge (the ostraciiform mode). Instead, the angle of inclination is altered as it moves from side to side so that the fin always has a backward-facing component even when moving away from the midline (Lighthill, 1969). The details of how the stiff fin is manipulated in this way by the muscles of the trunk and peduncle have not been measured.

Carangiform swimmers in the family Clupeidae (*Clupea*, *Sardinops*) have only moderate narrow necking of the caudal peduncle (Fig. 1F), and so have some Characins. The more extreme carangiform swimmers have pronounced narrow necking, and sharply swept back tails scooped out with a deep V (Fig. 1E). These include the bluefish *Pomatomus*, many species in Carangidae including *Caranx*, and some species in Scombridae (*Scomber japonicous*, S. *scombrus*). Also within Scombridae are species, with lunate tails, verging on the thunniform (*Sarda chiliensis*), as well as the tunas which exemplify the thunniform mode.

The unique Mormyridae, freshwater African fishes with electric sensory organs, possess paired longitudinal bones below the caudal electric organs which provide rigidity to that region of the body (Lissmann, 196la). The tendons which operate the tail fin run over this region and have their origin on more anteriorly placed myotomes. Lissmann writes that their swimming movements appear to be "of the normal carangiform type." The system of tendons running pulleylike past a rigid secton are suggestive of the thunniform mode, but the whole physiognomy of mormyrids, and the speed attained, is quite unlike the tunas. The mormyrid system has probably evolved to provide a stable base for the electric sensory system.

5. *Thunniform Mode*

In thunniform swimmers the thrust is generated exclusively by a high stiff caudal fin mounted on an extremely narrow peduncle (Fig. 1H). Significant lateral movement occurs only in the peduncle and tail fin (Figs. 2D and 4). The propulsive force is delivered from the massive body muscles to the caudal fin by a system of tendons which run like pulleys past two joints in the posterior of the vertebral column. The wavelength is long, and its amplitude is wide at the trailing edge (Fig. 4). The body is heavy toward the anterior, sometimes almost circular in cross section, and is beautifully streamlined. The massive anterior, the extreme narrow necking at the peduncle, and the high aspect-ratio of the tail, all combine to minimize sideways recoil despite the power

Fig. 4. Thunniform swimming by kawakawa *Euthynnus affinis*. Length unknown, perhaps about 40 cm. Cinephoto intervals 0.06 sec. Symbols as in Fig. 3. (Based on Fierstine and Walters, 1968.)

of the caudal thrusts. In tunas, the swiftest of all fish, "the quick and powerful strokes of [the tail fin] can be understood from the quick and high-pitched sound produced by the fish in its death-struggle on the deck of a boat" (Kishinouye, 1923).

The vertebrae of the caudal peduncle form a rigid unit, strengthened by lateral keels which make the peduncle wider than it is deep. Immediately in front is the prepeduncular joint which allows the peduncle to swing in a wide arc. Immediately behind is the postpeduncular joint, with the last three vertebrae shortened and providing a hinge on which the stiff caudal fin can swing.

This double-jointed system is operated by several sets of tendons. A deep-seated row of tendons lies on either side of the backbone, each tendon running backward from the anterior-facing cones of one myomere to insert on a vertebra from three to seven segments behind it. The most posterior of these, the "posterior oblique tendon" inserts on the first peduncular vertebra behind the prepeduncular joint. A pull on the posterior oblique tendon will swing the peduncle to that side.

Flexure at the postpeduncular joint is accomplished by the "great lateral tendon" lying outside the posterior oblique tendon and running out along the peduncle on either side to insert on the bases of the caudal fin rays. A cross section through this tendon reveals it as a series of nested cones representing contributions from the myosepta of successive myomeres. Also included are smaller deep-lying tendons which originate from particular parts of the posterior myomeres and insert on particular groups of caudal fin rays. In various scombroid fishes, other tendons and muscle bands have been described in the caudal complex (Fierstine and Walters, 1968).

The wide bony keel of the peduncle serves as a pulley which increases the angle of pull of the great lateral tendons which run on either side of it to flex the tail. The skin surrounding the peduncle is reinforced by collagenous fibers; it forms a strong sleeve which keeps the tendons from bowstringing away from the vertebral column during flexure of both the peduncular joints. The keels on either side of the peduncle, as well as its reduced span, probably also serve to streamline its extremely rapid sideways beats. Sometimes well-developed keels are present as purely external structures (e.g., in Carangidae) which do not separate the tendons but which provide transverse streamlining of the peduncle.

The angle of inclination of the caudal fin is altered during each phase of the beat, even more effectively than in the carangiform mode, so as to develop a maximum thrust at all times. The elaborate plan of tendons obviously allows fine control of the tail movements, but again the details are as yet undescribed.

The caudal fin is high, short antero-posteriorly, and lunate (Fig. 1H). It therefore has a very high aspect-ratio. The tips may project so far up and down as to operate in water relatively undisturbed by passage of the body. The analogy with a high aspect-ratio wing of a bird is rather complete, and the hydromechanical models used to analyze the thunniform mode are based on lifting-wing theory for oscillating aerofoils (Lighthill, 1969; Webb, 1975).

The intrinsic muscles of the caudal fin are reduced in contrast to those in subcarangiform fish, and the rays overlap the skeletal base widely. The span can probably be altered only very slightly in most scombrid fishes. In response to the powerful thrusts, even the stiff caudal fin rays show some bending. Here, unlike the subcarangiform mode, the center of the fin leads as it beats from side to side, and the tips follow.

In Scombridae there are from five to eleven small separate nondepressible saillike finlets in a row running from the dorsal and anal fins out onto the peduncle (Fig. 1H). Small anal and second dorsal fins occupy much the same positions in Istiophoridae, in Xiphiidae, and in some of the fast pelagic sharks (Fig. 1G). These finlets probably contribute little direct propulsive force; they probably serve to deflect water along the peduncle so as to prevent separation of the boundary layer and so reduce drag, working in the same fashion as multiple wing-tip slots. Scombroids (with the exception of Xiphiidae) also have a pair of short fleshy horizontal keels, slightly converging toward the rear, on either side of caudal fin base; these may direct a jet which reduces cross flow and boundary layer separation (Walters, 1962).

Swimming in the thunniform mode, the scombrid *Euthynnus affinis* continues to increase the amplitude of the tail beat so long as speed increases (Fierstine and Walters, 1968). This contrasts with the observation by Hunter and Zweifel (1971) that in the carangid *Trachurus symmetricus* the tail beat amplitude is a constant proportion of body length regardless of speed. Although Webb (1975) lists *T. symmetricus* as swimming in the thunniform mode, its shape suggests that it may in fact adopt the carangiform mode.

Striking convergence toward the thunniform mode of propulsion can be seen in four unrelated vertebrate groups: among the bony fish, several families of Perciform fishes (most or all being in the suborder Scombroidei); among the sharks (suborder Lamnoidei); among the marine mammals (whales and dolphins); and in the extinct marine reptiles, the ichthyosaurs (Lighthill, 1969). The mammals have a large lunate tail, narrow necking, and a massive streamlined anterior. The fact that their tails lie in a horizontal instead of vertical plane does not alter their basic similarity to the thunniform mode in tunas. Ichthyosaurs also had narrow necking, and lunate tails which, however, lay in a vertical plane; the vertebral column extended into the ventral lobe of the tail, rather than into the dorsal lobe as it does in the heterocercal tail of sharks.

The convergence between some sharks and scombroid fish shapes is remarkable (Fig. 1G, H). The caudal fin of a thunniform shark, although slightly heterocercal, has a high aspect-ratio. The narrow caudal peduncle carries lateral keels. Anteriorly the body is wide, heavy, and streamlined. A high stiff first dorsal fin reduces recoil. The small second dorsal fin lies far back and opposite to the small anal fin, suggesting similar hydrodynamic function to the finlets described in teleosts. Further parallels between the sharks *Lamna* and *Isurus* and the tunas are that all are heavier than water, have respiration geared to continuous swimming, achieve high velocities, and have counter-current blood systems which maintain their body temperatures well above that of the water (see Chapter 4).

The whale shark *Rhineodon* has a shape suggestive of the thunniform swimmers (Lighthill, 1969). Possibly this and some other sharks may use the thunniform mode, but the majority of sharks have highly flexible bodies (with many vertebrae) and swim in the anguilliform or subcarangiform modes.

6. Ostraciiform Mode

Ostraciiform swimmers have a body incapable of lateral flexure. Propulsion in the ostraciiform mode is by pendulumlike oscillation of the tail, which pivots on the caudal peduncle. The wigwag motion is induced by a nearly simultaneous contraction of all myomeres involved, on each side alternately. The body shape is variable, but poorly streamlined, and only low speeds are attained.

Curiously, the ostraciiform mode does not seem to have been described in detail in any living fish, but it has been simulated repeatedly by diligent builders of working models (Breder, 1926; Oehmichen, 1958; Kramer, 1960; Smith and Stone, 1961; Hertel, 1966; Gray, 1968). Breder's model, with a rigid oscillating tail, was used to disprove Pettigrew's (1874) contention that such a system would not produce net forward motion (see Section II). The closest approach to an analysis of ostraciiform mode in fish is suggested by Webb (1975) to be the description by Gray (1933c) of the propulsive cycle of a whiting *Gadus merlangus* in which the caudal fin had been amputated.

Probably a perfectly rigid tail is never found in live fish. Even in the boxfishes, family Ostraciidae, while the body is encased in an inflexible bony armor (Fig. 1I), there are a few peduncular myomeres which act "almost as a unit in sweeping this flexible tail from side to side" (Breder, 1926). So also in the puffers, family Tetraodontidae, and porcupinefishes, family Diodontidae, there is some bending of the peduncle and tail as it is oscillated in the ostraciiform fashion. In all these fish, alternative means of locomotion are available by swimming with the dorsal, anal, or pectoral fins. Norman and Greenwood (1975) write that in *Ostracion* the dorsal and anal fins normally form the chief propelling agents, but when greater speed is required the fish swings the tail vigorously from side to side.

The electric ray *Torpedo nobiliana* cannot bend the body laterally because of its wide expansion (Fig. 5). The caudal fin is quite well developed and almost symmetrical and beats from side to side while the edges of the expanded pectorals are held so as to provide some lift (Roberts, 1969c). Although the caudal fin is somewhat flexible and bends sideways during the

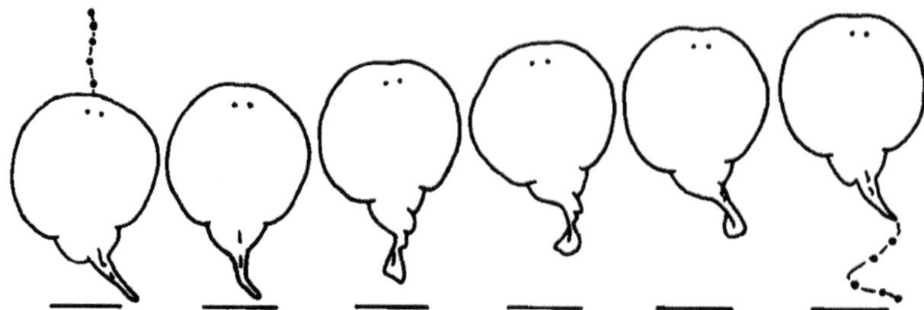

Fig. 5. Swimming by electric ray *Torpedo nobiliana*, approximating the ostraciiform mode. Length about 150 cm. Cinephoto intervals 0.5 sec. Symbols as in Fig. 3. (Based on Roberts, 1969c.)

tail beats (Fig. 5), this provides an approximation to the ostraciiform mode. Breder (1926) points out that, as demonstrated by his working model, the more flexible the tail in ostraciiform motion, the less the nose oscillates. The anterior of *Torpedo* can be seen in Fig. 5 to move very little from side to side, probably in part because the tail is not perfectly rigid.

The scabbardfish *Aphanopus carbo* is an elongate midwater species of the family Trichiuridae which may also use what can be considered an ostraciiform mode of swimming when slowly approaching its prey (Bone, 1971). Its small, deeply indented caudal fin, mounted on a narrow peduncle, can be sculled from side to side using only the musculature of the caudal region. The degree of sweep-back of the tail lobes is varied occasionally. The long body is held rigid, with the dorsal and anal fins retracted. This mode is evidently used in order to produce minimum disturbance to the well-developed lateralis system which senses the prey which it is stalking, and to provide minimum warning to the prey. In the final attack, the scabbardfish switches to a rapid swimming pattern involving powerful undulation of the entire body, with the median fins held erect. Thus, its rapid swimming is anguilliform, its slow stalking is ostraciiform. Bone (cited in Webb, 1975) suggests that two such swimming modes may also occur in the crestfish *Lophotus*, and in the trichiurid genera *Diplospinus*, *Benthodesma*, *Lepidopus*, and *Assurger* (Bone, 1971).

Although the ostraciiform mode involves only the peduncle and caudal fin, and so does the thunniform and to some extent the carangiform mode, nevertheless the hydrodynamics are probably entirely different. Species using ostraciiform swimming are poorly streamlined, and scull at very low speeds compared with carangiform or thunniform swimmers. The ensemble of features so characteristic of thunniform swimmers (fusiform body, narrow necking, lunate tail, and fixed caudal rays) are missing in whole or in part in ostraciiform swimmers. The term ostraciiform is intended for pendulumlike oscillation of an essentially rigid blade, unlike the active alteration of the angle of attack characteristic of the tails of carangiform and thunniform modes. Although ostraciiform waving of a somewhat elongate and flexible tail may produce undulations (comparable to flag waving), these undulations if passive are not moving faster than the water, and hence provide no thrust, unlike the anguilliform mode. It must be admitted that the ostraciiform category is a mixed bag of diverse forms, the details of whose swimming is not yet understood.

C. Propulsion by Undulation of Median or Pectoral Fins

Fish can propel themselves by passing undulations along longbased fins, in a manner analogous to the anguilliform undulations of the body. Instead of myomeres contracting to bend the body, muscles inserted on the sides of the fin rays at their bases (see Section V,B,2) contract to deflect the ray in an appropriate direction. Each fin ray can be moved about independently on a universal joint; successive rays are connected by a flexible membrane. The possible amplitude of the undulation is limited (compared to body flexures) because each ray is attached at its base to the body. Wavelengths can be very short, with several complete wavelengths present on a fin at once. Frequency of fin waves may be much higher than of body waves, reaching 70 Hz in the dorsal fin of a seahorse. The net effect of propulsion by undulation of extended fins rather than of the body is to attain only low or moderate speeds, but to achieve more precise control and maneuverability. Fin undulation usually allows both forward and backward movement, rapid reversal of direction without turning, and the ability to hover and to "drift" into confined apertures with precision. Fin propulsion also allows the body axis to remain straight; this is unavoidable in fish whose body shape or armor allows no flexure, and it may be desirable as an operating base for certain electrosensory or acoustico-lateralis systems.

Breder's (1926) compartmentalization of swimming modes was based on the particular fins involved. As explained earlier, his nomenclature is followed here and in Fig. 1 and Table III as a convenience, but it is to be supposed that essentially the same hydrodynamic processes may be involved regardless of whether the undulations progress along the dorsal fin, the anal fin, or both. The following discussion may be supplemented by reference to Table III and to the publications cited there.

1. *Amiiform Mode*

Breder (1926) applied the name "amiiform" to swimming by undulations of the long dorsal fin, as exemplified by the bowfin *Amia calva* (Fig. 1J). This fish, whose body is jacketed by a rather heavy armor of scales, usually employs its dorsal fin for locomotion, although to attain rapid movement it can also undulate its body in the subcarangiform mode. It has a full complement of developed paired and median fins. The swimming of *Amia* has not been studied in detail.

An amiiform swimmer that has been better described is the African freshwater *Gymnarchus niloticus* [for which Lissmann (1961a) suggested the term

Table III Swimming Modes Involving the Median or Pectoral Fins

Swimming mode and principal fin	Examples of fish groups	Remarks	Reference
Primarily undulations Amiiform			
Long-based dorsal	*Amia* *Gymnarchus*	May also use all other fins. No anal or caudal fins, ca. 200 dorsal rays. Up to 7 wavelengths visible. Wave velocity 0.63 ℓ/sec produces body velocity 0.5 ℓ/sec. Body held straight. Electrosensory	Breder (1926)Lissmann (1961a); Alexander (1967)
	Mormyrus	Short anal, propulsion primarily by dorsal. Body held straight. Electrosensory	Bennett (1971a)
	Trachipterus, Regalecus	Very long dorsal, highly compressed. Little body flexure. At slow speeds, inclination of body axis compensates for lack of gas bladder	Nishimura (1964); Nishimura and Hirosaki (1964)
	Trichiuridae: *Eupleurogrammus, Trichiurus, Lepturacanthus*	No anal or caudal fins. Body held straight. Detect prey by lateral line (decurved to avoid disturbance from dorsal)	Bone (1971)
Short-based dorsal	*Syngnathus,* *Nerophis*	Uses dorsal and pectorals, and largely ineffectual body curvature.	Breder (1926; Aleev (1963)
	Hippocampus	No caudal fin. Pectorals also used. Body axis inclined or vertical. Dorsal fin with 19 rays, 3.5 waves, frequency 70 Hz, amplitudes 0.16–0.24 times fin base, adjacent rays diverge by 17°	Breder and Edgerton (1942); Alexander (1967)

Gymnotiform Long-based anal	Gymnotoidel (South American electric fishes): *Gymnotus, Eigenmannia, Electrophorus*	Dorsal and caudal very small or absent. Pectorals small. Several half-waves visible on dorsal. Body held straight. Electrosensory. *Electrophorus* can also undulate body using wavelengths 4× fin waves	Bennett (1971a); Hertel (1966)
	Hyperopisus (Mormyridae)	Anal fin base 5× dorsal locomotion primarily Gymnotiform. Electrosensory	Bennett (1971a)
	Notopteridae: *Xenomystus*	Dorsal tiny or absent. Body held straight, but no special sensory system known. "Idles" with waves on anal passing in both directions.	Hertel 1966)
Balistiform Long-based dorsal and anal	*Monacanthus, Balistapus, Zeus*	Body deep, compressed, inflexible. Several halfwaves visible on dorsal and anal. Fins inclined to each other. Can reverse direction of undulations. Ray bases can be bent 90°. May augment by caudal or trunk undulation	Hertel (1966); Aleev (1963)
	Balistes	Anterior rays thickened, fins diverge in angle. Can also flap fins suggesting Tetraodontiform mode	Breder (1926)
	Isichthyes (Mormyridae)	Body elongate, dorsal and anal undulate in synchrony, 3 wavelengths visible. Electrosensory	Lissmann (1961a)
	Cichlidae: *Pterophyllum, Symphysodon;*	Deep-bodied, undulate hind edges of soft dorsal and anal, whose margins nearly vertical. Augment by pectoral	Bergmann (1968); Blüm (1968)
	Chaetodontidae	Sculling	
	Ophichthyidae, Branchiostegidae, Mastacembelidae, *Muraena*	Elongate; swim either by body undulation or localized undulations of dorsal and anal, or both	Breder (1926); Magnan (1930)
	Pleuronectiformes	Flatfish swim in midwater by body undulation plus synchronized flexures of dorsal and anal of wider amplitude	Aleev (1963); Ling and Ling (1974); Kruuk (1963)

Continued

Table III Swimming Modes Involving the Median or Pectoral Fins—Cont'd

Swimming mode and principal fin	Examples of fish groups	Remarks	Reference
Short-based dorsal and anal	Fistulariidae, Aulostomidae	Body slender, rather inflexible. Undulations in dorsal and anal. May augment by caudal	Lighthill (1969)
	Centriscidae	Body compressed, armored, stiff. Soft dorsal, anal and caudal close. Rapid vibration of these and pectorals move fish in any direction, usually with body vertical	Atz (1962); Klausewitz (1963)
Rajiform Wide horizontal pectorals	*Rajidae*, Dasyatidae	Pectorals expanded, undulate vertically, amplitude greatest at middle rays. No anal; caudal reduced; dorsals reduced or spinelike or absent. More than one wavelength visible. Winglike disc provides lift	Marey (1895); Magnan (1930)
	Myliobatidae, Mobulidae	Pectorals further expanded, width exceeds main body length. Mobile pectoral motion analogous to bird flight. Less than 1 wavelength visible. Pectoral base provides lift	Lighthill (1969); Klausewitz (1964)
	Pristidae, Rhinobatidae	Undulate pectorals, but also use moderate caudal fin and body undulation. Trunk and pectoral motions under separate nervous control	Campbell (1951); Alexander (1967)
Diodontiform Moderate pectorals with nearly vertical bases	Diodontidae, Tetraodontidae, some Balistidae	Can very plane of pectoral base, and angle of divergence from body. Rays diverge, partially cancelling vertical components of thrust. Fin base raised, versatile rays operated by tendons. May have 2 wavelengths at a time. May also flap fins (Labriform mode)	Breder (1926); Herald (1961); Schneider (1964); Webb (1975)
	Esocidae, Umbridae, Gasterosteidae, Gobiidae	Pectoral undulation is combined with other modes in many teleosts	

Caudal undulation	Serranidae, Pomacentridae	Vertical undulations of caudal fin translated into forward thrust by convergence of rays. Or upper and lower lobes may be moved in opposite directions, producing figure-8 sculling. Tail may open and shut scissorlike, or single lobe vibrated while maneuvering. Always accessory to other modes	DuBois-Reymond (1914); Breder (1926); Emery (1973); McCutchen (1970); Arita (1971)
Primarily oscillations Tetraodontiform Short-based dorsal and anal	Tetraodontidae, Diodontidae, Ostraciidae, some Balistidae	Paddlelike dorsal and anal flapped side-to-side. These occasionally alternate Anteriors of fins often stiff. "Erector" and "depressor" muscles modified. Augmented by pectoral or caudal propulsion	Schneider (1964); Breder (1926)
	Mola	High dorsal and anal fins flapped synchronously	Lighthill (1969)
Labriform Narrow pectorals with nearly vertical bases	Some Labridae (eg., *Tautoga*), Scaridae, Gasterosteidae, Umbridae	Pectorals flapped rapidly and synchronously, often fanlike and rounded. May also undulate (Diodontiform mode)	
	Some Labridae, Embiotocidae, Pomacentridae, Centrarchidae, Chaetodontidae, Acanthuridae, Serranidae, Sciaenidae, Cyclopteridae, Characinidae, Lamprididae	Row with pectorals, moving them forward horizontally and backward broadside. Base may rotate slightly. *Cymatogaster* achieves speed of 3.9 ℓ/sec, with pectoral frequency 4/sec. For high speeds, augment with caudal	Webb (1973b); Brett and Sutherland (1965); Harris (1953); Emery (1969, 1973); Rosenblatt and Johnson (1976)

"gymnarchiform"]. The dorsal fin of *Gymnarchus* extends along most of the length of the body, which tapers to a posterior point. There is no anal or caudal fin. When swimming, the body axis is usually held straight. Lissmann (196la) shows from cinematographs that locomotor waves may pass in either direction along the dorsal fin, and may show widely varying amplitude, particularly during turning or braking. The straight body and naked posterior provide a stable base line for a highly developed electrosensory system.

A related group of African freshwater electric fish, the Mormyridae, can also hold their bodies straight and propel themselves by undulations of the median fins. Most use the dorsal and anal fins together (balistiform mode) or the caudal, but in the genus *Mormyrus* the anal is very short compared to the dorsal (Bennett, 1971a), and locomotion is probably primarily amiiform.

The ribbonfishes, Trachypteroidei, have a dorsal fin running the whole length of the elongate and highly compressed body. Three species have been observed to swim slowly in the amiiform mode, passing undulations along the dorsal fin with little or no body flexure. At the usual low speeds they swim with the head inclined upward (even with the whole body vertical), so that the undulations of the fin furnish lift to compensate for the lack of a gas bladder. At increased swimming speed the body axis rotates to approach the horizontal, and in *Regalecus* the body as well as the dorsal fin may contribute to intermittent undulations.

Reference was made in the preceding section to certain genera or trichiurid fish which stalk their prey with a rigid body, sculling with the small caudal fin. In these the lateral line passes along the body in the usual position midway between the dorsal and ventral surfaces. In other trichiurid genera (Table III), the caudal and anal fins are absent, and locomotion is probably in the amiiform mode, using the enlarged dorsal fin and again holding the body straight. Here the lateral line, which is apparently used to detect prey, runs close to the ventral surface and is thus remote from disturbances caused by undulations of the dorsal fin.

The seahorse *Hippocampus* swims characteristically in an upright position, using both dorsal and pectoral locomotion, sometimes aided by the anal fin. There is no caudal fin. Breder and Edgerton (1942) measured dorsal fin waves in *Hippocampus* with remarkably high frequencies. Variations in wavelength, amplitude, and to a lesser extent frequency may occur in the versatile dorsal fin all at the same time. Maximum swimming speeds in seahorses are low; highest speeds are attained by inclining the body axis from the vertical toward the horizontal.

2. Gymnotiform Mode

The swimming of the South American electric fish *Gymnotus carapo*, after which the gymnotiform mode is named, is comparable to an upside-down *Gymnarchus* swimming in the amiiform mode (Fig. 1K). There is no dorsal fin. The body is held straight, and forward or backward locomotion is produced by passing rapid undulations of short wavelength in either direction along the elongate anal fin. The related electric eel *Electrophorus* can supplement the thrust from its anal fin oscillations by undulating the body; the waves in the body are about four times as long as those in the fin (Hertel, 1966).

The South American gymnotoid fishes described above probably combine a straight body and locomotion by fin undulation for the same reasons as do the unrelated gymnarchid and mormyrid fishes of Africa; all possess an electrosensory system (Bennett, 1971b) which requires a stable base undisturbed by body undulation.

Another tropical freshwater group, the featherbacks, family Notopteridae, have a long anal fin extending from just behind the head to the tip of the tail. The dorsal fin is tiny or absent. Propulsion is in the gymnotiform mode, with numerous waves moving along the anal fin to drive the fish either forward or backward. The body is usually held straight, although no special sensory system has been described in this group.

3. *Balistiform Mode*

The term balistiform refers to simultaneous undulation of the dorsal and anal fins. In trigger fish and file fish, family Balistidae, their fins are somewhat inclined to each other (Fig. 1L), so their propulsive waves produce a net forward thrust. Fin spines which lie ahead of the rays are detached from them and do not impede flexure of the fin rays. According to Hertel (1966), *Balistapus aculeatus* can swim backward by reversing the usual direction of wave generation on both fins. It can also move downward in the water by passing waves backward on the anal and forward on the dorsal fin, or can move upward by reversing both these wave directions. The caudal can be spread like a fan from the folded position to a span 2.5 times as wide, and is used to add propulsive strokes at maximum speeds.

The John Dory, Zeus *faber*, also has great transverse mobility of its posterior dorsal and anal rays which are the sole locomotor organs for slow swimming. For abrupt spurts, the entire trunk is thrown into undulations, and the median fins revert to stabilizers (Aleev, 1963). Other similarly deep-bodied fish such as the cichlids, and the butterfly fish, family Chaetodontidae, undulate the hind edges of their soft dorsal and anal fins.

The soft dorsal and anal fins are without anterior thickening in the file fish *Monacanthus,* and display conventional fore-and-aft undulations. In the trigger fish *Balistes* the anterior rays are thickened and elongated, and they can on occasion be flapped from side to side as a unit instead of undulating each, so as to produce a rowing motion approaching the tetraodontiform mode. Such thickening in the anterior rays of several forms tends to be associated with angular divergence of the fins away from the vertical.

Several different groups of highly elongate fish which are capable of anguilliform undulation of the whole body can also move by localized undulations passed along the dorsal and anal fins (Table III).

In flatfish also, anguilliform flexures of the body while swimming in midwater may be supplemented by synchronous flexures of the dorsal and anal fin rays which attain a wider amplitude. The sole *Solea vulgaris* can also move about on the bottom or dig into the sand without flexing the body, by means of pronounced undulations passed along the dorsal and anal fins (Kruuk, 1963).

The slender but rather inflexible fishes in the suborder Aulostomoidei swim by undulations of the soft dorsal and anal fins. Their swimming mode could be described as balistiform; although the fins have shorter bases than those of the previous species, their motion is typically undulation rather than oscillation. In this group the cornetfishes, family Fistulariidae, and trumpetfishes, family Aulostomidae, supplement balistiform swimming with caudal fin movement when hard pressed. The snipefishes, family Macrorhamphosidae, which are deeper-bodied, are armored and stiff, and depend largely on balistiform locomotion; they swim backward as easily as forward, usually with the body vertical and the head down (Herald, 1961). The shrimpfishes, family Centriscidae, also swim with the body vertical, either head-up or head-down. The soft dorsal, caudal, and anal fins are close together near the posterior and almost at right angles to the rigid body axis. By rapid vibrations of these three fins, and of the pectorals, shrimpfish can move in almost any direction with astonishing agility. Although the three fins can be closely coordinated, each is short-based, and the swimming shares features of the ostraciiform and tetraodontiform modes.

4. *Rajiform Mode*

The primary locomotor organs in most rays, order Rajiformes, are the greatly enlarged pectorals which form a wide lateral expansion of the body. Waves of undulation in a vertical plane are passed backward along the mobile fin margins ("rajiform mode," Fig. 1N). In the majority of rays the only organs other than the pectoral fins which may contribute at all to forward locomotion are the pelvic fins. These lie at the posterior of the disc created by the snout and pectoral margins and may be used to kick back as the ray is taking off from the bottom.

Two families, the eagle rays, Myliobatidae, and the mantas, Mobulidae, have the pectorals, even further expanded than in the skates, Rajidae, so that their width is greater than the length of the main part of the body. These animals are secondarily adapted for a free-swimming rather than a sedentary existence. Their tremendously mobile pectorals produce a graceful rajiform swimming motion (Fig. 6) analogous to the flight of birds (Lighthill, 1969). Longitudinal sections through the body show it as a series of hydrofoils. The base of the pectorals assumes a sharp angle of attack during the upstroke of the wing tips (Klausewitz, 1964), providing lift to overcome the excess of weight over buoyancy. *Mobula* also possesses free cephalic flippers modified from the anterior margins of the pectorals; these are operated by a separate musculature to capture food, and have not been shown to take part in forward locomotion.

Fig. 6. Lateral view of rajiform swimming by manta *Mobula diabolis*. Length unknown. Successive cinephotos (at 0.64 sec intervals), of fish swimming from left to right, are displaced upward. Vertical broken line represents a fixed position on background. (Based on Klausewitz, 1964.)

Chimaeras (Holocephali) are said by Breder (1926) to have two types of pectoral locomotion. They may row through the water (labriform mode), or they may hold the pectorals out at right angles to the body and undulate them in the rajiform mode. Their pectorals are not so wide-based as in Rajiformes, and the pectoral locomotion described by Breder is perhaps closer to the diodontiform mode.

5. *Diodontiform Mode*

The diodontiform mode of swimming practiced by porcupinefishes involves localized anguilliform undulation of the moderately broad pectoral fins. Unlike the horizontal pectoral fins of Rajiformes, the bases of the pectorals in diodontiform swimmers are held in a variable plane which may be nearly vertical (Fig. 1O). The undulatory waves produced by movement of successive pectoral rays on their bases may therefore have a vertical component, partly cancelled out by the fanlike divergence of the rays. The two opposing pectoral fins also diverge from the body laterally at complementary angles which can be altered. Undulatory waves may also be combined with flapping of the whole fin in the labriform mode. Combinations of these variables can generate thrust in almost any direction, allowing slow but very precise maneuvering.

The families Diodontidae, Tetraodontidae, and some Balistidae exemplify diodontiform swimming. The versatile pectorals of *Diodon* can show as many as two wavelengths at a time. Schneider (1964) describes how in *Tetraodon* the well-developed abductor muscles are attached to each pectoral ray by long tendons. In all these fish the base of the pectoral fin is raised, and the rays can be bent over at a sharp angle. Undulation of the pectorals may be an important locomotory component, always used in combination with other modes, in many other teleost groups (Table III).

6. *Caudal Undulation*

A category of propulsive techniques not included in the preceding survey and not formally named as a mode involves the propagation of vertical undulations in the caudal fin. This is always an accessory to other locomotory modes; it is not shown in Fig. 1, but could be entered in the center of the diagram.

Vertical undulations of the tail fin may be translated into forward thrust because the rays are convergent rather than parallel. "Considering each ray separately, in waving from side to side, it naturally has a forward reaction of the ostraciiform type" (Breder, 1926). A slightly different type of caudal thrust, described by DuBois-Reymond (1914), results when the upper lobe of the tail is moved to one side and the lower lobe to the other side, creating two backward diagonal thrusts. In this case the central caudal ray remains still, while in the former case all rays may pass through a complete cycle of oscillation. Diagonal thrusts of the opposing caudal lobes have been observed in the sea basses *Morone* (= *Roccus*) and *Epinephalus*, and may correspond to the figure-8 sculling of the caudal in young Pomacentridae (Emery, 1973). In other fish the tail is sometimes opened and shut in a scissorlike motion; one lobe may be vibrated while maneuvering; and individual rays may perhaps be capable of active bending (McCutchen, 1970; Arita, 1971).

D. Propulsion by Oscillation of Median or Pectoral Fins

At the opposite extreme from the undulations of extended longitudinal fins (dorsal, anal, or pectoral), which may be compared to anguilliform propulsion, lie very short-based fins whose oscillation may be compared with ostraciiform propulsion. Various intermediate fin modes may be compared roughly with carangiform or other body modes, depending on the length of the fin and the number of contained wavelengths. The analogy is not complete, because short-based fins may also be capable of rotation on the base so as to produce locomotory strokes without parallel in the body/caudal fin series of modes. Ventral segments of the pectoral fin may be rotated relative to dorsal segments, and single fin rays may be capable of active bending (Arita, 1971). These and other complexities, most of which have scarcely been investigated, suggest that this section of the Breder's classification of swimming modes is crude at best.

1. *Tetraodontiform Mode*

Puffers in the family Tetraodontidae use their short paddlelike dorsal and anal fins as a unit, simply flapping them from side to side (Fig. 1M). Breder, who named this the tetraodontiform mode, writes that these fins may be thought of as an ostraciiform tail in two parts moved slightly forward dorsally and ventrally. While *Lagocephalus laevigatus* usually fans its dorsal and anal in unison in this fashion, it may occasionally alternate them. *Tetraodon fluviatilis* swims with its dorsal and anal fins and its pectorals; the caudal fin is used for steering. The "erector" and "depressor" muscles of the dorsal and anal are enlarged and modified in this species so as to move the fin rays laterally rather than vertically (Schneider, 1964).

As mentioned under Balistiform Mode (Section III,C,3), some triggerfish, family Balistidae, which have stiffened anterior rays may flap the dorsal and anal fins from side to side instead of undulating the rays. This type of tetraodontiform swimming reaches an extreme in the ocean sunfish, family Molidae, where the body musculature, and the tail, are virtually lost, and propulsion depends on synchronized sideto-side paddling of the immensely high dorsal and anal fins. "These operate in the bird-flight mode of the Myliobatidae (eagle-rays) but, turned through 90°; apparently a unique case within the animal kingdom of a disjoint pair of high aspect-ratio wings deployed in a vertical plane" (Lighthill, 1969).

2. *Labriform Mode*

Fish which swim by oscillations of their narrow-based pectoral fins are said to swim in the. labriform mode (named from the wrasse family Labridae). The fins are often fanlike and rounded, and may be flapped synchronously. In the stickleback *Gasterosteus* and the mudminnow *Umbra* the pectorals may move in this way so rapidly as to appear as a blur. Breder (1926) likens this form of labriform motion to the action of a pair of anteriorly placed ostraciiform caudal fins, each producing a large forward thrust, plus a smaller lateral thrust which is cancelled out by the opposite fin.

A more complex style of labriform swimming is displayed in some other wrasse species, and in surfperches, family Embiotocidae, and several other families (Fig. 1P). Here, the fish "rows" with its pectorals, bringing them far forward almost edgewise, and forcing them back broadside. Webb (1973b) analyzed the pectoral movements of the shiner surfperch *Cymatogaster aggregata*. The fin base is inclined down and back at 35° from the horizontal. As the stiff leading edge swings forward it also moves slightly downward; the more flexible following rays lag behind, and do not complete their forward motion until the leading edge starts its backward swing. At the end of the back swing, the leading edge tip comes to rest against the body slightly below its point of departure; the fin is then rotated slightly upward while still pressed against the body, so that the tip starts the next stroke in its original position. The forward and backward phases of the cycle each generate thrust; they also produce components of positive and negative lift, respectively, which combined with slight negative buoyancy cancel out over a complete fin-cycle. At higher speeds the pectoral movements described above are replaced by a slightly different pattern in which the phase difference between posterior and anterior fin-rays is lessened and the fin tends to work more as an ostraciiform unit. The prolonged speed attained by a *Cymatogaster* using labriform propulsion compares favorably with similar activity levels for fish which use body and caudal propulsion. At this speed, pectoral beat frequencies are of the same order as caudal beat frequencies at similar speeds.

The damselfishes, family Pomacentridae, rely to various degrees on rowing with the pectorals. The fins are usually moved in synchrony, but may occasionally be alternated (as also in some Labridae). Emery (1969, 1973) describes different combinations of pectoral and caudal propulsion in the many species of damselfish inhabiting a coral reef, often involving progress in the form of a series of vertical loops. The pectoral girdles are massively developed in those species using their pectorals as the main locomotor organ, and poorly developed in those scarcely depending on labriform propulsion.

The pectoral fins, often augmented by other fins or by the trunk, are prominent in propulsion in several other families (Table III). Many other fish use their pectorals for slow swimming or for holding position, by means of subtle motions of the fins which have yet to be analyzed.

IV. NONSWIMMING LOCOMOTION

The preceding section dealt with straightforward horizontal motion free from the bottom. Obviously many fish spend much of their time in activities other than swimming in a straight line. Indeed, their abilities to perform other short-term maneuvers required for feeding, reproduction, or escape may be crucial in determining their survival. Such activities do not usually lend themselves as well as does straight swimming to kinematic analysis, physiological quantification, or calculation of energy budgets. Their brief discussion here is intended as a guide to further reading.

A. Jet Propulsion

Breder (1924, 1926) believed that water exhaled from the gill orifices of fish is sometimes an important locomotor asset. He recorded that the porcupinefish *Chilomycterus schoepfi* and the whale shark *Rhineodon* ejected strong currents through the gill slits. So do banjo catfish, family Aspredinidae, and the sargassum fish, family Antennariidae (Gradwell, 1971; Gregory, 1928). A posterior view of a swimming batfish (family Ogcocephalidae) shows the large circular exhalent apertures which bear a remarkable resemblance to jet aircraft engines.

Despite these observations, it is now generally believed that in most fish the small volume of water expelled during respiration makes the thrust which it could generate negligible (Bainbridge, 1961). Jet propulsion, although important in cephalopods (Packard, 1972), is most feasible only at high Reynold's numbers when inertia dominates viscosity (Lighthill, 1969). In fish it may be important most often in providing impetus during acceleration, particularly from a standing start (e.g., when a ray takes off from the bottom).

The slight forward thrust resulting from water ejected through the gill slits during normal respiration requires some locomotory compensation in a fish holding position in midwater. Stationary fish in aquaria are often seen to back-paddle gently with the pectorals or with some versatile fin tip such as the end of the soft dorsal or the caudal lobes.

Exhaled water may have another effect in fish locomotion. Aleev and Ovcharov (1973) showed by cinematography that dye ejected from the gill slits of a swimming fish forms a sleeve which envelops the body. It has been suggested that water from the gill slits may keep the boundary-layer fluid surrounding the body charged with enough kinetic energy to avoid separation (Lighthill, 1969). Exhalation has been shown to be coordinated with swimming movements (Satchell, 1968). Further references are given by Webb (Chapter 3, Section II, C, 1, d).

B. Terrestrial Locomotion

Elongate fish which swim in water by anguilliform motion may progress over land in the same way, although, as noticed by Aristotle, usually with undulations of wider amplitude (Breder, 1926). Eels can move overland in this way, and so can the mullet *Mugil corsula* (Ganguly and Mitra, 1962). The cyprinodont *Fundulus notti* moves on land by a series of "flips" which are apparently directed, the body being aligned before each jump by sun-compass so as to return a stranded fish to the water (Goodyear, 1970).

The air-breathing catfish *Clarias* moves on land by locomotory contractions which alternately affect the whole of the lateral musculature of the body halves. The anterior is raised and the body rolled so as to bring the tip of the erect pectoral spine on one side in firm contact with the ground. The body is thrown into a semicircle, and the spine serves as a lever which heaves the body forward in one of a series of steps. At the next step the body is bent in the opposite way. Boulenger (1907) believed that *Clarias* progressed by independent motions of its pectoral spines, but Johnels (1957) observed in *Clarias senegalensis* that both spines were held rigid. Nawar (1955) describes a large erector muscle, and a locking system, in the pectoral spine of *Clarias lazera*.

The "climbing perch" *Anabas testudineus*, family Anabantidae, moves over land at up to 3 m/min by spreading its gill covers and placing their ratchetlike spiny lower edges to the ground, and then rocking forward by vigorous action of the paired fins and tail (Herald, 1961). Fishes in this family have accessory air-breathing organs allowing extended aerial excursions. So have the snakeheads, family Channidae, which can move over land using rowing movements of their pectoral fins.

The most competently amphibious fish are the mudskippers *Periophthalmus* and their relatives in the family Gobiidae. Eggert (1929), Harris (1960), and Klausewitz (1967) give extensive anatomical and behavioral details. The pectoral fin consists of two functional parts, a rigid platelike proximal region, hinged to the shoulder girdle, which on land is functionally equivalent to the upper arm of a tetrapod, and a fanlike distal part equivalent to the lower arm and plantar surface (Gray, 1968). In water, the mudskipper can paddle slowly with the pectorals out forward, the bucco-pharynx air-filled, and the dorsal fin erect, or it can swim rapidly by vigorous tail movements, the air expelled, and the fins close against the body except for four to five of the pectoral rays which are flexed to form hydrofoils. It can also skim over the water in a series of bounds alternating with brief swimming. (A 14 cm fish covered 2.5 m/sec in two bounds.) On land the progression is usually ambipedal or "crutching"; the pectorals are swung forward in unison while the pelvic fins support the body; the pectorals are then pressed down and backward so as to lift the body and draw it forward. If these actions are speeded up the whole body leaves the ground in a series of small leaps. An even more rapid skipping movement is achieved by digging in the tail and straightening it violently, while also thrusting upward with the pectorals. (A 14 cm fish can cover 30–40 cm in a single skip.) Mudskippers can also climb, using the crutching motion. A general account of mudskippers in mangrove environments is given by Macnae (1968).

The amphibious clinid *Mnierpes macrocephalus* inhibits steep rocky shores, and makes terrestrial excursions for up to 30 min (Graham, 1970, 1973; Graham and Rosenblatt, 1970). Its lower surface is covered by thick protective pads, and its pectoral, pelvic, and anal fins are heavily padded and incised. On land, it curls its tail up toward its head, extends its pectorals forward, and raises itself up on its pelvic fins. It then simultaneously extends its tail and adducts its pectorals, thrusting the body forward. The tail is alternately curled to either side of the head. One fish was seen to jump 2 m from the shore to the sea. In this species several other physiological and anatomical adaptations to terrestrial excursion have been recorded, including cutaneous respiration, and optical modifications for acute vision in both air and water.

The catfish *Astroblepus* (=*Arges*) *marmoratus* was observed to use adhesion by its mouth and by its pelvic region to climb 6 m up a smooth slightly overhanging wall (Johnson, 1912). Suctorial discs formed by the mouth or the paired fins may also be used in combination with inching movements of the body in Petromyzontidae, Loricariidae, and Gobiidae.

C. Moving on the Bottom and Burrowing

Many fish which are heavier than water habitually rest on the bottom, but move forward by conventional body and tail movements that carry them on slight excursions away from the substrate. A few move forward by "walking" in continuous contact with the bottom. In *Lophius*, *Antennarius*, and *Ogcocephalus* (order Lophiiformes) the pectoral base is thick and muscular, and the rays are handlike. The frogfish *Histrio* climbs among sargassum weeds, moving each pectoral and pelvic fin independently, and actually clasping the branches with its prehensile fingerlike dermal rays (Gregory, 1928; Herald, 1961).

The goby *Gobius* and scorpionfish *Scorpaena* have less highly modified pectorals, but can move along the bottom either by drawing the whole fin in toward the body, or by passing a wave of successive contractions along the adductors of the fin rays (Aleev, 1963). In the family Synancejidae, the goblinfish *Choridactylus multibarbis* has the three lower rays of each pectoral fin detached from the web, protected by conical horny caps, and capable of independent movement (Samuel, 1961). The sea robins, family Triglidae, also have the lower rays of the large fanlike pectorals separate and capable of independent movement as the fish "feels" its way forward over the bottom.

Various cypriniform fish with suckerlike mouths modified for algae-scraping or bottom feeding can pull themselves along the bottom by rapid motions of their jaws. The hillstream loaches, family Homalopteridae, adhere to rocks with a large sucking disc formed by the pectoral and pelvic fins; they can creep using the anterior fin rays (Wickler, 1971).

The lungfish *Protopterus* often rests on the bottom supported on the pectoral and pelvic fins. Larvae 2–3 cm long will walk on firm bottom using alternating movements of the paired fins "highly resembling the movements of a lower tetrapod" (Johnels and Svensson, 1954). The lungfish *Neoceratodus*, whose paired fins are much thicker, also holds both the pectorals and pelvics down below the belly, and uses them actively during restricted locomotion in an aquarium.

Burrowing into soft bottom may be achieved by the same anguilliform mode used in swimming [e.g., in *Myxine* (Foss, 1968), Cobitidae, Ammodytidae, and Synbranchidae]. Fish that can swim backward to burrow include the Indian snake eel *Ophichthys* (=*Pisodonophis*) *boro* (Tilak and Kanji, 1967), the cusk eel *Otophidium scrippsi* (Greenfield, 1968), and the cucumber fish *Carapus acus* (Arnold, 1953). Nonanguilliform fish such as the mudminnow *Umbra* and several wrasses, family Labridae, "swim" into the soft bottom. Killifish, family Cyprinodontidae, bury themselves by diving into sediment headfirst at 45° (Minckley and Klaassen, 1969). Many different kinds of fish, including rays and flatfish, cover themselves by throwing up sand with undulatory fin movements; some burrow by ejecting water from the buccal cavity (Trueman and Ansell, 1969).

D. Jumping, Gliding, and Flying

A fish usually jumps into air by swimming rapidly upward through the surface of the water; its momentum alone carries it forward after the tail has left the water entirely. A basking shark *Cetorhinus* about 9 m long was seen to jump at least 2 m clear of the water (Matthews and Parker, 1951). Tarpon leap

7–8 ft, salmon 1–2 ft; a *Manta*, over 5 ft long and weighing over 1000 pounds, can leap over 5 ft with a noise audible a mile away (Norman and Greenwood, 1975). *Thunnus* leaps several feet, *Mugil* and *Labidesthes* leap 10–20 times their lengths in a low arc (Hubbs, 1933). Gray (1968) discusses the ascent by fish of waterfalls, either by swimming *in* the water stream or jumping over it. Aleev and Ovcharov (1969, 1971) published photographs of inked fish jumping through the water surface.

Swanson (1949) and Gunter (1953) summarize accounts of a variety of fish (including shark, herring, needlefish, halfbeak, and silversides) which have been seen to hurdle repeatedly over a floating object or a line in the water. No satisfactory explanation has been offered for this curious behavior, which at least in some cases cannot be attributed to removal of ectoparasites, and may fall into the category of "play."

Some fish skim or skip along the surface in a series of small leaps or a hydroplanelike precursor to gliding flight. The mullet *Mugil corsula* either swims rapidly along the surface with its head projecting, or skips along using its muscular caudal peduncle, tail fin, and paired fins (Ganguly and Mitra, 1962). *Mugil* and the halfbeak *Hemiramphus* have been seen to spread their pectoral fins when in the air (Hubbs, 1933). The needlefish *Tylosurus* rushes forward as much as 200 m, the body held rigidly at an angle of 30° or more above the surface while the strengthened lower lobe of the caudal fin remains submerged and vibrates rapidly (Breder, 1926). The halfbeak *Euleptorhamphus longirostrus*, with large pectoral fins, can skim in somewhat the same way and then rise entirely free from the water into a stiff wind and glide for over 50 m in the manner of a flying fish (Myers, 1950).

The most proficient aerial gliders are the flying fishes, family Exocoetidae, that despite their name do not flap their wings in active flight but rather use them as sail-planes. In *Cypselurus*, flight is initiated when the fish accelerates underwater and then emerges partially from the water, spreads its pectoral fins, and skims along with the body obliquely in the air while the elongate lower caudal lobe beats in the water. This taxi may average 9 m in length, with tail beats reaching nearly 70/sec (Hubbs, 1933). Franzisket (1965) calculated that from an underwater speed of 28 km/hr, one fish reached 61.7 km/hr at the end of its taxi. The fish then rises into the air by spreading its pelvic fins, which elevates the posterior part of the body. The pectorals and pelvics are held out to serve as aerofoils during the ensuing glide. After a flight of up to 13 sec, as speed through the air drops, the lower caudal lobe may reenter the water, and by vigorous beats provide increased velocity for another flight, often with a change in direction. As many as twelve consecutive flights have been recorded. Distances covered by one flight may reach 400 m. Hertel (1966) discusses some aerodynamic aspects of this type of flight. Details of the flights and of the relative development of the pectorals and pelvics vary among species of flying fish. Some fish may burst directly out of the water without a preliminary taxi.

A modification in some Asian cyprinids, not previously described, involves a curious "neck-bending" which is probably associated with escape by skittering along the surface. *Chela maassi* (Weber and de Beaufort) can snap its head back so that the dorsal surface is almost at right angles to the contour of the back. The action swings the large pectoral fins downward, so as to thrust the whole fish upward. The pivot for this action is between the first vertebra and the skull, and on either side between the upper end of the cleithrum and a large rounded expansion of the transverse processes of the first two vertebrae. A tough fibrous sheath protects the nerve cord at the site of bending. When fish in an aquarium, having body lengths of 5 cm, were alarmed, the neck-bending and pectoral thrust shot them abruptly upward in the water about 8 cm in 0.08 sec. If they were close to the top, they would break through the water surface. In their natural habitat, swampy pools in the Malaysian region, they elude capture by skittering away over the surface in what may be a rapid sequence of neckbending. Several other Asian cyprinids have a similar physiognomy, with large horizontal pectorals, depressed lateral line, and sharpedged thorax. Among them I have observed the neck-bending only in *Chela maassi* and *C. caeruleostigmata*; other genera, including Esomus, may be suspected of using similar escape patterns (Hertel, 1966) even though the neck-bending may be poorly developed. The action of *Chela maassi* when alarmed suggests a Mauthner reflex such as that of *Gasteropelecus* (see Section V,A,5), but the neural system has not been investigated.

A different neck-bending type of locomotion may have evolved in some South American characoids. The large, elongate, predatory *Rhaphiodon* (family Cynodontidae) has a sharp-edged thorax and the ability to bend the neck upward, perhaps coupled with downward thrusts of the strong pectoral fins. The anatomy and behavior of these fish, and of the Asian species, offer intriguing research topics.

The South American hatchetfishes, family Gasteropelecidae, have large highly placed pectorals operated by massive muscles originating on a greatly expanded pectoral girdle which again forms a sharp leading edge or "bow." A hatchetfish can taxi over the surface with the tail immersed and the pectorals beating; after about 12 m it may break entirely clear of the water and fly for about 3 m. Unlike the flying fish, hatchetfish apparently beat their wings in flight, emitting an audible buzz. Unlike the pectoral muscles of the Exocoetidae, which are not notably well developed, those in hatchetfish may comprise one quarter of the total weight. The anatomy (Weitzman, 1954) and innervation, (Auerbach, 1967) have been studied, but accurate analysis of the flight of hatchetfish is lacking.

A quite unrelated freshwater African osteoglossoid, the butterfly-fish *Pantodon buchholzi*, can also flap its wings. It has large fanlike pectoral fins, with powerful muscles which can move them up and down but cannot fold them against the body (Greenwood and Thomson, 1960). Butterfly-fish can make leaps of up to 2 m, but whether the fin muscles are used in "flight" or whether they only assist in a taxiing take-off is uncertain.

The "flying gurnards," family Dactylopteridae, have generated heated debate concerning their flying ability. They have large winglike pectorals, but these are delicately built, while the body is heavy and clumsy, and the caudal is small. Some authors have attributed to them grasshopperlike flight, but others have expressed disbelief. Hubbs (1933) pointed out that the pectorals are "assuredly utterly insufficient to hold the fish in the air, even for a moment" and their outer portions are "positively raglike." He concluded that their flight was a myth. On the other hand, these fish often turn up on the decks of small boats at sea. Moreover, if aquarists do not cover tanks containing flying gurnards, the fish are liable to escape and land several meters away in other tanks or on the ground (Bertin, 1958c). Just how they do this is not quite clear.

V. PROPULSIVE ANATOMY

A. Trunk

1. Segment Number and Flexibility

A fish is a chain of metameres, with a head on one end and a tail on the other. More than in most other vertebrates, locomotion in fish depends on the propagation of lateral propulsive waves by the sequential action of basically similar units arranged in a series. Each unit has a set of skeletal elements, a set of muscles, a set of nerves, and a set of arteries and veins. Hydrodynamically the chain could be indefinitely long and could still swim in the anguilliform mode; earlier attempts at mathematical analysis of undulatory locomotion indeed assumed an animal with no ends (Taylor, 1952).

The number of vertebrae (and hence of metameres) in living fish species ranges from 16 (*Mola*) to over 600 (*Nemichthys*). While many selective factors must converge to determine the optimum number for each species, locomotory ability is likely to be prominent. Such a striking range in segment number between very divergent forms could be expected to show some correlations with locomotor patterns. Perhaps more significantly, the considerable variation between races of the same species, and between populations, and even within populations, is likely to have locomotory and hence ecological importance. Few investigations so far have considered the significance of segment number with respect to locomotion.

The greater the number of myotomes, the more flexible the body is likely to be. Magnan (1929) demonstrated this by determining the arc of a circle into which various types of fish could be bent. Fish (such as eels) which could form more than one circle had 56–120 myomeres; fish (such as tuna) in which the head could not touch the tail had only 19–30. There does not seem to be a simple correlation between speed and segment number; the very fast thunniform swimmers have rather low counts (about 31), but fast sharks may have well over 100, and so may relatively slow congrid eels. Species which do not depend on trunk or tail propulsion may have very few (*Mola*) or very many (*Gymnarchus*) vertebrae.

There is a definite inverse correlation between vertebral number and robustness of the body. Among related species, those with wider or deeper bodies tend to have fewer vertebrae. The generalization is less valid between widely differing groups of fishes; here, also, narrow bodies tend to go with high vertebral count, but there are elongate blennies with rather few vertebrae, and fat sharks with many.

The thickness of the body rather than its segment number may be dominant in determining flexibility. Aleev (1963) measured flexibility along five equal sections of the body in a number of species. In the eel, which is fairly uniform in width (Fig. 2A), flexibility was high in all sections, and only moderately greater in the most posterior than in the most anterior section. In the cod, whose width tapers smoothly (Fig. 2B) the flexibility was zero in the most anterior region and rose evenly to a high value at the posterior. In a scombrid (Fig. 2C, D), flexibility was zero at the head and low in the next two sections (where the body is robust); it rose sharply in the region of the peduncle, and fell again in the tail region (a reflection of the double peduncular joint followed by stiff caudal rays). These differences along the length of the body do not, except in the case of the scombrid, reflect any notable difference in vertebral spacing, but they closely parallel the width of the visceral and muscle mass.

There is suggestion of a phyletic decrease in vertebral number in the course of evolution. No primitive fish (cyclostomes, elasmobranchs, lower teleosts) have low counts. Strikingly low counts occur irregularly among some but not all the higher teleosts. The forms retaining the most cartilage have many vertebrae and high flexibility. Ontogenetically, the young of all species have high flexibility quite uniformly distributed along the body; rigidity of the anterior may set in with ossification coupled with increased body width. Most young fish swim in the anguilliform mode (Fig. 3) regardless of their adult habits.

A widespread tendency for fish species with larger adult size to have more vertebrae than smaller related species may have locomotory significance. This phenomenon, termed pleomerism (Lindsey, 1975), occurs within families of differing shapes (sharks, scombrids, seahorses, and sand lances). It exists within genera, and sometimes between races, between populations, and even between the sexes. It holds only for comparisons between fish with roughly the same form. The average increase in vertebrae is about 10% for each doubling in maximum adult length, but ranges widely between groups.

The ubiquity of pleomerism is surprising. If, as is suggested, there is some functional correlation between body size and segment number, it may well be related to locomotion. The size upon which selection operates with respect to vertebral number is not likely to be the maximum size the species ever attains; it might more probably be the size when the larva starts to swim. Selection may not be operating directly on vertebral number, but possibly on some correlate such as optimal length of muscle fiber. Pleomerism seems to be so general a phenomenon that its explanation, if forthcoming, is likely to shed some light on the functional significance of segment number in fish.

Numbers of vertebrae (and of fin rays and other meristic series) are subject to several other puzzling sorts of variation which may have locomotory implications. Vertebral counts tend to be higher in forms from cooler waters, or from higher latitudes, than in related forms from the tropics. This phenomenon, known as Jordan's Rule, is not explainable through pleomerism, although it is reinforced by a tendency toward a higher proportion of large species at high latitudes (Lindsey, 1966, 1975). Aleev (1963) suggests that high vertebral counts in cool regions are functionally associated with larval adaptation to swimming under conditions of low Reynold's numbers, but experimental data are lacking. Many authors have implied that high vertebral counts at high latitudes are due to direct phenotypic modification by low developmental temperatures. But laboratory rearing of fish of one genetic type at different controlled temperatures does not produce a consistent negative correlation between meristic counts and rearing temperature; in half of the fourteen fish species which have been studied experimentally the curve of vertebral number against rearing temperature has been V-shaped (Lindsey and Harrington, 1972). Latitudinal gradients are more probably the result of selective factors which have produced a genetic cline. Meristic variation is surprisingly high within as well as between wild populations (Fowler, 1970). Some of this is phenotypic, some genotypic. Evidently the precise number of parts in a fish is not critical, but the range of variation which can be tolerated may depend on locomotory performance.

2. *Skeleton*

The vertebral column provides a series of incompressible blocks, separated by joints. Contraction of the body muscles on one side therefore produces bending instead of telescoping, and the bending is translated into backward thrust against the water in the manner which has been described.

Each vertebra is a complex of several firmly joined pieces (Bertin, 1958a), composed of cartilage in elasmobranchs, and of bone in most adult teleosts. Strength against compression is provided by a spoolshaped cylinder, the centrum, that articulates front and back with adjacent centra. In most fishes, both the anterior and posterior faces of the centra are concave, although in a few (*Lepisosteus*, and the blenny *Andamia*) they are convex in front, and in some eels both surfaces may be flat, or even convex in front (Norman and Greenwood, 1975). The centrum may be supported internally by longitudinal struts of stronger material, in chondrichthyes (Bertin, 1958a), and in teleosts

(Hübner, 1961). The relative lengths and widths of centra may vary widely between species, and also along the vertebral column of one fish.

Above each centrum is the neural arch which protects the nerve cord, and which (in teleosts but not in elasmobranchs) carries a median neural spine. Below each centrum in the caudal region is the hemal arch which protects an artery and vein, and in teleosts carries a median hemal spine; in the trunk region the transverse processes corresponding to the hemal arch carry pleural ribs which extend down into the abdominal wall. Overlapping extensions (zygopophyses) from the tops and bottoms of adjacent vertebrae maintain the alignment between vertebrae; they allow lateral flexure, but minimize dorsoventral bending. The neural and hemal spines, lying in the same sagittal plane as the vertebrae, can be rigidly connected to them, since almost no dorsoventral movement is required. In contrast, those bony structures which project laterally (ribs, intermuscular bones) must either be hinged at their point of attachment to the vertebrae or be completely free-floating.

Vertical flexibility of the axial skeleton does occur in some fish at the junction between the skull and the vertebral column. Gosline (1971) describes how the girdle is connected to the skull via a twoplane hinge system (involving the forked posttemporal and the supracleithrum) that allows some head movement without dislocating the lateral line sensory system where it passes forward onto the skull. In many teleosts the whole cranium tilts when the mouth is opened during feeding (Alexander, 1973). Some myctophids can bend their necks upward (Gosline, 1971). The clingfish *Lepadichthys lineatus* has a dorsal gap between the first two vertebrae and can bend its head up while attached to the substrate by its ventral sucker (Fishelson, 1968). The neck-bending of some species of cyprinids and characoids has been described in Section IV,D. The coelacanth *Latimeria* has a jointed cranium, the anterior part of which tilts up when the mouth is opened (Alexander, 1973). Some dorsoventral flexion of the vertebral column may be possible in sticklebacks (Nursall, 1956) and in tunas (Gosline, 1971). But these are limited abilities; the skeletal architecture of fish generally suits them for lateral, not vertical, undulation.

Flexibility may vary markedly along the vertebral column, controlled in part by the degree that the zygapophyses of one vertebra overlap the next. The prepeduncular and postpeduncular joints of some thunniform swimmers were described earlier. At each of these joints the zygapophyses are reduced and the adjacent vertebral faces form a lateral hinge. Between the two joints, the zygapophyses greatly overlap so as to lock several vertebrae into a rigid unit, further strengthened by a wide bony lateral keel (Fierstine and Walters, 1968).

Between adjacent vertebrae there are fluid-filled cushions, which are evidently remnants of the notochord. Gosline (1971) points out that were it not for these cushions the entire thrust when adjacent vertebrae are bent would have to be transmitted across the point where their lateral rims touched. The hydrostatic cushion, which fills the hollow between facing centra, receives the force from one rim of a vertebral pair and distributes it over the full face of the next vertebra.

The cephalochordate *Branchiostoma lanceolatum* lacks vertebrae, and retains a liquid-filled notochord as a hydrostatic skeleton. Guthrie and Banks (1970a,b) have found that the resistance of the notochord to bending can be altered at different swimming speeds, through altering its turgidity by means of muscle fibers in the curved lamellae which divide the notochord into compartments. Although flexibility of the notochord has not been shown to be controllable in vertebrates, Guthrie and Banks (1970a) note that parallel fibers have been reported in the notochord walls of the hagfish *Myxine*.

In addition to the pleural ribs, there are highly variable series of paired intermuscular bones which develop in the myocommata and are often separated from the vertebral bones, although they may be connected to them by a ligament (Nursall, 1956). The bones may be straight, C-shaped, or Y-shaped. The epipleural intermuscular bones extend out into the horizontal septum; the epineural intermuscular bones extend up and back in the epaxial myocommata. In seventeen freshwater species, Lieder (1961) found the number of epipleural bones (branched plus unbranched) varied from 0 (*Perca*) to 25 (*Aspius*), and the epineurals from 7 (*Acerina*) to 71 (*Aspius*). Intermuscular bones are exceedingly numerous and complex in some flounders (Pleuronectiformes); they may occur in four rows; each bone may be brushlike at both ends (Amaoka, 1969).

The nomenclature of the ribs and intermuscular bones is confusing (Emelianov, 1935; Bertin, 1958a; Harder, 1964). So is their function. Nursall (1956) concluded "These appear to have developed in reponse to stresses and strains set up by the musculature." Jarman (1961) pointed out that inelastic intermuscular bones might actually impede the contraction of adjacent muscles if they were parallel to the muscle fibers, and also if they were at right angles (since contracting muscles must increase in cross section). Jarman calculated the angle which must exist between the bone and the muscle fibers in order that there be no restriction. Unfortunately, this angle (55°) has not been found in any living animal. Perhaps if the intermuscular bones lie exclusively in the inextensible myocommata, onto which the muscle fibers insert at a considerable angle, they do not impede contraction, but serve somewhat the same function as the tendons discussed in the next section. Intermuscular bones are also discussed in Chapter 5.

Other bones unattached to the vertebral column may lie in the midline, often extending inward to interdigitate with the neural (or rarely, the hemal) processes of the vertebrae. These are usually serially continuous with the bones bearing spines or rays in the median fins, the pterygiophores, and are probably homologous with them. In addition to these rayless or spineless pterygiophores, there may be a series of detached median rods ahead of the first dorsal fin, the supraneurals, which usually correspond one-to-one with the neural vertebral processes (Goodrich, 1930; Eaton, 1945; Lindsey,

1955). All these free-floating median elements, which are very variable, are bound tightly into the median septum, and seem to lend strength without detracting from lateral flexibility.

3. *Myomeres*

The great lateral bands of muscles running along either side of the body are divided transversely into successive segments, the myomeres. Myomeres correspond in number with the vertebrae (over most of the vertebral column), but alternate with them, so that in the midline each myomere lies opposite the front half of one vertebra and the back half of the next. Within each myomere the muscle fibers are short and tend to run roughly parallel to the long axis of the body, although some may depart by as much as 35° (Alexander, 1969). Therefore most muscle fibers do not attach to skeletal parts, but instead to tough sheets of connective tissue, the myocommata (or "transverse septa"), which separate adjacent myomeres. The myocommata are anchored in the median plane to the vertebral column, and to its neural and hemal spines, and to the tough median septum. Within the myocommata there may lie the segmentally arranged ribs and intermuscular bones already referred to.

Embryonically, the myomeres develop as simple vertical bands, but these then fold in zigzag patterns whose complexity varies widely. The various grades of complexity are illustrated in Chapter 5 (Fig. 1). Myomeres of *Branchiostoma* are simply chevron-shaped, with the apex pointing forward. In lampreys and hagfish they are W-shaped with the center of the W, which points forward, low and rounded, and each myomere sloping forward and inward so that its attachment to the median line is ahead of its exposed lateral surface. The angles of the W-flexures become more acute toward the posterior, and additional dorsal flexures may exist in the region of the median fins. Hagfish also have a specialized ventral musculature (Nishi, 1938).

In elasmobranchs and teleosts the zigzag folding is complex in three dimensions (Chevrel, 1913; Shann, 1914). The salmonid shown in Fig. 7A illustrates a relatively simple arrangement of myomeres, which are W-shaped on the surface. The forwardly directed flexure of each myomere lies halfway down the flank, with backward flexures above and below it. In some forms an additional one or even two flexures may occur at the dorsal extremity of the myomeres, and occasionally one at the ventral extremity, producing up to six zigzag arms (Nishi, 1938). Beneath the surface, each flexure becomes sharper and assumes the shape of a plough. Internally the forwardly directed flexure often becomes divided into two, most pronouncedly in the more posterior myomeres (Fig. 7A).

Each myomere folds forward or backward under its neighbor, so that successive myomeres nest together. The apices are sometimes applied to the vertebrae or to the median septum, forming "halfcones" (Alexander, 1969) or "pyramids" (Shann, 1914). These have traditionally been referred to as

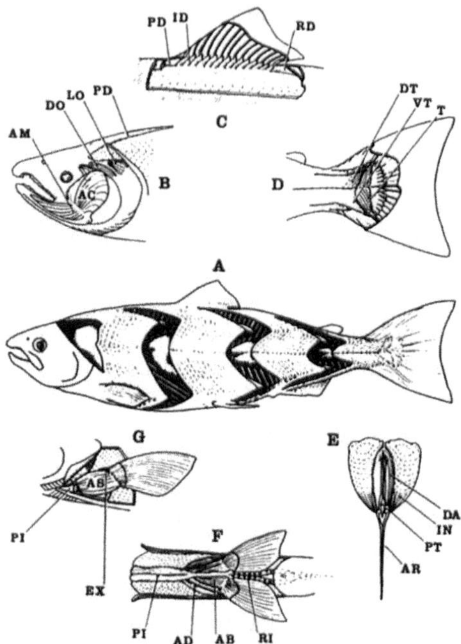

Fig. 7. Muscle systems in chinook salmon *Oncorhynchus tshawytscha.* (Based on Greene and Greene, 1914.) (A) Lateral view with some myomeres removed. (B) superficial head muscles after removal of the skin and part of the jaws. (C) Dorsal fin after removal of skin. (D) Superficial muscles of the caudal fin. (E) Section across anal fin in the plane of the pterygiophores. (F) Ventral view of pelvic region after removal of skin. (G) Oblique ventral view of pectoral fin muscles, with part of protractor ischii removed. AB, abductor ventralis superficialis; AC, adductor mandibulae, cephalic portion; AD, adductor ventralis profundus; AM, adductor mandibulae, mandibular portion; AR, anal fin ray; AS, abductor superficialis; DA, depressor analis; DT, dorsal tendon of lateralis superficialis; DO, dilator operculi; EX, extensor pectoralis; ID, inclinator dorsalis; IN, inclinator analis; LO, levator operculi; PD, protractor dorsalis; PI, protractor ischii; PT, pterygiophore; RD, retractor dorsalis; RI, retractor ischii; T, terminal tendons of lateralis profundus; VT, ventral tendon of lateralis superficialis.

complete cones (presumably considering the left and right halves together) or as deep cones. leDanois (1958) likens this arrangement to a series of babouches (oriental slippers) placed side by side with the toes pointed in alternate directions; the median skeleton and septum form a sort of common sole to all the slippers (Fig. 7A).

Often the apices of the forward or backward projecting zigzag do not touch the median plane, but instead lie in the midst of the muscle mass lateral to the midline. The myotomes then form nesting cones, which have been called incomplete cones or superficial cones. The nests of these cones will appear in transverse section as concentric circles of successive myomeres (Fig. 7E). Because of their sharper flexures, the more posterior myomeres show greater overlap of cones. In many teleosts the muscle cones extend past

4 or 5 vertebrae; in the swift lamnid sharks and in *Thunnus* they may extend past as many as 11 and 19 myomeres, respectively (leDanois, 1958).

If the apices of the cones lie away from the median plane, the myocommata which wrap the cones become attenuated at the apices, to form connective tissue extensions of varying thickness among different species (Nursall, 1956). In most higher teleosts the extensions are weak and disappear in the myomeres into which they project, except in the caudal region. *Amia* has no tendons; *Anguilla* and *Salmo* have tendons much less prominent than those in elasmobranchs (Alexander, 1969). In the shark *Scyliorhinus* there is a broad tendon which projects, lateral to the apex of each cone, in the direction in which the cone points. The scombroid fishes have very distinct tendons which attach to the vertebral column or to the pectoral girdle anteriorly, or to caudal structures posteriorly (Kishinouye, 1923). The origin and attachments of tendons which operate the prependuncular and postpenduncular joints in tuna have been described earlier (Section III, B,5). In nonscombroid teleosts the cones of the last few myomeres may also give rise to tendons which insert on the caudal base (Fig. 7D).

In sharks and bony fish (but not in lampreys or hagfish) the myomeres are divided into dorsal epaxial and ventral hypaxial portions by a horizontal septum of connective tissue. The horizontal septum runs inward to the centra usually slightly below the level of the forwardly directed myomere flexure. Kafuku (1950) concluded from examination of fifty-two species that the horizontal septum consists of two sheets of tendons which can slide over each other; an anterior oblique set runs forward and inward and attaches to the front half of each centrum, while a posterior oblique set runs backward and inward and attaches to the back half of the centra three or more vertebrae toward the posterior. In fish other than scombroids the distal (outer) end of each tendon in the septum, in the form of connective tissue, attaches to the superficial red muscle which lies as a thin strip or sheet just beneath the skin. In scombroids the comparable tendons attach to the deep-seated red muscles, which Kafuku believes are homologous with the superficial red muscles in nonscombroids. The absence in lampreys and hagfish of both a horizontal septum and a lateral band of red muscle supports the likelihood of the functional relationship between the two implied by Kafuku's observations. The significance of the red muscle system in swimming is referred to again later.

An additional ligament, parallel to the vertebral column, may be found fused to the median septum, where successive myocommata come together (thereby excluding all muscle fibers) to form a "deep longitudinal ligament" (Nursall, 1956). It commonly lies between the main anterior flexure and the epaxial posterior flexure. From species to species it varies in width and density. Longitudinal ligaments are particularly well developed in elasmobranchs, where they may occur in the hypaxial as well as the epaxial regions.

The muscle system contains two functionally different groups, designated red and white because of their usual color. Red muscle in most fish is superficial, lying in a relatively thin sheet along either flank, with its greatest thickness forming a wedge along the outer margin of the horizontal septum (which divides it into dorsal and ventral bands). In salmon the red muscles (musculus lateralis superficialis of Greene and Greene, 1914) run together at the posterior to form tendons attached to the caudal fin base (DT and VT in Fig. 7D). In some fish red muscle is scattered in a mosaic of red fibers among the more numerous white. In scombroids there is a substantial block of red muscle straddling the horizontal septum and in some species reaching in to the midline (Kafuku, 1950; Graham, 1975). The relative development of red and white muscle in different kinds of fish may be correlated roughly with their mode of life (Boddeke *et al.*, 1959). Red muscle is usually slow, with low contractile power, and is used for prolonged activity sustained by aerobic metabolism. White muscle is faster, more powerful, and capable of burst activity which may be anaerobic. Literature on the physiological and histological differences between red and white muscle is reviewed by Patterson and Goldspink (1972) and Johnston and Goldspink (1973a) and in Chapter 6.

Although individual muscle fibers extend only from one myocomma to the next, it is possible to tease out successive fibers which lie end-to-end on opposite sides of myocommata, and to follow this thread over as many as fifteen myomeres (Kashin and Smolyaninov, 1969). The paths followed by these threads are called muscle-fiber trajectories by Alexander (1969), who provides the most complete analysis so far available.

The fibers of red muscle tend to run parallel to the body axis, but the fibers in white muscle are arranged in complex three-dimensional patterns. There are two basic patterns, one found in all myomeres of elasmobranchs, primitive teleosts, *Anguilla* and *Salmo*. In higher teleosts this pattern is restricted to the last few myomeres. The patterns differ in the direction of slope of the muscle fibers relative to the myocommata. In the chondrichthyean pattern some of the resultant muscle-fiber trajectories run between a tendon of an anterior cone and a tendon of a posterior cone, while others run between the tendon of a cone and either the vertebral column or the median septum. In the teleostean pattern the trajectories form segments of helices with axes roughly parallel to the body. The helices are in coaxial bundles, with four bundles on each side in a typical teleost. They are illustrated in Chapter 6 (Fig. 3) of this volume. Most trajectories seem to begin and end at the median plane.

Probably the lengths and angles of the white fibers are disposed so that all contract to a similar extent when the fish bends (Alexander, 1969). The interposing of inextensible tendons (and perhaps intermuscular bones?) between some muscle trajectories and their ultimate attachments may serve to unify the rate of muscle contraction in different regions. In teleosts where tendons are poorly developed, myocommata running obliquely laterally from the points of attachment of small tendons to cones may act as lateral parts of the tendons. An additional reason for the transformation of muscles to tendons in the slender caudal peduncle of thunniform swimmers is that the tensile strength of the tendons, which transmit immense forces, is much greater than the maximum isometric tension of muscle.

Just where the body muscles attach to the axial skeleton has been the subject of debate. Nursall (1956) observed an "arch of muscle" visible in a near-horizontal section of a whole fish, which gave the impression of being attached to the vertebral column both anteriorly and posteriorly. The contraction of the arch on one side would bend the vertebral column. Nursall was aware that the pattern might be misleading, since it depended on the precise depth and plane at which the horizontal section was cut. Willemse (1959) published figures showing that this is so, and that no arch appears in sections cut at some other levels. He further dismissed the "arch of muscle" theory on the grounds that for a wave of flexure to pass backward the morphological structure would have to do likewise. Willemse also pointed out the myocommata are not designed to transmit bending forces to the vertebrae because of their angles of insertion.

Willemse (1959) proposed instead a "bimetal theory," whereby contractions of the muscles on one side of the body cause it to bend toward that side, after the manner of a bimetal strip with halves of different contractability. In this view, the myosepta should not be considered as tendonlike structures transmitting muscular forces to the vertebrae. He pointed out the extra advantage that "In our theory the intricate and rather confusing form of the myosepta is not used."

Szarski (1964) accepted Willemse's theory, emphasizing that if the myocommata were actually transmitting contracting forces to the vertebrae, one would expect the myosepta to be strongest near the body axis "whereas exactly the opposite is true." Szarski went on to seek a reason for the complex folding of the myomeres. He concluded that it is to ensure the smooth passage of contractile waves down the body. If the myomeres were flat transverse plates, a smooth passage of contraction could be achieved only by a complicated pattern of nervous impulses from the successive metameric spinal nerves. But with the overlapping myomere folds, even if all fibers in a single myomere contract and relax simultaneously, a gradual increase and then decrease is achieved in the total number of contracted fibers across any cross section of the body. Hence any type of myomere folding is advantageous, and works equally well at all swimming speeds.

Alexander (1969) argued against Willemse, who, he wrote, was "apparently forgetting that a bimetal strip will only be bent by differential expansion if its components are fastened together in such a way that they cannot slide past each other." Lund (1967) was also critical of some of Willemse's views, but he noted that fish could swim according to the bimetal theory if the lateral muscles were held together at the median septum. He also noted that leptocephalus larvae swim without any skeletal or cartilaginous elements to resist telescoping.

Alexander's analysis of the lateral muscles as a series of muscle fiber trajectories has been given above. Most of these trajectories probably run from bone to bone, bone to tendon, tendon to tendon, or in a helical path out from the median septum and then back to it farther along the body. If these trajectories are the functional unit's, they are suggestive of a whole series of Nursall's "muscle arches"; the problem of moving the arch no longer remains, if successive trajectories can be contracted in sequence. But since each "trajectory" is the sum of many fibers, each of which belongs to a separately innervated myomere, there remains the large question of how nervous coordination could contract each trajectory in sequence.

If the connection of the superficial strip of red muscle to the vertebral column via a double set of diagonal tendons in the horizontal septum (Kafuku, 1950) is of general occurrence, then the red muscle locomotory system seems to be mechanically independent of the white muscle system (and easier to understand). In salmon, the red muscle band inserts on to the caudal fin base by its own tendons. The red muscle system can function independently; for example, trout swimming slowly use only their red muscle, and switch on the much larger white (mosaic) muscle system only at intermediate and high speeds (Hudson, 1973). It seems that to some extent fish may have two alternate sets of engines whose wiring, fueling, and propulsive systems are separate.

It will be apparent from the foregoing that myomeres, septa, muscle fibers, tendons, and bones interlock in a three-dimensional jigsaw puzzle which is difficult to comprehend. Sir James Gray (1968) wrote "This complex arrangement is almost certainly of functional significance, but so far no completely convincing analysis seems available." At present, the emphasis on the trajectories traced by threads of end-to-end muscle fibers (Kashin and Smolyaninov, 1969; Alexander, 1969), rather than on the geometry of the fibrous walls (myocommata) which those trajectories pierce, seems to promise an improvement in the functional analysis of the jigsaw puzzle.

4. *Other Trunk Muscles*

The myomeres described above usually make up the great bulk of the lateral musculature, and provide almost all the propulsive force. Other muscle systems or modifications in the trunk will be described briefly. In addition to the references given below, Takahashi (1917) describes the structures and homologies of the carinal muscles in fish, and Winterbottom (1974) gives an extensive compilation of literature on the striated muscles in teleosts.

In the region of the visceral cavity the hypaxial trunk muscles of right and left are separated, and the myomere arms are flattened and modified where they form the abdominal wall. Below the lateral line there may be a patch of superior oblique muscles (with fibers sloping forward and up); ventral to it, a patch of inferior oblique muscles (with fibers sloping forward and down); and sometimes, on the ventral surface of the belly, a sheet of straight rectus muscles (Maurer, 1913; Nishi, 1938; leDanois, 1958).

The epaxial trunk muscles are fastened anteriorly to the back of the skull and to the upper pectoral girdles (Fig. 7B). The muscles expand and anchor into large concave spaces on the posterior face of the skull; in higher teleosts they may reach forward and attach to three longitudinal crests of bone, the supraoccipital and two frontal-parietals (Gosline, 1971).

Lying along the extreme dorsal margins of the lateral muscles are separate longitudinal muscles, the supracarinals. They run from head to tail, but may be interrupted by the dorsal fin(s). The anterior supracarinals (or protracter dorsalis) run from the skull and upper part of the pectoral girdle back to the pterygiophores of the first dorsal fin (PD in Fig. 7B and C). In salmon they have segmentally arranged septa which are in step with the myomeres but which have their own irregular and complex folding (Greene and Greene, 1914). Despite their name, they serve more to produce dorsal flexion of the body than erection of the dorsal fin.

Behind the dorsal fin, the posterior supracarinals (or retractor dorsalis) run from the posterior pterygiophores of the dorsal fin back to the neural elements of the caudal base (RD in Fig. 7C). In salmon the muscles are cylindrical, and are continuous past the base of the adipose fin as a pair of slender tendons. In species with two or three rayed or spinous dorsal fins the supracarinals may be interrupted. In scombroids the posterior supracarinal may insert on each finlet. In some eels with a long continuous dorsal fin the supracarinals run the whole length of the body with separate insertions on each fin ray (leDanois, 1958).

Lying along the mid ventral line is another set of paired longitudinal muscles, the infracarinals, that are interrupted by the pelvic and anal fins. The anterior infracarinals (or protractor ischii) run from the branchiostegal plate or the pectoral girdle, backward on either side of the midline to the pelvic girdle (PI in Fig. 7F and G). In salmon they are segmented, comparatively simple at the anterior but spirally folded toward the posterior (Greene and Greene, 1914). Although their contraction pulls the pelvic girdle forward, their more important functions may be to flex the body ventrally, and possibly to extrude eggs from the abdominal cavity. The anterior infracarinals are best developed in soft-rayed fishes whose pelvic fins are far back on the body; they may be small or absent in advanced forms whose pelvics are thoracic or jugular in position. In eels and brotulids which lack pelvic fins there are no anterior infracarinals, but in the sand lance *Ammodytes* the infracarinals are continuous from the throat to the anal fin (leDanois, 1958).

Behind the pelvic fin the retractor ischii pass backward from the pelvic girdle, running on either side of the anus to insert on an interhemal bone at the anterior of the anal fin (R1 in Fig. 7F). Their contraction may contribute to ventral flexion of the body, or to retraction of the pelvics, or to protraction of the anal fin.

A pair of slender muscles, the retractor analis, run from the posterior pterygiophores of the anal fin back to connective tissue and to the hemal spines at the caudal base. These muscles produce retraction of the anal fin, but are only slightly developed (Greene and Greene, 1914).

5. Muscles of Respiration

The possibility of jet propulsion, and of avoidance of boundary layer separation, by the ejection of water through the gill apertures has been discussed in Section IV,A. Water is pumped into the mouth and out through the gill apertures by a coordinated expansion and contraction of the oral chamber and the branchial chamber (Hughes and Shelton, 1962). An outline of the pump musculature and skeleton, in teleosts, elasmobranchs, and cyclostomes, is given by Shelton (1970). In Fig. 7B only one of the three muscles involved in inspiration (the dilator operculi), and one of the seven involved in expiration (the levator operculi) are labeled. Much of the muscle mass of the head functions to operate the jaws in the capture of food rather than in respiration; in Fig. 7B the cephalic and mandibular parts of the mandibular adductors are prominent and are labeled. The branchial muscles are described by Greene and Greene (1914), Nishi (1938), Thomas (1956), leDanois (1958), and Harder (1964). The mechanics of breathing by fishes are described by Hughes and Shelton (1962), Alexander (1967), Gosline (1971), and Shelton (1970).

6. Innervation and Coordination of Trunk Muscles

The anatomy of the central nervous system in fishes has been reviewed in an earlier volume of this series (Bernstein, 1970). Patterns of peripheral innervation of the muscles are reviewed by Barets (1961) and Bone (1964). Nishi (1938) discusses innervation of the trunk musculature. The anatomy of the spinal nerves of elasmobranchs is described by Norris and Hughes (1920), and Roberts (1969a). Innervation, proprioception, and coordination of the muscles is discussed in Chapter 6.

Of significance in the generation of waves of muscular contraction is the observation that while each spinal nerve provides branches mainly to its own myomere, the sensory fibers and probably the motor fibers also send branches across several adjacent segments (Nishi, 1938; Bertin, 1958d). "Because of the extent of sensory innervation any one segment will be provided with information about the activity of neighbouring segments, whilst the overlap of motor innervation, leading to the simultaneous contraction of many muscle fibres in different segments, will ensure a smooth transition from the contracting to the relaxed zone of the musculature" (Roberts, 1969a).

As would be expected from their ability to function independently, the red muscles in each segment receive motor nerves separate from those going to the white muscles. Roberts (1969a) found this in dogfish, and McMurrich (1884) observed a distinct superficial nerve plexus supplying the superficial lateral muscles in catfish. But the two muscle systems do not seem to have separate sensory nerves. Fish muscles lack the spindles which in tetrapod muscles are length detectors. Evidently the proprioceptors in fish are not deep among the muscles, but lie in the skin or in the very outer layers of the myomeres (Roberts, 1969a). Sensory discharges from these receptors in the body wall of dogfish are capable of providing information on the frequency and angle of bending of the trunk during swimming (Roberts, 1969d). Sense organs of the lateral line system (reviewed by Disler, 1960) may also provide information useful in locomotory coordination (Roberts, 1972; Roberts and Russell, 1972).

Regarding locomotory coordination, there has been much experiment and discussion as to whether the rhythmic muscular activity during swimming is generated (a) by the spinal neurons (with proprioceptive feedback being capable of modifying but not of initiating the rhythm) or (b) by impulses from the proprioceptors when the muscles develop tension or are passively stretched. The subject is well reviewed by Healey (1957). In teleosts, if the spinal cord is cut it is not possible to initiate locomotory activity by passive stretch of body muscles posterior to the cut. This suggests theory (a), in which spinal neurons in the central nervous system are essential (Gray, 1936a, 1968). Experiments on elasmobranchs give different results (Lissmann, 1946a,b). In dogfish the timing of the locomotory rhythm is dependent on proprioceptars which are excited by swimming movements—that is, theory (b). The spinal neurons are capable of intrinsic activity, but this is normally overriden by proprioceptive input (Roberts, 1969b).

The mechanism which controls the tonic posture, and difference in phase of the body waves, is different from that which controls rhythmic movements (Gray, 1968). Moreover, very young elasmobranchs show rhythmic activity which must be myogenically induced because it persists even after total destruction of the whole brain and spinal cord (Harris and Whiting, 1954).

Apparently the control of rhythmic movement may have input from several sources, whose relative importance may vary in different phyletic groups and also in different developmental stages. A table showing theoretical ways in which spinal activity waves might be propagated is presented by Riss (1972). A summary of current views on the role of proprioception in muscular control is given in Chapter 6.

For quick avoidance reaction when a fish is startled, the usual reflex pathways between sensory centers and motor nerves may be shortcircuited by the Mauthnerian system. This is a pair of large nerve cells in the medulla of the brain, each connected to a long axon running the whole length of the spinal cord. The Mauthner cells receive input from the optic and acoustico-lateralis centers in the brain; their axons give off branches to the motor nerves of the muscles. The small number of synapses, and the large diameter and heavy sheathing of the Mauthner fibers, allow for sudden and coordinated response of the swimming muscles. The result is usually a violent "startle response," usually in a direction unpredictable (by predators), followed by rapid directed swimming as the usual motor pathways take over from the Mauthnerian system.

The Mauthnerian system is present in Cyclostomes, Chimaeras, and Teleosts, but not in adult Elasmobranchs. Among Teleosts it is best developed in species relying on quick movement for escape. It is poorly developed in bottom dwellers and in some anguilliform swimmers, and is absent in scombroid fishes and in fishes lacking tails. The anatomy and physiology of the Mauthnerian system have been described earlier in this series (Diamond, 1971). Its differential development in fish with different habits is discussed by Marshall (1971).

B. Fins

1. *Anatomy of Rays and Spines*

The mechanical plan of the fin rays and fin spines is easier to comprehend than that of the myomeres. The flexible web of each fin is a double sheet supported by stiff rods like spokes of a fan. Each rod is attached at its base through a hinge or swivel joint to internal supports buried in the body. Each rod is operated by sets of muscles which can swing it in one or more planes. The whole complex of web, rods, internal supports, and muscles which make up a fin is usually inserted into the body as a largely self-contained unit without serial correspondence to the adjacent myomeres.

Teleost fin rays and spines arise embryonically as thickenings in the basement membrane of the dermal fin fold. The opposed pairs become fused along most of their length to form the shaft of the ray or spine. The halves remain separated at the base, where they are expanded to form a pair of tongs which in fin rays grasp the outer edge of their internal support (Fig. 7E). Fin rays may branch fore-and-aft several times as they approach the outer edge of the web, and their shafts are divided into segments by regularly spaced rings which probably contribute flexibility. Fin spines, in contrast, come to a point at their tips, rather than branching; the halves of the shaft are fused almost to the base to form a rigid, unsegmented spear, and the base typically forms a strong hinge on the internal support which can swing in the median plane but not laterally.

In adult teleosts the spines, rays, and internal supports are ossified; the bony rays are called lepidotrichia, and are thought to be homologous with the scales. The fins of elasmobranchs are supported by ceratotrichia composed of collagen; they are more closely packed then teleost lepidotrichia but are also paired structures, each grasping a median cartilaginous support. Very fine horny filaments, the actinotrichia, also occur in the fin fold of larval teleosts, and may persist in the outer edge of the adult fin membrane.

Rays in the dorsal and anal fins of most teleosts correspond one-to-one with their internal supports, and can be erected, depressed, or inclined freely on a universal joint. In the other teleost fins, and in elasmobranches, the rays are more numerous than their supports and are usually more restricted in their movements. The sets of muscles producing fin movement are described in the following sections. In addition to movement about the ray base, there is increasing evidence that sometimes individual rays are capable of active bending (McCutchen, 1970; Arita, 1971). This may depend on the fact that a ray consists of two halves whose tips are attached and whose bases are operated by separate muscles on each side (Gosline, 1971); it may also be that there are fine muscle fibers in or between the rays.

General discussions on the anatomy of fish fins may be found in Bridge (1896), Goodrich (1906, 1930), Schmalhausen (1912, 1913), Grenholm (1923), Eaton (1945), Lindsey (1955), Bertin (1958b), François (1959), Alexander (1967), and Gosline (1971).

2. *Dorsal and Anal Fins*

In elasmobranchs the internal support to the dorsal or to the anal fin is a flat median mosaic of cartilaginous plates called radials. The outer of the two or three rows of radials (which are joined by ligaments) projects out into the fat fin base, sometimes extending almost to the fin margin (François, 1959). The tightly packed ceratotrichia greatly outnumber the underlying radials, which they grasp (Roberts, 1969b). There is a single muscle on each side of each radial, originating on the outer margin of the myotomes and inserting by a broad tendon on the bases of all the adjacent ceratotrichia. Contraction of the radial muscles bends the fin to the side, along the longitudinal joints between the rows of radial elements. The fin of a shark cannot be collapsed, nor apparently can its area be altered.

In the Chondrostei, the most primitive living teleosts, including *Acipenser* and *Polypterus*, the median fin rays also outnumber their internal supports, but in all other teleosts the two correspond serially except at the fin extremities. The rays or spines are more widely spaced, and so are their internal supports. Each of these median supports, the pterygiophores, has two or three elements: a large proximal dagger-shaped element called a basal (whose inner end may overlap the tips of the neural or hemal spines), a small mesial element which slopes backward to form a "spacer" between successive pterygiophores and which during ontogeny usually becomes fused as part of the basal, and a distal radial element which may be grasped by the ray base.

Each fin ray rides on a double joint. Its tonglike base can pivot in the median plane on the distal radial beneath, so the fin can be erected or depressed. Each distal radial, in turn, can rotate from side to side at its junction with the more proximal pterygiophore. Spines, however, usually cannot be inclined to the side, although they can be depressed. Their bases may be massively constructed to resist forces from the side, and may have devices to lock them in the erect position (Hoogland, 1951; Bertin, 1958c). Useful diagrams of the joints in dorsal fin rays and spines of *Tilapia* are given by Geerlink and Videler (1974).

A fin ray is typically provided with three pairs of muscles, one of each pair inserted on each half of the ray base. The erectors and depressors originate on the basals and the median septum and are inserted on the front and back surfaces of the ray base, respectively. Contraction of one of these sets will cause the ray either to rise or to fall. The inclinator muscles originate on the inner side of the skin or from the myomeres, and run diagonally backward to insert on the sides of the ray bases (ID in Fig. 7C, IN in Fig. 7E). Their contraction will bend the ray to one side.

There is much variation in the inclinator muscles of the median fins. Spines, which cannot be bent sideways, usually lack inclinators. In *Tilapia* the spines have small inclinators and also a fourth set of "interinclinators" alternating with these; they may give support to the pterygiophores when strong lateral pressure is exerted on the fin (Geerlink and Videler, 1974). Catfish may lack inclinators but be able to swing the dorsal fin sideways, perhaps either by contracting the erector and depressor on one side simultaneously (Alexander, 1967), or by using the epaxial trunk musculature (Mahajan, 1967). The puffer *Tetraodon* has no inclinators, but uses its large erectors and depressors, which reach the ray bases via tendons, to swing the dorsal or anal fin sideways (Schneider, 1964).

There are many other variations in the origin and insertion of all these median fin muscles, which in various teleosts have become attached to the skull, or to the cleithrum, or to the neural or the hemal spines of the vertebrae. Both red and white muscles can occur; the dorsal fin of dogfish has an outer layer of red and a much thicker layer of white fibers (Roberts, 1969b). Bergman (1964) found that red fibers in the dorsal fin of a seahorse actually contracted faster than their associated white fibers.

The foregoing description applies equally to dorsal and to anal fins, which are essentially alike in their anatomy. Both may be composed only of soft rays, or, in teleosts, or spines anteriorly and rays posteriorly. If spines are present either in a separate fin or confluent with a soft-rayed fin, they almost invariably lie ahead of the soft rays. If no spines are present, the leading edge of the dorsal or anal may be stiffened by close-packed rays. This pattern of a rigid anterior and mobile posterior seems to conform to hydrodynamic requirements. The anterior spinous part can be erected rigidly for stabilization, defense, or display, and it can sometimes be folded flat. The posterior soft part provides the mobility. The evolution of amiiform, gymnotiform, balistiform, and tetraodontiform modes of swimming (Fig. 1) has depended on the development of versatile propellors provided by fin rays moving on universal joints.

In salmoniform, cypriniform, and a few other teleost groups there is an adipose fin lying between the dorsal rayed fin and the tail. It is often merely a small tab in the adult fish, but in some catfishes the adipose is both long and high. The adipose almost always lacks supporting structures other than fine actinotrichia in the membrane (San don, 1956). Muscles are absent, or rarely form inclinators inserted on a median membrane (Grenholm, 1923). Gosline (1971) suggests that the adipose may be significant chiefly for hydrodynamic reasòns during the juvenile stages.

3. Caudal Fin

Heterocercal caudal fins, found in most elasmobranchs and in chondrosteans (sturgeons and their relatives), have the posterior section of the vertebral column bent upward and continued almost to the tip of the fin. The hemal processes of the vertebrae are elongated, and project down and back to provide a stiff base to the expanded lower (or hypochordal) lobe of the fin. Some anterior hemal processes involved in the fin base may bear detached radial elements. The neural processes are shorter than the hemal, and carry an outer row of more numerous radials. These radials support a tightly packed row of fin rays (ceratotrichia in elasmobranchs, lepidotrichia in sturgeons) which occupy the long low ridge which is the upper (or epichordal) fin lobe. The hypochordal lobe is much deeper, and is strengthened by longer rays which grasp the expanded hemal processes. It is often indented on its trailing edge to form a ventral hypochordal lobe with long rays, and a longitudinal hypochordal lobe with shorter rays (nomenclature of Thomson, 1971). The posterior end of the latter lobe may be separated, by a subterminal notch, from a small flap carried beneath the tip of the vertebral column (called the inferior lobe of the epichordal lobe by Simons, 1970). In some swift pelagic sharks the vertebral column is bent up more sharply, the ventral lobe is extended, and the resultant high aspect-ratio outline resembles the symmetrical tail of a tuna (Fig. 1G).

The myomeres in elasmobranchs extend out along the posterior section of the vertebral column and retain their zigzag pattern almost to its tip. They are capable of resisting bending of the column. In the hypochordal lobe there are numerous stratified muscle bundles, running diagonally down and forward from the skin which covers the ventral edge of the myotomes, and inserting on the hemal processes (Nishi, 1938).

It is commonly stated that when a shark tail is swung sideways during swimming the hypochordal lobe lags behind the vertebral column and so contributes lift. This may not be so. Simons (1970) has shown that the hypochordal lobe may actually move in advance of the rest of the tail. The radial muscles in the lobe are so placed as to be able to alter the contribution it makes to the action of the whole tail (Alexander, 1965a). Simons believes that the hypochordal lobe is a device for altering the "horizontal trim" of the fish. Thomson (1971) points out that even though the hypochordal lobe used in this way opposes lift, the net effect of the whole tail is still to produce lift. Apparently muscular control of the lower lobe allows precise orientation of the direction of thrust from the tail.

In higher teleosts, the fin is superficially symmetrical top and bottom (homocercal). Internally, the vertebral column turns up sharply, but does not reach the hind edge of the fin as it does in elasmobranchs. On the dorsal side of the upturned vertebrae are a few epural elements probably derived from the neural processes. None of the major caudal fin rays arises here, and only a few minor rays above and ahead of the main caudal fin. Almost all the caudal fin arises from the lower surface of the upturned column, which may retain several centra or which may be reduced to a single rod, the urostyle. The hemal processes of the last few centra are greatly expanded, and together form a fan-like plate which retains several distinguishable elements (hypurals) in more primitive fish but which may become almost wholly fused to each other and to the urostyle in higher fish. On the dorsal side of the upturned terminal vertebrae are a few epural elements probably derived from the neural processes. Anteriorly, the caudal skeleton grades almost imperceptably into the backward sloping hemal and neural processes from the vertebrae of the caudal peduncle.

The caudal fin rays of teleosts are structurally similar to the dorsal and anal fin rays. Their split bases are widened and flattened to grasp the hind edge of the hypural bones. The major rays (whose number is largely fixed and characteristic for each species) bifurcate dorsoventrally several times as they approach the trailing fin edge; the series of minor rays, which do not bifurcate, are continuous with the major rays and run forward onto the upper and lower surfaces of the peduncle (Fig. 7D). The caudal rays may all be nearly horizontal, or they may fan out very widely, particularly in caudals with high aspect ratio. The rays of the upper and lower lobes may be separated by a slight gap at their bases, continuous with a gap in the underlying hypural elements (Nursall, 1963, Fig. 1), facilitating the operation of the two lobes as separate units (Gosline, 1971).

In most teleosts the bases of the rays near the center of the caudal fin usually overlap the hypural supports only slightly, providing a hinge along which the fin can bend somewhat from side to side. Nursall (1963) illustrates the joint, which also allows the rays to rotate in the sagittal plane (for expanding of the "fan"). Rays toward the upper and lower edges overlap their supports more. The result is a fin stiffer at its margins than its center; typically in the subcarangiform mode of swimming, as the tail is swept sideways the tips lead and the center bellies out behind. In contrast, tunas have extremely stiff and close packed caudal rays whose bases overlap the hypural base greatly and permit almost no bending there. The rays reaching the extreme upper and lower tips of the tuna tail are more than twice as long as the center rays, and they can be bent more. Consequently, during the very powerful sideways thrusts, the center leads and the tips follow.

In those teleosts which swim primarily with the pectoral, dorsal, or anal fins (Sections III,C and II,D), the propulsive importance of the caudal fin is often reflected in the number of the caudal fin rays. These are typically 19 in soft-rayed fishes and 17 in spiny-rayed groups, but in groups relying on the dorsal, anal, or pectoral fins as principal propellers the caudal rays may be reduced to 12–15 (Zeiformes, Labridae, Scaridae), 9–12 (most Tetraodontiformes), or even none in Molidae (Marshall, 1971).

The caudal musculature of teleosts ranges from a comparatively simple series of extensions from the posterior myomeres onto each fin ray base

(Polypterus), to a complex of seven or more sets of muscles in higher fishes. Nursall (1963) provides a valuable series of drawings and descriptions of the caudal muscles of the perciform *Hoplopagrus guntheri*, and a table of synonyms of muscles names by earlier authors. Other useful references on the caudal muscles are Schmalhausen (1913), Greene and Greene (1914), Grenholm (1923), and Nishi (1938).

The posterior myomeres of both the red muscle [lateralis superficialis of Greene and Greene (1914)] and white muscle (lateralis profundus) may be continued backward as tendons which insert on the caudal fin bases (DT and VT in Fig. 7D). Between these, at the surface at the horizontal septum, is a prominent deltoid flexor tendon [the terminal tendon of lateralis superficialis, of Greene and Greene (1914)] which originates in tendons of the myosepta and inserts on the central caudal rays (T in Fig. 7D). All the foregoing seem to be involved primarily in lateral flexion of various groups of caudal fin rays.

Deeper than the deltoid flexor is the hypochordal longitudinal muscle, running up and back from a prominent projection on the terminal hypural called the hypurapophysis. The muscle inserts on the upper three to five caudal rays. Marshall (1971) believes that the hypochordal pulls these rays sideways and down to produce a caudal twist which may be important in the locomotion of some groups.

Three more sets of muscles lie within the fin base: the deep dorsal flexor, the deep ventral flexors, and the dorsal adductors. These originate variously on the posterior vertebrae, epurals, or hypurals, and insert on caudal ray bases. They can adduct, abduct, or flex particular groups of rays.

Finally, a thin mat of interradial muscles knits together the proximal portions of the major caudal rays. These are of three sorts, forming fans overlying each other diagonally and connecting neighboring rays at various angles. Red fibers occur in the interradials, although the rest of the caudal musculature in *Hoplopagrus* is white (Nursall, 1963). Interradials can probably contract the fin, and produce some lateral bending of individual rays. Some of the fine caudal fin movements observed by Bainbridge (1963), and the caudal undulation described in Section III,D, 4 may be produced by the interradial muscles.

4. *Pectoral Fins*

The base of the elasmobranch pectoral fin is usually two flat cartilage plates, the mesopterygium and metapterygium, which articulate with the pectoral girdle. In skates a third member, the propterygium, supports the anterior of the fin; in some sharks only the metapterygium persists. Continuous with the distal margin of the pterygia is a row of triserial radial rods. The pattern of segmentations and fusions of pterygia and radials varies, but the whole forms a tightly knit mosaic, rigid proximally and somewhat flexible distally, which extends well out into the fin. As in the dorsal and anal fins, the outer web is supported by closely spaced unbranched unsegmented ceratotrichia, completely covered by skin and muscles so as to be externally invisible.

The muscles of the pectoral fin in the shark *Scyllium* consist of a deep and a superficial abductor, and a deep and a superficial adductor, each originating on the girdle and inserting on the fin supports; a tongue from the lateral musculature also recurves ventrally to insert on the lower fin base (leDanois, 1958).

The pectoral fins are less mobile in sharks than in most teleosts, but are more mobile than the dorsal or anal. The articulation with the pectoral girdle principally allows the fin to swing forward and back in its own plane. The leading edge turns slightly down as it swings forward, so that the amount of lift provided by the fin alters (Alexander, 1967). The low and horizontal position of the pectorals allows them to act as hydroplanes, which counteract the negative buoyancy and the pitch produced by the heterocercal tail. Marshall (1971) makes the interesting observation that in a few mesopelagic sharks which have achieved nearly neutral buoyancy the pectorals have shifted upward and rotated, coming to resemble the paddlelike pectorals of higher teleosts. Except perhaps in these aberrant forms, sharks do not use their pectorals as brakes the way teleosts do, and sharks are incapable of making sudden stops. In skates the pectorals are highly modified and become the principal locomotor organ (Section III,C,4).

In most teleosts the pectoral fin base is a row of hourglass-shaped radials (or actinosts) which articulate with the vertical posterior margin of the scapula and coracoid of the pectoral girdle. Their number is most often four, but may be higher or lower. Unlike the elasmobranch radials, they do not extend out into the fin, which is supported entirely by bony rays (lepidotrichia). These articulate along the distal margins of the radials, their split (and curved) bases sometimes encompassing a row of small round distal cartilages borne by the radials. The ray shafts are usually jointed, and they may branch toward the tips. The anterior (dorsal) edge of the fin is usually the stiffest.

The radials are progressively longer from top to bottom, and the fin therefore tends when it is extended to swivel about the anteriormost ray as an axis (Gosline, 1971). The pectorals of the more primitive teleosts are low on the body and largely horizontal, like those of sharks. In higher teleosts the pectorals have moved up on the sides, and their bases have changed to an essentially vertical alignment. The axis of rotation when the pectoral is extended is usually maintained, so that the outer ray swings out ahead of the lower (tending to push the head upward).

A nearly vertical hinge for the fin base allows fanning with the pectoral in order to maintain a stationary position. It also allows fish which habitually rest on the bottom to clap their pectorals back against the body to accelerate from a standing start (Gosline, 1971). In most fish with highly placed pectorals which are used as paddles, the lateral line curves upward as it approaches the front, presumably to avoid water disturbances from the pectorals (Marshall, 1971). The fins can also now be used as a brake; the upward shift of the fin origin ensures that braking by extending the pectorals will impart less of a turningcouple to the head.

The muscles operating the pectoral fin of the salmon were identified by Greene and Greene (1914) as follows: abductor superficialis running from the coracoid of the pectoral girdle to the ventral half of each ray base, bends fin forward and down and closes the rays (AS in Fig. 7F); abductor profundus, from the coracoid to the ray base inner margins, draws fin down; extensors, from the cleithrum to the outer surface of the first ray, spreads fin out in the horizontal position; adductor superficialis and adductor profundus, both from the coracoid–cleithrum junction to the ray bases, draw fin back against body; and interfilamentous, a network of delicate fibers running diagonally between the ray bases on their ventral surface, capable of closing up the rays.

Variations on this basic scheme of pectoral fin musculature have been discussed by Grenholm (1923), Shann (1924), and Sewertzoff (1926). The anatomy of particular species, including some highly modified ones, is described by McMurrich (1884), Howell (1933), Nawar (1955), Samuel (1961), Schneider (1964), Alexander (1965b), and Tilak and Kanji (1967). Correlations between relative development of the pectoral musculature and mode of life are shown by Ganguly and Nag (1964), Keast and Webb (1966), Emery (1973), and Horn (1975). Some references to occurrence of red muscle in the pectoral fins are given by Johnston and Goldspink (1973b) and Rosenblatt and Johnson (1976).

5. *Pelvic Fins*

The contribution of the pelvic fins to locomotion is usually minimal. They seldom contribute to forward propulsion, but may serve as stabilizers and maneuvering vanes. They are frequently modified for nonlocomotory purposes, and become copulatory organs, defensive spines, or adhesive discs. In quite a few groups the pelvic fins are absent.

The pelvic girdle of elasmobranchs is generally a single transverse cartilaginous bar. It bears a backwardly directed basipterygium on either side. The lateral margin of the basipterygium carries a row of triserial radial rods, and these in turn support the closely spaced ceratotrichia. The girdle is imbedded in the ventral myotomes. In advanced sharks there are more or less distinct muscle bundles evidently derived from myotomes and running between the basipterygium and the rays. In skates there is a rectus muscle attached to the pelvic girdle and to the first caudal vertebra, and a pelvic subspinal muscle (leDanois, 1958). In male elasmobranchs there are additional pelvic muscles associated with complex axial and terminal cartilages projecting back from the basipterygium.

In lower teleosts the two halves of the pelvic girdle are fused in the midline, and there are a few skeletal nodules, equivalent to the radials in the pectoral base, along the posterior margins where the rays insert. The innermost is the largest. This allows some rotation of the fin about the outermost ray as an axis. In higher teleosts the girdle halves are usually separate, there are no radials, and the fins can be extended or retracted only along an essentially single plane (Gosline, 1971). There is a concomitant reduction in the number of pelvic fin rays. In higher fish where the pectoral fins have moved upward the pelvics often move forward, and the pelvic girdle becomes strongly attached by ligaments to the cleithra of the pectoral girdle. Forms having such a firm pelvic base may develop strong pelvic spines, which may articulate with the girdle through a joint allowing movement only in one plane. In the anterior position, the locomotory role of the pelvic fins is largely to provide braking or turning, and to counteract the pitch produced when the pectoral fins are extended.

The muscles operating the pelvic fin in salmon include the two infracarinal muscles already referred to (Section V,B,3), and the following: abductor superficialis (AB in Fig. 7F), bends the fin downward away from the body, and closes up the rays; abductor profundus, bends the fin outward; adductor superficialis, draws the fin inward; adductor profundus (AD in Fig. 7F), rotates the fin inward and spreads the rays. The foregoing muscles originate at various places on the pelvic girdles and insert on appropriate parts of the fin ray bases.

The anatomy of pelvic fins was reviewed by Sewertzoff (1934). Sheldon (1937) presented an extensive study of catfish pelvic girdles. The morphology of the pelvics is highly modified in forms with adhesive or other specialized pelvics (Bertin, 1958c). Some peculiar instances are described by Tyler (1962) and Saxena and Chandy (1966a,b).

6. *Innervation of the Fins*

All the median and paired fins are innervated by spinal nerves from body segments in the vicinity of the fin. Because the fin rays and their muscles are almost always more tightly packed than the adjacent body segments, there is usually an anastomosis of the nerves, and the nervous supply to a particular fin muscle cannot be allocated to a particular spinal nerve. Even in the dorsal fin of the stickleback *Gasterosteus aculeatus*, in which a single set of fin muscles lies opposite each vertebral segment, branches of the spinal nerves anastomose to form a longitudinal collector. It has been shown experimentally that excitation of a single spinal nerve can cause contraction in two radial muscles in the pectoral fin of a dogfish, and in six to eight in a skate (Bertin, 1958d).

The pectoral muscles of the ray *Rhina* are served by a brachial plexus receiving contributions from the first ten spinal nerves plus three occipital–spinal and three occipital nerves. The pelvic plexus contains a large collector nerve with contributions from spinal nerves of the twenty-fifth to thirty-eighth segments (Bertin, 1958d). Similarly in the sturgeon *Acipenser* the pelvic fin is served by a plexus from thirteen spinal nerves.

Longitudinal collector nerves connecting the spinal nerves and fin muscles have been found under the dorsal and anal fins of catfish (McMurrich, 1884) and of *Polyodon* (Danforth, 1913). The path of nerves into the fin muscles is figured by Howell (1933) and Schneider (1964). Nishi (1938) summarizes the innervation of fish fins. Nursall (1963) provides a diagram and chart showing distribution of spinal nerves from the last five vertebrae into the various caudal muscles of *Hoplopagrus*. For literature on the experimental study of fin innervation and coordination see Roberts (1969b).

7. Functional Topography of Fins

Among the more than 20,000 species of fishes, there is spectacular variation in the number, shape, size, and position of the fins. Codfish can have ten paired and median fins; some eels have virtually none. Some fins are soft-rayed and some are spiny, and these perform different roles. A gymnotid anal fin may have over 500 rays, while a scombrid finlet contains only one.

Many different and sometimes conflicting selective forces must affect fin development and topography, but some general trends are observable when large numbers of species of widely different habit are compared. Magnan (1929) measured the size and position of the fins, along with many other parameters, in 171 fish species. Several characters showed trends with respect to the estimated speeds of the fish. Greenway (1965) presents calculations based on Magnan's data. The relative surface area of dorsal, anal, and pectoral fins all decrease regularly between "very slow" and "very fast" swimmers. The dorsal and anal fins also tend to shift toward the posterior. The caudal fin shows no clear trend in surface area, although its shape varies markedly (see Section III,B); the aspect-ratio increases strikingly between slow and swift species (Aleev, 1963). Kramer (1960) shows diagramatically how typical tail shapes alter from long and low in abyssal forms to short and high in pelagic forms. The decreased area and backward shift of the dorsal and anal fins with increasing speed are probably related to increasing emphasis on the caudal propellor, which reaches its acme in thunniform locomotion.

Webb (1975) discusses the hydrodynamic consequences of fins (or hydrofoils) having different geometric plans. Kramer (1960) measured the performance in a working model of cutout "fins" of various shapes. The outline shapes of different types of fins were categorized by Magnan (1929). Gregory (1928), as part of an ambitious analysis of body forms of fishes, distinguished twenty-four categories of shapes and positions of the fins, under each of which a species could be placed in one of three or four grades.

Among adult fish, broad and rounded pectoral, dorsal, anal, and caudal fins are typical of slower swimmers. Narrow, falcate pectorals and median fins often accompany high aspect-ratio tails in swift pelagic species. Changes in fin shape and size also occur during ontogeny; simple rounded fins typify the early stages even in forms which as adults acquire angular fins. Aleev (1963) suggests that rounded fins are adaptive to anguilliform movement under conditions of low Reynold's numbers, and hence characterize small fish, slow fish, and, he states, also fish adapted to colder (i.e., more viscous) water.

The effects of fin position on swimming and maneuvering have been discussed by Harris (1953). The positions which fins occupy on the body will depend on the allocation of the functions of generating forward movement, stabilizing that movement, changing direction, and stopping. One fin can fulfill more than one of these functions (and of course, other nonlocomotory functions such as display or defense). In anguilliform swimmers the requirements are fairly well distributed along the body, and the fins may be little differentiated. But in carangiform and thunniform swimmers the propulsive thrust is concentrated in the caudal fin, and the body is stout and rigid; the fins are more specialized, and each is most effective if located in a particular region of the body.

Aleev recognizes four functional regions: (1) an anterior zone of rudders and lifting surfaces, (2) a zone of keels, (3) a zone of stabilizers, and (4) a posterior zone of rudders and locomotor organs. The first zone involves the pectoral fins, and in higher fins the pelvic fins which are directly beneath the pectorals. The farther these lie ahead of the center of gravity the more effective they can be as rudders. The second zone may involve part of the dorsal fin, and of the anal fin, if they are situated far enough forward, and the pelvics if they are posteriorly placed. The third zone involves parts of the dorsal and anal fins behind the center of gravity; their effectiveness as stabilizers is greater when they are farther back, except that they should receive minimal lateral displacement from the propulsive waves. The fourth zone involves the caudal fin, and also sometimes the dorsal and anal fins if they are well back.

The relative lengths of these zones differ widely. For example, zones 2 and 3 are farther forward in a tuna than in a dogfish (Fig. 1). During ontogeny the positions of the fins often shift, the anterior of the first dorsal often moving forward and the posterior of the second moving back, both in elasmobranchs and in teleosts. Aleev (1963) has made extensive calculations of the dynamic stability resulting from fins of different sizes variously placed with respect to the center of gravity. He has also examined the relative body height, the location of the greatest height along the body, and the cross-sectional shape, all of which vary widely between species and sometimes during ontogeny. His conclusion is apt for the diffuse subject of functional topography of fins:

> It is clear that a certain change in the position and structure of a particular fin will affect the position and structure of the other fins and other hydrodynamically sensitive structural features merely through the resultant change of the total activity of all the fins and body parts. This example clearly illustrates that any morphological features of an organism, in both ontogeny and phylogeny, are connected with the environment not as independent parts of the structure but as parts of a whole. Here the effect is felt of the unity of all parts of the organism, in both the functional and morphological sense.

VI. LOCOMOTORY HABITS OF WILD FISH

Measurements of locomotion and metabolism of fish in the laboratory are steadily improving in precision, but the relevance of such data to wild fish is largely unknown. The gap will have to be closed before energy budgets can be drawn up for natural populations, in which food intake, metabolic costs, and growth can be balanced. Most of the discussion in this volume is of necessity based on laboratory observation, but first, knowledge will be summarized concerning the locomotory habits of fish in natural surroundings. Although current information is diffuse and anecdotal, substantial improvements can be expected soon through advancing technology. Methods of recording activity of fish, in the wild or, more often, in confinement, are discussed in Chapter 2.

A. Records of Long Distance Movements

The best evidence concerning long distance movements of fish comes from the recapture of individuals which previously had been marked in a distinctive manner at a known time and place. While there are important weaknesses in this type of data, tag recaptures offer positive evidence of movement, and allow the calculation of minimum distances covered and minimum rates of travel. A few examples of long distance movements are shown in Table IV, arranged roughly in descending order of the sizes of the fish involved. See also Table I in Chapter 2 for more detailed observations.

The distances covered by some fish are spectacular. Two large bluefin tuna *Thunnus thynnus* tagged at Cat Cay, Florida, were recaptured 7800 km away at Bergen, Norway, 118 and 119 days later (Mather, 1962). The same species in the Pacific has crossed from west to east—Japan to Baja, California (9700 km, Table IV), or east to west—San Diego to Japan (8800 km, Clemens and Flittner, 1969), and Guadalupe Island to Japan (10,750 km, Orange and Fink, 1963). By tagging albacore *Thunnus alalunga*. Clemens (1961) obtained evidence that they move in schools which probably make the two-way trip to California to the mid-Pacific or to Japan and then back to California. Other species of scombroid fishes may also migrate thousands of kilometers; Table IV gives a few examples. Diverse groups of fish, including clupeoids, salmonids, cods, flatfish, and sharks, may move long distances; a spiny dogfish tagged off the Washington coast turned up in Japan 8700 km away (Table IV), and the same species has undertaken various transatlantic movements (Holden, 1967). Many more examples could be given from tagging returns, and many species which have not been tagged may be suspected to migrate. The European eel *Anguilla anguilla* probably migrates over 6000 km between its freshwater feeding streams and its spawning ground south of Bermuda, but there are no tagging data.

All the distances quoted are the shortest possible distance by water between points of marking and recapture. The dogfish which crossed the Pacific may have covered substantially more than the 8700 km direct distance to Japan, if it traveled at its accustomed depth north around the Continental Shelf (Holland, 1957). Oceanic currents may also have deflected some of the courses followed.

Calculated rates of travel of tagged fish may be appreciably hastened or slowed by water movements. Mather (1962) points out that the two bluefin tuna swimming from Florida to Norway may have been assisted by the Gulf

Table IV Some Records of Long-Distance Movements by Tagged Fish

Species	Distance (km)	Time (days)	Rate (cm/sec)	Length[a] (cm)	Weight[a] (kg)	Author
Thunnus thynnus (bluefin tuna)	7800 Fla.–Bergen	118	77	–	ca. 170	Mather (1962)
Istiophorus platypeterus (saifish)	4440 Windward I.-Pensacola	98	52	228	20.9	Mather *et al.* (1974b)
Hippoglossus stenolepis (Pacific halibut)	2370 Alaska–Cape Blanco	175	16	–	–	Manzer (1946)
Tetrapturus albidus (white marlin)	5000 Bahamas–off mouth Amazon R.	516	11	–/185	16/24	Mather *et al.* (1974a)
Galeorhinus zyopterus (soupfin shark)	2040 Calif.–Q. C. I.	103	23	140/145	–	Herald and Ripley (1951)
Oncorhynchus tshawytscha (chinook salmon)	1610 Q. C. I. –Ore.	60	31	–	–	Manzer (1946)
Squalus acanthias (spiny dogfish)	8700 Wash-Honshu	2699	3.7	94/–	–	Holland (1957)
Gadus morhua (Atlantic cod)	2100 W. Greenland–Iceland	147	17	86	–	Meyer (1965)
Thunnus alalunga (albacore)	8534 Calif.–Japan	196	51	75.5/76	–	Clemens (1961)
Salmo salar (Atlantic salmon)	1100 Norway	11	116	85	–	Dahl and Sømme (1936, cited in Bainbridge, 1958)
Salmo salar (Atlantic salmon)	4270 Devon–Greenland	554	9	/73	/4	Allan and Bulleid (1963)
Thunnus thynnus (bluefin tuna)	9700 Japan–Baja Calif.	323	35	36/68	1.2/7.3	Clemens and Flittner (1969)
Oncorhynchus nerka (sockeye salmon)	1670 N. Pac.–Bristol Bay	35	55	–	–	Neave (1964)

[a]At tagging/at recapture.

Stream and the North Atlantic Drift. But many of the migrations cannot be attributed to water movement. Bluefin tuna and albacore cross the Pacific in both directions. So too tagged Atlantic cod have been recorded moving between Greenland and the Icelandic spawning grounds in either direction; Tåning (1937) noted however that the swim from Iceland to Greenland (roughly 2000 km), going with the current, takes only 90 days, while the reverse journey takes at least 164 days. The fastest daily rate of cod in West Greenland (length roughly 90 cm) was 32 cm/sec, which is close to a daily rate of 28 cm/sec, recorded earlier for cod leaving the spawning grounds at Lofoten, Norway.

Atlantic salmon *Salmo salar* also move both ways in the North Atlantic; some tagged fish have moved from several points on the British Isles to Greenland (Menzies and Shearer, 1957; Swain, 1963; Allan and Bulleid, 1963); others have moved to Greenland from New Brunswick (Kerswill and Keenleyside, 1961). Oceanic circulation may have modified these migrations, but it cannot be their sole cause.

The question of dependence on water currents in ocean migrations by Pacific salmon *Oncorhynchus* is ably discussed by Neave (1964), who synthesized results of massive tagging experiments in the North Pacific. His conclusions are as follows.

> The velocity and direction of recorded currents are often entirely inadequate to account for the rate of travel. In the central and eastern parts of the Gulf of Alaska, where current velocities are commonly of the order of 1–4 miles per day, the minimum travel speeds of tagged sockeye and pink salmon over distances of several or many hundred miles has frequently exceeded 25 and sometimes 40 miles per day. Moreover, these journeys were made in a wide variety of directions relative to known currents . . . . It is concluded that salmon are able to reach distant ocean areas which offer favourable conditions for survival and growth, and also to return from these areas, by relatively rapid journeys which are not closely controlled by currents.

Adult Pacific salmon of various species, usually with body lengths of roughly 50 cm, maintain average speeds of from 0.8 to 1.5 ℓ/sec for many days or even weeks; one pink salmon *O. gorbuscha* about 44 cm long achieved an average of 90 cm/sec (over 2 ℓ/sec) for 6 days (Hartt, 1966; Shepard *et al.*, 1968).

The same order of magnitude of travel rates (in absolute terms) has been recorded for a soupfin shark *Galeorhinus zyopterus* (length 142 cm) which averaged 43 cm/sec over 2 days (Herald and Ripley, 1951), and for a striped marlin *Tetrapturus audax* (length unknown) which averaged 75 cm/sec over 90 days swimming from Baja, California, to Hawaii (Squire, 1974).

A strikingly higher rate was recorded, over shorter periods, in schools of albacore *Thunnus alalunga* (Clemens, 1961). One school covered 428 km in 1 day, achieving the average rate of nearly 500 cm/sec. (The rates are calculated from recoveries at different times of fish all tagged in one school at the same time; they rest on the assumption that the fish remained together as a school, which is probable but not certain.) If these fish were about 80 cm long, the highest rate was about 6 ℓ/sec, well above the 3 ℓ/sec sec usually considered as normal cruising speed in fishes. Possibly harassment by the fishing boats which made the tag recaptures produced abnormally high speeds, comparable to the high speed of a skipjack tuna (5 ℓ/sec) observed by Yuen (1970) after he had chased it a long way from its home bank (see Section V,B,3).

Daily rates of travel, if calculated for a long period, may fall far short of the swimming speed displayed during the periods of each day when the fish was actively migrating. Sockeye salmon smolts migrating out of Babine Lake spent only about 8 hr per day traveling; much time was spent on feeding (Johnson and Groot, 1963). The daily period allocated to active travel by Pacific salmon in the ocean is uncertain, but they are known to feed extensively during migration; daily rate of movement might represent only a third of the actual active swimming speed (Neave, 1964). On this basis, maturing pink salmon (body length perhaps 44 cm) maintain travel rates of 46–55 km/day over many days, but if this is compressed into an 8 hr working day it represents swimming speeds of 155–180 cm/sec.

Migration rates of young fish are harder to obtain because tagging is usually inappropriate. Even so, rough measures can be made of the time and extent of the seaward migration of young Pacific salmon, capitalizing on their known life cycles and strong homing tendency. Many half-grown fish have been captured in the open North Pacific and tagged. Subsequent recapture of these as spawning adults in Asian or North American streams established where they themselves had originally come from as fry; the data and distance from home at their first capture coupled with their known age enabled Neave (1964) to estimate their movement rates since first entering the ocean. Thus, over periods of from 9 to 12 months, young pink salmon must have moved at daily rates of about 10.7 cm/sec, young sockeye at from 3.2 to 6.5 cm/sec, and young chum at about 8.6 cm/sec. These rates agree with Johnson and Groot's (1963) observations that young sockeye on their migration out of Babine Lake swam at daily rates of 6.5–13 cm/sec over a distance of 50–117 km.

Although the absolute speeds of young salmon referred to above are small compared with those of adult salmon, their "specific speeds" (i.e., body lengths/sec) are in the same range. In fact small fish tend to do better than large fish in their specific cruising speeds (Webb, 1975), as measured over short periods in the laboratory. Data on long distance travel given above are too rough for such a generalization concerning wild fish, but at least it can be said that, when related to body lengths, the performances of small fish seem to be at least comparable to those of large fish.

Much information has been accumulated about fish ascending rivers to spawn. In the confines of the river, tagging and recaptures are relatively easy. The ground distance covered, and also the elevation attained, are known more precisely than in marine migrations. What is not known with any precision is the length of the water column through which the fish have passed in stemming the current. Hence speed relative to the water, and work done, are imprecise. In general, the rates of movement upriver by Pacific salmon resemble the rates described earlier for their migrations in the sea (Idler and Clemens, 1959; Osborne, 1961; see also Chapter 2). Swimming against the flow of the river must require some additional outlay of energy, but just how much is a matter of conjecture.

B. Short-Term Components of Long-Term Movements

1. *Upstream Migration*

Distances of upstream fish migration have sometimes been calculated by adding to the ground distance covered a measure of the water column which has moved downstream during the time of the ascent. The calculation is illusory, since fish are cunning at seeking paths with unknown although probably low velocities. It has even been suggested (Osborne, 1961) that migratory salmon may extract energy from turbulent velocity fluctuations of the river; the possibility has not been disproven.

Observations on fish moving up rivers show that progress is irregular. During daylight, sockeye and coho salmon ascending a river progressed through slow stretches (less than 1.0 m/sec) in steadily swimming schools (Ellis, 1962, 1966). In somewhat faster water (1.0–1.5 m/sec) they interspersed their progress by rest periods. To navigate the rapids (more than 1.5 m/sec), they abandoned their schools and moved individually, alternating between bursts of high activity and short holding periods. Paths chosen by all fish represented only a small part of the total water space available. Migration through pools and past obstructions showed a diel pattern. Some migration occurred at night. Ellis (1966) gives estimates of their burst and sustained speeds, but it would not be practical under such conditions to measure water velocities and elapsed time along the precise path followed by each fish.

Upstream migration of sturgeons also shows complex patterns. *Acipenser gueldenstaedti* were found by Gayduck *et al.* (1971) to swim usually at a depth of 10–15 m (at 2-4 m above the bottom of the river), but they would occasionally rise and swim for 5–20 min only 1–2 m beneath the surface. Studies by Shubina (1971) on *Acipenser stellatus* in the Volga also demonstrated complex variation. The largest upstream-migrating fish rested for periods of 5–31 min; downstream-migrants did not rest, but moved slower than the flow of the river. Fish ascended the shallow east bank, and descended on the opposite side. The largest fish migrated fastest. Both in upstream and downstream migration, the distribution of males and females in the flow was different, and the distribution of fish of different sizes over the stream bed was different. Such data defy simple calculations of distance covered or energy expended.

Much of the literature on upstream migration of salmonids is reviewed by Banks (1969). Additional sustained cruising speeds of migrating fish, mostly measured over shorter time periods, are given in Chapter 2.

2. *Migration in the Ocean*

The problems of recording motion of a swimming fish with respect to the surrounding water are less in a nonturbulent ocean or lake than in a river. An eel, about 80 cm long, outbound in the North Sea on its spawning migration, was tracked for 14.25 hr, during which time it covered 25 km (Tesch, 1972). It swam in midwater so long as the depth exceeded 20 m. Its speed through the water did not vary greatly, with an average of about 51 cm/sec and a maximum for 1 hr of 77 cm/sec. Its ground speed was of course greater when moving with the tide than against it, but the effects of tidal drift on migrant eels apparently were largely cancelled out by alternating directions of flow.

Less regular locomotory habits have been observed in flatfish. A 43 cm plaice, *Pleuronectes platessa*, tracked for 15 hr in water 27 m deep (Greer Walker et al., 1971) spent 70% of the time off the bottom. During strongest tide flow it moved a ground distance of 7.41 km in 137 min, or 90 cm/sec (about the same speed and direction as the tidal current). At night its average height off the bottom increased.from 2 to 6.7 m. As the tide slackened, it settled on the bottom and buried itself. These direct observations are in agreement with earlier sightings of another flatfish species, the sole *Solea vulgaris*, often seen at the surface during night. deVeen (1967) analyzed fishermen's reports of such sightings, usually made in March, April, and May when the spawning migration is under way. He suggested that sole seen at the surface are those at the top of the main body of migrants which have left the bottom to use the tidal current for passive transport. Those seen at the surface are nearly always still, and drift with the tide. "It is not known how the diurnal vertical movements of the sole are synchronized with the tidal cycle so that they leave the bottom, at night, on an easterly tide" (Jones, 1968). The rise off the bottom at night, observed directly in the tagged plaice and inferred from sightings in soles, is in agreement with aquarium experiments on plaice by Verheijen and deGroot (1966), and on sole by Kruuk (1963); these flatfish were active at night and dug into the bottom during daylight. There is a concomitant diel rhythm in trawl catches of sole and other species (Woodhead, 1966).

From the viewpoint of calculating locomotory output for these types of migration, much of the lateral displacement is apparently passive, but this must be achieved by expending energy in vertical movement since the flatfish are heavier than water. To compute an energy budget, one would need to know how much work is required simply to remain off the bottom.

3. *Migration of Scombroids*

Large scombroid fishes range widely, often over water so deep that contact with the bottom is lost. A Pacific blue marlin 365 cm long, tracked for 22.5

hr, spent half the time within 10 m of the surface, one-sixth at from 10 to 30 m, and one-third at depths over 30 m, once diving to 80 m (Yuen *et al.*, 1974). The depth selected showed no diel pattern. During the period, speeds ranged from 31 to 228 cm/sec, with an average of 83 cm/sec. Another marlin remained at depths of 115–185 m during a 5.5 hr tracking period, but was probably suffering from effects of capture.

Longer tracking periods of skipjack tuna in Hawaiian waters are reported by Yuen (1970), who maintained contact with one fish (length 44 cm) for 12 hr and another for 7 days. His summary reads as follows.

> The following picture of the behavior of skipjack tuna that are associated with banks can be drawn from these results. They have a general daily pattern. They usually spend the day at the bank, where they swim to and fro and are away from the surface a good part of the time. This type of swimming is probably associated with searching for food and feeding. Later in the day, within a couple of hours of sunset, they leave the bank and swim with few changes in direction until approximately 2:00 AM, when they seem to adopt a more erratic swimming pattern. Although they leave the bank by different routes from day to day, they are usually back at the bank by sunrise. They are presumed to be close to the surface throughout the night. The repeated returns to the same spot by the fish at Kaula Bank after journeys of 25–106 km by various routes imply that skipjack tuna can navigate. Their consistent arrival times suggest that they have a sense of time. It is interesting to note that from 3:00 AM to 6:00 AM of the first morning after it was tagged, when it was unusually far from the banks [having been pursued for 28 km before tagging], the tuna at Kaula Bank averaged 8 km/hr (223 cm/sec), seven times its average speed for that time of day, as if it were compelled to arrive at the bank by a certain time.

Tracking of a fish carrying an acoustic tag usually involves fixing the position at regular intervals, and calculating speeds in terms of time elapsed and straight-line distance between successive fixes. Calculated speeds are therefore minima, since the method is open to the same objection, on a smaller scale, as is the determination of migration speeds from tag recoveries: The fish may very well have *not* swum in a straight line, and hence may have swum farther and faster than shown by its net movement.

There is reason to believe that swimming behavior of many wild fish includes important fine variations which may radically alter their energy expenditures. Fish heavier than water may alternate short periods of active swimming with periods of downhill "gliding." (The technique is not available to very small fish because of their relatively high drag and small momentum.) The energy saving could in theory amount to 50% over a given horizontal distance (Weihs, 1973a). Not only scombroids, but sharks, rays, flatfish, and others may be suspected of employing two-stage swimming modes. The subject of such locomotion in negatively buoyant midwater fishes is explored in Chapter 4.

C. Activity Cycles in Wild Fish

Levels of swimming activity alter rhythmically in many kinds of fish, timed to tidal, diel, or seasonal cycles. Locomotory changes may be spectacular in species which undergo long migrations. Young sockeye salmon, after a year or more of lake residence in Babine Lake, suddenly switch to a well-oriented directional swimming at speeds of 20–30 cm/sec (body length 8.3 cm), traversing 100 km of the lake and then proceeding down-river to the ocean (Johnson and Groot, 1963). The same species, homeward-bound from the high seas 2 years later, may cut across prevailing water currents for hundreds of kilometers at average speeds of 58 cm/sec (body length about 65 cm) (Neave, 1964). Increase in swimming speed are common at spawning time in many other species. They are particularly evident when reproduction involves the ascent of swift streams. Levels of locomotor activity also alter with season independent of reproduction. Swimming speed of yellow perch in Lake Mendota was much higher in summer than in winter, and was linearly related to water temperature (Hergenrader and Hasler, 1967). Temperature seems to be the dominant control of seasonal changes in activity (Andreasson, 1969; Gibson, 1969), although photoperiod may also be effective (Woodhead and Woodhead, 1955; Olla and Studholme, 1972).

Tidally controlled cycles in movement of fish in the littoral zone have been reviewed by Gibson (1969). Diel changes in swimming activity have been found in many species of fish (reviewed by Woodhead, 1966). Some species, such as the flatfishes, are largely nocturnal in their activity. Others, such as the pelagic clupeids, are much more active at dusk and at dawn. Activity peaks at dawn and dusk occur in freshwater (Spencer, 1939; Davis, 1964) as well as in marine species (Gibson, 1969; Stickney, 1972). Many sight-predators are active only in daylight.

Diurnal–nocturnal changes are particularly striking in the variegated fish fauna of coral reefs (Hobson, 1965, 1968, 1973, 1974; Collette and Talbot, 1972; Emery, 1973). The changeover between day-active and night-active fish groups is also clear in many freshwater communities. Lissmann (1961b) found in South American rivers that numerous gymnotid fishes, which have electrosensory systems, hid quietly in vegetation during daylight, but at night they swam actively throughout the open water, emitting a chorus of electrical signals. Temperate freshwater communities may also show exchanges between day and night occupants, with complex vertical exchanges in the open water, and onshore–offshore exchanges, between different species (Northcote *et al.*, 1964) and sometimes between young and adults of the same species. Young brown bullheads *Ictalurus melas* are active by day with peaks at twilight, while adults are active at night (Darnell and Meierotto, 1965). The diel changes in locomotor output of individual fish may therefore be pronounced. The immediate stimulus seems usually to be light intensity, but it is difficult to allocate the causes of diel activity patterns in wild fish among pursuit of food, avoidance of predation, response to light or other exogenous factors, and endogenous rhythm.

Diel vertical migration is widespread among diverse kinds of fish. Although most fish are reported to be deeper during daylight (Yuen, 1970; Cushing, 1973; Badcock, 1970), some show the reverse diel pattern (Northcote *et al.*, 1964; Hasler and Villemonte, 1953).

Vertical migration may be important in bringing about lateral transport from water currents. Examples have been given of flatfish rising off the bottom into a tidal flow destined to carry them toward their spawning ground. In many other species, too, diel vertical migration must expose them to changes in the actions of water currents. In fact, by appropriate vertical movements, fish in some regions might achieve most of their horizontal migratory movements by passive transport. Jones (1968), however, cautions that for no species are there data supporting the drift hypothesis which could be regarded as critical or conclusive.

Vertical migrations of fish must be achieved at the cost of some energy expenditure. The daily migration of small mesopelagic fish from the deep scattering layer involves an upward swim of 500 m lasting for an hour or more (Marshall, 1971). The metabolic cost, following calculations for zooplankton, may be less than 1% of the organic matter of the body. The descent needs little energy as the fish is likely to be slightly heavier than water. Vertical migration costs in other groups will depend partly on their buoyancy. Species which can adjust their gas bladder volumes ride up and down with less cost, but adjustments have been too slow in those (shallow water) species which have been studied, and a gas bladder imposes limits on the range of rapid depth change (Jones, 1952).

With respect to the adaptive advantages which must exist for vertical migrations in fish, Woodhead (1966) writes:

> It seems unlikely that the limited excursions from the sea bed of flatfishes, the nocturnal dispersions of single cod into midwater, the twilight migrations of large schools of many thousands of herring, both feeding and non-feeding, the extensive vertical movements of mesopelagic fish, and the deep scattering layer, can all be accommodated within a single simple theory. Vertical migration varies greatly in its nature and extent and is likely to hold different significance for different species of fish.

D. Schooling

Schooling has been observed in over 4000 species of fish (Shaw, 1962). The advantages of schooling may relate to reproduction, predator avoidance, feeding, learning, or energy conservation (Breder, 1959; Radakov, 1973). Models have been developed, based on theories of search, which show that there may be an advantage in schooling for prey, and in some circumstances for predators (Cushing and Jones, 1968).

Schools of fish can be very large indeed; Cushing and Jones (1968) refer to schools of spawning herring continuous in the Straits of Dover for 17 miles.

Within large groups are smaller more homogenous formations which behave in closely concerted ways (Zuyev and Belyayev, 1970). Schools usually contain fish all of the same size, swimming at the same rate and changing direction more or less as a unit. The range in relative body lengths within a school is usually from 1.0 to 0.6 (Breder, 1965). There are no permanent "leaders."

The sustained swimming speed of fish in a school may be different from that of isolated individuals. Single fish have been reported as faster than those in schools by some authors (Breder and Nigrelli, 1938; Escobar *et al.*, 1936; Ohlmer and Schwartzkopff, 1959; Schuett, (1934) and slower by others (Hergenrader and Hasler, 1967; Kleerekoper *et al.*, 1970). These observations were made under a variety of laboratory and field conditions, and on different species. Fish cannot school readily if an experimental chamber is too confining (Zuyev and Belyayev, 1970). Therefore, the performance of nonschooled specimens in the laboratory may give a misleading view of those species which school in the wild.

Important hydromechanical advantages may arise from swimming in a school. Belyayev and Zuyev (1969) stated that the endurance of fish may be increased two to six times by swimming in schools. Breder (1965) suggested that there may be effects of vortices in the wake of swimming fish on the fish following behind, but he did not analyze them. The Karman vortex sheet (the row of alternating eddies) may be used to lower the hydrodynamic resistance of followers; Zuyev and Belyayev (1970) offered experimental support of this hypothesis. Weihs (1973b) demonstrated, by calculation, the advantages of fish swimming in a diamond-shaped configuration such as is actually observed; he calculated that there would be reductions in relative speed of up to 30% between the best and worst positions a fish might adopt with respect to the wake of those fish ahead. Arguments for the reality of this effect are strong. The diamond pattern occurs in fish when they are swimming but disappears when they stop (Keenleyside, 1955). Dimensions of the diamonds observed in actual schools (van Olst and Hunter, 1970; Pitcher, 1973) agree with hydrodynamic calculations.

From the hydrodynamic advantages of schooling which have been proposed, it follows that the lead fish gets no advantage and must work harder. Observations by Zuyev and Belyayev (1970) are in agreement; the horse-mackerel swimming at the head of a grouping, although the same size as its followers, oscillated its caudal fins with up to 1.5 times the frequency. They concluded that the lead fish is in the least favorable circumstances, and it expends more energy on motion than the rest.

Distance between fish (or packing density) is of interest to those calculating fish numbers from echo records of fish schools. Interfish distances in various genera, reviewed by van Olst and Hunter (1970), suggest that mean distance to the nearest neighbor for most pelagic schooling species is generally about half a body length. But Cushing (1973) notes from observations of interfish distances that larger fish move at a greater distance apart than might be expected merely from linear differences in size. He suggests the increase is in proportion to their volumes. Perhaps an explanation may be found from hydrodynamic analysis.

E. Some Pitfalls in Locomotory Studies

Physiologists can easily make fools of themselves when generalizing about wild fish on the basis of observing laboratory fish. In most experiments the artificial constraints are so great that fish are likely to behave in a highly unnatural manner. Much of the repertoire of behavior patterns characteristic of a species in the wild may be physically impossible under laboratory conditions.

The capture and handling of a fish, even with all possible gentleness, has been shown to sometimes impose extreme stress and even death, without any apparent morphological damage. Fright stress simply from movements of the observer may induce prolonged physiological changes in a fish exposed without cover in an experimental chamber. Appropriate cover may radically alter locomotory performance, but just what cover is appropriate may not be obvious to the experimenter. Jones (1956) found that minnows *Phoxinus laevis* in an open aquarium were active by day and inactive at night, but provision of a simple shelter completely reversed their behavior so that they remained in the shelter by day and swam about by night. Flounders studied by Bregnballe (1961) swam both day and night in a tank, until sand was provided, whereupon they buried themselves all day and swam only in the dark.

Light intensity, and also temperature, can reverse the response of young fish to running water (Northcote, 1962; Pavlov *et al.*, 1968). Even the tone or color of the walls of the holding chamber is commonly observed to affect locomotor activities. This may be due to the light level which ought to be no more intense in the laboratory tank than that normally encountered by wild fish (Woodhead, 1966). Light intensities convenient to the physiologist may be acutely uncomfortable to the fish, for the latter, unlike the former, cannot close his eyes.

Size of the chamber in which a fish is confined imposes constraints on its locomotory possibilities, particularly for large fish species. Although big fish (and aquatic mammals) can be kept alive and apparently healthy in aquaria, their locomotory repertoire is severely limited. Studies on captive tuna have been of only limited value in predicting their wild behavior (Tester, 1959). Even small species, if they habitually cruise for long distances in the wild, may exhibit aberrant locomotion and metabolism in a small aquarium. Only the smallest fish in the largest tanks can achieve top forward speed even for a moment, before they have to turn. Moreover, wild fish are likely to switch frequently from one activity level to another, perhaps including important alternation between swimming and gliding (Weihs, 1973a). Their spectrum of activities is far wider than can be observed even in experimental tunnels or treadmills. The significant hydrodynamic advantages of swimming in a school are also denied by the size limitation of almost all experimental devices.

Many of the foregoing sources of error vanish when observations are made on free rather than on captive fish. Extended underwater fish-watching is now possible with free diving gear, underwater habitats, research submarines, or underwater television. Detailed accounts can be compiled of the minute-to-minute and daily locomotory patterns in natural fish communities. Such descriptions are essential precursors to drawing up metabolic budgets of wild fish, but they must be followed by quantification.

Miniaturization now offers the possibility of transmitting many types of information from a wild swimming fish. One limitation is the restricted distance over which the underwater signals carry [currently from a few hundred meters, up to 2.3 km for large tags (Yuen, 1970)]. More important are the possible effects which the shock of catching the fish and attaching the transmitting device may have on subsequent behavior. Yuen *et al.* (1974) found that the trauma of capturing marlin in order to attach a tag usually resulted in aberrant behavior and even early death. The device is likely also to have an effect on locomotor activity after the fish is released. Tags fastened outside the fish will have a hydrodynamic effect. Tags placed either inside or out may affect buoyancy; Gallepp and Magnuson (1972) concluded that bluegills provided with transmitters might, if released over deep water without adequate time (300 min) for buoyancy adjustment, sink and remain for a long time on the bottom.

The further that technology of miniaturization develops, the greater is the temptation to load animals with measuring devices, with the ever present danger that the behavior then recorded is not natural. With due precautions, new technology is certainly going to allow great strides in understanding fish locomotion. But zoologists should never forget that they suffer from the same problem as quantum physicists; they cannot observe the course of nature without disturbing it.

REFERENCES

Aleev, Y. G. (1963). "Function and Gross Morphology in Fish." Izd. Akad. Nauk SSSR. (Transl. from Russian, Isr. Program Sci. Transl. No. 1773, Jerusalem, 1969.)

Aleev, Y. G., and Ovcharov, O. P. (1969). On the development of processes of vortex formation and the character of border layer in the movement of fishes. *Zool. Zh.* **48,** 781–790.

Aleev, Y. G., and Ovcharov, O. P. (1971). The role of vortex formation in locomotion of fish and the influence of the boundary between two media upon the flowing pattern. *Zool. Zh.* **50,** 228–234.

Aleev, Y. G., and Ovcharov, O. P. (1973). The three-dimensional pattern of flow round a moving fish. *J. Ichthyol.* (*USSR*) **13,** 933–936.

Alexander, R. M. (1965a). The lift produced by the heterocercal tails of Selachii. *J. Exp. Biol.* **43,** 131–138.

Alexander, R. M. (1965b). Structure and function in the catfish. *J. Zool.* **148,** 88–152.

Alexander, R. M. (1967). "Functional Design in Fishes." Hutchinson, London.

Alexander, R. M. (1968). "Animal Mechanics." Sidgwick & Jackson, London.

Alexander, R. M. (1969). The orientation of muscle fibres in the myomeres of fishes. *J. Mar. Biol. Assoc. U.K.* **49,** 263–290.

Alexander, R. M. (1971). "Size and Shape." Arnold, London.

Alexander, R. M. (1973). Jaw mechanisms of the coelacanth Latimeria. *Copeia* pp. 156–158.

Allan, I. R. H., and Bulleid, M. J. (1963). Long-distance migration of Atlantic salmon. *Nature (London)* **200,** 89.

Amaoka, K. (1969). Studies on the sinistral flounders found in the waters around Japan—taxonomy, anatomy and phylogeny. *J: Shimonoseki Univ. Fish.* **18,** 1–340.

Andreasson, S. (1969). Locomotory activity patterns of *Cottus poecilopus* Heckel and *C. gobio* L. (Pisces). *Oikos* **20,** 78–94.

Arita, G. S. (1971). A re-examination of the functional morphology of the soft-rays in teleosts. *Copeia* pp. 691–697.

Arnold, D. C. (1953). Observations on *Carapus acus* (Brunnich), (Jugulares, Carapidae). *Pubbl. Staz. Zool. Napoli* **24,** 152–166.

Atz, J. W. (1962). Does the shrimpfish swim head up or head down?. *Anim. Kingdom* **65,** 175–179.

Auerbach, A. A. (1967). The synchronous effector system of *Gasteropelecus. Diss. Abstr. B.* **28,** 1367.

Badcock, J. (1970). The vertical distribution of mesopelagic fishes collected on the Sond cruise. *J. Mar. Biol. Assoc. U.K.* **50,** 1001–1044.

Bainbridge, R. (1958). The speed of swimming of fish as related to size and to the frequency and amplitude of the tail beat. *J. Exp. Biol.* **35,** 109–133.

Bainbridge, R. (1961). Problems of fish locomotion. *Symp. Zool. Soc. London* **5,** 13–32.

Bainbridge, R. (1963). Caudal fin and body movement in the propulsion of some fish. *J. Exp. Biol.* **40,** 23–56.

Banks, J. W. (1969). A review of the literature on the upstream migration of adult salmonids. *J. Fish Biol.* **1,** 85–136.

Barets, A. (1961). Contribution a l'étude des systèmes moteurs "lent" et "rapide" du muscle latéral des Teléostéens. *Arch. Anat. Microsc. Morphol. Exp.* **50,** 91–187.

Belyayev, V. V., and Zuyev, G. V. (1969). Hydrodynamic hypothesis of school formation in fishes. *Probl. Ichthyol.* **9,** 578–584.

Bennett, M. V. L. (1971a). Electric organs. *In* "Fish Physiology" (W. S. Hoar and D. J. Randall, eds.), Vol. 5, pp. 347–491. Academic Press, New York.

Bennett, M. V. L. (1971b). Electroreception. *In* "Fish Physiology" (W. S. Hoar and D. J. Randall, eds.), Vol. 5, pp. 493–574. Academic Press, New York.

Bergman, R. A. (1964). Mechanical properties of the dorsal fin musculature of the marine teleost, *Hippocampus hudsonius. Bull. Johns Hopkins Hosp.* **114,** 344–353.

Bergmann, H. H. (1968). Eine deskriptive Verhaltensanalyse des Segelflossers (*Pterophyllum scalare* Cuv. u. Val., Chichlidae, Pisces).*Z. Tierpsychol.* **25,** 559–587,

Bernstein, J. J. (1970). Anatomy and physiology of the central nervous system. *In* "Fish Physiology" (W. S. Hoar and D. J. Randall, eds.), Vol. 4, pp. 1–90. Academic Press, New York.

Bertin, L. (1958a). Squelette axial. *In* "Traité de Zoologie" (P. P. Grassé, ed.), Vol. 13, Part 1, pp. 688–709. Masson, Paris.

Bertin, L. (1958b). Squelette appendiculaire. *In* "Traité de Zoologie" (P. P. Grassé, ed.), Vol. 13, Part 1, pp. 710–747. Masson, Paris.

Bertin, L. (1958c). Modifications des nageoires. *In* "Traité de Zoologie" (P. P. Grassé, ed.), Vol. 13, Part 1, pp. 748–782. Masson, Paris.
Bertin, L. (1958d). Système nerveux. *In* "Traité de Zoologie". (P. P. Grassé, ed.), Vol. 13, Part 1, pp. 854–922. Masson, Paris.
Blüm, V. (1968). Das Kampfverhalten des braunen Diskusfisches, *Symphysodon aequifasciata axelrodi* L. P. Schultz (Teleostei, Cichlidae). *Z. Tierpsychol.* **25,** 395–408.
Boddeke, R., Slijper, E. J., and Van Der Stelt, A. (1959). Histological characteristics of the body—musculature of fishes in connection with their mode of life. *Proc. K. Ned. Akad. Wet., Ser. C* **62,** 576–588.
Bone, Q. (1964). Patterns of muscular innervation in the lower chordates. *Int. Rev. Neurobiol.* **6,** 99–147.
Bone, Q. (1971). On the scabbard fish *Aphanopus carbo. J. Mar. Biol. Assoc. U.K.* **51,** 219–225.
Boulenger, G. A. (1907). "Zoology of Egypt: The Fishes of the Nile." Published for the Egyptian Government, London.
Breder, C. M. (1924). Respiration as a factor in locomotion of fishes. *Am. Nat.* **58,** 145–155.
Breder, C. M. (1926). The locomotion of fishes. *Zoologica (N.Y.)* 4, 159–297.
Breder, C. M. (1959). Studies on social groupings in fishes. *Bull. Am. Mus. Nat. Hist.* **117,** 393–482.
Breder, C. M. (1965). Vortices and fish schools. *Zoologica (N.Y.)* **50,** 97–114.
Breder, C. M., and Edgerton, H. E. (1942). An analysis of the locomotion of the seahorse, *Hippocampus*, by means of high speed cinematography. *Ann. N.Y. Acad. Sci.* **43,** 145–172.
Breder, C. M., and Nigrelli, R. F. (1938). The significance of differential locomotor activity as an index to the mass physiology of fishes. *Zoologica (N.Y.)* **23,** 1–29.
Bregnballe, F. (1961). Plaice and flounder as consumers of the microscopic bottom fauna. *Medd. Dan. Fisk.-Havunders.* **3,** 133–182.
Brett, J. R., and Sutherland, D. B. (1965). Respiratory metabolism of pumpkinseed (*Lepomis gibbosus*) in relation to swimming speed. *J. Fish. Res. Bd. Can.* **22,** 405–409.
Bridge, T. W. (1896). The mesial fins of ganoids and teleosts. *J. Linn. Soc. London, Zool.* **25,** 530–602.
Bridge, T. W. (1904). Fishes (exclusive of the systematic account of Teleostei) *In* "The Cambridge Natural History" (S. F. Harmer and A. E. Shipley, eds.), Vol. 7, pp. 139–537. Macmillan, London.
Campbell, B. (1951). The locomotor behavior of spinal elasmobranchs with an analysis of stinging in *Urobatis. Copeia* pp. 277–284.
Chevrel, R. (1913). Essai sur la morphologie et la physiologie du muscle latéral chez les poissons osseux. *Arch. Zool. Exp. Gen.* **52,** 473–607.
Clemens, H. B. (1961). The migration, age and growth of Pacific albacore (*Thunnus germo*), 1951–1958. *Fish. Bull., Calif.* **115,** 1–128.
Clemens, H. B., and Flittner, G. A. (1969). Bluefin tuna migrate across the Pacific Ocean. *Calif. Fish Game* **55,** 132–135.
Collette, B. B., and Talbot, F. H. (1972). Activity patterns of coral reef fishes with emphasis on nocturnal–diurnal changeover. *Nat. Hist. Mus. Los Angeles Cty. Sci. Bull.* **14,** 98–124.
Cushing, D. H. (1973). "The Detection of Fish." Pergamon, Oxford.
Cushing, D. H., and Jones, F. R. H. (1968). Why do fish school? *Nature (London)* **218,** 918–920.
Dahl, K, and Sømme, S. (1936). Experiments in salmon marking in Norway, 1935. *Skr. Nor. Vidensk.-Akad. Oslo, 1* 1935, No. 12, 1–27.
Danforth, C. H. (1913). The myology of Polyodon. *J. Morphol.* **24,** 107–146.

Darnell, R. M., and Meierotto, R. R. (1965). Diurnal periodicity in the black bullhead, *Ictalurus melas* (Rafinesque). *Trans. Am. Fish. Soc.* **94,** 1–8.

Davis, R. E. (1964). Daily "predawn" peak of locomotion in fish. *Anim. Behav.* **12,** 272–283.

Dean, B. (1895). "Fishes, Living and Fossil." Columbia Univ. Press, New York.

deGroot, S. J. (1970). Some notes on an ambivalent behaviour of the Greenland halibut *Reinhardtius hippoglossoides* (Walb.) Pisces: Pleuronectiformes. *J. Fish Biol.* **2,** 275–279.

deVeen, J. F. (1967). On the phenomenon of soles (*Solea solea* L.) swimming at the surface., *J. Cons., Cons. Perm. Int. Explor. Mer* **31,** 207–236.

Diamond, J. (1971). The Mauthner cell. *In* "Fish Physiology" (W. S. Hoar and D. J. Randall, eds.), Vol. 5, pp. 265–346. Academic Press, New York.

Disler, N. N. (1960). "Lateral Line Sense Organs and Their Importance in Fish Behaviour." Izd. Akad. Nauk. SSSR. (Transl. from Russian, Isr. Program Sci. Transl. No. 5799, Jerusalem, 1971.)

DuBois, A. B., Cavagna, G. A., and Fox, R. S. (1974). Pressure distribution on the body surface of swimming fish. *J. Exp. Biol.* **60,** 581–591.

DuBois-Reymond, R. (1914). Physiologie der Bewegung. *In* "Handbuch der Vergleichenden Physiologie" (H. Winterstein, ed.), Vol. 3, Half 1, Part 1, pp. 1–248. Gustav Fischer, Jena.

Eaton, T. H. (1945). Skeletal supports of the median fins of fishes. *J. Morphol.* **76,** 193–212.

Eggert, B. (1929). Die Gobiidenflosse und ihre Anpassung an das Landleben. *Z. Wiss. Zool.* **133,** 411–439.

Ellis, D. V. (1962). Preliminary studies on the visible migrations of adult salmon. *J. Fish. Res. Board Can.* **19,** 137–148.

Ellis, D. V. (1966). Swimming speeds of sockeye and coho salmon on spawning migration. *J. Fish. Res. Board Can.* **23,** 181–187.

Emelĭanov, S. W. (1935). Die Morphologie der Fischrippen., *Zool. Jahrb., Abt. Anat. Ontog. Tiere* **60,** 133–262.

Emery, A. R. (1969). Comparative ecology of damselfishes (Pisces: Pomacentridae) at Alligator Reef, Florida Keys. *Diss. Abstr. B* **29,** 2962.

Emery, A. R. (1973). Comparative ecology and functional osteology of fourteen species of damselfish (Pisces: Pomacentridae) at Alligator Reef, Florida Keys. *Bull. Mar. Sci.* **23,** 649–770.

Escobar, R. A., Minahan, R. P., and Shaw, R. J. (1936). Motility factors in mass physiology: Locomotor activity of fishes under conditions of isolation, homotypic grouping, and heterotypic grouping. *Physiol. Zool.* **9,** 66–78.

Fierstine, H. L., and Walters, V. (1968). Studies in locomotion and anatomy of scombroid fishes. *Mem. South. Calif. Acad. Sci.* **6,** 1–31.

Fishelson, L. (1968). Structure of the vertebral column in *Lepadichthys lineatus*, a clingfish associated with crinoids. *Copeia* pp. 859–861.

Foss, G. (1968). Behaviour of *Myxine glutinosa* L. in natural habitat. Investigation of the mud biotope by a suction technique. *Sarsia* **31,** 1–13.

Fowler, J. A. (1970). Control of vertebral number in teleosts—an embryological problem. *Q. Rev. Biol.* **45,** 148–167.

François, Y. (1959). La nageoire dorsale anatomie comparée et évolution. *Année Biol.* **35,** 81–113.

Franzisket, L. (1965). Beobachtungen und Messungen am Flug der fliegenden Fische. *Zool. Jahrb., Abt. Allg. Zool. Physiol. Tiere* **70,** 235–240.

Gallepp, G. W., and Magnuson, J. J. (1972). Effects of negative buoyancy on the behavior of the bluegill, *Lepomis macrochirus* Rafinesque. *Trans. Am. Fish. Soc.* **101,** 507–512.

Ganguly, D. N., and Mitra, B. (1962). On the vertebral column of the teleostean fishes of different habits and habitats. I. *Mugil corsula*, *Pama pama*, *Triacanthus brevirostris* and *Andamia heteroptera*. *Anat. Anz.* **110,** 289–311.

Ganguly, D. N., and Nag, A. C. (1964). On the organisation of the myomeric musculature of some benthonic teleostean fishes with special reference to the functional morphology. Anat. Anz. **115,** 418–446.

Gayduck, V. V., Malinin, L. K., and Poddubnyy, A. G. (1971). Determination of the swimming depth of fish during the hours of daylight. *Ichthyologica* **11,** 140–143.

Geerlink, P. J., and Videler, J. J. (1974). Joints and muscles of the dorsal fin of *Tilapia nilotica* L. (Fam. Cichlidae). *Neth. J. Zool.* **24,** 279–290.

Gibson, R. N. (1969). The biology and behaviour of littoral fish. *Oceanogr. Mar. Biol.* **7,** 367–410.

Goodrich, E. S. (1906). Notes on the development, structure and origin of the median and paired fins of fish. *Q. J. Microsc. Sci.* **50,** 333–376.

Goodrich, E. S. (1930). "Studies on the Structure and Development of Vertebrates." Macmillan, London.

Goodyear, C. P. (1970). Terrestrial and aquatic orientation in the starhead topminnow, *Fundulus notti. Science* **168,** 603–605.

Gosline, W. A. (1971). "Functional Morphology and Classification of Teleostean Fishes." Univ. of Hawaii Press, Honolulu.

Gradwell, N. (1971). Observations on jet propulsion in banjo catfishes. *Can. J. Zool.* **49,** 1611–1612.

Graham, J. B. (1970). Preliminary studies on the biology of the amphibious clinid *Mnierpes macrocephalus. Mar. Biol.*, **5,** 136–140.

Graham, J. B. (1973). Terrestrial life of the amphibious fish *Mnierpes macrocephalus. Mar. Biol.* **23,** 83–91.

Graham, J. B. (1975). Heat exchange in the yellowfin tuna, *Thunnus albacares*, and skipjack tuna, *Katsuwonus pelamis*, and the adaptive significance of elevated body temperatures in scombrid fishes. *U.S. Fish Wildl. Serv., Fish. Bull.* **73,** 219–229.

Graham, J. B., and Rosenblatt, R. H. (1970). Aerial vision: Unique adaptation in an intertidal fish. *Science* **168,** 586–588.

Gray, J. (1933a). Studies in animal locomotion. I. The movement of fish with special reference to the eel. J. *Exp. Biol.* **10,** 88–104.

Gray, J. (1933b). Studies in animal locomotion. II. The relationship between waves of musculature contraction and the propulsive mechanism of the eel. *J. Exp. Biol.* **10,** 386–390.

Gray, J. (1933c). Studies in animal locomotion. III. The propulsive mechanism of the whiting (*Gadus merlangus*). *J. Exp. Biol.* **10,** 391–400.

Gray, J. (1936a). Studies in animal locomotion. IV. The neuromuscular mechanism of swimming in the eel. *J. Exp. Biol.* **13,** 170–180.

Gray, J. (1936b). Studies in animal locomotion. VI. The propulsive powers of the dolphin. *J. Exp. Biol.* **13,** 192–199.

Gray, J. (1968). "Animal Locomotion." Weidenfeld & Nicolson, London.

Greene, C. W., and Greene, C. H. (1914). The skeletal musculature of the king salmon. *Bull. U.S. Fish. Bur.* **33,** 21–59.

Greenfield, D. W. (1968). Observations on the behavior of the basketweave cusk-eel *Otophidium scrippsi* Hubbs. *Calif. Fish Game* **54,** 108–114.

Greenway, P. (1965). Body form and behavioural types in fish. *Experientia* **21,** 489–498.

Greenwood, P. H., and Thomson, K. S. (1960). The pectoral anatomy of *Pantodon buchholzi* Peters (a freshwater flying fish) and the related Osteoglossidae. *Proc. Zool. Soc. London* **135,** 283–301.

Greer Walker, M., Mitson, R. B., and Storeton-West, T. (1971). Trials with a transponding acoustic fish tag tracked with an electronic sector scanning sonar. *Nature* (*London*) **229,** 196–198.

Gregory, W. K. (1928). Studies on the body-forms of fishes. *Zoologica* (*N.Y.*) **8,** 325–421.

Grenholm, A. (1923). Studien über die Flossenmuskulatur der Teleostier. Inaug. Diss., Uppsala Univ. *Arsskr*., (*Matem. Naturvet. Uppsala*) **2,** 1–296.

Gunter, G. (1953). Observations on fish turning flips over a line. *Copeia* pp. 188–190.

Guthrie, D. M., and Banks, J. R. (1970a). Observations on the function and physiological properties of a fast paramyosin muscle—the notochord of amphioxus (*Branchiostoma lanceolatum*). *J. Exp. Biol.* **52,** 125–138.

Guthrie, D. M., and Banks, J. R. (1970b). Observations on the electrical and mechanical properties of the myotomes of the lancelet (*Branchiostoma lanceolatum*). *J. Exp. Biol.* **52,** 401–417.

Harder, W. (1964). Anatomie der Fische. *In* "Handbuch Binnenfischerei Mitteleuropas, "(R. Demoll, H. N. Maier and H. H. Wundsch, eds.) Vol. 2a, pp. 1–308, Figs. 1–96. Schweizerbart'sche Verlags., Stuttgart.

Harris, J. E. (1953). Fin patterns and mode of life in fishes. *In* "Essays in Marine Biology" (S. M. Marshall and A. P. Orr, eds.), pp. 17–28. Oliver & Boyd, Edinburgh.

Harris, J. E., and Whiting, H. P. (1954). Structure and function in the locomotory system of the dogfish embryo. The myogenic stage of movement. *J. Exp. Biol.* **31,** 501–524.

Harris, V. A. (1960). On the locomotion of the mud-skipper *Periophthalmus koelreuteri* (Pallas): (Gobiidae). *Proc. Zool. Soc. London* **134,** 107–135.

Hartt, A. C. (1966). Migrations of salmon in the North Pacific Ocean and Bering Sea as determined by seining and tagging, 1959–1960. *Int. North Pac. Fish. Comm. Bull.* **19,** 1–141.

Hasler, A. D., and Villemonte, J. (1953). Observations on the daily movements of fishes. *Science* **118,** 321–322.

Healey, E. G. (1957). The nervous system. *In* "The Physiology of Fishes" (M. E. Brown, ed.), Vol. 2, pp. 1–119. Academic Press, New York.

Herald, E. S. (1961). "Living Fishes of the World." Doubleday, Garden City, New York.

Herald, E. S., and Ripley, W. E. (1951). The relative abundance of sharks and bat stingrays in San Francisco Bay. *Calif. Fish Game* **37,** 315–329.

Hergenrader, G. L., and Hasler, A. D. (1967). Seasonal changes in swimming rates of yellow perch in Lake Mendota as measured by sonar. *Trans. Am. Fish. Soc.* **96,** 373–382.

Hertel, H. (1966). "Structure, Form and Movement." Reinhold, New York.

Hobson, E. S. (1965). Diurnal–nocturnal activity of some inshore fishes in the Gulf of California. *Copeia* pp. 291–302.

Hobson, E. S. (1968). Predatory behavior of some shore fishes in the Gulf of California. *Bur. Sport Fish. Wildl.* (*U.S.*)., *Res. Rep.* **73,** 1–92.

Hobson, E. S. (1973). Diel feeding migrations in tropical reef fishes. *Helgol. Wiss. Meeresunters.* **24,** 361–370.

Hobson, E. S. (1974). Feeding relationships of teleostean fishes on coral reefs in Kona, Hawaii. *U.S. Fish Wildl, Serv., Fish. Bull.* **72,** 915–1031.

Holden, M. J. (1967). Transatlantic movement of a tagged spurdogfish. *Nature* (*London*) **214,** 1140–1141.

Holland, G. A. (1957). Migration and growth of the dogfish shark, Squalus acanthias (Linnaeus), of the eastern North Pacific. *Wash. Dep. Fish., Fish. Res. Pap.* **2**(1), 43–59.

Hoogland, R. D. (1951). On the fixing-mechanism in the spines of *Gasterosteus aculeatus* L. *Proc. K. Ned. Akad. Wet. C* **54,** 171–180.

Hora, S. L. (1935). Ancient Hindu conception of correlation between form and locomotion of fishes. *J. Asiat. Soc. Bengal, Sci.* **1,** 1–7.

Horn, M. (1975). Swim-bladder state and structure in relation to behavior and mode of life in Stromateoid fishes. *U.S. Fish Wildl. Serv., Fish. Bull.* **73,** 95–109.

Houssay, F. (1912). "Forme, Puissance et Stabilité des Poissons," Collection de Morphologie Dynamique, Vol. 4. Herman, Paris.

Howell, A. B. (1933). The architecture of the pectoral appendage of the codfish. *Anat. Rec.* **56,** 151–158.

Hubbs, C. L. (1933). Observations on the flight of fishes, with a statistical study of the flight of the Cypselurinae and remarks on the evolution of the flight of fishes. *Pap. Mich. Acad. Sci., Arts Letts.* **17,** 575–611.

Hudson, R. C. L. (1973). On the function of the white muscles in teleosts at intermediate swimming speeds. *J. Exp. Biol.* **58,** 509–522.

Hübner, H. (1961). Die Wirbelsäule des Karpfens (*Cyprinus carpio* L.). *Z. Fisch. Deren Hilfswiss.* **10,** 429–505.

Hughes, G. M., and Shelton, G. (1962). Respiratory mechanisms and their nervous control in fish. *Adv. Comp. Physiol. Biochem.* **1,** 275–364.

Hunter, J. R., and Zweifel, J. R. (1971). Swimming speed, tail beat frequency, tail beat amplitude, and size in jack mackerel, *Trachurus symmetricus*, and other fishes. *U.S. Fish Wildl. Serv., Fish. Bull.* **69,** 253–266.

Idler, D. R., and Clemens, W. A. (1959). The energy expenditures of Fraser River sockeye salmon during the spawning migration to Chilko and Stuart Lakes. *Int. Pac. Salmon Fish. Comm. Prog. Rep.* pp. 1–80.

Jarman, G. M. (1961). A note on the shape of fish myotomes. *Symp. Zool. Soc. London* **5,** 33–35.

Johnels, A. G. (1957). The mode of terrestrial locomotion in *Clarias*. *Oikos* **8,** 122–129.

Johnels, A. G., and Svensson, G. S. O. (1954). On the biology of *Protopterus annectens* (Owen). *Ark. Zool.* **7,** 131–164.

Johnson, R. D. O. (1912). Notes on the habits of a climbing catfish (*Arges marmoratus*) from the republic of Colombia. *Ann. N.Y. Acad. Sci.* **22,** 327–333.

Johnson, W. E., and Groot, C. (1963). Observations on the migration of young sockeye salmon (*Oncorhynchus nerka*) through a large, complex lake system. *J. Fish. Res. Board. Can.* **20,** 919–938.

Johnston, I. A., and Goldspink, G. (1973a). A study of the swimming performance of the Crucian carp *Carassius carassius* (L.) in relation to the effects of exercise and recovery on biochemical changes in the myotomal muscles and liver. *J. Fish Biol.* **5,** 249–260.

Johnston, I. A., and Goldspink, G. (1973b). Quantitative studies on muscle glycogen utilization during sustained swimming in Crucian carp (*Carassius carassius* L.). *J. Exp. Biol.* **59,** 607–615.

Jones, F. R. H. (1952). The swimbladder and the vertical movements of teleostean fishes. II. The restriction to rapid and slow movements. *J. Exp. Biol.* **29,** 94–109.

Jones, F. R. H. (1956). The behaviour of minnows in relation to light intensity. *J. Exp. Biol.* **33,** 271–281.

Jones, F. R. H. (1968). "Fish Migration." Arnold, London.

Kafuku, T. (1950). "Red muscles" in fishes. I. Comparative anatomy of the scombroid fishes of Japan. *Gyoruigaku Zasshi* **1,** 89–100.

Kashin, S. M., and Smolyaninov, V. V. (1969). Concerning the geometry of fish trunk muscles. *J. Ichthyol. (USSR)* **9,** 923–925.

Keast, A., and Webb, D. (1966). Mouth and body form relative to feeding ecology in the fish fauna of a small lake, Lake Opinicon, Ontario. *J. Fish. Res. Board Can.* **23,** 1845–1874.

Keenleyside, M. H. A. (1955). Some aspects of the schooling behaviour of fish. Behaviour **8,** 183–248.

Kerswill, C. J., and Keenleyside, M. H. A. (1961). Canadian salmon caught off *Greenland. Nature (London)* **192,** 279.

Kishinouye, K. (1923). Contributions to the comparative study of the so-called scombroid fishes. *J. Coll.* Agric. *Tokyo Imp. Univ.* **8,** 293–475.

Klausewitz, W. (1963). Wie schwimmt der Schnepfenfisch? *Natur Mus.* **93,** 69–73.

Klausewitz, W. (1964). Der Lokomotionsmodus der Flügelrochen (Myliobatoidei). *Zool. Anz.* **173,** 110–120.

Klausewitz, W. (1967). Über einige Bewegungsweisen der Schlammspringer (*Periophthalmus*). *Natur Mus.* **97,** 211–222.

Kleerekoper, H., Timms, A. M., Westlake, G. F., Davy, G. F., Malar, T., and Anderson, V. M. (1970). An analysis of locomotor behaviour of goldfish (*Carassius auratus*). *Anim. Behav.* **18,** 317–330.

Kramer, E. (1960). Zur Form und Funktion des Lokomotionsapparates der Fische. Z. *Wiss. Zool. Abt. A* **163,** 1–36.

Kruuk, H. (1963). Diurnal periodicity in the activity of the common sole, *Solea vulgaris* Quensel. *Neth. J. Sea Res.* **2,** 1–28.

leDanois, Y. (1958). Système musculaire. *In* "Traité de Zoologie" (P. P. Grassé, ed.), Vol. 13, Part 1, pp. 783–817. Masson, Paris.

Lieder, U. (1961). Untersuchungsergebnisse über die Grätenzahlen bei 17 Süsswasser-Fischarten. *Z. Fisch. Deren Hilfswiss.* **10,** 329–350.

Lighthill, M. J. (1960). Note on the swimming of slender fish. *J. Fluid Mech.* **9,** 305–317.

Lighthill, M. J. (1969). Hydromechanics of aquatic animal propulsion. *Annu. Rev. Fluid Mech.* **1,** 413–446.

Lindsey, C. C. (1955). Evolution of meristic relations in the dorsal and anal fins of teleost fishes. *Trans. R. Soc. Can.* **49** (Ser. 3, Sect. 5), 35–49.

Lindsey, C. C. (1966). Body sizes of poikilotherm vertebrates at different latitudes. *Evolution* **20,** 456–465.

Lindsey, C. C. (1975). Pleomerism, the widespread tendency among related fish species for vertebral number to be correlated with maximum body length. *J. Fish. Res. Board Can.* **32,** 2453–2469.

Lindsey, C. C., and Harrington, R. W. (1972). Extreme vertebral variation induced by temperature in a homozygous clone of the self-fertilizing cyprinodontid fish *Rivulus marmoratus. Can. J. Zool.* **50,** 733–744.

Ling, S. C., and Ling, T. Y. J. (1974). Anomalous drag-reducing phenomenon at a water/fish-mucus or polymer interface. *J. Fluid Mech.* **65,** 499–512.

Lissmann, H. W. (1946a). The neurological basis of the locomotory rhythm in the spinal dogfish (*Scyllium canicula, Acanthias vulgaris*). I. Reflex behaviour. *J. Exp. Biol.* **23,** 143–161.

Lissmann, H. W. (1946b). The neurological basis of the locomotory rhythm in the spinal dogfish (*Scyllium canicula, Acanthias vulgaris*). II. The effect of de-afferentiation. *J. Exp. Biol.* **23,** 162–176.

Lissmann, H. W. (1961a). Zoology, locomotory adaptations and the problem of electric fish. In "The Cell and the Organism" (J. A. Ramsay and V. B. Wigglesworth, eds.), pp. 301–317. Cambridge Univ. Press, London and New York.

Lissmann, H. W. (1961b). Ecological studies on gymnotids. *In* "Bioelectrogenesis" (C. Chagas and A. P. deCarvalho, eds.), pp. 215–226. Elsevier, Amsterdam.

Lund, R. (1967). An analysis of the propulsive mechanisms of fishes with reference to some fossil actinopterygians. *Ann. Carnegie Mus.* **39,** 195–218.

McCutchen, C. W. (1970). The trout tail fin: A selfcambering hydrofoil. *J. Biomech.* **3,** 271–281.

McMurrich, J. P. (1884). The myology of *Amiurus catus* (L.) Gill. *Proc. R. Can. Inst.* **2,** 311–351.

Macnae, W. (1968). A general account of the fauna and flora of mangrove swamps and forests in the Indo-West-Pacific region. *Adv. Mar. Biol.* **6,** 73–270.

Magnan, A. (1929). Les caractéristiques géométriques et physiques des poissons. Première partie. *Ann. Sci. Nat. Zool.* **12,** 1–133.

Magnan, A. (1930). Les caractéristiques géométriques et physiques des poissons. Deuxième partie. *Ann. Sci. Nat. Zool.* **13,** 134–269.

Mahajan, C. L. (1967). *Sisor rabdophorus*—a study in adaptation and natural relationship. III. The vertebral column, median fins and their musculature. *J. Zool.* **152,** 297–318.

Manzer, J. I. (1946). Interesting movements as shown by the recoveries of certain species of tagged fish. *Progr. Rep. Pac. Biol. Sta.* **67,** 31.

Marey, E. J. (1895). "Movement." Heinemann, London.

Marshall, N. B. (1971). "Explorations in the Life of Fishes." Harvard Univ. Press, Cambridge, Massachusetts.

Mather, F. J. (1962). Transatlantic migration of two large bluefin tuna. *J. Cons. Cons. Perm. Int. Explor. Mer* **27,** 325–327.

Mather, F. J., Mason, J. M., and Clark, H. L. (1974a). Migrations of white marlin and blue marlin in the western North Atlantic Ocean—tagging results since May, 1970. *NOAA (Natl. Oceanic Atmos. Adm.) Tech. Rep. NMFS (Natl. Mar. Fish. Serv.) SSRF (Spec. Sci. Rep.—Fish.)* No. **675,** pp. 211–225.

Mather, F. J., Tabb, D. C., Mason, J. M., and Clark, H. L. (1974b). Results of sailfish tagging in the western North Atlantic Ocean. *NOAA (Natl. Oceanic Atmos. Adm.) Tech. Rep. NMFS (Natl. Mar. Fish. Serv.) SSRF (Spec. Sci. Rep.—Fish.)* No. **675,** pp. 194–210.

Matthews, L. H., and Parker, H. W. (1951). Basking sharks leaping. *Proc. Zool. Soc. London* **121,** 461–462.

Maurer, F. (1913). Die ventrale Rumpfmuskulatur der Fische (Selachier, Ganoiden, Teleostier, Crossopterygier, Dipnoer). *Jen. Z. Naturwiss.* **49,** 1–118.

Menzies, W. J. M., and Shearer, W. M. (1957). Long distance migration of salmon. *Nature (London)* **179,** 790.

Meyer, A. (1965). Results of cod tagging by the Federal Republic of Germany in the Greenland area from 1959 to 1964. *Int. Comm. Northwest Atl. Fish. Redbook Part III* pp. 148–152.

Minckley, C. O., and Klaassen, H. E. (1969). Burying behavior of the plains killifish, *Fundulus kansae. Copeia* pp. 200–201.

Myers, G. S. (1950). Flying of the halfbeak, *Euleptorhamphus. Copeia.* 320.

Nawar, G. (1955). On the anatomy of *Clarias lazera.* II. The muscles of the head and the pectoral girdle. *J. Morphol.* **97,** 23–58.

Neave, F. (1964). Ocean migrations of Pacific salmon. *J. Fish. Res. Board Can.* **21,** 1227–1244.

Nikolsky, G. V. (1963). "The Ecology of Fishes." Academic Press, New York.

Nishi, S. (1938). Muskelsystem II. Muskeln des Rumpfes. *Bolks Handb. Vgl. Anat. Wirbeltiere* **5,** 351–446.

Nishimura, S. (1964). Additional information on the biology of the dealfish, *Trachipterus ishikawai* Jordan and Snyder. *Bull. Jpn. Sea Reg. Fish. Res. Lab. (Nihonkai-Ku Suisan Kenkyusho Kenkyo)* **13,** 127–129.

Nishimura, S., and Hirosaki, Y. (1964). Observations on the swimming behavior of some taeniosomous fishes in aquaria and in nature. *Publ. Seto Mar. Biol. Lab.* **12,** 165–171.

Norman, J. R., and Greenwood, P. H. (1975). "A History of Fishes," 3rd Ed. Benn, London.

Norris, H. W., and Hughes, S. P. (1920). The cranial, occipital, and anterior spinal nerves of the dogfish, *Squalus acanthias. J. Comp. Neurol.* **31,** 293–404.

Northcote, T. G. (1962). Migratory behaviour of juvenile rainbow trout, *Salmo gairdneri*, in outlet and inlet streams of Loon Lake, British Columbia. *J. Fish. Res. Board Can.* **19,** 201–270.

Northcote, T. G., Lorz, H. W., and MacLeod, J. C. (1964). Studies on diel vertical movement of fishes in a British Columbia lake. *Verh. Int. Verein Limnol.* **15,** 940–946.

Nursall, J. R. (1956). The lateral musculature and the swimming of fish. *Proc. Zool. Soc. London* **126,** 127–143.

Nursall, J. R. (1963). The caudal musculature of *Hoplopagrus guntheri* Gill (Perciformes: Lutjanidae). *Can. J. Zool.* **41,** 865–880.

Oehmichen, E. (1958). Locomotion des poissons. *In* "Traité de Zoologie" (P. P. Grassé, ed.), Vol: 13, Part 1, pp. 818–853. Masson, Paris.

Ohlmer, W., and Schwartzkopff, J. (1959). Schwimmgeschwindigkeiten'von Fischen aus stehenden Binnengewässern. *Naturwissenschaften* **46,** 362–363.

Olla, B. L., and Studholme, A. L. (1972). Daily and seasonal rhythms of activity in the bluefish (*Pomatomus saltatrix*). In "Behavior of Marine Animals" (H. E. Winn and B. L. Olla, eds.), Vol. 2, pp. 303–326. Plenum, New York.

Orange, C. J., and Fink, B. D. (1963). Migration of a tagged bluefin tuna across the Pacific Ocean. *Calif. Fish Game* **49,** 307–309.

Osborne, M. F. M. (1961). The hydrodynamical performance of migratory salmon. *J. Exp. Biol.* **38,** 365–390.

Packard, A. (1972). Cephalopods and fish: the limits of convergence. *Biol. Rev. Cambridge Philos. Soc.* **47,** 241–307.

Patterson, S., and Goldspink, G. (1972). The fine structure of red and white myotomal muscle fibres of the coalfish (*Gadus virens*). *Z. Zellforsch. Mikrosk. Anat.* **133,** 463–474.

Pavlov, D. S., Sbikin, Y. N., and Mochek, A. D. (1968). The effect of illumination in running water on the speed of fishes in relation to feature of their orientation. *Probl. Ichthyol.* **8,** 250–255.

Pettigrew, J. B. (1874). "Animal Locomotion." Int. Sci. Ser., London and New York.

Pitcher, T. J. (1973). The three-dimensional structure of schools in the minnow, *Phoxinus phoxinus* (L.). *Anim. Behav.* **21,** 673–686.

Radakov, D. V. (1973). "Schooling in the Ecology of Fish." (Transl. from Russian, Isr. Program Sci. Transl. No. 22076, Jerusalem, 1973.)

Riss, W. (1972). Overview of the design of the central nervous system and the problem of the natural units of behavior. IV. Locomotion and spinal organization. *Brain, Behav., Evol.* **4,** 439–462.

Roberts, B. L. (1969a). The spinal nerves of the dogfish (*Scyliorhinus*). *J. Mar. Biol. Assoc. U.K.* **49,** 51–75.

Roberts, B. L. (1969b). The co-ordination of the rhythmical fin movements of dogfish. *J. Mar. Biol. Assoc. U.K.* **49,** 357–378.

Roberts, B. L. (1969c). The buoyancy and locomotory movements of electric rays. *J. Mar. Biol. Assoc. U.K.* **49,** 621–640.

Roberts, B. L. (1969d). The response of a proprioceptor to the undulatory movements of dogfish. *J. Exp. Biol.* **51,** 775–785.

Roberts, B. L. (1972). Activity of lateral-line sense organs in swimming dogfish. *J. Exp. Biol.* **56,** 105–118.

Roberts, B. L., and Russell, I. J. (1972). The activity of lateral-line efferent neurones in stationary and swimming dogfish. *J. Exp. Biol.* **57,** 435–448.

Rosenblatt, R. H., and Johnson, G. D. (1976). Anatomical considerations of pectoral swimming in the opah, *Lampris guttatus. Copeia pp.* 367–370.

Rosenthal, H. (1968). Schwimmverhalten und Schwimmgeschwindigkeit bei den Larven des Herings *Clupea harengus. Helgol. Wiss. Meeresunters.* **18,** 453–486.

Samuel, C. T. (1961). On the ambulatory mechanism in *Choridactylus multibarbis* Richardson. *Bull. Cent. Res. Inst. Univ. Kerala, Trivandrum* **8,** 79–83.

Sandon, H. (1956). An abnormal specimen of *Synodontis membranaceus* (Teleostei, Siluroidea), with a discussion on the evolutionary history of the adipose fin in fish. *Proc. Zool. Soc. London* **127,** 453–459.

Satchell, G. H. (1968). A neurological basis for the co-ordination of swimming with respiration in fish. *Comp. Biochem. Physiol.* **27,** 835–841.

Saxena, S. C., and Chandy, M. (1966a). The pelvic girdle and fin in certain Indian hill stream fishes. *J. Zool.* **148,** 167–190.

Saxena, S. C., and Chandy, M. (1966b). Adhesive apparatus in certain Indian hill stream fishes. *J. Zool.* **148,** 315–340.

Schmalhausen, J. J. (1912). Zur Morphologie der unpaaren Flossen. I. Die Entwicklung des Skelettes und der Muskulatur der unpaaren Flossen der Fische. *Z. Wiss. Zool.* **100,** 509–587.

Schmalhausen, J. J. (1913). Zur Morphologie der unpaaren Flossen. II. Bau und Phylogenese der unpaaren Flossen und insbesonders der Schwanzfosse der Fische. *Z. Wiss. Zool.* **104,** 1–80.

Schmidt-Nielsen, K. (1972a). "How Animals Work." Cambridge Univ. Press, London and New York.

Schmidt-Nielsen, K. (1972b). Locomotion: Energy cost of swimming, flying, and running. *Science* **177,** 222–228.

Schneider, H. (1964). Untersuchungen zur Schwimmweise der Kugelfische. I. Die Flossenmuskulatur der Flusskugelfisches (*Tetraodon fluviatilis*) in vergleich zu der der Schleie (*Tinca tinca*). *Z. Morphol. Oekol. Tiere* **54,** 414–435.

Schuett, F. (1934). Studies in mass physiology: The activity of goldfishes under different conditions of aggregation. *Ecology* **15,** 258–262.

Sewertzoff, A. N. (1926). Die Morphologie der Brustflossen der Fische. *Jena. Z. Naturwiss.* **62,** 343–392.

Sewertzoff, A. N. (1934). Evolution der Bauchflossen der Fische. *Zool. Jahrb., Abt. Anat. Ontog. Tiere* **58,** 415–500.

Shann, E. W. (1914). On the nature of the lateral muscle in Teleostei. Proc. *Zool. Soc. London* pp. 319–337.

Shann, E. W. (1924). Further observations on the myology of the pectoral region in fishes. *Proc. Zool. Soc. London* pp. 195–215.

Shaw, E. (1962). The schooling of fishes. Sci. *Am.* **206,** 128–138.

Sheldon, F. F. (1937). Osteology, myology and probable evolution of the nematognath pelvic girdle. *Ann. N.Y. Acad. Sci.* **37,** 1–96.

Shelton, G. (1970). The regulation of breathing. *In* "Fish Physiology" (W. S. Hoar and D. J. Randall, eds.), Vol. 4, pp. 293–359. Academic Press, New York.

Shepard, M. P., Hartt, A. C., and Yonemori, T. (1968). Salmon of the North Pacific Ocean—Part 8. Chum salmon in offshore waters. *Int. North Pac. Fish. Comm. Bull.* **25,** 1–69.

Shubina, T. N. (1971). Spawning and postspawning migrations of the sevryuga (*Acipenser stellatus* (Pallas)) in the lower Volga—routes and speeds. *J. Ichthyol.* (*USSR*) **11,** 88–97.

Simons, J. R. (1970). The direction of the thrust produced by the heterocercal tails of two dissimilar elasmobranchs: the Port Jackson shark, *Heterodontus portusjacksoni* (Meyer), and the piked dogfish, *Squalus megalops* (Macleay). *J. Exp. Biol.* **52,** 95–107.

Smith, E. H., and Stone, D. E. (1961). Perfect fluid forces in fish propulsion. *Proc. R. Soc., Ser. A* **261,** 316–328.

Spencer, W. P. (1939). Diurnal activity rhythms in fresh-water fishes. *Ohio J. Sci.* **39,** 119–132.

Squire, J. L. (1974). Migration patterns of Istiophoridae in the Pacific Ocean as determined by cooperative tagging programs. *NOAA* (*Natl. Oceanic Atmos. Adm.*) *Tech. Rep. NMFS (Natl. Mar. Fish. Serv.) SSRF (Spec. Sci. Rep.—Fish.)* No. 675, pp. 226–237.

Stickney, A. P. (1972). The locomotor activity of juvenile herring (*Clupea harengus harengus* L.) in response to changes in illumination. *Ecology* **53,** 438–445.

Swain, A. (1963). Long distance migration of salmon. *Nature* (*London*) **197,** 923.

Swanson, P. L. (1949). "Hurdling" by the needlefish. *Copeia* p. 219.

Szarski, H. (1964). The functions of myomere folding in aquatic vertebrates. *Bull. Acad. Pol. Sci., Ser. Sci. Biol.* **12,** 305–306.

Takahashi, N. (1917). On the homology of the median longitudinal muscles—supracarinalis and infracarinalis—with the fin muscles of the dorsal and anal fins, and their functions. *J. Coll. Agric., Tokyo Imp. Coll.* **6,** 199–213.

Tåning, Å. V. (1937). Some features in the migration of cod. *J. Cons. Cons. Perm. Int. Explor. Mer* **12,** 3–35.

Taylor, G. (1952). Analysis of the swimming of long and narrow animals. *Proc. R. Soc., Ser. A* **214,** 158–183.

Tesch, F.-W. (1972). Versuche zur telemetrischen Verfolgung der Laichwanderung von Aalen (*Anguilla anguilla*) in der Nordsee. *Helgol. Wiss. Meeresunters.* **23,** 165–183.

Tester, A. L. (1959). Summary of experiments on the response of tuna to stimuli. In "Modern Fishing Gear of the World" (H. Kristjonsson, ed.), pp. 538–542. Fishing News (Books), London.

Thomas, A. M. (1956). The cranial muscles of teleosts-1. *J. Anim. Morphol. Physiol.* **3,** 13–24.

Thomson, K. S. (1971). The adaptation and evolution of early fishes. *Q. Rev. Biol.* **46,** 139–166.

Tilak, R., and Kanji, S. K. (1967). Studies on the morphology of the pectoral girdle of *Pisoodonophis boro* (Ham.) in relation to its habit. *Anat. Anz.* **120,** 404–408.

Trueman, E. R., and Ansell, A. D. (1969). The mechanisms of burrowing into soft substrata by marine animals. *Oceanogr. Mar. Biol.* **7,** 315–366.

Tucker, V. A. (1975). The energetic cost of moving about. *Am. Sci.* **63,** 413–419.

Tyler, J. C. (1962). The pelvis and pelvic fin of plectognath fishes; a study in reduction. *Proc. Acad. Nat. Sci. Philadelphia* **114,** 207–250.

van Olst, J. C., and Hunter, J. R. (1970). Some aspects of the organization of fish schools. *J. Fish. Res. Board Can.* **27,** 1225–1238.

Verheijen, F. J., and deGroot, S. J. (1966). Diurnal activity of plaice and flounder (Pleuronectidae) in aquaria. *Neth. J. Sea Res.* **3,** 383–390.

Walters, V. (1962). Body form and swimming performance in the scombroid fishes. *Am. Zool.* **2,** 143–149.

Webb, P. W. (1973a). Effects of partial caudal-fin amputation on the kinematics and metabolic rate of underyearling sockeye salmon (*Oncorhynchus nerka*) at steady swimming speeds. *J. Exp. Biol.* **59,** 565–581.

Webb, P. W. (1973b). Kinematics of pectoral fin propulsion in *Cymatogaster aggregata. J. Exp. Biol.* **59,** 697–710.

Webb, P. W. (1975). Hydrodynamics and energetics of fish propulsion. *Bull., Fish. Res. Board Can.* **190,** 1–159.

Weihs, D. (1973a). Mechanically efficient swimming techniques for fish with negative buoyancy. *J. Mar. Res.* **31,** 194–209.

Weihs, D. (1973b). Hydromechanics of fish schooling. *Nature (London)* **241,** 290–291.

Weitzman, S. H. (1954). The osteology and the relationships of the South American characid fishes of the subfamily Gasteropelecinae. *Stanford Ichthyol. Bull.* **4,** 211–263.

Wickler, W. (1971). Verhaltensstudien an einem hochspezialisierten Grundfisch, *Gastromyzon borneensis* (Cyprinoidea, Gastromyzonidae). *Z. Tierpsychol.* **29,** 467–480.

Willemse, J. J. (1959). The way in which flexures of the body are caused by muscular contractions. *Proc. K. Ned. Akad. Wet., Ser. C* **62,** 589–593.

Winterbottom, R. (1974). A descriptive synonomy of the striated muscles of the Teleostei. *Proc. Acad. Nat. Sci. Philadelphia* **125,** 225–317.

Woodhead, P. M. J. (1966). The behaviour of fish in relation to light in the sea. *Oceanogr. Mar. Biol.* **4,** 337–403.

Woodhead, P. M. J., and Woodhead, A. D. (1955). Reactions of herring larvae to light: A mechanism of vertical migration. *Nature (London)* **176,** 349–350.

Yuen, H. S. H. (1970). Behavior of skipjack tuna, *Katsuwonus pelamis*, as determined by tracking with ultrasonic devices. *J. Fish. Res. Board Can.* **27,** 2071–2079.

Yuen, H. S. H., Dizon, A. E., and Uchiyama, J. H. (1974). Notes on the tracking of the Pacific blue marlin, *Makaira nigricans*. NOAA (*Natl. Oceanic Atmos. Adm.) Tech. Rep. NMFS* (*Natl. Mar. Fish. Serv.*) *SSRF* (*Spec. Sci. Rep. —Fish.*) No. **675,** pp. 265–268.

Zuyev, G. V., and Belyayev, V. V. (1970). An experimental study of the swimming of fish in groups as exemplified by the horsemackerel (*Trachurus mediterraneus ponticus* Aleev). *J. Ichthyol. (USSR)* **10,** 545–549.

Chapter 7

Fish locomotor muscle: Beginnings of mechanistic research on how muscle powers swimming

Robert E. Shadwick*
Department of Zoology, University of British Columbia, Vancouver, BC, United States
**Corresponding author: e-mail: shadwick@zoology.ubc.ca*

Chapter Outline

Robert Shadwick discusses the impact of Quentin Bone's chapter "Locomotor Muscle" in Fish Physiology, Volume VII, published in 1978.

Quentin Bone's review chapter in Fish Physiology Volume VII was a remarkably comprehensive summary of what was understood about the structure and function of fish locomotor muscle up to 1978. By today's standard, the knowledge of fish muscle structure was good, but that of its function was quite rudimentary. Despite these limitations Bone's review provided an important account of the structure and innervation of locomotor muscle fibers, and insights into how these are used in swimming. He integrated across biochemistry, metabolism, structure, development, and connections with the nervous system. He also described simple experiments using electromyography that showed that so-called "red" and "white" fiber types had different contractile properties that matched their roles in slow and burst swimming, respectively. Furthermore, Bone pointed out the paucity of species in which locomotor muscle had been studied and emphasized the need for a more comparative approach to increase the diversity of species studied. His goal was to guide readers toward important areas of future study that could ultimately result in understanding the complexity of fishes as swimming machines.

Fish Physiology, Vol. 40B. https://doi.org/10.1016/bs.fp.2024.08.001

1 Introduction

For researchers in the field of fish physiology, the structure and function of muscles used in swimming has been a major focus since the 1950s. William Hoar and David Randall introduced the 1978 volume on Locomotion by stating that recent research on a small number of species had "greatly increased our understanding of how fish move." In terms of the status of fish locomotor muscle research they pointed out that "… we are still at the stage of describing the types of nerve and muscle fibers present, how they are arranged and function to initiate and generate movements." This is the essence of the Locomotor Muscle chapter in that volume written by Quentin Bone (Fig. 1). Although comprehensive, in retrospect the chapter reveals how remarkably little was understood about contractile mechanics or muscle performance in swimming fish in 1978, compared to today. Bone begins modestly by remarking that the knowledge of fish muscle was relatively scanty, and that the chapter was more a progress report than a definitive account of the topic. In spite of the dearth of information about fish muscle at that time, he did provide remarkable insights and ideas, and added a significant comparative perspective to the burgeoning field. Thus, Bone's chapter has been highly influential, as indicated by its high and continued level of citation in modern research papers. It was written not just to summarize this area of research, but to inform and inspire future biologists to understand the whole body complexity of a swimming fish. In this sense Bones review chapter has been an unqualified success.

FIG. 1 Dr. Quentin Bone, FRS, in his laboratory at the Marine Biological Association (MBA) in Plymouth, U.K. Reproduced by permission from the MBA Archives, reference number UM 211.18.

2 Fish muscle research up to 1978

Bone's review chapter, with over 500 citations, is an engaging and well-illustrated, comprehensive, 64-page summary of the structure and function of fish skeletal muscle, as was known in 1978. The primary focus is on the microscopic anatomy of muscle cells, their innervation patterns, and their electrical activity and metabolism during locomotor activity, topics that Bone had established his expertise in previously (Bone, 1964, 1966, 1972; Bone and Chubb, 1975, 1976, 1978; Bone et al., 1978). Bone begins with a comparative description of the anatomical organization of the segmented lateral muscle blocks and the collagenous myoseptal sheets that separate them and connect to the skeleton. It was obvious that waves of muscle contraction traveling toward the tail created lateral undulations that power swimming. Superficial red muscle fibers are predominantly axially aligned, but white muscle fibers are organized into folded, nested myotomes with anterior and posterior projecting cones containing fibers whose orientation deviates from the body long axis. This makes the force trajectory from muscle to vertebral column appear to be complex and variable depending on body shape. The physical separation of these fiber types in fishes is unlike that in other vertebrates, as is their complex geometry.

In the 1970s our understanding of muscle kinematics and contractile dynamics powering teleost and elasmobranch swimming was limited. Only a few studies had used electromyography (EMG), an in vivo technique that could discern when muscle groups were active at different swim speeds, and these foreshadowed the major advances to come in the following decades. Interpretations of the complex muscle anatomy and force trajectories of fish locomotor muscle had been made by Alexander (1969) and others, using microscopy to map out what were revealed as helical trajectories in the white muscle that, it was argued, provided equal degrees of shortening and shortening velocity, regardless of distance from the backbone (the body bending axis). This work predicted that all white fibers will operate at the same point on their force-velocity curve and produce equal power, but there was no way to measure this. In most fishes, red fibers lie in thin subcutaneous bands at the midline between hypaxial and epaxial white fibers, separated by the horizontal septum. Fish physiologists interested in swimming performance had begun to use swim tunnels to measure oxygen consumption and determine what fuelled and put limits on sustained swimming in commercially important species (e.g., Bainbridge, 1960; Gordon, 1968; Greer-Walker and Pull, 1973; Hochachka, 1961; Hudson, 1973; Hunter, 1971; Johnston et al., 1977; Rayner and Keenan, 1967; Webb, 1971), but more sophisticated technology to probe muscle function was yet to come.

Bone's chapter provides a detailed discussion of the current knowledge of muscle fiber anatomy, primarily cell size, abundance, and innervation patterns. It contains excellent quality micrographs that demonstrate neural variations, and some that show histochemical differentiation of cell types. It was well recognized that the high range of swim speeds requires a separation of fast and slow fiber types. Fish are unusual among vertebrates in that these two types of muscle cells are readily distinguished by their color. In general, red fibers are smaller, have more mitochondria and vascularization, and have higher oxidative enzyme activity than white fibers. However, this bimodal segregation is simplistic, as other fibers intermediate in properties have been demonstrated in many species. Summary tables list the anatomical and biochemical distinctions of fast and slow fibers, with the caveat that there are areas of mixed fiber types and also some fibers that appear intermediate in properties. A considerable portion of the Locomotor Muscle chapter deals with the variety of intermediate-type fibers in different species, and there is a very good account of muscle fiber ontogeny with the general impression that slow fibers develop initially while fast fibers arise later. Questions on how this process changes with growth and differences between teleosts and sharks are raised as areas for future embryologists to address. Fig. 12 in Chapter "Locomoter muscle" by Quentin Bone is an excellent summary of the distribution of fiber types in a range of fishes.

In the lateral aspect of the body, the simplest V-shape myotome found in Amphioxus and larval fishes gives rise to a folded sideways W-shape in adults, reflecting the presence of anterior and posterior pointing nested cones. The benefits of this complex arrangement of muscle cells are considered in detail, but without the necessary understanding of how the tendinous myoseptal system connects contractile fibers to the skeleton that would come much later (e.g., Gemballa et al., 2003; Vogel and Gemballa, 2000). But Bone does comment on the difficulty of attaching the myoseptal connective tissue to a recording apparatus for in vitro experiments, and that fin muscles with discrete tendons are easier to use for mechanical studies.

The anatomical separation of red and white fibers is convenient for conducting in vivo experiments on each separately. For example, EMG electrodes can be placed in red or white fibers to reveal at which swim speeds each type is activated. In more recent in vitro studies of contraction dynamics this separation has also made it possible to use fiber bundles rather than single fibers. The studies of Boddeke et al. (1959) and of Bone (1966) revealed that the red fibers in fishes are primarily active in sustained swimming at low or moderate speed, while the white fibers are active in "sprinting" movements. Bone (1966) had also shown that these two muscle fiber types were innervated independently. To put the technological limitations of that time into perspective, in this 1966 experiment, muscle EMGs were recorded from sharks that were not actually swimming voluntarily, but pithed and immobilized on a board in a tank with continuous irrigation of the gills. Involuntary body oscillations were regarded as simulating slow swimming (35–40 tail beats/min) while fast "swimming" was achieved in response to pinching the tail. By the next decade Hudson (1973) was able to recorded EMGs from trout swimming at controlled speeds in a water tunnel, and he concluded that the white muscle mass of teleosts might also be involved at intermediate swimming speeds, while the red fibers may be active at all swimming speeds. These observations matched histochemical evidence that red muscle metabolized fats while white muscle was powered by glycolysis, implying function in sustained and burst swimming, respectively (Bone, 1966; Hudson, 1973).

Bone suggests that multiply innervated fast fibers in higher teleosts are probably active at swim speeds below the maximum sustainable. His extensive survey of innervation characteristics among different muscles of various species is replete with suggestions for future research. For example, Bone writes: "The richness of the innervation of teleost multiple innervated fast fibers is astonishing (Chapter Fig. 6B and E); the nexus of axons resembles the capillary bed of slow fibers in its abundance and nerve terminals are scattered all along the muscle fibers. The contrast with focal terminal innervation as seen in other fish fast fiber systems is very great and invites functional interpretation."

Additional studies on tuna had demonstrated differential contractile and metabolic activity by red and white fibers related to swim speed, (Gordon, 1968; Rayner and Keenan, 1967) but contractile properties of different muscle fiber types had not been measured in vitro. Wardle (1975) devised an apparatus to measure isotonic twitch times of bulk white muscle in whole unanesthetized fish and used these to predict maximum swim speeds. A similar approach was used by Brill and Dizon (1979a) to predict maximum swim speeds of skipjack tuna.

3 Significance of this chapter

This chapter is important for what it summarized but also for what was implied as important areas of study to be addressed. Apart from showing that different fiber types are activated differently according to swimming demands, the chapter also points out that control and coordination of locomotor muscle activity along the body, as required for generating lateral motion and propulsive force, was essentially unknown. In concluding, Bone reiterates that the field of fish muscle biology is very much open and awaiting new investigations Thus, he sets the stage for future researchers with emerging new technology to make important advances in this area.

4 The impact of new experimental methods post-1978

Efforts to address muscle contractile properties continued after Bone's review was published as new technology became available. Brill and Dizon (1979b) measured red and white muscle EMGs in captive skipjack tuna swimming in a gravity-fed flume and showed that the white fibers can be active at speeds below maximum sustainable. Johnston and Brill (1984) measured the effect of temperature on tension and shortening velocity of isolated red and white fibers from tunas and proposed that the internalized red fibers, operating at 8–10 °C higher than the white may be able to generate power for high speed swimming. Other researchers also probed force-velocity properties of isolated fiber types in relation to temperature and acclimation (Altringham and Johnston, 1982; Bone et al., 1986; Curtin and Woledge, 1988, 1991, 1993; Johnston, 1983).

Starting in the 1990s a more whole-body approach questioned how activation timing and contractile properties may vary along the body and how this might contribute to overall production of power for swimming. A boom in the use of swim tunnels with modern design (Steffensen et al., 1984) and digital computers to analyze kinematics and energetics of fish swimming allowed investigators to look at the coordination of locomotor muscle activation during swimming and ultimately to discover that muscle contractile properties and activation timing varied with axial location for both red and white fiber types (e.g., Ellerby et al., 2000; Gillis, 1998; Hammond et al., 1998; Hess and Videler, 1984; Jayne and Lauder, 1995; Rome et al., 1993; Shadwick et al., 1998, 1999; van Leeuwen et al., 1990; Wardle and Videler, 1993; Williams et al., 1989). Although much variation among species has been observed, in general, a wave of body undulation travels caudally with increasing lateral amplitude. This is driven by a wave of muscle activation that also travels posteriorly along each side, alternately, faster than the undulation and often with shorter duration at more posterior sites. Thus, as these waves travel caudally the activation onset occurs earlier relative to muscle shortening (Fig. 2). In some cases, the posterior muscles may be active while being lengthened by contralateral shortening (i.e., producing negative power)

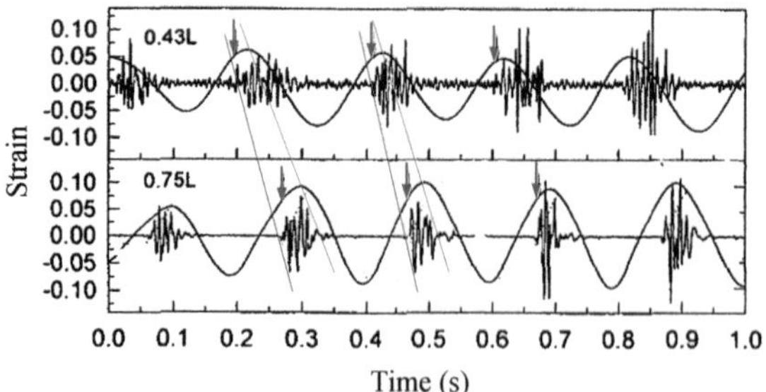

FIG. 2 Plots of EMG and red muscle strain, calculated from midline curvature, at axial locations of 0.43 *L* and 0.75 *L* of a 26.5 cm mackerel (*Scomber japonicus*) swimming at 4.9 Hz. (2.0 Ls^{-1}), showing that waves of muscle activation and contraction travel caudally. At both locations the onset of muscle activation (red arrows) occurs before maximum strain, i.e., peak muscle length, and the offset occurs during muscle shortening. The velocities of activation and contraction waves are represented by the slopes of the red and blue lines, respectively. These show that muscle activation travels faster than contraction and a phase advance occurs at the more posterior location. *From Shadwick, R.E., Steffensen, J.F., Katz. S.L., Knower, T., 1998. Muscle dynamics in fish during steady swimming. Amer. Zool. 38, 755–770.*

at the same time that the anterior muscles are producing positive power (Altringham et al., 1993; Altringham and Ellerby, 1999; van Leeuwen et al., 1990).

Muscle strain (i.e., degree of shortening) during swimming was initially predicted from midline curvature by treating the body as a bending beam with the vertebral column as the neutral axis, or using images acquired by high speed film cameras (Rome et al., 1990; van Leeuwen et al., 1990) or high speed VHS video (Jayne and Lauder, 1995), both of which required quite expensive instruments at that time. Quantification of these analog images required digital conversion for computer analysis and manually digitizing landmarks frame by frame. An alternative approach used by Lawrence Rome and colleagues determined muscle strain in swimming fish by calculating body curvature and then matching it to sarcomere lengths measured microscopically in sections made from fish frozen while bent into shapes that matched what was observed during swimming (Rome and Sosnicki, 1991; Rome et al., 1992).

A more direct but invasive technique called sonomicrometry was introduced to study fish muscle contractions by James Covell, working at UBC (Covell et al., 1991), who had developed this approach in his research on mammalian cardiac mechanics (Kirkpatrick et al., 1973). Covell used a home-made analog instrument, later marketed by Triton Technology Inc. Subsequently a more user-friendly digital sonomicrometer was introduced by Sonometrics Corp. and is now widely in use. This technique employs ultrasonic dimension gauges, consisting of pairs of small piezoelectric crystals implanted into muscle bundles, that transmit and receive ultrasound pulses. The time delay between the pair is converted to a distance, based on a known speed of sound in muscle. This gives a continuous real-time record of muscle segment length changes in a swimming fish. Coupled with EMG records at the same locations, it is possible to make direct determinations of activation timing relative to shortening timing and amplitude at different locations along the body (Fig. 3) (Coughlin et al., 1996; Hammond et al., 1998; Knower et al., 1999; Shadwick et al., 1999).

In vivo muscle strains measured in steady swimming have demonstrated, for the most part, that cyclic length changes are approximately sinusoidal. Direct measurements of in vivo muscle force in a swimming fish have not been made, as no suitable technology has yet been developed, leaving muscle power output to be determined in vivo by a work-loop technique (Josephson, 1985). In this approach cyclic contractions are imposed on isolated bundles of muscle fibers while an electrical stimulus is applied. Force and length changes are recorded and together give a work-loop which defines the positive, negative and net work done in each cycle. Multiplying by cycle frequency yields power which can be expressed as net or instantaneous. This approach can be used to explore the influence of strain amplitude, activation timing and cycle frequency (a proxy for tail beat frequency, and swim speed) and

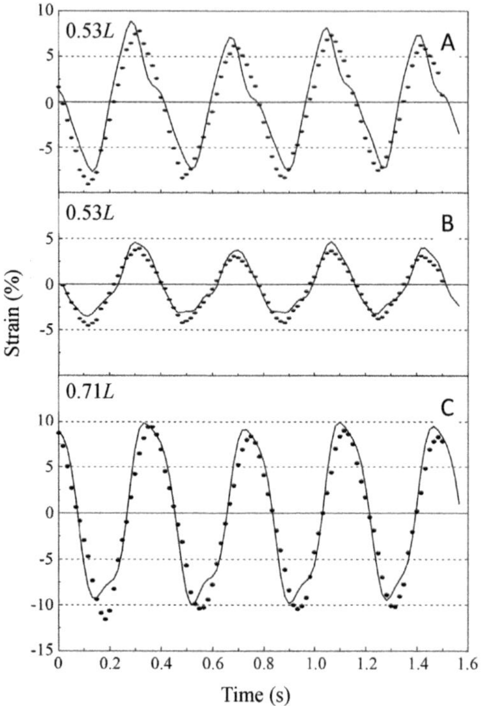

FIG. 3 Muscle strain calculated from midline curvature (filled symbols) and from sonomicrometry (continuous lines) at two locations on the body of a milkfish (*Chanos chanos*) (0.53 L and 0.71 L, where L is fork length) swimming at 2.3 FL s^{-1}, and at two depths in the myotomes at *0.53 L*. At this speed, the fish is using a steady gait. (A) Muscle strain calculated at 0.53 L for lateral red muscle located in close apposition to the skin. (B) Muscle strain calculated at 0.53 L for medial white muscle. At this swimming speed, the white muscle is not expected to be recruited, but is being passively deformed. (C) Muscle strain calculated at 0.7 L for lateral red muscle. *From Katz, S.L., Shadwick, R.E., Rapoport, H.S., 1999. Muscle strain histories in swimming milkfish in steady as well as burst swimming. J. Exp. Biol. 202, 529–541.*

temperature on the muscle performance. Plotting power against cycle frequency gives an indication of the muscle ability to function effectively at different swim speeds (Fig. 4). This approach was first used by Altringham and Johnston (1990) on red and white muscle from a sculpin and by Moon et al. (1991) on cod white fibers activated to maximize power output while strain amplitude and cycle frequency were varied. Rome and colleagues then used work-loops with parameters derived from measured EMGs and calculated muscle strain amplitudes in red muscle of scup swimming at maximum sustained speed (Rome and Swank, 1992; Rome et al., 1992, 1993). Their results showed that muscle performance was enhanced by elevated temperature and differed along the body, with the majority of power being produced by the posterior muscle. Additional work-loop studies on teleosts followed, in which

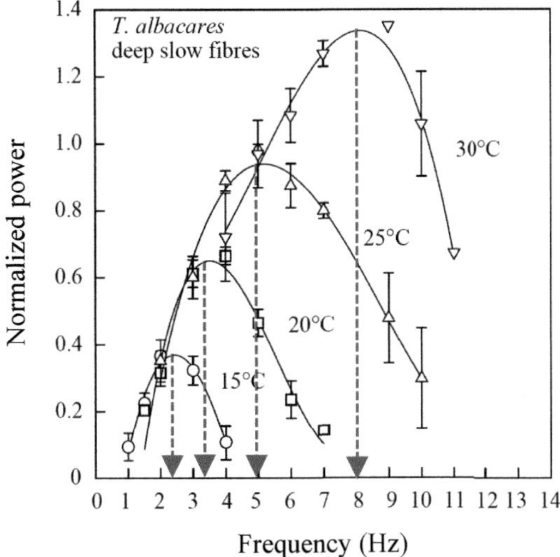

FIG. 4 Relative power vs frequency curves for deep red muscle fibers from yellowfin tuna (*Thunnus albacares),* a heterothermic species. Means ± S.E.M. Power is normalized to maximum power at 25 °C. These curves show the range of tail beat frequencies (a proxy for swim speed) that could be achieved by the red muscle at different temperatures. *Adapted from Altringham, J.D., Block, B.A., 1997. Why do tuna maintain elevated slow muscle temperatures? Power output of muscle isolated from endothermic and ectothermic fish. J. Exp. Biol. 200, 2617–2627.*

the influence of temperature and anatomical position were considered (Fig. 5; Altringham and Block, 1997; Altringham and Ellerby, 1999; Coughlin, 2000; Coughlin et al., 1996; Hammond et al., 1998; Johnson and Johnston, 1991; Rome and Swank, 1992; Shadwick and Syme, 2008; Syme and Shadwick, 2002). This technique has also been used to study red muscle function in sharks, including heterothermic species (Bernal et al., 2005, 2010; Donley et al., 2007, 2012; Stoehr et al., 2020).

In summary, since Bones chapter was published we now have methods to make direct and simultaneous measurements of muscle strain and EMG patterns in free-swimming fish, providing insights into the dynamics of muscle function along the length of the fish. We also know much more about contractile physiology of different muscle fiber types which includes power output in relation to axial position and temperature, and more sophisticated understanding of how muscles power swimming in a broad range of fish species.

5 A personal note

As a beginning graduate student I became aware of Quentin Bone when he came to UBC to collaborate with David Jones and Joe Kiceniuk, at about the time he would have been writing for Fish Physiology. Their research used EMG recordings to explore differences in activation of muscle fiber types

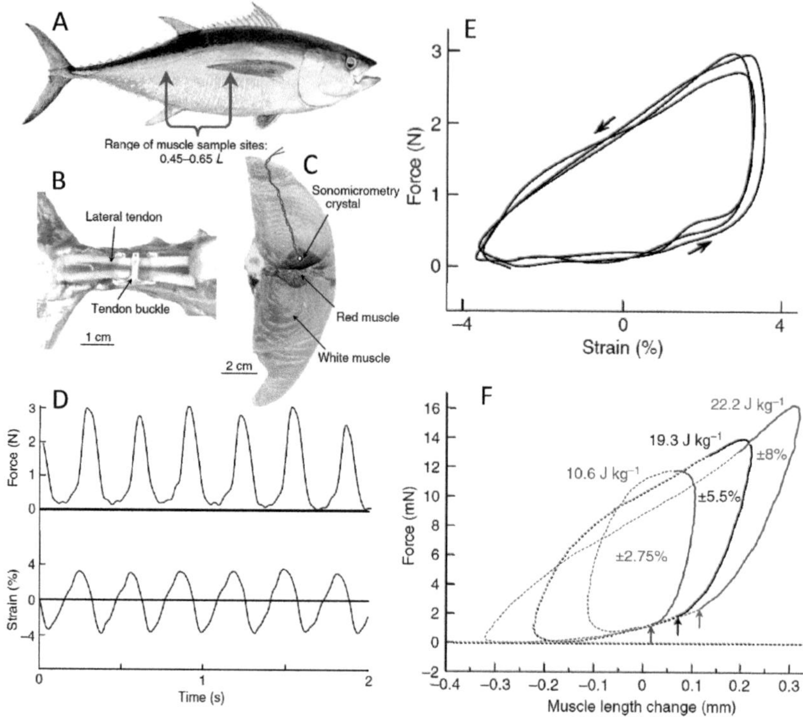

FIG. 5 (A) Lateral view of a yellowfin tuna (*Thunnus albacares*), indicating the region where sonomicrometry and muscle fiber sample sites were located (0.45–0.65 L, where L is fork length). (B) Lateral view of caudal peduncle with skin removed post-mortem, showing lateral tendons that transmit forces from the muscle cones to the caudal fin, and the tendon buckle transducer used to measure force in the tendons. (C) View of one side of a yellowfin in cross-section at 0.55 L showing the deep location of the aerobic red muscle and illustrating the positioning of a sonomicrometer crystal in this red muscle. (D) Example of force measured in the caudal tendons and red muscle length change (expressed as strain) measured at 0.5 L in a 45 cm yellowfin tuna swimming at 2.2 Ls^{-1}. (E) In vivo work loop resulting from plot of tendon force and muscle length for three cycles of traces in (D). (F) The effect of strain amplitude on in vitro work loops from red muscle taken from 0.5 L. Cycle frequency was 3 Hz, strains used were ±2.75%, ±5.5%, ±8%, and temperature was 25 °C. Network for each strain is shown in J kg^{-1}. Solid line portions of each cycle represent the period when activation stimulus was on, broken portions are the period when stimulus was off, and arrows indicate stimulus onset. *From Shadwick, R.E., Syme, D.A., 2008. Thunniform swimming: muscle dynamics and mechanical power production of aerobic fibres in yellowfin tuna (Thunnus albacares). J. Exp. Biol. 211, 1603–1611.*

during swimming in herring, carp and trout, and was referenced in the chapter (Bone et al., 1978). Although Dr. Bone was already well known for his studies on fish muscle, he had also published on the nervous system of Amphioxus and other lower chordates. A few years later when I was doing my PhD on cephalopod biomechanics with John Gosline, Bone and colleagues published a paper relevant to our work on squid mantle structure that revealed two distinct muscle fiber types used separately for slow and burst jet propulsion

(Bone et al., 1981). He also described the geometry of collagen fibers in the mantle wall that we used to model the elastic properties of the mantle (Gosline and Shadwick, 1983). In 1991 I met Dr. Bone at the Plymouth Laboratory for an international conference, broadly themed on animal swimming, which he helped to organize and publish (Maddock et al., 1994). He certainly had, as others have described (Mackie, 2022), a warm and welcoming nature and a broad interest in biology that served him well as an enthusiastic and energetic conference moderator.

Turning my attention to fish locomotor muscle in the 1990s, I was aided by having Bone's 1978 review as a guide to the important questions to be studied. Colleagues and I at the Scripps Institution of Oceanography began to explore the mechanical design of muscle and tendon complexes in tunas, a group of warm-bodied fishes regarded as supreme swimming machines. With a focus on red muscle located on the deep anterior pointing cones, we wanted to understand how contraction of these fibers resulted in motion at the caudal fin. Combining EMG and sonomicrometry (initially with help from Covell and his home-made sonomicrometer) with video analysis of body bending of yellowfin and skipjack tunas swimming in a water tunnel revealed this functional connection (Shadwick et al., 1999). In short, deep red muscle fibers are connected to long posterior pointing myoseptal tendons that fuse together and link directly onto the caudal fin rays. Consequently, these contractions cause midline bending 8–10 vertebral segments more posterior, rather than locally. Applying tendon force gauges (Biewener et al., 1988) to the lateral caudal tendons we showed peak thrust force occurs just as all ipsilateral muscle is deactivated and the fin is moving at its highest lateral velocity (Knower et al., 1999). Armed with activation and shortening data we used in vitro work loops to simulate steady swimming and found that red muscles produced near maximal power under those conditions (Katz et al., 2001; Shadwick and Syme, 2008; Syme and Shadwick, 2002).

Parallel studies we undertook on warm bodied lamnid sharks revealed strong convergence with tunas in their muscle morphology and physiology (Bernal et al., 2005; Donley et al., 2004, 2005, 2007). In mako and salmon sharks red muscle is located medially on anterior facing myomere cones, as in tunas, and connects to the skin and caudal fin through long tendons (Gemballa et al., 2006). In these lamnid species the power produced by red muscle is enhanced by elevated body temperature that can be up to 20 °C higher than the ambient water. These muscles are so well adapted to work at such high temperatures that they could not actually contract fast enough or with enough power to swim if they had to operate at the ambient ocean temperatures where they live (Bernal et al., 2005).

6 Looking ahead

In the study of fish muscle function since the 1960s electromyography has been invaluable. In spite of new technologies that allow other performance indicators to be recorded, knowledge of muscle activation timing will continue to be essential. One new direction has been to develop telemetry tags that record muscle EMGs in free ranging fish as an indicator of energy expenditure for swimming in natural conditions (Cooke et al., 2004). Others have characterized fish locomotor muscle performance in more diverse species, many of which are pelagic and possess some form of heat exchanger that elevates red muscle temperature (Bernal et al., 2010; Donley et al., 2012; Wegner et al., 2015). Studies of muscle function in intermittent or variable speed swimming reflect a more natural situation than older research with forced steady swimming (Coughlin et al., 2022). New techniques in imaging are being used to reveal the 3D organization of fish muscle fibers. van Meer and van Leeuwen (2024) used a transparent zebra fish line with green fluorescent protein in the fast muscle fibers to visualize the orientation of these fibers, and how this changes during development and along the body. Impacts of environmental change on metabolic rate and performance of locomotor muscle have also been considered (Gamperl and Syme, 2021; Joyce, 2023; McDonnell and Chapman, 2016). In the area of biomimetics, interest in using muscle analogs as actuators in fish-like robots has grown (Li et al., 2022). In all these cases, and in future endeavors, a fundamental understanding of the structural, electrical and contractile properties of fish locomotor muscle will continue to be as important as demonstrated by Quentin Bone in 1978.

Acknowledgments

The author thanks the volume editors and Dr. Doug Syme for helpful discussions and suggestions.

References

Alexander, R.M., 1969. The orientation of muscle fibres in the myomeres of fishes. J. Marine Biol. Assoc. 49, 263–290.

Altringham, J.D., Block, B.A., 1997. Why do tuna maintain elevated slow muscle temperatures? Power output of muscle isolated from endothermic and ectothermic fish. J. Exp. Biol. 200, 2617–2627.

Altringham, J.D., Ellerby, D.J., 1999. Fish swimming: patterns in muscle function. J. Exp. Biol. 202 (23), 3397–3403.

Altringham, J.D., Johnston, I.A., 1982. The pCa-tension and force velocity characteristics of skinned fibres isolated from fish fast and slow muscles. J. Physiol. (Lond.) 333, 421–449.

Altringham, J.D., Johnston, I.A., 1990. Modeling muscle power output in a swimming fish. J. Exp. Biol. 148, 395–402.

Altringham, J.D., Wardle, C.S., Smith, C.I., 1993. Myotomal muscle function at different locations in the body of a swimming fish. J. Exp. Biol. 182, 191–206.

Bainbridge, R., 1960. Speed and stamina in three fish. J. Exp. Biol. 37, 129–153.

Bernal, D., Donley, J.M., Shadwick, R.E., Syme, D.A., 2005. Mammalian-like muscle powers swimming in a cold water shark. Nature 437, 1349–1352.

Bernal, D., Donley, J.M., McGillivray, D.G., Albers, S.A., Syme, D.A., Sepulveda, C., 2010. Function of the medial red muscle during sustained swimming in common thresher sharks: contrast and convergence with thunniform swimmers. Comp. Biochem. Physiol. Part A: Molec. Integ. Physiol. 155, 454–463.

Biewener, A.A., Blickhan, R., Perry, A.K., Heglund, N.C., Taylor, C.R., 1988. Muscle forces during locomotion in kangaroo rats: force platform and tendon buckle measurements compared. J. Exp. Biol. 137, 191–205.

Boddeke, R., Slijper, E.J., van der Stelt, A., 1959. Histological characteristics of the body musculature of fishes in connection with their mode of life. *Proc. K. Ned. Akad. Wet.*, Ser. C 62, 576–588.

Bone, Q., 1964. Patterns of muscular innervation in the lower chordates. Int. Rev. Neurobiol. 6, 99–147.

Bone, Q., 1966. On the function of the two types of myotomal muscle fiber in elasmobranch fish. J. Mar. Biol. Assoc. UK 46, 321–334.

Bone, Q., 1972. The dogfish neuromuscular junction: dual innervation of vertebrate striated muscle fibers? Cell Sci. 10, 657–665.

Bone, Q., Chubb, A.D., 1975. The structure of stretch receptor endings in the fin muscles of rays. J. Mar. Biol. Assoc. UK 55, 939–943.

Bone, Q., Chubb, A.D., 1976. On the structure of corpuscular proprioceptive endings in sharks. J. Mar. Biol. Assoc. UK 56, 925–928.

Bone, Q., Chubb, A.D., 1978. The histochemical demonstration of myofibrillar ATPase in elasmobranch muscle. Histochem. J. 10, 489–494.

Bone, Q., Kiceniuk, J., Jones, D.R., 1978. On the role of the different fibre types in fish myotomes at intermediate swimming speeds. Fish. Bull. 76, 691–699.

Bone, Q., Pulsford, A., Chubb, A.D., 1981. Squid mantle muscle. J. Mar. Biol. Assoc. UK 61, 327–342.

Bone, Q., Johnston, I.A., Pulsford, A., Ryan, K.P., 1986. Contractile properties and ultrastructure of three types of muscle fibre in the dogfish myotome. J. Muscle Res. Cell Motil. 7, 47–56.

Brill, R.W., Dizon, A.E., 1979a. Effect of temperature on isotonic twitch of white muscle and predicted maximum swimming speeds of skipjack tuna, *Katsuwonus pelamis*. Environ. Biol. Fishes 4, 199–205.

Brill, R.W., Dizon, A.E., 1979b. Red and white muscle fibre activity in swimming skipjack tuna, *Katsuwonus pelamis* (Linnaeus). J. Fish Biol. 15, 679–685.

Cooke, S.J., Thorstad, E.B., Hinch, S.G., 2004. Activity and energetics of free-swimming fish: insights from electromyogram telemetry. Fish Fish. 5, 21–52.

Coughlin, D.J., 2000. Power production during steady swimming in largemouth bass and rainbow trout. J. Exp. Biol. 203, 617–629.

Coughlin, D.J., Valdes, L., Rome, L.C., 1996. Muscle length changes during swimming in scup: Sonomicrometry verifies the anatomical technique. J. Exp. Biol. 199, 459–463.

Coughlin, D.J., Chrostek, J.D., Ellerby, D.J., 2022. Intermittent propulsion in largemouth bass, *Micropterus salmoides*, increases power production at low swimming speeds. Biol. Lett. 18 (5). 20210658.

Covell, J.W., Smith, M., Harper, D.G., Blake, R.W., 1991. Skeletal muscle deformation in the lateral muscle of the intact rainbow trout *Oncorhynchus mykiss* during fast start maneuvers. J. Exp. Biol. 156, 453–466.

Curtin, N.A., Woledge, R.C., 1988. Power output and force velocity relationship of live fibres from white myotomal muscle of the dogfish, *Scyliorhinus canicula*. J. Exp. Biol. 140, 187–197.

Curtin, N.A., Woledge, R.C., 1991. Efficiency of energy conversion during shortening of muscle fibres from the dogfish *Scyliorhinus canicula*. J. Exp. Biol. 158, 343–353.

Curtin, N.A., Woledge, R.C., 1993. Efficiency of energy conversion during sinusoidal movement of red muscle fibres from the dogfish *Scyliorhinus canicula*. J. Exp. Biol. 185, 195–206.

Donley, J.M., Sepulveda, C.A., Konstantinidis, P., Gemballa, S., Shadwick, R.E., 2004. Convergent evolution in mechanical design of lamnid sharks and tunas. Nature 429, 61–65.

Donley, J.M., Shadwick, R.E., Sepulveda, C.A., Konstantinidis, P., Gemballa, S., 2005. Patterns of red muscle strain/activation and body kinematics during steady swimming in a lamnid shark, the shortfin mako (*Isurus oxyrinchus*). J. Exp. Biol. 208, 2377–2387.

Donley, J.M., Shadwick, R.E., Sepulveda, C.A., Syme, D.A., 2007. Thermal dependence of contractile properties of the aerobic locomotor muscle in the leopard shark and shortfin mako shark. J. Exp. Biol. 210, 1194–1203.

Donley, J.M., Sepulveda, C.A., Albers, S.A., McGillivray, D.G., Syme, D.A., Bernal, D., 2012. Effects of temperature on power output and contraction kinetics in the locomotor muscle of the regionally endothermic common thresher shark (*Alopias vulpinus*). Fish Physiol. Biochem. 38, 1507–1519.

Ellerby, D.J., Altringham, J.D., Williams, T., Block, B.A., 2000. Slow muscle function of Pacific bonito (*Sarda chiliensis*) during steady swimming. J. Exp. Biol. 203, 2001–2013.

Gamperl, A.K., Syme, D.A., 2021. Temperature effects on the contractile performance and efficiency of oxidative muscle from a eurythermal versus a stenothermal salmonid. J. Exp. Biol. 224 (15), jeb242487.

Gemballa, S., Ebmeyer, L., Hagen, K., Hannich, T., Hoja, K., Rolf, M., Treiber, K., Vogel, F., Weitbrecht, G., 2003. Evolutionary transformations of myoseptal tendons in gnathostomes. Proc. Roy. Soc. Lond. B. 270, 229–1235.

Gemballa, S., Konstantinidis, P., Donley, J.M., Sepulveda, C., Shadwick, R.E., 2006. Evolution of high-performance swimming in sharks: transformations of the musculotendinous system from subcarangiform to thunniform swimmers. J. Morphol. 267, 477–493.

Gillis, G.B., 1998. Environmental effects on undulatory locomotion in the american eel *Anguilla rostrata*: kinematics in water and on land. J. Exp. Biol. 201, 949–961.

Gordon, M.S., 1968. Oxygen consumption of red and white muscles from tuna fishes. Science 159, 87–90.

Gosline, J.M., Shadwick, R.E., 1983. Molluscan collagen and its mechanical organization in squid mantle. In: Hochachka, P.W. (Ed.), The Mollusca, Vol. 1, Metabolic Biochemistry and Molecular Biomechanics. Academic Press, New York, pp. 371–398.

Greer-Walker, M., Pull, G.A., 1973. Skeletal muscle function and sustained swimming speeds in the coalfish *Gadus oirens L.* Comp. Biochem. Physiol. A 44, 495–501.

Hammond, L., Altringham, J.D., Wardle, C.S., 1998. Myotomal slow muscle function of rainbow trout *Oncorhynchus mykiss* during steady swimming. J. Exp. Biol. 201, 1659–1671.

Hess, F., Videler, J.J., 1984. Fast continuous swimming of saithe (*Pollachius virens*): a dynamic analysis of bending moments and muscle power. J. Exp. Biol. 109, 229–251.

Hochachka, P.W., 1961. The effect of physical training on oxygen debt and glycogen reserves in trout. Can. J. Zool. 39, 767–776.

Hudson, R.C.L., 1973. On the function of the white muscles in teleosts at intermediate swimming speeds. J. Exp. Biol. 58, 509–522.

Hunter, J.R., 1971. Sustained speed of jack mackerel, *Trachyurus symmetricus. U.S. Fish Wildl. Serv., Fish. Bull.* 69, 267–271.

Jayne, B.C., Lauder, G.V., 1995. Speed effects on midline kinematics during steady undulatory swimming of largemouth bass, *Micropterus salmoides*. J. Exp. Biol. 198, 585–602.

Johnson, T.P., Johnston, I.A., 1991. Power output of fish muscle fibres performing oscillatory work: effects of acute and seasonal temperature change. J. Exp. Biol. 157, 409–423.

Johnston, I.A., 1983. Dynamic properties of fish muscle. In: Webb, P.W., Weihs, D. (Eds.), Fish Biomechanics. Praeger, New York, pp. 36–67.

Johnston, I.A., Brill, R.W., 1984. Thermal dependence of contractile properties of single skinned muscle fibres from antarctic and various warm water marine fishes including skipjack tuna (*Katsuwonus pelamis*) and kawakawa (*Euthynnus affinis*). *J. Comp. Physiol. B. Biochem. Syst., Environ. Physiol.* 155, 63–70.

Johnston, I.A., Davison, W., Goldspink, G., 1977. Energy metabolism of carp swimming muscles. J. Comp. Physiol. 114, 203–216.

Josephson, R.K., 1985. Mechanical power output from striated muscle during cyclic contraction. J. Exp. Biol. 114, 493–512.

Joyce, W., 2023. Muscle growth and plasticity in teleost fish: the significance of evolutionarily diverse sarcomeric proteins. Rev. Fish Biol. Fisher. 33, 1311–1327.

Katz, S.L., Syme, D.A., Shadwick, R.E., 2001. High-speed swimming: enhanced power in yellowfin tuna. Nature 410, 770–771.

Kirkpatrick, S.E., Covell, J.W., Friedman, W.F., 1973. A new technique for the continuous assessment of fetal and neonatal cardiac performance. Am J. Obst. 116, 963–972.

Knower, T., Shadwick, R.E., Katz, S.L., Graham, J.B., Wardle, C.S., 1999. Red muscle activation patterns in yellowfin (*Thunnus albacares*) and skipjack (*Katsuwonus pelamis*) tunas during steady swimming. J. Exp. Biol. 202, 2127–2138.

Li, Y., Xu, Y., Li, Y., 2022. A comprehensive review on fish-inspired robots. Int. J. Adv. Robot. Syst. https://doi.org/10.1177/17298806221110370.

Mackie, G., 2022. Quentin Bone. 17 August 1931—6 July 2021. *Biogr. Mems Fell. R. Soc.* 72, 55–76.

Maddock, L., Bone, Q., Rayner, J.M.V. (Eds.), 1994. Mechanics and Physiology of Animal Swimming. Cambridge University Press, Cambridge. 250 pp.

McDonnell, L.H., Chapman, L.J., 2016. Effects of thermal increase on aerobic capacity and swim performance in a tropical inland fish. Comp. Biochem. Physiol. A: Molec. Integ. Physiol. 199, 62–70.

Moon, T.W., Altringham, J.D., Johnston, I.A., 1991. Energetics and power output of isolated fish fast muscle fibres performing oscillatory work. J. Exp. Biol. 158, 261–273.

Rayner, M.D., Keenan, M.J., 1967. Role of red and white muscles in the swimming of the skipjack tuna. Nature 214, 392–393.

Rome, L.C., Sosnicki, A.A., 1991. Myofilament overlap in swimming carp. II. Sarcomere length changes during swimming. Am. J. Physiol. Cell Physiol. 260, C289–C296.

Rome, L.C., Swank, D., 1992. The influence of temperature on power output of scup red muscle during cyclical length changes. J. Exp. Biol. 171, 261–281.

Rome, L.C., Funke, R.P., Alexander, R.M., 1990. The influence of temperature on muscle velocity and sustained performance in swimming carp. J. Exp. Biol. 154, 163–178.

Rome, L.C., Choi, I., Lutz, G., Sosnicki, A.A., 1992. The influence of temperature on muscle function in the fast swimming scup: I. Shortening velocity and muscle recruitment during swimming. J. Exp. Biol. 163, 259–279.

Rome, L.C., Swank, D., Corda, D., 1993. How fish power swimming. Science 261, 340–343.

Shadwick, R.E., Syme, D.A., 2008. Thunniform swimming: muscle dynamics and mechanical power production of aerobic fibres in yellowfin tuna (*Thunnus albacares*). J. Exp. Biol. 211, 1603–1611.

Shadwick, R.E., Steffensen, J.F., Katz, S.L., Knower, T., 1998. Muscle dynamics in fish during steady swimming. Am. Zool. 38, 755–770.

Shadwick, R.E., Katz, S.L., Korsmeyer, K.E., Knower, T., Covell, J.W., 1999. Muscle dynamics in skipjack tuna: timing of red muscle shortening in relation to activation and body curvature during steady swimming. J. Exp. Biol. 202, 2139–2150.

Steffensen, J.F., Johansen, K., Bushnell, P.G., 1984. An automated swimming respirometer. Comp. Biochem. Physiol. A 79, 437–440.

Stoehr, A.A., Donley, J.M., Aalbers, S.A., Syme, D.A., Sepulveda, C., Bernal, D., 2020. Thermal effects on red muscle contractile performance in deep-diving, large-bodied fishes. Fish Physiol. Biochem. 46, 1833–1845.

Syme, D.A., Shadwick, R.E., 2002. Effects of longitudinal body position and swimming speed on mechanical power of deep red muscle from skipjack tunas (Katsuwonus pelamis). J. Exp. Biol. 2052, 189–200.

van Leeuwen, J.L., Lankheet, M.J.M., Akster, H.A., Osse, J.W.M., 1990. Function of red axial muscles of carp (*Cyprinus carpio*): recruitment and normalized power output during swimming in different modes. J. Zool. 220, 123–145.

van Meer, N., van Leeuwen, J.L., 2024. Quantification of three-dimensional architecture of axial muscle fibres in larval fish. In: Society for Integrative and Comparative Biology, Annual meeting Seattle USA 2024. https://www.xcdsystem.com/sicb/program/I1Kr23t/index.cfm?pgid=1343#V.

Vogel, F., Gemballa, S., 2000. Locomotory design of 'cyclostome' fishes: spatial arrangement and architecture of myosepta and lamellae. Acta Zool. 81, 267–283.

Wardle, C.S., 1975. Limit of fish swimming speed. Nature 255, 725–727.

Wardle, C.S., Videler, J.J., 1993. The timing of the electromyogram in the lateral myotomes of mackerel and saithe at different swimming speeds. J. Fish Biol. 42, 347–359.

Webb, P.W., 1971. The swimming energetics of trout. I. Thrust and power output at cruising speeds. J. Exp. Biol. 55, 489–520.

Wegner, N.C., Snodgrass, O.E., Dewar, H., Hyde, J.R., 2015. Whole-body endothermy in a mesopelagic fish, the opah, *Lampris guttatus*. Science 348, 786–789. https://doi.org/10.1126/science.aaa8902.

Williams, T., Grillner, S., Smoljaninov, V., Wallen, P., Rossignol, S., 1989. Locomotion in lamprey and trout: the relative timing of activation and movement. J. Exp. Biol. 143, 559–566.

Chapter 8

LOCOMOTOR MUSCLE☆

QUENTIN BONE

Chapter Outline

I. INTRODUCTION

Muscular tissue forms a larger part of the mass of the fish body than it does of other vertebrates: Some 40–60% of the total body mass in most fish is locomotor musculature (Table I). In part this is because economy in weight is not mandatory as it is for terrestrial and aerial forms, and in part because stringent demands are placed on the locomotor system by the density of the medium, so that a large amount of muscle is needed to generate sufficient power for rapid swimming.

In addition to sheer mass (which of course gives fish their culinary and economic importance), there are several design features of the muscular system that are not commonly found in the muscles of other classes of vertebrates. For example, in many fish, twitch fibers are multiply innervated; again, no fish muscles contain muscle spindles: Such features give fish muscle an especial comparative interest. It must be emphasized at the outset, however, that despite the efforts of a number of investigators (beginning with Lorenzini, 1678, who first described different muscle fiber types), knowledge of fish muscle is as yet relatively scanty. This chapter can thus

☆This is a reproduction of a previously published chapter in the Fish Physiology series, "1978 (Vol. 7)/Locomotor Muscle: ISBN: 978-0-12-350407-4; ISSN: 1546-5098".

Fish Physiology, Vol. 40B. https://doi.org/10.1016/bs.fp.2024.07.003

Table I The Amount of Locomotor Muscle as a Proportion of Total Body Weight in Different Fish

Species	% Muscle	Comments.	Authority
Galeocerdo arcticus	37	Incl. vertebrae	Warfel and Clague (1950)
Prionace glauca	36		Stevens (1976)
Scyliorhinus canicula	45		Bone and Roberts (1969)
Centroscymnus owstoni	55–60		Higashi *et al.* (1953)
Katsuwonus pelamis	68		Fierstine and Walters (1968)
Salmo irideus	55–67		Bainbridge (1960, 1962)
Carassius auratus	33–45	Incl. skin	
Leuciscus leuciscus	42–60		

only serve as a progress report, rather than as a definitive account of the muscular systems employed in locomotion. In several important respects, fish muscle is well-suited for physiological study (despite the disadvantages inherent in the myotomal muscle preparation), and certain puzzling features of vertebrate muscle fibers are perhaps most likely to be understood by work on fish.

The great majority of fishes swim using the segmental myotomal musculature, so most of this chapter will deal with the arrangement of the myotomes and of the muscle fibers within them. A few fish (e.g., some Trachyurids, many labrids, rays, and holocephali) swim by using the musculature of the paired or unpaired fins: These muscles will be considered in less detail. The nomenclature of fish muscles has been bedeviled by synonymy, but fortunately Winterbottom (1974) has provided a helpful guide to the nomenclature of the locomotor (and other) muscles.

In limiting this chapter only to locomotor muscles, it is important to observe that although this may yield a more uniform view of sets of fibers designed for similar ends in different forms, interesting work on other kinds of fish muscle fibers has to be omitted. For example, fish eye muscles have provided valuable material for experimental study of selective re-innervation (e.g., Mark and Marotte, 1972); the very rapid sonic muscles of the swimbladder in some forms have proven interesting (Skoglund, 1961); the gill muscles may yield useful mechanical preparations (Levin and Wyman, 1927); the muscular systems of fish barbels offer peculiar problems of coordination and control. Fascinating as these muscles all are, they fall outside the scope of this chapter.

II. THE ORGANIZATION OF THE MYOTOMES

The lateral musculature in all fish groups is subdivided into serially arranged myotomes of complex shape, delimited by connective tissue myosepta into which the myotomal muscle fibers insert. Figure 1 shows something of myotomal shape in different fish groups. There is plainly a phylogenetic increase in complexity of shape, from the simple V-shape in amphioxus via the shallow W-shape of the Agnatha, to the deep W-shape of gnathostomes. This phylogenetic change is reflected in the ontogeny of the higher fish groups where the somite first forms a V-shape, before folding further to yield the adult W-shape.

A number of workers have attempted to provide a functional basis for the complex shape of the fish myotome, and for its internal arrangement. These

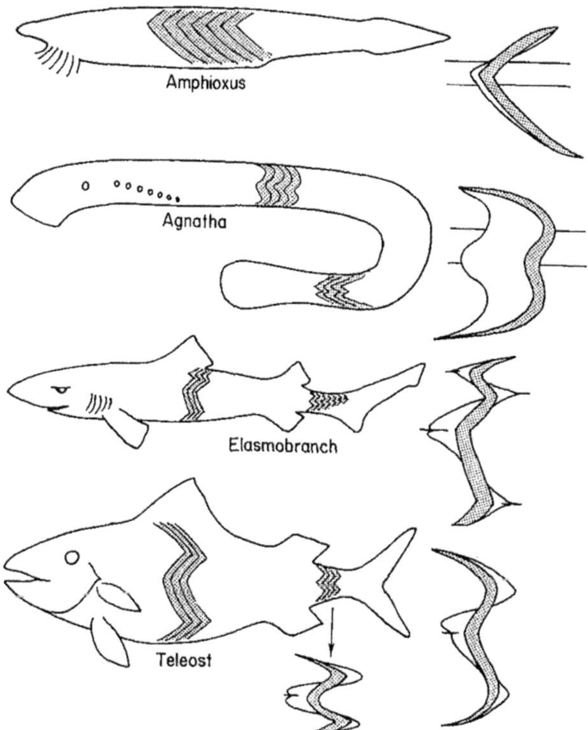

Fig. 1. Simplified diagrams of myotomal shape in various fish groups. An enlarged view of a midtrunk myotome in each is shown at right; a teleostean caudal myotome at bottom. Not to scale. (Redrawn from Nursall, 1956, *Proc. Zool. Soc. London* **126,** 127–143.)

attempts have been more successful in explaining the orientation of muscle fibers within the myotomes than the shape of the myotomes themselves; yet both presumably reflect the properties of the contractile units of the system and the properties of their insertions. The important properties insofar as myotomal design is concerned would seem to be the following.

1. Contraction takes place without volume change.
2. The muscle fibers insert onto deformable but inextensible partitions which are attached to the flexible but incompressible notochord or vertebral column.
3. Deformation of the myosepta is often (but not always) restrained by intermuscular bones.
4. Flexion of the body is required only in the lateral plane.
5. During flexion, the radius of curvature will be least next to the vertical column, largest just under the skin.
6. Both frequency and amplitude of flexions may vary.

The end result of the operation of the myotomal units with these properties is, of course, the lateral oscillations of the body brought about by transferring the contraction forces to the central strut. In the simple myotomes of amphioxus, the muscle lamellae run nearly parallel to the long axis, and the distance between inner and outer surface of the myotome in these animals is small. A single muscle lamella may span the whole of this distance (see Fig. 8), but the difference between the radii of curvature of its inner and outer edges is insignificant. The V-folding of the myotome is probably related to the need to avoid dorsoventral flexion. The notochord lies dorsally to accommodate the viscera below it, and if the myotomes were simple inclined blocks (the greater part of which lay below the strut), contraction would lead to ventral flexions of head and tail. With the V-shaped myotome arranged so that the arms of the V are unequal in length, and that the apex lies at the level of the notochord, solely lateral flexions are possible.

It is significant that with increase in scale, all other fishes have their myotomes arranged so that the muscle fibers of the greater part of the myotome do not run parallel to the long axis. Indeed, as van der Stelt (1968) and Alexander (1969) point out, the muscle fibers of the white or fast part of the myotome may make large angles with the long axis. Alexander (1969) found that some muscle fibers in the myotomes of both sharks and teleosts were oriented at nearly 40° to the long axis. These orientations are not random; Fig. 2 shows the two patterns of orientation found by Alexander in sharks and in higher teleosts. Alexander was able to show that these two patterns of orientation of the fast or white fibers of the myotome were a consequence of the requirement for all of these fibers to contract at about the same rate, whatever their position in the myotome. In other words, although during flexion the radius of curvature of the fish as a whole will be greatest next to the vertical column, and least superficially, suitable orientation of the fibers in these positions in

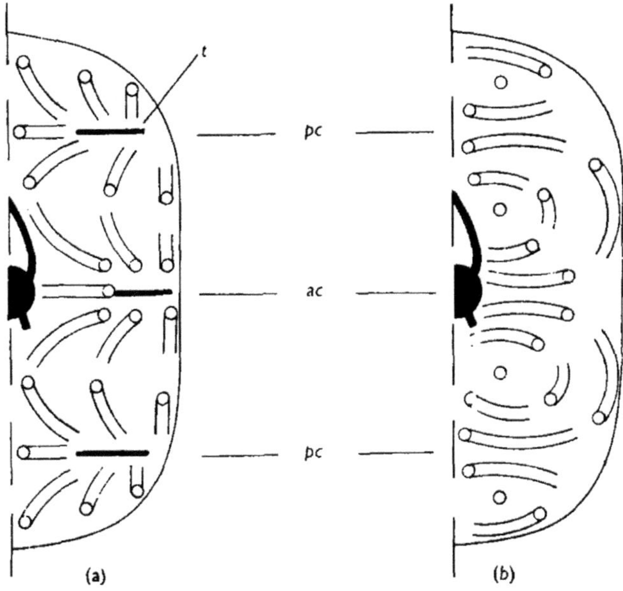

Fig. 2. Thick transverse slices across (a) selachian, and (b) teleostean body showing orientation of myotomal muscle fibers as viewed from behind. ac, anterior cone; pc, posterior cone; t, tendon. (From Alexander, 1969, *J. Mar. Biol. Assoc. U.K.* **49**, 263–290.)

the myotomes will enable each to contract to the same extent as the body flexes. The importance of this result is that, as Hill (1950) has pointed out, the shape of the force/velocity curve for muscle fibers means that maximum power will be produced at a particular rate of contraction. We do not know the shape of the force/velocity curve for the myotomal muscles of any fish, but we can assume that for most fishes, the white or fast portion of the myotome will be designed for maximum power (see Section IV). The orientation of the muscle fibers directly reflects this requirement, for if they were not able to contract at about the same rate throughout the myotome, the power extractable from the fast portion of the myotome would be much lessened. Note that this argument does not apply to the superficial sheet of slow (red) muscle fibers, for it is thin enough that the fibers which make it up can run more or less parallel to the long axis and yet all contract at about the same rate.

It is interesting that sharks and higher teleosts adopt different fiber orientations in the myotomes. Alexander has shown that the helical teleost arrangement (Fig. 3) will give a faster rate of body flexion for the same rate of contraction of the muscle fibers than will the selachian arrangement (at the expense of a weaker bending moment), and suggests that this is a better compromise for the teleosts since they swim faster than sharks of similar size. It is difficult to know whether this proposition is always applicable, since there are

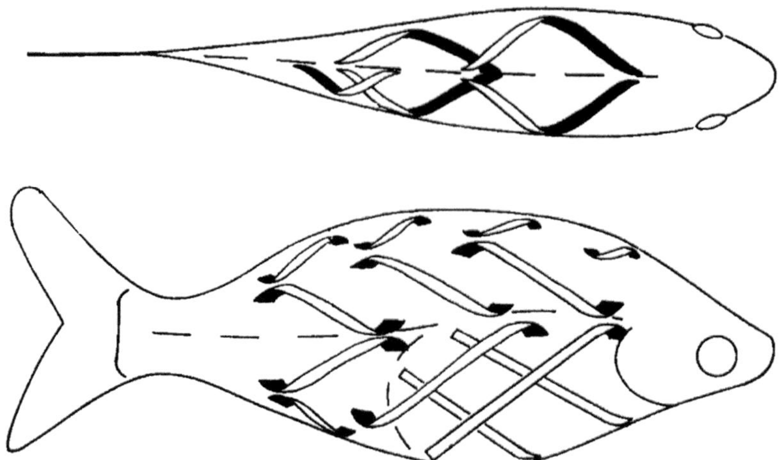

Fig. 3. Schematic dorsal and lateral views of typical teleost showing course of myotomal muscle fibers in successive myotomes along the body. The helices shown were found by taking the origin of one muscle fiber from the point at which the muscle fiber in the myotome next anterior inserts onto the common myoseptum, and so on along the fish. (From Alexander, 1969, *J. Mar. Biol. Assoc. U.K.* **49,** 263–290.)

no measurements of the speed of Isurids; it would certainly be very interesting to know if the selachian arrangement found by Alexander in *Hexanchus*, *Scyliorhinus*, and *Squalus* also obtains in *Carcharodon* or *Isurus*. In the caudal peduncles of higher teleosts, there is a transition to the selachian pattern of fiber orientation, and Alexander suggests that this is to enable the fish to cope with the larger stresses in this region while maintaining a similar rate of contraction of muscle fibers throughout the body.

Alexander's analysis was notable, for it gave the first convincing functional explanation of the fiber orientation of the myotomes. What is more, it seems likely that this fiber orientation underlies the complex shape of the gnathostome myotomes. The myotomal folding is so arranged that the muscle fibers insert into the myosepta at approximately the same angle, despite their very different orientations with respect to the long axis of the body. It would evidently be interesting to examine fiber orientations during ontogeny as the myotomes gradually achieve the W-shape, to provide a test of this view.

The larger scombroids have the most specialized teleost myotomes. In these fish, the inertia of the caudal peduncle and foil has been reduced as much as possible (just as in the legs of fast-running terrestrial vertebrates) and the foil is oscillated by tendons passing over a flexible joint in the vertebral column (Fierstine and Walters, 1968). These tendons are formed by a series of nested cones derived from successive myosepta (an analogous arrangement is found in the caudal region of certain rays), similar smaller tendons also run anteriorly to insert upon the pectoral girdle.

The result of this arrangement is that these fish have achieved the hydrodynamically desirable separation of the power source from the caudal propellor. Body flexions are less than those of fish swimming in less efficient and less rapid ways (see Chapter 3), and since the muscles used during cruising form the inner portion of the myotome (Fig. 4), it is possible that these contain muscle fibers oriented parallel to the long axis of the body. Fiber orientations have not yet been studied in the larger scombroids, nor in the similarly adapted Isurid sharks.

Little work has been done upon the myosepta themselves. Some of the problems of their design have been considered by Willemse (1975), who describes the organization of the connective tissue fibers in teleost myosepta. In many teleosts intermuscular bones of different forms lie in the myocommata, limiting the directions in which they may be deformed. Jarman (1961) pointed out that these bones could be arranged so that they did not interfere with contraction of the myotomal muscle fibers, but a functional analysis of the different patterns and numbers of these elements in different groups is yet to be made. They are absent from sharks.

III. FIN MUSCLES

The arrangement of the muscles of paired and unpaired fins is illustrated diagrammatically for many groups by Winterbottom (1974). In most, small and large fiber portions of the fin musculature are anatomically similar in arrangement, the small-diameter fibers lying superficially to the larger-diameter fibers. In rays, the pectoral fin muscles are divided into superficial small

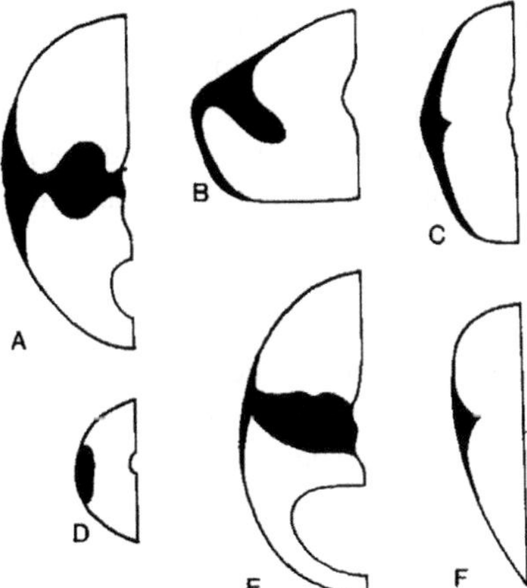

Fig. 4. Diagrammatic transverse sections across the body of various teleosts and elasmobranchs to show disposition of slow (red) muscle fibers in the myotomes (shaded). A, *Katsuwonus*; B, *Rhina*; C, *Squalus*; D, *Atherina*; E, *Euthynnus*; F, *Alosa*. A, redrawn from Rayner and Keenan (1967); remainder original.

bundles of small-diameter fibers, which arise on the pectoral girdle, and pass along the fin rays, to insert about halfway across the "wing" of the ray. These muscle fibers are, therefore, very long. By contrast, the larger-diameter fibers of the main portion of the fin ray musculature are arranged obliquely to the axis of the fin ray, inserting on the fin ray, and upon a connective tissue sheet underlying the superficial slow fibers. In this way, these fibers are virtually pinnate in arrangement. Calow and Alexander (1973) have shown how more power can be extracted from a pinnate fiber organization. In those teleosts where the fins are used for locomotion, e.g., *Cymatogaster* (Webb, 1975), or *Gasteropelecus*, the fibers are not pinnate, so far as is known. The arrangement in Holocephali is probably similar to that of the rays (see Kryvi and Totland, 1978).

IV. FIBER TYPES

A. General Considerations

The muscle fibers of the myotomes or fins of few fishes have been examined histologically, yet fewer physiologically, but the different muscle fiber types found are remarkably similar in those fish where they have been studied. Although there are great differences in general morphology between, say, acraniates and dipnoi, their locomotor muscle fibers are arranged in fundamentally the same way. It is reasonable therefore to generalize from the few cases which have been examined. At first sight this may seem surprising, but it is a consequence of the density of the medium in which fish swim.

The greater part of the drag opposing forward motion in most fishes is skin friction drag (see Chapter 3), approximately proportional to V^2, so that the thrust required at constant speed will also be proportional to V^2, the power required from the locomotor muscles proportional to V^3. There are complications of detail in the power dependence of the velocity (owing to uncertainties about the force/velocity curves of fish muscle and the nature of the boundary layer, see Webb, 1975; Bone, 1975) but this is an order of magnitude argument: Increased speed in water requires a great increase in power output. It is common observation that in individual species speed ranges of at least four times, and sometimes up to twenty times are found; such performances demand a wide range of power from the locomotor system, and have led to a highly specialized arrangement of the locomotor muscle fibers. Faced with the conflicting demands of low speed cruise economy, and short bursts of maximum speed, all fishes have devised the same solution, and have divided the locomotor musculature into two very different parts, each specialized for one of these functions. Naturally enough, as we should expect, the muscle fibers composing each of the two contrasting parts are entirely different in design, differentiated by a whole spectrum of histological and ultrastructural features, as well as by biochemical and physiological criteria. For example, the "cruising" muscle fibers have a high content of mitochondria and high oxidative enzyme activity, as compared with the fiber type used during burst locomotion. The remainder of this chapter will mainly be concerned with considering these differences, which are summarized in Table II and Fig. 5.

Most fish use the same locomotor organs (caudal region and caudal fin, or paired pectoral fins) whether they are cruising slowly or swimming rapidly for short bursts, and the two different parts of the locomotor system are therefore adjacent to each other in the myotomes or in the fin muscles. In rays, for

Table II A Comparison of Fast and Slow Muscle Fibers in Fish

Slow	Fast
Smaller diameter (20–50% of fast)	Larger diameter (may be more than 300 μm)
Well vascularized	Poorly vascularized
Usually abundant myoglobin, red color	No myoglobin, usually white
Abundant large mitochondria	Few smaller mitochondria with fewer cristae
Oxidative enzyme systems	Enzymes of anaerobic glycolysis
Low activity Ca^{2+}-activated myofibrillar ATPase	High activity of enzyme
Little low molecular weight protein	Rich in low molecular weight protein
Stored lipid and glycogen	Glycogen stored, usually little lipid
Myosatellite cells abundant	Fewer myosatellite cells
Sarcotubular system usually less in volume than in fast fibers	Relatively larger sarcotubular system
Z-lines broader than fast fibers in some cases	Z-lines usually thinner than in slow fibers
Distributed cholinergic innervation	Focal or distributed cholinergic innervation
Subjunctional folds usually absent	Subjunctional folds usually present
Lower resting potentials than fast fibers	Higher resting potentials
No propagated muscle action potentials, except under experimental conditions	Propagated muscle action potentials usual; may not always occur during activity of multiply innervated fibers
Long-lasting contractions evoked by depolarizing agents	Brief contractions evoked by depolarizing agents

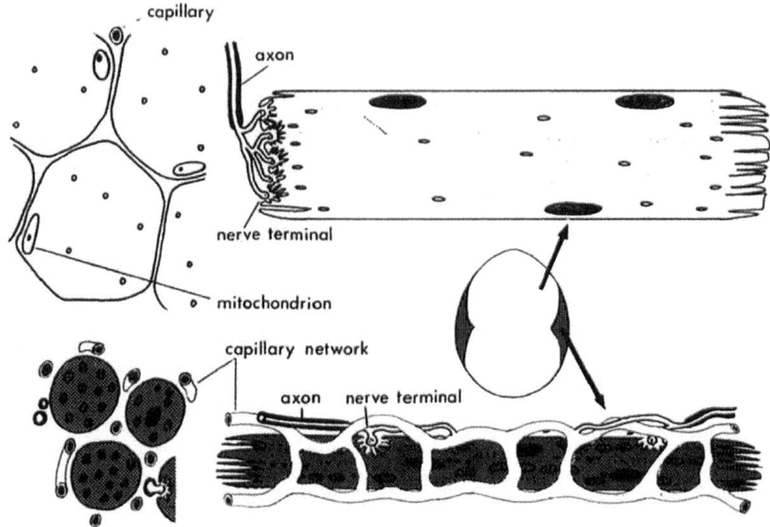

Fig. 5. Schematic comparison between slow fiber (above) and fast fiber. The fast fibers of some teleosts (see text) are multiply innervated. On the left, a transverse section showing comparative capillary and mitochondrial density.

example, the paired fins are employed during slow *and* fast swimming. In Holocephali, however, and such teleosts as labrids, many carangids, notopterids, young stromateoids, and alepocephalids, the paired fins only are used during slow swimming, and the myotomal musculature only during bursts of rapid swimming so that the two main muscle fiber types are physically divorced from each other.

In a crude way, the two main fiber types can be recognized macroscopically, even where they are both found together, because in most fish, fiber types are segregated so that a region of fibers rich in myoglobin with a large vascular bed appears as a red or pink zone as compared with the adjacent pale zone of fibers of contrasting type. The vascular bed of teleost red and white muscle has been described by Mosse (1978) and by Bone (in press). The striking difference between the vascularization of the two main fiber types can be seen by the tenfold difference in the capillary to muscle fiber ratio in favor of the red fibers. In the myotomal musculature of many fish, a superficial red layer of fibers covers the main mass of myotomal white fibers (Fig. 4). The proportions of these two fiber types differ in different species (Table III). There are some difficulties in comparing data given by different workers, for example, Greer-Walker and Pull (1975) examined the proportion of red and white fibers at a given level in different fish, whereas other workers have calculated the total amounts of the two types of fiber within the myotomes; moreover the proportions of the two fiber types may change with fish size (Magnuson, 1973).

Table III Relative Amounts of Slow and Fast Myotomal Muscle Fibers in Different Fish[a]

Species	Percentage of slow fibers in caudal region	Percentage of slow fibers overall
Scyliorhinus canicula	18	8
Prionace glauca	22	10–11
Scomber colias	30	
Gadus virens	11	
Gadus pollachius	3	
Chimaera monstrosa[b]	0.6	
Capros aper[b]	0.5	
Raia spp.[b]	0.0	

[a]Data for teleosts and *Chimaera* from Greer-Walker and Pull (1975).
[b]The last three fish swim slowly by means of the paired fins.

In 84 species examined by Greer-Walker and Pull (1975), red muscle fibers never constituted more than a quarter of the total myotomal musculature, and in most less than 10%. If only from consideration of the power output required from the musculature at different speeds, this distribution would suggest that the red fibers were employed during cruising, the much larger mass of white fibers during bursts of speed (see Section V).

In few fish is this procrustean division of fiber types into red and white an absolute one, and although it is convenient to begin by considering the locomotor apparatus as composed of two main fiber types, it is important to avoid the temptation of falling into what Austin (1962) called the deeply ingrained worship of tidy-looking dichotomies. In most fish there are more than two fiber types and, apparently, some overlap in function between these different types. It is for these reasons that the original simple division of myotomal muscle fibers by their color [as, for example, by Lorenzini (1678) or by Arloing and Lavocat (1875)] is probably best abandoned. Nevertheless, the proportion of visibly red muscle in the locomotor apparatus gives some kind of an indication of the habits of the fish, and various workers have commented on this relationship (Boddeke *et al.*, 1959; Bone, 1966; Greer-Walker and Pull, 1975); color is still a useful macroscopic guide to the fiber types which are found. This is, naturally, of particular importance in biochemical or electromyographic investigations. It is difficult to decide on an acceptable alternative to color as a description of the two main fiber types that does not have unsupported biochemical or physiological implications; but on the whole, it seems most appropriate to contrast slow and fast fibers.

It is admittedly unsatisfactory to categorize muscle fibers by their speed when in very few instances only has this been determined, but other suggested terms, such as anaerobic/aerobic, or twitch/non-twitch are still less satisfactory.

In higher vertebrates, for example, in mammals (e.g., Burke *et al.*, 1971), or in Anura (Smith *et al.*, 1974), good correlations have been established between muscle fiber histochemistry, morphology, and speed of contraction; the method of glycogen depletion has proven invaluable in distinguishing the scattered fibers belonging to a single motor unit. The situation is very different in fish, for although morphologically and histochemically different fiber types are arranged in discrete zones or layers so that they can be distinguished by their *position*, the inconvenient physiological "preparation" of the myotome has not been investigated in any detail, so that no correlations have been established between fiber types and contraction velocity. Apart from crude experiments such as those of Ranvier (1873), Barets (1961), and Bone (1966) differentiating the contractions of slow and fast muscle fibers, nothing has been done in this regard for locomotor muscle fibers, apart from the elegant work of Teräväinen and Rovainen (1971) on lamprey slow and fast myotomal units. The significance of the various morphologically different slow or fast fiber types is therefore not clear, and the problem is complicated by the continuous growth of fish, for this may mean that some fiber types should be regarded simply as stages in the development of others, rather than as types of especial contraction velocity and function. Thus Barker (1968) suggests that certain elasmobranch fibers are "slowtwitch" fibers, instead of "immature" twitch fibers (Bone, 1966), by analogy with the fiber types of higher forms, but this analogy may be invalid. Davies (1972) has shown that in mammals substantial changes take place in fiber types (based on histochemical criteria) as the muscle fiber population adapts to increase in body weight during growth; it is not yet known whether similar changes take place in fish muscle fibers, for detailed developmental studies are not available for any fish group (see Section IV,D). However, there is biochemical evidence, for example, in *Anguilla anguilla* (Bostrom and Johansson, 1972), that there are significant changes in enzyme activity profiles during development and growth, and these are probably reflected in such morphological parameters as mitochondrial content and vascularization.

In most fish the slow fibers form a superficial sheet (the Seitenlinie) covering the main mass of myotomal fast fibers. In rays, the slow fibers form a superficial zone in the fin ray muscle bundles. Presumably in these peripheral positions they are at a better mechanical advantage. They are only found deep in the myotome in secondarily flattened fish such as *Rhina*, or in fish where the slow fibers are operated above ambient temperature (Carey and Teal, 1966) (Fig. 4). There is some evidence that superficial and "deep" slow fibers in tuna are different in properties (see Section IV,F,5). It may be that this reflects the effects of different fiber contraction velocity on different positions within the myotome, and myotomal muscle fiber arrangement and physiology in tuna should repay further study (see Sharp and Dizon, in press).

B. Histology

Myotomal muscle fibers are often very long (in large fish, several centimeters) and insert at both ends into connective tissue sheets; fingers of connective tissue push into the ends of the muscle fibers in a complex interdigitation. At these points, there are couplings between the inpocketed tubes containing collagen fibers and the sarcoplasmic reticulum. These terminal couplings (first observed in anuran and lamprey muscle fibers by Nakao, 1975) are also found in elasmobranch fibers, and probably in all fish fibers. Nakao suggested that the couplings represented sites of calcium transfer and were in some way related to growth in length of myofibrils, but there is no evidence for this function. It is perhaps more probable that the couplings represent the response of the sarcoplasmic reticulum to an ingrowing portion of the sarcolemma, analagous to the SR response to the T-system. The organization of the sarcoplasmic reticulum and T-system is similar in most fish groups, triads occurring at Z-line level (e.g., Patterson and Goldspink, 1972; Kryvi, 1977; Kryvi and Totland, 1978). Exceptions to this general arrangement are seen in the Agnatha and Acrania.

Few quantitative studies of SR and T-systems in fish locomotor muscle have been carried out. Most of these (e.g., Hidaka and Toida, 1969; Nag, 1972; Korneliussen and Nicolaysen, 1975; Kryvi, 1977) agree with studies on other vertebrates where it is found that in slow fibers these systems concerned with activation are of lesser extent than in fast fibers. However, Patterson and Goldspink (1972) were unable to find significant differences between these systems in fast and slow myotomal fibers of *Gadus virens*, in agreement with Kilarski's (1967) original observations. It is, perhaps, probable that in these cases where no obvious quantitative difference between the SR of slow and fast fibers can be discerned, the fibers may overlap in function, and the dichotomy of function may not be so clear as once assumed. For example, there is indirect evidence for *Gadus virens* that the "fast" fibers may be active at slow speeds (Greer-Walker and Pull, 1973).

Morphological correlates of the metabolism of the different fiber types (manifested by their oxygen consumption, e.g., Gordon, 1968; Modigh and Tota, 1975) are seen in the different abundance, size, and cristal complexity of mitochondria (see Section IV,C), and also in the abundance of stored metabolites. Glycogen particles are found in all fiber types, usually as small units, but occasionally (e.g., in *Scyliorhinus* white fibers) in chain formations. Lipid is chiefly found in slow fibers, though in some species, such as *Squalus, Cetorhinus*, and *Ruvettus*, buoyancy lipid may be stored in fast fibers (Bone and Roberts, 1969; Bone, 1972b). In "fatty" fish, such as herring or mackerel (Bone, in press), not only is lipid stored in the slow fibers, but there are fat cells among the muscle fibers (Fig. 6G). Metabolic lipids of this kind, as distinct from buoyancy lipid, are often found in close association with mitochondria as pointed out by Nishihara (1967) and Nag (1972). Ultrastructural investigations of fish storing wax esters in muscle fibers for buoyancy would be of some interest; it is possible that "metabolic" and "buoyancy" lipid stores may be differently compartmented within the muscle fibers. During starvation (Greer-Walker, 1971; Johnston and Goldspink, 1973a,b), the different fiber types are affected differently; glycogen and protein diminish in the fast fibers although lipid and protein are little affected in the slow fibers. The result of protein loss is to increase the water content of the fibers to a maximum around 90%. A fairly large group of mesopelagic teleosts belonging to different families have the myotomal musculature naturally very watery as part of a general reduction of dense components in the absence of a swimbladder (Denton and Marshall, 1958; Blaxter *et al.*, 1971). There has been no systematic investigation of the locomotor muscles in such fish; the few that I have examined (alepocephalids, stromateoids, stomatoids) have achieved economy of the dense myofibrillar protein components by reducing muscle fiber diameter and increasing interfiber spacing, as compared with normally dense fish of the same size. Greer-Walker (1971) found that after 130 days starvation, white (fast) fibers of cod were reduced in diameter by 40%, red (slow) fibers by 15%. Ultrastructural studies by Johnston and Goldspink (1973b) on carp starved experimentally for 16 weeks showed that protein decline in the white fibers due to starvation chiefly represented loss of myofibrillar protein. In the red fibers, mitochondria degenerated and disappeared. There are then some similarities between fish where reduction of dense components is a consequence of starvation, and the mesopelagic fish where it is part of the normal life of the fish.

It has been found that both slow and fast fibers increase significantly in diameter during exercise (Greer-Walker and Pull, 1973). Bainbridge (1962) found no appreciable change in total muscle mass (in trout) after 1 year of continuous swimming, but his measurements did not take account of the (small) amount of slow muscle involved in this exercise. Slow fibers are smaller in diameter than fast fibers [between 22 and 53% on a series of marine teleosts examined by Greer-Walker and Pull (1975)] and are usually rich in myoglobin. Wittenberg (1970) has given grounds for supposing that myoglobin is equally distributed throughout the slow fibers, despite earlier attempts to localize it histochemically in specific regions of the sarcomeres in formalin-fixed material.

In some groups, for example, Anura (Smith and Ovalle, 1973), several distinct fiber types are recognizable by their myofibrillar patterning [recalling the

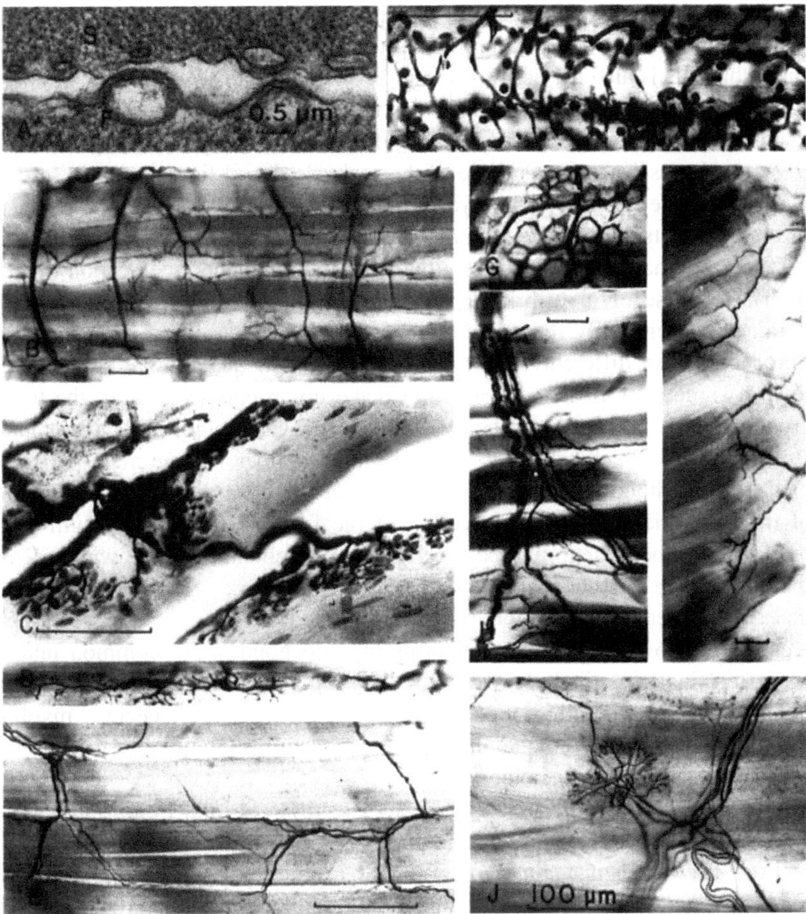

Fig. 6. (A) Transverse section of the edges of a slow (S) and a fast (F) myotomal muscle fiber of amphioxus, showing the larger subsarcolemmal cisternae (representing the sarcoplasmic reticulum) in the fast fiber. (B) Multiply innervated myotomal fast fibers in *Euthynnus*. (C) Focal innervation of pectoral fin fast fibers in *Scyliorhinus*. (D) Single end formation from multiply innervated slow fibers of pectoral fin of *Scyliorhinus*. (E) Multiply innervated fast fibers of pectoral fin of *Periopthalmus*. (F) Rich vascular bed of slow fibers from myotome of *Esox*. (G) Lipid-filled cells adjoining capillaries in the fast myotomal musculature of *Gempylus serpens*; these cells probably contain wax esters to provide static lift. (H) Large-diameter axon passing to focally innervated fast fiber portion of pectoral fin muscle in *Clupea*. Note division of axon at arrows. (I) Terminal pattern of innervation in the myotomal musculature of *Alepocephalus*. The myoseptum runs vertically at the right, muscle fibers pass from their insertions on the right obliquely to the left. (J) Focal innervation of fast fiber from pectoral fin of *Raia*; several nerve processes pass to motor endplate. All except (A) from whole mounts of silver-impregnated material. Scale bars: 100μm except for (A) 0.5μm.

crude but useful distinction made by Kruger (1950) between Feldenstruktur and Fibrillenstruktur fibers], and it is probable that closer investigations will show that this is also the case for fibers in different fish groups. For example, superficial and slow fibers in sharks are recognizable in this way (Fig. 7) but distinctions between the different slow fibers have not yet been quantified. Very rapid muscles, such as those of the seahorse dorsal fin, are rich in sarcoplasm and have relatively few myofibrils (Bergman, 1964a). A similar paucity of myofilaments occurs in certain mesopelagic fishes where the dense myofilament proteins have been reduced for buoyancy requirements (see Section IV,E).

The obverse of myofibrillar patterning is the mitochondrial distribution within the fibers. Here there is a rather wide range in size and abundance of mitochondria in different fiber types, and although quantitative studies are few, in some slow fibers mitochondria make up around 15–25% of the total cross-sectional area of the fiber (Patterson and Goldspink, 1972; Best and Bone, 1973). In *Scomber*, up to 45% of the slow fiber volume may be occupied by mitochondria (Bone, in press). In most fiber types, the distribution of mitochondria seen in cross section of the fiber is more or less uniform, but in some fibers there may be a peripheral mitochondrial zone, with a central zone where mitochondria are absent or rare. In sharks, Kryvi (1977) found that subsarcolemmal mitochondria make up about 10% of the crosssectional area of the slow fibers. Accumulations of mitochondria are common under nerve terminals, particularly where these are not embedded within indentations of the sarcolemma (see Section IV,E). In some species, myelin figures derived from mitochondria are abundant in different fiber types, and occasional multilamellar bodies are also observed. The possible significance of these organelles is discussed by Kordylewski (1974) and Bone (in press).

Individual fiber types in the different fish groups are considered in the next section. The reader is again warned to bear in mind that the general description of slow and fast fibers given above (summarized in Table II and Fig. 5) represent in most fish the different ends of a discontinuous spectrum of fiber types, so that in a given fish there may be several morphologically distinct slow fibers, or two fast fiber types. The essential plasticity of the muscular system, well shown by the changes between silver and yellow eels (Bostrom and Johansson, 1972), naturally means that fiber types may differ more in related fishes of different habit than between, say, elasmobranchs and teleosts of the same habit, but Acrania and Agnatha have peculiar arrangements of their myotomal fibers that are not seen in other fish.

C. Ultrastructure and Histochemistry in Different Fish Groups

1. *Acrania*

In amphioxus (*Branchiostoma*), Flood (1966, 1968) has shown that there are three distinct fiber types, each only a few micrometers thick (Fig. 8). A

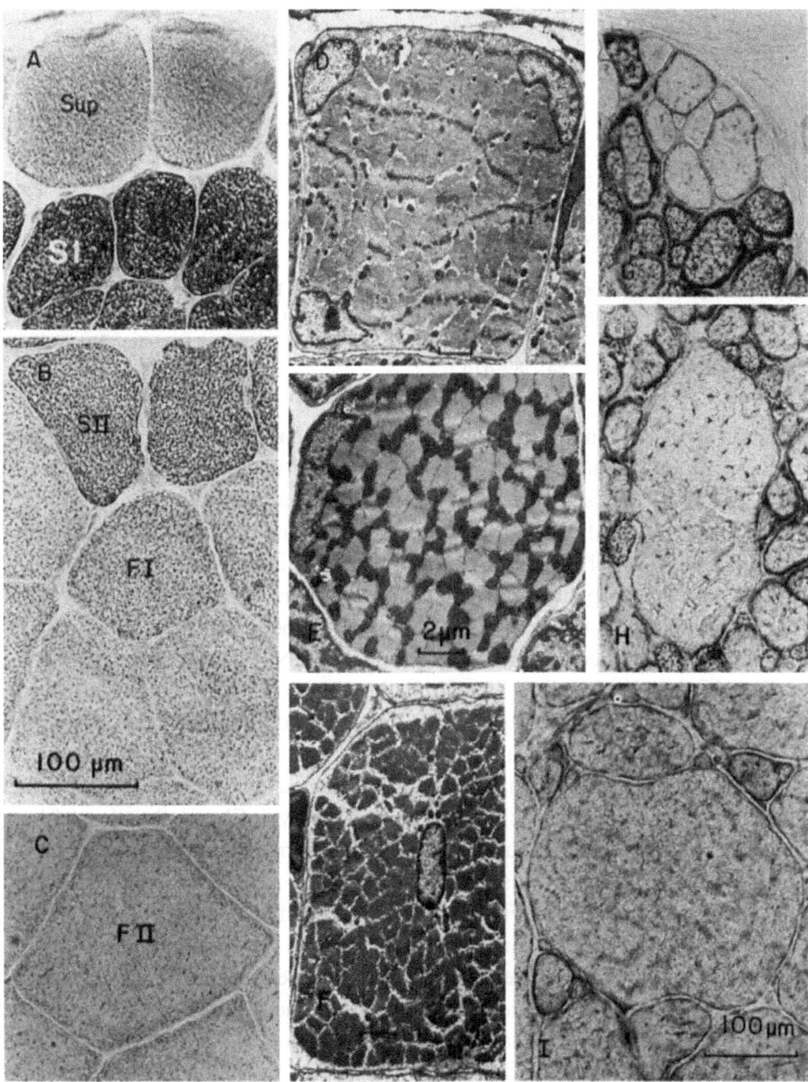

Fig. 7. (A, B, and C) Histochemical differentiation of myotomal fiber types in *Scyliorhinus*. Stained for malate dehydrogenase activity. Sup, superficial fibers; SI and SII, type I and type II slow fibers; FI and FII, type I and type II fast fibers. All to same scale. (D, E, and F) Low power electron micrographs of superficial, type I slow, and type II fast fibers respectively; from young specimen of *Scyliorhinus*. Note differences in mitochondrial size and abundance, and in myofibrillar patterning. All to same scale. (G, H, and I) Different fiber types in the myotome of *Lophius*, differentiated by succinic dehydrogenase staining. Note SDH-negative small diameter fibers in (G) (the superficial zone of the myotome), large SDH-negative fibers in (H) surrounded by SDH-positive fibers, and in (I), great variation in size of SDH-negative fibers from the fast zone of the myotome. All to same scale. Scale bars, 100 μm except for D, E, and F, 2 μm.

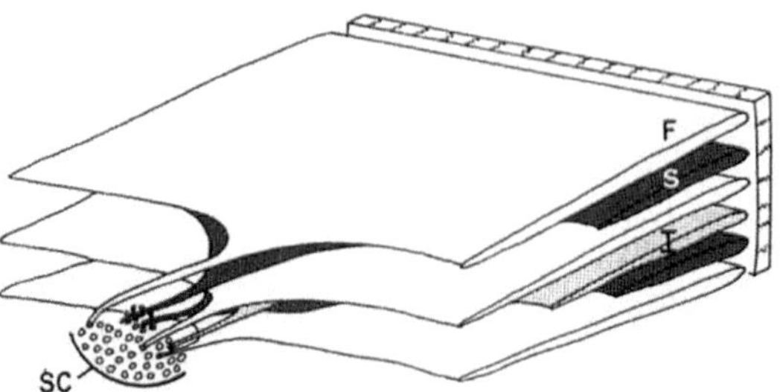

Fig. 8. Diagrammatic view of amphioxus myotome as seen from an internal aspect, showing three types of muscle fibers, and their connection (at SC) with the motor neurons at the surface of the spinal cord. F, fast fiber; S, slow fiber; I, intermediate fiber. (Redrawn after Flood, 1968, *Z. Zellforsch. Mikrosk. Anat.* **84**, 389–416.)

T-system is absent, and in these exceedingly flattened fibers, the sarcoplasmic reticulum is represented only by subsarcolemmal cisternae (Hagiwara *et al.*, 1971; Flood, 1977) (see Fig. 6A). Similar cisternae (in addition to the more usual triad couplings) have been reported in various muscles of higher vertebrates (Spray *et al.*, 1974); I have observed them in slow fibers of *Scyliorhinus* and they are probably figured in lamprey fibers by Teräväinen (1971, his Fig. 21). Superficial fibers contain abundant mitochondria and glycogen, as compared with the deeper mitochondria-poor fibers. The two fiber types send processes to the spinal cord (forming the ventral root "nerves"); each is in connection with a different region of the spinal cord. A third intermediate type of fiber was found to share this innervation region with the deep fibers. By analogy with other fish, it may be tacitly assumed that the superficial fibers (succinic dehydrogenase-positive) represent "slow" fibers, the deep fibers "fast" fibers. The unusual subsarcolemmal cisternae representing the sarcoplasmic reticulum support this identification, since these are larger in the deep than the superficial fibers (Fig. 6A). The status of the intermediate fibers is not clear. Flood suggests that they may perhaps be immature deep fibers, but the ontogeny of the system has not yet been examined at the ultrastructural level. Hagiwara and Kidokoro (1971) obtained physio-logical evidence for electrotonic coupling of adjacent fibers, but as they pointed out, no gap junctions have been observed between fibers.

2. *Agnatha*

In lampreys, the myotomal muscle fibers are arranged in compartments (Fig. 9) built from more or less cylindrical small-diameter slow parietal fibers, enclosing several layers of flattened fast central fibers. In different species the arrangement of the central fibers differs slightly, but in all, the central fibers closest to the parietal slow fibers are different in several respects from those in the middle of the compartment. Thus these "juxtaparietal" fibers are richer in mitochondria and oxidative enzymes than are the interior central fibers and

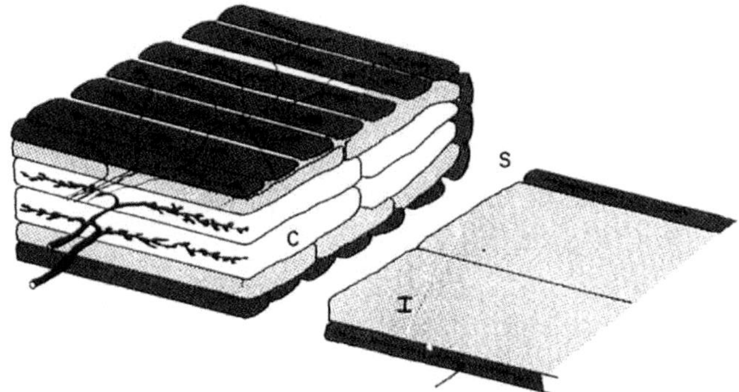

Fig. 9. Diagrammatic view of lamprey myotome from an internal aspect. Note absence of innervation of intermediate fibers (I); innervation of deep central fibers (C); and multiple innervation of slow fibers bordering muscle unit (S).

have a richer capillary bed (Lie, 1974). On histological or histochemical grounds they might be considered to be an intermediate fiber type. However, ultrastructural and physiological investigations by Teräväinen (1971) have shown that all central fibers within a compartment are coupled together electrically, and that all contract together, although only the most central are innervated. There are desmosomal connections between central fibers and between central and parietal fibers, but (as in amphioxus) the usual gap junctions correlated with electrical coupling were not observed by Teräväinen, or by Jasper (1967). SR and T-systems triads occur at Z-line level, but the T-tubules may run longitudinally along one sarcomere (Jasper, 1967).

Teräväinen's work showed conclusively that, despite some morphological diversity, the central fast fibers all operated as a unit. This result is of importance, for it shows that so-called intermediate fiber types recognized by virtue of ultrastructural or histochemical properties, need not have different functional characteristics from neighboring fibers of similar innervation pattern. We should be wary of multiplying functional fiber types on purely morphological grounds. Lie (1974) noted that in ammocoetes of *Lampetra fluviatilis* the compartments contained only a single central fiber, flanked by two intermediate fibers, whereas the adult compartments contained one or two central fibers. He concluded that the new central fibers had differentiated from intermediate fibers and that the latter could be considered as stages of development of central fast fibers.

In hagfish, detailed ultrastructural studies by Korneliussen and Nicolaysen (1973, 1975) have established three morphologically and histochemically distinct fiber types, first defined histochemically by Flood and Storm-Mathisen (1962) and by Dahl and Nicolaysen (1971). As in lampreys, the *Myxine* myotome is divided into muscle units composed of a regular arrangement of the different fiber types (see Fig. 10), and the three fiber types are segregated in

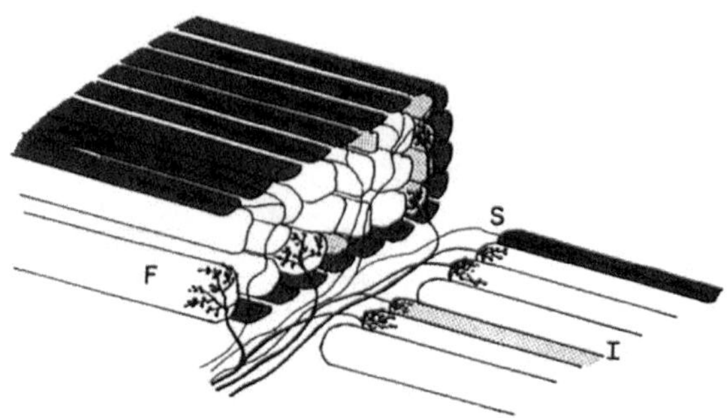

Fig. 10. Hagfish myotome viewed from an internal aspect. Fast fibers (F) terminally innervated, slow fibers (S) innervated by axons passing onto fibers from both myoseptal ends. Occasional intermediate fibers (I) scattered around borders of muscle unit next to slow fibers; these are apparently terminally innervated.

these units in a regular way. Ultrastructurally, the three fiber types are distinguished by their content of glycogen, lipid, and mitochondria, by the pattern of the myofilament fields, and by the organization of the Z-lines. All three fiber types possess M-lines; all three possess a T-system with triads at the A–I junction (unlike those of other fishes); in each fiber type there is a different relation between fiber volume and volume of the T-system. The familiar arrangement of the T-system as a collar around the muscle fiber is not found in all fiber types in *Myxine*, and Korneliussen and Nicolaysen (1975) ingeniously analyzed the density of triads to show that slow fiber myofilament bundles are only rarely encircled by T-tubules, whereas fast fibers are rarely devoid of them.

Regular elegant views of the T-system and its segmental relations with the sarcoplasmic reticulum have been demonstrated in many teleosts (e.g., Kilarski, 1965, 1967), so that it may seem unusual to find such diversity in arrangement as demonstrated for *Myxine*. Nevertheless, as will be seen, a similar less ordered system is found in elasmobranch myotomal muscle, and is probably common during muscle fiber ontogeny in different fish groups. The ultrastructural differences between the three fiber types briefly considered above are similar to those of the elasmobranch, where they will be considered in more detail. It is important to notice that physiological investigation (Andersen *et al.*, 1963) has only distinguished two functional fiber types, and the significance of the (morphologically) intermediate fibers is not yet clear, although myosin ATPase activity (Dahl and Nicolaysen, 1971) suggests that they may be intermediate in contraction speed.

3. *Elasmobranchs*

In the dogfish, *Scyliorhinus*, which has been studied in the most detail, four fiber types were initially recognized in the myotomes, on the basis of color, lipid content, and innervation (Bone, 1966). However, closer investigation

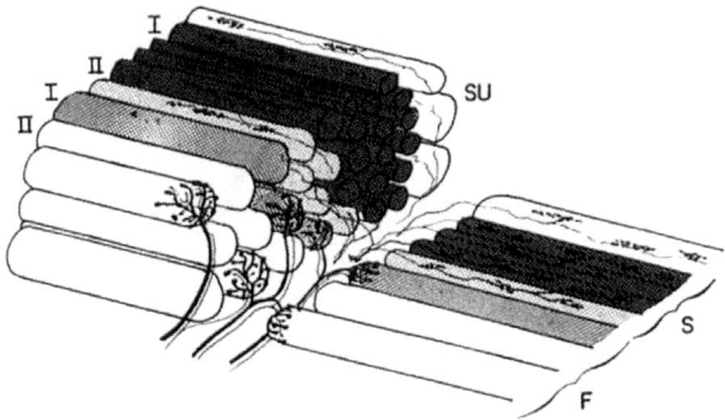

Fig. 11. *Scyliorhinus* myotome from an internal aspect. Superficial fibers (SU) overlie slow fibers (I and II); fast fibers are terminally innervated and mainly of type I. Note dual innervation of fast fibers.

(Bone *et al.*, 1978b) has shown that while the distinction between fast, slow, and superficial fibers seems to be an absolute one, the fast and slow fiber types should each be subdivided further into two types. There are, then, in the dogfish myotome five types of muscle fibers recognizable upon structural and histochemical grounds [just as there are in the anuran limb (Smith and Ovalle, 1973)], but this is not to say that these all play different functional roles (see Section IV,F,4). The general arrangement of the myotome is seen diagrammatically in Fig. 11; Figure 7A–C shows the distribution of fiber types seen after malate dehydrogenase staining. Similar results are obtained when sections are incubated for other oxidative enzymes such as citrate or succinate dehydrogenase (SDH). After incubation for Ca^{2+}-activated myofibrillar ATPase (Bone and Chubb, 1978), the same five fiber types are distinguishable; the ATPase activity increases from the outer surface of the myotome inward and is lowest in the superficial fibers and highest in the inner fast fibers.

The outer border of the myotome consists of a single, sometimes interrupted, layer of SDH-negative *superficial* fibers, forming a sort of thin skin over the myotome. Such superficial fibers are not found in all sharks. I have not observed them in the spurdog *Squalus acanthias* (Fig. 12), for example, and they do not occur in *Galeus* or *Etmopterus* (Kryvi, 1977). Immediately below (internal to) the superficial fibers are the typical *slow* fibers, or type I slow fibers, which are strongly SDH-positive. There may be a number of layers of these type I slow fibers, but they then merge into a second type of slow fiber which is less strongly SDH-positive, and usually larger in diameter than the type I fibers. There are two or three layers of these type II slow fibers, and an abrupt discontinuity where fibers that are nearly completely SDH-negative begin and make up the remainder of the myotome. At the zone of contact with the type II slow fibers, these *fast* fibers are only a little larger than the slow fibers in diameter, and faintly SDH-positive. As the interior of

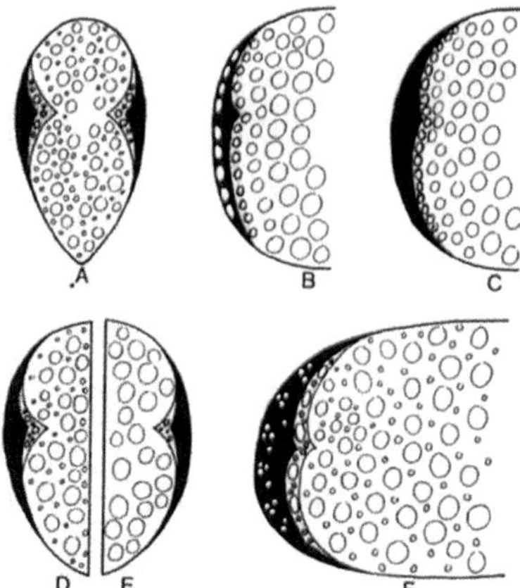

Fig. 12. Diagrammatic transverse sections of postanal region showing fiber types in myotomal musculature. Upper row, fast fibers focally innervated; lower, fast fibers multiply innervated. A, Clupeids, probably also eels; B, *Scyliorhinus*; C, *Squalus*, probably also Dipnoi; D, Salmonids, some cyprinodonts; E, Some cyprinodonts, gadoids; F, *Lophius*. Slow fibers, dark hatching; intermediate fibers, light hatching; fast fibers, unhatched.

the myotome is approached, after three or four rows of fibers, the type I fast fibers merge into larger SDH-negative type II fast fibers. Apart from the mosaic arrangement of salmonids, or the little known *Lophius* pattern, this myotomal organization is as complex as any and hence may be taken as a paradigm of the fiber types in fish locomotor musculature.

The ultrastructure of these different fiber types are shown in Fig. 7D-F. It is evident that the superficial fibers are entirely distinct from the adjacent type I slow fibers, and from the type I fast fibers with which Brotchi (1969) equated them. The inconspicuous M-line, virtual absence of lipid, the paucity and simplicity of the mitochondria, and the few triads of T-system and SR set these fibers aside as a particular type rather different from either slow or fast fibers. Fortunately, their position in the myotome allows us to categorize them by a noncommittal term! Interestingly enough, the superficial fibers at hatching are characterized by a simple myofibrillar arrangement recalling that of embryonic muscle cells, although the other fiber types are similar to their homologues in the adult fish.

The slow fibers of both types are rather similar ultrastructurally, differing in details of myofibrillar arrangement and mitochondrial density, but they are clearly separated histochemically when myotomal sections are incubated for glucose phosphoisomerase, since the type II slow fibers are positive, while the type I fibers are virtually negative.

The fast fibers of both types differ from both types of slow fiber in the virtual absence of mitochondria, in a thinner Z-line, more extensive sarcoplasmic reticulum, and abundant triads with the T-system. It is not yet clear whether the T-system is significantly different in slow and fast fibers: Identical values of C_m (membrane capacitance) were obtained by Stanfield (1972) for both types of fibers (Table IV). In *Galeus* and *Etmopterus*, Kryvi (1977) found similar values for T-tubule volume in slow, fast, and intermediate fibers, but the volumes of sarcoplasmic reticulum were higher in fast than in slow fibers. There are relatively more mitochondria in the type I fast fibers, and these fibers are weakly SDH-positive (they were originally termed intermediate fibers). Again, there seems to be a transition from type I to type II fast fibers over several fiber rows, the general impression gained is that the type I fibers are in some respects like the adjacent type II slow fibers, but that divorced from this propinquity, the fast fibers are uniformly of type II, evidently highly specialized functionally. In *Torpedo*, but not in *Scyliorhinus*, type I fast fibers are differently innervated to type II fast fibers.

Although ray fin muscle was long ago investigated by Ranvier (1873), it has received little attention since that time. As in the myotomes of sharks, there are slow and fast fibers distinguished by their color, diameter, and SDH activity (Bone and Chubb, 1975); superficial fibers are apparently absent, but there are both SDH-positive and negative small-diameter "slow" fibers. These fibers form a superficial strap in each fin ray muscle, unlike the "pinnate" fast fibers (see Section IV,A).

4. *Teleosts*

Slow and fast fibers (red and white fibers) were first characterized morphologically in teleosts (Arloing and Lavocat, 1875), and subsequent investigations at the ultrastructural level have apparently confirmed the existence of only two fiber types in several teleosts (Buttkus, 1963; Kilarski, 1965, 1967; Nishihara, 1967; Patterson and Goldspink, 1972; Nag, 1972). As usual, the two fiber types are distinguished by diameter, mitochondrial and lipid content, and sometimes by Z-line thickness and by difference in volume of the tubular systems. In some cases, e.g., in the fin muscle of goldfish, *Carassius*, or in the myotomal musculature of pike, *Esox*, no differences were observed between the sarcotubular systems of the two fiber types. As Nag (1972) remarks, it is possible that earlier workers did not appreciate the difficulties of making accurate estimates of the T-system. Since a large part of the membrane capacitance of the fiber resides in the T-system, it is desirable to know values of C_m to check morphological estimates of the extent of the T-system. In *Carassius*, fin muscles, for example, C_m for fast fibers is 7.23 $\mu F/cm^2$, as compared to 2.55 $\mu F/cm^2$ for slow fibers (Hidaka and Toida, 1969); although Nishihara (1967) did not note any difference in extent of the sarcotubular systems in the two fiber types, it is probable that such could only be detected by quantitative electron microscopy. It seems to be the case that the volume of the sarcoplasmic reticulum (that is, the internal tubular system excluding the T-system) may be very similar in both slow and fast fibers, even if the volume of the

Table IV A Comparison of Membrane Properties of Different Fish Muscle Fibers[a]

Species		Resting poten tial (mV)	Diameter (μm)	Input resis tance (MΩ)	R_i[b] (Ω/cm)	R_m (kΩ/cm^2)	τ (msec)	C_m (μF/cm^2)	λ (mm)	Authority
Scyliorhinus										
Myotomal	fast	–85.2	150	0.14	108	1.59	15.9	10.3	2.36	
	Slow	–71.7	50	0.75	136	5.4	47	10.2	2.27	Stanfield (1972)
Carassius										
Fin	fast	–82.4	60	1.01	294	10.9	48.4	7.23	1.9	
	Slow	–73.1	63	1.4	389	7.0	26.6	2.55	2.1	Hidaka and Toida (1969)
Lampetra										
Central	fast	–87.8		0.19		5.0	4.4	2–6	250 μm	Teräväinen (1971)
	Slow	–74.5		8.5		30.0	123	4.0	4.0	
Myxine										
Myotomal	fast	–75–85	56–155		40	5.5		3.68		
	Slow	–46	56.4	10.0	40	39.6		2.25		Nicolaysen (1976a,b)
Amphioxus		–53	1 – 2 μm thick	0.8–1.2			1.2–2.0			Hagiwara and Kidokoro (1971)

[a]Most values are the means of a number of experiments. Different experimental techniques and assumptions complicate direct comparisons between different species.
[b]R_i, specific internal resistance; R_m, membrane resistance; τ, time constant; C_m, membrane capacitance; λ, space constant.

T-system may be different. It is possible that Patterson and Goldspink (1972) were unable to demonstrate sarcotubular differences between slow and fast fibers in *Gadus virens* because SR and T-system were considered together.

In many teleosts (e.g., catfish, Barets, 1952; *Mollienesia*, Franzini-Armstrong and Porter, 1964), the muscle fibers have peculiar ribbonlike myofibrillar bundles round the edges of the fiber. In species with fibers of smaller diameter, these elongate bundles may occupy the entire fiber which then resembles a wheel with the myofibrillar bundles arrayed like spokes from a small central sarcoplasmic hub.

This ribbonlike myofibrillar arrangement [which was utilized by Lansimäki (1910) to categorize several muscle fiber types] is apparently unique to teleosts. However, sufficient investigation has not been made in such groups as Holostei and Dipnoi to exclude these: Certainly ribbon-myofibrils are not found in elasmobranchs, holocephali, or higher vertebrates, so far as is known. Lansimäki's different categories have not been found valuable in distinguishing functionally different muscle fiber types, but the simple duality of teleost muscle fiber types described so far (even if true in the cases mentioned), is in many teleosts complicated by the existence of "intermediate" fibers, and by the mosaic organization of the myotomes in different species. Barets (1952) described type I fast fibers (his aberrant fibers) in the catfish *Ameiurus*, and Greer-Walker (1970), using lipid staining noted in cod (*G. morrhua*), that there was an apparent mosaic arrangement among the type II fast fibers, as well as an intermediate fiber type lying between the slow and fast fibers. The situation in cod has been examined by Korneliussen *et al.* (1978).

Brotchi (1968) described intermediate fibers in carp (*C. carassius*) on the basis of SDH staining [they make up 7% of the myotome (Johnston, 1977)], and both SDH and myofibrillar ATPase activity differentiate several sorts of fiber in the herring *Clupea harengus* (Bone *et al.*, 1978a), in the angler (*Lophius*), and in various other species (Johnston and Tota, 1974; Patterson *et al.*, 1975; Mosse and Hudson, 1977). Figure 13 summarizes these various results. As yet, insufficient ultrastructural investigations have been made to allow comparison of these different fiber types with those of elasmobranchs. It seems evident, however, that the superficial red fibers correspond in all to type I slow fibers, but it is not clear whether "pink" or intermediate muscle fiber types should be considered as part of the slow or fast systems. In carp, Johnston *et al.* (1977) have shown intermediate fibers to be active at intermediate swimming speeds (see Section IV,G).

The mosaic arrangement reported in cod by Greer-Walker (1970) is similar to that observed in the deep lateral (fast) muscles of herring and *Lophius* where larger diameter fast fibers (equivalent to type II fast fibers of elasmobranchs) are surrounded by a regular array of smaller diameter fibers which are richer in mitochondria, slightly SDH-positive, and contain more glycogen than the type II fibers. These fibers (Fig. 7I) are probably to be equated with the type I fast fibers of elasmobranchs. Korneliussen *et al.* (1978) have

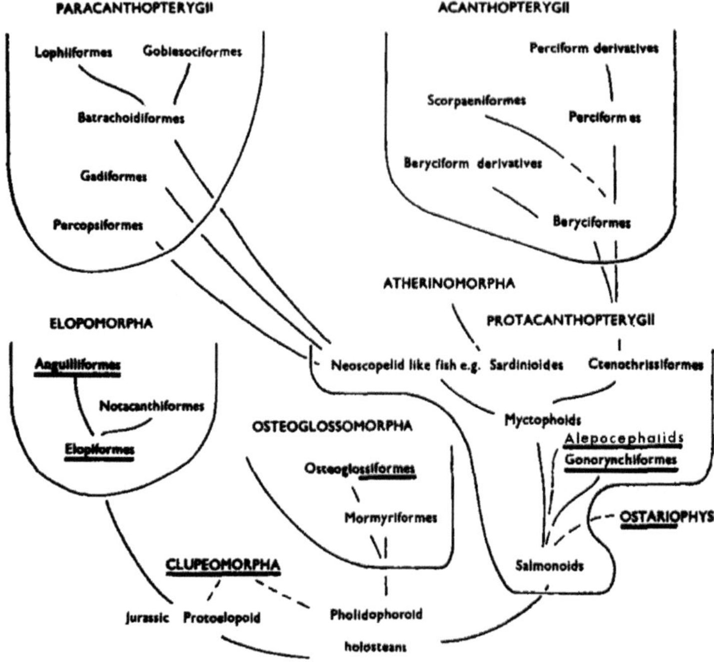

Fig. 13. Distribution of terminally innervated fast fibers (underlined) in teleost groups. All higher teleosts have multiply innervated fast fibers. Only *Hiodon* of the Osteoglossiformes has terminally innervated fast fibers; a few catfish of the Ostariophysi also show terminal innervation. (Modified from Bone, 1970; redrawn from Greenwood *et al.*, 1966, *Bull. Am. Mus. Nat. Hist.* **131**, 339–456.)

categorized no less than seven different fiber types in cod (on the basis of cryostat sections incubated for different enzymes), and three of these are found in the deep "fast" motor system. As these authors imply, it is an open question whether histochemical and morphological diversity of this kind in the teleost fast motor system (indeed in elasmobranchs and agnatha also) reflects functional diversity, or whether it is a simple concomitant of the extensive pattern of growth in fishes. In lampreys, the accident of coupling between the central and intermediate fibers (Teräväinen, 1971, see Section IV,C,2) taken together with Lie's observations on growth strongly suggest that intermediate fibers should properly be considered as growth stages in the formation of type II fast fibers. We do not know whether this is the case in other groups of fish, but the results of Korneliussen *et al.* (1978) certainly are most simply interpreted in this way.

The problem of the status of intermediate fiber types distinguished by morphology and histochemistry is particularly acute in salmonids. Here, it has long been recognized that there are small-diameter "red" fibers scattered among the deeper portion of the myotome, as well as segregated in a superficial lateral zone, as in other fish. The mosaic of the deep portion of the musculature (Boddeke *et al.*, 1959) was at first interpreted as a mixture of fast fibers with slow fibers that were the same as those of the lateral zone, i.e., that salmonids possessed "extra" slow fibers (e.g., Webb, 1971). More recently, Johnston *et al.* (1975) have examined the histochemistry of fiber types in *Salmo gairdneri*, and conclude that the mosaic portion of the myotome is composed of large-diameter fast fibers, and of small-diameter fibers which differ from those of the superficial zone in SDH and myofibrillar ATPase activity. Their conclusions are summarized in Fig. 12D. The similarity with the pattern found in the herring or in *Lophius* (Fig. 12F) is striking, and it is equally striking that there is in the mosaic muscle an almost continuous distribution of fiber size (from 15 to 95μm). Evidence from electromyography of salmonids (Hudson, 1973; Bone *et al.*, 1978a) considered in Section IV,G, is conflicting, and as yet, insufficient to decide whether slow fibers are intermingled with fast fibers, or whether the morphologically and histochemically similar mosaic of clupeids and gadoids indicates that the small-diameter fibers of the salmonid myotome are growth stages in the development of the larger fibers.

A special case is presented by *Lophius* (Bone and Chubb, 1978) for in the myotomal musculature of this relatively inactive fish, there are a great variety of fiber types. As well as the mosaic fibers of the deep (white) portion of the myotome, already referred to, which resemble those of salmonids or clupeids, *Lophius* is remarkable in having among the usual SDH-positive, lipid-rich, small-diameter fibers of the superficial lateral zone (the type I slow fibers), both bundles of SDHnegative, lipid-poor small-diameter fibers, and also occasional similar fibers of large diameter. This complex situation is seen in Fig. 7G–I. These fiber types are found in *Lophius* of different sizes; their significance is not understood, for although the mosaic arrangement in the deep portion of the myotome may be viewed as a consequence of a pattern of continued growth, it is less easy to interpret the arrangement in the superficial zone in such a way. It is obvious that further investigations are needed, both of the physiology of fiber types, and of their development in the mosaic muscle.

In the fins, so far as is known, slow and fast fibers are segregated, as they are in the fin muscles of other teleost groups (e.g., Nishihara, 1967).

D. Ontogeny of Fiber Types

There are two linked problems in the differentiation of muscle fibers in fish. First, there is differentiation in embryonic development giving rise to the fiber types and relative numbers found on hatching or metamorphosis; second, there are the separate problems raised by the continuous growth in all fish and, in some, the changes occurring prior to a migratory phase (as in *Anguilla*).

As the fish grows, increasing in length and mass, the relative amounts of slow and fast fibers change (at least in some species), in addition to actual increase in numbers of each fiber type. In other species, such as some of the pelagic stromateoids, there is a complete change in locomotor behavior during adolescence, which is probably again reflected in changes in the relative amounts of slow and fast fibers. How are the various fiber types differentiated, and what are the mechanisms for the changes observed in the fiber populations (e.g., Magnuson, 1973) as fish grow?

Although it is manifest that (as Nag and Nursall, 1972, put it) "fish muscle fibers offer an interesting problem in differentiation because of the presence of two types of fibers (white and red) which are involved in two kinds of swimming activities, which in their turn appear in different stages of development of the fish," there are few recent studies of these problems. In teleosts, Nag and Nursall studied the histogenesis of fiber types in the myotomes of *Salmo gairdneri* up to the free-swimming fry stage, and Waterman (1969) examined the development of the same system in the cyprinid *Brachydanio rerio*. Fortunately both accounts are in substantial agreement, strongly indicating that differentiation begins from the inner (medial) face of the myotome outward and that (in the way that Vialleton, 1902, had suggested) deep myoblasts differentiate into deep fast fibers, whereas surface myoblasts later differentiate into the superficial slow fibers. As Waterman emphasized "the two main fiber types exhibit structural differences from the time of their formation and differentiate along separate pathways leading to dissimilar adult configurations."

The somites are initially made up of more or less rounded cells with large intercellular spaces between them; as development continues these cells become more closely apposed and either develop specialized embryonic or focal intermediate junctions (*Brachydanio*) or adjacent cell membranes show indications of fusion (*Salmo*). The medial cells in each somite then elongate and begin fibrillogenesis to become myoblasts, containing bundles of myofilaments and the beginnings of a patterned SR. At this stage, the superficial presumptive myoblasts of the somite are not much elongated, nor do they contain myofilaments. These develop at a later stage, by which time the deep myoblasts have developed myofilament fields and may be considered as young fast muscle cells. During this process, the myoblasts destined to become fast muscle fibers become relatively less rich in mitochondria; indeed, the early myoblasts are, perhaps, more similar to the developed slow fibers than they are to the developed fast fibers. This is not to say that fast fibers pass through an "embryonic" slow stage, or that all myotomal fibers commence as slow fibers; development of the two seems to be quite separate. Waterman found that fibrillogenesis could occur either in mononucleate myoblasts or in the multinucleate later stages of the deep myoblasts, but he did not observe nuclear division or cytoplasmic fusion in his material. On the other hand, Nag and Nursall observed appearances indicating cytoplasmic fusion at different stages and suggest that multinucleated fibers are derived by coalescence of myoblasts and presumptive myoblasts. It is certainly possible that there are species differences and perhaps more significantly, differences between embryos of different dimensions, but further investigations are needed before a "norm" of development can be established.

In any event, considering only the broad categories of slow and fast fibers, these are apparently "fixed" from an early stage, and it is natural to ask whether this dichotomy is regulated by the nervous system. Waterman (1969) observed the first axons in the myotomes close to the medial ends of young superficial muscle cells, and concluded that this fiber type was the first to come under nervous control. Previous large irregular movements of the embryo were held to be due to myogenic contractions of the deep fast fibers, as yet not in connection with the nervous system. In a similar way, Nag and Nursall observed initial twitching movements of advanced embryos within the eggs, and stimulation of early larvae produced only short bursts of vigorous tail beats. Slow tail movements were not observed until a later stage, as Waterman had found. The early development of swimming movements is probably best known in dogfish (Harris and Whiting, 1954), and a preliminary study of the development of fiber types in dogfish ontogeny has shown the sequence to be similar to that in teleosts. That is, deep fast fibers develop first and seem to become innervated later than the slow fibers which appear after the fast fibers in ontogeny. The last fibers to appear are the superficial fibers which even after hatching are notably "embryonic" in appearance. Later development of *Etmopterus* slow and fast fibers has been studied by Kryvi and Eide (1977), who found that postembryonic growth takes place largely by muscle fiber hypertrophy; evidently this process gives rise to the notably uniform diameter of fibers in the fast region of the adult myotome, as compared to that in such teleosts as carp or herring, whose differentiation of fibers from myosatellite cells presumably continues throughout growth. Willemse and van den Berg (1978) examined the growth of myotomal fibers in *Anguilla* over a 2-year period and found that initial growth of red fibers takes place by increasing in diameter; later, new fibers are added, as they are in the white zone of the myotome at all stages, probably from myosatellite cells.

Underlying the early myogenic activity of the deep fast fibers of the myotomes in both dogfish and teleosts are longitudinal connections between the muscle cells of adjacent somites, first observed in teleosts by Waterman. These intermyotomal connections are transient in both groups, presumably disappearing as the deep fibers receive innervation from the segmental nerves.

It does not seem then, that the fiber types differentiate from an uniform population of late myoblasts following outgrowth of motor axons of different motoneuron classes, but it is obvious that further work is needed before the regulation of myogenesis is understood. In this field, fish muscle offers interesting material to the experimental embryologist, for the general agreement that fiber types are fixed from an early stage, and the segregation or zonation of fiber types at once suggests possibilities of experimental alteration to determine the effects of propinquity of one fiber type upon another, or the influence of innervation upon the different fiber types. More detailed histological studies are needed too, particularly upon species which have "intermediate" fibers of various sorts, and upon scombroids with a medial mass of slow fibers.

It will not have escaped notice that Nag and Nursall's description of the development of the salmonid myotomal fiber types up to the free-swimming stage only dealt with superficial slow fibers and deep fast fibers. Yet in the adult salmonid, the deep portion of the myotome is typically "mosaic" (Section IV,C,4), containing a mixture of fiber types. It is not yet clear how this mosaic fiber population arises, nor (in other species) how the relative amounts of slow and fast muscle fibers are altered. The continuous growth shown by fish is reflected in some species, for example, *Lophius* (Figs. 12F and 7G–I), by patterns of muscle fiber types arranged in such a way that the generation of the arrangement is evident. In most fish which have been examined, myosatellite cells are abundant and show in young specimens all stages of development of myofilament fields. In the shark *Galeus melastomus* Kryvi (1975) found that myosatellite cells were about twice as abundant on the slow fibers as on the fast fibers. As he observed, the presence of peripheral microtubules, of extensions deep into the body of the muscle fiber, as well as lysosomes and dense bodies in the satellite cell cytoplasm, all suggest that the functions of fish myosatellite cells have not yet been fully determined and the possibility of some trophic influence on adjacent cells has been suggested (Flood, 1971). No detailed study has yet been made of the organization of satellite cells in fish of different ages (see, however, Kryvi and Eide, 1977), but preliminary examination of *Scyliorhinus* material indicates that they become less abundant as growth proceeds, in accord with observations in teleosts by Nag and Nursall (1969). It is agreed that myosatellite cells are persistent myoblasts, and that increase in fiber number during growth is attributable to differentiation of myosatellite cells, but how this process is regulated is unknown. The view that the mosaic nature of the myotomal population in many teleosts is simply a reflection of continued differentiation of myosatellite cells has certainly the merit of simplicity. If it is true, then growth in sharks seems to occur rather differently, since their myotomes do not show a mosaic of fiber size. Nor, however, does there seem to be a special growth zone (for example, between the deepest slow fibers and the most superficial fast fibers) so that it is not known how the system increases in fiber number. Plainly, slight changes in the rates of differentiation of myosatellite cells in the slow and fast portions of the myotome during growth will result in changes in the relative amounts of slow and fast fibers in the myotomes, such as are found as fish increase in size.

E. Innervation

The innervation of fish muscle fibers presents unusual features since many fish have multiply innervated twitch fibers, and the pattern of innervation is of taxonomic value. There are, further, hints in some fish motoneurons of transmitter substances other than acetylcholine, and of dual innervation of the muscle fibers.

In all fish groups, myotomal superficial slow fibers are multiply innervated by small-diameter myelinated fibers that terminate in en grappe endings (Barets, 1961; Bone, 1964, 1966, 1970; Best and Bone, 1973). Where two types of slow fiber are recognized, as in dogfish, both are innervated in a similar way. In teleosts, the nerve terminals are usually embedded in the sarcolemma, and subjunctional folds are absent (Nishihara, 1967; Barets, cited in Barker, 1968); in dogfish, there are subjunctional folds under terminals on superficial and type I and type II slow fibers (Bone, 1972a). The slow fibers of the hagfish myotome are innervated by two axons only, passing onto the fiber at each of its myoseptal insertions (Andersen *et al.*, 1963) but in other groups, more than two axons probably innervate each slow fiber. In the fins, slow fibers are also multiply innervated, although the endformations may be larger than in the myotomal slow fibers (Fig. 6D and J). The interval between terminals along slow muscle fibers has been measured in a number of preparations and where the membrane constants of the fiber have also been measured there is a normal safety factor of around 10 times. For example, Stanfield (1972) found λ to be around 2.27 mm for slow fibers in dogfish myotomes; motor terminals along these fibers are some 150–200μm apart.

Acetylcholinesterase is demonstrable at the terminals on all slow fibers investigated; in teleosts Pecot-Dechavassine (1961) found that other esterases (e.g., butyrylcholinesterase) are absent. However, an interesting possibility has been raised in hagfish, by Korneliussen's (1973) suggestion that slow fibers may have monoaminergic innervation. He showed that dense core vesicles were especially abundant in nerve terminals on slow fibers of the hagfish myotome and craniovelar muscle, although formaldehyde-induced fluorescence proved inconclusive. Earlier work by Andersen *et al.* (1963) showed that acetylcholine was probably the transmitter at the neuromuscular junctions on slow and fast fibers in the myotome; Korneliussen's observations await further investigation. To judge from pictures by Teräväinen (1971) and Nakao (1976), dense-core vesicles are not found in unusual number in lamprey slow fiber terminals.

The innervation of fast fiber types is, however, of more interest than that of slow fibers, for there are large differences between different fish groups, such that the pattern of innervation may serve as a taxonomic character (Bone, 1970). In all groups except most of the teleosts, the fast fibers are focally innervated at their myoseptal ends (Fig. 6I); sometimes at both ends of the fiber (according to Barets, in *Ameiurus*), but more usually, at one end of the fiber only (in hagfish and elasmobranchs). This terminal innervation is also found in amphibia (at both ends of the fiber) and seems to be the original innervation pattern in lower chordate myotomal fast fibers. Best and Bone (1973) have suggested (see also Bone, 1975) that this innervation pattern is part of the various specializations of the fast fibers to increase power output by devoting the maximum space to myofilaments.

In elasmobranchs, in urodele amphibia, and in certain teleost groups, each fast fiber is apparently innervated by two separate axons, which both contribute to the formation of a single motor endplate. In most of these forms, there is no recognizable difference between the two motor terminals, but in some elasmobranchs (Bone, 1972a) they are recognizable by their different vesicle content. One type of terminal at the endplate contains "typical" cholinergic electron-lucent vesicles 50 nm diameter, while the other contains a high proportion of much larger dense core vesicles (up to 100 nm). As in the hagfish slow fibers, formaldehyde-induced fluorescence studies have so far proved inconclusive, and only acetylcholinesterase has been found in the subsynaptic folds. The significance of this dual innervation is therefore unresolved.

In the dogfish, as in most focally innervated species examined, every type of fast fiber is innervated terminally, even if (as in herring) they may be diverse in size. So both type I and type II fast fibers are terminally innervated in dogfish, and no difference has been recognized between the motor terminals on each fiber type. But in two cases, the "aberrant" fast fibers of *Ameiurus* (Barets, 1952), and the most superficial fast fibers of the myotomes in *Torpedo* (Bone, 1964), there is focal innervation in the midregion of the fiber. Again, the significance of this situation (rather unusual in fish groups with terminal innervation since it has not been observed in clupeids, eels, holostei, dipnoi, sharks, or other rays) is not understood. No intermediate types of innervation are observed, and it does not seem very probable that the terminal innervation pattern is derived from the midregion en plaque type. In *Ameiurus* (and possibly also in *Torpedo*), these peculiarly innervated fast fibers lie external to the connective tissue fascia separating the slow fibers from the main mass of fast fibers. It is premature therefore to equate them with the type I fast fibers in dogfish, and to suppose that their different innervation precludes the type I fast fibers of other fishes from being developmental stages in the formation of type II fibers.

Terminal innervation is found in certain teleost families (Barets, 1961; Bone, 1964, 1970); these have an interesting taxonomic distribution (Fig. 13). It is apparent from Fig. 13 that no acanthopterygians have this pattern and that, of the families regarded as primitive on other grounds, only the salmonids lack the terminal innervation pattern. The systematic position of a number of the groups possessing terminal innervation (for example, the alepocephalids) has not been agreed upon. The pattern of innervation of other taxonomically vagrant groups (e.g., many deep-sea families) has not been examined. The "innate conservatism" of soft parts should allow us to employ

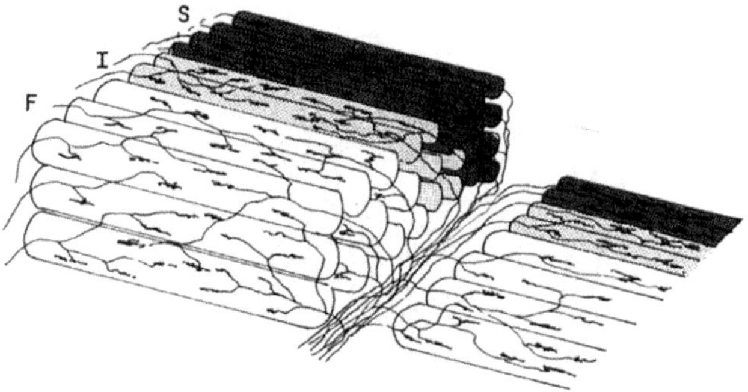

Fig. 14. Higher teleost myotome viewed from an internal aspect showing multiple innervation of all fiber types. It is not known whether any overlap of innervation takes place between fast (F) and intermediate (I) fibers.

muscle innervation as a character in determining the relationships of teleosts where skeletal structures give conflicting evidence for their affinities.

With the exception of the groups indicated in Fig. 13, all teleost families so far investigated (species in 55 families have been examined) have an entirely different innervation of the fast fibers. Each muscle fiber is multiply innervated (Barets, 1961; Hudson, 1969); the innervation is similar to that of the superficial slow fibers, distributed and punctate (Fig. 14). According to Barets the fast fiber innervation is more distributed than that of the superficial slow fibers. As in the slow fibers, the motor terminals are embedded in the sarcolemma and subjunctional folds are absent. The richness of the innervation of teleost multiply innervated fast fibers is astonishing (Fig. 6B and E); the nexus of axons resembles the capillary bed of slow fibers in its abundance and nerve terminals are scattered all along the muscle fibers. The contrast with focal terminal innervation as seen in other fish fast fiber systems is very great and invites functional interpretation (see Section IV,F,5).

Thus far, myotomal muscle fiber types have been considered. In the fin muscle, a similar division exists between the multiply innervated fast fiber fin muscles of higher teleosts (Fig 6E), and the focally innervated fast fibers of other groups (Fig. 6C). In the fin fast muscle fibers, however, innervation is not terminal; large en plaque endings are found in the midregion of the fibers. These end-formations may be very large, and derived from several branches of the same axon which have divided earlier along its course and come into proximity again at the motor endplate. Figure 15 illustrates the complexity of motor terminations in *Torpedo*, perhaps directed to prolonging the period of subjunctional permeability in these large diameter muscle fibers.

The ultrastructure of focally innervated fast fibers in the fins has not been examined, but that of multiply innervated fibers has been studied in several families (e.g., Nishihara, 1967; Bergman, 1964a), and resembles that of the body muscles, viz., in the absence of subjunctional folds.

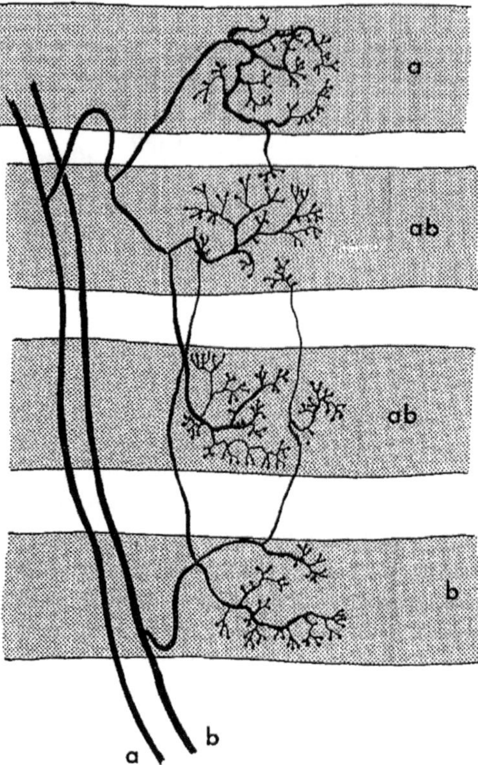

Fig. 15. Drawing from whole mount preparation of fast fibers of pectoral fin in *Torpedo* showing overlap of innervation between muscle fibers; a and b are two axons supplying muscle fibers labeled according to which axons innervate them.

F. Electrical and Mechanical Properties

Several workers have investigated the electrical properties of myotomal muscle fibers in different fish groups (e.g., Barets, 1961; Teräväinen, 1971; Stanfield, 1972), and, in Agnatha, Teräväinen and Rovainen (1971) have been able to obtain simultaneous intracellular records from myotomal muscle fibers and the motoneurons supplying them; but the mechanical properties of myotomal fibers have been little studied, probably because the myoseptal insertions of these fibers are inconvenient for attachment to recording apparatus. The recent work of Wardle (1975) shows that such investigations are likely to prove rewarding. Fin muscle fibers provide better physiological preparations, and their mechanical properties are therefore easier to study (Bergman, 1964b; Hidaka and Toida, 1969; Yamamoto, 1972). In elasmobranchs, fin muscle fibers are often of large diameter, and thus offer suitable material for studies on membrane properties (e.g., Hagiwara and Takahashi, 1974).

1. *Amphioxus*

The interesting work of Hagiwara and his colleagues (Hagiwara and Kidokoro, 1971; Hagiwara *et al.*, 1971) on the myotomal muscle lamellae (Fig. 8) of amphioxus has shown that there are two independent mechanisms for permeability increase; the normal action potential is mainly the result of increase in sodium conductance, but sufficient calcium ions enter during the spike to play a significant role in excitation–contraction coupling. Hagiwara and his colleagues consider that in practice what happens is that the sodium spike is important in conducting an impulse from the central motor endplate along the thin nervelike portion of the muscle fiber to its expanded contractile region where the calcium influx is concerned with excitation of the contractile apparatus. They also present evidence indicating that the subsarcolemmal cisternae of the SR are concerned with sequestering calcium rather than releasing it during contraction (as in higher forms). These unusual properties are of course related to the unique morphology of the system, presumably also responsible for the low values for the membrane resistance and time constant (Table IV).

Electrical coupling between separate muscle cells was detected in half of the tests made; the morphological basis for this has not yet been established (Section IV,C,1) and it would be interesting to know more of the properties of the (morphologically) different fiber types.

2. *Agnatha*

Lampreys and hagfish offer favorable material for determination of the electrical and mechanical properties of muscle fibers. In lampreys, Teräväinen's (1971) analysis has shown that the slow and fast (lateral and central) fibers have different input resistances and time constants, as expected from their differences in dimensions (Table IV); only fast fibers showed propagated overshooting action potentials (Fig. 16). Miniature endplate potentials were

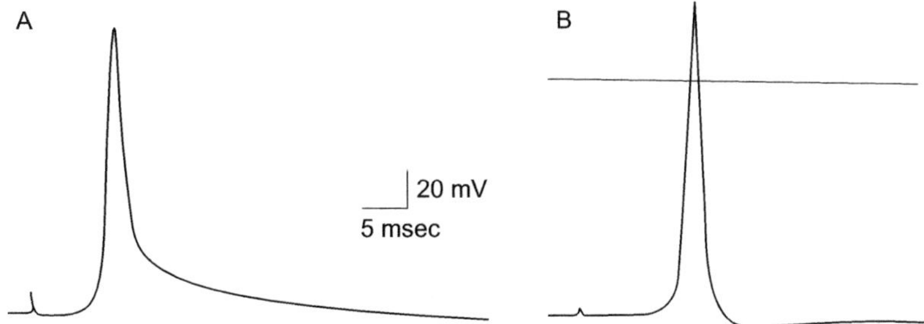

Fig. 16. Overshooting spikes recorded from fast (central) muscle fibers of lamprey. Note that A (an innervated fiber) shows a prolonged after potential, whereas B is a record from a noninnervated fiber coupled to an innervated fiber and thus does not show this endplate potential. (From Teräväinen, 1971, *J. Neurophysiol.* **34**, 954–973.)

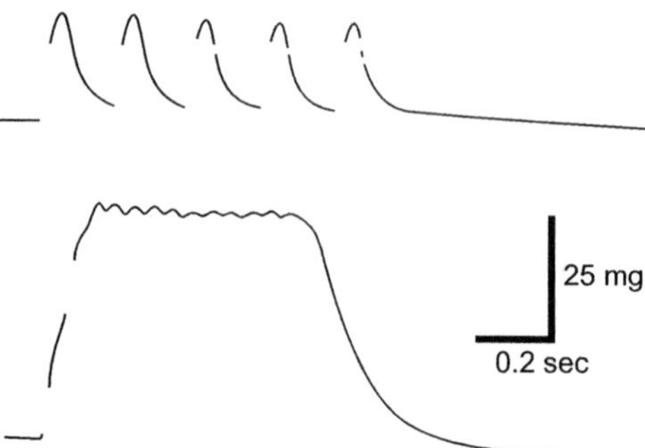

Fig. 17. Tension records from abdominal surface of myotome after nerve stimulation at 6/sec (upper) and 20/sec. Early and late relaxation seen in lower record presumably result from activity of fast and slow fibers, respectively. (From Teräväinen, 1971, *J. Neurophysiol.* **34**, 954–973.)

observed in both slow fibers and in fast central fibers (sometimes of two different rise times in the same fiber), but were not seen in lateral fast fibers. These last are electrically coupled to the central fast fibers, and are not directly innervated themselves. Again, as in amphioxus, gap junctions have not been observed; the morphological basis for electrical coupling has not been determined (Section IV,C,2). Recording tension directly from the abdominal surface of the myotome, Teräväinen was able to show (Fig. 17) that nerve stimulation gave a tetanic plateau at 20/sec, and that relaxation followed a dual time course, presumably because the slow and fast fibers relaxed at different rates.

In hagfish, Andersen *et al.* (1963) obtained similar mechanical results from the myotomes, i.e., single stimuli to the myotomal nerve gave evidence for the existence of slow and fast components, the slow component having a much longer relaxation time. As in lampreys, fast fibers gave propagated overshooting action potentials, from slow fibers only junctional potentials were elicited. But these junction potentials are unusual in their size, up to 30 mV, which is near to zero membrane potential! They have been studied by Alnaes *et al.* (1963). Jansen and his colleagues conclude from records such as that of Fig. 18 where discontinuities are found on the junction potentials, that the slow fibers in hagfish may produce abortive spikes when stimulated indirectly. Stanfield (1972) suggests from his study of slow fibers in the dogfish *Scyliorhinus* (see next section) that the observations in hagfish may be explained in terms of some sodium conductance in these fibers. The ionic basis of electrical activity in hagfish fibers has not been investigated fully, and would seem to be of interest in view of the more recent observations of "spikes" in slow fibers when these are immersed in experimental solutions (e.g., Hidaka and Toida, 1969). Most recently, Nicolaysen (1976a,b) has examined the spread of potential in the T-system of hagfish fast and slow fibers using sinusoidal transmembrane currents.

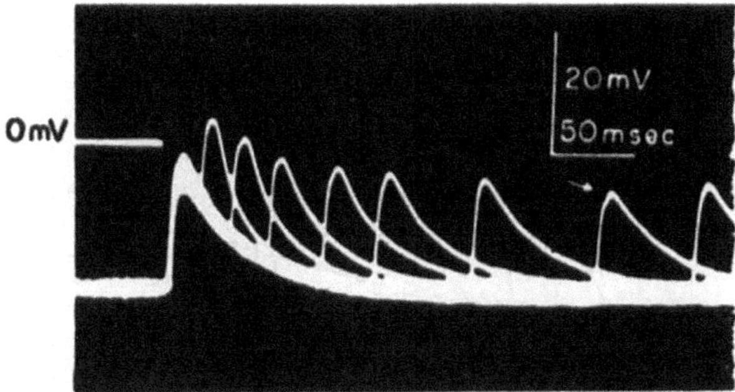

Fig. 18. Junction potentials from hagfish slow fibers. Note discontinuity on rising phase (arrow) interpreted as an abortive spike (see text). (From Andersen *et al.*, 1963).

3. *Elasmobranchs*

Hagiwara and Takahashi (1967, 1974) and Stanfield (1972) have examined the membrane properties of the fin muscles of several species of tropical stingrays, and of the myotomal muscle fibers of the dogfish, *Scyliorhinus.* The cable properties of the myotomal fibers investigated are seen in Table IV. Fast fibers in fins and myotomes gave overshooting propagated action potentials as expected from their focal innervation; slow fibers never gave propagated potentials, but an abortive spike was seen by Stanfield on one occasion. Using a twoelectrode clamp, Stanfield found that eight of twenty-seven myotomal slow fibers showed sufficiently large inward sodium currents as soon as depolarized to suggest that they were capable of propagating action potentials. Six other slow fibers showed no inward sodium current on depolarization, and others showed a small inward sodium current. Figure 19 illustrates active current–voltage relations in the extreme cases. No significant difference in cable properties was found between slow fibers capable of showing a marked conductance change to sodium and those in which no change was found. These results are interesting, for they suggest that reports of abortive spikes and small spike potentials at the break of strong inward currents are to be explained in terms of some sodium conductance, as Stanfield pointed out.

Since Stanfield wrote, there have been further investigations of morphology of the two fiber types, and it is known that the slow fibers can be divided into two types according to their histochemistry and position within the slow fiber portion of the myotome. Yet Stanfield emphasized the absence of sequestration of fibers with and without sodium currents in different parts of the slow fiber portion, so that it is not clear whether type I and type II slow fibers are different in this respect. Although there exists the possibility of propagated events among the slow fibers, no evidence has yet been adduced to suggest that these are found during normal swimming; crude records suggest that the slow fibers do indeed have the mechanical properties of slow fibers, contrasting with the twitch fast fibers.

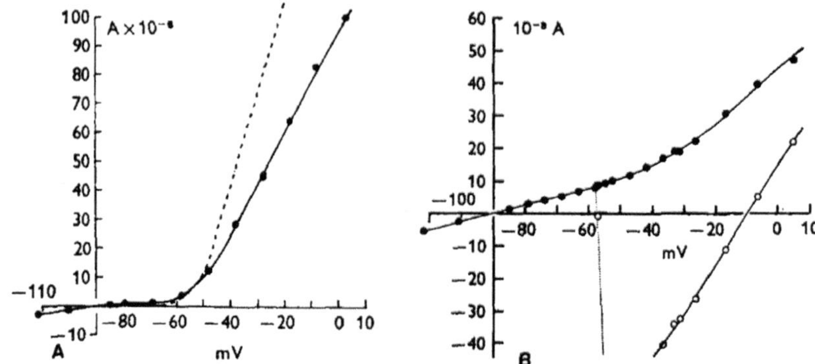

Fig. 19. Active voltage–current relations obtained from dogfish slow fibers using a two-electrode voltage clamp technique. In A, the dashed line shows the voltage–membrane current relation obtained from Cole's theorem. Note delayed rectification and absence of inward sodium current. In B, also a slow fiber, filled points indicate currents flowing at end of 100 msec pulse, open points initial currents flowing at about 2 msec. In this fiber, delayed rectification is again found, but significantly, there are large inward sodium currents, resembling those of the fast fibers. Threshold for the sodium conductance was around 60 mV. [From Stanfield, 1972, *J. Physiol.* (*London*) **222**, 161–186.]

4. *Teleosts*

Barets (1961) and Hudson (1969) have examined myotomal muscle fibers, but most workers have used fin muscle preparations. In a variety of freshwater and marine teleosts there is good agreement between the results obtained from fibers that are innervated in a comparable way. All authors agree that slow (red) fibers of fins or myotomes do not propagate action potentials; the properties of these slow fibers are similar to those of other fish groups; as in elasmobranchs (Stanfield) there are hints of a sodium conductance mechanism in some teleost slow fibers.

Where fast fibers are focally innervated, as in the myotomes of the catfish *Ameiurus* (Barets, 1961), or the fins of *Conger* (Hagiwara and Takahashi, 1967), typical overshooting propagated action potentials are found, similar to those in other vertebrate groups.

Many teleosts, however, possess multiply innervated fast fibers, and here the situation may be different. Most workers have found that muscle action potentials from fibers of this kind are characterized by failure to overshoot, or by overshoots close to zero membrane potential (Takeuchi, 1959; Barets, 1961; Hagiwara and Takahashi, 1967; Hidaka and Toida, 1969). Hudson (1968) found in the marine teleost *Cottus* that 20 mV overshoots (Fig. 20) were obtained provided care was taken to experiment using a Ringer solution containing appropriate values of Ca^{2+} and Mg^{2+} ions. Reduction of these ions by 50 and 30%, respectively, gave overshoots close to zero membrane potential. It is certainly tempting to suppose that appropriate ionic adjustments to the Ringer solutions used by previous workers would have allowed them to

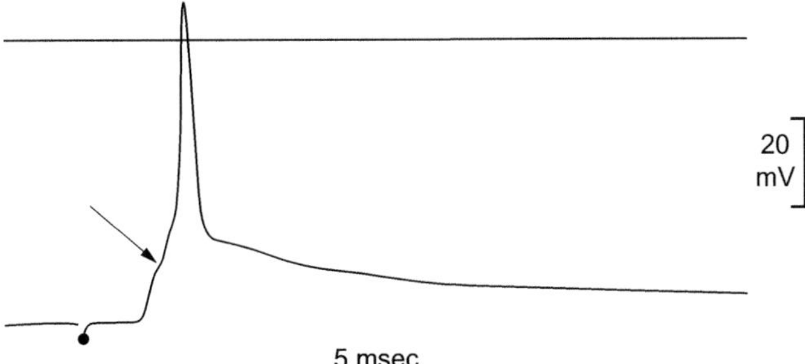

Fig. 20. Overshooting spike from multiply innervated fast myotomal fiber of *Cottus*. The arrow indicates the second of two summated junction potentials on the rising phase. (From Hudson, 1969.)

observe overshooting spikes, as Hudson suggests, but this has yet to be demonstrated. Alternative explanations of nonovershooting spikes are possible (e.g., in some fish high internal sodium may lower the sodium equilibrium potential), and as Hagiwara and Takahashi observe, multiply innervated fibers may be able to afford a lower safety factor than focally innervated fibers since contraction is not uniquely dependent upon propagated action potentials. In this way they could effect economies in ion transfer across the sarcolemma. However this may be, Hudson's (1969) work revealed several interesting points about teleost multiply innervated fast fibers. By simultaneous recording from the nerve which was stimulated, and a single muscle fiber (Fig. 21), Hudson showed that each muscle fiber was innervated by at least two axons in a single spinal nerve and by a similar number of axons in each of four neighboring spinal nerves, a remarkably high degree of polyneuronal innervation.

Until tension records are taken simultaneously with electrical records from multiply innervated fast fibers, it will not be known for sure if such fibers are capable of local contractions and also of twitches following propagated action potentials. Observational data (e.g., Barets, 1961; Takeuchi, 1959; Bone *et al.*, 1978a) suggest that this is the case; the functional advantage of such an arrangement is considered in the next section.

The number of muscle fibers innervated by single axons (i.e., the size of the motor unit) is not well known in either myotomal or fin muscles. Teräväinen and Rovainen (1971) suggest that in lamprey myotomal muscle, about ten to twelve fast motoneurons on each side of the spinal cord innervate an equal number of muscle units on that side, each muscle unit consisting of the central fibers and accompanying electrically coupled (noninnervated) lateral central fibers. In such focally innervated systems it is of course in principle possible to count motor axons in the nerve passing to the muscle, and divide the number obtained into the number of muscle fibers (making due allowance for the presence of different muscle fiber types where necessary).

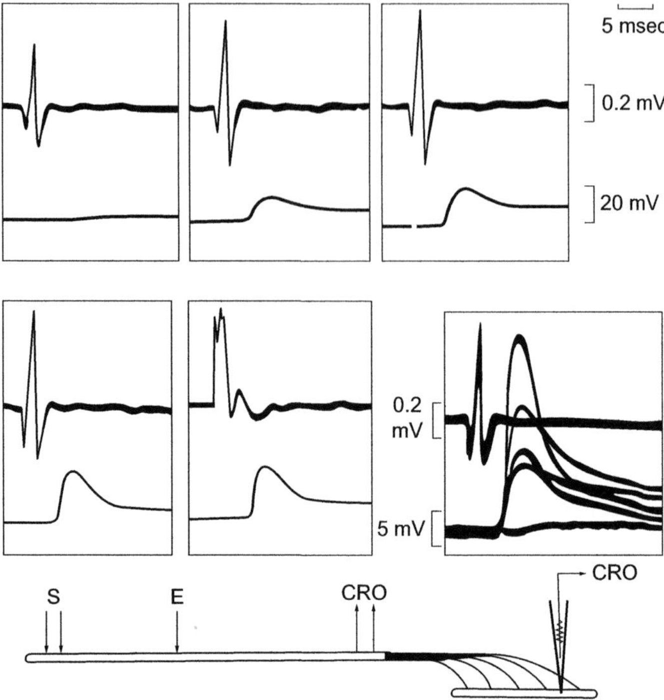

Fig. 21. Simultaneous recordings of the compound action potential from spinal nerve (upper) and concomitant electrical activity of a single muscle fiber (lower), when stimulus intensity is varied. Zero membrane potential indicated in all except (f) by baseline of nerve record. In (f) small changes in stimulus intensity give different (superimposed) muscle responses at similar latency. The experimental situation is shown at the bottom. S, stimulating electrodes; E, earth electrode. CRO, recording electrodes to oscilloscope. (From Hudson, 1969.)

On this basis, the myotomal fast motor unit in *Scyliorhinus* consists of some 50–100 muscle fibers. Since this fiber type is apparently used only during burst swimming, it would not be surmised that it was finely graded. The focally innervated fast fibers of the pectoral fin in herring are supplied by very few large axons which branch repeatedly (Fig. 6H): Gradation must be relatively coarse in this case. It is manifestly more complicated to unravel the possibilities of gradation in the multiply innervated fin or myotomal systems. Roberts (1969b) observed in slowly swimming spinal dogfish that there was good correlation between the duration of the muscle bursts (cf. his Fig. 21), the number of impulses in the burst, and the swimming frequency, indicating that the frequency and composition of the discharges of the motoneurons were controlled by a single mechanism. That is to say, the slow motor system is graded (as in amphibia) by variations in the amplitude and frequency of the junction potentials. Observations on the slow motor system of the unpaired fins led to the same conclusion.

These are slow fibers which apparently do not exhibit propagated muscle action potentials. Gradation in the multiply innervated *fast* fibers of teleosts is considered by Hudson (1973), who suggests that during sustained swimming activity, motor units may be rotated in the fast motor system.

The speed of fish locomotor fibers varies widely, from fusion frequencies up to 120 Hz for seahorse fin muscles (Bergman, 1964b) to 20 Hz for lamprey fast fibers (Teräväinen, 1971). Isometric muscle contraction time for the seahorse fin fibers was as little as 10 msec. Wardle (1975) has given comparative isotonic muscle contraction times for fast myotomal muscle fibers in various teleosts, ranging from 20 to 45 msec, contraction time being related, as expected, to the size of the fish (increasing as fish length increases). Little is known in fish of the factors determining contraction velocity of different fiber types from a single fish. It is possible that the low molecular weight proteins studied by Hamoir and his colleagues (e.g., Syrovy *et al.*, 1970; Hamoir *et al.*, 1972), which are known to be present in fish fast fibers but are much less abundant in slow fibers, may be concerned in some way in regulating muscle contraction velocity, perhaps by their effects on calcium-activated myosin ATPase, acting as calcium buffers in a situation where there are few mitochondria.

G. Functional Role of Different Fiber Types

A variety of suggestions have been made for the functions of the two main fiber types in fish muscle. These are summarized in Bone (1966), but it is now generally agreed that the superficial slow fibers are utilized by the fish for sustained slow-speed swimming or cruising and the deeper fast fibers for bursts of higher speed. Several lines of evidence point to this conclusion. First, direct electromyographic recording from teleosts swimming freely or in tunnel respirometers, or from spinal sharks, has shown that electrical activity is found within the zone of the superficial fibers when the fish is swimming slowly and within the deeper zone of the fast fibers during rapid swimming (Bone, 1966; Rayner and Keenan, 1967; Hudson, 1973; Bone *et al.*, 1978a). Second, biochemical and metabolic studies (reviewed in Bilinski, 1974) have shown that slow fibers operate mainly by aerobic glycolysis and lipolysis, fast fibers by anaerobic glycolysis. Examination of fish after exercise of different kinds has shown the expected utilization of metabolites by slow and fast fibers (see Chapter 8).

It was perhaps natural to follow Arloing and Lavocat (as did Boddeke *et al.*, 1959) in supposing that red and white fibers (or slow and fast fibers) were distinct and separate systems, utilized for different patterns of swimming at different speeds. The earlier electromyographic work on spinal dogfish showed that muscle action potentials could only be recorded from fast fibers in the deep portion of the myotome when the fish was strongly stimulated and swam very vigorously for a few tailbeats (see Fig. 22). At the usual slow

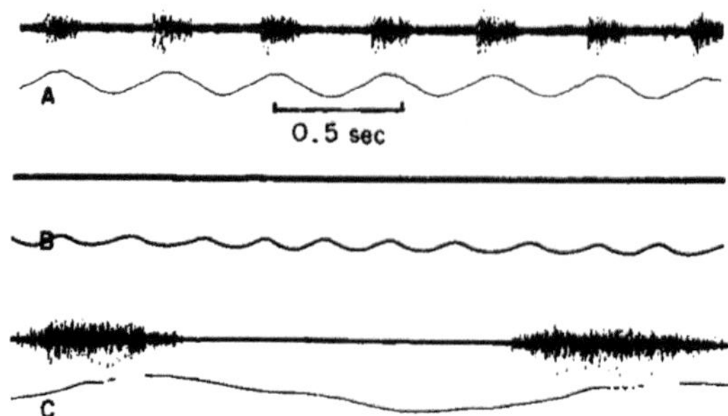

Fig. 22. Records obtained from swimming spinal dogfish using concentric needle electrodes. Upper in each, electrical activity; lower, record of movement of fish. (A) Activity of slow myotomal fibers; (B) absence of activity from fast fibers during slow spinal swimming; (C) prolonged bursts of action potentials from fast fibers during movements evoked by pinching tail. Note that electrical activity in C is recorded at a lower amplification than in A. Time bar: 0.5 sec.

spinal swimming rate of some 35 strokes/min, the fast fibers are silent, as they are even if the rhythm is speeded up by deliberately oscillating the fish to higher tailbeat frequencies. There seems to be no question that in sharks, the deep fast fibers are only utilized during fast bursts of swimming. Calculations based upon the rate of depletion of the fast fiber glycogen reserves suggest these fibers could only operate for around 2 min of sustained activity (the time actually observed). This is less disadvantageous than might appear, for it would allow (in *Scyliorhinus*) a distance of some 600 m to be covered if the fish swam continuously. Of course, the fast fiber system is normally used by the fish to swim for a few rapid tailbeats and then glide to rest or to slow sustained swimming using the slow fiber system.

Rather few teleosts have focally innervated fast fibers such as the dogfish, the herring is the only one in which fiber function has been investigated directly (Bone *et al.*, 1978a). By observing herring swimming in a tunnel respirometer at different water velocities it is a simple matter to show that muscle action potentials from the deep fibers are only obtained when the fish is swimming in rapid bursts, and that up to 5ℓ/sec, herring must utilize only the slow fibers, since action potentials are not found. It is important to realize that the mosaic arrangement in the herring fast fiber portion of the myotome does not represent a mixture of larger focally innervated fibers with small multiply innervated fibers. All are focally innervated; hence muscle action potentials must be a concomitant of their activity.

A certain amount of evidence has accumulated from different lines of investigation that where fast fibers *are* multiply innervated (i.e., in higher teleosts), the simple duality of slow and fast portions of the myotome is an oversimplification. There is now rather convincing support for the idea that in these fish, fast fibers are active during continuous swimming at speeds well below that which the fish is capable of sustaining for long periods. Thus,

Greer-Walker (1971) and Greer-Walker and Pull (1973) found hypertrophy of both slow fibers *and* fast fibers when coalfish (*Gadus virens*) were swimming for long periods at 2 and 3 ℓ/sec. Using the same species, and also the crucian carp (*C. carassius*) Johnston and Goldspink (1973a,b,c) were able to demonstrate from measurements of muscle lactate, that after continuous swimming at 2 ℓ/sec and above, the fast fiber system was active. Direct evidence for this conclusion was obtained in carp by Johnston *et al.* (1977) and by Bone *et al.* (1978a). A similar conclusion was drawn by Hunter (1971) in an interesting study of a fast swimming species (*Trachurus symmetricus*).

Hudson (1973) placed electrodes in lateral superficial slow fibers and in deep mosaic fast fibers in trout (*S. gairdneri*), and swam the fish in a tunnel respirometer at different speeds. He found that electrical activity was recorded from the superficial slow fibers at all swimming velocities, but the mosaic muscle was silent until the fish reached around 75% of the maximum sustainable swimming speed. At this swimming speed, the fast fibers showed electrical activity similar to that of the superficial slow fibers, but at burst speed, much larger electrical events were observed. Hudson's interesting results suggested that at intermediate and high cruising speeds, fibers in the fast portion of the myotome were operating without the production of muscle action potentials, whereas during bursts, the fast fibers produced action potentials. Similar conclusions were drawn from electromyography in carp (Bone *et al.*, 1978a), except that in this species, the fast fiber portion of the myotome was active even at the lowest speeds the fish would swim in the respirometer (Fig. 23).

It is not known whether these different patterns of activity from the fast motor system during cruising and during bursts of rapid swimming represent

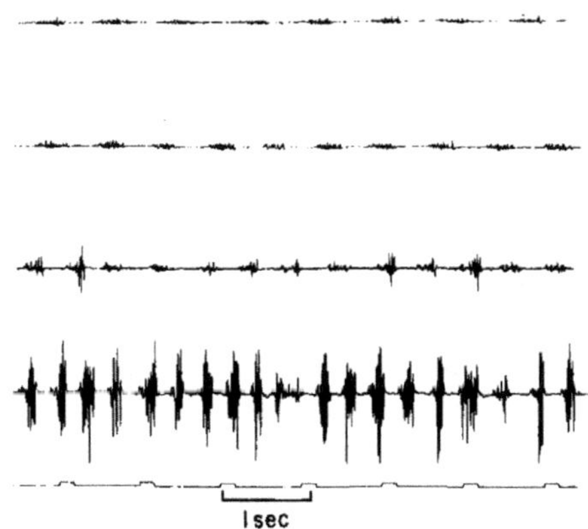

Fig. 23. Similar records to those of Fig. 22 obtained from fast myotomal muscle fibers of carp swimming in tunnel respirometer at different water velocities. At low swimming speeds (upper two records), only low amplitude electrical activity is obtained, but at higher speeds, (lower records) large events appear, probably muscle action potentials recorded extracellularly. (From Bone *et al.*, 1978a, *Fish Bull.*, in press.)

the activity of different fast fiber types. Functionally, it would be a neat trick if the fish were able either to operate the fast motor system by local contractions of the fibers so that the fibers contracted slowly when the fish swam slowly, or could contract them rapidly (with propagated muscle action potentials) during bursts of speed. The same fiber could then operate at the appropriate point on different force/velocity curves in both situations for maximum efficiency. Until single fiber activity records are obtained, the question must remain open, but results such as those obtained by Barets (1961) on tench (*Tinca*) do not rule out the possibility of two kinds of electrical (and mechanical) activity from single muscle fibers.

It is something of a paradox to find that it is apparently common in higher teleosts for muscle fibers specialized for short periods of anaerobic burst swimming to play a part in sustained long-term activity. During sustained swimming the requirement must be for efficient and economical utilization of metabolites, and anaerobic glycolysis yielding lactate is a relatively inefficient source of ATP (Bilinski, 1974; see also Chapter 8). Since little lactate is excreted during sustained swimming (Bilinski, 1974), an obvious solution to the paradox is that lactate is oxidized at various sites outside the fast motor system of the deep portion of the myotome. The fish would then be in overall oxygen balance, and no oxygen debt would have to be repaid after the period of swimming (as it has to be repaid after short bursts of swimming at high speed).

Two alternative sites have been proposed as capable of complete oxidation of lactate: the gills and the superficial slow muscle fibers. The discovery by Bilinski and Jonas (1972) that gill tissue in trout had a high capacity for lactate oxidation suggested that the energy required for exchange processes there is at least in part supplied in this way, so that lactate produced by the operation of the fast fibers at maintained high cruising speeds is oxidized to supply energy to drive the ion pumps of the gills. It is known that trout gill oxidizes lactate, but whether this mechanism is used to keep the fish in overall oxygen balance during sustained higher speed cruising is not known. It would be interesting to compare the capacity of the gill tissue to oxidize lactate in focally innervated forms (e.g., herring) which do not utilize the fast motor system during the cruise condition, with those from fish which are known to use the fast motor system during cruising (e.g., gadoids).

Various workers have suggested that lactate produced by the activity of fast fibers can be oxidized by the superficial slow fibers. Some, (e.g., Braekkan, 1956) have supposed that the superficial muscle fibers are indeed not concerned with thrust generation at all, acting solely as a sort of peripheral liver, accessory to the metabolism of the fast fibers. It is plain that this view is untenable, but some evidence suggests that the superficial (red) fibers may play a part in the glycolytic metabolism of the deep fast (white) fibers. In a series of papers Wittenberger (earlier references in Wittenberger *et al.*, 1975) has examined the metabolic interrelationships of the two muscle types after simple experimental procedures, and has concluded that the superficial fibers store glycogen for subsequent transfer to the fast fibers as well as oxidizing lactate derived from the fast fibers. Similarly, Smit and his colleagues (Smit *et al.*, 1971), observing that goldfish were able to swim at sustained fast speeds (up to 8.5 ℓ/sec for over 3 hr) without incurring an oxygen debt, assumed that slow fibers oxidized the lactate produced by the fast fibers, which must presumably have been active at these high speeds.

As Bilinski (1974) emphasizes, further experimental evidence is needed before this concept of "cooperative metabolism" between the two main fiber types can be accepted. The idea is in some ways an attractive one; what is needed are not only more experimental biochemical data but also some simple physiological data about diffusion pathways and capillary exchange between the two zones of the myotome. The poor vascularization of the deep fast fibers, and their distance from the superficial slow fiber zone in most fishes, would seem to make cooperative metabolism a very long-term process except in fish such as scombroids or carangids, where the deep fibers are better vascularized. Interestingly enough, the study by Pritchard *et al.* (1971), taken with that by Hunter (1971), on the carangid *Trachyurus*, and that by Johnston *et al.* (1977) on carp suggested sustained fast fiber activity over a wide speed range. If it is supposed that the mosaic portion of the myotomes in salmonids are composed of a mixture of slow and fast fibers (as, e.g., by Webb, 1971), then such an arrangement would certainly allow efficient exchange of metabolites between slow and fast fibers. Johnston (1977) has investigated the glycolytic enzyme profiles in slow and fast fibers of trout and mirror carp and has shown that these are similar in the carp, but differ in trout. As he points out, either there is something lacking in our understanding of anaerobic pathways in carp (but see Chapter 8) or there must be noncirculatory transfer of metabolites from white to red fibers which would require novel transport phenomena.

The experimental myography, biochemistry, and histology considered in this section, have on the whole been at a relatively crude level, so that it has only been possible to consider the roles of slow or fast fibers. The diverse other fiber types have been conflated in one or other of these two categories, and until more detailed studies are carried out, the functions of such fibers as the superficial fibers in the dogfish are unknown. An early hint of different roles for two types of slow fibers was obtained by Rayner and Keenan's (1967) electromyographic work on tuna where it was found that the superficial slow fibers and the deep, elevated temperature "chiai" slow fibers could operate under different conditions. Curiously enough, as Graham (1975) points out, the reason for maintainance of an elevated (slow) muscle temperature in larger scombroids and isurids is not entirely clear.

There is no direct evidence for the function of the different fiber types in the fins of fishes, so far as the author is aware, but since there are two main fiber types in fin muscle, very similar in most respects to the slow and fast fiber types of the myotomal musculature, it is natural to suppose that they function in a similar way. Nishihara (1967) points out that the pectoral fin muscles in goldfish are mainly red, slow fibers, whereas the pelvic fin muscles are chiefly composed of white fast fibers, and relates this to the different function of the two sets of fins.

V. PROPRIOCEPTION

Despite careful histological search by a number of workers, neuromuscular spindles have never been observed in the muscles of any fish. It seems extremely probable that they are indeed lacking, and that the few reports of their presence are mistaken. Either fish differ from other vertebrates in not requiring proprioceptors to regulate muscular contraction, or their proprioceptors are different to those of higher vertebrates. At present, except for elasmobranchs, there is insufficient evidence available to rule out the first possibility, but in some fish groups proprioceptors of different kinds are known, and on the whole it seems likely that they will eventually be found in all groups.

In two groups, the sharks and the rays, there is good histological and physiological evidence for proprioceptors associated with the locomotor musculature. In rays elongate beaded endings (Fig. 24B) among the fin ray muscles were found in the last century [they have been most recently investigated histologically by Barets (1956) and by Bone and Chubb (1975)]. Their position between muscle fibers, i.e., in parallel with the muscle fibers, naturally suggested that they were stretch receptors, and Fessard and Sand (1937) demonstrated that the static sensory discharge from the fin ray nerves was dependent upon the tension imposed upon the fin ray muscles. More recently, Ridge (1977) has investigated the dynamic properties of these endings, finding that they resemble neuromuscular spindles in certain respects (Fig. 25). There are some interesting features in the morphology of these elongate endings. First, they are only found between smalldiameter multiply innervated muscle fibers which form the superficial zone of the fin ray muscle bundles. The main mass of the fin ray muscle, consisting of larger focally innervated muscle fibers, does not possess these endings. Second, the coupling between the sensory neurites and the muscle fibers themselves is relatively loose. For most of the length of the ending the neurites are only coupled to the muscle fiber by loose strands of collagen, but at certain points, the sarcolemma is invaginated and the coupling is more direct. Because there is no capsule, and the endings are large and visible by means of Nomarski optics in the living state, it seems likely that the ray endings may be of interest to physiologists examining general aspects of stretch receptor function. It is certainly significant that these endings lie only among small-diameter muscle fibers which probably do not propagate muscle action potentials. In this respect they are similar to the nonencapsulated stretch receptors of urodele muscle (Bone *et al.*, 1976).

On the surface of the caudal myotomes just internal to the dermal connective tissue sheet, there are beaded brushlike endings derived from large

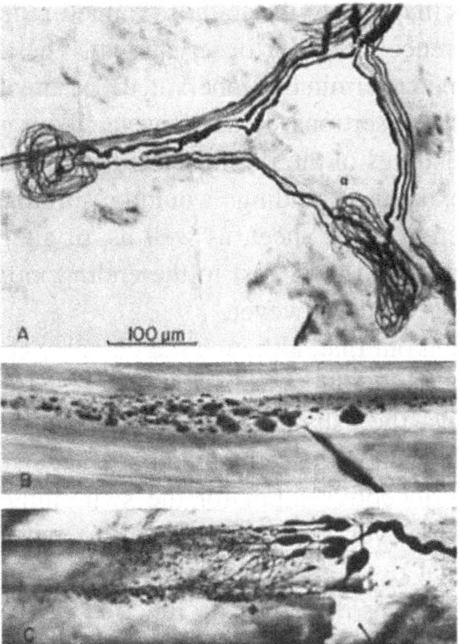

Fig. 24. The three proprioceptive endings known from elasmobranchs. (A) Wunderer corpuscles from flank of body in *Scyliorhinus*. Note division of parent fiber at arrow and complex coiling within corpuscles. (From Bone and Chubb, 1976, *J. Mar. Biol. Assoc. U.K.* **56**, 925–928.) (B) Stretch receptive ending from among slow fibers of pectoral fin in ray. (C) Similar receptor from caudal myotome surface in *Raia*. In this case, the ending is apparently more closely associated with a muscle fiber than are the endings of the pectoral and pelvic fins. All from whole mounts of silver-impregnated material. Scale bar: 100 μm.

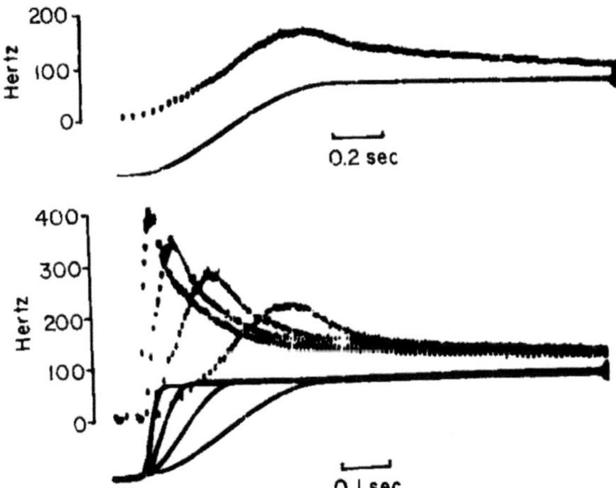

Fig. 25. Discharge frequency of stretch receptor endings from pectoral fin of *Raia clavata* in response to ramp stretches at different velocities. Lower records (solid lines), record of displacement (1 mm); upper record, discharge frequency (instantaneous frequency meter record). (A) Stretch at 1.25 mm/sec; (B) stretch at 2.5, 5, 10, and 20 mm/sec. (From Ridge, unpublished.)

diameter nerve fibers that are similar to those found in the fin ray muscles (see Bone, 1964, for reference to earlier observations). These endings lie superficial to the large-diameter terminally innervated fast muscle fibers of the caudal region, near to the insertions of the muscle fibers (Fig. 24C). They are therefore unlike the endings of the fin muscles (between slow fibers). Presumably, as the tail is flexed, these endings would be stimulated by pressure from the overlying connective tissue sheet, as well as, or alternatively, by the tension exerted by the muscle fibers next to the ending; no physiological investigations have yet been made however.

It should be pointed out that, although the elongate endings of rays are the best known, both histologically and physiologically, of fish proprioceptors, it is not yet clear what use the fish makes of the information which they provide. Since spinal rays do not show the same kind of reflex swimming as do sharks, the necessity of proprioceptive input for the swimming rhythm has not yet been demonstrated. However, having entered this caveat, it is obviously reasonable to suppose that (as Fessard and Sand suggested) the elongate receptors regulate the swimming rhythm. The function of the superficial endings of the tail is less easy to apprehend, since in most rays, the tail is relatively immobile.

In sharks, these elongate receptors are either absent or very rare. Fessard and Sand observed similar responses to those given by the elongate endings of rays in certain (unspecified) muscles of *Scyliorhinus*; Barets (personal communication) has occasionally observed them among the fin muscles. They are not found in the myotomal musculature, so far as I have been able to observe. Instead, endings of rather different morphology are found. These endings lie just superficial to the myotomes at the level of the myosepta. They are derived from large-diameter axons which form coiled corpuscles embedded in the connective tissue of the outer edge of the myospetum. Occasionally they are found among the superficial muscle fibers in the partition between two muscle stacks. These endings were first carefully described by Wunderer (1908) from the bases of the fins, where they were later examined physiologically by Lowenstein (1956), who showed them to be slowly adopting mechanoreceptors. In morphology they resemble most the coiled corpuscular endings of higher forms (e.g., Munger, 1961) being formed of a twisted skein of neurites surrounded by connective tissue elements.

Roberts (1969a) recorded from portions of the body wall as it was flexed, and showed that receptors existed which were sensitive to the frequency and amplitude of flexure. In all probability these receptors are the coiled corpuscles shown in Fig. 24A, but this has not yet been definitely proven. If so, they function during swimming as second-order proprioceptors, being stimulated by the alternate compression and release of the endings as the fish oscillates its body, just as at the bases of the fins they signal the bending of the fins. Spinal sharks appropriately set up (i.e., ventilated and held free from the bottom of the tank) swim continuously with a regular slow rhythm. This rhythmic activity is dependent upon sensory input; surgical removal of input abolishes the spinal swimming rhythm (Lissmann, 1946). It is very probable that it is the alternating proprioceptive input of the corpuscular endings of the myoseptal margins which maintains this swimming rhythm. There is some controversy (Grillner, 1974) whether the spinal swimming is dependent upon central oscillators, but whatever view is taken of the central organization, it is clear that proprioceptive input is of importance.

The occasional occurrence of the corpuscular endings between muscle fascia rather than superficial to the muscle fibers is of some interest, for in this position the endings (more elongate than usual) are presumably stimulated by pressure from contracting myotomal muscle fibers themselves rather than less directly; perhaps we have a hint here of the way in which proprioceptors linked to muscle fibers may have arisen from mechanoreceptors.

During investigation of the corpuscular endings of sharks, Roberts (1969a) found similar activity from the nerves passing to the body wall of the gurnard *Trigla cuculus* (L.), but the endings responsible have not been examined histologically. On the whole it seems unlikely that these endings (or indeed, any possible proprioceptive endings of teleosts) can be morphologically very noteworthy. Many histologists have examined teleost muscles without discovery of associated corpuscular or beaded endings that could be proprioceptive, so that such endings are probably simple branching endings in the myosepta, as are found in the hagfish, *Myxine* (Bone, 1963). In the hagfish it was possible to recognize the endings as sensory, since the nerve fibers giving rise to them could be traced back to their cells of origin in the dorsal root ganglia, but this is rarely likely to be possible. Since spinal teleosts do not normally exhibit steady swimming rhythms of the shark kind, it has not been shown whether proprioceptive input is required during myotomal locomotion.

A number of teleosts swim by means of the paired fins, as also do holocephali, and I have examined both groups without observing special sensory terminations associated with the fin musculature. In *Trigla*, the bases of the free fin rays of the pectoral fins are innervated by branching fibers resembling the endings found in the joints of higher vertebrates. Similar endings are found between the joints of the fin rays of elasmobranchs, between the vertebrae and fin ray joints of dipnoi, (personal observations) and are probably found in most teleosts. They have not received any attention physiologically, so far as I am aware. Holmes (cited in Barker, 1974) has found complex branching endings superficial to the musculature in the mobile, barbell-like. fins of the dipnoan *Protopterus*; again, these endings are somewhat similar to the joint receptors of amphibia.

Despite these rather scattered examples of proprioceptors associated with locomotor muscle fibers in different fishes, it seems to be correct to suppose that the innervation of any locomotor muscle in any fish is set apart from that of higher vertebrates, not only by the lack of spindles, but also by the poverty of the sensory component. In cat hindlimb muscles, for example

(Barker, 1974), some 75% of all the axons in the nerves passing to the muscle are sensory; if fusimotor axons are included, around 81% of the axons in the nerves pass to or from receptors. This is indeed very different from the arrangement in fish. Even in the fin muscles of rays, the large diameter axons supplying the elongate sensory terminals make up at most some 15% of the total muscular nerve supply. Partly, perhaps, this is because in fish, where the body weight is wholly or almost entirely supported by the water, postural problems are of little account (Bone, 1966), and so a rich sensory innervation giving a detailed pattern of information about muscle length and tension is not required. It is notable that the aquatic urodeles resemble fishes in that they are devoid of neuromuscular spindles in the myotomal locomotor musculature; they do have, however, sensory endings in the limb muscles which are sensitive to stretch (Bone *et al.*, 1976).

The absence of neuromuscular spindles from fish muscle is striking to the physiologist accustomed to higher forms, but to the fish physiologist what seems remarkable is that only in rays are there proprioceptive endings which seem analogous to spindles in that they are directly associated with muscle fibers, rather than being "secondorder" proprioceptors as are all the other sensory endings assumed to be proprioceptive in function. It may be that the ray method of swimming demands very delicate control of the fin ray musculature, unobtainable by indirect proprioception, but anyone who has watched the barbels of feeding mullet or the dorsal fins of *Notopterus* or gymnotids, will be aware of remarkably precise muscular movements apparently without benefit of direct proprioception. It is probably in delicately controlled muscles such as these that morphologically specialized proprioceptive endings will first be found in teleosts. Ono (personal communication) has found, however, only simple branched sensory endings in the barbels of *Mullus*.

From the account above it will be at once evident that further investigations are required before the role of proprioception in the control of fish muscles is understood; only in elasmobranchs have we a reasonable understanding of the system.

VI. FISH MUSCLE AND THE MUSCLES OF HIGHER FORMS

The preceding sections have indicated some of the differences between the muscles of fishes and those of terrestrial vertebrates. The differences are real, and they are differences not only of degree but also of kind. For example, neuromuscular spindles are absent, and they occur in all terrestrial vertebrates so far as is known; again, different fiber types are segregated or zoned to a much greater degree than they are in terrestrial forms. What is more, slow and fast fibers are more different from each other than they are even in amphibia. Not only are slow fibers apparently non-twitch fibers, but the fast twitch fibers are normally highly specialized for anaerobic operation, which is to say, for maximum power at the expense of sustained operation. On the whole, these differences between the muscle fibers of fishes and of terrestrial forms can be understood in terms of the rather stringent conditions set by the density of the water in which the fish lives. Fortunately for fish, the density of the water which imposes a requirement for a large power increase for small increment of swimming speed, also provides the possibility of achieving neutral buoyancy by storing small amounts of gas or lipid, and, so, with the possibility of greatly increasing the mass of locomotor muscle. The myotomal mass of fast fibers only used occasionally during escape or predatory movements is but a light penalty for the fish to carry around, since it is buoyed up by the water; such an arrangement of a mass of muscles used only as an emergency power pack would be quite unsuitable for a terrestrial animal subject to gravity.

Curiously enough, although all terrestrial vertebrates (with the exception of mammals) seem to have both multiply innervated slow fibers and focally innervated fast fibers in their locomotor muscles (see review by Barker, 1968), the function of the different fiber types is only known clearly in mammals. In amphibia, reptiles, and birds, it is possible that slow fibers are used for slow movements, perhaps additionally or alternatively, for isometric postural contractions, but this awaits investigation. Part of the difficulty resides in the mosaic arrangement of fiber types in most muscles of terrestrial forms, as opposed to the zonal fish arrangement.

It seems most appropriate to end this chapter by reminding the reader of two points. First, the study of fish muscle is as yet in a preliminary state; so far, few fishes out of the many different groups have been examined. Much remains to be done. There are a variety of fish groups still living, from acrania to dipnoi, and comparative studies are likely to prove fruitful in interpreting the functional roles of different fiber types. Second, fish muscle is very suitable experimental material (as Bilinski has already noted) for a wide variety of problems.

REFERENCES

Alexander, R. McN. (1969). The orientation of muscle fibers in the myomeres of fishes. *J. Mar. Biol. Assoc. U.K.* **49,** 263–290.

Alnaes, E., Jansen, J. K. S., and Rudjord, T. (1963). Spontaneous junctional activity of fast and slow parietal muscle fibers of the hagfish. *Acta Physiol. Scand.* **60,** 240–255.

Andersen, P., Jensen, J. K. S., and Løyning, Y. (1963). Slow and fast muscle fibers in the Atlantic hagfish (*Myxine glutinosa*). *Acta Physiol. Scand.* **57,** 167–179.

Arloing, S., and Lavocat, A. (1875). Recherches sur l'anatomie et la physiologie des muscles striés pâles et foncés. *Mem. Acad. Sci. Belles Lett. Toulouse* **7,** 177–194.

Austin, J. L. (1962). "Sense and Sensibilia" (G. J. Warnock, reconstr.). Oxford Univ. Press, London and New York.

Bainbridge, R. (1960). Speed and stamina in three fish. *J. Exp. Biol.* **37,** 129–153.

Bainbridge, R. (1962). Training, speed and stamina in trout. *J. Exp. Biol.* **39,** 537–555.

Barets, A. (1952). Différences dans le mode d'innervation des diverses portions du muscle latéral et leur rapports avec la structure musculaire chez le poisson-chat. (*Ameiurus nebulosus* Les.). *Arch. Anat. Microsc. Morphol. Exp.* **41,** 305–331.

Barets, A. (1956). Les récepteurs intra-musculaires des nageoires chez les sélaciens. *Arch. Anat. Microsc. Morphol. Exp.* **45,** 254–260.

Barets, A. (1961). Contribution à l'étude des systèmes moteurs lent et rapide du muscle latéral des téléostéens. *Arch. Anat. Morphol. Exp.* **50,** Suppl., 91–187.

Barker, D. (1968). L'innervation motrice du muscle strié des vertébrés. *Actual. Neurophysiol.* **8,** 23–71.

Barker, D. (1974). The morphology of the muscle receptors. *In* "Handbook of Sensory Physiology" (C. C. Hunt, ed.), Vol. III/2, pp. 1–190. Springer-Verlag, Berlin and New York.

Bergman, R. A. (1964a). The structure of the dorsal fin musculature of the marine teleosts, *Hippocampus hudsonius* and *H. zosterae. Bull. Johns Hopkins Hosp.* **114,** 325–343.

Bergman, R. A. (1964b). Mechanical properties of the dorsal fin musculature of the marine teleost *Hippocampus hudsonius. Bull. Johns Hopkins Hosp.* **114,** 344–353.

Best, A. C. G., and Bone, Q. (1973). The terminal neuromuscular junctions of lower chordates. *Z. Zellforsch. Mikrosk. Anat.* **143,** 495–504.

Bilinski, E. (1974). Biochemical aspects of fish swimming. *In* "Biochemical and Biophysical Perspectives in Marine Biology" (D. C. Malins and J. R. Sargent, eds.), Vol. 1, pp. 239–288. Academic Press, New York.

Bilinski, E., and Jonas, R. E. E. (1972). Oxidation of lactate to carbon dioxide by rainbow trout (*Salmo gairdneri*) tissues. *J. Fish. Res. Board. Can.* **29,** 1467–1471.

Blaxter, J. H. S., Wardle, C. S., and Roberts, B. L. (1971). Aspects of the circulatory physiology and muscle systems of deep-sea fish. *J. Mar. Biol. Assoc. U.K.* **51,** 991–1006.

Boddeke, R., Slijper, E. J., and van der Stelt, A. (1959). Histological characteristics of the body musculature of fishes in connection with their mode of life. *Proc. K. Ned. Akad. Wet., Ser. C* **62,** 576–588.

Bone, Q. (1963). Some observations upon the peripheral nervous system of the hagfish, *Myxine glutinosa. J. Mar. Biol. Assoc. U.K.* **43,** 31–47.

Bone, Q. (1964). Patterns of muscular innervation in the lower chordates. *Int. Rev. Neurobiol.* **6,** 99–147.

Bone, Q. (1966). On the function of the two types of myotomal muscle fiber in elasmobranch fish. *J. Mar. Biol. Assoc. U.K.* **46,** 321–349.

Bone, Q. (1970). Muscular innervation and fish classification. *Simp. Int. Zoofl., 1st Univ. Salamanca* pp. 369–377.

Bone, Q. (1972a). The dogfish neuromuscular junction: Dual innervation of vertebrate striated muscle fibers? *J. Cell Sci.* **10,** 657–665.

Bone, Q. (1972b). Buoyancy and hydrodynamic functions of integument in the castor oil fish, *Ruvettus pretiosus* (Pisces : Gempylidae). *Copeia* No. 1, pp. 78–87.

Bone, Q. (1975). Muscular and energetic aspects of fish swimming. *In* "Swimming and Flying in Nature" (T. Y.-T. Wu, C. J. Brokaw, and C. Brennen, eds.), Vol. 2, pp. 493–528. Plenum, New York.

Bone, Q. (in press). Myotomal muscle fiber types in *Scomber* and *Katsuwonus. In* "The Physiological Ecology of Tunas" (G. Sharp and A. Dizon, eds.). Academic Press, New York.

Bone, Q., and Chubb, A. D. (1975). The structure of stretch receptor endings in the fin muscles of rays. *J. Mar. Biol. Assoc. U.K.* **55,** 939–943.

Bone, Q., and Chubb, A. D. (1976). On the structure of corpuscular proprioceptive endings in sharks. *J. Mar. Biol. Assoc, U.K.* **56,** 925–928.

Bone, Q., and Chubb, A. D. (1978). The histochemical demonstration of myofibrillar ATPase in elasmobranch muscle. *Histochem. J.* (in press).

Bone, Q., and Roberts, B. L. (1969). The density of elasmobranchs. *J. Mar. Biol. Assoc. U.K.* **49,** 913–937.

Bone, Q., Ridge, R. M. A. P., and Ryan, K. P. (1976). Stretch receptors in urodele limb muscles. *Cell Tissue Res.* **165,** 249–266.

Bone, Q., Kiczniuk, J., and Jones, D. R. (1978a). On the role of the different fiber types in fish myotomes at intermediate swimming speeds. *Fish. Bull.* (in press)

Bone, Q., Moore, M. H., and Ryan, K. P. (1978b). Myotomal muscle fiber types in dogfish. In preparation.

Bostrom, S.-L., and Johansson, R. G. (1972). Enzyme activity patterns in white and red muscle of the eel (*Anguilla anguilla*) at different developmental stages. *Comp. Biochem. Physiol.* **42B,** 533–542.

Braekkan, O. R. (1956). Function of the red muscle in fish. *Nature* (*London*) **178,** 747–748.

Brotchi, J. (1968). Identification histoenzymoloqique des fibres lentes et rapides dans les muscles squelettiques des vertébrés. *Arch. Int. Physiol. Biochim.* **76,** 299–310.

Brotchi, J. (1969). Identification histo-enzymologique des types de fibres musculaires striées squelettiques chez la Roussette. *C. R. Soc. Biol.* **163,** 1457–1458.

Burke, R. E., Engel, W. K., Levine, D. N., Tsairis, P., and Zajac, F. E. (1971). Mammalian motor units: Physiological–histochemical correlation in three types in cat gastrocnemius. *Science* **174,** 709–712.

Buttkus, H. (1963). Red and white muscle of fish in relation to rigor mortis. *J. Fish. Res. Board Can.* **20,** 45–58.

Calow, L. J., and Alexander, R. Mc. N. (1973). A mechanical analysis of a hind leg of a frog (*Rana temporaria*). *J. Zool.* **171,** 293–321.

Carey, F. G., and Teal, J. M. (1966). Heat conservation in tuna fish muscle. *Proc. Natl. Acad. Sci. U.S.A.* **56,** 1464–1469.

Dahl, H. A., and Nicolaysen, K. (1971). Actomyosin ATPase activity in Atlantic hagfish muscles. *Histochemie* **28,** 205–210.

Davies, A. S. (1972). Postnatal changes in the histochemical fiber types of porcine skeletal muscle. *J. Anat.* **113,** 213–240.

Denton, E. J., and Marshall, N. B. (1958). The buoyancy of bathypelagic fishes without a gas-filled swimbladder. *J. Mar. Biol. Assoc. U.K.* **37,** 753–767.

Fessard, A., and Sand, A. (1937). Stretch receptors in the muscles of fishes. *J. Exp. Biol.* **14,** 383–404.

Fierstine, H. L., and Walters, V. (1968). Studies in locomotion and anatomy of scombroid fishes. *Mem. South. Calif. Acad. Sci.* **6,** 1–31.

Flood, P. R. (1966). A peculiar mode of muscular innervation in amphioxus. Light and electron microscopic studies of the so-called ventral roots. *J. Comp. Neurol.* **126,** 181–217.

Flood, P. R. (1968). Structure of the segmental trunk muscle in amphioxus. With notes on the course and "endings" of the so-called ventral root fibers. *Z. Zellforsch. Mikrosk. Anat.* **84,** 389–416.

Flood, P. R. (1971). The three-dimensional structure and frequency of myosatellite cells in trunk muscle of the axolotl (*Siredon mexicanus*). *J. Ultrastruct. Res.* **36,** 523–524.

Flood, P. R. (1977). The sarcoplasmic reticulum and associated plasma membrane of trunk muscle lamellae in *Branchiostoma lanceotatum* (Pallas). *Cell Tissue Res.* **181,** 169-196.

Flood, P. R., and Storm-Mathisen, J. (1962). A third type of muscle fibre in the parietal muscle of the Atlantic hagfish, *Myxine glutinosa* (L.)? *Z. Zellforsch. Mikrosk. Anat.* **58,** 638–640.

Franzini-Armstrong, C., and Porter, K. R. (1964). Sarcolemmal invaginations constituting the T system in fish muscle fibers. *J. Cell Biol.* **22,** 675–696.

Gordon, M. S. (1968). Oxygen consumption of red and white muscles from tuna fishes. *Science* **159,** 87–90.

Graham, J. B. (1975). Heat exchange in the yellow tuna, *Thunnus albacares*, and skipjack tuna, *Katsuwonus pelamis*, and the adaptive significance of elevated body temperatures in scombrid fishes. *U.S. Fish Wildl. Serv., Fish. Bull.* **73,** 219–229.

Greenwood, P. H., Rosen, D. E., Weitzman, S. H., and Myers, G. S. (1966). Phyletic studies of teleostean fishes, with a provisional classification of living forms. *Bull. Am. Mus. Nat. Hist.* **131,** 339–456.

Greer-Walker, M. (1970). Growth and development of the skeletal muscle fibers of the cod (*Gadus morhua* L.). *J. Cons., Cons. Perm. Int. Explor. Mer* **33,** 228–244.

Greer-Walker, M. (1971). Effect of starvation and exercise on the skeletal muscle fibers of the cod (*Gadus morhua* L.). *J. Cons., Cons. Perm. Int. Explor. Mer* **33,** 421–427.

Greer-Walker, M., and Pull, G. A. (1973). Skeletal muscle function and sustained swimming speeds in the coalfish *Gadus virens* L. *Comp. Biochem. Physiol. A* **44,** 495–501.

Greer-Walker, M., and Pull, G. A. (1975). A survey of red and white muscle in marine fish, *J. Fish Biol.* **7,** 295–300.

Grillner, S. (1974). On the generation of locomotion in the spinal dogfish. *Exp, Brain Res.* **20,** 459–470.

Hagiwara, S., and Kidokoro, Y. (1971). Na and Ca components of action potential in amphioxus muscle cells. *J. Physiol. (London)* **219,** 217–232.

Hagiwara, S., and Takahashi, K. (1967). Resting and spike potentials of skeletal muscle fibers in salt-water elasmobranch and teleost fish. *J. Physiol. (London)* **190,** 499–518.

Hagiwara, S., and Takahashi, K. (1974). Mechanism of ion permeation through the muscle fiber membrane of an elasmobranch fish, *Taeniura lymma. J. Physiol. (London)* **238,** 109–128.

Hagiwara, S., Henkart, M. P., and Kidokoro, Y. (1971). Excitation–concentration coupling in amphioxus muscle cells. *J. Physiol. (London)* **219,** 233–251.

Hamoir, G., Focant, B., and Distèche, M. (1972). Proteinic criteria of differentiation of white, cardiac and various red muscles in carp. *Comp. Biochem. Physiol. B* **41,** 665–674.

Harris, J. E., and Whiting, H. P. (1954). Structure and function in the locomotory system of the dogfish embryo. The myogenic stage of movement. *J. Exp. Biol.* **31,** 501–524.

Hidaka, T., and Toida, N. (1969). Biophysical and mechanical properties of red and white muscle fibers in fish. *J. Physiol. (London)* **201,** 49–59.

Higashi, H., Kaneko, T., and Sugii, K. (1953). Studies on utilization of the liver oil of deep sea sharks—IV. Hydrocarbon contents in "Ynmezame," *Centroscymnus owstoni* Garman. *Nippon Suisan Gakkaishi* **19,** 836–850.

Hill, A. V. (1950). The dimensions of animals and their muscular dynamics. *Sci. Prog. (London)* **38,** 209–230.

Hudson, R. C. L. (1968). A ringer solution for *Cottus* (teleost) fast muscle fibers. *Comp. Biochem. Physiol.* **25,** 719–25.

Hudson, R. C. L. (1969). Polyneuronal innervation of the fast muscles of the marine teleost *Cottus scorpius* L. *J. Exp. Biol.* **50,** 47–67.

Hudson, R. C. L. (1973). On the function of the white muscles in teleosts at intermediate swimming speeds. *J. Exp. Biol.* **58,** 509–522.

Hunter, J. R. (1971). Sustained speed of jack mackerel, *Trachyurus symmetricus. U.S. Fish Wildl. Serv., Fish. Bull.* **69,** 267–271.

Jarman, G. M. (1961). A note on the shape of fish myotomes. *Symp. Zool. Soc. London* **5,** 33–35.

Jasper, D. (1967). Body muscles of the lamprey. Some structural features of the T system and sarcolemma. *J. Cell Biol.* **32,** 219–227.

Johnston, I. A. (1977). A comparative study of glycolysis in red and white muscles of the trout (*Salmo gairdneri*) and mirror carp (*Cyprinus carpio*). *J. Fish Biol.* **11,** 575–588.

Johnston, I. A., and Goldspink, G. (1973a). Some effects of prolonged starvation on the metabolism of the red and white myotomal muscles of the plaice *Pleuronectes platessa. Mar. Biol.* **19,** 348–353.

Johnston, I. A., and Goldspink, G. (1973b). A study of the swimming performance of the crucian carp *Carassius cavassius* (L.) in relation to the effects of exercise and recovery on biochemical changes in the myotomal muscles and liver. *J. Fish Biol.* **5,** 249–260.

Johnston, I. A., and Goldspink, G. (1973c). A study of glycogen and lactate in the myotomal muscles and liver of the coalfish (*Gadus virens* L.) during sustained swimming. *J. Mar. Biol. Assoc. U.K.* **53,** 17–26.

Johnston, I. A., and Tota, B. (1974). Myofibrillar ATPase in the various red and white trunk muscles of the tunny (*Thunnus thynnus* L.) and the tub gurnard (*Trigla lucerna* L.). *Comp. Biochem. Physiol. A* **49,** 367–373.

Johnston, I. A., Ward, P. S., and Goldspink, G. (1975). Studies on the swimming musculature of the rainbow trout. 1. Fiber types. *J. Fish Biol.* **7,** 451–458.

Johnston, I. A., Davison, W., and Goldspink, G. (1977). Energy metabolism of carp swimming muscles. *J. Comp. Physiol.* **114,** 203–216.

Kilarski, W. (1965). Organizacja siateczki sarkoplazmatycznej miéni szkieletowych ryb. Cześć II. Okoń (*Perca fluviatilis* L.) [The organization of the sarcoplasmic reticulum in skeletal muscles of fishes. Part II. The perch (*Perca fluviatilis* L.)]. *Acta Biol. Cracov., Ser. Zool.* **8,** 51–57.

Kilarski, W. (1967). The fine structure of striated muscle in teleosts. *Z. Zellforsch. Mikrosk. Anat.* **79,** 562–580.

Kordylewski, L. (1974). Some observations on mitochondria in muscle fibers of *Salamandra salamandra* (L.). *Z. Mikrosk.-Anat. Forsch.* **88,** 937–947.

Korneliussen, H. (1973). Dense-core vesicles in motor nerve terminals. Monoaminergic innervation of slow non-twitch muscle fibers in the Atlantic hagfish (*Myxine glutinosa* L.). *Z. Zellforsch. Mikrosk. Anat.* **140,** 425–432.

Korneliussen, H., and Nicolaysen, K. (1973). Ultrastructure of four types of striated muscle fibers in the Atlantic hagfish (*Myxine glutinosa*, L.). *Z. Zellforsch. Mikrosk. Anat.* **143,** 273–290.

Korneliussen, H., and Nicolaysen, K. (1975). Distribution and dimension of the T-system in different muscle fiber types in the Atlantic hagfish (*Myxine glutinosa*, L.). *Cell Tissue Res.* **157,** 1–16.

Korneliussen, H., Dahl, H. A., and Paulsen, J. E. (1978). Histochemical definition of muscle fibre types in the trunk musculature of a teleost fish (cod, *Gadus morhua*, L.). *Histochemistry* **55,** 1–16.

Kruger, P. (1950). Ueber das Vorkommen von zweirlei Fasem in der muskulatur von Haien. *Z. Naturforsch.* **56,** 218–220.

Kryvi, H. (1975). The structure of the myosatellite cells in axial muscles of the shark *Galeus melastomus. Anat. Histol., Embryol.* **147,** 35–44.

Kryvi, H. (1977). Ultrastructure of the different fibre types in axial muscles of the sharks *Etmopterus spinax* and *Galeus melastomus. Cell Tissue Res.* **184,** 287–300.

Kryvi, H., and Eide, A. (1977). Morphometric and autoradiographic studies on the growth of red and white axial muscle fibres in the shark *Etmopterus spinax. Anat. Embryol.* **151,** 17–28.

Kryvi, H., and Totland, G. K. (1978). Fibre types in locomotory muscles of the cartilaginous fish *Chimaera monstrosa*. *J. Fish Biol.* **12,** 257–265.

Lansimäki, T. A. (1910). Ueber die Anordnung der Fibrillenbündel in den quergestreiften Muskeln einiger Fische. *Anat. Hefte, Abt. 1* **42,** 251–279.

Levin, A., and Wyman, J. (1927). The viscous elastic properties of muscle. *Proc. R. Soc., Ser. B* **101,** 218–243.

Lie, H. R. (1974). A quantitative identification of three muscle fiber types in the body muscles of *Lampetra fluviatilis*, and their relation to blood capillaries. *Cell Tissue Res.* **154,** 109–119.

Lissmann, H. W. (1946). The neurological basis of the locomotory rhythm in the spinal dogfish (*Scyllium canicula, Acanthias vulgaris*). II. The effect of de-afferentation. *J. Exp. Biol.* **23,** 162–176.

Lorenzini, S. (1678). "Osservazioni intorno alle Torpedini." Onofri, Florence.

Lowenstein, O. (1956). Pressure receptors in the fins of the dogfish *Scyliorhinus canicula*. *J. Exp. Biol.* **33,** 417–421.

Magnuson, J. J. (1973). Comparative study of adaptations for continuous swimming and hydrostatic equilibrium of scombroid and xiphoid fishes. *U.S. Fish Wildl. Serv., Fish. Bull.* **71,** 337–356.

Mark, R. F., and Marotte, L. R. (1972). The mechanism of selective re-innervation of fish eye muscles. IV. Identification of repressed synapses. *Brain Res.* **46,** 149–157.

Modigh, M., and Tota, B. (1975). Mitochondrial respiration in the ventricular myocardium and in the white and deep red myotomal muscles of juvenile tuna fish (*Thunnus thynnus* L.). *Acta Physiol. Scand.* **93,** 289–294.

Mosse, P. R. L. (1978). The distribution of capillaries in the somatic musculature of two vertebrate types with particular reference to teleost fish. *Cell Tissue Res.* **187,** 281–303.

Mosse, P. R. L., and Hudson, R. C. L. (1977): The functional role of different muscle fibre types identified in the myotomes of marine teleosts: A behavioural, anatomical and histochemical study. *J. Fish Biol.* **11,** 417–430.

Munger, B. L. (1961). Patterns of organization of peripheral sensory receptors. *In* "Principles of Receptor Physiology (W. R. Loewenstein, ed.), Handbook of Sensory Physiology, Vol. 1, pp. 523–556. Springer-Verlag, Berlin and New York.

Nag, A. C. (1972). Ultrastructure and adenosine triphosphatase activity of red and white muscle fibers of the caudal region of a fish, *Salmo gairdneri*. *J. Cell Biol.* **55,** 42–57.

Nag, A. C., and Nursall, J. R. (1972). Histogenesis of white and red muscle fibers of trunk muscles of a fish *Salmo gairdneri*. *Cytobios* **6,** 227–246.

Nakao, T. (1975). Fine structure of the myotendinous junction and "terminal coupling" in the skeletal muscle of the lamprey, *Lampetra japonica*. *Anat. Rec.* **182,** 321–327.

Nakao, T. (1976). An electron microscope study of the neuromuscular junction in the myotomes of larval lamprey, *Lampetra japonica*. *J. Comp. Neurol.* **165,** 1–16.

Nicolaysen, K. (1976a). The spread of the action potential in the T-system in hagfish twitch muscle fibers. *Acta Physiol. Scand.* **96,** 29–49.

Nicolaysen, K. (1976b). Spread of the junction potential in the T-system in hagfish slow muscle fibers. *Acta Physiol. Scand.* **96,** 50–57.

Nishihara, H. (1967). Studies on the fine structure of red and white fin muscles of the fish (*Carassius auratus*). *Arch. Histol. Jpn.* (*Niigata, Jpn.*) **28,** 425–447.

Nursall, J. R. (1956). The lateral musculature and the swimming of fish. *Proc. Zool. Soc. London* **126,** 127–143.

Patterson, S., and Goldspink G., (1972). The fine structure of red and white myotomal muscle fibers of the coalfish (*Gadus virens*). *Z. Zellforsch. Mikrosk. Anat.* **133,** 463–474.

Patterson, S., Johnston, I. A., and Goldspink, G. (1975). A histochemical study of the lateral muscles of five teleost species. *J. Fish Biol.* **7,** 159–166.

Pecot-Dechavassine, M. (1961). Étude biochemique, pharmacologique et histochemique des Cholinestérase des muscles striés chez les poissons, les batraciens et les mammifères. *Arch. Anat. Microsc. Morphol. Exp.* **50,** Suppl., 341–438.

Pritchard, A. W., Hunter, J. R., and Lasker, R. (1971). The relation between exercise and biochemical changes in red and white muscle and liver in the jack mackerel, *Trachurus symmetricus. U.S. Fish Wildl. Serv., Fish. Bull.* **69,** 379–386.

Ranvier, L. (1873). Propriétés et structures différentes des muscles rouges et des muscles blancs chez les lapins et chez les Raies. *C. R. Acad. Sci.* **77,** 1030–1034.

Rayner, M. D., and Keenan, M. J. (1967). Role of red and white muscles in the swimming of the skipjack tuna. *Nature (London)* **214,** 392–393.

Ridge, R. M. A. P. (1977). Physiological responses of stretch receptors in the pectoral fin of the ray *Raia clavata. J. Mar. Biol. Assoc. U.K.* **57,** 535–541.

Roberts, B. L. (1969a). The response of a proprioceptor to the undulating movements of dogfish. *J. Exp. Biol.* **51,** 775–785.

Roberts, B. L. (1969b). The co-ordination of the rhythmical fin movements of dogfish. *J. Mar. Biol. Assoc. U.K.* **49,** 357–425.

Sharp, G., and Dizon, A. (eds.)(in press). "The Physiological Ecology of Tunas." Academic Press, New York.

Skoglund, C. R. (1961). Functional analysis of swimbladder muscles engaged in sound production of the toadfish. *J. Biophys. Biochem. Cytol.* **10,** 187–200.

Smit, H., Amelink-Koutstall, J. M., Vijverberg, J., and von Vaupel-Klein, J. C. (1971). Oxygen consumption and efficiency of swimming goldfish. *Comp. Biochem. Physiol. A* **39,** 1–28.

Smith, R. S., and Ovalle, W. K. (1973). Varieties of fast and slow extrafusal muscle fibers in amphibian hind limb muscles. *J. Anat.* **116,** 1–24.

Smith, R. S., Blinston, G., and Ovalle, W. K. (1974). Organization of skeletal muscle in the amphibia. *In* "Control of Posture and Locomotion" (R. B. Stein, K. G. Pearson, R. S. Smith, and J. B. Redford, eds.), pp. 1–13. Plenum, New York.

Spray, T. L., Waugh, R. A., and Sommer, J. R. (1974). Peripheral couplings in adult vertebrate skeletal muscle. *J. Cell Biol.* **62,** 223–227.

Stanfield, P. R. (1972). Electrical properties of white and red muscle fibers of the elasmobranch fish *Scyliorhinus canicula. J. Physiol. (London)* **222,** 161–186.

Stevens, J. (1976). The ecology of the blue shark *Prionace glauca*. Ph. D. thesis, Univ. of London.

Syrovy, I., Gaspar-Godfroid, A., and Hamoir, G. (1970). Comparative study of the myosins from red and white muscles of the carp. *Arch. Int. Physiol. Biochim.* **78,** 919–934.

Takeuchi, A. (1959). Muscular transmission of fish skeletal muscles investigated with intracellular microelectrode. *J. Cell. Comp. Physiol.* **54,** 211–220.

Teräväinen, H. (1971). Anatomical and physiological studies on muscles of lamprey. *J. Neurophysiol.* **34,** 954–973.

Teräväinen, H., and Rovainen, C. M. (1971). Fast and slow motoneurons to body muscle of the sea lamprey. *J. Neurophysiol.* **34,** 990–998.

van der Stelt, A. (1968). Spiermechanica en myotoombouw Bij Vissen. Ph. D. Thesis, Univ. of Amsterdam.

Vialleton, L. (1902). Le développement des muscles rouges. *C. R. Assoc. Anat.* Part 4, pp. 47–53.

Wardle, C. S. (1975). Limit of fish swimming speed. *Nature (London)* **255,** 725–727.

Warfel, H. E., and Clague, J. A. (1950). Shark fishing potentialities of the Philippine seas. *Fish Wildl. Serv. (U.S.), Res. Rep.* No. **15,** pp. 1–19.

Waterman, R. E. (1969). Development of the lateral musculature in the teleost *Brachydanis serio*: A fine structural study. *Am. J. Anat.* **125,** 457-494.

Webb, P. W. (1971). The swimming energetics of trout. I. Thrust and power output at cruising speeds. *J. Exp. Biol.* **55,** 489–520.

Webb, P. W. (1975). Hydrodynamics and energetics of fish propulsion. *Bull., Fish. Res. Board Can.* No. 190, 158 pp.

Willemse, J. J. (1975). Some remarks on the structure and function of musculus lateralis in the European eel, *Anguilla anguilla* (L.) (Pisces, Teleostei). *Z. Morphol. Tiere* **81,** 195–208.

Willemse, J. J., and van den Berg, P. G. (1978). Growth of striated muscle fibres in the M. lateralis of the European eel *Anguilla anguilla* (L.) (Pisces, Teleostei). *J. Anat.* **125,** 447–460.

Winterbottom, R. (1974). A descriptive synonymy of the striated muscles of the Teleostei. *Proc. Acad. Nat. Sci. Philadelphia* **125,** 225–317.

Wittenberg, J. B. (1970). Myoglobin-facilitated oxygen diffusion: Role of myoglobin in oxygen entry into muscle. *Physiol. Rev.* **50,** 559–636.

Wittenberger, C., Coprean, D. C. and Morar, L. (1975). Studies on the carbohydrate metabolism of the lateral muscles in carp (influence of phloridzin, insulin and adrenaline). *J. Comp. Physiol.* **101,** 161–172.

Wunderer, H. (1908), Über terminal korperchen der Anamnien. *Arch. Mikrosc. Anat. Entwicklungsmech.* **71,** 504–569.

Yamamoto, T. (1972). Electrical and mechanical properties of the red and white muscles in the silver carp. *J. Exp. Biol.* **57,** 551–567.

Chapter 9

Fish swimming capacity: Keeping it current!

Jim Kieffer*
Department of Biological Sciences, University of New Brunswick, Saint John, NB, Canada
**Corresponding author: e-mail: jkieffer@unb.ca*

Chapter Outline

Jim Kieffer discusses the impact of F.W.H. Beamish's chapter "Swimming Capacity" in Fish Physiology, Volume 7, published in 1978.

Beamish's (1978) review was among the first to summarize concepts related to fish swimming performance. Beamish presents and discusses a large collection of data on swimming velocities for various fishes with consideration for differences in methodology (e.g., swimming tests), swimming behavior (types of swimming: endurance, prolonged, burst), fish lifestyle and ecology, and/or their locomotory strategies (e.g., schooling, high performance vs lower performance fish). The title of Beamish's paper "Swimming Capacity" is broad, encompassing many connected concepts related to swimming and biological and environmental constraints, and the importance of understanding how fish swimming performance guides management practices. Beamish's review represents the hallmarks of a seminal paper—it is widely cited, has had a significant impact on the field of fish swimming research and has catalyzed many new areas of investigation. It has shaped the basic and applied research programs of many researchers (including the author of this review) and continues to serve as a foundation for fish exercise studies. Though the chapter was published nearly 50 years ago, it continues to engage researchers with interests in fish swimming, and the ideas that Beamish proposed many decades ago still resonate strongly within the field.

The study of swimming capacity of fishes has a long and rich history, one which has continued to grow over the past 50–60 years (see Beamish, 1978; Brett, 1964; Hammer, 1995; Kieffer, 2000, 2010; Milligan, 1996; Plaut, 2001; Randall and Brauner, 1991; Wood, 1991). In 1978, F.W.H. Beamish

Fish Physiology, Vol. 40B. https://doi.org/10.1016/bs.fp.2024.05.006

published a review that summarized the field of fish swimming research to date. The studies outlined within this 1978 review, and the context Beamish placed around these studies, have catalyzed many research programs over the succeeding decades. Beamish's review remains relevant today, having been extensively cited, and continues to shape many different sub-fields of physiological and ecological research (see Figs. 1–3).

1 A seminal review!

As is the case with many seminal papers, the observations and comments presented by Beamish within his review have shaped our current thought, and the relationships it explores between fish swimming capacity and physiological, ecological, and behavioral factors continue to direct how biologists undertake new and exciting research. Beamish's (1978) paper was among the first papers in the field to summarize concepts related to fish swimming performance. It has not only contributed to our understanding of basic mechanisms involved in fish swimming but continues to influence technological and methodological advances. It has provided direction for subsequent research programs and the ideas and methodology presented are still influential within the field.

I first encountered Beamish's review in the early 1990s at the beginning of my PhD. Admittedly, I read it as part of my comprehensive exam preparation for my thesis topic: "Exhaustive exercise in fish." Its abundance of swimming data (e.g., *UCrit*) and discussion of the impacts that a changing environment had on fish swimming performance was valuable for my research. Having access to this enormous and detailed database on fish swimming capacity

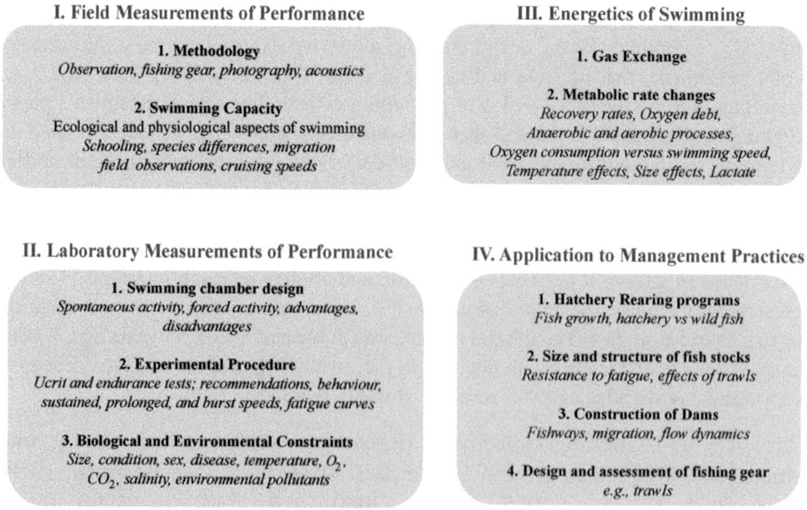

FIG. 1 General categories of research outlined in Beamish (1978).

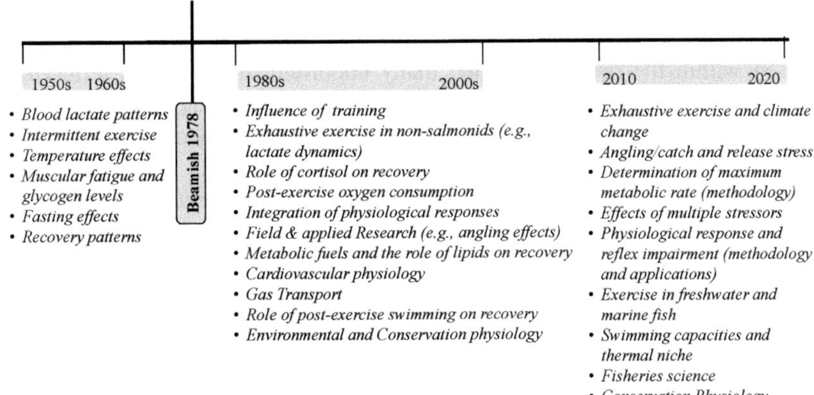

FIG. 2 Timeline depicting some advances in the study of fish (burst) exhaustive exercise.

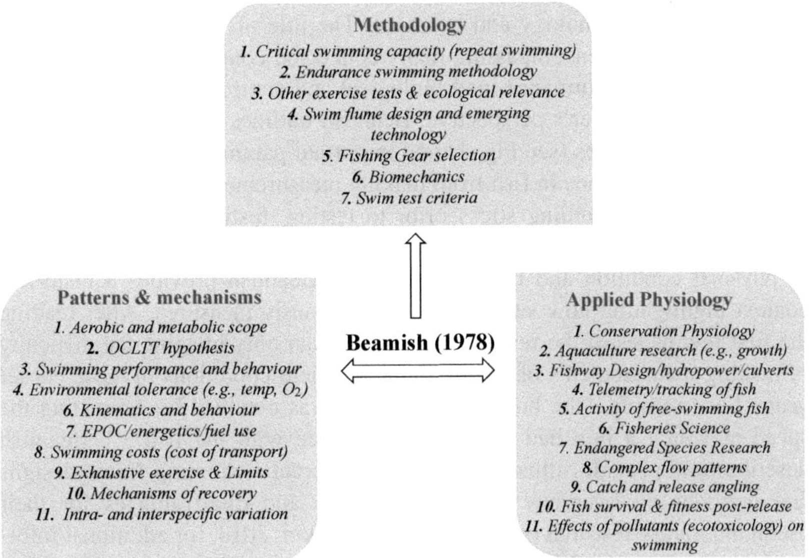

FIG. 3 Representative research areas influenced by Beamish (1978).

early in my career cannot be overstated. At the time, there were no (or at least limited) electronic database/repositories, no internet, and most publications could only be accessed either by writing to the author (who would hopefully respond and send a paper copy) or by photocopying the paper from the journal/text that was available at the campus library. I was fortunate to have access to the hard copied book!! During my PhD, post doc, and early career years, Beamish's review naturally provided me with much of the theoretical and practical framework for quantifying swimming capacity, a richness of swimming data for multiple fish species, conceptual figures, and suggestions for many proposed studies. The breadth and detail of the subject areas discussed within the review provided many opportunities for me to reflect on my own research trajectory and situate my work in the wider research field.

Beamish contextualizes his review into four main subject areas, which I highlight in Fig. 1. These include (I) Field measurements of performance, (II) Laboratory measurements of performance, (III) Energetics of swimming, and (IV) Application to management practices. These topic headings identify the many categories of fish swimming research and, even after nearly 50 years, still apply (Figs. 2 and 3). Beamish's review is novel because it presents and discusses a large collection of data on swimming velocities for various fishes; with consideration for differences in methodology (e.g., swimming tests), swimming behavior (types of swimming: endurance, prolonged, burst; effects of schooling), fish lifestyle and ecology, and/or their locomotory strategies (e.g., schooling, high performance vs lower performance fish). Beamish draws attention to the important relationship(s) between swimming physiology (performance), morphology and behavior. The title of his paper "Swimming Capacity" is broad, encompassing many connected concepts related to swimming, including performance, and biological and environmental constraints.

From a practitioner's perspective, Beamish outlines and comments on the various methodologies (see Fig. 1) and important parameters used to quantify swimming performance in fish from making measurements on individual fish, to recovery from handling stress prior to testing, fasting regimes, speed of current initiation, the methods used to promote swimming, and the importance of physical condition and body morphology. Beamish provides a historical context highlighting how swimming was previously measured. After reading the first few pages of the review chapter, one can only imagine the difficulty of making measurements of swimming under field conditions in those earlier years. For example, "the burst speed of pike was obtained by estimating the speed at which a boy had to run to keep pace with the fish…." Beamish describes and provides illustrations of the different swimming flume designs and provides an overview of the advantages and disadvantages for their use in laboratory research (see Ellerby and Herskin, 2013, for additional information on swimming flumes). He describes the two main methods to evaluate fish swimming performance: (1) increasing velocity or critical swimming (*UCrit*) test (Brett, 1964) and (2) fixed velocity or endurance test (see Beamish, 1978, for details). The *UCrit* test has been an important benchmark test for several reasons: (1) to compare swimming capabilities of different fish species, (2) to evaluate how environmental factors (e.g., pollutants, temperature) and other parameters impact swimming performance, and (3) to provide a general assessment of (mainly) aerobic swimming capacity. In contrast, fixed velocity (endurance) tests evaluate both the aerobic and anaerobic swimming capacities of fishes and allow researchers to construct "fatigue curves" based on the fish's modes of swimming behavior (i.e., sustained, prolonged, and burst swimming behavior; Beamish, 1978). Beamish recommends important parameters and guidelines for the magnitude and duration of velocity increments for subsequent *UCrit* studies. For the most part, these recommendations have been adhered to by researchers, but deviations from them have been noted in the literature for various practical and research-based reasons (see Hammer, 1995; Farrell, 2007, 2008; Nelson et al., 2002; Downie and Kieffer, 2017; for examples and discussion). Both the critical swimming and endurance protocols have stood the test of time and are still widely used today. However, since this seminal 1978 review, further discussions around the ecological relevance and repeatability of the *UCrit* test have emerged (Hammer, 1995; Nelson et al., 2002; Plaut, 2001). A cursory survey of the literature shows that the *UCrit* test remains a relevant and preferred methodology for assessing swimming capacity, but other tests have been developed (Nelson et al., 2002) including gait transition speed tests (Peake and Farrell, 2004) and volitional swimming tests (see Farrell, 2007, for a review). Regardless of the methodology used, these tests demonstrate the strong connection that exists between performance and behavior in fish, and that individual fish can use different behaviors when challenged under different flow conditions (e.g., Kieffer and May, 2020; Peake and Farrell, 2004).

Within the framework of fish exercise performance, Beamish expresses how abiotic and biotic factors impact (mainly aerobic) swimming in fish (see Fig. 1; Constraints on Performance), and specifically how fish size, water temperature, disease, water oxygen, temperature, CO_2 levels, and pollutants, among other variables, constrain swimming performance. In a broader context, the review underscores the relationship between physical environment and fish swimming performance, a subfield of research now referred to as *Environmental physiology* and/or *Conservation physiology*. These research areas have been active for several decades and reflect the growing impacts of human activities on aquatic environments. Beamish describes the relationship between swimming speed (and other metrics) and temperature and the *Scope for activity*, a term employed earlier by Fry (1947). Beamish further summarizes how metabolic scope for activity varies in fish with different life histories and thermal requirements, and how these relationships are impacted by thermal acclimation. He uses the terms "environmental constraints" and "the disturbed environment" in various sections of his review paper.

Beamish provides a broad summary of swimming energetics in fish (Fig. 1; Energetics of Swimming). He acknowledges the importance of measuring CO_2 production to fully understand the overall energy expenditure during exercise. At the time, however, there were limited techniques available to make CO_2 measurements (Beamish, 1978). Since then, metabolic fuel use approaches have become more available and the incorporation of CO_2 into metabolic energy use calculations (i.e., to determine the contributions of lipid, protein, and carbohydrates) is becoming more common (e.g., see Wang et al., 2021). Beamish cites numerous examples of the relationship between metabolic costs and swimming speeds in different fish species. He provides data figures outlining the relationships between swimming speeds and the metabolic costs associated with anaerobic processes at higher speeds—*the concept of oxygen debt* (now referred to as excess post-exercise oxygen consumption; EPOC) and the repayment of this debt (as per Brett, 1964). Research in this area continues to be widespread, and researchers have since addressed the effects of training, feeding status, exercise intensity, and temperature on oxygen consumption profiles and EPOC in a variety of fishes.

In the closing section of the review, Beamish comments on the Application to Management Practices (see Fig. 1). This section, in my opinion, is foretelling when one considers what is currently being studied and discovered in these applied research areas (Castro-Santos et al., 2022). For example, Beamish raises important points about how an understanding of swimming performance can address the survival rates of hatchery fish and contribute to growing a "fitter" fish. Beamish also explores the general impacts of dam construction and design (flows, fish passage) on the success of fish reaching their spawning grounds, and applications of fish swimming in the design and assessment of fishing gear (e.g., trawls).

2 The shaping of many research programs!

In 1978, when Beamish summarized the state of the field of fish swimming, many areas of research were in their infancy. The *UCrit* testing procedure and the important criteria for endurance swimming were established, and researchers had a general understanding of metabolic scope and oxygen debt. Over the subsequent decades, researchers, and their students(!), have addressed knowledge gaps, and new and significant fields of study have been established (see Figs. 2 and 3 for representative research areas). Beamish's review provided a backdrop for innovative studies and entire research programs on swimming performance.

The natural flow of my 30+ year research program closely mirrors the categories Beamish uses in his review. Throughout this time, my students, colleagues, and I have worked on diverse research topics related to fish swimming capacity and exercise, including the effect(s) of abiotic and biotic factors on burst exercise and aerobic exercise, aspects of anaerobic and aerobic metabolic fuel use, applied fish exercise (e.g., aspects of catch and release, use of triploid fish in aquaculture), methodologies used to test swimming performance, and the exercise capacities and metabolic scope. My earlier work focused almost exclusively on exhaustive exercise in salmonids and helped to address knowledge gaps related to the factors limiting exercise (reviewed in Kieffer, 2000). Being at the right place at the right time, I had an opportunity in the late 1990s to study the swimming performance of shortnose and Atlantic sturgeons. A few of our earlier studies showed that shortnose sturgeon had a muted response to exhaustive exercise (e.g., Kieffer et al., 2001) and these ancient fish showed interesting behaviors (e.g., station holding and substrate skimming) when swimming in a flume (e.g., Kieffer et al., 2009). However, regardless of the model species used in our basic and applied research programs, our work continues to focus on the whole organism (behavior and performance) and the physiological mechanisms related to fuel use in fish under various conditions.

Upon examination of the broader fish swimming research community, the past 40–50 years have been dynamic with respect to exhaustive exercise research (Fig. 2). Since the early and influential studies of the 1950s and 1960s on post exhaustive exercise stress and lactate dynamics in fish (Black, 1955), important benchmarks and significant advances in exhaustive exercise research occurred in the 1980s and 1990s, especially in the areas of exercise training (Pearson et al., 1990) and inter-specific research comparing high performance swimmers to sluggish fish (lactate releasers vs non-releasers, e.g., Milligan and McDonald, 1988). Research in the 1990s concentrated on intra-species comparisons (see Milligan, 1996; Kieffer, 2000, for reviews) including the factors that limited exhaustive exercise (reviewed in Kieffer, 2000, 2010), mechanisms of recovery and integration of processes (Milligan, 1996, 2003; Wood, 1991), and the effects of abiotic factors on recovery (reviewed in Milligan, 1996; Kieffer, 2000). During the 2000s, research examined the relationship between behavior and metabolic recovery following exhaustive exercise (e.g., Milligan et al., 2000; Peake and Farrell, 2004; Kieffer et al., 2011; see Fig. 2). Many of these areas continue to be studied, and mechanisms are still being worked out. From a comparative perspective, emphasis shifted, in part, from studies focusing largely on salmonid swimming physiology in the 1980s and 1990s, to studies utilizing other species (Kieffer, 2000), including Centrarchids (e.g., Kieffer and Cooke, 2009) and sturgeon (e.g., Kieffer et al., 2001), and multiple marine species. As research progressed into the 2010s and beyond, additional emphasis has been placed on applied aspects of exhaustive exercise physiology. This includes aspects of catch and release angling (reviewed in Cooke et al., 2013; Brownscombe et al., 2017), fish migration and the impacts of fish ladders and fishway design on fish movement around barriers (Katopodis et al., 2019), and aspects of exhaustive exercise under field conditions (Casselberry et al., 2023). Current information around exhaustive exercise has also re-ignited the decade old question "Why do fish die after severe exercise"? originally hypothesized by Chris Wood and his colleagues in the early 1980s (reviewed in Holder et al., 2022).

Our understanding of the aerobic swimming capacity of fish has advanced from the development and commercial availability of swimming flumes/respirometers. The number of studies and the quantity of data regarding *UCrit* have steadily grown in recent years, particularly in the context of practical applications of fish swimming (see Katopodis et al., 2019, see Fig. 3 for other representative examples). During those earlier years (Brett, 1964), researchers began to examine the effects of both abiotic and biotic factors on swimming speed (see Farrell, 2008; Randall and Brauner, 1991). Interestingly, while researchers continued to examine the swimming capacity of salmonids (e.g., Farrell, 2007, 2008), more research emerged on the aerobic swimming capacity of non-salmonids, and the relationship between fish behavior and swimming capacity (station holding and substrate skimming; see references within Kieffer et al., 2009 for sturgeon), and aerobic scope. From a methodological perspective, Farrell (2008) systematically compared fatigue velocities for constant acceleration tests and ramp-*UCrit* tests (and evaluates different temperatures, oxygen levels and recovery times; repeat swimming challenges). Farrell's study also provides valuable information on the design of repeat swim tests—an area of research Farrell's lab has extensively studied (see references in Farrell, 2008). Other complementary research around aerobic swimming methodology addressed the utility of the *UCrit* test (Farrell, 2007; Nelson et al., 2002; Plaut, 2001) and the repeatability of swimming in fish (see Farrell, 2008; Jain et al., 1997; Kieffer and May, 2020). Broadly speaking, many researchers continue to maintain research programs on the impact(s) of environmental factors, particularly temperature and oxygen, on oxygen consumption, metabolic scope, and swimming performance (see Clark et al., 2013). Since the late 2000s, researchers have debated the concept of oxygen and capacity-limited thermal tolerance (OCLTT) (see Ern et al., 2023 for a review). Information about the swimming performance of larval fish continues to grow, and the abiotic factors that affect their performance (metanalyses by Downie et al., 2020). This is just one example of how researchers are applying basic swimming/exercise data to investigate broader questions related to conservation biology, and the impacts of changing climate on fish locomotion (see Castro-Santos et al., 2022; Fig. 3).

3 Fish performance and swimming capacity: A bright future!

Beamish's (1978) review provided a snapshot of the research at that point in time. Overall, the review generated great interest in the subject (field) and provided a springboard for many subsequent studies. As we continue to add research to the boxes outlined in Fig. 1 (and create new boxes), our understanding of the complexities of fish swimming capacity will become clearer. New and/or improved technologies have allowed scientists to address research that was remarkably difficult to conduct in the past. Flume design continues to develop to accommodate fish with different body morphologies and swimming strategies. Flume respirometers units have become more portable allowing researchers to study wild fish *in situ* (for example, see Farrell et al., 2003; McKenzie et al., 2007; Holder et al., 2022). Technology to measure oxygen more precisely (and faster!) (Clark et al., 2013) and to modify swim flume speed has become automated, and standard criteria for measuring oxygen consumption rates are being adopted (Killen et al., 2021). Better capacities to measure carbon dioxide and other respiratory gases (e.g., ammonia) in swimming fish has allowed researchers to make better predictions related to fuel use in swimming fish (e.g., Wang et al., 2021), and tracking equipment enables researchers to explore fish swimming in both freshwater and seawater environments. With these more readily available technologies, physiologists and ecologists can evaluate the effects of multiple stressors on fish swimming and physiology (e.g., elevated temperature, combined with low oxygen, for example). Continued study in these broad areas will allow us to better understand how fish swimming ability will evolve in response to global change.

Some of the original research outlined by Beamish (1978) focused on salmonids, and "higher performance fish." Swimming studies continue to use salmonids as a model group of fish, but other researchers have focused on other freshwater and marine fish species, fish with unique morphologies, and larval fishes. Addressing swimming performance in larval fish is a difficult undertaking but it is an important research area because the larval phase is highly dynamic and influenced by many factors that control growth and development that can affect swimming capacity (see Downie et al., 2020). Scaling down technology to address swimming in larval fish continues to be challenging but there have been considerable advances (Downie et al., 2020). Continued work in larval fish swimming performance is paramount to our understanding of predator-prey relations, individual and population level dispersal processes, and/or recruitment of larval fish into the general population (Downie et al., 2020).

The methodology developed to evaluate swimming performance in fish has subsequently been used by researchers to address specific, targeted practical and industry related problems. Recent technologies have enabled research teams to investigate fish swimming under more natural (wild) conditions, and to study the energetics of free-swimming fish (that are not confined to flumes). New and updated remote tracking technology(ies), such as telemetry and underwater video systems will continue to guide this research. In addition, there has been wide interest in whether providing active species with a current to swim against can promote growth and improve feed conversion under aquaculture conditions (see McKenzie et al., 2021; Palstra and Planas, 2013). Other broad based, applied questions related to fish swimming requiring more research attention include the effects of water diversion, flow reduction, and river fragmentation as barriers to passage, and the role of fish swimming in fish capture by trawls (e.g., Castro-Santos et al., 2022; Hershey, 2021; Katopodis et al., 2019).

4 Concluding remarks

Many of the major advances and developments around fish swimming can be traced back to Beamish's seminal review on fish swimming (Fig. 3). As human activities continue to modify our aquatic environments, researchers are increasing their effort to understand how multiple stressors affect fish swimming (and physiology in general). Current and improved technology will allow researchers to study fish swimming with greater precision, and under field conditions. Revisiting the work of the many "fish swimmers" has provided me with a greater appreciation of the creative ideas these researchers have proposed, and how they have contributed to this robust research field. While technology may provide some tools to advance certain aspects of this research, it is these forward-thinking ideas that will continue to push the field forward. If you have not read Beamish's review, I encourage you to; if you have, consider reading it again—it is chock full of excellent insights and a rich historical context of the field. Even though the review is now close to 50 years old, it will continue to be a blueprint for future fish swimming studies.

Acknowledgments

I thank all the researchers, past and present, who have contributed their energy to this exciting field of fish swimming capacity. I thank my many mentors and colleagues, especially Dr. Bruce Tufts, Dr. Chris Wood, Dr. Mike Wilkie, and the late Dr. Gord McDonald for many, many excellent (and encouraging!!) discussions on fish swimming and energetics, and applied fish swimming research. I have been fortunate to have interacted with many great "fish swimming biologists" over the years (too many to list here) and thank all of them for the productive discussions around fish swimming! Lastly, thanks to all my students—current and past!

References

Beamish, F.W., 1978. Swimming capacity: fish physiology. Locomotion 7, 101–187.

Black, E.C., 1955. Blood levels of hemoglobin and lactic acid in some freshwater fishes following exercise. J. Fish. Board Can. 12 (6), 917–929.

Brett, J.R., 1964. The respiratory metabolism and swimming performance of young sockeye salmon. J. Fish. Board Can. 21 (5), 1183–1226.

Brownscombe, J.W., Danylchuk, A.J., Chapman, J.M., Gutowsky, L.F., Cooke, S.J., 2017. Best practices for catch-and-release recreational fisheries–angling tools and tactics. Fish. Res. 186, 693–705.

Casselberry, G.A., Drake, J.C., Perlot, N., Cooke, S.J., Danylchuk, A.J., Lennox, R.J., 2023. Allometric scaling of anaerobic capacity estimated from a unique field-based data set of fish swimming. Physiol. Biochem. Zool. 96 (1), 17–29.

Castro-Santos, T., Goerig, E., He, P., Lauder, G.V., 2022. Applied aspects of locomotion and biomechanics. Fish Physiol. A. 39, 91–140.

Clark, T.D., Sandblom, E., Jutfelt, F., 2013. Aerobic scope measurements of fishes in an era of climate change: respirometry, relevance and recommendations. J. Exp. Biol. 216 (15), 2771–2782.

Cooke, S.J., Donaldson, M.R., O'Connor, C.M., Raby, G.D., Arlinghaus, R., Danylchuk, A.J., Hanson, K.C., Hinch, S.G., Clark, T.D., Patterson, D.A., Suski, C.D., 2013. The physiological consequences of catch-and-release angling: perspectives on experimental design, interpretation, extrapolation and relevance to stakeholders. Fish. Manag. Ecol. 20 (2–3), 268–287.

Downie, A.T., Kieffer, J.D., 2017. Swimming performance in juvenile shortnose sturgeon (*Acipenser brevirostrum*): the influence of time interval and velocity increments on critical swimming tests. Conserv. Physiol. 5 (1), 1–12. cox038.

Downie, A.T., Illing, B., Faria, A.M., Rummer, J.L., 2020. Swimming performance of marine fish larvae: review of a universal trait under ecological and environmental pressure. Rev. Fish Biol. Fish. 30, 93–108.

Ellerby, D.J., Herskin, J., 2013. Swimming flumes as a tool for studying swimming behavior and physiology: current applications and future developments. In: Palstra, A.P., Planas, J.V. (Eds.), Swimming Physiology of Fish: Towards using exercise to farm a fit fish in sustainable aquaculture. Springer Press, pp. 345–376.

Ern, R., Andreassen, A.H., Jutfelt, F., 2023. Physiological mechanisms of acute upper thermal tolerance in fish. Physiology 38 (3), 141–158.

Farrell, A.P., 2007. Cardiorespiratory performance during prolonged swimming tests with salmonids: a perspective on temperature effects and potential analytical pitfalls. Philosoph. Trans. Royal. Soc. B: Biolog. Scien. 362 (1487), 2017–2030.

Farrell, A.P., 2008. Comparisons of swimming performance in rainbow trout using constant acceleration and critical swimming speed tests. J. Fish Biol. 72 (3), 693–710.

Farrell, A.P., Lee, C.G., Tierney, K., Hodaly, A., Clutterham, S., Healey, M., Hinch, S., Lotto, A., 2003. Field-based measurements of oxygen uptake and swimming performance with adult Pacific salmon using a mobile respirometer swim tunnel. J. Fish Biol. 62 (1), 64–84.

Fry, F.E.J., 1947. Effects of the environment on animal activity. Public. Ont. Fish. Res. Lab. 68, 1–63.

Hammer, C., 1995. Fatigue and exercise tests with fish. Comp. Biochem. Physiol. Part A 112 (1), 1–20.

Hershey, H., 2021. Updating the consensus on fishway efficiency: a meta-analysis. Fish Fish. 22 (4), 735–748.

Holder, P.E., Wood, C.M., Lawrence, M.J., Clark, T.D., Suski, C.D., Weber, J.M., Danylchuk, A.J., Cooke, S.J., 2022. Are we any closer to understanding why fish can die after severe exercise? Fish Fish. 23 (6), 1400–1417.

Jain, K.E., Hamilton, J.C., Farrell, A.P., 1997. Use of a ramp velocity test to measure critical swimming speed in rainbow trout (Onchorhynchus mykiss). Comp. Biochem. Physiol. Part A 117 (4), 441–444.

Katopodis, C., Cai, L., Johnson, D., 2019. Sturgeon survival: the role of swimming performance and fish passage research. Fish. Res. 212, 162–171.

Kieffer, J.D., 2000. Limits to exhaustive exercise in fish. Comp. Biochem. Physiol. Part A: Molecul. Integ. Physiol. 126, 161–179.

Kieffer, J.D., 2010. Perspective—exercise in fish: 50+ years and going strong. Comp. Biochem. Physiol. Part A: Molecul. Integ. Physiol. 156 (2), 163–168.

Kieffer, J.D., Cooke, S.J., 2009. In: Cooke, S.J., Philipp, D.P. (Eds.), Physiology and Organismal Performance of Centrarchids. Centrarchid Fishes: Diversity, Biology, and Conservation. Wiley-Blackwell, West Sussex, UK, pp. 207–263.

Kieffer, J.D., May, L.E., 2020. Repeat UCrit and endurance swimming in juvenile shortnose sturgeon (Acipenser brevirostrum). J. Fish Biol. 96 (6), 1379–1387.

Kieffer, J.D., Wakefield, A.M., Litvak, M.K., 2001. Juvenile sturgeon exhibit reduced physiological responses to exercise. J. Exp. Biol. 204 (24), 4281–4289.

Kieffer, J.D., Arsenault, L.M., Litvak, M.K., 2009. Behaviour and performance of juvenile shortnose sturgeon Acipenser brevirostrum at different water velocities. J. Fish Biol. 74 (3), 674–682.

Kieffer, J.D., Kassie, R.S., Taylor, S.G., 2011. The effects of low-speed swimming following exhaustive exercise on metabolic recovery and swimming performance in brook trout (*Salvelinus fontinalis*). Physiol. Biochem. Zool. 84 (4), 385–393.

Killen, S.S., Christensen, E.A.F., Cortese, D., Závorka, L., Norin, T., Cotgrove, L., Crespel, A., Munson, A., Nati, J.J.H., Papatheodoulou, M., McKenzie, D.J., 2021. Guidelines for reporting methods to estimate metabolic rates by aquatic intermittent flow respirometry. J. Exp. Biol. 224 (18), jeb242522. https://doi.org/10.1242/jeb.242522.

McKenzie, D.J., Garofalo, E., Winter, M.J., Ceradini, S., Verweij, F., Day, N., Hayes, R., Van der Oost, R., Butler, P.J., Chipman, J.K., Taylor, E.W., 2007. Complex physiological traits as biomarkers of the sub-lethal toxicological effects of pollutant exposure in fishes. Philosoph. Trans. Royal Soc. B: Biolog. Sci. 362, 2043–2059.

McKenzie, D.J., Palstra, A.P., Planas, J., MacKenzie, S., Bégout, M.L., Thorarensen, H., Vandeputte, M., Mes, D., Rey, S., De Boeck, G., Domenici, P., 2021. Aerobic swimming in intensive finfish aquaculture: applications for production, mitigation and selection. Rev. Aquacult. 13 (1), 138–155.

Milligan, C.L., 1996. Metabolic recovery from exhaustive exercise in rainbow trout. Comp. Biochem. Physiol. Part A 113 (1), 51–60.

Milligan, C.L., 2003. A regulatory role for cortisol in muscle glycogen metabolism in rainbow trout Oncorhynchus mykiss Walbaum. J. Exp. Biol. 206 (18), 3167–3173.

Milligan, C.L., McDonald, D.G., 1988. In vivo lactate kinetics at rest and during recovery from exhaustive exercise in coho salmon (Oncorhynchus kisutch) and starry flounder (*Platichthys stellatus*). J. Exp. Biol. 135 (1), 119–131.

Milligan, C.L., Hooke, G.B., Johnson, C., 2000. Sustained swimming at low velocity following a bout of exhaustive exercise enhances metabolic recovery in rainbow trout. J. Exp. Biol. 203 (5), 921–926.

Nelson, J.A., Gotwalt, P.S., Reidy, S.P., Webber, D.M., 2002. Beyond Ucrit: matching swimming performance tests to the physiological ecology of the animal, including a new fish 'drag strip.'. Comp. Biochem. Physiol. Part A: Molecul. Integ. Physiol. 133, 289–302.

Palstra, A.P., Planas, J.V. (Eds.), 2013. Swimming Physiology of Fish: Towards Using Exercise to Farm a Fit Fish in Sustainable Aquaculture. Springer Science & Business Media.

Peake, S.J., Farrell, A.P., 2004. Locomotory behaviour and post-exercise physiology in relation to swimming speed, gait transition and metabolism in free-swimming smallmouth bass (*Micropterus dolomieu*). J. Exp. Biol. 207, 1563–1575.

Pearson, M.P., Spriet, L.L., Stevens, E.D., 1990. Effect of sprint training on swim performance and white muscle metabolism during exercise and recovery in rainbow trout (*Salmo gairdneri*). J. Exp. Biol. 149, 45–60.

Plaut, I., 2001. Critical swimming speed: its ecological relevance. Comp. Biochem. Physiol. Part A: Molecul. Integ. Physiol. 131, 41–50.

Randall, D., Brauner, C., 1991. Effects of environmental factors on exercise in fish. J. Exp. Biol. 160 (1), 113–126.

Wang, S., Carter, C.G., Fitzgibbon, Q.P., Smith, G.G., 2021. Respiratory quotient and the stoichiometric approach to investigating metabolic energy substrate use in aquatic ectotherms. Rev. Aquacult. 13, 1255–1284.

Wood, C.M., 1991. Acid-base and ion balance, metabolism, and their interactions, after exhaustive exercise in fish. J. Exp. Biol. 160, 285–308.

Chapter 10

SWIMMING CAPACITY☆

F.W.H. BEAMISH

Chapter Outline

I. INTRODUCTION

Progress of fish maneuvering in water, the density of which is approximately equal to that of the animal itself, can be classified into three major categories: sustained, prolonged, and burst swimming speeds. Each reflects not only on the constraints imposed by time, but also on the biochemical processes which supply the fuel for their application.

Sustained swimming performance is applied to those speeds which can be maintained for long periods (greater than 200 min) without resulting in muscular fatigue. Included within sustained performance under the subcategory of cruising are those speeds achieved by migrating fish as well as the velocities which negatively bouyant species such as the scombroid and xiphoid fishes must achieve to maintain hydrostatic equilibrium. A second subcategory is sustained schooling which includes speeds displayed by groups of fish distributed in a regular array, examples of which include many of the clupeids and tunas. Also within sustained performance is routine activity which represents the daily movements of fish including foraging and station holding. Routine activity thus includes periods of steady and unsteady swimming.

☆This is a reproduction of a previously published chapter in the Fish Physiology series, "1978 (Vol. 7)/Swimming Capacity: ISBN: 978-0-12-350407-4; ISSN: 1546-5098".

Fish Physiology, Vol. 40B. https://doi.org/10.1016/bs.fp.2024.07.004

The quantitative measurement of routine activity in the field entails enormous practical difficulties which have yet to be overcome. In laboratory studies, routine activity is generally equated to the locomotion displayed by fish whose only movements are spontaneous which again is highlighted by periods of steady and unsteady swimming.

Prolonged swimming speed is of shorter duration (20 sec–200 min) than sustained and ends in fatigue. In field studies it is often difficult to separate sustained and prolonged, not only because of the practical difficulties imposed when attempting to track fish for long periods but also on account of the variability in swimming speed expressed by fish even when migrating or in schools. Rarely, if ever, is it possible to assess fatigue in the field. Prolonged speeds are most accurately measured in the laboratory in swimming flumes.

Critical swimming speed, a special category of prolonged, was first defined and employed by Brett (1964) to designate the maximum velocity fish could maintain for a precise time period. It is measured by interpolation for those fish that do not fatigue exactly at the beginning or end of a prescribed period. The details of its calculation are described elsewhere in this chapter. The application of critical swimming speed is confined to laboratory investigations.

The highest speeds of which fish are capable are organized under the category of burst swimming. These high speeds can be maintained only for short periods (less than 20 sec) and are characterized by an initial acceleration phase of unsteady swimming followed by a steady phase hereafter termed sprint. The capacity for short-term high performance is essential to the survival of many species as it facilitates the capture of prey, avoidance of predators, or the negotiation of rapid currents as may be encountered in rivers during spawning migration.

This chapter attempts to synthesize measurements of swimming performance within the categories described above in relation to environmental and biological factors. Some guidelines are provided for the measurement of performance both in the field and in the laboratory. Finally, the application of information on swimming performance to fishery management programs is discussed.

II. FIELD MEASUREMENTS OF PERFORMANCE

A. Methodology

1. Direct Observation and Conventional Tags

Perhaps the simplest procedure employed to measure swimming speed is to estimate by direct observation the time required for fish to swim a gauged distance (Lane, 1941) or to relate swimming velocity to an estimated rate of movement on land. For example, the burst speed of pike was obtained by estimating the speed at which a boy had to run to keep pace with the fish in still water (Stringham, 1924). Where migrating fish pass upstream through shallow channels or culverts of known length and velocity, direct estimates of burst, prolonged, or sustained cruising speed have been made (Wales, 1950; Dow, 1962; Ellis, 1966). Wantanabe (1942) measured the time required by skipjack tuna, *Euthynnus pelamis*, to swim a marked distance of 2 m alongside a ship but ignored currents in his calculation. Davidson (1949) observed the time required by young Atlantic salmon, *Salmo salar*, to cover specific distances in circular ponds in which the current pattern was well described.

Conservative estimates of sustained cruising or schooling swimming speeds have been made from conventional tagging studies, assuming that the fish travel the distance between the points of capture and recapture in a straight line.

2. *Sports Fishing Gear*

Modified sports fishing gear has been applied to the measurement of swimming speed (Magnan, 1929; Lane, 1941; Gero, 1952; Walters and Fierstine, 1964). Lane (1941) attached a motorcycle speedometer to a rod and reel to measure the burst swimming speed of a tuna caught by hook. Gero (1952) developed an instrument he termed a piscatometer, for measuring swimming velocity of hooked fish. The piscatometer was equipped with conventional sports fishing tackle. After a fish was hooked, a remote camera recorded swimming activity for subsequent analyses. Tension in the line was measured indirectly by a hydraulic strut and pressure was transmitted to an instrument where it was registered on a meter. Line velocity was determined by running the fishing line over a pulley which activated a tachometer. Iron particles, evenly spaced on a fishing line, in conjunction with a magnetic tape recorder head and an oscilloscope further facilitated measurement of the speed at which a hooked fish pulled line from a reel (Walters and Fierstine, 1964). Measurements of swimming speed made in this way assume that once hooked, fish swim along a straight course. Walters and Fierstine (1964) suggested that reliable estimates of burst speed could be made only within the first 10–20 sec after a fish was hooked. Thereafter changes in velocity occurred which they attributed to fish altering their direction of swimming. Other factors may also reduce line velocity including fishing line drag and probable turbulance induced in the boundary layer of the fish by the line and lure (Walters and Fierstine, 1964). The procedure of attaching a line to fish has been applied also in laboratory studies on the swimming capacity of fish (Ohlmer and Schwartzkopff, 1959).

3. *Photography and Television*

Motion picture records have been employed to determine swimming capacity of tunas at sea or in large outdoor tanks. (Magnuson and Prescott, 1966; Yuen, 1966; Magnuson, 1967, 1970, 1973; Dixon, Chang, Byles, and Neill, personal communication). High speed movie cameras are positioned so that the direction of swimming is at right angles to the long axis of the camera lens.

Swimming speed, estimated from photographs taken at sea, can be corrected for the vertical component of motion caused by rolling and pitching and for the horizontal component when the ship is moving forward, by realignment of the images in the manner described in detail by Yuen (1966).

Television was used by Brawn (1960) to estimate swimming performance of Atlantic herring, *Clupea harengus*. The television camera was attached at one end of a towed cage in which herring were forced to swim. The camera provided a visual record of swimming behavior at depths to 30 m. A current meter suspended within the cage measured the main current velocity, and a small streamer tag attached at the center of the chamber, where it was easily seen on the television screen, gave visual indication of the direction of minor currents.

4. Acoustics

The development of echo sounders, ultrasonic pulse sonar (Nishimura, 1963; Komarov, 1971), and tags containing radio or ultrasonic transmitting devices has greatly improved measurements of sustained cruising and schooling speeds, and to a lesser extent, routine velocities (Trefethen, 1956; Johnson, 1960; Novotny and Esterberg, 1962; Bass and Rascovich, 1965; Henderson *et al.*, 1966; McCleave *et al.*, 1967; Poddubny, 1967; Hasler *et al.*, 1969; McCleave and Horrall, 1970; Yuen, 1970; Dodson *et al.*, 1972; Madison *et al.*, 1972; Tesch, 1974).

Echo sounders, particularly those which operate in the higher range of frequencies (above 38 kHz), have been used to measure the sustained cruising and schooling speeds of fish with only moderate success. This method suffers in that, in general, it is not possible to identify fish to species without representative catches within the area under investigation. Further, it is extremely difficult to follow the swimming activity of individual fish for extended periods of time. The technique of echo sounding is described in numerous articles which should be consulted for a comprehensive understanding of the method (Cushing, 1957; Harden-Jones and McCartney, 1962; Dowd, 1967; Midttun and Nakken, 1968; Craig and Forbes, 1969; Dowd *et al.*, 1970).

Ultrasonic tags, which in water may approach neutral buoyancy, are customarily attached to the dorsal musculature (Tesch, 1974), dorsal fin (McCleave *et al.*, 1967; McCleave and Horrall, 1970), or placed in the stomach of fish (Henderson *et al.*, 1966; Hasler *et al.*, 1969). The effective transmission range of sonic tags varies from several hundred meters to over 2 km which allows for the tracking of fish with a minimum of interference to their behavioral pattern. The life expectancy of sonic tags is dependent on battery size, which in turn is dictated, at least in part, by the size of the fish to be tagged. Thus tags applied to white bass, *Morone chrysops*, transmitted sound for about 20 hr (Hasler *et al.*, 1969), considerably less than the 3-week expectancy from sonic devices attached to the larger American shad, Alosa sapidissima (Dodson *et al.*, 1972) and European eel, *Anguilla anguilla* (Tesch, 1974).

The pulsed sound emitted from the tags is received by a hydrophone housed in the tracking vessel. Navigational instruments facilitate measurement of the distance covered by the tracking vessel. Currents which may assist or hinder the movement of fish are difficult to determine with the precision necessary to correct for actual swimming speed. Further, with the methods presently available, it is not yet practical to clearly define distances covered by routinely active fish, such as when foraging. However, among migrating fish, sonic tags have contributed greatly to the present understanding of their speed of movement.

The development of pressure sensing ultrasonic transmitters by Stasko and Rommel (1974) has facilitated the measurement of swimming depth. Gayduk *et al.* (1971) measured swimming depth by attaching transmitters containing high-frequency photoresistors to fish. The photoresistors provide a measure of intensity of illumination which can be translated to vertical location assuming a constant attenuation of intensity with depth.

Swimming speed can be measured in the field by any of several methods. However, few allow for the identification of the category of swimming either through their failure to record a fish's progress for sufficient time or to provide a measure of fatigue. Among the methods employed, ultrasonic tags allow for a quantitative description of sustained swimming speed of migrating and, often, schooling fish. The shortcomings of the acoustic method is that it fails to define clearly the distance moved by routinely active fish. Methods for the identification and quantitative measurement of prolonged and burst swimming are not well developed. Pressure transducers have not been applied to the measurement of locomotion but offer an optimistic alternative to those methods presently in use.

B. Swimming capacity

Quantitative categorization of performance in free swimming fish is frustrated by environmental and biological constraints. Sustained and prolonged speeds are difficult to distinguish because fish seldom maintain a given speed for as long as 200 min. Actual swimming speed may be enhanced or hindered by currents that often are not measured. Thus migrations may be achieved largely on the strength of currents with little expenditure of energy for locomotion (Radakov and Solovyev, 1959; Zaitsev and Radakov, 1960; Harden-Jones, 1968; Royce *et al.*, 1968; Hasler *et al.*, 1969).

Most field measurements, exclusive of burst, have been made for migrating or schooling fish. Until quantitative procedures become available, it is convenient to consider the performance of migrating and schooling fish under sustained swimming with the realization that the range of recorded velocities may well extend to prolonged. Speeds achieved and maintained by individuals or a portion of a group of fish for under 20 sec and resulting either in fatigue or greatly reduced performance are classified within burst.

1. Sustained Swimming Speed

a. Cruising. Estimates of swimming performance by migrating fish based on tag and recapture studies suffer in that they assume a straight line course from the point of tagging to the place of recapture. Further, they fail to provide any information on the variation in swimming speed during the migration. The literature on tag and recapture studies is large, and in the light of recent studies with ultrasonic tags will not be dealt with in this chapter. Several of the more excellent references on tag and recapture studies include Dahl (1937), Dahl and Sømme (1938), Huntsman (1942), Pritchard (1944), Määr (1947), Dannevig (1953), Rasmussen (1959), Trout (1957), Fridriksson (1958), Lühmann and Mann (1958), Marty (1959), and Maslov (1960).

Ultrasonic tagging experiments have shown that the speed of migration depends on numerous factors including the negotiation of currents, topography, shoreline, temperature, meteorological conditions, and the physiological status of the fish (Malinin, 1973). Migratory fish seldom follow a strictly linear course although directionality in their pattern of movement is generally apparent (Poddubny, 1967; Hasler *et al.*, 1969; McCleave and Horrall, 1970; Yuen, 1970; Madison *et al.*, 1972; Dodson *et al.*, 1972).

Swimming speed among some migratory species appears related to the proximity of the land, which may serve as a navigational aid, while in others differences are not demonstrable. Cutthroat trout, *Salmo clarki*, in open water swam at an average speed of 22.9 cm sec^{-1}, appreciably less than the 36.6 cm sec^{-1} recorded when swimming near shore (McCleave and Horrall, 1970; see also Table I). In contrast, the migratory speed of adult sockeye salmon, *Oncorhynchus nerka*, did not differ with distance from shore (Madison *et al.*, 1972). The migratory speeds of sturgeon, *Acipsenser nudiventris*, sevryuga, *A. guldenstadtii*, and Atlantic salmon in freshwater rivers approximated those in the sea (Malinin, 1973).

Among some anadromous species, entry into freshwater may temporarily impair progress in the upstream migration. American shad tracked from the sea to freshwater exhibited considerable wandering and passive drift within the confines of the estuary (Dodson *et al.*, 1972). Further, mean swimming speed was reduced from 75 cm sec^{-1} in the saline environment to about 39 cm sec^{-1} in the estuary (Fig. 1). On entry into the estuary of the Miramichi River from the Atlantic Ocean, Stasko (1975) found Atlantic salmon tended either to drift with the tidal currents or hold station for as much as 14 days. Fish that achieved overall upstream progress did so by drifting with the flood tide and stemming ebb currents. Both Dodson *et al.* (1972) and Stasko (1975) considered this reduction in performance a manifestation of the physiological stress encountered in the transition from salt to freshwater.

Light conditions appear also to influence the speed and direction of migrants (Hasler *et al.*, 1958). Pronounced diel fluctuations in migratory swimming speeds of sockeye salmon were reported by Madison *et al.*

Table I Sustained Cruising Speeds for Migrating Fish

Species	Length (cm)	Number	Swimming speed						Time	Temperature (°C)	Location	Reference
			Range		Mean		Maximum[a]					
			cm sec^{-1}	ℓ sec^{-1}	cm sec^{-1}	ℓ sec^{-1}	cm sec^{-1}	ℓ sec^{-1}				
Carcharhinus leucas	ca. 110	9	18–202						1–11 days		Lake Nicaraguia and Caribbean Sea	Thorson (1971)
Acipenser	ca. 110	6	130–500	1.2–4.5	165	1.5			1 hr	20	Volga River, U.S.S.R.	Malinin *et al.* (1971)
Acipenser	ca. 110				15	0.14			1 day	20	Volga River, U.S.S.R.	Malinin *et al.* (1971)
Acipenser	ca. 110		25–58	0.2–0.5	33	0.3			1 day	20	Saratov HES, U.S.S.R.	Malinin *et al.* (1971)
Acipenser nudiventris					35				Many hr		Volga River,U.S.S.R.	Batchykov (1963)[b]
Acipenser nudiventris					25				Many hr		Volga River estuary, U.S.S.R.	Pavlov (1969)[b]
Acipenser nudiventris	80–130				15	0.1–0.2	70	0.5–0.9	Many hr		Volga River, U.S.S.R.	Malinin (1973)
Acipenser guldenstadtii					23				Many hr		Kura River, U.S.S.R.	Derzhavin (1922)[b]
Acipenser guldenstadtii					27		300		Many hr		Volga River, Rybinsk, and Volgogradsk Reservoir, U.S.S.R.	Poddubny (1967)
Acipenser guldenstadtii	70–120				33	0.3–0.5			Many hr		Kuban River, U.S.S.R.	Malinin (1973)
Anguilla anguilla	69–96	11	13–173	0.2–2.0	55–72	0.6–0.9	173	2.2	1–19 hr		North Sea	Tesch (1974)

Continued

Table I Sustained Cruising Speeds for Migrating Fish—Cont'd

Species	Length (cm)	Number	Swimming speed						Time	Temperature (°C)	Location	Reference
			Range		Mean		Maximum					
			cm sec^{-1}	ℓ sec^{-1}	cm sec^{-1}	ℓ sec^{-1}	cm sec^{-1}	ℓ sec^{-1}				
Alosa sapidissima		886			75				Many hr		Long Island Sound, Atlantic Ocean	Dodson *et al.* (1972)
Alosa sapidissima		5			39				Many hr		Connecticut River estuary, U.S.A.	Leggett and Jones (1973)
Alosa sapidissima		401			64				Many hr		Connecticut River U.S.A.	Leggett and Jones (1973)
Oncorhynchus gorbuscha					55				Many hr	5–6	Amur River, U.S.S.R.	Krykhtin (1964)[b]
Oncorhynchus gorbuscha					29				Many hr		Amur River estuary, U.S.S.R.	Krykhtin (1964)[b]
Oncorhynchus gorbuscha	64	11	23–78	0.4–1.2					3–50 hr		Pacific Ocean	Stasko *et al.* (1973)
Oncorhynchus keta					54				Many hr		Amur River, U.S.S.R.	Soldatov (1912)[b]
Oncorhynchus keta					7–8				Many hr		Amur River estuary, U.S.S.R.	Krykhtin (1964)[b]
Oncorhynchus kisutch	58	1			31	0.5			1–4 hr		Columbia River, U.S.A.	Johnson (1960)
Oncorhynchus kisutch	55	228	30–162	0.5–2.9	52–96	0.9–1.7	162	2.9	<30 sec	15.4–23.3	Somass River, Canada	Ellis (1966)

Oncorhynchus nerka	56	834	40–158	0.7–2.8	53–97	0.9–1.7	248	4.4	<30 sec	15.4–23.3	Somass River, Canada	Ellis (1966)
Oncorhynchus nerka	65	18	33–99	0.5–1.7	46–59	0.7–0.9	133–170	2.0–2.6	2–66 hr		North Pacific Ocean	Madison *et al.* (1972)
Oncorhynchus tshawytscha	84	37	4–176	0.05–1.8	154–176	0.8–1.8			<1–16 hr		Columbia River, U.S.A.	Johnson (1960)
Salmo clarki		42	0–82		23–37		82		1–13.5 hr		Yellowstone Lake, U.S.A.	McCleave and Horrall (1970)
Salmo clarki	34.5	7	0–44.7	0–1.3					3.5–18.5 hr		Yellowstone Lake, U.S.A.	McCleave and LaBar (1972)
Salmo gairdneri	72	2	9–44	0.1–0.6					1–3 hr		Columbia River, U.S.A.	Johnson (1960)
Salmo gairdneri	42–59		0.2–26	<0.01–0.7					10 hr	5	Lake Chuzenji, Japan	Kazihara *et al.* (1969)[b]
Salmo salar					54						Pechora River, U.S.S.R.	Letovaltseva (1967)[b]
Salmo salar			44–69						Many hr		Baltic Sea	Carlin (1968)[b]
Salmo salar	50–70		12–25		225				Many hr		N-Tulemskove Reservoir, U.S.S.R.	Malinin (1973)
Salmo salar	53–87	15	0–17	0–0.2			104	1.3	3–86 hr		Miramichi River, Canada	Stasko (1975)
Esox lucius	ca. 80				5.5	0.07	300	3.8	24 hr	0.5	Rybinsk Reservoir, U.S.S.R.	Poddubny *et al.* (1970)
Abramis brama	ca. 40		5–13	0.1–0.3					22–168 hr		Witham River, U.K.	Langford (1974)
Lota lota	ca. 50			16.7	0.3				48 hr		Sogozha River, U.S.S.R.	Malinin (1971)

Continued

Table I Sustained Cruising Speeds for Migrating Fish—Cont'd

Species	Length (cm)	Number	Swimming speed						Time	Temperature (°C)	Location	Reference
			Range		Mean		Maximum					
			cm sec^{-1}	ℓ sec^{-1}	cm sec^{-1}	ℓ sec^{-1}	cm sec^{-1}	ℓ sec^{-1}				
Morone chrysops	26–38	26	5–21	0.2–0.7	13	0.4			41–10.6 hr		Lake Mendota, U.S.A.	Hasler *et al.* (1969)
Tautoga onitus	47	1			60	1.3			2.9 hr	21.7	Atlantic Ocean, U.S.A.	Olla *et al.* (1974)
Makaira nigricans	ca. 215–360				61–95	0.2–0.5	86–228	0.3–0.7	1–22.5 hr		Pacific Ocean	Yuen *et al.* (1974)
Pleuronectes platessa	43	1	27.5–90	0.6–2.1					2.3 hr		North Sea	Greer Walker *et al.*, 1971

[a]Above true sustained speed.
[b]Data reported in Malinin (1973).

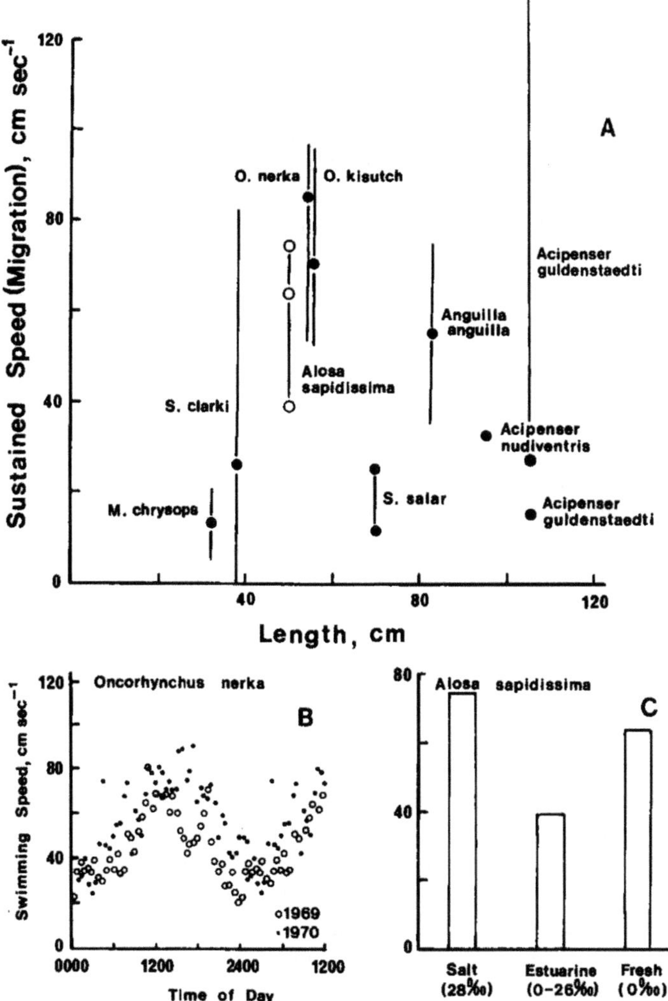

Fig. 1. Sustained cruising speeds of migrating fish. (A) Mean and range of sustained speed in relation to length (see text). (B) Diel fluctuations (redrawn from Madison *et al.*, 1972, *J. Fish. Res. Board Can.*). (C) Mean speed on entry into freshwater from the estuarine and marine environment (Dodson *et al.*, 1970, 1971, 1972).

(1972). Maximum speeds of about 70 cm sec^{-1} (60–90 cm sec^{-1}; Fig. 1) occurred at midday (1200 hr) with minimum speeds averaging near 35 cm sec^{-1} (20–50 cm sec^{-1}) during the hours of darkness (2300–0200 hr). Periods of greatest change in swimming speed coincided approximately with sunrise and sunset, which together with other information on angular change prompted the hypothesis of a daily reorientation process related to light conditions. Migrating adult chinook salmon in accord with sockeye also exhibit a reduction in swimming speed during the hours of darkness (Johnson, 1960). Differences between day and night migratory behavior of the European eel were not demonstrable; however, the prevailing influence of the tides precluded a valid comparison of the effects of light and darkness (Tesch, 1974). Among migrating schools of skipjack tuna, variation occurred in the spatial distance covered, although not necessarily in swimming speed (Yuen, 1970). During the hours of darkness, skipjack traveled from 25 to 106 km, equivalent to speeds approaching 100 cm sec^{-1}. By day the tuna remained in a reasonably confined area where they engaged in feeding. Under conditions such as these, ultrasonic devices often fail to provide a precise description of the actual distance covered. In fact, in the case of Yuen's tuna, the calculated speeds while feeding during the daylight hours were less than those necessary to maintain hydrostatic equilibrium (Magnuson, 1970).

Mean sustained cruising speeds of fish migrating in a horizontal plane lie mainly from 0.5 to 0.2 $\ell\ sec^{-1}$ (Fig. 1 and Table I). A low mean speed of swimming may have adaptive ecological significance by increasing that energetic efficiency of migration. Wiehs (1973a) calculated that the maximum distance that can be accomplished for a given expenditure of energy occurs when the energy utilized in propulsion is equal to the resting metabolic rate. In rainbow trout, *Salmo gairdneri*, the velocity for which this is the case approximates 1 $\ell\ sec^{-1}$.

Considerable data are available on swimming speeds of scrombroid fishes (Tables II and III), many of which are negatively buoyant (Magnan, 1929; Watanabe, 1942; Aleev, 1963) and include bonitos, mackerels, and tunas. To overcome negative buoyancy scrombroids must swim continuously (sustained cruising) with pectoral fins extended which produces a lift that balances their weight in water (Aleev, 1963; Alexander, 1967, 1968; Magnuson, 1970, 1973). Observations by Magnuson (1973) on the sustained cruising speeds necessary to overcome negative bouyancy for seven scrombroid species indicate a range from 0.3 to 3.3 $\ell\ sec^{-1}$ (Table II). The three species with the fastest speeds necessary to counter negative buoyancy, skipjack, kawakawa, *Euthynnus affinis*, and Pacific bonito, *Sarda chiliensis*, are without a swim bladder. Wahoo, *Acanthocybium solanderi*, the species with the lowest predicted speed has the largest bladder and the lowest density. Among the three species without swim bladders, skipjacks are the heaviest for their length, have the smallest lift area of pectoral fins, and have the highest range of minimum speeds.

Many fishes display regular diel vertical migrations which are of considerably smaller magnitude than those displayed in the horizontal direction. Patterns of vertical migration are best described for marine demersal and some pelagic species (Hjort, 1914; Saetersdal, 1956; Konstantinov, 1958; Richardson *et al.*, 1959; Brunel, 1964; Parrish *et al.*, 1964; Sundnes, 1963; Woodhead, 1964; Beamish, 1966a). Observations on vertical migration have been made almost entirely with echo sounders. This method generally

Table II Sustained Cruising Speeds Necessary to Overcome Negative Buoyancy among the Scombroid Fish[a]

			Swimming speed			
			Range		Mean	
Species	Number	Length (cm)	cm sec^{-1}	ℓ sec^{-1}	cm sec^{-1}	ℓ sec^{-1}
Acanthocybium						
solanderi	12					0.3
Auxis rochei		31	55–76	1.8–2.5	68	2.2
Euthynnus affinis		36	71–89	2.0–2.5	76	2.1
Euthynnus pelamis	12	38–48	39–125	1.0–3.3	70	1.6
Sarda chiliensis	12	52	47–83	0.9–1.6	83	1.6
Thunnus albacares	12	35	41–62	1.2–1.8	46	1.3
Thunnus obesus		55	55–94	1.0–1.7	66	1.2
Thunnus obesus		36	43–60	1.2–1.7	48	1.3

[a]Data from Magnuson (1973). Observations were made in large tanks 23°–26°C over a period of 1–4 hr.

Table III Sustained Schooling Speeds of Fish

			Swimming speed							
			Range		Mean					
Species	Number	Length (cm)	cm sec^{-1}	ℓ sec^{-1}	cm sec^{-1}	ℓ sec^{-1}	Observation time	Temperature (°C)	Location	Reference
Clupea harengus harengus		ca. 25–30			9–12	0.5			Norwegian Sea	Fridriksson and Aasen (1952)
Clupea harengus harengus		ca. 25–30	50–100	2–5					North Sea	Jones (1957)
Clupea harengus harengus		ca. 25–30			150	5–6			North Sea	Schärfe (1960)
Clupea harengus harengus		ca. 25–30			90	3			North Sea	Harden-Jones (1962)
Clupea harengus harengus		Juvenile				4–5	Hours		Tank	Hempel[a]
Clupea harengus pallasi		ca. 28			200	7				Chestonoy[b]
Clupea harengus pallasi		15			113	7.5	20 min		Pacific Ocean	High and Lusz (1966)
Engraulis encrasicholus					65–80				Tank	Lebedev (1936)[b]
Engraulis encrasicholus					75				Tank	Radakov (1962)[a]

Seriola quinqueradiata	1	80	<25–178	0.3–2.3	60	0.8	8 hr	18.5	Pacific Ocean	Ichihara *et al.* (1972)
Euthynnus affinis		35–42			82	2.0–2.3	1–4 hr	23–26	Tank	Magnuson (1967)
Euthynnus affinis		40			200	5		25	Tank	Dizon *et al.* (personal communication)
Euthynnus pelamis	10	49–57	500–1000	9.7–19.4	770	14.8		27–29	Pacific Ocean	Watanabe (1942)
Euthynnus pelamis	510	48–79	30–690	0.4–14.3	200	2.5–4.2	<3 sec	24–26	Pacific Ocean	Yuen (1966)
Euthynnus pelamis	2	44	20–435	0.5–9.9			7 days		Pacific Ocean	Yuen (1970)
Euthynnus pelamis		ca. 40	650–850	16–21			2–3 hr	25	Pacific Ocean	Neill (personal communication
Thunnus albacares		ca. 50			1100	12			Pacific Ocean	Kishinouye (1923)
Thunnus albacares		ca. 50			100–150	2–3			Pacific Ocean	Nishimura (1963)
Thunnus albacares	33	52	160–540	3.1–10.4			<3 sec	24–26	Pacific Ocean	Yuen (1966)

[a]Data reported in Blaxter (1967).
[b]Data reported in Radakov (1964).

precludes accurate information on the actual size of the migrants or on the rate of migration by individuals. Rather, it does provide information on the general rate of movement by concentrations of fish. In addition to the practical problems of obtaining quantitative measurements, the extent of the migration itself depends very much on factors such as season and depth (Verwey, 1960; deVeen, 1964; Beamish, 1966a; Hawkins *et al.*, 1974).

The vertical distance migrated by haddock, *Melanogrammus aeglefinus*, at night may be as much as 1000 m (Hjort, 1914) but more frequently is around 100 m (Beamish, 1966a). Atlantic cod, *Gadus morhua*, display diel vertical migrations of under 70 m (Richardson *et al.*, 1959). Redfish, *Sebastes marinus*, usually remain within 50 m from bottom (Beamish, 1966a). While the exact rate of ascent or descent is unknown, most appear to be completed within an hour. Assuming steady swimming over a 1-hr period, the mean rate of swimming in most cases is under 5 cm sec^{-1}. Echo sounder records made at sea from a stationary vessel indicate that periods of steady swimming alternate with periods of unsteady activity; however, the relative proportion of each is not clear.

The swimming depth of American eel, *Anguilla rostrata*, was monitored in Passamaquoddy Bay, Bay of Fundy, using pressure sensing ultrasonic transmitters (Stasko and Rommel, 1974). Eels made frequent dives from the surface to bottom during the hours of daylight and darkness at speeds of $0.8-1.1\ \ell\ sec^{-1}$. The maximum rate of ascent was $0.6-0.8\ \ell\ sec^{-1}$.

b. Schooling. Schools of fish are recognizable by the regular array of the component individuals and the synchrony displayed in their swimming speeds (Shaw, 1960). Belyayev and Zuyev (1969) and Wiehs (1973b) have suggested that the organized spatial distribution of fish within a school may, in fact, reduce the hydrodynamic resistance of individuals, thus providing for increased swimming efficiency.

Swimming speeds for the schooling clupeoid fish have been reported from under 1 to more than $7\ell sec^{-1}$ (Lebedev, 1936, reported in Radakov, 1964; Fridriksson and Aasen, 1952; Jones, 1957; Schärfe, 1960; Radakov, 1962, reported in Blaxter, 1967; Chestnoy, reported in Radakov, 1964; High and Lusz, 1966; see also Table III, Fig. 2). The higher velocities were recorded for herring swimming ahead of towed nets, for short periods of time, and probably represent prolonged speeds. Blaxter (1967) reported that a relative speed of $3-4\ \ell sec^{-1}$ is a valid estimate of sustained schooling performance among the clupeoid fish.

Swimming speeds of schooling scombroids, including the velocity necessary to counter negative buoyancy, have been reported from 2 to $21\ell sec^{-1}$ (Table III). Yuen (1966) calculated speeds of 30–690 cm sec^{-1} for feeding skipjack tuna from photographic records and found no demonstrable dependence on length (48–79 cm). Watanabe (1942) observed skipjack tuna as they swam by a measured distance of 2 m alongside a ship. The mean speed

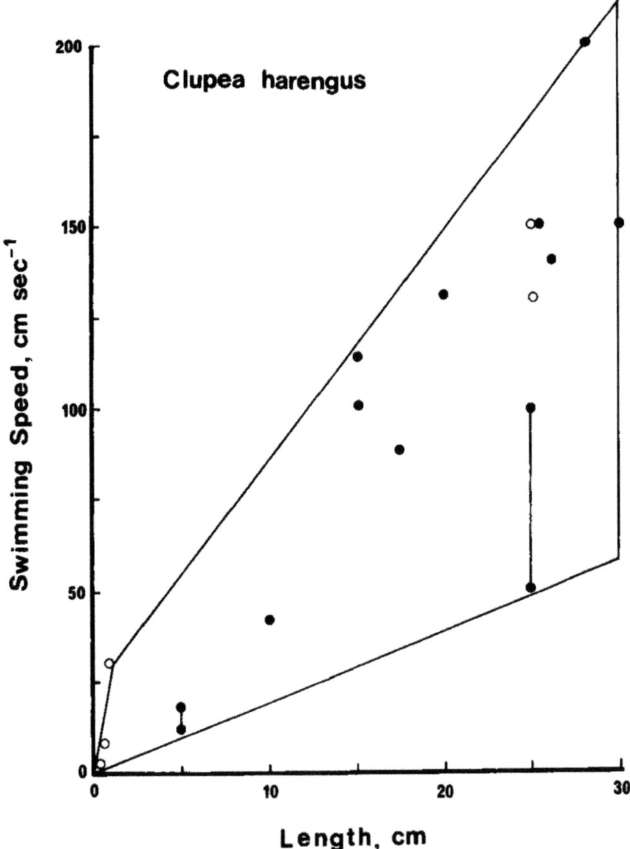

Fig. 2. Field and laboratory observations on the swimming capacity of herring, *Clupea harengus*, in relation to length. Sustained schooling and prolonged speeds are indicated by closed circles. Speeds reported as burst are recorded as open circles. [From Fridriksson and Aasen, 1952; Jones, 1957; Blaxter and Dickson, 1959; Bishai, 1960; Brawn, 1960; Schärfe, 1960; Boyar, 1961; Blaxter, 1962; Chestnoy (cited in Radakov, 1964); Blaxter and Parrish, 1966; High and Lusz, 1966.]

recorded for ten individuals (52 cm) was 770 cm sec^{-1} with a range of 500–1000 cm sec^{-1}. Yuen (1970) provides information from which it may be calculated that a 44 cm skipjack tuna maintained for 107 min a minimum average speed of 435 cm sec^{-1}(9.9 ℓ sec^{-1}). Such a speed is, according to Neill (personal communication), consistent with reports of skipjack schools "disappearing over the horizon after being pursued by fishing vessels at 650 to over 850 cm sec^{-1} for 2 or 3 hours." Whether, in fact, tuna fatigue at these speeds is as yet unanswered.

Among the other tunas for which speeds attributable to sustained schooling have been reported, Yuen (1966) calculated velocities of 160–540 cm sec^{-1} for a school of yellowfin tuna (52 cm) higher than the 100–150 cm sec^{-1} reported by Nishimura (1963) for this species. Kishinouye (1923) estimated the speed of yellowfin tuna at 1100 $cmsec^{-1}$. Dizon, Chang, Byles, and Neill (personal communication) observed speeds of 200 cm sec^{-1} for kawakawa (40 cm) feeding in large outdoor pools.

The hydrodynamic and vascular achievements in tunas (Barrett and Hester, 1964; Stevens and Fry, 1971; Linthicum and Carey, 1972; Graham, 1973, 1975; Magnuson, 1973; Stevens *et al.*, 1974) are manifested in their relative swimming performances which exceed by severalfold those recorded for any other schooling species. It is, of course, not possible to categorize tuna performance, without question, into prolonged and sustained schooling speeds. However, that relative velocities of 2–21 ℓ sec^{-1} can be maintained for at least several hours by skipjack tuna without fatiguing, supports the suggestion of high sustained schooling speeds. The lower speeds reported for tuna undoubtedly represent sustained routine and cruising speeds, rather than schooling performance.

2. *Burst Swimming Speed*

In an entertaining article, Lane (1941) described the maximum speeds which free swimming or tethered fish could achieve but based his measurements on estimated distances or movements in streams with little attention to currents. In a quantitative laboratory study on burst performances, Bainbridge (1958) supported the view that fish up to 1 m in length should be able to swim up to 10 times their own body length but only for a brief period of about 1 sec, beyond which velocity would decrease exponentially. Recent investigations suggest variability in burst swimming among species but that for at least some the relative performance maximum of 10 ℓ sec^{-1} is a conservative measure. Speeds of more than 20 ℓ sec^{-1} may better represent the upper threshold for steady or sprint burst swimming (Table IV). Thus the speed at which alewives, *Pomolobus pseudoharengus*, migrated upstream through a fishway ranged from 14 to 20 ℓ sec^{-1} (Dow, 1962; see also Table IV). Fierstine and Walters (1968) recorded burst acceleration velocities of 11.9 ℓ sec^{-1} for a 0.19 sec period by a yellowfin tuna and 18.4 ℓ sec^{-1} for a wahoo over the short time span of 0.05 sec. Skipjack tuna have been recorded in pursuit of food at speeds of over 20 ℓ sec^{-1} (Dizon *et al.*, personal communication). Komarov (1971) obtained speeds of 18.4 and 21.6 ℓ sec^{-1} for golden mullet, *Mugil auratus*, and mullet, *Mugil saliens*, respectively, but it is not clear whether the measurements were made in the field or in the laboratory. Of the species examined by Komarov (1971), carp, *Cyprinus carpio*, and green wrasse, *Creilabrus tinca*, were among the poorest swimmers but still registered relative velocities of 7–8 ℓ sec^{-1}.

Table IV Burst Swimming Speeds of Fish

Species	Number	Length (cm)	Velocity		Time (sec)	Temperature (°C)	Comments	References
			cm sec^{-1}	ℓ sec^{-1}				
Caraharhinus leucas	1	152	522	3.4	—		Hook and line	Gero (1952)
Negaprion brevirostris	1	184	244	1.3	—		Hook and line	Gero (1952)
Albula vulpes		<90	1000	>11.1			Hook and line	Lane (1941)
Anguilla vulgaris	1	60	114	1.9	2–5	10.0–15.0	Timed over measured distance (swimming tunnel)	Blaxter and Dickson (1959)
Alosa sapidissima			>350–402				Timed over measured distance (fishway)	Weaver (1965)
Clupea harengus	55	1.0–26	6–150	5.8–6.0	2–5	5.0–18.0	Timed over measured distance (swimming tunnel)	Blaxter and Dickson (1959)
Clupea harengus	1551	6.0–21.9	67–131	6.3–9.7	30	1.4–11.2	Annular trough	Boyar (1961)
Clupea harengus	1551	6.0–21.9	40–104	5.1	30	1.4–11.2	Annular trough	Boyar(1961)
Clupea sprattus	4	>75	24–36	3.2–4.8	2–5	<10–> 15	Timed over measured distance (swimming tunnel)	Blaxter and Dickson (1959)
Pomulobus pseudoharengus	21	27.1–31.2	415–485	13.6–15.7	2.6–6.2		Timed over measured distance (fishway)	Dow (1962)

Continued

Table IV Burst Swimming Speeds of Fish—Cont'd

Species	Number	Length (cm)	Velocity		Time (sec)	Temperature (°C)	Comments	References
			cm sec^{-1}	ℓ sec^{-1}				
Pomulobus pseudoharengus			304				Timed over measured distance (fishway)	Stringham (1924)
Engraulis encrasicholus pontious	—	12.1	162	13.4	<1		Distance covered by one tailbeat	Komarov (1971)
Oncorhynchus kisutch	9	44–67	161–216	2.7–4.4	10–15	10	Annular trough	Paulik and DeLacy (1957)
Oncorhynchus kisutch	5	51–76	220–372	4.3–6.4	0.1		Annular trough	Paulik and DeLacy (1957)
Oncorhynchus kisutch	4	36–61	287–533	6.2–9.2	1.5		Timed over measured distance (fishway)	Weaver (1963)
Oncorhynchus nerka	15	55–69	155–203	2.2–3.1	10–15	15	Annular trough	Paulik and DeLacy (1957)
Oncorhynchus nerka	—	—	268–313	—	—			Napier (1914)[a]
Oncorhynchus tshawytscha	6	51–97	543–668	6.9–10.7	1.5		Timed over measured distance (fishway)	Weaver (1963)
Salmo fario	1	25	350	10.0				Denil (1937)
Salmo gairdneri			550–1050				Hook and line	Lane (1941)
Salmo gairdneri	10	58–67	186–226	2.8–3.7	10–15	7	Annular trough	Paulik and DeLacy (1957)

Salmo gairdneri	6	61–81	536–817	7.5–13.4	1.5		Timed over measured distance (fishway)	Weaver (1963)
Salmo gairdneri		14.3	30–250	2.0–17.5	0.08			Webb (personal communication)
Salmo gairdneri		14.3	30–180	2.0–12.6	0.04			Webb (personal communication)
Salmo irideus	4	10.3–28.0	105–270	7.6–10.2	1	14	Annular trough	Bainbridge (1960)
Salmo irideus	4	10.3–28.0	32–73	2.6–3.1	20	14	Annular trough	Bainbridge (1960)
Salmo salar		35	347	—				Denil (1909)
Salmo salar		—	300					Lavollee (1902)[a]
Salmo salar		75–85	429–600	5.8–8.4				Denil (1937)
Salmo salar			805				Timed over measured distance (fishway)	Lane (1941)
Salmo trutta	19	13–37	137–305	8.2–10.5	2–5	10–15	Timed over measured distance (swimming tunnel)	Blaxter and Dickson (1959)
Salvelinus fontinalis	1	11.2	93	8.3	10	15	Swimming tunnel	Peterson (1974)
Euthynnus pelamis	10–20	40–50		13.5–18.6	0.2	25	Timed over measured distance (annular tank)	Neill *et al.* (personal communication)
Esox			590–1370				Hook and line	Lane (1941)

Continued

Table IV Burst Swimming Speeds of Fish—Cont'd

Species	Number	Length (cm)	Velocity		Time (sec)	Temperature (∘C)	Comments	References
			cm sec^{-1}	ℓ sec^{-1}				
Esox lucius	1		360–450	—				Stringham (1924)
Carassius auratus	7	9.0	138	15.3	2–5	<10.0	Timed over measured distance (swimming tunnel)	Blaxter and Dickson (1959)
Carassius auratus	8	6.7–21.3	74–200	9.4–11.0	1	14	Annular trough	Bainbridge (1960)
Carassius auratus	8	6.7–21.3	42–80	3.8–6.3	20	14	Annular trough	Bainbridge (1960)
Carassius auratus gibelio	—	23.0	226	9.8	<1		Distance covered by one tailbeat	Komarov (1971)
Chalcalburnus chaleoides montoides	—	12.5	187	15.0	<1		Distance covered by one tailbeat	Komarov (1971)
Cyprinus carpio	—	35.0	236	8.2	<1		Distance covered by one tailbeat	Komarov (1971)
Cyprinus carpio	3	—	—	5.2			Annular trough	Regnard (1893)
Hypophthalmichthys molotrix	—	27.0	248	9.2	<1		Distance covered by one tailbeat	Komarov (1971)

Leuciscus leuciscus			410				Tank observation	Lane (1941)
Leuciscus leuciscus	7	10.0–21.4	110–240	11.0–11.2	1	14	Annular trough	Bainbridge (1960)
Leuciscus leuciscus	7	10.0–21.4	46–90	4.2–4.4	20	14	Annular trough	Bainbridge (1960)
Rutilus rutilus			455				Tank observation	Lane (1941)
Tinca tinca	—	25.5	138	7.5	<1		Distance covered by one tailbeat	Komarov (1971)
Catastomus occidentalis	10	37–43	305	7.1–8.3	1.5–2.5		Timed over measured distance (culvert)	Wales (1950)
Gadus aeglefina	6	24–41	90–180	3.2–4.4	2–5	11.5–17.0	Timed over measured distance (swimming tunnel)	Blaxter and Dickson (1959)
Gadus morhua	17	12–57	90–240	4.7–7.5	2–5	9.5–12.0	Timed over measured distance (swimming tunnel)	Blaxter and Dickson (1959)
Gadus merlangus	13	14–20	70–180	5.0–9.0	2–5	9.0–17.0	Timed over measured distance (swimming tunnel)	Blaxter and Dickson (1959)
Gadus virens	17	12–19	60–122	5.0–6.4	2–5	9.0–12.0	Timed over measured distance (swimming tunnel)	Blaxter and Dickson (1959)
Odontogadus merlangus auxinas	—	18.0	19.5	10.8	<1		Distance covered by one tailbeat	Komarov (1971)

Continued

Table IV Burst Swimming Speeds of Fish—Cont'd

Species	Number	Length (cm)	Velocity		Time (sec)	Temperature (°C)	Comments	References
			cm sec^{-1}	ℓ sec^{-1}				
Zoarces vivi parous	1	6.4	18–21	2.8–3.3	2–5	<10.0–>15.0	Timed over measured distance (swimming tunnel)	Blaxter and Dickson (1959)
Gasterosteus spinachia	1	10.0	72	7.2	2–5	<10	Timed over measured distance (swimming tunnel)	Blaxter and Dickson (1959)
Pomicrops itajara	1	—	174	—	—		Hook and line	Gero (1952)
Serranus seriba		18.5	227	12.3	<1		Distance covered by one tailbeat	Komarov (1971)
Gymnocephalus cernua		10.5	133	12.7	<1		Distance covered by one tailbeat	Komarov (1971)
Lepomis cyanellus		8	10–70	1.3–8.8	0.04			Webb (personal communication)
Lepomis ctjanellus		8	10–150	1.3–18.8	0.08			Webb (personal communication)
Micropterus							Tank observation	Lane (1941)
Perca							Tank observation	Lane (1941)
Perca fluviatilis		11.5	145	12.6	<1		Distance covered by one tailbeat	Komarov (1971)

Pomatomus salatrix		16.0	204	12.8	<1		Distance covered by one tailbeat	Komarov (1971)
Trachura mediterraneus pontious		15.7	258	16.4	<1		Distance covered by one tailbeat	Komarov (1971)
Spicara smaris		17.0	235	13.8	<1		Distance covered by one tailbeat	Komarov (1971)
Mullus barbatus pontious		11.8	109	9.2	<1		Distance covered by one tailbeat	Komarov (1971)
Creilabrus tinca		15.0	107	7.1	<1		Distance covered by one tailbeat	Komarov (1971)
Mugil curatus		24.0	442	10.1	<1		Distance covered by one tailbeat	Komarov (1971)
Mugil saliens		18.5	400	21.6	<1		Distance covered by one tailbeat	Komarov (1971)
Sphyraena barracuda	2	121–130	1220	9.4–10.1	—		Hook and line	Gero (1952)
Pholis gunnelus	1	10	30	3.0	2–5	>15.0	Timed over measured distance (swimming tunnel)	Blaxter and Dickson (1959)
Gobius minutus	2	6.5	27	4.2	2–5	>15.0	Timed over measured distance (swimming tunnel)	Blaxter and Dickson (1959)
Acanthocybium solanderi	3	92–113	1204–2140	12.2–21.4	10–20		Hook and line	Walters and Fierstine (1964)

Continued

Table IV Burst Swimming Speeds of Fish—Cont'd

Species	Number	Length (cm)	Velocity		Time (sec)	Temperature (°C)	Comments	References
			cm sec^{-1}	ℓ sec^{-1}				
Acanthocybium solanderi	1	113.1	2082	18.4	<0.1		Hook and line (ocean)	Fierstine and Walters (1968)
Scomber scombrus	13	33.0–38.0	189–300	—	2–5	>15	Timed over measured distance (swimming tunnel)	Blaxter and Dickson (1959)
Thunnus albacares	5	53–98	523–2072	6.0–21.1	10–20		Hook and line	Walters and Fierstine (1964)
Thunnus albacares	1	98	1165	11.9	<0.2		Hook and line (ocean)	Fierstine and Walters (1968)
Thunnus thynnus	1	147		3.4			Hook and line	Lane (1941)
Istiophorus			3000				Hook and line	Lane (1941)
Trigla spp	2	18.0	129–135	7.2–7.5	2–5	10–>15	Timed over measured distance (swimming tunnel)	Blaxter and Dickson (1959)
Pleuronectes flesus	1	27.5	105	3.8	2–5	<10.0	Timed over measured distance (swimming tunnel)	Blaxter and Dickson (1959)

Pleuronectes microcephalus	1	8.0	15	1.9	2–5	>15	Timed over measured distance (swimming tunnel)	Blaxter and Dickson (1959)
Pleuronectes platesea	14	0.7–1.0	3.7–15.0	4.9–16.0	1–4	6.5–7.5	Timed over measured distance (swimming tunnel)	Ryland (1963)
Pleuronectes platesea	6	6.0–15.0	6–90		2–5	<10	Timed over measured distance (swimming tunnel)	Blaxter and Dickson (1959)

[a]Data from Lane (1941).

Maximum swimming speed was calculated by Wardle (1975) on the basis of contraction time of white lateral muscle and the relation between tailbeat frequency and forward motion, summarized by the equation:

$$U_{max} = \frac{S_L|\ell}{2t_M}$$

where U_{max}, maximum swimming speed (m sec^{-1}); S_L, stride length [0.6–0.8 times fish length (Bainbridge, 1958)]; ℓ, fish length (m); t_M, muscle contraction time (sec). Application of the formula indicates maximum velocities for a 10 cm fish of $8 - 15\,\ell\,\text{sec}^{-1}$. Maximum speeds for fish of 50 cm were estimated to lie between 6.4 and $10.6\,\ell\,\text{sec}^{-1}$. Actual measurement of swimming speeds were calculated from videotape recordings of fish stimulated into movement or conditioned to race between two feeding points by association of underwater flashing lights with the appearance of food. Haddock and sprats, *Clupea sprattus*, both 10 cm in length, achieved velocities of $26\,\ell\,\text{sec}^{-1}$ while the maximum speed recorded for salmon (25–28 cm) was $10\,\ell\,\text{sec}^{-1}$. A burst acceleration speed (0.06 sec) of $16.9\,\ell\,\text{sec}^{-1}$ was estimated for a sprat of 8.3 cm.

The capacity for burst swimming is, for many species, a prerequisite for their continued well being and even existence. The successful completion of the spawning migration by those species that ascend fast flowing waterways may depend on their burst performance capacity. Burst swimming is important also in escaping predators and in capturing food. Recently, Wiehs (1974) offered the intriguing suggestion that by alternating periods of fast swimming and motionless gliding, fish can reduce the expenditure of energy required to cover a given distance by over 50%.

III. LABORATORY MEASUREMENTS OF PERFORMANCE

The role of environmental and biological factors is perhaps most easily investigated in the laboratory. The advantages offered by the laboratory, however, are countered to at least a considerable extent by the difficulties experienced in the design of a suitable swimming chamber.

A. Swimming Chamber Design

1. Spontaneous Activity

Swimming activity exhibited by fish that are neither confined nor forced to perform frequently has been measured in conjunction with oxygen consumption, often with the specific intent of estimating basal (standard or resting) and intermediate levels of metabolism.

The first qualitative measures of spontaneous activity were made by Spoor (1946). He equated activity to the number of deflections of an aluminum paddle located within an enclosed metabolic chamber to movements of a fish. Corti and Weber (1948) designed an activitymeasuring apparatus which consisted of a delicately suspended experimental tank which moved with the fish's activity. Deflections of a glass plate suspended in a respiration chamber caused by water movements which were created when a fish was active were equated to activity by Fontaine (1956). Beamish and Mookherjii (1964) used a thermostatic heater probe to measure activity. Heat produced by the Joule effect is lost at a fixed rate if the water temperature is constant and the water is motionless. Disturbances within the water such as might be created by fish movement increase the heat loss from the thermistor; the temperature decreases and its electrical resistance increases. Recent investigators have improved the design of this apparatus (Heusner and Enright, 1966; Mathur and Shrivastava, 1970).

A manometric method was employed by Ruhland and Heusner (1959) in which swimming by fish caused waves on the water surface of the metabolic chamber. One electrode was immersed in the chamber water, and the second was suspended immediately above the surface so that the generation of waves allowed for the periodic completion of the electrical circuit. The length of time the circuit was completed was recorded and equated to spontaneous activity. Locomotory activity of the American eel, *Anguilla rostrata*, generated water currents, causing electrical contact between an armature wire and rods suspended in the experimental tank (Bohun and Winn, 1966). von Kausch (1968) recorded spontaneous activity by transformation of turbulence created by fish movement into electric impulses. Recently Spoor *et al.* (1971) developed a technique whereby water currents, generated by the activities of fish, by their influence on equilibrium potentials cause measurable changes in the potential difference between electrodes contained within the water bath. Subsequent application of multiple electrodes by Spoor and Drummond (1972) allowed a rather precise description of the movements of fish within a large gradient tank.

These methods while often extremely sensitive provide only a relative measure of activity. Several methods have been employed in recent years to provide absolute estimates of spontaneous activity. The application of a permanent magnet to a fish was used by Lillelünd (1967), DeGroot and Schuyf (1967), and Schuyf and DeGroot (1971), together with a Hall effect probe, to measure unrestrained movement within a tank. Smit (1965), Kleerekoper *et al.* (1969, 1970), Beamish (1973), and others have used photocells to record and quantify spontaneous movements. High-frequency sound has received attention as a technique to record fish movements in an annular trough (Cummings, 1963; Muir *et al.*, 1965; Meffert, 1968). Movement by fish within the ultrasonic transmission field shifts the audio frequency of the reflected signal, thereby providing a method for estimating activity.

2. Forced Activity

The design of swimming chambers can broadly be grouped into two categories: one in which the chamber is moved (Fry and Hart, 1948) and, second, where water flows through a stationary chamber (Katz *et al.*, 1959; Blažka *et al.*, 1960). Within each category there are numerous variations in design, performance, and cost.

An annular open trough, rectangular in cross section, mounted on a rotating turntable (Fig. 3) was first described by Fry and Hart (1948) and subsequently adopted in modified form by Graham (1949), Gibson and Fry (1954), Bainbridge and Brown (1958), Brett *et al.* (1958), Boyar (1961), Smit (1965), Hammond and Hickman (1966), Kutty (1969), and Fry and Cox (1970). The trough is rotated at a speed equal to that of the fish so that the animal remains stationary relative to the observer. The swimming response is attributable to the effect of the short radius of rotation (Gray, 1937) and is usually elicited by a fixed visual cue such as a lighted table lamp. The speed of the water within the trough in relation to the speed of rotation of the chamber was determined by Fry and Hart (1948) by measuring the rate at which a loose ball of absorbent cotton was carried about the chamber. Slippage measured in this way was independent of temperature. A thorough examination of the velocity profile has not been described but it is

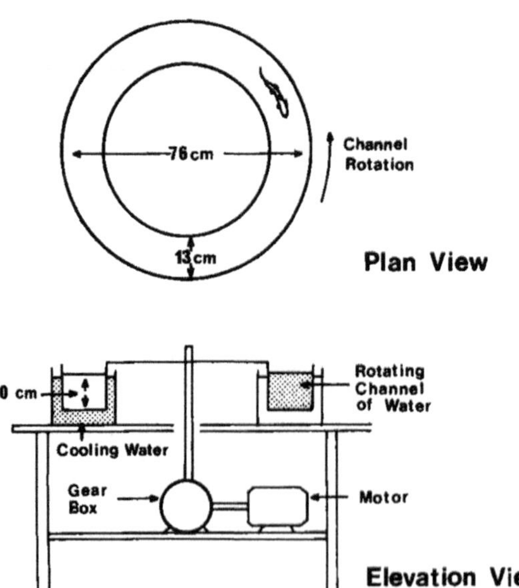

Fig. 3. Diagram of an annular swimming chamber, modified by Hammond and Hickman (1966) after that described and employed by Fry and Hart (1948). (From Hammond and Hickman, 1966, *J. Fish. Res. Board Can.*)

likely to be complex. For a given speed of rotation, a velocity differential is to be expected between the boundaries of the inner and outer radius. Further, vertically oriented clockwise spiral currents generated in the direction of flow alter the drag relations at all surfaces. Along the lower edges of the trough cumulative drag from both the walls and bottom would exceed that at other points along the trough surface, thereby reducing the velocity in this region. To measure active metabolic rate the trough was closed (Job, 1955; Fry, 1957; Wohlschlag, 1957; Basu, 1959).

A second broad category of swimming chambers includes those in which the chamber itself remains stationary and a current of water against which fish are encouraged to swim is generated by gravity flow or a pump. Chambers of this general design often take the shape of a simple tube or a series of tubes with or without expansion, and reduction cones and other devices such as screens appropriately constructed and positioned to encourage a rectilinear velocity profile within that portion of the chamber in which one or more fish are required to swim.

A current generated by gravity flow through a series of small diameter tubes was employed by Bishai (1960) to measure the swimming capacity of larval fish (Fig. 4) and subsequently adopted by Ryland (1963) and Houde (1969). A reservoir located above the swimming chamber provided the head necessary to generate a current. Water from the reservoir is conducted to the swimming portion of the chamber through a small bore tube. The rate and direction of flow are controlled by appropriately located needle valves. A description of the velocity profile was not provided. However, it can be anticipated that it would not follow a rectilinear pattern in profile as the relatively large internal boundary surface together with the low velocities against which larvae could swim would encourage the formation of layers of different velocity, resulting in a parabolic velocity profile. When laminar flow exists, a precise description of swimming capacity can be made only with great difficulty as it is imperative not only to continuously monitor the positions of the swimming fish, but also to calibrate the chamber sufficiently well that a velocity can be assigned to each location.

A stationary oval chamber in which water is circulated by a paddle wheel was first described by Lemke and Mount (1963) and subsequently modified and employed by MacLeod and Smith (1966), MacLeod (1967), and Oseid and Smith (1972). A description of the velocity profile was not provided; however, no specific attempt was made to encourage rectilinear flow. On the basis of the chamber design which was open at the top and rectangular in cross section, variations in drag are to be expected which would contribute to uneven flow. Recently, Lett (1975) used a paddle wheel to generate a current in an annular trough which was semi-circular in cross section. Even with the addition of a series of screens which acted as flow straighteners, a detailed description of the velocity profile indicated a 10% cross-sectional variation in current speed.

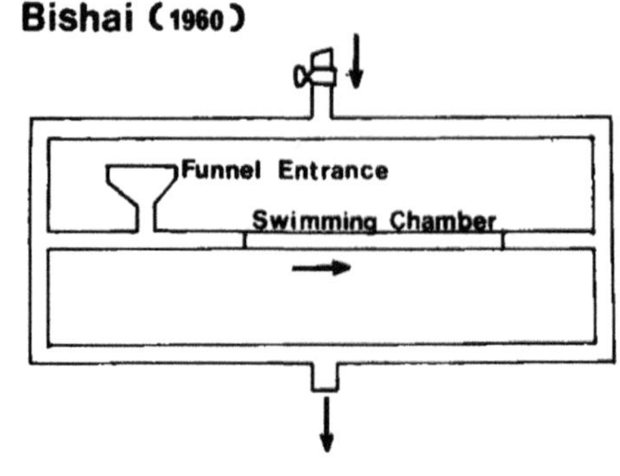

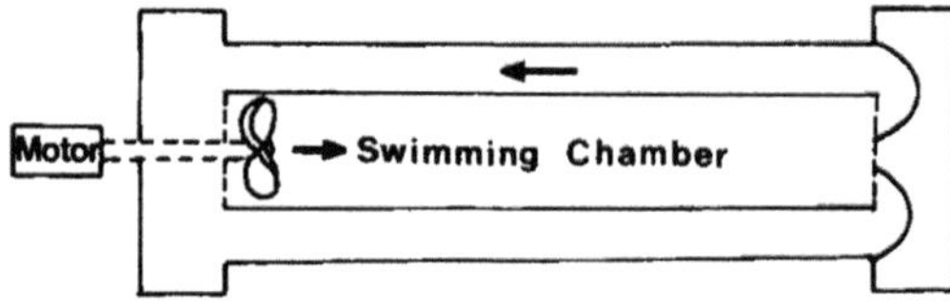

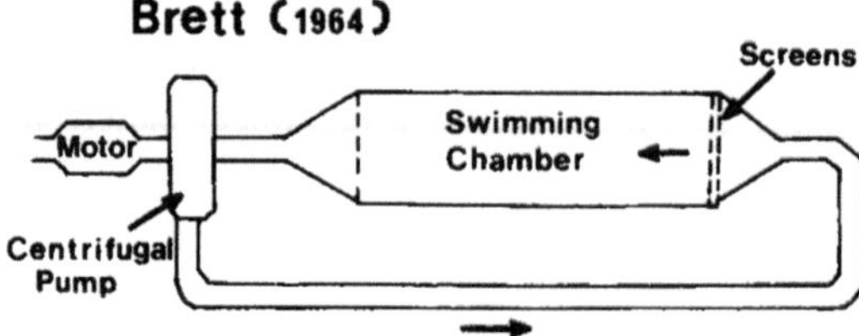

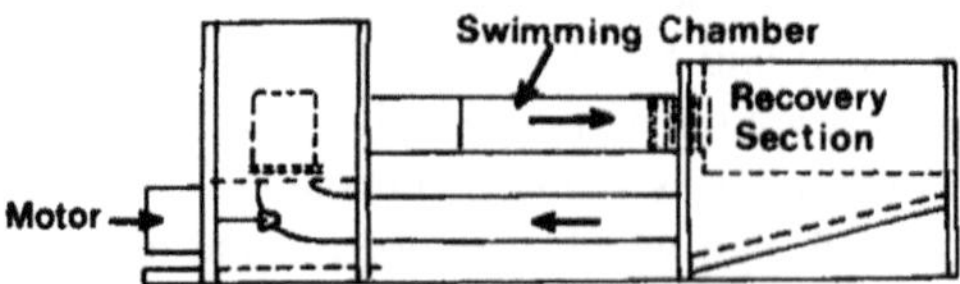

Fig. 4. Swimming flumes. The direction of flow is indicated by arrows. (Redrawn from Bishai, 1960, Thomas *et al.*, 1964, Smith and Newcomb, 1970.)

Swimming chambers in which the current is generated either with a pump, propeller, or impellor are of three basic types (Smith and Newcomb, 1970): small volume, low horsepower; first described by Blažka *et al.* (1960), (Fig. 4) and later modified and employed by Smit (1965), Beamish (1966b), Rao (1968), Kutty (1969), Smith and Newcomb (1970), Hunter and Zweifel (1971), Smit *et al.* (1971), Shazkina (1972a), Kutty and Saunders (1973), and Peterson (1974); large volume, low horsepower, (Fig. 4), independently designed and used by Thomas *et al.* (1964), Arnold (1969), and Griffiths and Alderdice (1972); small volume, high horsepower. In the latter category several chambers have been independently designed (Katz *et al.*, 1959; Mar, 1959; Brett, 1964; Farmer and Beamish, 1969). Recently, Bell and Terhune (1970), in collaboration with Brett (Brett and Glass, 1973) enlarged and further refined the earlier design.

Swimming chambers in this latter category have received the greatest engineering input and the velocity profiles closely approximate rectilinear flow. Each has advantages and disadvantages. A swimming chamber with small water volume provides the greatest sensitivity for measurement of changes in dissolved gases and consequently is recommended where the objective is not only concerned with swimming performance but also in the determination of metabolic rate. Chambers with large volumes tend to provide the least turbulent flow at lowest head; however diffusion grids greatly assist in smoothing the flow pattern. Low horsepower systems reduce both construction and operating costs as well as the input of heat into the flume. This is usually accomplished at the expense of a uniform cross-sectional velocity.

Blažka's apparatus (Blažka *et al.*, 1960) consists of two horizontal concentric cylinders in which water is driven through the inner cylinder by a propeller and returns between the outer and inner cylinders. The flow characteristics of this apparatus are regarded by Smith and Newcomb (1970) as the least desirable among the three basic types of tunnel chambers described above. In part this is attributable to the spiral movement of water encouraged by the propellor. Smith and Newcomb (1970) suggest this can be reduced by the use of a jet outboard impellor. Vanes located between the outer and inner chambers and within the inner chamber at the upstream end further reduce spiral movement. A characteristic of water flowing through a tunnel of fixed diameter is the growth of a boundary layer with length resulting ultimately in a parabolic velocity profile. Excessive turbulence may be avoided by the addition of appropriately machined end plates (Smith and Newcomb, 1970).

Chambers of large volume and low horsepower are used in swimming studies not involving the determination of metabolic rate and often in dealing with large numbers of individuals at the same time. The flume described by Thomas *et al.* (1964) consists of a head tank into which water is forced from an axial flow pump and from which water enters the swimming portion. Fluid acceleration between the large head tank and the entrance of the chamber, circular in cross section, encouraged a nearly uniform velocity profile throughout the length of the swimming portion of the flume. Provision is made for individual fish, unable to maintain a given speed, to collect in a trap immediately downstream from the swimming chamber.

Energy which is converted to heat is added continually to the water of a swimming chamber in operation mainly as the result of fluid friction. This problem is most severe when the flume volume is low and horsepower high, but can be avoided by the addition of a heat exchange system. To assure continuous swimming it is common practice to apply an electrical stimulus at the downstream end of the chamber, using metal electrodes in freshwater (Fry and Hart, 1948) or graphite rods in saltwater (Beamish, 1966b; Griffiths and Alderdice, 1972).

Intermediate between chambers of a large volume and low horsepower and those of small capacity and high horsepower is that described by Katz *et al.* (1959), which consists of a horizontal tube in which fish swim against a current generated by a centrifugal pump and regulated by a gate valve. Wire mesh screens located near the upstream end of the swimming chamber serve as baffles for dissipating gross turbulence or reducing it to finer patterns. The flow pattern based on the motion of particles suspended in water was reported as rectilinear. However, without some mechanism to impair boundary layer growth, a parabolic velocity profile would be anticipated toward the downstream portion of the swimming chamber.

The application of sophisticated engineering principles to flume design was applied by Mar (1959) and Brett (1964) and more recently by Farmer and Beamish (1969). These flumes, which were designed for metabolic studies, incorporated reduction and expansion cones to impair boundary layer growth in the swimming portion of the chamber. The cones were appropriately designed to discourage the formation of a venus contracta, manifested as an area of "dead" water, immediately downstream from a reduction cone. Grids of various mesh sizes further facilitated the production of a rectilinear plane of uniform micro turbulence.

Where the velocity profile does not approach rectilinearity, quantitative estimates of swimming performance must account for the fish's location within the swimming chamber. This is often sufficiently difficult to support the additional effort required in the design and construction of a more refined flume. For precise measurement of swimming speed it is necessary to correct also for the effect of the fish's body on current velocity by causing a narrowing of the available water channel resulting in an acceleration of flow over the body. This error can be corrected by the equation given in Smit *et al.* (1971):

$$U_c = U_s(1 + A_i/A_{ii})$$

where U_c, corrected velocity; U_s, velocity in the absence of a fish; A_{ii}, cross-sectional area of the swimming chamber; A_i, cross-sectional area of the fish, which is assumed to approximate an ellipse and thus equal $\Pi/\frac{1}{2}d/\frac{1}{2}w$, where d and w represent the maximum body height and width, respectively.

Drag experienced by fish in an enclosed flume is higher than that expected at comparable freestream velocities. Extra drag arises from horizontal buoyancy and solid blocking effects (Pope and Harper, 1966). The former effect results from the growth of the boundary layer along the chamber walls, which tends to decrease the effective crosssectional area of the tunnel through which water can flow. A pressure gradient is set up along the chamber, which tends to draw the fish toward the exit and, hence, increase the drag. The solid blocking effect arises from the increase in water velocity around the fish which results from the presence of that fish in an enclosed chamber. These corrections have been described by Webb (1971a). The horizontal buoyancy effect can be accounted for by a correction to the freestream velocity of about 1%. The solid blocking effect can similarly be accounted for by a correction of 7.5–15%, depending on the size of the fish. Consideration should be given also to the pressure effects on the propellor jet which results in an apparent thrust higher than expected. The correction applied to the freestream velocity is about 1% and opposite in effect to the horizontal buoyancy correction (Webb, 1971a). Thus the horizontal buoyancy and propellor corrections tend to cancel out.

B. Experimental Procedure

Swimming performance may be measured for single fish (Fry and Hart, 1948; Brett, 1964, 1965, 1967; Houde, 1969; Farmer and Beamish, 1969; Beamish, 1970, 1974; Oseid and Smith, 1972; Hocutt, 1973) or for groups of fish (Katz *et al.*, 1959; Boyar, 1961; Green, 1964; Thomas *et al.*, 1964; MacLeod and Smith, 1966; Dahlberg *et al.*, 1968; Griffiths and Alderdice, 1972; Johnston and Goldspink, 1973; Otto and Rice, 1974). In most cases groups consist of not more than ten individuals, although Thomas *et al.* (1964) ran performance tests on groups of one-hundred fish.

Fish are customarily deprived of food for a sufficient period prior to testing to assure the postabsorptative state. For some species and environmental conditions this is achieved within 24 hr (Davis *et al.*, 1963), while for others a longer interval is required (Molnár and Tölg, 1962; Farmer *et al.*, 1975).

After transfer of fish from the holding tank to the swimming chamber some investigators have followed the practice of allowing fish a sufficient period to recover from the effect of handling (Black, 1955, 1957a,b,c, 1958a,b; Black *et al.*, 1961). The usual period allowed is 12–16 hr, during which time a slow flow is generated which serves to provide orientation for the fish and to avoid oxygen depletion (Brett, 1964; Dahlberg *et al.*, 1968; Rao, 1968). The longest posthandling period followed appears to be the 3 days applied by Smit *et al.* (1971). Glova and McInerney (1977) suggest this extended period may not be necessary, at least for the measurement of critical swimming speed. They were unable to demonstrate significant differences in performance between young coho salmon allowed 1 and 12 hr recovery in the swimming chamber.

Often a period of recovery is not allowed, rather only the few minutes normally required for fish to orientate to a low velocity (Fry and Hart, 1948; Blaxter and Dickson, 1959; Bainbridge, 1960; Beamish, 1966b; MacLeod and Smith, 1966; MacLeod, 1967; Hocutt, 1973).

When a current is initiated most fish react by swimming against the flow. However, in most studies the objective is not simply to induce fish to swim, but rather to confine their activity within a limited portion of the chamber. This is facilitated by providing one or more visual cues together with a mild electrical stimulus. Alternating black and white stripes on the outer surface of the upstream end of the swimming chamber are frequently used to provide visual cues for orientation of swimming fish (Griffiths and Alderdice, 1972). At the downstream end an electric barrier often in association with a beam of light is employed to discourage fish (Fry and Hart, 1948; Brett, 1964). A small voltage (6–20 V, ac or dc) may be applied across the electrodes either continuously or as required (Brett, 1964; Thomas *et al.*, 1964; Griffiths and Alderdice, 1972). Larimore and Duever (1968) assisted young smallmouth bass, *Micropterus dolomieu*, in maintaining station by suspending a small open cylinder within the chamber. The avoidance displayed by bass for bright light encouraged their entry into the darkened cylinder. Where the chamber consists of an open channel fish are sometimes prodded into swimming by gently tapping the caudal fin with a rod (Fry and Hart, 1948). Particularly noticeable among some fish including the salmonids (Byrne *et al.*, 1972; Kutty and Saunders, 1973), centrarchids (Beamish, 1970), and scorpaenids (Beamish, 1966b) is their ability to hold station against a current using their large pectoral fins as depressors. This is usually effective only at the lower velocities and can be avoided by initiating tests at higher swimming speeds or rapidly alternating velocity until fish begin to swim. Occasionally individual fish do not perform well in swimming chambers despite the presence of visual cues, electrical stimuli, and repeated efforts. Where this occurs it is expedient to discard that individual from further tests. With experience an investigator can often eliminate within a few minutes after their introduction those individuals that are unlikely to perform.

Physical conditioning prior to measuring swimming performance can be an important experimental procedure and was early recognized by Gray (1953, 1957). Fatigue times of unconditioned hatchery rainbow trout forced to swim in a flume were much less than those for stream-conditioned trout (Reimers, 1956). Unexercised young sockeye and coho salmon tended to fatigue earlier than physically conditioned individuals (Brett *et al.*, 1958). Further, prolonged swimming speeds were higher among the conditioned salmon. Hammond and Hickman (1966) found that conditioning rainbow trout resulted in a marked increase in time required to fatigue fish subjected to strenuous exercise. The method employed by Brett (1964) of generating a current of water in oval or circular holding tanks against which the fish must swim offers a practical and efficient means of physically conditioning fish prior to experimentation.

The prolonged swimming speed of young largemouth bass, *Micropterus salmoides* (5.7 cm), subjected to a daily conditioning program increased from 18 to 30 cm sec^{-1} (3.1 – 5.3 ℓ sec^{-1}) after four trials (MacLeod, 1967). Similarly, the maximum prolonged speed of largemouth bass (22.5 cm) estimated by elevating velocity at 30 min intervals by increments of 10 cm sec^{-1} (30 min 10 cm sec^{-1}) increased with exercise (Beamish, 1970). Prolonged speed for bass exercised once every 3 days increased from 50 to 58 cm sec^{-1}(2.2–2.6 ℓ sec^{-1}) after three trials beyond which no further improvement in performance was demonstrable. In contrast, Bainbridge (1962) failed to demonstrate significant differences in burst speeds between rainbow trout exercised in a low current for up to 12 months and unexercised fish, although the best performances by the former were better than those by the latter group.

Concordant with an elevation in prolonged swimming performance with physical conditioning is a general increase in metabolic efficiency. Hochachka (1961) found exercised rainbow trout had higher levels of blood hemoglobin and relatively larger hearts than unexercised trout. Further, physically conditioned trout developed a greater oxygen debt and utilized more of their restricted glycogen reserves than untrained fish for comparable levels of exhaustion. Hochachka (1961) suggested stamina was limited not by the amount of energy reserves but by the excessive accumulation of lactic acid during exercise together with a limited capacity of the muscles to buffer this acid. High lactate levels in both muscle and plasma of conditioned rainbow trout have been demonstrated by Hammond and Hickman (1966). The high hemoglobin levels in conditioned fish would provide both for greater buffering and oxygen carrying capacity. That the primary effect of physical conditioning may be on enzyme systems involved in the mobilization of energy supplies for strenuous activity and in the rapid recovery from fatigue was suggested by Hammond and Hickman (1966).

Adaptation within the muscle fibers of fish exposed to physical conditioning was demonstrated for coalfish, *Gadus virens*, and Atlantic cod, by Greer Walker (1970, 1971) and Greer Walker and Pull (1973). Both red and white muscle fibers hypertrophied with physical conditioning, although the white fibers which are customarily associated with anaerobic metabolism increased in diameter only at the higher sustained swimming speeds. Hypertrophy of the muscle fibers would, on contraction, yield an increase in power output per unit effort. Production of anabolic steroids, although not experimentally demonstrated for fish, might accompany hypertrophy of the white muscle fibers and those of the red that are recruited to function in a similar manner to white fibers (George, personal communication). Anabolic steroids in mammals are known to elevate muscular activity by enhancement of the anaerobic metabolism which is in accord with the earlier observations of Hochachka (1961) and Hammond and Hickman (1966) on rainbow trout.

Procedural uniformity in the rate and magnitude of velocity increment is generally lacking from performance studies, particularly at the level of sustained, prolonged, and burst swimming. Velocity may be increased in a stepwise progression or gradually until either fatigue occurs or the prescribed swimming speed is realized. Where a stepwise progression is followed, the time required of fish to swim at each velocity plateau may vary from a few minutes (Fry and Hart, 1948; MacLeod, 1967; Larimore and Duever, 1968; Oseid and Smith, 1972) to 1 hr or more (Brett, 1964; Houde, 1969; Rao, 1968; Griffiths and Alderdice, 1972; Kutty and Saunders, 1973). Not always is the time period fish must swim at a given speed, nor the magnitude of the velocity increase held constant (Fry and Hart, 1948; MacLeod, 1967; Oseid and Smith, 1972).

The variation in swimming speed attributable to the method employed has received a minimum of investigation. To this end Jones (1971) was unable to detect differences in critical speed of rainbow trout where velocity increments of one-sixth to one-ninth of the maximum performance were applied. Moreover, demonstrable differences in performance were not apparent for stepwise intervals of 20–40 min between velocity increments. Beamish (unpublished data) compared critical speed of rainbow trout in relation to intervals of 30 and 60 min between velocity increments of 5 and 10 cm sec^{-1} after allowing fish to recover overnight from the effects of handling (Table V). For an increment of 5 cm sec^{-1} the critical speed was 36.1 ± 2.91 (95% confidence intervals) and 39.5 ± 5.90 cm sec^{-1} (4.1 and 4.1 ℓ sec^{-1}) when the interval between velocity increases was 30 and 60 min, respectively (Table V). When the increment was 10 cm sec^{-1}, the critical speed was 42.4 ± 8.81 and 39.9 ± 3.87 cm sec^{-1} (4.1 and 3.9 cm sec^{-1}) for intervals of 30 and 60 min, respectively. Thus, there is little indication that the critical speed of rainbow trout is influenced by velocity increments between one-fourth and one-ninth of the critical speed or the time interval, within the range of at least 20–60 min.

Recently, Farlinger and Beamish (1977) examined the influence of the magnitude of and interval between velocity increments on the critical swimming speed of largemouth bass (Fig. 5). Critical performance decreased curvilinearly with increasing intervals of time for a given velocity increment. When time was fixed, critical speed increased with velocity between 2.5 and 10 cm sec^{-1} and with further increases changed little or declined. The highest critical speeds were achieved when velocity increment was 10 cm sec^{-1} and the interval, low. Ideally the relationship between interval and increment on swimming performance should be established before an experiment is begun. If this is impractical, a velocity increment of 10 cm sec^{-1} appears a satisfactory choice. In the absence of quantitative information, the selection of the time interval may be determined by the objectives of the study.

Ultimately, if the velocity increments are continued, a speed is reached against which fish are unable to swim for the prescribed time period. When fatigue occurs in tunnel chambers, fish are forced through the electric field when present, against the downstream retaining screen (Boyar, 1961;

Table V Critical Swimming Speeds of Fish

Species	Number	Length (cm)	Weight (g)	Velocity increments (cm sec^{-1})	Time between increments (min)	Critical velocity		Temperature (°C)		Comments	Reference
						cm sec^{-1}	ℓ sec^{-1}	Acclimation	Experimental		
Hiodon alosoides	2	22.5		10	10	60	2.7	12	12		Jones *et al.* (1974)
Coregonus autumnalis	4	42.1		10	10	80	1.9	12	12		Jones *et al.* (1974)
Coregonus clupeaformis	159	6–51	2–1500	10	10	34.1–72.1	1.4–5.7	7–19	7–19		Jones *et al.* (1974)
Coregonus nasus	33	6–33	1–500	10	10	21.7–46.8	1.4–3.6	7–19	7–19		Jones *et al.* (1974)
Coregonus sardinella	2	29.5		10	10	60	2.0	12	12		Jones *et al.* (1974)
Oncorhynchus kisutch	1340	7.5–9.5		⅛ U_{crit}	60	7.5–55.1	1.0–5.8	2–26	2–26	Acclimation and acute temperatue exposure	Griffiths and Alderdice (1972)
Oncorhynchus kisutch	290	9.7		⅛ U_{crit}	60	38.6–57.9	4.0–6.0	5–19	5–19	Acclimation to temperature and salinity (0–20%)	Glova and McInerney (1977)
Oncorhynchus nerka	5	16.9	49.3	10	60	50.0	2.9	15	5	Acute temperature exposure	Brett (1967)
Oncorhynchus nerka	5	15.7	40.5	10	60	61.5	3.9	15	10	Acute temperature exposure	Brett (1967)

Continued

Table V Critical Swimming Speeds of Fish—Cont'd

Species	Number	Length (cm)	Weight (g)	Velocity increments (cm sec^{-1})	Time between increments (min)	Critical velocity		Temperature (°C)		Comments	Reference
						cm sec^{-1}	ℓ sec^{-1}	Acclimation	Experimental		
Oncorhynchus nerka	5	16.2	43.0	10	60	65.8	4.1	15	15		Brett (1967)
Oncorhynchus nerka	5	17.2	51.3	10	60	67.7	3.9	15	20	Acute temperature exposure	Brett (1967)
Oncorhynchus nerka	5	16.2	44.1	10	60	52.5	3.2	15	24.5	Acute temperature exposure	Brett (1967)
Oncorhynchus nerka	5	17.2	52.1	10	60	33.5	2.0	15	26	Acute temperature exposure	Brett (1967)
Oncorhynchus nerka	5	18.4	62.2	10	60	21.9	1.2	15	27.5	Acute temperature exposure	Brett (1967)
Oncorhynchus nerka		9.2–16.6	8.9–36.7	10	60	40.5–54.2	3.3–4.4	5	5		Brett and Glass (1973)
Oncorhynchus nerka	10	16.2	32.8	10	60	59.0	3.7	10	10		Brett and Glass (1973)
Oncorhynchus nerka		7.7–53.9	3.4–1432.0	10	60	51.5–178.0	3.3–6.7	15	15		Brett and Glass (1973)
Oncorhynchus nerka		5.6–61.4	1.1–1962.0	10	60	39.7–128.0	2.1–7.1	20	20		Brett and Glass (1973)
Oncorhynchus nerka	5	18.5	52.2	10	60	69.4	3.8	24	24		Brett and Glass (1973)

Oncorhytichus nerka	90	5.3–6.0		5	60	38.7–42.6	7.1–7.3	15	15	Sodium pentachlorophenate present (0–50 pph)	Webb and Brett (1973)
Prosopium williamsoni	9	30.4		10	10	42.5	1.4	12	7–12		Jones *et al.* (1974)
Salmo gairdneri	15	10.9	13.2	9	20	65.9	6.1	11.9	11.9		Jones (1971)
Salmo gairdneri	18	12.5	23.4	9	20	43.4	3.5	14.1	14.1		Jones (1971)
Salmo gairdneri	73	29.2	26.4	6	60	58.1	2.0	15	15		Webb (1971b)
Salmo gairdneri	15	11.8	18.1	9	20	52.3	4.4	22.4	22.4		Jones (1971)
Salmo gairdneri	14	12.5	21.0	9	20	79.3	6.3	22.6	22.6		Jones (1971)
Salmo gairdneri	6	30.6		10	10	66.6	2.2			Wild stock, 1 hr recovery prior to test	Jones *et al.* (1974)
Salmo gairdneri	6	32.8		10	10	91.0	2.8			Hatchery stock, 1 hr recovery prior to test	Jones *et al.* (1974)
Salmo gairdneri	10	9.3	7.5	5	60	39.5	4.2	15	15	Overnight recovery prior to test	Beamish (unpublished)
Salmo gairdneri	10	10.4	11.7	10	30	42.2	4.1	15	15	Overnight recovery prior to test	Beamish (unpublished)
Salmo gairdneri	10	10.3	10.2	10	60	39.9	3.9	15	15	Overnight recovery prior to test	Beamish (unpublished)

Continued

Table V Critical Swimming Speeds of Fish—Cont'd

Species	Number	Length (cm)	Weight (g)	Velocity increments (cm sec^{-1})	Time between increments (min)	Critical velocity		Temperature (°C)		Comments	Reference
						cm sec^{-1}	ℓ sec^{-1}	Acclimation	Experimental		
Salmo gairdneri	10	8.9	6.4	5	30	36.1	4.1	15	15	Overnight recovery prior to test	Beamish (unpublished)
Salmo gairdneri	10	9.2	7.6	5	60	39.2	4.3	15	15	Overnight recovery prior to test	Beamish (unpublished)
Salmo gairdneri	10	9.5	7.9	10	30	35.2	3.7	15	15	Overnight recovery prior to test	Beamish (unpublished)
Salmo gairdneri	10	9.3	8.1	10	60	41.9	4.4	15	15	Overnight recovery prior to test	Beamish (unpublished)
Salvelinus alpinus	11	35.5		10	10	100.2	2.8	12	12		Jones *et al.* (1974)
Stenodus leucichthys	22	8–41	11–700	10	10	144–490	12.0–18.0	12–19	12–19		Jones *et al.* (1974)
Thymallus articus	94	7–34	1.5–800	10	10	52–72	1.9–7.5	12–19	12–19		Jones *et al.* (1974)
Esox lucius	192	12–62	7–1800	10	10	19–47	0.8–1.6	12	12		Jones *et al.* (1974)
Notropis atherinoides	4	6.5		10	10	59	9.1	12	12		Jones *et al.* (1974)
Notropis spilopterus	15	7.5–8.4		6	20	23.2–67.2	3.0–8.6	30	15–35	Acute temperature exposure	Hocutt (1973)

Platygobio gracilis	28	17–30	40–300	10	10	42.9–62.7	2.1–2.5	12–19	12–19		Jones *et al.* (1974)
Catostomus catostomus	169	4–53	0.5–2200	10	10	23–91	1.7–5.8	7–19	7–19		Jones *et al.* (1974)
Catostomus commersoni	20	17–37	50–550	10	10	48–73	2.0–2.8	12–19	12–19		Jones *et al.* (1974)
Ictalurus punctatus	25	14.0–15.4		6	20	31.7–61.4	2.1–4.2	30	15–35	Acute temperature exposure	Hocutt (1973)
Percopsis omiscomaycus	3	7.2		10	10	55	7.6	12	12		Jones *et al.* (1974)
Lota lota	56	12–62	7–1800	10	10	36–41	0.7–3.0	7–12	7–12		Jones *et al.* (1974)
Pollachius viens	91	14.8–17.1	26.9–43.9	5	60	61.0–68.9	3.6–4.2	10	10	Trained and untrained	Greer Walker and Pull (1973)
Lepomis gibbosus	6	12.7	44.9	6	60	37.2	3.0	20	20		Brett and Sutherland (1965)
Micropterus salmoides	15	5.2–6.4		6	20	30.6–50.0	5.2–8.1	30	15–35	Acute temperature exposure	Hocutt (1973)
Micropterus salmoides	10	10.2		10	10	45.7	4.5	25	25		Farlinger and Beamish (1977)
Micropterus salmoides	10	10.0		10	60	35.1	3.5	25	25		Farlinger and Beamish (1977)

Continued

Table V Critical Swimming Speeds of Fish—Cont'd

Species	Number	Length (cm)	Weight (g)	Velocity increments (cm sec^{-1})	Time between increments (min)	Critical velocity		Temperature (°C)		Comments	Reference
						cm sec^{-1}	ℓ sec^{-1}	Acclimation	Experimental		
Perca flavescens	20	9.5		5	15	1.55–210	1.6–2.2	10	10		Otto and Rice (1974)
Perca flavescens	20	9.5		5	15	25.2–33.0	2.7–3.5	20	20		Otto and Rice (1974)
Perca flavescens	30	9.5		5	15	33.5	3.5	10	20	Acute temperature exposure	Otto and Rice (1974)
Perea flavescens	30	9.5		5	15	15.5	1.6	20	10	Acute temperature exposure	Otto and Rice (1974)
Stizostedion vitreum vitreum	54	8–38	4–500	10	10	38–84	2.2–4.7	19	19		Jones *et al.* (1974)

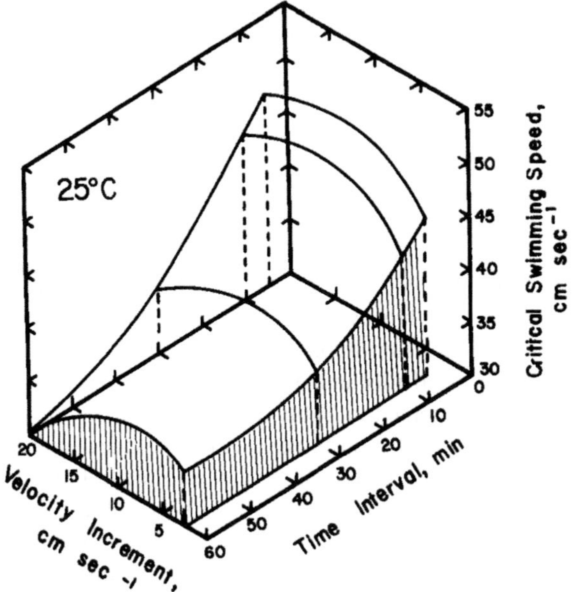

Fig. 5. Critical swimming speed of largemouth bass, *Micropterus salmoides* (10 cm), in relation to the interval between and magnitude of velocity increments. (Redrawn from Farlinger and Beamish, 1977.)

Brett, 1964) or into a recovery section of reduced water velocity (Thomas *et al.*, 1964; Griffiths and Alderdice, 1972). Where retaining screens are present the criteria for fatigue vary but fatigue is usually accepted when a fish by repeated efforts and despite the application of electrical stimulus or prodding can no longer hold itself off the screen. One variation of this procedure is that followed by Smit *et al.* (1971). When goldfish fell against the downstream screen, the velocity was reduced. If fish continued to swim even after the "fatigue" velocity was resumed, the first failure was ignored and the experiment was continued. However, when the second failure was recorded the experiment was terminated. Where oval or annular chambers are employed and a retaining screen is not used, fatigue is usually presumed when fish begin to loose laps. Larimore and Duever (1968) presumed young smallmouth bass to be fatigued when they left a darkened cylinder suspended in the chamber.

Critical swimming speed is measured by interpolation for those fish that do not fatigue exactly at the beginning or end of a prescribed swimming period. The formula described by Brett (1964) is as follows.

$$U_{\text{crit}}\ (\text{critical swimming speed}) = u_{\text{i}} + (t_{\text{i}}/t_{\text{iii}} \times u_{\text{ii}})$$

where u_{i}, highest velocity maintained for the prescribed period (cm sec^{-1}); u_{ii}, velocity increment (cm sec^{-1}); t_{i}, time (min) fish swam at the "fatigue" velocity; t_{ii}, prescribed period of swimming (min). Thus, a fish successfully swimming for a prescribed period of 60 min at 50 cm sec^{-1} but fatiguing at 60 cm sec^{-1} after 22 min would have a critical speed of

$$50 + \left(\frac{22}{60} \times 10\right) = 53.7 \text{ cm sec}^{-1}$$

Critical swimming speed is usually represented by the median performance of the fish used. This is determined graphically by plotting the logarithm of critical speed against the cumulative percentage of fish fatigued on a probit scale. As a rule this transformation allows the application of a linear regression. In some cases, however, the distribution is best described by more than a single linear regression implying mixed fatigue effects or compound responses to velocity (Griffiths and Alderdice, 1972). A summary of critical swimming speeds is presented in Table V.

Proper differentiation of sustained and prolonged swimming speed requires a description of the relationship between time to fatigue and velocity such as that presented by Brett (1964) for sockeye salmon (Fig. 6). In this figure there appears an obvious distinction between prolonged and sustained swimming. The logarithm of time to fatigue at prolonged speeds increased linearly as velocity was decreased. With a continued decrease in swimming speed, sockeye salmon did not fatigue. The separation between sustained and prolonged swimming is surprisingly sharp and represents only a few cm sec^{-1}. Where it is not practical to determine the response between velocity

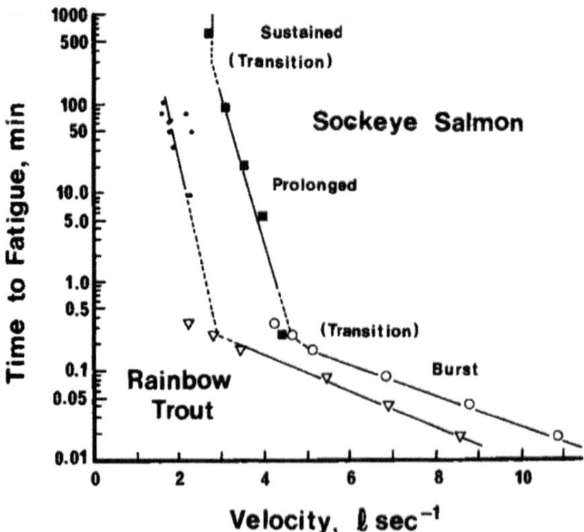

Fig. 6. Identification of sustained, prolonged, and burst speeds for rainbow trout, *Salmo gairdneri*, and sockey salmon *Oncorhynchus nerka*, on the basis of their fatigue time at different swimming velocities. Results for rainbow trout obtained from Bainbridge (1960, 1962), and for sockeye salmon from Brett (1964). (Redrawn from Brett, 1964, *J. Fish. Res. Board Can.*)

and time to fatigue it is assumed those speeds which can be maintained for a minimum of 200 min represent sustained. Just exactly how long fish will swim at sustained velocities has not been well investigated. Johnston and Goldspink (1973) were able to force coalfish to swim in a flume for 16 days at low velocities. Largemouth bass swam continuously at slightly in excess of $1 \ell \sec^{-1}$ for periods of 2 weeks (Beamish, 1975). Swimming endurance studies on redfish suggest some individuals were able to swim steadily at over $2 \ell \sec^{-1}$ for 10 days or more (Beamish, 1966b). Smit *et al.* (1971) forced goldfish to swim at low speeds for periods of 1 week as a general routine.

Distinct alteration in the coefficient of the linear relation between velocity and time to fatigue for burst and prolonged swimming speed implies physiological differences in the availability and mobilization of the fuel for muscular activity. Again, where it is not possible to determine the point of inflection between burst and prolonged speed, a period of 20 sec is presumed a reasonable approximation. The procedure generally applied in the measurement of burst speed in the laboratory is to prod, by mechanical or electrical stimuli, fish swimming steadily at a moderate velocity. This causes the fish to dart forward with an initial expression of accelerated or unsteady swimming followed by steady or sprint swimming. Photographic or electronic devices have been applied to improve precision (Bainbridge, 1960; Komarov, 1971).

C. Biological Constraints on Performance

1. Size

a. Length. Of the constraints on performance capacity, size is among the most important. As early as 1917, Thompson argued that sustained or prolonged speeds should be proportional to the length of fish raised to the power of 0.5. This conclusion was based not on measurements of swimming speed, but on the assumption that fish volume and the proportionate amount of muscle increases as the square of length. Assuming further that power is limited not by the volume of muscle but rather the surface area of the gills, Thompson (1917) suggested that maximum or burst speed was independent of length. This conclusion was reached also by Hill (1950) on the basis of heart capacity and blood flow through the vessels whose cross-sectional area increases as the square of length.

Most frequently the relationship between length of fish and performance is described by the equation

$$\log u = a + b(\log \ell)$$

where u is swimming speed (cm $\sec^{-1}$) and ℓ, length (cm) (Blaxter and Dickson, 1959; Bainbridge, 1960; Brett, 1965; Brett and Glass, 1973). The relation has been described also by a linear regression without logarithmic transformation (Glova and McInemey, 1977) or after logarithmic transformation of swimming speed but not length (Beamish, 1970).

Concordant with the earlier views of Thompson (1917), Brett (1965) found the (60 min, 10 cm sec^{-1}) critical speed of sockeye salmon (8–55 cm) was proportional to a fractional power of length equal to $1^{0.5}$ (Fig. 7A). At burst swimming speeds, the regression coefficient appears to approach unity. Blaxter and Dickson (1959) found the (2–5 sec) burst speed of Atlantic herring to increase linearly with length (1–26 cm) after logarithmic transformation, the coefficient of which was unity (Fig. 7B). Similarly, Bainbridge (1960) found a coefficient of unity for the (1–20 sec) burst swimming speed of the dace, *Leuciscus leuciscus*, (10.0–21.4 cm). Houde (1969) found a coefficient of approximately 1 for (0.5–5.0 sec) burst speed for larval yellow perch, *Perca flavescens*, (0.9–1.4 cm), and walleye, *Stizostedion vitreum*, (1.0–1.6 cm) following absorption of the yolk sac (Fig. 7C). Pavlov *et al.* (1968) reported a coefficient of unity for several species of minnows and yellow perch.

Methods have been described to correct swimming speed for variation in length where it is not practical to determine the precise relationship. Relative performance as ℓ sec^{-1} sometimes allows for comparison of fish of different length. Bainbridge (1960) found the relative burst speed (ℓ sec^{-1}) for dace, rainbow trout, and goldfish, *Carassius auratus*, was independent of length and equal to about 10ℓ sec^{-1}. Accordingly, relative burst speeds of larval plaice changed little with length (0.7–1.4 cm) from 10ℓ sec^{-1} (Ryland, 1963).

Relative burst speeds of longer duration generally favor the smaller individuals. Bainbridge (1960) found that when burst swimming speed was extended from 1 to 20 sec, relative performance by the smaller individuals displayed an improvement over that of larger fish. Thus over the size range of 7–20 cm, relative performance of goldfish decreased from about 6.3 to 4.0ℓ sec^{-1} and that for dace (10–22 cm) declined from 4.8 to 4.0ℓ sec^{-1}. Burst speeds (2–5 sec) determined for a number of teleosts by Blaxter and Dickson (1963) followed a similar pattern of decline in relative performance with size. Burst speed of brown trout, *Salmo trutta* (10–40 cm) decreased from approximately 17.5 to 1.5ℓ sec^{-1} while that for Atlantic herring (1–20 cm) was reduced from 10.1 to 5.6ℓ sec^{-1}. Among the many species examined only coalfish displayed an increase in relative speed with size, 5.2–5.7 ℓ sec^{-1} for individuals of 13–20 cm, which may be at least partly attributable to their narrow length range.

Relative performance at prolonged and critical swimming speeds generally favors the smaller individuals of a species. Thus, Brett and Glass (1973) found the (60 min, 10 cm sec^{-1}) critical speed of sockeye salmon decreased form 4.5 to 2.0ℓ sec^{-1} as length increased from 10 to 90 cm. Moreover, Brett and Glass (1973) demonstrated a similar pattern over the entire ecological range of temperatures for the species. Beamish (1970) found that relative prolonged performance of largemouth bass favored the smaller individuals at temperatures approximating their physiological optimum (Niimi and Beamish, 1974) but

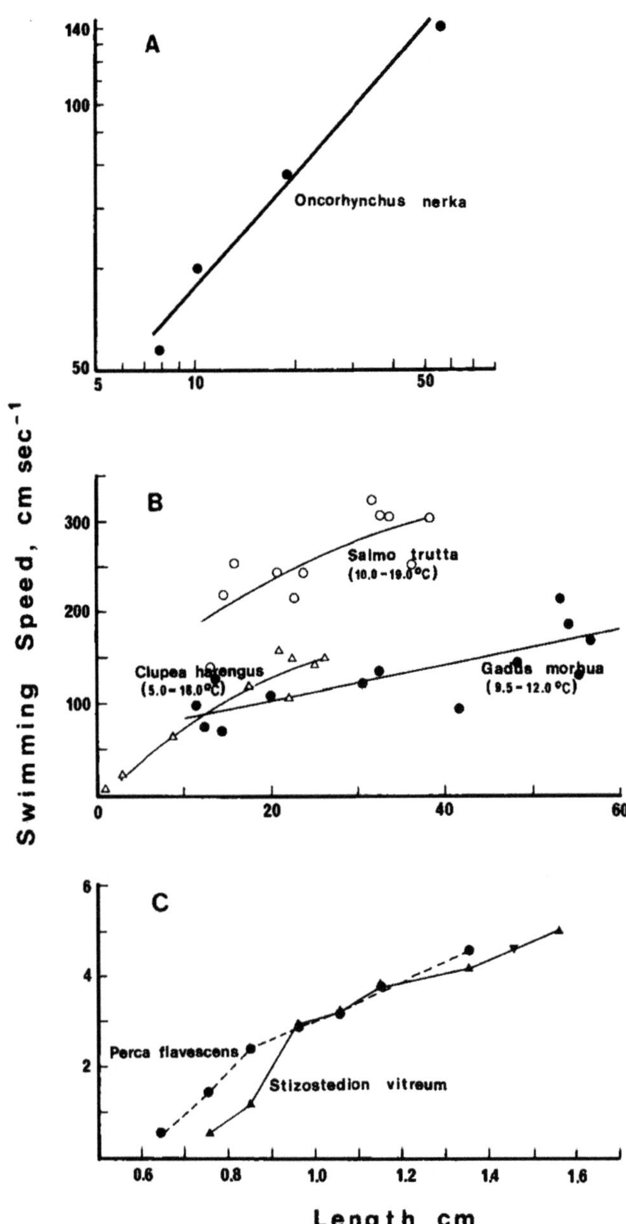

Fig. 7. Swimming speed and length. (A) Critical swimming speed of sockeye salmon (Brett, 1964). (B and C) Burst (sprint) swimming speeds (Blaxter and Dickson, 1959; Houde, 1969).

that at lower temperatures there was little evidence of this difference. More recently Glova and McInerney (1977) corroborated the dependency of the relation between length and performance on temperature for coho salmon but found it to be independent of salinity.

Smit *et al.* (1971) converted prolonged speeds of goldfish to relative velocity by the equation

$$u_r = \frac{\ell_s u^2}{\ell}$$

where u_r, relative velocity for the standard fish (cm sec^{-1}); ℓ_s, length of standard fish (cm); u, measured velocity of fish (cm sec^{-1}); ℓ, measured length of fish (cm).

The distance or length of time (endurance or stamina) fish are able to swim against a particular current is also dependent on length. Boyar (1961) found that as Atlantic herring increased in length their endurance increased, and that this relationship was best described by a linear regression after logarithmic transformation. Over the range of prolonged speeds applied Boyar found endurance to be a function of approximately the fourth power of length (Fig. 8). The distance sea lamprey, *Petromyzon marinus*, were able to swim at a fixed velocity and temperature increased approximately as the square root of their weight (Beamish, 1974; see also Fig. 9).

The percentage of muscle in at least some species of fish tends to increase with length (Bainbridge, 1960). However, hydrodynamic drag increases also with length. Sustained and prolonged swimming in contrast to burst is limited by the rate at which muscles can be supplied with the raw materials for contraction and relieved of waste products (Bainbridge, 1958; Jones, 1971). To this end large fish are able to provide for a higher relative metabolic scope for activity (Fry, 1947; Basu, 1959), although for some species this relationship is temperature dependent (Brett and Glass, 1973). After consideration of the pertinent factors, Brett (1965) attributed reduced relative performance by large sockeye salmon to an increase in hydrodynamic drag which he suggested outweighed the advantage of increased metabolic scope and body musculature. Burst speed is reliant on a store of raw materials such as glycogen within the muscle cells or possibly oxygen bound in muscle hemoglobin. The relative store of glycogen appears to be independent of size for at least Atlantic cod (Beamish, 1968). Hence in burst swimming it is likely the influence of increased drag in large fish is countered by proportional elevations in muscle development and metabolic fuel.

b. Weight. Swimming performance is most often expressed on the basis of length but has been described also for weight. Fry and Cox (1970) found the (1 min, 11–18 cm sec^{-1}) prolonged speed of rainbow trout to increase with weight (4–100 g) raised to the power of 0.13.

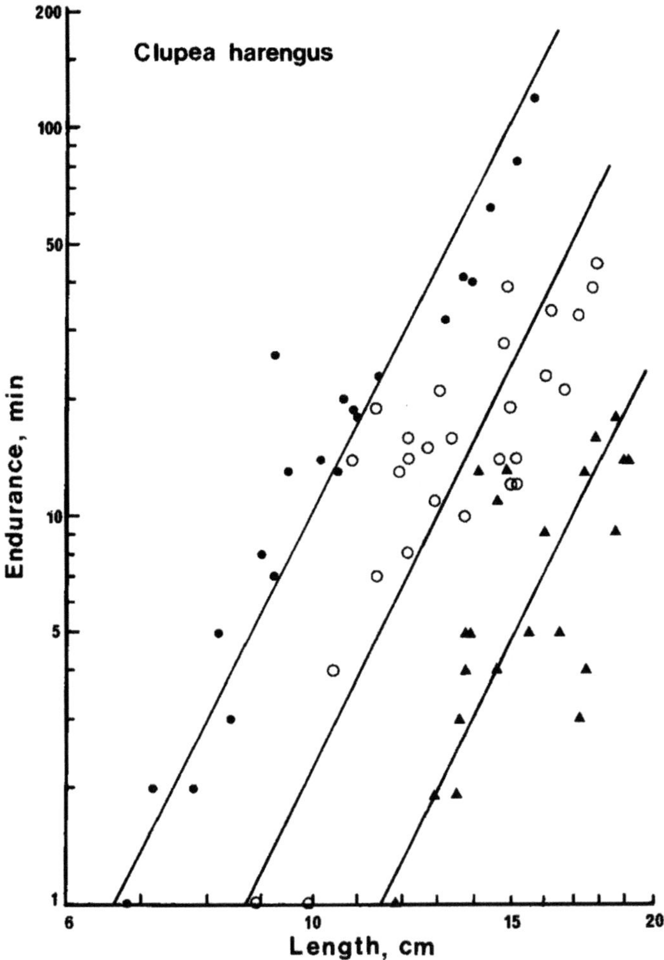

Fig. 8. The endurance of Atlantic herring, *Clupea harengus harengus*, of different lengths at three swimming speeds. (Redrawn from Boyar, 1961, Swimming speed of immature Atlantic herring with reference to the Passamaquoddy Tidal Project, *Trans. Am. Fish. Soc.* 90, 21–26.)

A method of swimming performance rating based on the ratio of useful work done to the muscle power available for fish of different weight but of the same species was developed by Thomas *et al.* (1964) and is described by the equation:

$$\text{Performance rating} = C_3\sqrt{\frac{v}{1.3\times10^5}\frac{1}{M}}\;u_i^{5/2t_i}$$

where u_i, the relative velocity between the fish and water; t_i, time interval; v, water viscosity; M, weight of fish; C_3 is a constant.

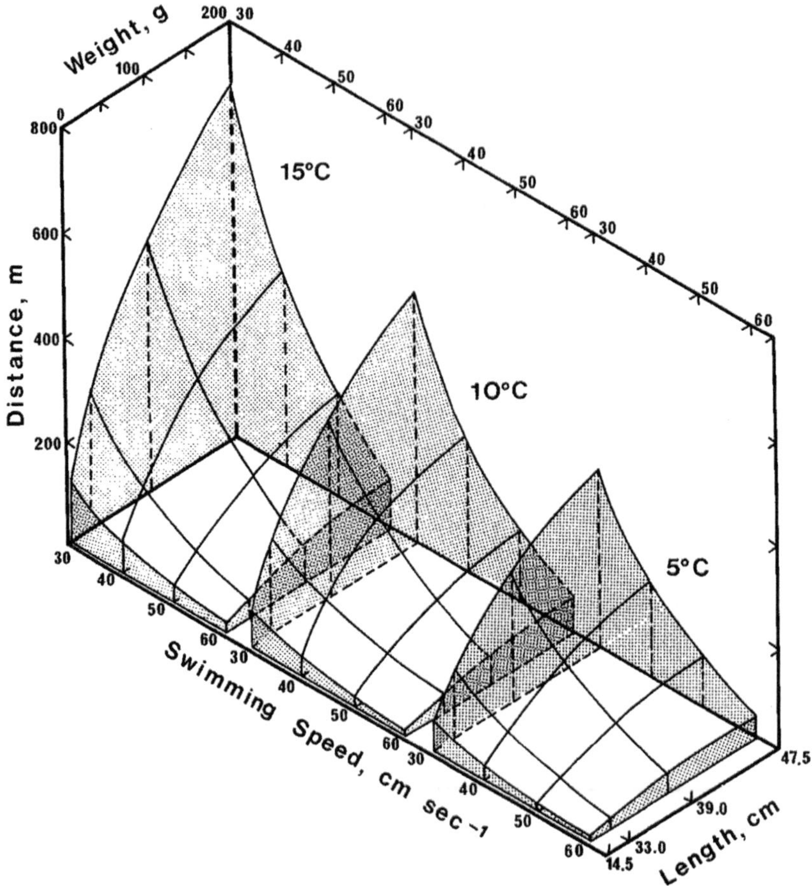

Fig. 9. Distance sea lampreys, *Petromyzon marinus*, swam in relation to speed, size, and temperature. (From Beamish, 1974.)

c. Condition. Swimming capacity is influenced also by the weight of fish relative to their length, most often described by a condition factor such as that computed by Fulton (1911, reported in Ricker, 1975):

$$M/\ell^3$$

where M is weight and ℓ, length. Bams (1967) expressed the condition of unfed sockeye migrant fry by the factor

$$\frac{10M^{1/3}}{\ell}$$

Ryland (1963) equated size of larval plaice, *Pleuronectes platessa*, to their length times the height of their musculature midway along the animal's length in describing the relationship with burst swimming speed.

The importance of condition to swimming capacity has been explored most thoroughly for the salmonids. Generally domestic stocks of trout not only grow faster than wild but are heavier for a given length. Vincent (1960) noted the chemical composition of wild and domestic stocks of brook trout, *Salvelinus fontinalis*, were similar except in the fat content which was higher in domestic stocks even when both groups were reared in the hatchery from the egg stage under experimentally similar conditions. Wild stock brook trout consistently out-performed domestic in stamina tests. In a comparison of stamina among three stocks of brook trout, Green (1964) first reared eggs and the young stages under similar hatchery conditions. Stamina tests were conducted at two velocities, 45 and 57 cm sec^{-1}. The number of fish able to swim for 2 min against the prescribed velocity increased consistently with length in all three stocks (Fig. 10). However, for fish of a given length, a greater proportion of fish from the wild stocks (Long Pond Outlet and Honnedaga Lake, New York State) were able to sustain the respective velocities for 2 min than was found for domestic stocks. Both Vincent (1960) and Green (1964) attributed the poorer performance by domestic stocks of brook trout to their higher fat content. Additionally, hydrodynamic drag would be expected to increase with condition factor. On the assumption that gill area and the efficiency of the pumping mechanism and gaseous exchange are similar for fish of a given species and length, the metabolic scope for activity would decline with an increase in weight.

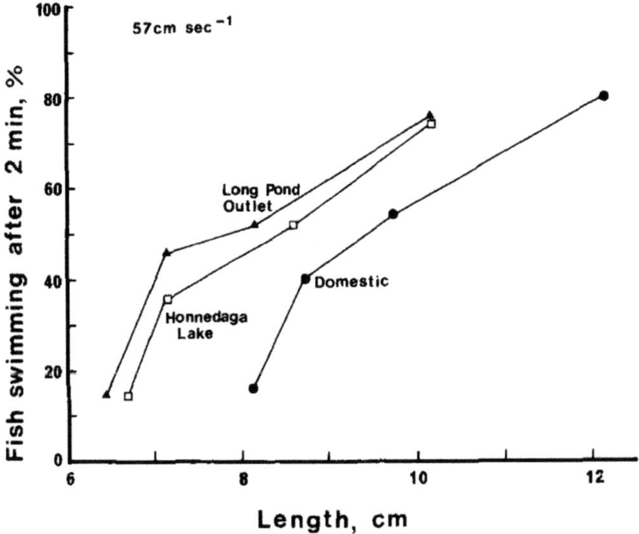

Fig. 10. Comparison of swimming performance among three stocks of brook trout, *Salvelinus fontinalis*. Performance is expressed as a percentage of the trout swimming after 2 min at 58 cm sec^{-1}. (From Green, 1964.)

Relative stamina of unfed migrant sockeye salmon fry was examined by Bams (1967) based on ranking order of fatigue. Since salmon were not fed, their condition factor decreased with progressive absorption of the yolk. Once the yolk was absorbed stored energy reserves were catabolized so that the fry became increasingly emaciated. Bams (1967) found relative swimming performance was optimal at the stage of almost total yolk absorption. With a well-developed yolk and hence a high condition factor, relative performance declined, which Bams (1967) attributed to the high drag associated with the protruding yolk sac. Performance was poor also, when the condition factor was low, which reflected the depletion of energy reserves among unfed fry after absorption of the yolk.

2. *Sex*

Little information is available on the influence of either sex or stage of maturity on swimming capacity. Brett (1965) compared the (60 min, 10 cm sec^{-1}) critical speed of adult male and female sockeye salmon for which relative performance favored the slightly shorter males. The critical speed for males (41.8 cm) was 125 cm sec^{-1} (3.0 ℓ sec^{-1}) while for females (53.9 cm) it was 143 cm sec^{-1} (2.7 ℓ sec^{-1}). Males were tested at their temperature optimum of 15°C (Brett, 1964), whereas females were examined at 17°C, suggesting that at comparable temperatures differences in performance would be minimal.

3. *Disease*

Restriction in capacity for swimming imposed by parasitic infection has received some consideration. However, there appears to be no information on the influence of bacterial or viral infection on swimming performance.

Relatively heavy infections in rainbow trout (11.0 cm) with metacercariae of the trematode, *Bolbpophorus confusus*, which locates in the body muscles, reduced the prolonged speed of 75 cm sec^{-1} by about 35% when compared with control fish (Fox, 1965; Butler and Milleman, 1971). Olson (1968, cited in Butler and Milleman, 1971) was unable to demonstrate differences in the swimming capacity of rainbow trout infected with metacercariae of the trematode, *Cotylurus erraticus*, which lodges in the pericardial cavity of its host. Swimming endurance of rainbow trout infected with metacercariae of the intestinal fluke, *Crepidostomum farionis*, and the cestode, Proteocephalus sp., was not significantly different from fish containing only *Proteocephalus sp.* (Klein *et al.*, 1969). However, while differences were not significantly different, the length of time *C. farionis*-infected fish swam, 32.5 min, was appreciably less than the 47.7 min recorded for control fish. Unfortunately the velocity at which tests were conducted was not reported nor was the diet fed control fish the same as that offered to *C. farionis*-infected trout. Smith and Margolis (1970) measured the (30 min,6 cm sec^{-1}) prolonged speed of sockeye salmon (2–4 g) free from and infected with the cestode, *Eubothrium salvelini*. Infected salmon, which were about 1 g lighter in weight, fatigued after swimming about 66% of the distance covered by control fish. It was estimated by Smith and Margolis (1970) that a reduction in swimming capacity of this magnitude was sufficient to reduce the success with which infected salmon are able to descend long rivers in their seaward migration, realizing that fish must procure food and escape from predation along the route.

Performance of rainbow trout and coho salmon infected with cercariae of the parasitic trematode, *Nanophytes salmincola*, was measured and compared with control fish by Butler and Milleman (1971) (Table VI). Two methods were employed to evaluate swimming performance. In one case, velocity was gradually increased to the prescribed velocity of 38.1 and 32.3 cm sec^{-1} for rainbow trout and coho salmon, respectively, at which endurance time was recorded. The second method employed a stepwise progression of velocity increments to delineate the maximum prolonged speed. Velocity was increased at 3 cm sec^{-1} every 20 min to 22.9 cm sec^{-1} and thereafter, 3 cm sec^{-1} at 10 min intervals until fish were fatigued. Immediately after exposure to 1500 cercariae, the maximum prolonged speed of infected trout was 32% below that found for control fish. With time this differential decreased so that after 15 days, infected and control fish displayed similar swimming capacity. Endurance was more markedly impaired among infected fish, being reduced by 36–54% over the first 96 hr after infection when the parasites were migrating through the tissues or had not yet completed development as metacercariae. After 15 days when the parasites had encysted, the differential in endurance time was only 3%. The pattern among coho salmon was similar to that for trout; however, the absolute reduction in endurance was the more marked among the former species. When trout and salmon were exposed to 100 cercariae daily for 15 days, which more closely approximated the natural rate of infection than a single large dose, endurance was reduced by 51 and 34% in trout and salmon, respectively. Hemorrhagic areas developed in the infected fish but not in the controls, to which Butler and Milleman (1971) attributed impairment of relative swimming performance.

The incidence of parasitism among fish used in swimming studies is seldom reported. The recent findings strongly suggest that closer attention to the species, stage, and abundance of parasites present in experimental fish would significantly reduce variability in measurements of performance, and better facilitate quantitative comparisons among the results of swimming speed studies.

Table VI Prolonged Swimming Speeds of Fish

Species	Number	Total length (cm)	Weight (g)	Velocity increments (cm sec_{-1})	Time between increments (min)	Swimming time at maximum velocity (min)	Maximum velocity		Temperature (°C)			Comments	Reference
							cm sec^{-1}	ℓ sec^{-1}	Acclimation		Experimental		
Petromyzon marinus	53	14.5–39.0	5–100	Gradually to prescribed velocity		10	16.6–33.6	0.9–1.2	5		5	Endurance	Beamish (1974)[a]
Petromyzon marinus	53	14.5–39.0	5–100	Gradually to prescribed velocity		10	16.8–34.7	0.9–1.2	10		10	Endurance	Beamish (1974)[a]
Petromyzon marinus	53	14.5–39.0	5–100	Gradually to prescribed velocity		10	24.2–41.3	1.1–1.7	15		15	Endurance	Beamish (1974)[a]
Alosa finita	30.0	29.7					75	2.5					Magnan (1929)
Clupea harengus		15.2–26.0		Gradually to prescribed velocity		1	91–143	5.4–6.0		12			Brawn (1960)[c]
Clupea harengus	22	6.8–15.7		Gradually to prescribed velocity		1–120	36.6	2.3–5.3		1.4–5.6		Endurance	Boyer (1961)[b]
Clupea harengus	16	12.9–20.5		Gradually to prescribed velocity		1–17	97.5	4.8–7.6		1.4–5.6		Endurance	Boyer (1961)[b]
Oncorhynchus kisutch	10	4.6–8.8	0.7–7.8	Gradually until laps lost		5	14.9–26.5	3.0–3.2	5–6		5–6		Brett *et al.* (1958)[b]
Oncorhynchus kisutch	11	4.9–9.1	1.4–8.1	Gradually until laps lost		5	22.0–35.4	3.9–4.5	10		10		Brett *et al.* (1958)[b]
Oncorhynchus kisutch	10	5.7–8.9	1.7–8.1	Gradually until laps lost		5	25.3–37.5	4.2–4.4	15		15		Brett *et al.* (1958)[b]

Oncorhynchus kisutch	10	5.6–9.0	1.9–8.3	Gradually until laps lost		5	30.5–41.2	4.6–5.4	20		20		Brett *et al.* (1958)[b]
Oncorhynchus kisutch	5	6.3	2.5	Gradually until laps lost		5	28.4	4.5	24		24		Brett *et al.* (1958)[b]
Oncorhynchus kisutch	20	7.9–9.0		Gradually to prescribed velocity			30–49		10		10	Endurance, 20%; O_2, 3–19 mg/liter	Davis *et al.* (1963)[a]
Oncorhynchus kisutch	26	6.7–8.9		Gradually to prescribed velocity			49–55		15		15	Endurance, 20%; O_2 3–19 mg/liter	Davis *et al.* (1963)[a]
Oncorhynchus kisutch	23	7.6–8.9		Gradually to prescribed velocity			23–55		20		20	Endurance, 20%; O_2, 3–19 mg/liter	Davis *et al.* (1963)[a]
Oncorhynchus kisutch	[315]	7.6–9.3	3.5–6.9	2.3	10	10	28–64	3.3–7.8	20		20	O_2, 2–26 mg/liter; CO_2, 1–120 mg/liter	Dahlberg *et al.* (1968)[a]
Oncorhynchus kisutch	[150]	5.7–6.0	1.4–1.7	2.3–3.0	10–20	10	22.9–34.3			18–19		Control	Butler and Milleman (1971)[a]
Oncorhynchus kisutch	[150]	5.7–6.0	1.4–1.7	2.3–3.0	10–20	10	19.8–29.7			18–19		Infected; *Nanophyetus salmincola*	Butler and Milleman (1971)[a]
Oncorhynchus nerka	3	6.2	1.7	Gradually until laps lost		5	14.3	2.3	1		1		Brett *et al.* (1958)[b]
Oncorhynchus nerka	5	6.6	2.9	Gradually until laps lost		5	22.9	3.5	6		6		Brett *et al.* (1958)[b]
Oticorhynchus nerka	15	6.2–15.2	1.9–35.9	Gradually until laps lost		5	26.8–43.3	4.3–2.8	10		10		Brett *et al.* (1958)[b]
Oncorhynchus nerka	10	7.1–13.9	2.8–26.9	Gradually until laps lost		5	32.6–46.4	3.3–4.6	15		15		Brett *et al.* (1958)[b]
Oncorhynchus nerka	5	7.4	3.4	Gradually until laps lost		5	27.1	3.7	20		20		Brett *et al.* (1958)[b]

Continued

Table VI Prolonged Swimming Speeds of Fish—Cont'd

Species	Number	Total length (cm)	Weight (g)	Velocity increments (cm sec_{-1})	Time between increments (min)	Swimming time at maximum velocity (min)	Maximum velocity		Temperature (°C)			Comments	Reference
							cm sec^{-1}	ℓ sec^{-1}	Acclimation		Experimental		
Oncorhynchus nerka	5	7.4	3.4	Gradually until laps lost		5	21.7	2.9	24		24		Brett *et al.* (1958)[b]
Oncorhynchus nerka	7	14.5		Gradually to prescribed velocity		>300	46.4	3.2	15		15	Endurance, 50% fatigue	Brett *et al.* (1958)[b]
Oncorhynchus nerka	17	14.3		Gradually to prescribed velocity		>300	51.1	3.6	15		15	Endurance, 50% fatigue	Brett *et al.* (1958)[b]
Oncorhynchus nerka	29	13.4		Gradually to prescribed velocity		66	53.6	4.0	15		15	Endurance, 50% fatigue	Brett *et al.* (1958)[b]
Oncorhynchus nerka	19	12.6		Gradually to prescribed velocity		21	55.4	4.4	15		15	Endurance, 50% fatigue	Brett *et al.* (1958)[b]
Oncorhynchus nerka	14	13.8		Gradually to prescribed velocity		6	66.2	4.8	15		15	Endurance, 50% fatigue	Brett *et al.* (1958)[b]
Oncorhynchus tshawytscha	160	3.8		Gradually to prescribed velocity		10	40	10.5		[22]		Endurance, 50% fatigue	Kerr (1953)[a]
Oncorhynchus tshawytscha	9	8.1–12.6		Gradually to prescribed velocity			23–67		11.5		11.5	Endurance, 20% fatigue; O_2, 2–10 mg/liter	Davis *et al.* (1963)[a]
Oncorhynchus tshawytscha	11	5.1–7.3		Gradually to prescribed velocity			29–53		15		15	Endurance, 20% fatigue; O_2 2–10 mg/liter	Davis *et al.* (1963)[a]
Oncorhynchus tshawytscha	16	5.7–7.6		Gradually to prescribed velocity			23–53		19.5		19.5	Endurance, 20% fatigue; O_2, 2–10 mg/liter	Davis *et al.* (1963)[a]

Salmo gairdneri	129		23–196	12–27	60	60	19–73		5–15		5–15	Salinity, 0–30%	Rao (1968)[a]
Salmo gairdneri	10	20.2		Gradually to prescribed velocity			30–70	1.5–3.5	15		15	O_2, 2.0–2.3 mg/liter	Kutty (1968)[a]
Salmo gairdneri	128	7.5–24	5–100	11–18	1	1	48–70	5.5–6.4	10		10		Fry and Cox (1970)[b]
Salmo gairdneri	[150]	5.7–6.0	1.4–1.7	2.3–3.0	10–20	10	25.1–43.3			18–19		Control	Butler and Milleman (1971)[a]
Salmo gairdneri	[150]	5.7–6.0	1.4–1.7	2.3–3.0	10–20	10	13.7–34.3			18–19		Infected; *Nanophyetus salmincola*	Butler and Milleman (1971)[a]
Salmo irideus		20.0					170	8.5				Photography	Gray (1953)
Salmo salar	>100	15–20		Gradually to prescribed velocity		20	70–100	3–4	1–14		1–14	Salinity, 0–30%	Byrne *et al.* (1972)[a]
Salmo salar	5	23.4	110.3	Gradually to prescribed velocity			50–76	2.1–3.2	15		15	O_2, 3.8–5.0 mg/liter	Kutty and Saunders (1973)[a]
Salmo trutta		34.0	34.1				92	2.7					Magnan (1929)
Salvelinus fontinalis	6	9.9		3	30	30	34.7	3.50	15		15	Exposure to 1.5 mg/liter fenitrothion	Peterson (1974)
Salvelinus fontinalis	6	9.6		3	30	30	37.4	3.90	15		15	Exposure to 0.5 mg/liter fenitrothion	Peterson (1974)
Salvelinus fontinalis	6	10.2		3	30	30	47.6	4.67	15		15	Exposure to 0.15 mg/liter fenitrothion	Peterson (1974)
Salvelinus fontinalis	18	10.3		3	30	30	49.5	4.82	15		15	0 fenitrothion	Peterson (1974)

Continued

Table VI Prolonged Swimming Speeds of Fish—Cont'd

Species	Number	Total length (cm)	Weight (g)	Velocity increments (cm sec_{-1})	Time between increments (min)	Swimming time at maximum velocity (min)	Maximum velocity		Temperature (°C)			Comments	Reference
							cm sec^{-1}	ℓ sec^{-1}	Acclimation		Experimental		
Salvelinus fontinalis	6	10.4		3	60	60	48.2	4.6	15		15	Metabolism	Peterson (1974)[a]
Salvelinus fontinalis	28	10.1		3	30	30	48.9	4.8	15		15	Metabolism	Peterson (1974)[a]
Salvelinus fontinalis	6	10.9		3	15	15	55.9	5.1	15		15	Metabolism	Peterson (1974)[a]
Salvelinus fontinalis	6	11.0		3	1.5	15	67.9	6.2	15		15	Metabolism	Peterson (1974)[a]
Salvelinus fontinalis	6	11.6		3	0.5	15	88.7	7.7	15		15	Metabolism	Peterson (1974)[a]
Salvelinus namaycush			27.7	4–12	2	2	35–54		8–23		8–23		Gibson and Fry (1954)[b]
Salvelinus namaycush			82.8	4–12	2	2	48–83		10–22		10–22		Gibson and Fry (1954)[b]
Esox lucius		16.5					210	12.7				Photography	Gray (1953)
Esox lucius		37.8			17		148	3.9					Magnan (1929)
Carassius auratus	3		4.4	4–12	2	2	22.4–34.7		5		5–25	Acute exposure to temperature	Fry and Hart (1948)[b]
Carassius auratus	3		4.4	4–12	2	2	29.1		10		10		Fry and Hart (1948)[b]
Carassius auratus	3		4.4	4–12	2	2	22.4–40.3		15		5–25	Acute exposure to temperature	Fry and Hart (1948)[b]

Carassius auratus	3		4.4	4–12	2	2	51		20		20		Fry and Hart (1948)[b]
Carassius auratus	3		4.4	4–12	2	2	28–51		25		15–35	Acute exposure to temperature	Fry and Hart (1948)[b]
Carassius auratus	3		4.4	4–12	2	2	50		30		30		Fry and Hart (1948)[b]
Carassius auratus	3		4.4	4–12	2	2	15.3–38.8		35		20–38	Acute exposure to temperature	Fry and Hart (1948)[b]
Carassius auratus	10	18.2		Gradually to prescribed velocity			15–85	1–3.2	20		20	O_2, 0.8–1.9 mg/liter	
Kutty (1968)[a]	*Carassius auratus*	7	15–17	50–60	110	40	40–300	60–126	3.8–8.4	15–30		15–30	Thermal acclimated, metabolism
Smit *et al.* (1971)[a]													
Cyprinus carpio		13.5					170	12.6				Photography	Gray (1953)
Leuciscus leuciscus		18.2					170	9.2				Photography	Gray (1953)
Pimephales promelas	465	4.8		Gradually until laps lost		3	19.6	4.1	15		15		McLeod (1967)[b]
Scardinius erythrophthalmus		19.0	18.8		13		114	6.0					Magnan (1929)
Scardinius erythrophthalmus		18.2					130	5.9				Photography	Gray (1953)
Gadus luscus		16.7	16.5				55	3.3					Magnan (1929)
Gadus merlangus		17.7	17.7				23	1.3					Magnan (1929)
Gadus morhua	34	35.5	560	Gradually to prescribed velocity		>5.6–240	75–135	2.1–3.8	8		8	Endurance	Beamish (1966b)[a]
Gadus morhua	40	35–37	580–635	Gradually to prescribed velocity		>4.2–240	75–135	2.1–3.7	5		5	Endurance	Beamish (1966b)[a]

Continued

Table VI Prolonged Swimming Speeds of Fish—Cont'd

Species	Number	Total length (cm)	Weight (g)	Velocity increments (cm sec_{-1})	Time between increments (min)	Swimming time at maximum velocity (min)	Maximum velocity		Temperature (°C)			Comments	Reference
							cm sec^{-1}	ℓ sec^{-1}	Acclimation		Experimental		
Merluccius vulgaris		22.6	23.7	19			79	3.5					Magnan (1929)
Macrozoarces americanus	12	33.6–38.4	156–237	Gradually to prescribed velocity		1.1–6.6	90–120	2.4–3.2	8		8	Endurance	Beamish (1966b)[a]
Morone saxatilis	1090	2–14		Gradually to prescribed velocity		10	35–87	7.6–12.6		22		Endurance, 50% fatigue	Kerr (1953)[a]
Morone saxatilis	340	8.9–11.4		Gradually to prescribed velocity		30	67	5.9–7.5		22		Endurance, 50% fatigue	Kerr (1953)[a]
Lepomis macrochirus		4.5–5.7	1.9–3.7	2–3	2–5	31–201	22.5	4.0–5.0	21		21	Endurance, O_2, 6.5 mg/liter; H_2S, 0–0.15 mg/liter	Oseid and Smith (1972)[b]
Lepomis macrochirus		5.1–5.4	2.9–3.4	2–3	2–5	22–28	28.0	5.2–5.5	21		21	Endurance, O_2 6.5 mg/liter; H_2S, 0–0.01 mg/liter	Oseid and Smith (1972)[b]
Micropterus salmoides		21.3					88	4.1					Magnan (1929)
Micropterus salmoides	30	5.7		Gradually until laps lost		3	18.8–30.7	3.3–5.4	20		20		MacLeod (1967)[b]
Micropterus salmoides	105	8.0–8.5	4.8–6.4	2.3	10	10	20–41	2.4–5.0	25		25	O_2, 1–24 mg/liter	Dahlberg *et al.* (1968)[a]
Micropterus salmoides	65	8.0–8.6	5.6–7.4	2.3	10	10	24–43	2.8–7.8	25		25	O_2, 1.2–8.1; CO_2 3–54 mg/liter	Dahlberg *et al.* (1968)[a]
Micropterus salmoides	45	2.0–2.2		2.2	3	3	4.8–14.6	2.2–6.5	5		5–20	Acute exposure to temperature	Larimore and Duever (1968)[b]

Micropterus salmoides	53	2.0–2.2		2.2	3	3	5.2–16.8	2.3–7.8	10		5–25	Acute exposure to temperature	Larimore and Duever (1968)[b]
Micropterus salmoides	51	2.0–2.2		2.2	3	3	7.2–23.9	3.3–10.1	15		10–30	Acute exposure to temperature	Larimore and Duever (1968)[b]
Micropterus salmoides	51	2.0–2.2		2.2	3	3	11.1–27.0	5.1–12.4	20		10–30	Acute exposure to temperature	Larimore and Duever (1968)[b]
Micropterus salmoides	45	2.0–2.2		2.2	3	3	8.5–29.2	3.6–13.0	25		10–30	Acute exposure to temperature	Larimore and Duever (1968)[b]
Micropterus salmoides	30	2.0–2.2		2.2	3	3	17.7–31.2	7.8–13.6	30		20–30	Acute exposure to temperature	Larimore and Duever (1968)[b]
Micropterus salmoides	30	5.7		Gradually until laps lost		3	18.8–30.7	3.3–5.4	20		20		Larimore and Duever (1968)[b]
Micropterus salmoides	32	15–27	45–270	10	30	30	24–55	1.6–2.0	10		10		Beamish (1970)[a]
Micropterus salmoides	45	15–27	45–270	10	30	30	3.3–58	2.2	15		15		Beamish (1970)[a]
Micropterus salmoides	45	15–27	45–270	10	30	30	45–63	2.3–3.0	20		20		Beamish (1970)[a]
Micropterus salmoides	48	15–27	45–270	10	30	30	47–64	2.4–3.1	25		25		Beamish (1970)[a]
Micropterus salmoides	51	15–27	45–270	10	30	30	48–66	2.4–3.2	30		30		Beamish (1970)[a]
Micropterus salmoides	38	15–27	45–270	10	30	30	40–60	2.2–2.7	34		34		Beamish (1970)[a]
Perca flavescens	137	0.6–1.4		Gradually to prescribed velocity		60	0.6–4.6	1.0–3.3	13		13	Velocity, 50% fatigue	Houde (1969)[a]

Continued

Table VI Prolonged Swimming Speeds of Fish—Cont'd

Species	Number	Total length (cm)	Weight (g)	Velocity increments (cm sec_{-1})	Time between increments (min)	Swimming time at maximum velocity (min)	Maximum velocity		Temperature (°C)			Comments	Reference
							cm sec^{-1}	ℓ sec^{-1}	Acclimation		Experimental		
Perca perca		18.3	18.4				66	3.6					Magnan (1929)
Stizostedion vitreum vitreum	181	0.7–1.5		Gradually to prescribed velocity		60	0.5–5.0	0.7–3.3	13		13	Velocity, 50% fatigue	Houde (1969)[a]
Sciaena aguila		29.7	29.5				113	3.8					Magnan (1929)
Mugil capito		26.5	26.0				61	2.3					Magnan (1929)
Trachurus symmetricus	15	14.6		Gradually to prescribed velocity		3.4	160	10.8	18.5		18.5	Endurance, 50% fatigue	Hunter (1971)[a]
Trachurus symmetricus	8	11.2		Gradually to prescribed velocity		4.5	139	12.4	18.5		18.5	Endurance, 50% fatigue	Hunter (1971)[a]
Gobius fluviatilus	13	8.8	28.6	Gradually to prescribed velocity		10–15	48–50	5.5–5.7		19–21			Shazkina (1972b)[a]
Gobius syrman	11	10.6	25.2	Gradually to prescribed velocity		8–10	44	4.2		19–21			Shazkina (1972b)[a]
Gobius melanostoma	22	9.0	29.3	Gradually to prescribed velocity		3–4	34	3.8		19–21			Shazkina (1972b)[a]
Scomber scombrus		25.3	34.1				81	3.2					Magnan (1929)
Sebastes dactylopterus		26.8					98	3.6					Magnan (1929)
Sebastes marinus	40	17–19	127–166	Gradually to prescribed velocity		0.7–14.2	52–135	3.1–8.0	5		5	Endurance	Beamish (1966b)[a]

Sebastes marinus	40	16–17	113–137	Gradually to prescribed velocity	0.8–13.2	52–135	3.1–8.0	8		8	Endurance	Beamish (1966b)[a]
Sebastes marinus	70	16–17	114–140	Gradually to prescribed velocity	0.6–12.7	52–135	3.1–8.0	11		11	Endurance	Beamish (1966b)[a]
Hemitripterus americanus	16	18.8–22.7	234–271	Gradually to prescribed velocity	0.4–0.7	105–120	4.6–6.4	8		8	Endurance	Beamish (1966b)[a]
Myoxocephalus octodecim spinosus	50	19.8–21.3	130–162	Gradually to prescribed velocity	0.5–6.7	60–120	2.8–5.7	8		8	Endurance	Beamish (1966b)[a]
Pseudopleuron ectes americanus	30	19–21	135–193	Gradually to prescribed velocity	1.9–14.1	75–135	4.0–6.5	5		5	Endurance	Beamish (1966b)[a]
Pseudopleuron ectes americanus	30	19–21	149–190	Gradually to prescribed velocity	1.7–10.3	75–135	3.6–6.6	8		8	Endurance	Beamish (1966b)[a]
Pseudopleuron ectes americanus	30	19–20	130–171	Gradually to prescribed velocity	1.4–10.0	7.5–135	3.9–6.6	11		11	Endurance	Beamish (1966b)[a]
Pseudopleuron ectes americanus	20	22–23	215–219	Gradually to prescribed velocity	5.3–25.4	75–135	3.3–6.1	14		14	Endurance	Beamish (1966b)[a]

[a]Tunnel swimming chamber.
[b]Oval or annular swimming chamber.
[c]Towed swimming cage.

D. Environmental Constraints on Performance

1. *Temperature*

Swimming capacity is regulated by the metabolic capacity of fish to convert chemical energy into propulsive thrust through muscular contraction. Adenosine triphosphate (ATP) generated by the stepwise degradation of carbohydrate and lipid is an essential prerequisite for muscle contraction. In sustained swimming the formation of ATP by aerobic processes would be expected to be more important than that by anaerobiosis. The contribution by both aerobic and anaerobic reactions is important in prolonged swimming (Bilinski, 1974). Burst swimming relies heavily on the anaerobic mobilization of metabolites from carbohydrate sources (Drummond and Black, 1960; Black *et al.*, 1961; Dean and Goodnight, 1964; Beamish, 1966c, 1968; Drummond, 1967, 1971; Dando, 1969). Since the physiological mechanisms associated with swimming vary with the category of locomotion, it is not surprising that they should differ in their response to temperature.

a. Sustained and Prolonged Swimming Speed. At swimming speeds where aerobic processes contribute significantly to the production of ATP the influence of temperature can perhaps best be understood from its relationship with oxygen consumption (see Brett, 1970b; Fry, 1971, for comprehensive reviews). In the metabolic processes oxygen serves as a final acceptor in the electron transport system rather than participating directly in the enzyme reactions involved in biological oxidation. Of the oxygen consumed, a portion serves to meet the basal metabolic requirements while the remainder provides for other activities including swimming, digestion, excretion, and growth. The amount of oxygen that can be extracted is, within limits, related to or dependent on the environmental conditions (Beamish and Dickie, 1967; Fry, 1967), each factor acting independently or interacting with others to alter the potential expression of maximum metabolic rate. The term "scope for activity" was employed by Fry (1947) to illustrate the effect of environmental identities on the oxygen available to vertebrate poikilotherms for activities excluding basal metabolism. Changes in scope for activity in relation to temperature are illustrated in Fig. 11 for brook trout (Graham, 1949), lake trout, *Salvelinus namaycush* (Gibson and Fry, 1954) and largemouth bass (Beamish, 1970).

Within the thermal range of tolerance for a species, prolonged speeds typically increased with temperature to a maximum and thereafter decline. Examples of prolonged swimming speeds in relation to temperature of acclimation are presented in Fig. 12. Temperature compensation, customarily applied to adaptive evolution of metabolic performance (Bullock, 1955; Fry, 1958; Roberts, 1966; Brett, 1970b) applies also to prolonged swimming speeds. Among the eurythermal temperate species such as goldfish (Fry and Hart, 1948) and large and smallmouth bass (Beamish, 1970; Larimore and Duever, 1968) the temperature for maximum prolonged performance lies within the range of 25°–30°C (Table VI). Stenothermal temperate species such as the lake trout, *Salvelinus namaycush* (Gibson and Fry, 1954), display maximum

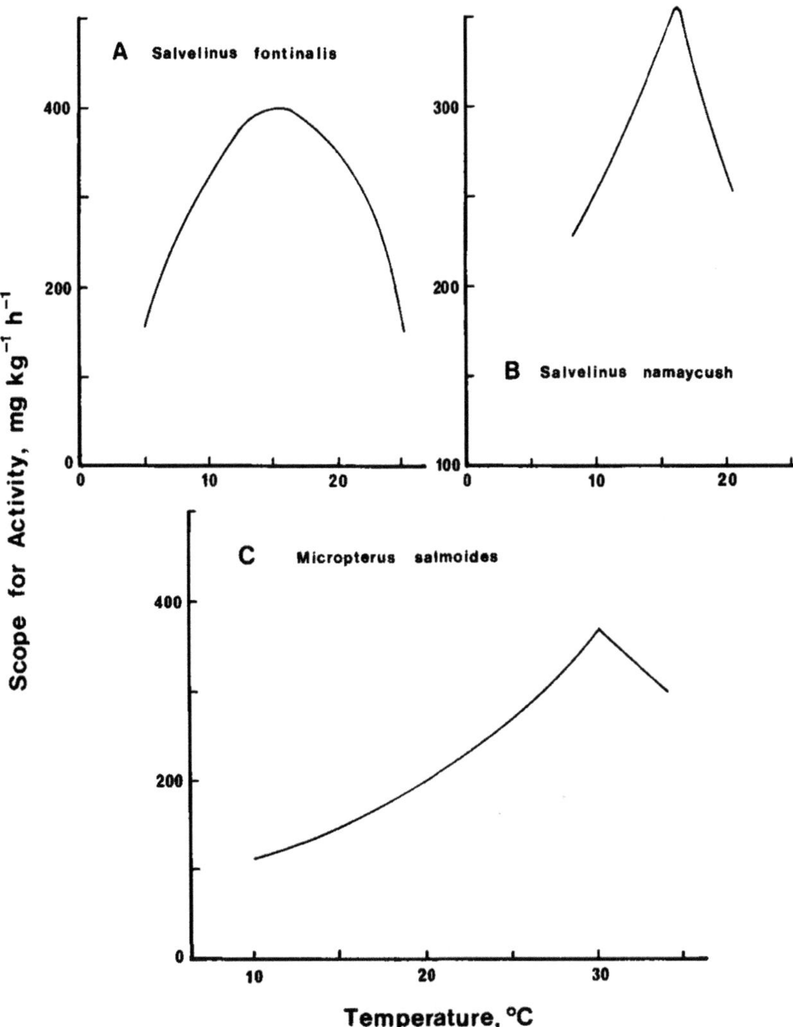

Fig. 11. Metabolic scope for activity of (A) brook trout, *Salvelinus fontinalis* (Graham, 1949); (B) lake trout, *Salvelinus namaycush* (Gibson and Fry, 1954); (C) largemouth bass, *Micropterus salmoides* (Beamish, 1970), in relation to temperature.

prolonged speeds between 15° and 20°C. The prolonged performance of the antarctic stenothermal, *Trematomus borchgrevinki* (Wohlschlag, 1964), in contrast to the temperate species was highest at −0.8°C and fish were unable to swim at temperatures above 2°C.

Prolonged performance may vary severalfold within the range of thermal tolerance for a species. The (3 min, 2.2 cm sec^{-1}) prolonged speed for smallmouth bass (2.2 cm) increased from 4.8 to 31.2 cm sec^{-1} (2.2–14.2 ℓ sec^{-1})

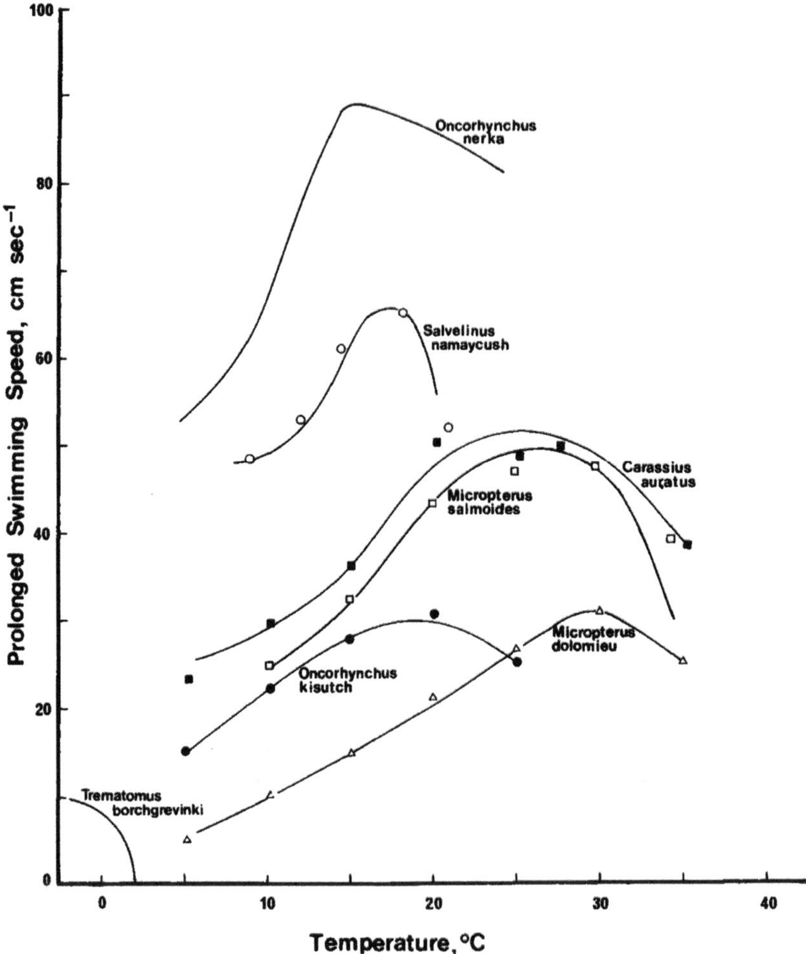

Fig. 12. Prolonged swimming speed and temperature. (From Fry and Hart, 1948; Gibson and Fry, 1954; Brett *et al.* 1958; Wohlschlag, 1964; Larimore and Duever, 1968; Beamish, 1970; Brett and Glass, 1973.)

between 5° and 30°C (Larimore and Duever, 1968). The prolonged (60 min, 10 cm sec^{-1}) critical for sockeye salmon (20 cm) varied from just over 50–90 cm sec^{-1} (2.5 – 4.5 $\ell\, sec^{-1}$) between 1° and 15°C.

The response of prolonged swimming speed to thermal acclimation is presented for three species in Fig. 13. In each, maximum performance progressively shifts toward a higher exposure temperature as acclimation is increased. This, of course, suggests that the capacity for activity over much of the range of thermal tolerance for a species is greatest at environmental temperatures equal to or above those to which it is acclimated. For example, goldfish acclimated to 5° and 25°C performed best at 18° and 28°C,

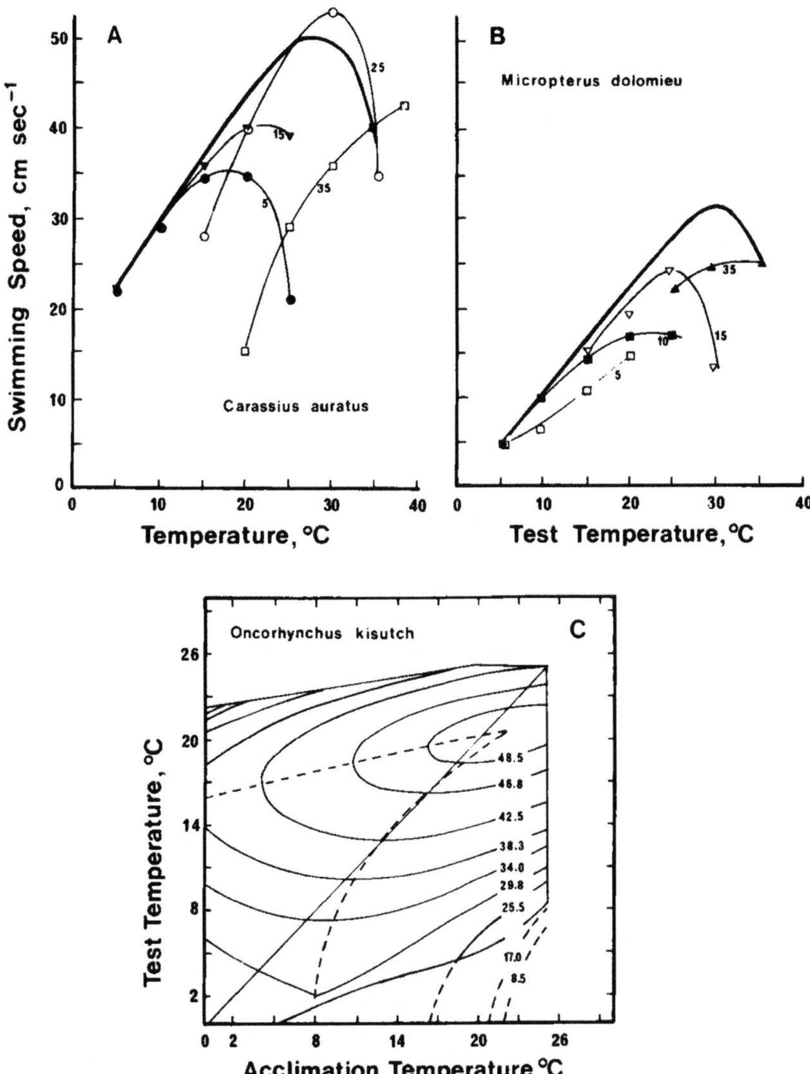

Fig. 13. Prolonged swimming speed and temperature. The heavy lines in panels (A) and (B) denote prolonged for fish acclimated to the test temperatures. Light lines indicate the response between prolonged and test temperature for a particular acclimation temperature (Fry and Hart, 1948; Larimore and Duever, 1968). In panel (C) the *U* critical speed isopleths (cm sec^{-1}) for coho salmon, *Oncorhynchus kisutch*, are presented in relation to acclimation and test temperature (Griffiths and Alderdice, 1972).

respectively (Fry and Hart, 1948). Similarly, the (3 min,2.2 cm sec^{-1}) prolonged speed of smallmouth bass, acclimated to 10° or 15°C occurred at 22° and 25°C, respectively (Larimore and Duever, 1968).

Griffiths and Alderdice (1972) thoroughly investigated the influence of acute temperature exposure on the (60 min, one-eighth critical) critical speeds of young coho salmon (7.9–9.5 cm). Maximum critical speeds occurred at test temperatures above the acclimation, denoted in Fig. 1.3 by line A. At a given test temperature, maximum performance coincided closely with acclimation, in the figure described by ridge B. Optimum performance occurred at a combination of acclimation and test temperatures near 20°C. Fry (1967) concluded, based on earlier evidence, that maximum swimming performance for a given temperature occurs when acclimation and exposure are identical which is generally concordant with the observations of Griffiths and Alderdice (1972).

b. Burst Swimming Speed. Temperature appears to exert little influence on burst speed although information at this level is particularly scarce. Blaxter and Dickson (1959) measured the burst speeds for a number of marine and freshwater teleosts (Table IV) and were unable to demonstrate any correlation between performances and temperature. Based on metabolic studies of sockeye salmon in relation to performance, Brett (1964) anticipated temperature independence for burst swimming. The swimming endurance of redfish, Atlantic cod, and winter flounder, *Pseudopleuronectes americanus*, at speeds approaching burst velocities did not vary appreciably over the ecological range of temperatures experienced throughout much of the year in the northwest Atlantic (Beamish, 1966b). More recently Groves (reported in Brett, 1970b) reported a temperature independence in the burst speed achieved but a dependence in terms of endurance by sockeye salmon.

2. Oxygen

In aquatic organisms which use oxygen for their respiration, the ambient oxygen consumption itself can limit swimming performance. There appears for most fish a threshold oxygen concentration below which swimming performance is reduced (Dizon, 1977).

a. Sustained and Prolonged Swimming Speed. Kutty (1968) and Kutty and Saunders (1973) introduced the term "critical oxygen concentration" to describe the concentration at which fish are unable to maintain sustained or prolonged speeds. Thus Atlantic salmon (23.4 cm) sustained speeds of 50 and 70 cm sec^{-1} (2.1 and 3.0ℓ sec^{-1}) for several hours until ambient oxygen was reduced to 4.0 and 4.8 mg O_2liter^{-1}, respectively (Kutty and Saunders, 1973; see also Fig. 14). Critical oxygen concentrations of goldfish (18.5 cm) at 59.2 and 18.6 cm sec^{-1} (3.2 and 1.0ℓ sec^{-1}) were 1.8 and 0.8 mg O_2liter^{-1}, respectively (Kutty, 1968). Similar reductions in sustained and prolonged swimming performance in the presence of low oxygen have been demonstrated for a number of species (Graham, 1949; Katz *et al.*, 1959; Davis *et al.*, 1963; Whitworth and Irwin, 1964; McLeod and Smith, 1966; Dahlberg *et al.*, 1968; see also Table VI; Fig. 14). Above air saturation, prolonged performance of coho salmon and largemouth bass was independent of oxygen (Dahlberg *et al.*, 1968; see also Fig. 14).

In contrast to the earlier observations by Prosser *et al.* (1957) on goldfish, Kutty (1968) found acclimation to low ambient oxygen did not alter the critical oxygen concentrations for a given sustained swimming speed. Failure to

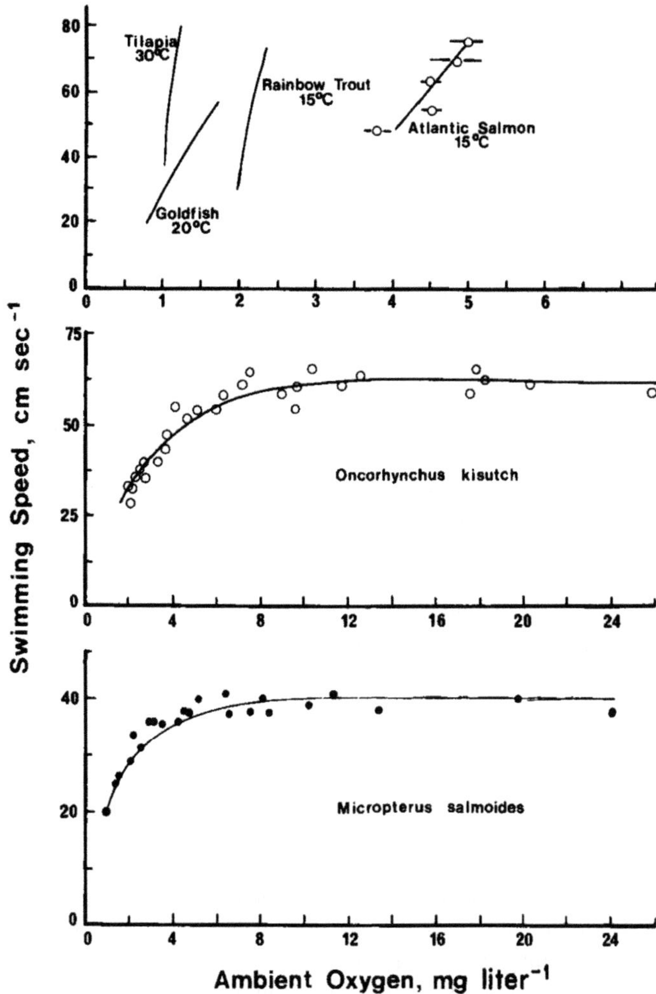

Fig. 14. Swimming speed and ambient oxygen concentration. In the upper panel, critical oxygen concentrations at which fish are unable to maintain a sustained speed (Kutty, 1968; Kutty and Saunders, 1973). In the lower two panels, the relationship between prolonged speed and oxygen (Dahlberg *et al.*, 1968).

swim at low ambient oxygen was not attributed by Kutty (1968) to fatigue as fish began to swim as soon as oxygen levels were increased, but rather to an oxygen sensing mechanism such as the peripheral or central oxygen receptors reported by Saunders and Sutterlin (1971).

b. Burst Swimming Speed. The effect of dissolved oxygen on burst swimming has not been measured. However, burst speed depending as it does on anaerobic energy sources may be expected to be largely independent of ambient oxygen except that between swimming events the accumulated metabolic debt must be repaid before the next burst of swimming can realize its full potential. The mobilization of energy resources for repeated bursts and therefore the frequency of rapid swimming may well be restricted by moderate oxygen deficiency.

3. *Carbon Dioxide*

Carbon dioxide has long been known to reduce the affinity of blood for oxygen (Root, 1931) and to influence the metabolic rate of fish (Basu, 1959; Beamish, 1964b). Particularly little information is available on the effect of carbon dioxide on swimming, a notable exception being the research of Dahlberg *et al.* (1968). They measured the prolonged speed of largemouth bass and coho salmon in response to dissolved oxygen and free carbon dioxide. The prolonged speeds of largemouth bass did not change in response to concentrations of carbon dioxide to 48 mg liter^{-1}. The performance of coho salmon in contrast to that observed for bass declined on exposure to concentrations of carbon dioxide between 2–61 mg liter^{-1}. In low concentrations of oxygen the influence of carbon dioxide was less pronounced. For an oxygen concentration of 10 mg liter^{-1}, the prolonged speed of salmon decreased from about 60 to fractionally above 50 cm sec^{-1} when carbon dioxide increased from 2 to 61 mg liter^{-1}. In contrast, when ambient oxygen was about 2 mg liter^{-1}, prolonged speed did not change with increase in free carbon dioxide.

4. *Salinity*

Salt concentration in the blood of fish is less than that of seawater. In a marine environment water is lost at the gills and other body surfaces (Potts, 1954; Black, 1951). Conversely freshwater homeostasis is dependent on the elimination of absorbed water, the concentration of the body fluids being greater than that of the environment (Black, 1957). The mechanism by which osmoregulation is achieved may vary among species (Parry, 1958; Gordon, 1963; Threadgold and Houston, 1964) but each requires the expenditure of energy.

Few measurements have been made on the relationship between salinity and swimming performance. A consistent pattern of change in the swimming speed of skipjack and yellow fin tuna did not occur in response to a salinity decrease from 34 to 29‰ (Dizon, 1977). Critical speeds (60 min, one-eighth critical) of coho salmon in relation to salinities and temperatures between 0–20‰ and 3°–23°C, respectively, were measured by Glova and McInerney (1977) (Fig. 15, Table V). The combined effects of salinity and temperature indicated critical swimming performance of underyearling coho was predominantly a temperature-dependent response during the premigratory stages of development. Swimming performance of fry was almost independent of salinity as reflected by the flat configuration of the performance isopleths in Fig. 15. Coho smolts achieved maximum critical speeds at salinities ranging from just under 8 to about 19‰. Relative to this salinity

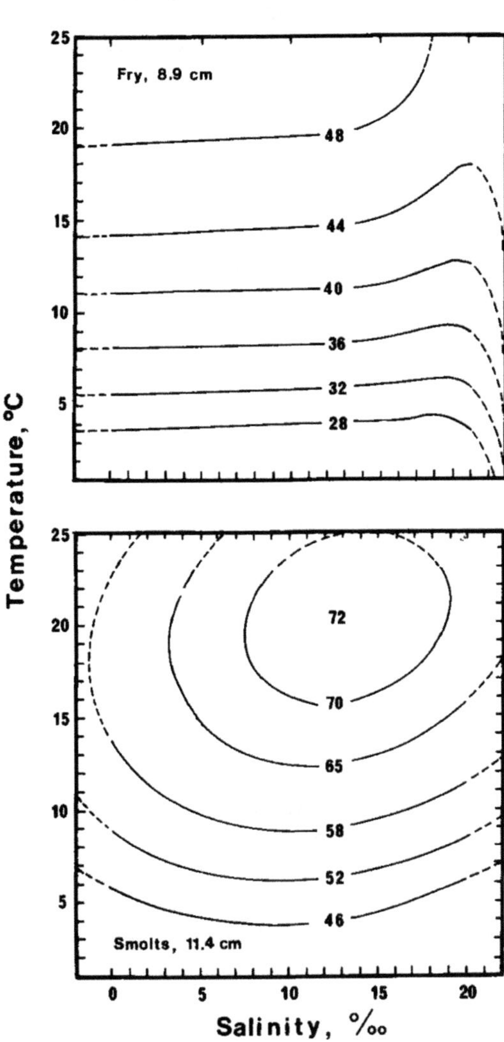

Fig. 15. Critical swimming speed isopleths (cm sec^{-1}) for coho salmon fry and smolts, *Oncorhynchus kisutch*, in relation to salinity and temperature (Glova and McInerney, 1977).

optimum, critical, performance declined by about 6 and 4 cm sec^{-1} at 0 and 20‰, respectively. Just prior to, or concurrent with smoltification, coho appear to lose their euryhaline capacity to function efficiently over a range of salinities which is concordant with observations on salinity tolerance for this species (Alderdice, 1963).

Performance restrictions attributable to salinity have not been measured for other species. However, fluctuations in metabolic expenditure in association with prolonged speeds have been measured for rainbow trout (Rao, 1968) and *Tilapia nilotica* (Farmer and Beamish, 1969) in salinities of 0–30‰. In both studies the energy actually expended in swimming was independent of salinity. Nevertheless, changes in metabolic rate did occur, suggesting that performance capacity would be reduced in proportion to the energy expended in osmotic regulation.

5. The Disturbed Environment

The influence of those identities introduced into the environment either directly or indirectly by man or through his activities has received attention only in recent years with most of the effort expended in the determination of their lethal concentrations. Of particular concern to the swimming capacity of fish are those identities which influence the exchange of respiratory gases or the metabolic pathways involved in the mobilization of energy. From among the many factors that may contribute to the disturbed environment, only a few have been examined with respect to their influence on swimming performance.

Conifer pulpwood fiber in suspension impairs the removal of oxygen from water by physically clogging the gill lamellae and interrupting the respiratory flow during gill cleaning reflexes (MacLeod and Smith, 1966). On exposure to suspensions of pulpwood fiber equivalent to 200 mg liter^{-1} the endurance of fathead minnows, *Pimephales promelas*, forced to swim at a low velocity was significantly less than that of fish in freshwater under comparable concentrations of dissolved oxygen (Fig. 16A). The influence of the pulpwood suspension was most pronounced at the higher temperatures (Fig. 16B) which is consistent with its impairment of gaseous exchange. In contrast to the influence on endurance at prolonged swimming, burst swimming speed was independent of the concentration of pulpwood fiber.

Hydrogen sulfide is found not infrequently in the aquatic environment and results from the decomposition of material either naturally occurring or present through the activities of man. Endurance of bluegills (3.2 cm) forced to swim at 22.5 cm sec^{-1} increased from just over 200 min in the absence of hydrogen sulfide to 240 min at 0.4 μg liter$^{-1}H_2S$, and with further increases, decreased so that at 14.6 μg liter^{-1} fish swam only for 30 min (Oseid and Smith, 1972). Coincident with the long term exposure to H_2S was an increase in the rate of gill irrigation which undoubtedly lead to an appreciable reduction in the mobilization of energy through aerobic processes.

Sodium pentachlorophenate (PCP), used as a defoliant or for the protection of timber from wood-boring insects and fungal infection, is considered a general metabolic poison for fish (Webb and Brett, 1973). However, the (60 min, 5 cm sec^{-1}) critical speed of sockeye salmon (5.3–6.0 cm) did not change significantly from about 40 cm sec^{-1} (7.3 ℓ sec^{-1}) on exposure to concentrations of PCP between $0-50\mu l^{-1}$ (Webb and Brett, 1973). Similarly, Krueuger *et al.* (1966) found that swimming performance of *Cichlasoma bimaculatum* was not reduced by pentachlorophenol until the concentration

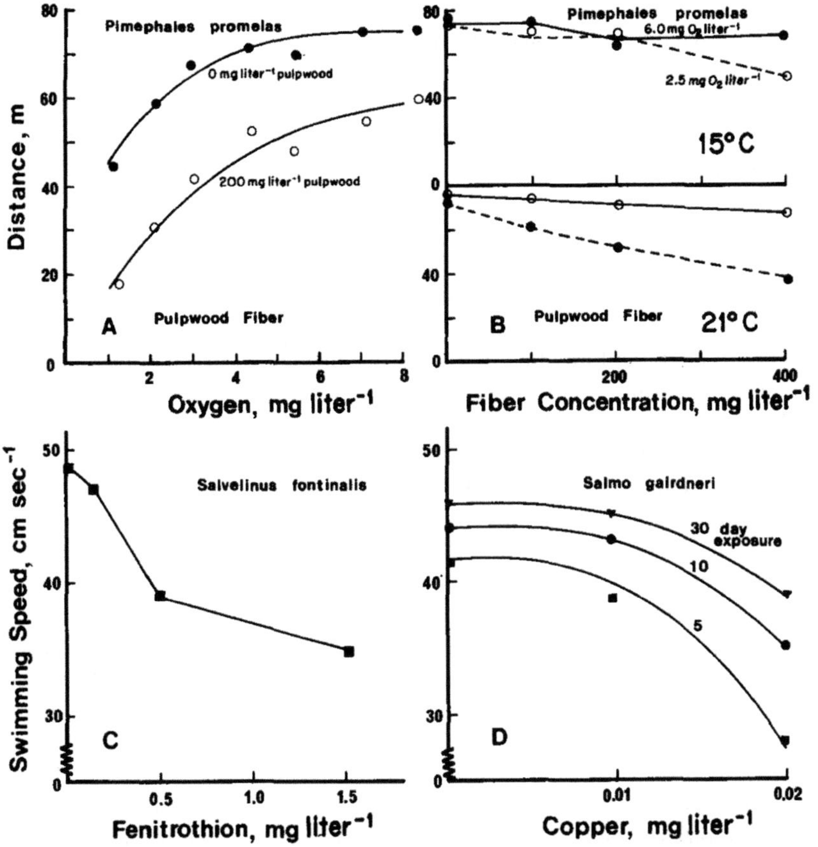

Fig. 16. (A) The influence of pulpwood fiber and dissolved oxygen on the distance swum by fathead minnows, *Pimephales promelas* (MacLeod and Smith, 1966). (B) Temperature, oxygen, and pulpwood fiber effect on swimming distance by fathead minnows (MacLeod and Smith, 1966). (C) Effect of fenitrothion on prolonged speed of brook trout, *Salvelinus fontinalis* (Peterson, 1974). (D) Effect of copper on critical speed of rainbow trout, *Salmo gairdneri* (Waiwood, personal communication).

approached lethal levels. Webb and Brett (1973) proposed that a general metabolic poison such as PCP should not "preferentially" affect the gas exchange system particularly where excitement, as included in fish forced to swim, serves as a "stressor."

The influence of fenitrothion, an organophosphate insecticide used in the control of spruce budworm, on prolonged swimming speed of brook trout was determined by Peterson (1974). Prolonged swimming speed decreased from 5.0 ℓ sec^{-1} for controls to 3.5 ℓ sec^{-1} for trout exposed to 1.5 mg $liter^{-1}$, the highest concentration applied (Fig. 16C). While the mechanism through which fenitrothion reduces swimming performance is unknown, Peterson (1974) suggested it may cause impairment of those areas of the nervous system concerned with muscle activity or alternately by causing indirect effects through "motivational" disturbances.

In a comprehensive study on the influence of bleached kraft mill effluent (BKME), Howard (1975) measured the (60 min, 5 cm sec^{-1}) critical speed of coho salmon to concentrations equivalent to 90% of the level at which 50% of the fish died within 96 hr (96 hr LC_{50}). Exposure for 18 hr to a concentration of 0.9 LC_{50} resulted in a 72% reduction in swimming capacity. Further critical speed for a given concentration of BKME was independent of exposure time beyond 18 hr and returned to control levels within 6–12 hr after being placed in effluent-free water. In swimming fish, Howard (1975) suggests BKME retards gaseous exchange either by absorption of the effluent on the gill surface or through the formation of a weak chemical bond to the gill epithelium. In addition, Javaid (1973) observed ventilatory irregularities among sockeye salmon exposed to BKME.

Effluents from mining operations discharged into waterways may also exert a pronounced influence on the capacity of fish to perform. Waiwood (personal communication) measured the influence of total copper and pH in relation to water hardness on the (60 min 5 cm sec^{-1}) critical swimming speed of rainbow trout (Fig. 16D). He found that for a given hardness, critical speed was reduced by increasing concentrations of copper but that the effect diminished with time of exposure to about 10 days (Fig. 16D). Further, the influence of a given concentration of copper on performance decreased inversely with water hardness. Copper is known to have a deleterious effect on the composition of blood (McKim *et al.*, 1970) and to damage various tissues including the kidney, liver, intestine, and cephalic lateral canals (Baker, 1969; Gardner and Laroche, 1973). Tissue damage would undoubtedly cause an elevation in basal metabolism and a decline in the scope for activity.

Swimming performance, depending as it does on the immediate recruitment of energy, has been recommended for use as a criterion in the determination of the sublethal effects of pollutants on fish (Brett, 1967; Sprague, 1971). However, the proper application of swimming speed, as well as the category of performance to be tested as a criterion of sublethal effect, requires a comprehensive prior knowledge of the pharmacological effects of the pollutants concerned. Impairment of the gaseous exchange of mechanism might, for example, be masked in burst swimming speeds which depend on anaerobic processes. Further, the capacity exhibited by some fish to acclimate in part or even fully to a given pollutant should be respected by the serious investigator.

IV. ENERGETICS OF SWIMMING

The expenditure of energy during swimming is reflected in gaseous exchange and should include measurements of both oxygen consumption and carbon dioxide production. However, due to limitations imposed by the techniques available, measurements of carbon dioxide production are infrequently made, the researches of Kutty (1968) being a notable exception. More generally, calculation of the energy expenditure for swimming is made from units of oxygen consumption and converted to units of energy on the basis of an oxycalorific coefficient derived for domestic homeothermic animals (Brody, 1945). This coefficient assumes not only the complete oxidation of catabolized substrates but also a normal balance of the sources such as is implied by an average respiratory quotient of 0.8. Winberg (1956) suggested that irrespective of the components oxidized, the oxycalorific coefficient will not vary more than 1.5%. It is generally agreed an oxycalorific coefficient of 3.36–3.44 is most applicable for teleosts (Warren and Davis, 1967; Brett, 1973; Beamish *et al.*, 1975). In contrast, Krueger *et al.* (1968) reported that calorific loss based on lipid depletion in strenuously exercised salmon was substantially greater than that estimated from respiratory rates, and questioned the method of evaluation of energy production indirectly from oxygen consumption. Brett (1973) on the other hand, concluded an oxycalorific coefficient of 3.36 cal mg ${O_2}^{-1}$ consumed is acceptable for teleosts and thatunder carefully regulated experimental conditions, estimates of energy expenditure made from the oxygen consumed by sockeye salmon are not at variance with those based on the utilization of body components.

In sustained swimming the mobilization of energy is achieved through aerobic processes so that the quantity of oxygen consumed is proportional to the amount of work performed. Fish swimming at prolonged speeds derive energy from both aerobic and anaerobic processes, the contribution from the latter increasing with the severity of exercise. At prolonged speeds, utilization of glycogen stores was reported by Pritchard *et al.* (1971) as the principal cause of swimming failure in the jack mackerel. The evolution of respiratory gases has not been measured for burst swimming because of the practical difficulties imposed by the short duration of muscular activity. It is presumed, however, that at burst swimming, fish consume some oxygen and that the remainder of the energy requirement is met through anaerobic processes. The latter results in an oxygen debt which is repaid subsequent to the termination of exercise. The allocation of aerobic and anaerobic processes in relation to the category of swimming is summarized in Fig. 17. Swimming energetics is the subject of several reviews which should be consulted (Fry, 1957; Fry and Hochachka, 1970; Brett, 1962, 1970a, 1972; Beamish and Dickie, 1967; Randall, 1970; Doudoroff and Shumway, 1970; Schmidt-Nielsen, 1972; Bilinski, 1974).

Present evidence indicates that swimming may elevate the total metabolic rate by as much as 15-fold (Beamish, 1964a; Brett, 1964). The oxygen consumed at sustained and prolonged swimming speeds is presented for a number of species in Fig. 18. Subtraction of standard or basal metabolism from the total oxygen uptake has been used to approximate the expenditure of energy associated with a particular swimming speed. Where anaerobic processes do not contribute significantly this would appear a satisfactory procedure. The rate of increase in the logarithm of oxygen uptake with relative swimming speed in Fig. 18 is surprisingly similar among species despite obvious variation in methodology, size, and temperature and is reasonably well represented

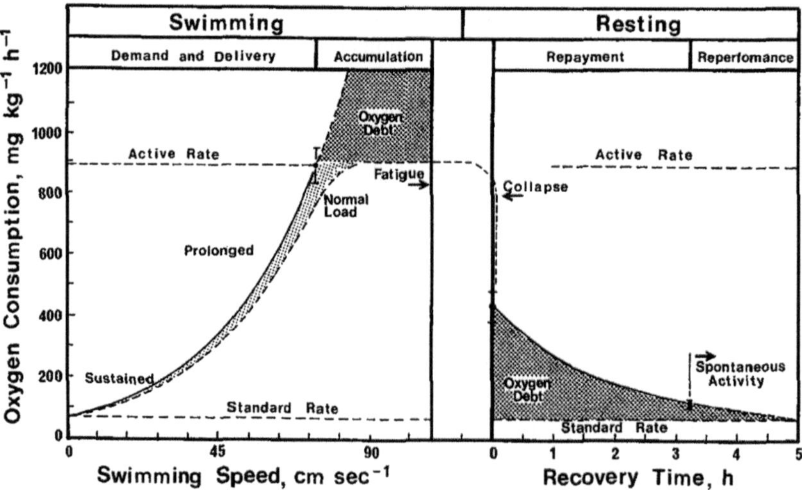

Fig. 17. Oxygen consumption and debt for sockeye salmon, *Oncohynchus nerka* (18 cm), in relation to swimming speed and recovery at 15°C. (Redrawn from Brett, 1964, *J. Fish. Res. Board Can.*)

by a coefficient of 0.36. Thus for each increase in relative swimming speed of ℓ sec^{-1} there is a corresponding 2.3-fold elevation in metabolic rate.

In severe prolonged and burst swimming, caution must be exercised in not accounting for energy expenditure by anaerobic processes. The anaerobic contribution is perhaps most conveniently assessed by continued measurement of oxygen consumption on completion of swimming until it returns to preexercise levels at which time the oxygen debt is presumably repaid (Heath and Pritchard, 1962; Brett, 1964; Smit *et al.*, 1971): This procedure assumes the products of anaerobic metabolism such as lactate are not excreted but subsequently oxidized during the recovery phase following exercise. Recently, Karuppannan (1972, reported in Kutty and Peer Mohamed, 1975) has shown that *Tilapia mossambica* excrete some lactate after strenuous exercise, corroborating the earlier observations by Blažka (1958) on the anaerobic metabolism of crucian carp, *Carassius carassius*.

The maximum rate of oxygen consumption among fish species appears to vary at least 5 -fold with maximum values in excess of 2000 mg kg^{-1}hr^{-1} (Stevens, personal communication). The metabolic capacity of the higher vertebrates is generally one or two orders of magnitude above that demonstrated for teleosts (Bartholomew and Tucker, 1963, 1964; Bartholomew *et al.*, 1965; Tucker, 1970; Brett, 1972). This discrepancy is compensated for, in part by a greater tolerance by teleosts to oxygen debt but from which recovery is slow. In sockeye salmon the rate of replacement of oxygen debt following fatigue was in excess of 3 hr and independent of temperature (Brett, 1964). The magnitude of the debt accumulated at the time of fatigue was influenced by temperature and increased 2 -fold between 5° and 15°C. Similarly, Heath and Pritchard (1962) found that bluegill sunfish, after strenuous exercise,

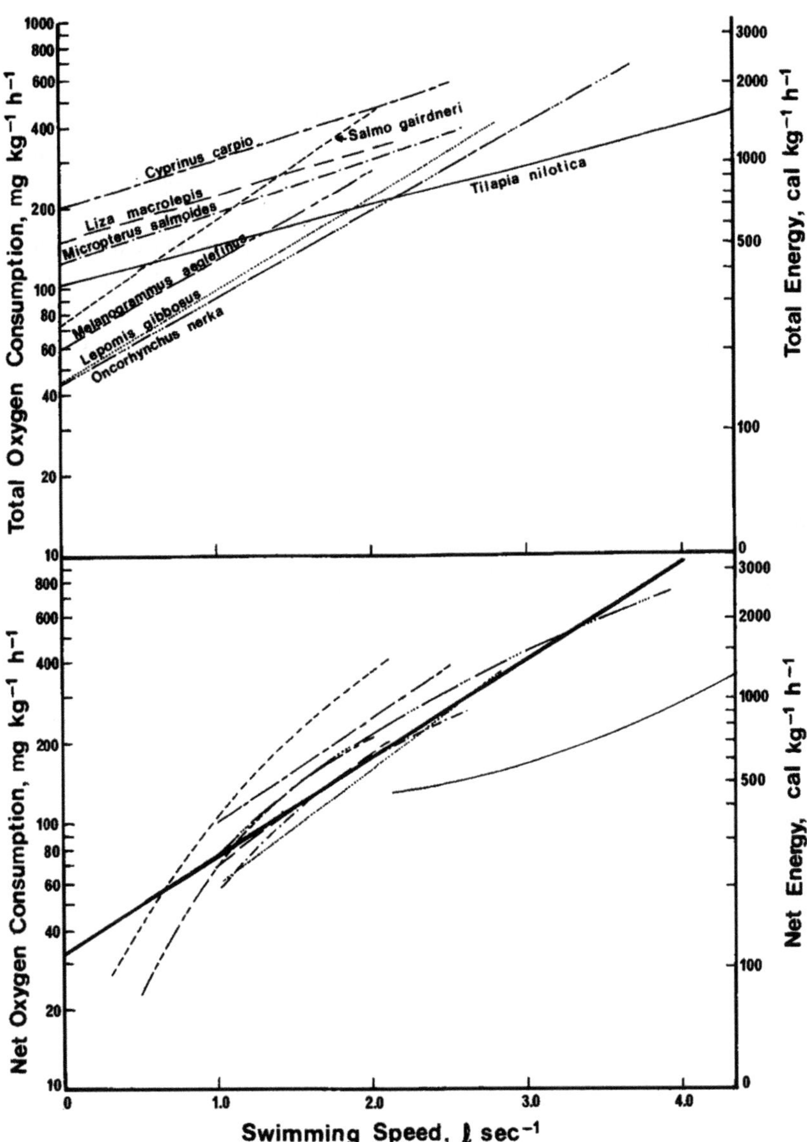

Fig. 18. Oxygen consumption, energy utilization, and swimming speed. The total oxygen consumed is presented in the upper panel. Subtraction of the basal from total metabolic rate provides a measure of the energy required for a given speed of swimming. The heavy line in the lower panel denotes the general rate of increase in net oxygen consumption and was fitted by eye. (From Basu, 1959; Beamish, 1964a, 1970; Brett, 1964; Brett and Sutherland, 1965; Farmer and Beamish, 1969; Kutty, 1969; Webb, 1971b.)

maintained a high consumption of oxygen followed after 1 hr by a gradual decline to preexercise levels 10–24 hr later.

Schmidt-Nielson (1972) expressed the energy cost for locomotion independently of swimming speed as the caloric expenditure to transport 1 unit of body mass 1 km. A reanalysis of Brett's data on prolonged swimming speeds of sockeye salmon by Schmidt-Nielsen showed a logarithmic linear decrease in energy expenditure with increase in body weight over a range of three orders of magnitude (Fig. 19). The application of this expression of the energetic cost of locomotion to other species exercised at sustained and prolonged speeds under different environmental conditions shows a remarkable similarity to the relation described for salmon in Fig. 19. Closer examination of the comparative energy cost at low and high prolonged speeds based on measurements of oxygen uptake indicates a reduction as velocity is increased. Thus, the energy expenditure for mullet, *Liza macrolepis*, declined from 2.09 to 1.48 cal g^{-1} km^{-1} between 10 and 22.5 cm sec^{-1} (Kutty, 1969), which presumably reflects a greater contribution of anaerobic metabolism at the higher speeds. Refinement in the measurement of total metabolism of swimming fish, while desirable, is unlikely to alter significantly the linear relationship described by Schmidt-Nielsen. Following Schmidt-Nielsen's interesting hypothesis, Gold (1973, 1974) and Calder (1974) expressed the energy cost of swimming one body length in terms of the mass of propulsive muscles relative to total weight and multiplied by the number of muscle contractions or tailbeats required to transport the animal

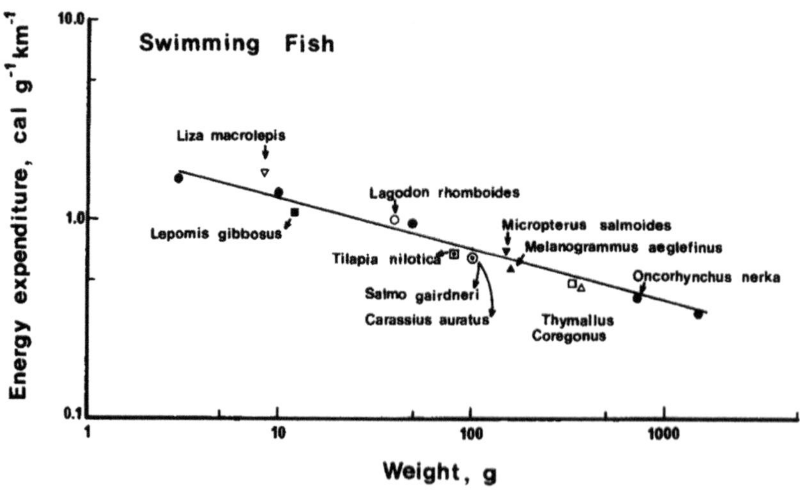

Fig. 19. Energy cost of swimming relative to body size calculated by SchmidtNielsen (1972) from data collected by Brett (1964), Wohlschlag *et al.* (1968), Matyukhin and Stolbow (1970, reported in Schmidt-Nielsen, 1972), Rao (1971), and Smit *et al.* (1971). In addition, measurements made by Brett and Sutherland (1965), Farmer and Beamish (1969), Kutty (1969), Tytler (1969), and Beamish (1970) have been recalculated and added to the figure.

one length. This assumed a constancy in the quantity of energy available per contraction per unit of muscle mass equal to 1 cal kg^{-1}.

Energy for muscular contraction is derived from the hydrolysis of adenosine triphosphate to adenosine diphosphate and inorganic phosphate. The evolution of adenosine triphosphate through the catabolism of organic compounds may occur under both aerobic and anaerobic conditions, the former being the more efficient in terms of yield but each offering distinct advantages to the swimming teleost. In sustained swimming where the amount of oxygen consumed is proportional to the work performed, the main source of energy is from long-chain fatty acids and to a lesser extent protein and glycogen (Greene, 1926; Idler and Tsuyuki, 1958; Drummond and Black, 1960). When the capacity for aerobic metabolism is exceeded as in severe prolonged or burst swimming, adenosine triphosphate is synthesized by anaerobic glycolysis of stored muscle glycogen. Lactic acid, the end product of glycolysis, diffuses from the muscle into the bloodstream (Nakatani, 1957; Black, Connor *et al.*, 1962; Driedzic and Hochachka, 1975). Both swimming and the accumulation of lactic acid may continue until the glycogen deposits are depleted or the end product of anaerobic glycolysis exerts a detrimental effect on activity. Where exercise is extreme in its severity, death may result during the recovery period (Parker *et al.*, 1959; Beamish, 1966c; Caillouet, 1967). The actual cause of death is uncertain but may result from interference with the acid base equilibrium, coupled with reduced affinity of hemoglobin for oxygen, and, in the presence of excess acid, lowered affinity for carbon dioxide (Black, 1958a). When death does not follow strenuous exercise, the elevated rate of oxygen consumption serves not only to meet the routine metabolic requirements but also to replace muscle supplies of adenosine triphosphate, creatine phosphate, and glycogen (Bilinski, 1974).

V. APPLICATION TO MANAGEMENT PRACTICES

Hatchery breeding programs have tended to select for qualities such as rapid growth, early maturity, high fecundity, and disease resistance which, of course, are of obvious benefit to the culturist. However, when the objective is to stock desirable waterways with the view of generating a sustainable population, selective breeding programs may have overlooked qualities essential to the continued well being of the population. Swimming performance is an important component of viability as it relates to a fish's capacity to maintain station against current, avoid predators, and acquire food. Bams (1967) proposed that unless severe environmental conditions impose a serious constraint, the most important component of survival is stamina. Vibert (1956) used the ability of fish to swim against a current as a test of adaptability for stocking. The importance of swimming performance is reflected by the higher survival of fish which were conditioned to a stream habitat prior to stocking (Shuck and Kingsbury, 1948; Miller, 1957).

The stamina of hatchery and stream-conditioned rainbow trout was investigated by Reimers (1956). Hatchery rainbow trout were able to swim against a current of 90 cm sec^{-1} for 5–10 min before fatigued, considerably less than the 30 min recorded for stream-conditioned trout. The performance of wild stocks of brook trout, even though reared under hatchery conditions, was consistently superior to that recorded for domestic stocks of the same species (Vincent, 1960; Green, 1964). The size at which fish are stocked may also influence their success. Survival of planted chinook salmon suggest a greater success among fingerlings than fry (Junge and Phinney, 1963). Thomas *et al.* (1964) attributed the greater survival of fingerlings to a number of factors including performance capacity. The role of nutrition on swimming performance and the ultimate capacity of planted fish to cope with the environment has not been examined but represents a potentially profitable area of research.

Hatchery procedure in the incubation of eggs may also regulate the ultimate capacity of the species to perform. Bams (1967) examined different methods of incubating eggs of sockeye salmon on the ultimate relative performance of fry. Naturally propagated salmon demonstrated the best relative performance followed by fry reared in gravel from the time of hatching and held, prior to the advanced alevin stage, in baskets or trays without a substrate. The poorest stamina was registered by fish reared in hatchery troughs without gravel at any stage.

Investigations aimed at determining size and structure of fish stocks can be influenced by the species' resistance to fatigue. This may result when fish are tagged and released subsequent to capture by any method which involves severe muscular exertion on the part of the fish. For example, marine demersal species are frequently captured for tagging purposes by otter trawls which are towed along the seabed at speeds of 140–200 cm sec^{-1} for 30 min or more. Many fish are unable to swim at these speeds for long and fatigue. This imposes a severe metabolic load, manifested by an oxygen debt, a depletion of glycogen reserves, and elevation in lactate, as well as a number of other physiological changes. Mortalities following capture by otter trawl have been reported and attributed to muscular fatigue. Among haddock, *Melanogrammus aeglefinus*, mortalities ranged between 7 and 78% of those captured by otter trawl (Beamish, 1966c). Fatigue deaths in ocean troll-caught chinook and coho salmon were observed by Milne and Ball (1956, 1958), Parker and Black (1959) and Parker *et al.* (1959). Barrett and Connor (1962) attributed some of the deaths of hook and line-caught yellowfin and skipjack tuna during recovery to fatigue.

The steadily increasing demand for greater utilization of waterways has resulted in the construction of dams on rivers and the location of electrical generating plants near rivers, lakes, and oceans (Kerr, 1953). One of the problems associated with the construction of dams is that of preserving fish populations indigenous to the waters. On the west coast of North America, particular concern has been expressed for the well being of valuable stocks and anadromous trout and salmon. This entails providing safe passage for fish through waterways and over obstacles. One of the considerations in the construction and operation of a fishway is to provide flows at the entrance which will attract the desired species. Weaver (1963) conducted a series of velocity preference studies at the site of the Bonneville Dam on the Columbia River. The experiments were conducted in large dual channels and compared the frequency of fish passing through each in relation to water velocity. The results suggested the proportion of rainbow trout and chinook and coho salmon selecting the channel with the highest current speed applied, 240 cm sec^{-1}, was appreciably greater than that at any other velocity, the lowest of which was 60 cm sec^{-1}. Information on the critical length of the passageway was provided from performance studies designed to measure the distance salmonids could swim at velocities to 500 cm sec^{-1}. Mean maximum speed which rainbow trout (68.5 cm) could maintain for 9.14 m (1.5 sec) was 642 cm sec^{-1} (9.5 ℓ sec^{-1}) with one individual (61 cm) achieving 817 cm sec^{-1} (13.4 ℓ sec^{-1}). Maximum speeds for chinook (75.3 cm) and coho salmon (51.0 cm) were 604 and 421 cm sec^{-1} (8.2 and 8.2 ℓ sec^{-1}), respectively. Earlier, Paulik and DeLacy (1957) measured the swimming speed of rainbow trout, coho, and sockeye salmon to provide information needed in the design of fishways. They found, in laboratory studies, the maximum prolonged speed for rainbow trout (63.6 cm) was 213 cm sec^{-1} (3.4 ℓ sec^{-1}), well below that found by Weaver (1963). Similarly the maximum prolonged speed of coho (56.2 cm) 190 cm sec^{-1}(3.5 ℓ sec^{-1}) fell short of the subsequent measurements made by Weaver (1963).

Another of the basic problems in the design of fishways is the location, number, and size of resting pools. Recovery of coho salmon (65.6 cm) from an exhaustive swimming effort at 100 cm sec^{-1} was found to be 31% complete after 1 hr rest and 67% after 3 hr. All fish recovered when allowed 18-24 hr (Paulik *et al.*, 1957). From these data the investigators concluded the necessity for adequate resting facilities along a fishway when velocities exceeded 100 cm sec^{-1} for more than a few minutes. Swimming performance of salmonids has been found to decline slightly as adult fish migrate upstream (Paulik and DeLacy, 1958).

In Passamaquoddy Bay of the Bay of Fundy, a study was initiated to examine the possible effects of the construction of a series of dams on the fishery. Of major importance were the Atlantic herring which accounted for the vast majority of the total fish landings. Movements of herring indicated they entered the bay through narrow passages in which water velocity occasionally reached 300 cm sec^{-1}. With the construction of the dams currents would have exceeded this speed. Laboratory measurements of swimming endurance at prescribed velocities indicated that had the dams been constructed, high currents together with the periods during which the dam gates were closed would have denied herring access to the bay for all but about 20 min every 24 hr.

Observations on the swimming capacity of the western sucker, *Catostomus occidentalis*, as they moved upstream through a culvert prompted Wales (1950) to note the possibility of excluding undesirable species from portions of a river by regulating current speed.

Swimming speed studies have been applied also in the design and assessment of fishing gear. The efficiency of otter trawls has received considerable attention as they supply much of the total annual harvest of fish from the marine environment. Reports from divers and from photographic observations (Blaxter and Parrish, 1966; Beamish, 1967) have indicated the orientation of fish swimming ahead of the trawl. With this information and the swimming capacity of the species the probability of escape can be estimated, assuming a straight line course and a fixed speed of swimming. Such estimates have been made for a number of demersal species in the northwest Atlantic by Beamish (1967) and the North Sea by Blaxter (1967; see also Fig. 20).

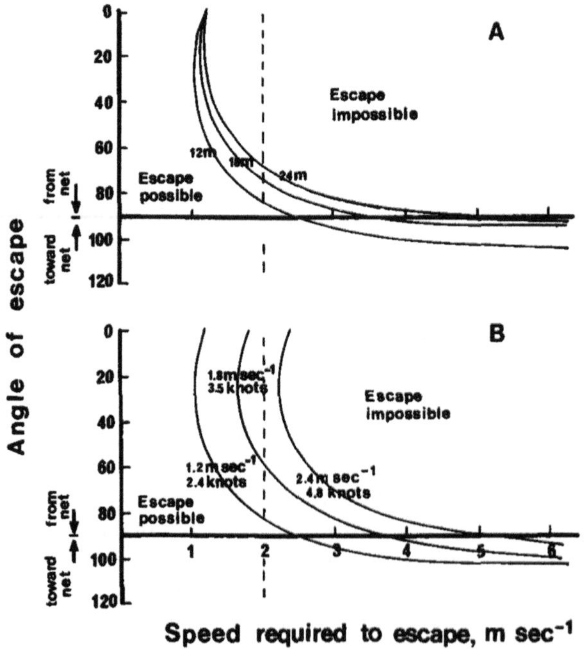

Fig. 20. Swimming speed required to avoid capture by otter trawl at different angles of escape when the fish is in the center of the path of the net and reacts 3 m away (Blaxter, 1967). Arrow at 2 m sec^{-1} indicates maximum burst speed for adult clupeoids. Panel (A) illustrates the change when nets of 12, 18, and 24 m in width are employed at speeds of 1.2 m sec^{-1}. Panel (B) indicates change when towing speed is varied from 1.2 to 2.4 m sec^{-1} and net width is held at 12 m.

ACKNOWLEDGMENTS

I am most grateful to Dr. E. D. Stevens for his comments on the manuscript and to Mrs. E. Thomas for her assistance with the figures.

REFERENCES

Alderdice, D. F. (1963). Some effects of simultaneous variation in salinity, temperature and dissolved oxygen on the resistance of juvenile coho salmon (*Oncorhynchus kisutch*) to a toxic substance. Ph.D. Thesis, Univ. of Toronto.

Aleev, Y. G. (1963). "Function and Gross Morphology in Fish," 245 pp. Izd. Akad. Nauk SSSR, Moscow. (Transl. by Isr. Program Sci. Transl., Jeruselam, 1969; available as TT67-51391, NTIS, Springfield, Virginia.)

Alexander, R. M. (1967). "Functional Design in Fishes," 160 pp. Hutchinson, London.

Alexander, R. M. (1968). "Animal Mechanics," 346 pp. Univ. of Washington Press, Seattle.

Arnold, G. P. (1969). A flume for behaviour studies of marine fish. *J. Exp. Biol.* **51**, 671–679.

Bainbridge, R. (1958). The speed of swimming of fish as related to size and to the frequency and amplitude of the tailbeat. *J. Exp. Biol.* **35**, 109–133.

Bainbridge, R. (1960). Speed and stamina in three fish. *J. Exp. Biol.* **37**, 129–153.

Bainbridge, R. (1962). Training, speed and stamina in trout. *J. Exp. Biol.* **39**, 537–555.

Bainbridge, R., and Brown, R. H. J. (1958). An apparatus for the study of the locomotion of fish. *J. Exp. Biol.* **35**, 134–137.

Baker, J. T. P. (1969). Histological and electron microscopical observations on copper poisoning in the winter flounder *Pseudopleuronectes americanus*. *J. Fish. Res. Board Can.* **26**, 2785–2793.

Bams, R. A. (1967). Differences in performance of naturally and artificially propagated sockeye salmon migrant fry, as measured with swimming and predation tests. *J. Fish. Res. Board Can.* **24**, 1117–1153.

Barrett, I., and Connor, A. R. (1962). Blood lactate in yellow fin tuna *Neothunnus macropterus* and skipjack *Katsuwonus pelamis* following capture and tagging. *Bull. Inter-Am. Trop. Tuna Comm.* **6**, 233–280.

Barrett, I., and Hester, F. J. (1964). Body temperature of yellow fin and skipjack tunas in relation to sea surface temperature. *Nature (London)* **203**, 96–97.

Bartholomew, G. A., and Tucker, V. A. (1963). Control of changes in body temperature, metabolism, and circulation by the agamid lizard, *Amphibolurus barbatus*. *Physiol. Zool.* **36**, 199–218.

Bartholomew, G. A., and Tucker, V. A. (1964). Size, body temperature, thermal conductance, oxygen consumption, and heart rate in Australian varanid lizards. *Physiol. Zool.* **37**, 341–354.

Bartholomew, G. A., Tucker, V. A., and Lee, A. K. (1965). Oxygen consumption, termal conductance and heart rate in the Australian skink, *Tiliqua scincoides*. *Copeia No.* 2, pp. 1969–1973.

Bass, G. A., and Rascovich, M. (1965). A device for the sonic tracking of large fishes. *Zoologica (N.Y.)* **50**, 75–82.

Basu, S. P. (1959). Active respiration of fish in relation to ambient concentrations of oxygen and carbon dioxide. *J. Fish. Res. Board Can.* **16**, 175–212.

Beamish, F. W. H. (1964a). Respiration of fishes with special emphasis on standard oxygen consumption. III. Influence of weight and temperature on respiration of several species. *Can. J. Zool.* **42**, 177–188.

Beamish, F. W. H. (1964b). Respiration of fishes with special emphasis on standard oxygen consumption. IV. Influence of carbon dioxide and oxygen. *Can. J. Zool.* **42**, 847–856.

Beamish, F. W. H. (1966a). Vertical migration by demersal fish in the Northwest Atlantic. *J. Fish. Res. Board Can.* **23**, 109–139.

Beamish, F. W. H. (1966b). Swimming endurance of some Northwest Atlantic fishes. *J. Fish. Res. Board Can.* **23**, 341–347.

Beamish, F. W. H. (1966c). Muscular fatigue and mortality in haddock, *Melanogrammus aeglefinus*, caught by otter trawl. *J. Fish. Res. Board Can.* **23**, 1507–1521.

Beamish, F. W. H. (1967). Photographic observations on reactions of fish ahead of otter trawls. *FAO Conf. Fish Behav. Relation Fish. Tech. Tactics, Bergen, Norway* Exp. Pap. No. 25, pp. 1–11.

Beamish, F. W. H. (1968). Glycogen and lactic acid concentrations in Atlantic cod (*Gadus morhua*) in relation to exercise. *J. Fish. Res. Board Can.* **25**, 837–851.

Beamish, F. W. H. (1970). Oxygen consumption of largemouth bass, *Micropterus salmoides*, in relation to swimming speed and temperature. *Can. J. Zool.* **48**, 1221–1228.

Beamish, F. W. H. (1973). Oxygen consumption of adult *Petromyzon marinus* in relation to body weight and temperature. *J. Fish. Res. Board Can.* **30**, 1367–1370.

Beamish, F. W. H. (1974). Swimming performance of adult sea lamprey, *Petromyzon marinus*, in relation to weight and temperature. *Trans. Am. Fish. Soc.* **103**, 355–358.

Beamish, F. W. H. (1975). Apparent specific dynamic action of largemouth bass, *Micropterus salmoides*. *J. Fish. Res. Board Can.* **31**, 1763–1769.

Beamish, F. W. H., and Dickie, L. M. (1967). Metabolism and biological production in fish. *In* "The Biological Basis of Fresh Water Fish Production" (S. D. Gerking, ed.), pp. 215–242). Blackwell, Oxford.

Beamish, F. W. H., and Mookherjii, P. S. (1964). Respiration of fishes with special emphasis on standard oxygen consumption. I. Influence of weight and temperature on respiration of goldfish, *Carassius auratus* L. *Can. J. Zool.* **42**, 161–175.

Beamish, F. W. H., Niimi, A. J., and Lett, P. F. K. T. (1975). Bioenergetics of teleost fishes: Environmental influences. *In* "Comparative Physiology—Functional Aspects of Structural Materials" (L. Bolis, H. P. Maddrell, and K. Schmidt-Nielson, eds.), pp. 187–209. North-Holland Publ., Amsterdam.

Bell, W. H., and Terhune, L. D. B. (1970). Water tunnel design for fisheries research. *Fish. Res. Board Can., Tech. Rep.* No. 195, pp. 1–69.

Belyayev, V. V., and Zuyev, G. V. (1969). Hydrodynamic hypothesis of school formation in fishes. *J. Ichthyol.* (*USSR*) **9**, 578–584.

Bilinski, E. (1974). Biochemical aspects of fish swimming. *In* "Biochemical and Biophysical Perspectives in Marine Biology" (D. C. Malins and J. R. Sargent, eds.), pp. 239–288. Academic Press, New York.

Bishai, H. M. (1960). The effect of water currents on the survival and distribution of fish larvae. *J. Cons., Cons. Perm. Int. Explor. Mer* **25**, 134–146.

Black, E. C. (1955). Blood levels of haemoglobin and lactic acid in some freshwater fishes following exercise. *J. Fish. Res. Board Can.* **12**, 917–929.

Black, E. C. (1957a). Alterations in the blood level of lactic acid in certain salmonid fishes following muscular activity. I. Kamloops trout, *Salmo gairdneri*. *J. Fish. Res. Board Can.* **14**, 117–134.

Black, E. C. (1957b). Alterations in the blood level of lactic acid in certain salmonid fishes following muscular activity. II. Lake trout, *Salvelinus namaycush*. *J. Fish. Res. Board Can.* **14**, 645–649.

Black, E. C. (1957c). Alterations in the blood level of lactic acid in certain salmonid fishes following muscular activity. III. Sockeye salmon, *Oncorhynchus nerka*. *J. Fish. Res. Board Can.* **14**, 807–814.

Black, E. C. (1958a). Hyperactivity as a lethal factor in fish. *J. Fish. Res. Board Can.* **15**, 573–586.

Black, E. C. (1958b). Energy stores and metabolism in relation to muscular activity in fishes. *In* "The Investigation of Fish Power Problems" (P. A. Larkin, ed.), H. R. MacMillan Lectures in Fisheries, pp. 51–67. Univ. of British Columbia, Vancourver.

Black, E. C., Robertson, A. C., and Parker, R. R. (1961). Some aspects of carbohydrate metabolism in fish. *In* "Comparative Physiology of Carbohydrate Metabolism in Heterothermic Animals" (A. W. Martin, ed.), pp. 89–124. Univ. of Washington Press, Seattle.

Black, E. C., Connor, A. R., Lam, K., and Chiu, W. (1962). Changes in glycogen, pyruvate and lactate in rainbow trout (*Salmo gairdneri*) during the following muscular activity. *J. Fish. Res. Board Can.* **19**, 409–436.

Black, V. S. (1951). Changes in body chloride, density and water content of chum and coho salmon fry when transferred from freshwater to seawater. *J. Fish Res. Board Can.* **8**, 164–177.

Black, V. S. (1957). Excretion and osmoregulation. *In* "The Physiology of Fishes" (M. E. Brown, ed.), Vol. 1, pp. 163–205. Academic Press, New York.

Blaxter, J. H. S. (1962). Herring rearing. 4. Rearing beyond the yolk-sac stage. *Mar. Res.* pp. 1–18.

Blaxter, J. H. S. (1967). Swimming speeds of fish. *FAO Conf. Fish Behav. Relation Fish. Tech. Tactics, Bergen, Norway* Rev. Pap. No. 3, pp. 1–32.

Blaxter, J. H. S., and Dickson, W. (1959). Observations on swimming speeds of fish. *J. Cons., Cons. Perm. Int. Explor. Mer* **24**, 472–479.

Blaxter, J. H. S., and Parrish, B. B. (1966). The reactions of marine fish to moving netting and other devices in tanks. *Mar. Res.* **1**, 1–15.

Blažka, P. (1958). The anaerobic metabolism of fish. *Physiol Zool.* **31**, 117–128.

Blažka, P., Volf, M., and Cepela, M. (1960). A new type of respirometer of the determination of the metabolism of fish in an active state. *Physiol. Bohemoslov.* **9**, 553–558.

Bohun, S., and Winn, H. W. (1966). Locomotor activity of the american eel (*Anguilla rostrata*). *Chesapeake Sci.* **7**, 137–147.

Boyar, H.C. (1961). Swimming speed of immature Atlantic herring with reference to the Passamaquoddy Tidal Project. *Trans. Am. Fish. Soc.* **90**, 21–26.

Brawn, V. M. (1960). Underwater television observations of the swimming speed and behaviour of captive herring. *J. Fish. Res. Board Can.* **17**, 689–698.

Brett, J. R. (1962). Some considerations in the study of respiratory metabolism in fish, particularly salmon. *J. Fish. Res. Board Can.* **19**, 1025–1038.

Brett, J. R. (1964). The respiratory metabolism and swimming performance of young sockeye salmon. *J. Fish. Res. Board Can.* **21**, 1183–1226.

Brett, J. R. (1965). The relation of size to rate of oxygen consumption and sustained swimming speed of sockeye salmon (*Oncorhynchus nerka*). *J. Fish. Res. Board Can.* **23**, 1491–1501.

Brett, J. R. (1967). Swimming performance of sockeye salmon (*Oncorhychus nerka*) in relation to fatigue time and temperature. *J. Fish. Res. Board Can.* **24**, 1731–1741.

Brett, J. R. (1970a). Fish—The energy cost of living. *In* "Marine Agriculture" (W. J. McNeil, ed.), pp. 37–52. Oregon State Univ. Press, Corvallis.

Brett, J. R. (1970b). 3. Temperature, 3.3 Animals, 3.32 Fishes. *In* "Marine Ecology, Vol. 1, Environmental Factors" (O. Kinne, ed.), Part 1, pp. 513–560. Wiley (Interscience), New York.

Brett, J. R. (1972). The metabolic demand for oxygen in fish, particularly salmonids and a comparison with other vertebrates. *Respir. Physiol.* **14**, 151–170.

Brett, J. R. (1973). Energy expenditure of sockeye salmon, *Oncorhynchus nerka*, during sustained performance. *J. Fish. Res. Board Can.* **30**, 1799–1809.

Brett, J. R., and Glass, N. R. (1973). Metabolic rates and critical swimming speeds of sockeye salmon, *Oncorhynchus nerka*, in relation to size and temperature. *J. Fish. Res. Board Can.* **30**, 379–387.

Brett, J. R., and Sutherland, D. B. (1965). Respiratory metabolism of pumpkinseed (*Lepomis gibbosus*) in relation to swimming speed. *J. Fish. Res. Board Can.* **22**, 405–409

Brett, J. R., Hollands, H., and Alderdice, D. F. (1958). The effect of temperature on the cruising speed of young sockeye and coho salmon. *J. Fish. Res. Board Can.* **15**, 587–605.

Brody, S. (1945). "Bioenergetics and Growth," 1023 pp. Reinhold, New York.

Brunel, P. (1964). Food as a factor or indicator of vertical migrations of cod in the western Gulf of St. Lawrence. *ICNAF Environ. Symp., Rome, 1964* Contrib. No. C-2, pp. 1–16.

Bullock, T. H. (1955). Compensation for temperature in the metabolism and activity of poikilotherms. *Biol. Rev. Cambridge Philos. Soc.* **30**, 311–342.

Butler, J. A., and Milleman, R. E. (1971). Effect of the "salmon poisoning" trematode, *Nanophyetus salmincola*, on the swimming ability of juvenile salmonid fishes. *J. Parasitol.* **57**, 860–865.

Byrne, J. M., Beamish, F. W. H., and Saunders, R. L. (1972). Influence of salinity, temperature, and exercise on plasma osmolality and ionic concentration in Atlantic salmon (*Salmo salar*). *J. Fish. Res. Board Can.* **29**, 1217–1220.

Caillouet, C. W., Jr. (1967). Hyperactivity, blood lactic acid, and mortality in channel catfish. *Iowa State Univ. Sci. Tech. Res. Bull.* No. 551, pp. 897–915.

Calder, W. A., III (1974). Energy cost of animal locomotion. *Science* **184**, 1098.

Corti, U. A., and Weber, M. (1948). Die Matrix der Fische. II. Intersuchungen über die Vitaliät von Fischen. *Schweiz. Z. Hydrol.* **11**, 297–300.

Craig, R. E., and Forbes, S. (1969). A sonar for fish counting. *Fiskeridir. Skr., Ser. Havunders.* **15**, 210–219.

Cummings, W. C. (1963). Using the Doppler effect to detect movements of captive fish in behavior studies. *Trans. Am. Fish. Soc.* **92**, 178–180.

Cushing, D. H. (1957). The number of pilchards in the channel. *Fish. Invest. Minist. Agric. Fish. Food (G.B.) Ser. II Salmon Freshwater Fish.* **21**, 1–27.

Dahl, K., (1937). Salmon migrations off Norway. *Salmon Trout Mag.* **88**, 229–234.

Dahl, K., and Sømme, S. (1938). Salmon markings in Norway 1937. *Skr. Nor. Vidensk. Akad. Oslo 1* **2**, 1–45.

Dahlberg, M. L., Shumway, D. L., and Doudoroff, P. (1968). Influence of dissolved oxygen and carbon dioxide on swimming performance of largemouth bass and coho salmon. *J. Fish. Res. Board Can.* **25**, 49–70.

Dando, P. R. (1969). Lactate metabolism in fish. *J. Mar. Biol.* Assoc. *U.K.* **49**, 209–223.

Dannevig, G. (1953). Tagging experiments on cod, Lofoten 1947–1952: Some preliminary results. *J. Cons., Cons. Perm. Int. Explor. Mer* **19**, 195–203.

Davidson, V. M. (1949). Salmon and eel movement in constant circular current. *J. Fish. Res. Board Can.* **7**, 432–448.

Davis, G. E., Foster, J., Warren, C. E., and Doudoroff, P. (1963). The influence of oxygen concentration on the swimming performance of juvenile Pacific salmon at various temperatures. *Trans. Am. Fish. Soc.* **92**, 111–124.

Dean, J. M., and Goodnight, C. J. (1964). A comparative study of carbohydrate metabolism in fish as affected by temperature and exercise. *Physiol. Zool.* **37**, 280–299.

DeGroot, S. J., and Schuyf, A. (1967). A new method for recording the swimming activity in flatfishes. *Experientia* **23**, 574–576.

Denil, G. (1909). Les échelles à poissons et leur application aux barrages des Meuse et d'Ourthe. *Ann. Trav. Publ. Belg.* **2**, 1–152.

Denil, G. (1937). La mécanique du poisson de rivière. Chapitre X. Les capacités mécaniques de la truite et du saumon. *Ann. Trav. Publ. Belg.* **38**, 412–423.

deVeen, J. F. (1964). On the phenomenon of soles swimming near the surface of the sea. *Rapp. P.-V. Reun., Cons. Explor. Mer* No. 155, p. 51.

Dizon, A. E. (1977). Effect of dissolved oxygen concentration and salinity on swimming speed of two species of tunas. *Fish. Bull.* **75**, 649–653.

Dodson, J. J., Leggett, W. C., and Jones, R. A. (1970). A study of the orientation and migration of American shad in Long Island Sound and the Connecticut River Estuary. *U.S. Bur. Commer. Fish. Prog. Rep.* Contract AFC 6-1, 16 pp.

Dodson, J. J., Leggett, W. C., and Jones, R. A. (1971). A study of the orientation and migration of American Shad in Long Island Sound and the Connecticut River Estuary. *U.S. Bur. Commer. Fish. Prog. Rep.* Contract AFC 6-2, pp. 1–26.

Dodson, J. J., Leggett, W. C., and Jones, R. A. (1972). The behaviour of adult American shad (*Alosa sapidissima*) during migration from salt to freshwater as observed by ultrasonic tracking techniques. *J. Fish. Res. Board Can.* **29**, 1445–1449.

Doudoroff, P., and Shumway, D. L. (1970). Dissolved oxygen requirements of freshwater fishes. *FAO Fish. Tech. Pap.* No. 86, 297 pp.

Dow, R. L. (1962). Swimming speed of River herring *Pomolobus pseudoharengus* (Wilson). *J. Cons., Cons. Perm. Int. Explor. Mer* **27**, 77–80.

Dowd, R. G. (1967). An echo counting system for demersal fishes. *FAO Conf. Fish Behav. Relation Fish. Tech. Tactics, Bergen, Norway* Exp. Pap. No. 7, pp. 1–6.

Dowd, R. G., Bakken, E., and Nakken, O. (1970). A comparison between two sonic measuring systems for demersal fish. *J. Fish. Res. Board Can.* **27**, 737–747.

Driedzic, W. R., and Hochachka, P. W. (1975). The unanswered question of high anaerobic capabilities of carp white muscle. *Gan. J. Zool.* **53**, 706–712.

Drummond G. I. (1967). Muscle metabolism. *Fortschr. Zool.* **18**, 360–429.

Drummond, G. I. (1971). Microenvironment and enzyme function: Control of energy metabolism during muscle work. *Am. Zool.* **11**, 83–97.

Drummond, G. I., and Black, E: C. (1960). Comparative physiology: Fuel of muscle metabolism. *Annu. Rev. Physiol.* **22**, 169–190.

Ellis, D. V. (1966). Swimming speeds of sockeye and coho salmon on spawning migration. *J. Fish. Res. Board Can.* **23**, 181–187.

Farlinger, S., and Beamish, F. W. H. (1977). Effects of time and velocity increments on the critical swimming speed of largemouth bass. *Trans. Am. Fish. Soc.* **106**, 436–439.

Farmer, G. J., and Beamish, F. W. H. (1969). Oxygen consumption of *Tilapia nilotica* in relation to swimming speed and salinity. *J. Fish. Res. Board Can.* **26**, 2807–2821.

Farmer, G. J., Beamish, F. W. H., and Robinson, G. A. (1975). Food consumption of the adult landlocked sea lamprey, *Petromyzon marinus, L. Comp. Biochem. Physiol.* A **50**, 753–757.

Fierstine, H. L., and Walters, V. (1968). Studies in locomotion and anatomy of scombroid fishes. *Mem. South. Catif. Acad. Sci.* **6**, 1–31.

Fontaine, M. (1956). Appareil pour la mésure de l'activité motrice d'un poisson. *In* "Notice sur les Travaux Scientiques de Maurice Fontaine," p. 81. F. Paillart, Abbeville.

Fox, A. C. (1965). Some effects of strigeid metacercariae on rainbow trout (*Salmo gairdneri*). *Trans. Am. Microsc. Soc.* **84**, 153.

Fridriksson, A. (1958). The tribes in the north coast herring of Iceland with special reference to the period 1948–1955. *Rapp. P.-V. Reun., Cons. Explor. Mer* **143**, 36–44.

Fridriksson, A., and Aasen, O. (1952). The Norwegian–Icelandic herring tagging experiments. Report No. 2. *Rit Fiskideildar* **1**, 1–54.

Fry, F. E. J. (1947). Effects of the environment on animal activity. *Univ. Toronto Stud., Biol. Ser.* **55**, 1–62.

Fry, F. E. J. (1957). The lethal temperature as a tool in taxonomy. *Ann. Biol.* **33**, 205–218.

Fry, F. E. J. (1958). Temperature compensation. *Annu. Rev. Physiol.* **20**, 207–224.

Fry, F. E. J. (1967). Responses of vertebrate poikilotherms to temperature. *In* "Thermobiology" (A. H. Rose, ed.), pp. 375–409. Academic Press, New York.

Fry, F. E. J. (1971). The effects of environmental factors on the physiology of fish. *In* "Fish Physiology" (W. S. Hoar and D. J. Randall, eds.), Vol. 6, pp. 1–98. Academic Press, New York.

Fry, F. E. J., and Cox, E. T. (1970). A relation of size to swimming speed in rainbow trout. *J. Fish. Res. Board Can.* **27**, 976–978.

Fry, F. E. J., and Hart, J. S. (1948). Cruising speed of goldfish in relation to water temperature. *J. Fish. Res. Board Can.* **7**, 169–175.

Fry, F. E. J., and Hochachka, P. W. (1970). Fish. *In* "Comparative Physiology of Thermoregulation, Vol. 1, Invertebrates and Nonmallian Vertebrates" (G. C. Whittow, ed.), pp. 79–134. Academic Press, New York.

Gardner, G. R., and Laroche, G. (1973). Copper-induced lesions in estuarine teleosts. *J. Fish. Res. Board Can.* **30**, 363–368.

Gayduk, V. V., Malinin, L. K., and Poddubnyy, A. G. (1971). Determination of the depths of fishes during the hours of daylight. *J. Ichthyol.* **11**, 140–143.

Gero, D. R. (1952). The hydrodynamic aspects of fish propulsion. *Am. Mus. Novit.* No. 1601, pp. 1–32.

Gibson, E. S., and Fry, F. E. J. (1954). The performance of the lake trout, *Salvelinus namaycush*, at various levels of temperature and oxygen pressure. *Can. J.Zool.* **132**, 252–260.

Glova, G. J., and McInerney, J. E. (1977). Critical swimming speeds of coho salmon (*Oncorhynchus kisutch*) fry to smolt stages in relation to salinity and temperature. *J. Fish. Res. Board Can.* **34**, 151–154.

Gold, A. (1973). Energy expenditure in animal locomotion. *Science* **181**, 275–276.

Gold, A. (1974). Comment on paper by Calder. *Science* **184**, 1098.

Gordon, M. S. (1963). Chloride exchange in rainbow trout (*Salmo gairdneri*) adapted to different salinities. *Biol. Bull.* **124**, 45–54.

Graham, J. B. (1973). Heat exchange in the black skipjack, and the blood–gas relationship of warm-bodied fishes. *Proc. Nat. Acad. Sci. U.S.A.* **70**, 1964–1967.

Graham, J. B. (1975). Heat exchanges in the yellow fin tuna, *Thunnus albacares*, and skipjack tuna, *Katsuwonus pelamis*, and the adaptive significance of elevated body temperatures in scombrid fishes. *U.S. Fish. Wildl. Serv., Fish. Bull.* **73**, 219–229.

Graham, J. M. (1949). Some effects of temperature and oxygen pressure in the metabolism and activity of the speckled trout, *Salvelinus fontinalis. Can. J. Res., Sect. D.* **27**, 270–288.

Gray, J. (1937). Pseudorheotropism in fishes. *J. Exp. Biol.* **14**, 95–103.

Gray, J. (1953). The locomotion of fishes. *Essays Mar. Biol.* pp. 1–16.

Gray, J. (1957). How fishes swim. *Sci. Am.* **192**, 48–54.

Green, D. M. (1964). A comparison of stamina of brook trout from wild and domestic parents. *Trans. Am. Fish. Soc.* **93**, 96–100.

Greene, C. W. (1926). The physiology of the spawning migration. *Physiol. Rev.* **6**, 201–241.

Greer Walker, M. (1970). Growth and development of the skeletal muscle fibres of the cod (*Gadus morhua* L.) *J. Cons., Cons. Perm. Int. Explor. Mer* **33**, 228–244.

Greer Walker, M. (1971). Effect of starvation and exercise on the skeletal muscle fibres of the cod (*Gadus morhua* L.) and the coal fish (*Gadus virens* L.) respectively. *J. Cons., Cons. Perm. Int. Explor. Mer* **33**, 421–427.

Greer Walker, M., and Pull, G. (1973). Skeletal muscle function and sustained swimming speeds in the coal fish (*Gadus virens* L.). *Comp. Biochem. Physiol. A* **44**, 495–501.

Greer Walker, M., Mitson, R. B., and Storeton-West, T. (1971). Trials with a transponding acoustic tag tracked with an electronic sector scanning sonar. *Nature (London)* **229**, 196–198.

Griffiths, J. S., and Alderdice, D. F. (1972). Effects of acclimation and acute temperature experience on the swimming speed of juvenile coho salmon. *J. Fish. Res. Board Can.* **29**, 251–264.

Hammond, B. R., and Hickman, C. P., Jr. (1966). The effect of physical conditioning on the metabolism of lactate, phosphate, and glucose in rainbow trout, *Salmo gairdneri. J. Fish. Res. Board Can.* **23**, 65–83.

Harden–Jones, F. R. H. (1962). Further observations on the movements of herring (*Clupea harengus* L.) shoals in relation to the tidal current. *J. Cons., Cons. Perm. Int. Explor. Mer* **27**, 52–76.

Harden-Jones, F. R. H. (1968). "Fish Migration," 325 pp. Arnold, London.

Harden-Jones, F. R. H., and McCartney, B. S. (1962). The use of electronic sector-scanning sonar for following the movements of fish shoals: Sea trials on R.R.S. Discovery II. *J. Cons., Cons. Perm. Int. Explor. Mer* **27**, 141–149.

Hasler, A. D., Horrall, R. M., Wisby, W. J., and Braemer, W. (1958). Sun-orientation and homing in fishes. *Limnol. Oceanogr.* **3**, 353–361.

Hasler, A. D., Gardella, E. S., Horrall, R. M., and Henderson, H. F. (1969). Open-water orientation of white bass, *Roccus chrysops*, as determined by ultrasonic tracking methods. *J. Fish. Res. Board Can.* **26**, 2173–2192.

Hawkins, A. D., MacLennan, D. N., Urquhart, G. G., and Robb, C. (1974). Tracking cod *Gadus morhus* L. in a Scottish sea. *Loch J. Fish Biol.* **6**, 225–236.

Heath, A. G., and Pritchard, A. W. (1962). Changes in the metabolic rate and blood lactic acid of bluegill sunfish, *Lepomis macrochirus*, Raf., following severe muscular activity. *Physiol. Zool.* **35**, 323–329.

Henderson, H. F., Hasler, A. D., and Chipman, G. C. (1966). An ultrasonic transmitter for use in studies of movements of fishes. *Trans. Am. Fish. Soc.* **95**, 350–356.

Heusner, A. A., and Enright, J. T. (1966). Long-term activity recording in small aquatic animals. *Science* **154**, 532–538.

High, W. L., and Lusz, L. D. (1966). Underwater observations on fish in an off-bottom trawl. *J. Fish. Res. Board Can.* **23**, 153–154.

Hill, A. V. (1950). The dimensions of animals and their muscle dynamics. *Sci. Prog. (London)* **38**, 209–230.

Hjort, J. (1914). Fluctuations in the great fisheries of Northern Europe. *Rapp. P.-V. Reun., Cons. Explor. Mer* **20**, 1–228.

Hochachka, P. W. (1961). The effect of physical training on oxygen debt and glycogen reserves in trout. *Can J. Zool.* **39**, 767–776.

Hocutt, C. H. (1973). Swimming performance of three warmwater fishes exposed to a rapid temperature change. *Chesapeake Sci.* **14**, 11–16.

Houde, E. D. (1969). Sustained swimming ability of larvae of walleye (*Stizostedion vitreum vitreum*) and yellow perch (*Perca flavescens*). *J. Fish. Res. Board Can.* **26**, 1647–1659.

Howard, T. E. (1975). Swimming performance of juvenile coho salmon (*Oncorhynchus kisutch*) exposed to bleached kraft pulpmill effluent. *J. Fish. Res. Board Can.* **32**, 789–793.

Hunter, J. R. (1971). Sustained speed of jack mackerel *Trachurus symmetricus*. *U.S. Fish. Wildl. Serv., Fish. Bull.* **69**, 267–271.

Hunter, J. R., and Zweifel, J. R. (1971). Swimming speed, tail beat frequency, tail beat amplitude, and size in jack mackerel, *Trachurus symmetricus*, and other fishes. *U.S. Fish Wildl. Serv., Fish. Bull.* **69**, 253–266.

Huntsman, A. G. (1942). Return of a marked salmon from a distant place. *Science* **95**, 381–382.

Ichihara, T., Soma, M., Yoshida, K., and Suzuki, C. (1972). An ultrasonic device in biotelemetry and its application to tracking a yellowtail. *Bull. Far Seas Fish. Res. Lab.* **7**, 27–49.

Idler, D. R., and Tsuyuki, H. (1958). Biochemical studies on sockeye salmon during spawning migration. I. Physical measurement, plasma, cholesterol, and electrolyte levels. *Can. J. Biochem. Physiol.* **36**, 783–791.

Javaid, M. Y. (1973). Effect of DDT on temperature selection in some salmonids. *Pak. J. Sci. Ind. Res.* **15**, 171–176.

Job, S. V. (1955). The oxygen consumption of *Salvelinus fontinalis*. *Univ. Toronto Studies Biol. Ser.* 61, pp. 1–39.

Johnson, J. (1960). Sonic tracking of adult salmon at Bonneville Dam, 1957. *U.S. Fish. Wildl. Serv., Fish. Bull.* **176**, 471–485.

Johnston, I. A., and Goldspink, G. (1973). A study of glycogen and lactate in the myotomal muscles and liver of the coalfish (*Gadus virens* L.) during sustained swimming. *J. Mar. Biol. Assoc. U.K.* **53**, 17–26.

Jones, D. R. (1971). The effect of hypoxia and anaemia on the swimming performance of rainbow trout (*Salmo gairdneri*). *J. Exp. Biol.* **55**, 541–551.

Jones, D. R., Kiceniuk, J. W., and Bamford, O. S. (1974). Evaluation of the swimming performance of several fish species from Mackenzie River. *J. Fish. Res. Board Can.* **31**,1641–1647.

Jones, F. R. H. (1957). Movements of herring shoals in relation to the tidal currents. *J. Cons., Cons. Perm. Int. Explor. Mer* **22**, 322–328.

Junge, C. O., Jr., and Phinney, L. A. (1963). Factors influencing the return of fall chinook salmon (*Oncorhynchus tshawytscha*) to Spring Creek Hatchery. *U.S. Fish Wildl. Ser., Spec. Sci. Rep.—Fish.* No. 445, pp. 1–32.

Katz, M., Pritchard, A., and Warren, C. E. (1959). Ability of some salmonids and a centrarchid to swim in water of reduced oxygen content. *Trans. Am. Fish. Soc.* **88**, 88–95.

Kerr, J. E. (1953). Studies of fish preservation at the Contra Costa steam plant of the Pacific Gas and Electric Company. *Calif. Dep. Fish Game Fish, Bull.* **92**, 1–66.

Kishinouye, K. (1923). Contributions to the comparative study of the so-called scombroid fishes. *J. Coll. Agric., Imp. Univ. Tokyo* **8**, 293–475.

Kleerekoper, H. A., Timms, M., Westlake, G. F., Davy, F. B., Malar, T., and Anderson, V. M. (1969). Inertial guidance system in the orientation of the goldfish (*Carassius auratus*). *Nature (London)* **223**, 501–502.

Kleerekoper, H. A., Timms, M., Westlake, G. F., Davy, F. B., Malar, T., and Anderson, V. M. (1970). An analysis of locomotor behaviour of goldfish (*Carassius auratus*). *Anim. Behav.* **18**, 317–330.

Klein, W. D., Olsen, O. W., and Bowden, D. C. (1969). Effects of intestinal fluke, *Crepidostomum farionis*, on rainbow trout, *Salmo gairdneri*. *Trans. Am. Fish. Soc.* **98**, 1–6.

Komarov, V. T. (1971). Speeds of fish movement. *Zool. Herald* **4**, 67–71. [Transl. by Fish. Res. Board Can., Transl. Ser. No. 2030 (1972).]

Konstantinov, K. G. (1958). Diurnal vertical migrations of the cod and haddock. *Tr. VNIRO* **36**, 62–82.

Krueger, H. M., Liu, S. D., Chapman, G. A., and Chang, J. T. (1966). Effects of pentachlorophenol on the fish *Cichlasoma bimaculatum*. *Int. Pharmacol. Congr., 3rd, Sao Paulo*, Abstr. No. 649.

Krueger, H. M., Sadler, J. B., Chapman, G. A., Tinsley, I. J., and Lowry, R. R., (1968). Bioenergetics, exercise, and fatty acids of fish. *Am. Zool.* **8**, 119–129.

Kutty, M. N. (1968). Influence of ambient oxygen on the swimming performance of goldfish and rainbow trout. *Can J. Zool.* **46**, 647–653.

Kutty, M. N. (1969). Oxygen consumption in the mullet (*Liza macrolepis*) with special reference to swimming velocity. *Mar. Biol.* **4**, 239–242.

Kutty, M. N., and Peer Mohamed, M. (1975). Metabolic adaptations of mullet *Rhinomugil corsula* (Hamilton) with special reference to energy utilization. *Aquaculture* **5**, 253–270.

Kutty, M. N., and Saunders, R. L. (1973). Swimming performance of young Atlantic salmon (*Salmo salar*) as affected by reduced ambient oxygen concentration. *J. Fish. Res. Board Can.* **30**, 223–227.

Lane, F. W. (1941). How fast do fish swim? *Country Life, London* 534–535.

Langford, T. E. (1974). Trials with ultrasonic tags for the study of coarse fish behaviour and movements around power stations outfalls. *J. Inst. Fish. Mgmt.* **5**, 61–62.

Larimore, R. W., and Duever, M. J. (1968). Effects of temperature acclimation on the swimming ability of smallmouth bass fry. *Trans. Am. Fish. Soc.* **97**, 175–184.

Leggett, W. C., and Jones, R. A. (1973). A study of the rate and pattern of shad migration in the Connecticut River–utilizing sonic tracking apparatus. *U.S. Natl. Mar. Fish. Serv., Rep.* AFG 5, pp. 1–118.

Lemke, A. E., and Mount, D. I. (1963). Some effects of alkyl benzene sulfonate on the bluegill, *Lepomis machrochirus*. *Trans. Am. Fish. Soc.* **92**, 372–378.

Lillelünd, K. (1967). Versuche zur Erbrütung der Eier vom Hecht, *Esox lucius* L., in Abhängigkeit von Temperatur und Licht. *Arch. Fischereiwiss.* **17**, 95–120.

Linthicum, D. S., and Carey, F. G. (1972). Regulation of brain and eye temperatures by the bluefin tuna. *Comp. Biochem. Physiol. A* **43**, 425–433.

Lühmann, M., and Mann, H. (1958). Wiederfänge markierter Elbaale vor der Küste Dänemarks. *Arch. Fischereiwiss.* **9**, 200–202.

McCleave, J. D., and Horrall, R. M. (1970). Ultrasonic tracking of homing cutthroat trout (*Salmo clarki*) in Yellowstone Lake. *J. Fish. Res. Board Can.* **27**, 715–730.

McCleave, J. D, and LaBar, G. W. (1972). Further ultrasonic tracking and tagging studies of homing cutthroat trout (*Salmo clarki*) in Yellowstone Lake. *Trans. Am. Fish. Soc.* **1**, 44–54.

McCleave, J. D., Jahn, L. A., and Brown, C. J. D. (1967). Miniature alligator clips as fish tags. *Prog. Fish-Cult.* **29**, 60–61.

McKim, J. M., Christensen, G. M., and Hunt, E. P. (1970). Changes in the blood of brook trout *Salvelinus fontinalis* after short-term and long-term exposure to copper. *J. Fish. Res. Board Can.* **27**, 1883–1889.

MacLeod, J. C. (1967). A new apparatus for measuring maximum swimming speeds of small fish. *J. Fish. Res. Board Can.* **24**, 1241–1252.

MacLeod, J. C., and Smith, L. L., Jr. (1966). Effect of pulpwood fiber on oxygen consumption and swimming endurance of the fathead minnow, *Pimephales promelas*. *Trans. Am. Fish. Soc.* **95**, 71–84.

Madison, D. M., Horrall, R. M., Stasko, A. B., and Hasler, A. D. (1972). Migrating movements of adult sockeye salmon (*Oncorhynchus nerka*) in coastal British Columbia as revealed by ultrasonic tracking. *J. Fish. Res. Board Can.* **29**, 1025–1033.

Määr, A. (1947). Über die Aalwanderung im Baltischen Meer auf Grund der Meerbusen Wanderaalmarkierungsversuche im Finnischen und Livischen Meerbusen in den Jahren 1937–1939. *Meed. Statens Unders.- O. Foers. Anst. Soetvatt Fiskeridir.* **27**, 1–56.

Magnan, A. (1929). Les caracteristiques geometriques et physiques des poissons. *Ann. Sci. Nat., Zool.* 13, 355–489.

Magnuson, J. J. (1967). Swimming activity of the scombroid fish *Euthynnus affinis* as related to search for food. *FAO Conf. Fish Behav. Relation Fish. Tech. Tactics, Bergen, Norway* Exp. Pap. No. 3, pp. 1–13.

Magnuson, J. J. (1970). Hydrostatic equilibrium of *Euthynnus affinis*, a pelagic teleost without a gas bladder. *Copeia* pp. 56–85.

Magnuson, J. J. (1973). Comparative study of adaptations for continuous swimming and hydrostatic equilibrium of scombroid and xiphoid fishes. *U.S. Fish Wildl. Serv., Fish. Bull.* **71**, 337–356.

Magnuson, J. J., and Prescott, J. H. (1966). Courtship, locomotion, feeding and miscellaneous behaviour of Pacific bonito (*Sarda chiliensis*). *Anim. Behav.* **14**, 54–67.

Malinin, L. K. (1971). Behaviour of burbot. *Piroda* **No. 8**, 77–79. (Transl. Ser. No. 2171. *J. Fish. Res. Board Can.*, 1972, 8 pp.)

Malinin, L. K. (1973). Speed of fish migration. *Fish. Ind.* **8**, 16–17. (Transl. Ser. No. 3146. Fish. Mar. Serv. Can., 1974.)

Malinin, L. K., Poddubny, A. G., and Gaiduk, V. V. (1971). Stereotypes of the Volga sturgeon behaviour in the region of the Saratov hydro-electric power station before and after the river regulation. *Zoologicheskii Zhurnal* **50**, 847–857. (Transl. Ser. No. 2155. *J. Fish. Res. Board Can.*, 1972, 27 pp.)

Mar, J. (1959). A proposed tunnel design for a fish respirometer. Tech. Memo. 59–3. Pac. Nav. Lab., D.R.B., Esquimalt, B.C.

Marty, J. J. (1959). The fundamental stages of the life cycle of Atlantic–Scandinavian herring. *U.S. Fish Wildl. Serv., Spec. Sci. Rep.—Fish.* No. 327, pp. 5–6a.

Maslov, N. A. (1960). Soviet investigations on the biology of the cod and other demersal fish in the Barents Sea. *In* "Soviet Fishery Investigations in North European Seas," pp. 185–231. VNIRO/PINRO, Moscow.

Mathur, G. B., and Shrivastava, B. D. (1970). An improved activity meter for the determination of standard metabolism in fish. *Trans. Am. Fish. Soc.* **99**, 602–603.

Meffert, D. (1968). Ultrasonic recorder for locomotor activities. *Trans. Am. Fish. Soc.* **97**, 12–17.

Midttun, L., and Nakken, O. (1968). Counting of fish with an echo-integrator. *Int. Counc. Explor. Sea Coop. Res. Rep. Ser. B* **17**, 1–7.

Miller, R. B. (1957). Permanence and size of home territory in stream-dwelling cutthroat trout. *J. Fish Res. Board Can.* **14**, 687–691.

Milne, D. J., and Ball, E. A. R. (1956). The mortality of small salmon when caught by trolling and tagged or released untagged. *Fish. Res. Board Can., Pac. Prog. Rep.* No. 106, pp. 10–13.

Milne, D. J., and Ball, E. A. R. (1958). The tagging of spring and coho salmon in the Strait of Georgia in 1956. *Fish. Res. Board Can., Pac. Prog. Rep.* No. 111, pp. 14–18.

Molnár, G., and Tölg, I. (1962). Relation between water temperature and gastric digestion of largemouth bass (*Micropterus salmoides* Lacépède). *J. Fish. Res. Board Can.* **19**, 1005–1012.

Muir, B. S., Nelson, G. J., and Bridges, K. W. (1965). A method for measuring swimming speed in oxygen consumption studies on the aholehole, *Kuhlia sandvicensis*. *Trans. Am. Fish. Soc.* **94**, 378–382.

Nakatani, R. E. (1957). Changes in the inorganic phosphate and lactate levels in blood plasma and muscle tissue of adult steelhead trout after strenuous swimming. *Univ. Wash., Sch. Fish. Tech. Rep.* 30, pp. 1–14.

Niimi, A. J., and Beamish, F. W. H. (1974). Bioenergetics and growth of largemouth bass (*Micropterus salmoides*) in relation to body weight and temperature. *Can. J. Zool.* **52**, 447–456.

Nishimura, M. (1963). Investigation of tuna behaviour by fish finder. *FAO Fish. Rep.* **3**, 1113–1123.

Novotny, A. J., and Esterberg, G. F. (1962). A 132–kilocycle sonic fish tag. *Prog. Fish-Cult.* **24**, 139–141.

Ohlmer, W., and Schwartzkopff, J. (1959). Schwimmgeschwindigkeiten von Fischen aus stehenden Binnengewässern. *Naturwissenschaften* **10**, 362–363.

Olla, B. L., Bejda, A. J., and Martin, A. D. (1974). Daily activity, movements, feeding and seasonal occurrence in the tautog, *Tautoga onitis*. *Fish. Bull.* **72**, 27–35.

Oseid, D., and Smith, L. L., Jr. (1972). Swimming endurance and resistance to copper and malathion of bluegills treated by long-term exposure to sublethal levels of hydrogen sulfide. *Trans. Am. Fish. Soc.* **101**, 620–625.

Otto, R. G., and Rice, J. O'H. (1974). Swimming speeds of yellow perch (*Perca Flavescens*) following an abrupt change in environmental temperature. *J. Fish. Res. Board Can.* **31**, 1731–1734.

Parker, R. R., and Black, E. C. (1959). Muscular fatigue and mortality in troll-caught chinook salmon (*Oncorhynchus tshawytscha*). *J. Fish. Res. Board Can.* **16**, 95–106.

Parker, R. R., Black, E. C., and Larkin, P. A. (1959). Fatigue and mortality in troll-caught Pacific salmon (*Oncorhynchus*). *J. Fish. Res. Board Can.* **16**, 429–448.

Parrish, B. B., Blaxter, J. H. S., and Hall, W. B. (1964). Diurnal variations in size and composition of trawl catches. *Rapp. P.-V. Reun., Cons. Int. Explor. Mer* **155**, 27–34.

Parry, G. (1958). Size and osmoregulation in salmonid fishes. *Nature* (*London*) **181**, 1218–1219.

Paulik, G. J., and DeLacy, A. C. (1957). Swimming abilities of upstream migrant silver salmon, sockeye salmon and steelhead at several water velocities. *Univ. Wash., Sch. Fish., Tech. Rep.* No. 44, pp. 1–40.

Paulik, G. J., and DeLacy, A. C. (1958). Changes in the swimming ability of Columbia River sockeye salmon during upstream migration. *Univ. Wash., Sch. Fish., Tech. Rep.* No. 46, pp. 1–67.

Paulik, G. J., DeLacy, A. C., and Stacy, E. F. (1957). The effect of rest on the swimming performance of fatigued adult silver salmon. *Univ. Wash., Sch. Fish., Tech. Rep.* No. 31, pp. 1–24.

Pavlov, D. S., Sbikin, Y. N., and Mochek, A. D. (1968). The effect of illumination in running water on the speed of fishes in relation to features of their orientation. *Vopr. Ikhtiol.* **8**, 318–324. (Transl. in *Probl. Ichthyol.* Am. Fish. Soc., Washington, D.C. Vol. 8, pp. 250–255, 1968.)

Peterson, R. H. (1974). Influence of fenitrothion on swimming velocity of brook trout (*Salvelinus fontinalis*). *J. Fish. Res. Board Can.* **31**, 1757–1762.

Poddubny, A. G. (1967). Sonic tags and floats as a means of studying fish response to natural environmental changes and to fishing gear. *FAO Conf. Fish Behav. Relation Fish. Tech. Tactics, Bergen, Norway* Exp. Pap. No.46, pp. 1–8.

Poddubny, A. G., Malinin, L. K., and Gaiduk, V. V. (1970;. Experiment in telemetric observations under ice of the behaviour of winter fish. *Biologiya Unutrehnikh. Vod.* **6**, 65–70. (Transl. Ser. No. 1817. *J. Fish. Res. Board Can.*, 1971.)

Pope, A., and Harper, J. J. (1966). "Low-Speed Wind Tunnel Testing." Wiley, New York.

Potts, W. T. W. (1954). The energetics of osmotic regulation in brackish and freshwater animals. *J. Exp. Biol.* **31**, 618–630.

Pritchard, A. L. (1944). Return of two marked pink salmon (*Oncorhynchus gorbuscha*) to the natal stream from distant places in the sea. *Copeia* pp. 80–82.

Pritchard, A. W., Hunter, J. R., and Lasker, R. (1971). The relation between exercise and biochemical changes in red and white muscle and liver in the jack mackerel, *Trachurus symmetricus*. *Fish. Bull.* **69**, 379–386.

Prosser, C. L., Barr, L. M., Pinc, R. D., and Lauer, C. Y. (1957). Acclimation of goldfish to low concentrations of oxygen. *Physiol. Zool.* **30**, 137–141.

Radakov, D. V. (1964). Velocities of fish swimming (in Russian). *Pamphlet A.N. Severtsov Inst. Anim. Morphol., Moscow* pp. 4–28.

Radakov, D. V., and Solovyev, B. S. (1959). First attempts to employ a submarine for observing the behaviour of herring. *Rybn. Khoz.* (*Moscow*) **35**, 16–21. (Fish. Res. Board Can., Transl. Ser. No. 338.)

Randall, D. J. (1970). Gas exhange in fish. *In* "Fish Physiology" (W. S. Hoar and D. J. Randall, eds.), Vol. 4, pp. 253–291. Academic Press. New York.

Rao, G. M. M. (1968). Oxygen consumption of rainbow trout (*Salmo gairdneri*) in relation to activity and salinity. *Can. J. Zool.* **46**, 781–786.

Rao, G. M. M. (1971). Influence of activity and salinity on the weight-dependent oxygen consumption of the rainbow trout *Salmo gairdneri*. *Mar. Biol.* **8**, 205–212.

Rasmussen, B. (1959). On the migration pattern of the West Greenland stock of cod. *Ann. Biol.* **4**, 123–124.

Regnard, M. P. (1893). Sur un dispositif qui permet de mesurer la vitesse de translation d'un poisson se mouvant dans l'eau. *C.R. Soc. Biol.* **5**, 81–83.

Reimers, N. (1956). Trout stamina. *Prog. Fish-Cult.* **18**, 112.

Richardson, I. D., Cushing, D. H., Harden-Jones, F. R., Beverton, R. J. H., and Blacker, R. W. (1959). Echo sounding experiment in the Barents Sea. *Fish. Invest. Minist. Agric., Fish. Food (G.B.) Ser. II Salmon Freshwater Fish.* **22**(9), 7–16.

Ricker, W. E. (1975). Computation and interpretation of biological statistics of fish populations. *Bull., Fish. Res. Board Can.* **191**, 1–382.

Roberts, J. L. (1966). Systematic versus cellular acclimation to temperature by poikilotherms. *Helgol. Wiss. Meeresunters.* **14**, 451–465.

Root, R. W. (1931). The respiratory function of the blood of marine fishes. *Biol. Bull.* (*Woods Hole, Mass.*) **61**, 427–456.

Royce, W. F., Smith, L. S., and Hartt, A. C. (1968). Models of oceanic migrations of Pacific salmon and comments on guidance mechanisms. *U.S. Fish Wildl. Serv., Fish. Bull.* **66**, 441–462.

Ruhland, M. L., and Heusner, A. (1959). Chambre respiratoire pour la détermination simultané de l'activité et de la comsommation d'oxygene par une méthode manométrique chez des poissons de 3-10g. *C.R. Soc. Biol.* **153**, 161–164.

Ryland, J. S. (1963). The swimming speeds of plaice larvae. *J. Exp. Biol.* **40**, 285–299.

Saetersdal, G. (1956). Fisheries research in northern waters—study items and results. Dep. Rep. No. 1. Deep Sea Research Inst., Fish. Directorate, Bergen. (Transl. by Cent. Off. Inf., London.)

Saunders, R. L., and Sutterlin, A. M. (1971). Cardiac and respiratory responses to hypoxia in the sea raven, *Hemitripterus americanus*, and an investigation of possible control mechanisms. *J. Fish. Res. Board Can.* **28**, 491–503.

Schärfe, J. (1960). A new method for aimed one-boat trawling in midwater and on the bottom. *Stud. Rev. Gen. Fish. Counc. Mediterr.* **13**, 1–38.

Schmidt-Nielsen, K. (1972). Locomotion: Energy cost of swimming, flying, and running. *Science* **177**, 222–228.

Schuyf, A., and DeGroot, S. J. (1971). An indictive locomotion detector for use in diurnal activity experiments in fish. *J. Cons., Cons. Perm. Int. Explor. Mer* **34**, 126–131.

Shaw, E. (1960). The development of schooling behaviour in fishes. *Physiol. Zool.* **33**, 79–86.

Shazkina, E. P. (1972a). Energy metabolism and food rations of steelhead trout under conditions of the Chernaya Rachka trout industry. *Tr. USES Nauchno. Issled. Inst. Morsk. Rubn. Khoz. Okeanogr.* **76**, 130–134. (Transl. for *Ref. Zh. Biol.*, No. 61203, pp. 138–144, 1972.)

Shazkina, E. P. (1972b). Active metabolism in Azov goby. *Proc. All-Union Res. Inst. Mar. Fish. Oceanogr.* **85**, 138–144. (Transl. Ser. No. 3040. Fish. Mar. Serv. Can., 1974.)

Shuck, H. A., and Kingsbury, O. R. (1948). Survival and growth of fingerling brown trout (*Salmo fario*) reared under different hatchery conditions and planted in fast and slow water. *Trans. Am. Fish. Soc.* **75**, 147–156.

Smit, H. (1965). Some experiments of the oxygen consumption of goldfish (*Carassius auratus* L.) in relation to swimming speed. *Can. J. Zool.* **43**, 623–633.

Smit, H., Amelink-Koutstaal, J. M., Vijverberg, J., and von Vaupel-Klein, J. C. (1971). Oxygen consumption and efficiency of swimming goldfish. *Comp. Biochem. Physiol.* A **39**, 1–28.

Smith, H. D., and Margolis, L. (1970). Some effects of *Eubothrium salvelini* (Schrank, 1790) on sockeye salmon, *Oncorhynchus nerka* (Walbaum), in Babine Lake, British Columbia. *J. Parasitol.* **56**, Sect. II, 321–322.

Smith, L. S., and Newcomb, T. W. (1970). A modified version of the Blazka respirameter and exercise chamber for large fish. *J. Fish. Res. Board Can.* **27**, 1321–1324.

Spoor, W. A. (1946). A quantitative study of the relationship between the activity and oxygen consumption of the goldfish, and its application to the measurement of respiratory metabolism in fishes. *Biol. Bull.* (*Woods Hole, Mass.*) **91**, 312–325.

Spoor, W. A., and Drummond, R. A. (1972). An electrode for detecting movement in gradient tanks. *Trans. Am. Fish. Soc.* **101**, 714–715.

Spoor, W. A., Neiheisel, T. W., and Drummond, R. A. (1971). An electrode chamber for recording respiratory and other movements of free-swimming animals. *Trans. Am. Fish. Soc.* **100**, 22–28.

Sprague, J. B. (1971). Measurement of pollutant toxicity of fish. III. Sublethal effects and "safe" concentrations. *Water Res.* **5**, 245–266.

Stasko, A. B. (1975). Progress of migrating Atlantic salmon (*Salmo salar*) along an estuary observed by ultrasonic tracking. *J. Fish Biol.* **7**, 329–338.

Stasko, A. B., and Rommel, S. A., Jr. (1974). Swimming depth of adult American eels (*Anguilla rostrata*) in a saltwater bay as determined by ultrasonic tracking. *J. Fish. Res. Board Can.* **31**, 1148–1150.

Stasko, A. B., Horrall, R. M., Hasler, A. D., and Stasko, D. (1973). Coastal movements of mature Fraser River pink salmon (*Oncorhynchus gorbuscha*) as revealed by ultrasonic tracking. *J. Fish. Res. Board Can.* **30**, 1309–1316.

Stevens, E. D., and Fry, F. E. J. (1971). Brain and muscle temperatures in ocean caught and captive skipjack tuna. *Comp. Biochem. Physiol.* **A38**, 203–211.

Stevens, E. D., Lam, H. N., and Kendall, J. (1974). Vascular anatomy of the counter-current heat exchanger in skipjack tuna. *J. Exp. Biol.* **61**, 145–153.

Stringham, E. (1924). The maximum speed of freshwater fishes. *Am. Nat.* **58**, 156–161.

Sundnes, G. (1963). Swimming speed of fish as a factor in gear research. *Fiskeridir. Skr., Ser. Havunders.* **13**, 126–132.

Tesch, F. W. (1974). Speed and direction of silver and yellow eels, Anguilla anguilla, released and tracked in the open North Sea. *Ber. Dtsch. Wiss. Komm. Meeresforsch.* **23**, 181–197.

Thomas, A. E., Burrows, R. E., and Chenoweth, H. H. (1964). A device for stamina measurement of fingerling salmonides. *Bur. Sport Fish. Wildl. (U.S.), Res. Rep.* No. 67, pp. 1–15.

Thompson, D. W. (1917). "On Growth and Form," 793 pp. Cambridge Univ. Press, London.

Thorson, T. B. (1971). Movement of bull sharks, *Carcharhinus leucas* between Caribbean Sea and Lake Nicaragua demonstrated by tagging. *Copeia* No. 2, pp. 336–338.

Threadgold, L. T., and Houston, A. H. (1964). An electron microscope study of the "chloride cell" of *Salmo salar* L. *Exp. Cell Res.* **34**, 1–23.

Trefethen, P. S. (1956). Sonic equipment for tracking individual fish. *U.S. Fish Wildl. Serv. Spec. Sci. Rep.—Fish.* **179**, 1–11.

Trout, G. C. (1957). The Bear Island cod: Migrations and movements. *Fish. Invest. Minist. Agric. Fish. Food (G.B.) Ser. II Salmon Freshwater Fish.* **21**, 1–51.

Tucker, V. A. (1970). Energetic cost of locomotion in animals. *Comp. Biochem. Physiol. A* **34**, 841–846.

Tytler, P. (1969). Relationship between oxygen consumption and swimming speed in the haddock, *Melanogrammus aeglefinus*. *Nature (London)* **221**, 274–275.

Verwey, J. (1960). Über die Orienteir ung wandern der Meerestiere. *Helgol. Wiss. Meeresunters.* **7**, 51–58.

Vibert, R. (1956). Methode pour l'étude et l'amelioration de la survie des alevins de repeuplement (triuites et saumons). *Ann. Sta. Cent. Hydrobiol. Appl.* **6**, 347–439.

Vincent, R. E. (1960). Some influences of domestication upon three stocks of brook trout (*Salvelinus fontinalis* Mitchill). *Trans. Am. Fish. Soc.* **89**, 35–52.

von Kausch, H. (1968). Der Einfluss der Spontanaktivität auf die Stoffwechselrate junger Karpfen (*Cyprinus carpio* L.) im Hunger und bei Fütterung. *Arch. Hydrobiol., Suppl.* **33**, 263–330.

Wales, J. H. (1950). Swimming speed of the western sucker *Catostomus occidentalis* Ayres. *Calif. Fish Game* **36**, 433–434.

Walters, V., and Fierstine, H. L. (1964). Measurements of swimming speeds of yellow fin tuna and wahoo. *Nature (London)* **202**, 208–209.

Wardle, C. S. (1975). Limit of fish swimming speed. *Nature (London)* **255**, 725–727.

Warren, C. E., and Davis, G. E. (1967). Laboratory studies on the feeding, bioenergetics and growth of fish. *In* "The Biological Basis for Freshwater Fish Production" (S. Gerking, ed.), pp. 175–214. Blackwell, Oxford.

Watanabe, N. (1942). Measurements on the bodily density, body temperature, and swimming velocity "katuwu", *Euthynnus vagans* (Lesson). *Nippon Suisan Gakkaishi* **11**, 146–148.

Weaver, C. R. (1963). Influence of water velocity upon orientation and performance of adult migrating salmonids. *U.S. Fish Wildl. Serv., Fish. Bull.* **63**, 97–121.

Weaver, C. R. (1965). Observations on the swimming ability of adult American shad (*Alosa sapidissima*). *Trans. Am. Fish. Soc.* **94**, 382–385.

Webb, P. W. (1971a). The swimming energetics of trout. I. Thrust and power output at cruising speeds. *J. Exp. Biol.* **55**, 489–520.

Webb, P. W. (1971b). The swimming energetics of trout. II. Oxygen consumption and swimming efficiency. *J. Exp. Biol.* **55**, 521–540.

Webb, P. W., and Brett, J. R. (1973). Effects of sublethal concentrations of sodium pentachlorophenate on growth rate, food conversion efficiency, and swimming performance in underyearling sockeye salmon (*Oncorhynchus nerka*). *J. Fish. Res. Board Can.* **30**, 499–507.

Whitworth, W. R., and Irwin, W. H. (1964). Oxygen requirements of fishes in relation to exercise. *Trans. Am. Fish. Soc.* **93**, 209–212.

Wiehs, D. (1973a). Optimal fish cruising speeds. *Nature* (*London*) **245**, 48–50.

Wiehs, D. (1973b). Hydromechanics of fish schooling. *Nature* (*London*) **241**, 290–291.

Wiehs, D. (1974). Energetic advantages of burst swimming of fish. *J. Theor. Biol.* **48**, 215–229.

Winberg, G. G. (1956). "Rate of Metabolism and Food Requirements of Fishes," 251 pp. Belorussian State Univ., Minsk (In Russian). (Fish. Res. Board Can., Transl. Ser. No. 194.)

Wohlschlag, D. E. (1957). Differences in metabolic rates of migrating and resident freshwater forms of an arctic whitefish. *Ecology* **38**, 502–510.

Wohlschlag, D. E. (1964). Respiratory metabolism and ecological characteristics of some fishes in McMurdo Sound, Antarctica. *Antarct. Res. Ser.* **1**, 33–62.

Wohlschlag, D. E., Cameron, J. N., and Cech, J. J., Jr. (1968). Seasonal changes in the respiratory metabolism of the pinfish (*Lagodon rhomboides*). *Contrib. Mar. Sci.* **13**, 89–104.

Woodhead, P. M. J. (1964). Diurnal changes in trawl catches of fishes. *Rapp. P.-V. Reun., Cons. Int. Explor. Mer* **155**, 35–44.

Yuen, H. S. H. (1966). Swimming speeds of yellowfin and skipjack tuna. *Trans. Am. Fish. Soc.* **95**, 203–209.

Yuen, H. S. H. (1970). Behaviour of skipjack tuna, *Katsuwonus pelamis*, as determined by tracking with ultrasonic devices. *J. Fish. Res. Board Can.* **27**, 2071–2079.

Yuen, H. S. H., Dizon, A. E., and Uchiyama, J. H. (1974). Notes on the tracking of the Pacific blue marlin, *Makaira nigricans*. *In* "Proceedings of the International Billfish Symposium, Kailua-Kona, Hawaii, 9–12 August 1972" (R. S. Shomura and F. Williams, eds.), Part 2, Review and Contributed Papers, pp. 265–268 (NOAA Tech. Rep. NMFS SSRF-675). NOAA, Honolulu.

Zaitsev, V. P., and Radakov, D. V. (1960). Submarines in fishery research. *In* "Soviet Fisheries Investigations in North European Seas" (J. J. Marty *et al.*, eds.), pp. 463–466. VNIRO and PINRO, Moscow.

Chapter 11

45 years of "The respiratory and circulatory systems during exercise" in *Fish Physiology*, as per David R. Jones and David J. Randall

Jodie L. Rummer*

College of Science and Engineering, James Cook University, Townsville, QLD, Australia

**Corresponding author: e-mail: jodie.rummer@jcu.edu.au*

Chapter Outline

Rummer discusses the impact of Jones and Randall's chapter "The respiratory and circulatory systems during exercise" in *Fish Physiology*, Volume VII published in 1978.

This classic contribution established a foundational framework in the field of fish physiology and the series of books of the same name and, as a result, has been pivotal in shaping research trajectories for decades. In this chapter, Jones and Randall define exercise as the quintessential, unifying stressor that is encountered by all animals and a principle that has become a focal point for extensive research across diverse taxa, habitats, populations, and environmental conditions. From a personal perspective, Jones and Randall's cornerstone work in exercise physiology in fishes has been instrumental in guiding my own research into the remarkable parallels between the athletic prowess of fish and human athletes. Indeed, it has catalyzed a career-long pursuit to understand

Fish Physiology, Vol. 40B. https://doi.org/10.1016/bs.fp.2024.05.001

the endurance, performance, and adaptations of these aquatic athletes, uncovering similarities that resonate beyond their aquatic environments and reflecting broader principles of resilience and vulnerability. This approach has not only shaped my research focus but has also helped me enhance my public engagement and deepen my understanding of the intricate dynamics at play within aquatic ecosystems.

1 What is exercise in fish?

Jones and Randall describe exercise as work performed by the locomotory muscles and emphasize that exercise leads to an increase in the rate of energy conversion above resting rates. It can range from sustained swimming against a current for long periods of time (i.e., sustained, endurance; also see Webb, 1998) to burst-type activities associated with predator avoidance or prey capture (also see Domenici and Blake, 1997). This heightened activity associated with exercise necessitates a surge in energy conversion rates to fuel not just the muscles responsible for movement but also the cardiac and respiratory systems. Fish increase the rate of gas exchange at the gills and tissues during exercise to meet the heightened oxygen demand. The heart and respiratory muscles work in tandem to augment oxygen uptake to fulfill the increased demand brought on by exercise.

In this seminal contribution, Jones and Randall focus intently on the physiological adaptations that enable fish to boost gas exchange rates at the gills and tissues, ensuring an adequate supply of oxygen to support the energetic costs of exercise. This exploration is crucial as it sheds light on the intricate balance of the respiratory and circulatory systems in meeting the metabolic demands of fish during periods of increased activity, providing insights into their overall fitness and survival. In response to the demands of exercise, fish exhibit a remarkable adaptation by significantly increasing their ventilation volume. This physiological adjustment is imperative for enhancing oxygen uptake from the water. During periods of increased activity, such as sustained swimming or escape responses, the oxygen demand of the fish's muscles escalates. To meet this heightened demand, fish increase the flow of water over their gills, thereby augmenting the volume of water ventilated. This increase in ventilation volume is not just about moving more water; rather, it is intricately tied to the efficiency of oxygen utilization. By passing more water over the gills, fish maximize the extraction of available oxygen, ensuring a steady supply to their tissues. This adaptation is particularly vital in environments where oxygen levels may be variable or limited. The increased ventilation volume, therefore, plays a dual role: it meets the immediate oxygen needs of the fish during exercise and enhances the overall efficiency of oxygen utilization, which is a critical factor in the fish's ability to sustain prolonged periods of physical activity. This adaptation underscores the fish's remarkable ability to modulate its respiratory system in response to varying metabolic demands, reflecting an intricate balance between energy expenditure and efficient oxygen uptake. Complementing this foundational perspective,

Claireaux et al. (2005) link swimming performance directly to cardiac pumping ability and cardiac anatomy in rainbow trout. Their research offers empirical evidence on how the cardiovascular system adapts to support the energetic costs of exercise, thereby deepening our understanding of the physiological mechanisms that underlie fitness and survival. This integration of cardiovascular function with swimming performance exemplifies the complex interplay of physiological systems in response to exercise in fish.

2 How is exercise performance measured and assessed in fish?

Jones and Randall refer to various methodologies used to quantify the physical exertion of fish. They also note and cite another fellow English-turned-Canadian fish physiologist, F. E. J. Fry, also with some of the most classic work in the field, still today, to clarify standard, routine, and active metabolism as terms to refer to basal metabolism, metabolism associated to a degree of random activity, and the maximum sustained metabolic rate, respectively (Fry, 1971). This would later become a topic heavily investigated among fish physiologists in terms of experimentally measuring and estimating such limits and comparing between species, conditions, etc. (e.g., see Chabot et al., 2016; Killen et al., 2017; Norin and Clark, 2016; Rees et al., 2024; Rummer et al., 2016).

Indeed, Jones and Randall discuss the use of "water tunnels," now commonly referred to as swim tunnels, which allow for the control and measurement of swimming speed as a means to simulate natural exercise conditions. Essentially, fish are placed in a controlled flow of water, and their swimming endurance and speed are measured. This can be adjusted to simulate different environmental conditions. Cardiorespiratory responses are monitored by measuring oxygen uptake or consumption rates, blood lactate concentrations, and changes in heart rate, to name a few. These physiological indicators provide insights into the metabolic rate and the capacity for oxygen delivery to the muscles during activity. In line with this, Yousaf et al. (2023) recently reviewed the use of heart rate bio-loggers via their team's studies assessing stress during critical swimming speed tests on farmed Atlantic salmon to further emphasize the importance of such methodologies in understanding the physiological impacts of exercise on fish.

An array of data can be collected from swimming protocols. For example, depending on system design, oxygen uptake rates (MO_2) can be obtained at any swimming speed (U), and the relationship between MO_2 and U can be established. This relationship can depend on life stage (e.g., across ontogeny in sockeye salmon, *Oncorhynchus nerka*; Brett, 1964) and differ by species (e.g., comparisons to rainbow trout, *Oncorhynchus mykiss*; Rao, 1968). Cost of swimming can also be determined, and this information can be combined with knowledge of energy substrates used to define total energy expenditure during swimming. That said, U can be expressed in a variety of ways. The speed at which fish are able to swim in the flume is often related to their critical swimming speed (U_{crit}), which is the maximum speed they can maintain before fatigue. U_{crit} is a commonly used measure of sustained swimming performance and defined as the highest speed a fish can maintain for a prolonged period without fatigue, usually measured over a series of increasing velocities in a swim tunnel until the fish can no longer keep pace with the water flow. U_{crit} is used to assess the aerobic swimming capacity of fish and is an indicator of their overall cardiovascular and muscular fitness. It can be influenced by various factors, including water temperature, fish size, species, and the state of acclimation to different flow regimes. In contrast, burst swimming speed (U_{burst}) is the maximum speed a fish can achieve in short, intense bursts, typically used during escape responses or predator-prey interactions. Unlike U_{crit}, U_{burst} is anaerobic and can only be sustained for a few seconds to a couple of minutes, depending on the species and conditions. U_{burst} is measured by observing the fish during these high-intensity efforts, either in a swim tunnel with a sudden increase in flow or in an open area where the fish can swim freely without constraint. Both swimming measurements provide valuable information about different aspects of fish swimming performance. While U_{crit} is more relevant to migratory behavior, endurance swimming, and the ability to cope with steady currents, which is important for understanding how fish may navigate their environment, particularly in the context of riverine habitats or steady ocean currents, U_{burst} is crucial for understanding escape responses, predator evasion, and the capture of prey, which are critical for survival in the wild. This would also become a topic (Domeneci and Blake, 1997) heavily investigated at the University of British Columbia (UBC), where Jones and Randall were based, by then PhD student, Paolo Domenici, and his supervisor, Bob Blake, who was a dear colleague of Jones and Randall.

3 Other metrics derived from exercise studies

Regardless of U_{crit} or U_{burst} protocols, so many other metrics can be derived from these studies, which would catalyze several subsequent decades of research. Moreover, Jones and Randall stress that no matter how closely we aim to simulate stream, river, and ocean conditions and force fish to swim to maximal performance, none of these conditions are exactly what fish experience in their natural habitat, and therefore it is crucial to understand the relationship between these various measures and other parameters. For example, tail beat frequency and amplitude can also be calculated these studies and provide information on the swimming kinematics, efficiency, and the type of swimming mode used (e.g., steady vs. burst swimming). With these swimming protocols, electromyography (EMG) can be added to record muscle activity patterns during swimming to understand muscle function and fatigue. Indeed, studies have focused on power output at various U to find the highest speeds at which the cost of transport is lowest. Given the relationship between oxygen uptake and speed, the efficiency of swimming increases to a point and then decreases with speed (Webb, 1971), which is important information that translates to various ecological and applied scenarios for fish. Heart rate can be monitored to assess the cardiovascular response to different swimming speeds and exercise intensities, and the rate and depth of opercular movements, which indicate the ventilation effort required to maintain aerobic metabolism during swimming, can also be measured. Metabolites may be most telling, as Jones and Randall suggest. Blood lactate concentration can be measured in cannulated fish during exercise or post-exercise to assess the level of anaerobic metabolism during high-intensity swimming. Likewise, other blood parameters, including pH, hematocrit, and concentrations of metabolic substrates and hormones, can be measured to assess physiological stress and energy utilization. By varying the water temperature in the swim flume, researchers can assess the thermal tolerance and the effect of temperature on swimming performance, which comes with its own challenges (e.g., failure temperatures, and what cardiorespiratory mechanisms are at play, see Eliason et al., 2013 and others). Finally, the time it takes for a fish to become exhausted at a given swimming speed, which can be an indicator of endurance and overall fitness, as well as post-exercise recovery rates for all the physiological parameters mentioned can provide insights into the fish's ability to recover from exertion. This approach has certainly catalyzed a comprehensive understanding of the multifaceted nature of exercise in fish, encompassing the physiological, biomechanical, and ecological dimensions of their swimming capabilities.

4 Use of water tunnels to exercise fish over the years

The first water tunnels for fish, also known as swim flumes or swimming respirometers, were developed in the 1960s. A pioneer in this field was Brett (1964), from the Fisheries Research Board of Canada in Nanaimo, British Columbia, not too far from the UBC where Jones and Randall were based. Brett's seminal work on the metabolic rates and swimming performance of salmonids in these swim tunnels laid the foundation for much of the current research on fish exercise physiology. I still remember using one of these large Brett-type swim flumes in the basement of the Department of Zoology at UBC during my PhD. While the Brett swim tunnel is typically an open-flow system, which can more closely replicate natural conditions for swimming fish, the Blazka swim tunnel—also designed in the 1960s (Blazka et al., 1960)—is a closed respirometry system, ideal for measuring oxygen uptake rates and other metabolic parameters during exercise. The Blazka swim tunnel recirculates water in a loop, creating a uniform flow and reducing the amount of water needed. However, this design may require more frequent water quality management to ensure fish health. Initially designed for smaller fish species, the Blazka tunnel allowed precise control of water flow and oxygen uptake measurements. In contrast, Brett's tunnels could be scaled up for larger fish and studies requiring longer swimming distances, though they required more space and could be more costly. The open system is easier to maintain for water quality over longer periods, beneficial for certain behavioral studies. Both types of swim tunnels have been essential in advancing fish exercise physiology, despite logistical constraints of the times. To address these challenges, Bell and Terhune (1970) provided a comprehensive guide on water tunnel design, offering detailed insights into construction and optimization for studying fish swimming behavior and physiology. This guidance overcame early logistical challenges, paving the way for evolving swim flumes that accommodate a wide range of species and experimental conditions, enhancing our understanding of fish locomotion and its ecological implications.

Since these early days, swim flumes have progressed tremendously, accommodating the tiniest vertebrates (e.g., larval coral reef fishes; Downie et al., 2023) to 100 kg sharks in a "mega-flume" (Payne et al., 2015). Technological advancements have also improved these tools, now allowing a wider range of species to be investigated under diverse conditions. High-resolution sensors and computer models enable detailed analyses of fish swimming in complex flow conditions. Additionally, 3D printing allows researchers to create custom-designed structures and components for swim tunnels and respirometry chambers, essential for small-scale species (Huang et al., 2020). Customization helps create controlled environments to study fish swimming behavior and physiology. While swim flumes offer numerous benefits in measuring physiological, morphological, and behavioral aspects of exercise, they are time-consuming and often allow only one fish to be swum at a time. Contemporary studies have compared swimming respirometry with various chase protocols as proxies for labor-intensive methods. Swim flumes simulate natural conditions, providing insights into the ecological relevance of fish swimming performance and highlighting the importance of replicating natural flow conditions to understand adaptive swimming behaviors in different species. Studies using swim flumes to explore physiological responses to environmental stressors during exercise have been pivotal in understanding the impacts of climate change on marine life (Rummer et al., 2016). These studies show swim flumes are crucial for understanding broader ecological and environmental contexts. Additionally, research has investigated energy expenditure and metabolic rates during sustained swimming, providing a comprehensive view of fish physiology and its implications for survival and fitness in the wild (Roche et al., 2013). The evolution of swim flumes from rudimentary setups to sophisticated systems equipped with advanced technology has revolutionized the study of fish physiology and behavior, enabling deeper insights into fish locomotion complexities and its ecological implications.

While the development of swim flumes has been instrumental in studying fish exercise, it is the intricate physiological responses, particularly cardiovascular and respiratory adaptations, that truly illuminate the complexities of exercise in fish. Exploring fish exercise performance via swim flumes, especially sustained swimming (Webb, 1998), and other strategies has led to a re-evaluation of how exercise is measured and interpreted. Such approaches have highlighted the importance of both aerobic and anaerobic metabolic pathways in determining exercise capacity (Rees et al., 2024) and deepened our understanding of physiological limits and fatigue mechanisms in fish (Farrell, 1997).

5 Exercise and the respiratory system

Jones and Randall discuss how the capacity for oxygen uptake in fish is influenced by various factors. The focus is on how changes in these components—including the surface area of the gas exchanger, its permeability, the distance between water and blood, and the difference in oxygen partial pressure across the exchange surface—affect oxygen uptake during exercise. The maximum oxygen uptake in fish is primarily constrained by morphometric factors, like gill surface area and diffusion distance, rather than physiological ones. There is a general relationship between these morphometric factors and metabolism, often explored in relation to body weight or mass. Metabolism follows a power function in relation to body mass. For example, in tench (*Tinca tinca*), as body weight increases, the capacity of the gas exchanger becomes a limiting factor for both resting and active metabolic needs. In contrast, for salmonids, the relationship between gill surface area and body mass is nearly linear, suggesting their gas exchanger can support resting metabolism across different sizes. However, during maximum exercise, the gas exchanger in larger salmonids struggles to meet increased metabolic demands. Jones and Randall summarize that size limitations for oxygen uptake in fish like tench are set by resting metabolism, while in more active fish like salmonids, they may be set by the demands of active metabolism. Different fish species have varying capacities to increase oxygen uptake during exercise, influenced by their morphometric and physiological characteristics.

In addition to morphometric factors, physiological adjustments play a crucial role in enhancing oxygen uptake during exercise. A key adaptation observed in fish is the significant increase in ventilation volume, a direct response to escalated oxygen demand during activities like sustained swimming or escape responses. By increasing the flow of water over their gills, fish augment the volume of water ventilated, thereby maximizing oxygen extraction. This enhanced ventilation ensures a steady oxygen supply to tissues, vital in environments where oxygen levels may be variable or limited. Hemoglobin's role in oxygen transport becomes increasingly important during heightened activity, facilitating efficient oxygen transport from gills to muscles. The increased ventilation and effective oxygen transport by hemoglobin underscore the fish's ability to modulate its respiratory system in response to varying metabolic demands. This balance between energy expenditure and efficient oxygen uptake is critical for sustaining prolonged physical activity and survival in diverse aquatic environments.

Jones and Randall highlight how Brett's (1964) research and subsequent studies with Brett and Glass (1973) made significant contributions to our understanding of exercise physiology in fish, particularly sockeye salmon. Brett discovered that both active and standard oxygen uptake in sockeye salmon were temperature-dependent, with a notable increase in metabolic rate up to 15 °C, beyond which the rate plateaued due to respiratory system limitations in supplying sufficient oxygen. Increasing the water's oxygen concentration by 50% at 20 °C significantly increased active metabolism, indicating that oxygen supply was a limiting factor. Brett and Glass also found that active oxygen uptake in sockeye salmon was largely independent of body size, contrasting with standard metabolism, which is proportional to body weight to the power of 0.78 (i.e., standard metabolism decreases relative to body weight as the fish grows, while active metabolism remains constant). This independence from body size makes active oxygen uptake a valuable and consistent measure of exercise performance in fish. These foundational findings are crucial today, informing current studies on fish physiology related to aquaculture, fisheries management, and ecological research, where understanding energy expenditure and oxygen usage under varying conditions is essential.

Research on oxygen consumption in fish during exercise highlights the complexity of energy conversion under these conditions. Oxygen uptake rates alone may not fully represent total energy expenditure, as anaerobic metabolism contributes significantly during high-intensity activities. Studies by Black et al. (1960, 1962), Connor et al. (1964), Beamish (1968), and Driedzic and Kiceniuk (1976) indicate that at moderate swimming speeds, changes in muscle glycogen and blood lactate levels are minimal, suggesting predominantly aerobic energy metabolism. However, during burst activities, there is a rapid increase in both muscle and blood lactate levels, along with a decrease in muscle glycogen, indicating a shift to anaerobic metabolism. The extent to which anaerobic metabolism contributes to the energy budget during sustained exercise remains debated. Smit et al. (1971) argued that anaerobiosis plays a significant role during sustained swimming in goldfish, while Kutty (1968), based on respiratory quotient (RQ) determinations, did not find evidence of anaerobic metabolism in goldfish and trout during sustained swimming. An initial anaerobic phase at the start of sustained swimming was suggested by a higher mean RQ in the first hour of exercise. As fish approach their critical velocity, anaerobic energy contribution increases, leading to a cumulative oxygen debt, described by Brett (1964). This debt is repaid during the post-exercise recovery period. In tests where velocity is incrementally increased, anaerobic metabolism is present immediately following each increment and throughout the exercise period at speeds close to the fish's critical velocity. This understanding is crucial in current research, providing insights into physiological responses during different types of exercise, aiding in designing appropriate exercise regimes in aquaculture, understanding fish endurance limits in natural settings, and assessing the impact of environmental stressors on fish metabolism. The balance between aerobic and anaerobic metabolism influences fish behavior, growth, and survival.

6 Swimming muscles

The application of electromyography (EMG) in fish exercise physiology, as detailed in the chapter, has been a pivotal development. EMG allows for the direct measurement of muscle activity, providing insights into the muscular workload during various forms of exercise (Rome et al., 1990). This technique has been instrumental in understanding the balance between energy expenditure and locomotor activity, offering a window into the metabolic strategies fish employ during sustained physical activity (Jayne and Lauder, 1995).

Fatigue in fish during sustained exercise is influenced by various factors, including muscle glycogen depletion and metabolic by-product accumulation (Brett, 1964). The mechanisms of fatigue are complex and can vary significantly among species and types of activity. Examining respiratory adjustments during exercise has shed light on the intricate mechanisms fish use to meet increased oxygen demands. The role of hemoglobin in oxygen transport and the adjustments in gill ventilation have been key areas of study (Perry and Gilmour, 2006). These insights have been crucial in understanding how fish respond to environmental challenges such as hypoxia (Randall and Daxboeck, 1984). Modern research, employing advanced imaging and molecular techniques, provides in-depth understanding of the biochemical pathways leading to fatigue in fish. These methods enable the observation of real-time changes in muscle tissues and blood chemistry during exercise. Investigating genetic factors that influence fatigue resistance is particularly promising, as it could identify specific genes or molecular pathways to enhance performance in aquaculture species. Additionally, understanding these mechanisms is vital for conservation, especially for migratory species. For example, transcriptomics studies have been instrumental in revealing exercise-induced changes in muscle tissue gene expression, offering new perspectives on fish physiology (Palstra and Planas, 2011).

7 Cardiovascular dynamics in fish during exercise

In addition to the respiratory adaptations, the cardiovascular system plays a crucial role in supporting exercise in fish. As Jones and Randall highlighted, arterial blood pressure and total peripheral resistance (TPR) undergo significant changes during exercise. This is echoed in the findings of Johansen et al. (1966), who observed that arterial blood pressure increases in teleosts during exercise, with ventral aortic pressure rising markedly in the early part of a bout of activity. Kiceniuk and Jones (1977) further elaborated on these changes, noting that the magnitude of the increase in mean arterial blood pressure is consistent, regardless of whether the fish starts swimming from rest or during an incremental velocity test. This suggests that the cardiovascular response may be more related to disturbance and cardiac overcompensation than to exercise stress, *per se*. The role of venous pressure and venous return during exercise is equally critical. The increase in venous pressure, particularly in the subintestinal vein during moderate swimming speeds, as reported by Stevens and Randall (1967), indicates hepatic venomotor activity and a potential redistribution of blood away from visceral regions during exercise. Kiceniuk and Jones (1977) observed that while dorsal aortic pressure changes are smaller than ventral aortic pressure changes, they still play a significant role in the overall cardiovascular response to exercise. These cardiovascular adjustments, including the changes in arterial and venous pressures and the differential responses in the gill and systemic circulations, highlight the complex interplay of physiological systems during exercise in fish. The ability to maintain arterial blood pressure against varying levels of TPR, as well as the efficient management of venous return, underscores the sophistication of the fish cardiovascular system in responding to the demands of exercise.

8 Exercise and the circulatory system

Having explored the specific dynamics of arterial and venous pressures during exercise, it becomes clear that the circulatory system in fish, encompassing heart function and blood flow, plays a pivotal role in supporting sustained exercise. Key adjustments during exercise include changes in heart rate, stroke volume, and blood pressure, which are essential for meeting the increased metabolic demands. Beyond these adjustments, the circulatory system is intricately involved in the transport of oxygen from the gills to the tissues. This process is elegantly described by the Fick equation, which has become a staple in fish physiology and physiological research in general, serving as a foundational concept for students and researchers alike, spanning those just beginning their research careers to seasoned scientists. The Fick equation states that the rate of oxygen consumption is the product of cardiac output and the difference in oxygen content between arterial and venous blood.

$$VO_2 = Q(CaO_2 - CvO_2)$$

where:

- VO_2 is the rate of oxygen consumption (uptake).
- Q is the cardiac output, which is the volume of blood pumped by the heart per minute.
- CaO_2 is the arterial oxygen content, which is the amount of oxygen carried in the blood leaving the heart and going to the tissues.

- CvO_2 is the venous oxygen content, which is the amount of oxygen remaining in the blood after it has passed through the tissues.

This equation not only highlights the critical role of the cardiovascular system in ensuring efficient oxygen delivery to meet the heightened metabolic demands during exercise but also serves as a cornerstone in the education and research of fish physiology. Indeed, focusing on the circulatory system's response to exercise has led to a deeper understanding of cardiovascular physiology in fish.

The discovery of the heart's remarkable plasticity and its ability to adapt to varying exercise demands represents a significant advancement in the field (Anttila et al., 2014). This research has shed light on how fish maintain physiological balance during periods of increased activity and stress (Steinhausen et al., 2008). Modern research, leveraging telemetry and biologging technologies, has enabled the study of these circulatory adjustments in natural environments, providing invaluable insights into how fish respond to various environmental stressors (Farrell et al., 2009). Looking ahead, future research could pivot toward assessing the impact of environmental stressors on circulatory health. This focus is crucial for conservation efforts and for predicting the resilience of fish populations in the face of ecological changes (McKenzie et al., 2007). Such studies are not only pivotal for ecological and conservation strategies but also offer broader implications for understanding vertebrate physiology under environmental stress.

9 Integrating advanced technologies in fish physiology research

The integration of molecular biology and telemetry in studying fish in natural environments is paving the way for more comprehensive ecological and conservation strategies. For example, transcriptomics and genomics have allowed for a more detailed understanding of the genetic and cellular mechanisms underlying fish physiology (Gracey et al., 2004). The integration of telemetry in studying fish in natural environments is paving the way for more comprehensive ecological and conservation strategies (Cooke et al., 2004). Advancements in fish physiology research not only pave the way for future academic studies but also hold significant practical applications. This evolving body of work is essential in shaping sustainable fisheries management and enhancing aquaculture practices, while simultaneously deepening our understanding of the broader ecological impacts stemming from physiological adaptations in fish (Brett, 1995; Cooke et al., 2013, 2016).

10 Concluding thoughts

Reflecting on the past 45 years since the seminal work of Jones and Randall on "The respiratory and circulatory systems during exercise" in *Fish Physiology*, it is evident that this field has undergone significant evolution and expansion. Their foundational framework has not only advanced our understanding of fish physiology but also set the stage for future research directions. Key advancements in the field have been driven by the development of sophisticated methodologies, such as swim flumes, and the integration of molecular biology and telemetry. These tools have enabled precise measurements of fish exercise performance and provided deeper insights into the physiological responses of fish to various stressors. The exploration of cardiovascular and respiratory adaptations during exercise has been particularly illuminating, revealing the remarkable plasticity of these systems in fish.

The research has shown that fish, much like human athletes, undergo complex physiological changes to meet the demands of exercise. These changes, including increased ventilation volume and adjustments in heart rate and blood pressure, are crucial for efficient oxygen delivery and utilization. The role of hemoglobin in oxygen transport and the intricate balance between aerobic and anaerobic metabolism have been key areas of study, offering insights into the endurance and survival strategies of fish.

Looking ahead, the field of fish exercise physiology is poised to address new challenges and questions. The impact of environmental stressors, such as climate change and habitat degradation, on the physiological health of fish populations is an area of growing concern. For example, research on heart rate and cardiac function in Arctic fishes provides critical insights into how rising temperatures affect cold-adapted species. Findings reveal that Arctic char (*Salvelinus alpinus*) exhibit rapid compensatory cardiac plasticity, reducing maximum heart rate (f_{Hmax}) over intermediate temperatures and improving their ability to increase f_{Hmax} during acute warming, which enhances cardiac thermal tolerance (Gilbert et al., 2022b). Further studies demonstrated that Arctic char can adjust their cardiac performance with thermal acclimation, increasing their cardiac heat tolerance significantly with acclimation temperatures of up to 14 °C; although, prolonged exposure to 18 °C proved lethal (Gilbert and Farrell, 2022). Field-based research using a mobile aquatic-research laboratory showed that Arctic char's f_{Hmax} increases with temperature up to a critical point, beyond which it declines and becomes arrhythmic, highlighting the vulnerability of this species to extreme thermal events (Gilbert et al., 2021). Similar rapid cardiac plasticity has been observed in rainbow trout, indicating that such mechanisms might be widespread among fish species to mitigate acute thermal challenges (Gilbert et al., 2022a). This is an area that certainly warrants further investigation.

On a personal note, I had the honor of learning from Jones during the first year of my PhD, before he passed on, and luckily much more time with Randall, not only through my PhD and post-doctoral work, but well into my career as a university professor as well, until his passing in April 2024. Undoubtedly, the legacy of Jones and Randall's work continues to inspire and guide research in fish physiology, and both have left a lasting impression on our field.

Indeed, future research could focus on assessing the resilience of fish to such stressors that will become more frequent and severe into the future and exploring conservation strategies to protect diverse aquatic ecosystems. As we advance our understanding of the complex interplay between the environment and physiological adaptations in fish, we are better equipped to appreciate the intricacies of aquatic life and the challenges it faces in a rapidly changing world.

References

Anttila, K., Couturier, C.S., Øverli, Ø., Johnsen, A., Marthinsen, G., Nilsson, G.E., Farrell, A.P., 2014. Atlantic salmon show capability for cardiac acclimation to warm temperatures. Nat. Commun. 5, 4252. https://doi.org/10.1038/ncomms5252.

Beamish, F.W.H., 1968. Glycogen and lactic acid concentrations in Atlantic cod (*Gadus morhua*) in relation to exercise. J. Fish. Res. Board Can. 25, 837.

Bell, W.H., Terhune, L.D.B., 1970. Water tunnel design for fisheries research. Fish. Res. Board Can. Tech. Rep. 195, 1–69.

Black, E.C., Robertson, A.C., Hanslip, A.R., Chiu, W.G., 1960. Alterations in glycogen, glucose and lactate in rainbow and Kamloops trout (*Salmo gairdneri*) following muscular activity. J. Fish. Res. Board Can. 17, 487.

Black, E.C., Connor, A.R., Lam, K.C., Chiu, W.G., 1962. Changes in glycogen, pyruvate and lactate in rainbow trout (*Salmo gairdneri*) during and following muscular activity. J. Fish. Res. Board Can. 19, 409.

Blažka, P., Volf, M., Cepela, M., 1960. A new type of respirometer for the determination of the metabolism of fish in an active state. Physiol. Bohemoslov. 9, 553–558.

Brett, J.R., 1964. The respiratory metabolism and swimming performance of young sockeye salmon. J. Fish. Res. Board Can. 21, 1183.

Brett, J.R., 1995. Energetics. In: Groot, C., Margolis, L., Clarke, W.C. (Eds.), Physiological Ecology of Pacific Salmon. UBC Press, Vancouver, Canada, pp. 3–68.

Brett, J.R., Glass, N.R., 1973. Metabolic rates and critical swimming speeds of sockeye salmon (*Oncorhynchus nerka*) in relation to size and temperature. Fish. Res. Board Can. 30, 379.

Chabot, D., Steffensen, J.F., Farrell, A.P., 2016. The determination of standard metabolic rate in fishes. J. Fish Biol. 88, 81–121. https://doi.org/10.1111/jfb.12845.

Claireaux, G., McKenzie, D.J., Genge, A.G., Chatelier, A., Aubin, J., Farrell, A.P., 2005. Linking swimming performance, cardiac pumping ability and cardiac anatomy in rainbow trout. J. Exp. Biol. 208, 1775–1784.

Connor, A.R., Elling, C.H., Black, E.C., Collines, G.B., Gauley, J.R., Trevor-Smith, E., 1964. Changes in glycogen and lactate levels in migrating salmonid fish ascending experimental "endless" fishways. J. Fish. Res. Board Can. 21, 255.

Cooke, S.J., Hinch, S.G., Wikelski, M., Andrews, R.D., Kuchel, L.J., Wolcott, T.G., Butler, P.J., 2004. Biotelemetry: a mechanistic approach to ecology. Trends Ecol. Evol. 19 (6), 334–343. https://doi.org/10.1016/j.tree.2004.04.003.

Cooke, S.J., Sack, L., Franklin, C.E., Farrell, A.P., Beardall, J., Wikelski, M., Chown, S.L., 2013. What is conservation physiology? Perspectives on an increasingly integrated and essential science. Conserv. Physiol. 1 (1), cot001. https://doi.org/10.1093/conphys/cot001.

Cooke, S.J., Blumstein, D.T., Buchholz, R., Caro, T., Fernández-Juricic, E., Franklin, C.E., et al., 2016. Physiological, behavioral and ecological aspects of conservation physiology. In: Burness, G.P. (Ed.), Conservation Physiology: Applications for Wildlife Conservation and Management. Oxford University Press.

Domenici, P., Blake, R.W., 1997. The kinematics and performance of fish fast-start swimming. J. Exp. Biol. 200 (8), 1165–1178. https://doi.org/10.1242/jeb.200.8.1165.

Downie, A.T., Lefevre, S., Illing, B., Harris, J., Jarrold, M.D., McCormick, M.I., Nilsson, G.E., Rummer, J.L., 2023. Rapid physiological and transcriptomic changes associated with oxygen delivery in larval anemonefish suggest a role in adaptation to life on hypoxic coral reefs. PLoS Biol. https://doi.org/10.1371/journal.pbio.3002102.

Driedzic, W.R., Kiceniuk, J.W., 1976. Blood lactate levels in free-swimming rainbow trout (*Salmo gairdneri*) before and after strenuous exercise resulting in fatigue. J. Fish. Res. Board Can. 33, 173.

Eliason, E.J., Clark, T.D., Hinch, S.G., Farrell, A.P., 2013. Cardiorespiratory collapse at high temperature in swimming adult sockeye salmon. Conserv. Physiol. 1 (1), cot008. https://doi.org/10.1093/conphys/cot008.

Farrell, A.P., 1997. Effects of temperature on cardiovascular performance. In: Wood, C.M., McDonald, D.G. (Eds.), Global Warming: Implications for Freshwater and Marine Fish. Cambridge University Press, Cambridge, pp. 135–158.

Farrell, A.P., Hinch, S.G., Cooke, S.J., Patterson, D.A., Crossin, G.T., Lapointe, M., Mathes, M.T., 2009. Pacific salmon in hot water: applying aerobic scope models and biotelemetry to predict the success of spawning migrations. Physiol. Biochem. Zool. 82 (6), 697–708.

Fry, F.E.J., 1971. The effect of environmental factors on the physiology of fish. In: Hoar, W.S., Randall, D.J. (Eds.), Fish Physiology. vol. 6. Academic Press, New York, p. 1.

Gilbert, M.J.H., Farrell, A.P., 2022. The thermal acclimation potential of maximum heart rate and cardiac heat tolerance in Arctic char (*Salvelinus alpinus*), a northern cold-water specialist. J. Therm. Biol. 95, 102816. https://doi.org/10.1016/j.jtherbio.2020.102816.

Gilbert, M.J.H., Harris, L.N., Malley, B.K., Schimnowski, A., Moore, J.-S., Farrell, A.P., 2021. The thermal limits of cardiorespiratory performance in anadromous Arctic char (*Salvelinus alpinus*): a field-based investigation using a remote mobile laboratory. Conserv. Physiol. 8 (1), coaa036. https://doi.org/10.1093/conphys/coaa036.

Gilbert, M.J.H., Adams, O.A., Farrell, A.P., 2022a. A sudden change of heart: warm acclimation can produce a rapid adjustment of maximum heart rate and cardiac thermal sensitivity in rainbow trout. Curr. Res. Physiol. 5, 179–183.

Gilbert, M.J.H., Middleton, E.K., Kanayok, K., Harris, L.N., Moore, J.-S., Farrell, A.P., Speers-Roesch, B., 2022b. Rapid cardiac thermal acclimation in wild anadromous Arctic char (Salvelinus alpinus). J. Exp. Biol. 225 (17), jeb244055. https://doi.org/10.1242/jeb.244055.

Gracey, A.Y., Fraser, E.J., Li, W., Fang, Y., Taylor, R.R., Rogers, J., Brass, A., Cossins, A.R., 2004. Coping with cold: an integrative, multitissue analysis of the transcriptome of a poikilothermic vertebrate. Proc. Natl. Acad. Sci. U. S. A. 101 (48), 16970–16975. https://doi.org/10.1073/pnas.0403627101.

Huang, S.-H., Tsao, C.-W., Fang, Y.-H., 2020. A miniature intermittent-flow respirometry system with a 3D-printed, palm-sized zebrafish treadmill for measuring rest and activity metabolic rates. Sensors 20 (18), 5088. https://doi.org/10.3390/s20185088.

Jayne, B.C., Lauder, G.V., 1995. Red muscle motor patterns during steady swimming in largemouth bass: effects of speed and correlations with axial kinematics. J. Exp. Biol. 198 (7), 1575–1587. https://doi.org/10.1242/jeb.198.7.1575.

Johansen, K., Franklin, D.L., Van Citters, R.L., 1966. Aortic blood flow in free swimming elasmobranchs. Comp. Biochem. Physiol. 19, 151.

Kiceniuk, J.W., Jones, D.R., 1977. The oxygen transport system in trout (*Salmo gairdneri*) during sustained exercise. J. Exp. Biol. 69, 247.

Killen, S.S., Norin, T., Halsey, L.G., 2017. Do method and species lifestyle affect measures of maximum metabolic rate in fishes? J. Exp. Biol. 90, 1037–1046. https://doi.org/10.1111/jfb.13195.

Kutty, M.N., 1968. Respiratory quotients in goldfish and rainbow trout. J. Fish. Res. Board Can. 25, 1689.

McKenzie, D.J., Axelsson, M., Chabot, D., Claireaux, G., Cooke, S.J., Corner, R.A., De Boeck, G., Domenici, P., Guerreiro, P.M., Hamer, B., Jørgensen, C., Killen, S.S., Lefevre, S., Marras, S., Michaelidis, B., Nilsson, G.E., Peck, M.A., Perez-Ruzafa, A., Rijnsdorp, A.D., Shiels, H.A., Steffensen, J.F., Svendsen, J.C., Svendsen, M.B.S., Teal, L.R., van der Meer, J., Wang, T., Wilson, J.M., Wilson, R.W., Metcalfe, J.D., 2007. Conservation physiology of marine fishes: state of the art and prospects for policy. Conserv. Physiol. 5 (1), cox005.

Norin, T., Clark, T.D., 2016. Measurement and relevance of maximum metabolic rate in fishes. J. Fish Biol. 88, 122–151. https://doi.org/10.1111/jfb.12796.

Palstra, A.P., Planas, J.V., 2011. Fish under exercise. Fish Physiol. Biochem. 37, 259–272. https://doi.org/10.1007/s10695-011-9505-0.

Payne, N.L., Snelling, E.P., Fitzpatrick, R., Seymour, J., Courtney, R., Barnett, A., Watanabe, Y., Sims, D.W., Squire, L., Semmens, J.M., 2015. A new method for resolving uncertainty of energy requirements in large water breathers: the 'mega-flume' seagoing swim-tunnel respirometer. Methods Ecol. Evol. 6 (6), 668–677.

Perry, S.F., Gilmour, K.M., 2006. Acid–base balance and CO_2 excretion in fish: unanswered questions and emerging models. Respir. Physiol. Neurobiol. 54 (1–2), 199–215. https://doi.org/10.1016/j.resp.2006.04.010.

Randall, D.J., Daxboeck, C., 1984. Oxygen and carbon dioxide transfer across fish gills. In: Hoar, W.S., Randall, D.J. (Eds.), Fish Physiology. vol. XA. Academic Press, New York, pp. 263–314.

Rao, G.M.M., 1968. Oxygen consumption of rainbow trout (*Salmo gairdneri*) in relation to activity and salinity. Can. J. Zool. 46, 781.

Rees, B.B., Reemeyer, J.E., Binning, S.A., Brieske, S.D., Clark, T.D., De Bonville, J., Eisenberg, R.M., Raby, G., Rummer, J.L., Zhang, Y., 2024. Estimating maximum oxygen uptake of fish during swimming and following exhaustive chase—different results, biological bases, and applications. J. Exp. Biol. 227 (11), jeb246439. https://doi.org/10.1242/jeb.246439.

Roche, D.G., Binning, S.A., Bosiger, Y., Johansen, J.L., Rummer, J.L., 2013. Finding the best estimates for metabolic rates in a coral reef fish. J. Exp. Biol. 216, 2103–2110. https://doi.org/10.1242/jeb.082925.

Rome, L.C., Funke, R.P., Alexander, R.M., 1990. The influence of temperature on muscle velocity and sustained performance in swimming carp. J. Exp. Biol. 154 (1), 163–178. https://doi.org/10.1242/jeb.154.1.163.

Rummer, J.L., Binning, S.A., Roche, D.G., Johansen, J.L., 2016. Methods matter: considering locomotory mode and respirometry technique for estimating metabolic rate in fish. Conserv. Physiol. 4 (1), cow008. https://doi.org/10.1093/conphys/cow008.

Smit, H., Amelink-Koutstaal, J.M., Vuverberg, J., von Vaupel-Klein, J.C., 1971. Oxygen consumption and efficiency of swimming goldfish. Comp. Biochem. Physiol. A 39, 1.

Steinhausen, M.F., Sandblom, E., Eliason, E.J., Verhille, C., Farrell, A.P., 2008. The effect of acute temperature increases on the cardiorespiratory performance of resting and swimming sockeye salmon (*Oncorhynchus nerka*). J. Exp. Biol. 211, 3915–3926.

Stevens, E.D., Randall, D.J., 1967. Changes in blood pressure, heart rate, and breathing rate during moderate swimming activity in rainbow trout. J. Exp. Biol. 46, 307.

Webb, P.W., 1971. The swimming energetics of trout. 11. Oxygen consumption and swimming efficiency. J. Exp. Biol. 55, 521.

Webb, P.W., 1998. Swimming. In: Evans, D.H. (Ed.), The Physiology of Fishes, second ed. CRC Press, pp. 3–24.

Yousaf, M.N., Røn, Ø., Keitel-Gröner, F., McGurk, C., Obach, A., 2023. Heart rate as an indicator of stress during the critical swimming speed test of farmed Atlantic salmon (*Salmo salar* L.). J. Fish Biol., 1–14. https://doi.org/10.1111/jfb.15602.

Chapter 12

THE RESPIRATORY AND CIRCULATORY SYSTEMS DURING EXERCISE☆

DAVID R. JONES
DAVID J. RANDALL

Chapter Outline

I. GENERAL INTRODUCTION

Exercise is the stress which animals most frequently experience and may be defined as work performed on the environment by the locomotory muscles. Exercise is accompanied by an increase in the rate of energy conversion from the resting rate. This increase provides for the energy requirements of the locomotory muscles as well as for extra work performed by the heart and respiratory muscles in supplying oxygen demanded in exercise.

☆This is a reproduction of a previously published chapter in the Fish Physiology series, "1978 (Vol. 7)/The Respiratory and Circulatory Systems During Exercise: ISBN: 978-0-12-350407-4; ISSN: 1546-5098".

Fish Physiology, Vol. 40B. https://doi.org/10.1016/bs.fp.2024.07.005

In this chapter the emphasis is placed on the ability of fish to increase the rate of gas exchange at the gills and tissues and the changes which occur in the components of the respiratory and circulatory systems facilitating this increase in gas exchange. On many occasions circulatory and respiratory adjustments have been measured along with indirect estimates of metabolism during exercise. Obviously metabolism, power output, and swimming speed are related but, for virtually all fish species, the relationship is too imprecisely known to use any one in calculations of the others. Consequently, before discussing respiratory and circulatory compensations in exercise, we have reviewed the literature on the metabolic adjustments to exercise in an attempt to establish criteria which could be used in assessing the exercise performance of a particular fish. These data have been more extensively reviewed by Beamish (see Chapter 2) and Fry (1971) in this and an earlier volume of this series, respectively.

Finally, a word is in order outlining the terminology used in this chapter. For metabolic rate, the terms standard, routine, and active refer to basal metabolism, metabolism associated and related to a degree of random activity, and the maximum sustained metabolic rate, respectively (Fry, 1971). Values for swimming speed are given in the Terminology to Describe Swimming Activity in Fish (p. xiii), while all physiological variables are described in the text in full.

II. ASSESSMENT OF EXERCISE PERFORMANCE

A. Introduction

Even the most cursory review of the literature leads to the conclusion that defining exercise performance in fishes is a difficult problem. In many experiments fish are forced to exercise up to a maximum sustained swimming speed which may seldom, if ever, be attained in nature. For instance, Brett (1972) suggests that the cardiovascular and respiratory systems in salmonids have evolved in response to the demands of upstream migration which are obviously very different from the demands placed on these systems in a fish being forced to swim as fast as it can in a water tunnel. Furthermore, disparate measures of exercise performance have been used by various investigators and it is necessary to know the relationship between them and metabolism since cardiovascular and respiratory compensations are best evaluated in terms of supply of metabolites. Therefore, if absolute or relative U has been used as a measure of performance, then the relationship between U and gas exchange for that particular species is required along with an estimate of whether anaerobic contributions to power output, and consequently U, are significant. Therefore, in this section a review of the measures of "effort" made by a fish during exercise is made as background against which the respiratory and circulatory adjustments, discussed in the subsequent sections, can be evaluated.

B. The Relation between Oxygen Metabolism and Swimming Speed, *U*

It is generally assumed that, in sustained exercise, respiratory and circulatory adjustments are adequate to meet increased energy demands aerobically. Unfortunately, there have not been many comprehensive surveys of oxygen metabolism in exercise and this lack of data has contributed to a controversy that surrounds not only the relationship between oxygen metabolism and *U*, but also whether oxygen metabolism is best described by absolute oxygen uptake or the difference between standard and active oxygen metabolism (metabolic scope) in exercise. The total amount of oxygen available for the locomotory muscles is generally assumed to be represented by the difference between the active and standard rate of uptake, although it is recognized that this represents the maximum estimate of aerobic energy available, since it includes increased energy demands of the heart and respiratory muscles as well as increased costs of osmoregulation (Brett, 1963, 1972; Webb, 1971a,b).

In view of this, Brett (1964) explored the relation between oxygen uptake and *U* and concluded that oxygen uptake increased exponentially with *U* in sockeye salmon (*Oncorhynchus nerka*) or, in other words, a linear relation was obtained when the logarithm of oxygen uptake was plotted against *U*, as determined in forced velocity tests using 75 min periods between velocity increments (Fig. 1). On the other hand, either a power or a linear function may describe the relationship between metabolic scope and *U*. Fry (1971) concluded, for a range of data, that the cost of swimming (as measured by the scope) increased approximately as the square of *U*. However, the data of Brett (1965a; Fig. 2), Rao (1968; Fig. 2), and Kutty (1968a) for salmon (*O. nerka*) and trout (*Salmo*

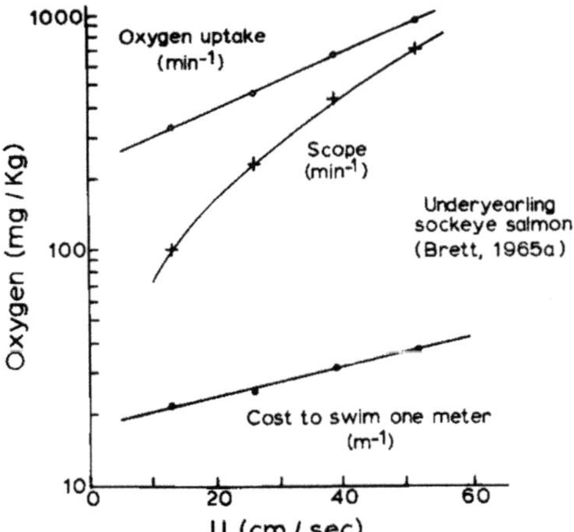

Fig. 1. Relationship between oxygen uptake (ml/kg/min), scope for activity (ml/kg/min) and cost to swim 1 m (ml/kg/m), and swimming speed (*U*) for underyearling sockeye salmon (*O. nerka*). (All data taken from Brett, 1965a.)

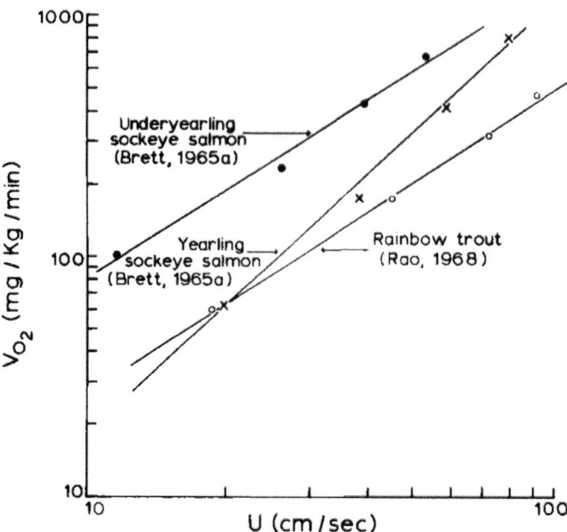

Fig. 2. Relationship between oxygen uptake (ml/kg/min) and swimming speed(U). For underyearling sockeye salmon and rainbow trout the slopes of the line are between 1.3 and 1.4, whereas for yearling sockeye salmon the slope is 1.8. (Data from Brett, 1965a; Rao, 1968.)

gairdneri) yield the relation that cost increases as $U^{1.35-1.8}$ and not U^2 as indicated by Fry (1971). Alternatively, Smit *et al.* (1971) claim that since power output is related to $U^{2.64}$ and swimming efficiency to $U^{1.64}$ (approximately) then scope must be related to power output divided by efficiency or $U^{1.0}$.

Undoubtedly, Fry (1971) correctly assesses the situation when he states that performance is qualitatively different from the power which produces it and that there need not be any simple proportionality between measures taken of the two. Many teleosts do not support their weight, being neutrally buoyant, so the power output of the locomotory muscles will be largely directed to overcoming the resistance to movement presented by the water. Since the drag on the fish will increase in proportion to U^2 [or, as Webb (1971a) suggested, $U^{1.8}$ for fish in tunnel respirometers] then a power relation between metabolic scope and U might be expected. This demands that the muscles, which produce the power to overcome the drag, be equally efficient at all power outputs, which is not the case. Muscular efficiency is not constant but usually increases with increasing power output until high levels are reached, when it becomes increasingly difficult and more costly in terms of energy input to augment the work output. This process can be better appreciated with reference to Fig. 3 taken from the work of Hill (1950) which shows that power output of muscle under a maximum load depends on the shortening speed, maximum power output being reached at a shortening speed of about 35% of maximum. On the other hand, efficiency of muscle contraction reaches its peak when the shortening speed is only 20% of maximum. Consequently, as the animal increases power output it does so at the expense of muscular efficiency, and an exponential

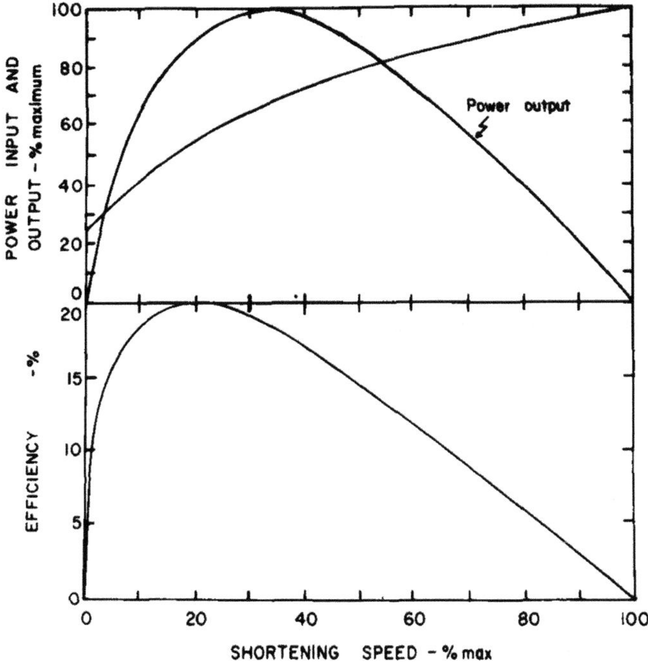

Fig. 3. The relationship between muscle power input, output, and efficiency as functions of muscle shortening speed, expressed in relation to maximum power, efficiency, and shortening speeds. [Redrawn from Hill, 1950, *Sci. Prog.* (*London*) **38,** 209.]

relationship of the type obtained by Brett (1964) might be expected between oxygen uptake and *U*. Of course a corollary of an exponential relationship between oxygen uptake and speed is that the efficiency of swimming decreases with speed. Webb (1971b) calculated the efficiency of swimming in trout and showed that a 20 cm/sec speed increase in the middle velocity range (25–45 cm/sec) was accomplished with a doubling in efficiency while a similar increase in the higher velocity range (45–65 cm/sec) caused swimming efficiency to increase only by one-quarter. In view of the above discussion it is difficult to find any justification for expecting a linear relation between scope and *U*, as suggested by Smit *et al.* (1971), unless the data are restricted to determinations in the midrange of both scope and *U*. In other words, the active oxygen consumption should increase sufficiently so that the standard uptake only contributes from 20 to 30% of the total, but not to the degree that the cost of increasing the power output becomes prohibitive.

In the absence of a clear relationship between *U* and oxygen uptake or metabolic scope for a particular species there are other measures which can be used to assess exercise performance. For instance, it seems plausible that the absolute rate of active oxygen metabolism could be a good indicator of the level of exercise stress to which a fish is subjected.

However, this measure of exercise stress is not without problems. Brett (1964) found that active, like standard, oxygen uptake in sockeye salmon was temperature dependent (Table I). Q_{10} for both active and standard metabolism over the range of 5°–15°C lay between 1.7 and 2, while at 20° and 24°C, the active metabolic rate was not significantly different from the value at 15°C, although standard rates were significantly elevated. In other words, salmon appeared to reach a maximum oxygen uptake of about 625 ml O_2/kg/hr at 15°C, under the conditions of Brett's experiments. Brett (1964) attributed the peaking of oxygen uptake at 15°C to a failure on the part of the respiratory system to supply enough oxygen at high temperatures and showed that, in one group of fish at 20°C, raising the oxygen concentration of the water by 50% caused active metabolism to reach over 900 ml O_2/kg/hr, yielding a Q_{10} between 15° and 20°C of 2.

Brett (1965a) and Brett and Glass (1973) also explored the relation between oxygen uptake during exercise and body size and found that the active level of oxygen uptake was virtually independent of body weight in sockeye salmon (Fig. 4), which is a considerable advantage to the investigator if active rates are used as a single measure of performance. Standard metabolism of sockeye salmon is, of course, proportional to weight to the power of 0.78, so standard metabolism, expressed on a unit weight basis, falls with weight increase ($W^{-0.22}$) while active metabolism does not.

It is generally assumed that maximum oxygen uptake is achieved just before fatigue in an incremental velocity test which should give some consistency to this variable. However, Webb (1971a,b) has shown that if drag loads are placed on a swimming trout (*S. gairdneri*), then maximum oxygen uptake falls along with U_{crit} (Table I). Webb (1971a,b) argued that the reason for this decline is that, although all fish are assumed to be making their maximum effort at U_{crit}, the rate at which they work falls as U_{crit} falls. Therefore the power output falls and hence maximum oxygen uptake. However, in terms of muscular effort expended, there would appear to be little difference between loaded and unloaded fish over the range of U_{crit}'s Webb investigated. For instance, in the majority of Webb's loaded fish both tailbeat frequency and specific amplitude (amplitude of tailbeat divided by wavelength) at U_{crit} either equaled or exceeded those in the unloaded group. The calculated thrust produced by an unloaded fish at U_{crit} was equal, in a loaded fish, to the calculated thrust plus the drag of the added load at a lower U_{crit}. There can be no doubt, however, that the oxygen uptake fell in loaded fish, so it must be concluded that, in loaded fish, the hydromechanics of the propulsive wave may not be adequately described by measurements of tailbeat frequency and amplitude. This effect has not been confirmed in other studies with salmonids. Kiceniuk and Jones (1977) found a poor relationship between the logarithm of oxygen uptake and absolute speed in trout (*S. gairdneri*) carrying various loads of instruments, whereas expressing U as a proportion of U_{crit} greatly improved the relationship.

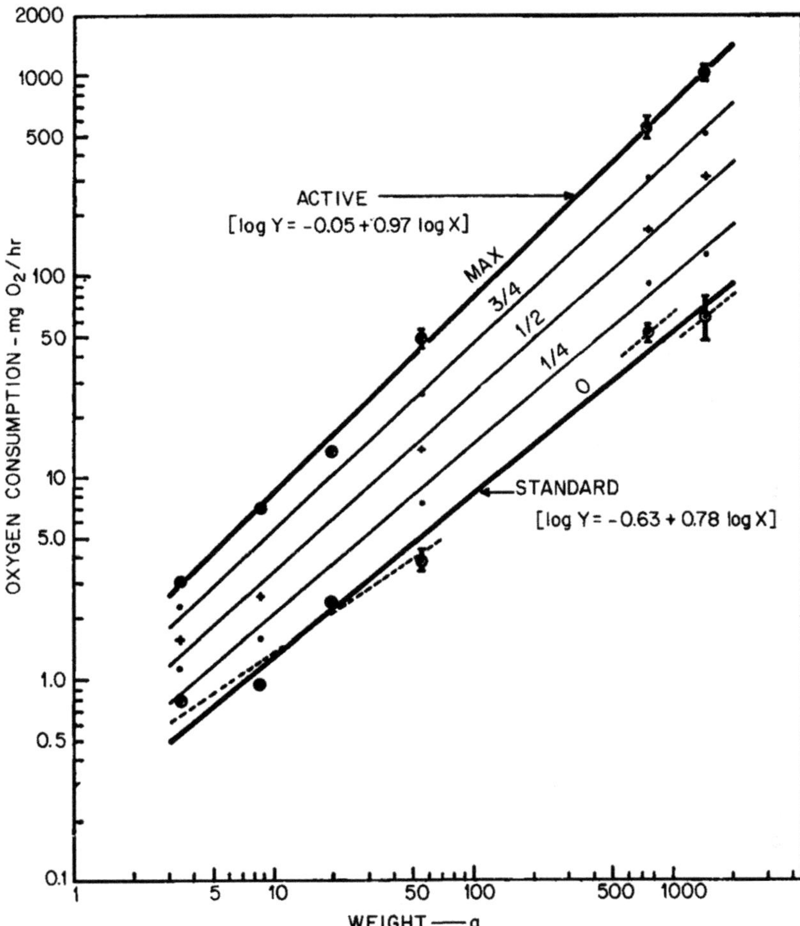

Fig. 4. The relationship between metabolic rate and weight for different levels of activity expressed as fractions of the maximum 60-min sustained speed (max) in *O. nerka*. Experiments performed in freshwater at 15°C. Broken lines represent possible relation of standard metabolism for immature freshwater stage and for mature fish of different sex. Whenever fish were tested singly limits of ±2 SE are indicated. (From Brett, 1965a, *J. Fish. Res. Board Can.* **22,** 1491.)

This observation suggests that fish attain their maximum oxygen uptake when making maximum effort regardless of absolute U. Furthermore, in Brett's (1964) experiment, when active oxygen consumption increased markedly after environmental oxygen concentrations were raised, U_{crit} was not significantly elevated, indicating a divorce between power output and oxygen uptake at maximum effort in this group of fish. Many fish are exercised under conditions that suggest there could be a restriction of power output, i.e., when swum to fatigue in a circular chamber when the muscles on one side may be held below their resting length, while on the other they may always be stretched above their

Table I Some Measures that May Be Used to Assess Maximum Exercise Performance in Teleosts[a]

Species	Temperature (°C)	Weight (g)	Critical velocity (cm/sec)	Active oxygen uptake (ml/kg/hr)	Change In oxygen uptake (active/standard)	Cost to swim 1 m (ml/kg/m)	Notes and references
Oncorhynchus nerka	5	36.7	53.7	360	12.5	0.17	Brett (1964)
	10	32.9	58	439	10.4	0.19	
	15	55.2	77	626	12.6	0.2	
	20	62.6	74	596	7.1	0.19	
	24	52.2	68	594	4.32	0.18	
	15	3.38	51.5	644	4	0.26	Brett (1965a) [1] Untrained fish [2] Speed quoted by Brett and Glass (1973) Cost = 0.2 ml/kg/m from data in Brett (1965a) [3] Corrected speeds from Brett and Glass (1973)
	15	3.47	59.3	581	7.5	0.23	
	15	19.1	53.2	490	5.5	0.2[1]	
	15	55.2	90.7	626	12.6	0.18[2]	
	15	746	150	511	10.28	0.085[3]	
		1432	178	502	16	0.07[3]	
	2	42	40	200	6.7	0.12	Brett and Glass (1973)
Lepomis gibbosus	20	45	37.2	285	9	0.19	Brett and Sutherland (1965)

Kuhlia sandvicensis	23	30[3]	54. 7[1]	407	8.8	0.18	Muir and Niimi (1972) [1]Freshwater [2]Saltwater [3]Scaled to 30 g weight
			54.3[2]	458	9.3	0.21	
Salmo sp (rainbow trout)	15	264	58.1	460	8.7	0.19	Webb (1971b) [1] Loaded fish Group 1 [2] Loaded fish Group 3
	15[1]	271	45.1	336	6.47	0.17	
	15[2]	258	28.6	304	5.76	0.24	
Salmo gairdneri	15	100[1]	90	408	5.2	0.1	Rao (1968) [1] Scaled to 100 g weight
Tilapia nilotica	25	80[1]	60	320	4.4	0.11	Farmer and Beamish (1969) Freshwater [1] Scaled to 80 g weight
Melanogrammus aeglefinus	10	156	52	193	5.5	0.08	Tytler (1969); fish unfit

[a]Active and standard rates of oxygen consumption are extrapolated.

resting length due to body curvature. Consequently, in view of the above conflicting data, it would be virtually impossible to predict how maximum aerobic metabolism under these conditions would compare with that obtained from fish swimming in a straight line.

It is possible to obtain an immediate idea of the increased demands for oxygen and carbon dioxide transport during sustained swimming by expressing activity in terms of a ratio of active to standard metabolism. Brett (1964) found that this ratio was about 10–12 for 33–65 g sockeye salmon over a temperature range of 5°–15°C, although it fell markedly at higher temperatures (Table I) due to the temperature-induced increase in standard metabolism in the face of no increase in active metabolism with temperature. However, sockeye salmon at 20°C in high oxygen-saturated water showed a marked increase in maximum oxygen uptake and the ratio was restored to 10. Obviously this ratio, as an estimate of exercise performance, is sensitive to temperature. Furthermore, since standard metabolism relates to $W^{0.78}$ while active metabolism is size independent, then the ratio is very sensitive to body size (Table I), increasing from 4 to 8 in small fish (3–4 g) to 16 in large fish (1.4–1.5 kg) (Fig. 4; Brett, 1965a).

The outcome of sustained swimming in still water is covering a distance between two points and what is probably most important to the fish is how much energy it uses doing this. Consequently, it might be more realistic to express exercise capability in terms of the metabolic cost to traverse a unit distance. This has been done previously in an attempt to establish an "optimum" speed, that is, when metabolic cost per unit distance traveled is minimal (Fig. 5; Brett, 1965b; Weihs, 1973a,b; Webb, 1975a). The shape of the curve

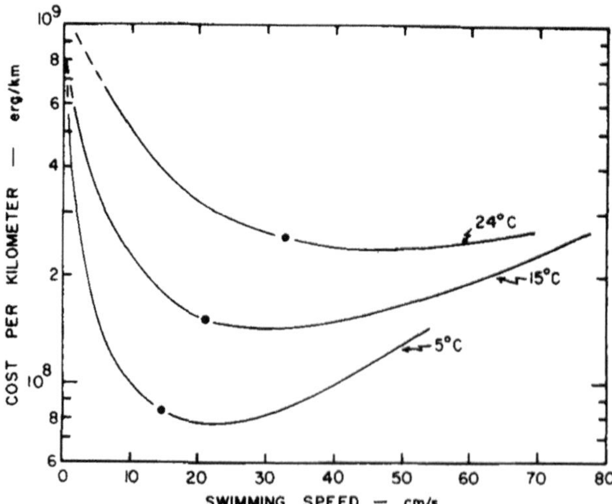

Fig. 5. Relation between energetic costs (erg/km) and swimming speed of yearling sockeye salmon at three temperatures. Optimum swimming speeds occur at the lowest point of the curve. Predicted optimum speeds (Weihs, 1973a) are solid circles. (Based on data in Brett, 1964; after Brett, 1965b; by Webb, 1975a, *Bull. Fish. Res. Board Can.* No. 190.)

in Fig. 5 results from the fact that when swimming speeds are low, standard metabolic costs are high, relative to costs of locomotion. Since speed is low then the time required to cover unit distance is long and therefore total cost per unit distance is high. Obviously cost per unit distance falls as oxygen uptake increases, but, since the latter has an exponential relation with *U*, cost rises again at the highest *U*'s. Weihs (1973a) derived an expression to predict the optimum speed for yearling sockeye salmon, finding that it occurred at two times the standard metabolic rate (Fig. 5). Since standard metabolism increases with temperature, then Weihs' (1973a) expression predicts that the optimum speed will increase with temperature, which fits well with Brett's (1964) data. At optimum speed and low temperature, cost of locomotion in fish is low, requiring less than one-fifth the amount of energy that a running mammal or flying bird of the same weight would require to cover the same distance (Tucker, 1970; Schmidt-Nielsen, 1972).

Assessment of exercise capability in terms of cost to traverse a unit distance assumes that anaerobic energy contributions are negligible. If there are no obvious discontinuities in the relation between oxygen uptake and sustained *U*, then probably only one group of muscles has been used throughout the exercise period. This appears to hold for trout (Kiceniuk and Jones, 1977) and salmon (Brett, 1964), although Brett and Sutherland (1965), Webb (1971a), and Hudson (1973) have indicated that at the lowest speeds these species may change their mode of propulsion. A change in the relation between oxygen uptake and *U* has been monitored in the goldfish (*Carassius auratus*) by Smit (1965), who found that around an imposed water velocity of 10 cm/sec goldfish seemed to switch to a "thriftier use of oxygen" (Fig. 6).

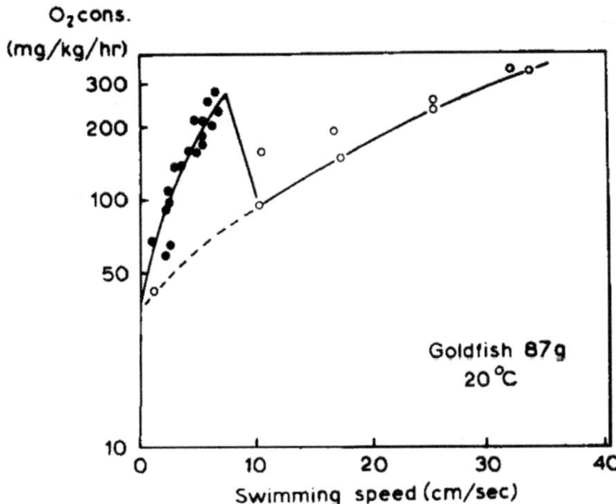

Fig. 6. Relationship between oxygen consumption and swimming speed of an 87-g goldfish (*C. auratus*) when spontaneously swimming in a nonrotating annular chamber (closed circles) and when forced to swim at various speeds in a rotating chamber (open circles). Throughout the experiments oxygen concentrations were at ambient. When forced to swim at 10 cm/sec the fish seemed to switch to a "thriftier use of oxygen." (From Smit, 1965, *Can. J. Zool.* **43,** 623.)

Assessment of energy cost per unit distance is complicated by a change in the mode of propulsion or by a shift from red to white (or mosaic) muscle at a particular speed, for in that case, muscle recruitment would not involve the same type of muscle throughout the exercise. Finally, the ideal mode of progression suggested by Weihs (1974) of alternate bursts and glides is difficult to compare, on a basis of oxygen cost per unit distance, with sustained forward progression as is observed in swimming tests.

Since there is an exponential relationship between oxygen uptake and U then, obviously, cost to cover a unit distance increases with speed; but in Table I cost is calculated using scope for activity (active minus standard metabolism) so an exponential increase in cost with U would not be expected. Furthermore, the faster the fish swims then the shorter the time it takes to cover a unit distance and any increase in the cost factor is offset somewhat. For instance, underyearling salmon (*O. nerka*) monitored by Brett (1965a) displayed a 1.7 times inorease in cost to swim a meter (0.15 to 0.26 ml/kg/m) while speed increased 4 times (12.8 to 51.5 cm/sec) (Fig. 1). Since the cost to swim a meter at maximum sustained swimming speed is temperature-independent it is a less variable measure of exercise capability compared with maximum or relative oxygen uptake (Table I). Unfortunately, "cost" is inversely proportional to weight so that low values are obtained from larger fish (Table I; Brett, 1965a) so, as Table I confirms, "cost" is far too variable to use as a single measure of whether fish performed maximally or not before fatigue. Nevertheless, it represents an estimate which, allied to absolute or relative oxygen uptake, is useful for making an assessment of exercise performance (Table I).

C. Anaerobic Contribution to Exercise Metabolism

Oxygen consumption, even in steady state conditions, may not be a good indication of the total energy conversion during exercise, since some of the energy budget may be provided by anaerobiosis. Consequently, during anaerobiosis, the relation between oxygen metabolism and U would alter markedly. At moderate swimming speeds depletion of muscle glycogen or build up of blood lactic acid is slight (Black *et al.*, 1960, 1962; Connor *et al.*, 1964; Beamish, 1968; Driedzic and Kiceniuk, 1976) and it is only during burst activity that both muscle and blood lactate levels increase rapidly and muscle glycogen falls (Black, 1958; Stevens and Black, 1966; Beamish, 1968), which is indicative of anaerobiosis.

The portion of the total energy budget contributed anaerobically in sustained exercise is a matter of some dispute. Smit *et al.* (1971) claim that anaerobiosis contributes markedly to the energy budget during sustained swimming in goldfish (*C. auratus*), although Kutty's (1968a) R.Q. determinations provide no evidence for anaerobic metabolism over a period from 3 to 12 hr after the start of sustained swimming at 20°C in goldfish (*Carassius sp.*) and trout (*S. gairdneri*), since the mean R.Q. in his fish was below unity.

Both species, however, may display an initial anaerobic phase at the start of sustained swimming because mean R.Q. in the goldfish is 1.66 in the first hour of exercise (Kutty, 1968a). In this first hour aerobic metabolism is also higher than in the remaining 3–12 hr swimming period, and it is likely that the stress of handling and being forced to swim raises both aerobic and anaerobic metabolism. As swimming fish approach critical velocity, then the anaerobic energy contribution increases forcing the fish into a cumulative oxygen debt (Brett, 1964). This "debt" is repaid in the postexercise recovery period. Hence, in an increasing velocity test, even one conducted on aerobic fish such as trout, anaerobic metabolism would be expected to be present immediately following each increment, and also throughout the exercise period at speeds of around 80–100% of critical velocity (Fig. 7; Webb, 1971b; Driedzic and Kiceniuk, 1976).

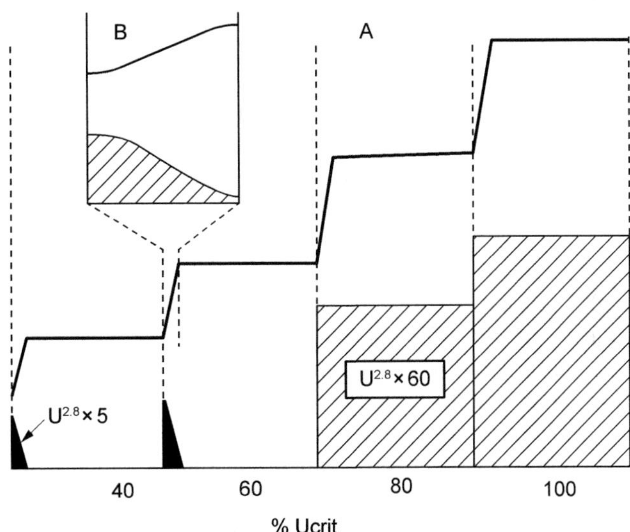

Fig. 7. (A) Diagram illustrating the proposed use of anaerobic energy sources in an increasing-velocity test. Four speeds are represented, corresponding to 40, 60, 80, and 100% of the critical swimming speed. The solid line represents the energy made available from aerobic energy sources. The shaded areas represent the energy made available continuously from anaerobic energy sources, and the solid shading represents the anaerobic energy requirements at a velocity increment. The total amount of anaerobic energy is distributed in proportion to the time for which the system operates and the rate at which energy is dissipated. This is proportional to $U^{2.8} \times t$ where t is 5 min at speeds less than 80% of U_{crit} and 60 min at speeds greater than this. At speeds greater than 80% U_{crit}, the anaerobic energy contribution at a velocity increment is assumed negligible in comparison with the continual anaerobic energy contribution. (B) Proposed energy changes over a 5 min period after a velocity increment. Note: the scales are not the same for the aerobic and anaerobic contributions. (From Webb, 1971b, *J. Exp. Biol.* **55,** 521.)

D. Limitations on Maximum Performance; Fatigue

Although all enforced exercise is terminated by fatigue it is surprising that there appears to be no consensus about what precipitates swimming failure in fish. Brett (1964) suggested that "burst" swimming was terminated by exhaustion of cellular energy supplies while metabolite supply limited steady (or sustained) performance. Studies by Black *et al.* (1962) and Stevens and Black (1966) indicate that in salmonids short-term exhausting exercise (being chased for up to 10 min) leads to severe depletion of muscle glycogen and marked elevation in muscle and blood lactate, whereas with moderate exercise muscle glycogen is hardly depleted at all (Black *et al.*, 1960, 1962). Obviously "burst" swimming is largely anaerobic, independent of the roles of circulatory and respiratory systems, and muscle lactate concentrations may be some 9–10 times higher than blood lactate levels (Stevens and Black, 1966). This suggests that the failure of muscle activity may be due to the fact that the lowest tolerable pH for anaerobic metabolism is attained rather than total depletion of cellular energy supplies. Unfortunately, even in increasing velocity tests, fish indulge in some "burst" type swimming, particularly when the imposed water velocity is increased or as U_{crit} is approached, which must complicate any metabolic analysis of fish swum to fatigue in this manner. Brett (1964, 1972) envisages fatigue occurring in increasing velocity tests due to energy demand exceeding the supply of oxygen at higher and higher speeds, so products of anaerobic metabolism gradually accumulate. However, changes in blood or tissue pH are unlikely to cause swimming failure in the absence of a profligate bout of burst swimming since lactate accumulation in increasing velocity tests is much less than occurs after violent chasing (Driedzic and Kiceniuk, 1976).

The failure to supply sufficient oxygen to contracting muscle during exercise has been variously attributed to limitations in the rate of oxygen transport in the body or oxygen exchange at the tissues or gills. When intracellular oxygen tension falls below the critical level for aerobic metabolism, anaerobic energy production increases greatly, which precipitates fatigue. This is obviously a different process from the swimming failure described by Kutty (1968b) in trout and goldfish. Kutty (1968b) found that fish failed to swim when the oxygen tension in the water fell below a critical level, even if the fish could normally swim at that speed for many hours. When the oxygen tension was raised, swimming restarted and was maintained for hours until the critical oxygen level was again attained. Unfortunately, since raising the water oxygen tension took a finite time and, in view of the fact that even after fatigue at U_{crit} many fish will swim again at low speed after a recovery period, these observations are not conclusive with respect to the nature of the fatigue. Nevertheless, both Smit *et al.* (1971) and Fry (1971) have suggested that failure to swim in hypoxic water is due to hypoxic depression of the central nervous system.

Although a reduction in environmental oxygen below ambient levels reduces swimming speed or active metabolism in many fish (Basu, 1959; Davis *et al.*, 1963; Silver *et al.*, 1963; Dahlberg *et al.*, 1968; Jones, 1971b), increases above ambient levels do not increase swimming speed (Brett, 1964; Dahlberg *et al.*, 1968) and appear to have variable effects on active metabolism (Brett, 1964). Therefore Smit *et al.* (1971) suggest that oxygen uptake rate is not limited by oxygen extraction from the medium or transport within the fish but probably by the aerobic capacity of the mitochondria or oxygen tension in muscle. Oxygen supply to the tissues depends on the provision of metabolic energy to power the branchial and cardiac muscles. Obviously, the costs of branchial and cardiac pumping increase as oxygen demand increases and ultimately an optimum level of oxygen uptake will be attained when supply to the tissues is maximal (Fig. 8; Jones, 1971a). Further increase in oxygen uptake requires more oxygen to be used by the heart and branchial muscles so that even less oxygen reaches the tissues, and the animal enters a "vicious circle" with regard to aerobic metabolism (Fig. 8; Jones, 1971a). Critical velocity of rainbow trout (*S. gairdneri*) is reduced by hypoxia or anemia, both of which might be regarded as restricting tissue oxygen supply by increasing the oxygen demand of the branchial and cardiac pumps

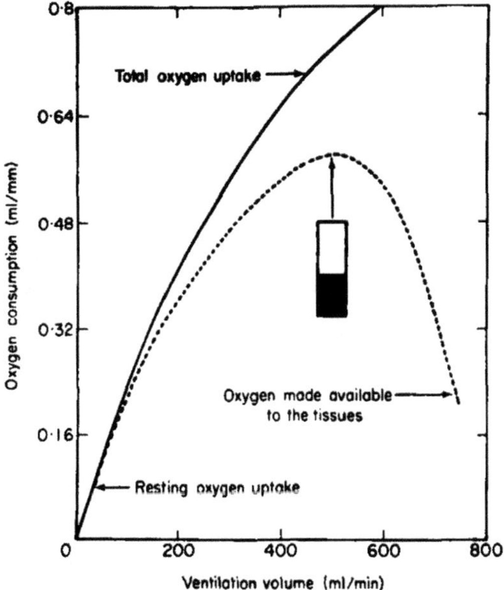

Fig. 8. The amount of oxygen made available to the tissues in an exercising salmonid-type fish at 15°C. The maximum oxygen uptake is predicted to occur when oxygen supply to the tissues is optimal. The arrow marks the optimum oxygen uptake at 15°C and the boxes the amount of oxygen consumed by the cardiac (□) and branchial (■) pumps. (From Jones, 197la, *J. Theor. Biol.* **32,** 341.)

(Jones, 1971b). This is probably a somewhat simplistic interpretation of the effects of hypoxia and anemia but nevertheless these results are not at variance with the theoretical analysis (Jones, 1971a).

There are also other metabolic costs, such as that for ionosmoregulation, which require some portion of the total aerobic energy available to the fish (Rao, 1968; Farmer and Beamish, 1969). However, whether the cost of ion-osmoregulation can be regarded in the same light as that of oxygen transport, that is, as a factor restricting tissue oxygen supply, is not clear. Certainly, the cost increases exponentially with U (Webb, 1975a), but so does the oxygen uptake and only if the cost of ion-osmoregulation increased at a substantially greater rate than the rate of oxygen uptake could this cost produce a point of optimum uptake where tissue supply was maximal. Support for this interpretation is provided by data not only on rainbow trout (Rao, 1968), but also on *Tilapia nilotica* (Farmer and Beamish, 1969) and *Kuhlia sandvicensis* (Muir and Niimi, 1972), for in these species there is no reduction in maximum performance or active levels of oxygen uptake in fresh or salt water or iso-osmotic media.

III. THE RESPIRATORY SYSTEM DURING EXERCISE

A. Introduction

The ability to exchange gases is limited by the surface area of the gas exchanger(s), A, the permeability of the surface to gases, k, the distance from the water to the blood, d, and the mean difference in oxygen partial pressure in the fluids on either side of the exchange surface, ΔPg. Therefore,

$$\text{oxygen uptake} = \frac{kA(\Delta Pg)}{d}$$

The main purpose of this section is to discuss how changes in the various components of the above equation facilitate an increase in oxygen uptake during exercise. It is obvious that a fish's maximum oxygen uptake will ultimately depend upon morphometric rather than physiological limitations for

$$\text{diffusing capacity} = \frac{\text{oxygen uptake}}{\Delta Pg} = \frac{kA}{d}$$

so some general relation between metabolism (M), A, and d is to be expected. Most commonly a relationship of this type is explored with respect to increase in body weight, W (Ultsch, 1973), rather than differences in exercise performance in animals of the same size. It is well established that M is not a linear function of W but rather a power function, W^b, where for standard M, b lies between 0.7 and 0.8. A similar relationship would be expected between A and W if d is weight independent. Hughes (1972) has presented data on both A and d for ten tench (*Tinca tinca*) varying in weight from 24.7 to 376 g and our analysis of the relation to W expressed in the allometric form is

$$A = KW^{0.72}$$
$$d = KW^{0.14}$$

so that diffusing capacity $= KW^{0.58}$. Consequently, as body weight increases in tench, the capacity of the gas exchanger will limit even the resting metabolic needs and will be unable to supply any extra oxygen for activity (Ultsch, 1973). However, for salmonids the relation of A to W approaches linearity, viz.,

$$A = KW^{0.91}$$

(derived from trout data given by Hughes, 1970; Hughes and Morgan, 1973). Hence, if d is relatively size-independent (interestingly d appears to be twice as large in trout as tench) then the gas exchanger will always be able to provide for resting metabolism regardless of the size of the fish, since in resting salmonids (Brett, 1965a)

$$M = KW^{0.78}$$

On the other hand, this relation changes during exercise to

$$M = KW^{0.95}$$

at maximum swimming speed (Brett, 1965a), which means that the gas exchanger is increasingly unable to cope with the active metabolic requirements of larger and larger fish. Hence, it can be argued that the size limitation for tench, which are relatively inactive fish, is set by resting M, while for salmonids it is set by active M.

B. Respiratory Adjustments to Exercise

1. Water Flow over and Blood Flow through the Gills

The amount of water flowing over the gills per unit time greatly increases during exercise (Saunders, 1962; Stevens and Randall, 1967a; Randall *et al.*, 1967; Heath, 1973; Roberts, 1975a; Kiceniuk and Jones, 1977). In resting fish, gill ventilation is achieved by alternate contractions and expansions of a buccal force pump and an opercular suction pump (Shelton, 1970) and both the rate and amplitude of these breathing movements increase during exercise (Stevens and Randall, 1967a; Davis, 1968; Sutterlin, 1969; Webb, 1971a; Heath, 1973; Kiceniuk and Jones, 1977) (Fig. 9). In rainbow trout (*S. gairdneri*) breathing rate falls to 71/min at intermediate U, from the resting rate of 83/min, although it rises to reach 146/min at maximum U (Webb, 1971a). However, Heath (1973) reported that the relative increase in breathing rate in trout is small compared to the marked increase in amplitude of both buccal and opercular breathing movements.

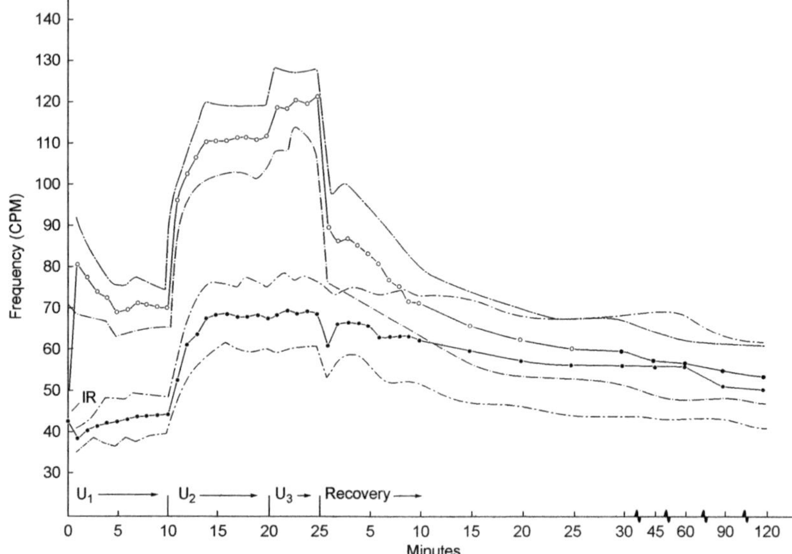

Fig. 9. The effects of various *U*'s ($U_1 = 0.8\ell$/sec; $U_2 = 1.5\ell$/sec; $U_3 = 2.2\ell$/sec) on the heart and ventilation rates of brown trout (*S. trutta*) at 8°C in a water tunnel. The trout were between 24 and 26 cm in length. ●, mean heart rate; ○, ventilation rate; dots and dashes, ±1 standard deviation. (From Sutterlin, 1969, *Physiol.Zool.* **42,** 36. Copyright 1969 by The University of Chicago Press.)

The forward motion of the fish will augment the action of the buccal and opercular pumps in irrigating the gills. In fact, when the speed of the fish reaches the range of 50–80 cm/sec then the kinetic energy of the flowing water is equivalent to a pressure of between 1 and 3 cm H_2O, which is adequate to provide a ventilation volume 10–15 times that obtained at rest. It is around this speed that many fishes cease making rhythmic breathing movements and rely on ram ventilation of the gills (Fig. 10). Among fishes which have been observed to change from rhythmic to ram ventilation in the 50-80 cm/sec velocity range are salmon (*O. nerka*) (Brett, 1964; Davis, 1968), remora (*Remora remora*) (Muir and Buckley, 1967), bluefish (*Pomatomus saltatrix*), Atlantic mackerel (*Scomber scombrus*), northern scup (*Stenotomus crysops*), and blue runner (*Caranx crysos*) (Roberts, 1975a) (Fig. 10). Some other fishes such as the skipjack tuna (*Katsuwonus pelamis*) and mackerel (S. *scombrus*) have largely lost the ability to pump water over the gills and are obligate ram ventilators (Hall, 1930; Magnuson, 1963; Brown and Muir, 1970). However, not all fast swimming fishes totally cease rhythmic breathing. Both rainbow trout and mullet (*Mugil cephalus*) often continue breathing in water tunnels at *U*'s which would appear to be adequate to maintain ventilation (Roberts, 1975b; Kiceniuk and Jones, 1977), yet resting rainbow trout will stop making breathing movements when force ventilated with volumes similar to those pumped by unrestrained resting fish (Jones and Schwarzfeld, 1974).

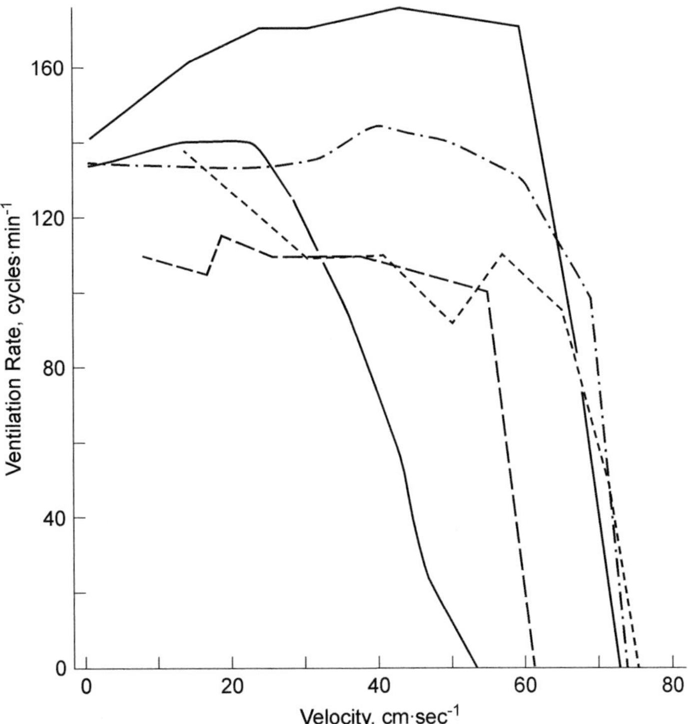

Fig. 10. The relationship between active gill ventilation rate and enforced *U* in five Atlantic mackerel (S. *scombrus*). When *U* reaches between 50 and 80 cm/sec the fish stop active gill ventilation and ram ventilate. [From Roberts, 1975a, *Biol. Bull.* (*Woods-Hole, Mass.*) **148,** 85.]

The most graphic demonstration of the pressure differences along the body of a fish, due to its forward *U*, has been given by Dubois *et al.* (1974, 1976). In bluefish (*P. saltatrix*), swimming at 1.8 m/sec, there is a pressure difference of more than 20 cm H_2O between the tip of mouth and the back of the operculum. However, small fish would never be able to attain the *U* necessary to generate sufficient upstream pressure to give adequate gill ventilation, while bottom feeders, with mouths at right angles or even directed away from the flowing water stream, would gain no assistance in gill ventilation. Small Atlantic mackerel (2–12 cm) actively ventilate their gills (Roberts, 1975b) which indicates that the capacity to ram ventilate is a consequence of the ability to swim fast, which only comes with increased size.

Even in those fish that do not rely on ram ventilation, forward motion may augment gill ventilation. There is evidence that respiratory and fin movements are coordinated, although in many fish the relationship between these activities is quite labile (von Holst, 1937; Satchell, 1968). However, in *Cymatogaster aggregata*, a fish which swims solely by means of pectoral fin movements, ventilation and fin beat are well synchronized so that the mouth-open phase of breathing coincides with the maximum rate of forward motion (Webb, 1975b). A 1:1 ratio between ventilation and fin movement was always observed at rest and was unchanged during exercise although the ratio rose to 2:1 during hypoxia. Satchell (1968) recorded discharges from medullary reticulomotor neurons in *Squalus acanthias* in time with respiration and suggested that these neurons may constitute a basis for this type of coordination.

Energetically there may be two effects of switching from rhythmic to ram ventilation. First, the drag characteristics of a swimming fish may be improved since the opercular clefts will no longer rhythmically protrude into and consequently disturb the flow profile over the fish. Furthermore, since the gills make a significant contribution to the total drag of the body there will no longer be sudden increases and decreases in drag with mouth opening, although the effect on U may be minimized by the coupling between locomotory movements and breathing so that maximum thrust coincides with maximum drag. Second, the work of gill ventilation will be transferred from the branchial to the swimming musculature and since the efficiency of contraction of the tail muscles is much higher than that of the branchial muscles (Webb, 1975a; Jones and Schwarzfeld, 1974), a saving in metabolic energy would be indicated. However, the situation is complicated by the fact that the branchial muscles are not relaxed but are tonically active during ram ventilation. The mouth of an anesthetized fish is forced closed or wide open if held in a water current, depending on the shape of the head and initial mouth gape (Roberts, 1975a), whereas mouth gape is adjusted with oxygen demand in swimming fish. Consequently, the net saving in energy for the fish is difficult to estimate.

Jones and Schwarzfeld (1974) calculated that the cost of breathing in resting, restrained rainbow trout (*Salmo gairdneri*) was about 10% of the total oxygen uptake and that the efficiency of the process was about 1%. Hughes and Saunders (1970) estimated a similar cost of breathing for the same species at ventilation volumes of 300–500 ml/kg/min, but at 1–2 liters /kg/min, during hypoxia, they estimated that oxygen cost of breathing could be as high as 50% of the total oxygen uptake. If 1% is taken as resting efficiency it is possible to construct a curve describing the efficiency of breathing with increases in work performed in breathing (Jones, 1971a). This relationship is shown in Fig. 11 and although it is a theoretical derivation the mean value of efficiency in trout (9.8%) given by Hughes and Saunders (1970), at what we calculate to be a 60–70 times increase in power output, fits it surprisingly well (Fig. 11). Hence, if efficiency increases to a maximum of 10% and then falls off once more as power output by the branchial muscles increases (Fig. 10) a portion of the increased cost of breathing at higher ventilation volumes will be offset by increased efficiency of breathing. It seems unlikely that power output by the branchial muscles will ever increase by more than 100 times in trout during sustained exercise so, due to the increase in efficiency, the oxygen cost of breathing would be the same proportion of the total oxygen uptake as it is at rest. In fact, at ventilation volumes reported for exercising trout (Kiceniuk and

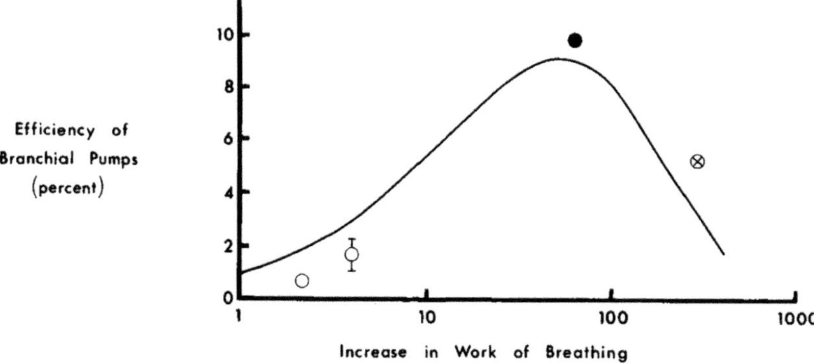

Fig. 11. A theoretical derivation of the change in efficiency of the branchial pumps of a salmonid fish with the number of times work performed in breathing increases. The curve is similar in shape to that derived by Jones (1971a) although the resting value is taken as 1% and not 4%. ○ represents values for efficiency obtained by Jones and Schwarzfeld (1974), while ● represents the mean efficiency value obtained by Hughes and Saunders (1970), at a power output of 62 times resting, during hypoxia. ⊗ is the efficiency value and power output calculated from Table II of Hughes and Saunders (1970).

Jones, 1977), applying Hughes and Saunders' (1970) data on oxygen cost of breathing gives an oxygen cost which is always less than 10% of the total oxygen metabolism up to U_{crit}.

The gills of skipjack tuna (*Katsuwonus pelamis*) (an obligate ram ventilator) contribute about 9% of the total drag [as calculated by Webb (1975b) from Brown and Muir (1970)] and Webb (1975a) suggested that gill drag might be of the order of 12% in neutrally buoyant fish because drag associated with the pectoral fins makes a smaller contribution to the total drag of the fish. Thus, it seems reasonable to assume that as the gills contribute about 10% of the drag, about 10% of the total energy expended in forward motion will be used in ram ventilating the gills. But, as Brown and Muir (1970) point out, gill drag increases as the square of the velocity of respiratory water flow. Consequently, doubling ventilation volume in tuna, at a given U, elevates gill resistance to 27% of the total swimming resistance and cost of swimming goes up. Brown and Muir (1970) showed that mouth gape in mackerel exposed to hypoxia increases at a given swimming speed and so does oxygen consumption, reflecting the additional work being performed to overcome the extra gill drag. However, under normoxic conditions, increased ventilation volume can be achieved in step with U and since ram pressure head generation and total body drag have the same power relation to U then it might be expected that, as in rhythmic breathers, the oxygen cost of ram ventilation, as a proportion of total oxygen uptake, would remain close to resting values (less than 10% of total oxygen uptake).

The control of breathing movements during exercise and the regulation of the switch from rhythmic breathing to ram ventilation are not understood.

Breathing rate changes instantaneously with increased *U* (Fig. 9; Sutterlin, 1969) which indicates central nervous drive or reflex excitation of the medullary respiratory neurons by receptors detecting body movements. However, in ram ventilators there must be another set of receptors whose input inhibits the output of rhythmic medullary respiratory neurons when ram ventilating U's are attained (Ballintijn and Roberts, 1976). Roberts (1975b) has suggested that the receptors whose input inhibits breathing may be those flow-sensitive elements of the acousticolateralis system. Alternatively, mechanoreceptors on the gill arches (Sutterlin and Saunders, 1969) could monitor water flow velocity and shut down breathing at a predetermined speed. Other possible candidates for initiating apnea during ram ventilation are stretch-sensitive receptors in the skin (Roberts, 1969, 1972), whose input could presumably be integrated to indicate tailbeat frequency. In any event, there does not appear to be a "master" switch in the sense that either the animal ram ventilates or not. In some instances the transfer to ram ventilation can be a graded process, rhythmic cycles falling out for longer and longer periods when *U* increases slowly (Fig. 12; Roberts,

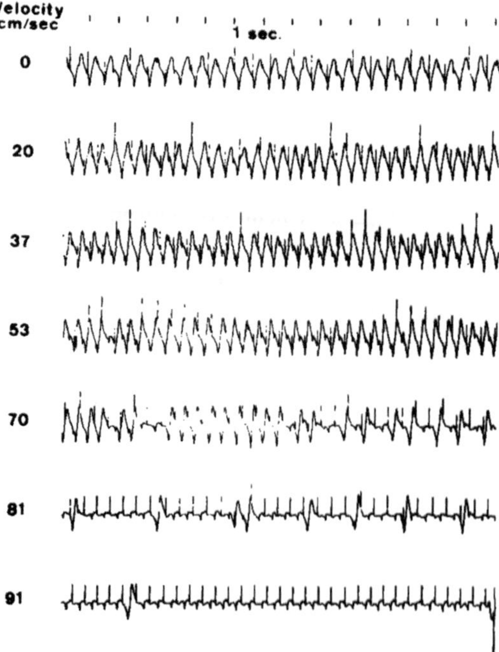

Fig. 12. Breathing movements, indicated by slow wave "muscle potentials" from the *adductor mandibularis*, and electrocardiogram of a bluefish (*P. saltatrix*) during acceleration to and above *U* sufficient to support ram ventilation. Breathing movements increase in frequency at low *U*'s and slow again as ram ventilation ensues at higher *U*'s, stopping at *U*'s of 91 cm/sec. The time marks denote 1 sec intervals. [From Roberts, 1975a, *Biol. Bull.* (*Woods Hole, Mass.*) **148,** 85.]

1975a). On the other hand, the latency of the "switch over" can be quite short, for when sudden velocity changes, from above to below ram ventilation speed, are imposed on swimming fish then breathing movements can be initiated within 0.2 sec.

Chemoreceptors may play some role in regulating ram ventilation even though they do not appear to be involved in either the switch from rhythmic breathing to ram ventilation or in causing the initial increase in rhythmic breathing at the start of exercise. Ram ventilating fish can adjust gill water flow by changing mouth gape at a given swimming speed. Mackerel (*S. scombrus*) increase mouth gape in response to increased oxygen demand after feeding and during hypoxia (Brown and Muir, 1970) but the nature and location of the receptors mediating this response are unknown.

Gill ventilation volume has never been directly measured in swimming fish although measurements have been made on a variety of restrained fishes by modifications of the van Dam (1938) technique or dye dilution method (Millen *et al.*, 1966). In exercising or unrestrained fishes, ventilation volume has been calculated by application of the Fick principle to measurements of oxygen uptake and oxygen content of mixed inspired and expired water (Saunders, 1962; Stevens and Randall, 1967b; Kiceniuk and Jones, 1977). Saunders (1962) observed a decrease in oxygen extraction per unit volume of water (percentage utilization) with exercise in a number of fish species, and his calculated increases in ventilation volume for a given rise in oxygen uptake were higher than those of Stevens and Randall (1967b) and Kiceniuk and Jones (1977), who observed little change in percentage O_2 utilization in the trout (*Salmo gairdneri*) with exercise. However, Davis and Watters (1970) criticized the sampling method used by both Saunders (1962) and Stevens and Randall (1967b) stating that small water samples taken via a polyethylene tube inserted into the opercular cavity were not representative of mixed expired water. Kiceniuk and Jones (1977) modified the technique of Davis and Cameron (1971), placing a rubber curtain over the operculum in order to create a collecting and mixing area for expired water, from which samples were taken as the fish swam. They observed that resting utilization of 33% was little changed in exercise and ventilation volume therefore increased linearly with oxygen uptake (Fig. 13). An increase in ventilation without a fall in percentage utilization was also recorded by Davis and Cameron (1971) who measured gill water flow directly in resting fish. They observed no change from resting utilization of 40–50% as the nonswimming fish voluntarily increased ventilation volume from 44 to 120 ml/min. However, these changes were only one-tenth those recorded by Kiceniuk and Jones (1977).

The percentage utilization of oxygen from water flowing over the gills is a measure of the amount of water actually involved in gas exchange (the respiratory volume) and an inverse measure of the size of the combined residual volume and water shunt. The latter has been characterized as a series of dead spaces (Randall, 1970b), namely:

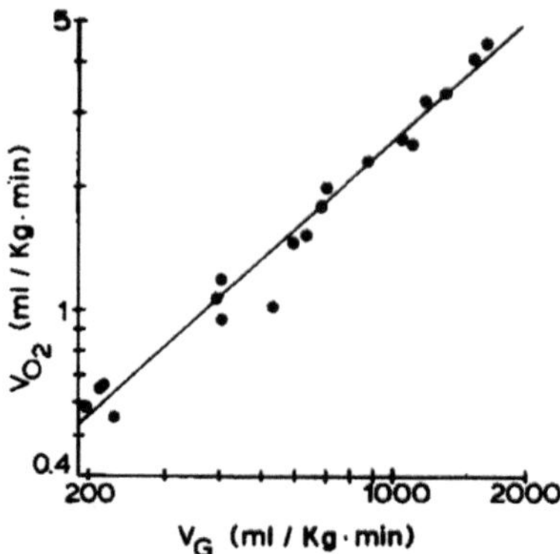

Fig. 13. The relationship between oxygen consumption and ventilation volume in rainbow trout at rest and during exercise of increasing intensity up to U_{crit}. (After Kiceniuk and Jones, 1977, *J. Exp. Biol.* **69,** 247.)

1. Diffusion dead space, where water and blood remain in contact with the gill epithelium for too short a time for blood and water gas tensions to equilibrate
2. Distribution dead space, associated with poor matching between ventilation and blood perfusion so that too much or too little oxygen is delivered either to portions or the whole gill sieve than is required to saturate the blood
3. Anatomical dead space, where water takes a nonrespiratory path through the gill sieve, for instance, between the tips of adjacent filaments

Increases in ventilation volume during exercise will tend to increase the magnitude of the diffusion dead space because of the reduced transit time for water (Randall, 1970b). If oxygen utilization remains constant as ventilation increases, however, then there must be a concomitant reduction in the magnitude of either the distribution or anatomical dead space. It is unlikely that the magnitude of the anatomical dead space will decrease at high ventilation rates; if anything, one might expect an increase in the anatomical dead space due to disruption of the gill sieve. Thus any increase in diffusion dead space, resulting from a reduction in transit time, is most probably offset by a reduction in distribution dead space in the gills during exercise which could be achieved by changes in flow characteristics of water or blood at the level of the gill filament, primary, or even secondary lamella.

The small leaflike secondary lamellae are the basic unit for gas transfer between blood and water (Fig. 14). Each secondary lamella has a small afferent arteriole which widens into twenty or thirty channels through the body of the lamella, narrowing to a small arteriole connected to the efferent circulation (Hughes and Morgan, 1973; Vogel *et al.*, 1973, 1976; Laurent and Dunel, 1976; Gannon *et al.*, 1978a,b,c). The channels are formed by pillar

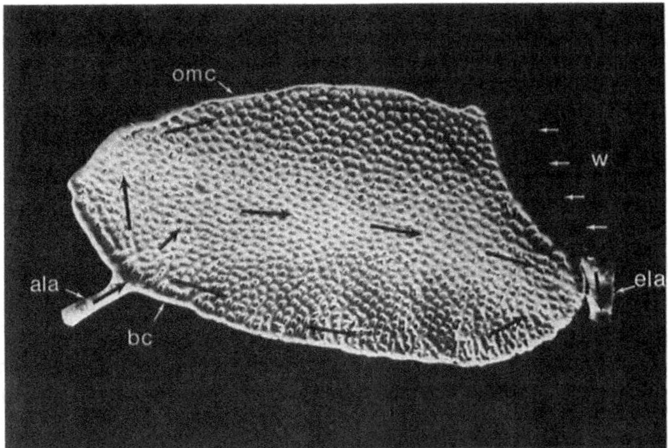

Fig. 14. Cast of the secondary lamellar vascular space dissected from a complete cast of the branchial vasculature. The black arrows indicate the probable paths of the erythrocytes. The direction of water flow (w) is shown by the white arrows. Blood enters through the afferent lamellar arteriole (ala) and leaves through the efferent lamellar arteriole (ela). The outer marginal channel (omc) and basal channel (bc) may serve as shunts or to improve the distribution of blood across the surface of the lamella (×140). (From Smith, 1976.)

cells which completely surround the blood spaces and function to hold the walls of the lamellae together, the latter being composed of a layer of epithelial cells. The thickness of the blood spaces within the secondary lamellae could conceivably be altered by contraction of the pillar cells (Hughes and Grimstone, 1965; Newstead, 1967; Bettex-Galland and Hughes, 1973) which could result in changes in flow distribution through the lamellae. Rankin (1976) has concluded, from direct visual observation on eels (*Anguilla anguilla*), that most of the blood flows through the somewhat larger peripheral vessels of the secondary lamellae (Fig. 14). A more even distribution of flow within the secondary lamellae during exercise would effectively increase the total surface area involved in gas exchange. Contraction of pillar cells would increase the area of the water channel correspondingly and water velocity past the lamellae would be reduced, favoring gas exchange, although diffusion distances in water would be increased.

The secondary lamellae are arranged in a ladderlike series along the gill filaments and in trout the walls of the arterioles leading to the proximal lamellae are devoid of smooth muscle, and these lamellae are probably perfused with blood at all times. The arterioles at the inlet and outlet of more distal lamellae, however, have a thick smooth muscle coat (Gannon *et al.*, 1978a) and it seems likely that perfusion of these lamellae with blood could be regulated according to requirements of gas exchange. Booth and Holeton (1977) have shown in resting trout (S. *gairdneri*) that only about 60% of the lamellae are perfused. Intravascular injection of acetylcholine further reduced lamellar perfusion to 40%, while after adrenaline injection at least 95% of all secondary lamellae are perfused. Dilatation of distal lamellar arteries could occur during exercise due to adrenergic nerve stimulation or increased levels of circulating catecholamines, exciting adrenergic β-dilator receptors associated with the smooth muscle coats of these arteries (Smith, 1976).

If lamellar recruitment occurs during exercise then the total gill area involved in gas transfer will increase and certainly change the magnitude of the distribution dead space. In resting fish, water flows over regions of the gill filament where there is little or no perfusion, creating a large distribution dead space. Perfusion of these regions of the filament during exercise could reduce the magnitude of the distribution dead space, which will tend to offset any increase in diffusion dead space associated with the increase in gill ventilation and maintain a constant oxygen utilization from water as it passes over the gills.

2. *Respiratory Gas Transport in the Blood and Peripheral Blood Flow*

Hemoglobin in arterial blood remains almost fully saturated with oxygen during exercise in salmonid fish. Kiceniuk and Jones (1977) found that oxygen saturation was 97% at rest in trout (*Salmo gairdneri*) and varied between 96 and 100% during exercise up to 92% of critical velocity. Stevens and Randall (1967b) observed no change in either arterial or venous oxygen tension during exercise; however, they used a different exercise pattern from Kiceniuk and Jones (1977), who observed a marked reduction in venous oxygen tension from 33 mm Hg at rest to 16 mm Hg during exercise. This was associated with a reduction in percentage saturation of hemoglobin in mixed venous blood from 70% at rest to 15% at exercise levels approaching the critical velocity for these fish. Thus the arterial–venous oxygen difference increased by a factor of between 2 and 3 during exercise in trout. Secondat (1950) also measured a reduction in venous oxygen content in carp (*Cyprinus carpio*) after exercise. Blood was collected by cardiac puncture and the mean venous oxygen content was 9.33 vol% ($n = 6$) before and 5.84 vol% ($n = 6$) after exercise. The reduction in venous content recorded in both carp and trout reflects increased tissue extraction of oxygen. The reduced venous oxygen tension will also increase oxygen differences across the gills and augment oxygen uptake.

Cole and Miller (1973) have suggested that, for mammals, the product of heart rate, stroke volume, and venous oxygen content is constant during submaximal work. This is also true for fish swimming at up to 90% of U_{crit}, whereas at maximal work outputs this product falls from values around 120 to 75 (Table III; Kiceniuk and Jones, 1977). It may be that the constancy of the quantity of oxygen returning to the heart reflects, as Cole and Miller (1973) point out, a peripheral regulation of oxygen delivery whereby oxygen extraction or blood flow rates are adjusted to match increases in supply with demand. However, one requisite of such a regulatory mechanism would be a means of determining venous oxygen tension, or preferably, content. In fact, Taylor *et al.* (1968) from a computer simulation of the cardiovascular and respiratory responses to exercise in trout (*Salmo gairdneri*) concluded that these responses must be controlled by a venous oxygen sensor.

It is clear that exercise in fish results in an increase in peripheral blood flow. associated with a reduction in the resistance to flow. Although there is an increase in the metabolic cost of circulation, gill ventilation, and osmoregulation during exercise, most of the increased oxygen uptake must be utilized by the muscles involved in locomotion. In mammals there is a marked shift in blood flow away from the gut and most of the increase in cardiac output is directed toward the exercising muscles. Stevens (1968), using radio-iodinated human serum albumin as a marker, could find no change in blood volume in any organ except the spleen (blood volume reduced) during exercise in trout. The method used, however, measured volumes and not flows. There might be marked changes in flow with only small changes in volume which would be undetected by this method. Clearly this question needs further examination.

The fish heart consists of an outer cortex and an inner trabeculated layer (Satchell, 1971; Cameron, 1975). The outer cortical layer has a rich coronary supply and in some species consists of about 20–30% of the ventricular mass. The inner spongy layer, like the atria, has few or no capillaries and must rely on the blood flowing through the chambers for nourishment and gas exchange (Voboril and Schiebler, 1970; Oštádal and Schiebler, 1971; Santer and Cobb, 1972; Cameron, 1975). The ratio of cortex to inner layer mass changes with the species. Sluggish species such as the toadfish (*Opsanus tau*) have a very small cortical layer whereas the skipjack tuna (*Katsuwonus pelamis*) has a well-developed cortical layer and coronary circulation. Changes in coronary blood flow during exercise have not been measured, but coronary flow presumably increases to meet the increased requirements of the cortex during exercise.

Based on data of Kiceniuk and Jones (1977), the oxygen uptake of the trout heart is calculated to be 3.5% of the total resting oxygen uptake, assuming that cardiac efficiency is 20% (Jones, 1971a) and that cardiac output multiplied by mean ventral aortic pressure is a measure of external work done by the heart in fish. At maximum exercise in rainbow trout, assuming a drop in cardiac efficiency to 10% (Jones, 1971a), the heart consumes 4.5% of the total oxygen uptake. That is, at maximum exercise levels oxygen uptake by the heart in trout increases by a factor of 9.6 whereas total oxygen uptake of the fish increases only 7.8 times above the resting level. Cardiac output triples and the large increase in calculated oxygen uptake by the heart is due to the assumption that cardiac efficiency is halved at maximum exercise. It is important to stress that nothing is known about efficiency of contraction in fish hearts and the above assumptions concerning cardiac efficiency are based on observations on mammalian hearts.

Around 70% of the ventricle of trout (*Salmo gairdneri*) is not supplied by coronary vessels but relies on blood flowing through the heart for nourishment. If one assumes that the inner spongy portion of the ventricle consumes 70% of the ventricular oxygen uptake then, as this oxygen comes from venous blood flowing through the ventricle, about 1% of the venous oxygen content will meet the ventricular requirements at rest. Venous blood flowing through the heart is 70% saturated with oxygen in resting trout (*Salmo gairdneri*). During exercise, however, the venous oxygen saturation falls to about 15% in rainbow trout (Kiceniuk and Jones, 1977) and 16% of the venous oxygen content must be removed to meet the requirements of the ventricle. In fact the above calculations indicate that at 92% critical velocity in trout more than three times as much oxygen must be removed from a unit volume of blood flowing through the heart; the blood remains within the ventricle for one-third as long and contains only about one-fifth as much oxygen as venous blood in the resting fish. It is possible that the myocardium becomes partially anaerobic under these conditions. Hearts from different species show different tolerances to anoxia (Gesser and Poupa, 1974), and it is possible that the anaerobic component of cardiac metabolism is elevated during exercise in fish. Alternatively, the ventricular myocardium may be able to extract oxygen from the venous blood even at very low blood oxygen tensions and contents.

Kiceniuk and Jones (1977) sampled blood before (cardinal sinus) or after (ventral aorta) it had passed through the heart. They observed no measurable differences in oxygen content between these two sites. This is to be expected in resting fish where only about 1% difference in content is predicted. A difference, however, should be apparent during maximum exercise. Unfortunately, at maximum exercise, Kiceniuk and Jones (1977) obtained only two samples from each site, and in no case were both sites sampled in a single fish. Their reported variability in venous oxygen content is in fact larger than the expected calculated difference in oxygen content in blood entering and leaving the ventricle. Thus this problem still needs further investigation.

Stevens (1968) showed a reduction in splenic volume in the trout (*Salmo gairdneri*) during exercise. This organ has both adrenergic and cholinergic innervation and nerve stimulation or application of α-adrenergic agonists will cause splenic contraction and the release of erythrocytes in cod (*Gadus morhua*) and tench (*Tinca tinca*) and to a lesser extent in dogfish (*Scyliorhinus canicula and Squalus acanthias*) (Nilsson *et al.*, 1975; Holmgren and Nilsson, 1975). The release of erythrocytes will cause an increase in circulating levels of hemoglobin.

Hematocrit, hemoglobin, and plasma protein levels increase during exercise in the rainbow trout (*Salmo gairdneri*) in freshwater (Stevens, 1968; Wood and Randall, 1973a; Kiceniuk and Jones, 1977). At maximum activity the increase in hematocrit is 9–14% above the resting level. This hemoconcentration is due not only to erythrocyte release from the spleen but also to a reduction in plasma volume caused by a marked diuresis (Wood and

Randall, 1973c). There are also longterm changes in hemoglobin levels with exercise. Hochachka (1961) found that rainbow trout made to swim in a current for 6 months had higher hemoglobin levels and larger hearts than untrained fish.

Both short and long-term hemoconcentration presumably augments oxygen transport and facilitates acid/base regulation during exercise. Cameron and Davis (1970) have shown in rainbow trout that variations in blood oxygen capacity are compensated for by changes in cardiac output. They argue that a given species has evolved a particular blood oxygen capacity so that the heart operates over a favorable efficiency range. In some species hemoglobin levels may have been selected so that maximum efficiency of the heart (external heart work divided by the total energy expended by the heart) occurs during prolonged exercise, as, for instance, in migrating salmon. In these fish a small increase in hemoglobin levels may result in a reduction in cost of cardiac work and be of enormous survival value. Cameron and Davis (1970) reported that anemic rainbow trout could hardly sustain any swimming activity and Jones (1971b) found that a decrease in hematocrit from 32 to 11% almost halved maximum swimming velocity in rainbow trout at 21°–22°C. Cameron and Davis (1970) also suggest that in some fish, for example, the toadfish (*Opsanus tau*), maximum cardiac efficiency may span much lower levels of activity. The toadfish is generally sluggish and shows only short bursts of activity and may simply tolerate a large oxygen debt during these bursts. In this case we would predict no change in hemoglobin levels during activity.

Blood volume of fishes lies in the range from 2 to 12% of the body weight (Thorson, 1959, 1960; Conte *et al.*, 1963; Smith and Bell, 1964, 1967; Smith, 1966; Hemmingsen and Douglas, 1970) with that of salmonids lying in the more restricted range of 4–10%. Resting cardiac outputs have been determined or estimated for a fairly large number of fishes but only in salmonids has cardiac output been determined during sustained exercise (Stevens and Randall, 1967a,b; Kiceniuk and Jones, 1977). In trout (*Salmo gairdneri*) resting cardiac output lies between 15 and 20 ml/kg/min for animals in the size range from 0.25 to 1.5 kg (Stevens and Randall, 1967b; Cameron and Davis, 1970; Kiceniuk and Jones, 1977). Hence one can expect long circulation times (blood volume divided by cardiac output) in these animals, ranging from 1 to5 min. In fact Davis (1970) obtained an estimate for mean circulation time in resting rainbow trout (approximately 200 g) of 64.1 sec by a dye injection technique, and similar circulation times were estimated by Itazawa (1970) for carp (*Cyprinus carpio*) and eel (*Anguilla japonica*) of about 200 g. When cardiac output increases during exercise, circulation time will be reduced, but, even so, in large fish (>1 kg) it is unlikely that mean circulation time will ever fall below a minute.

Exercise results in an increased production of CO_2 as well as oxygen uptake. The pattern of CO_2 excretion in resting fish is not well understood and even less is known about CO_2 excretion during exercise or its effect on oxygen transport. What is clear is that CO_2 can be transferred across the gills as either molecular CO_2 or bicarbonate ion and most of the excreted CO_2 comes from plasma bicarbonate. The transit time for blood flow through the gills is only a small fraction of that required for the uncatalyzed bicarbonate dehydration reaction, especially at temperatures of less than 20°C. Diamox (an inhibitor of carbonic anhydrase) reduces CO_2 excretion (Haswell and Randall, 1978) and increases CO_2 tensions in arterial blood in both normal (Hoffert and Fromm, 1966) and anemic fish (Haswell and Randall, 1978). The erythrocytic carbonic anhydrase is unavailable for plasma bicarbonate dehydration (Haswell and Randall, 1976) but there are similar levels of carbonic anhydrase in the gill epithelium. Plasma bicarbonate enters the gill epithelium where it is dehydrated to CO_2 before diffusing into the water. Presumably the fish is able to regulate plasma bicarbonate by controlling its movement through the gill epithelium (Randall *et al.*, 1976) and thus adjusts the CO_2: bicarbonate ratio and plasma pH. Haswell and Randall (1978) have proposed the following model for CO_2 excretion in fish (Fig. 15).

a. CO_2entering the blood from the tissues is rapidly dehydrated to form bicarbonate within the red blood cell (RBC). This reaction is catalyzed by RBC carbonic anhydrase and drives oxygen from hemoglobin.

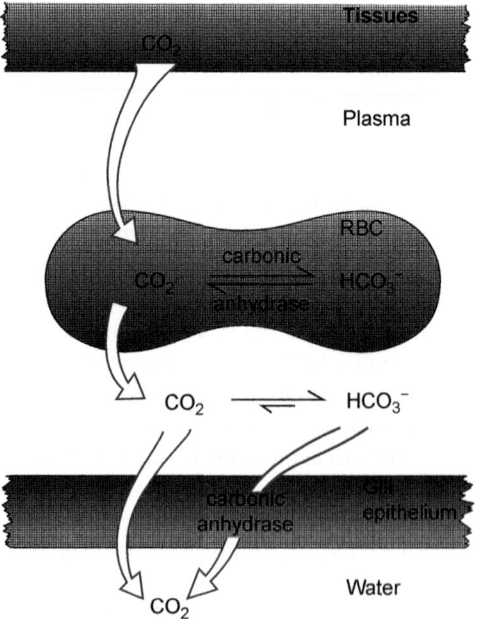

Fig. 15. A diagram depicting the pathway of carbon dioxide excretion in fish, based on the description of Haswell and Randall (1978).

b. The RBC swells due to water entry and, unlike mammals, there is no exchange of bicarbonate for chloride across the RBC membrane.
c. Bicarbonate is formed at the uncatalyzed rate from CO_2 in plasma; this reaction is slow and so bicarbonate is formed after blood has left the tissues and while it is in the veins.
d. The blood in the veins is a closed system as far as gas transfer is concerned so, as plasma bicarbonate levels increase, CO_2tension levels fall and RBC bicarbonate is dehydrated as CO_2 diffuses from the red cell into the plasma. Thus RBC bicarbonate formed while blood is in the tissue capillaries is dehydrated to CO_2, diffuses into the plasma, and forms bicarbonate while blood resides in the veins.
e. Blood entering the gills therefore has a higher plasma bicarbonate and a lower RBC CO_2 tension than blood leaving the tissues. The increase in RBC pH, resulting from the diffusion of CO_2 into the plasma, leads to an increased hemoglobin oxygen affinity, oxygen is bound to hemoglobin, and venous oxygen tension falls as blood flows from the tissues to the gills, augmenting oxygen transfer across the gills by decreasing venous oxygen levels.

It is not known how the pattern of CO_2 excretion is affected by exercise but even during exercise residence times for blood in the veins is probably still sufficient for adequate rates of plasma CO_2 hydration at the uncatalyzed reaction velocity. Stevens and Randall (1967a) observed increased venous CO_2 tension in trout, and Wood *et al.* (1977) measured a rise in total CO_2 as well as tension in venous blood after exhausting exercise in flounder (*Platichthys stellatus*). Obviously increased metabolic activity in the tissues results in increased CO_2 excretion to, as well as increased O_2 extraction from, blood. The elevated venous CO_2 tension is undoubtedly a contributing factor, for a given fall in venous oxygen tension, in causing the marked reduction in venous hemoglobin oxygen saturation.

Auvergnat and Secondat (1942) observed a reduction in plasma total CO_2 after exercise in the carp (*Cyprinus carpio*); they also measured a fall in blood pH. Wood *et al.* (1977) also recorded a fall in pH due to lactate accumulation during recovery from exercise in the flounder (*Platichthys stellatus*). Venous CO_2 content was elevated early in the recovery period but eventually fell below resting levels during recovery (Fig. 16) when the pH was reduced by lactate accumulation. The acid conditions result in dehydration of bicarbonate and a consequent rise in blood CO_2 tension. This increases the CO_2 tension difference between blood and water, enhances the excretion of CO_2 across the gills, and results in a reduction in plasma CO_2 content. Auvergnat and Secondat (1942) presumably missed the initial elevation in total CO_2 in blood observed by Wood *et al.* (1977) and reported only the secondary reduction in plasma total CO_2 after exercise.

Lactate is produced in large quantities in white muscle fibers during violent exercise; it enters the blood slowly, however, and the fish is usually

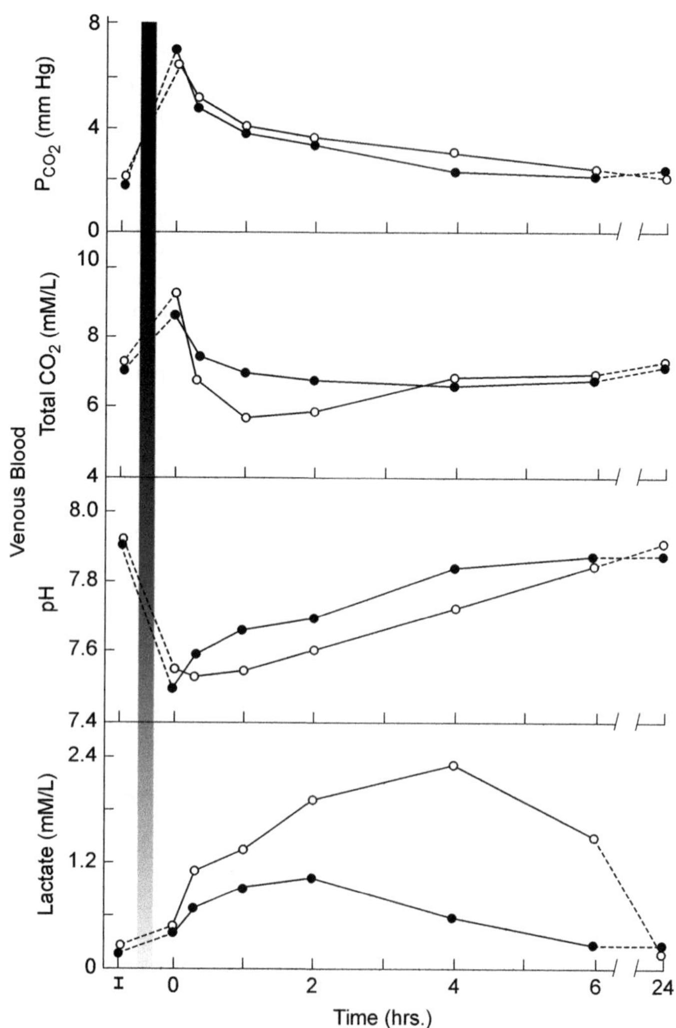

Fig. 16. Carbon dioxide tension (P_{CO_2}), calculated from the Henderson–Hasselbach equation, total CO_2 concentration, pH, and lactate concentration in the venous blood of two flounders (● = 800 g; ○ = 450 g) before and after 10 min of exhausting activity (bar). I = initial resting sample. Time 0 = immediately postexercise. (Modified from Wood *et al.*, 1977, *J. Exp. Biol.* **69,** 173.)

exhausted before there is any appreciable rise in blood lactate. Maximum levels of blood lactate occur 2–3 hr after the end of the exercise period (Black *et al.*, 1962). During exercise there may be a small acidosis related to elevated CO_2 levels, but reductions in pH due to lactate production do not appear until after the exercise period has ended. As Wood *et al.* (1977) have pointed out, the separation of the acidosis due to CO_2 and that due to lactate reduces the magnitude of the pH drop during exhausting activity, an important fact when one bears in mind the low buffering capacity of fish blood compared with that of mammals (Albers, 1970). The acidosis during exercise may also be ameliorated by the retention or uptake of H^+ in muscle tissue, as occurs during hypercapnia (Randall *et al.*, 1976), or by the movement of H^+, HCO_3^-, or OH^- ions across the gills (Maetz, 1973; Randall *et al.*, 1976).

A decrease in RBC pH results in a fall in hemoglobin oxygen capacity (Root shift) as well as hemoglobin oxygen affinity (Bohr shift). A sharp rise in lactate levels might affect arterial oxygen saturation via the Root effect, but trout (*Salmo gairdneri*) red cells are relatively impermeable to lactate (Randall *et al.*, 1978). There is a detectable loss of ^{14}C-labeled lactate from plasma, presumably into fish RBC's, *in vitro* at 11°C 8 hr after elevation of plasma lactate, but equilibrium is still not complete after 24 hr of exposure. Furthermore, trout RBC's have some capacity to regulate intracellular pH (Randall *et al.*, 1978) so a rise in blood lactate may have little effect on oxygen transfer.

3. *Ventilation: Perfusion Relationships*

Both blood flow through and water flow over the gills are pulsatile. During exercise the oscillations in water flow will be reduced as ram ventilation becomes more important at higher swimming speeds but blood flow pulsatility will increase if stroke volume increases markedly. Jones *et al.* (1974) have argued that pulsatile blood flow will only have a major effect on gas transfer if stroke volume is larger than gill volume. Under these circumstances the transit time for blood through the secondary lamellae at peak flow rates will be much less than that at minimum flow rates at the end of ventricular diastole. However, if stroke volume is less than lamellar blood volume, all blood will spend at least the equivalent time of one cardiac cycle during passage through the gills. Jones *et al.* (1974) reviewed estimates of gill blood volume and concluded that stroke volume is generally less than gill volume. However, from examination of vascular casts of trout gills (*S. gairdneri*), Gannon *et al.* (1978a) obtained values for total lamellar blood volume of about 0.7 ml/kg. If only 60% of these lamellae are perfused at rest (Booth and Holeton, 1977) then functional lamellar volume in resting trout is 0.42 ml, expanding to 0.7 ml when all lamellae are perfused during exercise. Kiceniuk and Jones (1977) estimated that stroke volume in a l-kg rainbow trout increased from 0.46 ml to 1.03 ml between rest and maximum exercise, so at rest or in exercise blood will only reside in the gills for one heart beat. In the resting sea raven (*Hemitripterus americanus*) blood takes four to seven heart beats to pass from the ventral to dorsal aorta (Stevens and Sutterlin, 1975). The gill lamellae in trout represent one-quarter (100% of lamellae perfused) to one-sixth (60% of lamellae perfused) of the total gill blood volume and if similar relationships exist in the sea raven then, as in trout, there appears to be an approximately one-to-one relationship between gill lamellar volume and stroke volume of the heart.

The ventilation/perfusion $(\dot{V}g/\dot{Q})$ ratio in fish is usually between 10 and 15 and reflects the oxygen content of the two media, blood and water (Rahn, 1966; Piiper and Baumgarten-Schumann, 1968; Holeton and Randall, 1967b; Cameron and Davis, 1970; Cameron *et al.*, 1971; Jones *et al.*, 1970; Kiceniuk and Jones, 1977). Thus, in fish, about ten times more water flows over the gills than blood perfuses them per unit time. Stevens and Randall (1967a) calculated a much higher $\dot{V}g/\dot{Q}$ ratio for trout (*Salmo gairdneri*), but this has not been substantiated by other workers and probably reflects either special experimental conditions (a water tunnel) or measurement inaccuracies. There are probably differences in $\dot{V}g/\dot{Q}$ ratios between species related to differences in blood hemoglobin oxygen affinity and capacity, the area of the gills, and venous oxygen levels (Jones *et al.*, 1970).

Kiceniuk and Jones (1977) found that exercise in rainbow trout resulted in an increase in the $\dot{V}g/\dot{Q}$ ratio from 12 to 32 (Table II). This increase in the $\dot{V}g/\dot{Q}$ ratio was the result of a much greater increase in Vg than $\dot{Q}$ during exercise. Ventilation volume in rainbow trout increased from 211 to 1700 ml/min/kg between rest and swimming speeds of up to 92% U_{crit}, that is, by a factor of 8. This must reduce the transit time for water flow through the gills and yet oxygen utilization from water did not change (Table II). At the same time cardiac output increased by a factor of over 3, decreasing residence time in the gills, and yet arterial blood remained fully saturated with oxygen. Although lamellar recruitment or a more even distribution of flow within a single secondary lamella would promote oxygen transfer, these effects will be offset by the decreased residence time for blood in the secondary lamellae. However, since cardiac output increases by a factor of 3, blood residence time in the secondary lamellae only decreases to one-half the resting value, and not one-third, due to lamellar recruitment. Consequently, since water flow increases by a factor of 8, there is a much more marked increase in the ventilation : perfusion ratio at the secondary lamellar level in terms of the flow velocities of the two media, than is apparent from analysis of the ratio of cardiac output and total gill water flow.

The carbon monoxide diffusing capacity of the gills can be used to indicate the ability of the gills to exchange oxygen. Fisher *et al.* (1969) concluded, from experiments in which carbon monoxide diffusing capacity was measured in catfish (*Ameiurus nebulosus*), that oxygen transport across the gills is primarily diffusion limited at ambient water oxygen levels. They observed that CO diffusing capacity doubled during hypoxia which they attributed to changes in conditions for diffusion between water and blood rather than diffusion and/or reaction velocity limitations within water or blood. The increase in diffusing capacity during hypoxia could have been related to lamellar recruitment, changing patterns of water and blood flow affecting the gill area

Table II Oxygen Transfer Factor (T_{O_2}) and Ventilation: Perfusion Ratio ($\dot{V}g/\dot{Q}$) for the Whole Gill in Resting and Exercising Rainbow Trout Compared with Values Obtained by Assuming that Only 60% of Secondary Lamellae Are Perfused at Rest and 100% at Maximum U[a]

Exercise level (U expressed as a proportion of maximum U)	Arterial Oxygen tension $P_{a_{O_2}}$ (mm Hg)	Mixed Venous oxygen tension $P_{\bar{v}_{O_2}}$ (mm Hg)	Inspired Water oxygen tension $P_{I_{O_2}}$ (mm Hg)	Exhaled Water oxygen tension $P_{E_{O_2}}$ (mm Hg)	Oxygen consumption $\dot{V}_{O_2}$ (ml/kg/min)	Mean Oxygen difference across gill ΔPg $\left[\frac{1}{2}\left(P_{I_{O_2}} + P_{E_{O_2}}\ \frac{1}{2}P_{a_{O_2}} - P_{\bar{v}_{O_2}}\right)\right]$ (mm Hg)	Mean[b] Oxygen difference across perfused lamellae ΔP_{lam} (mm Hg)	Oxygen transfer factor (ml/min/mm Hg): T_{O_2} Whole gill ($\dot{V}_{O_2}/\Delta Pg$)	Oxygen transfer factor (ml/min/mm Hg): T'_{O_2} Perfused lamellae ($\dot{V}_{O_2}/\Delta P_{lam}$)	Ventilation perfusion ratio $\dot{V}g/\dot{Q}$: Whole gill	Ventilation perfusion ratio $\dot{V}g/\dot{Q}$: Perfused lamellae
Resting	137	33	152	102	0.56	42	25	0.013	0.022	12	7.2
70–78% U_{crit}	123	23	155	102	1.9	53	—	0.036	—	21	—
81–91% U_{crit}	128	29	146	102	3.12	48	—	0.061	—	30	—
92–100% U_{crit}	126	16	151	102	4.34	56	56	0.078	0.078	32	32

[a] Data from Kiceniuk and Jones (1977).
[b] Anatomical dead space assumed to be zero.

available for gas transfer, or a change in the diffusion distance between water and blood. The catfish were "excited and exercising" so it is possible that all lamellae were perfused at all oxygen levels, and changing conditions for diffusion within each lamellae accounted for the increased CO diffusing capacity during hypoxia. Fisher *et al.* (1969) reported that the gill oxygen diffusing capacity changed in the same way as carbon monoxide diffusing capacity during hypoxia, indicating that increases in oxygen diffusing capacity are also related to changes in conditions for diffusion across the gill epithelium.

Another measure of the capacity of the gills to exchange oxygen is given by the transfer factor (T_{O_2}) which is defined as oxygen uptake divided by mean oxygen difference across the gills (ΔPg) between water and blood (Randall *et al.*, 1967). Randall *et al.* (1967) and Kiceniuk and Jones (1977) both observed a five- to sixfold increase in gill oxygen transfer factor during exercise in trout (*Salmo gairdneri*) (Table II), presumably reflecting either an increase in area available for diffusion or a decrease in the distance for diffusion between blood and water, since there was only a small increase in the mean oxygen difference across the gills (ΔPg) (Randall *et al.*, 1967; Kiceniuk and Jones, 1977). ΔPg increased from 42 mm Hg at rest to 56 mm Hg during exercise (Table II) which alone could not account for the sevenfold increase in oxygen uptake during exercise. However, it may be that resting ΔPg is not representative of O_2 differences across perfused lamellae, particularly if lamellar recruitment occurs during exercise. If only 60% of the lamellae are perfused at rest (Booth and Holeton, 1977), then assuming that (a) water is evenly distributed over the gill sieve and this distribution does not change with exercise, and (b) the anatomical dead space is constant and in the absence of any data is set at zero, then water passing over perfused lamellae will mix with water which has passed over unperfused lamellae to give a final mixed expired water O_2 tension of 102 mm Hg. The water passing over unperfused lamellae (40% of total flow) will have an O_2 tension the same as the inhaled O_2 tension, which in the experiments of Kiceniuk and Jones (1977) was 152 mm Hg. Thus, the O_2 tension in water that has passed over perfused lamellae (60% of total flow) is calculated to have been 67 mm Hg. The mean oxygen difference between water and blood across perfused lamellae (ΔP_{lam}) is 25 mm Hg, much lower than that determined for the whole gill in resting fish (Table II). If we assume that a[+]swimming speeds of 92% U_{crit} all lamellae are perfused, then ΔP_{lam} will equal ΔPg, which is 56 mm Hg (Table II). Thus, at the level of the perfused lamellae the mean oxygen difference between blood and water doubles between rest and exercise due to both a decrease in venous O_2 tension and a rise in exhaled water O_2 tension, and T_{O_2}', calculated using ΔP_{lam} rather than ΔPg, only increases 3.5 times between rest and exercise up to 92% of U_{crit} (Table II; Kiceniuk and Jones, 1977).

Doubling both the exchange surface area and the mean oxygen difference across the gills will quadruple oxygen uptake. Other subtle changes in water and blood flow or in conditions for diffusion across the gill epithelium must occur to account for the at least eightfold increase in oxygen uptake observed during exercise. These could include a redistribution of blood within each secondary lamella or a decrease in the thickness of the boundary layer of water or the mucus coat covering the gill epithelium. With respect to the former, it is noteworthy that a portion of each secondary lamella is buried in the surface of the filament and diffusion distances are larger between water and blood flowing in these basal channels. A redistribution of blood away from these basal channels would contribute to an increased gill oxygen transfer factor during exercise. With respect to the water boundary layer, this will presumably decrease with increasing water velocity.

4. *Ionic and Osmotic Regulation*

The permeability of gills is low compared with other portions of the vascular system and the tight coupling of epithelial cells restricts water diffusion across the gills of eels to a rate which is 3000 times less than the free diffusion coefficient of water (Steen and Stray Pedersen, 1975). Although water and ion permeability is relatively low, gill area is large and water influx rates may approach values of 1 μl/min/100 g for the eel (*Anguilla anguilla*) (Evans, 1969) and 50 μl/min/100 g for the trout (*S. gairdneri*) (Wood and Randall, 1973c). There is a net influx of water across the gills in freshwater fish which is slightly offset by the hydrostatic effect of blood pressure. In seawater the osmotic and hydrostatic forces act in the same direction contributing to a net water loss across the gills.

During exercise it is probable that lamellar recruitment results in an increase in functional gill area which facilitates the increase in oxygen uptake. This increase in functional area will also result in increased water and ion flux (Fig. 17). Farmer and Beamish (1969) observed an increase in plasma osmolality in *Tilapia nilotica* after exercise in saltwater and a decrease after swimming in freshwater. Byrne *et al.* (1972) observed little change in the ionic

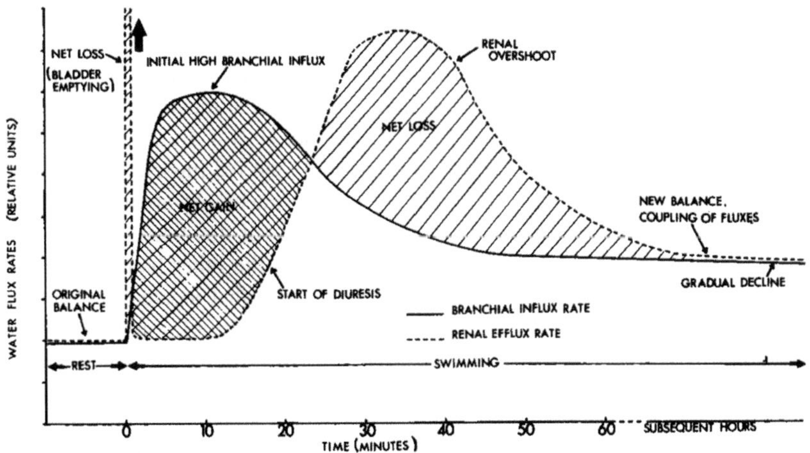

Fig. 17. Model of the temporal changes in branchial water influx and renal water efflux rates in rainbow trout during the course of swimming activity. (From Wood and Randall, 1973c.)

composition of the plasma of freshwater-adapted Atlantic salmon (*Salmo salar*) after exercise in freshwater while, after exercise in seawater, there was a marked increase in plasma Na^+, K^+, and Cl^- in postsmolt fish acclimated to seawater. Wood and Randall (1973a,b) found little change in plasma sodium in rainbow trout but a marked net sodium loss during exercise in freshwater, due to an increase in passive diffusion of sodium across the gills. Plasma sodium levels remained unchanged because the ion loss was ameliorated by reduction in blood volume due to water loss via the kidney (Fig. 17).

Wood and Randall (1973c) also observed a net uptake of water across the gills which was initially large and caused an increase in the weight of the fish. However after about 20 min of swimming, a large diuresis occurred, body weight was reduced below, and hemoglobin and plasma protein levels were raised above, the resting level. Gill water influx gradually fell but remained above resting levels for at least 8 hr of exercise. Kidney water loss exceeded gill water influx so blood volume was reduced (Fig. 17). Active sodium influx across the gills remained unchanged during exercise (Wood and Randall, 1973a,b). During the period after exercise loss of sodium by diffusion was below uptake levels and blood volume was expanded as sodium stores were replenished. Since the cost of osmoregulation is claimed to be at least 15% of oxygen uptake at all levels of exercise (Farmer and Beamish, 1969), then sodium cannot be typical of all ions because only the rate of passive loss changes with exercise, the rate of active uptake remaining unchanged.

IV. THE CIRCULATORY SYSTEM DURING EXERCISE

A. Introduction

The maximum oxygen consumption of most fish is an order of magnitude lower than in similarly sized mammals (Brett, 1972). The maximum uptake of an active fish (salmonid, weight 50 g 15°C) is about 1.0 ml O_2/100 g/min, representing an increase of 12 times above the resting level. Assuming that the locomotory muscles consume all of the increase in oxygen uptake, their rate of consumption will be 1.5 ml O_2/100 g/min, since muscle makes up about 63% of the body in adult salmonids (Brett, 1965a; Webb, 1971a). But, it is likely that only a small fraction of the total body musculature, the red muscle, is used during sustained swimming. In salmonids red muscle makes up about 4% of the myotomal mass (Webb, 1970) so the oxygen uptake of this tissue could be as great as 36 ml/100 g/min at maximum activity. On the other hand, if the red muscle scattered through the myotome is as active as that located in sheets at the sides of the fish, then the total red muscle mass quadruples and its oxygen consumption per unit weight is 9 ml/100 g/min. Nevertheless, it would appear that fish red muscle is capable of surprisingly high levels of oxygen consumption (in the mammalian range), and, since it is principally an aerobic tissue, the circulatory system must be capable of supplying it with sufficient oxygen. Since the capillary density of fish red muscle is about 1800/mm^2, some two to three times more than that of fish white muscle, and in the same range as that of mammalian skeletal muscle (Landis and Pappenheimer, 1963; Cameron and Cech, 1970), it follows that, at the tissue level, circulatory adjustments in fish red muscle will be far more pronounced than those at the whole organism level. Furthermore, they may even be more pronounced than those occurring in mammalian skeletal muscle in exercise.

B. Cardiac Adjustments to Exercise

Since oxygen is transported from the gas exchanger(s) to the tissues by the cardiovascular system it is convenient to discuss the exercise adjustments by using the Fick equation for oxygen and CO_2 transport:

$$\begin{aligned}\text{Oxygen uptake} &= \text{heart rate} \times \text{stroke volume}\\ &\quad \times \text{arterial-venous oxygen content difference}\end{aligned}$$

or

$$\begin{aligned}CO_2\ \text{elimination} &= \text{heart rate} \times \text{stroke volume}\\ &\quad \times \text{venous-arterial carbon dioxide content difference.}\end{aligned}$$

To meet an increase in oxygen demand the fish can increase one or more of the right-hand factors in the above equations. Since changes in $A - V_{O_2}$ difference have already been discussed, this section will concentrate on the cardiac adjustments.

1. *Heart Rate*

Heart rate is one of the easiest variables to monitor from exercising animals, and the relative abundance of heart rate determinations from swimming fish reflects this fact. Nevertheless, there is surprisingly little consensus about the heart rate response to exercise in fish. Bradycardia at the onset of exercise has been reported for lingcod (*Ophiodon elongatus*) (Fig. 18; Stevens *et al.*, 1972) and rainbow trout (Priede, 1974; Kiceniuk and Jones, 1977) but a decrease in heart rate with

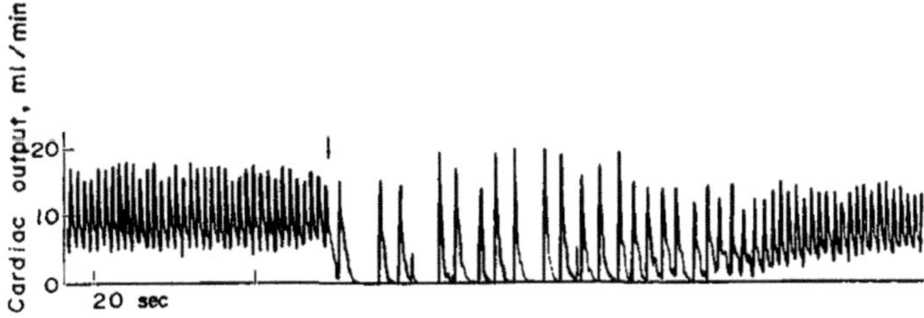

Fig. 18. Blood flow in the ventral aorta of an unrestrained lingcod. The start of spontaneous activity (at arrow) was associated with pronounced bradycardia, and diastolic flow fell to zero. (Modified from Stevens *et al.*, 1972, *Comp. Biochem. Physiol. A* **43,** 681. Copyright 1972 by Pergamon Press, Ltd.)

increasing intensity of sustained exercise has only been reported for "jack" sockeye salmon (Smith *et al.*, 1967). Tachycardia during sustained swimming has been reported for sucker (*Catastomus sp*), Atlantic mackerel (*S. scombrus*), blue runner (*C. crysos*), bluefish (*P. saltatrix*, Fig. 12), northern scup (*S. crysops*), mature sockeye salmon, pumpkinseed (*Lepomis gibbosus*), bullhead (*Ictalurus nebulosus*), brown trout (*S. trutta*, Fig. 9), and rainbow trout (Smith *et al.*, 1967; Sutterlin, 1969; Priede, 1974; Roberts, 1975a; Kiceniuk and Jones, 1977), while little (10–20% increase) or no change in heart rate with exercise has been reported for mullet (*Mugil cephalus*) (Roberts, 1975a), rainbow trout (Stevens and Randall, 1967a), cod (*Gadus morhua*) (Johansen, 1962), and a number of species of elasmobranchs (Johansen *et al.*, 1966; Hanson, 1967).

In those species in which heart rate increases during sustained swimming the relation between heart rate and speed (either absolute or relative) is not simple. Kiceniuk and Jones (1977) found for rainbow trout that below 50% of U_{crit} heart rate was little changed from resting levels whereas from 50 to 90% of U_{crit} heart rate increased linearly with U (Fig. 19). Sutterlin (1969) also noted in brown trout that heart rate was little affected by low swimming velocities. In fact, Priede (1974) took advantage of this relation and presented heart rate changes in rainbow trout as an exponential function of relative (or specific) velocity (Fig. 20a and b). However, since in salmonids the maximum heart rate is usually attained before maximum U an exponential relation is at best only a convenient approximation.

Specific variations in the response of the heart to exercise have never been rigorously investigated, so one cannot say whether active fish make larger or more rapid chronotropic adjustments than sluggish species. However, there are certainly some interesting indications in the literature which seem worthy of more concerted investigation, although the difficulties of making an assessment of the exercise performance in different species must be borne in mind.

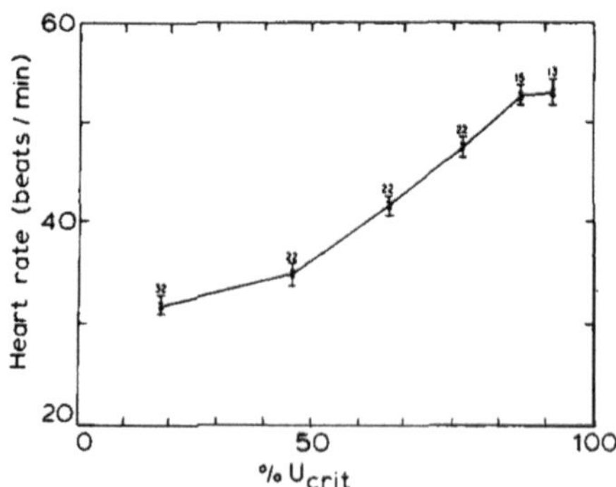

Fig. 19. Heart rate of rainbow trout at rest and during swimming (expressed as a percentage of each individual's U_{crit}. The value at 20% U_{crit} is for animals which were resting. Each point is a mean of *n* determinations (*n* being given above the points) and vertical bars denote one standard error. (Modified from Kiceniuk and Jones, 1977, *J. Exp. Biol.* **69**, 247.)

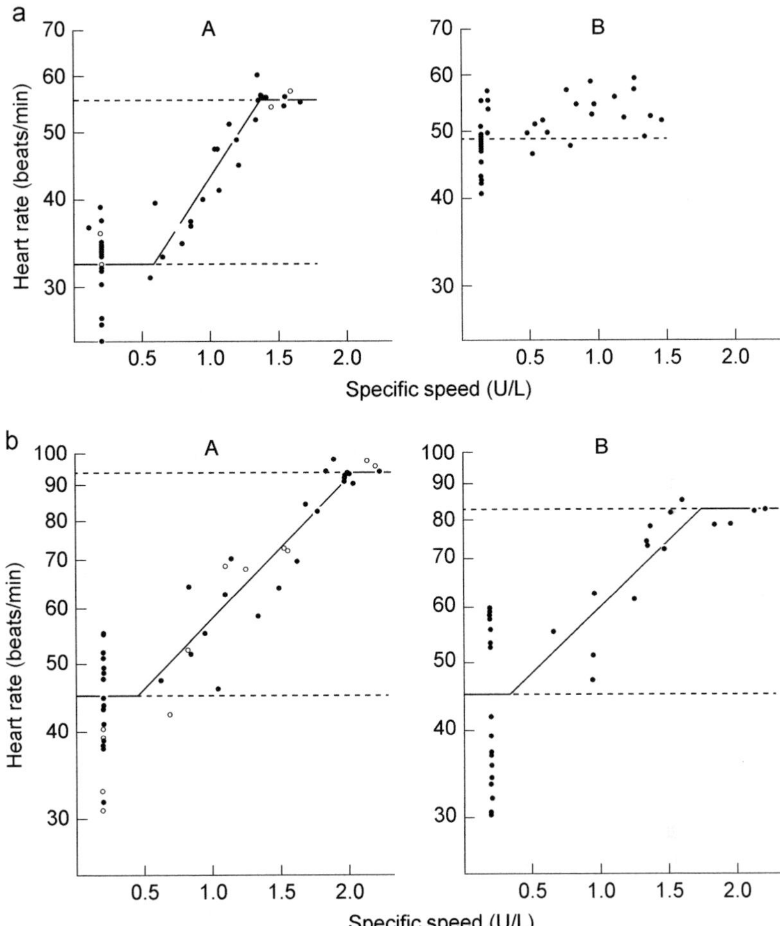

Fig. 20. (a) The relationship between heart rate and swimming speed of rainbow trout at 6.5°C. (A) Intact fish. Open circles are data from unilaterally vagotomized fish inserted for comparison but not used in the calculation of the lines shown. (B) Bilaterally vagotomized fish. The line shown is the mean basal heart rate; no significant relationship could be derived for the heart rates during swimming. (b) The relationship between heart rate and swimming speed of rainbow trout at 15°C. (A) Intact fish. Open circles are data derived from unilaterally vagotomized fish inserted for comparison but not used in the calculation of the lines shown. (B) Bivagotomized fish. (Figure and caption from Priede, 1974, *J. Exp. Biol.* **60**, 305.)

Exercise at one body length per second may be close to maximum effort for a bullhead while it may only be one-third or one-fifth maximum U for a trout. Sutterlin (1969) showed that exercise of one body length per second provoked an increase in heart rate in bullheads of 20–60% of the resting level over a period of 10 min, whereas in pumpkinseed heart rate increased 128% of the resting rate in about 2 min in response to the same relative U, both species being examined at 20°C. On the other hand, the active rainbow trout only shows a heart rate increase of 25–30% at one body length per second at 15°C, being about one-third of maximum heart rate at this temperature (Priede, 1974).

The effect of temperature on the cardiac chronotropic response to exercise has been investigated by Priede (1974). Raising the acclimation and exercising temperature from 6.5° to 15°C caused maximum exercise heart rate to increase from 56 to over 93 beats/min in rainbow trout. However, resting heart rate also rises from 32.5 to 45.5 so the maximum potential change in gas transport, as assessed from this term in the Fick equation, only increases from 1.7 to 2.0 times (Fig. 20a and b).

Changes in heart rate in fish can be provoked either neurally or aneurally. In fact, the hagfish heart, despite having an abundance of catecholamine-containing granules in specific cells (Augustinsson *et al.*, 1956; Bloom *et al.*, 1961; Hirsch *et al.*, 1964) and even a system of ganglion cells (Hirsch *et al.*, 1964), appears to be functionally aneural (Jensen, 1961, 1963, 1965). In hagfish (*Eptatretus stoutii*) changes in heart rate are provoked mechanically by means of a stretch-induced rate-controlling mechanism in the myocardium (Jensen, 1961), a mechanism which also operates in lamprey (*Petromyzon marinus*) and teleosts with innervated hearts (Jensen, 1969). Another aneural mechanism for change in heart rate is a direct effect on the heart muscle of circulating catecholamines which, it is suspected, are produced from chromaffin tissue. In elasmobranchs the chromaffin tissue is located close to the sinus venosus. The hearts of lampreys, elasmobranchs, and teleosts also receive vagal innervation which in lampreys is cholinergic and excitatory (Augustinsson *et al.*, 1956; Jensen, 1969), in elasmobranchs cholinergic and inhibitory, while in teleosts the vagal innervation is cholinergic and inhibitory (parasympathetic) along with a sympathetic adrenergic excitatory innervation (Gannon and Burnstock, 1969). Consequently an analysis of the cardiac chronotropic response to exercise in teleosts becomes quite complex, involving the interplay of four effector mechanisms (two aneural and two neural) with the multiple feedback systems which control the integrated cardiovascular response.

The neural control of heart rate during exercise has been investigated in both elasmobranchs and teleosts. In the former, heart rate appears to be unchanged during swimming (in a large aquarium tank) although atropine injection causes heart rate to increase by 30%, which implies that vagal tone is not reduced in exercise (Johansen *et al.*, 1966). However, in most teleosts atropinization or bilateral vagotomy increases resting heart rate at low temperature (6° – 10°C) although not at high temperature (15°C) (Stevens and Randall, 1967a; Stevens *et al.*, 1972; Priede, 1974) while maximum heart rates during exercise are either unaffected or reduced (Fig. 20a and b). This implies that vagal integrity is essential for the maximum expression of the heart rate response and is the best in vivo evidence for a vagal cardiac excitatory function.

The fall in heart rate which accompanies sudden movements of the fish or changes in imposed flow velocity is eliminated by atropinization or vagotomy in lingcod and trout (Stevens *et al.*, 1972; Priede, 1974), and Priede (1974) suggests that this is the type of vagal effect seen in approach reflexes (Labat, 1966). On the other hand, to what extent withdrawal of vagal parasympathetic activity contributes to the exercise heart rate adjustment is uncertain. In those fish which show no elevation in resting heart rate with vagotomy or atropinization it would appear to be negligible (Stevens and Randall, 1967a; Priede, 1974). On the other hand, in one individual sucker at 10°C (Stevens and Randall, 1967a) and perhaps rainbow trout at 6.5°C (Priede, 1974) withdrawal of vagal parasympathetic activity may account for the majority of the heart rate increase in exercise (Fig. 20a). However, in some bilaterally vagotomized trout at 6.5°C resting heart rate may, on occasion, fall back to that in intact animals, so in these fish vagal parasympathetic withdrawal cannot possibly effect the exercise tachycardia. Perhaps the best evidence for vagal withdrawal during exercise is the report of Stevens and Randall (1967a) describing heart rate response of rainbow trout to exercise during hypoxia. The bradycardia of hypoxia, a vagal parasympathetic effect (Holeton and Randall, 1967a,b; Randall and Smith, 1967), is transformed to tachycardia by exercise, the heart rate increase being much larger than normal, and must in part result from parasympathetic withdrawal.

The neural control of heart rate during exercise, discussed above, leaves the impression that a considerable portion of exercise tachycardia, when it occurs, may be effected either humorally or mechanically. Catecholamines may be secreted quite rapidly at the onset of exercise, for adrenaline in blood plasma increases some five times above the resting values after about 2 min of disturbance in rainbow trout (Nakamo and Tomlinson, 1967). After a period of extreme disturbance of 10–15 min, both adrenaline and noradrenaline are very high in rainbow trout, the former increases by 20–30 times the resting level and the latter by 10–12 times (Mazeaud *et al.*, 1978). What kind of cardiac chronotropic response would be provoked by these levels of circulating catecholamines? If the adrenergic receptors causing the cardiac chronotropic effect are of the β-type, as is found in plaice (*Pleuronectes platessa*) (Falck *et al.*, 1966), then it is to be expected that both adrenaline and noradrenaline would cause cardiac chronotropic effects. In the isolated trout heart, Bennion (1968) found that, at a given perfusion pressure at 6°C, heart rate was higher if the concentration of adrenaline in the perfusate was increased from 0.01 to 0.1 μg/ml, but at 15°C this was not the case and the lowest heart rates accompanied the highest adrenaline levels. On the other hand, in the isolated hearts of pike (*Esox lucius*) and carp (*Cyprinus carpio*) both adrenaline and noradrenaline at concentrations of 1 μg/ml cause marked cardiac acceleration (Cardot and Ripplinger, 1967). The cardiac effects of adrenaline *in vivo* are equally contradictory, only this is perhaps more to be expected since a large part of the cardiovascular response to sympathomimetic drugs is determined by compensatory reflexes. For instance, in most vertebrates adrenaline stimulates peripheral α-receptors and provokes a rise in blood pressure, the effect of which is lessened by the baroreceptor reflex causing simultaneous vagal bradycardia. Hence Randall and Stevens (1967) obtained bradycardia after intraarterial injection of 0.5 μg adrenaline in coho salmon (*O. kitsutch*) at 15°C (a barostatic reflex?) but tachycardia after atropinization. Increasing the dose level to 5 μg caused tachycardia and similar blood pressure increases in both control and atropinized fish (Fig. 21a and b). The existence of a barostatic reflex has never been confirmed in fish although there is certainly some suggestive evidence that one exists (Lutz and Wyman, 1932; Irving *et al.*, 1935; Mott, 1950; Satchell, 1968). However, even in Randall and Stevens' (1967) experiments, marked bradycardia only occurred some 90 sec after peak blood pressure, in fact when blood pressure had almost returned to the initial level. This suggests that these variables may not be reflexly related, a conclusion apparently confirmed by the injection of the larger doses of adrenaline. On the other hand, Stevens *et al.* (1972) found that increased stroke volume and bradycardia accompanied the arterial pressure rise provoked by dorsal

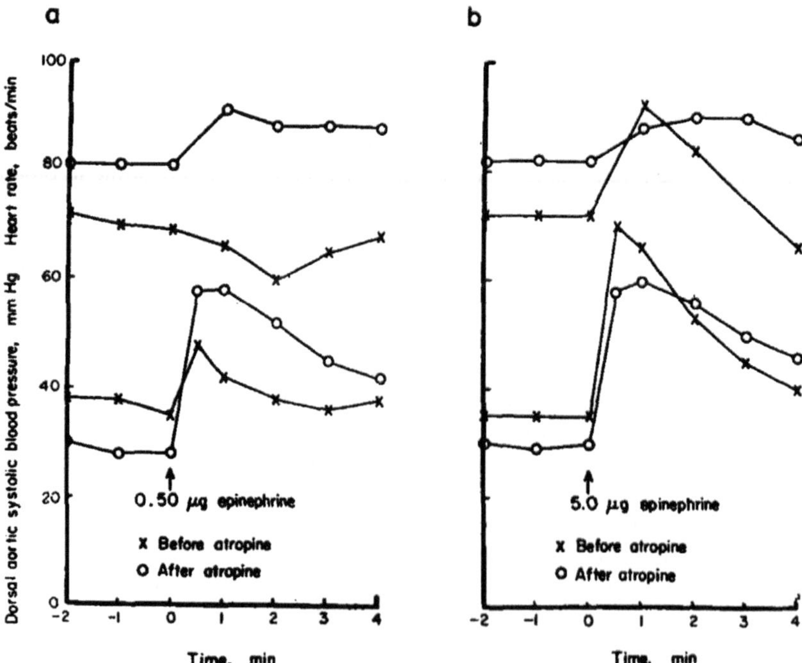

Fig. 21. The effect of intravenous injections of epinephrine on dorsal aortic blood pressure and heart rate in coho salmon before (a) and after (b) blockade of the efferent cholinergic fibers innervating the heart with atropine. (From Randall and Stevens, 1967, *Comp. Biochem. Physiol.* **21,** 415.)

aortic injections of 10 μg of adrenaline in lingcod (*O. elongatus*). Atropine eliminated the bradycardia and, after adrenaline injection, stroke volume fell and cardiac output was reduced by 15%. Using a similar dose level in carp (*Cyprinus carpio*) (intracardiac injection) Laffont and Labat (1966) obtained bradycardia at low temperatures ($1^{\circ}-8^{\circ}$C) and tachycardia at high temperatures ($9^{\circ}-20^{\circ}$C). The fact that the cardiac response to adrenaline is temperature-sensitive and perhaps dose-dependent may mean that the fish heart, like that of the frog, has α- and β-adrenoceptors which are allosteric conformations of the same structure, being predominantly β at high temperatures and α at low (Kunos and Nickerson, 1976). This suggestion is not confirmed by experiments with sole (*Solea vulgaris*) and plaice (*Pleuronectes flesus*) at 18°C since adrenaline injection caused bradycardia in both intact and vagotomized fish (Labat, 1964).

It has been suggested that the pacemaker cells of the fish heart are sensitive to stretch and depolarize more rapidly when extended (Jensen, 1961; Deck, 1964) although this has not been confirmed in all studies (Golenhofen and Lippross, 1969). The importance of this mechanism is that if the pacemaker is stretched during exercise due to an increase in venous pressure, then heart rate will increase (Pathak, 1972). This mechanical effect has been clearly demonstrated for the isolated ventricles of hagfish (Jensen, 1961, 1965), lamprey (Jensen, 1969), elasmobranchs (Jensen, 1970), carp (*Cyrpinus sp.*) (Harris and Morton, 1968), and rainbow trout (Jensen, 1969), and also in the isolated heart of rainbow trout (Bennion, 1968). A striking feature of all the experiments on isolated ventricles is that large changes in perfusion pressure are required to provoke any significant change in heart rate, much larger than occur *in vivo*. For instance, in the rainbow trout ventricle at 21°C, Jensen (1969) found that rate increased from 8 to 13 beats/min for an order of magnitude change in filling pressure. It could be argued that the ventricular pacemaker was much less stretchsensitive than that in the sinus venosus but Bennion (1968) obtained only slightly larger increases in rate at 15°C with an even bigger pressure increase in isolated trout hearts. Although high venous pressures have been reported in trout during exercise (Stevens and Randall, 1967a) they have not been confirmed by more recent work (Kiceniuk and Jones, 1977) so whether this mechanical factor plays any role in the chronotropic response to exercise is uncertain. However, saline infusion into the veins of brown trout (Sutterlin, 1969) and catfish (*Ameirus nebulosus*) (Labat *et al.*, 1961) caused elevation in heart rate but, even so, venous pressure reached 13 mm Hg before any effect was seen on trout heart rate. Despite reports to the contrary (Stevens and Randall, 1967a) the increases in heart rate in Labat *et al.*'s (1961) catfish (*sic.* bullheads) could not be blocked by atropine.

Obviously the control of the fish heart during exercise is still not understood, although it appears most likely that the tachycardia is a product of a vagal sympathetic contribution allied with a withdrawal of parasympathetic activity and a direct effect of circulating catecholamines. When occurring, exercise bradycardia would appear to be mediated by the vagal parasympathetic innervation, vagal inhibitory tone also being present in those species showing no cardiac chronotropic response to exercise. The afferent nervous link of the heart rate response has not been elucidated and, as in mammals, it might be expected to come from diverse inputs, although movement of water past the fish (Sutterlin, 1969) or passive tail beating (West and Jones, 1975) do not appear to affect heart rate.

2. *Stroke Volume*

Fishes are capable of large changes in stroke volume, and maximal values reported in the literature often exceed those given for similarly sized mammals. In the rainbow trout, for instance, maximum stroke volume seen in anemic animals is 2.67 ml/kg which is almost 10 times that obtained in similarly sized resting fish at the same temperature (Cameron and Davis, 1970; Davis and Cameron, 1971). Hence, the potential for change in gas transport due to changes in stroke volume in trout is equal to the maximum change in oxygen uptake or CO_2 excretion from resting to fully active metabolism at $8^\circ - 10^\circ C$.

It has been stated frequently that increased cardiac outputs in fish are achieved through adjustments in stroke volume rather than heart rate (Randall, 1970a). Virtually all exercise studies tend to confirm this except those on elasmobranchs where pulse flow in the ventral aorta (measured downstream of the first set of branchial arteries) appears to be unchanged by short periods of moderate activity, although pulse flow increases during the recovery period (Johansen *et al.*, 1966). Since the majority of the exercise periods monitored by Johansen *et al.* (1966) were probably less than one blood circulation time, the significance of the lack of adjustments is difficult to interpret. However, in lingcod (*O. elongatus*) stroke volume increases markedly along with the bradycardia at the onset of exercise (Fig. 18). An increase in pulse flow is also seen in the hemoglobin-free fish *Chaenocephalus aceratas* at the start of activity although heart rate is unchanged (Hemmingsen *et al.*, 1972). Although total heart output was not recorded in *Chaenocephalus*, if an allowance is made for flow through the last two pairs of branchial arches, stroke volume must have reached 10 ml/kg during activity, a testament to the remarkable stroke volumes in this fish, being the highest, on a unit weight basis, of any vertebrate.

Stroke volume determinations during sustained performance at a known work output have only been made on two occasions and both of these were by indirect methods. To determine stroke volume indirectly, simultaneous measures of oxygen uptake or carbon dioxide output, arterial and venous oxygen or CO_2 contents, and heart rate are required, preferably directly and from the same fish. Stevens and Randall (1967b) report a fivefold increase in stroke volume in rainbow trout but unfortunately none of the above conditions was fulfilled. In a recent series of experiments Kiceniuk and Jones (1977) used a 1-hr test period at each swimming speed and managed to obtain four simultaneous determinations of the required variables at speeds approaching U_{crit} (compared with nine resting determinations) (Table III). In these experiments stroke volume increased 2.24 times, while oxygen uptake increased by 7.75 times, but there was no indication that stroke volume had reached its limit, unlike the heart rate response (Fig. 22; Kiceniuk and Jones, 1977).

Changes in stroke volume occurring during exercise may be caused by all or some of the same factors that promote heart rate increases, but the effects of these factors are not linked since large changes in stroke volume can occur independent of rate changes and vice versa. Adrenergic compounds stimulate β-receptors in the fish heart and may have positive inotropic effects. In teleosts catecholamines may be released into the circulation or directly onto the heart from the sympathetic vagal innervation whereas in elasmobranchs there is no adrenergic innervation of the auricle and ventricle. Gannon *et al.* (1972) have suggested that in elasmobranchs amines are released from the anterior chromaffin masses into the blood in the posterior cardinal sinus and aspirated into the heart. These chromaffin cells are innervated by naked nerve endings and this indicates a neurohumoral control of the elasmobranch heart which may be much more specific than had been supposed. In the cod (*Gadus morhua*) sympathetic medullated fibers innervate the chromaffin cells lining the walls of the posterior cardinal veins, and following section of these nerves in acute preparations adrenaline release in response to stress is prevented (Nilsson *et al.*, 1976). Since release of adrenaline on nerve stimulation is abolished by mecamylamine (a ganglionic blocker) it is concluded by Nilsson *et al.* (1976) that chromaffin tissue is sympathetically controlled by preganglionic cholinergic fibers.

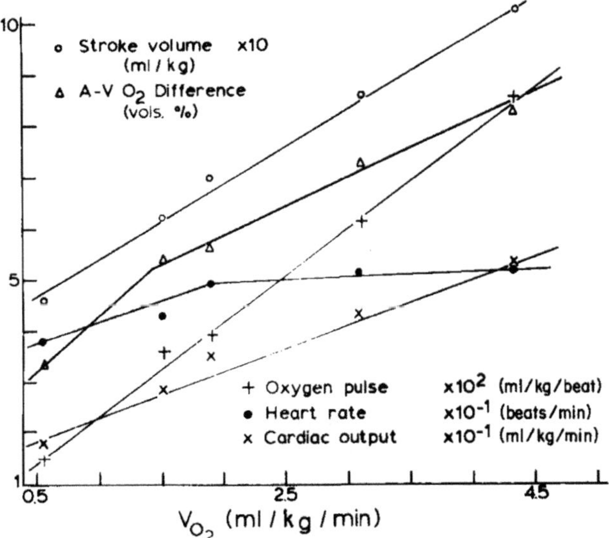

Fig. 22. The relationship between a number of cardiovascular variables and oxygen consumption in exercising rainbow trout. (All data from Kiceniuk and Jones, 1977.)

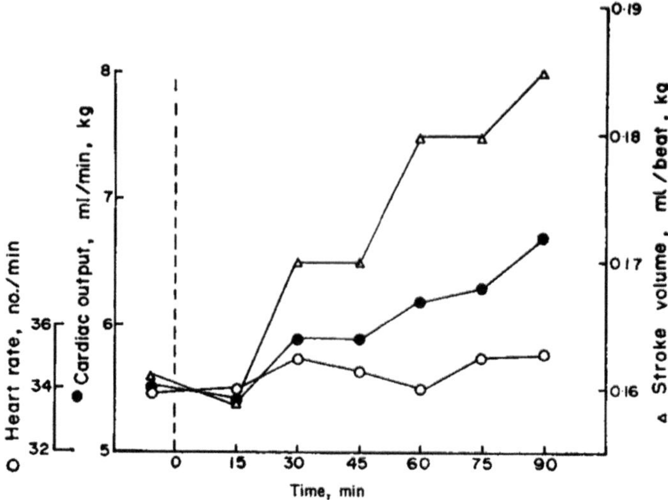

Fig. 23. The effect of the addition of 1.0 μg/ml adrenaline to the water in which the fish was situated on heart rate, stroke volume, and mean blood flow in the ventral aorta of a lingcod weighing 2.3 kg. Addition of adrenaline is indicated by the dashed line. (From Stevens *et al.*, 1972. Reprinted with permission from *Comp. Biochem.Physiol. A* **43,** 681. Copyright 1972 by Pergamon Press, Ltd.)

There seems to be general agreement that in virtually all fishes adrenergic compounds have marked inotropic effects in contrast to the weaker chronotropic effects. Falck *et al.* (1966) clearly demonstrated the positive inotropic effects of these compounds on the isolated hearts of lamprey (*Lampetra fluviatilis*) and plaice (*Pleuronectes platessa*) but, as Gannon (1971) pointed out, their technique did not allow them to differentiate between atrial and ventricular responses. However, in teleosts Gannon (1971) showed that both atrium and ventricle responded positively to adrenergic compounds and confirmed that the receptors were of the β-type. On the other hand, Bennion (1968) found that high perfusion pressures and temperatures were required for a marked increase in stroke volume of the isolated trout heart in response to an order of magnitude increase in adrenaline in the perfusate (0.01–0.1 μg/ml). When perfusion pressure was low (2 mm Hg) or at high perfusion pressure at low temperature (6°C) there was apparently little effect.

In intact fishes adrenaline injections cause both central and peripheral cardiovascular effects, which lead to the initiation of reflexes to counteract them, and complicate an assessment of the effect of adrenaline on a single target organ such as the heart. By placing lingcod (*O. elongatus*) in water containing 1 μg/ml adrenaline for a protracted period, Stevens *et al.* (1972) minimized compensatory reflexes and revealed the positive inotropic effect of adrenaline, for stroke volume increased by one-third (Fig. 23). In dogfish (*Squalus acanthias*) it has been clearly shown that some elements of a neuronal

Table III An Evaluation of the Oxygen Transport System in Trout during Exercise[a]

Swimming speed (% U_{crit})	Heart rate (beats/min)	Arterial O_2 content (vol%)	Venous O_2 content[b] (vol%)	Arterial O_2 tension (Torr)	Venous O_2 tension[b] (Torr)	Hematocrit (%)		pH_a	pH_v	Arterial–venous O_2 content (vol%)	Oxygen consumption (ml/kg/min)	Cardiac output (ml/kg/min)	Stroke volume (ml/kg)	Inspired water O_2 tension (Torr)	Arterial O_2 saturation (%)
						Hct_a	Hct_v								
Rest	37.8 ± 1.5	10.4 ± 0.5	7.1 ± 0.7	137 ± 4.2	33.2 ± 3.0	22.6 ± 1.0	24.2 ± 1.8	7.932	7.959	3.29 ± 0.26	0.56 ± 0.02	17.6 ± 1.1	0.46 ± 0.02	152.9 ± 1.9	97.0 ± 1.3
	$n = 9$	$n = 9$	$n = 9$	$n = 8$	$n = 8$	$n = 9$	$n = 8$	+7.991	+8.025	$n = 9$	$n = 9$	$n = 9$	$n = 9$	$n = 8$	$n = 9$
								−7.879	−7.902						
								$n = 5$	$n = 4$						
41–63%	42.7 ± 3.18	9.8 ± 0.74	4.4 ± 0.8	123.5 ± 7.5		22.7 ±1.4	24.45 ± 1.05			5.4 ± 0.1	1.52 ± 0.24	28.4 ± 5.0	0.62 ± 0.8	152.0 ± 2.31	96.0 ± 5.00
	$n = 3$	$n = 3$	$n = 3$	$n = 2$		$n = 3$	$n = 2$			$n = 3$	$n = 3$	$n = 3$	$n = 3$	$n = 3$	$n = 3$
70–78%	49.0 ± 1.00	9.02 ± 0.5	3.4 ± 0.4	123.0 ± 4.2	23.5 ± 2.1	20.34 ± 1.4	21.85 ± 2.4	7.924	7.988	5.6 ± 0.58	1.9 ± 0.27	34.8 ± 4.8	0.7 ± 0.09	155.75 ± 0.95	98.75 ± 1.0
	$n = 5$	$n = 5$	$n = 5$	$n = 4$	$n = 4$	$n = 5$	$n = 4$	+ 8.046	+ 8.081	$n = 5$	$n = 5$	$n = 5$	$n = 5$	$n = 5$	$n = 5$
								−7.829	−7.911						
								$n = 3$	$n = 3$						
81–91%	51.3 ± 4.6	10.2 ± 1.3	2.9 ± 1.47	128.0 ± 5.0	29.3 ± 6.2	22.5 ± 1.35	25.8 ± 0.9	7.859	7.883	7.3 ± 0.49	3.12 ± 0.38	42.9 ± 5.4	0.86 ± 0.16	146.7 ± 0.67	99.7 ± 0.67
	$n = 3$	$n = 3$	$n = 3$	$n = 3$	$n = 3$	$n = 3$	$n = 3$	+7.970	+7.950	$n = 3$	$n = 3$	$n = 3$	$n = 3$	$n = 3$	$n = 3$
								−7.770	−7.825						
								$n = 2$	$n = 2$						

Continued

Table III An Evaluation of the Oxygen Transport System in Trout during Exercise—Cont'd

Swimming speed (% U_{crit})	Heart rate (beats/min)	Arterial O_2 content (vol%)	Venous O_2 content (vol%)	Arterial O_2 tension (Torr)	Venous O_2 tension (Torr)	Hematocrit (%)				Arterial–venous O_2 content (vol%)	Oxygen consumption (ml/kg/min)	Cardiac output (ml/kg/min)	Stroke volume (ml/kg)	Inspired water O_2 tension (Torr)	Arterial O_2 saturation (%)
						Hct_a	Hct_v	pH_a	pH_v						
Maximum (92% +)	51.4 ± 2.48	9.7 ± 0.7	1.35 ± 0.4	126.0 ± 5.4	16.0 ± 2.1	25.7 ± 0.8	27.4 ± 1.2	7.610	7.548	8.3 ± 0.5	4.34 ± 0.17	52.6 ± 2.2	1.03 ± 0.07	151.8 ± 2.6	98.5 ± 0.87
	$n = 4$	$n = 4$	$n = 4$	$n = 4$	$n = 4$	$n = 4$	$n = 4$	+7.620	+7.630	$n = 4$	$n = 4$	$n = 4$	$n = 4$	$n = 4$	$n = 4$
								−7.600	−7.480						
								$n = 2$	$n = 2$						

[a]Data obtained from six trout and n = the number of determinations. All values given as means ± SEM except pH_a and pH_v, for which the values are means + and − SEM. Mean and standard errors were calculated on hydrogen ion concentration and reconverted to pH values.
[b]Values obtained from the common cardinal vein and ventral aorta were similar.

autonomic cardiovascular control system are present (Opdyke *et al.*, 1972) and in the skate (*Raja binoculata*) injection of $5\,\mu g/ml$ of adrenaline causes an elevation in ventral aortic blood velocity as well as an increase in blood pressure which again implies a positive inotropic role for adrenaline *in vivo* (Johansen *et al.*, 1966).

As has already been discussed, part of the cardiac chronotropic response to exercise may result from a withdrawal of vagal tone and since acetycholine has a negative inotropic effect on some teleost hearts (Falck *et al.*, 1966), cessation of cholinergic inhibition will tend to augment the force of contraction. However, there appears to be little or no cholinergic innervation of the teleost ventricle (Cobb and Santer, 1972) and correspondingly no inotropic effect of applied acetycholine (Belaud and Peyraud, 1970; Gannon, 1971). The negative inotropic effects of acetycholine on the teleost atrium are pronounced (Gannon, 1971) and since the single atrium feeds the single ventricle an increase in atrial stroke volume in exercise would tend to raise ventricular stroke volume by means of the passive Starling mechanism, since atrial contraction is the major, if not the sole, contributor to ventricular filling in teleosts (Randall, 1968, 1970a). Starling (1918) stated that "the energy of contraction (of a muscle) is a function of the length of the muscle fiber." Consequently, if a cardiac chamber is filled to a greater extent (end-diastolic volume increased) then the next contraction is stronger and a greater stroke volume is pumped. Both the aneural hagfish heart (*E. stoutii*) and the isolated trout heart appear to follow Starling's law (Chapman *et al.*, 1963; Bennion, 1968). However, it is not clear what happens in innervated fish hearts in response to increased filling pressure, when adjustment of stroke volume to the varying needs of the animal may be more potently influenced by nervous or hormonal systems. In the intact cod (*Gadus morhua*) experimentally increasing venous return causes stroke volume to almost double in a few beats (Johansen, 1962). However, the fact that the elevated stroke volume persisted for some time after the actual infusion, while ventricular pressures were relatively unaffected throughout, implies the involvement of neurohumoral regulation in this response, and therefore casts doubt on the role of a "pure" Starling mechanism in intact fishes, at least in teleosts.

In summary, there is no doubt that stroke volume in fishes is extremely labile and there is good evidence that it increases markedly during exercise in teleosts. However, the genesis of the increased cardiac contractility is not understood although neurohumoral factors are probably more important in this respect than passive mechanisms.

Oxygen uptake is a function of cardiac output, the product of heart rate and stroke volume, and the $A-V_{O_2}$ content difference which, on occasion, increases by 2.5 times (Kiceniuk and Jones, 1977) in exercise, adding markedly to the capacity for gas transport (Table III). In teleosts cardiac output has an approximately linear relation to oxygen uptake across the activity spectrum but since heart rate approaches its maximum value when oxygen uptake has increased by, at most, three times, then this linear relation is achieved due to a proportionately greater increase in stroke volume (Fig. 21; Kiceniuk and Jones, 1977). The product of stroke volume and $A - V_{O_2}$ content difference is the oxygen pulse which is more conveniently obtained by dividing oxygen uptake by heart rate. In exercising rainbow trout oxygen pulse increased by 5.67 times, from 0.0148 to 0.084 ml/kg/ beat, while oxygen uptake rose by 7.75 times, showing that heart rate changes only make a minor contribution to the increase in oxygen transport in trout during exercise (Fig. 21; Table III; Kiceniuk and Jones, 1977). In contrast, in mammals oxygen pulse would not be expected to change by more than three or four times for even a ten times increase in oxygen uptake above resting levels.

C. Arterial Blood Pressure and Total Peripheral Resistance during Exercise

In mammals maximum oxygen uptake during exercise may be limited either by the maximum cardiac output and oxygen extraction at the tissues or by the ability of the cardiovascular system to generate pressure against the minimum total peripheral resistance (TPR). It has been suggested that, in mammals, maximum oxygen uptake may be best expressed as the product of maximum $A - V_{O_2}$ content difference times the ratio of maximum blood pressure to minimum peripheral resistance (Rowell, 1974). In other words, cardiac output has been redefined as this ratio. Certainly in fish, where 50–60% of the body may be locomotory muscle, the potential exists for large changes in total peripheral resistance which may have to be controlled more with respect to maintenance of arterial blood pressure than total blood flow.

It is generally agreed that arterial blood pressure increases during exercise in teleosts although in elasmobranchs, during very short bursts of activity, there appears to be no pressure change (Johansen *et al.*, 1966). In teleosts ventral aortic pressure rises markedly in the early part of a bout of activity and mean pressure may attain values 10-15 mm Hg above the resting level (Johansen and Waage-Johannessen, 1962; Stevens and Randall, 1967a; Kiceniuk and Jones, 1977). It is interesting that Kiceniuk and Jones (1977) found that the magnitude of the increase in mean arterial blood pressure was the same whether the animal started swimming from rest or whether a new speed was imposed on a swimming fish in an incremental velocity test. Peak mean arterial pressure occurs between 5 and 15 min after the onset of swimming and then declines so that, after 1 hr of swimming, mean pressure is only elevated by 3–4 mm Hg (Fig. 24; Kiceniuk and Jones, 1977). Hence if four or five velocity tests are imposed, then mean ventral aortic pressure will increase by about 15–20 mm Hg (Fig. 25). The fact that the change in mean pressure early in exercise is independent of the imposed velocity implies that it may be a response to disturbance in terms of cardiac overcompensation, rather than exercise stress and further emphasizes the need to

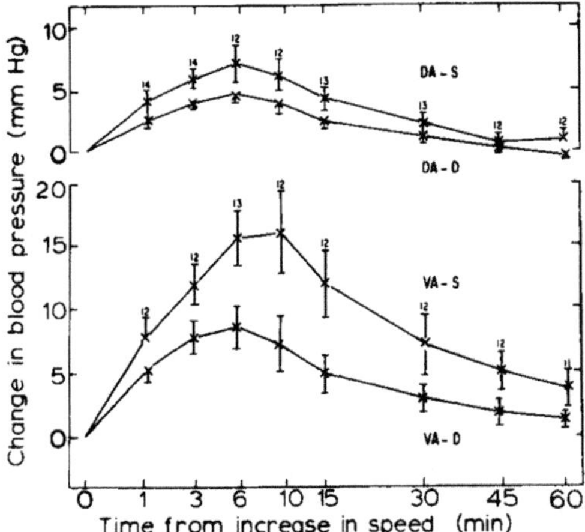

Fig. 24. Change in systolic (S) and diastolic (D) blood pressure in the dorsal aorta (DA) and ventral aorta (VA) following an increase in swimming speed. X's are means of (pressure at time t_η) – (pressure at time t_0), pressure at t_0 being the pressure immediately before a speed increase was imposed. Vertical bars denote one standard error and numbers above the points represent the number of determinations on five individuals. (From Kiceniuk and Jones, 1977, *J. Exp. Biol.* **69,** 247.)

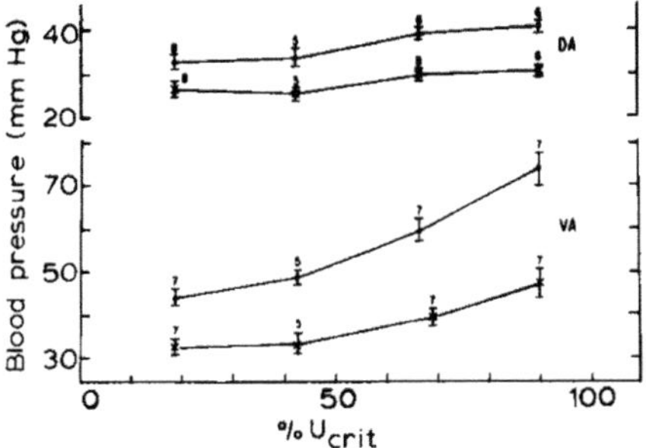

Fig. 25. Changes in diastolic (×) and systolic (●) blood pressures in both the dorsal aorta (DA) and ventral aorta (VA) of trout at rest and during swimming (expressed as a percentage of each individual's critical velocity). Values at less than 25% U_{crit} are for animals which were resting. The points are means of individual determinations, the vertical bars denote one standard error, and the numbers above the points indicate the numbers of determinations on eight animals. (From Kiceniuk and Jones, 1977, *J. Exp. Biol.* **69,** 247.

differentiate between shortterm (Stevens and Randall, 1967a,b) and steady state exercise (Kiceniuk and Jones, 1977).

On the other hand, dorsal aortic pressure changes are much smaller than ventral aortic pressure changes in both short-term and steady state exercise. In fact Smith *et al.* (1967) reported that in mature and some jack sockeye salmon dorsal aortic mean pressure only started to rise when the maximum speed was approached and was closely followed by fatigue of the fish, although they also report that blood pressures were usually higher in the first 30 min of an exercise period, before falling back to control levels. In the early period of swimming (10–15 min) at any velocity mean dorsal aortic pressure rises by 5 mm Hg in trout which is less than one-half the change in ventral aortic pressure (Stevens and Randall, 1967a; Kiceniuk and Jones, 1977). After 1 hr of swimming at any fixed speed dorsal aortic pressure is only 1 mm Hg above the starting level so an increasing velocity test with five increments may cause an elevation in dorsal aortic pressure of only 5 mm Hg. This is about one-quarter to one-third of the observed increase in mean ventral aortic pressure (Fig. 24; Kiceniuk and Jones, 1977).

As we have seen, cardiac output may increase from 3 to 6 times during strenuous activity so, in the absence of change in total peripheral resistance, mean blood pressure might be expected to increase by this amount. However, this does not occur in rainbow trout for, close to U_{crit}, ventral aortic pressure is usually only 20–25 mm Hg above the resting value of 35–40 mm Hg (Fig. 25; Kiceniuk and Jones, 1977). In Kiceniuk and Jones' (1977) experiments cardiac output only increased by 2.4 times so total peripheral resistance fell by about one-third. The pressure drop across the gills (ventral minus dorsal aortic pressure) increased from a resting level of 8 to 24 mm Hg at $\geq$92% U_{crit}, which slightly exceeded the flow changes, so gill resistance increased. On the other hand, the pressure drop through the systemic circuit (dorsal aortic minus venous pressure) only increased slightly so systemic resistance fell markedly. The differential changes in resistance for these two vascular beds drastically alter the relationship seen at rest between these resistances. At rest the resistance of the body circuit is 4 times that of the gills but it falls as maximum activity is approached and reaches a level only 1.5 times that of the gill circuit.

The largest fall in total peripheral resistance occurs, under steady state conditions, between rest and activity, and TPR is relatively unaffected by further speed increases up to U_{crit} (Kiceniuk and Jones, 1977). In fact, as Wood (1974a) has pointed out, further small decreases (5–10%) in resistance that accompany increasing U, in an incremental velocity test, could be due to passive dilatation caused by the increase in blood pressure.

The question arises as to what agents provoke these changes in resistance for the two vascular beds. Obviously a rise in blood pressure will cause passive vasodilatation of the vascular beds and reduce resistance, although this effect would appear to be more pronounced on the branchial than systemic circulation (Wood, 1974a; Wood and Shelton, 1975). Further, local auto-regulation may play a role on the systemic side but some changes in vascular resistance are known to be under neural or hormonal control. Kirby and Burnstock (1969) suggested that there was a gradual transition in the innervation of arteries from a predominance of cholinergic involvement in fish to one of adrenergic predominance in mammals. Certainly muscarinic cholinergic receptors, which cause contraction on stimulation, are present in pregill vessels of trout (Kirby and Burnstock, 1969; Klaverkamp and Dyer, 1974). Their existence in postgill vessels, however, is less conclusively proved (Holmgren and Nilsson, 1974). However, α-adrenergic excitatory receptors have been shown to exist in the major arteries of dogfish (*Squalus acanthias*) (Capra and Satchell, 1974), trout, and cod (*Gadus morhua*) (Holmgren and Nilsson, 1974), and β-adrenergic inhibitory receptors in dogfish and trout (Capra and Satchell, 1974; Klaverkamp and Dyer, 1974). In many instances only slight changes in resistance will be provoked by contraction and relaxation of smooth muscle in walls of large arteries so that any changes occurring during exercise must be attributed to contraction or relaxation of the smaller "resistance" vessels. The wall of the coeliac artery in some teleosts is particularly well invested with smooth muscle and it is conceivable that contraction of this segment could eliminate blood supply to the gut allowing blood to be redirected to the muscles. However, vascular control of this type is unlikely to make a large direct contribution to blood flow adjustments in exercise.

On the other hand, the compliance of the vessels may alter drastically on receptor stimulation, which could cause marked circulatory effects. For instance, an increase in systemic compliance alone would be expected to increase ventral aortic flow pulsatility while decreasing pressure pulsatility in both ventral and dorsal aortas if stroke volume is unaltered (Satchell, 1971; Jones *et al*., 1974; Fig. 26). A decrease in systemic compliance would have the reverse effects on pressure and flow pulsatility. Pressure pulsatility (pulse pressure/mean pressure) in both the dorsal and ventral aortae increases by 70%, from the resting level, in swimming trout in a steady state close to U_{crit} (Kiceniuk and Jones, 1977). Furthermore, the ratio between pressure pulsatility in the ventral and dorsal aortas changes little during exercise, from 1.66 at rest to 1.55 at 80–100% U_{crit} (Kiceniuk and Jones, 1977). Unfortunately, since heart stroke volume, heart rate, and gill and systemic resistances all change during exercise it is impossible to judge the relative contribution of arterial compliance changes to these effects on pulsatility.

With respect to resistance vessels in the gills and systemic circulation, a large number of studies have demonstrated that catecholamines cause branchial vasodilatation (Keys and Bateman, 1932; Ostlünd and Fänge, 1962; Kirschner, 1969; Steen and Kruysse, 1964; Richards and Fromm, 1969) and systemic vasoconstriction (Keys and Bateman, 1932; Reite, 1969; Wood and Shelton, 1975). The actual classification of the adrenoreceptors responsible for these actions is not entirely clear although β_1-dilatory receptors predominate in the gills (Wood, 1974a, 1975) and α-constrictor receptors in the body circulation (Wood and Shelton, 1975). Cholinergic receptors, on the other hand, cause vasoconstriction in both gills and systemic portions of the circulation (Davies and Rankin, 1973; Reite, 1969; Wood, 1974a, 1975).

During exercise gill resistance changes variably but there is a marked fall in systemic resistance. Circulating catecholamines, which are frequently invoked

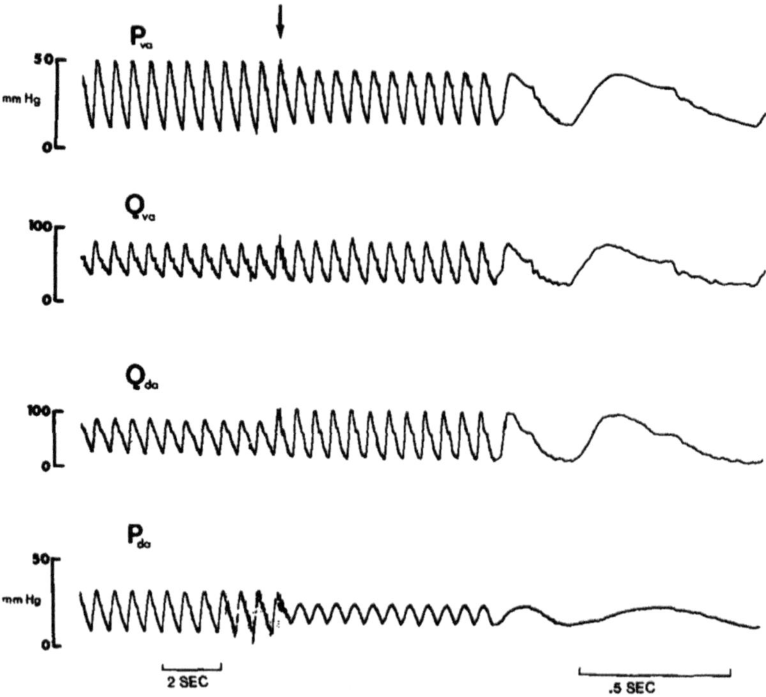

Fig. 26. Pressures and flows before and after "gill resistance" in a hydraulic model of the fish circulation. Traces (from top to bottom)—first: pressure proximal to "gills" in "ventral aorta" (P_{va}); second: flow proximal to "gills" in "ventral aorta" (Q_{va}) (arbitrary units); third: flow distal to "gills" in "dorsal aorta" (Q_{da}) (arbitrary units); fourth: pressure distal to "gills" in "dorsal aorta" (P_{da}). At arrow, a compliance was introduced into "dorsal aorta" distal to gills. (From Jones *et al.*, 1974, *Am. J. Physiol.* **226,** 90.)

to explain cardiovascular adjustments to exercise, would cause dilation in gill and constriction in systemic vessels. Certainly the levels of circulating catecholamines which occur in intact stressed dogfish (Mazeaud, 1969) and trout (Nakano and Tomlinson, 1967) are sufficient to cause gill vasodilatation (Davies and Rankin, 1973; Wood, 1974a) and presumably systemic vasoconstriction. There is evidence that β-inhibitory receptors in the gills are much more sensitive to circulating catecholamines than the α-excitatory receptors in the body circulation (Wood, 1974a; Wood and Shelton, 1975), the body resistance being primarily under direct neural control (Wood and Shelton, 1975). However, extreme stress (chasing and air exposure) is likely to cause far larger increases in circulating catecholamines than actual swimming activity. Consequently, this could explain a lack of gill resistance changes in steady state activity (Kiceniuk and Jones, 1977). On the other hand, any drop in gill resistance during exercise (Stevens and Randall, 1967a) could be due to neurally controlled vasodilatation. A decrease in vagal cholinergic tone (parasympathetic) could cause gill vasodilatation by eliminating excitatory muscarinic activity while an increase in vagal adrenergic activity (sympathetic) could promote gill vasodilatation through the agency of adrenergic α - and β-inhibitory receptors (Wood, 1974a; Johansen, 1972; Rankin and Maetz, 1971). On the body side of the circulation one might look for the enhancement of cholinergic and adrenergic neural activities in some beds while for others, such as locomotory muscles, these neural activities may be eliminated. There is some indirect evidence, due to the presence of "Mayer" type waves in resting fish after hemorrhage, that systemic vasomotor tone is neurally controlled through the agency of α-adrenergic receptors (Wood, 1974b) but in the absence of detailed knowledge on the balance between adrenergic and cholinergic controlling mechanisms, even in resting animals, it is difficult to predict what may occur during exercise.

D. Venous Pressure and Venous Return during Exercise

The veins are large vessels offering low resistance to flow and, as in the major arteries, energy gradients (kinetic and pressure energy) required for flow are small. Obviously, in the absence of pronounced venomotor activity, any increase in energy gradient between the periphery and heart will increase venous return. In exercise an increase in energy gradient may be caused by a rise in peripheral venous pressure, on a sustained or momentary basis, due to systemic vasodilatation or activities of the muscle pump. Also, in those fishes with a rigid pericardium, intrapericardial pressure may fall well below environmental pressure during activity which will expand the great veins and reduce central venous pressure, further enhancing the energy gradient for flow.

Venous pressures are higher at sites peripheral to the heart and in the caudal vein of resting elasmobranchs reach 1–3 mm Hg (Hanson, 1967; Birch *et al.*, 1969; Satchell, 1965) while close to the heart negative pressures are commonly obtained (Sudak, 1965a,b; Hanson, 1967; Birch *et al.*, 1969). A similar picture seems to hold in some teleosts for Stevens and Randall (1967a) recorded pressures of 8 mm Hg in the subintestinal vein of resting rainbow trout while Kiceniuk and Jones (1977) obtained a value of 1.4 mm Hg in the central common cardinal vein. Unfortunately no recordings of venous pressure have been made from free swimming elasmobranchs but in trout pressure in the subintestinal vein rises substantially at moderate swimming speed (Stevens and Randall, 1967a) while in the cardinal veins a small increase (0.5 mm Hg) is only observed close to U_{crit} (Kiceniuk and Jones, 1977). The subintestinal vein empties into the liver before reaching the heart, and Stevens and Randall (1967a) attributed the pressure rise to hepatic venomotor activity. Also, since the subintestinal vein drains an area not directly involved in locomotion, an increase in flow resistance in the vascular bed it drains might be expected in exercise and, in fact, Stevens and Randall (1967a) found the amount of blood they could drain from this vein (central end occluded) fell during swimming, while the pressure gradient from the dorsal aorta to the subintestinal vein was virtually unchanged. This observation is probably the best evidence for redistribution of blood away from visceral regions during exercise.

In many elasmobranchs cardiac contraction creates a negative pressure within the pericardium and great veins entering the ductus cuvieri (Johansen and Hanson, 1967; Hanson, 1967; Birch *et al.*, 1969). A structural prerequisite for negative pressure generation is a rigid pericardium, which is lacking in teleosts. Nevertheless, reports have appeared of negative pressure in the central veins and sinus venosus of teleosts (Brunings, 1899; Mott, 1951). Johansen (1965a,b) has shown that as heart rate slows the pericardial pressure approaches zero whereas tachycardia causes it to become more negative; tachycardia during exercise would therefore be expected to promote venous return. However, pronounced tachycardia during exercise has only been shown to occur in teleosts, and since there is no rigid pericardium in teleosts, there is consequently no good evidence that negative intrapericardial pressure plays any specific role in exercise.

Obviously venous return increases during exercise but the only certain agent of its promotion lies in the auxillary pumping activities of the waves of locomotory muscle contraction acting on veins or special venous reservoirs. In elasmobranchs these reservoirs are located in the tail and under the median fins (Birch *et al.*, 1969). However, for a muscle pump to be effective the veins must be compressible and possess valves to ensure that flow only travels unidirectionally (usually toward the heart). Although these criteria are well met in the veins associated with the median fins and tail (Fig. 29), all the major longitudinal veins in fishes, on the other hand, only possess valves where they enter the sinus venosus or atrium. In elasmobranchs the valves which prevent back flow from the ductus cuvieri into the major longitudinal vessels open during cardiac systole and close during diastole (Birch *et al.*, 1969). However, the long veins possess no other valves and in elasmobranchs they are incompressible tunnels surrounded by connective tissue (Satchell, 1971). Only the hepatic vein of elasmobranchs is a discrete tubular structure, similar to teleostean veins, but its opening into the sinus venosus is protected by a muscular sphincteric-type valve. In fact, there is evidence showing that this valve opens during swimming and may aid in the mobilization of the liver blood store during exercise (Johansen and Hanson, 1967). However, the lack of valves, combined with the noncollapsible nature of veins in elasmobranchs, or the protection of veins, such as the caudal vein, within the incompressible hemal canal, suggest that, if anything, these vessels are designed to be protected from the waves of muscular contraction rather than designed to be the propulsive part of a muscle pumping system. In fact it is the segmental veins of the postpelvic regions which form the muscle pump in both elasmobranchs and teleosts. The segmental intercostal veins, draining blood from the muscle blocks, possess valves where they enter the main longitudinal veins (Fig. 27a; Sutterlin, 1969) and, in elasmobranchs which retain

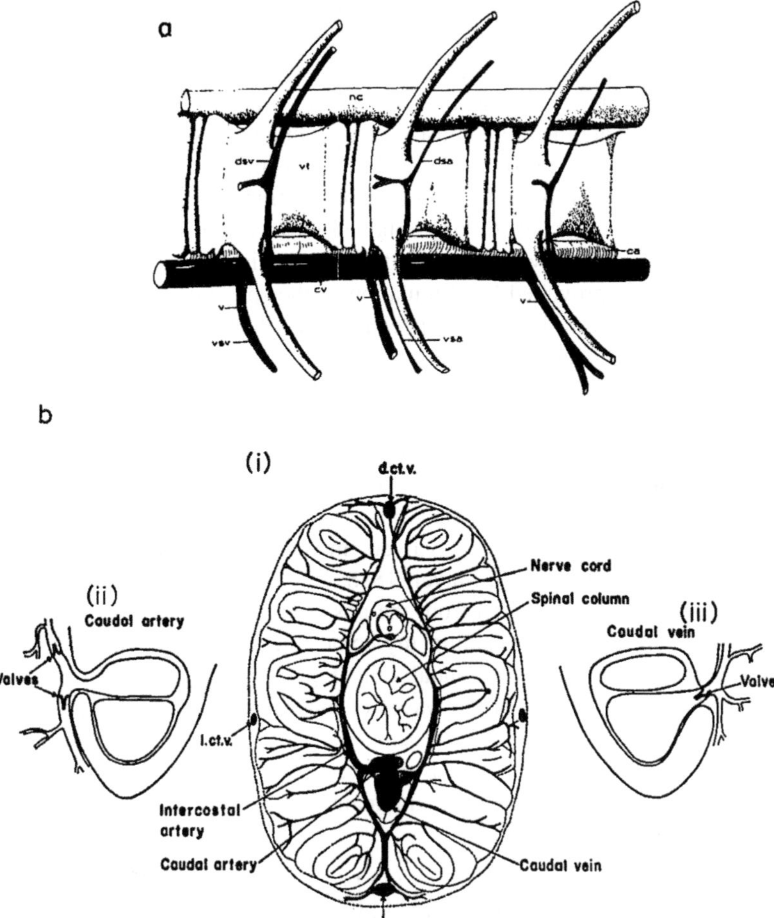

Fig. 27. (a) The arrangement of the blood vessels in the caudal region of a trout, showing the position of the valves in the segmental veins. The anterior end of the fish is toward the left.v, valves; nc, nerve cord; vt, vertebra; ca, caudal artery; cv, caudal vein; vsv, ventral segmental vein; dsv, dorsal segmental vein; vsa, ventral segmental artery; dsa, dorsal segmental artery. (From Sutterlin, 1969, *Physiol. Zool.* **42,** 36. Copyright 1969 by the University of Chicago Press.) (b) (i) Diagrammatic cross section of the trunk near the base of the tail. The intercostal artery and intercostal vein have their origins half a segment apart; they are here shown at the same level. (ii) The location of the arterial valves. (iii) The location of the venous valves. d.ct.v., dorsal cutaneous vein; l.ct.v., lateral cutaneous vein; v.ct.v., ventral cutaneous vein. [From Birch *et al.*, 1969, *J. Zool.* (*London*) **159,** 31.]

their tail as the major means of propulsion, the intercostal arteries also are valved where they arise from the dorsal aorta (Fig. 27b; Birch *et al.*, 1969). The orientation of these valves is such that back flow of blood is prevented. Consequently, as the muscular wave of contraction passes backwards along the body the capillaries, veins and perhaps even the arteries are squeezed and the blood is forced into the caudal vein. When active tailbeat flexions are caused experimentally, elevations (2–4 mm Hg) in caudal vein pressure occur in both teleosts and elasmobranchs (Satchell, 1965; Sutterlin, 1969), but Satchell (1965) found that passive rhythmic flexion in *Heterodontus portusjacksoni* only increased mean caudal vein pressure by 1 mm Hg. Hence, it seems unlikely that, due to the muscle pump, venous pressure will rise to such a level that the pressure difference for flow across the systemic vascular bed is substantially reduced. In fact, using an isolated trunk preparation perfused at constant pressure, Satchell (1965) obtained an increase in perfusate flow of 46% over the resting level (Fig. 28) during a 1-min period when swimming movements were provoked by stimulation of the spinal cord.

There seems little doubt that specialized auxillary pumping mechanisms will also augment venous return in exercise. In elasmobranchs valved venous reservoirs are associated with the median fins (Mayer, 1888; Marples, 1935–1936; Birch *et al.*, 1969) and fin movements probably pump blood from cutaneous vessels into the caudal vein. Mayer (1888) working with the dogfish [*Scyliorhinus* (*Scyllium*) *canicula*] observed that dorsal fins that had been congested with blood at rest paled as soon as movement occurred. Also in elasmobranchs there are paired, valved, interconnecting venous sinuses on each side of the tail (Fig. 29; Mayer, 1888; Birch *et al.*, 1969). These sinuses are filled with blood squeezed from the posterior ends of the cutaneous veins by contraction of the surrounding myotome, the valve orientation ensuring that the blood flows into the caudal vein (Fig. 29). The caudal vein itself is protected from the effects of the waves of muscular movement during swimming since it is enclosed within the hemal canal and therefore its effectiveness in conducting blood to the heart from these auxillary pumps, as well as the generalized muscle pump, will be unimpaired in exercise.

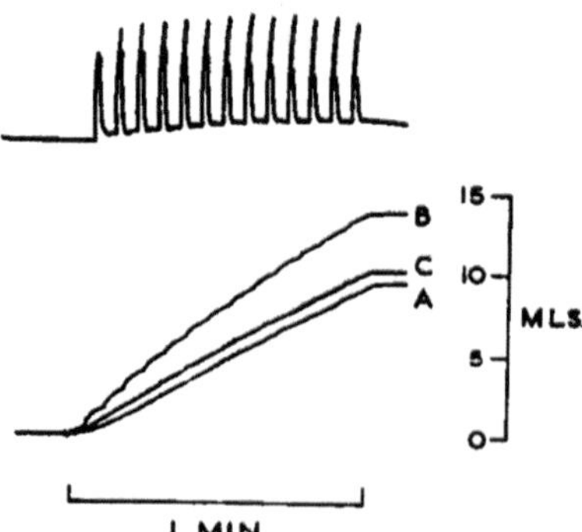

Fig. 28. The effect of trunk movements on venous flow in the elasmobranch *Heterodontus portusjacksoni*. Upper trace, movement of postpelvic trunk. Lower trace, outflow of perfusate from caudal vein in 1 min. A, at rest; B, during movement; C, immediately following B. The perfusate (dextran–Ringer solution) was fed into the aorta of an isolated trunk preparation from a constant pressure reservoir and outflow was collected from the cannulated caudal vein. (From Satchell, 1965.)

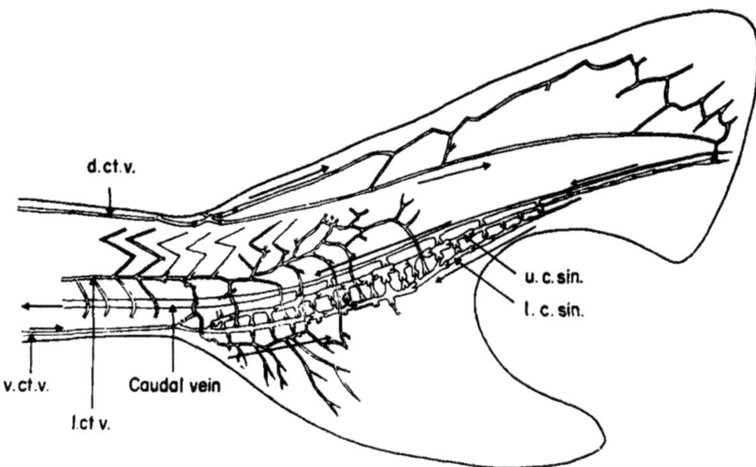

Fig. 29. The veins in the tail of an elasmobranch (*H. portusjacksoni*) with the blood flow pattern indicated by the arrows. d.ct.v., l.ct.v., v.ct.v., Dorsal, lateral and ventral cutaneous veins; u.c.sin., l.c.sin., upper and lower caudal sinuses. [From Birch *et al.*, 1969, *J. Zool.* (*London*) **159,** 31.]

Other external influences such as respiratory movements or the hydrodynamic pressure gradients along the body surface of swimming fish (Dubois *et al.*, 1974, 1976) may also influence venous return. The former might be expected to be more important in elasmobranchs where the location of the anterior cardinal, hyoidean, and inferior jugular sinuses indicate that they may be compressed in expiration and expanded in inspiration. In many teleosts the largest body crosssectional area occurs in the region of the heart and, during swimming, significant negative pressures may occur on the body surface of this region (Dubois *et al.*, 1974, 1976). If these pressures were transmitted by the body structures to the outside of the heart then the return of blood to the cardiac chambers would be augmented. Further, in posterior regions of the body the venous transmural pressure could be more or less doubled during rapid swimming. Fish achieve their highest swimming speeds in "bursts," which are effected anaerobically, and are therefore independent of circulatory sufficiency. Consequently these hydrodynamic surface effects are only likely to be significant in fish which can maintain high swimming speeds (approximately 100 cm/sec) for long periods of time and, even then, the effect on tissue fluid balance, by changing capillary filtration pressures, may be more important than the changes in venous transmural pressure.

ACKNOWLEDGMENTS

D. R. J. was supported by grants from the NRC of Canada. D. J. R. was supported by grants from the NRC of Canada and Environment Canada.

We are extremely grateful to Drs. N. H. West, W. Burggren, and C. Milliken for carefully reading early drafts of the manuscript and suggesting changes which have improved the final version and to Mrs. N. A. Virani for her patience in preparing the bibliography.

REFERENCES

Albers, C. (1970). Acid base balance. *In* "Fish Physiology" (W. S. Hoar and D. J. Randall, eds.), Vol. 4, p. 173. Academic Press, New York.

Augustinsson, K. B., Fänge, R., Johnels, A. and Östlund, E. (1956). Histological, physiological and biochemical studies on the heart of two cyclostomes, hagfish (*Myxine*) and lamprey (*Lampetra*). *J. Physiol. (London)* **131,** 257.

Auvergnat, R., and Secondat, M. (1942). Retentissement plasmatique de l'exercise musculaire chez la carpe (*Cyprinus carpio* L.). *C. R. Acad. Sci.* **215,** 92.

Ballintijn, C. M., and Roberts, J. L. (1976). Neural control and proprioceptive load matching in reflex respiratory movements of fishes. *Fed. Proc., Fed. Am. Soc. Exp. Biol.* **3,** 1983.

Basu, S. P. (1959). Active respiration of fish in relation to ambient concentrations of oxygen and carbon dioxide. *J. Fish. Res. Board Can.* **16,** 175.

Beamish, F. W. H. (1968). Glycogen and lactic acid concentrations in Atlantic cod (*Gadus morhua*) in relation to exercise. *J. Fish. Res. Board Can.* **25,** 837.

Belaud, A., and Peyraud, C. (1970). Actions de l'acetylcholine sur le coeur perfuse de congre (*Conger conger* L.) modifications de e.c.g. *C. R. Soc. Biol.* **146,** 405.

Bennion, G. R. (1968). The control of the function of the heart in teleost fish. M.S. Thesis Univ. of British Columbia, Vancouver.

Bettex-Galland, M., and Hughes, G. M. (1973). Contractile filamentous material in the pillar cells of fish gills. *J. Cell Sci.* **13,** 359.

Birch, M. P., Carre, C. G., and Satchell, G. H. (1969). Venous return in the trunk of the Port Jackson shark, *Heterodontus portus jacksoni*. *J. Zool. (London)* **159,** 31.

Black, E. C. (1958). Energy stores and metabolism in relation to muscular activity. *In* "The Investigation of Fish-power Problems" (P. A. Larkin, ed.), p. 51. Univ. of British Columbia Press, Vancouver.

Black, E. C., Robertson, A. C., Hanslip, A. R., and Chiu, W. G. (1960). Alterations in glycogen, glucose and lactate in rainbow and Kamloops trout (*Salmo gairdneri*) following muscular activity. *J. Fish. Res. Board Can.* 17, 487.

Black, E. C., Connor, A. R., Lam, K. C., and Chiu, W. G. (1962). Changes in glycogen, pyruvate and lactate in rainbow trout (*Salmo gairdneri*) during and following muscular activity. *J. Fish. Res. Board Can.* 19, 409.

Bloom, G., Östlund, E., von Euler, U. S., Lishajko, F., Ritzen, M., and Adams-Ray, J. (1961). Studies on catecholamine-containing granules of specific cells in cyclostome hearts. *Acta Physiol. Scand., Suppl. No.* 185, p. 1.

Booth, J., and Holeton, G. F. (1977). Unpublished data.

Brett, J. R. (1963). The energy required for swimming by young sockeye salmon with a comparison of the drag force on a dead fish. *Trans. R. Soc. Can., Sect. 1, 2, 3* 1, 441.

Brett, J. R. (1964). The respiratory metabolism and swimming performance of young sockeye salmon. *J. Fish. Res. Board Can.* **21,** 1183.

Brett, J. R. (1965a). The relation of size to the rate of oxygen consumption and sustained swimming speeds of sockeye salmon (*Oncorhynchus nerka*). *J. Fish. Res. Board Can.* **22,** 1491.

Brett, J. R. (1965b). The swimming energetics of salmon. *Sci. Am.* **213,** 80.

Brett, J. R. (1972). The metabolic demand for oxygen in fish, particularly salmonids, and a comparison with other vertebrates. *Respirat. Physiol.* **14,** 151.

Brett, J. R., and Glass, N. R. (1973). Metabolic rates and critical swimming speeds of sockeye salmon (*Oncorhynchus nerka*) in relation to size and temperature. *J. Fish. Res. Board Can.* **30,** 379.

Brett, J. R., and Sutherland, D. B. (1965). Respiratory metabolism of pumpkinseed, *Lepomis gibbosus*, in relation to swimming speed. *J. Fish. Res. Board Can.* **22,** 405.

Brown, C. E., and Muir, B. S. (1970). Analysis of ram ventilation of fish gills with application to skipjack tuna (*Katsuwonus pelamis*). *J. Fish. Res. Board Can.* **27,** 1637.

Brunings, W. (1899). Zur Physiologie des Kreislaufes der Fische. *Pfluegers Arch. Gesamte Physiol. Menschen Tiere* **75,** 599.

Byrne, J. M., Beamish, F. W. H., and Saunders, R. L. (1972). Influence of salinity, temperature, and exercise on plasma osmolality and ionic concentration in Atlantic salmon (*Salmo salar*). *J. Fish. Res. Board Can.* **29,** 1217.

Cameron, J. N. (1975). Morphometric and flow indicator studies of the teleost heart. *Can. J. Zool.* **53,** 691.

Cameron, J. N., and Cech, J. J. (1970). Notes on the energy cost of gill ventilation. *Comp. Biochem. Physiol.* **34,** 447.

Cameron, J. N., and Davis, J. C. (1970). Gas exchange in rainbow trout with varying blood oxygen capacity. *J. Fish. Res. Board Can.* **27,** 1069.

Cameron, J. N., Randall, D. J., and Davis, J. C. (1971). Regulation of the ventilationperfusion ratio in the gills of *Dasyatis sabina* and *Squalus suckleyi*. *Comp. Biochem. Physiol. A* **39,** 505.

Capra, M. F., and Satchell, G. M. (1974). Beta-adrenergic dilatory responses in isolated, saline perfused arteries of an elasmobranch fish, *Squalus acanthias*. *Experientia* **30,** 927.

Cardot, J., and Ripplinger, J. (1967). Action de l'adrenaline et de la noradrenaline sur la coeur lavé du brochet et de la carpe. *J. Physiol.* (*Paris*) **59,** 399.

Chapman, C. B., Jensen, D., and Wildenthal, K. (1963). On circulatory control mechanisms in the Pacific hagfish. *Circ. Res.* **12,** 427.

Cobb, J. L. S., and Santer, R. M. (1972). Excitatory and inhibitory innervation of the heart of plaice (*Pleuronectes platessa*); anatomical and electrophysiological studies. *J. Physiol.* (*London*) **222,** 42.

Cole, T. J., and Miller, G. J. (1973). Interpretation of the parameters relating oxygen uptake to heart rate and cardiac output during submaximal exercise. *J. Physiol.* (*London*) **231,** 12 P.

Connor, A. R., Elling, C. H., Black, E. C., Collines, G. B., Gauley, J. R., and Trevor-Smith, E. (1964). Changes in glycogen and lactate levels in migrating salmonid fish ascending experimental "endless" fishways. *J. Fish. Res. Board Can.* **21,** 255.

Conte, F., Wagner, H. H., and Harris, T. (1963). Measurements of the blood volume of the fish, *Salmo gairdneri*. *Am. J. Physiol.* **205,** 533.

Dahlberg, M. L., Shumway, D. L., and Doudoroff, P. (1968). Influence of dissolved oxygen and carbon dioxide on swimming performance of largemouth bass and coho salmon. *J. Fish. Res. Board Can.* **25,** 49.

Davies, D. T., and Rankin, J. C. (1973). Adrenergic receptors and vascular responses to catecholamines of perfused dogfish gills. *Comp. Gen. Pharmacol.* **4,** 139.

Davis, G. E., Foster, J., Warren, C. E., and Doudoroff, P. (1963). The influence of oxygen concentration on the swimming performance of juvenile Pacific salmon at various temperatures. *Trans. Am. Fish. Soc.* **92,** 111.

Davis, J. C. (1968). The influence of temperature and activity on certain cardiovascular and respiratory parameters in adult sockeye salmon. M.S. Thesis Univ. of British Columbia, Vancouver.

Davis, J. C. (1970). Estimation of circulation time in rainbow trout, *Salmo gairdneri*. *J. Fish. Res. Board Can.* **27,** 1860.

Davis, J. C., and Cameron, J. N. (1971). Water flow and gas exchange at the gills of rainbow trout, *Salmo gairdneri*. *J. Exp. Biol.* **54,** 1.

Davis, J. C., and Watters, K. (1970). Evaluation of opercular catheterization as a method for sampling water expired by fish. *J. Fish. Res. Board Can.* **27,** 1627.

Deck, K. A. (1964). Dehnungseffekte am spontanschlagenden, isoliertem Sinusknoten. *Pfluegers Arch. Gesamte Physiol. Menschen Tiere* **280,** 120.

Driedzic, W. R., and Kiceniuk, J. W. (1976). Blood lactate levels in free-swimming rainbow trout (*Salmo gairdneri*) before and after strenuous exercise resulting in fatigue. *J. Fish. Res. Board Can.* **33,** 173.

Dubois, A. B., Cavagna, G. A., and Fox, R. S. (1974). Pressure distribution on the body surface of a swimming fish. *J. Exp. Biol.* **60,** 581.

Dubois, A. B., Cavagna, G. A., and Fox, R. S. (1976). Locomotion of bluefish. *J. Exp. Zool.* **195,** 223.

Evans, D. H. (1969). Studies on the permeability to water of selected marine, freshwater and euryhaline teleosts. *J. Exp. Biol.* **50,** 689.

Falck, B., von Mecklenburg, C., Myhrberg, H., and Persson, H. (1966). Studies on adrenergic and cholinergic receptors in the isolated hearts of *Lampetra fluviatilis* (Cyclostomata) and *Pleuronectes platessa* (*Teleostei*). *Acta Physiol. Scand.* **68,** 64.

Farmer, G. J., and Beamish, F. W. H. (1969). Oxygen consumption of *Tilapia nilotica* in relation to swimming speed and salinity. *J. Fish. Res. Board Can.* **26,** 2807.

Fisher, T., Coburn, R., and Forster, R. (1969). Carbon monoxide diffusing capacity in the bullhead catfish. *J. Appl. Physiol.* **26,** 161.

Fry, F. E. J. (1971). The effect of environmental factors on the physiology of fish. In "Fish Physiology" (W. S. Hoar and D. J. Randall, eds.), Vol. 6, p. 1. Academic Press, New York.

Gannon, B. J. (1971). A study of the dual innervation of teleost heart by a field stimulation technique. *Comp. Gen. Pharmacol.* **2,** 175.

Gannon, B. J., and Burnstock, G. (1969). Excitatory adrenergic innervation of the fish heart. *Comp. Biochem. Physiol.* **29,** 765.

Gannon, B. J., Campbell, G. D., and Satchell, G. M. (1972). Monamine storage in relation to cardiac regulation in the Port Jackson shark, *Heterodontus portusjacksoni*. *Z. Zellforsch. Mikrosk. Anat.* **131,** 437.

Gannon, B. J., Campbell, G., Randall, D. J., and Smith, D. G. (1978a). The vasculature of the gill of rainbow trout, *Salmo gairdneri*. I. The respiratory vascular bed. In preparation.

Gannon, B. J., Campbell, G., Randall, D. J., and Smith, D. G. (1978b). The vasculature of the gill of the rainbow trout, *Salmo gairdneri*. II. Gill structures related to blood flow regulation and gas exchange in the respiratory vascular bed. In preparation.

Gannon, B. J., Campbell, G., Randall, D. J., and Smith, D. G. (1978c). The vasculature of the gills of the rainbow trout, *Salmo gairdneri*. III. Nonrespiratory blood channels. In preparation.

Gesser, H., and Poupa, O. (1974). Relations between heart muscle enzyme pattern and directly measured tolerance to acute anoxia. *Comp. Biochem. Physiol. A* **48,** 97.

Golenhofen, K., and Lippross, H. (1969). Mechanische Koppelungswirkungen der Atmung auf den Herzschlag. *Pfluegers Arch.* **309,** 156.

Hall, F. G. (1930). The ability of the common mackerel and certain other marine fishes to remove dissolved oxygen from sea water. *Am. J. Physiol.* **93,** 417.

Hanson, D. (1967). Cardiovascular dynamics and aspects of gas exchange in chondrichthyes. Ph.D. Thesis Univ. of Washington, Seattle.

Harris, W. S., and Morton, M. J. (1968). A cardiac intrinsic mechanism that relates heart rate to filling pressure. *Circulation, Suppl.* **6,** 95.

Haswell, M. S., and Randall, D. J. (1976). Carbonic anhydrase inhibition in trout plasma. *Respirat. Physiol.* **28,** 17.

Haswell, M. S., and Randall, D. J. (1978). The pattern of carbon dioxide excretion in the rainbow trout (*Salmo gairdneri*). *J. Exp. Biol.* **72,** 17.

Heath, A. G. (1973). Ventilatory responses of teleost fish to exercise and thermal stress. *Am. Zool.* **13,** 490.

Hemmingsen, E. A., and Douglas, E. L. (1970). Respiratory characteristics of the hemoglobin free fish, *Chaenocephalus aceratus. Comp. Biochem. Physiol.* **33,** 733.

Hemmingsen, E. A., Douglas, E. L., Johansen, K., and Millard, R. W. (1972). Aortic blood flow and cardiac output in the hemoglobin free fish, *Chaenocephalus aceratus. Comp. Biochem. Physiol. A* **43,** 1945.

Hill, A. K. (1950). The dimensions of animals and their muscular dynamics. *Sci. Prog. (London)* **38,** 209.

Hirsch, E. F., Jellinek, M., and Cooper, T. (1964). Innervation of the systemic heart of the California hagfish. Circ. Res. **14,** 212.

Hochachka, P. W. (1961). The effect of physical training on oxygen debt and glycogen reserves in trout. *Can. J. Zool.* **39,** 767.

Hoffert, J. R., and Fromm, P. O. (1966). Effect of carbonic anhydrase inhibition on aqueous humor and blood bicarbonate ion in the teleost (*Salvelinus namaycush*). *Comp. Biochem. Physiol.* **18,** 333.

Holeton, G. F., and Randall, D. J. (1967a). Changes in blood pressure in the rainbow trout during hypoxia. *J. Exp. Biol.* **46,** 297.

Holeton, G. F., and Randall, D. J. (1967b). The effect of hypoxia upon the partial pressure of gases in the blood and water afferent and efferent to the gills of rainbow trout. *J. Exp. Biol.* **46,** 317.

Holmgren, S., and Nilsson, S. (1974). Drug effects on isolated artery strips from two teleosts, *Gadus morhua* and *Salmo gairdneri. Acta Physiol. Scand.* **90,** 431.

Holmgren, S., and Nilsson, S. (1975). Effects of some adrenergic and cholinergic drugs on isolated spleen strips from the cod, *Gadus morhua. Eur. J. Pharmacol.* **32,** 163.

Hudson, R. C. L. (1973). On the function of the white muscles in teleosts at intermediate swimming speeds. *J. Exp. Biol.* **58,** 509.

Hughes, G. M. (1970). Morphological measurements on the gills of fishes in relation to their respiratory function. *Folia Morphol. (Prague)* **18,** 78.

Hughes, G. M. (1972). Morphometrics of fish gills. *Respirat. Physiol.* **14,** 1.

Hughes, G. M., and Grimstone, A. V. (1965). The fine structure of the secondary lamellae of the gills of *Gadus pollachius. Quart. J. Microsc. Sci.* **106,** 343.

Hughes, G. M., and Morgan, M. (1973). The structure of fish gills in relation to their respiratory function. *Biol. Rev. Cambridge Philos. Soc.* **48,** 419.

Hughes, G. M., and Saunders, R. L. (1970). Responses of the respiratory pumps to hypoxia in the rainbow trout (*Salmo gairdneri*). *J. Exp. Biol.* **53,** 527.

Irving, L., Solandt, D. T., and Solandt, O. M. (1935). Nerve impulses from branchial pressure receptors in the dogfish. *J. Physiol. (London)* **84,** 187.

Itazawa, T. (1970). Characteristics of respiration of fish considered from the arteriovenous difference of oxygen content. *Nippon Suisan Gakkaishi* **36,** 57.

Jensen, D. (1961). Cardioregulation in an aneural heart. *Comp. Biochem. Physiol.* **2,** 181.

Jensen, D. (1963). Eptatretin: A potent cardioactive agent from the branchial heart of the Pacific hagfish *Eptatretus stoutii*. *Comp. Biochem. Physiol.* **10,** 129.

Jensen, D. (1965). The aneural heart of the hagfish. *Ann. N.Y. Acad. Sci.* **127,** 443.

Jensen, D. (1969). Intrinsic cardiac rate regulation in the sea lamprey, *Petromyzon marinus*, and rainbow trout, *Salmo gairdneri*. *Comp. Biochem. Physiol.* **30,** 685.

Jensen, D. (1970). Intrinsic cardiac rate regulation in elasmobranch: The horned shark, *Heterodontus prancisci*, and thornback ray, *Platyrhinoidis triseriata*. *Comp. Biochem. Physiol.* **34,** 289.

Johansen, K. (1962). Cardiac output and pulsatile aortic flow in the teleost *Gadus morhua*. *Comp. Biochem. Physiol.* **7,** 169.

Johansen, K. (1965a). Dynamics of venous return in elasmobranch fishes. *Hvalradets Skr.* **48,** 94.

Johansen, K. (1965b). Cardiovascular dynamics in fishes, amphibians, and reptiles. *Ann. N.Y. Acad. Sci.* 127, 414.

Johansen, K. (1972). Heart and circulation in gill, skin, and lung breathing. *Respirat. Physiol.* **14,** 193.

Johansen, K., and Hanson, D. (1967). Hepatic vein sphincters in elasmobranchs and their significance in controlling hepatic blood flow. *J. Exp. Biol.* **46,** 195.

Johansen, K., and Waage-Johannessen, N. (1962). Unpublished observations.

Johansen, K., Franklin, D. L., and Van Citters, R. L. (1966). Aortic blood flow in free-swimming elasmobranchs. *Comp. Biochem. Physiol.* **19,** 151.

Jones, D. R. (1971a). Theoretical analysis of factors which may limit the maximum oxygen uptake of fish. The oxygen cost of the cardiac and branchial pumps. *J. Theor. Biol.* **32,** 341.

Jones, D. R. (1971b). The effect of hypoxia and anaemia on the swimming performance of rainbow trout (*Salmo gairdneri*). *J. Exp. Biol.* **55,** 541.

Jones, D. R., and Schwarzfeld, T. (1974). The oxygen cost to the metabolism and efficiency of breathing in trout (*Salmo gairdneri*). *Respirat. Physiol.* **21,** 241.

Jones, D. R., Randall, D. J., and Jarman, G. M. (1970). A graphical analysis of oxygen transfer in fish. *Respirat. Physiol.* **10,** 285.

Jones, D. R., Langille, B. L., Randall, D. J., and Shelton, G. (1974). Blood flow in dorsal and ventral aortas of the cod, *Gadus morhua*. *Am. J. Physiol.* **226,** 90.

Keys, A., and Bateman, J. B. (1932). Branchial response to adrenaline and pitressin in the eel. *Biol. Bull. (Woods Hole, Mass.)* **63,** 327.

Kiceniuk, J. W., and Jones, D. R. (1977). The oxygen transport system in trout (*Salmo gairdneri*) during sustained exercise. *J. Exp. Biol.* **69,** 247.

Kirby, S., and Burnstock, G. (1969). Comparative pharmacological studies of isolated spiral strips of large arteries from lower vertebrates. *Comp. Biochem. Physiol.* **28,** 307.

Kirschner, L. B. (1969). Ventral aortic pressure and sodium fluxes in perfused eel gills. *Am. J. Physiol.* **217,** 596.

Klaverkamp, J. F., and Dyer, D. P. (1974). Autonomic receptors in isolated rainbow trout vasculature. *Eur. J. Pharmacol.* **28,** 25.

Kunos, G., and Nickerson, M. (1976). Temperature-induced interconversion of α- and β-adrenoreceptors in the frog heart. *J. Physiol. (London)* **256,** 23.

Kutty, M. N. (1968a). Respiratory quotients in goldfish and rainbow trout. *J. Fish. Res. Board Can.* **25,** 1689.

Kutty, M. N. (1968b). Influence of ambient oxygen on the swimming performance of goldfish and rainbow trout. *Can. J. Zool.* **46,** 647.

Labat, R. (1964). Action de l'adrenaline sur la frequence cardiaque de Pleuronectes vagotomises. *C. R. Soc. Biol.* **158,** 371.

Labat, R. (1966). Electrocardiologie chez les poisson téléostéens: Influence de quelques facteur ecologiques. *Ann. Limnol.* **2,** 1.

Labat, R., Raynaud, P., and Serfaty, A. (1961). Réactions cardiaques et variations de masse sanguine chez les Téléostéens. *Comp. Biochem. Physiol.* **4,** 75.

Laffont, J., and Labat, R. (1966). Action de l'adrenaline sure la frequence cardiaque de la Carpe commune. Éffet de la temperature du milieu sur l'intensité de la reaction. *J. Physiol.* (*Paris*) **58,** 351.

Landis, E. M., and Pappenheimer, J. R. (1963). Exchange of substances through the capillary walls. *In* "Handbook of Physiology: Circulation" (W. F. Hamilton and N. F. Dow, eds.), Vol. 2, p. 961. Am. Physiol. Soc., Washington, D.C.

Laurent, P., and Dunel, S. (1976). Functional organization of the teleost gill. I. Blood pathways. *Acta Zool.* (*Stockholm*) **57,** 189.

Lutz, B. R., and Wyman, L. C. (1932). The effect of adrenaline on the blood pressure of the elasmobranch, *Squalus acanthias. Biol. Bull.* (*Woods Hole, Mass.*) **62,** 17.

Maetz, J. (1973). Na^+/NH_4^+, Na^+/H^+exchanges and NH_3 movements across the gill of *Carassius auratus. J. Exp. Biol.* **58,** 255.

Magnuson, J. J. (1963). Tuna behavior and physiology, a review. *Proc. World Sci. Meet., Biol. Tunas Relat. Species, La Jolla, Calif., 1962* **3,** 1057.

Marples, B. J. (1935-1936). The blood vascular system of the elasmobranch fish *Squatina squatina* (Linne). *Trans. R. Soc. Edinburgh* **58,** 817.

Mayer, P. (1888). Über Eigentumlichkeiten in den Kreislauforganen der Selachiern. *Mitt. Zool. Stat. Neapel., Leipzig* **8,** 307.

Mazeaud, M. M. (1969). Influence de stress sur les teneurs en catecholamines du plasma et des corps axillaires chez un Selacian, la Rousette (*Scyliorhinus canicula* L.). *C.R. oc. Biol.* **163,** 2262.

Mazeaud, M. M., Mazeaud, F., and Donaldson, E. M. (1978). Stress resulting from handling in fish: Primary and secondary effects. *Trans. Am. Fish. Soc.* (in press).

Millen, J. E., Murdaugh, H. V., Hearn, D. C., and Robin, E. D. (1966). Measurement of gill water flow in *Squalus acanthias* using the dye-dilution technique. *Am. J. Physiol.* **211,** 11.

Mott, J. C. (1950). Radiological observations on the cardiovascular system in *Anguilla anguilla. J. Exp. Biol.* **27,** 324.

Mott, J. C. (1951). Some factors affecting the blood circulation in the common eel (*Anguilla anguilla*). *J. Physiol.* (*London*) **114,** 387.

Muir, B. S., and Buckley, R. M. (1967). Gill ventilation in *Remora remora. Copeia* 3, 581.

Muir, B. S., and Niimi, A. J. (1972). Oxygen consumption of the euryhaline fish Aholehole (*Kublia sandvicensis*) with reference to salinity, swimming, and food consumption. *J. Fish. Res. Board Can.* **29,** 67.

Nakano, T., and Tomlinson, N. (1967). Catecholamine and carbohydrate concentrations in rainbow trout (*Salmo gairdneri*) in relation to physical disturbance. *J. Fish. Res. Board Can.* **24,** 1701.

Newstead, J. D. (1967). Fine structure of the respiratory lamellae of teleostean gills, Z. *Zellforsch. Mikrosk. Anat.* **79,** 396.

Nilsson, S., Holmgren, S., and Grove, J. D. (1975). Effects of drugs and nerve stimulation on the spleen and arteries of two species of dogfish, *Scyliorhinus canicula* and *Squalus acanthias. Acta Physiol. Scand.* **95,** 219.

Nilsson, S., Abrahamsson, T., and Grove, J. D. (1976). Sympathetic control of adrenaline release from the chromaffin tissue in a fish. *Acta Physiol. Scand.* **96,** C11.

Opdyke, D. F., McGreehan, J. R., Messing, S., and Opdyke, N. E. (1972). Cardiovascular responses to spinal cord stimulation and autonomically active drugs in *Squalus acanthias. Comp. Biochem. Physiol. A* **42,** 611.

Ošťádal, B., and Schiebler, T. H. (1971). The terminal blood bed in the heart of the fish and in the heart of the turtle. *Z. Anat. Entwicklungsgesch.* **134,** 101.

Ostlünd, E., and Fänge, R. (1962). Vasodilation by adrenaline and noradrenaline and the effects of some other substances on perfused fish gills. *Comp. Biochem. Physiol.* **5,** 307.

Pathak, C. L. (1972). Stretch sensitive intrinsic autoregulatory mechanisms for rhythmicity and contractility of the heart. *Experientia* **28,** 650.

Piiper, J., and Baumgarten-Schumann, D. (1968). Transport of O_2 and CO_2 by water and blood in gas exchange of the dogfish (*Scyliorhinus stellaris*). *Respirat. Physiol.* **5,** 326.

Priede, I. G. (1974). The effect of swimming activity and section of the vagus nerves on heart rate in rainbow trout. *J. Exp. Biol.* **60,** 305.

Rahn, H. (1966). Aquatic gas exchange: Theory. *Respirat. Physiol.* **1,** 1.

Randall, D. J. (1968). Functional morphology of the heart in fishes. *Am. Zool.* **8,** 179.

Randall, D. J. (1970a). The circulatory system. *In* "Fish Physiology" (W. S. Hoar and D. J. Randall, eds.), Vol. 4, p. 133. Academic Press, New York.

Randall, D. J. (1970b). Gas exchange in fishes. *In* "Fish Physiology" (W. S. Hoar and D. J. Randall, eds.), Vol. 4, p. 253. Academic Press, New York.

Randall, D. J., and Smith, J. C. (1967). The regulation of cardiac activity in fish in a hypoxic environment. *Physiol. Zool.* **40,** 104.

Randall, D. J., and Stevens, E. D. (1967). The role of adrenergic receptors in cardiovascular changes associated with exercise in salmon. *Comp. Biochem. Physiol.* **21,** 415.

Randall, D. J., Holeton, G. F., and Stevens, E. D. (1967). The exchange of O_2 and CO_2 across the gills of rainbow trout. *J. Exp. Biol.* **46,** 339.

Randall, D., Heisler, N., and Drees, F. (1976). Ventilatory response to hypercapnia in the larger spotted dogfish, *Scyliorhinus stellaris. Am. J. Physiol.* **230,** 590.

Randall, D. J., Milliken, C., and Haswell, M. S. (1978). Permeability of rainbow trout red blood cells to lactate. In preparation.

Rankin, J. C. (1976). Personal communication.

Rankin, J. C., and Maetz, J. (1971). A perfused teleostean gill preparation: Vascular actions of neurohypophysial hormones and catecholamines. *J. Endocrinol.* **51,** 621.

Rao, G. M. M. (1968). Oxygen consumption of rainbow trout (*Salmo gairdneri*) in relation to activity and salinity. *Can. J. Zool.* **46,** 781.

Reite, O. B. (1969). The evolution of vascular smooth muscle responses to histamine and 5-hydroxytryptamine. I. Occurrence of stimulatory actions in fish. *Acta Physiol. Scand.* **75,** 221.

Richards, B. D., and Fromm, P. O. (1969). Patterns of blood flow through filaments and lamellae of isolated-perfused rainbow trout (*Salmo gairdneri*) gills. *Comp. Biochem. Physiol.* **29,** 1063.

Roberts, B. L. (1969). The spinal nerves of the dogfish (*Scyliorhinus*). *J. Mar. Biol. Ass. U.K.* **49,** 105.

Roberts, B. L. (1972). Activity of lateral-line sense organs in swimming dogfish. *J. Exp. Biol.* **56,** 105.

Roberts, J. L. (1975a). Active branchial and ram gill ventilation in fishes. *Biol. Bull.* (*Woods Hole, Mass.*) **148,** 85.

Roberts, J. L. (1975b). Cardio-ventilatory interactions during swimming, and during thermal and hypoxic stress. *Respirat. Mar. Organisms, Proc. Mar. Sect., 1st Mar. Biomed. Sci. Symp. p.* **139.**

Rowell, L. B. (1974). Human cardiovascular adjustments to exercise and thermal stress. *Physiol. Rev.* **54,** 75.

Santer, R. M., and Cobb, J. L. S. (1972). The fine structure of the heart of the teleost, *Pleuronectes platessa L. Z. Zellforsch. Mikrosk. Anat.* **131,** 1.

Satchell, G. H. (1965). Blood flow through the caudal vein of elasmobranch fish. *Aust. J. Sci.* **27,** 241.

Satchell, G. H. (1968). A neurological basis for the coordination of swimming with respiration in fish. *Comp. Biochem. Physiol.* **27,** 835.

Satchell, G. H. (1971). "Circulation in Fishes." Cambridge Univ. Press, London and New York.

Saunders, R. L. (1962). The irrigation of gills in fishes. II. Efficiency of oxygen uptake in relation to respiratory flow, activity, and concentrations of oxygen and carbon dioxide. *Can. J. Zool.* **40,** 817.

Schmidt-Nielsen, K. (1972). Locomotion: Energy cost of swimming, flying and running. *Science* **177,** 227.

Secondat, M. (1950). Influence de l'exercise musculaire sur la valeur de la glycéme de la carpe (*Cyprinus carpio* L.). *C. R. Acad. Sci.* **231,** 796.

Shelton, G. (1970). The regulation of breathing. *In* "Fish Physiology" (W. S. Hoar and D. J. Randall, eds.), Vol. 4, p. 293. Academic Press, New York.

Silver, S. J., Warren, C. E., and Doudoroff, P. (1963). Dissolved oxygen requirements of developing steelhead trout and chinook salmon embryos at different water velocities. *Trans. Am. Fish. Soc.* **92,** 327.

Smit, H. (1965). Some experiments on the oxygen consumption of goldfish (*Carassius auratus* L.) in relation to swimming speed. *Can. J. Zool.* **43,** 623.

Smit, H., Amelink-Koutstaal, J. M., Vuverberg, J., and von Vaupel-Klein, J. C. (1971). Oxygen consumption and efficiency of swimming goldfish. *Comp. Biochem. Physiol. A* **39,** 1.

Smith, D. G. (1976). The structure and function of the respiratory organs of some lower vertebrates. Ph.D. Thesis, Vols. 1 and 2. Zool. Dep., Univ. of Melbourne, Melbourne.

Smith, L. S. (1966). Blood volumes of three salmonids. *J. Fish. Res. Board Can.* **23,** 1439.

Smith, L. S., and Bell, G. R. (1964). A technique for prolonged blood sampling in free-swimming salmon. *J. Fish. Res. Board Can.* **32,** 711.

Smith, L. S., and Bell, G. R. (1967). Anesthetic and surgical techniques for Pacific salmon. *J. Fish. Res. Board Can.* **24,** 1579.

Smith, L. S., Brett, J. R., and Davis, J. C. (1967). Cardiovascular dynamics in swimming adult sockeye salmon. *J. Fish. Res. Board Can.* **24,** 1775.

Starling, E. H. (1918). "The Linacre Lecture on the Law of the Heart, Given at Cambridge, 1915." Longmans, Green, London.

Steen, J. B., and Kruysse, A. (1964). The respiratory function of teleostean gills. *Comp. Biochem. Physiol.* **12,** 127.

Steen, J. B., and Stray-Pedersen, S. (1975). The permeability of fish gills with comments on the osmotic behavior of cellular membranes. *Acta Physiol. Scand.* **95,** 6.

Stevens, E. D. (1968). The effect of exercise on the distribution of blood to various organs in rainbow trout. *Comp. Biochem. Physiol.* **25,** 615.

Stevens, E. D., and Black, E. C. (1966). The effect of intermittent exercise on carbohydrate metabolism in rainbow trout, *Salmo gairdneri*. *J. Fish. Res. Board Can.* **23,** 471.

Stevens, E. D., and Randall, D. J. (1967a). Changes in blood pressure, heart rate, and breathing rate during moderate swimming activity in rainbow trout. *J. Exp. Biol.* **46,** 307.

Stevens, E. D., and Randall, D. J. (1967b). Changes in gas concentrations in blood and water during moderate swimming activity in rainbow trout. *J. Exp. Biol.* **46,** 329.

Stevens, E. D., and Sutterlin, A. M. (1975). A technique for measuring heat exchange across the gills of a teleost, Hemitripterus americanus. *Respirat. Mar. Organisms. Proc. Mar. Sect., 1st Mar. Biomed. Sci. Symp.* p. 195.

Stevens, E. D., Bennion, G. R., Randall, D. J., and Shelton, G. (1972). Factors affecting arterial pressures and blood flow from the heart in intact, unrestrained lingcod, *Ophiodon elongatus*. *Comp. Biochem. Physiol. A* **43,** 681.

Sudak, F. N. (1965a). Intrapericardial and intracardiac pressures and events of the cardiac cycle in *Mustelus canis* (Mitchill). *Comp. Biochem. Physiol.* **14,** 689.

Sudak, F. N. (1965b). Some factors contributing to the development of subatmospheric pressure in the heart chambers and pericardial cavity of *Mustelus canis* (Mitchill). *Comp. Biochem. Physiol.* **15,** 199.

Sutterlin, A. M. (1969). Effects of exercise on cardiac and ventilation frequency in three species of freshwater teleosts. *Physiol. Zool.* **42,** 36.

Sutterlin, A. M., and Saunders, R. L. (1969). Proprioceptors in the gills of teleosts. *Can. J. Zool.* **47,** 1209.

Taylor, W., Houston, A. H., and Horgan, J. D. (1968). Development of a computer model simulating some aspects of the cardiovascular-respiratory dynamics of the salmonid fish. *J. Exp. Biol.* **49,** 477.

Thorson, T. B. (1959). Partitioning of body water in sea lamprey. *Science* **130,** 99.

Thorson, T. B. (1960). The partitioning of body water in Osteichthyes: Phylogenetic and ecological implications in aquatic vertebrates. *Biol. Bull. (Woods.Hole, Mass.)* **120,** 238.

Tucker, V. A. (1970). Energetic cost of locomotion in animals. *Comp. Biochem. Physiol.* **34,** 345.

Tytler, P. (1969). Relationship between oxygen consumption and swimming speed in the haddock, *Melanogrammus aeglefinus*. *Nature* (*London*) **221,** 274.

Ultsch, G. R. (1973). A theoretical and experimental investigation of the relationships between metabolic rate, body size, and oxygen exchange capacity. *Respirat. Physiol.* **18,** 143.

van Dam, L. (1938). On the utilization of oxygen and regulation of breathing in some aquatic animals. Ph.D. Thesis, Univ. of Groningen, Groningen, Netherlands.

Voboril, Z., and Schiebler, T. H. (1970). The blood supply of fish hearts. *Z. Anat. Entwicklungsgesch.* **130,** 1.

Vogel, W., Vogel, V., and Keemers, H. (1973). New aspects of the intrafilamental vascular system in gills of a euryhaline teleost *Tilapia mossambica*. *Z. Zellforsch. Mikrosk. Anat.* **144,** 573.

Vogel, W., Vogel, V., and Pfautsch, M. (1976). Arterio-venous anastomoses in rainbow trout gill filaments. *Cell Tissue Res.* **167,** 373.

von Holst, E. (1937). Vom Wesen der Ordnung im Zentralnervensystem. *Naturwis senschaften* **40,** 641.

Webb, P. W. (1970). Some aspects of the energetics of swimming of fish with special reference to the cruising performance of rainbow trout. Ph.D. Thesis Univ, of Bristol, Bristol, England.

Webb, P. W. (1971a). The swimming energetics of trout. I. Thrust and power output at cruising speeds. *J. Exp. Biol.* **55,** 489.

Webb, P. W. (1971b). The swimming energetics of trout. II. Oxygen consumption and swimming efficiency. *J. Exp. Biol.* **55,** 521.

Webb, P. W. (1975a). Hydrodynamics and energetics of fish propulsion. *Bull., Fish. Res. Board Can. No.* **190,** 158 pp.

Webb, P. W. (1975b). Synchrony of locomotion and ventilation in *Cymatogaster aggregata*. *Can. J. Zool.* **53,** 904.

Weihs, D. (1973a). Optimal cruising speed for migrating fish. *Nature* (*London*) **245,** 48.

Weihs, D. (1973b). Mechanically efficient swimming techniques for fish with negative buoyancy. *J. Mar. Res.* **31,** 194.

Weihs, D. (1974). Energetic advantages of burst swimming of fish. *J. Theor. Biol.* **48,** 1.

West, N. H., and Jones, D. R. (1975). Unpublished observations.

Wood, C. M. (1974a). A critical examination of the physical and adrenergic factors affecting blood flow through the gills of the rainbow trout. *J. Exp. Biol.* **60,** 241.

Wood, C. M. (1974b). Mayer waves in the circulation of a teleost fish. *J. Exp. Zool.* **189,** 267.

Wood, C. M. (1975). A pharmacological analysis of the adrenergic and cholinergic mechanisms regulating branchial vascular resistance in the rainbow trout (*Salmo gairdneri*). *Can. J. Zool.* **53,** 1569.

Wood, C. M., and Randall, D. J. (1973a). The influence of swimming activity on sodium balance in the rainbow trout (*Salmo gairdneri*). *J. Comp. Physiol.* **82,** 207.

Wood, C. M., and Randall, D. J. (1973b). Sodium, balance in the rainbow trout (Salmo gairdneri) during extended exercise. *J. Comp. Physiol.* **82,** 235.

Wood, C. M., and Randall, D. J. (1973c). The influence of swimming activity on water balance in the rainbow trout (*Salmo gairdneri*). *J. Comp. Physiol.* **82,** 257.

Wood, C. M., and Shelton, G. (1975). Physical and adrenergic factors affecting systemic vascular resistance in the rainbow trout: A comparison with branchial vascular resistance. *J. Exp. Biol.* **63,** 505.

Wood, C. M., McMahon, B. R., and McDonald, D. G. (1977). An analysis of changes in blood pH following exhausting activity in the starry flounder (*Platichthys stellatus*). *J. Exp. Biol.* **69,** 173.

Chapter 13

Understanding the effects of environmental factors on fish performance and tolerance in an era of climate change

Patricia M. Schulte*
Department of Zoology and Biodiversity Research Centre, The University of British Columbia, Vancouver, BC, Canada
**Corresponding author: e-mail: pschulte@zoology.ubc.ca*

Patricia Schulte discusses the impact of Fred Fry's chapter "The effects of environmental factors on the physiology of fish" in Fish Physiology, Volume 6 published in 1971.

We live in an era when anthropogenic activities are causing dramatic changes to our environment. Carbon emissions have led to increases in global mean temperature and increases in the frequency, intensity, and duration of heat waves (Frölicher et al., 2018). Changes in agricultural practices and land use are causing eutrophication that is increasing the frequency, magnitude and extent of exposure to hypoxia, and this interacts with elevated temperatures with sometimes devastating effects on fish (Deutsch et al., 2015; Earhart et al., 2022). At the same time, changes in precipitation and runoff are affecting the salinity of freshwater and coastal ecosystems (Jeppesen et al., 2020; Trenberth, 2011), and chemical pollution is now reaching dangerous levels in many areas (Persson et al., 2022; Wang et al., 2020). In the sweeping chapter "The effects of environmental factors on the physiology of fish" (Fry, 1971), Fry addresses how all of these factors influence the metabolism and activity of fish, outlining a conceptual framework that came to be known as the "Fry Paradigm" (Kerr and Lawrie, 1976). There are many lessons in this chapter that are directly applicable to fish physiologists attempting to understand and predict the likely responses of fish populations to changing conditions in the Anthropocene.

Fry's world view was fundamentally shaped by his conception of the pre-eminence of energetics. He viewed organisms as systems through which

Fish Physiology, Vol. 40B. https://doi.org/10.1016/bs.fp.2024.05.007

energy flows, and his key insight is summed up in his own words that: "Energy comes from the environment, and… the environment sets to a large degree the conditions under which the organism uses the energy it has assimilated" (Fry, 1971). Fry's insight was that the effects of environmental factors on fish can best be understood by examining their effects on energetics.

Fry had direct experience with fish from his earliest days, as his father operated a wholesale fish business in Toronto, Canada (Evans and Neill, 1990). These experiences led him to study aquatic ecology at the University of Toronto (B.A., M.A., Ph.D.), and he subsequently joined the faculty there, where he remained throughout his long career (Kerr and Lawrie, 1976). During the Second World War, Fry served with the Royal Canadian Airforce, where he worked in the area of aviation medicine, where he used his technical acumen and insight into physiology and energetics to help develop equipment that allowed fighter pilots to breathe at high altitudes (Kerr and Lawrie, 1976; McCauley, 1990). These experiences provide a window into some of the characteristics of Fry's approach to science, including his interest in methods development, his focus on the role of energetics and aerobic metabolism, and his passion for both theoretical and practical approaches in scientific research. Indeed, as stated during a symposium at the time of his retirement from the University of Toronto "The question of whether his scientific research was 'pure' or 'applied' never has had much meaning for Fry" (Loftus, 1976), and he made major contributions both to practical problems in fisheries management and fundamental aspects of fish physiology. It is thus appropriate that Fry's ideas are now being applied in the quest to understand the effects of anthropogenic environmental change on fish.

Fry thought of himself as an ecological physiologist (Evans and Neill, 1990), and was fundamentally interested in how fishes interact with their environment and how they perform the activities that allow them to survive and thrive in a changing environment. He felt that this question could be best understood in the context of the underlying mechanisms that support animal activity. Fig. 1 illustrates the conceptual schema with which Fry introduced his chapter (Fry, 1971), emphasizing that his main focus was on "what fish can do in relation to their environment," which today we would term their activity or performance. His key insight was that these organismal-level processes are constrained and influenced by processes at lower levels of biological organization, which Fry thought of in two categories: "how it works" and "what makes it go." Today we would call these physiological and biochemical processes, respectively. Fry was particularly focused on the physiological process associated with aerobic metabolism, and the underlying biochemistry that supports this metabolic activity. In this wide-ranging chapter, he summarizes and extends some of the ideas outlined in his earlier monograph "Effects of the environment on animal activity." (Fry, 1947), covering everything from methods of measuring metabolism and activity, to

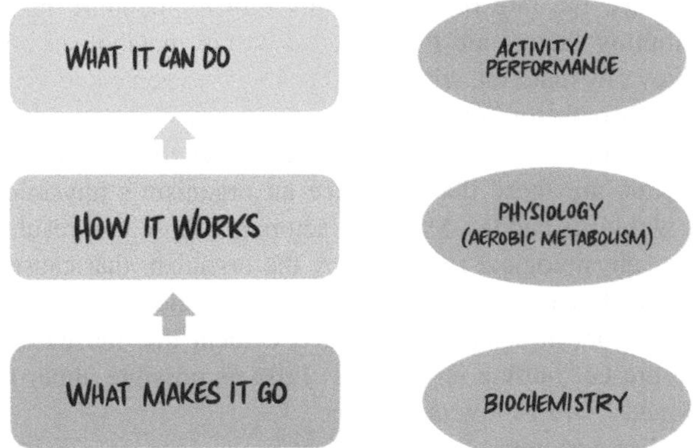

FIG. 1 Fry's conceptual schema for the relationship of animal activity to the underlying metabolic processes. On the left are Fry's descriptions and on the right are the modern analogs of these processes. Processes at lower levels of organization shape and constrain those at higher levels.

providing a theoretical and conceptual schema for understanding fish performance and tolerance in relation to the environment, as well as comprehensively summarizing the state of the literature to date. The deep conceptual insights presented in this chapter were developed at a time when the available data to address these issues were still limited, which makes the clarity and insight of Fry's thinking even more remarkable. Although the chapter addresses many subjects, at least from my perspective, it presents two major concepts that have had long-lasting impacts on our understanding of the environmental physiology of fishes:

(1) the classification of types of environmental factors
(2) the Fry paradigm emphasizing the importance of aerobic metabolic scope.

Fry's perspective stemmed from his view of the primacy of energetics, and thus he considered that fundamental insights into the impacts of environmental factors on organisms can be gained by considering their impacts on metabolism. This understanding led him to propose that we can best understand the interaction between an organism and its environment by classifying environmental factors not by the type of factor (e.g., temperature, oxygen, photoperiod etc.), but instead by grouping environmental factors based on their impacts on the organism and its metabolism. He therefore proposed dividing environmental factors into five classes: lethal, controlling, limiting, directive and masking (Fry, 1947, 1971). Note that in the 1947 monograph he also proposed "accessory" factors, which he did not discuss in later work.

Lethal factors are those that destroy the integrity of the organism, and thus its ability to perform metabolic functions. Controlling factors are those that directly influence the rate of metabolism, with temperature being a prime example. Limiting factors are those that constrain metabolism because of a shortage of key raw material, with environmental oxygen being a prime example. The final two factors (directive and masking) are more challenging to think about, as they involve regulated responses of the organism itself. Directive factors are those that influence an organism's physiology and its movements within its habitat. Masking factors are those that result in homeostatic or other physiological responses by the organism that cause additional metabolic work. Because of their complexity Fry did not discuss these last two types of factors in detail, stating "it is evident that the wisest course to adopt in the present outline is to say as little as possible about the matter" (Fry, 1947), and I will follow this advice.

Note that Fry was very clear that these five classes were neither comprehensive nor mutually exclusive. For example, temperature has the potential to act as a controlling factor under some circumstances and a lethal factor under others. Similarly, Fry was careful to point out that organisms have substantial capacity to alter the impacts of these factors on their metabolism and activity through various types of physiological regulation, acclimation and acclimatization. Fry also considered the possible interactions between these factors, indeed in his earlier work (Fry, 1947) he referred to "accessory factors" which he considered to be "a useful heading under which to bring together various interactions of the [factors] which result in the death of the animal." However, in his later work (Fry, 1971), he no longer used a separate term to address these interactions, and simply emphasized that "death often ensues because of interactions between controlling and limiting factors."

Fry considered temperature to be the prime example of a controlling factor that directly influences metabolic rate. He was very aware of the work of Eyring and Arrhenius on the effects of temperature on biological reactions (Arrhenius, 1889; Eyring, 1935). This work demonstrated that the rate of chemical reactions increases with temperature as an inevitable result of thermodynamics. Thus, temperature acts as a controlling factor by increasing metabolic rate through its effects at the chemical and biochemical levels. Fry's understanding of these biochemical effects is an illustration of his ability to easily move across levels of organization to gain new insights into animal function (Kerr, 1976).

When thinking about the effects of environmental factors, Fry defined three important metabolic states. The first is the standard metabolic rate (SMR), which he considered to be an approximation of the minimum rate of aerobic metabolism for the intact organism. By contrast, he defined routine metabolic rate (RMR) as the mean rate of oxygen consumption observed in fish that exhibit some level of random activity under experimental conditions in which they are protected from most outside stimuli and where movements are somewhat restricted. Finally, Fry discussed the active metabolic rate, or what we now call maximum metabolic rate (MMR), which he felt could best be measured as the maximum sustained aerobic metabolic rate for a fish swimming steadily. In the chapter, Fry spent considerable time discussing the difference between SMR and RMR, noting that RMR can briefly approach MMR under some circumstances, and he emphasized the importance of estimating SMR to allow calculation of the scope for activity (MMR-SMR), or what we now term aerobic scope. Fry hypothesized that the scope for activity (aerobic scope) was an indicator of the capacity for fish to perform important activities such as foraging, evading predators, and competing with conspecifics. Note that Fry was largely focused on aerobic metabolism, mentioning the possibility of anaerobic metabolism only briefly, although he acknowledged its potential importance, particularly for hypoxia tolerant species such as goldfish.

Although the chapter covers many topics, there was a strong emphasis on the effects of temperature. Fry used the available data at that time to show that that SMR increased essentially exponentially with temperature while MMR reached a plateau at high temperature. This resulted in aerobic scope exhibiting a bell-shaped curve with a clear optimum temperature (T_{opt}), which Fry suggested was the temperature at which activity could be maximized (Fig. 2). Above this temperature, SMR continues to increase exponentially, while MMR reaches a plateau or even begins to decrease, and at some point SMR reaches MMR, resulting in no available scope for activity at the T_{crit}. In recent years Fry's hypothesis that the scope for activity (aerobic scope) is a key index of how temperature impacts organismal performance and fitness has been extended to form the basis of the hypothesis of oxygen and capacity limited thermal tolerance (OCLTT), which is currently a widely used approach for forecasting the effects of climate change on species distributions (Clark and Mark, 2017; Pörtner, 2010; Pörtner and Farrell, 2008).

The high profile of the OCLTT hypothesis has inspired a great resurgence of interest in the interactions between temperature and metabolic rate in fish in the context of climate change (Claireaux and Lefrançois, 2007; Eliason et al., 2011; Nilsson et al., 2009; Pörtner and Peck, 2010). As a result, there are now abundant data on the relationship between temperature and metabolic rate for many species of fish. Meta-analysis of these data indicate that aerobic scope does not always follow the predicted pattern of a bell-shaped curve, and that this shape varies among species, largely because in some species MMR does not plateau at high temperatures, and instead continues to increase with temperature up to close to the lethal temperature for the organism (Lefevre, 2016). This suggests that constraints on aerobic scope at high temperature are not a universal mechanism that limits the performance of fish at high temperature. On the other hand, the temperature at which aerobic scope is maximized is strongly correlated with the temperature at the warm range edge for many fish species (Payne et al., 2016), supporting the potential of a strong link between the thermal optimum for aerobic scope and fitness. The evidence linking temperature effects on aerobic scope to performance, fitness and

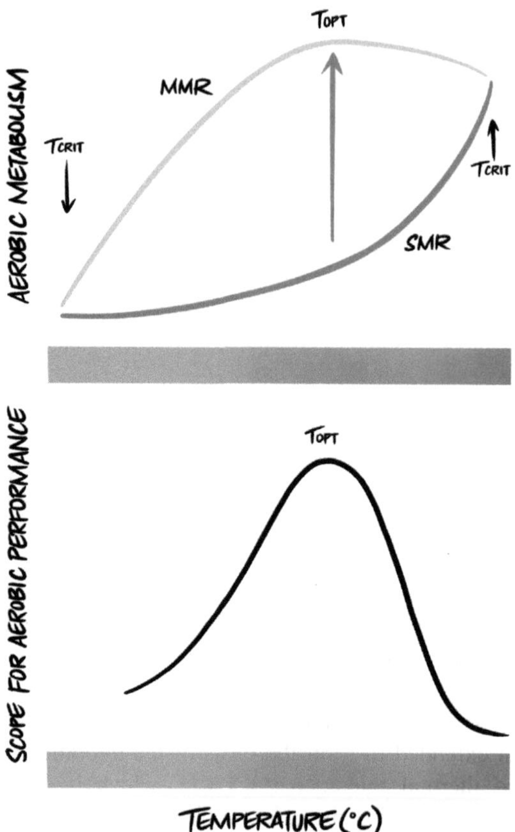

FIG. 2 The effects of temperature on standard metabolic rate (SMR), maximum metabolic rate (MMR) and absolute aerobic scope (AS), or the difference between MMR and SMR. The upper panel shows SMR and MMR and the key metrics that can be derived from these curves. T_{opt} is the temperature at which aerobic scope is maximized. T_{crit} (the critical temperature) is the temperature at which aerobic scope falls to zero. Note that some authors label the T_{crit} as CT_{max}. See the text for additional discussion of the confusion that has ensued by mixing these terms. The lower panel shows AS (or the scope for aerobic performance, which Fry called the scope for activity).

habitat temperature is particularly compelling for salmonid fishes (Eliason et al., 2011), the group that Fry used to develop his paradigm.

Using the data that he had available, Fry also pointed out an extremely interesting pattern that provides insights in to the mechanisms that set the bell-shaped curve for aerobic scope in salmonids (and possibly other fish species). Recall that the bell-shape of the aerobic scope curve in salmon is due largely to the plateau in MMR at high temperature. Using data from Sockeye salmon collected by his former student Roly Brett (Brett, 1964), Fry observed that the plateau in MMR at high temperature could be relieved by providing supplementary oxygen. This strongly suggested that oxygen becomes a limiting factor on activity at high temperature. Indeed, Fry considered the limiting effects of oxygen levels on both MMR and SMR, showing that fish are able to maintain their SMR across a fairly wide range of oxygen levels, until a point which he called the "level of no excess activity," and which we now term the P_{crit}, below which the rate of oxygen consumption declines with ambient oxygen levels. By contrast, MMR is gradually reduced by decreasing oxygen. These ideas have been greatly developed and extended in subsequent years (Claireaux and Chabot, 2016), but the fundamental pattern remains robust.

Despite the clear relevance of the OCLTT concept, it has been remarkably controversial (Clark et al., 2013; Gräns et al., 2014; Jutfelt et al., 2018; Norin et al., 2014). From my perspective, part of this confusion has arisen through the attempt of the OCLTT to unify the Fry paradigm with the concept of acute thermal tolerance, which is more directly related to temperature's role as a lethal factor (see below) rather than its role as a controlling factor—which is the central focus of Fry's conception of aerobic scope. Problematically, in the OCLTT framework the term CT_{max} is often used to describe the upper endpoint of the aerobic scope curve. This has resulted in confusion (Ørsted et al., 2022), because in the broader literature, CT_{max} is a term referring to a measure of acute thermal tolerance determined using a dynamic temperature ramp. In this context, CT_{max} is derived using a method termed critical thermal methodology (CTM), which was originally developed for reptiles (Cowles and Bogert, 1944) and later extended to fish (Becker and Genoway, 1979; Beitinger et al., 2000). In this approach, temperature is acutely increased (or decreased) until the fish is no longer able to maintain the appropriate dorso-ventral orientation (i.e., it loses equilibrium). At this point, the fish is considered to have reached "ecological death," as it will quickly die unless removed to a more permissive temperature. Thus, in this framework the CT_{max} (or CT_{min} for low temperatures) is an index of the lethal effects of temperature exposure. This is not necessarily the same temperature as the T_{crit} of an aerobic scope curve (Fig. 2), which is not a lethal measure. Indeed, T_{crit} would be expected to be lower than CT_{max} in many species. The net result of the confusion in the literature between CT_{max} and T_{crit} is that a substantial literature has developed that shows that (at least in some species) CT_{max} is not oxygen dependent (Ern et al., 2015; McArley et al., 2021, 2022), suggesting that it is not necessarily determined by the same mechanisms that shape the aerobic scope curve. These results have often been used to call into question the generality of the OCLTT and (by extension) the Fry paradigm, but it is critically important to remember that T_{crit} and acutely lethal temperatures are not the same thing. Fry was well aware of this distinction, and in his chapter, he states that "the relation of maximum scope to lethal temperature depends on whether and where the normal oxygen content begins to exert a limiting effect" (Fry, 1971). He then goes on to discuss how species with very similar thermal optima for aerobic scope may have very different lethal temperatures, which suggests that Fry was clear that these two processes can be decoupled.

Little was known about the mechanisms that led to thermal death at the time that Fry was developing his ideas on the subject. Indeed, he described the state of understanding of the underlying mechanisms at that time as "our still profound ignorance of the nature of thermal death," stating that "the whole problem is still obscured by our ignorance of the specific sites of breakdown in thermal death" (Fry, 1971). However, he considered that high temperature could potentially exert its lethal effects via its direct effects on metabolic processes, or alternatively temperature might cause lethality via mechanisms that are independent of its direct impacts on metabolism.

There have been several recent syntheses that outline the current state of our understanding of the mechanisms that cause fish to die at high temperatures (Ern et al., 2023). In general, we have returned to the place where Fry began, as the current data suggest that there is strong evidence for oxygen limitation of thermal tolerance in some species and strong evidence against it in other species (Ern et al., 2023). Of course, we now have much greater insight into the molecular processes that fail at these temperatures (e.g., effects on protein structure, membrane fluidity, mitochondrial function) and while our understanding of the "nature of thermal death" remains incomplete, we are coming close to a synthetic understanding of the issue (Ern et al., 2023). These issues are critically important as we think about the effects of climate change on fishes.

Lethal factors destroy an organism's ability to perform integrative physiological functions and ultimately lead to death. These factors thus set limits on the environments in which a fish can live, and can shape species' distribution and abundance. Therefore, having appropriate methods for determining these limits is very important. In this chapter, Fry addresses this technical question at some length, and although he felt there were a number of unresolved issues, he was quite optimistic that this problem was largely solved, stating in his earlier monograph: "There appears to be hope that methods for the laboratory measurement of the effects of temperature as a lethal factor will soon be perfected" (Fry, 1947). Unfortunately, more than 50 years later, we are still debating the appropriateness of various measures of thermal tolerance for estimating lethal temperatures (Cooper et al., 2008; Jørgensen et al., 2021; Kilgour and McCauley, 1986; Lutterschmidt and Hutchison, 1997; Ørsted et al., 2022; Rezende et al., 2014). In essence, the problem derives from the fact that thermal injury is the result of an interaction between the level of thermal exposure (how hot it is) and the length of time that an animal is exposed to those temperatures (duration of exposure). Fry drew a very useful distinction between the "zone of tolerance," which is the range of temperatures over which a fish can maintain functional capacity and survive over long periods of time period of time (months to years), and a time-limited "zone of resistance." In the zone of resistance, fish is experiencing temperatures that will kill them over a relatively short period of time (hours to days), and this lethal exposure is determined by the length of time spent at temperatures likely to cause thermal damage.

This distinction is important because, since the time of Fry, thermal tolerance has been estimated using two rather different classes of methods. The first group of methods are typically called static methods. When using a static method, fish are exposed to one or more levels of a constant temperature and either time to heat failure or percent mortality after a given time is reported. By contrast, with a dynamic method, the fish is exposed to a controlled rate of temperature increase until the failure temperature is reached (this is the typical method for determining CT_{max} and CT_{min}). Fry preferred to use static methods, as he felt that dynamic methods confound time and temperature, and thus were of limited physiological utility. Fry thus placed great importance on estimating the upper incipient lethal temperature (UILT), or the maximum temperature at which at least 50 % of a sample of fish would be able to survive indefinitely, which he suggested could not be easily estimated using dynamic methods. However, the vast majority of studies have estimated thermal tolerance using dynamic methods, reporting CT_{max} as the endpoint, because, as Fry himself stated, these methods are "extremely economical of material," requiring far fewer fish. In addition, dynamic methods do not use death as an endpoint, and thus they are now preferred from an animal welfare perspective. But the physiological and environmental relevance of CT_{max} is far from clear (Terblanche et al., 2011), as fish are seldom exposed to heating at the rates typically used in laboratory studies, and CT_{max} is generally substantially higher than the temperatures that are encountered in the environment.

Recently, there have been attempts to develop models that will allow upper incipient lethal temperatures to be estimated from measures of CT_{max} (Jørgensen et al., 2021) using the fundamental principles of cellular physiology. In essence, this model (Fig. 3) considers the balance between the damaging effects of exposure to high temperatures (e.g., protein denaturation, changes in membrane fluidity, exposure to oxidative stress), and the action of homeostatic processes that repair this damage (e.g., heat shock proteins) (Ørsted et al., 2022). At some temperature (termed the critical temperature (T_c) in the model), repair processes can no longer keep up with the rate of damage, and homeostasis and integrated organismal function is lost. Note that T_c in this model is conceptually equivalent to the upper incipient lethal temperature, as it is the highest temperature that can be sustained indefinitely, and is not necessarily the same as the T_{crit} of an aerobic scope curve. This model follows the thinking of Fry, as he felt that "the incipient lethal level should be looked on as the boundary of the immediate direct lethal effects, …. on a site of metabolism so as to destroy it more rapidly than the organism can keep it in repair" (Fry, 1971). The advance made in the more recent work is to show mathematically that it is possible to predict T_c using data from a

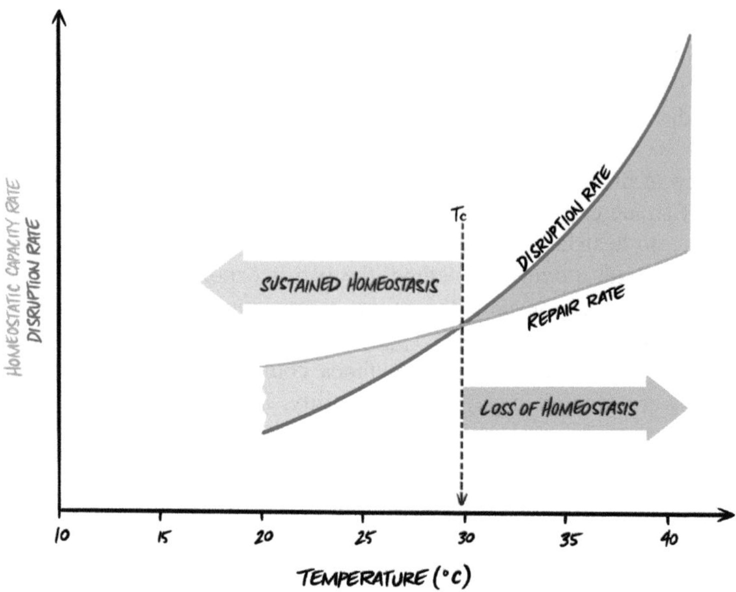

FIG. 3 A unifying model of the lethal effects of temperature. This model suggests that the lethal effects of temperature are the result of the relative activity of processes that negatively affect cellular function and the homeostatic mechanisms that limit, repair, or restore this damage. At the critical temperature (T_c), which is conceptually equivalent to Fry's upper incipient lethal temperature, the rate of thermally induced disruption becomes too rapid for homeostatic and repair mechanisms to keep up. Over time, with exposure to temperatures above T_c, damage will accumulate, and thermal death will occur. The time to thermal death will be shorter at high temperatures because the difference between the rate of damage and the rate of repair is larger than at low temperatures. Note that the central assumption of this model is that damage and repair processes have different thermal sensitivities, such that the rate of damage accelerates more rapidly with increasing temperature than does the rate of repair (Jørgensen et al., 2021; Ørsted et al., 2022).

dynamic ramping assay such as a CT_{max} trial, and to demonstrate that this approach is consistent with published data (including some data presented by Fry), showing that this model can replicate real-world observations (Jørgensen et al., 2021; Ørsted et al., 2022).

T_c (which is conceptually equivalent to the upper incipient lethal temperature) thus separates a thermal zone that is dominated by processes determining "the rate of life" in the permissive temperature range from a zone that is dominated by processes that are related to "the rate of death" in the stressful temperature range (Jørgensen et al., 2021; Ørsted et al., 2022). Note that Fig. 3 only captures the upper lethal temperatures, but these ideas can easily be extended to address lethality at low temperatures by taking into account the observation that the rate of processes related to cellular damage start to increase at low temperatures (Peck, 2016), whereas repair process likely operate very slowly at low temperatures, resulting in accumulation of cold-induced damage.

One of the critical questions that remains, however, is the extent to which the temperature at which aerobic scope falls to zero is aligned with the upper incipient lethal temperature (or T_c in the model). In essence, this addresses the question: to what extent are the controlling effects of temperature and the lethal effects of temperature acting on the same processes? The authors of this unifying model (Jørgensen et al., 2021; Ørsted et al., 2022) propose that the complete aerobic scope curve should fall within the zone of sustained homeostasis, and thus aerobic scope will fall to zero at temperatures below the upper incipient lethal temperature. However, in my opinion this question is largely unresolved at present, as there is currently rather limited data to address this question across fish taxa. There may be some species where the upper bound of the aerobic scope curve may correspond to the upper incipient lethal temperature, whereas there may be others where they are decoupled. In essence, measures of thermal performance (scope for activity, aerobic scope) may be correlated in some species, but there is no underlying physiological constraint that indicates that this must be the case. Indeed, as mentioned above, this idea was quite apparent to Fry, and much debate in the literature could have been avoided had the field placed the lessons of his chapter more clearly in focus. The lethal, controlling and limiting effects of environmental factors on organisms are all important, and we gain great clarity from making the distinction among them.

Here, I have focused on the impacts of temperature, but Fry's chapter is much more comprehensive, and looks in detail at the effects of oxygen, salinity and pollutants as well as temperature, placing all of these types of environmental stressors into an overarching framework united by the idea of the primacy of energetics in determining animal activity. There are many lessons to be learned from this work that are directly applicable to studies of fishes in the Anthropocene, when they are being exposed to a novel suite of multiple interacting stressors that affect their ability to assimilate energy and use it to perform the activities of their daily life.

Acknowledgments

This work was funded by a Canada Research Chair (CRC-2021-00040) and a Natural Sciences and Engineering Council of Canada (NSERC) Discovery Grant (RGPIN-2017-04613) to Patricia Schulte. Thanks to Rush Dhillon for developing the illustrations presented in the Figures.

References

Arrhenius, S., 1889. Über die Dissociationswärme und den Einfluss der Temperatur auf den Dissociationsgrad der Elektrolyte. Z. Phys. Chem. https://doi.org/10.1515/zpch-1889-0408.

Becker, C.D., Genoway, R.G., 1979. Evaluation of the critical thermal maximum for determining thermal tolerance of freshwater fish. Environ. Biol. Fish 4, 245–256. https://doi.org/10.1007/BF00005481.

Beitinger, T., Bennett, W., McCauley, R., 2000. Temperature tolerances of North American freshwater fishes exposed to dynamic changes in temperature. Environ. Biol. Fish 58, 237–275.

Brett, J.R., 1964. The respiratory metabolism and swimming performance of young sockeye salmon. J. Fish. Res. Board Can. 21, 1183–1226. https://doi.org/10.1139/f64-103.

Claireaux, G., Chabot, D., 2016. Responses by fishes to environmental hypoxia: integration through Fry's concept of aerobic metabolic scope. J. Fish Biol. 88, 232–251. https://doi.org/10.1111/jfb.12833.

Claireaux, G., Lefrançois, C., 2007. Linking environmental variability and fish performance: integration through the concept of scope for activity. Philos. Trans. R. Soc. Lond. Ser. B Biol. Sci. 362, 2031–2041. https://doi.org/10.1098/rstb.2007.2099.

Clark, T.D., Mark, F.C., 2017. An introduction to the special issue: "OCLTT: a universal concept?". J. Therm. Biol. 68, 147–148. https://doi.org/10.1016/j.jtherbio.2017.08.003.

Clark, T.D., Sandblom, E., Jutfelt, F., 2013. Aerobic scope measurements of fishes in an era of climate change: respirometry, relevance and recommendations. J. Exp. Biol. 216, 2771–2782. https://doi.org/10.1242/jeb.084251.

Cooper, B.S., Williams, B.H., Angilletta, M.J., 2008. Unifying indices of heat tolerance in ectotherms. J. Therm. Biol. 33, 320–323. https://doi.org/10.1016/j.jtherbio.2008.04.001.

Cowles, R.B., Bogert, C.M., 1944. A preliminary study of the thermal requirements of desert reptiles. Bull. Am. Mus. Nat. Hist. 83, 265–296.

Deutsch, C.A., Ferrel, A., Seibel, B., Pörtner, H.-O., Huey, R.B., 2015. Climate change tightens a metabolic constraint on marine habitats. Science 348, 1132–1135. https://doi.org/10.1126/science.aaa1605.

Earhart, M.L., Blanchard, T.S., Harman, A.A., Schulte, P.M., 2022. Hypoxia and high temperature as interacting stressors: will plasticity promote resilience of fishes in a changing world? Biol. Bull. 243, 23. https://doi.org/10.1086/722115.

Eliason, E.J., Clark, T.D., Hague, M.J., Hanson, L.M., Gallagher, Z.S., Jeffries, K.M., Gale, M.K., Patterson, D.A., Hinch, S.G., Farrell, A.P., 2011. Differences in thermal tolerance among sockeye salmon populations. Science 332, 109–112. https://doi.org/10.1126/science.1199158.

Ern, R., Huong, D.T.T., Phuong, N.T., Madsen, P.T., Wang, T., Bayley, M., 2015. Some like it hot: thermal tolerance and oxygen supply capacity in two eurythermal crustaceans. Sci. Rep. 5, 10743. https://doi.org/10.1038/srep10743.

Ern, R., Andreassen, A.H., Jutfelt, F., 2023. Physiological mechanisms of acute upper thermal tolerance in fish. Phys. Ther. 38, 141–158. https://doi.org/10.1152/physiol.00027.2022.

Evans, D.O., Neill, W.H., 1990. Introduction to the proceedings of the symposium "from environment to fish to fisheries: a tribute to F.E.J. Fry": transactions of the American fisheries society. Trans. Am. Fish. Soc. 119 (4), 567–570.

Eyring, H., 1935. The activated complex in chemical reactions. J. Chem. Phys. 3, 107–115. https://doi.org/10.1063/1.1749604.

Frölicher, T.L., Fischer, E.M., Gruber, N., 2018. Marine heatwaves under global warming. Nature 560, 360–364. https://doi.org/10.1038/s41586-018-0383-9.

Fry, F.E.J., 1947. Effects of the environment on animal activity. Univ. Tor. Stud. Biol. Ser. 55, 1–62.

Fry, F.E.J., 1971. The effects of environmental factors on the physiology of fish. Fish Physiol. 6, 1–98. https://doi.org/10.1016/S1546-5098(08)60146-6.

Gräns, A., Jutfelt, F., Sandblom, E., Jönsson, E., Wiklander, K., Seth, H., Olsson, C., Dupont, S., Ortega-Martinez, O., Einarsdottir, I., Björnsson, B.T., Sundell, K., Axelsson, M., 2014. Aerobic scope fails to explain the detrimental effects on growth resulting from warming and elevated CO_2 in Atlantic halibut. J. Exp. Biol. 217, 711–717. https://doi.org/10.1242/jeb.096743.

Jeppesen, E., Beklioğlu, M., Özkan, K., Akyürek, Z., 2020. Salinization increase due to climate change will have substantial negative effects on inland waters: a call for multifaceted research at the local and global scale. The Innovation 1, 100030. https://doi.org/10.1016/j.xinn.2020.100030.

Jørgensen, L.B., Malte, H., Ørsted, M., Klahn, N.A., Overgaard, J., 2021. A unifying model to estimate thermal tolerance limits in ectotherms across static, dynamic and fluctuating exposures to thermal stress. Sci. Rep. 11, 12840. https://doi.org/10.1038/s41598-021-92004-6.

Jutfelt, F., Norin, T., Ern, R., Overgaard, J., Wang, T., McKenzie, D.J., Lefevre, S., Nilsson, G.E., Metcalfe, N.B., Hickey, A.J.R., Brijs, J., Speers-Roesch, B., Roche, D.G., Gamperl, A.K., Raby, G.D., Morgan, R., Esbaugh, A.J., Gräns, A., Axelsson, M., Ekström, A., Sandblom, E., Binning, S.A., Hicks, J.W., Seebacher, F., Jørgensen, C., Killen, S.S., Schulte, P.M., Clark, T.D., 2018. Oxygen- and capacity-limited thermal tolerance: blurring ecology and physiology. J. Exp. Biol. 221, jeb169615. https://doi.org/10.1242/jeb.169615.

Kerr, S.R., 1976. Ecological analysis and the Fry paradigm. J. Fish. Res. Board Can. 33, 329–335. https://doi.org/10.1139/f76-051.

Kerr, S.R., Lawrie, A.H., 1976. Natura naturans: a symposium on the Fry paradigm. J. Fish. Res. Board Can. 33, 296–299. https://doi.org/10.1139/f76-044.

Kilgour, D.M., McCauley, R.W., 1986. Reconciling the two methods of measuring upper lethal temperatures in fishes. Environ. Biol. Fish 17, 281–290. https://doi.org/10.1007/BF00001494.

Lefevre, S., 2016. Are global warming and ocean acidification conspiring against marine ectotherms? A meta-analysis of the respiratory effects of elevated temperature, high CO_2 and their interaction. Conserv. Physiol. 4, cow 009. https://doi.org/10.1093/conphys/cow009.

Loftus, K.H., 1976. A new approach to fisheries management and F.E.J. Fry's role in its development. J. Fish. Res. Board Can. 33, 321–325. https://doi.org/10.1139/f76-049.

Lutterschmidt, W.I., Hutchison, V.H., 1997. The critical thermal maximum: history and critique. Can. J. Zool. 75, 1561–1574. https://doi.org/10.1139/z97-783.

McArley, T.J., Sandblom, E., Herbert, N.A., 2021. Fish and hyperoxia—from cardiorespiratory and biochemical adjustments to aquaculture and ecophysiology implications. Fish Fish. 22, 324–355. https://doi.org/10.1111/faf.12522.

McArley, T.J., Morgenroth, D., Zena, L.A., Ekström, A.T., Sandblom, E., 2022. Prevalence and mechanisms of environmental hyperoxia-induced thermal tolerance in fishes. Proc. R. Soc. B Biol. Sci. 289, 20220840. https://doi.org/10.1098/rspb.2022.0840.

McCauley, R., 1990. Frederick Earnest Joseph Fry, M.B.E., M.A., Ph.D., D.Sc., F.R.S.C.—1908–1989. Environ. Biol. Fish 27, 241–242. https://doi.org/10.1007/BF00002742.

Nilsson, G.E., Crawley, N., Lunde, I.G., Munday, P.L., 2009. Elevated temperature reduces the respiratory scope of coral reef fishes. Glob. Chang. Biol. 15, 1405–1412. https://doi.org/10.1111/j.1365-2486.2008.01767.x.

Norin, T., Malte, H., Clark, T.D., 2014. Aerobic scope does not predict the performance of a tropical eurythermal fish at elevated temperatures. J. Exp. Biol. 217, 244–251. https://doi.org/10.1242/jeb.089755.

Ørsted, M., Jørgensen, L.B., Overgaard, J., 2022. Finding the right thermal limit: a framework to reconcile ecological, physiological and methodological aspects of CTmax in ectotherms. J. Exp. Biol. 225, jeb244514. https://doi.org/10.1242/jeb.244514.

Payne, N.L., Smith, J.A., van der Meulen, D.E., Taylor, M.D., Watanabe, Y.Y., Takahashi, A., Marzullo, T.A., Gray, C.A., Cadiou, G., Suthers, I.M., 2016. Temperature dependence of fish performance in the wild: links with species biogeography and physiological thermal tolerance. Funct. Ecol. 30, 903–912. https://doi.org/10.1111/1365-2435.12618.

Peck, L.S., 2016. A cold limit to adaptation in the sea. Trends Ecol. Evol. 31, 13–26. https://doi.org/10.1016/j.tree.2015.09.014.

Persson, L., Carney Almroth, B.M., Collins, C.D., Cornell, S., de Wit, C.A., Diamond, M.L., Fantke, P., Hassellöv, M., MacLeod, M., Ryberg, M.W., Søgaard Jørgensen, P., Villarrubia-Gómez, P., Wang, Z., Hauschild, M.Z., 2022. Outside the safe operating space of the planetary boundary for novel entities. Environ. Sci. Technol. 56, 1510–1521. https://doi.org/10.1021/acs.est.1c04158.

Pörtner, H.-O., 2010. Oxygen- and capacity-limitation of thermal tolerance: a matrix for integrating climate-related stressor effects in marine ecosystems. J. Exp. Biol. 213, 881–893. https://doi.org/10.1242/jeb.037523.

Pörtner, H.-O., Farrell, A.P., 2008. Physiology and climate change. Science, 690–692.

Pörtner, H.-O., Peck, M.A., 2010. Climate change effects on fishes and fisheries: towards a cause-and-effect understanding. J. Fish Biol. 77, 1745–1779. https://doi.org/10.1111/j.1095-8649.2010.02783.x.

Rezende, E.L., Castañeda, L.E., Santos, M., 2014. Tolerance landscapes in thermal ecology. Funct. Ecol. 28, 799–809. https://doi.org/10.1111/1365-2435.12268.

Terblanche, J.S., Hoffmann, A.A., Mitchell, K.A., Rako, L., le Roux, P.C., Chown, S.L., 2011. Ecologically relevant measures of tolerance to potentially lethal temperatures. J. Exp. Biol. 214, 3713–3725. https://doi.org/10.1242/jeb.061283.

Trenberth, K.E., 2011. Changes in precipitation with climate change. Clim. Res. 47, 123–138. https://doi.org/10.3354/cr00953.

Wang, Z., Walker, G.W., Muir, D.C.G., Nagatani-Yoshida, K., 2020. Toward a global understanding of chemical pollution: a first comprehensive analysis of national and regional chemical inventories. Environ. Sci. Technol. 54, 2575–2584. https://doi.org/10.1021/acs.est.9b06379.

Chapter 14

THE EFFECT OF ENVIRONMENTAL FACTORS ON THE PHYSIOLOGY OF FISH☆

F.E.J. FRY

Chapter Outline

☆This is a reproduction of a previously published chapter in the Fish Physiology series, "1971 (Vol. 6)/The Effect of Environmental Factors on the Physiology of Fish: ISBN: 978-0-12-350406-7; ISSN: 1546-5098".

Fish Physiology, Vol. 40B. https://doi.org/10.1016/bs.fp.2024.07.006

I. INTRODUCTION

The study of animal function is organized more or less under three heads which in everyday language are, as applied to a machine, what it can do, how it works, and what makes it go. Insofar as fields of study can be classified in biology these divisions of the subject are ordinarily considered to be autecology, physiology, and biochemistry, with a great deal of individual taste governing the label any particular worker may choose for himself. The subject of this chapter is what fish can do in relation to their environment and therefore largely autecology.

The organism can be taken to be an open system (von Bertalanffy, 1950), suitably walled off from its milieu, through which energy flows by appropriate entrances and exits. The organism uses this energy to maintain and extend its being. The energy comes from the environment, and further the environment sets to a large degree the conditions under which the organism uses the energy it has assimilated, but all organisms have regulatory powers and bargain with the environment in regard to the extent they make use of the energy they have gained. Such bargaining involves the use of some energy for regulation against the environment to free the rest for the organism's other activities.

Thus the prime subject of this chapter will be the action of the environment on metabolism and the effects of this action on the activity of the organism.

A. Metabolism and Activity

A careful distinction will be made here in the usage of the terms metabolism and activity. Metabolism as used here is catabolism as ordinarily understood, that is, the sum of the reactions which yield the energy the organism utilizes. Activities are what the organism does with the energy derived from metabolism. Thus activities are such processes as running or fighting or other manifestations of the energy released by metabolism. These manifestations are not all movements; growth is activity and so is excretion. By this definition anabolism is an activity.

While the influence of the environment is on metabolism, the effect of that influence is displayed through the activity of the organism whose metabolism has been so affected.

The purpose of belaboring the distinction between metabolism and activity here is not to introduce a novel thought, for these generalities are what we all recognize, but rather to provide a consistent treatment of the whole organism in relation to its biochemical basis. Activity is fundamentally the result of transformation of energy from one form to another and the application of that energy to a given performance. Two generalizations arise from these circumstances. First, all the energy released will not likely be applied to the final outcome which is the object of its release. The organism will take its levy for its maintenance as a system, and there will be the ancillary costs of supply and disposal of the metabolites which pass through the system. Second, performance is qualitatively different from the power which produces it and there need not be any simple proportionality between the measures taken of the two.

These circumstances will be recognized here by considering the difference between resting and active metabolic rates, which will be termed "scope for activity," as being the power available for activity, and, where appropriate, relations will be sought between activity and scope. These concepts are, of course, regularly applied to homoiotherms by those interested in animal production (e.g., Brody, 1945) where the costs and consequences of thermoregulation are so prominent and have been simply transferred to poikilotherms over the past quarter century.

B. Measurement of the Metabolic Rate

The metabolic rate of fish has almost universally been measured by determining oxygen consumption. The fundamental method of measuring heat production has been applied (e.g., Davies, 1966) but probably never will be suitable for measurements required for environmental physiology.

It cannot be assumed that all fish are obligate aerobes and that a measure of oxygen consumption is always a measure of the metabolic rate. Coulter (1967) reported that extensive catches of fish are regularly taken in oxygen-poor water in Lake Tanganyika under circumstances which suggest they are resident there. The goldfish (Kutty, 1968a) can live for months with a respiratory quotient of 2, and there are the dramatic reports of Blažka (1958) and Mathur (1967) on extensive survival of fish under completely anaerobic conditions.

The newer methods of easy determination of carbon dioxide in water should soon be rapidly applied to the determination of the respiratory quotient (e.g., R. W. Morris, 1967) although, as yet, the margin of error in them requires to be narrowed. At present the error inherent in the new methods is of the order of twice that for determinations with the Van Slyke apparatus or by distillation (e.g., Maros *et al.*, 1961).

Three levels of metabolism will be distinguished here. Following the usage now current among a number of fisheries workers, these will be termed "standard," "routine," and "active" levels of metabolism. Standard metabolism is an approximation of the minimum rate for the intact organism. It is preferably determined as the value found at zero activity by relating metabolic rate to random physical activity in fish in the postabsorptive state (e.g., Beamish and Mookherjii, 1964; Spoor, 1946). The fish should be able to swim freely in the respiration chamber while protected from outside disturbance and should have been in the chamber long enough to recover from the effects of transfer to it. It may also be important that the chamber is supplied with water from the aquarium in which the fish was living. Foreign water may provide disturbing chemical stimuli or perhaps more importantly may lack the familiar chemical milieu of the home tank. Standard metabolism can also be determined by extrapolation to zero activity from determinations at various levels of forced activity (e.g., Brett, 1964). The routine rate of metabolism is the mean rate observed in fish whose metabolic rate is influenced by random activity under experimental conditions in which movements are presumably somewhat restricted and the fish protected from outside stimuli. The value has usually been given only for the normal working hours of the experimenter (e.g., Beamish, 1964a). Active metabolism is the maximum sustained rate for a fish swimming steadily.

Standard and active metabolism are determined to permit calculation of scope for activity. Routine metabolism is largely to be considered as a measure of the degree of random activity and is discussed in Section VI.

Various types of apparatus used in such determinations are discussed below, together with comments on experimental precautions. Most measures of metabolism have been measurements of routine metabolism. Standard and active metabolism have not yet often been measured, and the limits of these have been still less often well worked out.

Figure 1 shows determinations of the metabolic rate of the goldfish, *Carassius auratus*, at 20°C by various workers in the same laboratory at various times over a number of years. Figure 1A shows oxygen consumption and Fig. 1B CO_2 output. For the sake of clarity, Kutty's points for oxygen consumption are not shown but the number of his readings under forced activity can be inferred from the number of points in Fig. 1B since he made determinations of the respiratory quotient. His curve for routine metabolism is based on 35 points. Smit's data (1965) are illustrative material based on a single fish. There are three salient points to be considered in Fig. 1A:

(1) The dots which represent oxygen consumption during routine activity show how high the metabolic rate can go when a fish is randomly active within the confines of a respiration chamber. The routine respiration rate as shown by the mean of these values approaches half the active respiration rate.

(2) An extrapolation from either forced or random activity to zero activity gives a similar value for standard metabolism if the fish are not disturbed.

(3) A major problem in the measurement of metabolism of fish is the wide range of values that may be obtained for a given fish in the same state of overt physical activity. A fish resting quietly may be consuming oxygen at a given rate from its standard to well over half its maximum level. Ordinarily a fish need only be moved from its bath to the respiration chamber to elicit almost its active metabolic rate. Cook's data illustrate this phenomenon in Fig. 1. Her data are for the first 15 min after the fish were transferred from their acclimation tank to the respiration chamber. The effect of swimming is to somewhat increase oxygen uptake in an excited fish, possibly by facilitating venous return, but more than half the extrapolated maximum rate of oxygen

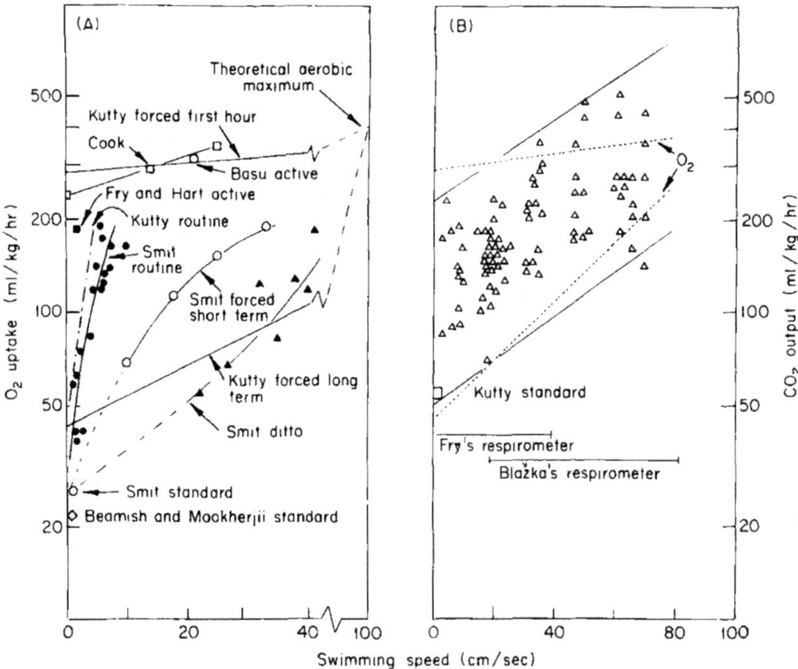

Fig. 1. Various measures of metabolism of the goldfish, *Carassius auratus,* under conditions of random and forced activity at 20°C. (A) Oxygen consumption; (B) carbon dioxide production. Data of Basu (1959), Beamish and Mookherjii (1964), C. N. Cook (personal communication in Fry, 1967), Kutty (1968a), and Smit (1965). Kutty's data for oxygen consumption under forced activity are omitted from A but are based on the same number of observations as are shown by the points in B. For further explanation see text.

consumption associated with the most vigorous sustained activity can be displayed by the goldfish with no overt movement at all. Kutty (1968a) found that under forced activity the metabolic rate of goldfish fell after about 3 hr from these initially high values to a steady state which more truly reflected the degree of swimming effort, but his data were for well-trained fish. Smit (1965), who did not train his fish, found the rate declining sometimes over at least 8 hr.

The active rate most frequently reported in the literature (e.g., Basu, 1959; shown in Fig. 1) has been an acute measurement at a moderate rate of swimming speed. Only the more recent authors (Brett, 1964; Kutty, 1968a; Smit, 1965) have pushed their measurements to the point where the maximum sustained rate could be estimated. Similarly, most earlier workers concerned with standard metabolism (e.g., Fry and Hart, 1948) approximated that value by taking the lowest point in the daily cycle. Beamish (1964a) and Fry and Hochachka (1970) gave comparisons of many of the remaining published values for the metabolic rate of the goldfish.

Routine metabolism has, of course, been the level of metabolism most frequently reported, usually with the implication that the fish were in a quiescent state when the measurements were made. Most fish show a daily cycle of activity, so that the degree of quiescence depends to a considerable extent on the time of day when the measurements are made in spite of all the usual precautions to confine the fish within an appropriately limited space and to protect it from disturbance (e.g., Kausch, 1968). The significance of measuring routine metabolism is that it is a reflection of random activity, the degree of which reflects response to the directive effects of the environment. Routine metabolism is unsuitable as an approximation of standard metabolism because of the high metabolic rate which may be achieved by a fish within a restricted space. In Fig. 1 the peak routine rate for the goldfish is approximately six times the standard rate. These rates reflect presumably the cost of continual small accelerations as the fish starts to swim and then checks its progress.

The emphasis above has been on physical activity as a variable influencing the metabolic rate, together with what has been termed "excitement." There are of course other well-known influences, the two major ones being the cost of assimilation and of ion-osmoregulation. The latter ugly term seems necessary because we have not distinguished the cost of transfer of ions from that of the transfer of water. The cost of assimilation has been eliminated by a fast of some 48 hr (Beamish, 1964d). In the species Beamish investigated assimilation accounted for about 50% of the inactive metabolism. In general, major costs of assimilation have been removed from most measurements since it is usual to fast the animals, if only to avoid feces in the respiration chamber. In fresh water or salt water the cost of ion-osmoregulation in the rainbow trout, as determined by subtracting the minimum rate in an isosmotic dilution of seawater, is 20–30% of the metabolic rate (Rao, 1968). Such a cost, however, is a proper fraction of standard metabolism.

In addition to such costs of regulation and activity as may be included in the resting metabolism, there may be changes in the residual (standard) metabolism with season, a subject which has been little explored, although Beamish (1964e) showed an approximate doubling of the standard metabolic rate of the eastern brook trout from its low to its high in the annual cycle.

Figure 1B shows the metabolism of the goldfish as carbon dioxide output. The chief purpose of the comparison is to show to what extent the metabolic rate of this species under these circumstances can be expressed by oxygen consumption alone. The points shown for carbon dioxide output were all obtained under forced activity. Goldfish which are randomly active in water high in oxygen have a respiratory quotient (*RQ*) of approximately unity (Kutty, 1968a). Under forced activity at air saturation the *RQ* is again unity or lower for the long term and also excitement alone can be satisfied by aerobic respiration. However, as the swimming speed is increased an increasingly large segment of the upper symbols (excitement plus activity) falls above the boundary of the area encompassed by the values for oxygen consumption. Thus, metabolism at some acute values in the goldfish cannot be estimated by oxygen consumption although all the long-term values at air saturation can be.

C. The Relation of Metabolism to Size and Physical Activity

Most discussions of metabolism here will be divorced from a consideration of the size of the fish involved by expression as rate per unit weight.

In general the relation of metabolism to body weight has been described by the equation $y = ax^b$ where y is the rate of metabolism and x is the body weight (often the formula is used in the form $y/x = ax^{1-b}$). The exponent b has usually been found to be of the order of 0.8 (Paloheimo and Dickie, 1966a; Winberg, 1956), and most examples have been for routine metabolism. There have been some notable exceptions to this rule. The cichlids, in particular (Ruhland, 1965; R. W. Morris, 1967), have shown some exponents of the order of 0.5 as has also been found for other species on occasion (e.g., Wells, 1935; Barlow, 1961). An exponent of unity, which indicates the metabolic rate is weight proportional, has also been found from time to time. Beamish (1964a) found the standard metabolic rate of brook trout, *Salvelinus fontinalis*, to be weight proportional. Both Brett (1965) and Rao (1968) found active metabolism to be essentially weight proportional in the two salmonids they investigated. Job (1955), however, found a decrease in the exponent for active metabolism with increasing temperature. Brett and Rao worked at or below the temperature optimum for their species and probably also stimulated their fish to greater activity. In cases where routine or standard metabolism have been compared at different temperatures the weight exponent has been temperature independent.

While the relative amount of energy the animal can produce or may require in relation to its size is of great importance in determining its relation to the environment there are no established explanations for the various differences found in the magnitude of the exponent b, and indeed the reality of many of these differences is in question. Glass (1969) questioned whether the best values for the exponent have been calculated in most instances. It has been the practice to fit a straight line to a logarithmic transformation of the data, like the treatment shown in Fig. 32. Glass demonstrated in a series of examples that a better fit to the points can be obtained by using the equation in its arithmetic form. The treatment here will be to use the exponent given by the author for any weight corrections, but such corrections will be largely ignored. It seems important however that the notion, now rather thoroughly fixed in the literature, that the general value for b is approximately 0.8 should not yet be allowed to become a dogma. This point is emphasized in Section V where a special case of a change in the exponent in relation to osmoregulation is dealt with (Fig. 32).

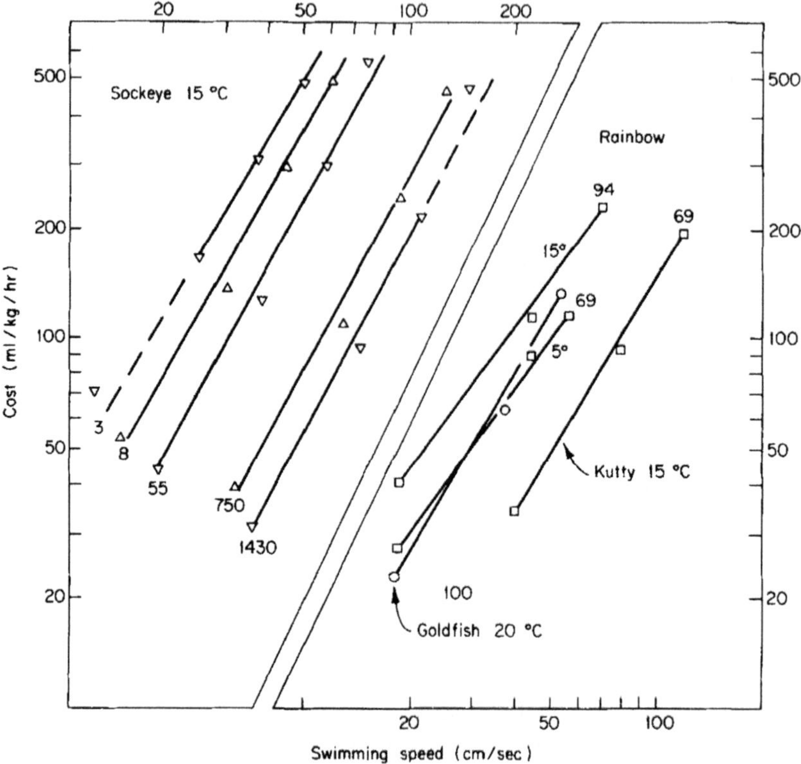

Fig. 2. The cost of swimming in the sockeye salmon, *Oncorhynchus nerka*, rainbow trout, *Salmo gairdneri*, and the goldfish. Data of Brett (1965), Kutty (1968a), and Rao (1968). The cost shown is scope for activity, i.e., metabolism at activity stated minus standard metabolism. The numbers associated with the curves are weights in grams unless indicated as degrees. Note Kutty's data show the same slope with a different intercept from Rao's.

There has been only one thorough investigation into the cost of swimming, that of Brett (1965) for the sockeye salmon. His data can be interpreted (Fig. 2) as showing that the metabolic cost of swimming increases approximately as the square of the swimming speed, a relation suggested earlier by fragmentary data (Fry, 1957). The restricted data of Kutty and Rao also support Brett (Fig. 2), as do the data for the had-dock, *Melanogrammus aeglefinus* (Tytler, 1969).

Brett's data (1964) for fingerling sockeye indicate that the cost of swimming is essentially independent of temperature, a conclusion which differs from that of Rao (1968) who indicated a somewhat greater cost at a higher temperature, so the matter is still in question. The relation of the metabolic rate to speed of swimming needs further examination before we can generalize. While the main example in Fig. 2 is statistically strong and there are subsidiary data of a similar nature in the literature, there is at least one contradictory series. From the power relation between scope and swimming speed in Fig. 2 and the relation of scope to size given in Brett (1965), it follows that the maximum speed of swimming at any size is related to approximately $L^{0.5}$, where L is the total length, as Brett showed empirically. However, Pavlov *et al.* (1968), working with small minnows, found that the maximum swimming speed was a simple multiple of the length of the fish over lengths of 8–35 mm, which infers either a completely different relation of scope for activity to size in these small fish or a different relation of scope to swimming. speed. These workers used a gravity head of water to provide their water flow rather than the more turbulent recirculating systems of Brett and Kutty and others whose data follow the power relation. Perhaps smooth-skinned fish swimming in still water have an economy of effort not to be found in rapid streams or fine turbulence activity apparatus, or perhaps the relation for small fish differs from that for larger ones.

D. Apparatus for the Determination of Metabolic Rate

It seems unlikely that any useful determination of the metabolic rate of fish can now be made which is not accompanied by a measure of physical activity. There are two broad approaches to this problem, one is to record the random activity of an individual free from outside disturbance while measuring its metabolic rate. The other approach is to force the fish to swim at certain constant speeds while making the measurement. Wohlschlag (1957) has somewhat combined the two approaches by driving a rotating chamber so as to counter the random swimming speed of the fish.

In the last decade several methods of measurement, or at least registration, of random activity have been reported following the pioneering work of Spoor (1946) and other workers cited in Fry (1958). The present most quantitative practical method, which under proper design should measure the energy output, is to measure the degree of turbulence of the water in the respiration chamber as affected by movements of the fish. Such measurements can be readily made by a heat loss flowmeter (Beamish and Mookherjii, 1964; Dandy, 1970; Heusner and Enright, 1966; Kausch, 1968). A convenient but less quantitative method is to equip the respiration chamber with photocells to measure the number of circuits the fish may make around it (e.g., Smit, 1965; Peterson and Anderson, 1969a). Finally, the entry of fish into an echo-sounder beam (Muir *et al.*, 1965) offers the possibility of integrating accelerations if proper geometry of the chamber and suitably refined recording apparatus can be combined. The method of de Groot and Schuyf (1967, which see also for other references) in which the fish carry a magnet offers similar possibilities.

Undoubtedly, the best form of apparatus in which to induce the active respiration of fish is a tunnel through which the water is recirculated. Probably the best-engineered example of such an apparatus is that described by Brett (1964, see also Mar, 1959) and illustrated by Phillips, Volume I, this treatise. Rotating chambers have been used extensively in the author's laboratory because of their great simplicity and convenience, but they are suitable only for approximating the active metabolic rate. A most interesting circumstance is that it is apparently impossible to induce a fish to swim as fast in a closed rotating chamber as in a tunnel or a flume. Presumably the discrepancy is a result of the Taylor effect (Taylor, 1923, and earlier). A body in rotating water is subject not only to form and surface drag as it is in a tunnel but carries with it a cell in which water circulates vertically from the body to the water surface. It does seem likely though that the Taylor effect can be eliminated by allowing a free meniscus at the inner and outer margins as indicated in Fig. 3.

The problem in the design of chambers for active respiration is to have a uniform cross-sectional flow through the section where the fish is held; thus, a measure of the speed of the water provides a measure of the counterspeed of the fish. A secondary problem is to keep the hydro-dynamic efficiency high for quietness, ease of speed control, and reduction of heat input while keeping the circulating volume low to minimize lag in measurement. Blažka's (Blažka *et al.*, 1960) apparatus, with appropriate modifications to improve the flow, promises to be a convenient form of apparatus for the study of active metabolism but has not yet been given appropriate engineering treatment.

Chambers for the study of routine and standard metabolism have taken a variety of forms, one even with an appropriate posterior upturn to accommodate the heterocercal tail of sturgeon (Pavlovskii, 1962), but the majority have been a horizontal cylinder (e.g., Halsband and Halsband, 1968). The author favors a chamber in which the bottom is flat so that a fish resting on the bottom has the least lateral restriction (e.g., Bullivant, 1961). Chambers which are circular in cross section do not have this feature. The top of the chamber is preferably arched to permit the easy removal of bubbles. If activity is to be measured as well as oxygen consumption, the geometry of the chamber should take that factor into consideration. Various chambers are shown in Fig. 3.

The various electrochemical methods now in vogue for measuring oxygen in water have greatly simplified the measurement of metabolism although, as mentioned above, the measurement of production of carbon dioxide cannot yet be carried out as conveniently with an equivalent accuracy. The Van Slyke apparatus still seems to be the basic tool for the measurement of carbon dioxide. The distillation method of Maros *et al.* (1961) has about the same precision as the Van Slyke and can probably be readily mechanized to become at least semiautomatic. No doubt gas chromatography and infrared analysis are on the verge of giving results precise enough so that the usual small difference between two relatively large values for total carbon dioxide in water will be determined to the same accuracy as can the comparable difference between

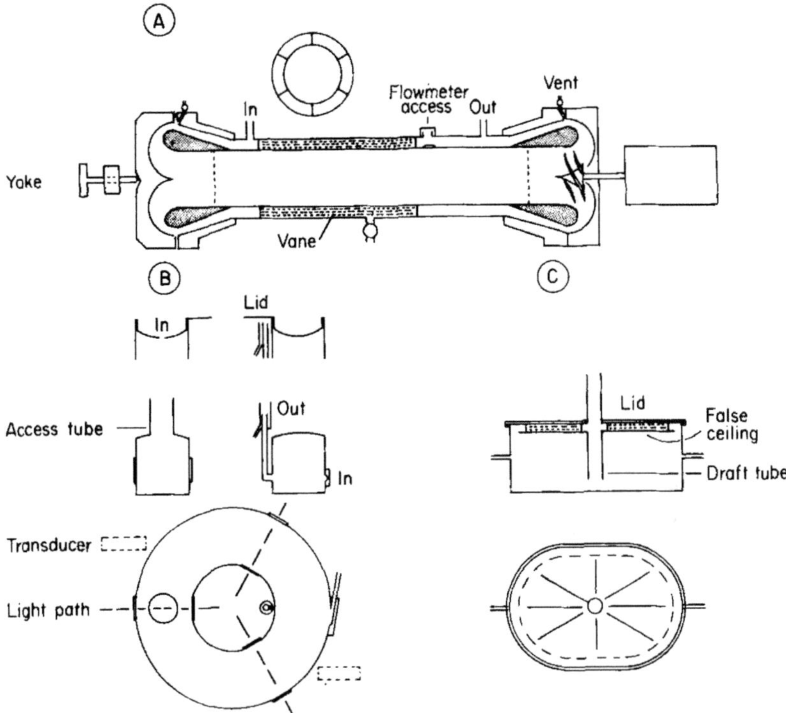

Fig. 3. Three types of apparatus for the measurement of standard and active metabolism. (A) Modified Blažka chamber for the determination of active metabolism (Blažka *et al.*, 1960) essentially as described by Smith and Newcomb (1970). The drive indicated is a variable speed, solid-state controlled motor. (B) Annular chamber for standard or active metabolism. In the configuration shown, standard metabolism is determined by monitoring activity by lights and photocells (e.g., Smit, 1965; Peterson and Anderson, 1969a). Muir *et al.* (1965) immersed their annular chamber in a second water bath in which echo sounder heads were placed as indicated, whereby activity was monitored by the Doppler effect. An annular chamber can be rotated for forced activity. In that case a lid allowing a small open water surface at the inner and outer peripheries (as indicated in the upper section) promotes the best reaction from the fish. (C) Chamber for use with heat loss flowmeter. Currents induced by the fish are in general parallel to the walls of the chamber and constrained by it. The resulting constraint to a curved path induces a current down the draft tube via the false ceiling (cf. Beamish and Mookherjii, 1964; Mathur and Shrivastava, 1970; Dandy, 1970). The draft tube is not essential (Kausch, 1968). Brett's apparatus for the measurement of active metabolism is illustrated in Fig. 1, p. 420, Vol. I, this treatise.

two determinations of oxygen content. The use of decarbonated water adds to the accuracy when determining the respiratory quotient (Kutty, 1968a).

One major nuisance in respiration chambers is the formation of gas bubbles which can provide a substantial reservoir for oxygen. It is well, therefore, to slightly undersaturate the water supply to a respiration chamber. If sufficient headroom is available in the laboratory, undersaturation can be easily achieved by passing the water down over Raschig rings in an exchange column subjected to a vacuum, say, of 50 cm H_2O. Otherwise the water supply can be heated 2° or 3° while it passes through an aerator and then be brought back to the desired temperature. Mount (1964) described a convenient apparatus for degassing water by vacuum which also can be used to control initial oxygen concentration.

Whenever the relation between activity and metabolism is the concern, it is necessary to pay attention to the lag in measurement resulting from the volume of the chamber. This problem can be met by using a closed system whereby activity and oxygen consumption are integrated over a fixed period, which in general is considerably longer than the lag between activity and oxygen consumption within the organism. However, there are limitations to the use of closed systems for the measurement of oxygen consumption, the greatest of which is probably that something like half the time for experiment must be wasted in flushing the chamber. There is the further problem that the action of the valves for opening and closing the chamber may be a source of disturbance to the fish. The objection sometimes put forward that products of metabolism accumulate in closed systems has no real validity since an open system is in effect merely an enlarged closed one.

Lag may be accounted for in an open system by the use of a factor to correct for changes in oxygen content in the water in the chamber during the period of measurement. The formula for calculating the oxygen consumption in a constant flow system for a period of time t is

$$F(y_1 - \bar{y}_2) + V(y_{2,0} - y_{2,t})$$

where F is the volume of flow through the chamber, V is the volume of the chamber, y_1 is the inlet oxygen concentration taken to be constant throughout the period, and y_2 is the outlet concentration. The subscripts 0, t, indicate readings at the beginning and end of the period.

The correction factor $V(y_{2,0} - y_{2,t})$ in this formula has unfortunately been overlooked by most authors (e.g., Kausch, 1968).

The problem of lag under the conventional method of maintaining a constant rate of flow and monitoring change in oxygen, while amenable to correction as indicated above, still requires time-consuming tabulation or an expensive digital output. With present methods of electronic control it appears feasible to reverse the approach so that oxygen is maintained constant in the respiration chamber by a control which monitors the level there and regulates the supply of water, analogous to the method so often employed to measure the respiration of air breathers. Oxygen consumption would then be recorded in terms of volume of water pumped and chamber lag would be eliminated. Such a chamber, of course, probably could not be provided with a turbulence meter, but activity could be monitored by one of the other methods available.

Systems such as those of Scholander *et al.* (1943) and Ruhland and Heusner (1959), where oxygen pressure is kept constant in a gas phase over water, still have at least the same lag as is found in the constant flow system but offer the advantage of a simple determination of oxygen added to the system. The correction term can be found by inserting an auxiliary oxygen electrode into the water, as indeed Ruhland (1967) has already done for another purpose.

E. Acclimation

It is well accepted that an organism is not the same organism, even from day to day, but that its physiological state is continuously being modified by its environmental history. These effects of the environment during the individual's life will be taken into account as far as possible—which can still, however, only be done in a somewhat rudimentary fashion. The two terms, "acclimation" and "acclimatization" will be applied to the conditioning of the individual by its experience. Acclimation will be used to designate the process of bringing the animal to a given steady state by setting one or more of the conditions to which it is exposed for an appropriate time before a given test. Such conditions may be fixed or cycled depending on the circumstances. A common practice is to maintain fish at a given constant temperature for such a purpose. The animal is then said to be acclimated to that particular temperature. Tests, say of its ability to swim at that temperature, will show constancy over some days or weeks after acclimation, whereas during acclimation there may have been considerable change. However, while there may be such constancy within a season, fish acclimated to the same temperature in summer may be constant at a different level from those acclimated to the same temperature in winter (Wohlschlag *et al.*, 1968). Thus there may be a major difference in the physiological state of an organism acclimated to a low temperature and one acclimatized to winter conditions, the latter term being reserved here for an organism whose history has been exposure to the total environmental complex throughout its life up to the time of test.

The most significant aspect of acclimatization as opposed to acclimation is that acclimatization allows the organism to acquire an adjustment, say to higher temperatures, in advance of the event if that event is appropriate to the seasonal cycle. Thus, acclimatization provides for anticipatory adjustment as well as reactive adjustment.

In drawing the distinctions above between acclimation and acclimatization, only reversible effects were considered. The modifications of most physiological responses by environmental history that have been investigated have been essentially reversible given sufficient time, but the possibility of irreversible changes remains, especially for influences at points of development where a given growth stanza may be prolonged or curtailed (e.g., Martin, 1949). Thus the rearing temperature has been found to have an influence on the lethal temperature of the guppy, *Poecilia reticulata* (Gibson, 1954), which cannot be eliminated by extensive thermal acclimation at a later date. To the ecologist (V. E. Shelford, personal communication), acclimatization may have also a phylogenetic implication on the subspecific level. In each locality, with its unique environment, a species is subject to different selection pressures as well as to any different ontogenetic influence which bears on the successful individuals.

The aim in the laboratory should be to duplicate the significant ontogenetic influences of acclimatization by suitable acclimations. The hope would be that residual differences then observed among populations would be the phylogenetic aspects of acclimatization.

A still unsolved problem in acclimation is how to condition animals to long-term cyclic changes such as the annual cycle of day length. Responses to such cycles appear to have an inertia which cannot be easily overcome. Moreover, the interactions between such cycles and, say, a constant temperature have not been adequately explored. On the whole it appears better at present to maintain an organism on its normal light cycle and state the season at which the work was done. Workers (e.g., Jankowsky, 1968) are now beginning to add the latter important information to their papers.

F. A Classification of the Environment

There are two fundamental bases for a classification of the environment, that is, either by its elements according to their identities such as light, heat, and oxygen or by the manner in which the identities may influence organisms. The second method has been chosen here following Fry (1947). The term "factor," apparently introduced by Blackman (1905), from whose powerful exposition it certainly gained its widespread currency, has been commonly employed to designate such a category of effect. The effects of the environment on organisms may be grouped into five categories. These will be designated lethal, controlling, limiting, masking, and directive factors. The first factor restricts the range of the environment in which the organism can exist; beyond this range metabolism is destroyed. The second and third factors govern metabolic rate. The remaining two are exploited by the organism to achieve and maintain its being through organic regulation. These categories are defined and discussed in an introductory fashion below.

While these categories are stated to be categories of effect it must be recognized, as is the case of most classifications, that they are also categories of convenience and imperfect to the degree that this is so. Thus the category of lethal factors, while undoubtedly dealing with lethal effects, is basically set up here to deal with the statistical aspects of mortality without reference to any specific cause of death. Death often ensues because of interaction between controlling and limiting factors and when this is so it is artificial to stop dealing with these factors at the verge of annihilation and to switch to another category. Again in the discussion of controlling factors the organism is considered as having no powers of organic regulation, which again is an evasion of reality. An organism cannot be without regulation.

1. *Lethal Factors*

An environmental identity acts as a lethal factor when its effect is to destroy the integration of the organism. Properly speaking such destruction should be independent of the metabolic rate to be the result of a lethal factor.

The lethal effect of any identity may be separated into two components: (a) the *incipient lethal level*, that level of the identity concerned beyond which the organism can no longer live for an indefinite period of time, and (b) the *effective time*, the period of time required to bring about a lethal effect at a given level of the identity beyond the incipient lethal level.

2. *Controlling Factors*

Controlling factors comprise one of two categories which govern the metabolic rate. What are considered here as controlling factors are what Blackman (1905) termed “tonic effects.” Controlling factors govern the metabolic rate by their influence on the state of molecular activation of the components of the metabolic chain.

Those not familiar with the general notion of normal and activated states of molecules in relation to rates of chemical reaction will find a recent general treatment of the subject in the introductory chapter in Johnson *et al.* (1954). Temperature is the most outstanding of the controlling factors.

Controlling factors place bounds to two levels of metabolism. They permit a certain maximum in the absence of a limiting factor through their influence on the rates of chemical reactions. The controlling factors also demand a certain minimum metabolic rate which, it is taken, is necessary to release the energy required for the repair reactions needed to keep the organism in being.

3. *Limiting Factors*

Limiting factors make up the second category of identities that govern the metabolic rate. They are Blackman’s “factors of supply” in his original treatment of “limiting factors” and the category to which Liebig’s “law of the minimum” applies. Both the term and the concept of the limiting factor have been widely used in this connection. The usage here is simply to restrict the definition to what was the major burden of Blackman’s exposition (1905).

Limiting factors operate by restricting the supply or removal of the materials in the metabolic chain. Thus a reduction in the supply of oxygen below a certain level can reduce the metabolic rate, and below that level it can be said that the oxygen supply is limiting.

The effect of a limiting factor is to throttle the maximum metabolic rate permitted by the existing level of controlling factors. Concentrations outside the limiting levels are to be considered as being neutral unless toxic levels are reached.

4. *Masking Factors*

A masking factor is an identity which modifies the operation of a second identity on the organism. An organism achieves all its physiological regulation by the exploitation of masking factors through the channeling of energy by some anatomical device.

For example, deep-sea fishes with swim bladders have pressures of gas in these bladders far in excess of the pressure that could be generated by releasing all the atmospheric gases held in the blood. To make this gas available at the higher pressure the fish exploits a second physical law, namely, the property of dissolved gases to diffuse down a pressure gradient. The rete mirabile that connects the swim bladder gas gland with the general circulation provides a countercurrent path for such diffusion and the circuit, arteriole → gas gland → venule, forms a regenerative loop which accumulates the gas in solution at the gas gland until the final chemical release overcomes the hydrostatic pressure (e.g., Steen, 1963). In this example the essential anatomy is the loop which brings the blood leaving the gas gland in close association with the blood about to enter it. Here the physical arrangement of the blood vessels permits a result which no chemical activity in the gas gland can achieve alone. The energy to drive the fraction of the system which permits the masking factor to operate is provided by the heart. (For details see chapter by Randall, Volume IV, this treatise.)

5. *Directive Factors*

These allow or require a response on the part of the organism directed in some relation to a gradient of the factor in space or in time.

The directive factors elicit the well-known forced movements (Loeb, 1913, 1918; Fraenkel and Gunn, 1961). They also provide for the animal's guidance in moving about in the environment in relation to physical obstacles and for its interactions with other organisms. The directive factors also trigger physiological responses without the mediation of the senses, as in the effect of photoperiod on the pituitary.

Directive factors operate by the impingement of energy on some appropriate target. The energy absorbed initiates a signal which appropriately channels metabolism into the appropriate response.

II. LETHAL FACTORS

Lethal factors as used here do not fall in a pure category of effect but constitute rather a section heading under which various common aspects of bioassay can be grouped. In dealing with lethal factors the primary approach will be one of description. Such a "blinkered" consideration is compulsory when dealing with lethal temperatures, which are taken as the main example here, because of our still profound ignorance of the nature of thermal death. In any event description should logically precede and lead to analysis. Description itself, of course, can be analytical and should be so; at least things should be described to the extent that questions can be raised as to the mechanisms underlying the phenomena observed. The causes of environmental death can, for example, be divided immediately into two fundamental categories by observing the rate of dying in relation to the metabolic rate of the organism. The rate of dying at a lethal temperature, for example, is signally independent of the metabolic rate while in lethal oxygen the metabolic rate is almost paramount. The former case where the lethal effect is independent of the metabolic rate is the pure lethal factor, and lethal temperature will be considered here as such a factor. Where the metabolic rate influences the rate of dying, death is usually brought about by the interaction of limiting and controlling factors. Whenever such a case can be recognized, such as the effects of decreased oxygen, an analysis of the interaction is infinitely more valuable than a determination of the lethal level (see Section IV, D). The importance of the distinction above lies in the so-called sublethal effects which might be seen in better perspective if they were termed "prelethal."

If there is no relation of the lethal effect to the metabolic rate, then the division of the effects into zones of resistance and tolerance defined below is meaningful. The incipient lethal, the boundary between the zones of resistance and tolerance, is then a real threshold, and the factor concerned can be taken to no longer exert any direct harmful effect. Indeed, in the case of temperature, the homoiotherms have found the successful evolutionary path to be the one which has led them to live within a few degrees of their upper lethal temperature.

The range of intensity of a given identity, which at some levels has a lethal effect, can be divided into a zone of resistance—over which it will operate to kill the organism in a determinate period of time—and a zone of tolerance—over which the life span of the organism is not influenced by the direct lethal effect of the identity concerned. Thus while the life span of a poikilotherm becomes progressively shorter as temperature increases, since increasing temperature speeds metabolism, temperature is not a lethal factor until the threshold is reached above which there is a drastic change in the length of life. Below that threshold the organism will be said to be in the zone of tolerance, above it to be in the zone of resistance (see Fig. 7, goldfish). The boundary between the two will be taken as being sharp and will be designated the "incipient lethal level," which is ordinarily expressed as the median lethal dose (LD_{50}). Dealt with according to the scheme above, events in the zone of resistance are measured according to the principles of time mortality (Bliss, 1935) while the incipient lethal level is expressed through the determination of dosage mortality (Bliss, 1937).

In most assays of lethality where the concept of a zone of resistance is not applied, all estimates are dealt with by dosage mortality. In Fig. 7A, the various crosses are determinations by dosage mortality, i.e., the median lethal temperature for exposure for the indicated period of time. The circles are estimates by time mortality, i.e., the median time to death at the temperature indicated. Time mortality and dosage mortality yield equivalent numerical values where the two determinations can be compared.

The importance of the distinction between the zones of resistance and tolerance comes largely because the two are not necessarily correlated. In the work of V. M. Brown *et al.* (1967) (Fig. 4) at a concentration of phenol in the mutual zone of resistance, the resistance time is longer at a lower temperature than at a higher one. However, the fish tolerate more phenol on a long-term basis at a higher temperature than a lower one, as can be seen from the way the assay lines cross each other in the figure.

While analysis into zones of resistance and tolerance as treated here gives a satisfying sense of completeness to the data, it must always be realized that there is no finality to the incipient lethal temperature short of maintaining a test throughout the whole life of the organism. The incipient lethal level

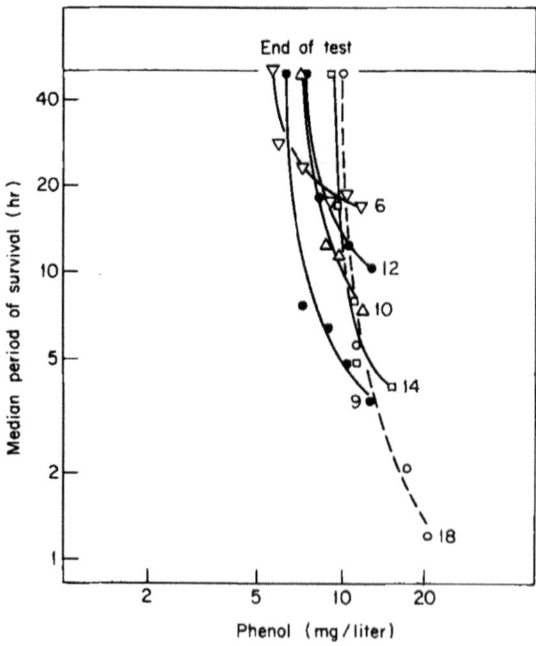

Fig. 4. Concentration vs. survival-time curves for rainbow trout, *Salmo gairdneri*, at different temperatures in phenol solutions. Slightly modified from Brown *et al.* (1967). The numbers associated with the curves indicate the test temperatures.

should be looked on as the boundary of the immediate direct lethal effects, "immediate" being taken as a matter of days or weeks and "direct" as the operation of the identity directly on a site of metabolism so as to destroy it more rapidly than the organism can keep it in repair. Allen and Strawn (1968) take this point of view when they accept heat death as being complete by 20,000 min although the fish were apparently not able to live indefinitely beyond that period since their food intake could not meet their maintenance requirements.

A. Determination of Lethal Effects

With the exception of certain determinations of lethal temperature, and to some extent the lethal effects of unsuitable concentrations of the respiratory gases, tests for lethality have been carried on by acute exposure of samples to various levels of the identity concerned until death or for a given period of time. There has been a tendency to extend the period as more experience has been gained; thus, now a 96-hr test period is probably the most widely approved, although 48 hr is probably most often used. These times have largely been set by rule of thumb and are essentially aimed at a period which would be long enough so that the LD_{50} would represent the tolerance level as defined here. With regard to the toxicity of phenol to rainbow trout (Fig. 4) it appears probable that a 48-hr test would barely provide incipient lethal levels but that the 96-hr test would give an adequate margin.

The lethal effects of temperature will be treated by time mortality here in the zone of resistance. In the case of temperature, much work also has been devoted to the determination of lethal temperatures by placing the organism at some intermediate temperature and then heating at some convenient rate, usually a Celsius degree every few minutes. The temperature at which the animal dies or becomes visibly incapacitated, often termed the "critical thermal maximum" (CTM), is taken as a measure of its lethal temperature. This particular technique will be treated below as a special case.

1. *Measurement of Thermal Resistance*

Time mortality curves for fish are usually surprisingly regular when the mortality in probits is plotted against the logarithm of time as Fig. 5 indicates. Here probit lines are simple, straight and parallel, indicating a statistical homogeneity both in the population and in the locus in the organism which breaks down under the influence of excessive temperature.

However, in many instances, particularly in determinations in the lower zone of thermal resistance, the regularity breaks down. Figure 6 shows four examples of such statistical heterogeneity. Figure 6A shows a case at the lower lethal which clearly displays the phenomena designated by Doudoroff

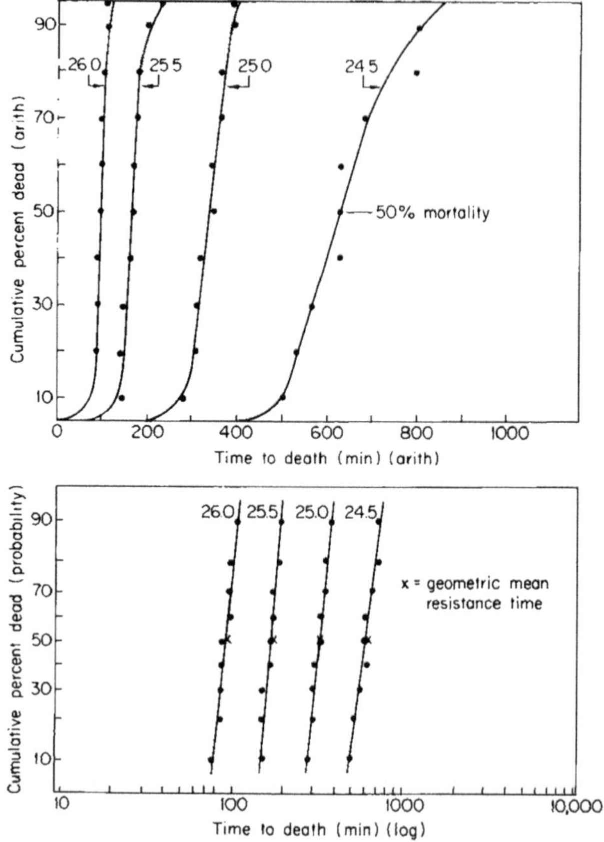

Fig. 5. Time mortality curves for death of fish exposed directly to various constant temperatures. Chinook salmon, *Oncorhynchus tshawytscha*, acclimated to 10°C. From Brett (1952).

(1945) as primary and secondary chill coma. All deaths at 0.0°C and the early deaths up to 0.7°C are the result of primary chill coma; the remaining deaths are the result of secondary chill coma. Pitkow (1960) considered primary chill coma to be the result of failure of the respiratory center while Doudoroff (1945), who showed that some of the lethal effect at the secondary coma point could be removed by using an isosmotic solution, considered the lethal action of temperature at that point was to suppress the ion-osmoregulatory mechanism. Brett's data illustrate that point also. Exposure in dilute seawater (nearly isosmotic) prolonged life in the secondary phase, slightly at 0.7°C and significantly at 3.2°C. It is probably better to speak of a breakdown in ionosmoregulatory regulation. Wikgren (1953) showed that carp have an excessive loss of ions at low temperatures and R. Morris (1960) made the same observation for the lamprey.

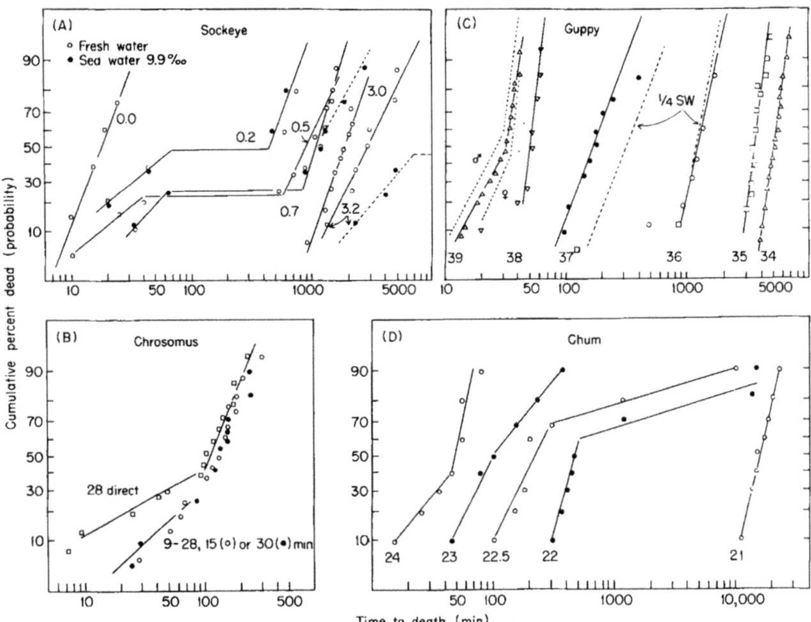

Fig. 6. Statistical heterogeneity in time mortality at various lethal temperatures. (A) Cold death in *Oncorhynchus nerka*, acclimation temperature 20°C [from Brett (1952)]. (B) Heat death in *Poecilia reticulata* acclimated to 25°C [from Arai *et al.* (1963 and unpublished observations)]. (C) Heat death in the minnow, *Chrosomus eos*, acclimated to 9°C [from Tyler (1966)]. (D) Heat death in *Oncorhynchus keta* acclimated to 5°C (Brett, 1952). The numbers associated with the curves indicate the test temperatures. All results are from direct transfer from the acclimation temperature except as indicated in B.

The remaining parts in Fig. 6 show statistical discontinuities at high lethal temperatures. Figure 6B shows a sudden "shock" effect which has been intuitively feared by practical fish culturists and has led to the practice of tempering the transfer of fish from one temperature to a different one by equalizing the two over a period of a fraction of an hour. In the case shown in panel B such equalization over 15 or 30 min does largely remove the first lethal effect on the section of population sensitive to it. However, at least in the laboratory, the shock effect is not a prominent feature.

Figure 6C shows two heterogeneities in the response of the guppy to high temperature. There is a sex difference in response at 39°C, while the major feature of the panel is the jump in the time–temperature sequence between 37° and 36°C where the guppy goes from a stage where exposure in 25% seawater lengthens life to one where there is no effect. The guppies used in these experiments were genetically homogeneous so that the shift in response represents a change in the locus of breakdown with a change in the intensity of the lethal factor. While not shown here (but see Fry, 1967) unselected stocks of guppies show statistical heterogeneity below 36°C. The pure line tested by Arai *et al.* (1963) were all sensitive to the first locus for temperature death at 35° and 34°C.

While the analysis of such discontinuities, as are mentioned above, is a most fruitful field for further research, the matter will be dropped at this point to take up the relation between survival at different temperatures in the zone of resistance. Figure 7 shows two typical series of determinations of thermal resistance for the upper lethal zone. The two species in the figure illustrate contrasting types of response. The goldfish has a very short zone of resistance and a very high incipient lethal temperature. The bullhead while having almost as high an incipient lethal temperature as the goldfish has a more normal zone of resistance. The salmonids (e.g., Brett, 1952) have low tolerance but high resistance.

Since the goldfish has such a short zone of thermal resistance, the data for the bullhead give a more typical picture (Fig. 7B). Both species, however, are typical in their response to thermal acclimation where both the height and the extent of the resistance curves increase with increasing acclimation temperature. It is typical too for thermal resistance to continue to increase with thermal acclimation beyond the point where there is no further change in thermal tolerance, a feature also illustrated in Fig. 8.

Dosage mortality will not be dealt with extensively since this is the customary method of bioassay of lethality (e.g., Bliss, 1952). The only point to be made is that the period of exposure for the determination of tolerance cannot be arbitrary but must be based on a knowledge of the extent of the zone of resistance. If in the case of the bullhead (Fig. 7B) samples of fish acclimated to 5°C were exposed to various constant temperatures from 26.5° to 28.5°C and observed for 1000 min, mortality would have been complete at 28.5°C while no mortality would have been observed at 27°C as indicated by the dotted line. By extrapolation of the relation between time and mortality at the higher temperatures, 50% mortality would have been expected at 500 min at 27°C. Since the latter did not take place, the total mortalities in such a series of baths would then be plotted against temperature to give the incipient lethal temperature since it could be expected that mortality was complete in the lowest temperature although only a fraction of the sample had died in the test.

For other species, or the bullhead at another acclimation temperature, another period of observation would be more suitable for the determination of dosage mortality so that an arbitrary period of determination is not practical. A good method of determining the period of exposure is thus to expose the samples in which mortalities are not complete for the period of time indicated by extrapolation from events at higher lethal temperatures in which 50% of the sample in the lowest test temperature might have died. There are still dangers to such extrapolation. Figure 6D illustrates the difficulty of being sure that an assay proceeds to the tolerance level when there

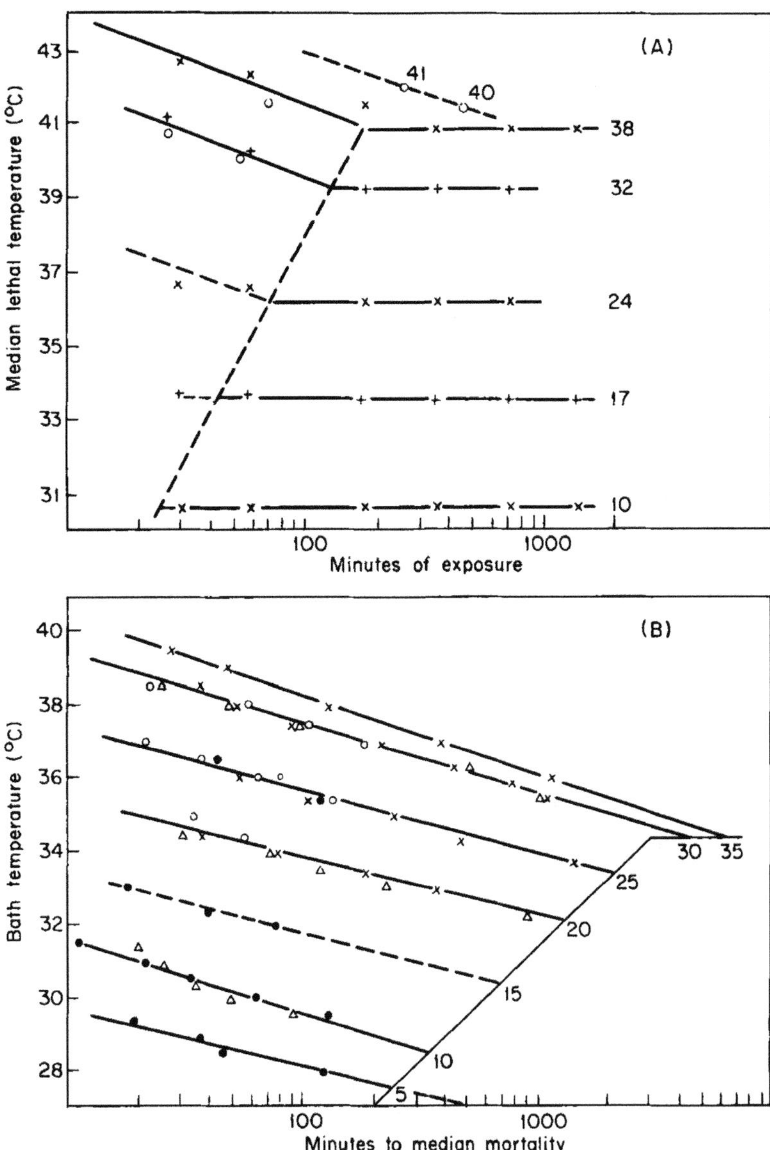

Fig. 7. Thermal resistance times for goldfish (A) and bullhead (B), *Ameiurus* nebulosus. From Fry *et al.* (1946) and Hart (1952). Circles in A represent rèsistance times at a given lethal temperature and thus correspond to the points in B. The various symbols for the bullhead represent samples from different localities over the range of the species. The numbers associated with the curves indicate the acclimation temperatures. The extension of the resistance line for bullheads acclimated to 5°C is discussed in Section II, A, 1.

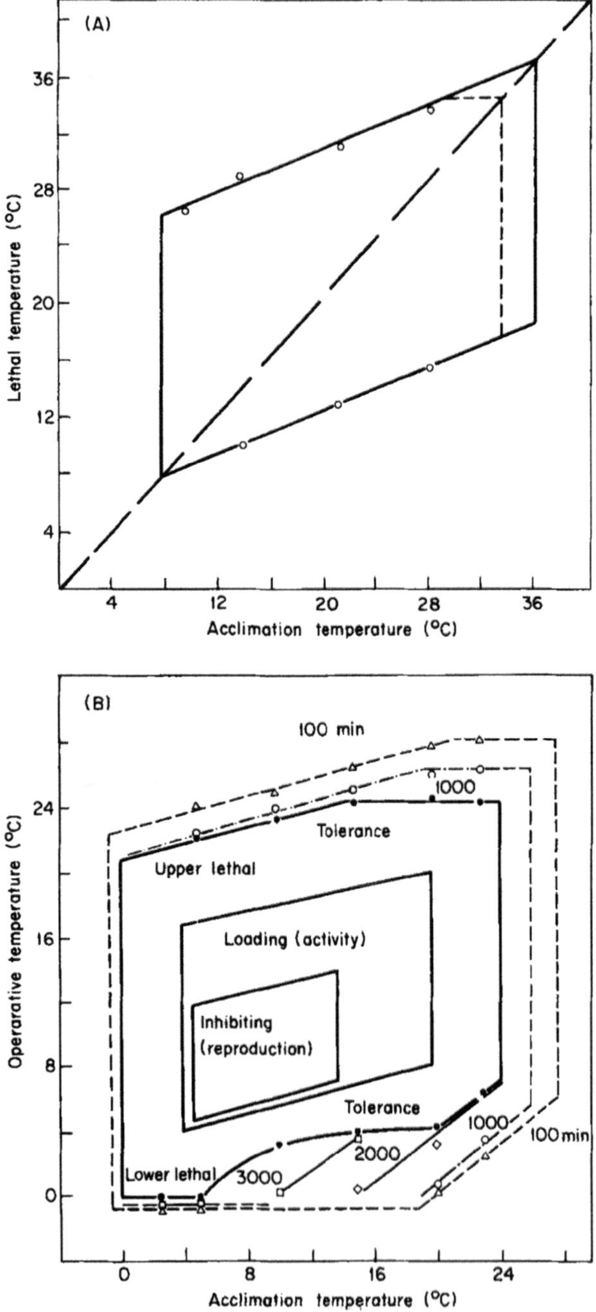

Fig. 8. Typical thermal tolerance diagrams. (A) Puffer, *Spheroides maculatus* [from Hoff and Westman (1966)]; (B) sockeye salmon, *Oncorhynchus nerka* [from Brett (1952, 1958) and Brett and Alderdice (1958)]. See text for definitions.

is a discontinuity in the time–temperature response. Here, suppose the assay had been carried out over a 5-day week, a most likely practical operation, which, when one considers the time at which the working week would start and stop, allows for a period of about 6200 min. In that case the mortality at 21°C would have been entirely overlooked.

2. *Thermal Tolerance*

The incipient lethal temperatures can be plotted as shown in Fig. 8. Here the various typical responses for a thermal tolerance diagram are to be seen. Typically the upper incipient lethal temperature changes approximately 1° for a 3° change in acclimation temperature. The lower incipient lethal shows a somewhat greater response, usually shifting 1° for about 2° change in acclimation temperature. The response of the lower incipient lethal temperature to acclimation temperature is not always linear. In the sockeye salmon (Fig. 8B) Brett found a sigmoid response in the lower incipient lethal to the acclimation temperature, the flex no doubt coming where the cause of cold death passes from primary to secondary chill coma.

The example Fig. 8A was chosen because it displays almost completely the simplest relation to be expected in the response of the incipient lethal to thermal history, a regular linear change in lethal temperature to acclimation temperature so that a trapezoidal figure bounds the zone of thermal tolerance. The lowest and highest incipient lethal temperatures which an organism can attain by extreme acclimation have been termed the "ultimate incipient lethal temperatures" (Fry *et al.*, 1942). In Fig. 8A these ultimate lethals have the ideal values where the acclimation temperature equals the lethal temperature. Figure 8B, on the other hand, shows the various modifications that have been encountered in the tolerance diagram: a plateau in the upper incipient lethal at high acclimation temperatures and a floor to the lower lethal at low acclimation temperatures together with a flexure in the course of the lower incipient lethal temperature to acclimation temperature, as mentioned above. The precise level of this latter floor has been little explored. In many freshwater species the ultimate lower lethal is indeterminate since the fish can still be active at the freezing point of water. Most marine species apparently will freeze in seawater before the latter itself freezes, and the floor there may be set by the freezing point of blood. In the case of the sockeye in Fig. 8, however, the floor comes a little above the freezing point of the blood and has no direct explanation. Death is obviously not a result of the formation of ice crystals in this case.

Some marine species have the ability to supercool, as is discussed in Chapter 3 by DeVries. Such species then may also have ultimate incipient lower lethal temperatures which are indeterminate in their normal habitat. There is some question as to whether the increases in salt content, noted in the bloods of some marine species (e.g., Fig. 9), are adjustments to low temperature. To some extent at least, they may be symptoms of lack of

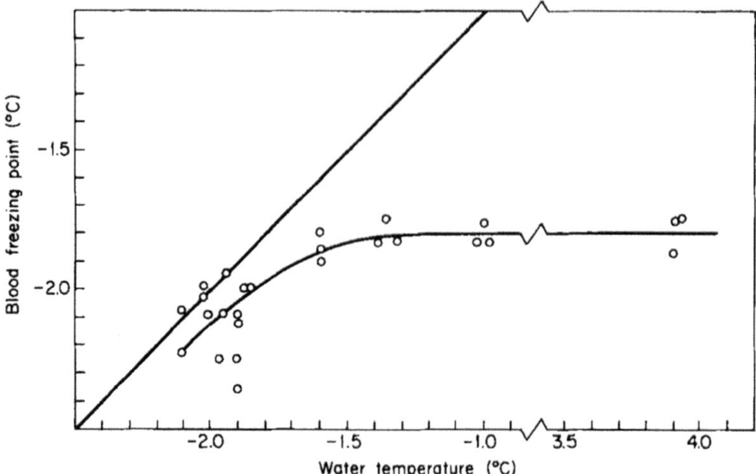

Fig. 9. Relation between water temperature and blood freezing point in *Trematomus bernachii*. Modified from Potts and Morris (1968). The diagonal passes through the points where blood and water would have identical freezing points.

acclimative capacity in the ion-osmoregulatory system and represent approaches of secondary chill coma. Thus, Woodhead (1964) shows a decided upward drift in the serum sodium concentration at about 3°C in the sole, *Solea vulgaris*, a species which he considers from field evidence to have its ultimate lower incipient lethal temperature at about 2°C, well above the freezing point of its blood.

The Antarctic species, for which data are shown in Fig. 9, apparently restricts the blood flow through its gills at −2°C, possibly to restrict loss of water or influx of salt.

The thermal tolerances of fish vary greatly. Antarctic species, for which unfortunately we do not have any tolerance diagrams, die at 5°C or a little above (Wohlschlag, 1964). Low temperate species like the goldfish have ultimate lethal temperatures of 0°C and in the vicinity of 40°C and tolerance diagrams that bound an area of some 1200°C^2. High temperate species have ultimate upper lethals ranging from 20°C to approximately 35°C. Tropical species appear to be distinguished by high, low lethal temperatures and as investigated have not shown higher upper lethal temperatures than some temperate species (e.g., Allanson and Noble, 1964). Indeed, the guppy cannot be carried through its life cycle above 32°C.

Brett (1958) has extended the concept of the tolerance diagram by considering that the lethal temperature is the ultimate response to thermal stress and recognized loading and inhibiting stresses, which terms he uses to designate limits within the zone of thermal tolerance where growth and activity and reproduction, respectively, are suppressed His suggested boundaries for these

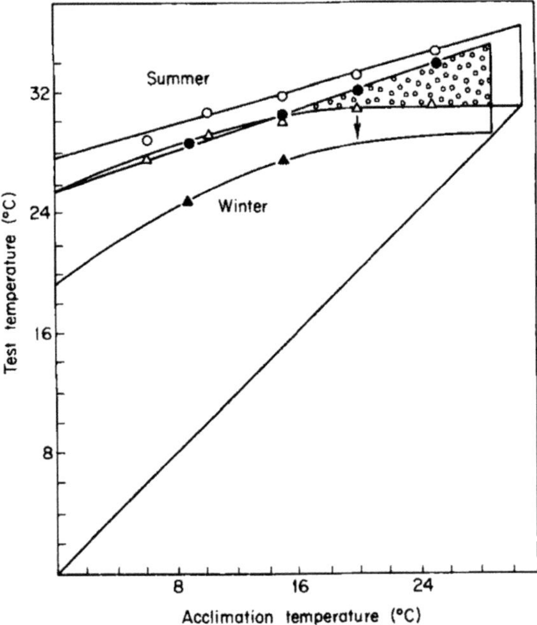

Fig. 10. Summer-winter differences in lethal temperature in the minnow *Chrosomus eos*. The upper boundary to each area is the 30-minute resistance level (circles), the lower, the incipient lethal level (triangles). Modified from Tyler (1966), Solid symbols are winter values; the arrow represents an incomplete winter determination. Stippling indicates the area of overlap.

levels are shown for the sockeye. However, these concepts may be modified, particularly in view of his recent work on growth (Brett *et al.*, 1969).

It has long been recognized that there can be differences in lethal temperature among fish acclimated to the same temperature at different seasons of the year. In this respect most attention has been paid to the upper lethal temperature. An example of the magnitude of such differences is given in Fig. 10. Some modification of the lethal temperature in the appropriate directions has been achieved by manipulation of the photoperiod, a long day bringing an increase in thermal resistance and in the incipient lethal temperature (Hoar and Robinson, 1959; Tyler, 1966), but neither the seasonal effect nor the extent to which manipulation of the photoperiod can modify the lethal temperature has yet been the subject of any exhaustive analysis. Recently (Johansen, 1967), it has been shown that the pituitary must be intact if the goldfish is to acclimate to a higher temperature, which is the first clear indication of endocrine involvement in the response to lethal temperature.

Ushakov (e.g., 1968) and his school have shown that the thermal resistance of muscle is relatively unaffected by acclimation temperature although some slight seasonal effects are to be seen. Interestingly the upper incipient lethal temperature of excised muscle (or tissue cultures) approximates the ultimate upper incipient lethal temperature in four species (Fry, 1967; Fry and Hochachka, 1970). However, the whole problem is still obscured by our ignorance of the specific sites of breakdown in thermal death and, indeed, by lack of unequivocal comparisons between the whole animal and its tissues.

3. Rates of Thermal acclimation

Brett (1944, 1946) early showed that the lethal temperature of various freshwater species adjusted so rapidly to changes in water temperature that changes in the weather were reflected as well as the seasonal cycle (Fig. 11). He and various workers, in particular Doudoroff (1942) and Cocking (1959), have addressed themselves to the measurement of the rate at which fish adjust their lethal temperature in relation to a change in acclimation temperature. Such changes can be very rapid when the temperature is adjusted upward while downward changes are much slower. The main feature of the observations of Brett and Doudoroff on rates of adjustment of the upper and the lower lethal temperature to changes in acclimation temperature are shown in Fig. 12. It would appear from these data that the change of heat resistance follows a different course from change in cold resistance indicating that the

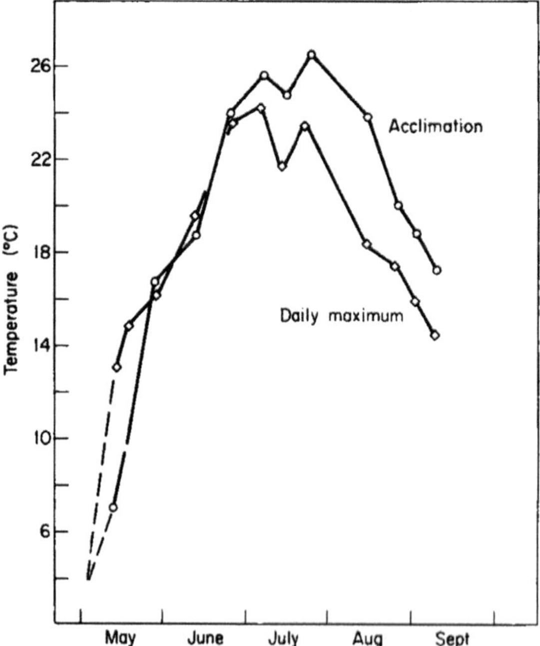

Fig. 11. Acclimation temperature of the brown bullhead, *Ameiurus nebulosus*, as determined from the lethal temperature, in relation to mean daily maximum water temperature. From Brett (1946).

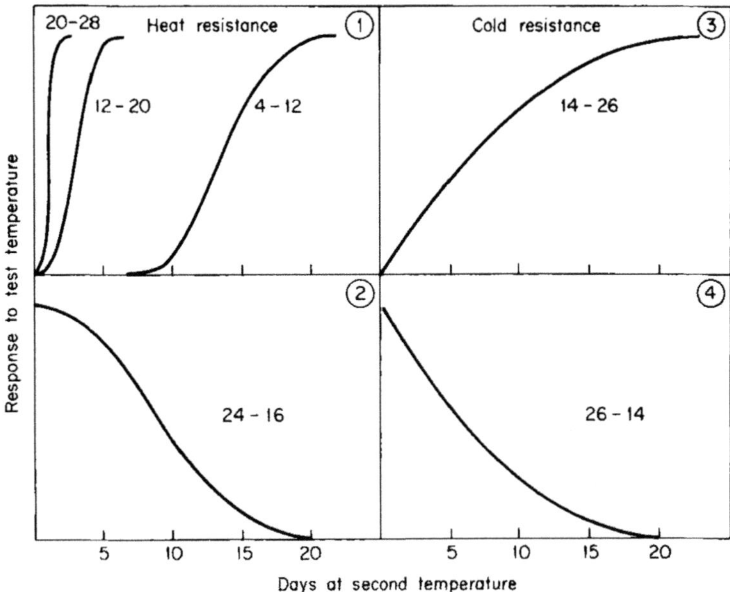

Fig. 12. The four basic courses of acclimation to lethal temperature. Heat resistance determined as average survival time; cold resistance as 24-hr median tolerance limit. (1) Goldfish (Brett, 1946). (2) The minnow, *Pimephales promelas* (Brett, 1944). (3 and 4) *Girella nigricans* (Doudoroff, 1942). Numbers associated with the curves indicate the magnitude and direction of shift of acclimation temperature.

two responses do not operate on the same site. Change in resistance to high temperature follows a sigmoid course and can show a long latent period which is about 1 week at 12°C for goldfish when moved from 4°C, while the curves for change in resistance to low temperature are simply convex or concave. Heinicke and Houston (1965) found a distortion in the plasma sodium:chloride ratio in goldfish transferred abruptly from 20° to 30°C which reached its extreme in the first 4 days of exposure. The ratio returned to normal by about 10 days. Another major difference made clear in the figure is that, as the data for *Girella* show, while adjustments in cold resistance are about equally rapid whether a given step is up or down over a given acclimation range, heat resistance is gained much more rapidly than it is lost. The upper lethal temperature of goldfish is adjusted within a day when they are shifted from 20° to 28°C, while the shift from 24° to 16°C requires over 2 weeks for the reciprocal adjustment in *Pimephales*. It is not likely that these differences are the result of different species being used in the two experiments. In consequence of this differential in rate, acclimation of the upper lethal temperature tends to follow the daily maximum with increasing temperatures, as Fig. 11 indicates, where the acclimation temperature rapidly catches up with the daily maximum temperature in approximately the first 3 weeks after breakup. Heath (1963) found that in the cutthroat trout, *Salmo clarki*, a 24-hr period of fluctuating temperature resulted in the highest acclimation temperature as compared to the mean temperature throughout the whole period. There is no information on the effect of fluctuating acclimation temperatures on the lower lethal temperature.

Rates of adjustment to acclimation temperature (neglecting the Q_{10} effect), vary from approximately 1°C/day in the goldfish and the roach (Cocking, 1959) to the same change in an hour or so in *Girella*, the bullhead, and various salmonids.

There are experimental difficulties in the determination of rates of acclimation which have not yet been thoroughly explored. Brett (1946) noted that the brown bullhead would not respond to a change in acclimation temperature if the oxygen in the acclimation bath were reduced to approximately 10% air saturation. Hart (1952) was not able to acclimate yellow perch, *Perca flavescens*, to high temperature in winter and noted that he was unable to get them to feed. Later, in our laboratory, perch which were feeding did acclimate. On the other hand, Brett found bullheads starved for a total of 40 days from June 16 acclimated precisely as did fish which were feeding up to a day or so of test.

4. *Death in Changing Temperatures*

Lethal temperatures in the environment are ordinarily temporary changes brought about by unusual seasons or extreme weather. Thus, time is a major element in lethality as well as is temperature, which indeed is the reason for segregating thermal resistance from thermal tolerance. It is of importance therefore to be able to integrate the temperature experience. The progress to death at any one lethal temperature can be taken as linear (e.g., Jacobs, 1919; Olson and Stevens, 1939). Thus fractions of the respective resistance times spent at each of a series of lethal temperatures can be summed to indicate total lethal experiences as Table I indicates. If the temperature changes continuously then the summation can be made by relating the time-temperature curve to the temperature-resistance time curve. This may be done graphically as in Fig. 13 where temperatures are plotted as equivalent to the reciprocals of resistance times appropriate to them, or by calculation based on the same principle, as is done by engineers interested in heat sterilization (Olson and Jackson, 1942).

To simplify the problem in the examples given above the experiment began with a sudden transfer from the acclimation temperature to the zone of resistance in order to eliminate the effects of acclimation during the heating period before the incipient lethal temperature was passed. There is acclimation in the zone of resistance, too, as can be demonstrated by subjecting a fish to a sublethal exposure at temperatures above the incipient lethal and then returning it to its original acclimation temperature for a period of recovery, after which its resistance to a given lethal temperature is tested again (Fry *et al.*, 1946). Cocking (1959) slowly heated from the acclimation temperature and continued to the death of the fish to get a measure of the rate of thermal acclimation.

Table I Summation of Lethal Experience in Various Temperatures in the Eastern Brook Trout, *Salvelinus fontinalis*[a,b]

Acclimation temp. (°C)	Thermal experience	Observed time to death (min)	Theoretical time to death (min)	Summed lethal dose
11	27.1, 33 min (0.50) to 26.3 till death (0.45)	102±18	107	0.95
11	26.3, 60 min (0.39) to 27.1 till death (0.80)	112±10	100	1.19
11	25.8, 90 min (0.34) to 26.1, 40 min (0.21) to 26.5 till death (0.48)	190±22	177	1.03
20	26.5, 230 min (0.48) to 28.0 till death (0.64)	291±16	281	1.10
20	27.0, 125 min (0.44) to 28.0 till death (0.60)	183±20	179	1.04
20	27.5, 75 min (0.46) to 28.0 till death (0.55)	128±10	122	1.01

[a]Error given is 2σ.
[b]Based on Fry *et al.* (1946), their Tables 4 and 6. Numbers in parentheses indicate calculated fraction of lethal experience corresponding to the period of exposure at the particular temperature.

As mentioned earlier, lethal temperatures (CTM) are often determined by steadily increasing the temperature a degree every few minutes and recording the temperature at which the sample dies or is incapacitated. Incapacity is taken as the equivalent of death for two reasons: first, the indicator must provide an unambiguous point under such rapidly changing conditions; second, it

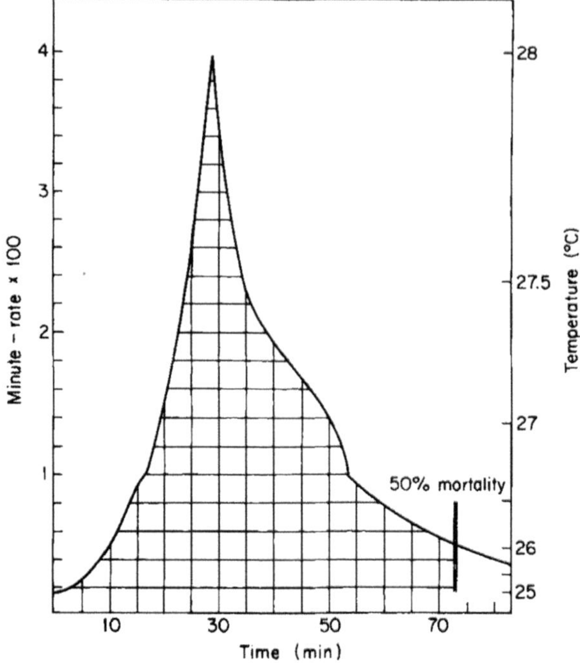

Fig. 13. Accumulative effects of exposure to changing lethal temperature in eastern brook trout, *Salvelinus fontinalis*, acclimated to 11°C. From Fry *et al.* (1946). The "minute-rate" is the fraction of the resistance time at any lethal temperature represented by an exposure time of one minute. The temperature scale on the right is based on the minute-rate scale. Thus, for example, the resistance time at 27° was 72.5 min so that the minute-rate $\times 100(1/72.5 \times 100)$ is 1.37. The median mortality point indicated was the observed one. Each square under the curve indicates 1% of the total theoretical lethal exposure. There are approximately 101 squares under the curve up to the median mortality point.

is assumed that if the animal becomes incapacitated it will not be able to escape from further stress and will be trapped in a lethal situation. From the narrow ecological point of view the CTM is useful and is above all extremely economical of material. It can even be used to determine the ultimate upper incipient lethal temperature if rate of heating is slow enough, say, 0.5°C/day day (Cocking, 1959; Spaas, 1959).

As a means of physiological analysis however the CTM has many failings. The determination confounds time and temperature, but in particular discontinuities of response such as are illustrated in Fig. 6 are lost. In practice when the rates of heating are of the order of minutes per degree only the most acute cause of death will be displayed. There is no time for the slower breakdowns. Moreover, the plateau often to be found in thermal tolerance diagrams is not likely to be displayed in the response of the CTM to acclimation temperature, as is well shown for the guppy, *Poecilia reticulata* (Fry, 1967, Fig. 3) in which the incipient lethal temperature is approximately 32°C at all acclimation temperatures above 20°C while the CTM increases steadily about 1° for every 3° change in acclimation temperature throughout the whole range of observations. Thus, whereas the ultimate incipient lethal temperature of this species is below 33°C, the maximum CTM is somewhat over 40°C.

Finally, it should be remarked that when a fish is in water warmed at the rate of a degree every few minutes its internal temperature will lag somewhat behind the ambient temperature. It can be calculated from Harvey's data (1964) that the lag for a 50-g sockeye, *Oncorhynchus nerka*, would be 0.4°C if the heating range were 1°C/5 min (Fry, 1967). A tenfold change in weight would bring about a twofold change in lag on the basis of present data on the size-thermal conductivity relation.

B. Toxicity Studies

This section is written almost with reluctance. It may be said that pollution biologists have backed into bioassay. It is proper for those interested in chemical control of pests to assay their biocides by determining the lethal dose with care and precision, for their purpose is to load the environment with the minimum proper dose. They wish to be sure they have killed their target at a minimum of cost and further damage to the organic community. The pollution biologist, on the other hand, wishes to protect the organism of his concern and to see it prosper—again with a minimum of interference with man's other interests. Thus, his aim is to protect the organism from damage, not from death alone. To deal then, even learnedly, with lethal levels of a pollutant is somewhat to serve orthodoxy at the expense of progress. Lethal levels are to be considered only as the boundaries of the zone within which the real work goes on.

Statistical response to toxic materials is similar to response to lethal temperatures. For example, Fig. 14 shows that the same statistical discontinuity can be found in stress from a poison as is found with the physical identity, temperature. With regard to acclimation—a prominent feature in temperature death, and presumably also for setting the lethal levels of toxic substances—there is little that can yet be said. Most assays are made with animals which have not been previously exposed to the lethal agent under test; thus, they present the most severe case.

One important difference between the harmful effects of toxic agents and the lethal action of temperature as discussed above is that the effects of incomplete exposure to a lethal level are by no means necessarily reversible. Sublethal exposure can lead to permanent destruction of critical tissue such as gill (Scheier and Cairns, 1966) or sensory epithelium (Bardach *et al.*, 1965).

An extensive review on toxic substances as they affect aquatic organisms with an exhaustive bibliography has been prepared by McKee and Wolf (1963). Here only a few remarks on the interactions of multiple materials in the environment will be offered.

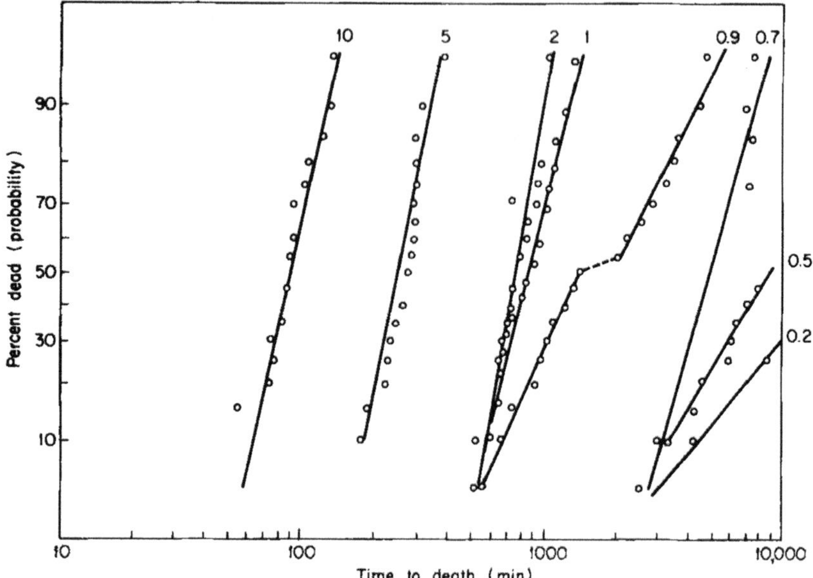

Fig. 14. Statistical response of the eastern brook trout, *Salvelinus fontinalis*, to various dosages of 20% dinitro-*o*-cyclohexylphenol, dicyclohexylamine salt, at approximately 9°C. Data of D. F. Alderdice (personal communication). Numbers associated with the probit lines indicate the dosage in milligrams per liter.

An effluent often contains a mixture of various toxic materials and it is necessary to consider whether each of these acts independently, in which case all that is necessary is to be certain that each is kept below its threshold level, or whether the effects are additive or even synergistic. Lloyd (1961c) demonstrated that fractions of the incipient lethal doses were additive in the case of copper and zinc. This rule also held for resistance times at higher doses in hard water, but mixtures of the two metals were relatively more toxic than either alone at high doses in soft water. Sprague and Ramsay (1965) reported similar findings. Such additivity has been reported for various other combinations (e.g., Herbert and Vandyke, 1964). Much of this work is reviewed compactly in Herbert (1965). Again, it has long been known that the toxicity of metals, for instance, is highly variable in natural waters when the concentration of the toxic agent is expressed as the total of that element present in solution. It has also long been well-recognized that such differences are in large part the result of differences in pH, but the principles of chemical dissociation appear to have first been applied extensively to the effects of toxic substances on fish by Wuhrmann and Woker (1948) who demonstrated that the amount of un-ionized ammonia in a given solution was the significant measure to take to determine its toxicity. Cyanide, which complexes with metals, offers another example (Fig. 15).

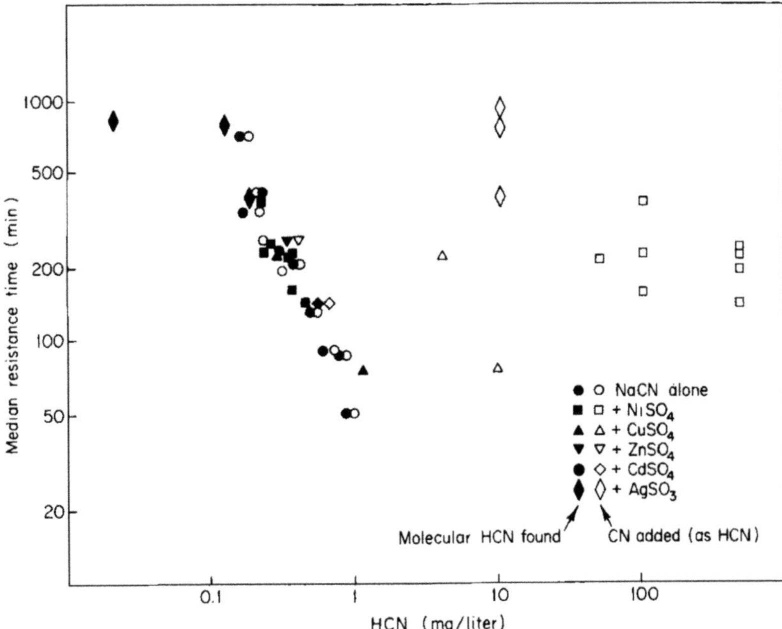

Fig. 15. Median resistance time of bluegill, *Lepomis macrochirus*, in relation to free molecular HCN and total cyanide concentrations (expressed as HCN equivalents) in various simple and complex cyanide solutions. Modified from Doudoroff *et al.* (1966).

Determination of the concentration of the toxic form in the environment may not entirely resolve the question of toxicity. The dose the fish absorbs may be further modified by the same agent which influences the state of the toxic agent. Thus, Lloyd and Herbert (1960) calculated that the interaction of various ambient concentrations of free CO_2 with the respiratory exchange so modified conditions within the interlamellar spaces that the same concentration of un-ionized ammonia (0.40 mg/liter) was presented to the gill under circumstances where the outside concentrations varied from 0.84 to 0.49 mg/liter.

An agent which changes respiratory flow will also modify the dosage received from a given ambient concentration, again through the agency of the countercurrent exchange system in the gill (Lloyd, 1961a, b; Herbert and Shurben, 1963).

III. CONTROLLING FACTORS

The operation of controlling factors will be illustrated exclusively by the effects of temperature. The effects of pressure, which must have a major controlling effect in the oceanic depths, have been reviewed in the chapter by Gordon, Volume IV, this treatise.

The thesis taken here, which seems to be the one generally implied, is that, other things being equal, the metabolic rate is a function of molecular activity while otherwise under the regulation of agents which operate in ways not yet much understood (but see Chapter 2 by Hochachka and Somero). On these premises, the controlling factor, setting a limit to molecular activity, sets an upper limit to the metabolic rate. By conferring a given level of molecular activity the controlling factor also imposes a given level of instability on the living system, which must be counteracted by repair through energy-yielding reactions. Thus a given level of controlling factor is taken to permit an upper limit to the metabolic rate and to require a lower one. As defined in the Introduction, the upper limit is active metabolism, the lower one standard metabolism. Activity, a transformation of the energy released by metabolism and a function of some fraction of the difference between these two levels, may or may not bear the same relation to temperature as either of them.

Controlling factors operate at the cellular level, the site of the metabolism yielding the energy. The potential for energy yield in the cells may not always be permitted full expression, even in the normal environment, because of restrictions imposed by the nature of the whole organism, as will be discussed in Section IV. There is also the complication that random activity varies with temperature. Thus, the relation of routine metabolism to temperature cannot be expected to fit directly any of the curves described by the temperature formulas. The relation of routine metabolism to temperature will be considered in Section VI.

A. Formulas Relating Temperature to Metabolism and Activity

Figure 16 gives a crude genealogy of the major formulas which have been applied to the effects of temperature on living processes. Briefly, the earliest formulation appears to have been the rule of thermal sums proposed by Réaumur and applied to the effect of temperature on the date of the appearance of such phenological events as the ripening of crops or the development of larvae in relation to whether the season is advanced or retarded in different climates. Much later Berthelot, studying the rate of fermentation, found the effect of temperature could be described as a process which increased geometrically as temperature increased arithmetically. From this point of view effects of temperature could be characterized by a coefficient comparing rates over a stated interval. This coefficient is the well-known Q_{10} and the rates of biological processes usually double or treble over an interval of 10°C ($Q_{10}=2$ to 3). Through van't Hoff to Arrhenius the effect of temperature on rates of chemical reaction was related to molecular theory and much later the coefficient μ (which represents e, the energy of molecular activation) came into widespread use in biology in America, largely through the work of Crozier and his colleagues. Bělehrádek proposed a third coefficient b, derived from the slope of the relation between log temperature and log time,

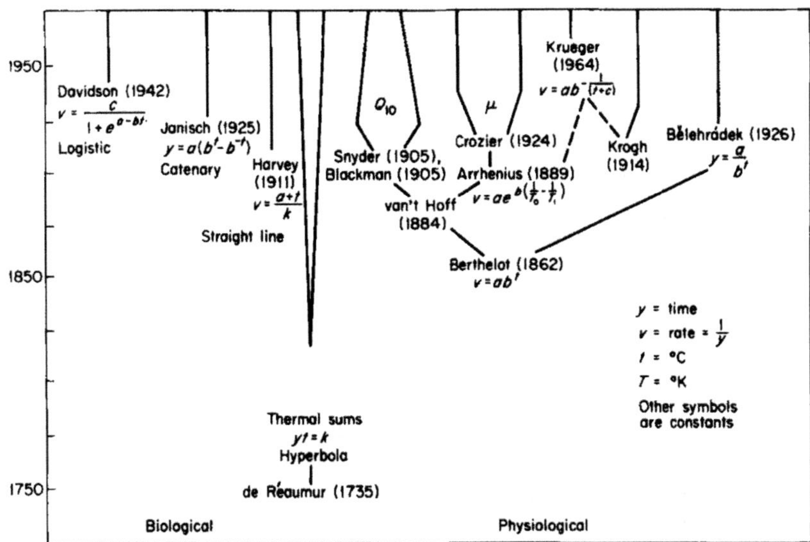

Fig. 16. Historical and algebraic relations between the various formulas applied to the effect of temperature on the rate of processes in organisms. The dates given indicate original or early references to each formula.

not being satisfied that the μ as found in whole organisms necessarily strictly reflected the e of the gaseous reactions of the physical chemists. Bělehrádek considered that the rates of reaction in organisms were more likely to be limited by diffusion than to be set by the level of molecular activation. The final item given on the right hand side of Fig. 16 is Krogh's curve for standard metabolism, which was not formulated but simply a descriptive curve of proportional change with temperature that fitted various observations that Krogh and his colleagues had gathered, among them the metabolic rate of the goldfish. Krueger later proposed a method to fit a formula to the curve.

The first formula given to the left of the rule of thermal sums is Harvey's fit of the straight line to various rates such as heartbeat, which is mathematically identical to the rule of thermal sums. The remaining two formulas were proposed by workers concerned with the rate of development of insect eggs to meet the long-recognized inadequacy of the rule of thermal sums as a fit over the whole range of temperature at which development can take place, by formulating curvilinear fits. Janisch fitted a catenary to the time curve to permit inclusion of the cases at high temperature where time increases over that required for development at a somewhat lower temperature. Davidson fitted the logistic formula to the rate curve to provide for the deviation from linearity at low temperatures where the rates are higher than would be predicted from extrapolation of the straight line which fits the central range. He ignored the response to the extreme upper temperatures which were the special concern of Janisch, as Janisch ignored the response to the extreme lower ones.

A good review through which to enter the earlier literature on this subject is that of Bělehrádek (1930). The essentially chemical approach is well presented in Precht *et al.* (1955) and Johnson *et al.* (1954). In particular, the latter authors state the general biochemical case for the controlling factors in their treatment of the interaction between temperature and pressure, which is not dealt with in the summary above.

The two series of formulas are segregated in the figure under the terms "biological" and "physiological." Andrewartha and Birch (1954) recognized the same segregations but preferred the terms "empirical" and "theoretical"; but while the terminology of these authors is correct concerning the origins of the formulas (except for Berthelot's), it misses the point of their application. Their terminology seems to imply that the "theoretical" series will ultimately prevail over the "empirical" one. However, the distinction is really in kind not in worth. The biological series relate directly to activity in relation to temperature. The physiological formulas relate to the chemical basis which yields the energy for activity.

The difference between the biological and physiological needs for formulation are well expressed by Booij and Wolvekamp (1944, p. 212) as follows:

> Whilst the chemist takes care that the form of the reaction vessels and the properties of the substances of which they are manufactured do not interfere with the processes under investigation, the engineer on the contrary will design special structures and make use of their physical properties in order to obtain a harmonious cooperation of physico-chemical processes. . . .
>
> It is to the segregated parts of the more complex processes taking place within the organism (just as those taking place within an engine) only that the fundamental laws of physical chemistry may be applied.

In the discussion below, concerned as it is with the whole organism, there will be little direct application of the physiological series of formulas, only the Q_{10} being used in general terms. The physiological relation looked for will be the effect of temperature on the relation of active and standard metabolism and on the relation of the difference between these values to activity.

B. Active and Standard Metabolism in Relation to Temperature

The most thorough determinations of active and standard metabolism have been made by Brett (1964) for his stock A. His data are summarized in Table II and were obtained in his apparatus which is illustrated by Phillips in Volume I of this treatise. Brett's determinations were the product of experiments each several hours long in which the fish were stimulated to swim faster and faster by small steps. He took the maximum rate so achieved as his measure of active metabolism while he extrapolated the rate determined at various speeds to zero speed as his measure of standard metabolism. Thus not only did he obtain a close approximation to the maximum continuous rate

Table II Swimming Speed and Active and Standard Metabolism of the Sockeye Salmon in Relation to Acclimation Temperature[a]

Temp.	O	Metabolism (ml/kg/hr)				Swimming speed	Ratio
(°C)	(mg/liter)	Active	Standard	Scope	Scope$^{0.55}$	(length/sec)	Active/Std.
A							
5	12.8	364	29	335	24.5	3.26	12.5
10	11.2	445	42	403	27.1	3.65	10.6
15	10.1	635	50	585	33.3	4.12	12.7
20	14.0	921	85	836	40.5	4.27	10.8
B							
20	9.1	604	85	519	31.2	3.94	—
24	8.5	601	139	462	29.2	3.75	—

[a]Data of Brett (1964) for stock A, from his Tables 1 and 6 and Fig. 16. "Scope" is the difference between standard and active metabolism. Fish weight, approximately 50 g.

of oxygen consumption but also probably largely eliminated the early effects of excitement from his data and thus obtained as well a close approximation for the minimum resting level.

Brett determined only oxygen consumption, but it probably is safe to infer from Kutty's work (1968a) on the respiratory quotients of goldfish and rainbow trout in similar experiments that the respiration of the salmon was essentially aerobic in these long-term determinations.

In general, Brett worked with the dissolved oxygen concentration in the water at approximately air saturation. At 20°C, however, a special experiment was performed in which the water was considerably enriched to14 mg/literO_2. Table II is divided into two parts: part A shows experiments to 15°C plus the one at increased oxygen at 20°C, and part B, shows experiments at air saturation at 20° and 24°C. Part B will be dealt with again in Section IV. All Brett's measurements were made with the fish acclimated to the test temperature.

Looking at the data in part A it is apparent, as would be expected, that increasing temperature accelerates both active and standard metabolism. The point of major interest here though is that within the limits of error, standard metabolism is the same fraction of active metabolism at the four temperatures concerned. Thus, standard metabolism reflects the possibilities for active metabolism.

The various data in the table are plotted on a semilogarithmic grid in Fig. 17. Here the data of part B are plotted as well as those of part A of Table II, again for further consideration below.

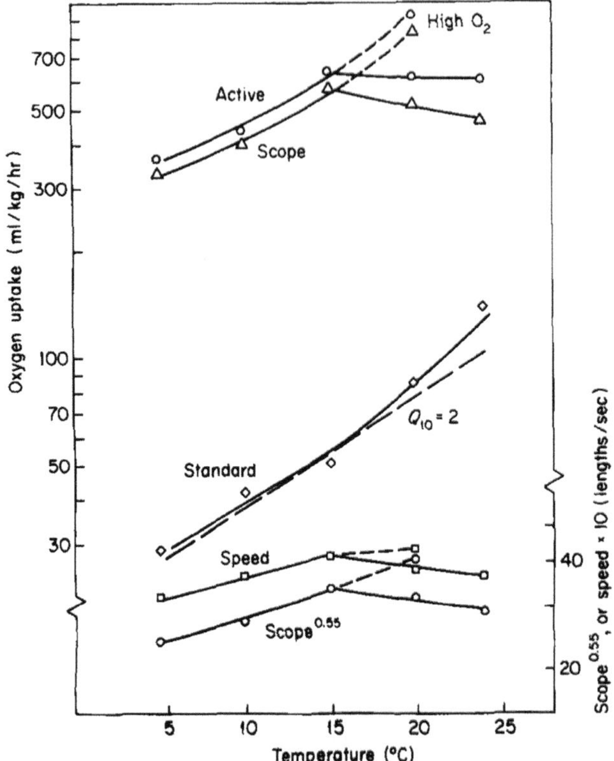

Fig. 17. The effect of temperature on metabolism and activity in the sockeye salmon, *Oncorhynchus nerka*. From Brett (1964), his stock A. For further explanation see text.

The curve for active metabolism with supplementary oxygen and that for standard metabolism are parallel and almost straight on the semilogarithmic plot and have a Q_{10} of approximately 2. The points for standard metabolism at 20° and 24°C with oxygen at air saturation continue the trend of the points at the lower temperatures. The corresponding points for active metabolism, of course, do not, because of the limiting effect of oxygen in air-saturated water. The curve for standard metabolism can therefore be taken as reflecting the potential for active metabolism even though the latter may not be attained. However, such a conclusion can only be highly tentative. In particular, the measurement of standard metabolism is only yet in its infancy as is, of course, also the measurement of active metabolism, most especially at high temperatures without oxygen being limiting.

1. The Q_{10} of Standard Metabolism

Few general statements can yet be made of the relation of standard metabolism to temperature except that different species are adapted to different temperature ranges [see Wohlschlag's (1964) summary diagram], a point the fish make well themselves without recourse to a respiration chamber. Another general point is that over the biokinetic range of a species the metabolic rate is approximately 75 ml/O_2/hr for a 100-g individual at the mid range. The relation of standard metabolism to temperature is often described by Krogh's "standard curve" (Ege and Krogh, 1914) which has been formulated by Krueger (1964) and which Winberg (1956) applied so effectively in his generalizations. However, Krogh's curve as an empirical formula applies much more generally to curves of routine metabolism than to standard metabolism, at least as the latter has been determined by extrapolation to zero physical activity.

The relations of the extrapolated values for standard metabolism to temperature (Beamish, 1964a; Beamish and Mookherjii, 1964; Brett, 1964; Rao, 1968) have been various, ranging from a response similar to the Krogh curve with constantly decreasing Q_{10} through a case of constant $Q_{10}=2$ in the goldfish—which was the fish species on which the Krogh curve was determined—to the case of the sockeye (Fig. 17) where Q_{10} shows a slight constant increase with increasing temperature (as the dotted line indicates). The convex course of Krogh's curve with decreasing Q_{10} with increasing temperature will be considered in Section VI.

2. *The Relation of Activity to Temperature*

The remaining curves in Fig. 17 are concerned with the effect of temperature on activity, directly or indirectly. The uppermost of these curves, labeled "scope" is the difference between active and standard metabolism and is shown both for the case where oxygen was not limiting and for the case of air saturation over all temperatures. The curve for scope where oxygen is not limiting is parallel to the corresponding curves for active and standard metabolism and therefore can be interpreted by the same temperature coefficient. Thus, under circumstances where the activity concerned has a linear relation to the metabolic scope available for that activity and no limiting factor intervenes, then the Q_{10} for the activity will be the same as the Q_{10} for standard metabolism. However, no specific example of such a relation is at hand.

The series of curves grouped within the central box on the figure represent the relation of swimming speed to temperature and the function of scope related to that activity derived from Fig. 2. The Q_{10} for the maximum sustained swimming speed up to 15°C is much less than 2 and approximates $\sqrt{2}$. Scope$^{0.55}$, the power relation derived from Fig. 2, has a similar temperature response, as would be expeted since Brett found that the increase in metabolism brought about by a given increase in swimming speed was the same at all temperatures. The Q_{10} therefore for the activity, maximum continuous swimming speed, when no limiting factor intervenes is approximately 1.45 but it is based on a Q_{10} of 2 for metabolism. In the example oxygen is limiting above 15°C and the swimming speed curve drops. The course however is still explained by the same function applied to the metabolic scope still available. Various other similar curves for the effect of temperature on swimming speed are collected in Fry (1967).

Larimore and Duever (1968) give a curve for the swimming speed of smallmouth bass fry, *Micropterus dolomieui*, where the Q_{10} approximates 2 from 5° to 25°C. These fish, about 20 mm long, may not show the same relation between speed and metabolism. Pavlov *et al.* (1968) and Houde (1969) showed that maximum sustained swimming speed increases directly with length in small smooth-skinned fish rather than as length$^{0.5}$, which was the case for the sockeye salmon (Brett, 1965). However, such evidence is not conclusive since the divergence may be in the capabilities for metabolism, which have not been measured.

In the example given the Q_{10} of active and standard metabolism is approximately 2, the commonly found relation for biochemical reactions. One example, cruising speed, is worked out to show a quantitative relation between temperature and metabolism. The example is perhaps deceptively simple and probably oversimplified (for example, ancilliary costs are ignored) so that only the general principle of activity being related to some function of some fraction of total metabolism (in the present case aerobic, but not necessarily so) should be retained after considering these paragraphs. In fact, while the relation of muscular activity to temperature can be justified in terms of metabolic cost in the straightforward fashion indicated above the response of other activities cannot yet be so easily analyzed. The rate of embryonic development (e.g., Krogh, 1914; Garside, 1966; Kinne and Kinne, 1962) has a mean Q_{10} ranging up to 6. Perhaps these high Q_{10}'s are a reflection of the multiplicative nature of growth which accelerates the oxygen supply as growth is faster.

3. *The Rule of Thermal Sums*

The rule of thermal sums, which states that time × temperature is a constant for a given developmental or phenological event, has widespread use in practical fish culture. Normally the "thermal unit" (cf. Embody, 1934) is employed which is the Fahrenheit expression ($^\circ\mathbf{F} - 32 \times$ days). As a practical tool it is highly useful but some biologists, particularly those concerned with the morphometric consequences, have endowed the rule with an undue constancy and an unproven physiological significance (e.g., Tåning, 1952). In such cases the confidence in the rule is often misplaced, and it is doubtful whether it should be used. Shelford (1929) is a useful reference for anyone to consult who wishes to apply this rule. Basically the relation between temperature and rate of development, as ordinarily observed, is sigmoid so that the central section about the point of flexure, which is rather gradual, is well approximated by a straight line. If the linear section is extrapolated to zero rate the intercept T_0 provides a correction so that $(T_1 - T_0)$ provides a corrected temperature, which when multiplied by the *time* for development gives a constant. The rule, with the correction when T_0 is not 0°C, therefore is good for a median range of temperatures within the total range over which a given species can develop. At extreme temperatures the linear approximation of rate breaks down and the thermal sum is no longer constant. Hence, while the rule is useful in phenology and in hatchery practice, where the normal annual fluctuations are not likely to have a mean far from the median range, in physiological investigations, where controlled temperatures are used over the total range for development, the rule will break down and should not be used. The formulas of Janisch (1925) and Davidson (1944) are empirical fits to the development curve. Since there is evidence the inflection in the curve is the result of the limiting effect of oxygen, as is discussed below, these formulations are likely to apply to only the specific case of air saturation and have no general application to aquatic organisms which often develop under various degrees of oxygen deficiency.

C. Acclimation to Controlling Factors

The relation of cruising speed to temperature given in Fig. 17 is for the fish acclimated to each test temperature before test. The general case for response at a given test temperature over all acclimation temperatures is given for the cruising speed of goldfish in Fig. 18 as a response surface (see Alderdice, 1971, for a general review).

The figure is to be viewed as a hillock rising to a peak at the point "S" with contours as shown by the curved lines. The contours are elliptical with

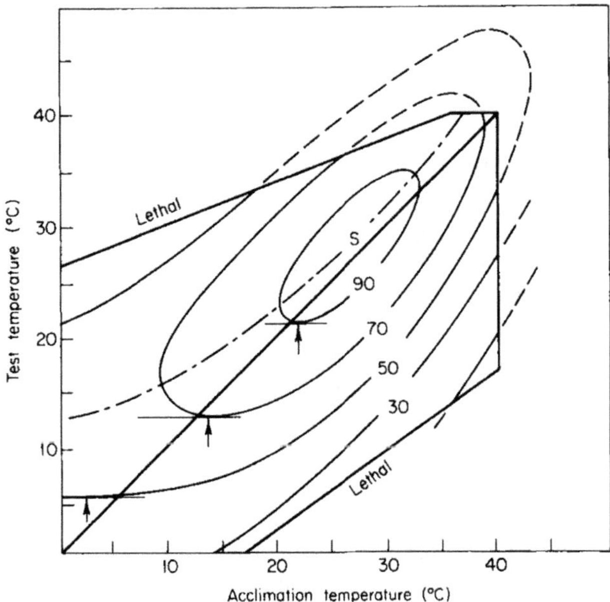

Fig. 18. The relation between acclimation temperature and test temperature as these affect the sustained swimming speed of 5-g goldfish within the zone of thermal tolerance. From Lindsey *et al.* (1970), their model B, and Fry *et al.* (1942). Numbers on isopleths indicate speed in feet per minute. For further explanation see text.

the major axis inclined at approximately 45° to the temperature axes. The major axes are joined by a broken line which represents the so-called ridge line, the path of most gradual ascent up the surface. In terms of adjustment to temperature the ridge line represents the combinations of test temperatures and acclimation temperatures which gives the least variation in swimming speed over the tolerance domain. The ridge line is above the 45° line (dotted) at low acclimation temperatures, indicating that performance is better at a temperature somewhat higher than the acclimation temperature when the acclimation temperature is low, perhaps an adaptation to favor activity in the spring warming period. The conditions for peak performance "S" come at a point where the ridge line, the acclimation, and the test temperature coincide, which may or may not have any significance.

The arrows impinging on the horizontal tangents extending to the test temperature axis indicate the conditions where performance is maximized for each test temperature concerned. These points come where test temperature is also the acclimation temperature. All in all, therefore, the process of thermal acclimation, at least as exemplified by this analysis of the response of the cruising speed of the goldfish, fits the organism to best meet the problem it faces at a given temperature. Similar studies of cruising speed on four other species (McCrimmon, 1949; Roots and Prosser, 1962; unpublished data of Ferguson, in Fry, 1964; Fry, 1967) indicate similar responses.

Accordingly, while it is most likely that the present skimpy data are overinterpreted in the paragraph above, it does seem clear that there are major and meaningful adjustments to temperature to make the organism able to perform more effectively, as indeed is the current general opinion.

However, it is not possible here to give a clear-cut analysis of the changes in the level of metabolism which bring about the change in activity, in spite of the number of contributions to the subject (e.g., see the review of Precht, 1968). The problem is that most shifts in metabolism of the whole organism observed with changing temperature have undoubtedly resulted from changes in random activity which have neither been observed nor controlled (see analysis in Fry and Hochachka, 1970). Except for the work of Kanungo and Prosser (1959), changes in active metabolism do not appear to have been followed in the course of thermal acclimation. Kanungo and Prosser, while finding values only about one-fourth of those reported by Basu (1959) and Kutty (1968a), got a shift such that up to 25°C active metabolism was higher for fish acclimated to 10°C than for those acclimated to 30°C. Above 25°C the positions were reversed. Both curves reached the same peak, about 200 ml/kg/hr, the curve for 10°C acclimation at 25°C, the other at 30°C. The two curves appear to be essentially parallel with a lateral shift when plotted on semilogarithmic paper. The authors themselves, by plotting logarithms of the rates on a logarithmic scale, suggested that the curves rotate, but they did not give the mathematical justification for their procedure.

It seems impossible to measure standard metabolism in fish at any but the acclimation temperature because of the stimulus brought about by temperature change which may not be reflected in overt movement and thus escape detection by current methods of accounting for departure from the standard state. Even changes in active metabolism will have to be viewed with the reservation that excitement metabolism may also enter in, and, in addition, that the cost of the various regulatory functions may, and probably does, vary with any departure from the acclimation temperature. Perhaps the problem will be seen more clearly after the consideration of the costs of regulation in Section V.

The rate and degree of adjustment to change of temperature as reflected in tissue metabolism and adjustments in the capacities of organ systems have not often been investigated but all the work done indicates appropriate changes in capacity to compensate for change in temperature. Thus, Prosser and his associates (Prosser and Fahri, 1965; Roots and Prosser, 1962) have found the activity of the nervous system of the goldfish to show almost complete compensation, i.e., to change almost to the same degree as the change in acclimation temperature. Four degrees' change in acclimation temperature brought about three degrees' change in cold block temperature for nervous activity. Compensatory changes in the function of the circulatory system are suggested by the work of Hart (1957), Das and Prosser (1967), and Jankowsky (1968). Smit (1967) showed temperature compensation in the digestive system, in the rate of secretion of pepsin and acid.

Because of the difficulties of interpretation (see, e.g., Peterson and Anderson, 1969b), the effect of acclimation temperature on the metabolic rate of various tissue slices, minces and breis will not be discussed. The question of cellular restructuring in relation to temperature adjustment will be dealt with by Hochachka and Somero in Chapter 2.

IV. LIMITING FACTORS

The limiting factors are first of all the metabolites, food, water, and the respiratory gases. Other identities operate as secondary limiting factors when they influence the rate of exchange of the metabolites between organism and environment.

The discussion of limiting factors will be largely confined to consideration of the effects of varying the concentration of the respiratory gases, oxygen and carbon dioxide. Basically a decrease in oxygen or an increase in carbon dioxide, over the ranges of concern here, operate in the same way. The supply of oxygen to the tissues is restricted.

The effect of a limiting factor is to restrict activity. Two examples of such restrictions are shown in Fig. 19. A feature common to both these examples, and indeed ordinarily to be found, is that the limiting factor (here oxygen) becomes operative at a relatively high value. Thus, in Fig. 19A, the ability to swim is first affected by oxygen concentration at about 6 mg/liter O_2 at $10^\circ C$

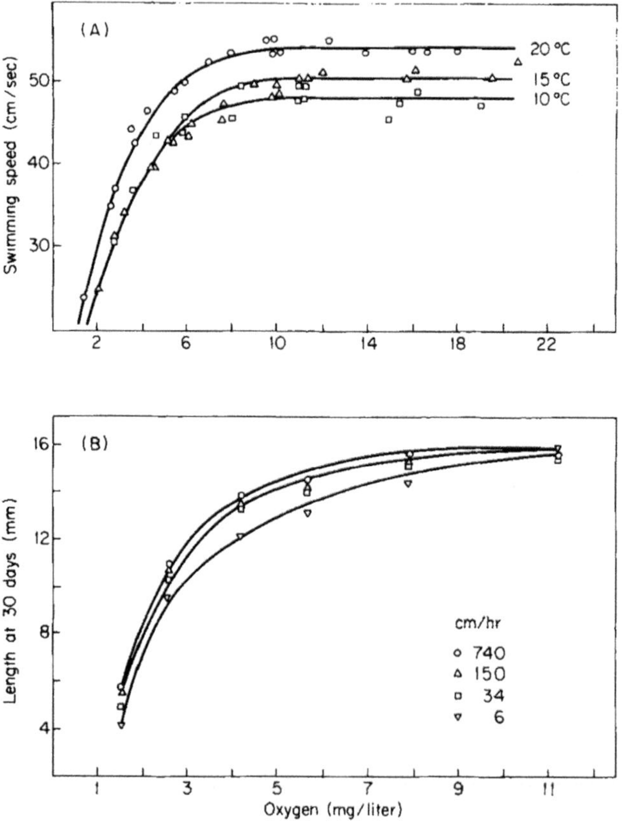

Fig. 19. (A) Swimming speed of underyearling coho salmon, *Oncorhynchus kisutch*, in relation to oxygen concentration. From Davis *et al.* (1963). (B) Effect of oxygen concentration and rate of percolation on the growth of embryos of steelhead trout, *Salmo gairdneri*, at 9.5°C. From Silver *et al.* (1963).

and 10 mg/liter at 20°C. The coho, a salmon, would of course be expected to be sensitive to relatively slight decreases in oxygen concentration, but the goldfish, which in contrast is expected to withstand low oxygen, surpasses another salmonid, the rainbow trout, only below 2.5 mg/liter (Fig. 22) and in this range probably does so because it can operate to a certain degree anaerobically (Kutty, 1968a) not because it can take up a great deal more oxygen (Basu, 1959).

Panel B not only shows an example of the operation of a limiting factor on another type of activity, namely, development, but emphasizes the problem of "supply." When the fish is in the egg, oxygen supply depends on three circumstances, one chemical—concentration, one physical—diffusion pressure, and one mechanical—rate of flow. A complete unit of oxygen supply has not yet been proposed. In the present section, as a compromise, the unit of

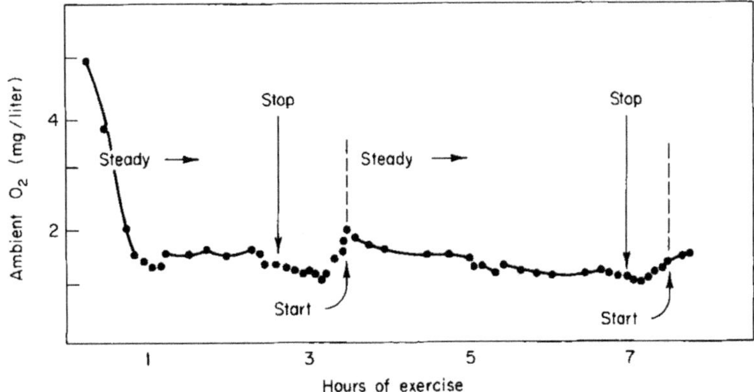

Fig. 20. Swimming response of an 18-cm goldfish to changing oxygen concentration while exposed to a water current of 60 cm/sec at 20°C. Arrows indicate times at which the fish stopped swimming and started again. From Kutty (1968b).

concentration will be used since the counter-current system by which the fish in general gains its oxygen when out of the egg seems to be more dependent on the mass of oxygen presented to the respiratory surface than on the partial pressure.

The response of swimming to the limiting effect of low oxygen is under the control of the central nervous system as Fig. 20 shows. In this example the fish was stimulated to swim steadily at a moderate speed (about one-half its maximum capacity for steady swimming, cf. Kutty, 1968a) and the oxygen allowed to fall gradually. After about 2½ hr the fish abruptly stopped swimming and fell back against the screen when the oxygen content fell a little below 2 mg/liter. The current in the chamber was maintained and the oxygen allowed to subside a little further for another half-hour. Meanwhile the fish remained on the screen. Then the oxygen content was raised again. Within about 5 min when the oxygen content had risen above the level at which swimming had stopped the fish was again breasting the current steadily and continued to do so for some 3 hr until the oxygen became critically low once more. Again the fish abruptly stopped swimming, and again it resumed swimming promptly when the oxygen content was slightly increased.

The indication from such an experiment as Kutty's above is that an organism may adjust to a limiting factor and restrict activity during the period when it is imposed. However, there can be circumstances where the organism becomes committed to a given resource when a given identity is not limiting and then a fluctuation produces limiting conditions. Daily fluctuations in the oxygen content of well-vegetated waters is a familiar case of this sort. Brook trout will not grow in an aquarium where the oxygen is restricted only for part of the daily cycle (Fig. 21).

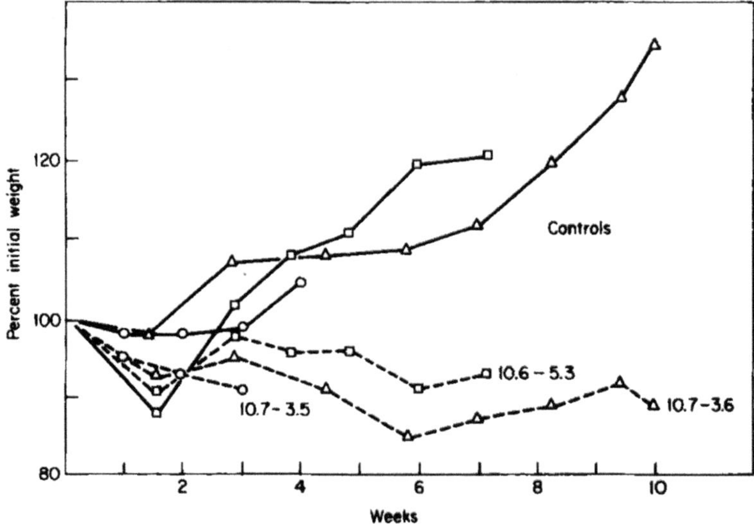

Fig. 21. Growth of yearling eastern brook trout, *Salvelinus fontinalis*, at constant high (solid lines) and various daily fluctuating levels of oxygen (broken lines). Numbers indicate upper and lower levels of oxygen in milligrams per liter. From Whitworth (1968).

The effect of small inert bodies, such as silt or pulp fines, which apparently operate by displacing an equivalent amount of water and the oxygen supply it contains and otherwise interfering with the respiratory flow, is a case of operation of a limiting factor of special interest to pollution biologists.

Some effects of limiting factors may be extremely obscure. For example, Kinne and Kinne (1962) observed a reduction in the rate of development of the desert pupfish, *Cyprinodon macularius*, in relation to increased salt content in the water in which they were incubated. These authors were able to correlate the change with the change in the saturation value of oxygen as influenced by the presence of the salt. There was apparently no other substantial effect of the widely differing salt content in the water in which the various samples were hatched.

A. Acclimation to Low Oxygen

A clear shift in the lethal level of oxygen in relation to acclimation level was shown by Shepard (1955) for the eastern brook trout, *Salvelinus fontinalis*, and for three warm-water species by Moss and Scott (1961). To explain the change, Shepard found an increase in the ability to extract oxygen from water when the concentration was low, as did Prosser *et al.* (1957) also, for goldfish. Similarly, MacLeod and Smith (1966) found that the hematocrit of the fathead minnow, *Pimephales promelas*, changed in response to lowered oxygen or increased content of pulp fiber. Thus changes have been found in the supply system, as is well known for mammals.

In the brook trout the relation between lethal and acclimation levels of oxygen was linear and can be expressed by the formula $y = 0.88 + 0.08x$, where y is the lethal level and x the acclimation level of oxygen, both being expressed in milligrams per liter. The lower limit of the formula is when $y = x$; the upper limit of Shepard's observations was air saturation. MacLeod and Smith also found that the response of the hematocrit was linear over the whole range of oxygen concentration they investigated and for concentrations of pulp fiber from zero to 800 mg/liter.

All adjustments, however, do not seem to be so simple. There appears to be a good deal of accommodation. Kutty (1968b) found no difference between goldfish acclimated to low oxygen and those acclimated to air saturation with respect to the speed at which they could be induced to swim at limiting levels of oxygen (Fig. 22), although Prosser *et al.* (1957) showed the potential for increased oxygen uptake. Furthermore, Kutty demonstrated (Fig. 23) that at a given swimming speed the oxygen consumption was reduced greatly in goldfish acclimated to low oxygen as compared with those acclimated to high oxygen. A similar reduction in the overt cost of running was found by Segrem and Hart (1967) for the white-footed mouse. Thus, assuming all these observations to be valid, it seems that acclimation to low oxygen involves both an enhancement of supply and a restraint in utilization. With regard to the reduction in consumption at a given speed, comparison of Figs. 1 and 23

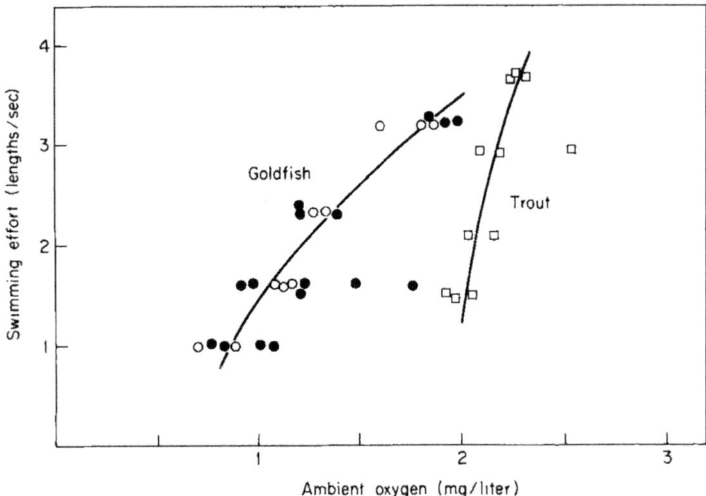

Fig. 22. Swimming effort of 18-cm goldfish at 20°C and 20-cm rainbow trout, *Salmo gairdneri*, at 15°C in relation to ambient oxygen. Open symbols denote fish acclimated to oxygen at air saturation, closed (goldfish only) acclimated to 15% air saturation. Both species acclimated to their respective test temperatures. From Kutty (1968b).

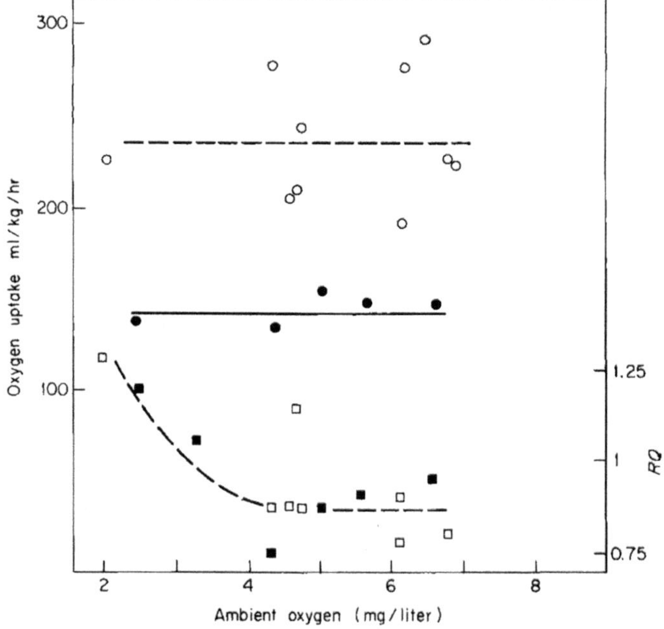

Fig. 23. Oxygen consumption (circles) and *RQ* determinations (squares) of goldfish acclimated to oxygen at air saturation (open symbols) and 15% air saturation (closed symbols). Fish approximately 18 cm long swimming continuously at 45 cm/sec. Determinations after the first two hours of swimming. From Kutty (1968a).

suggests, since the oxygen consumption of the fish acclimated to low oxygen lies on the lower boundary of the triangle relating oxygen consumption to speed, that when acclimated to low oxygen the fish is applying its energy only to the business at hand. Such a behavioral modification is also indicated in the response of fish to restricted food supply. Paloheimo and Dickie (1966b) showed the efficiency of food conversion was inversely related to the size of the ration, but it must be noted as Brett *et al.* (1969) state that the examples available to those authors did not have data for cases of severe restriction.

There are few data on the rate of acclimation to low oxygen. Shepard (1955) demonstrated that acclimation of the eastern brook trout, *Salvelinus fontinalis*, to a change of oxygen concentration was 95% complete in 100–200 hr at 10°C, being somewhat slower if the fish were in the dark and presumably thus not so active. The data of Moss and Scott (1961) support Shepard's findings.

It seems likely that a moderate imposition of a limiting factor does not involve the imposition of stress, although there will be restriction in activity. It is normal of course, as shown in Fig. 18, for air saturation to limit active metabolism at higher temperatures which otherwise are well within the normal range. Dahlberg *et al.* (1968) showed that while the growth rate of the largemouth bass was restricted in their experiments from an oxygen concentration of 8 mg/liter down, the food conversion ratio remained stable at least down to 4 mg/liter and perhaps 3 mg/liter. In this case, food consumption was progressively restricted throughout the total range of oxygen in question, presumably through some central control as in the case of swimming speed (Fig. 20). Their experiments were carried out at constant levels of oxygen in contrast to the fluctuating levels employed by Whitworth (1968) (Fig. 21).

B. Oxygen Concentration and Metabolic Rate

There is an extensive early literature on this subject which can now be considered to be of only historic interest and which can largely be found through the reviews of Tang (1933) and von Ledebur (1939). The confusion in the earlier literature lies in a lack of distinction between the various levels of metabolism. The point of view expressed in the present discussion has its origins in the work of Van Dam (1938) on resting metabolism and of Lindroth (1940), who appreciated that decreased oxygen limited active metabolism. Lindroth (1942) stated the concept designated below as the "level of no excess activity." Figure 24 shows the typical response of the standard and active

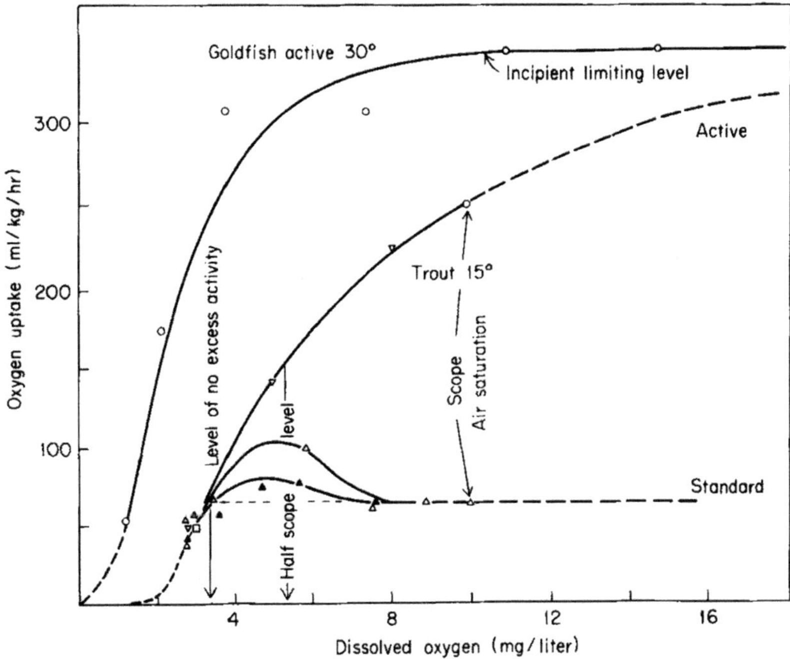

Fig. 24. Active metabolism of the goldfish, and active and standard metabolism of the eastern brook trout, *Salvelinus fontinalis*, in relation to oxygen concentration. Closed symbols, fish acclimated to test level of oxygen; open, acclimated to air saturation. From Basu (1959) (○), Beamish (1964c) (△▲), Graham (1949) (▽), and Job (1955) (□).

metabolic rates to oxygen concentration. Standard metabolism in the brook trout is relatively unaffected by oxygen concentration until the level of oxygen drops to approximately 50% air saturation. Below that point there is first an increase, as was well-demonstrated by Van Dam (1938), which it is presumed takes care of the increased needs of ventilation. The need for increased ventilation seems to vary with the species and with the temperature. In the goldfish (Beamish, 1964b) such cost of respiration can hardly be detected at 10°C when the standard oxygen consumption is only 15 ml/kg/hr for a fish of approximately 100 g, while there is an approximate doubling of the standard rate at approximately 40% air saturation if the goldfish are at 20°C. As the oxygen concentration drops still further the fish can finally no longer increase its oxygen consumption, even with heavy breathing, and the total rate of oxygen consumption then falls progressively with further lowering of the oxygen concentration. This is not to say that ventilation may not still increase, but the fish then has to resort to anaerobic support. In his study of respiratory efficiency in relation to respiratory flow, Saunders (1962) took advantage of the increased ventilation induced by lowering the ambient oxygen to stimulate the higher rates of ventilation. Acclimation to low oxygen possibly reduces the cost of ventilation, as is indicated by the solid triangles in Fig. 24.

In contrast to the course of standard metabolism in relation to oxygen concentration, which is relatively unaffected, active metabolism may be strongly influenced by oxygen concentration at all levels up to air saturation and even higher. The experimenter often must artificially increase the oxygen content of the water if he wishes to obtain the full extent of active metabolism (e.g., Fig. 17). In Fig. 24 the curve for the brook trout was carried only as far as air saturation but that for the goldfish was continued to higher levels of oxygen concentration. Both are for fish acclimated to air-saturated water. As mentioned above, acclimation to lower oxygen would have displaced the curves to the left (e.g., Prosser et al., 1957; Shepard, 1955). An earlier speculation of Fry (1947) that the maximum would be reduced still has not been tested, in spite of the apparent confirmation shown by Prosser *et al.*, since it is probable that they did not increase the swimming speed to the limit.

The terminology chosen here to describe the relations of metabolism to a limiting factor is indicated in Fig. 24. As was the usage in considering controlling factors, the scope for activity is taken as the difference between active and standard metabolism. Two restrictions of scope are designated: the "half-scope concentration"—that concentration of oxygen where the active metabolism is reduced to the point where the scope is one-half that at air saturation—and the "level of no excess activity"—the point where the active metabolism is reduced to the standard level. The half-scope level (Basu, 1959) is simply a convenient arbitrary point to take in discussing the restrictive effects of a limiting factor. The level of no excess activity approximates the lethal level. It will be noted in the diagram that both these points have been estimated from an extrapolation of the line for standard metabolism determined at concentrations of oxygen higher than those at which the cost of respiration increases. In the case of the half-scope value, it can be taken that the cost of respiration is often absorbed in the general cost of physical activity, at least within the limits of accuracy of the index, since swimming fish frequently, and perhaps ordinarily, passively irrigate their gills by their forward movement, except of course in start and stop activity. Swimming with the mouth open will, of course, contribute to drag so that some cost is still there, but the efficiency of irrigation is greatly increased since there is no longer the need to accelerate each mouthful of water to take it in nor to accelerate it again to expel it from the epibranchial cavity on exhalation. Only the friction to pass water steadily through the branchial sieve remains to be overcome in the work of respiration (see C. E. Brown and Muir, 1970, for a quantitative analysis). Similarly, it can be expected that the activity of the swimming muscles will promote the circulation and reduce the work of the heart. The position of the level of no excess activity is decidedly more arbitrary, for at that point the fish is breathing heavily and the cost of respiration must be maximal. Extrapolation in this case really relies on an overestimate in the determination of standard metabolism to compensate for a change in the cost of respiration and on the possibility of some anaerobic support for activity.

The final term indicated in Fig. 24 is the "incipient limiting level" shown for the goldfish, that is, the point where a further reduction in oxygen begins to restrict the active metabolic rate. The point is of little ecological interest as a datum since it does not necessarily appear under natural conditions, being often above air saturation. Moreover, it is not a precisely defined point.

C. Combinations of Oxygen and Carbon Dioxide

The effects of low oxygen and high carbon dioxide show the typical consecutive interaction of limiting factors (Fig. 25A). Here, to take the dashed line for the effect of 18 mg/liter CO_2 at 20 hr of acclimation, oxygen below 4 mg/liter is limiting and no effect is seen from the additional carbon dioxide present, while the effect of carbon dioxide as indicated by the horizontal position of the curve is complete at about 10 mg/liter O_2. At 61 mg/liter CO_2, carbon dioxide exerts a graduated limiting effect over the whole region of the data up to 15 mg/liter O_2. Thus, there is a wide transition phase between the operation of oxygen as the preponderant limiting factor to that of carbon dioxide. The data are imperfect here, and the effect is better shown in Fig. 28. There is, of course, a substantial early literature on the precise nature of the operation of limiting factors which is to a large degree nowadays irrelevant but which is admirably discussed in Booij and Wolvekamp (1944). There seems to be no point in dealing with the sharpness of the transition phase, which was the subject of most of the early work, nor indeed in the complete preponderance of one limiting factor over another. The best present approach appears to be empirical description of the effects of limiting factors on active metabolism.

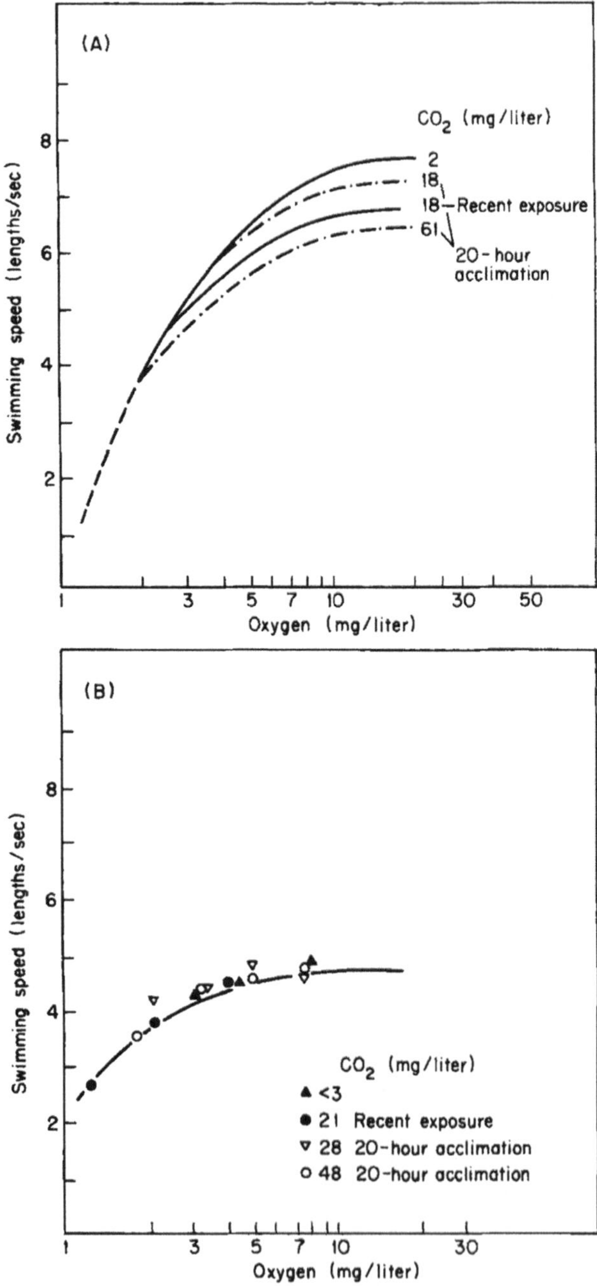

Fig. 25. Swimming speed of (A) coho salmon, *Oncorhynchus kisutch*, at 20°C and (B) largemouth bass, *Micropterus salmoides*, at 25°C. From Dahlberg *et al.* (1968). Fish approximately 8 cm long.

The major point of concern to the fisheries biologist with respect to the respiratory gases is that under natural conditions oxygen lack is a much more likely limiting factor than carbon dioxide excess, particularly since it is only under anaerobic conditions that free carbon dioxide can ordinarily reach major levels. It is only under rather special conditions that carbon dioxide becomes a limiting factor. The commonest such condition is when fish are transported (e.g., see review of Fry and Norris, 1962).

Different species display different sensitivities to carbon dioxide. The largemouth bass (Fig. 25B) showed no effect on exposure to 48 mg/liter free CO_2.

A final point to be noted in Fig. 25A is that there may be extensive acclimation to carbon dioxide in a comparatively short time, as is shown by the difference between the curve for acute exposure to 18 mg/liter free CO_2 and for 20-hr acclimation to that level. Similarly, Saunders (1962) found that the efficiency with which oxygen is taken up at the gills, which is reduced by the presence of increased carbon dioxide, is recovered in a few hours of continuous exposure to moderate levels of CO_2. Again, the data of Lloyd and White (1967) suggest that the change in blood bicarbonate (Lloyd and Jordan, 1964) in response to increased CO_2 is largely complete in 24 hr in rainbow trout at $12^\circ - 16^\circ C$.

Beamish (1964c) and Basu (1959) have carried out the most extensive researches to date on the interaction of various levels of oxygen and carbon dioxide on standard and active metabolic rates. Beamish (Fig. 26) showed that standard metabolism was essentially uninfluenced by the level of free carbon dioxide until the total uptake of oxygen was reduced below the requirements for standard metabolism. On the other hand, as Basu found, the active metabolic rate at any given level of oxygen concentration is progressively reduced by increase in carbon dioxide concentration (Figs. 26 and 27). There is, however, a great limitation in Basu's data in that his results are for acute exposure of animals acclimated to low CO_2. No one has yet, apparently, measured the active metabolism of fish acclimated to high CO_2.

The essential feature of Basu's results is that the response to increasing carbon dioxide is an exponential decrease in active oxygen consumption. The same proportionate effect was found at all oxygen concentrations down to a low value, which in the case of the carp shown is 12.5% air saturation; below this the rate of reduction of oxygen consumption with increasing carbon dioxide sharply increased. The increased slope shown by the response to CO_2 at low oxygen in the carp was found also in the bullhead, *Ameiurus nebulosus*, and in the goldfish, *Carassius auratus*. Basu had no explanation for this phenomenon but did show the change in response was statistically significant, whereas there was no significant difference in slopes at the various higher concentrations of oxygen.

A special feature in the data for the eastern brook trout is shown by the points enclosed in the dashed ellipse. Basu was unable to reduce the oxygen

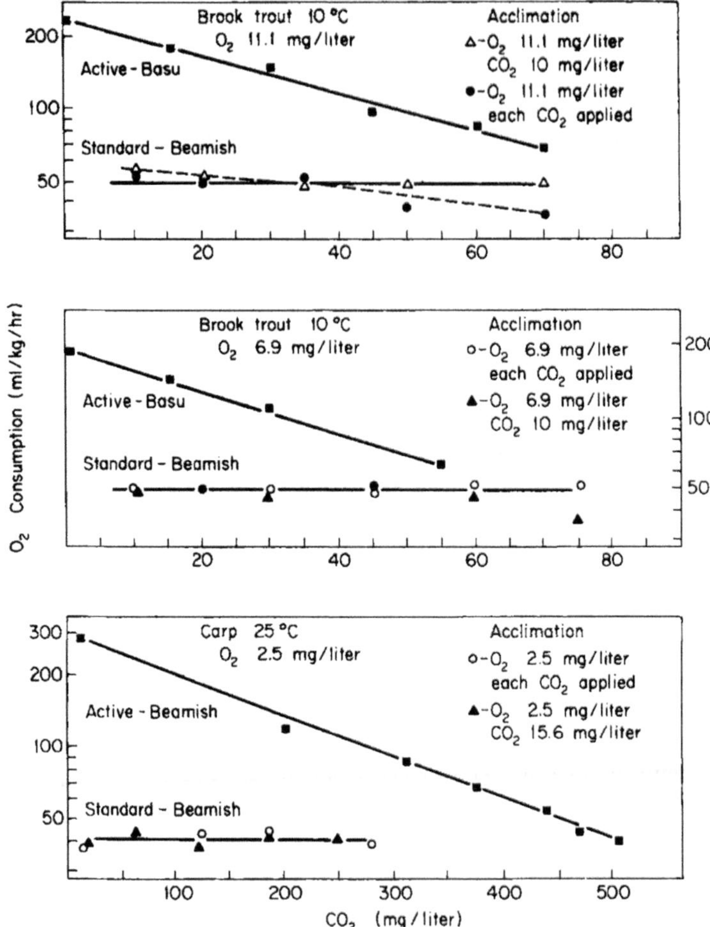

Fig. 26. Effect of various concentrations of oxygen and carbon dioxide on active and standard metabolism. From Basu (1959) and Beamish (1964c). Measurements with active metabolism made only with fish acclimated to air-saturated water.

consumption of this species below approximately 70 mg/kg hr O_2 under the conditions of the experiment and concluded that the species was able to transport that much oxygen by serum transport alone at the oxygen level indicated.

Figures 27C and D show the response to temperature. The characteristic shown here, namely, that the effect of a given concentration of carbon dioxide was least at the highest temperature investigated for a given species, is true of goldfish and the bullhead also (Basu, 1959).

Basu was able, on the assumption there would be no effect of carbon dioxide on standard metabolism, which Beamish (1964c) later demonstrated, to show that the curves for respiratory sensitivity as found, for example, by

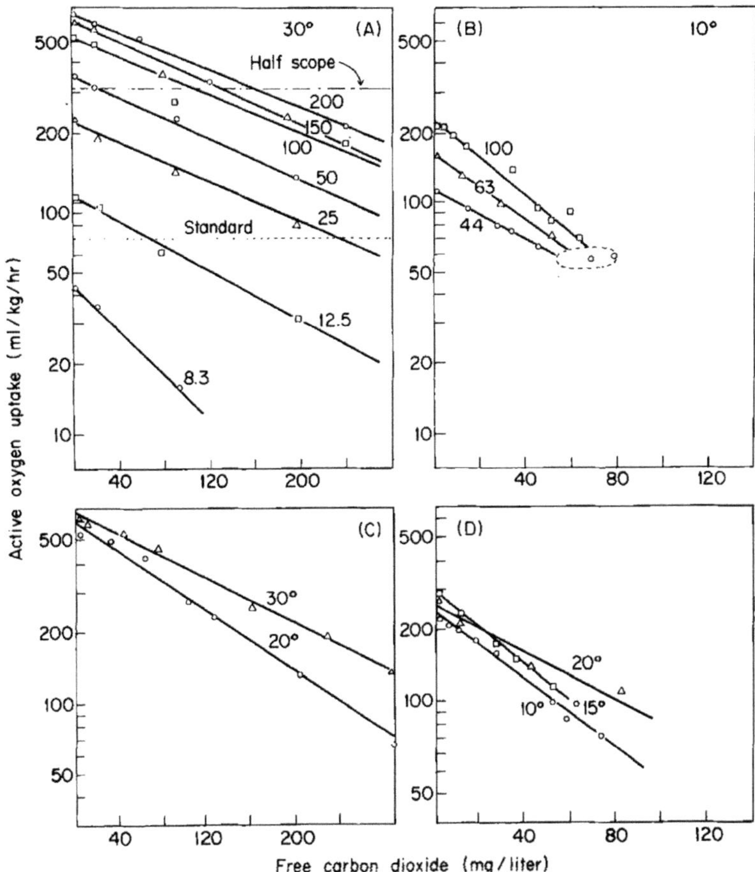

Fig. 27. Effect of various concentrations of oxygen and carbon dioxide on active metabolism of brook trout, *Salvelinus fontinalis* (B,D) and carp, *Cyprinus carpio* (A,C). From Basu (1959) and Beamish (1964a).

E. C. Black *et al.* (1954) could be calculated from the data on oxygen consumption with a high degree of approximation. Curves based on his calculations of the effects of the interaction between oxygen and carbon dioxide on various levels of activity are shown in Fig. 28.

The various isopleths in Fig. 28 are constructed by taking points at which the rate of oxygen consumption had been reduced to the same given level from such curves as are in Figs. 27A and B. Values for respiration at the half-scope level and the level of no excess activity based on Beamish's (1964a) more recent determinations of standard metabolism are included in Fig. 27A. Points determined from the intersection of these lines are also plotted in Fig. 28 as indicated.

The analysis presented above glosses over a number of points. First, as already mentioned, the active rates are for fish not previously acclimated to

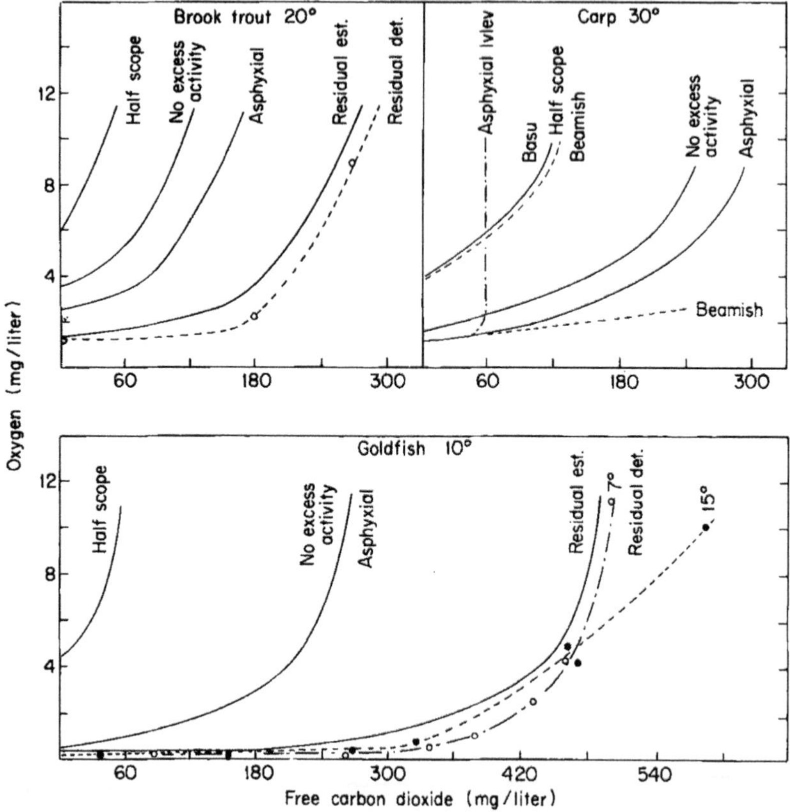

Fig. 28. Determined and calculated sensitivity of fish to various combinations of oxygen and carbon dioxide. From Basu (1959).

the test levels of carbon dioxide, except for the lowest. Second, Basu's calculations did not take into account any change in the cost of irrigation of the gills (Beamish, 1964c) or transport of oxygen in the blood. Finally, Basu's work was carried out with water of 270 mg/liter $CaCO_3$ hardness. Ivlev (1938), using water with a bicarbonate alkalinity of about 40 mg/liter, found the asphyxial level of CO_2 to be independent of oxygen concentration above 60 mg/liter CO_2, whereas Basu's respiratory data indicate a continuing interaction between oxygen and carbon dioxide up at least to air saturation. It is probable that the difference is due to the difference in water hardness between the two tests. Lloyd and Jordan (1964) reported an interaction between high CO_2 and low pH on the time of survival of rainbow trout, *Salmo gairdneri*, such that, with oxygen at air saturation, approximately 20 mg/liter free CO_2 was fatal to rainbow trout at approximately pH 5.5. In neutral water, trout could be expected to survive continuous exposure to at least twice that concentration of free CO_2. A free carbon dioxide concentration of 60 mg/liter may possibly therefore represent a lethal pH for carp in soft water.

D. Interaction of Limiting and Controlling Factors

Figure 18, as pointed out at the time it was introduced, could not be completely discussed from the point of view of controlling factors alone. In air-saturated water, the normal environment for the sockeye, the limiting effect of oxygen intervened above 15°C to suppress the full potentiality for active metabolism permitted by temperature. Thus the effect is for the sockeye to have an optimum for activity at 15°C, much below its lethal temperature. Oxygen at air saturation is often limiting for fish as the example shows. Such a temperature optimum is frequently called a "conditioned" optimum, e.g., an optimum temperature conditioned by the existing level of oxygen.

Figure 29 shows the generalized response of the eastern brook trout to various combinations of oxygen and temperature. Unfortunately, there is not sufficient information to provide a strictly quantitative picture. In particular, there are no data for active metabolism in relation to acclimation to oxygen except those of Shepard (1955). Figure 29A shows the typical effect of temperature on the rate of oxygen uptake when oxygen is limiting. At a higher

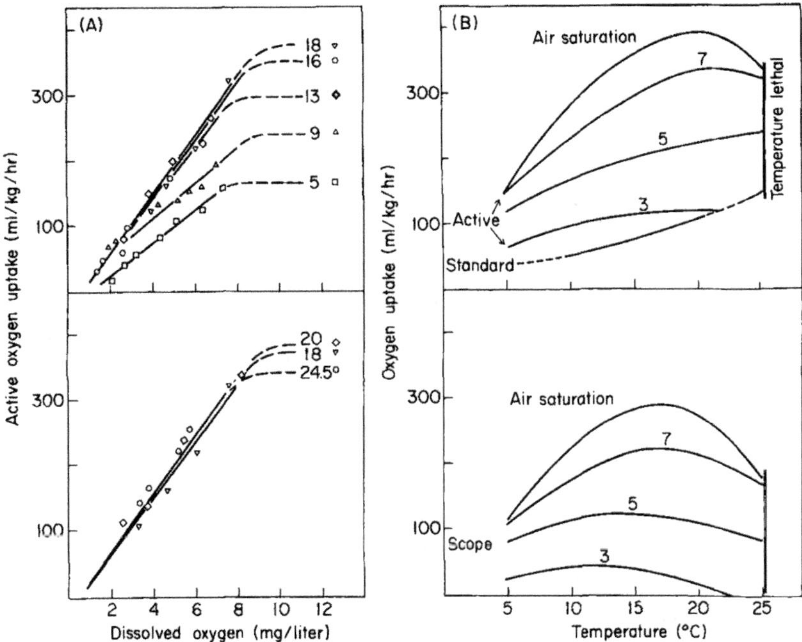

Fig. 29. (A) Limiting effect of oxygen on active respiration at various temperatures. (B) The interaction between oxygen concentration and temperature on the metabolic rate of the eastern brook trout, *Salvelinus fontinalis*. Based on the data of Basu (1959), Beamish and Mookherjii (1964), and Graham (1949). [Note that ppm in Fig. 9, Graham (1949) should have been ml/liter.] Numbers in A indicate temperature; in B, mg/liter O_2.

temperature more oxygen can be taken up at a given concentration of oxygen over the whole range, whether oxygen is limiting or not, although in the brook trout that effect is not seen above 133°C. In part the difference is the result of the imperfect expression of supply, as mentioned earlier, but also there is a change in regulation. There may be more irrigation and flow will be higher for a given pressure head at a higher water temperature, since the gill is essentially a capillary sieve and flow through it at a given pressure depends on viscosity. Utilization may also be more complete with increasing temperature because of an increase in blood flow, more rapid diffusion, and more rapid progress toward equilibration in the hemoglobin.

An important consequence of the interaction of oxygen concentration and temperature is that the optimum temperature in terms of scope for activity cannot be predicted from the lethal temperature. The relation of maximum scope to lethal temperature depends on whether and where the normal oxygen content begins to exert a limiting effect. Species with similar lethal temperatures may have quite different optima as conditioned by oxygen at air saturation. While the data must now be considered only semiquantitative, Fry (1957) showed that various salmonids have similar lethal temperatures but while two species of trout (*Salmo*) show increasing active oxygen consumption right up to the lethal temperature, two chars (*Salvelinus*) have their active oxygen consumption limited by air saturation above about 15°C.

Consideration of the previous section indicates how a second limiting factor, say, carbon dioxide, can be added to the interaction with temperature. Basu (1959) gave an example of such a plot in his Fig. 9.

V. MASKING FACTORS

The masking factors and the directive factors, discussed below, deal with the channeling of the energy available to the organism, which in the broad sense is all applied to organic regulation.

As an organized segment of the universe it goes without saying that all the independence an organism achieves comes through the interaction of the various identities in an appropriate matrix, itself fashioned from the environment or a pattern in it. Life is governed by all the laws of nature, but like a good corporation flourishes by playing one against another to its own advantage. Also, like the corporation it must pay its lawyer.

Organic regulation can be broadly classified as mechanical, physiological, and behavioral, the latter term being taken to include all the manifestations of the central nervous system, some of which as we know them ourselves may be internal but which in fish, if solely so, would not be accessible to us. Structure apparently always enters into regulation and the least costly regulation is achieved by some structural isolation which, after the investment has been made in ancestry and individual development, calls for next to no cost for operation and maintenance.

The present section will be brief since physiological regulation is the main subject of most of the treatise and will be confined to a summary of some recent work on the cost of regulation of the body fluids in the rainbow trout and an outline of our knowledge of the regulation of the body temperature in the tunas and the lamnid sharks. The latter is taken as an example of a regulation which is brought about almost entirely by the appropriate development of form.

A. Cost of Ion-Osmoregulation

The physiological details of water economy and ion exchange are now becoming clear (see Chapters 1–3, Volume I) but we still have only fragmentary information concerning the metabolic cost of such regulations. Various workers (e.g., Keys, 1931; Leiner, 1938; Veselov, cited by V. S. Black, 1951; Hickman, 1959; Job, 1959) have noted differences in the routine metabolic rate of fish in waters of various salinities. Recent preliminary studies by Rao (1968) and Farmer and Beamish (1969) of the relation between activity, metabolic rate, and the salinity of the medium appear to be the only ones to offer a quantitative estimate of the cost of regulation of the body fluids in fish. Rao's work is the basis of the present section.

Rao measured the metabolic rate of the rainbow trout, *Salmo gairdneri*, at rest and at various swimming speeds over a modest size range (approximately 140–120 g) in freshwater and various dilutions of seawater. He measured standard metabolism in Fry's apparatus and active metabolism in Blažka's chamber (Fig. 3). He acclimated his subjects to the test conditions of temperature and salinity, taking cognizance of the observation of Conte and Wagner (1965) that this species has a seasonal cycle in its tolerance to seawater. Thus, he worked with high salinities in late summer and autumn. However, he maintained his fish under a constant 12-hr light period.

Rao's data calculated for a 100-g fish are given in Table III. The boldface numbers are his observations, those in italics are the differences between the metabolic rate in the respective medium and that in 7.5‰ salinity, which is approximately isosmotic with fish blood. Rao obtained the result to be expected from the work of his predecessors (e.g., Job, 1959). The metabolic rate for a given level of activity was least in an isosmotic dilution of seawater. Further, he found that the cost of ionosmoregulation was proportional to the metabolic rate and hence presumably a function of the respiratory flow (Fig. 30). Thus, isolation plays a large part in osmotic regulation, the general body surface being largely impermeable to water, as was already considered to be the case. Another finding of considerable ecological significance was that the cost of regulation is little affected by temperature. Indeed, if the solubility of oxygen in water is taken into account and with the assumption that the efficiency of extraction is constant, then the cost of regulation bears the same relation to respiratory flow at both temperatures investigated. Further, the increase in cost of regulation in seawater is not proportional to the increase in osmotic gradient.

Table III Cost of Ion-Osmoregulation in 100-g Rainbow Trout, *Salmo gairdneri*, in Relation to Swimming Speed and Salinity[a,b]

	Oxygen uptake (ml/kg/hr)			
		Excess over uptake at 7.5‰		
Speed (cm/sec)	Uptake at 7.5‰	FW	15‰	30‰
			5°	
0	**38**	*2*	*4*	*14*
18.5	**62**	*11*	*12*	*20*
45.1	**94**	*29*	*32*	*43*
57.5	**123**	*36*	*50*	*65*
Max. speed	**186**	*62*	*66*	*92*
			15°	
0	**66**	*13*	*11*	*23*
18.5	**92**	*26*	*28*	*41*
45.1	**159**	*36*	*29*	*62*
72.7	**246**	*59*	*62*	*87*
Max. speed	**340**	*69*	*78*	*97*

[a]As indicated by excess in oxygen consumption at a given salinity over oxygen consumption at 7.5‰ salinity (approximately isosmotic).
[b]Based on Rao (1968), his Table 2.

The metabolic data confirm the findings of Houston (1959) and Gordon (1963) that there is a decrease in the permeability of the exposed membranes as an adjustive response to increased salinity.

The second major point of ecological interest in Rao's work is that under the conditions of his tests the systems for uptake and transport of oxygen could handle the cost of ion-osmoregulation in addition to the cost of physical activity; thus, no penalty with regard to the ability to physically compete was imposed at air saturation by the increased regulatory load. Under his circumstances oxygen was not limiting, and each organ system had full scope to carry on its appropriate activity as required and as permitted by the controlling factors.

We can only speculate concerning the result if oxygen were limiting, but such speculation at least leads to an interesting hypothesis. Figure 31 suggests that as oxygen becomes limiting the cost of internal regulation must compete with the scope for activity. In consequence, with increasing limiting conditions the scope for activity can be expected to be reduced more quickly than

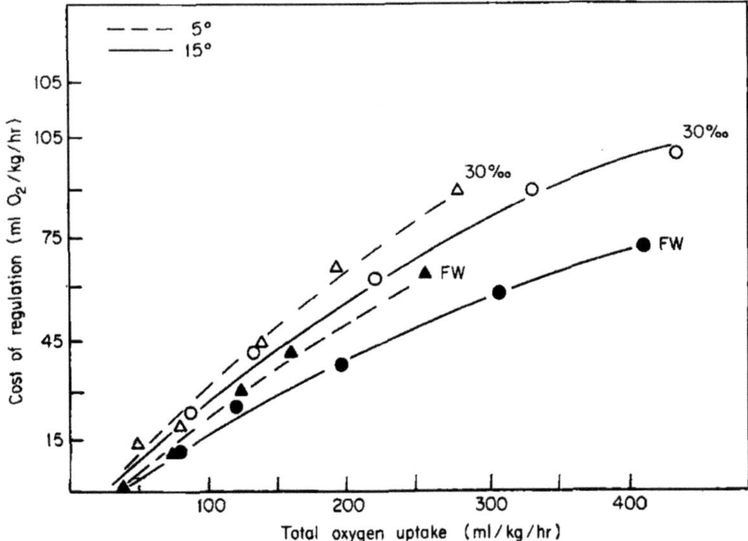

Fig. 30. Relation between total metabolism and the cost of ion-osmoregulation in fresh water and 30‰ salinity for 100-g rainbow trout. From Rao (1968).

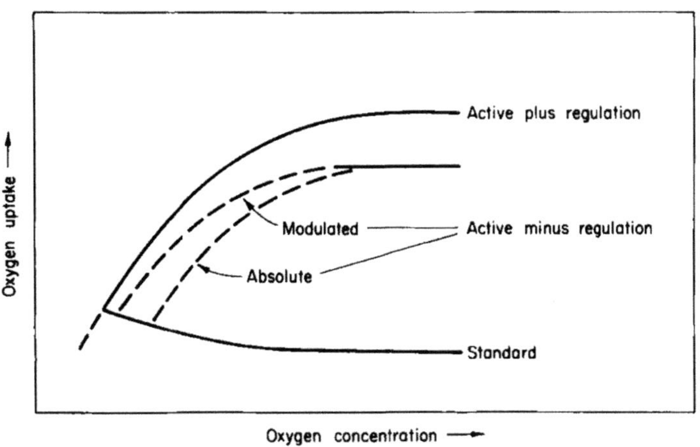

Fig. 31. Hypothetical effects of cost of physiological regulation on scope for activity in the presence of a limiting factor. See text for explanation.

the metabolic rate. Two courses for such a decrease are suggested in Fig. 31 labeled "absolute" and "modulated," respectively. The absolute curve postulates that all regulatory processes will have precedence over external activity; the modulated curve postulates that activity will compete with regulation so that regulation is not perfect and the internal condition can be allowed to drift within limits toward the external one. In the rainbow trout (Rao, 1969) there appears to be such modulation in osmotic concentration, but there is no profit at present in carrying the speculation further.

Since Rao has shown that the cost of ion-osmoregulation can be added to the scope for activity it can be presumed that other costs may be similarly added, at least up to some limit of the oxygen supply system we have not yet determined. Such an important cost is the cost of assimilation, and Fry's suggestion (1957) that the rate of oxygen consumption might be the limit to growth, which has been questioned by Swift (1964), needs examination. Measurements of active metabolism have been made till now with fasted fish so that the cost of assimilation has been largely removed from them. Again experiment needs to catch up with speculation.

Small rainbow trout (approximately 50 g) are difficult to maintain in seawater and probably do not fully adjust. Rao (1969) suggested that when the capacity to regulate is overreached then metabolism (irrigation) may be restricted. A selection of his data are plotted in Fig. 32. In the lowermost series of curves where the fish are swimming slowly (about 1 length per second for a 100-g fish), the cost of regulation is proportional to the metabolic rate up to a salinity of 15‰ (about halfstrength seawater) as shown by the parallel course of the log weight-log metabolism lines. At the highest salinities (30‰) the cost increases for smaller fish. The convergence of the lines at the upper ends may be taken as an artifact. It is probable that if a range of large fish had been examined the whole line for 30‰ salinity would have been a concave curve, with the slope at the upper weights being parallel to the slopes found at lower salinities.

In the next series of curves the fish were forced to swim about four times as fast. All sizes still seem capable of regulation up to 15‰ salinity, and the same relative increased cost of regulation for small fish is shown at 22.5‰ as was found at the lower speed. However, there is a decided change at 30‰ so that now the metabolism of the smaller fish is depressed. The same picture is found in the metabolism at maximum speed where there is also a suggestion that metabolism is beginning to be depressed in the smaller fish at 22.5‰.

Rao presented further data for standard metabolism and at an intermediate swimming speed which are concordant with the series shown. At 5°C he found little if any evidence of inability to regulate in the small fish within a similar size range but did find a relatively higher metabolic rate for standard metabolism in small fish at 30‰ salinity.

Unfortunately, Rao had to stop his investigations at this point so that an analysis of the response is not available. In particular, the osmotic pressure of the blood in the small fish is not known. Rao (1969) found a significant increase in the osmotic pressure of swimming fish weighing 100 g, but he did not examine either smaller or larger specimens. There may then be a reduction in cost of regulation because the fish allows the osmotic gradient to be reduced, although it would take a major departure in serum values from those found in 100-g fish for such an effect to account for the changes in

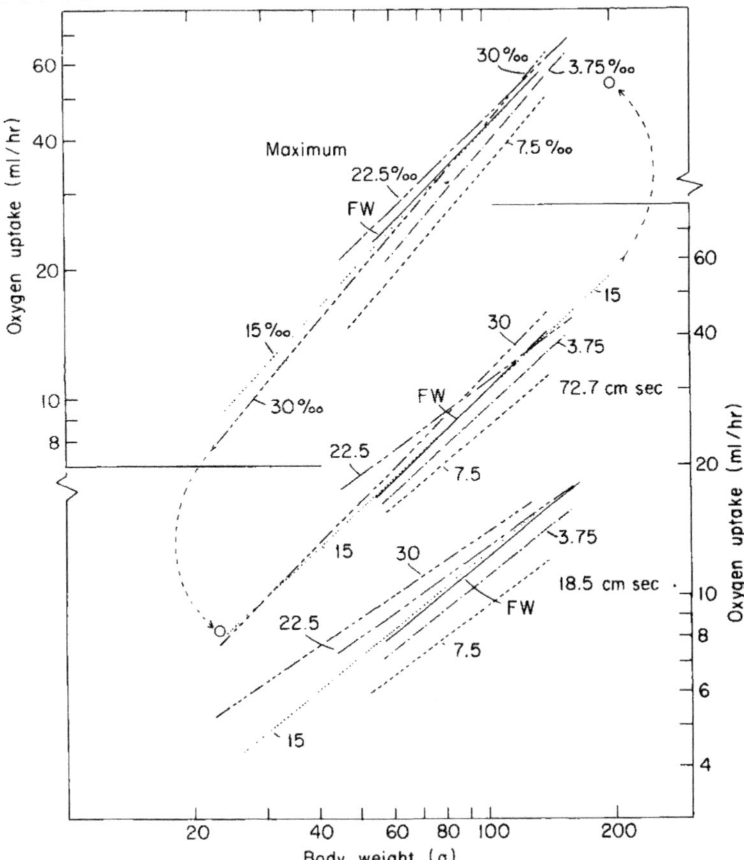

Fig. 32. Metaholic rate–body weight relation in rainbow trout, *Salmo gairdneri*, as related to swimming speed and salinity at 15°C. From Rao (1967, 1968). Note that the scale for the ordinate is displaced upward for curves relating to maximum swimming speed. The curved arrows give the positions of the indicated ends of the 30‰ and 15‰ curves on the respective scales. Numbers indicate salinity in parts per thousand; FW is fresh water.

metabolic rate. Another possibility is that the fish reduces its irrigation or perhaps increases mucus secretion at the gills. These various possibilities, however, are simply speculation.

An important consideration is that rainbow trout have little capacity for anaerobic metabolism (Kutty, 1968a). However, if the oxygen consumption data of Rao (1968) and Kutty (1968a) for this species are compared it will be found that Rao's data for fresh water are somewhat higher than Kutty's. Since he did not extend his readings at any one speed over as long a time as did Kutty, it is probable that he did not ordinarily find the minimum metabolic rate at a given swimming speed. Accordingly, part of the result of increasing ion-osmotic stress could also be an adjustment in behavior to conserve energy while swimming at a given speed, as appeared to be the case in the goldfish acclimated to reduced oxygen (Fig. 23). In any event, it appears likely that activity is accommodated to the limits of regulation in this activity as it may be in others.

B. Thermoregulation in Fish

In general the body temperature of fish is slightly above the ambient temperature, the difference being of the order of 0.5°C (Nicholls, 1931). Such a difference can be explained by the use of gills for respiration. The countercurrent system of exchange in the gill assures that the temperature of the blood will be reduced almost to that of the water on every circuit. The potential heat capacity of chemical transport is of the order of 0.5 cal/ml; thus, an excess temperature of the order of 0.5°C can be gained each circuit. Consequently, if the heat is also dissipated each circuit, the excess temperature at the site of metabolism will be of the order of 0.5°C. Lindsey (1968) has found excess deep muscle temperatures up to 2.6°C in very large fish (Fig. 34A). Since arteries and veins tend to run parallel in tissues, there is the probability that there may be some incidental local conservation of heat by countercurrent exchange, which can account for the excess temperatures these authors reported. The highest excess temperature found by Lindsey was in white muscle of fish which had recently struggled. Probably the anaerobic activity is not immediately balanced by increased circulation to drain away the heat. There do not appear to be any body temperature data for predominantly air-breathing fish.

Two groups of fish, the tunas (various authors to Carey and Teal, 1969b) and the lamnid sharks (Carey and Teal, 1969a), have deep muscle temperatures which may be greatly in excess of the water temperature (Fig. 33). These fish have retia mirabilia which in particular supply the red lateral muscles from the lateral artery and drain them through the lateral vein. These retia conserve the metabolic heat of the muscles they serve, which are those continuously active. The remainder of the muscle mass is presumably heated by conduction from the red muscle (Fig. 34B), although in some species there are retia associated with the dorsal and visceral as well as lateral blood vessels. The retia were figured by Kishinouye (1923) who also surmised their function. Aside from a diagram given by Carey and Teal (1966) there appears to be no recent description or specific detail with respect to the distribution from any one rete to a given muscle.

The data in Fig. 33 are for very large fish, and in these the muscle temperature can be extremely stable with respect to ambient temperature. However, when even these large fish become active their body temperature drops, presumably because the main body mass is supplied by blood from the dorsal aorta. Accordingly, the thermoregulation of these fish appears to be achieved by thermoconservation rather than by thermogenesis and thus does not call for the expenditure of extra metabolism. There is, of course, regulation, but the

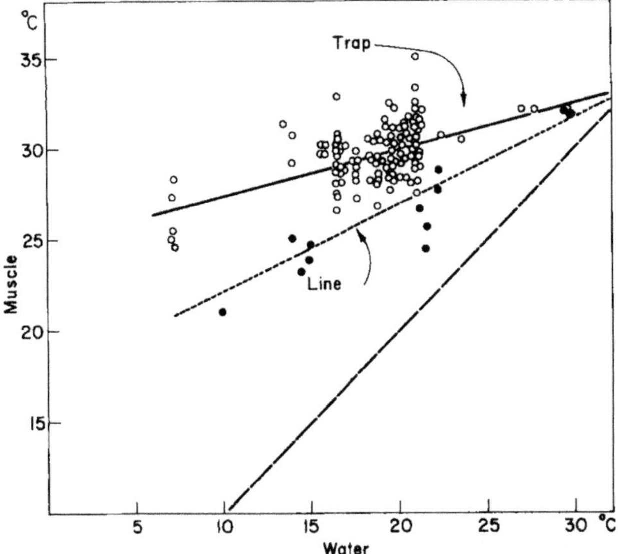

Fig. 33. Body temperature regulation in bluefin tuna, *Thunnus thynnus*. Data of Carey and Teal (1969b). From Fry and Hochachka (1970). Trap-caught fish were killed without a struggle. Line-caught fish were played for some time before landing.

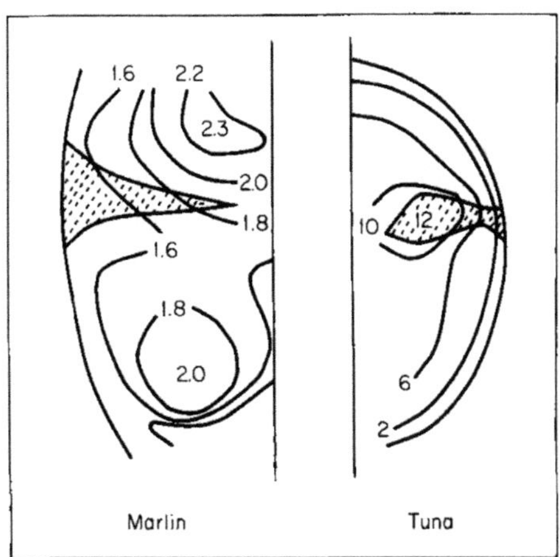

Fig. 34. Excess temperature profiles (°C) in muscles of newly landed bluefin tuna, *Thunnus thynnus* (Carey and Teal, 1969b) and marlin, *Makaira mitsukurii* (Lindsey, 1968). The marlin had been played for some time, the bluefin was shot in a trap. Hatched areas are red muscle. Note centers of high temperature are in the white muscle in the marlin and in the red in the tuna.

regulation is best expressed by the lowering of the differential in warm water, presumably by a relaxation of the countercurrent conservation system. In smaller fish, regulation is not as efficient as in the large tuna and the excess muscle temperature declines as the ambient temperature decreases although a substantial differential may still be maintained, e.g., 8°C at 20°C in the skipjack, *Katsuwonis pelamis* (Barrett and Hester, 1964).

A point of great interest on which there is at present virtually no information is whether the brains of these fishes are maintained at the temperature found in their muscles. Stevens and Fry (1971) found an excess temperature of about 4.5°C in the brain of skipjack in a sample of 20 fish in which the excess muscle temperature was 9°.

If it should transpire, as appears probable, that the tunas do not regulate the brain temperature to the degree they regulate the muscle temperature, the question then arises as to what is the utility of regulating the muscle temperature. Looking back to Fig. 17, which presumably shows the limits for acclimation of muscle, the utility of a high muscle temperature is clear. While the ratio of active to standard metabolism is constant, the difference between the two increases greatly with increasing temperature so that scope is much greater in that species at 15°C than at 5°C. Thus there are limits to temperature compensation with respect to total metabolism, which must be taken to be in large degree muscle metabolism. Brain metabolism however may compensate much more perfectly (Baslow, 1967; Baslow and Nigrelli, 1964) so that a cold brain may still be able to govern a warm muscle.

VI. DIRECTIVE FACTORS

A directive factor is an environmental identity which exerts its effect on the organism by stimulating some transductive response. The examples of the elaborate sense organs such as the eye or the ear are of course self-demonstrative from common knowledge. Other sense organs such as those which sense water temperature (see Murray, Volume V, this treatise) have been less obvious. Transduction may not necessarily lead to sensation. Signals from the environment initiate other important events, as, for example, the effect of day length on the pituitary (see chapter by Liley, Volume I, this treatise). It is assumed that the directive factors are the basis for all behavioral and psychical regulation and for anticipatory adjustments in physiological regulation. By anticipatory adjustment is meant, for example, a hormonal change elicited by the annual photoperiod cycle which prepares the organism for future seasonal events of temperature change. An example is given in Fig. 10 for lethal temperature. The subject of directive factors is therefore a large one, but comment here will be restricted to a brief consideration of the reactions of fish to dissolved substances and temperature gradients.

Such reactions are undirected movements which are ordinarily called kineses (Fraenkel and Gunn, 1961). The definition of a kinesis implies that such activity is purely random. However, Sullivan (1954; see also in Fry, 1964), dealing with temperature, and Hemmings (1966), concerned with an odor gradient, have both pointed out that the degree of movement at any one time may depend on the immediate past experience; thus, while the direction of movement may be random the degree is directed. It seems likely too that the distinction Fraenkel and Gunn made between the undirected movements (kineses) and the taxes, the directed movements, while highly convenient, does not point to a fundamental division. A kinesis is orientation by consecutive sampling of a gradient in an opaque environment. A taxis is alignment to the source of stimulus in a transparent environment. A nice analogy that has been used is that the relation between taxes and kineses is the same as between melody and harmony.

As in physiological regulation there is a great deal of acclimation and acclimatization in behavioral regulation. There may also be a large element of what can be called transferred adjustment which has not yet been well analyzed. Transferred adjustment is typified by the well-known conditioned reflex and can serve for a taxis where the environment permits only a kinesis as the primary response to the actual gradient. The surfacing of fish when oxygen is low, and particularly their rapid gathering around a hole cut in the ice of a snow-covered pond, is probably an example of such a transferred adjustment. Here the animal may be triggered to respond to the light gradient by the stress of low oxygen, but this possibility does not yet appear to have been put to an experimental test with fish.

Experiments with the directive factors, like all experiments in behavior, require more than mechanical excellence of the apparatus to assure good results, and they depend as much on interpretation of circumstances as on the calculation of the results. A good example is the work of Ozaki (1951) on the orientation of young fish to various colors of light. He observed that a single fish alone in his apparatus could not respond and that orientation was progressively more precise as he used 2 or 3 fish; but in particular he noted that orientation was most associated with those moments when two fish were aligned as in a school. One is tempted to say that these fish were not free to respond to the subtleties of their surroundings until their primary requirement of orientation in a school had been satisfied. Again Verheijen (1958) makes a good case for the point of view that a positive phototaxis to a point source of light is really a strong disorientation occasioned by glare. Thus, while good results may be achieved by fishing with lights, such light experiments may say little about the responses of fish to natural light in its various forms. Fry (1958) has assembled a number of other similar examples. Harden Jones (1968) has an interesting chapter on the reactions of fish to stimuli.

Up to now there has been little standardization of apparatus or refinement in approach in the study of these mechanical aspects of behavior. Fry (1958) compiled a bibliography which contained references to most types of gradient apparatus used up to that time. Kleere-koper (1967) has probably produced the most elaborate means of monitoring the paths taken by fish in responding to various stimuli. His largest tank to date had an area of 25 $meter^2$ with a grid of 2500 photocells linked to a computer.

A. Reactions to Dissolved Substances

1. Gradient Experiments

Figure 35 shows the results of two recent workers and relates the gradient experiments to the lethal levels. In each case the fish react to avoid a level far below the incipient lethal. There are not sufficient comparisons of this sort yet to generalize, but the reactions in the two examples have high statistical reliability. At present, however, such reactions are not recognized by a standard method of bioassay (McKee and Wolf, 1963; American Public Health Association, 1965). It would seem desirable that they should be since they afford a rapidly measured prelethal test. However, Sprague (1968) pointed out that in contrast to the sharp avoidance of low concentrations of metals (e.g., Fig. 35A) rainbow trout do not avoid phenol at 10 mg/liter, a near-lethal concentration, and apparently could not discriminate between a lethal concentration and clean water in his gradient although such a situation provoked high activity. A similar confusion resulted when an alkyl benzene sulfonate (ABS) detergent was presented at 10 mg/liter, which Sprague suggested may be the result of damage to the olfactory receptors (Bardach, 1956). When the fish were presented with a slowly lethal concentration of chlorine (0.1 mg/liter), they showed a net preference for it although they avoided higher or lower concentrations (0.01 and 1.0 mg/liter). The threshold

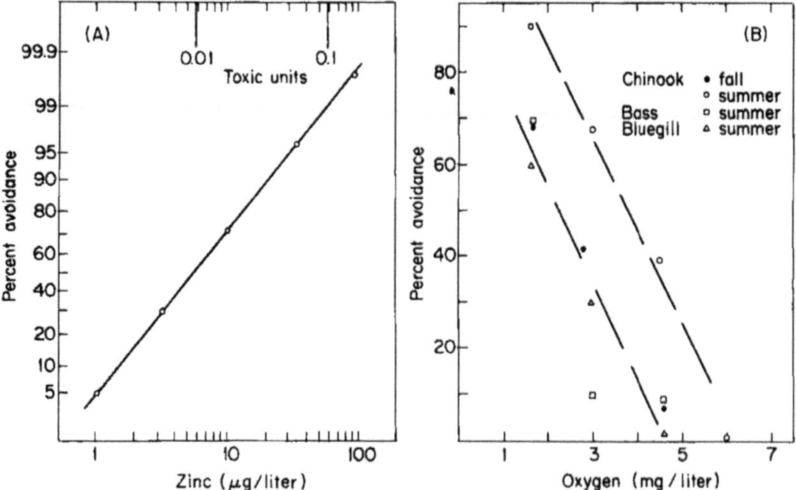

Fig. 35. (A) Avoidance of zinc sulfate solution by rainbow trout, *Salmo gairdneri*, at 9.5°C, water hardness 15 mg $CaCO_3$. From Sprague (1968). The avoidance is essentially complete at 0.1 toxic unit. One toxic unit is the incipient lethal concentration (Sprague and Ramsay, 1965). (B) Avoidance of low oxygen by chinook salmon, *Oncorhynchus tshawytscha*, largemouth bass, *Micropterus salmoides*, and the bluegill, *Lepomis macrochirus*, at existing river temperatures. Data from Whitmore *et al.* (1960), their Table 1, "periodic count." Avoidance of low oxygen is apparently not complete until the incipient lethal level is approached.

for avoidance of Kraft pulp mill effluent was also approximately the incipient lethal level. Thus, Sprague concluded that it cannot be arbitrarily assumed that any given pollutant will automatically repel fish.

There do not appear to be any reaction experiments in which there has been acclimation to the test substances so that the ecological meaning of such tests is not clear.

2. *Changes in Activity*

There have been other tests in which the reaction to dissolved substances has been monitored by changes in random activity (e.g., Dandy, 1967, 1970), opercular movements (e.g., Halsband and Halsband, 1968), or routine metabolic rate (e.g., Kutty, 1968a). In continuous exposure to 100 μg/liter Cu in Toronto tapwater (survival time $>$ 7 days) the increase in activity shown on the introduction of the metal subsided to the pre-introduction level in about 6 hr and continued at that level thereafter (Dandy, 1967). Similar behavior was shown in the response to H_2S down to the threshold found at 100 μg/liter, at which level no initial increase in activity was found. It appears from Dandy's data that such clear-cut selection as shown in the examples in Fig. 35 may only appear in acute experiments.

The subsidence of response described by Dandy may be habituation of the sense organ rather than acclimation in terms of the whole animal's increase in ability to resist the influence of the toxicant, but further work is needed, particularly with fish acclimated to a given level and then tested over the whole range of reaction.

B. Temperature Selection

A typical example of the response of fish to a temperature gradient is given in Fig. 36. Such responses usually have the sort of statistical precision shown in the figure, but the whole pattern of all responses for a given species still shows variations that have not all been explained. Figure 37 shows the observations available for the rainbow trout. While rainbow trout may not be genetically homogeneous, particularly with respect to the various domestic stocks in different parts of the world, three of the groups whose work is illustrated in Fig. 37 worked within a few hundred miles of each other. There is as yet no complete explanation for the differences in behavior these various workers have reported. Three different types of apparatus were used, but the differences found were not necessarily related to differences between apparatus.

Among the major sources of variability in temperature selection is time of year, as was first pointed out by Sullivan and Fisher (1953). Unfortunately, there are still no complete annual series of observations on the relation of the preferred temperature to acclimation temperature. The most complete are those of Zahn (1963) (Fig. 38) who, however, dampened the annual light

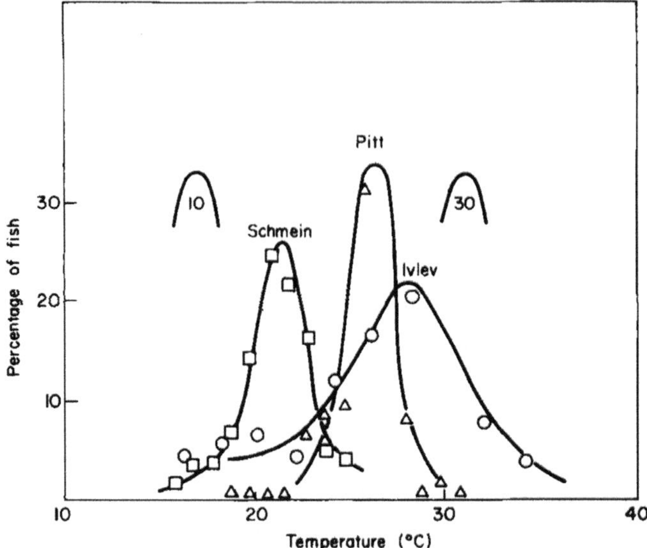

Fig. 36. Distribution of carp, *Cyprinus carpio*, in temperature gradients. Data of Ivlev (1960), Pitt *et al.* (1956), and Schmein-Engberding (1953). All fish acclimated to approximately 20°C prior to test. The numbers 10 and 30 indicate modes for fish acclimated to those temperatures. From Fry and Hochachka (1970).

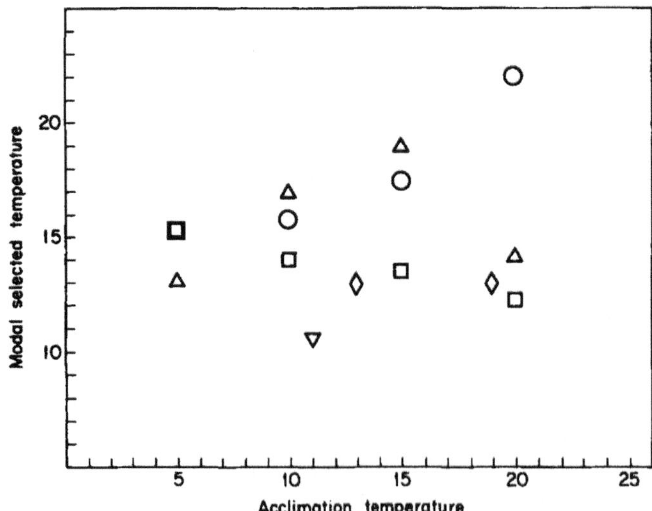

Fig. 37. Various modal selected temperatures in relation to thermal history for the rainbow trout, *Salmo gairdneri*. Data of W. J. Christie (personal communication) (△), Garside and Tait (1958) (□), Javaid and Anderson (1967) (○), Mantelman (1958) (◊), and Schmein-Engberding (1953) (▽).

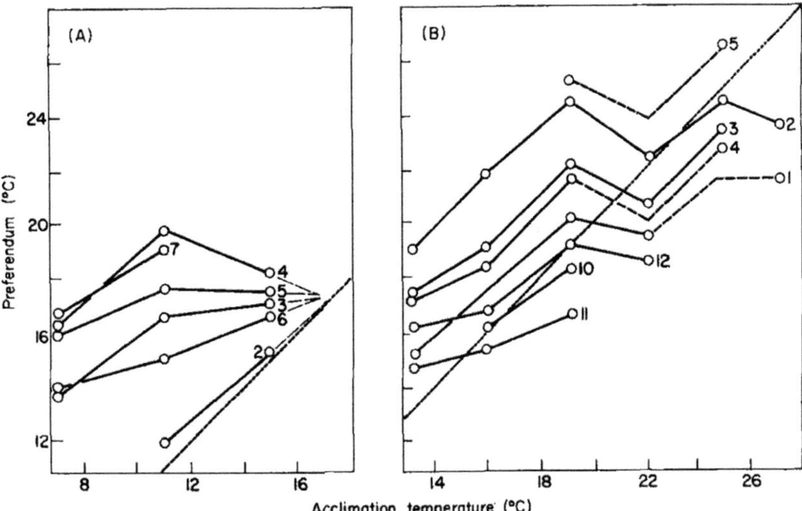

Fig. 38. Effect of season and acclimation temperature on temperature selection in the (A) plaice, *Pleuronectes platessa*, and (B) bitterling, *Rhodeus sericeus*. After Zahn (1963), from Fry and Hochachka (1970). Numbers indicate month of test.

cycle somewhat by superimposing a minimum day length in the case of the bitterling, while with the plaice he used a constant day length for the various relatively short periods of time that he maintained these fish after capture and before experiment. Despite this, the influence of season is profound in both cases in Fig. 38 and leads to the conclusion that any discussion of the relation between acclimation temperature and temperature preferendum without reference to season should be in the most general terms.

Both Ferguson (1958) and Zahn (1962) have noted that the response of the temperature preferendum to acclimation temperature has been most diverse, ranging from a slightly negative one such as shown by the data of Garside and Tait in Fig. 37 to cases like the bitterling (Fig. 38B, any single curve), where the temperature preferendum increases almost in step with increase in acclimation temperature. In the plaice, Zahn found almost the whole range of response in the one species at different times of the year. There is clear need for some patient description of the temperature preferendum in relation to season and latitude under various conditions of acclimation.

The response of the fish to the gradient apparatus itself also requires further analysis. There are three fundamentally different methods of presenting a temperature gradient to a fish. The first and most widely used method has been the longitudinal horizontal gradient (e.g., Norris, 1963; Schmein-Engberding, 1953; Alabaster and Downing, 1966) whereby the fish are placed in a tube or trough in which the water changes in temperature from one end to the other so that the gradient and the swimming path of the fish are constrained into the same plane and are parallel. The second method has been to establish a vertical thermal gradient (e.g., Brett, 1952) in a tank large enough to allow the fish some freedom in a horizontal path and in which temperature selection is effected by the fish swimming higher or lower. Thus, the fish has two degrees of freedom in its swimming path and the temperature gradient is at right angles to the longitudinal axis of the fish in its normal orientation. In this chamber the fish may react to depth as well as temperature (Javaid and Anderson, 1967). The third method has been to place the fish in a central chamber into which all choice chambers open individually. In its simplest form such apparatus is a divided trough (e.g., Collins, 1952). A more elaborate form is the rosette (e.g., Kleerekoper, 1969) in which the central chamber is surrounded by a number of choice chambers. In the third method both the direction and the magnitude of the change experienced by the animal in passing from the central condition to a given test condition is randomized. In the first and second methods the change is gradual as the animal passes up or down the gradient. The direction is random in the horizontal gradient but has a fixed association with gravity in the vertical gradient. In general, although no extensive comparisons have been made, these three types of apparatus yield similar results, or at least they are not consistently associated with any given result. A thorough statistical comparison of the three types in a single laboratory is highly desirable.

1. *Random Activity in Relation to Temperature Change*

Sullivan's distinction (1954; see also in Fry, 1964) cleared up a great deal of confusion with regard to temperature selection. In effect, she pointed out that certain activity is stimulated by recent temperature change and thus is a response to temperature as a directive factor, while activity at a given constant temperature is rather a facilitation of response by temperature as a controlling factor. The latter response has been widely but mostly unwittingly reported in the literature as a homeostatic response in the metabolic rate. Its general effect is to produce a central horizontal section or even a dip in the curve relating routine metabolism to temperature, or at least make that curve decidedly convex on a semilogarithmic plot. The clearest statement of such influence of random activity on the course of the temperature metabolism curve is probably that of Schmein-Engberding (1953) whose data are illustrated in Fig. 39 and who demonstrated that the anomaly could be removed by anesthesia. That author also pointed out there was a correspondence between the anomaly in the curve and the temperature preferendum.

Unfortunately, the works of Schmein-Engberding and Sullivan have been largely overlooked in the physiological literature, where discussion has followed the lines of workers such as Meuwis and Heuts (1957) whose data are summarized in Fig. 39 and who concluded there could be a "broad homeostatic zone of independence of the breathing frequency on temperatures" (Meuwis and Heuts, 1957, p. 107, 1.12). Such a notion of metabolic regulation is still

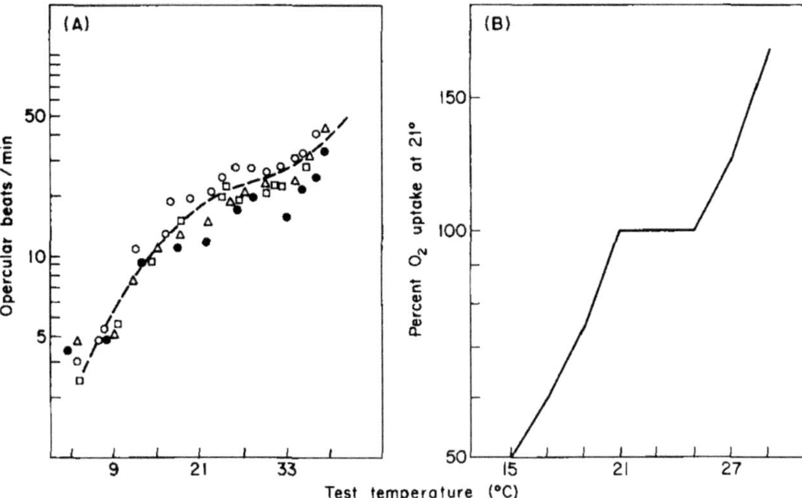

Fig. 39. Measures of routine metabolism of the carp, *Cyprinus carpio*, in relation to temperature. (A) From Meuwis and Heuts (1957) and (B) from Schmein-Engberding (1953). The various symbols in A represent different individuals.

widely held although Beamish (1964a) and Roberts (1966), for example, refer clearly to the effect of random activity on the level of metabolism as ordinarily measured (routine metabolism). The apparent homeostasis in the routine metabolism curve is brought about by a peak of random activity associated with the temperature that is the thermal preferendum, presumably for the state of thermal acclimation of the fish in question. Scope for activity appears to be ordinarily greatest at this temperature, and the animal presumably reacts most vigorously here to any stray stimuli. As the temperature increases beyond the preferendum, then increase in standard metabolism counteracts lessening random activity; thus, the routine metabolism curve rises again at higher temperatures. The interaction of these two effects is responsible for the so-called homeostasis. Figure 40 shows the relation of random movement to temperature when fish are equilibrated to successive temperature levels in turn.

The movements associated with thermoregulation are of two degrees. There is a gross response of activity (the response temperature of Rubin, 1935) whereby a fish resting quietly will show activity in warming water as its lethal temperature is reached. Fisher and Sullivan (1958) also showed the response temperature in the brook trout, together with the controlling effect. It is the second peak in Fig. 40.

There is also a tendency, much harder to demonstrate, for fish which are randomly active during a period of temperature change to be at first progressively more active as the temperature departs from their thermal preferendum. Figure 41 shows the latter phenomenon. Line A is the immediate response to temperature change which can be taken to be the response to temperature as a

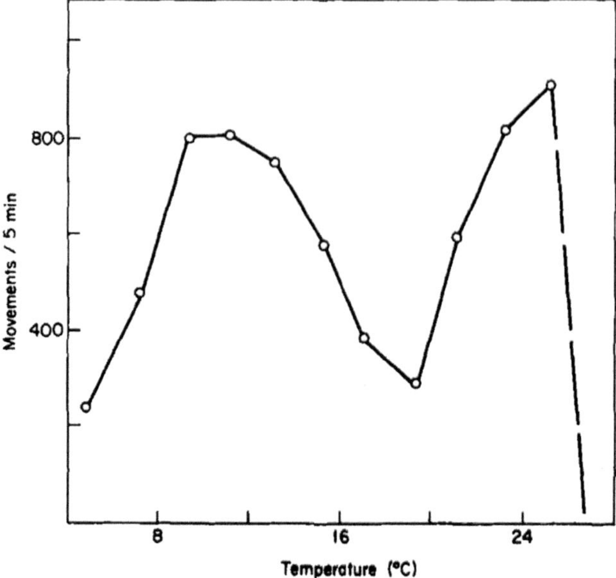

Fig. 40. Random activity of the eastern brook trout, *Salvelinus fontinalis*, in relation to temperature after approximately 30 min exposure to a given temperature. From Fisher and Sullivan (1958).

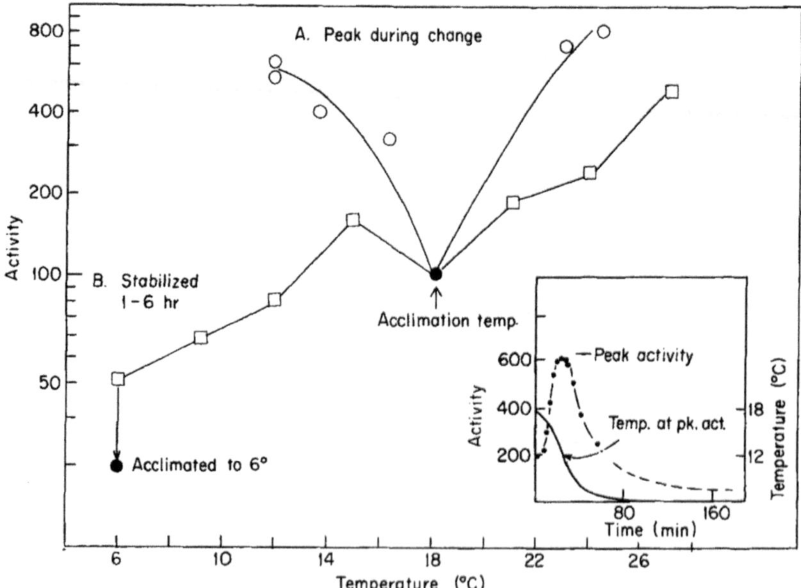

Fig. 41. Random activity of Atlantic salmon, *Salmo salar*, in relation to changing and stabilized temperatures. From Peterson and Anderson (1969a). The inset gives an example of the details of the short-term effects of temperature change on random activity.

directive factor. The final preferendum of the Atlantic salmon is close to 18°C (Javaid and Anderson, 1967) so that any departure from the preferred temperature evokes an immediate increase in activity of the fish.

Line B approximates the response to temperature as a controlling factor. It seems probable, both from the data of Peterson and Anderson and from Fisher and Sullivan (1958), who intentionally allowed time for thermal equilibrium, that what is "stabilized" to give the difference between curves A and B in Fig. 41 is the body temperature of the fish.

VII. RECAPITULATION

Figure 42 is presented as a summary of the various relations of the organism to its environment insofar as these can be pictured in one plane by quantitative expressions of metabolism.

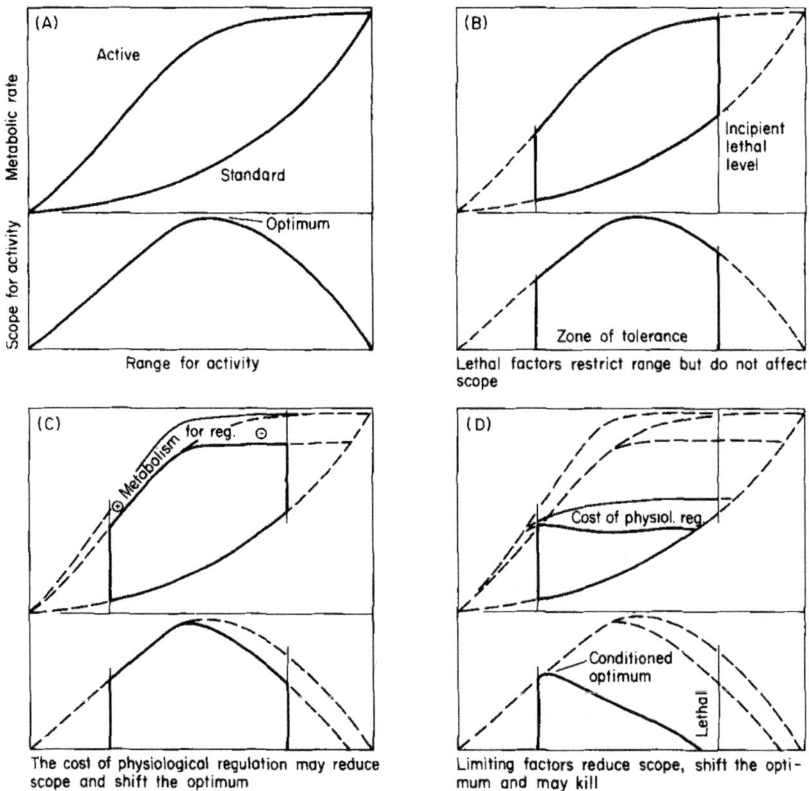

Fig. 42. Summary of relations of metabolism to environmental factors modified from Fry (1947). The solid lines represent the boundaries to scope for activity as various factors interact from A to D. For further explanation see text.

The pervading environmental factor is the controlling factor which sets the upper and lower limit to the metabolic rate. Thus the potential range for activity is set by the controlling factors and by the capacity of the organism to satisfy the requirements for metabolism that the controlling factors require or permit. The potential range can be envisioned as extending from a level where the state of activation, and with it the metabolic rate, is so low that the organism cannot respond at all, to a level where molecules are so activated that the standard metabolic rate absorbs the organism's total metabolic capacity (Fig. 42A).

The potential range over which the controlling factors can operate may be restricted by a controlling factor becoming lethal without any effect on scope for activity within the zone of tolerance (Fig. 42B).

The scope for activity within the limits set by the controlling factors is restricted by the operation of any limiting factor. Limiting factors operate to suppress the active metabolic rate and therefore to reduce scope for activity. The limiting factor will have its greatest effect on scope where the controlling factors require the highest level of standard metabolism; if severe enough, it will have a lethal effect down to the level of controlling factors where the limiting factor permits sufficient metabolism to satisfy the standard requirement. Thus, a limiting factor also reduces the range of controlling factors available for activity when the restriction to the active metabolic rate is sufficiently severe (Fig. 42D).

The organism regulates to maintain its own integrity and the continuation of the species. Such regulation involves mechanical barriers, physiological and biochemical activity, and behavioral responses. Regulation is achieved by channeling energy through appropriate form and is accomplished through the masking and directive factors. Standard metabolism probably represents the regulatory energy required by the quiescent animal and is related to the level of controlling factors which impinge on the organism. Beyond that the cost regulation is some function of activity (Fig. 42C).

Regulation mitigates the effects of the controlling factors and facilitates the uptake, discharge, and transport of metabolites. In the broad sense all physical activity is also regulation, but attention is usually focused on ancilliary activities which support growth or behavior. The masking factors provide the direct machinery of regulation by providing form and energy. The directive factors provide signals which permit the organism to respond both behaviorally and physiologically to its environment. The physiological response to a directive factor allows regulation to meet some future event by anticipatory adjustment linked to the precursor which acts as the directive factor in question.

Where no limiting factor operates, the immediate cost of the ancillary activity may be added to the cost of the behavioral activity or growth and so be shown as an increment to the curve for active metabolism. Under limiting conditions such costs become competitive to a greater and greater degree, but the interaction between these costs is not known. Certainly, as in Figs. 20 and 32, the behavioral activity is restricted to permit regulation to continue, either to lessen the requirement for regulation or to meet the lessened possibilities.

Organisms are continually adjusting to the fluctuating environment so that their reaction always depends on their history. Many aspects of their history may be stabilized by exposure to various constant conditions for various periods (acclimation). However, in a cyclic environment the organism may have evolved endogenous cyclic adjustments that cannot be entirely dampened by acclimation; moreover, structural characteristics determined during various critical periods of development may not be reversible at a later date. Hence, the ideal description in terms of reproducible response which can be evoked at any time by a stabilized environment is impossible to attain. Descriptions of metabolism require statements of history such as latitude and time of year that are not usually provided. Figure 42 is a compromise in which certain gross aspects of stability have been required.

REFERENCES

Alabaster, J. S., and Downing, A. L. (1966). A field and laboratory investigation of the effect of heated effluents on fish. *Min. Agr. Fish Food, U. K., Fish Invest., Ser. I.* **6,** No. 4, 1–42.

Alderdice, D. F. (1971). Factor combinations. *In* "Marine Ecology" (O. Kinne, ed.). Wiley, New York (in press).

Allanson, B. R., and Noble, R. G. (1964). The tolerance of *Tilapia mossambica* (Peters) to high temperature. *Trans. Am. Fisheries Soc.* **93,** 323–332.

Allen, K. O., and Strawn, K. (1968). Heat tolerance of channel catfish *Ictalurus punctatus. Proc. 21st Ann. Conf., Southeast. Assoc. Game Fish Comm., 1967* pp. 399–411.

American Public Health Association. (1965). "Standard Methods for the Examination of Water and Wastewater," 12th ed. Am. Public Health Assoc., New York.

Andrewartha, H. G., and Birch, L. C. (1954). "The Distribution and Abundance of Animals." Univ. of Chicago Press, Chicago, Illinois.

Arai, M. N., Cox, E. T., and Fry, F. E. J. (1963). An effect of dilutions of seawater on the lethal temperature of the guppy. *Can. J. Zool.* **41,** 1011–1015.

Arrhenius, S. (1889). Über die Reaktionsgeschwindigkeit bei der Inversion von Rohrzucker durch Säuren. *Z. Physik. Chem.* **4,** 226–248.

Bardach, J. E. (1956). The sensitivity of the goldfish (*Carassius auratus* L.) to point heat stimulation. *Am. Naturalist* **90,** 309–317.

Bardach, J. E., Fujiya, M., and Holl, A. (1965). Detergents: Effects on the chemical senses of the fish *Ictalurus natalis* (leSueur). *Science* **148,** 1605–1607.

Barlow, G. W. (1961). Intra- and interspecific differences in rate of oxygen consumption in gobiid fishes of the genus *Gillichthys. Biol. Bull.* **121,** 209–229.

Barrett, I., and Hester, F. (1964). Body temperature of yellowfin and skipjack tunas in relation to sea surface temperatures. *Nature* **203,** 96–97.

Baslow, M. H. (1967). Temperature adaptation and the central nervous system of fish. *In* "Molecular Mechanisms of Temperature Adaptation," Publ. No. 84, Am. Assoc. Advance. Sci., Washington, D. C.

Baslow, M. H., and Nigrelli, R. F. (1964). The effect of thermal acclimation on brain cholinesterase activity of the killifish, *Fundulus heteroclitus. Zoologica* **49,** 41–51.

Basu, S. P. (1959). Active respiration of fish in relation to ambient concentrations of oxygen and carbon dioxide. *J. Fisheries Res. Board Can.* **16,** 175–212.

Beamish, F. W. H. (1964a). Respiration of fishes with special emphasis on standard oxygen consumption. II. Influence of weight and temperature on respiration of several species. *Can. J. Zool.* **42,** 176–188.

Beamish, F. W. H. (1964b). III. Influence of oxygen. *Can. J. Zool.* **42,** 355–366.

Beamish, F. W. H. (1964c). IV. Influence of carbon dioxide and oxygen. *Can. J. Zool.* **42,** 847–856.

Beamish, F. W. H. (1964d). Influence of starvation on standard and routine oxygen consumption. *Trans. Am. Fisheries Soc.* **93,** 103–107.

Beamish, F. W. H. (1964e). Seasonal changes in the standard rate of oxygen consumption of fishes. *Can. J. Zool.* 42, 189–194.

Beamish, F. W. H., and Mookherjii, P. S. (1964). Respiration of fishes with special emphasis on standard oxygen consumption. I. Influence of weight and temperature on respiration of goldfish, *Carassius auratus* L. *Can. J. Zool.* **42,** 161–175.

Bělehrádek, J. (1926). Sur la formule généale exprimant l'action de la tempèrature sur les processus biologiques. *C. R. Soc. Biol.* **95,** 1449–1452.

Bělehrádek, J. (1930). Temperature coefficients in biology. *Biol. Rev.* **5,** 30–58.

Berthelot, M. (1862). Essai d'une theorie sur la formation des ethers. *Ann. Chim. Phys.* 3° *serie* **66,** 110–128.

Black, E. C., Fry, F. E. J., and Black, V. S. (1954). The influence of carbon dioxide on the utilization of oxygen by some freshwater fish. *Can. J. Zool.* **32,** 408–420.

Black, V. S. (1951). Osmotic regulation in teleost fishes. *Univ. Toronto Biol. Ser.* **59,** 53–89.

Blackman, F. F. (1905). Optima and limiting factors. *Ann. Botany* (*London*) **19,** 282–295.

Blažka, P. (1958). The anaerobic metabolism of fish. *Physiol. Zool.* **31,** 117–128.

Blažka, P., Volf, M., and Čepela, M. (1960). A new type of respirometer for the determination of the metabolism of fish in an active state. *Physiol. Bohemoslov.* **9,** 553–558.

Bliss, C. I. (1935). The calculation of the dosage mortality curve. *Ann. Appl. Biol.* **22,** 134–167.

Bliss, C. I. (1937). The calculation of the time-mortality curve. *Ann. Appl. Biol.* **24,** 815–852.

Bliss, C. I. (1952). "The Statistics of Bioassay," pp. 445–628. Academic Press, New York.

Booij, H. L., and Wolvekamp, H. P. (1944). Catenary processes, master reactions and limiting factors. *Bibliotheca Biotheor.* D1, 145–224.

Brett, J. R. (1944). Some lethal temperature relations of Algonquin Park fishes. *Univ. Toronto Studies Biol. Ser.* **52,** 1–49.

Brett, J. R. (1946). Rate of gain of heat-tolerance in goldfish (*Carassius auratus*). *Univ. Toronto Studies Biol. Ser.* **53,** 1–28.

Brett, J. R. (1952). Temperature tolerance in young Pacific salmon genus *Oncorhynchus. J. Fisheries Res. Board Can.* **9,** 265–323.

Brett, J. R. (1958). Implications and assessments of environmental stress. In "Investigations of Fish-power Problems" (P. A. Larkin, ed.), pp. 69-83. H. R. MacMillan Lectures in Fisheries, Univ. of Brit. Columbia.

Brett, J. R. (1964). The respiratory metabolism and swimming performance of young sockeye salmon. *J. Fisheries Res. Board Can.* **21,** 1183–1226.

Brett, J. R. (1965). The relation of size to rate of oxygen consumption and sustained swimming speed of sockeye salmon (*Oncorhynchus nerka*). *J. Fisheries Res. Board Can.* **22,** 1491–1501.

Brett, J. R., and Alderdice, D. F. (1958). The resistance of cultured young chum and sockeye salmon to temperatures below 0°C. *J. Fisheries Res. Board Can.* **15,** 805–813.

Brett, J. R., Shelbourn, J. E., and Shoop, C. T. (1969). Growth rate and body composition of fingerling sockeye salmon, *Oncorhynchus nerka*, in relation to temperature and ration size. *J. Fisheries Res. Board Can.* **26,** 2363–2394.

Brody, S. (1945). "Bioenergetics and Growth." Reinhold, New York.

Brown, C. E., and Muir, B. S. (1970). Analysis of ram ventilation of fish gills with application to skipjack tuna (*Katsuwonus pelamis*). *J. Fisheries Res. Board Can.* **27,** 1637–1652.

Brown, V. M., Jordan, D. H. M., and Tiller, B. A. (1967). The effect of temperature on the acute toxicity of phenol to rainbow trout in hard water. *Water Res.* **1**, 587–594.

Bullivant, J. S. (1961). The influence of salinity on the rate of oxygen consumption of young Quinnat salmon, *Oncorhynchus tschawytscha*. *New Zealand J. Sci.* **4,** 381–391.

Carey, F. G., and Teal, J. M. (1966). Heat conservation in tuna fish muscle. *Proc. Natl. Acad. Sci. U. S.* **56,** 1464–1469.

Carey, F. G., and Teal, J. M. (1969a). Mako and porbeagle: Warm-bodied shàrks. *Comp. Biochem. Physiol.* **28,** 199–204.

Carey, F. G., and Teal, J. M. (1969b). Regulation of body temperature by the bluefin tuna. *Comp. Biochem. Physiol.* **28,** 205–213.

Cocking, A. W. (1959). The effects of high temperatures on roach (*Rutilus rutilus*). II. The effects of temperature increasing at a known constant rate. *J. Exptl. Biol.* **36,** 217–226.

Collins, G. B. (1952). Factors influencing the orientation of migrating anadtomous fishes. *U. S. Fish Wildlife Serv., Fishery Bull.* **73** 52, 375–396.

Conte, F. P., and Wagner, H. H. (1965). Development of osmotic and ionic regulation in juvenile steelhead trout *Salmo gairdneri*. *Comp. Biochem. Physiol.* **14,** 603–620.

Coulter, G. W. (1967). Low apparent oxygen requirements of deep-water fishes in Lake Tanganyika. *Nature* **215,** 317–318.

Crozier, W. J. (1924). On the critical thermal increment for the locomotion of a diplopod. *J. Gen. Physiol.* **7,** 123–136.

Dahlberg, M. L., Shumway, D. L., and Doudoroff, P. (1968). Influence of dissolved oxygen and carbon dioxide on swimming performance of largemouth bass and coho salmon. *J. Fisheries Res. Board Can.* **25,** 49–70.

Dandy, J. W. T. (1967). The effects of chemical characteristics of the environment on the activity of an aquatic organism. Ph.D. Thesis, University of Toronto (National Library of Canada, Canadian theses on microfilm, No. CM. 68–616).

Dandy, J. W. T. (1970). Activity response to oxygen in the brook trout, *Salvelinus fontinalis* (Mitchill). *Can. J. Zool.* **48,** 1067–1072.

Das, A. B., and Prosser, C. L. (1967). Biochemical changes in tissues of goldfish acclimated to high and low temperatures. I. Protein synthesis. *Comp. Biochem. Physiol.* **21,** 449–467.

Davidson, J. (1942). On the speed of development of insect eggs at constant temperatures. *Aust. J. Exp. Biol. Med. Sci.* **20,** 233–239.

Davidson, J. (1944). On the relationship between temperature and rate of development of insects at constant temperatures. *J. Animal Ecol.* **13,** 26–38.

Davies, P. M. C. (1966). The energy relations of *Carassius auratus* L. II. The effect of food, crowding and darkness on heat production. *Comp. Biochem. Physiol.* **17,** 983–995.

Davis, G. E., Foster, J., Warren, C. E., and Doudoroff, P. (1963). The influence of oxygen concentration on the swimming performance of juvenile Pacific salmon at various temperatures. *Trans. Am. Fisheries Soc.* **92,** 111–124.

DeGroot, S. J., and Schuyf, A. (1967). A new method for recording the swimming activity of flatfishes. *Experientia* **23,** 1–6.

de Réaumur, R. A. F. (1735). *Mém. Acad. Roy. Sci. Paris* [cited in Shelford (1929)].

Doudoroff, P. (1942). The resistance and acclimatization of marine fishes to temperature changes. I. Experiments with *Girella nigricans* (Ayres). *Biol. Bull.* **83,** 219–244.

Doudoroff, P. (1945). II. Experiments with fundulus and Atherinops. *Biol. Bull.* **88,** 194–206.

Doudoroff, P., Leduc, G., and Schneider, C. R. (1966). Acute toxicity to fish of solutions containing complex metal cyanides, in relation to concentrations of molecular hydrocyanic acid. *Trans. Am. Fisheries Soc.* **95,** 6–22.

Ege, R., and Krogh, A. (1914). On the relation between temperature and the respiratory exchange in fishes. *Intern. Rev. Ges. Hydrobiol. Hydrog.* **7,** 48–55.

Embody, G. C. (1934). Relations of temperature to the incubation periods of eggs of four species of trout. *Trans. Am. Fisheries Soc.* **64,** 281–292.

Farmer, G. J., and Beamish, F. W. H. (1969). Oxygen consumption of *Tilapia nilotica* in relation to swimming speed and salinity. *J. Fisheries Res. Board Can.* **26,** 2807–2821.

Ferguson, R. G. (1958). The preferred temperature of fish and their midsummer distribution in temperate lakes and streams. *J. Fisheries Res. Board Can.* **15,** 607–624.

Fisher, K. C., and Sullivan, C. M. (1958). The effect of temperature on the spontaneous activity of speckled trout before and after various lesions of the brain. *Can. J. Zool.* **36,** 49–63.

Fraenkel, G. S., and Gunn, D. L. (1961). "The Orientation of Animals," 2nd ed. Dover, New York.

Fry, F. E. J. (1947). Effects of the environment on animal activity. *Univ. Toronto Studies Biol. Ser.* **55,** 1–62.

Fry, F. E. J. (1957). The aquatic respiration of fish. *In* "The Physiology of Fishes" (M. E. Brown, ed.), Vol. 1, pp. 1–63. Academic Press, New York.

Fry, F. E. J. (1958). Laboratory and aquarium research. II. The experimental study of behaviour in fish. *Proc. Indo-Pacific Fishery Council* pp. 37–42.

Fry, F. E. J. (1964). Animals in aquatic environments: Fishes. In "Handbook of Physiology" (Am. Physiol. Soc., J. Field, ed.), Sect. 4, pp. 715–728. Williams & Wilkins, Baltimore, Maryland.

Fry, F. E. J. (1967). Responses of vertebrate poikilotherms to temperature. *In* "Thermobiology" (A. H. Rose, ed.), pp. 375–409. Academic Press, New York.

Fry, F. E. J., and Hart, J. S. (1948). The relation of temperature to oxygen consumption in the goldfish. *Biol. Bull.* **94,** 66–77.

Fry, F. E. J., and Hochachka, P. W. (1970). Fish. *In* "Comparative Physiology of Thermoregulation" (G. C. Whittow, ed.), pp. 79–134. Academic Press, New York.

Fry, F. E. J., and Norris, K. S. (1962). The transportation of live fish. *In* "Fish as Food" (G. Borgstrom, ed.), Vol. 2, pp. 595–608. Academic Press, New York.

Fry, F. E. J., Brett, J. R., and Clawson, G. H. (1942). Lethal limits of temperature for young goldfish. *Rev. Can. Biol.* **1,** 50–56.

Fry, F. E. J., Hart, J. S., and Walker, K. F. (1946). Lethal temperature relations for a sample of young speckled trout, *Salvelinus fontinalis. Univ. Toronto Studies Biol. Ser.* **54,** 1–47.

Garside, E. T. (1966). Effects of oxygen in relation to temperature on the development of embryos of brook trout and rainbow trout. *J. Fisheries Res. Board Can.* **23,** 1121–1134.

Garside, E. T., and Tait, J. S. (1958). Preferred temperature of rainbow trout (*Salmo gairdneri* Richardson) and its unusual relationship to acclimation temperature. *Can. J. Zool.* **36,** 563–567.

Gibson, M. B. (1954). Upper lethal temperature relations of the guppy, *Lebistes reticulatus. Can. J. Zool.* **32,** 393–407.

Glass, N. R. (1969). Discussion of calculation of power function with special reference to respiratory metabolism in fish. *J. Fisheries Res. Board Can.* **26,** 2643–2650.

Gordon, M. S. (1963). Chloride changes in rainbow trout (*Salmo gairdneri*) adapted to different salinities. *Biol. Bull.* **124,** 45–54.

Graham, J. M. (1949). Some effects of temperature and oxygen pressure on the metabolism and activity of the speckled trout, *Salvelinus fontinalis. Can. J. Res,* **D27,** 270–288.

Halsband, E., and Halsband, I. (1968). Eine Apparatur zur Messung der Stoff-wechselintensität von Fischen und Fischnährtieren. *Arch. Fischereiwiss.* **19,** 78–82.

Harden Jones, F. R. (1968). "Fish Migration." St. Martin's Press, New York.

Hart, J. S. (1952). Geographic variations of some physiological and morphological characters in certain freshwater fish. *Univ. Toronto Biol. Ser.* **60,** 1–79.

Hart, J. S. (1957). Seasonal changes in CO_2 sensitivity and blood circulation in certain freshwater fishes. *Can. J. Zool.* **35,** 195–200.

Harvey, E. N. (1911). Effect of different temperatures on the medusa *Cassiopea*, with special reference to the rate of conduction of the nerve impulse. *Carnegie Inst. Wash. Publ., Pap. Tortugas Lab.* **3,** 27–39.

Harvey, H. H. (1964). Dissolved nitrogen as a tracer of fish movements. *Verhandl. Intern. Ver. Limnol.* **15,** 947–951.

Heath, W. G. (1963). Thermoperiodism in sea-run cutthroat trout (*Salmo clarki clarki*). *Science* **142,** 486–488.

Heinicke, E. A., and Houston, A. H. (1965). Effect of thermal acclimation and sublethal heat shock upon ionic regulation in the goldfish, *Carassius auratus* L. *J. Fisheries Res. Board Can.* **22,** 1455–1476.

Hemmings, C. C. (1966). The mechanism of orientation of roach, *Rutilus rutilus* L. in an odor gradient. *J. Exptl. Biol.* **45,** 465–474.

Herbert, D. W. M. (1965). Pollution and fisheries. Ecology and the industrial society. *Ecol. Ind. Soc., Symp.*, 1964 pp. 173–195.

Herbert, D. W. M., and Shurben, D. S. (1963). A preliminary study of the effect of physical activity on the resistance of rainbow trout (*Salmo gairdneri* Richardson) to two poisons. *Ann. Appl. Biol.* **52,** 321–326.

Herbert, D. W. M., and Vandyke, J. M. (1964). The toxicity to fish of mixtures of poisons. II. Copper-ammonia and zinc-phenol mixtures. *Ann. Appl. Biol.* **53,** 415–421.

Heusner, A., and Enright, J. T. (1966). Long-term activity in small aquatic animals. *Science* **154,** 532–533.

Hickman, C. P., Jr. (1959). The osmoregulatory role of the thyroid gland in the starry flounder *Platichthys stellatus*. *Can. J. Zool.* **37,** 997–1060.

Hoar, W. S., and Robinson, G. B. (1959). Temperature resistance of goldfish maintained under controlled photoperiods. *Can. J. Zool.* **37,** 419–428.

Hoff, J. G., and Westman, J. R. (1966). The temperature tolerances of three species of marine fishes. *J. Marine Res.* (*Sears Found. Marine Res.*) **24,** 131–140.

Houde, E. D. (1969). Sustained swimming ability of larvae of walleye (*Stizostedion vitreum*) and yellow perch (*Perca flavescens*). *J. Fisheries Res. Board Can.* **26,** 1647–1659.

Houston, A. H. (1959). Osmoregulatory adaptation of steelhead trout (Salmo gairdnerii Richardson) to sea water. *Can. J. Zool.* **37,** 729–748.

Ivlev, V. S. (1938). The effect of temperature on the respiration of fish. *Zool. Zh.* **17,** 645–660. (Engl. transl. by E. Jermolajev.)

Ivlev, V. S. (1960). Analiz mekhanizma raspredelniia ryb v usloviiakh temperaturnovo gradienta. *Zool. Zh.* **39,** 494–499 [for translation, see *Fisheries Res. Board Can., Transl. Ser.* **364,** (1961)].

Jacobs, M. H. (1919). Acclimatization as a factor affecting the upper thermal death points of organisms. *J. Exptl. Zool.* **27,** 427–442.

Janisch, E. (1925). Über die Temperaturabhängigkeit biologischer Vorgänge und ihre kurvenmässige Analyse. *Arch. Ges. Physiol.* **209,** 414–436.

Jankowsky, H.-D. (1968). Versuche zur Adaptation der Fische im normalen Temperaturbereich. *Helgolaender Wiss. Meeresuntersuch.* **18,** 317–362.

Javaid, M. Y., and Anderson, J. M. (1967). Thermal acclimation and temperature selection in Atlantic salmon, *Salmo salar*, and rainbow trout, S. *gairdneri*. *J. Fisheries Res. Board Can.* **24,** 1507–1513.

Job, S. V. (1955). The oxygen consumption of *Salvelinus fontinalis*. *Univ. Toronto Biol. Ser.* **61,** 1–39.

Job, S. V. (1959). The metabolism of *Plotosus anguillaris* (Bloch) in various concentrations of salt and oxygen in the medium. *Proc. Indian Acad. Sci.* **B50,** 267–288.

Johansen, P. H. (1967). The role of the pituitary in the resistance of the goldfish (*Carassius auratus* L.) to a high temperature. *Can. J. Zool.* **45,** 329–345.

Johnson, F. H., Eyring, H., and Polissar, M. J. (1954). "The Kinetic Basis of Molecular Biology." Wiley, New York.

Kanungo, M. S., and Prosser, C. L. (1959). Physiological and biochemical adaptation of goldfish to cold and warm temperatures. I. Standard and active oxygen consumptions of cold- and warm-acclimated goldfish at various temperatures. *J. Cellular Comp. Physiol.* **54,** 259–263.

Kausch, H. (1968). Der Einfluβ der Spontanaktivität auf die Stoffwechselrate junger Karpfen (*Cyprinus carpio* L.) im Hunger und bei Futterung. *Arch. Hydrobiol.* Suppl. 33, No. 3, 263–330.

Keys, A. B. (1931). A study of the selective action of decreased salinity and of asphyxiation on the Pacific killifish, *Fundulus parvipinnis*. *Bull. Scripps Inst. Oceanog. Univ. Calif., Tech. Ser.* **2,** 417–490.

Kinne, O., and Kinne, E. M. (1962). Rates of development in embryos of a cyprinodont fish exposed to different temperature-salinity-oxygen combinations. *Can. J. Zool.* **40,** 231–253.

Kishinouye, K. (1923). Contributions to the comparative study of the so-called scombroid fishes. *J. Coll. Agr., Tokyo Imp. Univ.* **8,** 293–470.

Kleerekoper, H. (1967). Some aspects of olfaction in fishes, with special reference to orientation. *Am. Zool.* **7,** 385–395.

Kleerekoper, H. (1969). "Olfaction in Fishes." Indiana Univ. Press, Bloomington, Indiana.

Krogh, A. (1914). On the influence of the temperature on the rate of embryonic development. *Z. Allgem. Physiol.* **16,** 163–177.

Krueger, F. (1964). Neuere mathematisch Formulierung der biologischen Temperaturfunktion und des Wachstums. *Helgolaender Wiss. Meeresuntersuch.* **9,** 108–124.

Kutty, M. N. (1968a). Respiratory quotients in goldfish and rainbow trout. *J. Fisheries Res. Board Can.* **25,** 1689–1728.

Kutty, M. N. (1968b). Influence of ambient oxygen on the swimming performance of goldfish and rainbow trout. *Can. J. Zool.* **46,** 647–653.

Larimore, R. W., and Duever, M. J. (1968). Effects of temperature acclimation on the swimming ability of smallmouth bass fry. *Trans. Am. Fisheries Soc.* **97,** 175–184.

Leiner, M. (1938). "Die Physiologie der Fischatmung." Akad. Verlagsges., Leipzig.

Lindroth, A. (1940). Sauerstoffverbrauch der Fische bei verschiedenem Sauerstoff-druck und verschiedenem Sauerstoffbedarf. *Z. Vergleich. Physiol.* **28,** 142–152.

Lindroth, A. (1942). Sauerstoffverbrauch der Fische. II. Verschiedene entwicklungsund altersstadien vom Lachs und Hecht. *Z. Vergleich. Physiol.* **29,** 583–594.

Lindsey, C. C. (1968). Temperatures of red and white muscle in recently caught marlin and other large tropical fish. *J. Fisheries Res. Board Can.* **25,** 1987–1992.

Lindsey, J. K., Alderdice, D. F., and Pienaar, L. V. (1970). Analysis of nonlinear models—the nonlinear response surface. *J. Fisheries Res. Board Can.* **27,** 765–791.

Lloyd, R. (1961a). The toxicity of ammonia to rainbow trout (*Salmo gairdneri* Richardson). *Water Waste Treat. J.* **8,** 278–279.

Lloyd, R. (1961b). Effect of dissolved oxygen concentrations on the toxicity of several poisons to rainbow trout (*Salmo gairdneri* Richardson). *J. Exptl. Biol.* **38,** 447–455.

Lloyd, R. (1961c). The toxicity of mixtures of zinc and copper sulphates to rainbow trout (*Salmo gairdneri* Richardson). *Ann. Appl. Biol.* **49,** 535–538.

Lloyd, R., and Herbert, D. W. M. (1960). The influence of carbon dioxide on the toxicity of un-ionized ammonia to rainbow trout (*Salmo gairdneri Richardson*). *Ann. Appl. Biol.* **40,** 399–404.

Lloyd, R., and Jordan, D. H. M. (1964). Some factors affecting the resistance of rainbow trout (*Salmo gairdneri* Richardson) to acid waters. *Air Water Pollution* **8,** 393–403.

Lloyd, R., and White, W. R. (1967). Effect of high concentration of carbon dioxide on the ionic composition of rainbow trout blood. *Nature* **216,** 1341–1342.

Loeb, J. (1913). Die Tropismen. *In* "Handbuch der Vergleichenden Physiologie" (H. Winterstein, ed.), Vol. 4, pp. 451–519. Fischer, Jena.

Loeb, J. (1918). "Forced Movements, Tropisms and Animal Conduct." Lippincott, Philadelphia, Pennsylvania.

McCrimmon, H. R. (1949). The survival of planted salmon (*Salmo salar*) in streams. Ph.D. Thesis, University of Toronto.

McKee, J. E., and Wolf, H. W. (1963). "Water Quality Criteria," 2nd ed., Publ. 3A. Res. Agency Calif. State Water Qual. Control Board, Sacramento, California.

MacLeod, J. C., and Smith, L. L., Jr. (1966). Effect of pulpwood fiber on oxygen consumption and swimming endurance of the fathead minnow, *Pimephales promelas. Trans. Am. Fisheries Soc.* **95,** 71–84.

Mantelman, I. I. (1958). O raspredelenii molodi nekotorykh vidov ryb v termogradientnykh usloviiakh. *Izv. Vses. Nauch-Issled. Inst. Ozer. Rechn. Ryb Khoz.* **47**(1), 1–63 [for translation, see *Fisheries Res. Board Can., Transl. Ser.* **257** (1960)].

Mar, J. (1959). "A Proposed Tunnel Design for a Fish Respirometer," Tech. Memo. 59-3. Pacific Naval Lab., D. R. B. Esquimalt, B. C.

Maros, L., Schulek, E., Molnar-Perl, I., and Pinter-Szakacs, M. (1961). Einfaches destillationsverfahren zur titrimetrischen bestimmung von Kohlendioxyd. *Anal. Chim. Acta* **25,** 390–399 [for translation, see *Fisheries Res. Board Can., Transl. Ser.* **596** (1965)].

Martin, W. R. (1949). The mechanics of environmental control of body form in fishes. *Univ. Toronto Biol. Ser.* **58,** 1–91.

Mathur, G. B. (1967). Anaerobic respiration in a cyprinoid fish *Rasbora daniconius* (Ham). *Nature* **214,** 318–319.

Mathur, G. B., and Shrivastava, B. D. (1970). An improved activity meter for the determination of standard metabolism in fish. *Trans. Am. Fisheries Soc.* **99,** 602–603.

Meuwis, A. L., and Heuts, M. J. (1957). Temperature dependence of breathing rate in carp. *Biol. Bull.* **112,** 97–107.

Morris, R. (1960). General problems of osmoregulation with special reference to cyclostomes. *Symp. Zool. Soc. London* **1,** 1–16.

Morris, R. W. (1967). High respiratory quotients of two species of bony fishes. *Physiol. Zool.* **40,** 409–423.

Moss, D. D., and Scott, D. C. (1961). Dissolved-oxygen requirements of three species of fish. *Trans. Am. Fisheries Soc.* **90,** 377–393.

Mount, D. I. (1964). Additional information on a system for controlling the dissolved oxygen content of water. *Trans. Am. Fisheries Soc.* **92,** 100–103.

Muir, B. S., Nelson, G. J., and Bridges, K. W. (1965). A method for measuring swimming speed in oxygen consumption studies on the aholehole *Kuhlia sandvicensis. Trans. Am. Fisheries Soc.* **94,** 378–382.

Nicholls, J. V. V. (1931). The influence of temperature on digestion in *Fundulus heteroclitus. Contrib. Can. Biol. Fisheries* **7,** 47–55.

Norris, K. S. (1963). The functions of temperature in the ecology of the percoid fish *Girella nigricans* (Ayres). *Ecol. Monographs* **33,** 23–62.

Olson, F. C. W., and Jackson, J. M. (1942). Heating curves: Theory and practical application. *Ind. Eng. Chem.* **34,** 334–341.

Olson, F. C. W., and Stevens, H. P. (1939). Thermal processing of canned foods in tin containers. II. Nomograms for graphic calculation of thermal processes for non-acid canned foods exhibiting straight-line semi-logarithmic heating curves. *Food Res.* **4,** 1–20.

Ozaki, H. (1951). On the relation between the phototaxis and the aggregation of young marine fish. *Rept. Fac. Fisheries Prefect. Univ. Mie* **1,** 55–66.

Paloheimo, J. E., and Dickie, L. M. (1966a). Food and growth of fishes. II. Effects of food and temperature on the relation between metabolism and body weight. *J. Fisheries Res. Board Can.* **23,** 869–908.

Paloheimo, J. E., and Dickie, L. M. (1966b). Food and growth of fishes. III. Relations among food, body size and growth efficiency. *J. Fisheries Res. Board Can.* **23,** 1209–1248.

Pavlov, D. S., Sbikin, Yu. N., and Mochek, A. D. (1968). The effect of illumination in running water on the speed of fishes in relation to features of their orientation. *Vopr. Ikhtiol.* **8,** 250–255 (for translation, see "Problems of Ichthyology." Am. Fisheries Soc., Washington, D. C., 1968).

Pavlovskii, E. N., ed. (1962). "Techniques for the Investigation of Fish Physiology." Izd. Akad. Nauk S.S.S.R. (Transl. No. OTS 64-11001. Off. Tech. Serv., U. S. Dept. Comm., Washington, D. C., 1964).

Peterson, R. H., and Anderson, J. M. (1969a). Influence of temperature change on spontaneous locomotor activity and oxygen consumption of Atlantic salmon, *Salmo salar*, acclimated to two temperatures. *J. Fisheries Res. Board Can.* **26,** 93–109.

Peterson, R. H., and Anderson, J. M. (1969b). Effects of temperature on brain tissue oxygen consumption in salmonid fishes. *Can. J. Zool.* **47,** 1345–1353.

Pitkow, R. B. (1960). Cold death in the guppy. *Biol. Bull.* **119,** 231–245.

Pitt, T. K., Garside, E. T., and Hepburn, R. L. (1956). Temperature selection of the carp (*Cyprinus carpio* Linn.) *Can. J. Zool.* **34,** 555–557.

Potts, D. C., and Morris, R. W. (1968). Some body fluid characteristics of the Antarctic fish, *Trematomus bernacchii. Marine Biol.* **1,** 269–276.

Precht, H. (1968). Der Einflβ "normaler" Temperaturen auf Lebensprozesse bei wechselwarmen Tieren unter Ausschluβ der Wachstums- und Entwicklungsprozesse. *Helgolaender Wiss. Meeresuntersuch.* **18,** 487–548.

Precht, H., Christophersen, J., and Hensel, H. (1955). "Temperatur und Leben." Springer, Berlin.

Prosser, C. L., and Farhi, E. (1965). Effects of temperature on conditioned reflexes and on nerve conduction in fish. *Z. Vergleich. Physiol.* **50,** 91–101.

Prosser, C. L., Barr, L. M., Pinc, R. A., and Lauer, C. Y. (1957). Acclimation of goldfish to low concentrations of oxygen. *Physiol. Zool.* **30,** 137–141.

Rao, G. M. M. (1967). Oxygen consumption of rainbow trout (*Salmo gairdneri*) in relation to activity, salinity and temperature. Ph.D. Thesis, University of Toronto (National Library of Canada, Canadian theses on microfilm, No. 1987).

Rao, G. M. M. (1968). Oxygen consumption of rainbow trout (*Salmo gairdneri*) in relation to activity and salinity. *Can. J. Zool.* **46,** 781–786.

Rao, G. M. M. (1969). Effect of activity, salinity, and temperature on plasma concentrations of rainbow trout. *Can. J. Zool.* **47,** 131–134.

Roberts, J. L. (1966). Systemic versus cellular acclimation to temperature by poikilotherms. *Helgolaender Wiss. Meeresuntersuch.* **14,** 451–465.

Roots, B. I., and Prosser, C. L. (1962). Temperature acclimation and the nervous system in fish. *J. Exptl. Biol.* **39,** 617–628.

Rubin, M. A. (1935). Thermal reception in fishes. *J. Gen. Physiol.* **18,** 643–647.

Ruhland, M. L. (1965). Etude comparative de la consommation d'oxygène chez différentes espèces de poissons teléostéens. *Bull. Soc. Zool. France* **90,** 347–353.

Ruhland, M. L. (1967). Controle des pressions partielles d'oxygène au cours des mésures de la consommation d'oxygène chez les poissons dans un appareil enregistreur continu. *Bull. Soc. Zool. France* **92,** 787–792.

Ruhland, M. L., and Heusner, A. (1959). Technique d'enregistrement de faibles consommations d'oxygène: Application aux poissons de petite taille. *Compt. Rend. Soc. Biol.* **153,** 161–164.

Saunders, R. L. (1962). The irrigation of the gills in fishes. II. Efficiency of oxygen uptake in relation to respiratory flow, activity and concentrations of oxygen and carbon dioxide. *Can. J. Zool.* **40,** 817–862.

Scheier, A., and Cairns, J., Jr. (1966). Persistence of gill damage in *Lepomis gibbosus* following a brief exposure to alkyl benzene sulfonate. *Notulae Naturae (Acad. Nat. Sct. Phila.)* **391,** 1–7.

Schmein-Engberding, F. (1953). Die Vorzugstemperaturen einiger Knochenfische und ihre physiologische Bedeutung. *Z. Fischerei* **2,** 125–155.

Scholander, P. F., Haugaard, N., and Irving, L. (1943). A volumetric respirometer for aquatic animals. *Rev. Sci. Instr.* **14,** 48–51.

Segrem, N. P., and Hart, J. S. (1967). Oxygen supply and performance in *Peromyscus*. Metabolic and circulatory responses to exercise, *Can. J. Physiol. Pharmacol.* **45,** 531–541.

Shelford, V. E. (1929). "Laboratory and Field Ecology." Williams & Wilkins, Baltimore, Maryland.

Shepard, M. P. (1955). Resistance and tolerance of young speckled trout (*Salvelinus fontinalis*) to oxygen lack, with special reference to low oxygen acclimation. *J. Fisheries Res. Board Can.* **12,** 387–446.

Silver, S. J., Warren, C. E., and Doudoroff, P. (1963). Dissolved oxygen requirements of developing steelhead trout and Chinook salmon embryos at different water velocities. *Trans. Am. Fisheries Soc.* **92,** 327–343.

Smit, H. (1965). Some experiments on the oxygen consumption of goldfish (*Carassius auratus* L.) in relation to swimming speed. *Can. J. Zool.* **43,** 623–633.

Smit, H. (1967). Influence of temperature on the rate of gastric juice secretion in the brown bullhead, *Ictalurus nebulosus. Comp. Biochem. Physiol.* **21,** 125–132.

Smith, L. S., and Newcomb, T. W. (1970). A modified version of the Blažka respirometer and exercise chamber for large fish. *J. Fisheries Res. Board Can.* **27,** 1321–1324.

Snyder, C. D. (1905). On the influence of temperature upon cardiac contraction and its relation to influence of temperature upon chemical reaction velocity. *Univ. Calif. Publ. Physiol.* **2,** 125–146.

Spaas, J. T. (1959). Contribution to the biology of some cultivated cichlidae. Temperature, acclimation, lethal limits and resistance in three cichlidae. *Biol. Jaarboek Konink. Natuurw. Genoot. Dodonaea Gent* **27,** 21–38.

Spoor, W. A. (1946). A quantitative study of the relationship between the activity and oxygen consumption of the goldfish and its application to the measurement of respiratory metabolism in fishes. *Biol. Bull.* **91,** 312–325.

Sprague, J. B. (1968). Avoidance reactions of salmonid fish to representative pollutants. *Water Res.* **2,** 23–24.

Sprague, J. B., and Ramsay, B. A. (1965). Lethal levels of mixed copper-zinc solutions for juvenile salmon. *J. Fisheries Res. Board Can.* **22,** 425–432.

Steen, J. B. (1963). Oxygen secretion in the swimbladder. *Proc. 5th Intern. Congr. Biochem., Moscow, 1956* pp. 621–630. Pergamon Press, Oxford.

Stevens, E. D., and Fry, F. E. J. (1971). Brain and muscle temperatures in ocean caught and captive skipjack tuna. *Comp. Biochem. Physiol.* **38A,** 203–211.

Sullivan, C. M. (1954). Temperature reception and responses in fish. *J. Fisheries Res. Board Can.* **11,** 153–170.

Sullivan, C. M., and Fisher, K. C. (1953). Seasonal fluctuations in the selected temperature of speckled trout, *Salvelinus fontinalis* (Mitchill). *J. Fisheries Res. Board Can.* **10,** 187–195.

Swift, D. R. (1964). The effect of temperature and oxygen on the growth rate of the Windermere char (*Salvelinus alpinus willughbii*). *Comp. Biochem. Physiol.* **12,** 179–183.

Tang, P. (1933). On the rate of oxygen consumption by tissues and lower organisms as a function of oxygen tension. *Quart. Rev. Biol.* **8,** 260–274.

Tåning, Å. V. (1952). Experimental study of meristic characters in fishes. *Biol. Rev.* **27,** 169–193.

Taylor, G. I. (1923). Experiments on the motion of solid bodies in rotating fluids. *Proc. Roy. Soc.* **A104,** 213–218.

Tyler, A. V. (1966). Some lethal temperature relations of two minnows of the genus *Chrosomus. Can. J. Zool.* **44,** 349–364.

Tytler, P. (1969). Relationship between oxygen consumption and swimming speed in the haddock, *Melanogrammus aeglefinus. Nature* **221,** 274–275.

Ushakov, B. P. (1968). Cellular resistance, adaptation to temperature and thermostability of somatic cells with special reference to marine animals. *Marine Biol.* **1,** 153–160.

van Dam, L. (1938). "On the Utilisation of Oxygen and Regulation of Breathing in Some Aquatic Animals," pp. I-143. Volharding, Gröningen.

van't Hoff, J. H. (1884). Études de dynamique chimique. Amsterdam.

Verheijen, F. J. (1958). The mechanism of the trapping effect of artificial light sources upon animals. *Arch. Neerl. Zool.* **13,** 1–107.

Veselov, E. A. (1949). Effect of salinity of the environment on the rate of respiration in fish. *Zool. Zh.* **28,** 85–98.

von Bertalanffy, L. (1950). The theory of open systems in physics and biology. *Science* **111,** 23–29.

von Ledeburg, J. F. (1939). Der Sauerstoff als ökologischer Faktor. *Ergeb. Biol.* **16,** 173–261.

Wells, N. A. (1935). Variations in the respiratory metabolism of the Pacific killifish *Fundulus parvipinnis*, due to size, season and continued constant temperature. *Physiol. Zool.* **8,** 318–336.

Whitmore, C. M., Warren, C. E., and Doudoroff, P. (1960). Avoidance reactions of salmonid and centrarchid fishes to low oxygen concentrations. *Trans. Am. Fisheries Soc.* **89,** 17–26.

Whitworth, W. R. (1968). Effects of diurnal fluctuations of dissolved oxygen on the growth of brook trout. *J. Fisheries Res. Board Can.* **25,** 579–584.

Wikgren, B. J. (1953). Osmotic regulation in some aquatic animals with particular respect to temperature. *Acta Zool. Fenn.* **71,** 1–93.

Winberg, G. G. (1956). Intensivnost obmena i pichchevye potrebnosti ryb. *Nauch. Tr. Belorussk. Gos. Univ.* [for translation, see *Fisheries Res. Board Can., Transl. Ser.* **194** (1960)].

Wohlschlag, D. E. (1957). Differences in metabolic rates of migratory and resident freshwater forms of an Arctic whitefish. *Ecology* **38,** 502–510.

Wohlschlag, D. E. (1964). Respiratory metabolism and ecological characteristics of some fishes in McMurdo Sound, Antarctica. *In* "Biology of the Antarctic Seas" (M. O. Lee, ed.), Antarctic Res. Ser. No. 1, pp. 33–62. Am. Geophys. Union.

Wohlschlag, D. E., Cameron, J. N., and Cech, J. J., Jr. (1968). Seasonal changes in the respiratory metabolism of the pinfish (*Lagodon rhomboides*). *Contrib. Marine Sci.* **13,** 89–104.

Woodhead, P. M. J. (1964). The death of North Sea fish during the winter of 1962-63, particularly with reference to the sole, *Solea vulgaris. Helgolaender Wiss. Meeresuntersuch.* **10,** 283–300.

Wuhrmann, K., and Woker, H. (1948). Beiträge zur Toxikologie der Fische. II. Experimentelle Untersuchungen über die Ammoniak- und Blausäurevergiftung. *Schweiz. Z. Hydrol.* **11,** 210–244.

Zahn, M. (1962). Die Vorzugstemperaturen zweier Cypriniden und eines Cyprinodonten und die Adaptationstypen der Vorzugstemperatur bei Fischen. *Zool. Beitr.* [N.S.] **7,** 15–25.

Zahn, M. (1963). Jahreszeitliche Veränderungen der Vorzugstemperaturen von Scholle (*Pleuronectes platessa* Linne) und Bitterling (*Rhodeus sericeus Pallas*). *Verhandl. Deut. Zool. Ges. Muenchen* pp. 562–580.

Chapter 15

From foundations to frontiers: Setting the stage for advances in fish physiological energetics

Shaun S. Killen[a,*] and Tommy Norin[b]

[a]*School of Biodiversity, One Health, and Veterinary Medicine, University of Glasgow, Glasgow, United Kingdom*

[b]*DTU Aqua: National Institute of Aquatic Resources, Technical University of Denmark, Lyngby, Denmark*

[*]*Corresponding author: e-mail: shaun.killen@glasgow.ac.uk*

Shaun Killen and Tommy Norin discuss the J. R. Brett and T. D. D. Groves chapter "Physiological Energetics" in *Fish Physiology*, Volume 8, published in 1979. The next chapter in this volume is the re-published version of that chapter.

In the 45 years since its publication, Brett and Groves seminal review chapter on fish physiological energetics has been enormously influential across multiple disciplines, providing a comprehensive overview of fish energy budgets, including methodologies, nutrition, thermal adaptation, metabolic scaling, and environmental effects on growth. Here, we highlight the chapter's enduring relevance and impact on contemporary research, noting how advancements in knowledge, technology, and methodological approaches that have built upon Brett and Groves' foundational principles. Notably, there are also numerous concepts discussed by Brett and Groves that have been largely forgotten but that could form the basis for valuable conceptual frameworks or analytical approaches in contemporary studies of fish ecophysiology. Using Brett and Groves chapter as a historical reference point, we highlight ongoing challenges in accurately partitioning energy budgets and the complexity of metabolic processes, including costs associated with maintenance, activity, and digestion. We also discuss the intricate relationship between metabolism and behavior, an intense focus of recent research, in the context of Brett and Groves original perspectives on this topic. Brett and Groves' work presciently identified key research areas, many of which continue to be explored today. We discuss the historical significance of their chapter and emphasize its continuing influence and our evolving understanding of fish bioenergetics in the face of environmental changes.

As we examine the enduring impact of Brett and Groves' chapter *Physiological Energetics*, it is clear that their review has been a foundation for inspiring

Fish Physiology, Vol. 40B. https://doi.org/10.1016/bs.fp.2024.08.002

a generation of ecophysiologists and advancements in fish physiology and related fields. During the PhD studies of the eldest author of this retrospective, for example, Brett and Groves' chapter was repeatedly photocopied from the university library, due to copies being worn by highlighting, note-scribbling, and coffee spills. Having been cited nearly 3000 times to date, including by studies of comparative physiology, behavioral ecology, aquaculture, and fisheries science, the chapter has been an invaluable reference across these fields for more than four decades. Structured with sections focussing on the primary components of the fish energy budget, the chapter covers an enormous range of subtopics including methodologies, nutrition, thermal adaptation, metabolic scaling, metabolic fuel use, and environmental effects on growth. The chapter acknowledges the difficulties in piecing together the multifaceted components of the fish energy budget, but nonetheless provides a comprehensive attempt to do so that elegantly ties together food intake and food composition with metabolism, digestibility, excretion, and growth that remains unrivaled today; it does so while drawing valuable (and sometimes forgotten) comparisons with endotherm energetics that reveal substantial differences in daily energy uses between fishes and mammals such as ourselves (e.g., Pontzer et al., 2021). The work also goes beyond examining physiological mechanisms by touching on how physiological traits and data can be used to understand behavioral ecology and responses to environmental factors. Given the recent explosion of research centred on how energetics and metabolism affect how fish behave and cope with ongoing environmental change, one wonders if Brett and Groves could have anticipated the direction of this field and the importance of fish bioenergetics in ecophysiological research.

While the ecophysiological toolbox has expanded substantially since the time this chapter was written, the foundational principles underlying estimation of metabolic rates of fishes remain essentially the same. Brett and Groves initiate their review with a summary of the basic theory underlying calorimetric methods for estimating metabolic rates. They lay out the distinction between direct and indirect calorimetry, and briefly discuss some of the challenges of direct calorimetry in aquatic systems. Despite the notable development of at least one system for aquatic direct calorimetry that can detect metabolic heat produced by fish (Regan et al., 2013), indirect calorimetry via respirometry remains by far the most popular method of estimating metabolic rates in fishes (Killen et al., 2021). Technological innovations, including the development of optodes for measuring dissolved oxygen, have greatly improved the versatility of respirometry systems, allowing estimates of oxygen uptake at various life-stages and under varying ecologically-relevant conditions (e.g., Norin et al., 2018). Rapidly advancing technologies and approaches are also improving our ability to piece together energy budgets for free-ranging fish in the wild (e.g., accelerometry, otoliths; Chung et al., 2019; Metcalfe et al., 2016a,b). Indeed, in the concluding paragraph of their chapter, Brett and Groves specifically comment on how, at that time, deriving estimates of energy budgets in free ranging fish was all but impossible without attempting to align behavioral observations over time with the estimated costs of those behaviors as derived using indirect calorimetry. It is important to note, however, that for many of the recently developed techniques, estimates of energy expenditure similarly need to be calibrated against measures of oxygen uptake that are still obtained using indirect calorimetry. For example, the estimated costs of activity derived from accelerometers must be calibrated against the costs of movement for given estimates of whole-body acceleration using swim-tunnel respirometry. Therefore, Brett and Groves' chapter serves as a reminder that although these advanced approaches allow estimations of energy use that were previously impossible, they are another step removed from the core assumptions and constraints that are already present when using any form of indirect calorimetry, and so appropriate caution is warranted when interpreting results. This is particularly true given that indirect calorimetry only estimates the aerobic component of metabolism within an animal, while anaerobic metabolism cannot be directly estimated using measurements of oxygen uptake (although the aerobic repayment of the oxygen debt incurred during anaerobic activity can be estimated; Brett, 1964; Svendsen et al., 2010). When using swim-tunnel respirometry, therefore, there may be an underestimation of activity costs, especially when fish are swimming at intermediate or high speeds (Svendsen et al., 2010).

The discussion of the principles of respirometry is a starting point for Brett and Groves to review the biochemical substrates for metabolism in fishes and the energetic consequences of their usage. Metabolic fuel use has been studied from a purely mechanistic perspective, and by comparing fuel use among different species, but has been surprisingly neglected in terms of the environmental and ecological effects that may affect fuel use within species. As discussed by Brett and Groves, the mix of molecules used to fuel metabolism—whether it be carbohydrates, lipids, or proteins—will ultimately determine the exact ratio between oxygen uptake and actual energy use by the animal. They make the point that—where carbohydrates and lipids are the primary energy substrates for mammals—fishes (at least the carnivorous or omnivorous ones that have been primarily studied) burn mainly lipids and proteins, and thus a comparatively lower oxycaloric coefficient applies to fishes. In addition, higher density food items that are richer in lipids are also beneficial for maximizing growth at warmer temperatures, because individuals can satisfy their increased metabolic demand while maximizing surplus energy for growth. This phenomenon has been noted in the aquaculture industry, with high-energy feeds being used to promote maximum food-conversion efficiency at warmer temperatures, but there could also be interesting ecological effects that we know little about. For example, to support growth at higher temperatures, fish may need to switch to prey with a higher energy density, or individuals and species that are more sensitive to warming may be most likely to display changes in prey selection that can negatively affect their growth if they are unable to appropriately adjust their diet (Guzzo et al., 2017). Aquaculture research has also sought to find ways of maximizing feed conversion efficiency, by adjusting feed formulation or stimulating some amount of swimming activity during rearing to promote increased carbohydrate metabolism and preserve protein for growth. Overall, there is mixed evidence as to the effectiveness of these strategies, because the exact feed intake and metabolic adjustments are often not directly measured, and so the exact mechanisms of any increased growth rate in response to manipulations are unclear (McKenzie et al., 2021). For example, increased food intake due to activity or changes in feed composition may increase growth rates but not necessarily increase the efficiency with which nutrients are converted to structural mass. Brett and Groves also draw attention to the substantial differences between fishes with different lifestyles and diets (e.g., herbivores vs carnivores) in the fraction of ingested energy that is lost in excretion—and thus the fraction that is available for metabolism and growth—and they underline the difficulties in assessing this, largely due to an inaccuracy of fecal (energy) measurements (disappearance of feces because they dissolve, or altered energy content due to bacterial colonization). Taken together, the interactions among food intake, food composition, fuel use, and growth reviewed by Brett and Groves remain an area of intense study in the field of aquaculture and could also provide novel insights into fish ecophysiology.

Brett and Groves then discuss factors that influence standard metabolic rate (SMR), defined as the minimum rate of energy throughput required to sustain life in a resting, non-growing and non-digesting animal at a specific temperature. The criterion that SMR can only be estimated in a non-growing animal is often relaxed in research on fishes, given that they are indeterminate growers and continue to grow on even low rations of food, making it nearly impossible to avoid some form of growth. However, Brett and Groves estimate the cost of growth to equal that of true (non-growing) SMR, each amounting to ~10% of the energy available for metabolism after excretory loses (or ~7% of gross energy intake) for an average carnivorous fish, providing an important reminder of the non-negligible cost of growth. It is interesting to note that, throughout the chapter, SMR is treated as this generally minor consideration for fish in terms of its importance for overall energy budgeting and its ecophysiological consequences (in comparison, basal metabolic rate of humans is 50–70% of their daily energy use; Pontzer et al., 2021). In contrast, large volumes of current research on fishes are aimed at understanding how inter- and intra-specific variation in SMR may be related to variation in individual behavior (Metcalfe et al., 2016b), species' morphology and lifestyle (Killen et al., 2016a), biogeography (Deutsch et al., 2020), and broad-scale ecological phenomena (Brandl et al., 2023). Within a given fish species, for example, it is now commonly observed and appreciated that the SMR can show approximately twofold variation among individuals that are the same size and measured under identical conditions, with consequences for behavior and overall performance (Metcalfe et al., 2016b).

Brett and Groves go on to discuss how SMR among species may be affected by metabolic cold adaptation (MCA), or the idea that cold-adapted species may display an elevated SMR when compared to warmer-adapted species measured at the same acclimation temperature. In their discussion of this topic, Brett and Groves side with Holeton, who had rightly pointed out that early examinations of MCA were likely biased by stress, spontaneous activity, and short acclimation times (e.g., Holeton, 1974). The debate surrounding MCA in fishes and ectotherms in general has continued for decades, with recent evidence from allopatric Icelandic populations of threespine sticklebacks suggesting that evolution of populations in geothermal hot pools can cause a compensatory reduction in SMR, providing support for the existence of MCA in fishes (Pilakouta et al., 2020). Furthermore, it has now also been observed that MCA occurs at the mitochondrial and enzyme levels, suggesting that MCA is not simply an artifact of handling stress during measurements of whole-animal oxygen uptake (White et al., 2012).

Brett and Groves conclude their discussion of SMR with a brief overview of the effects of allometric scaling, stating that the effects of body mass on SMR in fish is much different than the ⅔ scaling exponent that had been observed for endotherms. They not only propose a much higher scaling exponent for SMR of fishes (0.86), which we now know is more appropriate (e.g., Clarke and Johnston, 1999; White et al., 2006), but also point out the still overlooked phenomenon that the scaling exponent varies with taxonomic level, being different when investigated across vs within species—a difference that extends even further to scaling within individuals as they grow throughout ontogeny (Norin, 2022). Brett and Groves conclude their section on SMR by stating that "… the species-specific characteristic [of the scaling exponent] is sufficiently unique to justify separate determination for any given species, at the normal temperature experienced …". While research on the effects of body mass on maintenance metabolism in fishes has advanced since Brett and Groves' chapter, they were somewhat ahead of their time by recognizing that patterns of metabolic scaling in fishes can indeed be influenced by their species-specific characteristics, such as lifestyle, activity level, and temperature, which is well-supported today for both fishes (Killen et al., 2010) and animals in general (e.g., Glazier, 2005; Harrison et al., 2022).

In downplaying the importance of maintenance costs in fish bioenergetics, Brett and Groves provide an important reminder of just how critical activity costs can be for an animal's energy budget, often dwarfing maintenance metabolism over specific timeframes. Fig. 3 of their chapter presents a "time–speed" curve, showing the times taken for a 50 g sockeye salmon at 15 °C to consume 10 mg O_2 while performing different levels of swimming activity. Amazingly, within only a few seconds, an individual will use the same amount of energy while burst swimming (such as when involved in a predator–prey interaction) as they would in several minutes on maintenance costs while motionless. This time–speed analysis has been largely forgotten in contemporary ecophysiological research but could be applied at the level of the individual to understand intraspecific variation in costs associated with specific behaviors, and perhaps linking movement efficiency to consistent individual differences in behavior (i.e., animal personality). For example, it has been demonstrated that individuals within species show varying degrees of swimming efficiency and increases in energy expenditure with swimming speed (Killen et al., 2016b; Svendsen et al., 2013). Correspondingly, individuals that show higher integrated movement costs over time, due to a lower locomotor efficiency, may be more motivated to reduce activity and therefore display a slower, risk-averse lifestyle. As described by Brett and Groves, the upper limit for active metabolism is set by an individual's maximum rate of oxygen uptake, or maximum metabolic rate (MMR). Despite recognizing the importance of active metabolism for fish bioenergetics, there had been relatively little research into this aspect of whole-animal metabolism at the time Brett and Groves compiled their chapter. For example, only four fish species had been thoroughly studied for changes in MMR across a range of temperatures, with three of these showing an optimal temperature for MMR, beyond which performance declined with further warming. Today, dozens more species have been measured for changes in MMR across temperatures, and while many show a peak in MMR (and aerobic scope, AS) at some optimal temperature, many others do not show a clear optimum (Lefevre, 2016). Furthermore, while current evidence suggests that the decline in MMR with warming past the thermal optimum may be related to reductions in cardiac function or other aspects of oxygen supply capacity, nearly 45 years after Brett and Groves chapter, we still do not fully understand this adverse warming effect on the upper limits of aerobic metabolic capacity of many fish.

In addition to physical activity, elevated metabolism in fishes can also occur after feeding. The digestion and assimilation of nutrients causes a substantial increase in oxygen uptake in fishes, often referred to as specific dynamic action (SDA). Brett and Groves noted that, at that time, more information on this post-feeding increase in metabolic rate was needed for constructing energy budgets and for understanding feeding efficiency. This remains true today, as SDA has been relatively neglected in terms of research attention when compared to quantification of SMR, MMR, and aerobic scope. Far fewer species have been measured for oxygen uptake following feeding as compared to those that have been measured for these other metabolic traits, and we know little about the effects of environmental and intrinsic factors on post-feeding metabolic rates. Compared to maintenance metabolism or maximum metabolism, for example, we know very little about how the costs of digestion and assimilation scale with body mass (e.g., Secor, 2009). These short-comings are unfortunate given that feeding is obviously an ecologically-relevant task that all animals participate in on a frequent basis, and they are very likely to be digesting meals or assimilating nutrients at any given moment. Any metabolic effects on the efficiency of food conversion or digestion may have profound consequences on growth, fitness, behavior (McLean et al., 2018), and potentially the ability to respond to stressors or environmental change (Lefevre et al., 2021; Nuic et al., 2024). In addition, while Brett and Groves mainly discuss SDA in relation to the energy content of the ingested meal, more recent evidence suggests that the post-feeding rise in oxygen uptake in fish may occupy a large proportion of an individual's available aerobic scope (Sandblom et al., 2014), potentially constraining their ability to perform other oxygen-requiring tasks or behaviors after feeding. In fact, for some species, it seems likely that maximum metabolic rate—and therefore accurate estimates of aerobic scope—may only be reached after feeding and not during activity-induced increases in oxygen uptake (Fu et al., 2022). These effects may be especially relevant for selection or behavior given that individuals within species seem to differ in the nature of their post-feeding increase in metabolism, with some showing a higher peak in oxygen uptake following feeding but with a reduced duration, while others show a lower peak but more prolonged period over which metabolic rate is apparently elevated (McLean et al., 2018). Interestingly, despite their initial call for increased data on post-feeding metabolism, Brett and Groves conclude their section on feeding metabolism by stating that "It is safe to say … that unless serious attention to the effect of diet composition on heat increment is the aim, it is doubtful if the necessary effort to correctly partition feeding metabolism into its respective components is warranted." On the contrary, the most recent developments in fish bioenergetics suggest that increased general knowledge in this area could be invaluable for understanding fish behavior, ecology, and their ability to cope with climate change (Lefevre et al., 2021; Sandblom et al., 2014).

The continued paucity of data on post-feeding metabolism is probably related to the increased methodological complexity needed to acquire these data. As Brett and Groves describe, attention must be paid to the exact amount of food consumed by individuals and the effects of any disturbance or stress during feeding on the rise in oxygen uptake that is subsequently observed. Indeed, some species are reluctant to feed when contained in a respirometer or without the presence of conspecifics, which can make controlled feeding with minimal handling difficult. Brett and Groves describe an interesting experiment by Beamish that attempted to control for the metabolic effects of excitement and activity during feeding, in which fish were first trained to receive food with forceps, then control fish were offered empty forceps before being measured for oxygen uptake (Beamish, 1974). This clever design would help control for effects of the feeding process itself, but the exact approach to quantifying and controlling for these effects will likely depend on the species of interest and should be carefully considered on a case-by-case basis. Unfortunately, some contemporary studies of the post-feeding rise in metabolic rate provide scarce details of the control treatments that were used, and so in addition to there being relatively little data on post-feeding metabolism as compared to other whole-animal metabolic traits, the quality of the data is often also unclear. Finally, the SDA response has recently been argued to represent the energy cost of growth, rather than the traditional view of it being the energy cost of digestion (Goodrich et al., 2024), underlining how we still lack a good understanding of this post-feeding rise in metabolic rate.

Brett and Groves' consideration of how estimates of SMR and SDA may be affected by fish activity relate to their overall appreciation for the links between behavior and energetic demand. More specifically, Brett and Groves' discussion of how aggression, migration, and spawning are related to metabolic rates foreshadowed the surge in studies examining the interactions between behavior and metabolic rates that was to emerge during the subsequent decades. In fact, much of the early work examining links between energy use and behavior in fish focussed on aggression and dominance, mostly in salmonids (e.g., Metcalfe et al., 1995). Interestingly, Brett and Groves strictly focus on how behaviors can influence energy demand, for example, describing the pronounced increase in energy demand that can result during agonistic interactions. Within the next 10–15 years, however, there would be a strong research effort to understand how baseline energy requirements may influence the expression of behavior and dominance (Biro and Stamps, 2010; Metcalfe et al., 1995; Metcalfe et al., 2016b). Even more recently, these two streams of research have converged to highlight the complex interactions and feedbacks that exist between energy demand and behavior (Mathot et al., 2019). On one hand, for example, increased maintenance metabolism may drive increased foraging demand, risk-taking, and aggression in individuals, presumably due to an increased motivation to acquire resources. On the other hand, social behaviors, foraging history, and activity level can feedback to alter organ systems and whole-animal energy requirements. The most recent studies examining the interactions between behavior and energetics also emphasize the potential constraining effects that short-term increases in stress or activity can have on the aerobic scope available for other aerobic processes during that time period. Contemporary work examining links between energy use and behavior also highlight that these relationships are highly dependent on the environmental context (Killen et al., 2013), with implications for trait variation expressed within populations and the potential for correlated selection between metabolic and behavioral traits (Crespel et al., 2024). Indeed, it may be argued that of all the topics examined within Brett and Groves' comprehensive review, the study of how behavior and metabolism are related in an ecological context may have experienced the greatest increase in research attention since the time their chapter was published, with fishes acting as key model organisms in this line of research.

It has now been 45 years since Brett and Groves so comprehensively reviewed the literature and paved the way for future work in fish physiological energetics. Upon re-reading this classic chapter, one is struck by the apparent prescience expressed in their synthesis and their suggested priorities for future research. Incredibly, for example, most of the areas they highlight for further study in their "Concluding Commentary" remain the focus of intense contemporary research. This includes an increased understanding of the energetics of fish in the wild, the effects of fluctuating environmental conditions, changes in metabolic rates with body size, and the importance of feeding metabolism. In Brett and Groves' own final words of their chapter, this is likely because "The diversity of morphology of fishes is undoubtedly matched by a comparable diversity in physiology." Indeed, the ability of Brett and Groves to identify priority research avenues, along with providing a rich yet succinct overview of fish physiological energetics, has allowed their chapter to be influential for decades and be likely to remain a keystone contribution well into the future.

References

Beamish, F.W.H., 1974. Apparent specific dynamic action of largemouth bass, *Micropterus salmoides*. J. Fish. Res. Board Can. 31, 1763–1769.

Biro, P.A., Stamps, J.A., 2010. Do consistent individual differences in metabolic rate promote consistent individual differences in behavior? Trends Evol. Evol. 25, 653–659.

Brandl, S.J., Lefcheck, J.S., Bates, A.E., Rasher, D.B., Norin, T., 2023. Can metabolic traits explain animal community assembly and functioning? Biol. Rev. 98, 1–18.

Brett, J.R., 1964. The respiratory metabolism and swimming performance of young sockeye salmon. J. Fish. Res. Bd. Canada 21, 1183–1226.

Chung, M.-T., Trueman, C.N., Godiksen, J.A., Holmstrup, M.E., Grønkjær, P., 2019. Field metabolic rates of teleost fishes are recorded in otolith carbonate. Comm. Biol. 2, 24.

Clarke, A., Johnston, N.M., 1999. Scaling of metabolic rate with body mass and temperature in teleost fish. J. Anim. Ecol. 68, 893–905.

Crespel, A., Lindström, J., Elmer, K.R., Killen, S.S., 2024. Evolutionary relationships between metabolism and behaviour require genetic correlations. Philos. Trans. R. Soc. B 379. 20220481.

Deutsch, C., Penn, J.L., Seibel, B., 2020. Metabolic trait diversity shapes marine biogeography. Nature 585, 557–562.

Fu, S.J., Dong, Y.W., Killen, S.S., 2022. Aerobic scope in fishes with different lifestyles and across habitats: trade-offs among hypoxia tolerance, swimming performance and digestion. Comp. Biochem. Physiol. A 272, 111277.

Glazier, D.S., 2005. Beyond the '3/4-power law': variation in the intra- and interspecific scaling of metabolic rate in animals. Biol. Rev. 80, 611–662.

Goodrich, H.R., Wood, C.M., Wilson, R.W., Clark, T.D., Last, K.B., Wang, T., 2024. Specific dynamic action: the energy cost of digestion or growth? J. Exp. Biol. 227, jeb246722.

Guzzo, M.M., Blanchfield, P.J., Rennie, M.D., 2017. Behavioral responses to annual temperature variation alter the dominant energy pathway, growth, and condition of a cold-water predator. Proc. Natl. Acad. Sci. U. S. A. 114, 9912–9917.

Harrison, J.F., Biewener, A., Bernhardt, J.R., Burger, J.R., Brown, J.H., Coto, Z.N., Duell, M.E., Lynch, M., Moffett, E.R., Norin, T., Pettersen, A.K., Smith, F.A., Somjee, U., Traniello, J.F.-A., Williams, T.M., 2022. White paper: an integrated perspective on the causes of hypometric metabolic scaling in animals. Integr. Comp. Biol. 62, 1395–1418.

Holeton, G.F., 1974. Metabolic cold adaptation of polar fish: fact or artifact. Physiol. Zool. 47, 137–152.

Killen, S.S., Atkinson, D., Glazier, D.S., 2010. The intraspecific scaling of metabolic rate with body mass in fishes depends on lifestyle and temperature. Ecol. Lett. 13, 184–193.

Killen, S.S., Marras, S., Metcalfe, N.B., McKenzie, D.J., Domenici, P., 2013. Environmental stressors alter relationships between physiology and behaviour. Trends Ecol. Evol. 28, 651–658.

Killen, S.S., Glazier, D.G., Rezende, E.L., Clark, T.D., Atkinson, D., Willener, A., Halsey, L.G., 2016a. Ecological influences and morphological correlates of resting and maximal metabolic rates across teleost fish species. Am. Nat. 187, 592–606.

Killen, S.S., Croft, D.P., Salin, K., Darden, S.K., 2016b. Male sexually coercive behaviour drives increased swimming efficiency in female guppies. Funct. Ecol. 30 (4), 576–583.

Killen, S.S., Christensen, E.A.F., Cortese, D., Závorka, L., Norin, T., Cotgrove, L., Crespel, A., Munson, A., Nati, J.J.H., Papatheodoulou, M., McKenzie, D.J., 2021. Guidelines for reporting methods to estimate metabolic rates by aquatic intermittent-flow respirometry. J. Exp. Biol. 224, jeb242522.

Lefevre, S., 2016. Are global warming and ocean acidification conspiring against marine ectotherms? A meta-analysis of the respiratory effects of elevated temperature, high CO_2 and their interaction. Conserv. Physiol. 4.

Lefevre, S., Wang, T., McKenzie, D.J., 2021. The role of mechanistic physiology in investigating impacts of global warming on fishes. J. Exp. Biol. 224, jeb238840.

Mathot, K.J., Dingemanse, N.J., Nakagawa, S., 2019. The covariance between metabolic rate and behaviour varies across behaviours and thermal types: meta-analytic insights. Biol. Rev. 94, 1056–1074.

McKenzie, D.J., Palstra, A.P., Planas, J., MacKenzie, S., Bégout, M.L., Thorarensen, H., Vandeputte, M., Mes, D., Rey, S., De Boeck, G., Domenici, P., 2021. Aerobic swimming in intensive finfish aquaculture: applications for production, mitigation and selection. Rev. Aquac. 13, 138–155.

McLean, S., Persson, A., Norin, T., Killen, S.S., 2018. Metabolic costs of feeding predicatively alter the spatial distribution of individuals in fish schools. Curr. Biol. 28, 1144–1149.

Metcalfe, N.B., Taylor, A.C., Thorpe, J.E., 1995. Metabolic rate, social status and life-history strategies in Atlantic salmon. Anim. Behav. 49, 431–436.

Metcalfe, J.D., Wright, S., Tudorache, C., Wilson, R.P., 2016a. Recent advances in telemetry for estimating the energy metabolism of wild fishes. J. Fish Biol. 88, 284–297.

Metcalfe, N.B., Van Leeuwen, T.E., Killen, S.S., 2016b. Does individual variation in metabolic phenotype predict fish behaviour and performance? J. Fish Biol. 88, 298–321.

Norin, T., 2022. Growth and mortality as causes of variation in metabolic scaling among taxa and taxonomic levels. Integr. Comp. Biol. 62, 1448–1459.

Norin, T., Mills, S.C., Crespel, A., Cortese, D., Killen, S.S., Beldade, R., 2018. Heat-induced anemone bleaching increases the metabolic rate of symbiont anemonefish. Proc. R. Soc. B 285, 20180282.

Nuic, B., Bowden, A., Franklin, C.E., Cramp, R.L., 2024. Atlantic salmon *Salmo salar* do not prioritize digestion when energetic budgets are constrained by warming and hypoxia. J. Fish Biol. https://doi.org/10.1111/jfb.15693.

Pilakouta, N., Killen, S.S., Kristjánsson, B., Skulason, S., Lindstrom, J., Metcalfe, N.B., Parsons, K., 2020. Multigenerational exposure to elevated temperatures leads to a reduction in standard metabolic rate in the wild. Funct. Ecol. 34, 1205–1214.

Pontzer, H., et al., 2021. Daily energy expenditure through the human life course. Science 373, 808–812.

Regan, M.D., Gosline, J.M., Richards, J.G., 2013. A simple and affordable calorespirometer for assessing the metabolic rates of fish. J. Exp. Biol. 216, 4507–4513.

Sandblom, E., Gräns, A., Axelsson, M., Seth, H., 2014. Temperature acclimation rate of aerobic scope and feeding metabolism in fishes: implications in a thermally extreme future. Proc. R. Soc. B 281, 20141490.

Secor, S.M., 2009. Specific dynamic action: a review of the postprandial metabolic response. J. Comp. Physiol. B 179, 1–56.

Svendsen, J.C., Tudorache, C., Jordan, A.D., Steffensen, J.F., Aarestrup, K., Domenici, P., 2010. Partition of aerobic and anaerobic swimming costs related to gait transitions in a labriform swimmer. J. Exp. Biol. 213, 2177–2183.

Svendsen, J.C., Banet, A.I., Christensen, R.H., Steffensen, J.F., Aarestrup, K., 2013. Effects of intraspecific variation in reproductive traits, pectoral fin use and burst swimming on metabolic rates and swimming performance in the Trinidadian guppy (*Poecilia reticulata*). J. Exp. Biol. 216, 3564–3574.

White, C.R., Phillips, N.F., Seymour, R.S., 2006. The scaling and temperature dependence of vertebrate metabolism. Biol. Lett. 2, 125–127.

White, C.R., Alton, L.A., Frappell, P.B., 2012. Metabolic cold adaptation in fishes occurs at the level of whole animal, mitochondria and enzyme. Proc. R. Soc. B 279, 1740–1747.

Chapter 16

PHYSIOLOGICAL ENERGETICS[☆]

J.R. BRETT
T.D.D. GROVES

Chapter Outline

[☆]This is a reproduction of a previously published chapter in the Fish Physiology series, "1979 (Vol. 8)/ Physiological Energetics: ISBN: 978-0-12-350408-1; ISSN: 1546-5098".

Fish Physiology, Vol. 40B. https://doi.org/10.1016/bs.fp.2024.07.007

I. INTRODUCTION

Physiological energetics, or animal bioenergetics, concerns the rates of energy expenditure, the losses and gains, and the efficiencies of energy transformation, as functional relations of the whole organism. This distinguishes the use of "bioenergetics" as applied to the mechanisms of energy exchange within the cell, but still observing the same physical and chemical laws (e.g., Lehninger, 1965). It also distinguishes the study from "ecological energetics," which involves the transfer of energy from one trophic level to another (e.g., Odum, 1971; Winberg, 1970). All three are closely related.

The majority of such presentations commence with an energy flow diagram indicating the main steps that the energy of food intake follows through the organism, and the paths of energy distribution (e.g., Davies, 1964; Beamish *et al.*, 1975). Each of these steps with their appropriate values is subject to quantitative change, depending on many biotic and abiotic factors. With the thought that the basis of these energy exchanges needs to be elaborated first, it was deemed more fitting to conclude with a quantitatively expressed flow diagram (Fig. 18). It may be of help, however, to keep this format in mind, segments of which constitute the contents of this chapter. For the reader unfamiliar with the various components of expended energy, repeated reference to Fig. 18 is recommended.

The essential concept of biological energetics was captured by Kleiber (1961) with the title "The Fire of Life." Indeed, at the time when coal-fired steam engines were gaining acceptance, research on the comparative fuel requirements and work efficiency of domestic animals was greatly stimulated. Study of the energetic efficiencies of agricultural processes involving the production of such commodities as meat, milk, and eggs (e.g., Brody, 1945) has continued to provide an advanced background of knowledge from which the investigation of fish energetics has drawn insight and inspiration. An understanding of the physical, chemical, and biological basis on which the energetics is built, and the equivalents employed, constitutes the opening section of this chapter. Some necessary distinctions between mammalian and nonmammalian systems are made. An adequately nutritious diet is assumed; the basic source of fuel for the fire of life is solely derived from the food.

Since much of the energetics terminology must be used at the very outset, recurring in later passages where the meaning may be more explicit, a short glossary is included (Section VIII).

II. ENERGY: RELATIONS AND MEASUREMENTS

A. Thermodynamics and Biological Energy Flow

Fish, like other living systems, must conform to the laws of thermodynamics. Matter and energy may be converted but never destroyed. Fish gain matter and energy in food, and they lose absorbed matter and energy as a result of catabolism (which provides energy for maintenance and activity) and the elaboration of reproductive products. Catabolism of substrates results in the production of carbon dioxide and water (which are excreted), heat, and in some cases intermediate partially oxidized products (which may also be excreted).

During aerobic metabolism, 40–50% of the substrate chemical free energy is temporarily trapped in adenosine triphosphate and related labile compounds. These so-called "high energy" compounds provide the immediate driving energy for endergonic processes such as biosynthesis and membrane transport, and are the immediate fuel for conversion to mechanical work in muscle tissues. Chemical free energy is degraded to heat. In the case of homeotherms at temperatures below their thermal neutral range, this metabolic heat contributes to the maintenance of body temperature. With the possible exception of some warm-bodied sharks and tunas (Carey *et al.*, 1971), metabolic heat in the fish represents a complete loss of energy to the animal.

If body mass is to be maintained, absorbed dietary energy (exogenous sources) must equal energy loss for maintenance and activity. When exogenous sources exceed these requirements, growth can occur from the deposition of matter, which for fish is largely protein. Energy is also stored in growth as the chemical energy of covalent bonds in proteins, fats, and carbohydrates. If dietary energy is insufficient to cover catabolism, growth of some organs or body components may occur at the expense of internal (endogenous) sources previously stored in growth. In the absence of any dietary input all energy for maintenance and activity must be provided from endogenous sources.

A good example of the metabolic and growth alternatives available to fish according to the laws of thermodynamics and conservation of matter is provided in comparative data on systems for hatching salmonid eggs and holding alevins. Since the amount of yolk is fixed and hence the amount of matter and energy is fixed, systems that stimulate activity of embryos and alevins result in small "swim-up" fry having limited energy reserves. Systems that minimize such physical activity and hence minimize catabolism result in larger fry (Marr, 1966; Bams, 1970).

In living systems, the fate of absorbed food is more complicated than simple metabolic combustion for energy or deposition of matter in growth. One reason of major importance is that environmental factors strongly influence the biochemical state of the fish (see Chapters 1 and 10). A second concerns particularly the fate of protein. In catabolism, the total heat produced bioenergetically differs from that obtained by direct combustion of food because protein nitrogen is not fully oxidized. Therefore, protein catabolism produces less net energy than expected from the heat of combustion as determined by bomb calorimetry. In growth, only those proteins with an adequate amino acid balance can be utilized. Surplus dietary energy ingested as protein must be stored as fat. Efficient use of body fat at some later time however requires some degradation of carbohydrate and protein.

B. Calorimetry: Metabolic Heat Production

1. *Direct Calorimetry*

The basic principles of calorimetry date to the early experiments of Lavoisier and Laplace in 1780 and are summarized by Brody (1945), Kleiber (1961), and Blaxter (1965). In the case of direct calorimetry, heat loss is measured. Lavoisier and Laplace determined the amount of ice melting in a chamber which surrounded a guinea pig. Subsequent direct calorimeters have been based on the measurement of the heat increment in water circulated in the walls of the calorimeter, or of the air passing through the chamber. Another direct technique involves the measurement of heat input by a calibrated electric heater to maintain a given chamber temperature in the presence or absence of an animal. To obtain a more rapid response than that possible with heavy calorimeter chambers having high heat capacity, gradient layer calorimeters have been devised in which the temperature differential across a thin container wall of known thermal conductivity is measured. Calorimeters of this type have been described by Benzinger *et al.* (1958), Pullar (1958), and Mount (1963).

Direct calorimeters have been considered unsatisfactory for studies on fish because of the relatively low heat production by fish and the high heat capacity of the water in the system. Smith (1976), however, argues to the contrary, pointing out that problems of heat loss by vaporization and radiation, which complicate direct calorimetry in terrestrial animals, do not occur in aquatic forms. Using a modified adiabatic calorimeter, Smith (1976) determined the heat production of four species of salmonids. From the data presented, the sensitivity of metabolic response does not appear to be detected as well as by methods depending on measurements of oxygen consumption, due to the high heat buffering capacity of the direct calorimeter used.

2. *Indirect Calorimetry*

Indirect calorimetry, which also began with Lavoisier, involves measurement of gas exchange. The calorific equivalents for oxygen uptake and CO_2 production are based on heats of combustion of different energy substrates, measured in a direct bomb calorimeter. The type of physiological fuel involved can be established by determining the RQ, or respiratory quotient (defined as the ratio of moles of CO_2 produced to O_2 utilized), and the urinary or nonfecal nitrogen excretion. Under totally aerobic conditions, the values of RQ vary from 0.7 for fat to 1.0 for carbohydrate. In the classical literature on terrestrial ureotelic animal respirometry, the RQ for protein catabolism is approximately 0.82. In the case of ammonotelic fish, the RQ for protein is nearly 0.9. In either case, if the urinary or nonfecal nitrogen excretion is measured, an estimate can be made of heat production due to amino acid or protein breakdown. This amount, with appropriate adjustment of the oxygen and CO_2 volumes, allows an estimate of heat production from nonprotein energy sources. RQ values of greater than unity have been recorded in cases where there was active synthesis of body fat. In addition, particularly in the case of fish, anaerobic metabolism can result in RQ values of between 1 and 2 (Kutty, 1972). The caloric equivalent of oxygen uptake, Q_{ox}, consequently varies with the physiological state and substrate utilized.

In mammalian studies, the caloric equivalent of respired oxygen during catabolism of a mixed protein (RQ = 0.81) is 4.82 kcal/liter O_2 at standard temperature and pressure (Brody, 1945). The caloric equivalent of oxygen during aerobic carbohydrate catabolism (RQ = 1.0) is 5.04 kcal/liter O_2, and that for mixed fat (RQ = 0.71), 4.69 kcal/liter O_2. At a mean nonprotein RQ of 0.85, the equivalent is 4.86 kcal/liter O_2. Thus, for studies in mammalian metabolism where carbohydrate and fat are the principal energy substrates, a mean oxycalorific value of approximately 4.86 kcal/liter O_2 is appropriate.

For fish, the Q_{ox} for carbohydrate is also taken as 5.04 kcal/liter O_2. Oxycalorific equivalents for fat of 4.66 and 4.69 have been determined from the literature by Brafield and Solomon (1972) and Elliot and Davison (1975), respectively. The same authors record a Q_{ox} of 4.58 kcal/liter O_2 for the ammonotelic catabolism of protein. Since the principal energy sources in carnivorous fish appear to be lipid and protein rather than lipid and carbohydrate, a mean Q_{ox} of 4.63 kcal/liter O_2 may be more appropriate in aerobic, steady state metabolic studies of fish than the frequently applied values of 4.8–5.0 kcal/liter O_2.

Indirect calorimeters for animals have been operated on both the constant volume, open circuit principle, where differential composition of air entering and leaving the chamber is measured, and on the closed circuit principle in which the change of concentration of chamber oxygen is measured after the carbon dioxide produced is absorbed. In studies on fish, tunnel-type respirometers have been operated on a closed circuit system during sampling intervals (Blazka *et al*., 1960; Brett, 1964; Beamish, 1970). By determining the rates of decrease in dissolved oxygen concentration during closed circuit intervals, precise data on the energy expenditure of fish at different temperatures and at different levels of activity can be obtained. Since dissolved oxygen is more conveniently measured than dissolved carbon dioxide, the RQ of fish is not usually determined. Tunnel respirometers of the type described by Brett (1964) measure the energy exchange of *swimming* fish since movement of water through the apparatus is necessary for both temperature control and representative sampling for oxygen.

Because of some doubt cast by Krueger *et al*. (1968) on the oxycalorific value appropriate for fish respiration, Brett (1973) resorted to testing the basis by a technique first applied to fish by Pentegov *et al*. (1928) and subsequently developed by Idler and Clemens (1959). These authors determined energy expenditure by measuring the change in body composition of migrating salmon, which naturally starve as they proceed from the sea to distant spawning grounds. Thus, by direct determination of the change in energy content (bomb calorimetry) of the flesh and viscera of exercised fish, and by comparison of this with the value deduced from oxygen consumption measurements, an oxycalorific equivalent of 4.8* kcal/liter O_2 was shown to be an acceptable value for fish. The availability of large numbers of fish, providing an opportunity for sacrificing subsamples for determining energy content, provides an opportunity of study not possible with many other vertebrates.

C. Energy Sources

There are some major quantitative differences between the dietary energy sources of fish and land mammals. In contrast to terrestrial animals, fish utilize dietary carbohydrate poorly, both at the level of digestion and in their capacity to metabolize absorbed carbohydrate. Raw starches are only 30–40% digestible by salmonids, and digestibility appears to decrease markedly when carbohydrate levels exceed 25% of the ration. More omnivorous or herbivorous species such as catfish and carp have a higher capacity to utilize carbohydrates. A number of species have been shown capable of developing a cellulolytic intestinal microflora (Stickney, 1974). Even in herbivorous fish such as grass carp, the principal digestible energy available in the diet is in the form of protein and simple carbohydrates (e.g., disaccharides, oligosaccharides, and hemicelluloses).

Protein is a major energy source for all fish, and it appears that blood glucose may derive more readily through'gluconeogenesis than directly from dietary carbohydrates. This may account for the fact that the optimum protein/calorie ratio in prepared salmonid rations is in the vicinity of 120 mg protein/digestible kcal, as compared to a ratio of approximately 70 mg/kcal in mammalian rations.

Lipids are the principal nonprotein energy source in the natural diets of both carnivorous and omnivorous fish. Provided dietary fats are liquid at ambient temperatures, fats are both highly digestible and readily metabolized.

D. Available Energy

Phillips (1969) assigned caloric values to dietary protein, carbohydrate, and fat that allow an estimate of the available energy of a trout ration based on the proximate composition of the feed. The assigned caloric values take into consideration the average digestibility of the components together with the nonfecal excretory energy losses but not the heat increment, and are therefore an estimate of metabolizable energy rather than net energy (Fig. 18). The values given for dietary fats and carbohydrates are the gross energies (9.45 kcal/g of fat, and 4.0 kcal/g of raw starch) modified only by mean digestibilities of 0.85 for fat and 0.4 for carbohydrate. The resulting metabolizable energy values are 8.0 kcal/g of dietary fat and 1.6 kcal/g of dietary carbohydrate. The latter

---

*The accuracy of the method did not allow a precise measure, such as 4.63 kcal recommended now.

value is not valid for cooked starch, which has been shown to be approximately 80% digestible by trout (Nose, 1967) resulting in a metabolizable energy closer to 3.3 kcal/g. Phillips' value of 3.9 kcal "available" energy/g of protein was derived by applying an average digestibility factor of 0.9 and by deducting 1.3 kcal/g from the gross energy of digestible protein (5.66 kcal/g).

Brody (1945) refers to a urinary loss of about 1.3 kcal/g of protein catabolized in the form of nitrogenous excretory products, principally urea. This corresponds approximately to the heat of combustion of urea (151.6 kcal/mole—kleiber, 1961) equivalent to 0.86 kcal/g of protein, plus the energy cost of urea biosynthesis amounting to approximately 0.35 kcal/g of protein (or a total of 1.21 kcal/g of protein). The difference between 1.21 and 1.3 kcal/g of protein is due to the other nitrogenous wastes (uric acid and creatinine) which have higher heats of combustion than urea, but which constitute only about 10% of the total urinary nitrogen in mammals. The heat loss due to urea synthesis would be more properly included as part of the heat increment of dietary protein. The 0.35 kcal/g of protein in ureotelic animals does, however, represent a component of energy which is unavailable for purposes other than nitrogen excretion when protein is catabolized.

Since teleost fish excrete primarily ammonia in freshwater, the nonmetabolizable energy fraction of absorbed dietary protein will reflect the heat of combustion of ammonia—5.94 kcal/g of excreted ammonia nitrogen or 0.95 kcal/g of protein (Elliott and Davison, 1975). The metabolizable energy for *absorbed* protein in freshwater teleosts is therefore close to 4.70 kcal/g of protein catabolized. Assuming a mean digestibility of 0.9, the energy available to fish in dietary protein is approximately 4.23 kcal/g rather than the 3.9 of Phillips.

Although ammonia has been shown to contribute up to 75–90% of the nitrogenous excretion of marine teleosts (Wood, 1958; cf. Table V) a small proportion of the nitrogen excreted by marine species is in the form of trimethylamine oxide, some of which may have originated as such in the diet. Excretion of endogenous trimethylamine oxide, synthesized as a result of protein catabolism in the fish, would result in a significant reduction of the metabolizable energy from protein. Methyl carbon and hydrogen excreted as trimethylamine oxide represent a loss of approximately 42.7 kcal/g of trimethylamine nitrogen.

Thus, the amount of energy physiologically available to fish from *synthesized* body carbohydrate, fat, and protein appears to be approximately 4.10, 9.45, and 4.80 kcal/g, respectively. The available energy from protein for marine species excreting a significant amount of trimethylamine oxide would be less. The corresponding mean metabolizable energies of the *dietary* components for salmonids in fresh water, obtained by applying the mean digestibilities used by Phillips (1969), are 1.6, 8.0, and 4.2 kcal/g of raw starch, fat, and protein, respectively. The value for cooked starch consumed by salmonids or for raw starch consumed by herbivorous fish is closer to 3.3 kcal/g. The further energy losses due to the heat increments of the respective dietary components are discussed in Section III, D.

E. Energy Density of Feed

Fish, like other animals, tend to eat to meet their energy requirements (Rozin and Mayer, 1961). Assuming an adequate dietary nutrient balance, the fish can compensate for a low energy density (calories per gram) by eating more of the ration. Compensation of this sort can occur below the limits of the physical capacity of the gut. Up to that point, weight gain may be similar between groups of fish, but the fish on the high energy feed require less feed per unit of gain. However, since fish receiving a ration of high energy density can consume more nutrients at maximum physical intake they are able to grow at a higher rate. These principles are illustrated in data reported for trout by Phillips and Brockway (1959) and Ringrose (1971). Brett (1976a,b) has demonstrated how both the metabolic rates and maximum feed intakes of sockeye salmon receiving isocaloric diets increase with increasing ambient temperature (cf. Figs. 7 and 13). It follows that high energy rations can be important in supporting maximum growth and performance at high temperatures. This has been demonstrated in the experience with chinook fry raised at high ambient temperatures (Robertson Creek Salmon Hatchery, Port Alberni, British Columbia). In this case, the most successful rations were high fat, high energy rations.

F. Storage of Body Substance

The metabolizable energy values discussed in Section D above are the amounts of physiologically useful energy obtained when protein, fat, or carbohydrate is catabolized by the fish. When growth occurs, however, the total body energy increases according to the gross combustion energies of the new tissue components. The minimum energy cost of growth may be estimated from the moles of ATP required per mole of subunit polymerized as protein, fat, or carbohydrate, and by considering the total metabolic energy cost of forming and then breaking a labile phosphate bond. If glucose is the energy source, 1 mole of ATP formed and hydrolyzed represents approximately 18.7 kcal of metabolic heat loss. By this approach, the minimum energy costs of growth are approximately 0.23, 1.24, and 0.56 kcal/g of glycogen, fat, or protein synthesized. The corresponding chemical net energetic efficiencies are approximately 95, 87, and 90%, respectively. The biological net efficiencies are slightly lower than this due to transport and metabolic cycling of precursors prior to synthesis. Brody (1945) reports the net energetic efficiencies of poultry egg production, of milk production, and of prenatal growth in animals to be 77, 61, and 60%, respectively. The gross caloric efficiency of growth of young coho (55%) observed by Averett (1969) indicates a net efficiency of approximately 70% (assuming that the energy cost of maintenance and movement was of the order of 30% of the total energy intake).

For practical purposes it is of use to consider that the energy cost of new tissue synthesis is in the range of 15–30% in terms of metabolizable energy of the gross energy of the body components stored. Further discussion of the gross and net efficiencies of growth are given in Section V.

G. Summary of Equivalents

There has been some confusion and inconsistency in the literature on fish energetics according to the caloric equivalents used (mammalian or fish), the substrate being respired, and the amount of energy in question, that is, the absolute caloric content (heat of combustion), the metabolizable content of the food (physiological value), and the metabolically available energy from body resources. A frequent error has been the use of the relatively high caloric equivalents for saturated animal fats rather than the lower caloric content of highly unsaturated fats in fish meal. Table I summarizes the various equivalents, drawn from literature already cited.

Table I Energy Equivalents Used for Mammalian and Fish Bioenergetics[a]

		Energy of food and body resources			Respiratory energy equivalents Oxycalorific values		Respiratory quotient
		Food		Body			
Component or substrate		Heat of combustion (kcal/g)	Physiological value (kcal/g)	Metabolic energy (kcal/g)	Q_{ox} (kcal/liter (of O_2)	Q_{ox} (cal/mg of O_2)	Ratio, Co_2/O_2
Carbo hydrate	Mammal	4.10	4.0–3.2	4.10	5.04	3.53	1.0
	Fish	4.10	3.3–1.6	4.10	5.04	3.53	1.0
Fat	Mammal	9.45	9.0	9.45	4.69	3.28	0.71
	Fish	8.66	8.0	9.45–8.66[b]	4.69	3.28	0.70
Protein	Mammal	5.65	4.2–3.9	4.70	4.82	3.37	0.81
	Fish	5.65	4.5–3.9	4.80	4.58	3.20	0.9
Mixed	Mammal	5.95	—	—	4.86	3.40	0.83
	Fish	5.89	—	—	4.63	3.25	0.90

[a]*The range in physiological values is largely affected by digestibility.*
[b]*From Beamish* et al. *(1975).*

III. METABOLISM: RATES OF ENERGY EXPENDITURE

The prime demand for food, before any energy storage or somatic growth can be achieved, is to meet maintenance requirements. Within this need, the highest priority must be ascribed to *basal metabolism*—the minimum rate of energy expenditure to keep the organism alive. Among the various metabolic demands, the determination of this physiological minimum has received most attention (Winberg, 1956) undoubtedly influenced by the considerable documentation of basal metabolism for homeotherms (Brody, 1945; Kleiber, 1961). The case is perhaps not so justified for making comparable measurements on cold-blooded vertebrates, which are characterized by resting metabolic rates that are 10 to 30 times less than mammals, and up to 100 times less than small birds of the same weight (Brett, 1972). It is locomotion that is metabolically costly for fish. Because of such relatively low maintenance requirements fish can withstand long periods of starvation, and frequently cease feeding for many days when experiencing major environmental change (vertical and horizontal migrations) or if by chance they are denied food, as is witnessed by the not infrequent lack of any trace of digested material in the full length of the intestine of captured fish.

Because of the natural, ready elevation of the metabolic rate of fish, with concomitant technical difficulties of measuring resting states, the term *standard metabolism* was early applied to describe the minimum rates observed (Krogh, 1914). Cases involving normal, spontaneous activity were called *routine metabolism* (Fry, 1957; Beamish, 1964a). The introduction of activity meters and tunnel respirometers, relating the rate of doing work (power) to swimming speed (performance), allowed estimates of the minimum metabolic rate for complete rest by extrapolation to zero activity—a true basal rate. A functional distinction between basal metabolism (warm-blooded) and standard metabolism (cold-blooded) was no longer necessary. In accordance with custom however, the term standard metabolism will be used throughout for fish. While there is no doubt that this parameter of metabolism offers meaningful insight, particularly with regard to the response to environmental factors, nevertheless the animal's need for foraging and food-processing led Winberg (1956) to make a plea for research to be directed to higher levels of metabolic requirement, in the interest of contributing to problems of bioenergetics involving the feeding, digesting, growing, competing animal.

By measuring *active metabolism*, limits for the maximum rate of energy expenditure at maximum sustained activity could be established experimentally. But, short of knowing what these limits were, and that episodes of chase and escape would undoubtedly invoke active metabolic rates, knowledge of the average daily energy costs for fish under almost any natural conditions was not forthcoming. This led Paloheimo and Dickie (1966) to relate feeding levels to the apparent rate of metabolism that could be derived from energy equations, involving rates of food intake and growth. Assuming certain corrections for the metabolizable fraction of the food, these authors deduced from the literature that maximum food intake would be accompanied by maximum or active metabolic rate. However, ultimate support for this conclusion has not been forthcoming from laboratory experiments on the metabolism of feeding fish (Warren, 1971). It remains a challenge to biologists to devise some remote means of measuring the actual rates of energy expenditure under field conditions. The nearest experimental approach is to determine the metabolic costs associated with the daily processes and activities of normal living, and sum these costs according to closely documented behavior in nature. For captive fish in a hatchery or other aquaculture system (usually fed to near satiation) direct opportunity of metabolic measurement is afforded.

General consideration will be given first to standard metabolic rates, as the minimum energy costs, together with some comparison of maximum rates of energy expenditure (but see also Fry, 1971). The lesser known aspects of metabolic costs associated with feeding, and the apparent patterns of daily energy expenditure, will then be compared with the standard and active metabolic rates.

A. Standard Metabolism

1. *Level and Metabolic Compensation*

Despite the sensitivity of metabolic rate to many factors, particularly excitement, size, and temperature, some idea of the general level for fish can be obtained from the listing in "Biological Data Book" (Altman and Dittmer, 1974). Of the 365 records involving 34 species, 57 entries relate to adequately derived standard metabolism, that is, activity was accounted for and the fish were fully acclimated to the test temperature. These records have a mean of 89 ± 34 (SD) mg of O_2/kg/hr (0.29 ± 0.11 kcal/kg/hr), and an extreme range from 26 to 229 mg of O_2/kg/hr (0.08–0.74 kcal/kg/hr). With few exceptions, the records all apply to fish commonly found in temperate climatic zones. Studies comparing the temperature effect on standard metabolic rates for fish from different climates have been conducted, particularly by Scholander *et al.* (1953) and Wohlschlag (1960, 1964). Although some of the records appear to border on the high side, possibly because of uncontrolled or unaccounted activity (closer to routine metabolic rates), they clearly demonstrate a compensative, adaptive response of polar species to maintain a higher metabolic rate at low temperatures, above that which would be expected from studies on temperate species (Fig. 1). By contrast, only limited metabolic compensation appears to have taken place among tropical species, which have paced their standard rates downward, mostly in the lower range of their thermal tolerance. These fishes are consequently operating at a higher maintenance level in accordance with their higher environmental temperatures. Applying the same correction factor (× 0.56) developed for the temperate species (see Fig. 1), tropical fish would have a minimum energy expenditure of about 0.5 kcal/kg/hr in the midpoint of their temperature range (26°C)—an elevation of 70% over the mean value for temperate species.

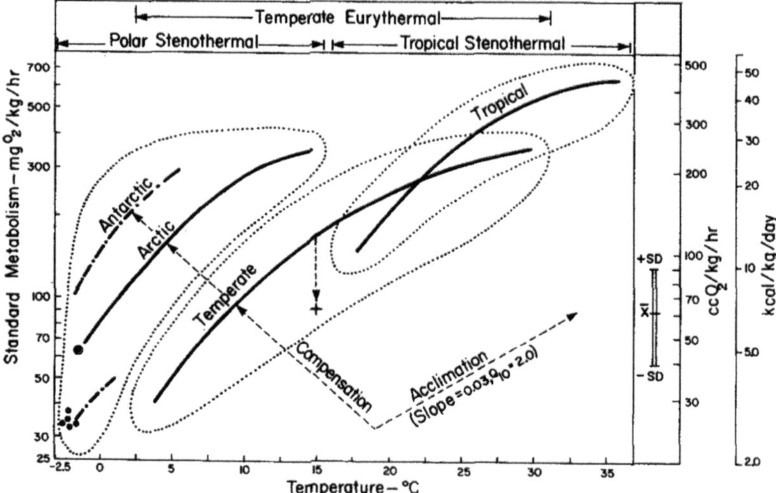

Fig. 1. Schematic representation of relation between temperature and "standard" metabolic rates (log scale) of fish from different climatic zones. Drawn from compilation of Wohlschlag (1964) with selected additions from Holeton (1974) (points for polar species). Dotted lines indicate general range of variability within each zone. Adaptive metabolic compensation is shown for polar species, following the construction line indicated. The effect of acclimation temperature is shown by the average slope (0.03), with a construction line for a temperature increase of 10°C ($Q_{10}=2.0$). The mean ± 1 SD of the standard values in Altman and Dittmer (1974) (column to right) indicates that the curves are somewhat higher than true standard metabolic rates, the midpoint for the temperate species being about 1.8 times the indicated mean for an average weight of 50–100 g (note the central arrow pointing to the + mark). (Modified from Brett, 1970b; reproduced by permission of Wiley–Interscience.)

Holeton (1973, 1974) has justly questioned the conclusions from earlier studies on metabolic cold adaptation of polar species, pointing out the likelihood of error from prolonged, elevated metabolic rates resulting from capture and handling, and the short acclimation times applied in some cases. However, Holeton's (1974) most extensive records on the Arctic cod, *Boreogadus saida*, were consistent with previously published values for arctic fish, with a mean standard metabolic rate of 70 mg of O_2/kg/hr at −1. 5°C, a temperature that would be rapidly lethal for most temperate species. Five species of arctic cottids had standard metabolic rates of about 38 mg of O_2/kg/hr at this temperature (shown in Fig. 1), whereas three species of zoarcids and two species of liparids occurred at lower "uncompensated" rates (not shown). It appears that the upper curve for antarctic fish (broken line) in Fig. 1 is unsupported as attributable to metabolic compensation and must represent the upper range for routine metabolic rates. For the balance of polar species investigated by Holeton (1974) there are obviously major exceptions, showing lack of any apparent metabolic compensation to low temperature. No standard rate below

20 mg of O_2/kg/hr has been obtained, which can be taken as the minimum metabolic rate to support life in free-living fish in the 10–100 g size range (approximately 0.07 kcal/kg/hr).

2. *Temperature Effect*

The foregoing presents the general range of standard metabolism as it varies among species and between climatic zones. The effect of temperature on standard metabolism *within* species has been considered by Fry (1971) in Vol. VI of Fish Physiology, and should be consulted. Some examples occur in the present text, starting with Fig. 1 in which the shape of each curve follows a generalized form, first defined for goldfish, *Carassius auratus*, by Ege and Krogh (1914), and called Krogh's "standard curve." This curvilinear relation was subsequently elaborated upon and supported by Winberg (1956). Since no simple mathematical transformation could be derived, a set of empirical multipliers based on the Q_{10}* values was developed by Winberg (1956) for temperate species. When plotted as the logarithm of standard metabolism against temperature, the curve follows the convex slope in Fig. 1. Variability in level (intercept) and shape of the nontransformed curve can be seen in Figs. 2, 10, and 13. The validity of Krogh's standard curve has been examined by Holeton (1974) who shows that the data points were considerably elevated above those obtained for the same species by Beamish and Mookherji (1964); also that the Q_{10} value for low temperatures ($0^\circ - 5^\circ$C) was undoubtedly in error as a measure of standard metabolism. In general it appears that a simple exponential transformation [$\log M_s = a + bT$ (M_s, standard metabolism; T, temperature)] comes close to linearizing the data for goldfish over the temperature range of $10^\circ - 30^\circ$C, with a mean Q_{10} of 2.3.

Once the level of standard metabolism has been established for a species at the midpoint of its normal, environmental temperature range, then the use of 2.3 as a multiplier would provide a near approximation for temperature effects within the span of $\pm 10^\circ$C. This can be shown to apply to brook trout, bullhead, and carp (Beamish, 1964b) and sockeye salmon (Brett, 1964) without introducing an error greater than 20% of the observed value. However, beyond noting the sort of exponential increase, as Fry (1971) comments, "few general statements can yet be made of the relation of standard metabolism to temperature except that different species are adapted to different temperature ranges" (pp. 44–45).

3. *Size Effect*

The effect of size on the standard metabolic rate of animals has been the focus of much research and conjecture (Zeuthen, 1953, 1970; Kleiber, 1961; Gordon, 1972). The general relation is described by the allometric equation,

---

*Q_{10} = the increase in rate for an increase of 10°C.

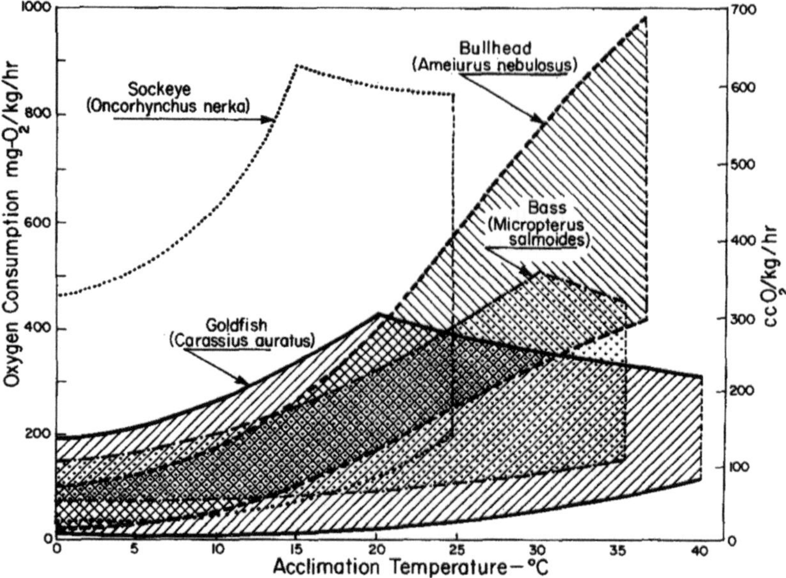

Fig. 2. Rate of oxygen consumption in relation to temperature for four species of fish. Active and standard metabolic rates are indicated in the upper and lower lines for each species, terminated by a vertical line at the upper lethal temperature. Data for sockeye from Brett (1964); for bullhead from Fry (1947); for bass from Beamish (1970); and for goldfish from Fry and Hochachka (1970). (From Brett, 1972; courtesy of North-Holland Publ. Co.)

$Y = aX^b$, where Y = rate of oxygen consumption, X = body weight, and a and b are constants characteristic of a given species. The rate per unit weight (Y/X) usually diminishes with increasing size both within and among species such that, for instance, in salmon the standard rate of a 3000-g adult is about one-fifth that of a l-g fry, or, in the popular elephant–mouse example, the fraction is more like one-fiftieth. Since the exponent b approximates 0.67 (i.e., two-thirds) in many warm-blooded animals it was thought to reflect the simple physical relation of the surface-to-volume law. However, among fishes the value for individual species has rarely been found this low, and would be better placed at 0.86 ± 0.03 (SE)* for the general case (Glass, 1969). This is further supported by the table of values compiled by Kausch (1972) where the mean was also 0.86 (± 0.04 (SE), n = 25). Kayser and Heusner (1964) reported an exponent of 0.70 ± 0.02 for four species *grouped* together; the species however are very unevenly distributed in weight dispersion over the full size range, depressing the separate slopes within species. In the interest of accurate assessment the species-specific characteristic is sufficiently unique to justify separate determination for any given species, at the normal temperatures experienced (see table in Glass, 1969).

*Recalculated from Glass (1969) using estimated rate for the killifish.

B. Active Metabolism

Over and above maintenance metabolism, the greatest energy demands for most animals derive from those activities which call on locomotion (attacking, escaping, migrating, jumping). When swimming, a major increase in demand for oxygen arises from the contraction of the large lateral muscles of fish. Burst speeds overtax the capacity of the respiratory–circulatory system to provide sufficient oxygen. Fatigue results. The lesser, marathonlike, sustained speeds are set by the maximum rate of oxygen consumption, defining the upper limit for the *active metabolic rate*; this is equivalent to the "aerobic capacity" of mammals.

1. Level and Temperature Effect

The effect of temperature on active metabolism has been reviewed by Fry (1957, 1971). A different set of temperature relations than those for standard metabolic rates apply. Four of the better known species are represented in Fig. 2, including their standard rates to provide an indication of the full scope for sustained metabolic rate. Three of the four species are characterized by an optimum temperature above which the active metabolic rate decreases. This phenomenon has not been explained although it has been hypothesized by Jones (1971) that the energy demands of ventilation and associated circulation become excessive beyond critical temperatures (e.g., 15°C for sockeye salmon) restricting increased supply of oxygen to the tissues. There is also provisionary evidence that oxygen becomes a respiratory limiting factor at high temperatures for some salmonids as the oxygen content of air-saturated water decreases with increasing temperature (Brett, 1964; Fry, 1971). It can be seen (Fig. 2) that the highest active metabolic rates approach 1000 mg of O_2/kg/hr (3.3 kcal/kg/hr). Fast swimming, streamlined fish can readily elevate their metabolic rate from standard levels by a factor of 8 to 10 times.

It would be of interest to have comparable observations on active metabolic rates of sluggish fish to compare with the more active species. As yet such are not available. However, Wohlschlag (1960) made records of the metabolic rate of *Trematomus bernacchii*, a benthic species which lives under a heavy, sea-ice cover in the Antarctic. Although the fish could not be induced to swim continuously, excitement apparently drove the metabolic rate up in a number of cases, rising to 140–180 mg of O_2/kg/hr (0.46–0.58 kcal/kg/hr). This is about one-fifth the active metabolic rate of sockeye salmon. With the exception of hemoglobinless fish (Holeton, 1970), the above records represent the sort of overall range of energy expenditure that may be expected among fishes with regard to their maximum oxygen consumption rates (0.5–3.3 kcal/kg/hr). It cannot of course include the energy expenditure of burst speeds, which are almost entirely anaerobic, relying on subsequent hyperventilation to pay off the accumulated oxygen debt. The extrapolation of power–performance curves reveals that the energy liberated during maximum bursts can reach 100 times the standard rate or 10 times the active rate (Brett, 1972). This may also be demonstrated by computing an "equal energy" curve, such as that in Fig. 3. It underscores the fact that just a few bursts of attack or escape each day would be equivalent to doubling the daily costs incurred from the standard metabolic rate—unlike mammals or birds. There is sufficient quantitative difference in the metabolic rates of cold-blooded vertebrates (ectotherms) compared to warm-blooded vertebrates (endotherms) to warrant one such example: at the same weight (about 2 kg) the highest level of active metabolic rate of a salmon just equals the basal metabolic rate of a rabbit.

2. *Size Effect*

Few systematic studies have been conducted on the effect of size on active metabolic rates of fish. This is a very important area of enquiry since it is a

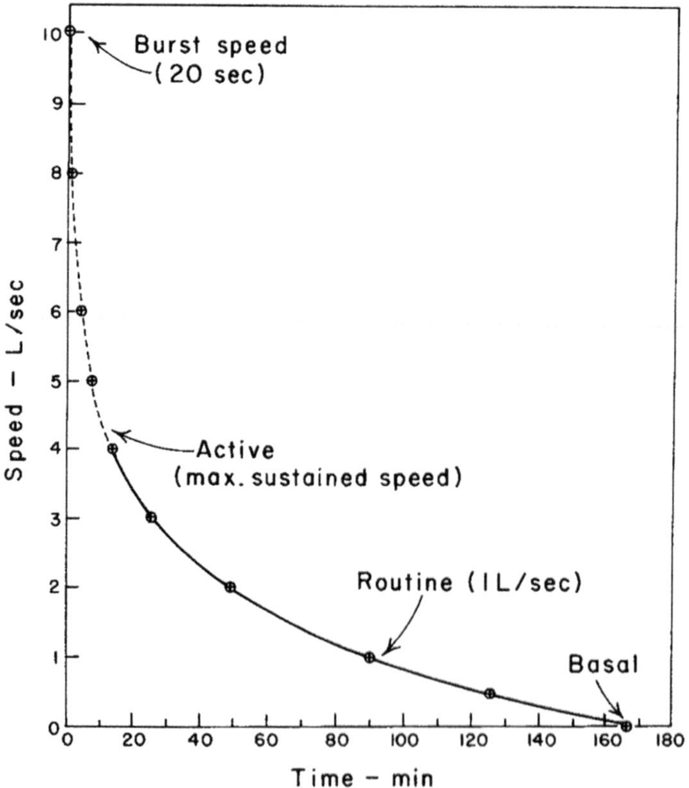

Fig. 3. Time–speed curve for equal energy expenditure of a 50 g, 18 cm sockeye salmon at 15°C. The calculated points are the times to consume 10 mg O_2 (about 35 cal), from a resting (basal) state to swimming at 10 body lengths/sec (L/sec). Routine metabolic rate shown as equal to 1 L/sec. Broken line for extrapolated points. (Derived from data in Brett, 1964.)

common error in most "biological production models" involving energy requirements to incorporate a scaling factor for size according to standard metabolic rates—which is clearly a false premise for growing, active fish. Studies on size effect of sockeye salmon (Brett, 1965) showed a *continuous* change in the weight exponent (W^b), from 0.78 to 0.97, with increasing levels of activity (at 15°C). The mean value of b for all temperatures (5°–20°C) for the active metabolic rate of this species was 0.98, indicating an almost insignificant effect of weight for most temperature circumstances (Brett and Glass, 1973). Because of the large percentage of muscle in a big fish (increases from about 35% at 10 g to about 65% at 1000 g in salmon) the expected decrease in tissue respiration rate with increasing size is largely offset by the relative increase in mass of "working" tissue.

3. *Combined Effect*

A composite graph of the combined effects of weight and temperature on rates of energy expenditure was developed for sockeye salmon by Brett and Glass (1973). It provides a means of approximating standard and active metabolism graphically from the plotted isopleths (Fig. 4). The slopes of the lines show how weight has a consistent interacting effect with temperature over the tolerable range, reducing the standard metabolic rate with increasing size at all levels of temperature (Fig. 4A). This contrasts with the circumstance for active metabolism (Fig. 4B) which progresses from almost complete independence of weight below 10°C (isopleth lines are almost horizontal) to increasing dependence for the combined circumstances of high temperature and large size.

C. Feeding Metabolism

The need for a better understanding of the metabolic requirements of *feeding* fish has been recognized by a number of investigators. Warren and associates

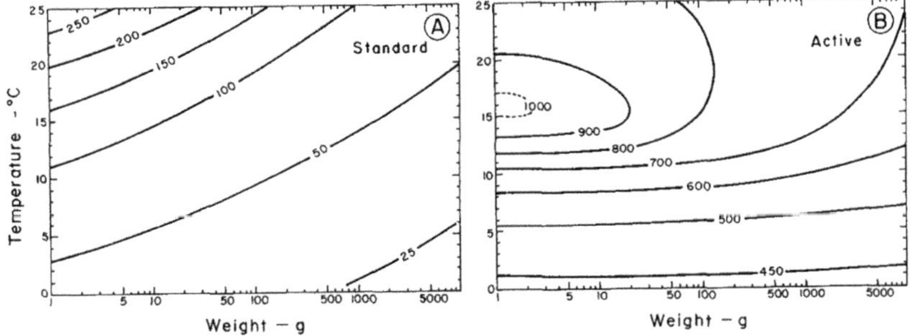

Fig. 4. Developed response surface for standard (A) and active (B) metabolic rates of sockeye salmon in relation to weight (log scale) and temperature. Rates shown in mg of O_2/kg/hr. Dotted isopleth line for 1000 mg of O_2/kg/hr obtained by extrapolation. Equivalents: 100 mg O_2 = 70 cc O_2 = 325 cal. (From Brett and Glass, 1973.)

(Warren and Davis, 1967; Warren, 1971) have repeatedly tried to develop complete energy budgets of growing fish, including measurements of the daily energy expenditure on food processing separate from the requirements of standard metabolism and those generated by the accompanying feeding activity (locomotor requirement and excitement). Such partitioning has its attendant difficulties since the animal does not conveniently separate the components of its total metabolic demands. Some of the metabolic pathways can be recognized if accompanying measurements of CO_2, NH_3, and urea are made. But for the present no such distinctions will be considered—just the increase in oxygen consumption accompanying different levels of ration and temperature.

Using a "mass respirometer" Saunders (1963) was among the first to study the effect of feeding on the metabolic rate of Atlantic cod, *Gadus morhua*, at 10°C. Routine metabolic rates of starving fish (wt = 1 kg) rose from 75 to 112 mg of O_2/kg/hr after feeding and remained at this elevated level for 1 to 2 days, falling gradually back to the fasting, routine rate by the seventh day. More extensive studies on this species were conducted by Edwards *et al.* (1972). Weighed daily rations of plaice fillets were provided to individual cod, and oxygen consumption measured at 12°C in a closed-circuit respirometer. In order to compare temperature relations, a Q_{10} value of 2.5 was applied to convert "basal rates" (determined under light anaesthetic) from 12°C to various levels of seasonal temperature. A maximum increase of 4.7 times the "basal rate" occurred when fed a high ration at 15°C.

The fairly extended metabolic response of cod to feeding was also observed for the subtropical reef fish, aholehole (*Kuhlia sandvicensis*), by Muir and Niimi (1972). When given a single ration of 4.5% body wt, the metabolic rate rose from a routine level of 76 mg of O_2/kg/hr(23°C) to 182 mg of O_2/kg/hr by the fourteenth hr, falling gradually back to the initial level after about 50 hr (Fig. 5A). This contrasts with the excited, more rapid metabolic responses of sockeye salmon, which anticipate a normal feeding time in a daily cycle,

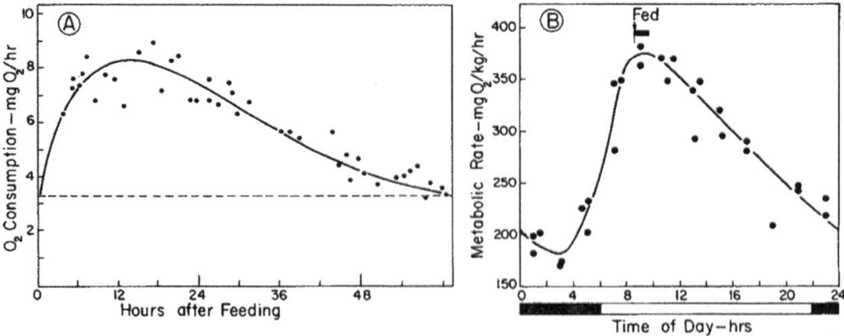

Fig. 5. Change in oxygen consumption following a single feeding for (A) a 44-g aholehole given a 4.7% ration at 23°C (Muir and Niimi, 1972) and (B) a group of sockeye salmon averaging 29 g weight and given a 3% ration at 15°C (Brett and Zala, 1975). Dark period shown cross-hatched in (B).

raising their metabolic rate from an early morning minimum (0300 hr) of 170 to 370 mg O_2/kg/hr* right at feeding time (0800 hr; Fig. 5B). This peak rate falls as digestion proceeds, reaching a minimum 19 hr later at 15°C.

The above examples serve to illustrate the differences in response between species and the varying approach of investigators to measuring metabolic rates. The computation of daily energy expenditure requires complete information on the hourly metabolic rates over the full 24 hr. To the extent that this was possible to determine, records for eleven species have been compiled in Table II, according to temperature and ration. In some cases only the lowest and highest feeding conditions have been included; not all cases have a true standard metabolic rate determination, and entries under routine metabolic rates are naturally subject to wide variation. Nevertheless some useful generalities can be extracted. Increasing ration has an almost direct, proportional effect on increasing daily metabolic rate. Temperature also increases the feeding metabolic rate both by stepping up the pace and by increasing the daily food intake. The ratio of feeding metabolic rate (F) to standard rate (S) ranges from less than 1.0 for submaintenance rations up to 5.8 for small fish on high rations. The ratio, F/S, for all cases involving a high ration ($n = 8$) averaged 3.7 ± 1.2 (SD). This may be compared with 1.7 ± 0.4 (SD) for the average ratio of feeding metabolic rate to routine metabolic rate (F/R).

From Winberg's (1956) review, a multiple of 2 times standard rate was given as a near approximation of the metabolic effect of feeding. This is now seen to be understated for cases where accurate estimates of the standard rate have been made, and instead is closer to the ratio for routine rates. However, such early generalities do not take into account the more recent and important finding that the daily metabolic expenditure is highly ration dependent, as may be determined from the entries in Table II. The relation is either a linear function of metabolic rate increasing with ration, up to maximum intake, as has been shown for flatfish, reef fish, and salmon (Fig. 6), or a linear increase at lower levels of ration tapering off at higher rations to form an upper plateau, as in the case of carp (Huisman, 1974).

From a compilation of data available on the feeding metabolic rates of fingerling sockeye salmon a "predictive model"** in the form of isopleths of daily oxygen consumption rate in relation to temperature and ration was computed by Brett (1976b). The response surface (Fig. 7) was developed by use of a set of regression lines, some of which are shown in Fig. 6A. Over the whole surface, from low temperature and low ration to high temperature and high ration, the metabolic rate ranges from 50 to 400 mg of O_2/kg/hr, a factor of 8 times.

*This relatively high "minimum" was partly due to water velocities of 10–14 cm/sec (1–1.2 length/sec) in the culture tanks.

**No data were available at temperatures above 20°C and very few for 5°C and below. By assuming similar slopes for temperature effects on the *metabolic rate x ration* relation, the full response surface was computed (Brett, 1976b).

Table II Records of Daily Metabolic Rates of Feeding Fish in Relation to Standard and Routine Metabolic Rates[a]

Species	Weight (g)	Temperature (◦C)	Ration (% wt)	Metabolic rates (mg of O_2/kg/hr					Time (hr)	Ratio		Reference and remarks
				Standard	Routine	Peak	Feeding 1	Feeding 2		*F/S*	*FIR*	
Gadus morhua Atlantic cod	1000	10	(Fed)	—	75	—	112	112	24–48	—	1.5	Saunders (1963); fed on fresh and frozen herring
	1000	15	(Fed)	—	75	—	120	120	24–48	—	1.6	
Cyprinus carpio Carp	10±5	10	2	80	—	—	120	116	24	1.5	—	Kausch (1969); standard determined for zero activity; ration determined from *ad libitum* supply
	10±5	15	7	136	—	—	—	230	24	1.7	—	
	10±5	20	6	214	—	—	—	240	24	1.1	—	
Salmo gairdneri Rainbow trout	30	15	(High)	—	220	290	264	264	24	—	1.2	Mann (1968); standard rate for sockeye salmon in brackets
	30	15	—	(100)	—	—	—	—	—	(2.6)	—	
Oncorhynchus kisutch Coho salmon	3.3	8	11.7	60	240	980	460	350	12	5.8	0.7	Averett (1969) and Warren (1971); ration is percentage fly larvae weight to fish weight; metabolic rates estimated from graphs; seasons were July to September
	3.9	11	5.3	120	220	415	320	270	12	2.2	1.2	
	4.7	14	3.4	170	220	370	350	280	14	1.6	1.3	
	2.6	20	7.7	230	540	1010	900	750	15	3.3	1.4	

Oncorhynchus tshawytscha Chinook salmon	18–20	11.7	Ex.	(75)	165	—	277	221	8	(2.9)	1.3	Elliot (1969); ration was excess (Ex.) feeding 2 times/day; standard rates for sockeye salmon; routine rates interpolated; see Fig. 7
	18–20	12.6	Ex.	(80)	180	—	286	228	8	(2.8)	1.3	
	18–20	15.8	Ex.	(96)	220	—	360	287	8	(3.0)	1.3	
	20	12.6	Ex.	(80)	170	366	289	230	8	(2.9)	1.4	
Cynoglossus sp. (5 species) *Sole*	9	28	(Low)	105	—	—	(105)	(105)	?	1.0		Edwards *et al.* (1971)
	9	28	(Max)	105	—	—	(390)	(390)	?	3.7	—	
Kuhlia sandvicensis Aholehole	71	23	1.4	57	78	118	118	95	72	1.7	1.2	Muir and Niimi (1972); metabolic rate increases proportional to ration; daily feeding rate in first 24 hr approximates 80% of peak rate
	71	23	2.8	57	78	158	158	126	72	2.2	1.6	
	71	23	4.2	57	78	195	195	156	72	2.7	2.0	
	44	23	2.3	62	76	121	121	97	72	1.6	1.3	
	44	23	4.5	62	76	182	182	146	72	2.4	1.9	
Gadus morhua Atlantic cod	500	15±2	(Low)	88	–	–	(88)	(88)	?	1.0	–	Edwards *et al.* (1972); standard = basal under anesthetic
	500	15±2	(Max)	88	–	–	(440)	(440)	?	5.0	–	
Perea fluviatilis Perch	12	14	(Low)	—	175	—	—	146	24	—	0.8	Solomon and Brafield (1972); low ration is approximately maintenance
	12	14	(Max)	—	175	350	—	296	24	—	1.7	

Continued

Table II Records of Daily Metabolic Rates of Feeding Fish in Relation to Standard and Routine Metabolic Rates—Cont'd

Species	Weight (g)	Temperature (∘C)	Ration (% wt)	Metabolic rates (mg of O_2/kg/hr					Time (hr)	Ratio		Reference and remarks
				Standard	Routine	Peak	Feeding 1	Feeding 2		*F/S*	*FIR*	
Lepomis macrochirus Bluegill	49	15	Ex.	—	48	93	—	62	24	—	1.3	Pierce and Wissing (1974); ration is excess of mayfly nymphs; nocturnal metabolic rate 26% higher than day
	92	20	Ex.	—	45	62	—	71	24	—	1.6	
	83	25	Ex.	—	89	140	—	120	24	—	1.3	
Cyprinus carpio Carp	31–47	17	10	48	87	—	—	141	24	2.9	1.6	Huisman (1974); standard rate from starving fish
	2–16	23	5	83	156	—	—	243	24	2.9	1.6	
Oncorhynchus nerka Sockeye salmon	10–20	10	4.5	60	100	—	—	190	24	3.2	1.9	Brett (1976b); maximum observed feeding rates for highest rations given; greater rates possible
	10–20	15	6.0	71	125	—	—	315	24	4.4	2.5	
	10–20	20	6.5	120	161	—	—	420	24	3.5	2.6	

[a]Entries in the column under Feeding I have heen translated into average daily rates in the column under Feeding 2 according to the times during which oxygen consumption rates were measured. This was done by assuming that routine metabolic rates applied for the balance of the 24 hr, if better information was not available in the reference noted. F/S and F/R are the respective ratios of Feeding 2 to standard and routine metabolic rates. Entries in order of year published. Bracketed values were estimated or obtained indirectly.

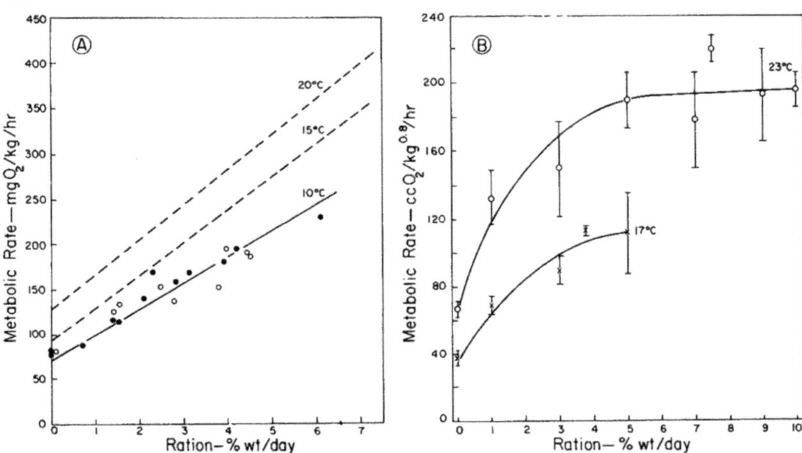

Fig. 6. Response of metabolic rate of three species of fish to different levels of daily ration, at a number of temperatures. (A) Sockeye salmon at 10°C, open circles, and aholehole, *Kuhlia sandvicensis*, at 23°C, solid circles (Muir and Niimi, 1972). Additional lines for sockeye shown for 15° and 20°C (Brett, 1976b). (B) Carp, *Cyprinus carpio*, at 17°C and 23°C; variance shown as ±1 SD (Huisman, 1974). Note that metabolic rate in B is expressed as cc O_2/kg$^{0.8}$/hr.

The outer encompassing periphery of the response surface is defined by the maximum daily food intake (as measured in Brett *et al.*, 1969), and an inner boundary set by the maintenance ration, below which loss of weight would occur. It applies to sockeye with a mean weight of about 20 ± 10 g (range). The strong interaction between temperature and ration is apparent; for a change in ration of 10% of the maximum intake the temperature would have to fall by an average of 3.3°C (13% of the tolerable range) to maintain the same metabolic rate.

Elliot (1969) determined the metabolic rate of hatchery-reared chinook salmon (*Oncorhynchus tshawytscha*) in experimental troughs and raceways. Fingerling fish varied in size from approximately 1 to 25 g and were tested over a temperature range of 6° – 16°C. Two conditions were examined: (1) normal hatchery procedure involving a morning cleaning and feeding, followed by an afternoon feeding, and (2) normal activity without feeding, defined as "a state in which oxygen consumption is fairly stable and the fish are in a relatively quiescent state."

The mean metabolic rate for the first condition, over the period of 0800 to 1600 hr, was 277 mg of O_2/kg/hr at 11.7°C, 289 mg of O_2/kg/hr at 12.16°C, and 360 mg of O_2/kg/hr at 15.8°C. This compares quite closely with the daily rate for sockeye of the same weight on maximum ration. Under quiescent, nonfeeding conditions, the mean rate for 6 g and 18 g chinook is also shown in Fig. 7. The rates occur about halfway between the maintenance and maximum values obtained for sockeye. Parallel but less extensive determinations made in raceways were about 16% higher than the above trough levels. If this

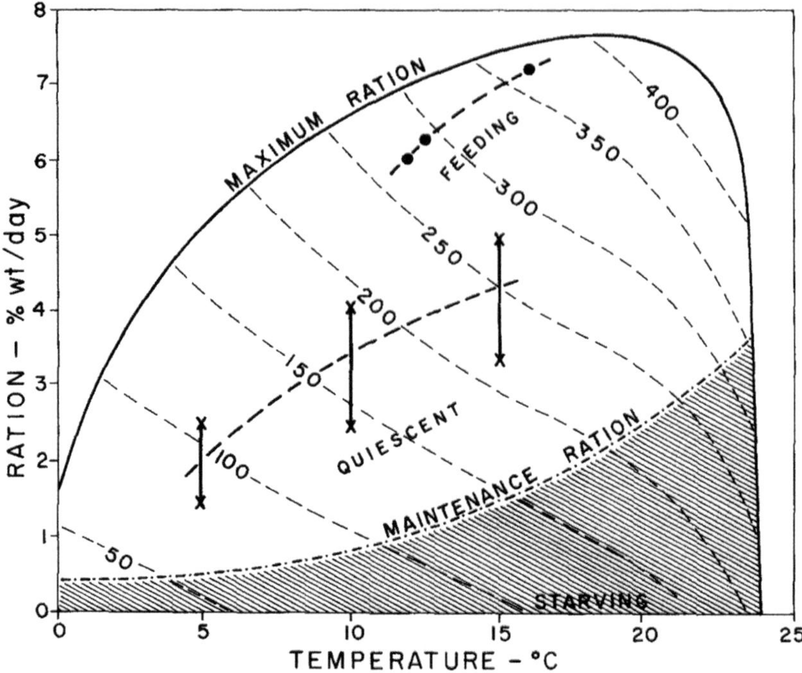

Fig. 7. Isopleths of feeding metabolic rate (mg of O_2/kg/hr) for fingerling sockeye salmon (20 ± 10 g) in relation to ration level and temperature. Ration is expressed in dry weight of food as a percentage of dry weight of fish. The peripheral boundary line of the response surface defines the maximum food intake; the inner maintenance line defines the boundary for positive growth. Confidence limits for any one value are approximately ± 20% of the rate depicted. Metabolic rates of juvenile chinook salmon in hatchery troughs are shown (Elliot, 1969). Upper points (●) are for 20 g chinook fed twice a day; lower points (×) relate to the range for fish weighing 6 and 18 g when not fed and considered quiescent under normal activity (routine metabolism). Energy equivalents as in Fig. 4. (From Brett, 1976b.)

correction is applied to the feeding metabolic rates of the chinook salmon, their 8-hr average falls almost right on the 24-hr average for sockeye.

In the studies on cod (Edwards *et al.*, 1972), evidence for correcting for weight effect by using the approximate power function ($W^{0.8}$) is presented. While this may be applied with some confidence to standard metabolic rates there is no convincing argument developed for applying such an exponent value to the metabolic rates of feeding fish. Two phenomena make the presumption questionable. First, there is nothing to support the physiological assumption that the metabolic cost of digesting a unit of food would be different in small and large fish. Second, maximum food intake (as a proportion of body weight) decreases with increasing size. This association confounds, but does not discredit, the weight exponents for fed fish reported by Saunders

(1963); the values for fed fish (0.76–0.83) were lower than those for starved fish (0.79–0.89), which could be a result of the above-noted confounding.

D. Heat Increment

Within the total energy expenditure associated with feeding there is a segment derived from the biochemical transformation of ingested food into a metabolizable, excretable form. A major contribution to this exothermic loss apparently comes from deamination of protein, mostly in the liver (Krebs, 1964; Buttery and Annison, 1973). Rubner (1902) first drew attention to this phenomenon in domestic animals. The nutritional quality of the food affected the magnitude of the resulting *heat increment*. Kleiber (1967) noted that the original German term was wrongly translated as "specific dynamic action" (SDA) instead of "specific dynamic effect." Further, the concept of "specificity," restricting the heat loss to the specific process of deamination, is no longer applicable since similar but smaller energy releases accompany lipid and carbohydrate catabolism. In homeotherms these latter losses can amount to 13% of the caloric content of lipid and 5% for carbohydrate, compared to 30% for protein (Harper, 1971). In consequence, preference is given to the use of the term *heat increment*, as favored by Kleiber (1967).

For fish, the first experimental determinations of metabolic loss ascribed to heat increment were reported by Warren and Davis (1967), presenting previously unpublished data of H. Sethi on cichlids (*Cichlasoma bimaculatum*). Greater elaboration of the phenomenon was provided by Averett (1969), following extensive studies on juvenile coho salmon, *Oncorhynchus kisutch*. Under conditions of varying ration and temperature, heat increments ranging from 4 to 45% of the caloric content of the food were reported, with most values occurring between 9 and 15%. In their reviews, Warren and Davis (1967) and Warren (1971) give special attention to this incremental heat loss. With subsequent insight these authors recognized that the upper values obtained by Averett (1969) were attributable in part to metabolic excitation accompanying feeding. It is absolutely necessary to separate the energy expenditure of excitability and increased activity (occurring in conjunction with food intake) by rigorous experimental technique, otherwise the partitioning is a useless exercise. Beamish's (1974) carefully conducted study on largemouth bass, *Micropterus salmoides*, is almost a model in this regard of appropriate energy partitioning. In one experiment, conditioned fish were forced to swim in a tunnel respirometer at each of three velocities (1.4, 1.9, and 2.5 body lengths/sec). Continuously monitored oxygen consumption rates ceased to fluctuate after 14–18 hr, as the fish became habituated. A single ration of 4% weight was fed after 24 hr, and monitoring continued until prefeeding metabolic rates were resumed. The diet consisted of freshly thawed shiners. In a second experiment a fixed swimming speed was imposed and four separate rations used (2, 4,6, and 8% weight). After training the fish to accept food from forceps, a final experiment was conducted to determine the metabolic expenditures associated with the feeding procedure itself. By withholding the food after the fish bit at the forceps, the ancillary excitation of feeding could be assessed and then deducted to obtain an unconfounded estimate of heat increment. This daily energy loss per unit weight did not differ significantly with the weight of the fish (mean, 65 g; range, 9–90 g; temperature, 25°C) but varied as a mean percentage of 14.2% ± 4.2 (SD) of the food energy ingested, or 17.2% of the metabolizable energy (Fig. 8). Peak oxygen consumption rates rose to 80–100% of the active metabolic rate, lasting from 1 to 2 hr. At the highest ration (8% weight) the average total daily metabolic rate can be computed as 339 mg of O_2/kg/hr, partitioned as 35% standard, 41% excitation, and 24% heat increment (standard metabolic rate, from Beamish, 1970).

Three other studies on heat increment have been conducted with sufficient attention to the problem of increased activity and excitability, that bear comparison with the largemouth bass. Aholehole, *Kuhlia sandvicensis*, were fed chopped tuna flesh at two ration levels (2.3 and 4.5% weight) (Muir and

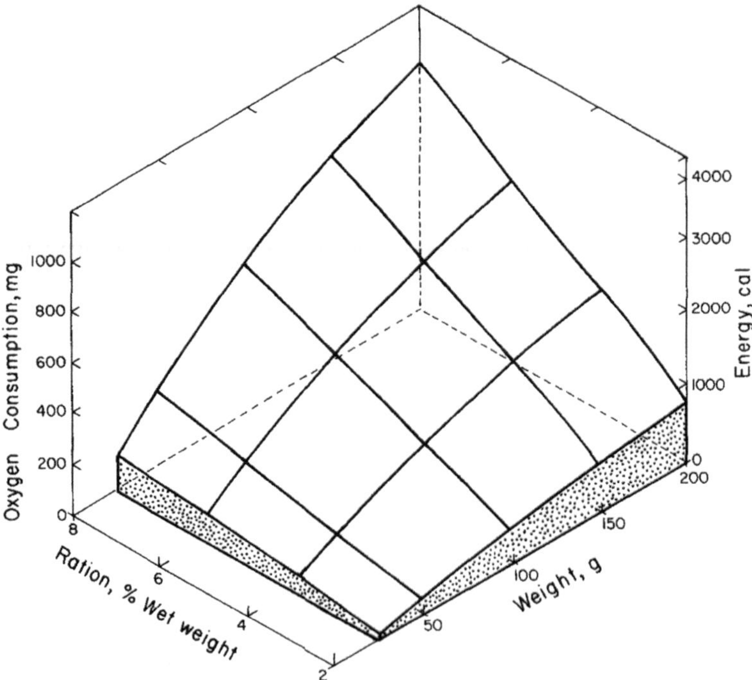

Fig. 8. Heat increment or specific dynamic action of ingested food (minnows) expressed in terms of oxygen consumption and equivalent energy expenditure of largemouth bass, *Micropterus salmoides*, in relation to ration and weight (Beamish, 1974). The results are shown in absolute values. In relative terms, according to the food intake, approximately 14% of the energy of the ration is lost to heat increment, fish weight having no significant effect on this percentage.

Niimi, 1972). Elevated metabolic rates, over baseline estimates for nonfeeding swimming fish, indicated that 76 mg of O_2/g of ration, or approximately 16% of the energy of the food, was appearing as heat increment. Energy expenditure for food utilization in bluegill, *Lepomis macrochirus*, fed to satiation on mayfly nymphs, ranged from 4.8 to 24.4% $(\overline{X} = 12.7 \pm 1.5\%)$ of the total intake, which was considered by Pierce and Wissing (1974) to be mostly heat increment. Satiation feeding (although with quite variable daily intake) of the sluggish sargassum fish, *Histrio histrio*, resulted in a range of heat increments from 15.2% for 1-g fish to 36.2% for 28-g fish; the overall average was 23.7% (Smith, 1973). This is in agreement with the mean value of 23.3% obtained by Miura *et al.* (1976) for biwamasu salmon, *Oncorhynchus rhodurus*, when fed chopped pieces of fish.

Drawing on the more refined experiments, it would appear that a loss of about 12–16% of the ingested food energy can be attributed to heat increment. This generalization applies under favorable environmental conditions to ration levels that are well above maintenance. More evidence is necessary for extreme conditions before further conclusions can be drawn. It is safe to say, however, that unless serious attention to the effect of diet composition on heat increment is the aim, it is doubtful if the necessary effort to correctly partition feeding metabolism into its respective components is warranted.

E. Other Metabolic Costs

As was pointed out, determination of energy expenditure of fishes has evolved and expanded in its centers of attention from standard and routine metabolic rates to active and feeding metabolic rates, including partitioning of the heat increment fraction. Earlier generalizations that standard metabolism was about one-quarter active metabolism, and that twice the standard rate was a fair approximation of average daily metabolic costs, are no longer tenable (Ware, 1975). The enquiry now stands at the threshold of breaking the barrier of knowledge on metabolic rates of free-roaming fishes, with all their diversity of behavior and energy-saving strategies. Unlike the captive fish, little of this is open to direct study, with but one notable exception—the spawning migration of nonfeeding fish such as the salmon and shad. Otherwise it is only possible to examine “components” of wild behavior in the walled-off compartments devised by experimenters, not excluding the artificial stream. Although information is most limited and spotty, this area of enquiry is highly deserving of attention. A few cases are presented.

1. *Aggression*

A chance to record the metabolic rate accompanying aggression in the pumpkinseed, *Lepomis gibbosus*, was afforded while studying group respiratory rates (Brett and Sutherland, 1965). Four fish of similar size had been selected. Within a few hours of being placed in the respirometer, periods of intense contesting for selected areas in the tunnel ensued. Energy expenditure rose to a rate of 180 mg of O_2/kg/hr (0.63 kcal/kg/hr) during peaks of attack and defense, equivalent to about one-half the active metabolic rate of this species.

On another occasion, under very similar circumstances, young sockeye salmon were being exercised as a group (Brett, 1973). Aggression that broke out at the start of one series reached peak metabolic demands of 450 mg of O_2/kg/hr—about one-half the active metabolic rate. Gradually over 6 days this diminished from a daily average of 360 mg of O_2/kg/hr (0.83 kcal/kg/hr) to 180 mg of O_2/kg/hr (0.58 kcal/kg/hr), slightly above the expected rate for 36-hr starved fish forced to swim at 1.25 lengths/sec, the actual imposed velocity in this case.

These two examples serve to illustrate the great energy drain that can occur when unresolved disputes for territory occur, amounting to one-third to one-half the active metabolic rate when contesting is intense.

2. *Migration*

Observations on the rate of swimming of migrating young sockeye in Babine Lake have been reported by Johnson and Groot (1963), and Groot and Wiley (1965). From a knowledge of the temperature, weight, and speeds of migration at near-surface and subsurface levels the likely energy requirements can be calculated. For the period of about 6 hr a day, when fish are on the move, from 1.3 to 2.1 kcal/kg/hr are expended, that is, average metabolic rates of 380–640 mg of O_2/kg/hr (equivalent rates of 30–50 kcal/kg/ day).

Precise records of energy drain exist for adult sockeye salmon (2–3 kg) migrating up the Fraser River, British Columbia (Idler and Clemens, 1959). By means of the method of caloric equivalents described previously, the change in body constituents, from estuary to spawning grounds, was traced by sampling particular races and determining the progressive depletion of body reserves. Over 90% of the fat and, in the female, 50–60% of the total protein may be utilized. This amounts to 2.1 kcal/kg/hr equivalent to 600 mg of O_2/kg/hr, or about three-quarters of the active metabolic rate for fish of this size.

Less precise estimates can be made of the energy expenditure of American shad, *Alosa sapidissima*, which enter the Connecticut River to spawn in the spring (Leggett, 1972). Adults do not normally feed in fresh water. The average weight loss of spent fish ranged from 48 to 55% of the weight prior to entering fresh water. Allowing for the weight of ovaries (13–15%) or testes (8–9%), the average somatic weight loss of females and males was 690 g (45%) and 613 g (48%), respectively, over an 83-day period, that is, 7.8 g/day. Using a mean weight of pre- and postmigrating fish, and allowing for an expected increase in water content of the spent fish, the energy expenditure for this species was about 10 kcal/kg/day.

3. Spawning

Although measurements have been made of the metabolic rates of salmon at a stage close to spawning time (Awakura, 1963; Brett, 1965) none are available for the cost of digging and ultimate spawning. The caloric content of the body and gonads, however, has been measured (Brett and Glass, 1973). When these values are applied to the body and ovary weights of female sockeye salmon on the spawning grounds (Stuart Lake race; see Idler and Clemens, 1959) the energy content of the ovary equals about 25% of the remaining body caloric content.

4. Daily Patterns: Vertical Migrations

Many fishes, particularly the mesopelagic fishes of the oceans such as the myctophids or lantern fishes, rise in the evening through several hundred meters of water column, descending at dawn. They feed on plankton near the surface at night, resting during the day at depth. Some clupeoids make similar vertical migrations. Alexander (1972) has analyzed the energetics for fish of different buoyancy, considering the daily energy cost of remaining at depth. Estimates of 30 mg of O_2/kg/hr (over standard metabolism) were calculated for fish with a swimbladder, and 13-20 mg of O_2/kg/hr for fish deriving their buoyancy from a high proportion of lipids. The swimbladder was considered the most economical mechanism for near-surface buoyancy, but not so at considerable depth.

Similar vertical migrations have been reported during the summer for the lake-dwelling stage of young sockeye salmon, involving descent from surface temperatures of 17°C to approximately 5°C at 30 m or deeper. The possibility of an energy-saving device through the mechanism of behavioral thermoregulation, which favorably balances daily metabolic expenditure, was considered by Brett (1971). Reduced metabolic rates at low temperatures would confer an "energy bonus" when food was limiting. Using the information on feeding times and amounts, and applying appropriate metabolic rates according to the temperatures experienced, a daily pattern of energy expenditure was computed (Fig. 9). This shows a remarkable resemblance to the daily metabolic patterns of the little brown bat, *Myotis lucifugus*, which undergoes daily torpor periods apparently to conserve energy (Gordon, 1972, p. 362).

F. Summary

The rate at which energy is expended by fish can be seen to vary greatly according to species, climatic zone, temperature, size, and level of activity. Standard metabolic rates (maintenance metabolism) for fish in the 10- to 100-g size range vary from an average of 0.2 kcal/kg/hr for arctic species at –1.5°C through the temperate species at 0.3 kcal/kg/hr (15°C) to the tropical fishes at 0.5 kcal/kg/hr (26°C). An elevation in temperature of 10°C increases

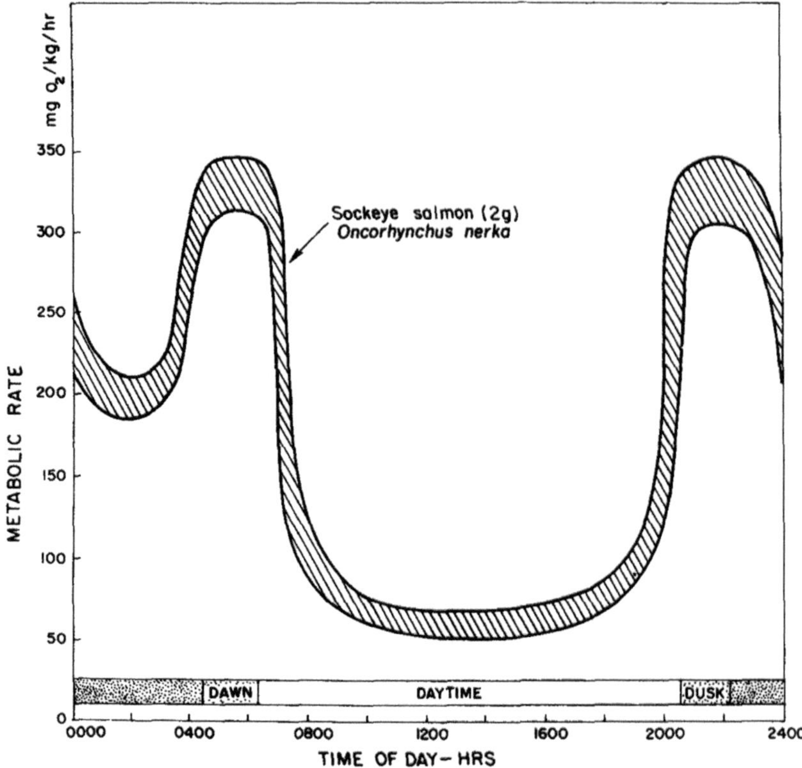

Fig. 9. Schematic representation of daily midsummer metabolic rates of young sockeye salmon in Babine Lake, British Columbia. Feeding occurs almost entirely near the surface at dawn and dusk, with descent from surface temperatures of 16 – 17°C to deep water temperatures of 5° – 6°C (Brett, 1971). The daily mean computes to 183 mg of O_2/kg/hr or 14.2 kcal/kg/day. Energy equivalents as in Fig. 4.

the metabolic rate by a factor of about 2.3 times; this varies considerably between species and over the tolerable temperature range.

Through increased activity, streamlined fast-swimming forms can elevate their metabolic rate on a sustained basis by as much as 10 times, whereas more sluggish, cold-adapted species are confined to a multiple of 2 to 3 times.

A loss of 14% of the energy value of the food occurs as a result of the costs of metabolic processing. This heat increment, along with the activity accompanying feeding, raises the daily metabolic rate of fish that are feeding heavily by a factor of approximately 4 times the standard rate.

The various metabolic rates of sockeye salmon have been compiled to show their relative magnitudes (Fig. 10, Table III). Upstream migration of the spawning adult salmon is the most costly *sustained* energy expenditure of all the activities recorded.

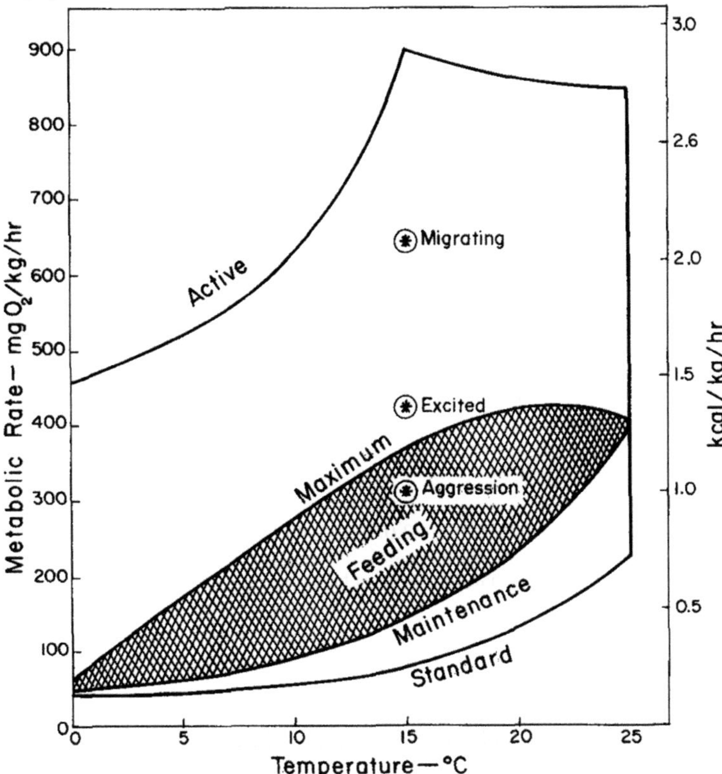

Fig. 10. Rate of energy expenditure of fingerling sockeye salmon showing standard and active metabolic rates (lower and upper lines) in relation to temperature. Metabolic rates associated with feeding (from maintenance to maximum ration) shown crosshatched. Maximum oxygen consumption rate observed for aggression and excitement, and calculated rate for lake migration shown for a temperature of 15°C.

Table III Metabolic Rates Associated with Various Activities of Sockeye Salmon[a]

	mg O_2/kg/hr	kcal/ kg/hr	kcal/ kg/day	Multiple of standard	Percentage of active
Standard	75	0.24	5.8	1.0	9.6
Active	790	2.52	60.5	10.5	0.0
Routine	120	0.39	9.3	1.6	15.2
Feeding (maximum)	320	1.03	24.8	4.3	40.5

Continued

Table III Metabolic Rates Associated with Various Activities of Sockeye Salmon—Cont'd

	mg O_2/kg/hr	kcal/ kg/hr	kcal/ kg/day	Multiple of standard	Percentage of active
Feeding (maintenance)	140	0.45	10.8	1.9	17.7
Heat increment (maximum)	110	0.35	8.5	1.5	13.9
Aggression (maximum)	310	1.00	24.0	4.1	39.2
Excitability (maximum)	420	1.36	32.6	5.6	53.2
Migration	640	2.07	49.6	8.5	81.0

[a]Data determined for a 100 g fish at *15°C*. See text for basis.

IV. EXCRETION: RATES OF ENERGY LOSS

A. Composition and Energy Loss in Feces

The nondigestible fraction of a diet, along with sloughed intestinal epithelial cells, mucus, catabolized digestive enzymes, and bacteria constitute the main components of feces. Smith and Thorpe (1976) also include metabolic fecal nitrogen secreted into the gut, which according to Nose (1967) could amount to 5–17% of the total fecal nitrogen. Diet formulation has a primary aim of using selected, highly digestible nutritious components with a balanced energy content such that utilization is maximized and fecal loss reduced (Hastings, 1969). These involve drawing on many available sources of feedstuff that are not necessarily normal to the diet of wild fish. Only the loss from natural foods will be considered here. Various terms are used in the literature to describe the first stage of incorporating food (less feces) into the body, all with the same general meaning: absorption, assimilation, digestibility, and availability. This brings the food to the second step or metabolizable level, a fraction of which is also excreted mostly as soluble nitrogenous wastes. This latter aspect is considered subsequently.

The opportunistic nature of feeding of many fishes presents a wide variety of organisms, which can change with season and alter as the fish grows, affecting fecal loss. An omnivorous stage in early life is frequently followed by a maturing, carnivorous stage. When the zooplankton feeding young ayu, *Plecoglossus altivelis*, leave the sea to enter freshwater they become grazing herbivores utilizing the diatoms and blue–green algae of streams (Kawanabe, 1969). Chitin and cellulose (fiber) are nondigestible fractions of such diets contributing to a more copious fecal production than the highly digestible flesh and bone of a fish diet. Hence the composition of the feces necessarily varies; it also changes significantly from the composition of the food. Since the ash fraction is often 2 to 3 times that of the food (Rosenthal and Paffenhöfer, 1972; Hickling, 1966; Kelso, 1972), the fecal caloric content on a dry weight, unit basis ranges from one-quarter to one-half that of the food (Table IV). Studies on fecal production that only measure dry weight miss this essential energetic assessment, leaving the impression of greater loss than the caloric values reveal.

In many cases the accuracy of fecal measurement leaves much to be desired. Suspended and soluble fractions may be lost. Allowing feces to accumulate over a number of days invites inaccuracy from bacterial action (Iwata, 1970; Smith and Thorpe, 1976), although this would not be of consequence if the total organic content of the water were analyzed (e.g., Davies, 1963, 1964). Blackburn (1968) gently agitated fresh fecal matter for 24 hr and showed that 16.8% of the feces was lost in suspension. Since the caloric content of the feces was low (2.7 kcal/g) this loss represented a 4% unaccounted reduction in original food energy content. It was further demonstrated that different fecal samples from the same fish had significantly different caloric content. A similar sort of loss (18%) in protein nitrogen content of fecal matter of bass was reported by Beamish (1972). Soluble organic material was shown by Elliott (1976a) to amount to only 1–4% of the total fecal energy, depending on ration level (fresh amphipods fed to brown trout).

Various methods of determination exist, beyond simply siphoning off the gross particulate matter and supplementing this with fine-pore filtration. The addition of an inert reference material to the diet (1% chromic oxide) is frequently used in nutritional studies (Furukawa and Tsukahara, 1966). Approximate homogeneity of the indicator chemical in the consumed ration, and a similar rate of passage through the gut, appear to be justifiable assumptions. Bryan (1974), quoting Bakus (1969), supported the method of using ^{14}C-labeled food—in this case taken up by algae through photosynthesis. Gross carbon assimilation was determined in short-term experiments that were not considered to be significantly affected by any respiratory loss of ^{14}C in 6–8 hr from uptake.

For caloric assessment Warren and Davis (1967) substituted and recommended the wet combustion method (Kanzinkin and Tarkovskaya, 1964) to determine the relative energy content of food and feces using powerful oxidants.

In the case of carnivorous fish, feeding on invertebrates with a hard exoskeleton (e.g., amphipods, prawns, midge larvae), the energy loss in the feces was 16.8 ± 5.9% (SD) ($n = 14$; Table IV). A diet of soft-bodied invertebrates (e.g., polychaetes, squid—but not tubifex) showed only a 4.5% loss for the two cases recorded. Tubifex, however, was relatively poorly digested, losing

$22.1 \pm 7.0\%$ (SD) ($n = 5$). A piscivorous diet had a mean, nonassimilated fraction of $6.1 \pm 3.4\%$ (SD).

For herbivorous fish, grazing many hours of the day, the digestibility of the selected plant life (algae, grasses) is comparatively low. Among grass carp, *Ctenopharyngodon idella*, Hickling (1966) analyzed the food and fecal ash content, which increased from 6 to 12% following passage through the gut. A fecal loss of 30–40% of the weight of the ingested food was considered normal. Stanley (1974a,b) recorded a 50% dry-weight loss, equivalent to a 42% energy loss for this species (Table IV). Milkfish, *Chanos chanos*, were found to lose from 50 to 65%, depending on the diet (Tang and Hwang, 1967). However, when three species of algae (Chlorophycae) were incorporated in an experimental feed for common carp, the algae *Mougeotia* sp. was highly digestible (Table IV) and provided best growth, with *Sirogonium* sp. being comparatively poor (Singh and Bhanot, 1970).

The above considerations take some account of the dietary quality of the food and the type of fish but not of the other potentially influential factors such as temperature, size of meal, frequency of feeding, weight of fish, and prestarvation. With the exception of Elliott's (1976a) work, none of these factors show significant effects except at extremes. Brocksen and Brugge (1974) reported a significant temperature effect, with higher temperatures conferring a greater efficiency on the assimilation of tubifex by rainbow trout (fecal caloric loss = 28.2% at 5°C, falling to 15.2% at 20°C. However, on a fixed 5% ration, as applied, the relative satiation would be greater at 5°C than 20°C possibly influencing the percentage loss. Blackburn (1968) found that the highly efficient digestive process of largemouth bass feeding on guppies improved from 3.2% loss for daily feeding to only 0.8% loss when fed once every 5 days. This is in some contrast to Pandian's (1967b) findings that starvation of 10-40 days had no significant effect on the efficiency of digestion by *Megalops cyprinoides*.

Elliott (1976a) has conducted the most searching analysis to date, determining the separate and combined effects of temperature, ration, and body weight on the fecal energy loss of brown trout feeding on *Gammarus pulex*. When ration was provided as a fixed proportion of the maximum intake ($0.1 - 1.0\ R_{max}$) for the particular temperature tested (e.g., 10°C), the percentage energy loss increased exponentially with increasing ration from a mean low of 14% ($0.1\ R_{max}$) to 23% ($1.0\ R_{max}$). Temperature had a significant effect; at a given ration (e.g., R_{max}) fecal loss decreased with increasing temperature, falling from approximately 29% at 4°C to 20% at 19°C (Fig. 11A). Weight, however, had no significant influence within the range studied (11–250 g).

B. Metabolizable Energy and Nitrogen Excretion Rates

The protein of natural diets (except the hard, cuticle-forming keratins) is usually assimilated to a greater degree than other components of the food.

Table IV Records of the Digestibility of Various Natural Foods in Terms of Percentage Loss in the Feces, as either Dry Weight or Caloric Content[a]

Species	Size (g)	Temperature (°C)	Food			Feces			Comments	Reference
			Diet	Calories (kcal/g/dry)	Ration (% weight)	Calories (kcal/g/dry)	Fraction % dry	Fraction % kcal		
Holacanthus bermudensis Angel fish	200+	19–28	Algae (e.g., *Enteromorpha* sp.)	2.5	*Ad lib.*	—	—	77.7	Feces siphoned daily and water filtered. Food composition was 50–70% carbohydrate	Menzel (1958)
Epinephelus guttatus Red hind	350	19–28	Fish (e.g., *Sardinella* sp.)	—	Excess	—	—	4.7	Feces siphoned off daily. Fed 3 species of fish	Menzel (1960)
Chanos chanos Milkfish	4–80	25–30	1. Diatoms	—	20–25	—	50	(42)	Feces occur in compact form. Siphoned without loss of dilution or suspension	Tang and Hwang (1966)
		(?)	2. Algae	1.2	20–25		65	(55)		
Cichlasoma bimaculatum Cichlid		20–32	Tubifex (*Tubifex* sp.)	5.5	*Ad lib.*	—	—	15	Refers to unpublished results of H. Sethi	Warren and Davis (1967)
		36		5.5	*Ad lib.*			30		
Ophiocephalus striatus Ophiocephalus	30	28	Prawn (*Metapenaeus* sp.)	—	3–8	—	—	9.4	Feces collected after 7–10 days by filtration. Chitin fraction subtracted from food and feces	Pandian (1967a,b)

Continued

Table IV Records of the Digestibility of Various Natural Foods in Terms of Percentage Loss in the Feces, as either Dry Weight or Caloric Content—Cont'd

Species	Size (g)	Temperature (°C)	Food			Feces			Comments	Reference
							Fraction			
			Diet	Calories (kcal/g/dry)	Ration (% weight)	Calories (kcal/g/dry)	% dry	% kcal		
Megalops cyprinoides Megalops	2–150	28	Prawn (*Metapenaeus* sp.)	—	Satiat^n	—	—	8.5	No size effect on absorption	Pandian (1967b)
						—				
Megalops cyprinoides Megalops	50	28	Fish (*Gambusia* sp.)	—	Satiat^n	—	—	7.3	Fish fed once per day to satiation	Pandian (1967c)
						—	—			
Micropterus salmoides Largemouth bass	81	21	Guppy (*Lebistes* sp.)	5.3	0.8	2.8	22	9.8	Three levels of ration used. All feces filtered and losses accounted for	Blackburn (1968)
				—	2.7	—	19	9.3		
				—	3.2	—	17	7.3		
Cottus perplexus Reticulate sculpin	1–5	10	Midge larvae (*Chironomus* sp.)	5.3	—	—	—	18.1	Food and feces caloric equivalent determined by wet combustion method	Brocksen *et al.* (1968)
Salmo clarkii Cutthroat trout	2–8	10	Midge larvae (*Chironomus* sp.)	5.3	—	—	—	14.5	Food and feces caloric equivalent determined by wet combustion method	Brocksen *et al.* (1968)

Oncorhynchus kisutch Coho salmon	3	5–17	Fly larvae (*Musca* sp.)	5.6	1–4	—	—	15	Nonassimilated food showed no trend with season or temperature	Averett (1969)
Ctenopharyngodon idella Grass carp	20–70	23	Lettuce (*Lactuca* sp.)	3.5	*Ad. lib.*	—	—	13.2	High fraction of nondigestable cellulose	Fischer (1970)
Cyprinus carpio Common carp	0.5	25?	Algae (Chlorophyceae)	—	10	—	—	15–20	Control fish fed on live plankton—nearly the same as formulated diets	Singh and Bhanot (1970)
Cynoglossus sp. Flatfish	10	28	Polychaete (*Diopatra* sp.)	5.3	Excess	1.4	19	5.0	Caloric content of feces only 26% of food	Edwards *et al.* (1971)
Perca fluviatilis Perch	12	14	Amphipods (*Gammarus* sp.)	4.3	0.3	—	—	13	Possibly missed small fraction of feces, collected daily	Solomon and Brafield (1972)
Gadus morhua Cod	400	14–16	Plaice (*Pleuronectes* sp.)	5.6	Various	3.1	2.3	1.3	Fed on fillets of plaice. Some feces lost in suspension	Edwards *et al.* (1972)
Micropterus salmoides Largemouth bass	7–91	25	Shiner (*Notropis* sp.)	6.5	2.8	—	—	10.4	Feces collected within 30 min	Beamish (1972)

Continued

Table IV Records of the Digestibility of Various Natural Foods in Terms of Percentage Loss in the Feces, as either Dry Weight or Caloric Content—Cont'd

Species	Size (g)	Temperature (°C)	Food			Feces			Comments	Reference
			Diet	Calories (kcal/g/dry)	Ration (% weight)	Calories (kcal/g/dry)	Fraction			
							% dry	% kcal		
Stizostedion vitreum Walleye	113–502	12	1. Amphipods		1–5	1–3.0	—	17.9	Four diets used. Feces collected within 1 hr by pipette and filtering. Ration level had no effect on assimilation; large fish showed some decrease	Kelso (1972)
		20	2. Crayfish		1–5	1–3.0	—	16.5		
			3. Perch		1–5	1–3.0	—	3.1		
			4. Shiners	4.96	1–5	1–3.0	—	2.1		
Blennius pholis Blenny	19	25	Squid	0.97 (wet)	High	—	4	4	Two temperatures tested. Feces collected by pipette each morning prior to feeding	Wallace (1973)
Histrio histrio Sargassum fish	1–13	21–24	Shrimp (*latreutes* sp.)	—	Satiat^{n}	—	—	27	Two sizes tested. Water filtered daily. Fecal loss greatest in smaller fish	Smith (1973)
	28			—	Satiat^{n}	—	—	18		

Ctenopharyngodon idella Grass carp	1100	23	Egeria (*Egeria* sp.)	—	1.2	—	50	42	Feces siphoned and suspended organics filtered. (Also called white amur)	Stanley (1974a)
Salmo gairdneri Rainbow trout	15	5	Tubifex (*Tubifex* sp.)	—	5	—	—	28.2	Three temperatures tested. All organic refuse determined by chemical oxidation	Brocksen and Brugge (1974)
		17		—	5	—	—	22.1		
		20		—	5	—	—	15.2		
Siganus spinus Rabbitfish	5.4 (cm)	—	Algae (*Enteromorpha* sp.)	—	—	—	64	(54)	Two sizes tested. Used ^{14}C-labeling of algae to determine gross carbon assimilation	Bryan (1974)
	12.9 (cm)	—		—	—	—	84	(70)		
Salmo trutta Brown trout	10–302	4	Amphipod (*Gammarus* sp.)	4.4	Maximum	—	—	15	All particulate feces and dissolved organic material analyzed. Six temperatures and five levels of ration used, within range recorded here	Elliott (1976b)
					0.1 Maximum	—	—	29		
	10–302	20			Maximum	—	—	11		
					0.1 Maximum	—	—	21		

[a]*Data placed in order of publication. Entries of fecal dry weight for herbivores converted to caloric content (in parentheses) by the ratio obtained by Stanley (1974a).*

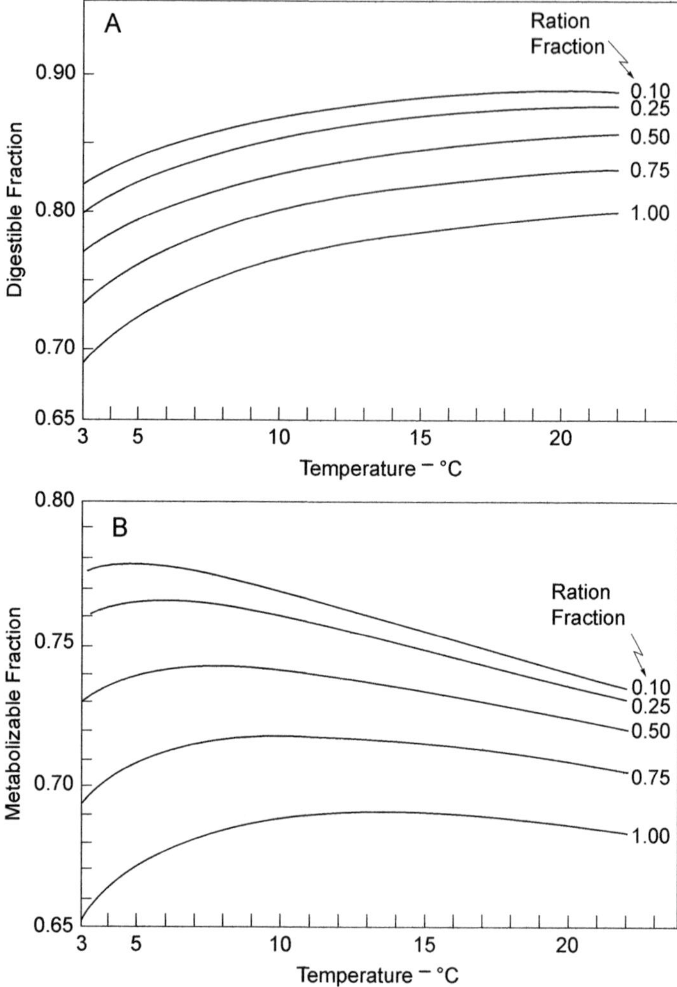

Fig. 11. The fraction of daily energy intake available to brown trout in relation to temperature at different levels of ration (R). Ration is expressed as a fraction of the maximum intake (R/R_{max}). (A) is the amount absorbed, the balance being lost in feces; (B) is the amount available after both fecal and nitrogenous excretion losses (ammonia-N and urea-N) have been subtracted. (From Elliott, 1976a.)

This assumes that the protein is available for assimilation. If the cell wall of plant material is not broken down, then an inability to digest cellulose can obviously prevent access to the protein content. Sunfish, feeding on meal-worms, absorbed 96–98% of the protein at all levels of ration (Gerking, 1955). Birkett (1969) obtained an average absorption of 92.3% for plaice, sole, and perch feeding on live invertebrates. Animal proteins in general have been shown to have high assimilation efficiencies, for example, beef heart, 96% (Morgulis, 1918), whitefish meal, 92%, casein, 99% (Nose, 1967), and fish protein concentrate, over 90% (Cowey and Sargent, 1972). This is also true for some plant proteins but with greater variability [e.g., 73–93% for algal diets fed to silver carp (Chiang, 1971) dropping to 54–63% for goldfish (Nose, 1960)]. Converted by enzymatic hydrolysis into their constituent amino acids, dietary protein is usually readily absorbed through the gut wall into the blood stream. However, if protein is in excess of the requirements of the organism, or the constituent amino acids are poorly balanced in relation to growth needs, deamination occurs, with excretion of nitrogen, mainly as ammonia and urea across the gills (Forster and Goldstein, 1969). This dietary or *exogenous* fraction of nitrogenous excretion represents an energy loss from the food, the balance of which is available for metabolic use, either as an energy source or as a growth increment. This balance is the *metabolizable* fraction (see Fig. 18), providing a measure of the *physiological value* of the diet. It can only be determined experimentally. It is not to be confused with the normal loss of energy associated with protein catabolism, which occurs through the excretion of the *energy-containing* end products of ammonia and urea (Krueger *et al.*, 1968). It is this latter, chemically determinable reduction in energy content, that defines a major portion of the remaining physiological value, which for protein is approximately 88% of the gross caloric content (as presented in Section I). The principle of physiological value applies equally well to carbohydrates and lipids as it does to proteins, but the former constituents tend to be fully utilizable once absorbed, mostly as sugars and fatty acids, respectively. There is great importance associated with determining the N excretion rates as a means of assessing the metabolizable protein available to the organism (Smith, 1971). The problem, however, is to distinguish the normal, maintenance waste fraction (endogenous) from the true metabolizable fraction (Iwata, 1970). This can be seen best by summarizing the expressions involved: Consumed N = Assimilated N + Fecal N; Assimilated N = Metabolizable N + Excreted N (mostly exogenous); Metabolizable N = Retained N+ Excreted N (mostly endogenous). This follows the sequential steps depicted in Fig. 18 (Birkett, 1969).

Unlike mammals which tend to use a large proportion of carbohydrate and lipid for energy (conserving proteins unless starving), fish normally exploit a portion of their protein stores as an energy source. In the case of the mullet, *Rhinomugil corsula*, Kutty and Mohamed (1975) showed this to amount to 14–15% of the total energy expended (in routine metabolism). The dynamic state of tissue proteins, involving catabolic and anabolic processes in an open system, led to the well-recognized concept of a metabolic pool of amino acids (see discussion in Cowey and Sargent, 1972). Exogenous and endogenous sources are common to this pool. To separate the relative contributions, endogenous nitrogen excretion was defined for domestic animals as "the lowest level of N excretion attained after an empirically defined time interval on a low nitrogen but otherwise complete diet" (Brody, 1945, p. 59). This maintains the normal energy requirements of the animal without involving it in any serious nutritional stress from complete lack of a nitrogen source, (i.e., just maintaining the status quo without sacrificing body protein). In accordance with the need for meeting energy expenditure, Savitz (1961,1971) force-fed just enough glucose to bluegill, *Lepomis macrochirus*, to meet their estimated maintenance metabolism following the method of Gerking (1955) (see excretion rates in Table V). However, according to Iwata (1970) it is very difficult to determine endogenous nitrogen excretion accurately because it is hard to give an amount of nonprotein food equivalent in calories to the maintenance metabolic requirement of fish.

Records of N excretion of starved animals provide a near approximation of the endogenous fraction, particularly for cold-blooded vertebrates that normally endure prolonged periods of starvation (Stover, 1967). Error may be minimized by extrapolating back to zero time (postdigestive state), where it is obvious that any decay in maintenance metabolism can be considered as insignificant. The best method is undoubtedly the use of ^{15}N-labeled protein, tracing the fraction that is excreted before any new dietary protein has been made available for energy.

Gerking (1955) demonstrated a weight-dependent relation for endogenous N excretion (approx. $W^{0.54}$), following the well-known relation of decreasing rate of metabolism with increasing size. A considerably higher value for the weight exponent ($W^{0.9}$) was reported for starving crucian carp (Iwata, 1970). It is undoubtedly important to take size into consideration when comparing various endogenous rates among fishes; however, some doubt has been cast on the correct exponent (Savitz, 1969; Iwata, 1970), and it can be expected that considerable species variation will occur, as indicated in the compilation of Table V. Davies (1963) determined the metabolizable fraction of live, white worms (*Enchytraeus albidus*) when fed to goldfish, *Carrasius auratus*, at various ration levels. The metabolizable portion of the food varied quite significantly, ranging from 72% at a ration of 1.5% dry weight/day to 86% for a ration of 4.5%.

Few studies on fish have been performed where the exogenous and endogenous fractions of nitrogen metabolism could be separated. By tracing the hourly pattern of ammonia and urea excretion of fingerling sockeye salmon, following a single daily meal (Fig. 12), it was shown that a strong pulse of ammonia excretion occurred, peaking 4–4.5 hr after the start of feeding (Brett and Zala, 1975; see also Durbin, 1976). A formulated diet based on fish meal was used (Oregon moist pellets). From comparison with the endogenous excretion of starving "control" fish, the exogenous fraction could be computed. This amounted to approximately 15 mg of N/kg/ day, or 27% of the nitrogenous intake. Assuming that 97% of the dietary protein would normally be absorbed in the gut, the metabolizable fraction would only amount to 70% of the intake, that is, a loss of 3% in feces and 27% excreted. This example serves to illustrate the areas of potential loss not usually determined for the

Table V Daily Rates of *Endogenous* Nitrogen Excretion[a]

					Nitrogen excretion (mg of N/kg/day)			
Species	Weight, average or range (g)	Temperature (°C)	Salinity (‰)	Starvation (days)	Ammonia	Urea	Total	Reference
Lepomis macrochirus	50	26	Fresh	3–8	—	—	154	Gerking (1955)
Lepomis macrochirus	144	26	Fresh	3–8	—	—	95	Gerking (1955)
Leptocottus armatus	165–391	12	30	1	41	14	64	Wood (1958)
Platichthys stellatus	310–335	12	30	1	61	9	73	Wood (1958)
Taeniotoca lateralis	360	12	30	1	14	13	30	Wood (1958)
Salmo gairdneri	129	13	Fresh	6–14	75	35	136	Fromm (1963)
Lepomis macrochirus	10–100	7	Fresh	3	—	—	58	Savitz (1969)
Lepomis macrochirus	10–100	31	Fresh	3	—	—	289	Savitz (1969)
Carassius auratus	1	20	Fresh	7	105	—	135	Iwata (1970)
Carassius auratus	10	20	Fresh	7	75	—	100	Iwata (1970)
Salmo gairdneri	50–100	13	Fresh	7	—	40	160	Olson and Fromm (1971)
Carassius auratus	1–4	22	Fresh	7	—	71	—	Olson and Fromm (1971)
Lepomis macrochirus	70–90	24	Fresh	3	—	—	114	Savitz (1971)

Continued

Table V Daily Rates of *Endogenous* Nitrogen Excretion—Cont'd

					Nitrogen excretion (mg of N/kg/day)			
Species	Weight, average or range (g)	Temperature (°C)	Salinity (‰)	Starvation (days)	Ammonia	Urea	Total	Reference
Lepomis macrochirus	70–90	24	Fresh	7–28	—	—	74	Savitz (1971)
Perca fluviatilis	12	14	Fresh	7+	170	—	—	Solomon and Brafield (1972)
Salmo gairdneri	900	10	Fresh	4	31	—	—	Nightingale (1974)
Salmo gairdneri	900	15	Fresh	4	55	—	—	Nightingale (1974)
Salmo gairdneri	900	20	Fresh	4	85	—	—	Nightingale (1974)
Oncorhynchus nerka	29	15	Fresh	1–22	175	46	221	Brett and Zala (1975)
Salmo gairdneri	30–40	12	Fresh (smolt)	10	—	—	92	Smith and Thorpe (1976)
Salmo gairdneri	30–40	12	Fresh (postsmolt)	10	—	—	60	Smith and Thorpe (1976)
Salmo gairdneri	30–40	12	Salt (smolt)	10	—	—	97	Smith and Thorpe (1976)
Salmo gairdneri	30–40	12	Salt (postsmolt)	10	—	—	93	Smith and Thorpe (1976)
Dicentrarchus labrax	5–235	16–20	Salt	7–10	72	12	84	Guérin-Ancey (1976)

[a]Listed in order of publishing date.

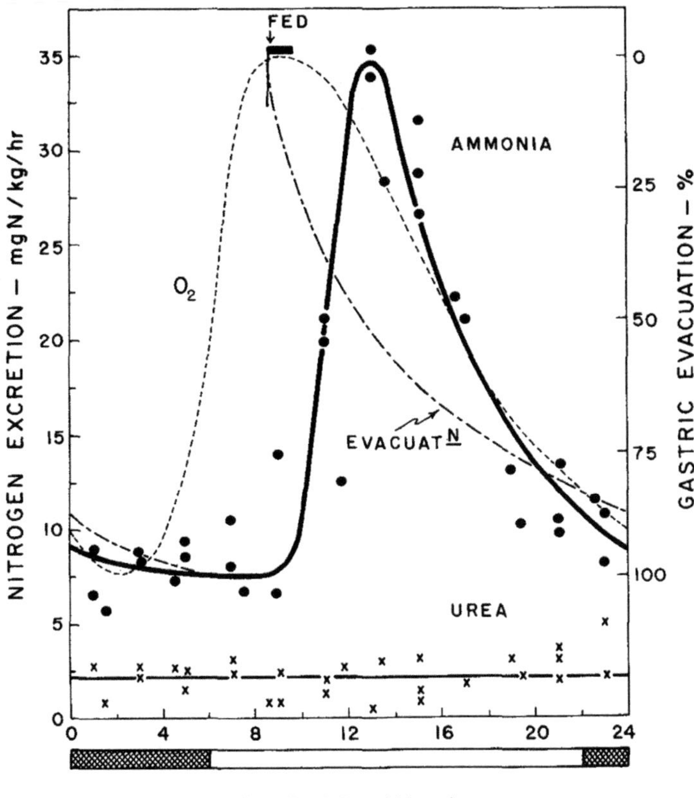

Fig. 12. Diurnal variation in rate of ammonia (●) and urea (×) excretion of fingerling sockeye salmon (*Oncorhynchus nerka*). Ration =3% weight/day; temperature =15°C; photoperiod =16 hr with dark period shown cross-hatched; freshwater pH=7.0-6.7. For comparison the associated oxygen consumption rate and expected gastric evacuation rate have been included, scaled to the peak and baseline excretion of ammonia. Note that the oxygen consumption rates, which peaked at 370 mg of O_2/kg/hr, are about 10 times the nitrogen excretion rates. (From Brett and Zala, 1975.)

components of fish food. The complication of separating the sources of the excretory products is further involved by posing the question of identifying the metabolic source for the accompanying heat increment, which in fish could be derived in part from protein catabolism.

The daily fluctuations of ammonia and urea excretion for heavily feeding fish have been studied in at least two hatcheries. Quite irregular patterns in the relative proportions of these nitrogenous products were recorded by Burrows (1964) for chinook and coho salmon. Stocking density, handling stress, and temperature levels were factors contributing to the variability. When ammonia did predominate, as in the sockeye experiments, it tended to occur over 14 hr of the day, with a peak at 1600 hr. In the studies of McLean and Fraser (1974) on hatchery-reared coho, ammonia nitrogen accounted for over 60% of the total nitrogen excreted, rising to over 90% on some days (at 40 mg of N/kg/hr). With fish of 12–15 g weight, on a dawn to dusk feeding regime, a daily pulse in ammonia excretion occurred, reaching a peak 9 hr after the start of feeding.

When applied to the maintenance ration, Brett (1976a) further attempted to account for the energy losses occurring over the full range of temperature tolerance (approximately $0^\circ - 25^\circ$C), using Phillips' (1969) physiological values for energy equivalents (Fig. 13). Despite the relatively low caloric value so computed (3.92 kcal/g, i.e., a factor of 0.72 times the total energy content) the appropriate feeding metabolic rate did not equate to the physiological

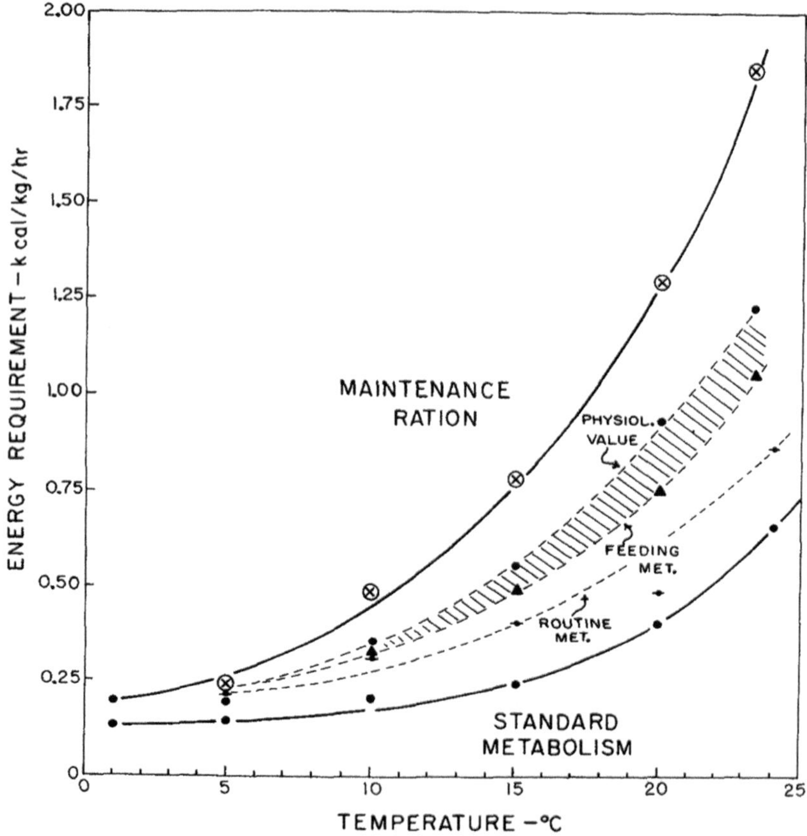

Fig. 13. Energy requirement of yearling sockeye for standard metabolism and maintenance ration, over the tolerable range of temperature. The metabolic rate associated with feeding at the maintenance level (F_{met}) is shown together with the amount normally expended in spontaneous activity (routine metabolism). The net energy available from the maintenance ration is shown as the physiological value. The shaded area is unaccounted energy, probably attributable to an increasing inefficiency of food conversion (i.e., a reduced physiological value at higher temperatures). (From Brett, 1976a.)

value at temperatures above 10°C. The unaccounted caloric difference, amounting to as much as 18% at 20°C, was considered to be the result of an increasing inefficiency of food conversion at higher temperatures, indicating lower physiological values than the ones applied.

While determining the effect of temperature, ration, and size on the energy loss in the feces of brown trout, Elliott (1976a) also examined the daily energy loss from excretion of nitrogenous wastes—ammonia and urea. The proportion of the daily food intake (gammarids) lost from these sources increased as temperature increased and the level of intake decreased (i.e., just the opposite to the relation for fecal loss). Thus, nitrogenous excretory products accounted for about 4–6% loss in energy at 4°C, and from 11 to 15% loss at 20°C. The greater percentage loss at each temperature occurred at the lowest intake (0.1 R_{max}). Endogenous excretory levels were not distinguished from exogenous sources, although nitrogen excretion rates were shown to decrease over a 4- to 6-day period. The fractional loss was independent of weight (4- to 300-g range). The combined energy losses (fecal plus soluble nitrogenous) resulted in a range from 22 to 35% on a low ration, and from 26 to 31% on a high ration (Fig. 11B).

By experimenting with different types of natural diets, Elliott (1976a) further showed that although there were some significant differences in the comparative energy losses according to pathway, the total energy losses from each diet were very similar. Consequently, the fraction of energy available for growth and metabolism remained remarkably constant at each ration level and temperature.

C. Summary

From the foregoing it can be seen that there is a wide range in the value of food energy available for metabolism and growth. Fecal energy alone accounted for over 50% in some herbivorous fish, leading to the belief that the metabolizable fraction could easily be as low as 40%. For omnivorous, invertebrate-feeding fish, such as the brown trout and perch, a range of 25–30% loss would apply for most combinations of temperature and ration. Carnivorous fish, feeding on other fish, could be expected to approach a 20% loss under favorable environmental conditions—the only category supporting the former generalization of Winberg (1956) on the average metabolizable fraction of the diet. Least information is available on the energy loss from nitrogenous excretion. An erroneous assumption by Winberg (1956), that the energy value of the chief excretory product (ammonia) was negligible, has contributed in the past to overrating the potential net energy from the food.

These cases have been followed through indicating how gaps in knowledge of fish energetics, frequently riding on assumptions from mammalian physiology, can lead to potential errors.

V. GROWTH: RATES OF ENERGY GAIN AND FOOD CONVERSION

The net energy derived from the food (metabolizable minus heat increment) is that portion available for all additional forms of metabolism and activity, of which the three major components—swimming, maintenance, and growth—make the greatest demands. Growth has the lowest immediate priority, but in the long run growth and reproduction dictate the species survival. The need to collect and convert enough food to meet the growth requirements is almost always pressing in nature. Records of large fish demonstrate the rare capacity afforded to particular members of populations (World Record Board, National Fresh Water Fishing Hall of Fame, Box 33, Hayward, Wisconsin 54843).

By developing the relation between experimentally controlled rations (of high diet quality) and growth rate, the full range of capacity to grown can be defined. The characteristics of this growth–ration relation (the GR curve) and its derivative, the conversion-efficiency relation (the KR curve), are dealt with in Chapter 10 (see Fig. 4 of that chapter). It is sufficient here to present a few of the main findings, and extend these to elaborate on the principles of energy conversion and the effects of environmental factors.

A. Ration and Fish Size Relations

In almost all cases recorded, the higher the ration the greater the growth rate. Only carp, *Cyprinus carpio*, has been shown to experience some rate reduction at maximum ration (Huisman, 1974, 1976). This case serves to illustrate an extreme of the general "glutton effect" of high rations, in which the efficiency of food utilization is decidedly less than at some submaximal point defining the optimum ration (R_{opt}). The position of R_{opt} in the growth–ration curve may range from a relatively low to a relatively high ration according to species and environmental conditions, especially temperature effect (e.g., Elliott, 1976b). Two cases of extreme difference are shown in Fig. 14A. A point of major significance is the consequence to the conversion-efficiency relation illustrated in Fig. 14B. In the first case (curve a) a long and well-defined reduction in efficiency accompanies increasing ration beyond R_{opt}, whereas in the second case (curve b) there may be little to no reduction of efficiency at any point as ration increases. This variation of pattern accounts for much of the controversy over how efficiency changes with ration, the so-called " *K*-line" of Paloheimo and Dickie (1966) being applicable to those cases resembling curves of type a (Fig. 14). *K* refers to the gross efficiency (K_1) used by Ivlev (1945). *K* was shown to decrease exponentially with increasing ration.

Relative growth rate* is greatest at the smallest size (see Chapter 10, Figs. 22 and 29). Some exception to this principle may occur at first feeding

*Equivalent to specific or instantaneous growth rate. See Chapter 11.

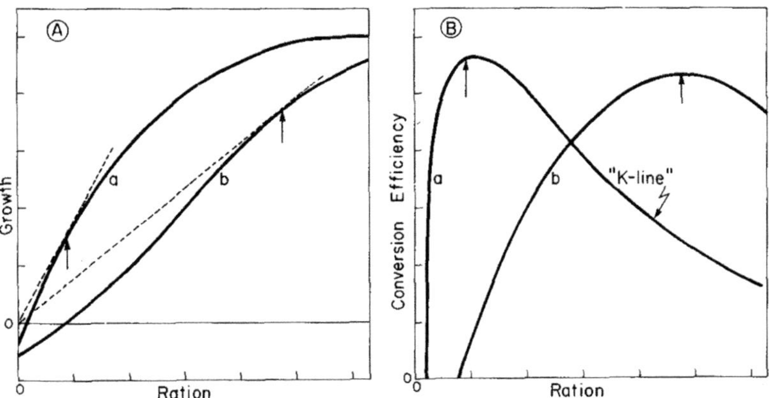

Fig. 14. (A) Variation in form of the relation between growth rate (G) and ration (R), both expressed in terms of % dry body weight/day. Broken lines are the tangents to the curves a and b with arrows indicating the defined position of optimum ration. (B) Corresponding conversion efficiency curves ($G/R \times 100\%$) with maximum efficiency (K_{max}) indicated by arrows also corresponding to the positions in A for R_{opt}. The segment of curve a in B, from optimum to maximum ration, is the K-line of Paloheimo and Dickie (1966), shown to be linearized by an exponential transformation.

of larval or fry stages, where effective feeding behavior is being learnt and the digestive system may not be fully differentiated or completely free of yolk. Once maximum feeding is established, growth rates can easily reach 8–10% body weight/day, as in young salmon (see Chapter 10), requiring feeding rates of 25–30% per day. Winberg (1956) compiled tables of average daily gain in weight during initial periods of development, giving examples for bream and carp where the growth rate exceeded 35% per day. Although for very brief periods, it is apparent that some of these early forms would have to consume more than their body weight in a day.

On maximum ration at a young stage the rate at which fish can accumulate body calories exceeds the rate at which metabolism expends calories. Both rates are relatively high initially. With increasing size and age these two relative rates decrease but at different declining slopes such that eventually the rate of metabolic energy expenditure is considerably higher than the associated capacity to deposit energy (Fig. 15). This fundamental circumstance is a basic cause for decreasing conversion efficiency accompanying increasing size, which would approach zero as size reached an upper limit. The relation is apparent in the studies of Kinne (1960) on *Cyprinodon macularius*, of Smith (1973) on *Histrio histrio*, and for salmonids (see Chapter 10, Fig. 22). Huisman (1974) has shown that conversion efficiency declines exponentially with increasing weight in carp (35–210 g). Such a decline with

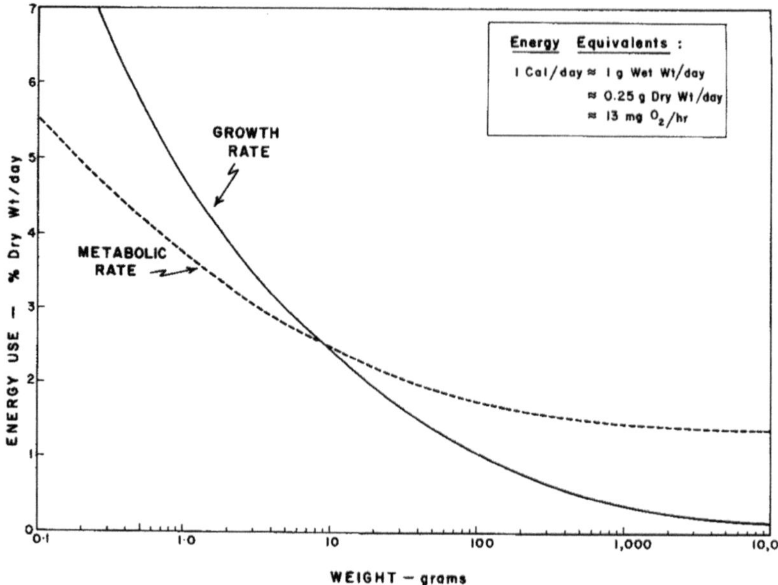

Fig. 15. General relation between the energy deposited in growth and that expended in total metabolism according to size (sockeye salmon). The ratio of the energy involved in the two systems (growth: metabolism) changes from >1.0 at approximately 10 g to <1.0 above 10 g, accounting largely for the decrease in food conversion efficiency accompanying increasing size. (From Brett, 1970a.)

size could easily confound the K-line relation of Paloheimo and Dickie (1966) where increased ration bringing decreased efficiency could well be associated with increased fish size.

B. Gross and Net Conversion Efficiencies

Possibly the most meaningful and simplest indicator of adequacy of diet, ration level, state of health, and environmental suitability for an organism is the capacity to convert food into flesh: the gross conversion efficiency,

$$K_1 = (G/R) \times 100\%$$

Growth rate (G) and ration (R) may be expressed in terms of wet weight, dry weight, or caloric content. Wet weight is only suitable when the moisture content of both the food and the fish are approximately the same. Determinations based on dry weight will vary from those in calories according to whether the level of ration (if of constant composition) induces weight gains that increase or decrease the relative fat content of the fish (e.g., Huisman, 1974).

Net conversion efficiency provides a measure of the capacity to convert that fraction of the ration in excess of the maintenance level into flesh,

$$K_2 = (G/R - R_{\text{maint}}) \times 100\%$$

Its measurement will depend greatly on the accuracy of determining the maintenance ration—not always accomplished too successfully. Its usefulness is unquestionable where partitioning of energy or nutritional adequacy are under investigation. Otherwise this demanding refinement hardly adds much to the acknowledged value of knowing the gross efficiency.

Many growth studies include determinations of gross conversion efficiencies. For juvenile fish up to maturity, gross efficiencies mainly range between 10 and 25%, depending on size, age, diet, ration, and environmental conditions (Pandian, 1967a,b; Yoshida, 1970; Chesney and Estevez, 1976). Highest efficiencies are associated with conversion of yolk in early development, for example, 65–70% in salmonids (Marr, 1966) and 75–80% in sardine (Lasker, 1962). Maximum efficiencies for post embryonic stages were earlier thought to have an upper ceiling of about 35% (Ivlev, 1945). However, in caloric equivalents considerably higher values have been recorded, for example, 55% for 1–2 gm coho salmon feeding on live fly larvae at $8^{\circ} - 14^{\circ}C$ (Averett, 1969), 52% for immature cichlids feeding on oligochaetes at $28^{\circ}C$ (Warren and Davis, 1967), and 60% for 3.8 g mackerel feeding on chopped anchovy at $15^{\circ}C$ (Hatanaka and Takahashi, 1956). Herbivorous fishes are characteristically on the low side of efficiency (10–20%) because of the large nondigested fraction of their normal diet (Welch, 1968). However, Stanley (1974a,b) has reported that grass carp feeding on green algae reach an efficiency of 40% or more by virtue of a compensating low metabolic rate (see also Stanley and Jones, 1976). Using a radioactive tracer, ^{137}Cs, Kevern (1966) was able to trace the annual efficiency of a population of carp, *Cyprinus carpio*, (150–200 g) feeding on algae and detritus. On a caloric basis, and a mean ration of 3.9% wet weight day, an efficiency of only 6.5% was determined.

C. Nitrogen Retention

Although growth must be considered as a net increase in any of the body constituents, not excluding water, continued elaboration of tissues cannot proceed in the absence of an adequate supply of protein. In consequence nitrogen retention (protein synthesis) rather than carbon or caloric retention has been considered as the fundamental unit of growth (Brody, 1945; Maynard and Lousli, 1962). There is an optimum nitrogen content of the diet, subject in particular to age and temperature effects, ranging from 35 to 55% of the diet composition (see Chapter 1).

Such an approach to growth was adopted by Gerking (1971) in a study of the effects of nitrogen consumption and body weight on the N retention of bluegill sunfish, *Lepomis macrochirus*. His findings are in keeping with the basic principles of growth and conversion efficiency already set forth, with the modifier that protein would be required to a lesser extent than calories as size and age increased, diminishing to a state of nitrogen maintenance for an old fish mainly metabolizing for energy and not for growth. With an increase in nitrogen consumption, the rate of N retention of sunfish increased linearly up to a maximum for the range of rations provided, following a conversion efficiency curve similar to that in Fig. 14B, type b. Maximum protein conversion efficiency decreased from 39% for a 14-g fish to 10% at 85 g.

Iwata's (1970) detailed studies on nitrogen retention in relation to absorption in the carp, *Carassius auratus cuvieri*, showed that efficiency of retention increased during the latter half of a 20-day experiment. The improvement was ascribed to recovery of the fish from the primary stress of handling and effects of anaesthetic. From the data obtained during the tenth to twentieth day of feeding it is apparent that the carp exhibited a relation for gross conversion efficiency of protein in relation to nitrogen absorption that falls in an intermediary position between the a- and b-type curves of Fig. 14A,B. Maximum gross efficiency was approximately 35%. This appears to be in agreement with Birkett's (1969) findings for plaice, sole, and perch which ranged from 27.5 to 49.0% of the "gross efficiency."*

D. Environmental Factors

The manipulation of environmental conditions has been shown to provide favorable combinations that significantly improve both growth and conversion efficiency. Growth rate of 20 g sockeye at an optimum temperature of 15°C in freshwater was increased from 1.4 to 2.4% per day by using isosmotic salinity, an increasing photoperiod, cover from direct light, and low water velocity (Brett and Sutherland, 1970). The salmon not only increased their daily food intake but also showed improved conversion efficiency at any given level of ration. As presented in Chapter 10, abiotic factors can be classified in terms of how they influence any activity, a classification first elaborated by Fry (1947) and subsequently extended (Fry, 1971). Thus, temperature governs the rates of metabolic reactions (Controlling Factor), salinity imposes a metabolic load on internal regulation (Masking Factor), daily light cycles affect endocrine activity (Directive Factor), and oxygen, size, and ration can each restrict growth through one mechanism or another (Limiting Factor). The consequences of these factors will be considered as they relate to conversion efficiency in a manner similar to that developed for the growth–ration curves (see Chapter 10, Figs. 18 and 26).

---

*Birkett's (1969) paper should be consulted for his use of "gross efficiency," which by definition would be equivalent to net efficiency as used in this text.

1. Temperature

In three species receiving most attention (coho, Averett, 1969; sockeye, Brett *et al.*, 1969; trout, Elliott, 1975a,b), at temperatures below 10°C conversion efficiency rises most rapidly from the base level (maintenance ration) reaching a peak efficiency at an intermediate ration (Fig. 16). This relation changes progressively with rising temperature such that at 17°C and above, the highest efficiency occurs near the maximum ration. The shift in the relation of conversion efficiency to ration with increasing temperature follows a pattern of change not unlike that from curve a to curve b depicted in Fig. 14B. Over the complete range of tolerable temperatures maximum efficiency occurred between 5 and 10°C (coho), at 9°C (trout), and at 11°C (sockeye).

In cases where growth–ration curves are established for a wide range of temperatures, the interaction between ration and temperature on growth rate can be determined, from which the full range of conversion efficiencies may be derived. Two such examples are shown, for trout (Fig. 16) and sockeye (Chapter 10, Fig. 27B). Although the central axes of the isopleths are not particularly different, the epicenter for the efficiency isopleths of trout lies just outside the boundary (R_{max}) of the figure, in contrast to that for the sockeye where it lies well inside the configuration (at 11.5°C and 4.0% per day ration).

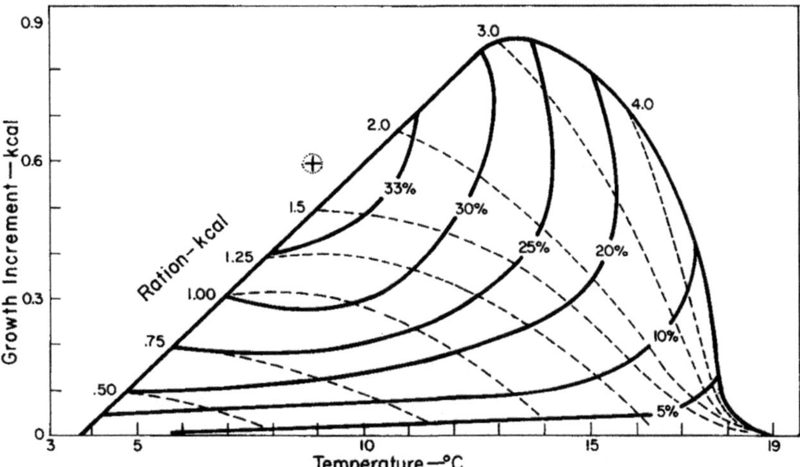

Fig. 16. The relation of temperature to daily growth increment (broken lines) according to fixed ration (shown for each curve), measured in terms of energy values for brown trout of 50 g initial weight. Gross conversion efficiencies (%) drawn as isopleths (solid lines) over the growth curves (Elliott, 1976b). The probable center of the isopleth configuration has been added (circled +) for comparison with Fig. 27B, Chapter 10. The interaction axis of the isopleths would run through this center approximately parallel to the outer diagonal boundary line (set by the maximum food intake).

The sockeye were more prone to feed well at low temperatures, accounting for much of the difference. Brown trout did not feed much above the maintenance level when at the lowest (3.8°C) and the highest (19.5°C) temperatures studied.

2. *Salinity*

When exposed to increasing salinity the capacity to regulate ionic balance of stenohaline freshwater fish decreases rapidly above the blood isosmotic level of $10 \pm 2\%$ (see Chapter 10, Fig. 13). The increasing ionic load of this Masking Factor becomes intolerable. This is reflected in the fall of conversion efficiency reported by Shaw *et al.* (1975) for Atlantic salmon *parr*, which decreased from a maximum of 22% in fresh water to 7% in salt water (Chapter 10, Fig. 14B). By contrast the euryhaline pupfish, *Cyprinodon macularius*, showed greatest efficiency at 15‰ with a maximum in excess of 30% conversion when on a high ration at temperatures from 17° to 22°C (Kinne, 1960) (see Chapter 10, Fig. 28B).

3. *Photoperiod*

The effect of this Directive Factor on conversion efficiency has received relatively little attention. By stimulating the production of growth hormone (Chapter 9) improved efficiency would likely follow the path of change noted earlier for sockeye salmon, following environmental manipulation (Brett and Sutherland, 1970). This conjecture is supported by the findings of Gross *et al.* (1965) for sunfish, *Lepomis cyanellus*, which had greatest consumption and maximum conversion efficiency (48%) after exposure to an increasing photoperiod of 8–16 hr light per day.

4. *Oxygen*

Factors such as oxygen concentration have a predictable effect on conversion efficiency by restricting the development of the normal GR curve, as discussed in Chapter 10. Below the critical O_2 level of 5 ppm, growth becomes dependent on oxygen concentration for many fish, for example, largemouth bass, *Micropterus salmoides* (Stewart *et al.*, 1967) (see Chapter 10, Fig. 16). A precipitous decline from an average of 27% efficiency above 5 ppm O_2 occurred among the bass, falling to 0% at approximately 2 ppm of O_2.

E. Summary Configurations

The above environmental relations have been depicted in general form in Fig. 17, and bear comparison with the recapitulation of growth × ration

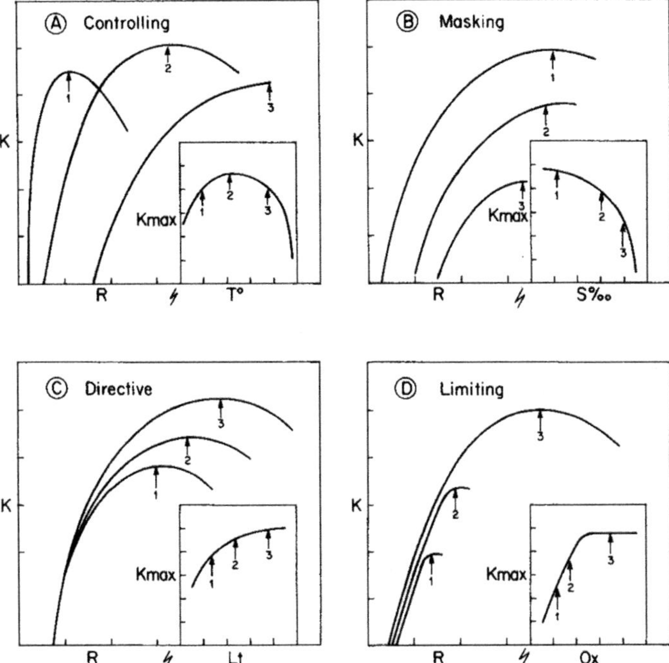

Fig. 17. Recapitulation of gross conversion efficiency × ration curves (*KR*) in relation to abiotic entities illustrating the basic forms for: (A) *temperature* (T°), a Controlling Factor; (B) *salinity* (S‰), a Masking Factor; (C) *light* (L_t= static photoperiods), a Directive Factor; and (D) *oxygen* (O_x), a Limiting Factor. Each *KR* curve is shown at three levels of each abiotic entity, with an arrow for the position of maximum efficiency (K_{max}) associated with optimum ration (R_{opt}). The insert box shows the path of change of K_{max} as the abiotic entity increases over the tolerable range. The intercept on the x-axis is defined by R_{maint}. *K* is represented in terms of % and *R* as % dry body weight/day. Compare with Chapter 10, Fig. 18.

(GR) curves presented in Chapter 10, Fig. 18. Maximum efficiency (K_{max}) usually occurs at an intermediate ration (R_{opt}) below R_{max}, though not always, as is apparent for extremes of temperature, salinity, and limiting levels of oxygen. The change in efficiency with increasing salinity is depicted for a freshwater fish; the relation would be reversed for a marine species, with maximum efficiency occurring at relatively high salinities, that is, the curve in "B-box" would be reversed. Least information is available for the influence of photoperiod. Limiting effects of reduced oxygen concentration probably impose some increased maintenance ration through increased ventilation; this speculation is shown by the separation of the curves in D, which would otherwise be superimposed.

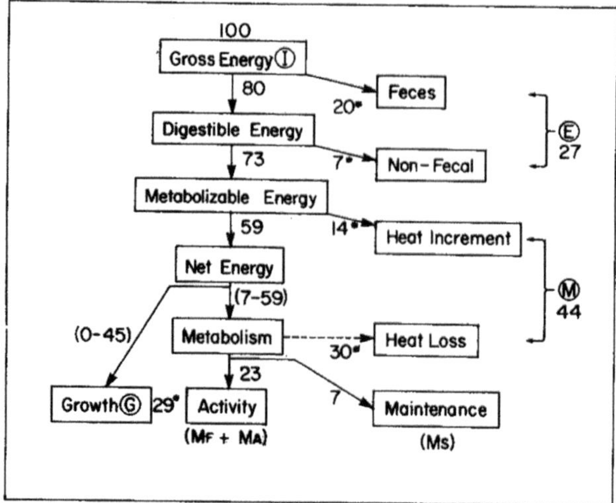

Fig. 18. Average partitioning of dietary energy for a carnivorous fish. A ration of 100 calories is equivalent to about 2% dry weight/day for a 1 kg fish. Nonfecal energy is mostly that excreted as ammonia and urea. Physiological value is equal to metabolizable energy. Circled lettering is according to equations for energy budgets (Table VI): $I=M+G+E$, where I, rate of ingestion; M, metabolic rate; G, growth rate; E, excretion rate. Amounts marked with asterisk total to 100; figures in brackets indicate the possible range of net energy distributed to metabolism and growth. Thus, of the net energy (59 cal) a minimum of 7 cal is required for maintenance of a non-feeding fish (standard metabolism); alternatively all 59 cal could be required by a very active fish for metabolism alone. A growing fish would require some increase in the metabolic rate over the standard rate, here shown as reducing the upper limit of energy available for growth by an additional 7 cal [i.e., $59-14(M_F)=45$]. Note that Metabolism gives rise to Heat Loss (30 cal) which is the sum of Maintenance (7 cal) and Activity (23 cal), which in turn are coupled with Heat Increment (14 cal) to make up the total metabolism (M = 44 cal).

VI. ENERGY BUDGETS

A. Balanced Equations

Since biological systems conform to the laws of thermodynamics, all the energy ingested (I) by a fish must turn up in one form or another through metabolism (M), growth (G) and excretion (E), where $I=M+G+E$.

As presented in Section III, total metabolism (M) may be divided into a variety of levels, namely, standard metabolism (M_S), routine metabolism (M_R), feeding metabolism (M_F) and active metabolism (M_A). If the *additional* energy demand (scope) explicit in each is considered to be additive, then

$$M = M_S + aM_{R-S} + bM_{F-S} + cM_{A-S}$$

where the constants (a, b, and c) apply to estimates of the fraction of time each day that routine, feeding, and active metabolism occur. Determination of any subcomponent helps in an estimate of M. In a similar manner growth (G, see Section V) may include a subcomponent related to production of gametes (G_G) to add to somatic or general body growth (G_S), where $G = G_G + G_S$. Finally, excretion (E, see Section IV) has subcomponents mainly as feces (E_F), urea and ammonia (E_U), and a small amount of mucus and sloughed epidermal cells from the skin (E_S). When all assembled,

$$I = (M_S + aM_{R-S} + bM_{F-S} + cM_{A-S}) + (G_S + G_G) + (E_F + E_U + E_S)$$

Each of these subcomponents has already been considered (except E_S), including the range of values and variations that can be expected, particularly from the effects of environmental factors. Because they form a balanced equation, determinations (or well-founded estimates) of any three of the main components provide a value for the fourth by difference. However, all errors in assessing the three components obviously become a pooled error in the fourth, unless by chance they cancel out.

B. Carnivores

Following the major review of metabolism and growth of carnivorous fish by Winberg (1956), and greatly influenced by Ivlev's (1939) pioneering studies in fish energetics, a general value of 20% of the food intake was assigned to excretion (E). A simplified form of the general equation ($0.8^{*}I = M + G$) was consequently accepted as applicable in many cases. Coupled with measurements of growth (G) and estimating total metabolism[†] (M), the amount of food consumed (I) was deduced by difference (Mann, 1965); or alternatively by knowing I and G, M could be deduced (Paloheimo and Dickie, 1966). Efforts have been made to measure most of the components and subcomponents directly, notably by Warren and associates (Warren, 1971). In addition, two of the most complete studies are those by Solomon and Brafield (1972) on perch, and by Elliott (1976b) on brown trout. These important contributions along with a selection of others have been assembled in Table VI. Although some degree of arbitrary selection has been made, they all represent advanced studies in the field and can be analyzed for an assessment of energy budgets. It must be noted that these are almost entirely for carnivorous fish, feeding mostly on stream invertebrates and fish (perhaps some distinction should be made between insectivores and carnivores, as in mammals).

---

*Here expressed as a fraction, as used by Winberg (1956), rather than as a percentage (80%) as used in the equations below.

†Frequently by relating established values of standard metabolism (M_s) with estimates of feeding (M_F) and activity effects (M_A), for the weight and temperatures concerned.

While noting the uniqueness of species and their individualistic response to different diets and environmental conditions, some average relations have been determined. Using only those energy budgets where the fish were fed a ration well above maintenance, and where temperatures were not extreme (i.e., cases where conversion efficiency was not less than 20%), and excluding the one herbivorous fish, a mean budget ($n = 15$) with 95% confidence limits computes to

$$99.5\ I = (43.8 \pm 7.0)M + (28.9 \pm 5.7)G + (26.8 \pm 3.2)E$$

or balances to

$$100\ I = (44 \pm 7)M + (29 \pm 6)G + (27 \pm 3)E$$

This general expression is different from the one developed by Winberg (1956) (Table VI), being reduced in the relative metabolic fraction (–16%) and greater in both the fractions for growth (+9%) and excretion (+7%). This does not mean that the absolute metabolic rates were lower; indeed, there is reason to believe these are generally higher (Solomon and Brafield, 1972), which in turn would mean a greater food requirement. As stated previously, Winberg (1956) did not ascribe any caloric value to nonfecal excretion, an oversight that accounts for most of the error in that component, necessitating a compensatory change in the others. Regardless of this change, it should be noted that the variability is least for the excretion fraction, a fact which tends to apply over a wide range of rations and natural diets (Elliott, 1976b). Hence, changes in rations affecting growth rate will likely have greater effect on the relative metabolic fraction. Turning to the growth fraction, a value of $29 \pm 6\%$ is indicative of the conversion efficiency of young, fast-growing, well-fed fish. Accepting the fairly consistent provision of 59% net energy from the food (Fig. 18, derived from Table VI), growth and metabolism are competing components for this remaining energy. Increased metabolism will inevitably result in a relative decrease in growth, particularly when food is limiting or the fish cannot compensate by increasing its food intake. Further, it is apparent that for some species efficient food conversion is only possible over the environmental range of temperature where the population is naturally abundant (e.g., for sockeye, Brett, 1971). Thus, given a good estimate of growth of a normal population the likely values for the other components of the energy budget can be determined.

C. Herbivores

The herbivorous fishes are undoubtedly the most important species of food-fish in the world for aquaculture, yet they have received far less attention in physiological studies than the carnivores. In particular, the families Cyprinidae (carps), Cichlidae (tilapias), Mugilidae (mullets), and Chanidae (milkfish) form the basis of much of the world production (Bardach *et al.*, 1972). In the foregoing considerations, they have been shown to be characterized by generally low total digestibility of their plant food (e.g., 60–70%) (Hickling, 1966) (Table IV). High amounts of fiber and ash contribute to this more copious fecal matter, since the digestibility of the protein, fat, and carbohydrate components of some algal feeds may be as high as 92, 97, and 80%, respectively (Singh and Bhanot, 1970). However, Chiang (1971) recorded a wide variability in protein digestibility, ranging from (a) 73–93% for silver carp, *Hypophthalmichthys molitrix*, feeding on phytoplankton, (b) 54–63% for goldfish (chlorophyceous diet), and (c) 23–69% for milkfish (miscellaneous plant material).

Stanley (1974a) attempted to determine an energy budget for large grass carp or white amur, *Ctenopharyngodon idella*, feeding on the aquatic macrophyte *Egeria densa* (Table VI). Although there was no net gain in nitrogen retention, it was determined that from a caloric intake of 96 kcal approximately 40 kcal were excreted, 74 kcal deposited in growth, and 8 kcal expended in metabolism. Within the acknowledged difficulties of measuring all components of the energy budget with equal accuracy, the divergence from a balanced equation appears greatest in growth (a remarkable mean conversion efficiency of 74%), and in metabolism at only 8%. Greatest support can be given for the fraction in excretion (42%) which agrees well with an overall mean rate of fecal production for herbivores of 40.5 ± 8.6 (SE) from Table IV. If a nonfecal total fraction of 2.5% is added to this (since the protein fraction of plant diets is frequently one-fourth to one-third that of meats) then a total of 43% excretion would be considered to characterize herbivores.

Proceeding to conversion efficiency (*G* in the balanced equation), the generally lower capacity of herbivores would likely not exceed 20% for young fish. This leaves 37% for metabolic rate. Any improvement by reducing the metabolic fraction would turn up as increased growth, and vice versa. Some herbivores are known to be fairly sluggish (carp) while others may become remarkably active (milkfish). Thus, any generalized equation for herbivores will leave considerably greater limits of confidence than that for carnivores. As recorded in Table VI, the general equation may be written: $100\,I = 37\,M + 20\,G + 43\,E$. Lack of sufficient data does not permit determining confidence limits for the values given, which would appear to be greater than those for carnivores, judging from the excretion records.

When grass carp were fed only on lettuce (*Lactuca sativa*), Fischer (1972) showed that they were hardly able to maintain their body weight. For fish ranging in body weight from 40 to 120 g an average balanced equation at 22°C worked out to $100I = 16M + 3G + 81E$. When switched to a diet of tubificidae, considerable improvement in growth occurred, resulting in an average equation of $100I = 23M + 17G + 60E$.

The omniverous nature of this species (see review in Fischer and Lyakhnovich, 1973) appears to put it part way between the strictly carnivorous and herbivorous forms or stages. Excretion remains exceedingly high (60% of energy intake) even

Table VI Energy Budgets Listed in Order of Increasing Assumptions (Values in Parentheses) Regarding One or More Components

Species	Starting weight (g)	Average temperature (°C)	Diet	Ingest$_a$ I (kcal)	Energy budget							Excretion		Remarks	Reference
					I (%)	=	M (%)	+	G (%)	+	E (%)	Fecal (%)	Non-fecal (%)		
Perca fluviatilis Perch	12.0	14	*Gammarus*	14.1	98.7*		54.4		20.5		23.9	16.3	7.6	All components monitored continuously. Three levels of ration selected, including a near maximum and near maintenance level	Solomon and Brafield (1972)
	12.8	14	*Gammarus*	9.8	103.3*		59.0		20.4		23.9	15.7	8.2		
	18.7	14	*Gammarus*	6.8	105.8		74.4		6.4		25.0	14.6	10.4		
Salmo trutta Brown trout	50.1	3.8	*Gammarus*	0.28	100		(62)		5		33	29	4	Four temperatures selected in relation to a variety of rations near maximum for each temperature. Metabolic rate determined by difference. Weight not significant for range studied (12–258 g)	Elliott (1976b)
	50.7	9.5	*Gammarus*	1.60	100*		(36)		33		31	24	7		
	50.1	15.0	*Gammarus*	3.84	100*		(49)		20		31	22	9		
	50.6	17.8	*Gammarus*	5.14	100		(66)		3		31	21	10		
Salmo gairdnerii Rainbow trout	3500	15	Prepared	112	99.2*		25.9		50.1		23.2	15.2	(8.0)	Domesticated stock of hatchery trout tested using a diet with 55% protein and 13% fat. Nonfecal estimated as 8%	Cho (1975)
	3500	15	Prepared	102	99.9*		22.5		48.0		29.4	2.14	(8.0)		
Histro histro Sargasum fish	1.0	7	Shrimp	1.2	100*		33.4		34.8		31.8	27.7	(4.1)	Use of organic carbon determinations for food and feces. Three size classes tested, with decreasing conversion efficiency as size increased	Smith (1973)
	13			9.5	100*		41.9		24.0		34.1	26.1	(4.1)		
	28			19.7	100*		55.1		15.8		29.1	17.7	(11.4)		

Oncorhynchus kisutch Coho salmon	1.2	17	Fly larvae	1.5	100	(43)	32	25	—	—	Approximated from data in Fig. 29. Feces was 25%. (Recalculated assuming an 8% nonfecal component)	Avertt (1969); Warren (1971)
	—	—		—	100*	(35)	32	(33)	25	(8)		
Ctenopharyngodon idella Grass carp	1200	23	*Egeria*	96	122	8	74	40	—	—	Fixed amount of food used—1 kg wet food per day. (Recalculated, metabolic rate least accurate determination; no net gain in protein)	Stanley (1974a)
					100	8.3	(50)	41.7	—	—		
Gadus morhua Atlantic cod	223	15	Chopped fish	415	100	(76.5)	22.2	1.3	—	—	Metabolic rate determined by difference. Nonfecal not measured. (Recalculated by assuming an 8% nonfecal component)	Edwards *et al.* (1972)
	—	—		—	100*	(68.5)	22.2	(9.3)	1.3	(8)		
Gasterosteus aculeatus Stickleback	1.0	15	Tubifex	—	100	46	4.5	(49.4)	20	(29.4)	Nonfecal portion assigned the balance, by difference (unusually high)	Walkey and Meakins (1969)
Esox lucius Pike	250	12	*Phoxinus*	—	100*	43	29	(28)	—	—	Metabolism is equivalent to maintenance requirement and is therefore on the low side	Johnson (1966); Welch (1968)
Lepomis macrochirus Bluegill sunfish	49	15	Mayfly nymphs	620	100	36	(44)	(20)	(17)	(3)	Fecal loss assumed as 20% with growth determined by difference	Pierce and Wissing (1974)
	—	—		—	100*	36	(36)	(28)	(20)	(8)	(Recalculated, assigning 8% to nonfecal loss)	

Continued

Table VI Energy Budgets Listed in Order of Increasing Assumptions (Values in Parentheses) Regarding One or More Components—Cont'd

Species	Starting weight (g)	Average temperature (°C)	Diet	Ingest *I* (kcal)	Energy budget							Excretion		Remarks	Reference
					I (%)	=	M (%)	+	G (%)	+	E (%)	Fecal (%)	Non-fecal (%)		
	92	20		1060	100		47		(33)		(20)	(17)	(3)	Same as above. Higher temperature	
	—	—		—	100*		47		(25)		(28)	(20)	(8)	(Recalculated)	
	63	25		1213	100		50		(30)		(20)	(17)	(3)	Same as above. Highest temperature	
	—	—			100*		50		(22)		(28)	(20)	(8)	(Recalculated)	
Carnivorous fish No. 1	—	—		—	100		60		20		20	20	—	General budget developed from pertinent literature published to 1955. Nonfecal energy considered insignificant	Winberg (1956); Welch (1968)
Carnivorous fish No. 2	—	—		—	100		44		29		27	20	7	General budget for young fish feeding well (selected data, to 1976)	This table
Herbivorous fish	—	—	—	—	100		37		20		43	41	2	General budget for young fish feeding well	See text

[a]*Values determined in calories consumed (I) and expressed as percentage of this amount. I, ingestion; fractions. Growth as a percentage of ingestion is equivalent to conversion efficiency (column G). Only those cases are included. If one or more values were determined by difference, ingestion (I) occurs as 100%; otherwise because different times and expressions of amounts were used. Only "recalculated" values were used for statistics, marked with an asterisk were used in deriving the "general budget" for carnivorous fish, including cases where M, total metabolism; G, growth; and E, total excretion. Excretion is further divided into fecal and nonfecal cases showing positive growth are presented. Where various temperatures, rations, or sizes were tested not all I is the sum of the components. The column for calories consumed is only useful for comparison within species The third to last entry is the general equation for carnivorous fish developed by Winberg (1956). The budgets good growth and/or recalculation were involved.*

on the tubificid diet. Indeed, there is obviously a general range in functional capacity, developmental stage, and feeding opportunity between carnivore, insectivore, omnivore, and herbivore. Experiments with single diets deny the benefits of mixed diets so essential for the subtle requirements of good growth.

VII. CONCLUDING COMMENTARY

Studies on the growth of fishes have been more concerned with the nutritional adequacy of the diet than the provision of an energy component suited to the life stage, environmental conditions, and routine activity of the species. By studying the metabolic rates of *feeding fish* in relation to *ration level* and *temperature*, energy requirements can be assessed with increased confidence—but only in the laboratory, or hatchery rearing-pond. For the wild fish, energy expenditure still defies direct measurement. Detailed records of daily activities in nature (e.g., Olla *et al.*, 1974) hold promise for assigning likely hourly energy expenditures, calculated from tables of metabolic rates corresponding to type activities. The nearest such table is that for sockeye salmon (Table III) from which the daily metabolic pattern in Fig. 9 was derived. However, the phenomenon of metabolic acclimation under conditions of fluctuating environmental temperature has yet to be investigated with any thoroughness.

Advances in the development and application of direct calorimeters for fish should result in a final conclusion to the refined but possibly still inaccurate oxycalorific equivalent for respired oxygen. Such expression of metabolic energy release for carnivores and herbivores alike needs greater attention. Further, it is possible that in some species the anaerobic fraction or the oxygen debt accumulated at times of active metabolism may not be fully replaced by respiration, leading to underestimation of this important parameter of energetics.

The variability between species in metabolic response to temperature and the effect of size on metabolic rate make any generalization subject to considerable inaccuracy where a particular species is involved. The mean values of weight exponents and temperature Q_{10}'s may offer a useful measure for the broad insights of the ecologist but hardly satisfy the more detailed interests of the physiologist. The assumption that the exponent, approximately 0.8, for weight relations derived for standard metabolism ("metabolic weight") can be applied directly to other levels of metabolism, as well as to digestion, excretion and growth, needs critical examination.

Improved techniques and procedural systems have allowed better assessment of the digestible, metabolizable, and net energy of the diet but all too few cases have been recorded, even among the better known carnivorous fishes. Comparable knowledge and detailed studies on herbivorous fishes are greatly needed, particularly in view of the expanding aquaculture interest for plant-eating species.

Bioenergetic studies are of fundamental importance to the advance of aquaculture. Manipulation of environmental factors in conjunction with improved diets holds great promise that gross conversion efficiencies of 50% and over can be achieved, at least for young fish.

Finally, among families of fishes, the Salmonidae appear to have received greater bioenergetic attention than any other family. It would be a mistake to apply the findings too generally. The diversity of morphology of fishes is undoubtedly matched by a comparable diversity in physiology.

VIII. GLOSSARY

absorption uptake of digested material across a membrane such as the gut wall

active metabolism mg of O_2/kg/hr; rate of oxygen consumption during maximum sustained activity

adiabatic calorimeter one in which heat flow through the wall is prevented by a water jacket maintaining similar temperature inside and out

ammonotelic organisms that excrete ammonia as the major end product of nitrogen metabolism

anabolism metabolic processes resulting in increased body substance (=growth)

assimilation pertaining to that part of the food intake which is utilized

availability as applied to diet, that fraction of the food that is capable of being absorbed

basal metabolism mg of O_2/kg/hr; as applied to mammals, the minimum metabolic rate of a resting animal in the postabsorptive state under thermoneutral conditions

catabolism the breakdown of metabolites resulting in loss of energy

calorie the amount of energy required to heat 1 kg of water by 1°C, at 15°C (=1000 cal)

cellulolytic inducing the hydrolysis of cellulose

digestible energy kcal; energy remaining from food *less* feces

direct calorimetry energy expenditure by an organism through direct measurement of heat loss

ectotherm an organism that does not maintain body temperature by internal heat production (cf. poikilotherm)

endergonic absorbing energy

endogenous arising from body substance (i.e., the metabolic energy source is internal)

endotherm pertaining to organisms that maintain body temperature by internal heat production (cf. homiotherm)

energy the term used to designate all forms of work and heat

energy budget a quantitative accounting of all the energy inputs and losses of an organism

energy density kcal/g; the energy per unit weight of food

exergonic producing or losing energy

exogenous arising from the food (i.e., having an external source from that of body substance)

feeding metabolism mg of O_2/kg/hr; the metabolic rate of an organism during feeding and absorption

gross conversion efficiency K or K_1, %; the percentage of total food intake converted into body substance

gross energy kcal; the total energy of food as determined by bomb calorimetry

gluconeogenesis formation of blood glucose as a result of the breakdown of amino acids

heat increment kcal/kg/hr; the metabolic heat loss from the digestion and transformation of food (=specific dynamic action, or SDA)

indirect calorimetry kcal/liter of O_2; measurement of the respiratory gas exchange of an organism for estimating energy exchange by oxycalorific equivalents

isocaloric the provision of equal amounts of dietary energy to different treatments in a feeding experiment

instantaneous growth rate the relative growth rate, as expressed in specific growth rate (see specific growth rate)

***K*-line** the rate of decrease in gross food conversion efficiency (K) with increasing ration (R) (i.e., the straight line obtained from the equation log $K=a+bR$)

maintenance metabolism the minimum rate of energy expenditure maintaining an organism at rest (=standard metabolism)

maintenance ration that level of food intake just maintaining a fixed weight (i.e., maintaining energy equilibrium)

metabolic compensation a shift in standard metabolic rate as an adaptation to living under extremes of climatic conditions

metabolizable energy the energy available from the food less total exogenous excretion (fecal + nonfecal)

net conversion efficiency K_2, %; the percentage of food intake *less* maintenance fraction converted into body substance

net energy metabolizable energy *less* heat increment (i.e., the food energy available for work and growth)

nonfecal excretion the excretion of soluble metabolites via the kidney and gills (principally nitrogenous)

oxycalorific equivalent kcal/liter of O_2; the number of calories released by a substrate per liter of oxygen consumed (also expressed as mg of O_2/cal)

routine metabolism mg of O_2/kg/hr; the metabolic rate of an organism during normal spontaneous activity

standard metabolism mg of O_2/kg/hr; minimum rate of oxygen consumption of an organism at rest in the postabsorptive state and thermally acclimated

specific dynamic action SDA; see heat increment

specific growth rate G, as % wt/day; the percentage increase in body weight (W) per unit time (T), determined as:

$$[(\ln W_2 - \ln W_1) \times 100]/(T_2 - T_1)$$

ureotelic organisms that excrete urea as the major endproduct of nitrogen metabolism

REFERENCES

Alexander, R. M. (1972). The energetics of vertical migration by fishes. *Symp. Soc. Exp. Biol.* 26, 273–294.

Altman, P. L., and Dittmer, D. S., eds. (1974). "Biological Data Book," Vol. 3, Part V, "Fishes," 2nd Ed. pp. 1624–1630. Fed. Am. Soc. Exp. Biol., Biol. Handbooks, *Bethesda, Maryland.*

Averett, R. C. (1969). Influence of temperature on energy and material utilization by juvenile coho salmon. Ph.D. Thesis, Oregon State Univ., Corvallis.

Awakura, T. (1963). Physiological and ecological studies on the fishes at spawning stage. I. On oxygen consumption of spawning chum salmon *Oncorhynchus keta* (Walbaum). *Sci. Rep. Hokkaido Fish Hatch.* 18, 1–10.

Bakus, G. J. (1969). Energetics and feeding in shallow marine waters. *In* "International Review of General and Experimental Zoology" (W. J. Felts and R. J. Harrison, eds.), pp. 275–369. Academic Press, New York.

Bams, R. A. (1970). Evaluation of a revised hatchery method tested on pink and chum salmon fry. *J. Fish. Res. Board Can.* 27, 1429–1452.

Bardach, J. E., Ryther, J. H., and McLarney, W. O. (1972). "Aquaculture. The Farming and Husbandry of Freshwater and Marine Organisms." Wiley (Interscience), New York.

Beamish, F. W. H. (1964a). Influence of starvation on standard and routine oxygen consumption. *Trans. Am. Fish. Soc.* 93, 103–107.

Beamish, F. W. H. (1964b). Respiration of fishes with special emphasis on standard oxygen consumption. II. Influence of weight and temperature on respiration of several species. *Can. J. Zool.* 42, 177–188.

Beamish, F. W. H. (1970). Oxygen consumption of largemouth bass, *Micropterus salmoides*, in relation to swimming speed and temperature. *Can. J. Zool.* 48, 1221–1228.

Beamish, F. W. H. (1972). Ration size and digestion in largemouth bass, *Micropterus salmoides Lacepede*. *Can. J. Zool.* 50, 153–164.

Beamish, F. W. H. (1974). Apparent specific dynamic action of largemouth bass, Micropterus salmoides. J. Fish. Res. Board Can. 31, 1763–1769.

Beamish, F. W. H., and Mookherji, P. S. (1964). Respiration of fishes with special emphasis on standard oxygen consumption. I. Influence of weight and temperature on respiration of goldfish, *Carassius auratus* L. *Can. J. Zool.* 42, 161–175.

Beamish, F. W. H., Niimi, A. J., and Lett, P. F. K. P. (1975). Bioenergetics of teleost fishes: Environmental influences. *In* "Comparative Physiology—Functional Aspects of Structural Materials" (L. Bolis, H. P. Maddrell, and K. Schmidt-Nielsen, eds.), pp. 187–209. North-Holland Publ., Amsterdam.

Benzinger, T., Huebscher, R. G., Minard, D., and Kitzinger, C. (1958). Human calorimetry by means of the gradient principle. *J. Appl. Physiol.* 12, S1–S28.

Birkett, L. (1969). The nitrogen balance in plaice, sole and perch. *J. Exp. Biol.* 50, 375–386.

Blackburn, J. M. (1968). Digestive efficiency and growth in largemouth black bass. M.A. Thesis in Zoology, Univ. of California, Davis.

Blaxter, K. L., ed. (1965). "Energy Metabolism." Academic Press, New York.

Blazka, P., Volt, M., and Cepela, M. (1960). A new type of respirometer for the determination of the metabolism of fish in an active state. *Physiol. Bohemoslov.* 9, 553–558.

Brafield, A. E., and Solomon, D. J. (1972). Oxycalorific coefficients for animals respiring nitrogenous substratus. *Comp. Biochem. Physiol.* A 43, 837–841.

Brett, J. R. (1964). The respiratory metabolism and swimming performance of young sockeye salmon. *J. Fish. Res. Board Can.* 21, 1183–1226.

Brett, J. R. (1965). The relation of size to rate of oxygen consumption and sustained swimming speed of sockeye salmon (*Oncorhynchus nerka*). *J. Fish. Res. Board Can.* 22, 1491–1501.

Brett, J. R. (1970a). Fish—The energy cost of living. *In* "Marine Aquiculture" (W. J. McNeil, ed.), pp. 37–52. Oregon State Univ. Press, Corvallis.

Brett, J. R. (1970b). 3. Temperature. 3.3 Animals. 3.32 Fishes. *In* "Marine Ecology," Vol. 1, "Environmental Factors" (O. Kinne, ed.), Part 1, pp. 515-560. Wiley (Interscience), New York.

Brett, J. R. (1971). Energetic response of salmon to temperature. A study of some thermal relations in the physiology and freshwater ecology of sockeye salmon (*Oncorhynchus nerka*). *Am. Zool.* 11, 99–113.

Brett, J. R. (1972). The metabolic demand for oxygen in fish, particularly salmonids, and a comparison with other vertebrates. *Respir. Physiol.* 14, 151–170.

Brett, J. R. (1973). Energy expenditure of sockeye salmon, *Oncorhynchus nerka*, during sustained performance. *J. Fish. Res. Board Can.* 30, 1799–1809.

Brett, J. R. (1976a). Scope for metabolism and growth of sockeye salmon, *Oncorhynchus nerka*, and some related energetics. *J. Fish. Res. Board Can.* 33, 307–313.

Brett, J. R. (1976b). Feeding metabolic rates of sockeye salmon, *Oncorhynchus nerka*, in relation to ration level and temperature. *Envir. Can., Fish. Mar. Serv.* Tech. Rep. No. 675, 18 pp.

Brett, J. R., and Glass, N. R. (1973). Metabolic rates and critical swimming speeds of sockeye salmon (*Oncorhynchus nerka*) in relation to size and temperature. *J. Fish. Res. Board Can.* 30, 379–387.

Brett, J. R., and Sutherland, D. B. (1965). Respiratory metabolism of pumpkinseed (*Lepomis gibbosus*) in relation to swimming speed. *J. Fish. Res. Board Can.* 22, 405–409.

Brett, J. R., and Sutherland, D. B. (1970). Improvement in the artificial rearing of sock-eye salmon by environmental control. *Fish. Res. Board Can., Gen. Ser. Circ.* No. 89.

Brett, J. R., and Zala, C. A. (1975). Daily pattern of nitrogen excretion and oxygen consumption of sockeye salmon (*Oncorhynchus nerka*) under controlled conditions. *J. Fish. Res. Board Can.* 32, 2479–2486.

Brett, J. R., Shelbourn, J. E., and Shoop, C. T. (1969). Growth rate and body composition of fingerling sockeye salmon, *Oncorhynchus nerka*, in relation to temperature and ration size. *J. Fish. Res. Board Can.* 26, 2363–2394.

Brocksen, R. W., and Brugge, J. P. (1974). Preliminary investigations on the influence of temperature on food assimilation by rainbow trout *Salmo gairdneri* Richardson. *J. Fish Biol.* 6, 93–97.

Brocksen, R. W., Davis, G. E., and Warren, C. E. (1968). Competition, food consumption, and production of sculpins and trout in laboratory stream communities. *J. Wildl. Manage.* 32, 51–75.

Brody, S. (1945). "Bioenergetics and Growth." Reinhold, New York.

Bryan, P. G. (1974). Food habits, functional digestive morphology, and assimilation efficiency of the rabbitfish *Siganus spinus* (Pisces: Siganidae) on Guam. M.S. Thesis, Univ. of Guam, Agana.

Burrows, R. E. (1964). Effects of accumulated excretory products on hatchery-reared salmonids. *Fish Wildl. Serv. (U.S.), Res. Rep.* 66, 1–12.

Buttery, P. J., and Annison, E. F. (1973). Considerations of the efficiency of amino acid and protein metabolism in animals. *In* "The Biological Efficiency of Protein Production" (J. G. W. Jones, ed.), pp. 141–171. Cambridge Univ. Press, London and New York.

Carey, F. G., Teal, J. M., Kanwisher, J. W., and Lawson, K. D. (1971). Warm-bodied fish. *Am. Zool.* 11, 137–145.

Chesney, E. J., Jr., and Estevez, J. I. (1976). Energetics of winter flounder (*Pseudopleuronectes americanus*) fed the polychaete, *Nereis virens*, under experimental conditions. *Trans. Am. Fish. Soc.* 105, 592–595.

Chiang, W. (1971). Studies on feeding and protein digestibility of Silver carp. *Hypophthalmichthys molitrix* (C. & V.). *Chin.-Am. Jt. Comm. Rural Reconstr., Fish. Serv.* No. 11, pp. 96–114.

Cho, C. Y. (1975). Aquaculture with emphasis on nutrition and diet formulation. Prog. Rep. to D.O.E., Fish. Mar. Serv., from Dep. Nutr., Guelph, Ontario.

Cowey, C. B., and Sargent, J. R. (1972). Fish nutrition. *Adv. Mar. Biol.* 10, 383–492.

Davies, P. M. C. (1963). Food input and energy extraction efficiency in *Carassius auratus. Nature (London)* 198, 707.

Davies, P. M. C. (1964). The energy relations of *Carassius auratus* L. I. Food input and energy extraction efficiency at two experimental temperatures. *Comp. Biochem. Physiol.* 12, 67–79.

Durbin, A. G. (1976). Oxygen consumption and ammonia excretion of adult Atlantic menhaden, *Brevoortia tyrannus*. Grad. School Oceanogr., Univ. Rhode Island, Kingston,16 pp.

Edwards, R. R. C., Blaxter, J. H. S., Gopalon, U. K., Mathews, C. V., and Finlayson, D. M. (1971). Feeding, metabolism, and growth of tropical flatfish. *J. Exp. Mar. Biol. Ecol.* 6, 279–300.

Edwards, R. C. C., Finlayson, D. M., and Steele, J. H. (1972). An experimental study of the oxygen consumption, growth, and metabolism of the cod (*Gadus morhua* L.). *J. Exp. Mar. Biol. Ecol.* 8, 299-309.

Ege, R., and Krogh, A. (1914). On the relation between the temperature and the respiratory exchange in fishes. *Int. Rev. Gesamten Hydrobiol. Hydrogr.* 1, 48–55.

Elliot, J. E. (1969). The oxygen requirements of chinook salmon. *Prog. Fish Cult.* 31, 67–73.

Elliott, J. M. (1975a). The growth rate of brown trout, *Salmo trutta* L., fed on maximum rations. *J. Anim. Ecol.* 44, 805–821.

Elliott, J. M. (1975b). The growth rate of brown trout (*Salmo trutta* L.) fed on reduced rations. *J. Anim. Ecol.* 44, 823–842.

Elliott, J. M. (1976a). Energy losses in the waste products of brown trout (*Salmo trutta* L.). *J. Anim. Ecol.* 45, 561–580.

Elliott, J. M. (1976b). The energetics of feeding, metabolism and growth of brown trout (*Salmo trutta* L.) in relation to body weight, water temperature and ration size. *J. Anim. Ecol.* 45, 923–948.

Elliott, J. M., and Davison, W. (1975). Energy equivalents of oxygen consumption in animal energetics. *Oecologia* 19, 195–201.

Fischer, Z. (1970). The elements of energy balance in grass carp (*Ctenopharyngodon idella* Val.). Part 1. *Pol. Arch. Hydrobiol.* 17, 421–434.

Fischer, Z. (1972). The elements of energy balance in grass carp (*Ctenopharyngodon idella* Val.). Part II. Fish fed with animal food. *Pol. Arch. Hydrobiol.* 19, 65–82.

Fischer, Z., and Lyakhnovich, V. P. (1973). Biology and bioenergetics of grass carp (*Ctenopharyngodon idella* Val.). *Pol. Arch. Hydrobiol.* 20, 521–557.

Forster, R. P., and Goldstein, L. (1969). Formation of excretory products. *In* "Fish Physiology." Vol. 1. (W. S. Hoar and D. J. Randall, eds.), pp. 313-350. Academic Press, New York and London.

Fromm, P. O. (1963). Studies on renal and extra-renal excretion in the freshwater teleost, *Salmo gairdneri*. *Comp. Biochem. Physiol.* 10, 121–128.

Fry, F. E. J. (1947). Effects of the environment on animal activity. *Univ. Toronto Stud., Biol. Ser.* 55, 1–62.

Fry, F. E. J. (1957). The aquatic respiration of fish. *In* "The Physiology of Fishes" (M. E. Brown, ed.), Vol. 1, pp. 1–63. Academic Press, New York.

Fry, F. E. J. (1971). The effect of environmental factors on the physiology of fish. *In* "Fish Physiology" (W. S. Hoar and D. J. Randall, eds.), Vol. 6, pp. 1–98. Academic Press, New York.

Fry, F. E. J., and Hochachka, P. W. (1970). Fish. *In* "Comparative Physiology of Thermoregulation" (G. C. Whittow, ed.), Vol. 1, pp. 79–134. Academic Press, New York.

Furukawa, A., and Tsukahara, H. (1966). On the acid digestion method for the determination of chromic oxide as an index substance in the study of digestibility of fish feed. *Bull. Jpn. Soc. Sci. Fish.* 32, 502–506.

Gerking, S. D. (1955). Endogenous nitrogen excretion of bluegill *sunfish*. *Physiol. Zool.* 28, 283–289.

Gerking, S. D. (1971). Influence of rate of feeding and body weight on protein metabolism of bluegill sunfish. *Physiol. Zool.* 44, 9–19.

Glass, N. R. (1969). Discussion of calculation of power function with special reference to respiratory metabolism in fish. *J. Fish. Res. Board Can.* 26, 2643–2650.

Gordon, M. S. (1972). "Animal Physiology: Principles and Adaptations," 2nd Ed. Macmillan, New York.

Groot, C., and Wiley, W. L. (1965). Time-lapse photography of an ASDIC echo-sounder PPI-scope as a technique for recording fish movements during migration. *J. Fish. Res. Board Can.* 22, 1025–1034.

Gross, W. L., Roelofs, E. W., and Fromm, P. O. (1965). Influence of photoperiod on growth of green sunfish, *Lepomis cyanellus*. *J. Fish. Res. Board Can.* 22, 1379–1386.

Guérin-Ancey, O. (1976). Étude expérimentale de l'excretion azotée du bar (*Dicentrar chus labuax*) en cours de croissance. II. Effects du jeûne usr l'excretion d'ammoniac et d'urée. *Aquaculture* 9, 187–194.

Harper, H. A. (1971). "Review of Physiological Chemistry," 13th Ed. Lange Med. Publ., Los Altos, California.

Hastings, W. H. (1969). Nutritional score. *In* "Fish in Research" (O. W. Neuhaus and J. E. Halver, eds.), pp. 263–292. Academic Press, New York.

Hatanaka, M., and Takahashi, M. (1956). Utilization of food by mackerel *Pneumatophorus japonicus*. *Tohoku J. Agric. Res.* 7, 51–57.

Hickling, C. F. (1966). On the feeding process in the white amur, *Ctenopharyngodon idella*. *Proc. Zool. Soc. London* 148, 408–419.

Holeton, G. F. (1970). Oxygen uptake and circulation by a hemoglobinless Antarctic fish (*Chaenocephalus aceratus Lonnberg*) compared with three red-blooded Antarctic fish. *Comp. Biochem. Physiol.* 34, 457–472.

Holeton, G. F. (1973). Respiration of Arctic char (*Salvelinus alpinus*) from a high arctic lake. *J. Fish. Res. Board Can.* 30, 717–723.

Holeton, G. F. (1974). Metabolic cold adaptation of polar fish: Fact or artifact. *Physiol. Zool.* 47, 137–152.

Huisman, E. A. (1974). A study on optimal rearing conditions for carp (*Cyprinus carpio* L.). Ph. D. Thesis, Agricult. Univ. Wageningen. (Spec. Publ., Organisatie ter Verbetering van de Binnenvisserij, Utrecht.)

Huisman, E. A. (1976). Food conversion efficiencies at maintenance and production levels for carp, *Cyprinus carpio* L., and rainbow trout, *Salmo gairdneri* Richardson. *Aquaculture* 9, 259–273.

Idler, D. R., and Clemens, W. A. (1959). The energy expenditure of Fraser River sockeye salmon during the spawning migration to Chilko and Stuart Lakes. *Int. Pac. Salmon. Fish. Comm., Prog. Rep.* 80 pp.

Ivlev, V. S. (1939). Energy balance of carps. *Zool. Zh.* 18, 303–318.

Ivlev, V. S. (1945). The biological productivity of waters. *Usp. Sovrem. Biol.* 19, 98-120. [Engl. transl., *J. Fish. Res. Board Can.* 23, 1727–1759 (1966).]

Iwata, K. (1970). Relationship between food and growth in young crucian carps, *Carassius auratus cuvieri*, as determined by the nitrogen balance. *Jpn. J. Limnol.* 31, 129–151.

Johnson, L. (1966). Experimental determination of food consumption of pike, Esox lucius, for growth and maintenance. *J. Fish. Res. Board Can.* 23, 1495–1505.

Johnson, W. E., and Groot, C. (1963). Observations on the migration of young sockeye salmon (*Oncorhynchus nerka*) through a large, complex lake system. *J. Fish. Res. Board Can.* 20, 919–938.

Jones, D. R. (1971). Theoretical analysis of factors which may limit the maximum oxygen uptake of fish: The oxygen cost of the cardiac and branchial pumps. *J. Theor. Biol.* 32, 341–349.

Kanzinkin, G. S., and Tarkovskaya, O. I. (1964). Determination of caloric values of small samples. *In* "Techniques for the Investigation of Fish Physiology" (E. N. Pavlovskii, ed.), pp. 122–124. IPST, Jerusalem.

Kausch, H. (1969). The influence of spontaneous activity on the metabolic rate of starved and fed young carp (*Cyprinus carpio* L.). *Verh. Ver. Limnol.* 17, 669–679.

Kausch, H. (1972). Stoffwechsel und Ernährung der Fische. *Handb. Tierernäehr., Band II* 8, 690–738. [Fish. Res. Board Can., Transl. Ser. No. 2489 (1973).]

Kawanabe, H. (1969). The significance of social structure in production of the "Ayu," *Plecoglossus altivelis*. *In* "Symposium on Salmon and Trout in Streams" (T. G. Northcote, ed.), pp. 243–251. Inst. Fish., Univ, of British Columbia, Vancouver, B.C.

Kayser, C., and Heusner, A. (1964). Étude comparative du métabolisme énergétique dans la série animale. *J. Physiol.* (*Paris*) 56, 489–524.

Kelso, J. R. M. (1972). Conversion, maintenance, and assimilation for walleye, *Stizostedion vitreum vitreum*, as affected by size, diet, and temperature. *J. Fish. Res. Board Can.* 29, 1181–1192.

Kevern, N. R. (1966). Feeding rate of carp estimated by a radioisotopic method. *Trans. Am. Fish. Soc.* 95, 363–371.

Kinne, O. (1960). Growth, food intake, and food conversion in a euryplastic fish exposed to different temperatures and salinities. *Physiol. Zool.* 33, 288–317.

Kleiber, M. (1961). "The Fire of Life. An Introduction to Animal Energetics." Wiley, New York. (Rev. Ed., R. E. Krieger Publ. Co., Huntington, New York, 1975.)

Kleiber, M. (1967). An old professor of animal husbandry ruminates. *Annu. Rev. Physiol.* 29, 1–20.

Krebs, H. A. (1964). The metabolic fate of amino acids. *In* "Mammalian Protein Metabolism" (H. N. Munro and J. B. Allison, eds.), Vol. 1, pp. 125–176. Academic Press, New York.

Krogh, A. (1914). The quantitative relation between temperature and standard metabolism in animals. *Int. Z. Phys.-Chem. Biol.* 1, 491–508.

Krueger, H. M., Saddler, J. B., Chapman, G. A., Tinsley, I. J., and Lowry, R. R. (1968). Bioenergetics, exercise, and fatty acids of fish. *Am. Zool.* 8, 119–129.

Kutty, M. N. (1972). Respiratory quotient and ammonia excretion in *Tilapia mossambica. Mar. Biol.* 16, 126–133.

Kutty, M. N., and Mohamed, M. P. (1975). Metabolic adaptations of mullet *Rhinomugil corsula* (Hamilton) with special reference to energy utilization. *Aquaculture* 5, 253–270.

Lasker, R. (1962). Efficiency and rate of yolk utilization by developing embryos and larvae of the Pacific sardine *Sardinops caerulea* (Girard). *J. Fish. Res. Board Can.* 19, 867–875.

Leggett, W. C. (1972). Weight loss in American shad (*Alosa sapidissima*, Wilson) during the freshwater migration. *Trans. Am. Fish. Soc.* 101, 549–552.

Lehninger, A. L. (1965). "Bioenergetics. The Molecular Basis of Biological Energy Transformations." Benjamin, New York.

McLean, W. E., and Fraser, F. J. (1974). Ammonia and urea production of coho salmon under hatchery conditions. *Envir. Prot. Ser. Pac. Region, Surveillance Rep.* EPS 5-PR-74-5.

Mann, H. (1968). Der Einfluss der Ernährung auf den Sauerstoffverbranch von Forellen. *Arch. Fischereiwiss.* 19, 131–133.

Mann, K. H. (1965). Energy transformation by a population of fish in the River Thames. *J. Anim. Ecol.* 34, 253–257.

Marr, D. H. A. (1966). Influence of temperature on the efficiency of growth of salmonid embryos. *Nature (London)* 212, 957–959.

Maynard, A. L., and Lousli, K. J. (1962). "Animal Nutrition," 5th Ed. McGraw-Hill, New York.

Menzel, D. W. (1958). Utilization of algae for growth by the angel fish, *Holacanthus bermudensis. J. Cons., Cons. Perm. Int. Explor. Mer* 24, 308–313.

Menzel, D. W. (1960). Utilization of food by a Bermuda reef fish, *Epinephelus guttatus. J. Cons., Cons. Perm. Int. Explor. Mer* 25, 216–222.

Miura, T., Suzuki, N., Nagoshi, M., and Yamamura, K. (1976). The rate of production and food consumption of the Biwamasu, *Oncorhynchus rhodurus*, population in Lake Biwa. *Res. Popul. Ecol. (Kyoto)* 17, 135–154.

Morgulis, S. (1918). Studies on the nutrition of fish. Experiments on brook trout. *J. Anim. Sci.* 2, 263 277.

Mount, L. E. (1963). The thermal insulation of the newborn pig. *J. Physiol. (London)* 168, 698–705.

Muir, B. S., and Niimi, A. J. (1972). Oxygen consumption of the euryhaline fish aholehole (*Kuhlia sandvicensis*) with reference to salinity, swimming, and food consumption. *J. Fish. Res. Board Can.* 29, 67–77.

Nightingale, J. W. (1974). Bioenergetic responses of nitrogen metabolism and respiration to variable temperature and feeding interval in Donaldson strain rainbow trout. Ph.D. Thesis, Univ. of Washington, Seattle.

Niimi, A. J., and Beamish, F. W. H. (1974). Bioenergetics and growth of largemouth bass (*Micropterus salmoides*) in relation to body weight and temperature. *Can. J. Zool.* 52, 447–456.

Nose, T. (1960). On the effective value of freshwater green algae, *Chlorella ellipsoidea*, as a nutritive source to goldfish. *Bull. Freshwater Fish. Res. Lab.* 10, 1–10.

Nose, T. (1967). On the metabolic fecal nitrogen in young rainbow trout. *Bull. Freshwater Fish. Res. Lab.* 17, 97–106.

Odum, E. P. (1971). "Fundamentals of Ecology," 3rd Ed. Saunders, Philadelphia, Pennsylvania.

Olla, B. L., Bejda, A. J., and Martin, A. D. (1974). Daily activity, movements, feeding, and seasonal occurrence in the tantog, *Tantoga onitis. U.S. Fish Wildl. Serv., Fish. Bull.* 72, 27–35.

Olson, K. R., and Fromm, P. O. (1971). Excretion of urea by two teleosts exposed to different concentrations of ambient ammonia. *Comp. Biochem. Physiol.* A 40, 999–1008.

Paloheimo, J. E., and Dickie, L. M. (1966). Food and growth of fishes. III. Relations among food, body size, and growth efficiency. *J. Fish. Res. Board Can.* 23, 1209–1248.

Pandian, T. J. (1967a). Food intake, absorption and conversion in the fish *Ophiocephalus striatus. Helgol. Wiss. Meeresunters.* 15, 637–647.

Pandian, T. J. (1967b). Intake, digestion, absorption and conversion of food in the fishes *Megalops cyprinoides* and *Ophiocephalus striatus. Mar. Biol.* 1, 16–32.

Pandian, T. J. (1967c). Transformation of food in the fish *Megalops cyprinoides*. I. Influence of quality of food. *Mar. Biol.* 1, 60–64.

Pentegov, B. P., Mentov, Y. N., and Kurnaev, E. F. (1928). Physicochemical characteristics of spawning migration fast of chum salmon. *Bull. Pac. Sci. Fish. Res. Stn.* (*Vladivostok*) 2, 47 pp.

Phillips, A. M., Jr. (1969). Nutrition, digestion, and energy utilization. *In* "Fish Physiology" (W. S. Hoar and D. J. Randall, eds.), Vol. 1, pp. 351–432. Academic Press, New York.

Phillips, A. M., Jr., and Brockway, D. R. (1959). Dietary calories and the production of trout in hatcheries. *Progr. Fish. Culturist* 21, 3–16.

Pierce, R. J., and Wissing, T. E. (1974). Energy cost of food utilization in the bluegill (*Lepomis macrochirus*). *Trans. Am. Fish. Soc.* 103, 38–45.

Pullar, J. D. (1958). Direct calorimetry of animals by the gradient layer principle. *Proc. Symp. Energy Metab., 1st, Copenhagen* pp. 95–98.

Ringrose, R. C. (1971). Calorie-to-protein ratio for brook trout (*Salvelinus fontinalis*). *J. Fish. Res. Board Can.* 28, 1113–1117.

Rosenthal, H., and Paffenhöfer, G. A. (1972). On the digestion rate and calorific content of food and feces in young gar fish. *Naturwissenschaften* 59, 274–275.

Rozin, P., and Mayer, J. (1961). Regulation of food intake in the goldfish. *Am. J. Physiol.* 201, 968–974.

Rubner, M. (1902). "Die Gesetze des Energieverbranchs bei der Ernahrung." Deuticke, Vienna.

Saunders, R. L. (1963). Respiration of the Atlantic cod. *J. Fish. Res. Board Can.* 20, 373–386.

Savitz, J. (1969). Effects of temperature and body weight on endogenous nitrogen excretion in the bluegill sunfish (*Lepomis macrochirus*). *J. Fish. Res. Board Can.* 26, 1813–1821.

Savitz, J. (1971). Nitrogen excretion and protein consumption of the bluegill sunfish (*Lepomis macrochirus*). *J. Fish. Res. Board Can.* 28, 449–451.

Scholander, P. F., Flagg, W., Walters, V., and Irving, L. (1953). Climatic adaptation in arctic and tropical poikilotherms. *Physiol. Zool.* 26, 67–92.

Shaw, H. M., Saunders, R. L., Hall, H. C., and Henderson, E. B. (1975). The effect of dietary sodium chloride on growth of Atlantic salmon (*Salmo salar*). *J. Fish. Res. Board Can.* 32, 1813–1819.

Singh, C. S., and Bhanot, K. K. (1970). Nutritive food values of algal feeds for common carp, *Cyprinus carpio* (Linnaeus). *J. Inland Fish. Soc. India* 2, 121–127.

Smith, K. L., Jr. (1973). Energy transformations by the Sargassum fish, *Histrio histrio* (L.). *J. Exp. Mar. Biol. Ecol.* 12, 219–227.

Smith, M. A. K., and Thorpe, A. (1976). Nitrogen metabolism and trophic input in relation to growth in freshwater and saltwater *Salmo gairdneri*. *Biol. Bull.* (*Woods Hole, Mass.*) 150, 139–151.

Smith, R. R. (1971). A method for determining digestibility and metabolizable energy of fish feeds. *Prog. Fish Cult.* 33, 132–134.

Smith, R. R. (1976). Studies on the energy metabolism of cultured fish. Ph.D. Thesis, Cornell Univ., Ithaca, New York.

Solomon, D. J., and Brafield, A. E. (1972). The energetics of feeding, metabolism and growth of perch (*Perca fluviatilis* L.). *J. Anim. Ecol.* 41, 699–718.

Stanley, J. G. (1974a). Energy balance of white amur fed *Egeria*. *Hyacinth Control J.* 12, 62–66.

Stanley, J. G. (1974b). Nitrogen and phosphorus balance of grass carp, *Ctenopharyngodon idella,* fed elodea, *Egeria densa*. *Trans. Am. Fish. Soc.* 103, 587–592.

Stanley, J. G., and Jones, J. B. (1976). Feeding algae to fish. *Aquaculture* 7, 219–223.

Stewart, N. E., Shumway, D. L., and Doudoroff, P. (1967). Influence of oxygen concentration on the growth of juvenile largemouth bass.

Stickney, R. R. (1974). Occurrence of cellulase activity in the stomachs of fishes. *J. Fish Biol.* 6, 779–782.

Stover, J. H. (1967). Starvation and the effects of cortisol in the goldfish (*Carassius auratus* L.). *Comp. Biochem. Physiol.* 20, 939–948.

Tang, Y.-A., and Hwang, T.-L. (1966). Evaluation of the relative suitability of various groups of algae as food of milkfish produced in brackishwater ponds. *Proc. World Symp. Warm-Water Pond Fish Cult., Rome* FAO Fish. Rep. No. 44, Vol. 3, pp. 365–372.

Walkey, M., and Meakins, R. H. (1969). Energy transformation in a host-parasite system. *Parasitology* 59, 1–26.

Wallace, J. C. (1973). Observations on the relationship between the food consumption and metabolic rate on *Blennius pholis* L. *Comp. Biochem. Physiol. A* 45, 293–306.

Ware, D. M. (1975). Growth, metabolism, and optimal swimming speed of a pelagic fish. *J. Fish. Res. Board Can.* 32, 33–41.

Warren, C. E. (1971). "Biology and Water Pollution Control." Saunders, Philadelphia, Pennsylvania.

Warren, C. E., and Davis, G. E. (1967). Laboratory studies on the feeding bioenergetics and growth of fishes. *In* "The Biological Basis of Freshwater Fish Production" (S. D. Gerking, ed.), pp. 175–214. Blackwell, Oxford.

Welch, H. E. (1968). Relationships between assimilation efficiencies and growth efficiencies for aquatic consumers. *Ecology* 49, 755–759.

Winberg, G. G. (1956). "Rate of Metabolism and Food Requirements of Fishes." Beloruss. State Univ., Minsk. [Fish. Res. Board Can., Transl. Ser. No. 194 (1960).]

Winberg, G. G. (1970). Energy flow in aquatic ecological system. *Pol. Arch. Hydrobiol.* 17, 11–19.

Wohlschlag, D. E. (1960). Metabolism of the Antarctic fish and the phenomenon of cold adaptation. *Ecology* 41, 287–292.

Wohlschlag, D. E. (1964). Respiratory metabolism and ecological characteristics of some fishes in McMurdo Sound, Antarctica. *In* "Biology of the Antarctic Seas" (M. O. Lee, ed.), Antarctic Research Series, Vol. 1, pp. 33–62. American Geophysical Union.

Wood, J. D. (1958). Nitrogen excretion in some marine teleosts. *Can. J. Biochem. Physiol.* 36, 1237–1242.

Yoshida, Y. (1970). Studies on the efficiency of food conversion to fish body growth. III. Total uptake of food and the efficiency of total food conversion. *Bull. Jpn. Soc. Sci. Fish* 36, 914–916.

Zeuthen, E. (1953). Oxygen uptake as related to body size in organisms. *Q. Rev. Biol.* 28, 1–12.

Zeuthen, E. (1970). Rate of living as related to body size in organisms. *Pol. Arch. Hydrobiol.* 17, 21–30.

Chapter 17

Establishing the principles by which the environment affects growth in fishes

David J. McKenzie[a,*] **and Peter V. Skov**[b]
[a]*UMR Marbec, CNRS, IRD, Ifremer, INRAE, Université de Montpellier, Montpellier, France*
[b]*Technical University of Denmark, Hirtshals, Denmark*
[*]*Corresponding author: e-mail: david.mckenzie@cnrs.fr*

Chapter Outline

David J. McKenzie and Peter V. Skov discuss the impact of J.R. Brett's chapter "Environmental Factors and Growth" in Fish Physiology, Volume 8, published in 1979.

This chapter reviews how abiotic and biotic environmental factors shape the growth of fishes, based upon how they influence "scope for growth", the difference in growth rate between maintenance ration and maximum ration. Brett develops a framework to classify environmental factors based on the mechanisms by which they influence scope for growth, by direct analogy to the highly influential Fry Paradigm which focusses on how the abiotic environment affects metabolism and scope for activity. Brett was Fry's student and his highly inspired scope for growth paradigm considers the anabolic component of animal energetics, how food is used to grow body tissues, and how abiotic and biotic conditions affect feed intake and utilization. Brett adopts Fry's categorization of "Factors," defined as being either Controlling, Limiting, Masking, or Directive, depending on whether they govern, restrict, modify, or cue a response, respectively. The main Controlling Factor is temperature whereas major Limiting Factors are food availability (ration) and oxygen availability (hypoxia). The chapter appeared just as the growth of fish farming began to take off and is still heavily cited in aquaculture journals because it covers issues that remain important in fish nutritional and bioenergetic research. Brett's paradigm has received some attention from fisheries biologists, but we believe that it deserves much greater attention from conservation physiologists

Fish Physiology, Vol. 40B. https://doi.org/10.1016/bs.fp.2024.05.002

and modelers that are interested in understanding and projecting how the multifaceted effects of global change will affect fish populations. In this context, Brett's careful consideration of how Factors can interact to influence growth is an exceptionally significant aspect of fish environmental physiology, since abiotic and biotic factors always occur in combinations. Overall, the chapter demonstrates the brilliance of Roly Brett as a scientist, his analysis of fish growth and bioenergetics was far ahead of his time and is still extremely relevant today.

Mankind must have contemplated for millennia about how environmental conditions influence fish growth but, as European culture evolved, there was a surge of interest. In the 19th century, the public became concerned that industrial activities were having negative impacts on water bodies and causing declines in fish populations (Stickney and Treece, 2012). This was a stimulus for development of hatcheries in Europe and the United States (Stickney and Treece, 2012), and marks the appearance of early recorded observations on how temperature affects growth of farmed brown trout *Salmo trutta* in Scotland (Maitland, 1887). By the 1900s, fish physiologists had become interested—Johansen and Krogh (1914) demonstrated that the rate of development of embryos of European plaice *Pleuronectes platessa* increased with warming over their viable temperature range. The earliest controlled feeding experiments, investigating effects of ration and water temperature on growth rate and feed efficiency, seem to date from the 1940s (Brown, 1946; Donaldson and Foster, 1941; Wingfield, 1940). By the 1970s it was recognized that ration and temperature were the environmental factors with greatest impacts on fish growth, and studies had also been performed on effects of dissolved oxygen, photoperiod, and salinity, among others (e.g. Brett, 1971; Brett et al., 1969; Elliott, 1975; Shelbourn et al., 1973). There was understanding of the physiological principles that underpinned the profound influences of temperature and oxygen on energetics and growth of fishes (Brett, 1956; Brett and Groves, 1979; Fry, 1957, 1971).

Roly Brett was a student of Fred (F.E.J.) Fry (of the "Fry Paradigm") at the University of Toronto in the 1940s, doing a Masters and then a PhD on thermal tolerance in fishes (Brett, 1944, 1952). In 1944 he was hired by the Fisheries Research Board of Canada, to work at the Pacific Station in Nanaimo on how environmental factors affected salmon populations and their fisheries. While we never met him, his brilliance and ingenuity are very evident in his research. He is especially famous for designing swim tunnels (the "Brett" type) and the critical swimming speed (U_{crit}) protocol (Brett, 1964). People that worked with him say that, beyond being an extraordinary scientist, he was also a gentleman and excellent mentor, patient, generous and kind. Dave (D.J.) Randall wrote to him as a graduate student, suggesting they measure blood gases in swimming fish, and Brett forwarded his letter to Edgar Black at the University of British Columbia. This led to Dave being offered a job in the UBC Department of Zoology in 1963 and, in terms of the Fish Physiology series, the rest is history. Eventually, Dave did go on to collaborate with Roly to swim cannulated fish and, with Don (E.D.) Stevens, used the Brett-type tunnels to perform classical fundamental studies of fish respiratory physiology (e.g. Stevens and Randall, 1967).

At the Pacific Station, Brett started working on physiological energetics (Brett, 1956) and, by the late 1970s, he was an authority in the field (Brett, 1979; Brett and Groves, 1979). Brett devoted a considerable effort to investigating how abiotic and abiotic factors influence growth in fishes, especially Pacific salmon, and this provided the depth of understanding and reflection that are so evident in this 1979 Fish Physiology chapter. It is an extremely impressive piece of scholarship, densely-written over 69 pages with 32 figures, an essential reference work for anybody studying the mechanisms underlying fish growth. We focus on the core concept of the chapter, the notion of "scope for growth," which demonstrates the influence that Fred Fry had on his brilliant student.

1 Environmental factors and scope for growth

Brett conceived the notion of "scope for growth" by direct analogy to the—now highly-influential—Fry Paradigm, which defines abiotic factors according to how they influence aerobic metabolic rates and consequent scope for activity (Claireaux and Lefrançois, 2007; Fry, 1947, 1971). Aerobic metabolism comprises the catabolic processes that use nutritional substrates and oxygen to create ATP, to meet the energetic demands of life. Fry classified abiotic factors according to how they influence the magnitude of basal costs of living—basal metabolic rate (called "standard metabolic rate" in ectotherms, in particular because it changes with temperature), and the maximum capacity for oxygen supply and ATP generation—maximum metabolic rate. The scope for activity is the difference between standard and maximum rates, it defines the capacity of an ectotherm to perform essential activities beyond simply staying alive. How environmental factors affect this is expected to influence ecological performance of the animal (Fry, 1947, 1971).

Scope for growth considers the anabolic side of animal energetics, where acquired nutritional resources are used to deposit tissue and accumulate energy reserves, measurable as an increase in body mass. Brett (1979) defines scope for growth in relation to two main rate functions, the rate of growth (G, in % increase in body mass per day) and its main "driving force"—the rate at which substrates can be consumed, ration (R, in % of body mass consumed per day). Understanding the relation of G to R is key to understanding the action of environmental factors on scope for growth. Brett adopts the categorization of "Factors" developed by Fry (1971), capitalized to avoid confusion with an environmental factor, and defined as being either Controlling, Limiting, Masking, or Directive, depending on whether they govern, restrict, modify, or cue a response, respectively. The scope for growth paradigm considers the effects of both abiotic factors such as temperature and biotic factors such as food availability.

Measuring scope for growth under controlled conditions requires rearing replicated groups of fish on a range of rations, so that a growth-ration (GR) curve can be developed. The GR curve has the form shown in Fig. 1A, redrawn from Brett's Fig. 4A (Brett, 1979). At zero ration (R_0), G has a minimum negative value for starvational mass loss (G_{starv}), and G then rises with R to a point of zero growth rate (G_0) at the maintenance ration (R_{maint}). As R increases, there is a concomitant increase in G that tends toward an asymptote as the animal reaches its maximum capacity for food consumption (R_{max}) and growth (G_{max}). The scope for growth is then defined as G_{max}—G_0, which is determined by R_{max}—R_{maint}. The relation of G to R is a measure of gross conversion efficiency ($K = G/R \times 100\%$) and, if a tangent is drawn from the origin, it meets the GR curve at the point where K is maximal. This represents the optimum ration (R_{opt}) in terms of feed utilization for growth. Fig. 1B (Brett's 4B) then demonstrates how the R_{opt} for K actually occurs below R_{max} and G_{max} (Brett, 1979).

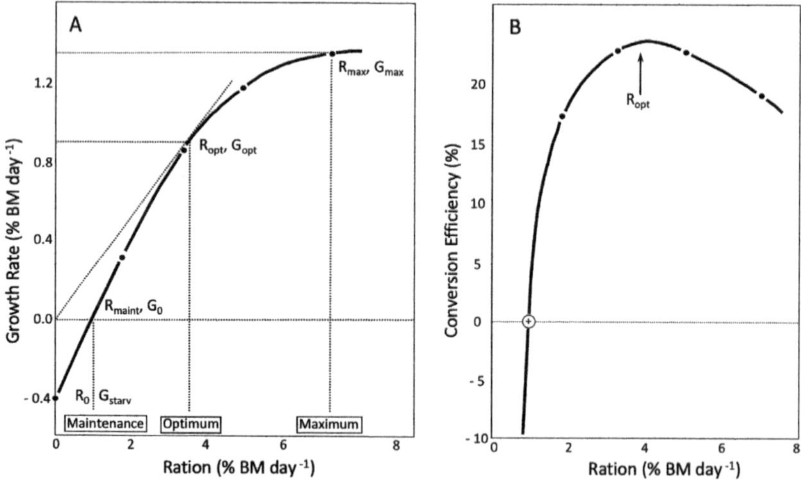

FIG. 1 The concept of scope for growth and its relation to feed conversion efficiency. A growth-ration (GR) curve (A) shows the specific growth rate (G) of fingerling sockeye salmon in relation to ration (R), at 10°C. Scope for growth is given by the difference between maintenance ration (R_{maint}) at which G is zero (G_0) and maximum ration (R_{max}) eliciting maximum growth (G_{max}). The optimum R for feed efficiency (R_{opt}) causing optimal growth (G_{opt}) is the point of intersection between the GR curve and a tangent drawn from the origin. A feed conversion efficiency curve (B) shows the ratio of G to R, revealing that optimal efficiency is achieved at an R below R_{max}. The circled point at zero feed efficiency is derived from A. Note that feed efficiencies have been improved very markedly for farmed fishes since this study was performed in the late 1960s. *Panel (A) data from Brett, J., Shelbourn, J.E., Shoop, C.T., 1969. Growth rate and body composition of fingerling sockeye salmon, Oncorhynchus nerka, in relation to temperature and ration size. J. Fish. Res. Board Can. 26, 2363–2394.*

Clearly, investigating how environmental factors affect scope for growth by performing multiple GR curves across a range of factor levels is a major undertaking in terms of manpower and facilities. Very fortunately, insight into how a factor or combination of factors affects scope for growth can be inferred just by measuring G_{max} at R_{max}—the growth rate obtained when feed is provided ad-libitum or to excess.

2 Abiotic factors

Here we concentrate on Controlling Factors, using temperature as the example, and Limiting Factors, using oxygen as an example. Controlling Factors operate at all levels of the environmental factor concerned, by governing biochemical and physiological rates. Temperature is the prime example that has direct thermodynamic effects on such rates in ectotherms (Fry, 1947, 1971). Using his own data for juvenile sockeye salmon, Brett explains why all fishes show a unimodal bell-shaped relationship of G_{max} to temperature, with a clear optimal temperature (T_{opt}) for growth (Fig. 2A, redrawn from fig. 18A in Brett (1979)). At low temperatures, R_{maint} is low but scope for growth is also low because the cold depresses R_{max} and G_{max} (Fig. 2A).

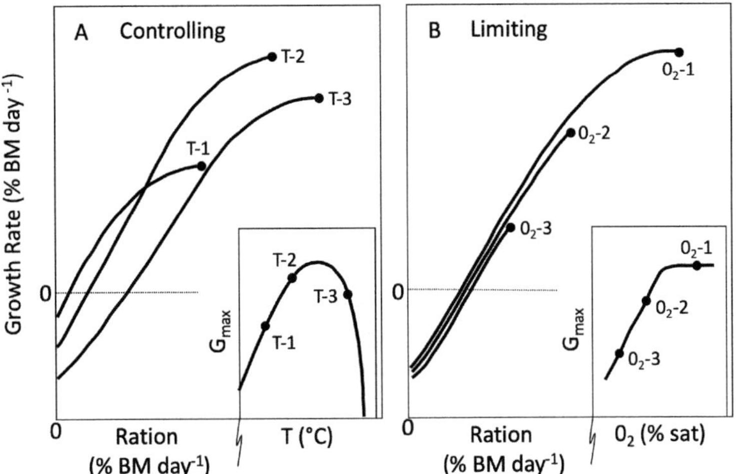

FIG. 2 The dependency of growth-ration (GR) curves on abiotic factors, illustrating the basic forms for (A) temperature (T), a Controlling Factor, and (B) oxygen (O_2), a Limiting Factor. Each GR curve is shown at three levels of each abiotic factor, with a point for the position of maximum growth rate at maximum ration (G_{max}) terminating the top of the curve. The insert box shows the consequent dependence of G_{max} on the abiotic factor, across the tolerable range. The consequences of the abiotic factors for maintenance ration (R_{maint}) and negative G in starvation (G_{starv}) can be seen by the points of intersection with the respective lines for zero growth (G_0, horizontal broken line) and zero ration (R_0 on y-axis). Curves were drawn by Brett (1979) based on data presented in his chapter.

At temperatures near T_{opt}, all biochemical and physiological rates are stimulated and there is an increase in R_{maint} but also in physiological capacity for food intake and processing, so increasing R_{max} and therefore permitting a high G_{max} and scope for growth (Fig. 2A). Beyond T_{opt}, R_{maint} starts to represent an ever-larger proportion of R_{max}, leading to a progressive decline in the proportion of R available for growth, hence a decline in G_{max} (Fig. 2A). This effect beyond T_{opt} is exacerbated by a drop in appetite, presumably under endocrine control (Volkoff and Rønnestad, 2020). Brett reviews data for a number of temperate fishes to show that a T_{opt} for G_{max} can be identified in each and is related to a species' thermal range, being higher in species that occupy warmer habitats.

Limiting factors "restrict the supply or removal of metabolites"; these "become operational at a particular level of the factor and comprise dependent and independent states." Availability of dissolved oxygen (DO) is the most well recognized abiotic Limiting Factor, also in direct homology to the Fry Paradigm. The relation of G to R is the same at all oxygen levels but, below a certain hypoxic DO threshold, fish will become limited in their ability to meet the oxygen demands of food digestion and assimilation—the so-called specific dynamic action response. As a result, the GR curve is truncated (Fig. 2B from fig. 18D in Brett (1979)), R_{max} and G_{max} decline in direct proportion to water DO. Brett points out that part of this response is also because hypoxia depresses appetite, through endocrine pathways.

Brett defines Masking Factors as "modifying or preventing the effect of an environmental factor through some regulatory device" by which he indicates that they incur a metabolic load for homeostasis that influences how an animal responds to other factors. He provides the example of salinity, where costs of active water and ion balance would raise maintenance costs and therefore R_{maint} for G_0 in the scope for growth paradigm. He notes that measuring costs of osmoregulation is not easy and this question is still being investigated (Little et al., 2023). Directive Factors cue or signal the animal to select or respond to particular characteristics of the environment, and Brett provides photoperiod as an example. He argues that any abiotic factor can be assigned to one of the four categories, Controlling, Limiting, Masking or Directive, through its specific effect on the GR curve.

3 Biotic factors

Having established the mechanism of action of environmental factors on growth, Brett considers the main biotic factors and argues that these all generally act as either Limiting or Masking. For example, R is unequivocally a Limiting Factor because it affects how much food energy, beyond R_{maint}, is available for growth. Competition is interesting because it is a Limiting Factor when it affects access to food but a Masking Factor if it causes agonistic interactions and stress.

4 Interactions among factors

Brett (1979) emphasized the importance of considering how biotic and/or abiotic factors might interact to influence growth. This was linked to the work of Alderdice (1972), which examined how marine poikilotherms respond to multiple environmental factors occurring simultaneously. Brett defined interacting effects elegantly as when the response to two variables does not move along a line parallel to a single axis. He exemplified this with how sockeye salmon perform better on restricted ration at low temperatures, because these reduce R_{maint} and therefore more of the limited food energy can be allocated toward G. As a consequence, T_{opt} for G declines with declining rations.

The classification of Factors, based upon how they act, provides a clear framework to predict how they will interact. For example temperature, possibly the only Controlling Factor in fishes, has a "pacing effect" that will then raise or lower the critical level of a Limiting Factor, such as R or DO. Masking Factors would place a "regulatory toll" upon the animal that is added to the prevailing pacing effects of a Controlling Factor. That is, a Controlling Factor will always act "in conjunction and never in sequence" with Limiting or Masking Factors. Brett then argues that "Limiting Factors together cannot result in any interaction; they operate in series, like links in a chain." For example, DO will limit G at a given threshold concentration. This concentration is then influenced by R—at lower R, the consequently lower G will not become limited until a lower DO.

5 Main strengths and influence

The chapter is a very rich source of information and reflection, it considers questions and subjects that are still being actively researched more than 40 years later. Although the prime importance of R and T for G was widely recognized at the time, the scope for growth paradigm provided a framework to categorize how abiotic and biotic factors affected growth through their effects on GR curves, and especially on G_{max} at R_{max}. The chapter has more than 1250 citations and, perhaps not surprisingly, the majority are indexed under key words relating to *growth* or *growth rate*. More than 200 citations deal with temperature as an environmental factor; studies investigating the effects of temperature on growth, survival, and the metabolism of fishes, have cited the chapter intensively during the past decade. The review is still heavily cited in aquaculture journals because it covers issues that remain significant in fish nutritional and bioenergetic research today.

Effects of temperature remain perhaps the most studied abiotic factor for fish growth. Brett's paper on energetic responses of sockeye salmon to temperature (Brett, 1971) has been cited repeatedly for the role of temperature as "the ecological master factor." The effects of temperature on growth has remained an active research area but has gradually considered more details over the years, in efforts to identify the mechanistic pathways involved in effects of temperature on the biotic factors. The two main effects of temperature on growth are in feed intake and utilization efficiency. Brett's approach of investigating how temperature affects voluntary feed intake and meal size has remained a common approach in aquaculture studies, but has been supplemented with other variables such as gene expression or assays of various orexigenic and anorexigenic factors, digestive performance, specific dynamic action (SDA), and SDA coefficient. Brett focused on energetics when classifying Controlling and Limiting Factors, but temperature and hypoxia may also be Directive Factors if they elicit neuroendocrine responses that effect, for example, appetite. Over the years, Brett's list of environmental factors has gradually been expanded to include variables such as dissolved gases other than oxygen, and physical and chemical water quality parameters. Although Brett (1979) adopted the categorization of "Factors" developed by Fry (1971), and this terminology is still in use (Claireaux and Lefrançois, 2007; Niklitschek and Secor, 2009), the Factors are often considered in the literature without necessarily being categorized and occasionally using other terminologies (loading, constraining).

A major strength of the scope for growth paradigm is its clear framework for considering interactions among factors. Despite Brett's obvious passion for such interacting effects, it did not appear to take hold among his contemporaries. The past decades have perhaps not seen a greater realization but, nonetheless, more interest and appreciation for how different water quality parameters interact to influence fish growth. This may have been enabled by technical advances in experimental physiology, particularly in sensor technology that allows for control of multiple water quality parameters in real-time, and automation of oxygen consumption measurements (Steffensen, 1989; Svendsen et al., 2016).

Brett acknowledged the importance of reduced DO for appetite, and probably inspired many of the researchers who in recent years have examined the interacting effects of temperature and oxygen saturation levels on feed intake. An example of this is the work by Remen et al. (2016), who demonstrated that increasing temperatures positively affected daily voluntary feed intake in Atlantic salmon, while at the same time increasing the sensitivity to critical oxygen saturation thresholds, below which fish would cease feeding (Fig. 3). Another example on interacting environmental factors is found in Hamad et al. (2023) who investigated tolerance toward nocturnal hypoxic and hypercapnic events in Nile tilapia (Fig. 4). Nile tilapia are considered quite resilient to hypoxia, and nocturnal hypoxia by itself had no effect on feed intake (DFI) the following day. Hypercapnia did significantly depress DFI (R), the effect became much more severe when hypoxia and hypercapnia occurred simultaneously, showing a very marked effect on SGR (G) driven both by changes in R, but also an uncoupling in the correlation between R and G. Beyond classic water quality parameters such as temperature and

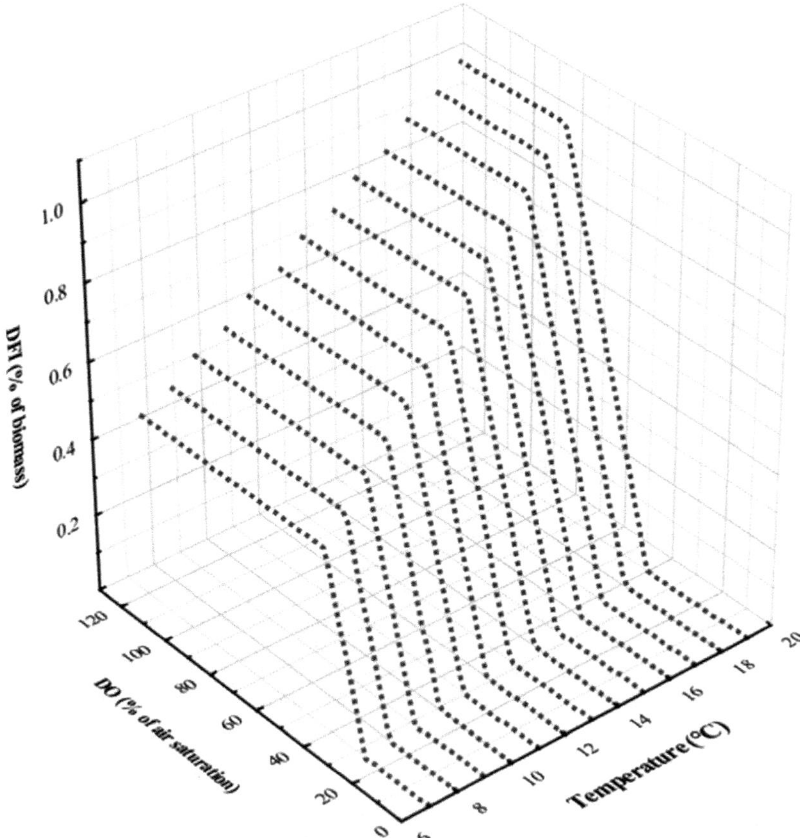

FIG. 3 A model of daily feed intake (DFI, % of biomass, BM) as a function of temperature (7–19 °C) and dissolved oxygen (DO, % of air saturation, 0–120% O_2), for post-smolt Atlantic salmon (0.3–0.5 kg), illustrating how DFI increases linearly with temperature, but decreases linearly with decreasing DO until zero is reached at the routine limiting oxygen saturation (LOS rout). Both DO max FI and LOS rout increase linearly with temperature. *From Remen, M., Sievers, M., Torgersen, T., Oppedal, F., 2016. The oxygen threshold for maximal feed intake of Atlantic salmon post-smolts is highly temperature-dependent. Aquaculture 464, 582–592.*

dissolved gases, Brett (1979) inspired work investigating the thermal dependence of chemical toxicity in fish, for example the occurrence of increased bioaccumulation of heavy metals under climate change (Kazmi et al., 2022), or delousing of Atlantic salmon in aquaculture (Walde et al., 2022).

Brett's concepts on interacting factors have also been influential in fisheries science, for growth modeling. Historically, growth has been considered a constant in stock assessment models, with most variability in stocks being derived from recruitment and mortality, whether natural or induced by fishing. Based much on Brett (1979), Lorenzen (2016) emphasized that the constant

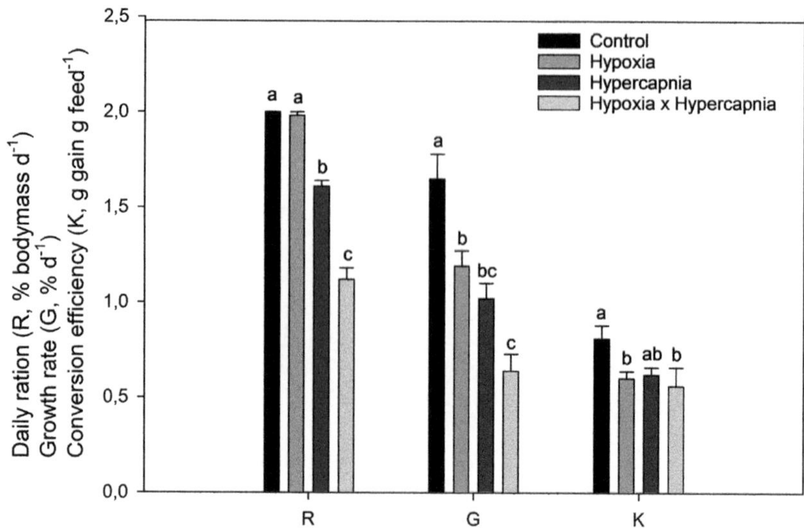

FIG. 4 Effects of nocturnal hypoxia (20% sat.) and hypercapnia (30 mg CO_2 l-1), individually and in combination, on the daily feed intake (R), growth rate (G), and conversion efficiency (K) for Nile tilapia *Oreochromis niloticus*. Columns in a group not sharing a common superscript are significantly different. Values are shown as the mean and standard error of the mean from 4 replicate tanks, each containing 10 fish. *Modified from Table 1 in Hamad, M.I., Lamtane, H.A., Munubi, R.N., Skov, P.V., 2023. Individual and combined effects of hypoxia and hypercapnia on feeding and feed utilization in Nile tilapia (Oreochromis niloticus). Aquaculture 567, 739239.*

growth paradigm is challenged by the fact that fish growth shows considerable plasticity in response to food availability and environmental factors, as evident from the observations that fish from the same cohort can vary by 90% in their weight-at-age, 40% in their length-at-age. However, Lorenzen also emphasizes that while the effects of environmental factors on growth may not be relevant for species or populations under stable conditions, an ecosystem-based approach to fisheries management should embrace the plasticity of growth (Lorenzen, 2016).

6 Time for a reappraisal

In fishes, there has been much focus on whether the Fry Paradigm, and its mechanistic elaboration the Oxygen and Capacity Limited Thermal Tolerance (OCLTT) hypothesis, are universal mechanisms to understand how global change, especially progressive warming, heatwaves and hypoxic events, will affect fish population distributions and abundance. These investigations have often failed to find that temperature controls aerobic scope as predicted by the Fry-OCLTT model, leading to much debate about its universality (Farrell, 2016; Jutfelt et al., 2018; Lefevre et al., 2021; Pörtner, 2021). Brett's scope for growth model is at least as intuitive as the Fry Paradigm, if not more so.

While experiments to measure G_{max} at R_{max} under different environmental conditions may be more expensive and time-consuming than respirometry to measure aerobic scope, thermal performance curves for growth have greater ecological relevance because of the potential link of population growth rate to fitness (Lefevre et al., 2021).

Brett's focus on interacting effects of factors (and Factors) is an exceptionally important aspect of fish environmental physiology, since abiotic and biotic factors almost always occur in combination. Logistically, it may be beyond the capacity of many laboratories to investigate the effects of several environmental factors acting in concert and so, for practical reasons, most studies on the effects of environmental factors on growth only apply a single factor. These results can, however, be used in modeling approaches to investigate potential interaction effects (Cuenco et al., 1985; Niklitschek and Secor, 2009). Therefore, as has already begun for fisheries science (Lorenzen, 2016), Brett's framework for how environmental factors interact to affect growth may have general applicability as a theoretical basis for incorporating such interactions into mechanistic models. This could improve our understanding of why some fish species are more vulnerable to climate change than others. Overall, therefore, Brett's highly innovative research and critical analysis of fish growth and bioenergetics is extremely relevant today, and clearly illustrates that he was much ahead of his time.

Acknowledgments

We are grateful to Dave Randall, Don Stevens and Paul Webb for information about Roly Brett as a colleague and mentor.

References

Alderdice, D.F., 1972. Factor combination: responses of marine poikilotherms to environmental factors acting in concert. In: Marine Ecology Vol. 1, Environmental Factors, Part III. Wiley, London, pp. 1659–1722.

Brett, J.R., 1944. Some lethal temperature relations of Algonquin Park fishes. Univ. Tor. Stud. Biol. Ser. 52, 5–41.

Brett, J.R., 1952. Temperature tolerance in young Pacific salmon, genus Oncorhynchus. J. Fish. Res. Board Can. 9, 265–323.

Brett, J.R., 1956. Some principles in the thermal requirements of fishes. Q. Rev. Biol. 31, 75–87.

Brett, J.R., 1964. The respiratory metabolism and swimming performance of young sockeye salmon. J. Fish. Res. Board Can. 21, 1183–1226.

Brett, J.R., 1971. Energetic responses of salmon to temperature. A study of some thermal relations in the physiology and freshwater ecology of Sockeye salmon (Oncorhynchus nerka). Am. Zool. 11, 99–113.

Brett, J.R., 1979. Environmental factors and growth. In: Hoar, W.S., Randall, D.J., Brett, J.R. (Eds.), Fish Physiology, *Vol.* 8. Acadmeic Press, New York, pp. 599–675.

Brett, J.R., Groves, T.D.D., 1979. Physiological energetics. In: Hoar, W.S., Randall, D.J., Brett, J.R. (Eds.), Fish Physiology *Volume 8*. Academic Press, New York, pp. 280–352.

Brett, J., Shelbourn, J.E., Shoop, C.T., 1969. Growth rate and body composition of fingerling sockeye salmon, Oncorhynchus nerka, in relation to temperature and ration size. J. Fish. Res. Board Can. 26, 2363–2394.

Brown, M.E., 1946. The growth of brown trout (Salmo trutta Linn.) I. Factors influencing the growth of trout fry. J. Exp. Biol. 22, 118–129.

Claireaux, G., Lefrançois, C., 2007. Linking environmental variability and fish performance: integration through the concept of scope for activity. Philos. Trans. R. Soc. Lond. B Biol. Sci. 362, 2031–2041.

Cuenco, M.L., Stickney, R.R., Grant, W.E., 1985. Fish bioenergetics and growth in aquaculture ponds: II. Effects of interactions among, size, temperature, dissolved oxygen, unionized ammonia and food on growth of individual fish. Ecol. Model. 27, 191–206.

Donaldson, L.R., Foster, F.J., 1941. Experimental study of the effect of various water temperatures on the growth, food utilization, and mortality rates of fingerling sockeye salmon. Trans. Am. Fish. Soc. 70, 339–346.

Elliott, J.M., 1975. The growth rate of Brown trout (Salmo trutta L.) fed on maximum rations. J. Anim. Ecol. 44, 805.

Farrell, A.P., 2016. Pragmatic perspective on aerobic scope: peaking, plummeting, pejus and apportioning. J. Fish Biol. 88, 322–343.

Fry, F.E.J., 1947. The effects of the environment on animal activity. Univ. Tor. Stud. Biol. Ser. 55, 1–62.

Fry, F.E.J., 1957. The aquatic respiration of fish. In: Brown, M.E. (Ed.), The Physiology of Fishes *Volume 1*. Academic Press, New York, pp. 1–63.

Fry, F.E.J., 1971. The effect of environmental factors on the physiology of fish. In: Hoar, W.S., Randall, D.J. (Eds.), Fish Physiology *Volume 6*. Academic Press, New York, pp. 1–98.

Hamad, M.I., Lamtane, H.A., Munubi, R.N., Skov, P.V., 2023. Individual and combined effects of hypoxia and hypercapnia on feeding and feed utilization in Nile tilapia (Oreochromis niloticus). Aquaculture 567, 739239.

Johansen, A.C., Krogh, A., 1914. The influence of temperature and certain other factors upon the rate of development of the eggs of fishes. ICES J. Mar. Sci. s1, 1–43.

Jutfelt, F., Norin, T., Ern, R., Overgaard, J., Wang, T., McKenzie, D.J., Lefevre, S., Nilsson, G.E., Metcalfe, N.B., Hickey, A.J.R., et al., 2018. Oxygen- and capacity-limited thermal tolerance: blurring ecology and physiology. J. Exp. Biol. 221, jeb169615.

Kazmi, S.S.U.H., Wang, Y.Y.L., Cai, Y.-E., Wang, Z., 2022. Temperature effects in single or combined with chemicals to the aquatic organisms: an overview of thermo-chemical stress. Ecol. Indic. 143, 109354.

Lefevre, S., Wang, T., McKenzie, D.J., 2021. The role of mechanistic physiology in investigating impacts of global warming on fishes. J. Exp. Biol. 224, jeb238840.

Little, A., Pasparakis, C., Stieglitz, J., Grosell, M., 2023. Metabolic cost of osmoregulation by the gastro-intestinal tract in marine teleost fish. Front. Physiol. 14, 1163153.

Lorenzen, K., 2016. Toward a new paradigm for growth modeling in fisheries stock assessments: embracing plasticity and its consequences. Fish. Res. 180, 4–22.

Maitland, J.R.G., 1887. The History of Howietoun: Containing a Full Description of the Various Hatching-houses and Ponds, and of Experiments which Have Been Undertaken There, from 1873 to the Present Time and Also of the Fish-cultural Work and the Magnificent Results Already Obtained. JR Guy, secy. Howietoun fishery.

Niklitschek, E.J., Secor, D.H., 2009. Dissolved oxygen, temperature and salinity effects on the ecophysiology and survival of juvenile Atlantic sturgeon in estuarine waters: II. Model development and testing. J. Exp. Mar. Biol. Ecol. 381, S161–S172.

Pörtner, H.-O., 2021. Climate impacts on organisms, ecosystems and human societies: integrating OCLTT into a wider context. J. Exp. Biol. 224, jeb238360.

Remen, M., Sievers, M., Torgersen, T., Oppedal, F., 2016. The oxygen threshold for maximal feed intake of Atlantic salmon post-smolts is highly temperature-dependent. Aquaculture 464, 582–592.

Shelbourn, J., Brett, J., Shirahata, S., 1973. Effect of temperature and feeding regime on the specific growth rate of sockeye salmon fry (Oncorhynchus nerka), with a consideration of size effect. J. Fish. Res. Board Can. 30, 1191–1194.

Steffensen, J.F., 1989. Some errors in respirometry of water breathers: how to avoid and correct for them. Fish Physiol. Biochem. 6, 49–59.

Stevens, E.D., Randall, D.J., 1967. Changes of gas concentrations in blood and water during moderate swimming activity in rainbow trout. J. Exp. Biol. 46, 329–337.

Stickney, R.R., Treece, G.D., 2012. History of aquaculture. In: Aquaculture Production Systems, pp. 15–50.

Svendsen, M.B.S., Bushnell, P.G., Steffensen, J.F., 2016. Design and setup of intermittent-flow respirometry system for aquatic organisms. J. Fish Biol. 88, 26–50.

Volkoff, H., Rønnestad, I., 2020. Effects of temperature on feeding and digestive processes in fish. Temperature 7, 307–320.

Walde, C.S., Stormoen, M., Pettersen, J.M., Persson, D., Røsæg, M.V., Jensen, B.B., 2022. How delousing affects the short-term growth of Atlantic salmon (Salmo salar). Aquaculture 561, 738720.

Wingfield, C., 1940. The effect of certain environmental factors on the growth of brown trout (Salmo trutta L.). J. Exp. Biol. 17, 435–448.

Chapter 18

ENVIRONMENTAL FACTORS AND GROWTH☆

J.R. BRETT

Chapter Outline

I. INTRODUCTION

Although readily observed and easily measured, growth is one of the more complex activities of the organism. It represents the net outcome of a series of behavioral and physiological processes that begin with food intake (the consummation of an appetitive behavior) and terminate in deposition of animal substance. The processes of digestion, absorption, assimilation, metabolic expenditure, and excretion all interplay to affect the final product. These individual functions are properly the content considered under digestion and

☆This is a reproduction of a previously published chapter in the Fish Physiology series, "1979 (Vol. 8)/ Environmental Factors and Growth: ISBN: 978-0-12-350408-1; ISSN: 1546-5098".

Fish Physiology, Vol. 40B. https://doi.org/10.1016/bs.fp.2024.07.008

bioenergetics (Chapters 4 and 6); however, to understand how the many environmental factors influence growth the basic rate functions should never be lost sight of, despite the general simplicity of growth measurements (Fig. 1). In so far as possible these individual functions will be introduced to provide some explanation of the response to environmental factors, without deviating from the major purpose of examining the recorded trends in the activity of growing.

In relation to environmental effects, quite obviously growth cannot be studied without involving food consumption, even though the latter may not necessarily be measured. In this respect growing, unlike such activities as swimming or respiring, is inseparably coupled with a powerful *biotic* factor, so that any *abiotic* factor is necessarily involved in some form of interaction between the two. For example, if temperature rises, the amount of food consumed usually increases as well as the rate of digestion. Growth rate, however, may either increase or decrease depending on the nature of the *food* × *metabolism* × *temperature* relation; on analysis it can be seen that the energy

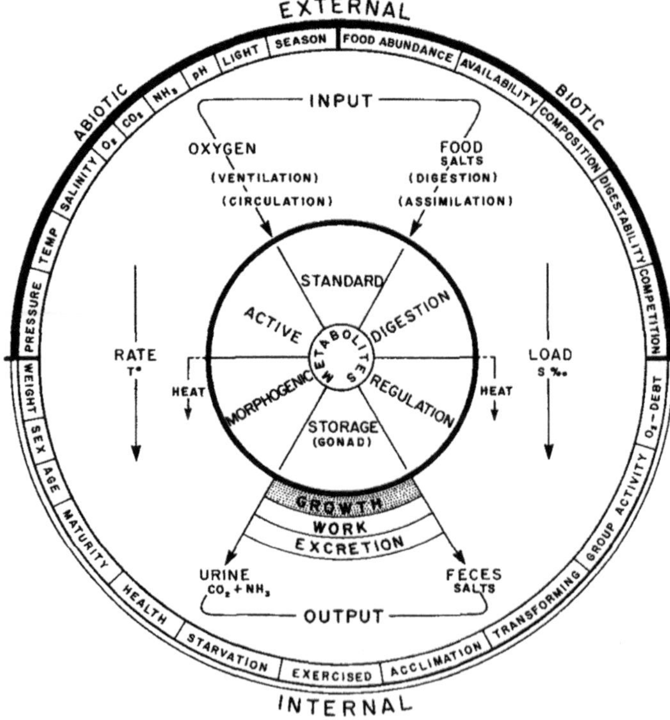

Fig. 1. Diagrammatic representation of the general metabolic system involved in growth, from input sources (food and oxygen) to output products (growth, work, excrements), showing some of the external and internal factors affecting the system (center circle). Morphogenic metabolism applies to the energy of internal work to produce body substance. Heat released from the system includes the heat increment or SDA. Temperature is shown as a major controlling factor of metabolism, and salinity as the chief energy-demanding factor in regulation.

demand of elevated metabolic rate could have exceeded the gain from increased food uptake resulting in reduced growth rate. Furthermore, the act of growing necessarily alters size so that another important biotic factor continually changes with time. An understanding of these inherent involvements is necessary before any clear insight of the environmental relation to the growth process can be achieved.

The general approach adopted is that elaborated by Fry (1947, 1971), who classified environmental factors not by simple typegrouping but by considering how the separate entities (temperature, oxygen, light, etc.) *acted* on any particular function or activity. Fry's extension of the classical concept of "limiting factors" (Blackman, 1905) was further enhanced by the recognition that the environment acts *through* metabolism *on* activity and not directly on the activity itself. Most of Fry's (1947, 1971) considerations were devoted to catabolic aspects of metabolism and the influence of abiotic factors. Such an approach was conceptually extended by Warren (Warren and Davis, 1967; Warren, 1971) to encompass the anabolic process of growth. Biotic factors as a group were never classified.

These basic concepts are applied and developed in the following considerations. At the outset it is necessary to define the environmental factors classified by Fry (1971) and to introduce the main terms (and abbreviations) appropriate to experimental growth studies. Fry adopted the label "factors," such as Lethal Factors, in keeping with their historical meaning as *effectors*—the influence any particular environmental entity had on the organism. Unfortunately this is now in conflict with the common use of "factors" to imply the environmental entities themselves, as used in the title of this chapter. To avoid misunderstanding, the use of "Factors" as a category of effects will be capitalized. The great advantage that such a system of classification offers comes from the small number of categories involved, and the clear distinction of action they normally imply. Any environmental factor may fall into one or more of the functional categories.

As an activity, growth must always come within the lethal limits of the environmental factor considered, so that this category (Lethal Factors) does not enter into further consideration here. Fry (1971) invokes four other categories, namely:

1. Controlling Factors, which *govern* the rates of reaction by influencing the state of molecular activation of the metabolites (e.g., temperature, pH); these operate at all levels of the environmental factor concerned
2. Limiting Factors, which *restrict* the supply or removal of metabolites, as links in the chain of metabolism [e.g., oxygen, light (as in photosynthesis)]; these become operational at a particular level of the factor, involving dependent and independent states
3. Masking Factors, which *modify* or prevent the effect of an environmental factor through some regulatory device (e.g., humidity influencing body

temperature by affecting evaporative heat loss, or temperature regulation by countercurrent heat flow as in warm-bodied fish)

4. Directive Factors, which *cue* or signal the animal to select or respond to particular characteristics of the environment (e.g., temperature preference, photoperiod-induced smoltification); these may be compared to the category of "releasing mechanisms" in animal behavior (Baerends, 1971), and hormonal response may be involved

In the formulation of a growth model for salmonids, Stauffer (1973) assessed the various factors influencing growth. He concluded that any attempt at modeling must include at least the three factors, *ration*, *size*, and *temperature* as the most important independent variables. Elliott (1975c,d) supports this view. Although not attempting any classification, ration was considered by Stauffer to be the sole driving force, temperature the major rate-controlling force, and weight was thought to act as a scaling factor that adjusts these rates to the size of the growing individual.

Starting with the fundamental growth–ration relationship as the basis, all other factors will be examined as they affect this key "driving force" relation.

II. BASIC GROWTH RELATIONS*

A. The Growth Curve

The growth of any organism is a multiplicative process in which cell number and cell volume increase (Needham, 1964). Given an unlimited source of nutrition, growth proceeds in an almost exponential burst. The burst, however, is continuously dampened as size and age increase. Nevertheless, over short periods of time the exponential relation

$$W = ae^{gt}$$

where W=weight, t=time, and a and g are constants, closely approximates the relation (Fig. 2; see also discussion in Weatherley, 1972). The absolute rate of increase in weight (dW/dt) is greatest just at the inflection point of the sigmoid curve. However, the *relative* rate of increase (dW/dtW) is usually greatest at the youngest, smallest stage. The latter derivative is the instantaneous or *specific growth rate* represented by the slope (g) of the lines in Fig. 2B; when multiplied by 100 it is equivalent to the percentage increase in weight per unit time (usually 1 day) and is labeled "G." The latter is used throughout this chapter as "rate of growth," involving a change in weight irrespective of length (see Chapter 11).

Growth rate decreases greatly at the onset of maturity. This phenomenon can result in initially fast-growing fish, which mature at an early age, being

*The notation used in this section and elsewhere is as follows: T, temperature (°C); L, hours of light per day (photoperiod); R, ration (% weight/day); S, salinity (‰); G, % weight per day.

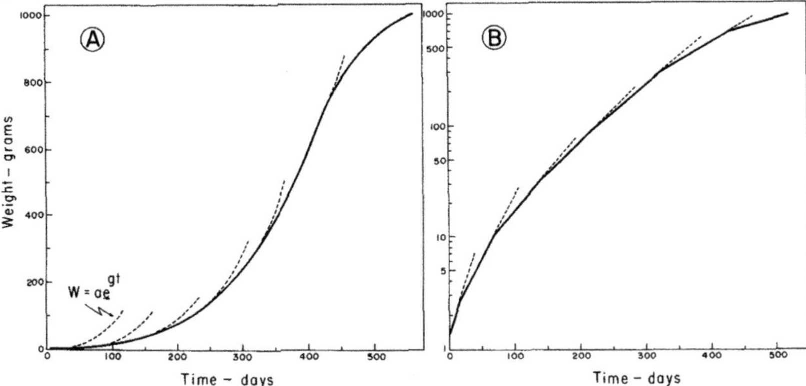

Fig. 2. (A) Generalized growth curve for a fish under constant environmental conditions, and (B) the transformed curve using a logarithmic scale for weight. Experimentally observed growth "stanzas" are indicated by straight lines in B, with their extrapolation shown as broken lines in A and B following the form of an exponential curve (Brett *et al.*, 1969—up to a weight of 100 g in B). W, weight; t, temperature, and a and g are constants. $G = 100 \times g$, the specific growth rate. The broken lines simply display what the path would be if a given specific growth rate persisted. When food is not limiting, size and age progressively depress the specific growth rate, undoubtedly following a continuous curve except when developmental stages alter the pattern.

surpassed in ultimate size by slower growing fish, demonstrating that the differences in growth rates established in young fish do not necessarily persist throughout life (Kinne, 1960).

B. Natural Environmental Relations

In nature, the timing of hatching and early growth (in temperate climatic zones) usually coincides with increasing daylength, rising temperature, and seasonal abundance of food. There is a natural compounding of these factors with a rapidly changing size of the organism. In addition, changes in prey organisms and a change in ability to seize ever larger prey introduce the variables of diet quality and quantity.

Add to this a possible factor of migrating into new environments that may involve a drastic change in salinity, it can be seen that no single, natural growth curve can reveal the underlying physiological responses to environmental factors. Nevertheless, the correlates of seasonal temperature (Fig. 3), and the ultimate damping of growth with age and maturity, show through this complexity in gross form (Larkin *et al.*, 1956; Gerking, 1966; Weatherley, 1972).

For the experimenter there is no small challenge to manipulating environmental factors in a design that is not confounded in one way or another. Kerr (1971) has elaborated on some of the inherent difficulties.

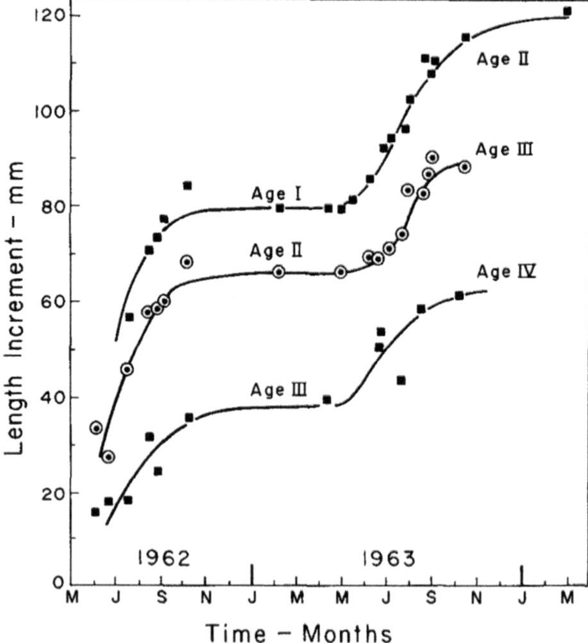

Fig. 3. Seasonal gain in length of a population of bluegill, *Lepomis macrochirus*, by age groups. The growth increment varies with season and decreases with age and size. (From Gerking, 1966.)

C. The Growth–Ration (GR) Curve

The overall relation of growth rate (G) to rate of food uptake (R) from minimum to maximum consumption was first depicted graphically for large-mouthed bass, *Micropterus salmoides*, by Thompson (1941). Although no actual data were presented, the figure clearly indicated some of the simple geometry of the relation, from which certain important parameters such as optimum and maintenance ration could be derived. Earlier work of Pentelow (1939) suggests something of this relation, but only with subsequent insight is it possible to see that the feeding experiments on brown trout, *Salmo trutta*, support the general configuration.

All too little attention appears to have been paid to this fundamental relation which, together with its derived conversion efficiency equivalents, holds much of the key to understanding the action of environmental factors on growth. Figure 4 depicts the relation for sockeye fingerlings. Starting at zero ration (R_0) the GR curve rises steeply from a minimum negative value (G_{starv}) to cross the point of zero growth rate (G_0) at the maintenance ration (R_{maint}). This continued steep rise begins to flex downward such that a tangent from the origin passes through the point where the ratio of G to R is maximal, providing a measure of the optimum ration (R_{opt}). With increasing ration the GR curve flexes

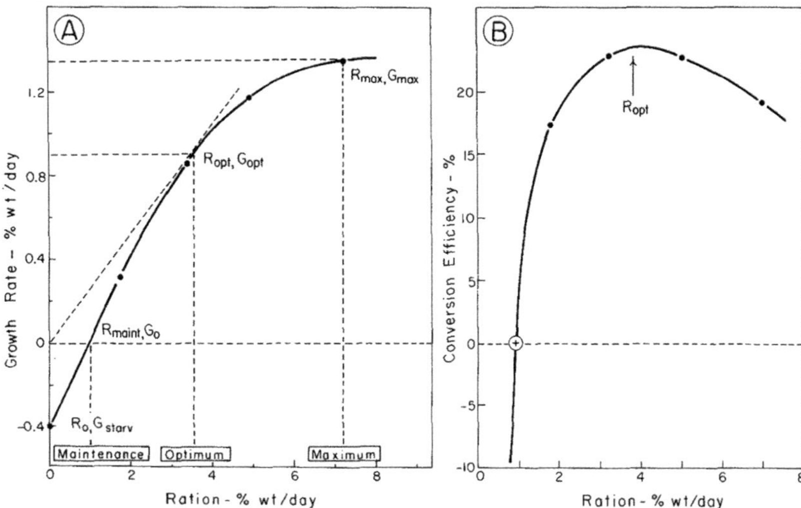

Fig. 4. (A) Specific growth rate of fingerling sockeye salmon (mean weight, 13 g) in relation to ration, at 10°C. The key parameters of the GR curve, with abbreviations, are indicated. (Data from Brett *et al.*, 1969.) (B) Conversion efficiency curve (KR) for the same data. Circled point (+) derived from A.

further, reaching a plateau of maximum growth rate (G_{max}) at the point of maximum ration (R_{max}). Figure 4A is just one example of the curve. Huisman (1974) has shown that no change in shape is manifest when the specific growth rate is determined in terms of either wet weight, dry weight, or caloric content. Some difference in the value of G for a given ration occurs because the body composition changes with the level of ration, altering the protein, fat, and water ratios and hence affecting relative dry weight and energy content.

Variations in the form of the curve do occur, according to environmental conditions and species; however, the general form and indicated parameters are indisputable. Elliott (1975d) tried unsuccessfully to find a suitable linear transformation for GR curves derived from rigorously performed experiments on brown trout, and resorted to smooth curves fitted by eye. A straight line has not infrequently been used to depict the relation, especially where a considerable scatter of points exists and the growth rate for R_{max} is not well defined (e.g., Hatanaka and Takahashi, 1960; Brocksen and Cole, 1972). Stauffer (personal communication) at first accepted this linear expression because of difficulty in finding a more appropriate transformation. Subsequently the curvilinear segment was found to be fitted more closely by a sine curve, $G = a[\sin(bR + c)]$, which significantly reduced the error component for the sockeye data (Stauffer, 1973).

It is a simple and illuminating step to derive the relation for gross conversion efficiency (K), which usually traverses a tapered, domeshaped curve (Fig. 4B).

When ration is expressed in the same units as growth rate* (% weight/day), $K = G/R \times 100\%$. At R_{maint}, K must obviously be zero; similarly it is apparent that R_{opt} must define the peak of the dome where maximum efficiency occurs. It is significant to note that in the range of the dome there will be a lower ration (below R_{opt}) and higher ration (up to R_{max}) for which one and the same conversion efficiency would be obtained. This fact has confused some reporting of results. Further, there is still debate on the particular path traced by the segment of the curve beyond R_{opt}. This has been stated to follow a logarithmic decay with increasing daily food uptake, according to the equation

$$K = ae^{-bR}$$

where K is gross conversion efficiency (Paloheimo and Dickie, 1966). Greater consideration of this phenomenon is provided in Chapters 6 and 11.

Insofar as possible, each environmental factor will be examined for its effect on the GR curve and the consequence to its parameters—G_{max}, R_{max}, R_{maint}. This is the basic, comparative theme throughout the chapter. It should be noted that growth in nature must be responding more or less at every point along the GR curve according to feeding opportunity, with not infrequent periods of food scarcity. Hatcheries, on the other hand, tend always to be dealing with the upper end of the GR curve, above R_{opt} and G_{opt}. Experimentally it is undeniably important to examine the full GR curve over a wide range of the environmental factor(s) concerned. If that is not possible then the effect on G_{max} and its corresponding R_{max} is the most important relation to determine.

III. ABIOTIC FACTORS

A. Temperature

Since temperature acts as a Controlling Factor pacing the metabolic requirements for food and governing the rate processes involved in food processing, the GR curve responds to temperature in a varied manner as each parameter shifts. The response of fingerling sockeye salmon (mean weight = 13 g) has been studied at temperatures from 1° to 24°C (Brett *et al.*, 1969); three cases are illustrated (Fig. 5). At a comparatively low temperature (5°C) the curve is far to the left by virtue of a low G_{starv} and small R_{maint}; it rises in a sweeping curve to a low plateau (G_{max}) in keeping with reduced daily food requirements and generally suppressed growth rate. As temperature increases to 10°C the whole curve moves to the right and is elevated with approaching

*Because maximum growth rate usually decreases with increasing size, this expression of relative food consumption can only be applied to fish of the same, limited weight range, or to fish that maintain the same specific growth rate over a wide range of weights.

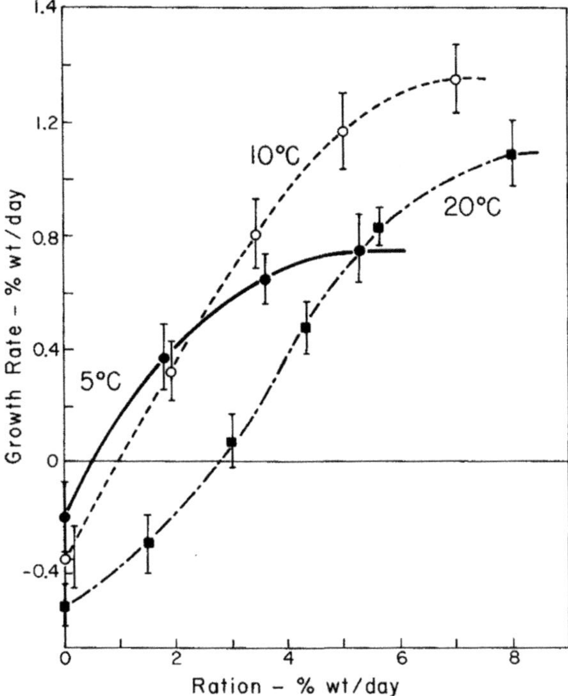

Fig. 5. The effect of temperature on the growth-ration (GR) curve for fingerling sockeye salmon. Mean weight, 13 g. Limits shown as ±2 SE. (Data from Brett *et al.*, 1969.) Maintenance requirements (R_{maint}) occur at the point of intersection of each curve with the baseline for zero growth (G_0). The highest ration at each temperature approximates R_{max}, providing a measure of the maximum growth rate (G_{max}).

optimum thermal conditions and consequent enhanced G_{max}; maintenance requirements (R_{maint}) are necessarily up. Higher temperatures, above the optimum, bring a further shift to the right as maintenance costs escalate accompanied by a lowering in G_{max} despite an increase in R_{max}; conversion efficiency has begun to decline.

The general shape of the GR curve for this species appears to shift gradually from a simple concave form at low temperatures to a sigmoid shape at high temperatures; fish that are underfed at a relatively high temperature (e.g., 20°C) exhibit a great deal of searching which would compound the expenditure of energy in the lower segment of the curve causing the convex upsweep. Some of the data for brown trout at 19.5°C suggest the same relation (Elliott, 1975d). Averett (1969) also presents a set of GR curves for young coho salmon, ranging from 5° to 17°C. Although subject to considerable variability these all show the features of the sine curve applied by Stauffer (1973). At the upper end of the GR curve for carp, *Cyprinus carpio*, at 23°C a distinct inflection downward occurs, which must reflect some adverse effects on G_{max}

with high ration, decreasing conversion efficiency considerably (Huisman, 1974). This is hardly apparent, if at all, for carp at 17°C.

It is these changing positions and shapes of the GR curve that dictate the temperature relations for each parameter. Most experiments on temperature effects have been involved with studying a given parameter (e.g., G_{max}, G_{starv}) without the insight of the complete GR relation. Since some parameters (e.g., G_{opt}, R_{maint}) are difficult to determine directly the advantages of interpolation from the GR curve are worth emphasizing.

1. *Maximum Growth Rate (G_{MAX}) × Temperature*

Most effort has been devoted to studying the controlling effect of temperature on this parameter of growth (i.e., changes in G_{max} for fish fed a maximum ration). Almost all species in the young stages show a typical rapid increase in growth rate as temperature rises passing through a peak (optimum temperature) and frequently falling precipitously as high temperatures become adverse. This is well illustrated for juvenile brook trout, *Salvelinus fontinalis*, by the studies of McCormick *et al.* (1972) who also included the change in sample biomass, resulting from the onset of some mortality at high

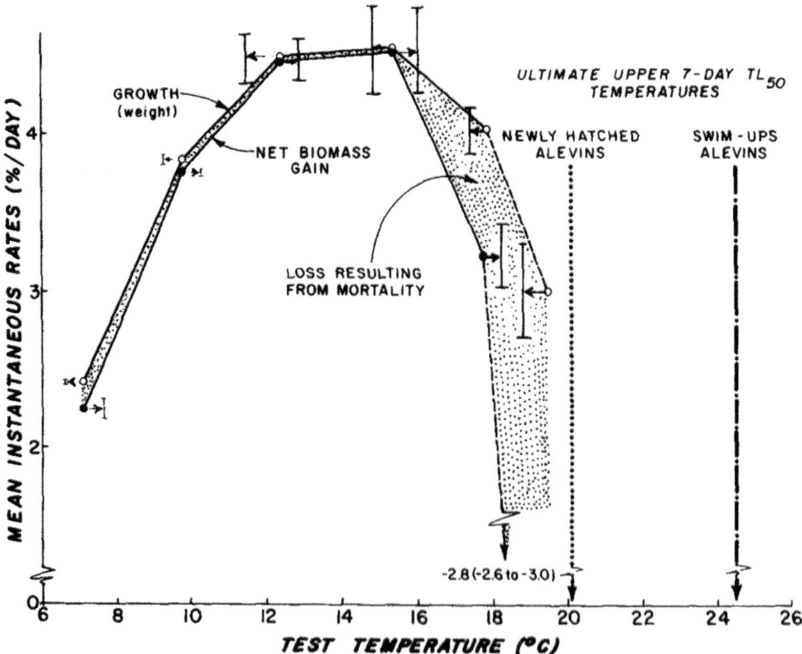

Fig. 6. Effects of temperature on mean instantaneous rates of growth, mortality, net biomass gain, and ultimate 7-day median temperature tolerance limits of brook trout, *Salvelinus fontinalis*, during their first 8 weeks after hatching. (From McCormick *et al.*, 1972.)

Table I Temperature Optima for Growth when on a High Ration[a]

Species	Optimum (°C)	Test interval (°C)	Salinity (‰)	Initial size (cm, g)	Comment	Reference
Salmo trutta	12.8	1.5	Fresh	10–300 g	Fed maximum ration individually	Elliott (1975c)
Oncorhynchu keta	13	3	35	Under-yearling	Size not given	Kepshire (1971)
Salvelinus fontinalis	14	3	Fresh	0.1–0.2 g	Alevin to juvenile stage; fed nauplii, liver and starter feed	McCormick *et al.* (1972)
Salvelinus alpinus	14	2	Fresh	Yearling	Diet of minced beef liver in gelatine	Swift (1964)
Pleuronectes platessa	14.2	2	Seawater	0.5–2.0 cm	Computed from observations	Jansen (1938); Ursin (1963)
Oncorhynchusnerka	15	5	Fresh	6–20 g	Moist pellet feed	Brett *et al.* (1969)
Oncorhynchus tshawytscha	15.5	2.5	Fresh	2–9 g	Moist and dry pellet feed	Banks *et al.* (1971)
Oncorhynchus gorbuscha	15.5	3	35	Under-yearling	Size not given	Kepshire (1971)
Salmo gairdneri	17.2	1.5	Fresh	0.3–3 g	Alevin to juvenile stage, brine shrimp and trout pellets	Hokanson *et al.* (1977)

Continued

Table I Temperature Optima for Growth when on a High Ration—Cont'd

Species	Optimum (°C)	Test interval (°C)	Salinity (‰)	Initial size (cm, g)	Comment	Reference
Coregonus artedii	18.1	3	Fresh	0.2 g	Larval stage; nauplii	McCormick *et al.* (1971)
Morone saxatilis	24–25	2	Fresh	90–100 g	Fed on live minnows	Nelson (1975); Cox and Coutant (1975)
Lebistes reticulatus	24–25	3	Fresh	0.7–1.8 cm	Reared from birth to adult (males); dry fish food	Gibson and Hirst (1955); Ursin (1963)
Micropterus salmoides	25	7	Fresh	8–140 g	Fed on shiners and beef liver	Niimi and Beamish (1974)
Etheostoma spectabilis	26	2	Fresh	0.6–2.1 cm	Larvae; feeding on brine shrimp	West (1966)
Micropterus salmoides	27	2	Fresh	30–240 g	Fed on live minnows	Nelson (1974); Coutant and Cox (1975)
Catostomus commersoni	27	3	Fresh	Larvae	Fed *ad libitum* on brine shrimp	McCormick *et al.* (1977)
Micropterus salmoides	27.5	2.5	Fresh	0.6–2.6 m	Fry; feeding on zooplankton	Strawn (1961)
Cichlasoma bimaculatum	28	4	Fresh	2–3 g	Fed on Tubifex	Warren and Davis (1967)

Cyprinodon macularius	28	5	15	20–30 cm	Fry; brine shrimp, white worms and fish food	Kinne (1960)
Perca flavescens	28	2	Fresh	0.4 g	Fry; brine shrimp, liver and yeast	McCormick (1976)
Ictalurus punctatus	29	2	Fresh	1.5–7.2 cm	Fry; feeding on *zooplankton*	West (1965)
Ictalurus punctatus	30	4	Fresh	4 g	Pelleted feed presented at 3 rates	Andrews and Stickney (1972); Andrews *et al.* (1972)
Cyprinodon macularius	30	5	35	20–30 cm	Fry; shrimp, worms and fish food	Kinne (1960)
Lepomis gibbosus	30	5	Fresh	24–34	Fed on oligochaetes	Pessah and Powles (1974)

[a]Temperature interval is indicated as a measure of experimental sensitivity (i.e., the smaller the interval the better the optimum will be defined). Arranged in general order of increasing temperature optimum.

temperatures (Fig. 6). As might be expected the optimum temperature for growth of different species ranges generally upward from the ecologically cold-adapted to the warm-adapted species (Table I). One remarkable case is that for larval ciscoe, *Coregonus artedi*, a particularly temperature-tolerant member of the coregonids, which at this very early stage displays a growth rate that continues to rise almost to the lethal temperature (19.8°C) despite the onset of some mortalities above 17°C (McCormick *et al.*, 1971).

Among the highest optima in temperate freshwater are those for the channel catfish, *Ictalurus punctatus* (29°–30°C), and the pumpkinseed, *Lepomis gibbosus* (30°C) (Table I). The lethal temperature for the latter species is 34.5°C (Altman and Dittmer, 1966). Pessah and Powles (1974) consider that the pumpkinseed shows a form of "growth homeostasis," between 15° and 30°C, as a result of stable and similar growth rates displayed in their test lots *following* about 5 weeks of constant environmental and feeding conditions. Although the conclusion appears to be supported by the data there are a number of potentially confounding problems including progressive differences in size between lots, differential and possibly temperature-correlated hierarchial effects (for which the species is renowned), growth depensation, and the fact that temperatures in the region of the optimum may show relatively little difference in growth rate—but are likely to be significantly different in food conversion efficiency. (All of these aspects of growth are considered further in the following text.)

The records in Table I apply to different *fixed* levels of temperature held constant within experimental limits. In early tests on growth, some authors (e.g., Pentelow, 1939) relied on using the normal seasonal change in environmental temperature to conduct their "controlled" experiments. Brown (1946a) attempted to evaluate the effect of increasing and decreasing temperatures (18° to 4°C and reverse) on fingerling brown trout, *Salmo trutta*, by manipulating tank temperatures at the rate of 0.5° and 1°C per week over a 1-year period. There was no indication that either rising or falling temperature at these slow rates of change had any different effect on growth than when held at equivalent, constant, mean temperatures. The rates of temperature change were apparently well within the acclimation rate for growth.

This circumstance applies to the varying temperatures imposed under laboratory and natural stream conditions on brown trout fed to satiation (Elliott, 1975c). The resulting weight at temperatures that rose from 6.8°C in March to 12.1°C in June, and decreased from 12.9°C in August to 7.2°C in November, were estimated quite accurately by applying the developed equations of growth determined for static temperatures.

Rapid temperature change, however, results in a more complex growth response. Daily fluctuating temperatures of ±4°C, oscillating around six mean levels from 12° to 22°C, were applied to juvenile rainbow trout, *Salmo gairdneri*, by Hokanson *et al.* (1977). When the growth rates were compared with results from constant temperatures (equivalent to the mean of the fluctuating temperatures) the growth–temperature curve was shifted to the left such that the optimum mean temperature occurred 2°C below that for the constant temperatures (Fig. 7). Fluctuation resulted in growth rates which were higher than expected when in the thermal range below the "static" optimum (17°C), somewhat lower when at and above the optimum (to 21°C), and highly depressed at a mean of 22°C. The explanation of these results does not lie in the simple multiplication of the instantaneous growth rates (according to the range of temperatures experienced) and then determining the mean of these rates. It is apparent that the daily peak temperatures have a carry-over, beneficial effect when the thermal range applied occurs below the static optimum. This benefit declines where higher temperatures are involved, and comes crashing down when the peak of fluctuation enters the fringe of temperature tolerance for the species. Similar effects of daily fluctuating temperatures were obtained by Wurtsbaugh (1973) who reported negative effects on growth of trout when relatively high, oscillating temperatures were involved.

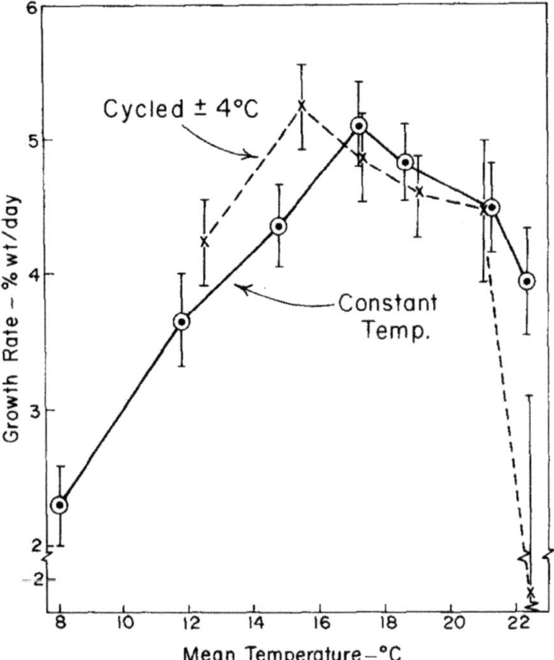

Fig. 7. Growth rate of juvenile rainbow trout in relation to constant and cycled temperatures (plotted at corresponding mean temperature). Limits shown as ±2 SE. (From Hokanson *et al.*, 1977.)

2. OPTIMUM GROWTH RATE (G_{OPT}) × TEMPERATURE

Few studies are complete enough to permit examining how the optimum growth rate changes with temperature. For young sockeye the course of this intermediate parameter of the GR curve was found to rise steadily from a position well below G_{max} at low temperatures to be almost superimposed on G_{max} at a high temperature (Fig. 8). The escalating maintenance cost with increasing temperature apparently forces G_{opt} to approach G_{max} at the upper end of the temperature tolerance scale.

3. MAXIMUM RATION (R_{MAX}) × TEMPERATURE

The need for meeting the increased appetite of fish with rising temperature has long been appreciated in hatchery practices. Haskell (1959) credits the introduction of the temperature-scaled feeding chart to Tunison and Deuel in 1933. In a comparison of six feeding tables for salmonids, Stauffer (1973) shows the rapid increase in recommended ration, up to a maximum of about 10% per day for 2-g salmon at 18°C (Fig. 9). Above this peak, one curve (EIFAC) inflects downward reflecting the loss of appetite that can occur at a relatively high temperature (shown to be true for brown trout above 19°C, Elliott, 1975a). Although these charts provide the fish-culturist with a good guideline they are not necessarily right up to the maximum ration, and in some cases may be as much as one-quarter below this parameter (Stauffer,

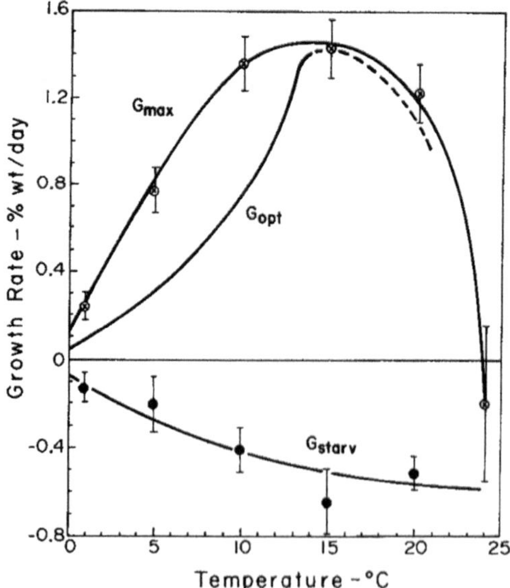

Fig. 8. Relations of various defined growth rates according to maximum ration (G_{max}, R_{max}), optimum ration (G_{opt}, R_{opt}), maintenance ration (G_0, R_{maint}), and starving (G_{starv}, R_0). Limits shown as ±2 SE. (Data from Brett *et al.*, 1969.)

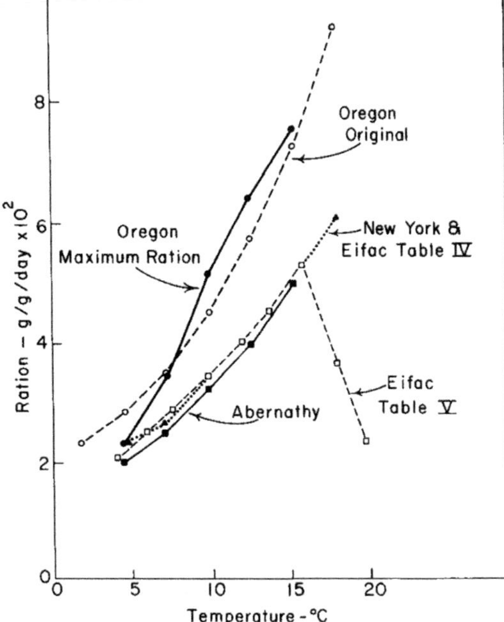

Fig. 9. The relation between recommended ration (as pelleted feed) and temperature for 1.5–2.5 g salmonids, proposed by six commercial feeding charts. Values approach R_{max}. (From Stauffer, 1973.)

1973). On a measured excess ration (i.e., an estimated amount over satiation) fingerling sockeye salmon increased their uptake from 3% weight/day at 1°C to 8% weight/day at 20°C, followed by a rapid decline at higher temperatures (Brett *et al.*, 1969). The rate of increase in R_{max} declined with increasing temperature (T) following a concave curve described by the equation $R = 2.68 + 1.76(\log_e T)$ (Fig. 10A).

4. *Maintenance Ration* (R_{MAINT}) × *Temperature*

One of the earliest critical studies on the effect of temperature on R_{maint} was that of Pentelow (1939) who held brown trout, *Salmo trutta*, individually in small troughs. Although troubled by the difficulties which Brown (1946a) experienced from the constant need to adjust the ration upward or downward as the fish lost or gained weight, the deviations were not great. Other factors such as random activity, changing size, and seasonal effects were of greater consequence to the observed variability. The data were reanalyzed by Stauffer (1973) and grouped according to size and seasonal temperature (Fig. 11). Except for the lowest temperatures, where little change was noted (between 5° and 8°C), there is an exponential increase in R_{maint} with rising temperature, reaching a requirement of 4% weight/day at 15°C for the

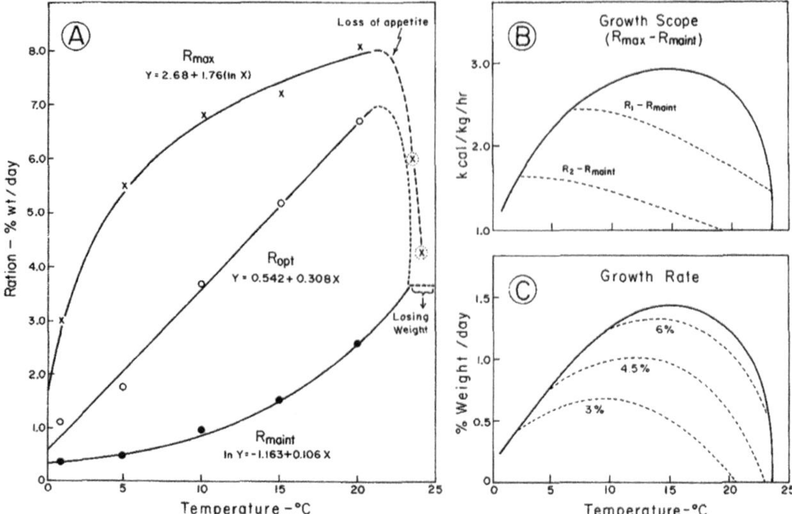

Fig. 10. (A) Relation of different levels of ration to temperature. Upper curve for R_{max}, middle line for R_{opt}, and lower curve for R_{maint}. (From Brett *et al.*, 1969.) (B) Scope for growth, derived from difference between R_{max} and R_{maint} (expressed here in kcal/kg/hr). (From Brett, 1976.) The growth scope for two hypothetical fixed rations, R_1 and R_2 (less than R_{max}), is shown by dotted lines. These may be compared with the observed growth of fingerling sockeye (C) on excess and restricted rations; much of the growth response obtained in C is predicted by the growth scope relations developed in B.

smallest size (1–2 g). A similar exponential increase was noted by Brett *et al.* (1969) for 13-g sockeye (Fig. 10A). At this weight the R_{maint} at 1°C rose from 0.3% weight/day to an estimated 3.7% weight/day at 23°C (Fig. 10). Applying the method of interpolation from GR curves, Elliott (1975d) determined the maintenance ration for brown trout. R_{maint} was found to increase exponentially from approximately 100 mg dry weight/day at 6.6°C to 350 mg dry weight/day at 19.4°C (50-g fish, Fig. 11).

5. *Scope For Growth* (G_{SCOPE}) × *Temperature*

The concept of *scope for activity* was first elaborated by Fry (1947) as the difference between active and standard metabolic rates; this "metabolic scope" defined the amount of energy available for activity, over the tolerable range of temperature. Fry (1947, 1971) demonstrated that the maximum sustained swimming speed of fish was closely correlated with metabolic scope and not active metabolic rate. Warren and Davis (1967) paid tribute to Fry's concept and elaborated a *scope for growth*, which they considered to be "the difference between the energy of the food an animal consumes and all other energy utilizations and losses" (p. 184). These authors recognized that the anabolic aspects of food consumption cannot be considered in the same

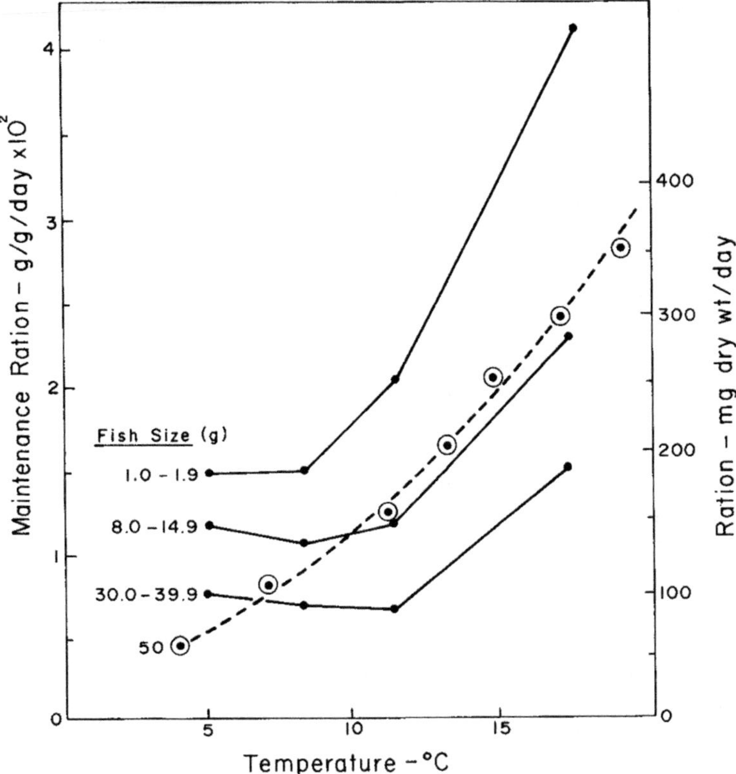

Fig. 11. Maintenance ration (R_{maint}) of three size groups of brown trout, *Salmo trutta*, in relation to temperature (solid lines). (Data of Pentelow, 1939; analyzed by Stauffer, 1973.) Values obtained by Elliott (1975d) for 50 g brown trout included for comparison (broken line), plotted as mg dry weight/day to approximately the same scale.

context with the catabolic relations of active and standard metabolic rates, although one system obviously influences the other. A number of examples of scope for growth, depicted as growth rates in relation to temperature, were given by Warren and Davis (1967) in conjunction with certain associated bioenergetic relations, namely: food consumed, fecal wastes, specific dynamic action, and starving metabolic rate. However, these do not by themselves involve a calculated scope (by difference) such as Fry applied to metabolic rates. Further, it is evident that environmental factors will not act directly on growth—the pathway will be through the mechanisms of energy supply and demand influencing the scope, in a way that Warren and Davis (1967) implied but did not actually determine. Warren (1971) further elaborated on this thesis, illustrating the theoretical effects of limiting food on the energy budget and the deduced scope for growth.

Brett (1976) was able to demonstrate that the difference between R_{max} and R_{maint} gave a simple measurement of scope for growth in relation to temperature (Fig. 10B). For sockeye salmon the reason for the rapid.decline in G_{max} above 8°C, in the face of increasing R_{max}, was due to the exponential increase in R_{maint} (Fig. 10A). Maximum growth is a consequence of the interplay between these two levels of ration over the tolerable range of temperature for the species. Only the food component in excess of R_{maint} is available for growth. The relation that R_{max} and R_{maint} bear to growth capacity as it is influenced by temperature is entirely analogous to the relation that standard and active metabolism bear to swimming capacity and temperature.

The effect of moderately restricted ration is to limit R_{max} at high temperatures without affecting R_{maint}. It can be deduced from such a growth-scope model (Fig. 10B) that the optimum temperature for growth would shift progressively to a lower temperature as ration was reduced—a phenomenon which has already been demonstrated for sockeye salmon (Brett *et al.*, 1969) and more recently supported by similar studies on brown trout (Elliott, 1975c).

B. Light

Studies on the influence of light on growth have not infrequently resulted in variable, complex, and confusing results. This appears to arise from the multiplicity of ways in which light can act (quality, quantity, and periodicity), its interaction with other environmental factors, particularly temperature, and the possible harmony or disharmony with internal rhythms of the fish—both daily and seasonal. Great discretion has to be exercised by the experimenter to avoid confounding the variables (light, temperature, size, season) and to avoid introducing serious uncontrolled or undocumented interfering responses involving activity levels, ion regulation, and possible induced prematuration; equal opportunity to feed must be presented in each experimental lot independent of light. Furthermore, past history and acclimation steps have an importance that has yet to be critically assessed and clarified.

Light usually acts as a Directive Factor stimulating brain–pituitary responses which radiate through the endocrine and sympathetic systems. Its natural periodicity undoubtedly induces the production of growth hormone (STH)* and anabolic steroids, and can influence locomotor activity in association with thyroid stimulation. Since injections of STH can produce great increases in growth rate (Chapter 9) light is a potentially powerful environmental factor, but this degree of influence has yet to be clearly shown experimentally despite results which are quite significant. More frequently than not, light periodicity has been manipulated in relation to the question of smolt transformation (salinity effect) or of inducing early maturation, with growth

---

*Somatotrophic hormone or growth hormone (GH).

responses additionally recorded (Henderson, 1963; Pyle, 1969; Wagner, 1974). The consequent interplay of Directive and Masking Factors has tended to obscure the particular role of light on growth. In consequence of the lack of definitive experiments and the absence of testing over a sufficient range of restricted rations, only the parameters G_{max} and R_{max} of the GR curve can be given any serious consideration. From these the effect on conversion efficiency can then be considered.

Evidence for the effect of light periodicity (fixed and variable) will be examined first.

1. Photoperiod and Maximum Growth Rate (G_{MAX})

In a study of seasonal variation in the growth rate of 3-year-old, hatchery brown trout, *Salmo trutta*, Swift (1955) drew attention to two striking anomalies: (1) Growth surged forward in spring while the temperature was still cold; (2) the rate fell in summer and again in autumn when the water was still warm. This occurred despite the fact that the fish were fed daily to satiation. The high correlation of increasing daylength with growth in the spring was subsequently confirmed as the stimulating agent for endocrine activity, enhanced by rising temperature as the season progressed (Swift, 1959, 1960, 1961). The decrease in growth rate during autumn could be shown to relate to the advance of maturation accompanied by an *inflection* to falling temperature and photoperiod.

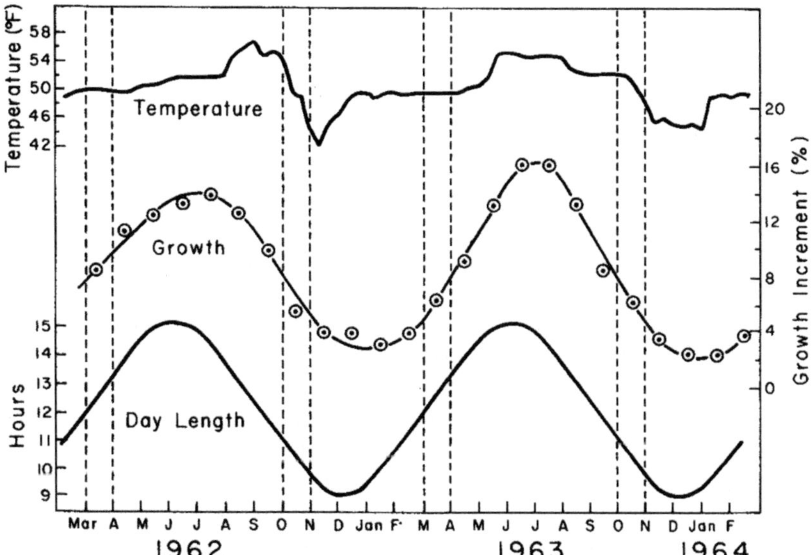

Fig. 12. Annual change in average monthly growth increments of lake whitefish, *Coregonus clupeaformis*, in relation to seasonal daylength and temperature. Fish were reared in tanks supplied with water from a natural spring. (From Hogman, 1968.)

But the growth slump in summer still remained open to speculation, in which the observation that change in spontaneous activity, thyroid depletion, and greater susceptibility to handling mortality might well provide some diagnostic clues.

During an investigation of the annulus formation on scales of four species of coregonids, Hogman (1968) also noted that seasonal change in growth rate was more closely related to daylength than to changes in partially controlled water temperature (Fig. 12). Rising in close harmony with increasing photoperiod, the peak in growth increment trailed about 1 month behind the maximum daylength; no such correlated phasing occurred in relation to some uncontrolled fluctuations in temperature (maximum change $=7^{\circ}$C).

These clear demonstrations of seasonal effects on growth in association with the *natural* photoperiod have only found moderate support from laboratory tests conducted under more rigorous environmental control. Indeed, early efforts by Brown (1946b) using static daylengths suggested an inverse relation of growth rate with daylength, whereas those of Anderson (1959) and Bjorklund (1958) appeared to show no relationship at all. Most studies indicate that, whatever the response, it takes at least 6–8 weeks to become manifest (e.g., Eisler, 1957).

One of the most searching enquiries was that by Gross *et al.* (1965) on the green sunfish, *Lepomis cyanellus*, using both fixed and variable photoperiods. From four 6-week experiments conducted during various seasons of the year, a comparison was made on each occasion between photoperiods of 16, 8, 8–16, and 16–18 L, all at $25^{\circ} \pm 1^{\circ}$C. A marginally significant difference in the weight gain occurred between 16 and 8 L ($P=0.10$) with a greater significant difference obtained between 8 and 16 L, compared with 16 and 8 L ($P<0.05$). The greatest weight gain accompanied 16 L (gain = 5.7 g) and the least with 16–8 L (gain = 4.0 g). Assuming the mean initial weight for each random sample of fish was 15 g (not given, but see Gross *et al.*, 1963) the greatest difference in achieved weight between treatments was only 8.3% in 6 weeks (difference in $G=0.2\%$ weight/day). Although not analyzed by the authors it is of interest to note that for the months selected the mean weight gains for all photoperiods were greatest in early spring (increase of 6.3 g) and least in late fall (increase of 2.7 g).

By extending the latter experiment another 3 weeks, and considering the previous 6-week experimental period as acclimation time, the difference between the growth increment of the 8 L and 16–8 L was significantly increased. This led the authors to conclude that not only did increasing daylength have a stimulating effect and decreasing daylength an inhibiting effect on growth, but also that prior photoperiod history was of demonstrable importance.

In the uniform time allotted for feeding, it was further found that the sunfish consumed more food on a longer photoperiod and were also more effective at conversion, the highest average efficiency (48%) occurring on 8–16 L. The seasonal influence on conversion efficiency was even more pronounced, with the early spring period topping each of the four test seasons with an amazing average of 72% K.

Some corroborative evidence for the growth-promoting effect of long day-length was obtained by Kilambi *et al.* (1970) for channel catfish, *Ictalurus lacustris*, using a 2 × 3 factorial experiment*. Over a period of 120 days, fry (0.07 g) of this species grew at the following mean rates† (G_{max}) according to the combination of temperature and photoperiod applied:

	26°C	28°C	32°C	Mean G_{max}
10 L	2.07	1.55	2.30	1.97
14 L	2.30	2.48	2.29	2.36
Mean G_{max}	2.18	2.02	2.30	

Although the average growth rate at the lowest temperature (26°C) for both photoperiods occupied an intermediate position between the other two temperatures (28° and 32°C) the mean rate for all temperatures at 14 L was 16% higher than that at 10 L. As reported, the relation between treatments was not without odd variability in the respective order of the growth rates (of intermediate stages) during the experiment. Chance variation resulting from small samples ($n = 5$) and the fact that the optimum temperature for growth occurs at 29°–30°C (Table I) would tend to confuse the issue. Temperatures beyond the optimum or near the limit of tolerance usually suppress growth, and may also suppress photoperiod effects. The immediate prehistory of the fish was chosen as 24 L, apparently to remove any periodicity difference between lots. The consequence of this fairly extreme pretreatment is worth investigating in itself.

There was a tendency for conversion efficiency to be higher on 14 L than 10 L, but this difference diminished with time and size of the catfish.

Turning to experiments on anadromous species, Saunders and Henderson (1970) investigated the effect of natural,‡ constant (13 L), and reciprocal photoperiod on 35-g Atlantic salmon smolts, *Salmo salar*. Growth and feeding rates (G_{max} and R_{max}) were followed over a 10- to 12-month period in two successive years. At the time of downstream migration (April–May), the smolts were transferred to full seawater (30‰); this was accompanied by an increase in temperature from 10° to 15°C after which temperature was held constant. Prior history was in freshwater on natural photoperiod and seasonal temperature increase. The prime interest was environmental control of the smolting process and the possible stimulation of rapid growth in the sea, hence the experimental design. In rigorous terms it is difficult to separate

*See also studies by Huh *et al.* (1976) on yellow perch and walleye.

†Specific growth rates recalculated from length–weight equations applied to the mean starting and terminal sizes, using a least squares projection from the last five samples at each temperature.

‡The authors have used a steady rate of increase rather than the accelerating and decelerating normal rates of increase as the season approaches the longest day.

seasonal, photoperiod, temperature, and size effects; however, segments of the experiment are comparable and lead to certain conclusions. At 15°C, *in seawater*, prior to June 21, the reciprocal (decreasing) photoperiod induced a small but significant increase in weight over the other photoperiod effects ($G=1.37\%$ weight/day versus 1.29% weight/day); when applied after June 21 a lesser weight increase resulted than on natural photoperiod, but not different from a constant 13 L.

When reciprocal photoperiod was applied during 2.5 months *in freshwater* subsequent growth in saltwater was comparatively retarded ($G=0.31\%$ weight/day). Under these circumstances no difference between natural and constant photoperiods occurred, both of which were accompanied by good growth for the size involved ($G=0.85\%$ weight/day). Although no differences in plasma chloride concentrations could be detected in any of the fresh- and saltwater comparisons, it was apparent that the reciprocal photoperiod had affected the success of the smolting process prior to saltwater transfer, and this negative effect carried through to affect growth in salt water.

Little difference in conversion efficiency between treatments could be detected. A decrease from an average of 31 to 16% K took place as the year progressed and fish weight rose to over 700 g. Some decrease in K with increasing weight can be expected.

Knutsson and Grav (1976) examined the effects of increasing photoperiods (6–19, 8–19, and 12–19 L) on yearling Atlantic salmon held in freshwater at three temperatures (7°, 11°, and 15°C). The increasing daylengths were applied in the fall–winter season, with the inflection to 12–19 L commencing 3 months ahead of the more normal 6–19 L (at Bergen, Norway, 60°N. Lat.). The temperature effect was considerably greater than the photoperiod effect. Greatest growth occurred at 15°C on the 12–18 L, which was also the regime with most advanced, seasonal inflection.

In general, for freshwater fish, the evidence indicates that long daylength, and more especially increasing daylength applied over a number of months (particularly in the right season) is stimulating to growth. But the effects at best are not large. Decreasing daylengths have an inhibiting effect on some freshwater fish. The lack of greater induced response, compared with natural seasonal effects on normal populations (separate from temperature effects), suggests that experimental designs are somehow inadequate, or that a circannular rhythm is present and not subject to displacement.

For anadromous fish such as salmonids, if the size and season are right, decreasing photoperiod improves the smolt transformation allowing greater growth in saltwater. But this is not so if applied in a freshwater environment at a time when the seasonal norm is the reverse photoperiod (i.e., increasing daylength).

It is possible to generalize that at the time of transformation the stimulation of neuroendocrine pathways related to either sodium excretion (ACTH, prolactin), or maturation (gonadotropin), are partially and temporarily antagonistic to growth hormone production. However, as intermediate developmental stages are achieved, growth can resume to a greater degree (see Chapter 9).

2. Light Intensity (without Periodicity) and Growth Rate

Although light intensity has received considerable attention in relation to both egg development (MacCrimmon and Kwain, 1969) and meristic characters (Lindsey, 1958), little research has been performed on the growth relations.

Eisler (1957) reviewed earlier literature noting that chinook salmon were reported to grow better under "dark-reared" than "light-reared" conditions, and that cod fry were attracted to artificial light. In a 12-week experiment involving exposure to four light intensities (0.02, 88, 116, and 157 f. c. fluorescent light, λ 3350 – 6000 Å; no photoperiod stated) Eisler reared chinook salmon fry on raw beef liver. Only the lowest intensity ("dark") showed a significant difference in growth, achieving only 68% of the weight of the "light" intensities. He concluded that high light conditions stimulated growth, apparently not conceiving that low light might have an inhibiting effect.

Kwain (1975) subjected hatched rainbow trout to three levels of light (0.2, 2, and 20 lx) at two temperatures (3° and 10°C). Significantly reduced growth rate was reported at the lowest intensity only, for both temperatures. No significant difference occurred between 2 and 20 lx. Trout reared at 0.2 lx were observed to be sluggish and not actively seeking food as at higher light intensities. If reduction in visual perception was the cause, as speculated, light intensity by itself can be dismissed as a factor impinging on growth directly.

C. Salinity

Fishes regulate their plasma ions such that the internal osmotic pressure of their body fluids is equivalent to approximately 10‰ salinity, with a range of ±2‰ depending on tolerance, regulating capacity, and environmental salinity (Holmes and Donaldson, 1969). Maintenance of internal balance in freshwater, where loss of ions and bodyflooding are the major problems, is associated with the hormone *prolactin* which serves to control membrane transport and kidney function (Bern, 1975). A highly dilute urine is excreted with a concomitant small loss of salts which are replaced by the diet. Strictly freshwater fish are stenohaline (here abbreviated to Fr–St), being subject to rapid death in normal seawater.

In the marine environment where the problem of ion control is reversed, fishes survive by drinking saltwater, secreting chlorides mostly via the gills, and excreting an isotonic urine. Prolactin is suppressed; the ATPase enzymes for sodium and potassium occur abundantly in the mitochondrial-rich cells of the gills. The majority of marine fishes cannot tolerate the brackish waters of estuaries indefinitely and are therefore also stenohaline (Sa–St). Some marine species, however, are very tolerant, remaining in estuaries and occurring part way up rivers during the freshwater cycle at low tide. It is of interest to note that two species of predatory sea basses, *Dicentrarchus* sp., have been used to control excessive fry production in freshwater reservoirs (Chervinski, 1975). Such tolerant species are labeled saltwater euryhaline (Sa–Eu).

Some freshwater species, such as those occurring among the Salmonidae, commence life in freshwater, where they are spawned, and subsequently pass through a transformation (smoltification) at the time of migrating to sea—either as fry (some *Oncorhynchus* sp.) or as smolts (some *Oncorhynchus* sp. and *Salmo* sp.). These pass from freshwater stenohaline to saltwater euryhaline and are here referred to as anadromous euryhaline (An–Eu).

A few species apparently of freshwater origin* occur in a wide range of salinities, even tolerating the highly saline conditions of inland seas where evaporative loss may result in salinities of 60‰ and greater. These euryhaline fishes have been labeled Fr/Sa–Eu.

Since internal ion regulation is involved, despite the rare natural possibility of osmotic equilibrium in certain brackish waters, salinity must be classed as a Masking Factor, constantly requiring some energy expenditure associated with active transport of ions to maintain the internal melieu.

Studies on the effect of salinity on growth tend to be scattered, somewhat conflicting, and frequently lacking in any record of the internal ionic state. Although some measure of salinity tolerance may be included, the ion regulatory capacity is almost never ascertained. These ancillary determinations are not essential to the basic observation of salinity–growth responses. But the variability in reported response has been such that uniformity of physiological state is apparently not present in many cases, and hence the explanation of some of the variability is denied. Because temperature as a Controlling Factor and photoperiod as a Directive Factor enter into the consideration, as well as prior exposure to some intermediate saline concentration (acclimation), there is need for wide-ranging systematic study of this environmental entity. As for the two previous environmental factors, the effect of G_{max} will be considered first.

1. *Maximum Growth Rate* (G_{MAX}) × *Salinity*

A wide variety of species have been tested for the effect of salinity on growth rate, when provided with an abundance of food (R_{max}). In particular, the anadromous fishes have received attention because of their capacity to transform from a fresh to a saltwater habitat, and their suitability for aquaculture. This is also true of at least one species of catadromous fish, the grey mullet (De Silva and Perera, 1976). Other interest has centered on the wide-ranging capacity for the estuarine and euryhaline species of freshwater origin to grow in saline environments. Since the number of searching enquiries is not great, the ecological diversity of the species studied has tended to diffuse the picture. From the compilation in Table II the greatest growth rates within species cluster either around zero salinity (freshwater), or 10 ± 2‰, or

---

*Or originally inhabiting seawater and penetrating freshwater secondarily (e.g., *Tilapia*, Chervinski, 1961).

Table II Rate of Growth (G_{max}) of Different Species of Fish Exposed to Various Salinities and Fed a High Ration[a]

Species	Type	Initial size		Salinity acclimation		Test temperature (°C)	Test duration (days)	Growth rate for salinity ranges (‰)						Remarks	Reference
		cm	g	‰	Days			0	1–6	7–12	13–18	19–35	35+		
Poecilia reticulata	Fr–St	0.7	—	0	—	24	40	1.25*	—	1.35[(8)]	1.25[(16)]	—	—	* Interpolated lengths (cm) achieved in 40	Gibson and Hirst (1955)
Carassius auratus	Fr–St	8.4	12.1	0	—	20	70	+	+[(6)]	—	Lethal[(15)]	—	—	Slight benefit in fresh at 35 days; none by 70 days	Canagaratnam (1959)
Oncorhynchus kisutch	Au–Eu	3.6	0.47	0	—	10	70	1.65	1.90[(6)]	2.41[(12)]	—	—	—	All fish fed 10% weight day; no acclimation	Canagaratnam (1959)
Oncorhynchus kisutch	An–Eu	4.2	1.01	9	2	10	70	1.13	—	—	1.58[(18)]	—	—	Initial weight doubled	Canagaratnam (1959)
Salmo salar	An–Eu	10.7	15	0–22	—	8–4*	126	0.23	—	0.22[(7)]	0.16[(15)]	0.16[(15)]	—	* Seasonal change (not controlled), Dec. to April	Saunders and Henderson (1969)
Oncorhynchus kisutch	An–Eu	3.0	1.0	0.25	1	10	32	1.60	1.65[(5)]	1.55[(10)]	1.15[(15)]	1.10[(20)]	—	Tested in Aug.–Sept. (presmolt)	Otto (1971)
Oncorhynchus kisutch	An–Eu	6.0	—	0.25	1	10	32	0.12	0.55[(5)]	0.60[(10)]	0.10[(15)]	0.10[(20)]	—	Tested in Jan.–Feb. with larger fish (smolt)	Otto (1971)
Oncorhynchus tshawytscha	An–Eu	—	0.5	0–24	80	11	70	2.7	—	—	2.6[(17)]	2.2[(33)]	—	Intermediate, increasing salinity acclimation applied	Kepshire and McNeil (1972)
Oncorhynchus tshawytscha	An–Eu	—	0.8	5–25	66	12	28	—	—	—	3.0[(18)]	2.4[(33)]	—	Reduced growth rate in full seawater despite acclimation	Kepshire and McNeil (1972)

Continued

Table II Rate of Growth (G_{max}) of Different Species of Fish Exposed to Various Salinities and Fed a High Ration—Cont'd

Species	Type	Initial size		Salinity acclimation		Test temperature (°C)	Test duration (days)	Growth rate for salinity ranges (‰)						Remarks	Reference
		cm	g	‰	Days			0	1–6	7–12	13–18	19–35	35+		
Salmo salar	An–Eu	—	40	0–20	21	10	56	$\underline{0.7}$	—	$0.5^{(10)}$	—	$0.5^{(20)}$	—	No significant difference reported (Oct.–Nov.)	Shaw *et al.* (1975*b*)
Oncorhynchus keta	An–Eu	—	0.35	0–28	7	15	35	5.7	—	—	—	$\underline{5.9^{(28)}}$	—	Not significantly different	Shelbourn (1976)
Morone saxatilis	An–Eu	Fry*		0.2–4.8	5+	(18)**	7	—	$9.6^{(6)}$	$\underline{10.7^{(12)}}$	—	$9.4^{(20)}$	—	* Fry aged 5 to 63 days; ** 3 test temperatures, 12°, 18°, 24°C	Otwell and Merriner (1975)
Mugil cephalus	Ca–Eu	—	0.2	<20*	—	25	360	4	—	$4^{(10)}$	$\underline{5^{(20)}}$	$0.3^{(30)}$	—	* Caught in coastal lagoons; conversion efficiency greatest at 10‰	De Silva and Perera (1976)
Cyprinodon macularius	Fr/Sa–Eu	0.8	0.02	0–55*	70	30	84	1.51**	—	—	$1.86^{(15)}$	$\underline{2.06^{(35)}}$	$1.78^{(55)}$	* Brood stock acclimated; ** length achieved	Kinne (1960)
Cyprinodon macularius	Fr/Sa–Eu	0.8	0.02	0–55*	70	15	84	$\underline{1.55^{*}}$	—	—	$1.30^{(15)}$	$1.02^{(35)}$	—	best in freshwater when at low temperature (15°C); * length achieved	Kinne (1960)
Tilapia mossambica	Fr/Sa–Eu	—	0.22	0–35	2	25*	56	2.24	—	$3.17^{(9)}$	$3.96^{(18)}$	$\underline{4.06^{(27)}}$	$3.23^{(35)}$	* Assumed room temperature (not given)	Canagaratnam (1966)

Trinectes maculatus	Fr/Sa–Eu	3.5	0.7	0–30	4	15	7–14	0.53	—	—	1.03$^{(15)}$	1.40$^{(30)}$	—	Only three fish per test used (usually)	Peters and Boyd (1972)
Trinectes maculatus	Fr/Sa–Eu	3.5	0.7	0–30	4	35	7–14	3.14	—	—	2.59$^{(15)}$	2.09$^{(30)}$	—	Large temperature effect in freshwater	Peters and Boyd (1972)
Paralichthys dentatus	Sa–Eu	—	0.02	25–35	—*	20	4–7	—	7.5**$^{(5)}$	8.0$^{(10)}$	8.7$^{(15)}$	10$^{(30)}$	10.5$^{(35)}$	* Held through post- larval stages; ** interpolated	Peters (1971)
Paralichthys dentatus	Sa–Eu	—	0.02	25–35	—	30	4–7	—	12.8*$^{(5)}$	14.8$^{(10)}$	15.2$^{(15)}$	17.4$^{(30)}$	18.2$^{(35)}$	* All values interpolated from graphs	Peters (1971)
Anisotremus davidsoni	Sa–Eu	—	1.5	29–45	14	25	14	—	—	—	—	2.2$^{(33)*}$	1.2$^{(45)}$	* Tested at 5 salinities from 29–45‰.	Brocksen and Cole (1972)
Bairdiella icistia	Sa–Eu		5.0	29–45	14	25	14	—	—	—	—	3.7$^{(29)}$	3.0$^{(45)}$	* Same as above	Brocksen and Cole (1972)

[a]Most cases are expressed as specific growth rate in weight/day, but some are in terms of length achieved (as indicated under Remarks). Selected salinity ranges are indicated, with the particular salinity shown as a superscript in parentheses. Each species is classified according to ecological type, using the abbreviations: Fr; freshwater; Sa, saltwater; An, anadromous; Ca, catadromous; St, stenohaline; Eu, euryhaline. Species that normally occur in freshwater but are found abundantly in saltwater have been labeled Fr/Sa–Eu. Because initial size, acclimation state, test temperature, and duration of experiment all vary greatly between species, and can be shown to interact within species, comparison is best made as a *relative* effect of salinity on growth, as in Fig. 13. The highest value for growth in any one species is underlined. Species are arranged in ecological order, from freshwater to marine. Asterisks refer to Remarks column.

28–35‰ (see references in Table II). These roughly distinguish between the ecologically separable freshwater stenohaline species, the anadromous species, and the euryhaline and strictly marine species (Fig. 13). On excess ration, greatest growth rate occurred at 20‰ in *Mugil cephalus*, whereas highest conversion efficiency was reported at 10‰; salinity concentration affected the level of maximum ration (De Silva and Perera, 1976).

The grouping of the freshwater species includes the presmolt stage of the anadromous fishes (e.g., *Salmo salar*). There may be some slight benefit in the isosmotic range, but above this concentration G_{max} falls off; indeed salinities in the region of 15‰ may be lethal (e.g., *Carassius auratus*) for both osmotic and ionic reasons. Such a euryhaline species as *Cyprinodon macularius* can grow comparatively well anywhere from 0 to 55‰; *Tilapia mossambica* almost matches this capacity. Chervinski (1961) reported that *Tilapia nilotica* would grow as well in 18‰ as in freshwater (avg. $G_{max} = 2.4\%$ weight/day; weight $= 25$ g; temperature $= 24 \pm 3^{\circ}$C), and would adapt to at least 50‰. No records for the minimum salinity permitting growth for strictly marine species has been found; there is a suggestion in the response of *Paralichthys dentatus* at 20°C that a falling off occurs at low salinities, as depicted in Fig. 13.

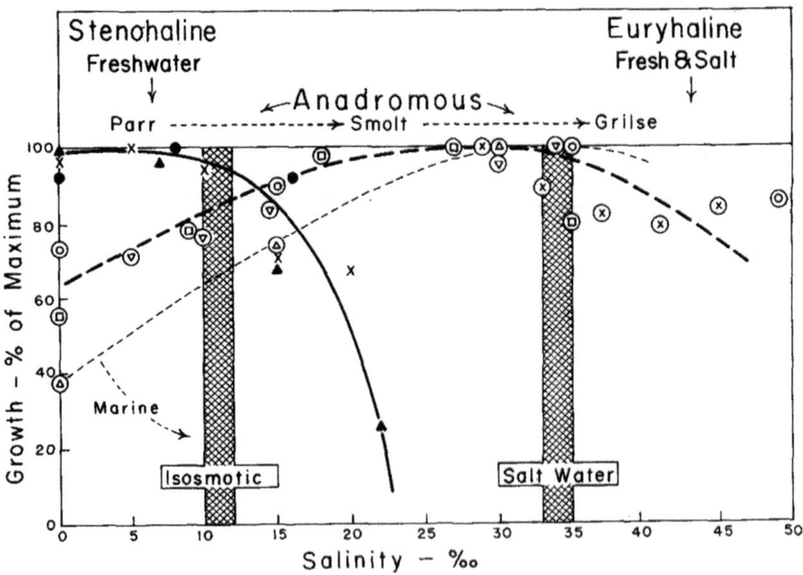

Fig. 13. Effect of salinity on growth of various fishes expressed as a percentage of the maximum rate determined for each species. Data and sources in Table II. Solid line is for stenohaline, freshwater fish, and parr stage of salmon (species: *P. reticulata*, *O. kisutch*, *S. salar*). Heavy broken line and all circled points apply to euryhaline fish distributed in freshwater and saltwater environments including the postsmolt anadromous salmonids (grilse) (species: *C. macularius*, *T. mossambica*, *T. maculatus*, *P. dentatus*, *B. icistia*). Light broken line is for an estuarine species, and indicates the likely shift in relation to a fully marine stenohaline species.

At the time of smolting the salmonid fishes shift to tolerating a comparatively high salinity, accompanied by good growth capacity. Such an enhanced ability to grow appears to characterize the marine stage, quite separate from any circumstance of abundance of food. This is supported by pen-rearing studies at sea (Falk, 1968; Novotny, 1975). These general relations, however, do not take into account the interacting effect of elevated temperature which, in the case of *Cyprinodon macularius*, favors a high salinity to enhance growth whereas the reverse is true of *Trinectes maculatus* (Table II). In this species high temperature will only support good growth in freshwater. These findings serve to reveal some of the basis for the great diversity between species recorded so far.

2. Restricted Rations × Salinity

In separate studies on the effect of salinity and of dietary sodium chloride on the growth of Atlantic salmon parr at 10°C, Shaw *et al.* (1975a, b) used daily rations from 0 to 3% weight/day for fish in the 40–60 g range. Surprisingly, no undue consequence of the greatly increased dietary loading with salt was observed at the selected environmental salinities of 0,10, and 20‰ (cf. Basulto, 1976; Zaugg and McLain, 1969). Possibly the normal saltwater drinking rate of 5–13% weight/day (Shehadeh and Gordon, 1969) reflects such a well-developed ion secretory capacity. Despite the age and size of the Atlantic salmon, some lack of tolerance above a salinity of 22‰ was reported, with a few deaths occurring at 30‰ (in May and June). No significant effect of salinity on the GR curve could be demonstrated up to 20‰. At 30% the curve was shifted to the right (Fig. 14) along the abscissa in

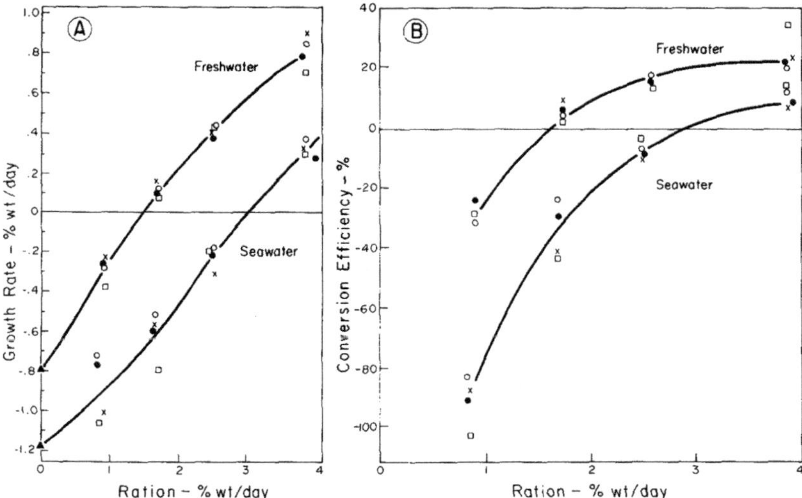

Fig. 14. (A) Growth–ration (GR) curves for Atlantic salmon parr raised in freshwater and seawater (30‰) at 10°C. (B) Derived gross conversion efficiency (K) curves. (From Shaw *et al.*, 1975a.)

accordance with an increase in maintenance ration from 1.3% weight/day (fresh) to 3% weight/day (salt). A great depression in conversion efficiency occurred at all ration levels in saltwater. The nature of the relation suggests that the highest ration provided was in the region of R_{opt}, not R_{max}. The level of R_{max} and therefore G_{max} was not defined, yet the growth rate in freshwater was comparable with sockeye salmon G_{max} for the same weight and temperature (Brett, 1974).

In contrast to the above results with Atlantic salmon, Smith and Thorpe (1976) obtained a significant improvement in the growth rate of yearling rainbow trout, *Salmo gairdneri* (over 40 g), when acclimated to saltwater and compared with freshwater controls at 12°C. The increased growth rate only occurred in late summer following parr-smolt transformation, and at maximum ration; no advantage occurred under restricted rations. Analysis indicated that the higher growth rates in saltwater resulted from an improved ability to retain assimilated nitrogen and not from increased consumption.

The progressive increase in the salinity of the Salton Sea (37‰)* brought Brocksen and Cole (1972) to examine the effect on the growth rate of two introduced sports fish—bairdiella, *Bairdiella icistia*, and sargo, *Anisotremus davidsoni*. Each species was tested at five salinities (29–45‰) and five rations (1–10% weight/day) at 25°C. Both species grew best at salinities lower than the existing environmental level (Table II); sargo, however, required a particularly high R_{maint} at all salinities (avg. = 7.4% weight/day) which would undoubtedly affect its success. Some interaction between ration and salinity occurred for *Bairdiella*, a salinity of 37‰ being worst at high rations and best at low rations.

Peters (1971) and Peters and Boyd (1972) have used restricted rations in experiments on the combined effects of temperature, salinity, and food availability on the growth of young flatfish, *Trinectes maculatus*. By examining the response at an intermediate temperature (20°C) it is apparent that the slope of the GR curve increases with increasing salinity such that, from a common pivot point of similar maintenance rations, an increase in salinity induced a higher G_{max} when on ad *libitum* feeding. When starved, at 15°C, *Trinectes maculatus* lost weight faster in freshwater than at higher salinities; the reverse was observed at 35°C.

These illustrations serve to point up the complexity and diversity of the growth response to salinity, a circumstance that is likely to require searching physiological analysis for adequate explanation. Whatever this may be, it is of interest to note that in nature these species of flatfish occur most frequently at salinities where they grow fastest (Peters, 1971).

It can be concluded that, with the possible exception of the smolting stage in salmonids, evidence for any substantial increase in growth accompanying

---

*Expected to be 40‰ by 1975, i.e., increasing at a rate of 1‰ per year.

isosmotic conditions is lacking (isionic conditions have not been examined). Pronounced decreases in G_{max} with increasing salinity, among freshwater adapted species, appear to be the result of large increases in R_{maint}. As the limit of salinity tolerance is approached and regulation becomes progressively inadequate, R_{max} falls to the level of R_{maint} blocking further growth.

D. Oxygen

At the time of feeding and for some hours thereafter the metabolic rate of fish is elevated (Chapter 6). Some authors deduced that for maximum rations the increased demand for oxygen could reach as high as the active metabolic rate (Paloheimo and Dickie, 1966). While this may be true in nature where the daily encounter of predator and prey can involve numerous bursts of activity, in the laboratory the maximum metabolic requirement does not appear to exceed two or three times the standard rate (about one-third active metabolism). The total daily increase in oxygen consumption is directly related to the size of the meal; like any other activity, as the load increases the energy expenditure goes up (Chapter 6). One aspect of the metabolic process, *specific dynamic action*, has not always been clearly defined and measured for fish (Warren, 1971; Beamish, 1974). Despite the sequence of complex digestion–absorption–transformation steps in food processing, environmental oxygen can be shown to act as a simple Limiting Factor, sharply curtailing growth and food conversion efficiency at critical oxygen levels, usually well below the air-saturation point. However, oxygen cannot be confined *a priori* to this single role. It is conceivable that reduced oxygen content could act as a cue (Directive Factor) for reduction of appetite—or more likely some associated change accompanying lowered oxygen, since the existence of O_2-sensors among fish has yet to be confirmed.

1. *Maximum Growth Rate (G_{MAX}) × Oxygen*

While many studies have been conducted on the effects of oxygen supply on the growth rate of fish, these are very varied in relation to duration, size, and survival of fish, nature of diet, temperature applied, and the level and precision of environmental oxygen control, such that useful systematic tabulation is difficult if the comparability is to be preserved (see Brungs, 1971; Swift, 1963, 1964; Doudoroff and Shumway, 1967, 1970; Warren, 1971; Ebeling and Alpert, 1966; Davison *et al.*, 1959). The studies involving late embryonic and early larval stages, where first feeding occurs, are complicated by the change in developmental state during the test period and the initiating of an adequate feeding response. Further, there is a bias introduced where only the size of the surviving fish can be used to determine growth. Some increased sensitivity to reduced oxygen is apparently present at the larval stage in comparison with the juvenile stage; the margin, however, is not great (Carlson and Siefert, 1974; Carlson *et al.*, 1974).

From three well-documented studies on juveniles it has been possible to determine the mean and variance of the "growth index" applied in each case (Chiba, 1966; Stewart *et al.*, 1967; Herrmann *et al.*, 1962). This was done by grouping according to concentration* intervals of 0.5 ppm O_2, and plotting the results as the mean of G_{max} against the mean value of the concentrations within each of the O_2 intervals (Fig. 15). Where only two values occurred in a given interval, the range was used. Despite the considerable deviation accompanying each mean determination (partly due to experimental variability in O_2 level) it is clear that an oxygen concentration of close to 5 ppm is critical for growth, below which increasing suppression of G_{max} is directly proportional to decreasing O_2 concentration—a drop of 1 ppm causes a 30% reduction in growth rate. The analysis bears further comment. None of the authors would dispute the O_2-dependent segment of the curves as depicted. However, the balance of the relation might be contested, here shown as

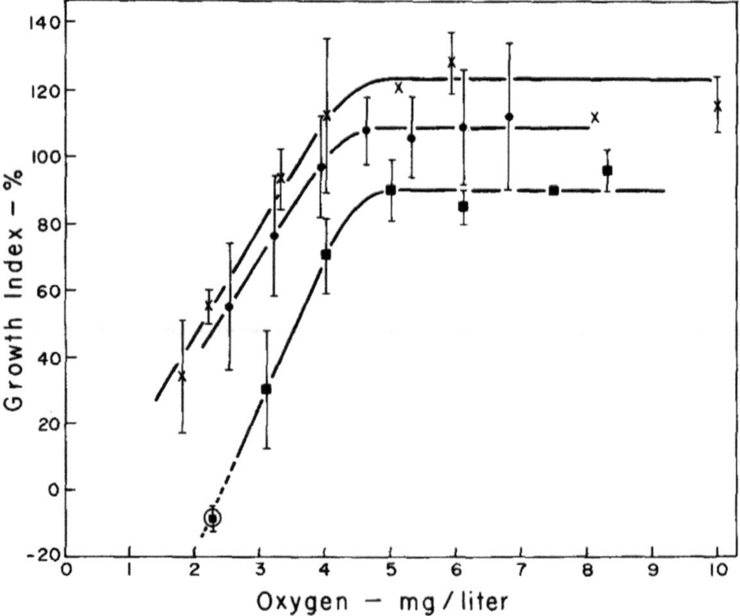

Fig. 15. Relation of oxygen concentration to growth rate expressed as the percentage of the "growth index" developed by each author. Limits $= \pm 1$ SD. Data for *Micropterus salmoides* (26°C; 2.5– 4.5 g—upper curve) from Stewart *et al.* (1967), for *Cyprinus carpio* (22°C; 0.5–3.4 g—middle curve) from Chiba (1966), and for *Oncorhynchus kisutch* (20°C; 2–6 g—lower curve) from Herrmann *et al.* (1962); circled point was accompanied by significant mortalities. The vertical positioning by species has no relative significance. Lines drawn according to interpretation presented in text.

*Since fish have a great ability to extract oxygen from water, concentration is a better indicator of available amount than saturation.

forming a plateau beyond the critical transition zone (depicting complete *independence* of O_2 concentration above 5 ppm, as in simple limiting cases). In two of the original sources (Stewart *et al.*, 1967; Herrmann *et al.*, 1962) and again in Doudoroff and Shumway (1970) the data for largemouth bass are interpreted by the authors as rising to a peak of growth close to 8 ppm (100% saturation) and falling off at concentrations far in excess of this, for example, 17 ppm at 26°C (212% saturation). It is known that excessive oxygen concentrations can even be lethal (Hubbs, 1930). However, from the variance of the small samples involved above the 5 ppm critical level, there is insufficient evidence to negate the present interpretation. Support for the latter can be gained from the fact that oxygen concentration would have to be reduced to the critical level indicated (approximately 5 ppm) in order to act as a Limiting Factor to the associated metabolic rate (see Chapter 6). It should be noted that *experimentally reduced* oxygen concentration is maintained in the complete absence of any change in the rest of the controlled environment. In nature, and in hatcheries, decreased oxygen would likely be accompanied by an increase in other environmental factors such as ammonia, urea, and nitrites which would act antagonistically to growth.

The presence of a definite upper plateau, supporting the Limiting Factor concept, is clearly depicted in the conversion efficiency relation in Fig. 16.

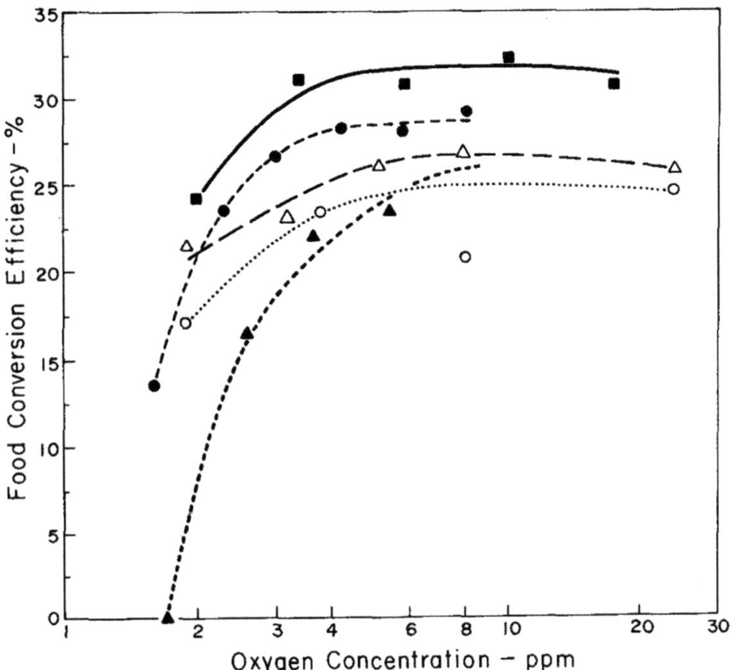

Fig. 16. Food conversion efficiency in relation to oxygen concentration. Results are for separate experiments conducted on *Micropterus salmoides* at 26°C. (From Stewart *et al.*, 1967.)

The values obtained above 5 ppm are in keeping with the high efficiency that usually accompanies good growth when young fish are fed a nutritious diet. Tests conducted on a small number of juvenile Northern pike, *Esox lucius*, by Adelman and Smith (1970) follow the same pattern as the three species illustrated in Fig. 15. Evidence for reduction in food consumption and conversion efficiency for this species suggests a critical level between 3 and 4 ppm O_2 (at 19°C).

2. Restricted Rations × Oxygen

Since ration is the "driving force," any restriction obviously reduces the growth opportunity to a lower level; the GR curve is truncated. As was pointed out, a restricted ration is also accompanied by a lowering of the daily metabolic rate and consequently a reduced demand for oxygen. Thus, it might be expected that the critical oxygen level would fall. Fisher (1963) demonstrated this clearly for underyearling coho salmon (Fig. 17). On a limited ration, which resulted in approximately 0.4 G_{max}, the critical oxygen level dropped to between 3 and 4 ppm. If the positions of inflection inferred by Fisher (1963) are used, intermediate levels of restricted ration would be expected to follow the paths indicated in Fig. 17 (dotted lines).

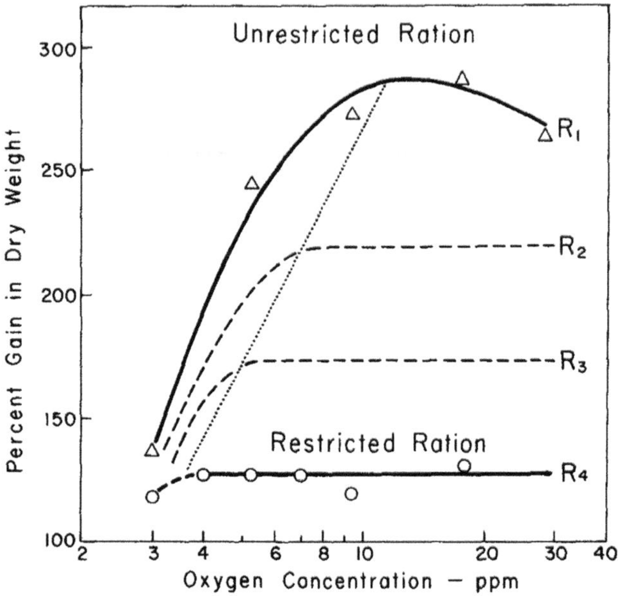

Fig. 17. Growth rate of coho salmon, *Oncorhynchus kisutch*, on unrestricted ration (R_1) and restricted ration (R_4) in relation to oxygen concentration. (Modified from Fisher, 1963; reproduced in Doudoroff and Shumway, 1967.) The dotted construction line has been added, joining the points of inflection for R_1 and R_4. The expected relations for intermediate restricted rations (R_2 and R_3) have been drawn as broken lines.

The general phenomenon witnessed under these circumstances is the interrelation of two Limiting Factors—food and oxygen—acting in series. Successive plateaus below G_{max} are set by the degree of restricted ration. But as oxygen is reduced, an O_2-dependent stage enters to depress the growth rate below that previously dictated by the limited ration (depicted by the slope of the line below the plateau, Fig. 17).

This phenomenon can be identified in the response of channel catfish exposed to three oxygen concentrations and two feeding regimes (Andrews *et al.*, 1973). At 26.6°C the saturations imposed convert to the following concentrations: 100% = 7.9 ppm O_2, 60% = 4.7 ppm O_2, and 30% = 2.8 ppm O_2. On *ad libitum* feeding (R_{max}) there was a significant reduction in growth rate between each O_2 level, which would be expected for concentrations below the critical O_2 level of 5 ppm (results: G = 3.1% weight/day at 7.9 ppm, 2.7% at 4.7 ppm, and 1.8% at 2.8 ppm). When fed a fixed ration of 3% weight/day there was no significant difference between the two higher O_2 concentrations (avg. G = 1.8% weight/day); only the growth rate at 2.8 ppm O_2 was depressed (G = 1.3% weight/day). Thus, the limiting effect of reduced oxygen was shifted to a lower concentration when on restricted ration.

3. *Varying Oxygen × Growth*

Daily oscillations in oxygen content are not unusual in nature, frequently produced through the light cycle on photosynthesis. Stewart *et al.* (1967) subjected largemouth bass to alternately low (2–4 ppm) and higher (4–8 ppm) oxygen concentrations and showed that growth was markedly impaired. The reduction was greater than that which would have occurred if kept at the mean O_2 level, showing the detrimental consequences of the low concentrations when these were below the critical level. When the variation was entirely below the critical value (e.g., 1.8–3.7 ppm) the growth was only 32% of that expected for a concentration equivalent to the mean daily O_2 level.

Whitworth (1968) demonstrated a similar inhibiting effect on the growth of brook trout, *Salvelinus fontinalis*. When subjected to fluctuating oxygen levels (10.6 ppm reduced to either 5.3 or 3.6 ppm) the fish lost weight and were approximately 75% of the weight achieved by a control group at 10.6 ppm (for 49 days). Diurnal fluctuations from 3.0 ppm O_2 rising to either 9.5 or 18 ppm O_2, applied by Fisher (1963) to underyearling coho, resulted in an almost equally depressed growth rate, similar to that which would have occurred at fixed levels of 3.5 and 3.9 ppm, respectively.

It is apparent from these few but revealing studies on fluctuating O_2 that high concentrations above air-saturation do not confer any substantial benefit compensating for the periods of low concentration. Further, an exposure to

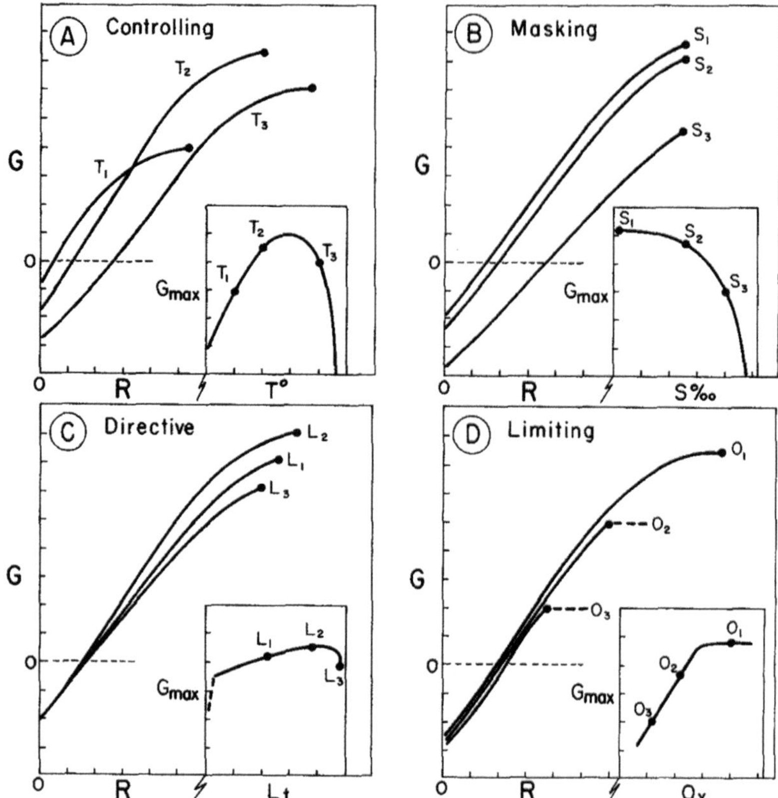

Fig. 18. Recapitulation of GR curves in relation to abiotic entities illustrating the basic forms for: (A) *temperature* (T°), a Controlling Factor; (B) *salinity* (S‰), a Masking Factor; (C) *light* (L_t= static photoperiods), a Directive Factor; and (D) *oxygen* (O_x), a Limiting Factor. Each GR curve is shown at three levels of each abiotic entity, with a point for the position of G_{max} terminating the top of the curve. The insert box shows the path of change of G_{max} as the entity increases from minimum to maximum values (e.g., as temperature increases over the tolerable range) in A. The paths of change for R_{maint} and G_{starv} can be seen by the points of intersection with the respective lines for G_0 (horizontal broken line) and R_0 (*y*-axis). Curves have been drawn where possible from data and figures presented; the circumstance for levels of fixed photoperiod remains speculative. *G*, the specific growth rate, and *R*, the ration, are both represented in terms of % body weight/day (or cal/kcal/day). Maximum values of *G* range up to 10% weight/day and of *R* to 30% weight/day for young fish.

subcritical levels of oxygen for only a portion of the day (e.g., 8–12 hr) is sufficient to depress the growth rate to that comparable with the constant low O_2 level. And this is despite feeding during the high O_2 period. There is obviously not a simple on–off effect without a carryover of serious consequences into the higher O_2 period.

E. Summary Configurations

More information is available on the effects of temperature on growth than any other abiotic factor; oxygen is next, with light and salinity not at all well documented. Within these limitations an attempt has been made to present the various patterns that the GR curves follow for each environmental factor, and the consequent path of the major parameter G_{max} (see Fig. 18). These illustrate the categories of effect (Factors) postulated by Fry (1947).

In the case of light (L) only the effects of static photoperiods are indicated, whereas dynamic states of changing photoperiod are undoubtedly more important. The consequences of the extremes (complete darkness or 24 hr light) are indicated as depressive in the former and suboptimal in the latter. Rasquin and Rosenbloom (1954) have recorded the long-term stress effects on fish held continuously in darkness.

By way of summary and recapitulation to this point, the GR curves can be seen to alter in four quite distinct patterns, according to the category of the abiotic factor (i.e., how the abiotic factor acts). Thus, in Fig. 18A, where temperature is acting as a Controlling Factor, the GR curve not only shifts to the right as R_{maint} increases but also rises and subsequently falls as G_{max} passes through the optimum temperature. In B, where salinity as a Masking Factor is imposing an increasing load on a freshwater species, the GR curve also shifts to the right as R_{maint} increases, but there is no optimum salinity; G_{max} falls off rapidly depressing the whole curve. In C, where light is depicted as a Directive Factor, there is a common origin of G_{starv}, little effect on R_{maint}, and an optimum daylength where the GR curve rises to a maximum in all respects. In D, where oxygen is acting as a Limiting Factor, the curves are superimposed (here shown slightly separated) and truncated at intermediate plateaus of growth according to the depression of ration induced by lowered oxygen concentrations. Any environmental factor acting within any of the four categories shown would cause the same sort of response of the GR curve.

IV. BIOTIC FACTORS

A. Ration

1. Quantity (R_{MAX})

Up to this point the use of maximum growth rate (G_{max}) obtained by providing a maximum ration (R_{max}), or something over and above that amount (*ad libitum*, or excess), has been used in a conceptual or assumed sense. Critical examination reveals a number of factors related to feeding frequency and quantity that bear on the maximum daily intake of fishes. The most important of these factors include (a) the duration of a given feeding (satiation time), (b) individual meal size (stomach capacity), (c) time between meals (feeding interval), and (d) the interaction of these. To these must be added the consequences of abiotic and biotic factors, among which temperature and fish size are of greatest importance. In practice it is not uncommon to feed very young fish almost continuously on a fine feed "mash," switching to interval feeding on progressively larger pellets as fish size increases.

It should be noted further that the conceptual approach to growth and feeding relations adopted in this chapter assumes that the prime demand for food is imposed by the maintenance requirements of the fish, with a further demand dictated by the potential growth capacity (influenced by growth hormone). These interlocking requirements set the limits of voluntary food intake, not the reverse, that is, not the case of appetite governing growth rate (compare Chapter 3).

As any experimenter will attest, when dealing with various feeding rates and groups of fish, the problem presented on restricted rations is that of providing a fair share to each ravenous member of the sample, never the total daily intake. However, as R_{max} is approached, the question of an ultimate level becomes more elusive, subject to many minor forms of disturbance, in which taste, size, shape, and movement of food particles are important as well as the elimination of any unusual sound or light stimuli. At this level of daily consumption it takes very little to put the fish off feeding to maximum capacity.

A number of species of fish have been shown to require up to an hour or more to reach satiation, for example, *Trachurus japonicus* and *Salmo gairdneri* (Ishiwata, 1968), *Oncorhynchus nerka* (Brett, 1971a), and *Salmo trutta* weighing 100 g or more (Elliott, 1975a). Other species such as the puffer, *Fugi vermiculatus*, the filefish, *Stephanolepis cirrhifer* (Ishiwata, 1968), and relatively small *Salmo trutta* (about 15 ± 10 g, Elliott, 1975a) were satiated in 15 min or somewhat less. The influence of weight and temperature on satiation time has been shown to be highly significant in the response of *Salmo trutta* (Elliott, 1975a), whereas weight did not appear to have a strong influence in *Oncorhynchus nerka* (Brett, 1971a). Differences in experimental technique may have some bearing on this apparent fundamental difference in response, namely the consequences of feeding individual fish a single food item at a fixed rate (Elliott, 1975a) and that of broadcast feeding of pellets to a school of fish (Brett, 197la).

By contrast, maximum meal size of sockeye salmon was greatly influenced by fish weight whether on a single or multiple feeding regime, the smallest fish consuming the largest relative amount (percentage of body weight, Fig. 19). This is similarly apparent from the hatchery feeding tables mentioned previously. The relation of fish weight to maximum meal size was shown by Elliott (1975a) to be proportional to $W^{0.75}$ for brown trout. A greater exponent of weight ($W^{0.95}$), approaching weight independence, derived by Kato (1970) for rainbow trout appears to be in some contradiction to the more general findings of considerable weight-dependence suggested by hatchery feeding tables.

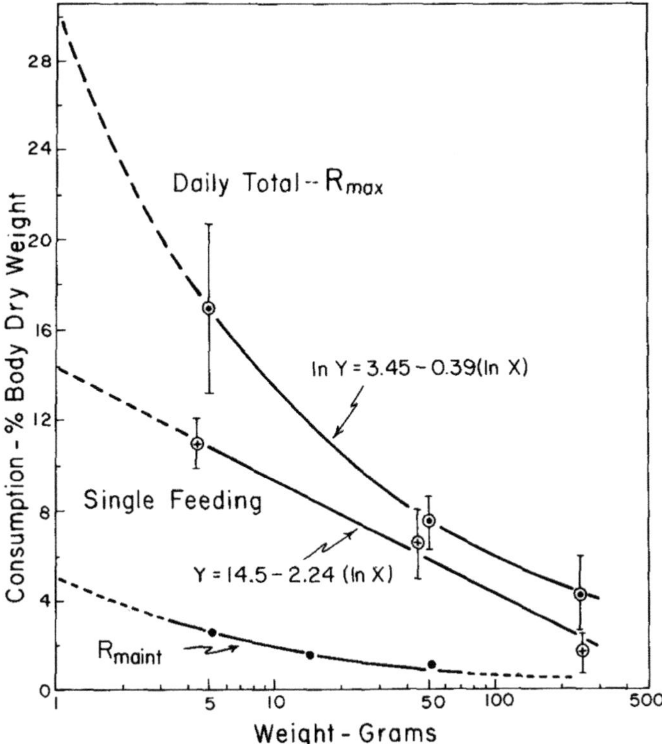

Fig. 19. Maximum food intake of sockeye salmon at 15°C on single and triple daily feedings, in relation to fish wet weight (log scale). Mean daily totals (±1 SD) are shown with calculated extrapolations according to the equations. (Modified from Brett, 1971a.) Values for maintenance ration (R_{maint}) have been plotted in the bottom line. The difference between R_{max} and R_{maint} is the scope for growth, which can be seen to decrease with increasing weight.

In addition to a weight influence, the effect of temperature on maximum meal size of brown trout was shown by Elliott (1975a) to follow four apparently distinct temperature ranges (Fig. 20). Temperatures between 13° and 18°C resulted in highest intake for a single meal. However, since rate of digestion increases with rising temperature the highest daily intake for *multiple* meals was reported at a slightly higher temperature, 18.4°C (Elliott, 1975b). Small salmonids (under 3 g) at a high temperature (15°C and above) can be shown to consume over 20% of their body weight per day (dry weight basis); large fish (over 1000 g) frequently require less than 1% body weight per day to meet their maximum consumption levels (cf. Fig. 19).

By studying the time sequence for return of full appetite in relation to deprivation time and rate of stomach evacuation, both Elliott (1975b) and Brett (1971a) showed that when the stomach was 75–95% empty voluntary food intake was not far from the maximum. Since this is a rate function

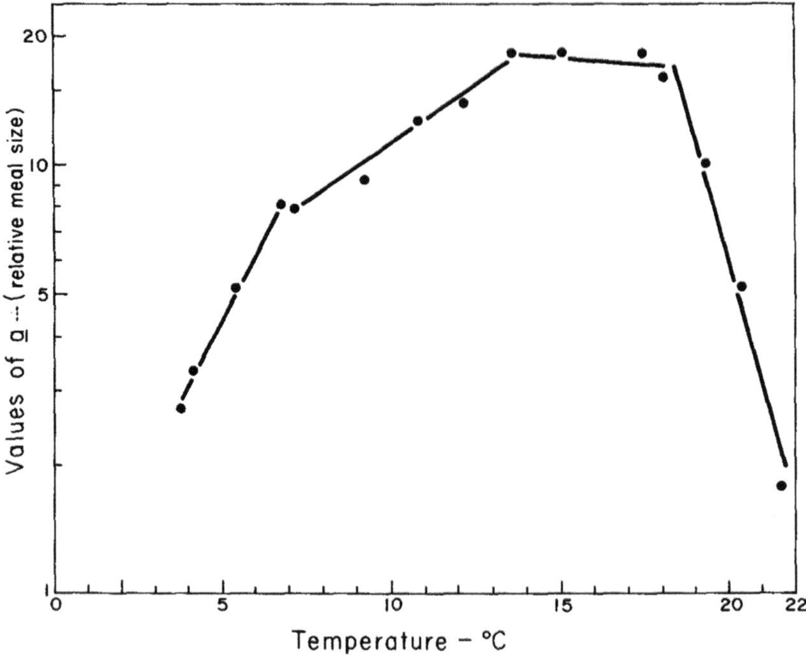

Fig. 20. Maximum meal size of brown trout in relation to temperature. The relation is shown in terms of the constant a from the general equation, $Q = aW^b$, relating meal size (Q) to weight (W) for single daily feedings. The relation of a to temperature is independent of fish weight (10–350 g). (From Elliott, 1975a.) The use of a set of straight lines was arbitrarily selected by Elliott who considered that there were four distinct temperature ranges suitable for mathematical modeling.

dependent on meal size and temperature, the time sequence of feeding can be strategically manipulated to increase daily intake. As an example, restricted rations in the morning designed to permit greatest intake at the *end* of an 8-hr day provide most use of the 16-hr digestive interval until next feeding opportunity. Brett (1971a) showed that on a selected 11-hr feeding interval at 15°C the sum of the two feedings was considerably greater when the first feeding was 3.4% and not the maximum intake of 4.4% (for the size of fish involved—50 g). The repeated 11-hr interval applied over a series of days was chosen to remove the possibility of endogenous or habitual rhythms of digestion interfering with the experimental design (offset from multiples of 12 hr).

Using automatic feeders that dispersed food from one to twentyfour times per day, the growth and conversion efficiency of catfish, *Ictalurus punctatus*, were discovered to be highest on satiation feeding twice per day at 28°C; the most frequent feeding gave the least benefit (Andrews and Page, 1975). Apparently, nibbling was not an effective feeding strategy for this species.

This is in contrast to the behavior of young salmon where continuous feeding for 15 hr/day at 20°C produced significantly greater growth rate than feeding to satiation three times daily (Shelbourn et *al.*, 1973).

An explanation for some of these differences has been offered by Kono and Nose (1971) who examined the effect of various feeding frequencies on six diverse species of fish. They concluded that the suitability of different time sequences was influenced by stomach size, with the smallest stomachs requiring most frequent feedings. This is exemplified by the continuous browsing behavior of surf fishes (Embiotocidae) which have a long intestine and an almost undifferentiated stomach region (De Martini, 1969).

2. *Quality*

The general form of the GR curve was presented earlier (Fig. 4), and the interrelations of restricted ration with individual abiotic factors examined. It is quite apparent that restricted ration simply cuts the curve off at the particular point of intersection for any level of ration below R_{max}. Growth would continue at a fixed rate, all things being equal. As will be seen this is not entirely the case, since increasing size would sooner or later exert a restricting effect depending on the level of reduced ration, the only exception being the maintenance ration (R_{maint}, G_0). This size involvement does not affect the selfevident conclusion that food acts as an unequivocal Limiting Factor.

The foregoing assumes that the ration is formulated as a wellbalanced diet (see Chapter 1). Few nutritional experiments on fish have been concerned with comparisons over the full span of rations, or planes of nutrition, from R_{maint} to R_{max}, except some of those involving protein conversion ratios (e.g., Nose, 1963). The question of quality difference among formulated and natural diets was posed during studies on environmental effects on the growth of young sockeye salmon (Fig. 21). Greatest growth at all feeding levels was obtained with Halver's test diet, formulated with casein as the major protein source; least growth occurred on frozen marine zooplankton (mostly *Calanus plumchrus*, Brett, 1971b). From Fig. 21 it is apparent that as the slope of the GR curve falls, from the most effective to the least effective diet, maximum intake increases. In the present case R_{max} shifted from about 7% weight/day on the best diet to 14% per day on the worst, while the associated G_{max} fell from approximately 3% per day to 1% per day. Quite obviously this is accompanied by a great decrease in conversion efficiency.

The associated increase in R_{maint} (from 2 to 6% per day) provides some insight for the plight of the fish on a nonnutritious, low-energy diet. Even when such a diet was present in abundance the fish were constantly hungry, packing their stomachs to the maximum at every opportunity.

Elliott (1975a) found that naturally occurring food organisms of six varieties were consumed by brown trout to the same extent (meal size) and over

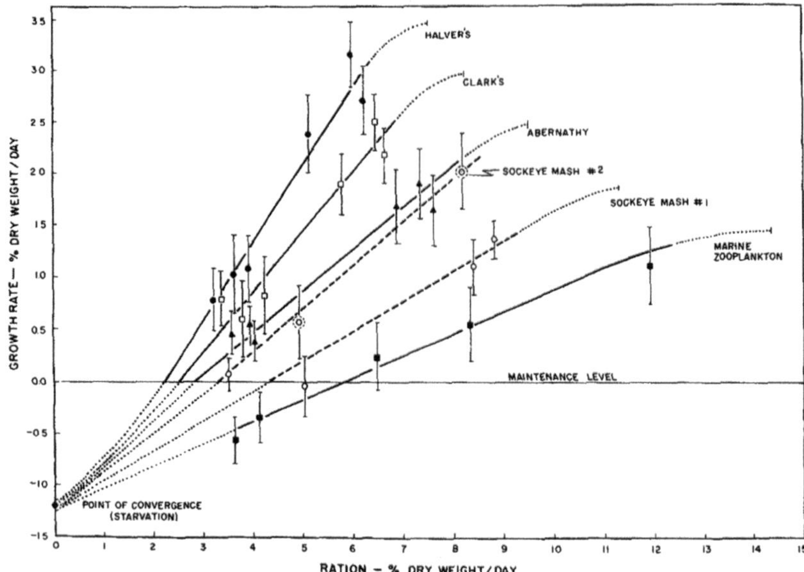

Fig. 21. Growth rate of sockeye salmon in relation to ration, for various diets. Limits: ±2 SE. All solid lines determined by least squares. Broken lines for sockeye "mash" were drawn by eye; two encircled points were for No. 2 formulation. The "point of convergence" was estimated by extrapolation. Dotted extensions of lines indicate expected shapes of curves terminating at maximum voluntary food intake. (From Brett, 1971b.)

the same time frame (satiation) with the exception of meal worms. The latter were much lower in water content with the result that fewer were eaten, involving a significantly shorter satiation time. All other parameters of weight relations and temperature effect on feeding remained within the 95% confidence limits developed for this species on a "standard" diet of amphipods.

B. Size

Few metabolic relations are independent of size. The anabolic process of growth is no exception. As an animal increases in size the metabolic activities are paced at a generally declining rate, frequently proportional to weight raised to the power 0.7 ($W^{0.7}$) in warm-blooded vertebrates (Kleiber, 1961). The weight relation of standard metabolism for fish is more varied with the weight exponent ranging from 0.65 to 0.85 (Beamish, 1964; Glass, 1969). The early concept that these might be explained by simple surface–volume proportionality as size increases (i.e., proportional to $W^{0.67}$) has too many exceptions to be a valid hypothesis for such a highly complex enzyme–substrate–transport–exchange system. Size has a greater restricting effect on growth rate than on metabolic rate, a difference which will be shown to account for a declining conversion efficiency with size (see Fig. 22).

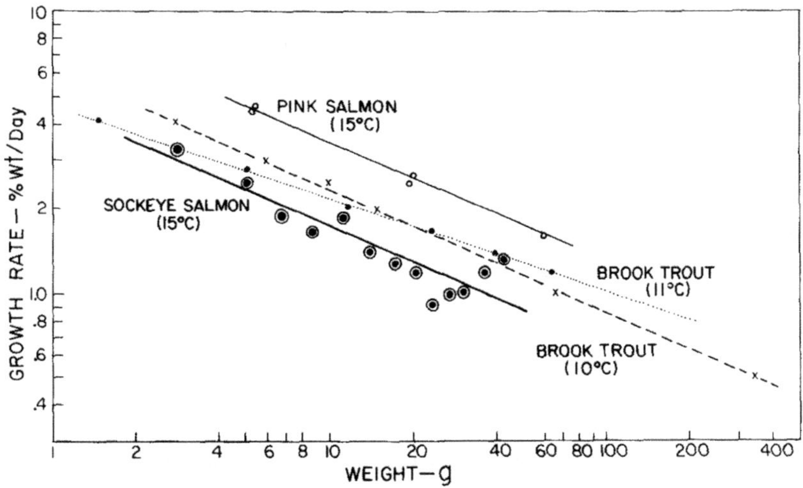

Fig. 22. The growth rate–weight relation for three salmonids. Brook trout (10°C) from Cooper (1961); brook trout (11°C) from Haskell (1959); pink salmon (15°C) from Brett (1974). (Figure from Brett and Shelbourn, 1975.)

The separate effects of age (not maturity) independent of size have yet to be defined. However, there appears to be reason to support an independent age effect of small magnitude (Brett, 1974).

Since size has a continuous effect on growth rate throughout life, *without* an independent phase, it is difficult to conceive of size as a normal Limiting Factor, restricting the supply of metabolites at some critical point of dependence. Also, since the weight influence does not conform to the natural, physical laws of surface relations it is tempting to classify size as a Controlling Factor possibly influencing the metabolic pacing through some form of size-dependent hormonal or enzymatic control of metabolism (the "scaling effect" of Stauffer, 1973). It is not the purpose here to attempt to find biochemical support for such an hypothesis. Rather, the manner in which the GR curve responds to size will be examined and compared with the types of response already defined for abiotic factors.

1. *Maximum Growth Rate* (G_{MAX}) × *Size*

In studies on the utilization of food by a number of marine species, Hatanaka and associates (Hatanaka and Takahashi, 1956; Hatanaka *et al.*, 1957; Hatanaka and Murakawa, 1958) showed that the growth rate of young mackerel, *Pneumatophorus japonicus*, and amberfish, *Seriola quinqueradiata*, decreased rapidly with size. Thus, at seasonal temperatures of 23 ± 2°C the growth rate of mackerel fed on anchovies (five to six times/day) fell from 9.5% weight/day at 4 g to 2.5% weight/day at 40 g. Kinne (1960) reported on the size relations affecting the increase in length of the euryhaline desert pupfish, *Cyprinodon macularius*, which displayed remarkably high growth rates soon after hatching, for example, 23% weight/day (5 mm, 20 mg, 20°C, 35‰) falling to 1.3% weight/day by the thirty-fourth week (27 mm, 220 mg).

Extensive size-effect studies have been conducted on members of the family Salmonidae, particularly influenced by the need for appropriate hatchery management practices. If the logarithm of growth rate (G_{max} as % weight/day) is plotted against the logarithm of weight (W in g) over a range of weights from 1 to 400 g, the growth rate can be seen to decrease at a fairly uniform rate (Fig. 22) (see also Kato and Sakamoto, 1969). From a compilation of these sources (Table III), Brett and Shelbourn (1975) concluded that the slope of the decreasing growth rate with size ($b=-0.41$) was characteristic of the family, with the intercept a taking on various values according to species and environmental factors (temperature, salinity). This generalization is equivalent to saying that for salmonids G_{max} is proportional to $W^{0.6}$.

Since the compilation in Table III, Elliott (1975c) has shown that brown trout are characterized by a slope of −0.33(5.6°C) to −0.28 (19.5°C) with a mean of −0.32 for all temperatures. This is close to the value for coho salmon (−0.34) determined by Stauffer (1973), and indicates a somewhat wider range of possible slopes.

2. Growth on Restricted Rations × Size

A few studies have been conducted where it is possible to develop more-or-less complete GR curves for different weight ranges under otherwise similar conditions. Hatanaka *et al.* (1957) grouped data on young mackerel into weight ranges of 7–24, 26–48 and 50 – 55 g. Depicting the GR relation as a straight line (no data plotted), the slope of each GR curve was shown to decrease with increasing weight. Maintenance ration (R_{maint}) fell, and the rate of loss of weight (G_{starv}) also decreased as size increased. The same general phenomenon is apparent in the studies of protein metabolism of bluegill sunfish, *Lepomis macrochirus* (Gerking, 1971). A weight-related decrease in slope between nitrogen retained versus nitrogen consumed (equivalent to GR curves) was shown to be accompanied by a decrease in the protein maintenance requirements (e.g., 0.36 mg of N/g/day for a 14-g fish decreasing to 0.26 mg of N/g/day for an 85-g fish).

By pooling the findings on sockeye salmon from a number of studies it was possible to display a set of curves for three mean sizes (Fig. 23). These confirm the general pattern of change in the GR curve as weight increases.

It can be seen that, in the interrelation of the curves, there is a similarity of patterns between size effect and temperature effect—increasing size shifts the curve down and to the left much as decreasing temperature does (below the optimum temperature). However, as will be developed under Section V, "Interaction and Optimizing," there is compelling evidence to classify size as a Limiting Factor rather than a Controlling Factor.

Table III Values of Parameters a and b for the Equation $\ln G = a + b(\ln W)^a$

Species	Weight range (g)	Temperature (°C)	Number (n)	Intercept (a)	Slope (b)	Source
Salvelinus fontinalis Brook trout	1.5–60	11.0	7	4.66	−0.33	Haskell (1959)
Salvelinus fontinalis Brook trout	2.5–350	10.0	8	6.49	−0.47	Cooper (1961)
Oncorhynchus nerka Sockeye salmon	1.0–30	15.0	9	5.58	−0.43	Brett *et al.* (1969); Shelbourn *et al.* (1973)
Oncorhynchus nerka Sockeye salmon	6.0–7.0	15.0	4	7.72	−0.49	Brett (1974)— wild stock
Oncorhynchus nerka Sockeye salmon	50–190	15.0	3	26.31	-0.69^b	Brett (1974)— cultured stock
Oncorhynchus gorbuscha Pink salmon	5.0–60	15.0	5	9.78	−0.45	Brett (1974)

Continued

Table III Values of Parameters a and b for the Equation $\ln G = a + b(\ln W)$—Cont'd

Species	Weight range (g)	Temperature (°C)	Number (n)	Intercept (a)	Slope (b)	Source
Oncorhynchus nerka Sockeye salmon	0.3–75	15.5	22	5.42	−0.40	Brett (1974)—in Table IV, column 15.5°C
Oncorhynchus nerka Sockeye salmon	0.3–75	10.5	22	3.46	−0.39	Brett (1974)—in Table IV, column 10.5°C
Oncorhynchus kisutch Coho salmon	0.3–75	15.5	22	5.53	−0.34	Stauffer (1973)
Oncorhynchus kisutch Coho salmon	0.3–75	10.5	22	3.94	−0.3	Stauffer (1973)
Oncorhynchus nerka Sockeye salmon	3.0–45	15.0	13	4.47	−0.42	Brett and Shelboum (1975)
Mean ± 2 SE					−0.43 ± 0.06 –	
Mean ± 2 SE[c]					0.41 ± 0.04	

[a]G, specific growth rate (% weight/day) on maximum ration; W, weight (g). Size and source of various salmonid species indicated. From Brett and Shelbourn (1975).
[b]This stock noted as possibly atypical in Brett (1974).
[c]Mean calculated without value noted in *b*.

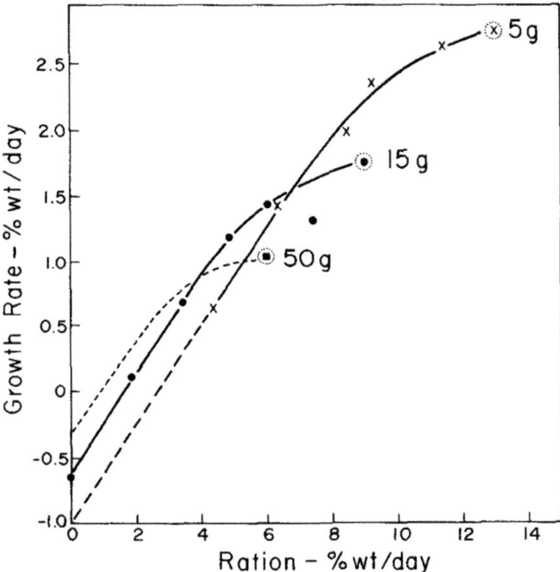

Fig. 23. The effect of size on the growth rate versus ration relation for young sockeye salmon. (Data for 5 and 15 g sizes from Brett *et al.*, 1969; Brett and Shelbourn, 1975.) Circled points representing G_{max} and R_{max} derived for each weight from Brett (1974). Only one point available for 50 g fish; expected curve drawn.

3. *Scope for Growth* (G_{SCOPE}) × *Size*

Sparse as the information is, nevertheless it is apparent that the same approach to scope for growth ($R_{max} - R_{maint}$) which was developed for temperature effect can be applied to weight effect. As size increases, R_{max} falls rapidly; R_{maint} also decreases but at a slower rate than R_{max} (see Fig. 19). This results in a converging, so that G_{scope} diminishes with size. An old mature fish is mainly eating for maintenance and gonad development; somatic growth is almost terminated. The diminution of growth promoting hormone that must accompany advanced age and size results in a decreased demand for food such that R_{max} approaches R_{maint}.

It is apparent that size is affecting both R_{max} and R_{maint} at the same time, and therefore operating to constrain *demand* rather than limit supply. This conveys a somewhat different mechanism than a simple Limiting Factor.

C. Competition

When considering the effects of abiotic factors on growth it was convenient to assume that each fish was responding as an independent entity, feeding and growing by itself without reference to the activities of any other member of the group. In truth, behavioral interaction can have great impact both masking and confounding the results of strictly physiological tests designed to seek the basic parameters of organismal growth. Perhaps the most extreme example of behavioral consequence is that of the Siamese fighting fish, *Betta splendens*, for which any close association is only resolved with the death of an opposing member. Less extreme cases of aggression have nevertheless imposed the necessity of raising each fish in a separate cell, out of visual or odor contact with its confrere (e.g., *Lepomis macrochirus*, Gerking, 1971; *Odontobutis obscurus*, Yamagishi *et al.*, 1974). The exact opposite may be true for some schooling species which are so restless and excitable away from the association of like members of similar size that feeding is disrupted and any determination of normal growth made impossible (e.g., *Clupea* sp., see Blaxter, 1970; Blaxter and Holliday, 1963). At this present stage in the elaboration of growth responses of fish it can be seen that behavioral involvements were undoubtedly the greatest obstacle in earlier research to achieving any clear understanding of environmental effects on growth. This is nowhere more apparent than in the treatise on fish physiology (Brown, 1957) where size hierarchial effects received prime attention and obviously entered into many reported experiments in a persistent, uncontrolled, if not unperceived fashion.

The interrelation of common members of a sample of fish is affected by *numbers*, *space*, *size*, and *species*. The relation that these factors bear to growth, and the extent to which they interact with each other is greatly influenced by *food availability*. In the face of the resulting competition for space and food, fishes have evolved various behavioral patterns involving defense and dominance, with their associated patrolling of territories and acts of aggression. Not only do these behavioral patterns require expenditure of energy, increasing daily maintenance requirements, but also there is evidence for suppression of growth in low-ranking members by intimidation (Wirtz, 1974) or possibly through water-borne inhibiting agents (cf. Richards, 1958; West, 1960; Yu, 1968).

When such behavioral interactions affect growth the relative difference in size of the members of a population usually increases—the large grow ever larger while the small lag further behind. This phenomenon has been called "growth depensation" and refers to the increase in variance of a size frequency distribution with time (see Chapter 11). In studies on growth variability, Yamagishi (1969) emphasized the importance of assessing the above parameter by the simple determination of the *coefficient of variation* [$CV = 100 \times (SD/mean)$, in %]. In this way a measure of the degree of interaction can be assessed permitting greater insight of the factors in play; no change in CV with time signifies no significant interaction, or completely random behavioral interrelations.

Among the many experiments performed it is frequently difficult to distinguish the separate effects of numbers, space, and feeding opportunity. These factors are not infrequently correlated with each other; as numbers decrease, space and feeding opportunity increase. If the relative space configuration is maintained then the significance of numbers can be examined distinct from density, space, and configuration (e.g., two fish in a 2 liter cube versus 200 fish in a 200 liter cube). However, in the following consideration the latter distinction is not attempted.

(1) Numbers (density). If it were not for behavioral relations, the number of fish that could be grown in a given volume would be directly proportional to the rate of exchange. Allen (1974) examined the growth obtained by stocking five densities (90–720 fish/m^3) of channel catfish at five different rates of flow. As the density increased, the mean weight achieved decreased. This was almost entirely due to decreasing oxygen concentrations which, as the numbers increased, fell progressively below the critical O_2 concentration level of 5 ppm.

When such obvious limitations are removed there appears to be an optimum density for some species of fish when on unrestricted rations. Brown (1946b) demonstrated this phenomenon among fingerling brown trout. At lowest densities (1 fish/50 liter) the fish did not feed as well and appeared to lack the social stimulation accompanying greater numbers. Highly crowded conditions (1 fish/3 liter) resulted in reduced food conversion efficiency and some physical interference between fish. Magnuson (1962) explored density-related growth depensation in populations of medaka, *Oryzias latipes*, after removal of the variability resulting from congenital differences. Increasing the population's size by four times, even when space was increased proportionately, reduced growth rate. This appeared to be associated with some reduction in food intake, leading Magnuson to conclude that neither a general depression in growth rate nor growth depensation occurred in this species if food was *effectively* present in excess.

More recent experiments have been conducted by Refstie and Kittelsen (1976), who followed the growth and mortality of Atlantic salmon reared from first-feeding fry to smolt at various levels of crowding. These authors noted that many studies on density effects have been performed at comparatively low concentrations, not representative of the conditions usually found in hatcheries. Using fry densities of five to 35 liters, only density effects on survival were apparent in the first 6 weeks, best survival occurring at highest density. Transferred to larger tanks at the parr stage for 30 weeks, the highest densities resulted in the lowest growth rates, without any mortalities. While it might be supposed that the main factor depressing feeding rates in this territorial species was social interaction involving dominance, no increase in the coefficient of variation occurred. It was concluded that free movement of the fish was inhibited, affecting food availability despite excess ration.

The question of how density affects growth rate when food is unrestricted appears to depend on the natural schooling relation of the species or the extent to which territorial behavior in some nonschooling species is suppressed by the sheer weight of numbers. Yamagishi (1963) observed that in a schooling race of crucian carp, *Carassius carassius*, growth depensation was significantly less than in two other nonschooling races when reared under similar conditions. This behavioral relation was further investigated in three contrasting marine species: (i) the halfbeak, *Hemirampus sajori*, a schooling fish, (ii) the red seabream, *Chrysophrys major*, a territorial fish, and (iii) the zebra sole, *Zebrias zebra*, which transforms from planktonic feeding to demersal or bottom'feeding. Among the halfbeak, growth depensation increased during earliest feeding but decreased as schooling developed. The seabream continued to increase in growth depensation, associated with aggression and cannibalism. The zebra sole displayed a sudden increase in the coefficient of variability on becoming bottom-feeders and establishing territories. These examples were considered by Yamagishi (1969) to be representative of three general types which are depicted in Fig. 24.

The shift with development to increased growth depensation of the zebra sole is reversed in the case of the ayu, *Plecoglossus altivelis*, a salmonlike fish that grazes on diatoms and algae growing on stones in rivers. As population density increases, the social structure of this species changes fron near-the-bottom territorial behavior to off-the-bottom schooling behavior (Kawanabe, 1969). Utilization of the limited food resource improves, and variability between individuals diminishes.

Density relations are of great importance in fish pond management in which carrying capacity defines the maximum weight of fish that can be sustained, and delimits the point at which no further growth is possible. Within

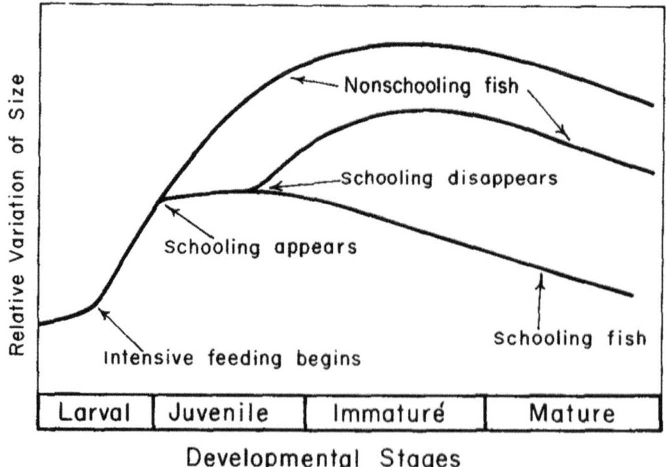

Fig. 24. General trend in relative variation of size (growth depensation) in fish characterized by different types of behavior, according to stages in development. (From Yamagishi, 1969, *Res. Popul. Ecol. (Kyoto)* **11**, 14–33.)

the carrying capacity there is a high correlation between increasing density and decreasing individual growth (Hepher, 1967). In addition, it can be shown in ponds that there is a direct relation between weight gain and initial weight (for carp, Wohlfarth and Moav, 1972) although this is apparently only true during the warm, growing season. Such observations, although derived from well-designed "plot" comparisons and most meaningful in aquaculture, are the result of many interacting mechanisms which defy strict physiological interpretation.

(2) Competition for space. That space, as a factor by itself, could affect growth was demonstrated in the case of female guppies, Poecilia reticulata, raised in aquaria of different sizes (Comfort, 1956; also cited in Brown, 1957). Greatest growth was achieved in the greatest space; if switched from small to large aquaria growth increased accordingly. Similarly, Allee *et al.* (1948) found that four green sunfish, *Lepomis cyanellus*, grouped in unobstructed aquaria containing 4 liters of water grew faster than single individuals in one-fourth that volume.

Keeping density constant (by volume), Yamagishi (1962) observed that growth rate of rainbow trout fry increased with an increase in the bottom area of the culture tanks. Fighting was intense, lasting from a fraction of a minute to 20 min. Growth was greatest when the total area exceeded the maximum size of territory that each fry could occupy.

In the presence of excess food distributed evenly, space per se was not found to be an influential factor for medaka (Magnuson, 1962). Growth depensation was no larger or smaller in populations of high density with less space per fish than among isolated control fish. By a comparison of treatments it was evident that aggressiveness in this species was a competitive mechanism for food and not for space. Even large fish had no competitive advantage over small fish.

(3) Size and hierarchical effects. Studies on the dominance interrelation between fish of different size and rank-order reveal two fundamentally different responses. Nagoshi (1967a,b) reared mixed populations of small and large guppies in 1-liter baskets using stratified rations of live zooplankton (*Cyclops vicinus*). On restricted rations a social hierarchy developed in which small fish were subordinate to large fish and comparatively suppressed in their growth rate. This difference persisted as ration increased, with only slight diminution up to an unrestricted level of feeding when size was no longer used to advantage (Fig. 25A). This may be compared with observations on the highly territorial and predacious goby, *Odontobutis obscurus*, which, when space is limited, attempt to dominate the competitor at all feeding levels (Yamagishi *et al.*, 1974). Starting with fish of almost equal size the lowest rank-order goby lost weight at the greatest rate when starving, and remained at half the G_{max} of the largest fish when preyfish were provided in excess (Fig. 25B). The intensity of attacks by the first-ranking fish on subordinates was such that the second-ranking fish eventually surpassed the growth of the first-ranking fish without deposing this "ruler."

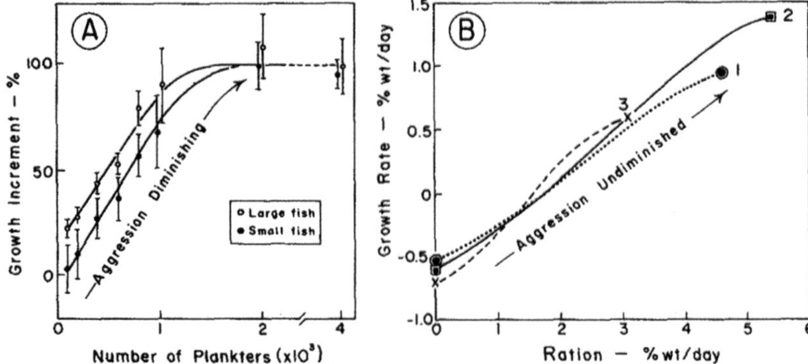

Fig. 25. (A) Relation of growth rate (% of maximum) to feeding rate in large and small guppies, *Poecilia reticulata*. (From Nagoshi, 1967a.) (B) Growth rate of gobies, *Odontobutis obscurus*, according to rank order indicated by numbers; fish No. 2 was smaller than No. 1 at start. (Data from Yamagishi *et al.*, 1974.) Aggression diminished with increased ration in A but not in B.

Kato and Sakamoto (1969) examined the effect of grading on growth rate and growth depensation among three groups of young rainbow trout [initial sizes: large, 3.8 ± 0.13 cm (SD); medium, 3.6 ± 0.02 cm; small, 2.8 ± 0.15 cm]. Growth rate among groups declined almost uniformly with increasing size without any compensatory shift of the smaller fish, indicating true genetic differences. Depensation, however, increased in all three groups with the smallest fish exhibiting the largest change.

When food supply was limited for medaka, the larger fish were socially dominant, chasing smaller fish away from food and growing faster (Magnuson, 1962). If food was spatially localized the freeroaming hierarchial societies changed into territorial societies with the dominant defending the food areas. Magnuson concluded that aggressive behavior is a competitive mechanism that can provide the dominant animal with an advantage when food and space are limited. Aggressive behavior will disperse the competitors throughout the habitat only if food is found in all areas. Such relations would be expected to occur among fish that exhibit aggressive behavior in connection with food, in habitats containing contagiously distributed food limited in supply, and among fishes which live on the substrate or among aquatic vegetation.

In general, it can be seen that competition acts to restrict the food intake of subordinate fish (Limiting Factor—reducing R_{max}). It may be sufficiently demanding of energy expenditure to invoke a metabolic cost that significantly reduces growth (Masking Factor—elevating R_{maint}). The behavior of some species is such that competition only occurs when food is restricted (e.g., *Poecilia reticulata*); in others both Limiting and Masking Factors can reduce the scope for growth (e.g., *Odontobutis obscurus*).

D. Summary Configurations

Classifying biotic factors within the same scheme as abiotic factors (in terms of how they act) has proved possible, and reveals some new insights. None appears as a Controlling Factor, although size is not without some similarities (the "scaling" effect of Stauffer, 1973). Ration, an obvious Limiting Factor, has a feature of diet quality to be applied to the quantitative expression, decreasing the slope and extending the terminal point (R_{max}). Competition may evoke either one or both of Limiting and Masking effects.

These relations are assembled in Fig. 26, and bear comparison with the abiotic series (see Fig. 18). Again, it will be noted that there are only a few basic types of response of the GR curve. Ration, as a Limiting Factor, *categorically* acts like oxygen concentration by setting intermediate limits to

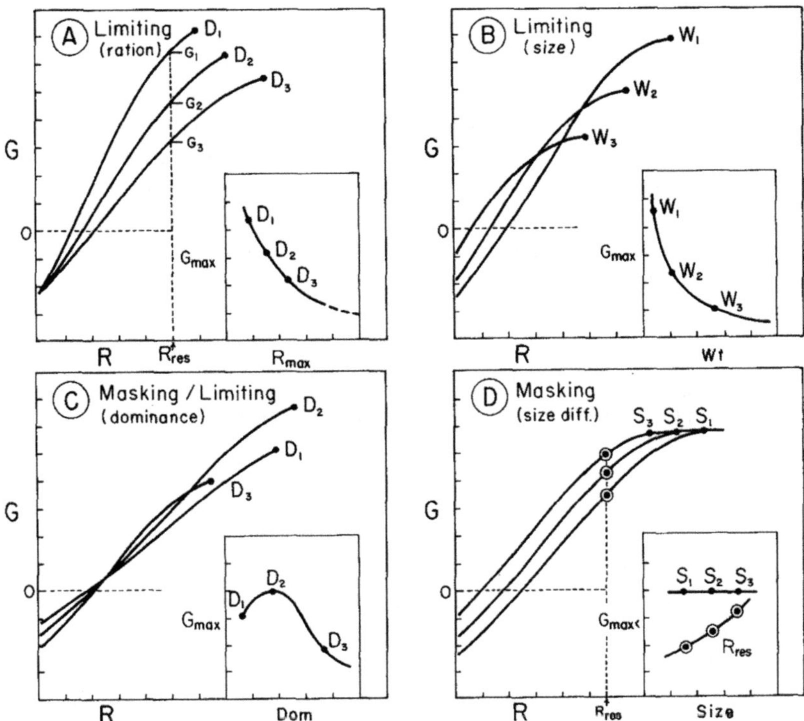

Fig. 26. Recapitulation of GR curves in relation to biotic entities, illustrating the basic forms for Limiting and Masking Factors, and the path of G_{max} according to the level of the biotic entity (boxed inserts). (A) illustrates the effects of diets of decreasing nutrient value (D_1–D_3) and the consequence of a restricted ration (R_{res}) limiting growth (G_1–G_3). (B) illustrates the effect of increasing weights (W_1–W_3) on the GR curve, and on G_{max}. (C) illustrates the effect of dominance order (D_1–D_3) requiring either increased energy expenditure (D_1), or limiting the food availability (D_3). (D) illustrates the case where size difference (S_1–S_3) between competing groups of fish only affects growth when food is limiting (R_{res}).

growth, when not present in excess. Size, as will be shown, also acts to set limits on G_{max} only when ration is not limiting. Despite the complexity of behavioral relations they tend to be either Limiting by denying food to a competing member, or to be Masking by placing a metabolic burden of increased activity (attacking and/or defending) in a feeding territory.

V. INTERACTION AND OPTIMIZING

As stated in the introduction it is impossible to consider growth in relation to any one of the abiotic factors without introducing the biotic factor of *ration*. When food is present in excess (undocumented) then no account of the consequences of a given abiotic factor in relation to R_{max} and R_{maint}, and their derivative G_{scope}, is provided. Since an abiotic factor such as temperature affects maintenance metabolism it might be expected that temperature and ration would interact, affecting growth in some progressively shifting relation. Other cases of interaction have been indicated previously, including the combined effects of salinity and photoperiod, salinity and temperature, oxygen and ration, and size and ration. These couplets can be extended to multifactor combinations involving complex interactions for which the growth response could be described but which, by their very complexity, could well defy physiological explanation—a black box of stimuli producing unlimited patterns of growth response. Fortunately, the system of classifying the action of environmental factors lends insight to their combined effects.

The term "interaction" has been used so far in a general sense without any special connotation; "interrelation" could be substituted. Computational methods such as analysis of variance and covariance allow assessing the extent to which *statistical interaction* accounts for a portion of observed variability of a response. Factor combinations may also be assessed by response surface analysis. The essence of this method has been described by Alderdice (1972). *Response surface interaction* can be detected when the path of maximum response in relation to changing levels of two variables (x and y) moves along a line that is not parallel to either the x- or y- axis. The extent of this rotation is a measure of the degree of interaction. Thus, if the temperature for maximum growth shifts as ration decreases (see Fig. 27) then an *interaction* is present. This is the meaning implied henceforth. The path of maximum growth represents the optimum condition in terms of the two or more environmental factors considered; where applicable, the center of the whole configuration is the ultimate optimum.

Consideration will be given to the effects of environmental entities on growth in terms of their classified Factor relations.

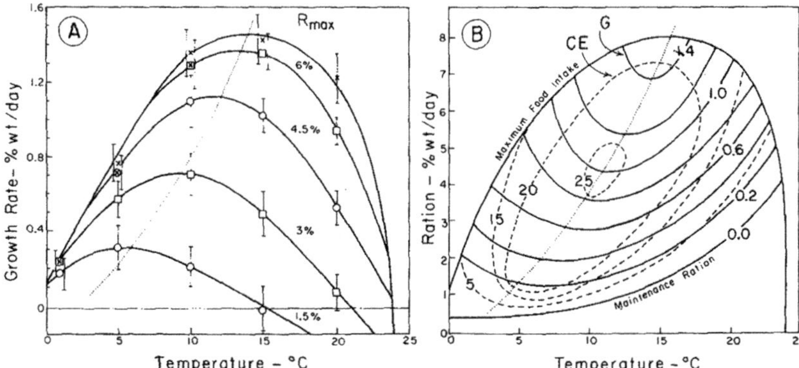

Fig. 27. Effect of restricted rations on growth rate and conversion efficiency in relation to temperature, for 13-g sockeye salmon. (From Brett *et al.*, 1969.) (A) illustrates the basic data as mean $G \pm 2$ SE for rations from maximum down to 1.5% weight/day, from which (B) the isopleths for growth rate (% weight/day) and for food conversion efficiency (CE* in %, broken lines) have been approximated. Note in A how the highest growth rate shifts to a lower temperature as ration is reduced, following the path of the dotted line (interaction), and in B that the optimum temperature for growth occurs at approximately 15°C on maximum ration, whereas the optimum temperature for the highest CE occurs close to 11°C on a ration of 4% per day (center of concentric broken lines). Dotted line in B forms the central axis of the isopleths passing through the respective centers for optimal growth rate and optimal conversion efficiency. This is comparable to the line for G_{opt} in Fig. 8.

A. Controlling × Limiting Factors

1. *Temperature × Ration* (R_{MAX} *to* R_{MAINT})

Previous reference has been made to the studies on fingerling sockeye salmon in which six levels of temperature were combined with five to six levels of ration* in a multifactor experiment (Brett *et al.*, 1969). A plot of the growth rate versus temperature (Fig. 27A) reveals that as ration is reduced the optimum temperature for growth moves from 15°C for excess ration to 5°C on a restricted ration of 1.5% weight/day. In terms of surface response analysis these data are plotted as in Fig. 27B, with growth rates shown as interpolated isopleths. Each isopleth is terminated by the limits of food uptake (R_{max}) at the respective lower and upper temperature levels involved. The center appears to fall right at the peak of maximum ration (8% R at 15°C) or just outside the observed limits. A similar interaction resulting in a downward shift of the optimum temperature as ration was restricted has been obtained in preliminary experiments on growth of striped bass, *Morone saxitalis*

*Conversion efficiency represented by CE in (B) has been referred to as Kin the text to conform with more common usage.

*Except at 1° and 24°C where meals were not accepted over such a full range of rations.

(Cox, 1975, personal communication); Elliott (1975c) has confirmed the same phenomenon for brown trout.

Because the highest rations at any temperature do not produce the highest conversion efficiencies, the **K** isopleths follow a different pattern with a different center from maximum growth, as indicated in Fig. 27B. The rotation and curvature of the interaction "line" must obviously remain the same for both parameters, G_{opt} and **K**. The reason for the shift in optimum temperature was ascribed to the reduced demand for maintenance ration at lower temperatures, allowing a greater fraction of the available ration to be converted into growth. It was possible to conceive that the conversion efficiency of this available fraction could be so reduced at lower temperatures that little growth benefit would result, or conversely that relatively high efficiency would be retained at *one* optimum temperature, suppressing any interaction. This appears to be the case for channel catfish which showed no change in optimum temperature (30°C) as ration was reduced although growth rate was greatly depressed ($3R \times 5T$ factorial experiment of Andrews and Stickney, 1972). This interpretation is supported by the generally high conversion efficiencies reported at all temperatures on a low ration of 2% weight/day, with the highest efficiency still remaining at 30°C.

2. *Temperature × Size*

The effect of these two factors acting together is taken into account in hatchery feeding charts (Haskell, 1959) which prescribe reduced ration (as % weight per day) with increasing size without altering the position of the optimum temperature. This interrelation is reflected in the growth rate table developed for sockeye (for sizes from 0.3 to 500 g), which depicts the highest growth rates remaining in the region of 15°–16°C *independent* of size (Brett, 1974). Brown trout display a similar relation, the optimum temperature for growth of all sizes tested occurring at 13°C (Elliott, 1975c). Lack of any interaction between temperature and size is, however, not the case for yellowtail, *Seriola quinqueradiata*, for which the optimum falls from 27°C for juveniles to 21°C for large adults (Oshima and Ihaba, 1969).

3. *Temperature × Oxygen*

Because the metabolic rate is reduced by both low temperatures and low rations, the critical level of O_2 concentration permitting the necessary energy expenditure can be expected to be reduced by their combined influence. Furthermore, although it was pointed out that O_2 concentration was a better measure of oxygen availability than O_2 saturation, the latter is not without some influence in terms of the O_2 supply that saturation provides at different temperatures (e.g., 8 ppm O_2= saturation at 25°C; 12 ppm O_2=saturation at 6°C). Hence it can be predicted that reduced oxygen would affect the temperature promoting maximum growth in a combined fashion; as O_2 concentration falls the critical O_2 level permitting growth decreases with decreasing temperature, but not greatly because the necessary minimum O_2 tension (% saturation) must be maintained. This conjecture would result in an interaction configuration for limiting oxygen not unlike that for reduced rations (Fig. 27).

The only experiments performed that bear on this hypothesis are those reported by Doudoroff and Shumway (1970) on the unpublished data of Trent. Largemouth bass were acclimated to 10°, 15°, and 20°C and their growth rate determined at reduced oxygen concentrations. At the two lower temperatures the critical O_2 level was reduced from 5 ppm to about 3 ppm, supporting the above prediction (i.e., that growth is still possible at either a lowered oxygen concentration, or a lowered ration, only when temperature is reduced).

Thus the pacing effect of temperature, a Controlling Factor, raises or lowers the critical level of any Limiting Factor. The processes of growth are never free from temperature, which, as is true of any Controlling Factor, acts always in *conjunction*, never in sequence, with another environmental factor.

B. Controlling × Masking Factors

TEMPERATURE × SALINITY

As stated earlier, this combination of two environmental factors, one governing the rates of reaction (*T*, °C), the other exacting a toll of energy required for regulation (s, ‰), has been studied by Kinne (1960) on the desert pupfish, and by Peters and Boyd (1972) on the hogchoker. Only *ad libitum* feeding will be considered here.

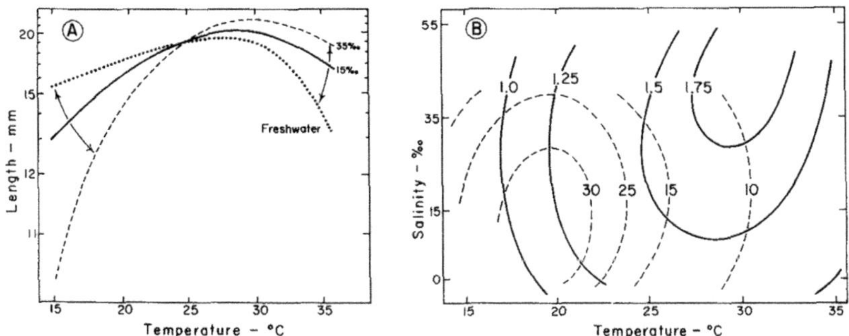

Fig. 28. Growth of the desert pupfish, *Cyprinodon macularius*, in relation to temperature at three salinities (35‰, 15‰, and freshwater). (A) from Kinne (1960). (B) illustrates isopleths at growth rates (% weight/day) with an overlay of conversion efficiencies (%, broken lines). (After Alderdice, 1972.)

For the pupfish, as temperature increased from a low level to a high level the accompanying salinity promoting best growth also increased from low to high, resulting in a "low-low/high-high" interaction. Alderdice (1972) analyzed these data showing that the center for maximum growth came at about 30°C at 40‰, while that for maximum conversion efficiency was close to 20°C and 15‰ (Fig. 28). Although the fish would voluntarily consume a large meal at 30°C and 40 ‰ (resulting in maximum growth) it was not converted with anywhere near the same efficiency as a smaller intake (R_{opt}) at lower temperature and salinity.

The hogchoker also displayed a significant temperature-salinity interaction, with the overall peak of maximum growth occurring at 25°C and 30‰, and the least recorded at 15°C and 0‰ (Peters and Boyd, 1972).

C. Limiting × Limiting Factors

Ration × Size

By definition, Limiting Factors cannot result in any interaction; they operate in series, like links in a chain. Optimum combinations do not exist; an ultimate maximum can occur where neither is limiting. By coupling these two entities together, ration and size, in an experiment at one temperature and one salinity it became obvious that size was acting in the role of a Limiting Factor when ration was not limiting (Brett and Shelbourn, 1975). An initial high ration of 12% weight/ day fed to sockeye fry (2.4 g initial weight) soon became excessive, the growth rate falling progressively from 3.6 % per day at 2.4 g to 1 % per day at 37 g. This was accompanied by reduced voluntary food intake (Fig. 29). When lower levels of ration were provided *at the start* (e.g., 6 and 4 % weight/day) growth rate remained constant within limits of normal variability* until reaching a limiting size at which point the growth rate followed the normal decline dictated by increasing size.

The same sort of interrelation of *ration* and *oxygen* acting as sequential Limiting Factors was depicted in Fig. 17.

D. Multifactor Effects

1. Controlling × Masking × Limiting [T · (°C) × S (‰) x R]

In an effort to determine how energy utilization was affected by temperature, salinity, and feeding in two estuarine fish, summer flounder, *Paralichthys dentatus*, and southern flounder, *Paralichthys lethostigma*, Peters (1971) conducted a multifactorial experiment involving fifteen different treatment combinations. Five levels of ration were prescribed, from 30 to 90% of R_{max}. Since R_{max} changes with size and environmental conditions, separate, parallel

---

*A case for possible slight increase in G can be made, which does not affect the conclusions.

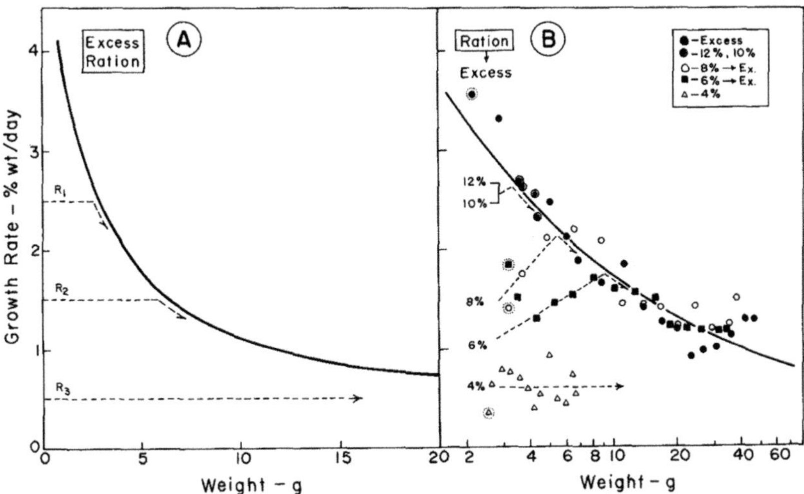

Fig. 29. The relation of growth rate to size (weight) on unrestricted and restricted rations (indicated as % weight/day). (A) depicts the model of expected responses where $R_1 - R_3$ are limited rations. (B) displays the observed growth rates for young sockeye in freshwater at 15°C and 16 L. (From Brett and Shelbourn, 1975.) Note that weight is shown on a logarithmic scale in B, to separate points for small weights.

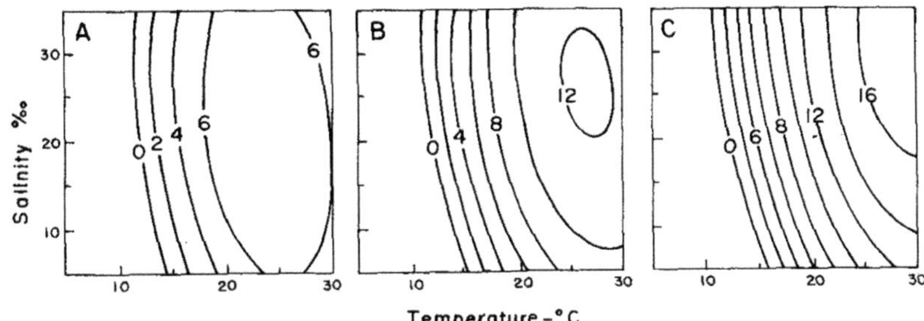

Fig. 30. Isopleths of growth rate (calories/kg/day) of *Paralychthys dentatus* in relation to salinity and temperature, for three levels of ration—(A) 60 % maximum, (B) 80% maximum, (C) *ad libitum*. (From Peters, 1971.)

determinations were made throughout the experiment to determine R_{max} at each of the temperature-salinity combinations. Fixed rations as % weight/day were therefore not used, but rather a sliding scale varying in *fixed proportion* to observed changes in R_{max}. This novel approach permitted maintaining the composite, three-variable design no matter how much or little the "parallel" control fish consumed; it does, however, add a complexity to interpretation in absolute values of ration.

Of the various configurations depicting the growth relations, one set of response surfaces has been selected (Fig. 30). These show that for *ad libitum* feeding, temperature as a Controlling Factor exerts most effect on growth. Salinity as a Masking Factor has relatively little effect but interacts in such a way that low salinities (5–15‰) at high temperatures (20°–30°C) have some suppressing effect on growth rate. The optimum combination lies outside the high temperature (30°C) × high salinity (35‰) corner of the "factor space" considered (Fig. 30C); G_{max} would be in excess of 16% weight/day. As ration is reduced, becoming limiting, growth rate decreases without altering the rotation of the temperature–salinity interaction axis. The center of optimum conditions can be seen to move such that at 60% of R_{max} it occurs at about 24°C and 21‰, with a G_{max} of close to 7% weight/day. In nature, as feeding opportunity was reduced, it would consequently be best for this species to seek intermediate temperature and salinity combinations.

Peters (1971) further notes that conversion efficiencies were generally higher near 80% of R_{max} than either 60 or 100%. This would be predicted from the nature of the basic GR curve where R_{opt} is usually somewhat less than R_{max}. It would also be predicted that the limiting effect of reduced ration would be to slice the dome off the isopleth "mound" such that the lower growth rates (below 6% weight/day) would remain in a similar relation to temperature and salinity for all three configurations—which is approximately the case.

A somewhat different factorial experiment involving the same Factors but with age (=*size* ?) as the Limiting Factor was conducted on postlarval striped bass, *Morone saxatilis*, by Otwell and Merriner (1975). Using a combination of three temperatures (T=12°, 18°, and 24°C), three salinities (S=4,12, and 20‰) and six ages (A=5–63 days), all factors and interactions were found to be significant ($P<0.01$). Temperature, as a main factor, accounted for 83% of the variance in growth (length), with 5% attributable to age and less than 1% to salinity; first-order interaction ($T \times A$) was 3%, and second-order interaction ($T \times \mathrm{S} \times \mathrm{A}$) was 4% of the variability. High temperature (24°C) gave best growth under all circumstances; at this temperature the best combination was the younger ages (less than 28 days) and a salinity of 12‰.

2. *Controlling × Masking × Directing × Limiting [T(°C) × S(%) × L × W]*

An experiment involving fixed levels of temperature × salinity × photoperiod × size has been attempted for sockeye salmon.* From an analysis of variance of the preliminary results, significant consequences of temperature on growth rate could be identified but salinity and photoperiod effects were not distinguishable from the error component, particularly at high temperatures. This was true for the two size ranges studied. It appeared that high temperature blocked the ability of photoperiod to induce smoltification, thereby eliminating potentially favorable effects of salinity.

---

*Unpublished records from annual reports of Pacific Biological Station, British Columbia.

The complexity of the system, the technical difficulties of exact control, and the need for constant attention to a large number of test tanks brought a temporary halt to this approach. Less involved combinations, using dynamic changes in photoperiod and temperature (increasing, decreasing, and static) were commenced. Preliminary results indicate that some combinations of temperature and photoperiod are synergistic; the magnitude of the photoperiod effect however is considerably less than that of temperature. For sockeye of 2 g, the difference in growth rates between an increasing and a decreasing daylength was 0.44% weight/day, whereas a temperature increase from 10° to 17.5°C stimulated growth by 1.1% per day. For 4.5 g sockeye, the corresponding effects were 0.40 and 0.78% per day. A more dramatic photoperiod effect was achieved by applying photoperiod control over a period of 5 months. In this experiment, an unnatural "seasonal" cycle depressed growth rates by 1% per day. The effect of photoperiod on growth is apparently not directly related to number of hours of exposure to daylight. The direction of change in daylength is the important cue; stimulation of growth apparently occurs when the photoperiod cycle is somewhat in advance of the normal seasonal cycle.

In conclusion, these various examples serve to illustrate the forms of growth response when more than one environmental factor is involved. They undoubtedly take physiological enquiry a few steps nearer the complexity of natural environments. Also, as an exercise in the extension of the "Factors" approach to anabolic systems, involving both abiotic and biotic effects, it is apparent that justification of this system of classification is supported by virtue of the insight provided.

VI. GOVERNING MECHANISMS

It was noted at the outset that an integrated series of rate functions—feeding, assimilating, metabolizing, transforming, excreting—were the underlying mechanisms providing net energy gain and governing the rate of growth. Within the span of abiotic and biotic factors affecting these rates, temperature was identified as the only Controlling Factor, setting the pace of each function. The supply of food, oxygen, hormone, or size restrictions could act as Limiting Factors, or salinity could place some regulatory burden on the organism, but, given the supply, the power of temperature remained the greatest determinant of growth.

It is, therefore, unnecessary to seek further than the temperature relations of these metabolic functions to reveal just how they combine to regulate growth. Drawing on the various studies on sockeye salmon, the feeding, digesting, converting, and growing rates have been plotted on a common scale equating the maximum rate for each function to 100, that is, a percentage scale (Fig. 31A). The energy demands dictated by standard metabolism, feeding metabolism, and their relation to maintenance ration are depicted in Fig. 31B. From these a composite picture of the mechanisms inducing the rise and decline of maximum growth (G_{max}) with increasing temperature can be assembled. For clarity of presentation three temperature ranges will be considered: low, intermediate, and high.

a. Low Temperatures (1°–5°C)

Rate of digestion is greatly depressed, approaching zero in the region of 0°C. However, digestion shows some compensatory increase; young sockeye will grow at 1°C, consuming 1.5% weight/day. Gross conversion efficiency is only one-half the maximum, which reflects either reduced digestive efficiency or high nitrogenous excretion, or both. Rate of digestion appears to be the chief limiting factor to food consumption and hence to growth at low temperatures, with a concomitant reduction in appetite (daily meal size). Some seasonal increase in food intake may occur subsequently, indicating an adaptive capacity to take on energy and retain it for gradual digestion. All levels of metabolism (standard, routine, feeding) are a small fraction of their maximum, with the result that maintenance ration is exceedingly low (Fig. 31B).

b. Intermediate Temperatures (13°–17°C)

With rising temperature all feeding and growth functions (Fig. 31A) increase in roughly the same proportion except growth rate, which is relatively more

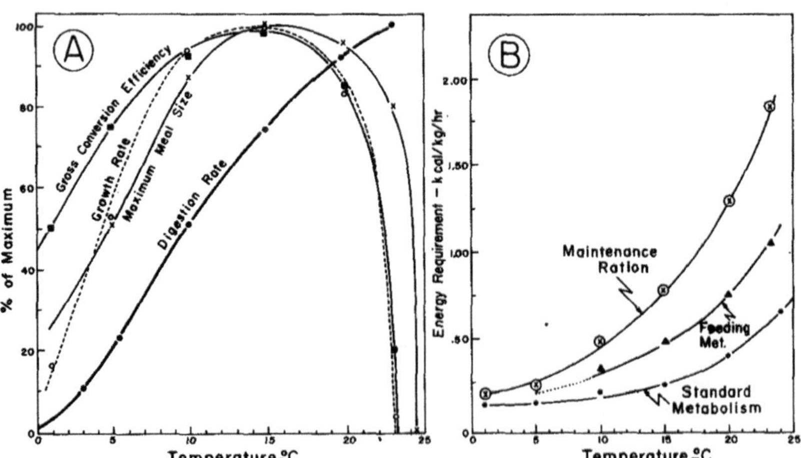

Fig. 31. The effect of temperature controlling various rate functions affecting the growth rate of young sockeye salmon. (A) illustrates the relative rates of feeding, digesting, growing and converting, plotted as a percentage of the maximum for each function. (From Brett and Higgs, 1970.) (B) illustrates the relation of upkeep to temperature, as standard metabolism, feeding metabolism, and maintenance ration, in terms of energy units (kcal/kg/hr). (From Brett, 1976.)

rapid owing to an increased food conversion efficiency coupled with higher intake per meal. At 15°C an optimum is reached for three systems; only rate of digestion continues to rise. Furthermore, at 15°C a significant threshold is reached with meals being fully digested within 24 hr. Supporting metabolic rates have risen in accordance with an exponential effect of temperature, maintenance ration rising at the greatest rate.

c. High Temperatures (20°–24°C)

Rate of digestion continues to increase, abetting the ability to deal with multiple daily meals. However, appetite is reduced and conversion efficiency declines precipitously. As a result the sustained capacity to digest is not accompanied by growth, which follows conversion efficiency in a parallel decline. At 23°C high levels of standard metabolism, maintenance ration, and excretion all combine to suppress growth to the point of extinction.

VII. CONCLUDING COMMENTARY

Over the past two decades significant advances have been made in determining the effects of single and multiple environmental factors on the capacity of fish to grow. By analyzing the growth–ration (GR) response curves (where these have been recorded) an understanding of how the major parameters of growth (G_{max}, G_{opt}, G_0, G_{starv}) and of ration (R_{max}, R_{opt}, R_{maint}) respond to environmental factors has been afforded.

Conducting precise, well-controlled experiments on growth requires close attention to many aspects of biological significance; the experimenter is faced with a choice of feeding strategies and culturing conditions which can profoundly affect the outcome of the research. Food quality, quantity, and timing of presentation influence appetite, feeding reponse, and digestability; water currents, hours of light, space, density, and cover affect excitability and swimming activity thereby influencing the level of daily energy expenditure; exact measurements of food intake, fish weights, and moisture content are required before conversion efficiencies can be determined accurately. Since these normal concomitants may interact with the prescribed environmental conditions (e.g., the higher the temperature, the greater the excitability), knowledge of both the way in which environmental and behavioral factors affect growth helps to avoid pitfalls in the design of growth experiments.

Differences in behavioral response between and within tanks contributes greatly to variance, and will continue to be a challenging problem both as an object of study and as a target for improved experimental control. Considerably increased statistical sensitivity and insight can be obtained by distinctively marking each individual in a group. Assessing growth depensation is essential.

It is apparent that factorial experiments involving small numbers of combinations can provide confusing results, particularly where values of one or more of the factors occur on either side of an optimum. Interaction, rather than interrelation, will be most likely where a Controlling Factor is involved (e.g., temperature or pH). The consequence of a Masking Factor, such as salinity, is complicated by the state of neuroendocrine activity which in turn is subject to cyclic or ontogenetic change (e.g., smolting) under the influence of a Directive Factor (e.g., photoperiod, temperature cycles). The immature, stable stages of development not subject to change through differentiation (e.g., metamorphosis) are undoubtedly the best period in which to study the basic mechanisms of growth.

Since the channeling of food energy is involved, and maintenance requirements have first priority, it is fitting to reemphasize that bioenergetics and growth are inseparable—advances in each go hand in hand.

REFERENCES

Adelman, I. R., and Smith, L. L. (1970). Effect of oxygen on growth and food conversion efficiency of northern pike. *Prog. Fish Cult.* **32,** 93–96.

Alderdice, D. F. (1972). Factor combinations. Responses of marine poikilotherms to environmental factors acting in concert. *In* "Marine Ecology," Vol. 1, "Environmental Factors" (O. Kinne, ed.), Part 3, pp. 1659–1722. Wiley (Interscience), New York.

Allee, W. C., Greenberg, B., Rosenthal, G. M., and Frank, P. (1948). Some effects of social organization on growth in the green sunfish, *Lepomis cyanellus. J. Exp. Zool.* **108,** 1–19.

Allen, K. O. (1974). Effects of stocking density and water exchange rate on growth and survival of channel catfish *Ictalurus punctatus* (Rafinesque) in circular tanks. *Aquaculture* **4,** 29–40.

Altman, P. L., and Dittmer, D. S. (1966). "Environmental Biology." Fed. Am. Soc. Exp. Biol., Biol. Handbooks, Bethesda, Maryland.

Anderson, R. O. (1959). The influence of season and temperature on growth of the bluegill, *Lepomis machrochirus* (Rafinesque). Ph.D. Thesis, Univ. of Michigan, Ann Arbor.

Andrews, J. W., and Page, J. W. (1975). The effects of frequency of feeding on culture of catfish. *Trans. Am. Fish. Soc.* **104,** 317–321.

Andrews, J. W., and Stickney, R. R. (1972). Interactions of feeding rates and environmental temperature on growth, food conversion and body composition of channel catfish. *Trans. Am. Fish. Soc.* **101,** 94–99.

Andrews, J. W., Knight, L. H., and Murai, T. (1972). Temperature requirements for high density rearing of channel catfish *Ictalurus punctatus* from fingerling to market size. *Prog. Fish Cult.* **34,** 240–241.

Andrews, J. W., Murai, T., and Gibbons, G. (1973). The influence of dissolved oxygen on the growth of channel catfish. *Trans. Am. Fish. Soc.* **4,** 835–838.

Averett, R. C. (1969). Influence of temperature on energy and material utilization by juvenile coho salmon. Ph.D. Thesis, Oregon State Univ., Corvallis.

Baerends, G. P. (1971). The ethological analysis of fish behavior. *In* "Fish Physiology" (W. S. Hoar and D. J. Randall, eds.), Vol. 6, pp. 279–370. Academic Press, New York.

Banks, J. L., Fowler, L. G., and Elliott, J. W. (1971). Effects of rearing temperature on growth, body form, and hematology of fall chinook fingerlings. *Prog. Fish Cult.* **33,** 20–26.

Basulto, S. (1976). Induced saltwater tolerance in connection with inorganic salts in the feeding of Atlantic salmon (*Salmo salar* L.). *Aquaculture* **8,** 45–55.

Beamish, F. W. H. (1964). Influence of starvation on standard and routine oxygen consumption. *Trans. Am. Fish. Soc.* **93,** 103–107.

Beamish, F. W. H. (1974). Apparent specific dynamic action of largemouth bass, Micropterus salmoides. *J. Fish. Res. Board Can.* **31,** 1763–1769.

Bern, H. A. (1975). Prolactin and osmoregulation. *Am. Zool.* **15,** 937–948.

Bjorklund, R. G. (1958). The biological function of the thyroid and the effect of length of day on growth and maturation of goldfish, *Carassius auratus* Linn. Ph.D. Thesis, Univ. of Michigan, Ann Arbor.

Blackman, F. F. (1905). Optima and limiting factors. *Ann. Bot.* (*London*) **19,** 282–295.

Blaxter, J. H. S. (1970). Sensory deprivation and sensory input in rearing experiments. *Helgol. Wiss. Meeresunters.* **20,** 642–654.

Blaxter, J. H. S., and Holliday, F. G. T. (1963). The behavior and physiology of herring and other cluepids. *Adv. Mar. Biol.* **1,** 261–393.

Brett, J. R. (1971a). Satiation time, appetite, and maximum food intake of sockeye salmon, *Oncorhynchus nerka. J. Fish. Res. Board Can.* **28,** 409–415.

Brett, J. R. (1971b). Growth responses of young sockeye salmon (Oncorhynchus nerka) to different diets and planes of nutrition. *J. Fish. Res. Board Can.* **28,** 1635–1643.

Brett, J. R. (1974). Tank experiments on the culture of pan-size sockeye (*Oncorhynchus nerka*) and pink salmon (*O. gorbuscha*) using environmental control. *Aquaculture* **4,** 341–352.

Brett, J. R. (1976). Scope for metabolism and growth of sockeye salmon, *Oncorhynchus nerka*, and some related energetics. *J. Fish. Res. Board Can.* **33,** 307–313.

Brett, J. R., and Higgs, D. A. (1970). Effect of temperature on the rate of gastric digestion in fingerling sockeye. *J. Fish. Res. Board Can.* **27,** 1767–1779.

Brett, J. R., and Shelbourn, J. E. (1975). Growth rate of young sockeye salmon, *Oncorhynchus nerka*, in relation to fish size and ration level. *J. Fish. Res. Board Can.* **32,** 2103–2110.

Brett, J. R., Shelbourn, J. E., and Shoop, C. T. (1969). Growth rate and body composition of fingerling sockeye salmon, *Oncorhynchus nerka*, in relation to temperature and ration size. *J. Fish. Res. Board Can.* **26,** 2363–2394.

Brocksen, R. W., and Cole, R. E. (1972). Physiological responses of three species of fishes to various salinities. *J. Fish. Res. Board Can.* **29,** 399–405.

Brown, M. E. (1946a). The growth of brown trout (*Salmo trutta* Linn.). III. The effect of temperature on the growth of 2-year-old trout. *J. Exp. Biol.* **22,** 145–155.

Brown, M. E. (1946b). The growth of brown trout (*Salmo trutta* Linn.). II. Growth of 2-year-old trout at a constant temperature of 11.5°C. *J. Exp. Biol.* **22,** 130–144.

Brown, M. E. (1957). Experimental studies of growth. *In* "Physiology of Fishes" (M. E. Brown, ed.), Vol. 1, pp. 361–400. Academic Press, New York.

Brungs, W. A. (1971). Chronic effects of low dissolved oxygen concentrations on the fathead minnow (*Pimephales promelas*). *J. Fish. Res. Board Can.* **28,** 1119–1123.

Canagaratnam, P. (1959). Growth of fishes in different salinities. *J. Fish. Res. Board Can.* **16,** 121–130.

Canagaratnam, P. (1966). Growth of *Tilapia mossambica* Peters in different salinities. *Bull. Fish. Res. Stn., Ceylon* **19,** 47–50.

Carlson, A. R., and Siefert, R. E. (1974). Effects of reduced oxygen on the embryos and larvae of lake trout (*Salvelinus namaycush*) and largemouth bass (*Micropterus salmoides*). *J. Fish. Res. Board Can.* **31,** 1393–1396.

Carlson, A. R., Siefert, R. E., and Herman, L. J. (1974). Effects of lowered dissolved oxygen concentrations on channel catfish (*Ictalurus punctatus*) embryos and larvae. *Trans. Am. Fish. Soc.* **103,** 623–626.

Chervinski, J. (1961). Laboratory experiments on the growth of *Tilapia nilotica* in various saline concentrations. *Bamidgeh* **13,** 8–14.

Chervinski, J. (1975). Sea basses, *Dicentronchus labrax* (Linne) and *D. punctatus* (Bloch) (Pisces, Serranidae), a control fish in fresh water. *Aquaculture* **6,** 249–256.

Chiba, K. (1966). A study on the influence of oxygen concentration on the growth of juvenile common carp. *Bull. Freshwater Fish. Res. Lab. Tokyo* **15,** 35–47.

Comfort, A. (1956). "The Biology of Senescence." Routledge & Kegan Paul, London.

Cooper, E. L. (1961). Growth of wild and hatchery strains of brook trout. *Trans. Am. Fish. Soc.* **90,** 424–438.

Coutant, C. C., and Cox, D. K. (1975). Growth rates of subadult largemouth bass, 24–33.5°C. Environ. Sci. Div., Oak Ridge Natl. Lab., Oak Ridge, Tennessee.

Cox, D. K. (1975). Growth rate of striped bass, *Morone saxatilis*, as a function of temperature and ration. Annu. Rep. 1975. Environ. Sci. Div., Oak Ridge Natl. Lab., Oak Ridge, Tennessee.

Cox, D. K., and Coutant, C. C. (1975). Growth–temperature response of striped bass, *Morone saxatilis*. Environ. Sci. Div., Oak Ridge Natl. Lab., Oak Ridge, Tennessee. (Also personal communication.)

Davison, R. C., Breese, W. P., Warren, C. E., and Doudoroff, P. (1959). Experiments on the dissolved oxygen requirements of cold-water fishes. *Sewage Ind. Wastes* **31,** 950–966.

De Martini, E. E. (1969). A correlative study of the ecology and comparative feeding mechanism morphology of the Embiotocidae (Surf-fishes) as evidence of the family's adaptive radiation into available ecological niches. *Wasmann J. Biol.* **27,** 177–247.

De Silva, S. S., and Perera, P. A. B. (1976). Studies on the young grey mullet, *Mugil cephalus* L. I. Effects of salinity on food intake, growth and food conversion. *Aquaculture* **7,** 327–338.

Doudoroff, P., and Shumway, D. L. (1967). Dissolved oxygen criteria for the protection of fish. *Am. Fish. Soc. Spec. Publ.* No. **4,** pp. 13–19.

Doudoroff, P., and Shumway, D. L. (1970). Dissolved oxygen requirements for freshwater fishes. *FAO Fish. Tech. Pap.* No. 86.

Ebeling, A. W., and Alpert, J. S. (1966). Retarded growth of the paradise fish *Macropodus opercularis* (L.) in low environmental oxygen. *Copeia* No. 3, pp. 606–610.

Eisler, T. (1957). The influence of light on the early growth of chinook salmon. *Growth* **21,** 197–203.

Elliott, J. M. (1975a). Weight of food and time required to satiate brown trout, *Salmo trutta* L. *Freshwater Biol.* **5,** 51–64.

Elliott, J. M. (1975b). Number of meals in a day, maximum weight of food consumed in a day and maximum rate of feeding for brown trout, *Salmo trutta* L. *Freshwater Biol.* **5,** 287–303.

Elliott, J. M. (1975c). The growth rate of brown trout, *Salmo trutta* L., fed on maximum rations. *J. Anim. Ecol.* **44,** 805–821.

Elliott, J. M. (1975d). The growth rate of brown trout (*Salmo trutta* L.) fed on reduced rations. *J. Anim. Ecol.* **44,** 823–842.

Falk, K. (1968). Versuche zur Forellenmast in Küsten-und Binnengewässern. *Fisch. Forsch. Wiss. Schriftenr.* **6,** 93–98.

Fisher, R. J. (1963). Influence of oxygen concentration and of its diurnal fluctuations on the growth of juvenile coho salmon. M.S. Thesis, Oregon State Univ., Corvallis.

Fry, F. E. J. (1947). Effects of the environment on animal activity. *Univ. Toronto Stud. Biol. Ser.* **55,** 1–62.

Fry, F. E. J. (1971). The effect of environmental factors on the physiology of fish. *In* "Fish Physiology" (W. S. Hoar and D. J. Randall, eds.), Vol. 6, pp. 1–98. Academic Press, New York.

Gerking, S. D. (1966). Annual growth cycle, growth potential, and growth compensation in the bluegill sunfish in northern Indiana lakes. *J. Fish. Res. Board Can.* **23,** 1923–1956.

Gerking, S. D. (1971). Influence of rate of feeding and body weight on protein metabolism of bluegill sunfish. *Physiol. Zool.* **44,** 9-19.

Gibson, M. B., and Hirst, B. (1955). The effect of salinity and temperature on the preadult growth of guppies. *Copeia* No. 3, pp. 241–243.

Glass, N. R. (1969). Discussion of calculation of power function with special reference to respiratory metabolism in fish. *J. Fish. Res. Board Can.* **26,** 2643–2650.

Gross, W. L., Fromm, P. O., and Roelofs, E. W. (1963). Relationship between thyroid and growth in green sunfish, *Leopmis cyanellus* (Rafinesque). *Trans. Am. Fish. Soc.* **92,** 401–408.

Gross, W. L., Roelofs, E. W., and Fromm, P. O. (1965). Influence of photoperiod on growth of green sunfish, *Lepomis cyanellus. J. Fish. Res. Board Can.* **22,** 1379–1386.

Haskell, D. C. (1959). Trout growth in hatcheries. *N.Y. Fish Game J.* **6,** 204–2.37.

Hatanaka, M. A., and Murakawa, G. (1958). Growth and food consumption in young amberfish, *Seriola quinqueradiata* (T. et S.). *Tohoku J. Agric. Res.* **9,** 69–79.

Hatanaka, M. A., and Takahashi, M. (1956). Utilization of food by mackerel *Pneumatophorus japonicus* (Houttuyn). *Tohoku J. Agric. Res.* **7,** 51–57.

Hatanaka, M. A., and Takahashi, M. (1960). Studies on the amounts of the anchovy consumed by the mackerel. *Tohoku J. Agric Res.* **11,** 83–100.

Hatanaka, M. A., Sekino, K., Takahashi, M., and Ichimura, T. (1957). Growth and food consumption in young mackerel, (*Pneumatophorus japonicus*, Houttuyn). *Tohoku J. Agric. Res.* **7,** 351–368.

Henderson, N. E. (1963). Influence of light and temperature on the reproductive cycle of the eastern brook trout, *Salvelinus fontinalis* (Mitchill). *J. Fish. Res. Board Can.* **20,** 859–897.

Hepher, B. (1967). Some biological aspects of warm-water fish pond management. *In* "The Biological Basis of Fresh Water Fish Production" (S. Gerking, ed.), pp. 417–428. Blackwell, Oxford.

Herrmann, R. B., Warren, C. E., and Doudoroff, P. (1962). Influence of oxygen concentration on the growth of juvenile coho salmon. *Trans. Am. Fish. Soc.* **91,** 155–167.

Hogman, W. J. (1968). Annulus formation on scales of four species of coregonids reared under artificial conditions. *J. Fish. Res. Board Can.* **25,** 2111–2112.

Hokanson, K. E. F., Kleiner, C. F., and Thorsland, T. W. (1977). Effects of constant temperature and diel fluctuation on growth, mortality, and yield of juvenile rainbow trout, *Salmo gairdneri* (Richardson). *J. Fish. Res. Board Can.* **34,** 639–648.

Holmes, W. N., and Donaldson, E. M. (1969). Excretion, ionic regulation, and metabolism. *In* "Fish Physiology" (W. S. Hoar and D. J. Randall, eds.), Vol. 1, pp. 1–89. Academic Press, New York.

Hubbs, C. L. (1930). The high toxicity of nascent oxygen. *Physiol. Zool.* **3,** 441–460.

Huh, H. T., Calbert, H. E., and Steiber, D. A. (1976). Effects of temperature and light on growth of yellow perch and walleye using formulated feed. *Trans. Am. Fish. Soc.* **105,** 254–258.

Huisman, E. A. (1974). A study on optimal rearing conditions for carp (*Cyprinus carpio* L.). Spec. Publ., Organisatie ter Verbetering van de Binnenvisserij, Utrecht.

Ishiwata, N. (1968). Ecological studies on the feeding of fishes. IV. Satiation curve. *Bull. Jpn. Soc. Sci. Fish.* **34,** 691–693.

Jansen, A. C. (1938). The growth of the plaice in the transition area. *Rapp. P.-V. Reun. Cons. Int. Explor. Mer* **108,** 104–107.

Kato, T. (1970). Studies on the variation of growth in rainbow trout, *Salmo gairdnerii*. II. Regression line of satiation amount on the body weight as an indicator of food amount. *Bull. Freshwater Fish. Res. Lab.* **20,** 101–107.

Kato, T., and Sakamoto, Y. (1969). Studies on the variation of growth in rainbow trout, *Salmo gairdnerii*. I. The effect of grading of body size on the course of growth. *Bull. Freshwater Fish. Res. Lab.* **19,** 9–16.

Kawanabe, H. (1969). The significance of social structure in production of the "ayu," *Plecoglossus altivelis. In* "Symposium on Salmon and Trout in Streams" (T. G. Northcote, ed.), pp. 243–251. Inst. Fish., Univ. of British Columbia, Vancouver.

Kepshire, B. M., Jr. (1971). Growth of pink, chum, and fall chinook salmon in heated seawater. *Proc. Annu. N.W. Fish Cult. Conf., 22nd* pp. 25–26.

Kepshire, B. M., Jr., and McNeil, W. (1972). Growth of premigratory chinook salmon in seawater. *U.S. Fish Wildl. Serv., Fish. Bull.* **70,** 119–123.

Kerr, S. R. (1971). Analysis of laboratory experiments on growth efficiency of fishes. *J. Fish. Res. Board Can.* **28,** 801–808.

Kilambi, R. V., Noble, J., and Hoffman, C. E. (1970). Influence of temperature and photoperiod on growth, food consumption, and food conversion efficiency of channel catfish. *Proc. Annu. Conf. Southeast. Assoc. Game Fish Comm., 24th* pp. 519–531. (*Aquat. Sci. Fish. Abstr.* **4,** 4Q5139F, p. 231.)

Kinne, O. (1960). Growth, food intake, and food conversion in a euryplastic fish exposed to different temperatures and salinities. *Physiol. Zool.* **33,** 288–317.

Kleiber, M. (1961). "The Fire of Life. An Introduction to Animal Energetics." Wiley, New York.

Knutsson, S., and Grav, T. (1976). Seawater adaptation in Atlantic salmon (*Salmo salar* L.) at different experimental temperatures and photoperiods. *Aquaculture* **8,** 169–187.

Kono, H., and Nose, Y. (1971). Relationship between the amount of food taken and growth in fishes. I. Frequency of feeding for a maximum daily ration. *Bull. Jpn. Soc. Sci. Fish.* **37,** 169–175.

Kwain, W.-H. (1975). Embryonic development, early growth, and meristic variation in rainbow trout (*Salmo gairdneri*) exposed to combinations of light intensity and temperature. *J. Fish. Res. Board Can.* **32,** 397–402.

Larkin, P. A., Terpenning, J. G., and Parker, R. R. (1956). Size as a determinant of growth rate in rainbow trout *Salmo gairdneri. Trans. Am. Fish. Soc.* **86,** 84–96.

Lindsey, C. C. (1958). Modification of meristic characters by light duration in kokane, *Oncorhynchus nerka. Copeia* No. 2, pp. 134–136.

McCormick, J. H. (1976). Temperature effects on young yellow perch, *Perca flavescens* (Mitchell). Ecol. Res. Ser., U.S. Environ. Prot. Agency, Duluth, Minnesota.

McCormick, J. H., Jones, B. R., and Syrett, R. F. (1971). Temperature requirements for growth and survival of larval ciscos (*Coregonus artedii*). *J. Fish. Res. Board Can.* **28,** 924–927.

McCormick, J. H., Hokanson, K. E. F., and Jones, B. R. (1972). Effects of temperature on growth and survival of young brook trout, *Salvelinus fontinalis. J. Fish. Res. Board Can.* **29,** 1107–1112.

McCormick, J. H., Jones, B. R., and Hokanson, K. E. F. (1977). White sucker (*Catostomus commersoni*) embryo development, and early growth and survival at different temperatures. *J. Fish. Res. Board Can.* **34,** 1019–1025.

MacCrimmon, H. R., and Kwain, W. -H. (1969). Influence of light on early development and meristic characters in the rainbow trout, *Salmo gairdneri* Richardson. *Can. J. Zool.* **47,** 631–63.

Magnuson, J. J. (1962). An analysis of aggressive behavior, growth, and competition for food and space in medaka, *Oryzias latipes* (Pisces, Cyprinodontidae). *Can. J. Zool.* **40,** 313–363.

Nagoshi, M. (1967a). Experiments on the effects of size hierarchy upon the growth of guppy (*Lebistes reticulatus*). *J. Fac. Fish. Prefect. Univ. Mie* **7,** 165–189.

Nagoshi, M. (1967b). On the effects of size hierarchy upon the growth of fishes. *J. Fac. Fish. Prefect. Univ. Mie* **7,** 191–198.

Needham, A. E. (1964). "The Growth Process in Animals." Pitman London.

Nelson, D. J. (1974). Temperature effects on growth of largemouth bass. Annu. Prog. Rep., pp. 28–29. Environ. Sci. Div., Oak Ridge Natl. Lab., Oak Ridge, Tennessee.

Nelson, D. J. (1975). Growth, consumption, and conversion rates as a function of temperature of subadult striped bass. Annu. Prog. Rep., pp. 45–47. Aquat. Stud., Environ. Sci. Div., Oak Ridge Natl. Lab., Oak Ridge, Tennessee.

Niimi, A. J., and Beamish, F. W. H. (1974). Bioenergetics and growth of largemouth bass (*Micropterus salmoides*) in relation to body weight and temperature. *Can. J. Zool.* **52,** 447–456.

Nose, T. (1963). Determination of nutritive value of food protein on fish. II. Effect of amino acid composition of high protein diets on growth and protein utilization of the rainbow trout. *Bull. Freshwater Fish. Res. Lab.* **13,** 41–50.

Novotny, A. J. (1975). Net-pen culture of Pacific salmon in marine waters. *Mar. Fish. Rev.* **37,** 36–47.

Oshima, Y., and Ihaba, D. (1969). "Fish Culture," Vol. **4,** "Yellowtail—Amber Jack." Publ. Midori-Shobo, Tokyo.

Otto, R. G. (1971). Effects of salinity on the survival and growth of pre-smolt coho salmon (*Oncorhynchus kisutch*). *J. Fish. Res. Board Can.* **28,** 343–349.

Otwell, W. S., and Merriner, J. V. (1975). Survival and growth of juvenile striped bass, *Morone saxatilis*, in a factorial experiment with temperature, salinity and age. *Trans. Am. Fish. Soc.* **104,** 560–566.

Paloheimo, J. E., and Dickie, L. M. (1966). Food and growth of fishes. III. Relations among food, body size, and growth efficiency. *J. Fish. Res. Board Can.* **23,** 1209–1248.

Pentelow, F. T. K. (1939). The relation between growth and food consumption in the brown trout (*Salmo trutta*)*J. Exp. Biol.* **16,** 446–473.

Pessah, E., and Powles, P. M. (1974). Effect of constant temperature on growth rates of pumpkinseed sunfish (*Lepomis gibbosus*). *J. Fish. Res. Board Can.* **31,** 1678–1682.

Peters, D. S. (1971). Growth and energy utilization of juvenile flounder, *Paralichthys dentatus* and *Paralichthys lethostigma*, as affected by temperature, salinity, and food availability. Ph.D. Thesis, Dep. Zool., North Carolina State Univ., Raleigh.

Peters, D. S., and Boyd, M. T. (1972). The effect of temperature, salinity, and availability of food on the feeding and growth of the hogchoker, *Trinectes maculatus* (Block and Schneider). *J. Exp. Mar. Biol. Ecol.* **9,** 201–207.

Pyle, E. A. (1969). The effect of constant light or constant darkness on the growth and sexual maturity of brook trout. *Fish. Res. Bull.* No. 31, pp. 13–19.

Rasquin, P., and Rosenbloom, L. (1954). Endocrine imbalance and tissue hyperplasia in teleosts maintained in darkness. *Bull. Am. Mus. Nat. Hist.* **104,** 361–425.

Refstie, T., and Kittelsen, A. (1976). Effect of density on growth and survival of artificially reared Atlantic salmon. *Aquaculture* **8,** 319–326.

Richards, C. M. (1958). The inhibition of growth in crowded *Rana pipiens* tadpoles. *Physiol. Zool.* **31,** 138–151.

Saunders, R. L., and Henderson, E. B. (1969). Survival and growth of Atlantic salmon parr in relation to salinity. *Fish. Res. Board Can. Tech. Rep.* No. 147.

Saunders, R. L., and Henderson, E. B. (1970). Influence of photoperiod on smolt development and growth of Atlantic salmon (*Salmo salar*). *J. Fish. Res. Board Can.* **27,** 1295–1311.

Shaw, H. M., Saunders, R. L., Hall, H. C., and Henderson, E. B. (1975a). The effect of dietary sodium chloride on growth of Atlantic salmon (*Salmo salar*) parr. *J. Fish. Res. Board Can.* **32,** 1813–1819.

Shaw, H. M., Saunders, R. L., and Hall, H. C. (1975b). Environmental salinity: its failure to influence growth of Atlantic salmon (*Salmo salar*) parr. *J. Fish. Res. Board Can.* **32,** 1821–1824.

Shehadeh, Z. H., and Gordon, M. S. (1969). The role of the intestine in salinity adaptation of the rainbow trout, *Salmo gairdneri*. *Comp. Biochem. Physiol.* **30,** 397–418.

Shelbourn, J. E. (1976). Early growth rates of chum salmon fry (*Oncorhynchus keta*) in the laboratory in fresh and salt water. Unpublished manuscript, Pacific Biological Station, Nanaimo, British Columbia.

Shelbourn, J. E., Brett, J. R., and Shirahata, S. (1973). Effect of temperature and feeding regime on the spcific growth rate of sockeye salmon fry (*Oncorhynchus nerka*), with a consideration of size effect. *J. Fish. Res. Board Can.* **30,** 1191–1194.

Smith, M. A. K., and Thorpe, A. (1976). Nitrogen metabolism and trophic input in relation to growth in freshwater and saltwater. *Biol. Bull.* (*Woods Hole, Mass.*) **150,** 139–151.

Stauffer, G. D. (1973). A growth model for salmonids reared in hatchery environments. Ph.D. Thesis, Univ. of Washington, Seattle.

Stewart, N. E., Shumway, D. L., and Doudoroff, P. (1967). Influence of oxygen concentration on the growth of juvenile largemouth bass. *J. Fish. Res. Board Can.* **24,** 475–494.

Strawn, K. (1961). Growth of largemouth bass fry at various temperatures. *Trans. Am. Fish. Soc.* **90,** 334–335.

Swift, D. R. (1955). Seasonal variations in the growth rate, thyroid gland activity, and food reserves of brown trout, (*Salmo trutta* Linn.). *J. Exp. Biol.* **32,** 751–764.

Swift, D. R. (1959). Seasonal variation in the activity of the thyroid gland of yearling brown trout (*Salmo trutta* Linn.). *J. Exp. Biol.* **36,** 120–125.

Swift, D. R. (1960). Cyclical activity of the thyroid gland of fish in relation to environment changes. *Symp. Zool. Soc. London* No. 1, pp. 17–27.

Swift, D. R. (1961). The annual growth rate cycle in brown trout (*Salmo trutta* Linn.) and its cause. *J. Exp. Biol.* **38,** 595–604.

Swift, D. R. (1963). Influence of oxygen concentration on growth of brown trout, *Salmo trutta* L. *Trans. Am. Fish. Soc*, **92,** 300–301.

Swift, D. R. (1964). The effect of temperature and oxygen on the growth rate of the Windermere char (*Salvelinus alpinus willughbii*). *Comp. Biochem. Physiol.* **12,** 179–183.

Thompson, D. H. (1941). The fish production of inland streams and lakes. *Symp. Hydrobiol., Univ. Wis., Madison* pp. 206–217.

Ursin, E. (1963). On the incorporation of temperature in the von Bertalanffy growth expression. *Medd. Dan. Fisk.- Havunders.* **4,** 1–16.

Wagner, H. H. (1974). Photoperiod and temperature regulation of smolting in steelhead trout (*Salmo gairdneri*). *Can. J. Zool.* **52,** 219–240.

Warren, C. E. (1971). "Biology and Water Pollution Control." Saunders, Philadelphia, Pennsylvania.

Warren, C. E., and Davis, G. E. (1967). Laboratory studies on the feeding bioenergetics and growth of fishes. *In* "The Biological Basis of Freshwater Fish Production" (S. D. Gerking, ed.), pp. 175–214. Blackwell, Oxford.

Weatherley, A. H. (1972). "Growth and Ecology of Fish Populations." Academic Press, New York.

West, B. W. (1965). Growth, food conversion, food consumption, and survival at various temperatures, of the channel catfish *Ictalurus punctatus* (Rafinesque). M.S. Thesis, Univ. of Arkansas, Fayetteville.

West, B. W. (1966). Growth rates at various temperatures of the orange-throat darter *Etheostoma spectabilis*. *Proc. Ark. Acad. Sci.* **20,** 50–53.

West, L. B. (1960). The nature of growth inhibiting material from crowded *Rana pipiens* tadpoles. *Physiol. Zool.* **33,** 232–239.

Whitworth, W. R. (1968). Effects of diurnal fluctuations of dissolved oxygen on the growth of brook trout. *J. Fish. Res. Board Can.* **25,** 579–584.

Wirtz, P. (1974). The influence of the sight of a conspecific on the growth of *Blennius pholis* (Pisces, Teleostei). *J. Comp. Physiol.* **91,** 161–165.

Wohlfarth, G. W., and Moav, R. (1972). The regression of weight gain on initial weight in carp. I. Methods and results. *Aquaculture* **1,** 7–28.

Wurtsbaugh, W. A. (1973). Effects of temperature, ration, and size on the growth of juvenile steelhead trout, *Salmo gairdneri*. M.S. Thesis, Oregon State Univ., Corvallis.

Yamagishi, H. (1962). Growth relation in some small experimental populations of rainbow trout fry, *Salmo gairdneri* Richardson, with special reference to social relations among individuals. *Jpn. J. Ecol.* (*Nippon Seitai Gakkaishi*) **12,** 43–53.

Yamagishi, H. (1963). Some observations on growth variation and feeding behavior in the fry of two races of Japanese crucian carp, *Carassius carassius* L. *Jpn. J. Ecol.* (*Nippon Seitai Gakkaishi*) **13,** 156–161.

Yamagishi, H. (1969). Postembryonal growth and its variability of the three marine fishes with special reference to the mechanism of growth variation in fishes. *Res. Popul. Ecol.* (*Kyoto*) **11,** 14–33.

Yamagishi, H., Maruyama, T., and Mashiko, K. (1974). Social relation in a small experimental population of *Odontobutis obscurus* (Temminck et Schlegel) as related to individual growth and food intake. *Oecologia* (*Berlin*) **17,** 187–202.

Yu, M.-L. (1968). A study on the growth inhibiting factors of zebra fish, *Brachydanio rerio*, and blue gourami, *Trichogaster trichopterus*. Ph.D. Thesis, Dep. Biol., New York Univ., New York.

Zaugg, W. S., and McLain, L. R. (1969). Inorganic salt effects on growth, saltwater adaptation, and gill ATPase of Pacific salmon. *In* "Fish in Research" (O. W. Newhaus and J. E. Halver, ed.), pp. 293–306. Academic Press, New York.

Chapter 19

"Growth rates and models": A classic to read and to use!

Céline Audet*
Institut des sciences de la mer de Rimouski (ISMER), Université du Québec à Rimouski, Rimouski, QC, Canada
**Corresponding author: e-mail: celine_audet@uqar.ca*

Céline Audet discusses the impact of William E. Ricker's "Growth Rates and Models" in Fish Physiology, Volume 8, published in 1979.

"Growth Rates and Models," written by W. E. Ricker, a renowned Canadian fisheries scientist, still fills a critical need in fish biology. It has been cited every year since its publication in 1979, which not only indicates how the subject is fundamental in fish biology, but also how scientifically sound it was to so successfully stand the test of time. Growth rate is such a key index that we tend to apply the formula without revisiting the principles supporting this approach, and this metric is used in fish physiology, developmental and environmental studies, aquaculture, and fisheries. Ricker's chapter presents basic equations, reminds one of the notions of allometry and isometry, how growth varies with age, how to fit growth curves, and so on. But more important, it also highlights the difficulties related to sampling in the wild, sampling bias, and pitfalls to avoid in the interpretation of growth fitting. Do not be intimidated by the math! The text is easy to understand and written with a sense of humor that, unfortunately, we do not see much these days in the scientific literature.

To any physiologist involved in fisheries or aquaculture research (and many others!), W. E. Ricker is a keystone reference, even if they are not aware of it. His chapter, "Growth rates and models," published in Fish Physiology in 1979, is cited every year and has received an average of 30 citations per year for the last 15 years. At the time of submitting this overview of his chapter, I found that the number of citations for the current year is already over 20, showing that his chapter is not only a classic, but that it fills a critical need in fish research for measuring and estimating growth.

Who was Bill Ricker? A Fellow of the Royal Society of Canada (1956) and Officer of the Order of Canada (1986), Bill Ricker was a fisheries scientist who produced close to 300 publications and was renowned for his work on biological statistics of fish populations and stock recruitment (Beamish et al., 2003). After completing a Ph.D. at the University of Toronto (1936), he began

Fish Physiology, Vol. 40B. https://doi.org/10.1016/bs.fp.2024.05.008

his career at the Fisheries Research Board in British Columbia and in 1939 he became professor of zoology at Indiana University (Power, 2001). In 1950, he came back to Canada and became editor of publications for the Fisheries Research Board. He was stationed at the Pacific Biological Station, where he became Chief Scientist of the Fisheries Research Board and continued to do research even after his retirement in 1973. The research vessel *W.E. Ricker*, named in his honor, was often moored outside his office window (Metcalfe, pers. comm.), which is quite unique and certainly meant a lot. Dr. Ricker passed away in 2001.

Why is it so important to measure growth? Growth is a function among others, so why is there such an interest in estimating it? I would be tempted to say that this is because it can represent a fish's general status. Measuring growth is a way to estimate the fish's general condition: a fish that grows well is perceived as healthy. In fisheries, evaluating growth-at-age provides important clues for determining the stock's evolution, and for assessing its durability. In periods of climate change, the same metrics allow the evaluation of tolerance to new environmental conditions and of geographic expansion or limitation of suitable habitat conditions; permit the prediction of new species assemblages and the vulnerability to predation, and so on. Growth is one of the main indicators revealing how fishes perform in the wild.

In the aquaculture industry, fish growth is a prerequisite to profitability. Aquaculture scientists not only use growth measurements to identify optimal rearing conditions, but also to compare genetic-based performance, to study characteristics of young developmental stages (which are the most difficult to optimize in production), to establish environmental conditions that are needed for successful transition from one developmental stage to the next, and to test the suitability of new feeds that will ensure durability of fish production. This is a non-exhaustive list, indicating just how integral growth measurement is in this field.

Indeed, growth measurement is a vital tool used to resolve a plethora of questions related to fish physiology. But let's come back to Ricker's chapter. Personally, I liked it for three main reasons. (1) In this chapter, Ricker reminds us of the principal equations that are needed to measure fish growth. While doing research to prepare this overview, I even learned that the origin of the appellation "Fulton's condition factor" should indeed be attributed to Ricker (Nash et al., 2006). (2) This chapter is a remarkable historical overview of the different early attempts made to evaluate fish growth. (3) The text is scattered with colorful remarks ("fanciful speculations"; "for the resulting abstruse speculations"; "we experience a feeling of disappointment"; "but fail to indicate how"; and the one I prefer, "it is difficult to picture how this formula can be anything more than a mathematical curiosity of no real utility"), lightening a mathematical text that could otherwise have been a grueling effort for many.

There are two semantic issues that need to be explained before describing this chapter. Ricker used the word "stanzas" to distinguish different phases of life history characterized by different growth characteristics (transformations relative to metamorphosis, physiological changes occurring in diadromous fishes and associated with trophic or reproductive migrations, diet shifts, seasonal growth cycles), and he made a distinction between "stanza" and "stage of development." I found the use of this term quite uncommon, but also found it interesting, biologically speaking. In his chapter, Ricker stated that he used mass, biomass, and weight "interchangeably." However, I prefer to use "mass," and will do so in the following lines.

The chapter published by Ricker in Fish Physiology is divided into six sections: (1) Measurement of fish length and weight; (2) Estimation of growth rate in nature; (3) Characteristics of fish growth; (4) Growth models related to age; (5) Growth in relation to temperature; (6) Growth in relation to ration. For fish physiologists, fisheries biologists, and ecologists, all sections are of interest. Do not be afraid of the equations, the text will guide you.

The first section begins with an overview of the different ways to measure growth. This may seem basic, but everyone who measures fish growth should know how standard length, fork length, and total length (among others) differ. Fork length, as stated by Ricker, "is widely used by fishery biologists." Fork length is certainly *my* favorite; which is yours?

Ricker then reminds us that when estimating mass from length or length from mass, "at each stage of the life cycle, mass varies as some power of length," and he reiterates the importance of getting as much data as possible to obtain a general mass–length relationship for a population. I loved reading "... all biological situations where variability is for the most part inherent in the material rather than a result of errors in measuring"; it reminded me of how many times I taught my students that what they called "outliers" were in fact part of the equation and should not be excluded. "Always keep everything as simple of possible" is something that must also be remembered. Finally, the notion that "any measure of growth must be referred to some definite interval of time" is so important. Yes, measuring growth rate is what we mean most of the time when we measure "growth."

Indeed, in most lab experiments, we are interested in measuring relative growth and relative growth rate. This section of the chapter indicates the main equations used in most papers along with equations allowing one to measure the instantaneous growth rate (for a particular instant of time). One thing is sure, he knew his math and equation transformations! Luckily, we do not have to remember how to do the calculations—the final equations are listed here.

As a fishery biologist, Ricker was well aware of the different issues that scientists face when interpreting length frequency distributions and what Ricker calls "humps" in these distributions. Age can sometimes—but not always—be determined by marks on scales or bones, so many fisheries studies need to calculate age distributions to study recruitment rates, displacements during life stages, or to monitor populations over the years. This is something Wahiba Ait Youcef (our Ph.D. student at the time), Yvan Lambert (DFO fishery biologist, Mont-Joli, QC), and I became acutely aware of while studying length frequencies in the estuary and Gulf of St. Lawrence of Greenland halibut, a fish for which age determination through scale or otolith measurement is not possible (Youcef et al., 2015). In Ricker's section on the estimation of the growth rate in nature, he points out the importance of time series to interpret successive humps in length measurements, the importance of mortality rates in the different age groups, the relationship between growth rate and susceptibility to predation, and sex differences in growth and mortality rates, not forgetting size selection by a fishery. Ricker thoroughly describes the difficulties of samplings in the wild and the different biases that can be encountered and that must be accounted for. It reminded me why I loved working with fishery scientists throughout my career: they certainly learned a lot from him, and they in turn taught me as much as Ricker did. It is in this section of his chapter that Ricker discusses isometric (body parts being compared growing proportionally) and allometric (body parts being compared growing at different rates) growth, presents growth models related to age, describes how growth varies during early life, and explains how to fit growth curves. This is probably the heart of his chapter. Importantly, Ricker insists that growth models should be seen as means to fit lines to data, and models should not be forced based on physiological interpretations.

The next two sections are about growth in relation to temperature and in relation to food ration. These two sections are of great interest to aquaculture or experimental biology scientists. Not only does temperature influence fish metabolism, but the relationship between growth and temperature evolves according to fish age. We just need to think about very young stages, which require a fixed number of degree days to reach the next step in their development, a relationship, which "within limits, is independent of the temperatures that actually prevail." The relationship between growth and temperature also presents other particularities including fish that tend to live longer and grow larger in the cooler parts of their geographical range, or variations of growth rates according to whether or not fish are exposed to their optimal temperature range for growth. In section 5, Ricker identified a growth–temperature relationship "flexible enough to take care of almost any set of observations."

When temperature varies, food consumption may also vary to meet the different energy requirements; this issue is one that biologists modeling growth are well aware of. Ricker presents the different curve models that existed at the time he wrote his chapter. I am not skilled enough in growth models to appreciate how much this biological field has evolved. Ricker went through five growth–food ration analysis and concludes, "It now seems safe to conclude that no such simple relationship exists." I will let the "growth model biologists" comment on this.

The summary is great and wrapped up all the essentials.

Reading the classics is not only still informative biologically speaking, but it also gives us a historical overview on how knowledge evolves.

Enjoy!

References

Beamish, R.J., Noakes, D.J., Noakes, D.L.G., Beamish, F.W.H., 2003. William Edwin Ricker-in memoriam. Can. J. Fish. Aquat. Sci. 60, iii–v. https://doi.org/10.1139/f03-900.

Nash, R.D., Valencia, A.H., Geffen, A.J., 2006. The origin of Fulton's condition factor—setting the record straight. Fisheries 31 (5), 236–238.

Power, M., 2001. William E. Ricker, 1908-2001. FishBytes. The newsletter of the Fisheries Centre–University of British Columbia 7(5), 1–2.

Youcef, W.A., Lambert, Y., Audet, C., 2015. Variations in length and growth of Greenland halibut juveniles in relation to environmental conditions. Fish. Res. 167, 38–47. https://doi.org/10.1016/j.fishres.2015.01.007.

Chapter 20

GROWTH RATES AND MODELS☆

W.E. RICKER

Chapter Outline

☆This is a reproduction of a previously published chapter in the Fish Physiology series, "1979 (Vol. 8)/ Growth Rates and Models: ISBN: 978-0-12-350408-1; ISSN: 1546-5098".

Fish Physiology, Vol. 40B. https://doi.org/10.1016/bs.fp.2024.07.009

I. MEASURING LENGTH AND WEIGHT OF FISH

A. Methods of Measuring Length

Fish lengths have been measured in many different ways (Fig. 1). The differences arise from choosing different reference points near the anterior end and near the posterior end of the fish, and from using different methods of making the measurement.

Anterior reference points (A) include (1) tip of the snout; (2) tip of the snout or of the lower jaw when the mouth is closed, whichever protrudes farther; (3) anterior margin of the orbit of the eye; (4) midline of the orbit; (5) posterior margin of the orbit.

Posterior reference points (P) include (1) end of the scale covering of the body; (2) hind margin of the hypural bone of the tail (usually located by the

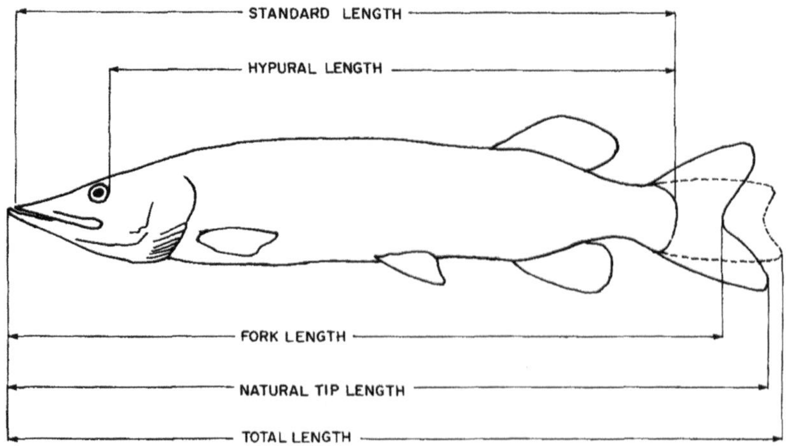

Fig. 1. Definition of five different measurements of the length of a fish.

position of a crease when the tail is bent a little); (3) tip of the shortest median ray of the tail fin; (4) tip of the longest lobe of the tail when held in a natural position; (5–7) tip of the upper (5), the lower (6), or the longer (7) lobe of the tail when squeezed into a position of maximum extension.

Methods of making the measurement (M) include (1) using calipers; (2) using a tape held along the curve of the body; (3) laying the fish on a measuring board with the front end pressed against an upright piece; (4) laying the fish on a board with a movable cross hair above it, attached to an indicator running along a scale.

In theory any combination of reference points and methods might be used, but practice is considerably more restricted. Some of the commoner combinations have special names; these are given below, together with the reference points used.

Standard length: A1, P2, and M1 or M4. Used mainly by systematists.

Median length or fork length (formerly often called total length): A2, P3, and M3 or M4. Widely used by fishery biologists for both marine and freshwater fishes.

Total length, extreme tip length: A2, P7, and M3 or M4. Widely used for freshwater fishes in the United States, and the usual "legal" length measurement there.

Postorbital–hypural length (also called simply hypural length): A5, P2, M4. Used recently by salmon biologists in British Columbia to avoid confusion caused by elongation of the snout and fraying of the tail at maturity. Anterior points A3 and A4 have been used elsewhere for the same purpose, but the hind margin of the orbit is solid bone and makes a more definite reference point.

Natural tip length: A2, P4, and M3 or M4. This measure is often used in Europe, but for every fish it is necessary to decide what tail position is natural, and different observers tend to make different choices.

In addition to the above methodological differences, length will vary with the condition of the fish, for example, whether it is alive, recently killed, after rigor mortis has set in, or at different intervals of time after preservation in formalin or alcohol.

B. Methods of Obtaining Fish Weights

In this chapter the terms mass, biomass, and weight are used interchangeably for the whole weight of a fish. There are differences in the methods and conditions under which weight is determined, just as for length. The usual procedure is to weigh the fish whole, either alive or after death. If it is alive, or preserved in liquid, some standard drip period or amount of blotting should be adopted in each experiment, but I know of no general rules about this. Also, weight can change somewhat after death with exposure to air, and also on preservation and afterward.

Stomach contents often contribute substantially to variability in weight, but usually no attempt is made to adjust for this. However, when cultured fish are weighed, greater uniformity can be achieved by always doing it at the same time of day, and the same time after the last feeding.

Another big source of variability in weight of adult fish is the seasonal sexual cycle. Particularly in females, gonad weight when the current year's eggs are nearing maturity is far greater than when the ovary is in the resting condition.

In weighing commercial catches it is sometimes necessary to use eviscerated fish, with or without the gills left in, and with or without the head left on.

C. Conversions between Length or Weight Measurements

Choice of a length must be governed by convenience and by custom, and as much uniformity as possible is desirable; nevertheless there will always be a need for conversion from one system to another. Pairs of measurements from the same fish are compared graphically or by means of a regression equation. Some important considerations here as follows.

1. Comparisons should include as wide a range of sizes as possible. Failing this, very large numbers of fish must be measured in order to obtain a reliable conversion line.
2. If X and Y represent the two measurements being compared, the regression line should pass through the point $\overline{X}, \overline{Y}$—the means of the two types of measurements—and its slope should be a functional regression, as defined in the next section. This line is symmetrical with respect to X and Y, and can be used to convert X measurements to Y, or Y to X.
3. If the intercept of the functional regression line does not differ significantly from the origin (as is usually the case), it is desirable to use the line that joins the origin to $(\overline{X}, \overline{Y})$ for conversion purposes. This means using a simple factor, which is a great computational convenience.

Conversions between different types of weights should follow the same procedure as for lengths.

D. Estimation of Weight from Length, and Length from Weight

It has been found that, within any stanza of a fish's life, the weight varies as some power of length:*

$$w = al^b \quad (1)$$

$$\log w = \log a + b(\log l) \quad (2)^*$$

*The symbol ln means natural logarithm; log is used when either natural or base-10 logarithms may be employed.

These expressions would apply best to an individual fish that is measured and weighed in successive years of its life. This of course is rarely possible. The value of b is usually determined for a population, by plotting the logarithm of weight against the logarithm of length for a large number of fish of various sizes, the slope of the fitted line being an estimate of b. As in converting between lengths, and indeed in all biological situations where variability is for the most part inherent in the material rather than a result of errors in measuring, a "functional" line should be used: that is, the two variates must be treated symmetrically (Ricker, 1973, 1975b). Putting $\log l = X$ and $\log w = Y$, a least-squares functional line can be obtained by first dividing the values of X and Y by their respective standard deviations; when these transformed data are plotted so that one unit occupies the same distance on both axes, the functional regression line minimizes the sum of squares of the distances from the observed points to itself, and it has a slope of 1. However, it is not necessary to go to the trouble of transforming the data in this way, for it was shown by Teissier (1948) that the slope of the functional line is simply the ratio of the standard deviations of Y and X—surprising as this may seem. It is also the geometric mean of the ordinary regression of Y on X and the reciprocal of the regression of X on Y (or vice versa) and for that reason it has been called the geometric mean (GM) functional regression; other names are standard major axis (Jolicoeur, 1975) and reduced major axis (Teissier, 1948; Kermack and Haldane, 1950; Imbrie, 1956). The GM functional regression is also equal to the ordinary regression of Y on X divided by the coefficient of correlation between Y and X.

The two regressions are compared in Fig. 2. When variability is small, and the range of lengths and weights available is large, there is not much difference between the functional line and an ordinary regression line. But when, as in Fig. 2, neither of these conditions obtains, the difference can be considerable. In any event, the ordinary regression is always numerically smaller than the functional regression, and it is well to avoid any kind of consistent bias, however small.

The functional regression can be expressed either as the change in log weight per unit change in log length, or as the change in log length per unit change in log weight, but exactly the same line is specified in either case; thus the same line is used to convert from weight to length or from length to weight. This contrasts with the ordinary regression of log weight on log length, which is a different line from the regression of log length on log weight.

When data are extensive they have sometimes been grouped into short length-classes, and the mean length and weight of each class have been used as the primary data in computing the regression line. The motive is to speed up computations, but with modern equipment this is no longer necessary, and when using means it is impossible to compute a representative standard error, nor can the functional regression be computed.

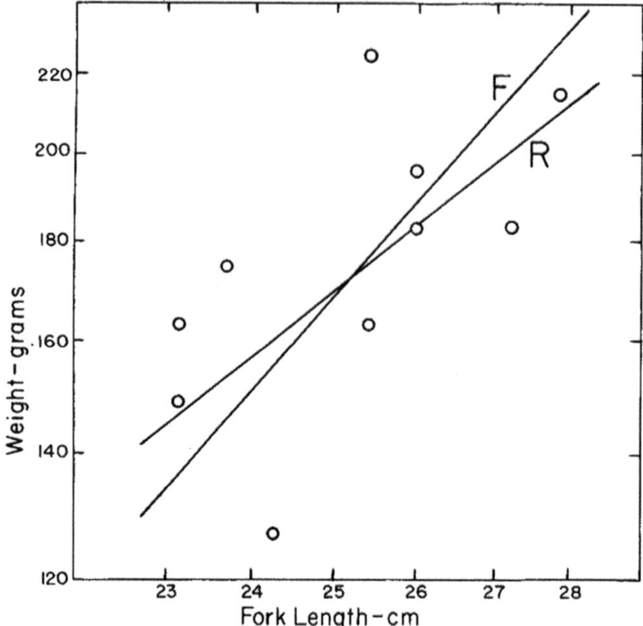

Fig. 2. Relation between weight and length of ten specimens of brook trout, plotted on logarithmic axes. The slope of the ordinary regression of weight on length (R) is 1.955, while that of the GM line (F) is 2.816. The latter is much nearer the true relationship, which is close to 3 in this species.

A more legitimate reason for grouping data is to distribute the observations more evenly among the range of sizes present, and so get a more representative relationship. This is best done by measuring some fixed number of fish within each of a series of short length intervals and weight intervals. The intervals used should preferably be in logarithms. However, to obtain a representative functional regression it is necessary to select half of the total sample on the basis of length and the other half on the basis of weight—otherwise there will be bias (Ricker, 1973).

When a general weight–length relationship for a population is desired, every effort should be made to obtain fish of a wide range of sizes, down to and including age 0 (unless of course the younger fish belong in a different growth stanza). When only a short range of fish sizes is available the parameters estimated can deviate importantly from the population values simply from sampling variability.

E. Numerical Representation of Growth

Growth may be described in terms of length (l) or weight (w), but for simplicity we will set down the expressions for weight only. Any measure of growth must be referred to some definite interval of time, either expressed or implied. It is when this interval is equal to one unit of whatever measure of time is being employed (days, months, years) that it becomes most appropriate to speak of a *growth rate*, although there has been no great consistency in this usage.

For an interval of time from t_1 to t_2 we may distinguish the following.

1. Absolute growth (or absolute increment), and *absolute growth rate*:

$$w_2 - w_1 \quad \text{and} \quad \frac{w_2 - w_1}{t_2 - t_1} \tag{3}$$

2. Relative growth and *relative growth rate* [often associated with the name of Minot (1891)]:

$$\frac{w_2 - w_1}{w_1} \quad \text{and} \quad \frac{w_2 - w_1}{w_1(t_2 - t_1)} \tag{4}$$

Special cases of the above rates occur when they refer to a particular instant of time rather than an interval. The absolute growth rate is then represented by dw/dt, and the relative growth rate becomes $(dw/dt)/w = dw/wdt = G$. The most appropriate name for G would be the instantaneous relative growth rate, but this is usually shortened to *instantaneous growth rate*, and is so used in this chapter. Other names are the *specific, intrinsic, exponential, logarithmic,* or *compound interest* rate. In biology this measure of growth is often associated with the name of Schmalhausen (1926). Its great advantage is that it is additive.

When growth of a fish continues at a constant instantaneous rate for a finite interval of time, its size at any time during that interval is described by an exponential curve having the formula

$$w = ae^{Gt} \tag{5}$$

where a is initial size (when $t = 0$). The absolute rate of growth at any time t during the interval is described by the slope of Eq. (5) at time t:

$$\frac{dw}{dt} = aGe^{Gt} \tag{6}$$

The constant instantaneous rate of growth is of course dw/wdt, or Eq. (6) divided by Eq. (5), which is G.

In actual work growth must be measured over an interval of time rather than at a particular instant. Let w_1 and w_2 be the weights of a fish at times t_1 and t_2; substitute each of these pairs of values in Eq. (5), take logarithms, subtract, and transpose; this gives

$$\frac{\ln w_2 - \ln w_1}{t_2 - t_1} = G \tag{7}$$

Equation (7) is the one that is usually used in practice to estimate the instantaneous rate of growth. Sometimes G is approximated by $(w_2 - w_1)/w_1$ or, somewhat more accurately, by $2(w_2 - w_1)/(w_2 + w_1)$, when $t_2 - t_1 = 1$ and the unit of time is short —1 day, for example.

For some purposes it is not even necessary that growth be exponential in order for a value of G computed from Eq. (7) to be useful. As an illustration, suppose w_2 and w_1 are measured 1 year apart; G is then the instantaneous rate of growth for that year, even though the fish's increase in weight actually followed a seasonal S-shaped cycle rather than an exponential curve. Although G cannot then be used to compute fish sizes within the year, it can be used for comparisons with other instantaneous rates on a yearly basis. For a population, for example, the instantaneous mortality rate for the year can be subtracted from the instantaneous rate of growth to give the net rate of increase or decrease in biomass for the year, regardless of how either growth or mortality are distributed throughout the year.

For a unit interval of time we can compute from Eq. (5)

$$\frac{w_2}{w_1} = e^G \quad (t_2 - t_1 = 1) \tag{8}$$

This is sometimes called the *finite rate of growth.*

The various definitions above can also be applied to length, although instantaneous rates are only rarely used for length. As a matter of fact the instantaneous rate of increase in length and instantaneous rate of increase in weight are very similar statistics, differing only by a constant. For a unit time interval we may combine Eq. (7) with Eq. (2), as follows.

$$\begin{aligned} G &= \ln w_2 - \ln w_1 \\ &= \ln a + b(\ln l_2) - \ln a - b(\ln l_1) \\ &= b(\ln l_2 - \ln l_1) \end{aligned} \tag{9}$$

This provides a convenient method of estimating G from length data, provided b is known.

Figure 3 compares some of the growth indices described above.

II. ESTIMATION OF GROWTH RATES IN NATURE

A. Age from Frequency Distributions (Petersen's Method)

Samples of fish taken in nature frequently exhibit well-marked humps in their length frequency distribution, especially near the lower end of the size range. These can have different interpretations.

1. A succession of humps among the smaller fish may represent a succession of spawnings that occurred at different times during the current year.
2. The humps may represent successive broods (year-classes) of fish, spawned in successive calendar years.

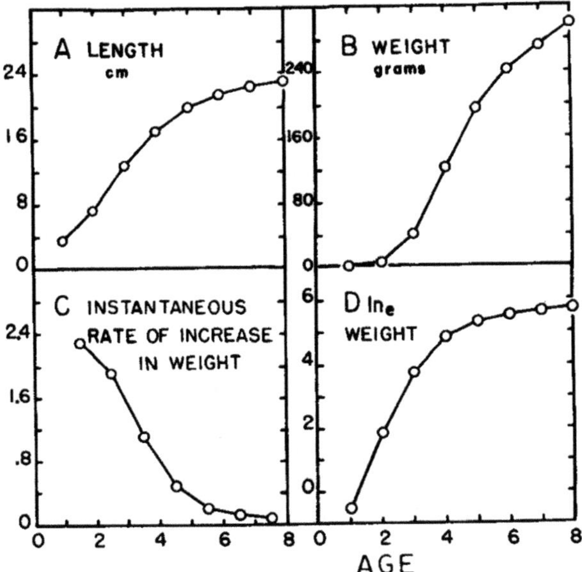

Fig. 3. Comparison of different measures of growth, for bluegill sunfish (*Lepomis macrochirus*). The instantaneous rates in panel C are the slopes of the line segments in panel D. (From Ricker, 1975a, *Bulletin of the Fisheries Research Board of Canada*, No. 191; reproduced by permission.)

3. The humps may represent year-classes spawned 2 or more years apart. This can occur when success of reproduction varies considerably from year to year.
4. The humps may represent merely random sampling variability.

If there is doubt about the nature of any series of humps, a decisive test is to take samples two or more times during a calendar year. The rate of progression of the humps with time will indicate their nature.

Frequency distributions of weight could be used in place of length, and might be superior among the older fish. However, lengths are easier to obtain, and are less subject to variation from environmental or biological causes.

To estimate growth rates, the simplest procedure, introduced by Petersen (1892), is to take the position of the mode of the successive humps (of Type 2 above) of a frequency distribution as the representative length of the year-class. If rate of growth varies from year to year, greater accuracy is achieved by sampling the same brood at the same time in successive years. After the first, second, third, or fourth years, depending on the species and the size of the samples available, year-class modes become indistinguishable from random fluctuation in the length distribution. In any event the mode is not the best measure of central tendency in length. An alternative is to fit normal distributions to the successive humps in the series. This is made easy by using probability paper, which transforms successive frequencies in a normal distribution to a straight line (Cassie, 1954). A series of humps (year-classes) gives points that describe a series of straight lines at different levels on the graph, but usually having much the same slope. For separating age groups it is usually sufficient to fit lines by eye to these points and divide the series at the middle of each transition from one straight segment to the next. By replotting cumulative percentage frequencies within each segment, the mean and standard deviation of the corresponding normal curve can be estimated, hence the curve itself if it is needed (Partlo, 1955; Tesch, 1971).

B. Age from Marks on Scales or Bones

In temperate and northern latitudes most species of fish carry a record of their age on the hard structures of their body. The most useful feature for this purpose varies with the species; scales, otoliths, fin rays, opercula and vertebrae are all used, more or less in order of decreasing frequency. In few cases is the record completely unambiguous: different biologists will often disagree as to what age should be assigned, especially for older fish. However there have been quite a number of cases where age determinations have been checked against fish of known age, or by the progression of a dominant yearclass over a period of years, and the method usually has proved sufficiently reliable for routine use (Chugunova, 1959; Graham, 1929; Tesch, 1971; Van Oosten, 1929).

Not only can the current age be determined from scales or other features, but usually also the length at the end of successive annual growing seasons. This is obtained by back-calculation from measurements of the scale, using an empirically determined relation between fish length and scale radius or diameter.

C. Effects of Bias in Sampling

Samples of fish taken from the wild very rarely include all age groups in proportion to their abundance. For the most part this is because the gear used to capture them is selective by size. Even if it is not, the fish themselves tend to inhabit different parts of their environment at different ages, or within a single environment they may assemble in schools containing individuals more or less uniform in length and different from other schools. The result is that the fish toward the lower end of the length range sampled will usually include only the larger individuals of the youngest ages present. There may be the opposite bias at the upper end of the sampling range. If the mean lengths of incompletely sampled age groups at either end of the range are compared directly with the representatively sampled middle group of ages, the result is that the rate of growth is underestimated. Thus great care should be taken to exclude such ages from an analysis. To obtain an unbiased picture of growth throughout life is no easy task, and almost always requires the use of several methods of sampling.

Ideally growth should be estimated by following a single yearclass of fish throughout its entire life, because growth rate may vary between broods, especially if they differ greatly in abundance. Usually, however, this is not possible, and the lengths of a succession of ages taken in a single year is used to give a general picture of growth over the time period represented.

D. Effects of Within-Age Size-Selective Mortality

Even when sampling is perfectly representative, there remains a major obstacle to an accurate appreciation of fish growth in nature. It frequently happens that the larger fish in an age group have a different mortality rate from the smaller ones: either greater or less, but usually greater. This can be detected when back-calculations of length at earlier ages are made from scales or otoliths, *using samples that are representative of the whole of each age group involved*. When a larger fraction of the larger fish die, the result is "Rosa Lee's phenomenon" (Sund, 1911; Lee, 1912), whereby the calculated average size of fish of younger ages is the smaller, the older the fish from whose scales they were calculated. Two recent reviews of this subject are by Jones (1958) and Ricker (1969).

1. *Natural* selection for size can bear more heavily on either the larger or the smaller fish. Faster-growing fish frequently tend to mature earlier and also become senile and die earlier than slowergrowing fish of the same brood (Gerking, 1957). This is the principal and perhaps the only cause of natural Lee's phenomenon in unfished populations. However there are at least two possible situations that act in the opposite direction. (a) There is considerable evidence that during the first year of life slower-growing individuals are more susceptible to predation. Such selective mortality during the first year cannot affect calculated growths differentially, because it is only after the first annulus is laid down that there can be any back-calculation. But if the same situation persists into the second or later years of life it means that, for example, the size at annulus 1 computed from fish of age 2 will tend to be *greater* than the same computed from fish of age 3. (b) The other situation occurs when fish of both sexes are sampled and are analyzed together, but there are in fact sex differences both in rate of growth and in natural mortality rate. Among most flatfishes, for example, females grow faster and live longer than males. If lengths are back-calculated from samples in which the sexes are not distinguished, the increasing representation of the faster-growing females at older ages tends to increase the calculated mean size at younger ages, and the result can be "reversed Lee's phenomenon."
2. Size selection by a *fishery* can also be important. The larger members of a year-class are the first to become vulnerable to a given type of gear, and it may be several years before the smallest members are fully vulnerable. In

sport fisheries there is often a minimum size limit for retention of fish caught. Obviously these can be major causes of Lee's phenomenon. It is also possible for the largest fish in a population to be less vulnerable to fishing than those of intermediate size, but in practice this is far less important.

III. CHARACTERISTICS OF FISH GROWTH

A. Growth Stanzas

Considering the whole life of a fish, its growth can conveniently be divided into a series of stages or stanzas,* a concept given formal development by Vasnetsov (1953). The change from one stanza to the next is characterized by some kind of crisis or discontinuity in development, such as hatching or maturation, or a change of habits or habitat. In order of decreasing severity, these may include the following.

1. A major reorganization of body structure, comparable to what occurs in the metamorphosis of moths or wasps. Such drastic changes are not very common among fishes, but they occur in a number of oceanic species. Eels (Anguillidae) are familiar examples, in which a flat, transparent leptocephalus is reorganized into a cylindrical pigmented elver. Somewhat less drastic is the change, by the various flatfishes, from a symmetrical pelagic fingerling to an adult form with both eyes on the same side of the head.
2. Any fairly abrupt change in body form, or in the relative lengths of appendages, or in the relative length and structure of the digestive tract. For example, Martin (1949) illustrated breaks in the slope of the plot of the logarithm of one or more linear measurements of the body or fins and the logarithm of standard length for ten species of fish, at lengths varying from 27 mm for a characin (*Brycon guatamalensis*) to 50 mm for herring (*Clupea harengus*). There was also a break at 35 mm in the slope of log weight against log length for rainbow trout (*Salmo gairdneri*), and Tesch (1971) illustrated a similar break for brown trout (*S. trutta*).
3. Major physiological changes, for example, in tolerance to temperature or salinity, accompanied by corresponding changes in endocrine and other internal organs. The diadromous fishes are the best known examples, and among them physiological changes are often accompanied by a change in form or color. For example, a young salmon, in adapting to marine life, changes the color of its sides from barred to silvery, its body becomes more elongate, and greater tolerance of salinity is acquired. Once in saltwater, growth rate increases greatly, and this is reflected in broader spaces between the circuli on the scales. Returning from the sea, salmon

*I use the word stanza rather than stage, to avoid confusion with developmental stages such as are described by Ahlstrom (1943) or Pelluet (1944).

enter a new stanza characterized by changed color, often to bright hues, and by tolerance of freshwater, thickening of the skin, partial resorption of the scales, reduction of the digestive tract, and marked changes in external form. In Pacific salmon (*Oncorhynchus*) these changes are irreversible and the fish dies after spawning, but Atlantic salmon and others of the genus *Salmo* may recover and return for another stanza of ocean life.

4. A sudden increase or decrease in rate of growth. This is a borderline case, and whether to consider such a change the start of a new growth stanza is largely a matter of individual preference. For example, at a certain size some perch (*Perca*) shift from an insect to a fish diet and increase their growth rate rather abruptly, which can be called the start of a new stanza if you feel inclined. If growth data were available for individual fish, it is probable that the onset of first sexual maturity would be recognized as the start of a new stanza in most cases. But when averages are used, as is customary, the change in growth rate at maturity becomes blurred by the fact that different members of a brood mature at different ages.

Within any stanza of growth of animals or plants, increase in size may follow an S-shaped curve. This was originally suggested for plants by Sachs (1874) and is often called a Sachs cycle. The lower part of the S may (or may not) approximate to an exponential curve, while the upper asymptotic part may reflect preparations for the next stanza—unless the fish is already in its final stanza. However the complete Sachs cycle need not always be present, and Hayes (1949) points out that the time at which the inflexion point occurs differs greatly, depending on whether it is length or weight that is under consideration (compare A and B in Fig. 3). This difference is of course a direct consequence of the relationship between length and weight described by Eq. (1).

B. The Seasonal Growth Cycle

Another characteristic of the growth of many fishes in nature is a marked seasonal variability. This is universal outside of tropical regions, and is by no means rare within them, where it is usually related to seasonal rainfall. One thinks of such extremes as the Alaska blackfish (*Dallia*) that hibernate like frogs at the bottom of tundra ponds, or the African lungfishes (*Protopterus*) that retreat into a cocoon far down in the mud during the dry season.

Under less severe conditions there have been a number of investigations showing that growth tends to follow the cycle of the seasons, usually faster in summer and slower in winter. An example at random is Alexander and Shetter's (1961) study of brook and rainbow trout (*Salvelinus fontinalis* and *Salmo gairdneri*) in a Michigan lake. Growth was very slow from late December to early April, but did not stop entirely. Bluegill sunfish (*Lepomis macrochirus*) ceased to grow during cool weather in several Indiana Lakes (Gerking,

1966), while in ponds carp (*Cyprinus carpio*) sometimes lose weight in winter. Similarly, a species that must endure summer temperatures considerably greater than its preferred temperature may feed little and stop growing at that season; indeed, growth may slow down even if the fish continue to feed at their maximum rate for the prevailing temperature (see Sections V and VI).

However, the seasonal cycle of growth is not always, perhaps never, wholly under temperature control. The whitefish (*Coregonus clupeaformis*) studied by Hogman (1968) provide an example (see Chapter 10, Fig. 12). Their growth was closely correlated (with a time lag of about 1 month) with the seasonal cycle of daylength, and was only weakly correlated with the temperature of the spring water in which they were held. It seems possible also that an internal seasonal rhythm may play a role in regulating growth rate, though I know of no experiments or observations that bear on this, apart from the seasonal sexual cycle of mature fish.

C. Shape and Variability of Frequency Distributions of Length and Weight

In temperate and northern latitudes the distribution of lengths within a single brood (year-class) of fish is usually unimodal, apart from random variation, and frequently it is reasonably close to a normal or Gaussian distribution. Exceptions occur when spawning takes place at intervals over a considerable period of time, as in anchovies (*Engraulis*); or when a few individuals in a brood start to grow exceptionally rapidly by reason of preempting favorable territory, or because they turn cannibal on their siblings, or both; the largemouth bass (*Micropterus salmoides*) is an example.

Very often, however, not only is the length frequency distribution approximately normal at the end of the first year, but it remains so throughout life. This might appear to conflict with the widespread occurrence of greater mortality rates among the larger members of a brood. However Jones (1958) showed that size-selective mortality does not necessarily change the shape of the length distribution of a year-class. When the gradient of instantaneous mortality rate within a year-class is linear with respect to length, an originally normal length frequency distribution will remain normal no matter how severe the mortality gradient (Fig. 4). In nature, of course, the length–mortality relation need not be exactly linear, but it requires a marked deviation from linearity to produce appreciable skewness in the derived length distribution (Ricker, 1969). Jones (1958) also showed that with any linear length–mortality relation the original variability of the frequency distribution is conserved (Fig. 4).

If the distribution of lengths of an age group of fish conforms to a normal curve, then its weight distribution will not be exactly normal, and vice versa (Fig. 5). This follows directly from the weight–length relationship of Eq. (1). The relation between variability in length and in weight is somewhat less obvious. For example, if all the fish in a year-class increase in length by

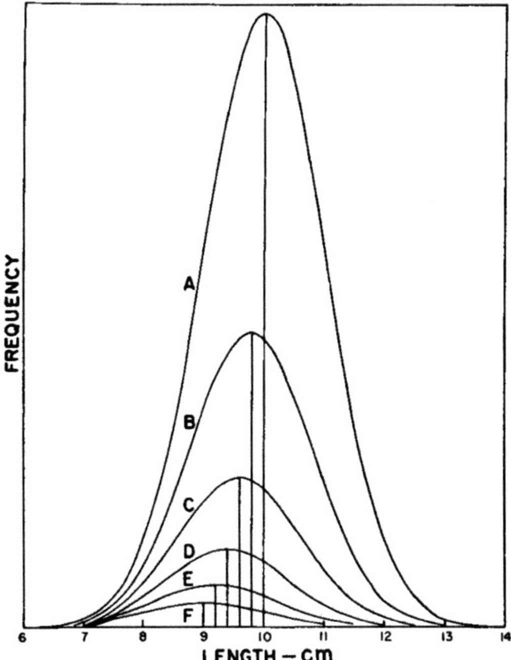

Fig. 4. Normal length distribution curves having a standard deviation of 10 mm. Each curve is obtained by subjecting the fish in the next larger curve to an instantaneous mortality rate that increases by 0.02 per millimeter of length, averaging 0.75. (From Ricker, 1969, *Journal of the Fisheries Research Board of Canada*; reproduced by permission.)

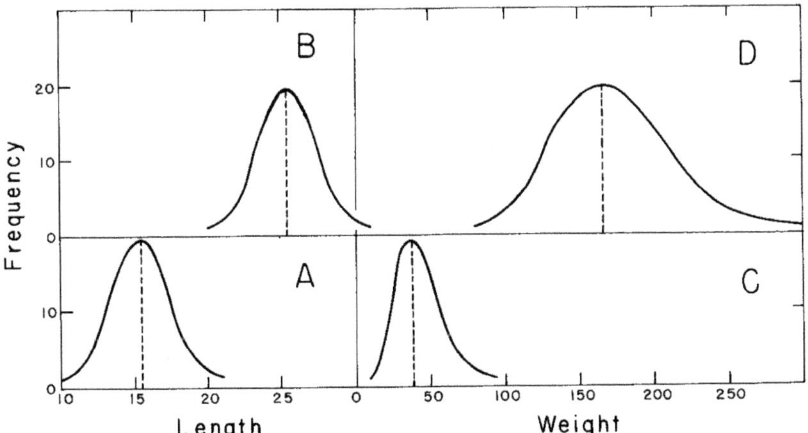

Fig. 5. Frequency distributions of length and weight in a model year-class of fish in successive years of life. From A to B the fish of all sizes have increased in length by 10 units, thus maintaining their original normal frequency distribution. The corresponding distributions of weight are shown in C and D.

the same absolute amount, their variability in length will remain the same, but their variability in weight will increase greatly (Fig. 5).

In practice, the variability in length of a year-class usually increases somewhat during the first few years of life; then it stabilizes, and it may decrease in old age. The phase of increasing variability is sometimes called *growth depensation*, because the longer fish are becoming even longer relative to their smaller congeners. The later stage is *growth compensation*, when the smaller fish start to catch up to the larger ones. Variability in weight, however, continues to increase for some time after variability in length has started to decrease. Thus growth compensation in terms of weight starts much later than the same phenomenon in terms of length, and sometimes does not occur at all.

D. Isometric and Allometric Growth

Even within a single growth stanza, different parts of the body of a fish may grow at different rates. This can frequently be described by an expression similar to Eq. (1) or (2):

$$l_2 = al_1^b \tag{10}$$

$$\log l_2 = \log a + b(\log l_1) \tag{11}$$

where l_1 and l_2, lengths of any two body parts (l_1 is often the standard length of the body); a, a constant equal to the value of l_2 when $l_1 = 1$; b, the exponent that indicates the direction and speed of any change in body form. When $b = 1$, growth is *isometric*: The body parts being compared are growing proportionally. When $b > 1$ the length l_2 is increasing faster than l_1, and vice versa; in either case growth is *allometric*.

Zar (1968) and others have pointed out that fitting Eq. (11) by least-squares is not the same thing as obtaining a least-squares fit to Eq. (10), and nowadays a least-squares fit to Eq. (10) can be obtained iteratively by computer. Nevertheless the latter procedure is not appropriate here, for two reasons. In the first place, there is a general tendency for natural variability to increase as size increases, so that the logarithmic transformation tends to stabilize variance and so make a least-squares fit more appropriate. Second, the variability in the data will be almost entirely natural, with very little contribution from errors in the measurements (assuming these are made with ordinary care), and the body parts measured cannot be categorized as "dependent" and "independent." Hence the line must be fitted symmetrically with respect to both variates, and this requires a functional regression, specifically, the GM functional regression described in Section I,D, which is always numerically greater than the ordinary regression.

For example, suppose l_2 is eye diameter and l_1 is head length. If a functional regression indicates that $b = 1$, eye and head are growing proportionally. But if an ordinary regression of $\log l_2$ on $\log l_1$ is fitted to the same data, b will be less than 1 and the eye will appear to be decreasing in size relative to the head. Because ordinary regressions have been used for this relationship up to recently, there must be many erroneous interpretations of this type in the literature. The magnitude of the correlation coefficient between l_2 and l_1 indicates the difference between the two regressions. The difference is small if the variability in the $l_2:l_1$ ratio between different fish of the same size is small, and also if the range of fish sizes being compared is large. But increasing the number of fish measured, within the same size range, will not tend to make the two regressions more alike.

When comparing weight and length the situation is similar. A functional slope or exponent $b = 3$ of the line relating $\log w$ and $\log l$ indicates isometric growth, in which weight increases as the cube of length. When $b > 3$ the fish is increasing in weight (presumably also in volume) at a greater rate than required to maintain constant body proportions, and vice versa. While many fishes grow approximately isometrically during their final growth stanza, values of b up to about 3.5 have been observed in some species. Sometimes, also, values less than 3 are observed, usually in populations where the larger individuals lack a suitable food supply.

Lumer (1937) pointed out that Eq. (10) implies that the instantaneous rates of increase in the lengths of two body parts must either remain constant or change at the same instantaneous rate. The initial growth rates of the two parts, however, must then differ if $b \neq 1$.

Although growth conforming to Eq. (10) is very common, it is not universal. Some internal organs reach their maximum size, or even regress, before adult body size is achieved, not to mention the seasonal development of the gonads. Actually there is some ambiguity as to whether the term "allometric" should apply only to cases that conform to Eq. (10) or whether it can be used for any kind of nonisometric growth.

E. Effects of Size-Related Mortality on Estimates of Growth Rate

In Section II,C mention was made of the effect of sampling bias in producing erroneous estimates of growth rate, and how this can be avoided if it is possible to take samples using gears of different selectivities. However, the effect of within-age selective mortality on the population, as distinct from the sample, is less easily disposed of. Whether it is a fishery or natural causes that make the faster-growing fish die sooner than slower-growing ones, the result is a serious complication of the growth picture. Figure 6 shows the kind of analysis that is possible when back-calculated lengths of fish of a series of ages are available. For the fish taken at each age there is a separate growth line, which typically lies below the line for the next younger age. (If at the younger ages the smaller fish of a brood experience the greater mortality, the picture can become quite complicated.) The growth rates that are most nearly representative of the population that exists during each year's interval are indicated by the final segment of each line, shown broken in Fig. 6.

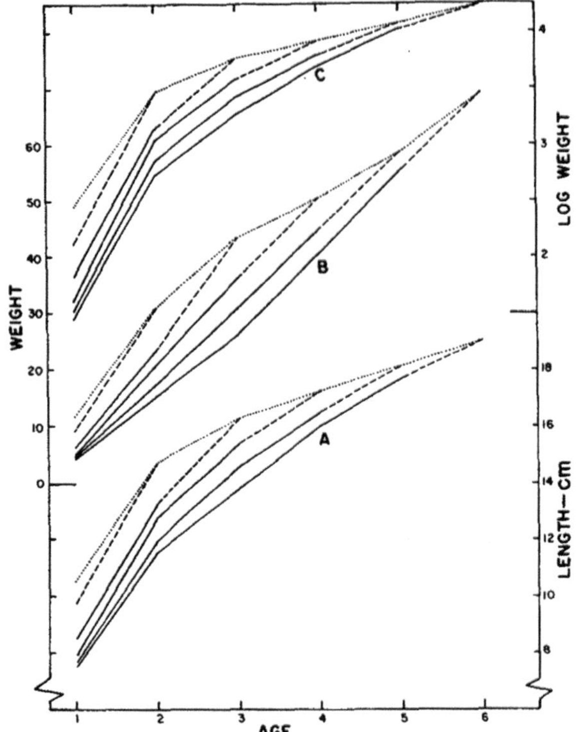

Fig. 6. Growth of a population of ciscoes (*Coregonus artedii*), somewhat idealized, in terms of length (A), weight (B), and natural logarithm of weight (C). Solid and broken lines represent back-calculated sizes from the age at which they terminate. The dotted line joins the observed sizes at successive ages. (From Ricker, 1969, *Journal of the Fisheries Research Board of Canada*; reproduced by permission.)

On the other hand, if back-calculated lengths are not available, and what is known is only the average size of the fish present in successive years, the computed growth rate is less than the actual at all ages, as shown by the fine dotted lines in Fig. 6. This latter has been called the *population growth rate*, in contrast to the true average growth rate of the fish themselves (Ricker, 1975a, p. 217), and the difference between the two can be quite large.

We are left with a very inconvenient conclusion, which has not yet been widely appreciated. Usually it is impossible to represent true growth rates and the true size of the fish at successive ages in natural populations by a continuous line or continuous series of lines on a graph. If fish sizes are plotted, the slopes of the lines joining the points will underestimate growth rate, as do the fine dotted lines in Fig. 6. But if a continuous series of lines were to be plotted with the correct slope for the prevailing growth rate each year, the fish sizes indicated would be increasingly greater than actual average size as age increases.

IV. GROWTH MODELS RELATED TO AGE

A. Growth during Early Life

The early development of fish eggs and young has been studied at constant temperature under laboratory conditions by several authors, usually in terms of weight. For short periods of time, growth can be described by the exponential curve of Eq. (5). Exponential curves can also be used to describe any growth sequence, by dividing the latter into short segments. These may represent either natural periods of exponential growth at different rates, or merely arbitrary time divisions. Hayes and Armstrong (1943) show data on growth of embryos of salmon (*Salmo salar*), plotted on a semilogarithmic scale (Fig. 7). With a little imagination four linear segments can be distinguished, and if these could be related to embryological events they might serve as the dividing points for as many growth stanzas. On the other hand, a single smooth curve would describe the data about as well.

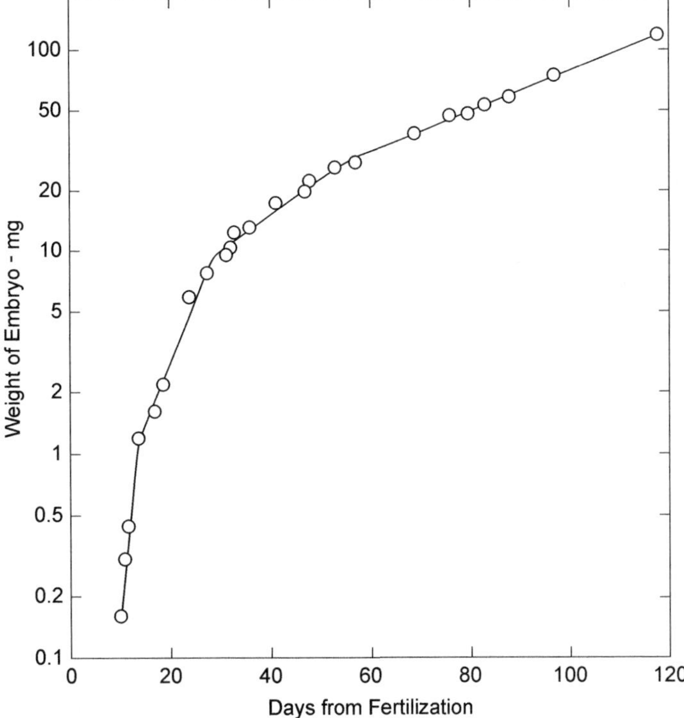

Fig. 7. Growth of embryos of Atlantic salmon, plotted with a logarithmic ordinate. The four straight lines indicate possible successive stanzas of simple exponential growth. (From Hayes and Armstrong, 1943. Redrawn by permission of the National Research Council of Canada from the *Canadian Journal of Research*, Volume 21, Section D, pp. 19–33.)

An alternative expression to describe early growth is

$$w = a(t - t_0)^b \quad (12)$$

$$\log w = \log a + b[\log(t - t_0)] \quad (13)$$

where w, weight at time t; t_0, an initial time chosen so as to provide best agreement to the formula; a and b, constants. This expression, without the t_0, was proposed by Friedenthal for the growth of mammalian embryos, and it was later applied to other organisms. The modified form shown above was proposed by MacDowell *et al.* (1927), and it was applied to salmon development by Hayes and Armstrong (1943). Figure 8 shows the same data

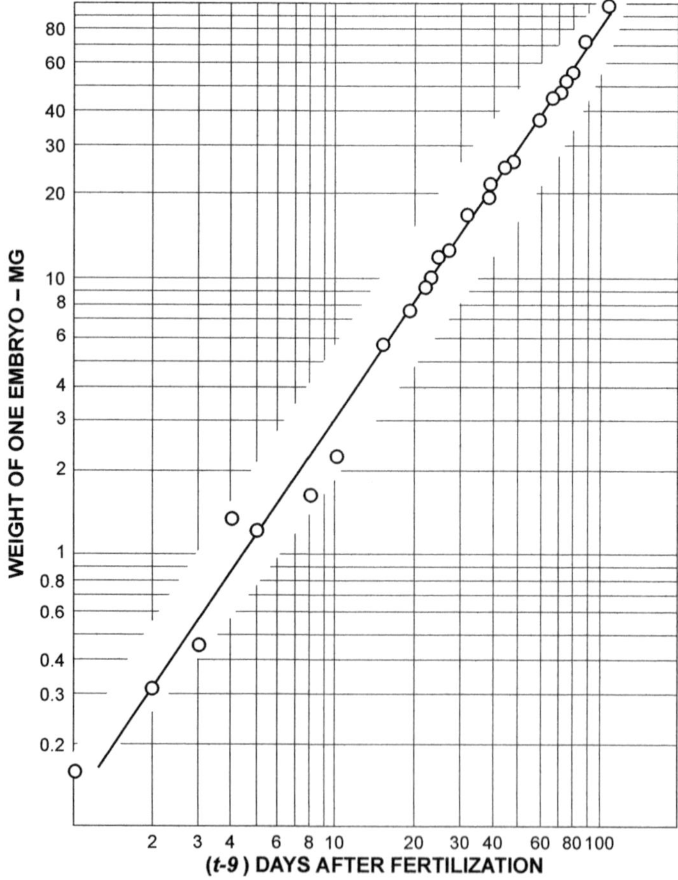

Fig. 8. Double logarithmic plot of the data of Fig. 7, with the time axis originating 9 days after fertilization, when the embryonic axis is established. (From Hayes, 1949, Fig. 7; reproduced by permission.)

as Fig. 7, plotted in this manner and fitted with the double logarithmic straight line.

Gray (1928), however, observed that growth of embryos of trout (*Salmo trutta*) declined greatly during the last 20 days before hatching, suggesting a complete S-shaped Sachs cycle before the next growth stanza began. To fit this with a single line would require one of the curves described in the sections below—probably a Richards curve.

The straight line of Eq. (13) was found by Allen (1951, p. 121) to describe the growth of brown trout in the wild during their first 450 days after hatching. Considering the changes in temperature and other conditions over so long a period, this is quite remarkable, and perhaps fortuitous, characteristic only of the Horokiwi Stream and others of its type.

B. Growth Modeled by Successive Exponential Segments

The most convenient way to model a fish's growth is to compute the instantaneous rate of growth for successive time intervals. In this form growth rate can be compared directly with mortality rate using simple subtraction. The shorter the intervals at which observations of size are available, the more accurate will be the resulting representation. During the phase of increasing rate of growth (in length or weight), the exponential curves will agree better with observation because they are concave upward; when growth rate starts to decrease, the graph of size against age is convex upward and so has the opposite curvature to that of the corresponding exponential segments. However, by using short time intervals the difference between the observed and computed lines can easily be made so small as to be of no consequence. If size is known only at yearly intervals, segments less than a year long are pointless because of the seasonal variation in growth rate.

It is not necessary to actually compute the exponential curves when using successive instantaneous rates of growth [G in Eqs. (5)–(7)]. For example, the increase in biomass of a fish during a unit time interval, or the production of a population, is equal to the instantaneous rate of growth times the mean biomass present (Ricker, 1946). In this way Ricker and Foerster (1948) used a series of instantaneous growth rates in making a computation of sockeye production in a small lake, and Ricker (1945, 1958, 1975a) used the same plan to compare growth and mortality rate and to compute the catch taken from a unit weight of recruits to a fishery.

In dealing with natural populations, rather than individual fish, a computation using successive exponential segments has the important advantage that the true growth rate, as defined in Section III,E, can be combined directly with actual population biomass in calculating the production of the population. This is not possible with the growth curves of Sections IV, D–IV, I, which are always based on the size of the surviving fish at each age, and accordingly yield only minimum estimates of growth rate and hence of production.

C. General Characteristics of Curves Applied to the Final Stanza of Life

Growth curves for fish in the wild are usually fitted to data on size at yearly intervals. Either length or weight can be fitted, but length is usually easier because the inflection point for length has usually (not always) been passed by age 1, so it is only the part of the curve having decreasing curvature that needs to be described by a formula. By contrast, the absolute rate of increase in weight often continues to increase for several years before decreasing; hence in order to fit the whole of a weight curve it is necessary to find an S-shaped curve having the correct curvature on both sides of the inflection point.

Whether length or weight data are to be fitted with a curve will depend on the quality of the data available, but if the weight : length relationship of Eq. (1) is known, the one can always be computed from the other. The curves most frequently fitted to size data are almost all *bipartite* in their general differential form (Fletcher, 1973, 1975). Rate of increase in size is proportional to the difference between a positive constant times size already achieved, ay, and some function of that size, $f(y)$. When $f(y) < ay$, this differential form is

$$\frac{dy}{dt} = ay - f(y) \tag{14}$$

When $f(y) > ay$, it becomes

$$\frac{dy}{dt} = f(y) - ay \tag{15}$$

All the curves described by Eqs. (14) and (15) have an upper asymptote, and most are either asymptotic to the time axis or are at some point tangent to it. One or more of the asymptotic curves described below may be fitted to at least the upper portion of a series of annual observations of either length or weight. The symbols l, L_∞ are used for length, end w, W_∞ for weight. Formulas are given in terms of one or the other, as seems most appropriate; but as a matter of observation *all these curves have given reasonable fits to both length and weight data*, but not necessarily on the same species, and not necessarily for the complete range of ages.

The only criteria for choosing a growth curve that have proved valid are goodness of fit and convenience. Historically, however, most of the curves in use have been proposed along with some mathematico–physiological theory as to how growth might be regulated, and much ingenuity has been expended in trying to relate them to growth processes (Pütter, 1920; Brody, 1927; von Bertalanffy, 1934; Parker and Larkin, 1959; Taylor, 1962; Laird *et al.*, 1965; Ursin, 1967; Zweifel and Lasker, 1976). Although none of these theories has been demonstrated to have any biological basis, it is of interest to mention them when introducing each of the curves in turn.

All of the curves in use may be written with a variety of different parameters, and in different ways. Rather than aiming at a uniform manner of presentation, the forms shown are those most commonly encountered in the literature on growth, particularly of fishes and fish populations. Relationships among the different parameters are indicated, and their derivation from the basic differential form.

D. Logistic Growth Curve

Figure 9 is an example of a logistic curve, which also represents the "autocatalytic law" of physiology and chemistry. It is one of several bipartite expressions proposed by Verhulst (1838) as possible descriptions of the succession of age frequencies in human populations. Its differential is of the type of Eq. (14), with $f(y) = by^2$. Using weight symbols, and putting $W_\infty = a/b$ and $g = a$, Eq. (14) becomes

$$\frac{dw}{dt} = gw - \frac{g}{W_\infty} w^2 = \frac{gw(W_\infty - w)}{W_\infty} \tag{16}$$

The first form shown above shows that the absolute rate of increase in weight, plotted against weight, will be an inverted quadratic parabola. The second form of Eq. (16) illustrates the rationale of this curve as applied to growth; divided by w, it becomes the instantaneous rate of increase in weight, and this is evidently proportional to the difference between the asymptotic weight W_∞ and the actual weight. Two integral forms of Eq. (16) are

$$w = \frac{W_\infty}{1 + e^{-g(t-t_0)}} \tag{17}$$

$$w = \frac{W_\infty}{1 + ce^{-gt}} \tag{18}$$

where w, weight at any time t; W_∞, asymptotic weight; g, instantaneous rate of growth when $w \rightarrow 0$; t_0, the time at which the absolute rate of increase in

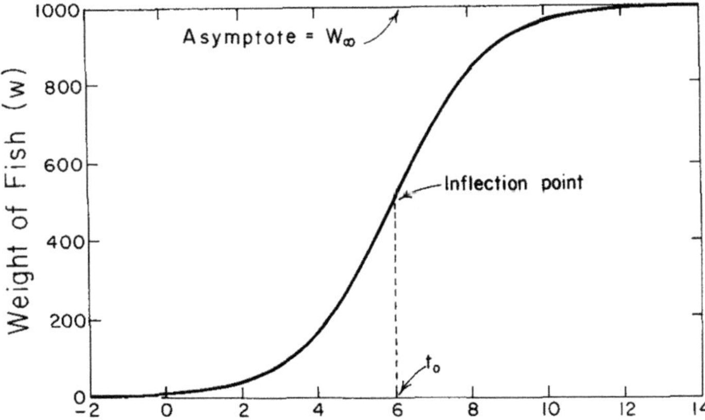

Fig. 9. Example of a logistic growth curve.

weight begins to decrease, that is, the inflection point of the integral curve, or the maximum of the differential Eq. (16); c, e^{gt_0}.

A logistic curve has the t-axis as its lower asymptote, when $t \rightarrow -\infty$. The point of inflection is at $t = t_0$, $w = W_\infty/2$, exactly half-way between the two asymptotes. The two halves of the curve are symmetrical, or rather, antisymmetrical. The constant t_0 adjusts the time scale so that time is in effect measured from the inflection point. The instantaneous rate of growth at the inflection point is $dw/wdt = g/2$.

The logistic curve of Fig. 9 has the parameters $W_\infty = 1000$, $g = 0.8$, $t_0 = 6$, and $c = 121.5$.

Ricklefs (1967) showed that the logistic curve described well the growth in weight of three species of birds. In ichthyology it has been used mainly to describe the increase in weight of *populations*, rather than individuals (Graham, 1935; Schaefer, 1954). In that context the differential form of the curve is called the Graham surplus production curve, or the Graham–Schaefer curve (Ricker, 1975a, p. 310).

E. Gompertz Growth Curve

Apart from seasonal variations, a fish's rate of increase in biomass typically decreases throughout life, or at any rate throughout its last growth stanza. If the instantaneous rate of decrease in the instantaneous rate of increase is constant, it leads to the type of curve shown in Fig. 10, which was originally proposed by Gompertz (1825) to describe a portion of the distribution of ages in human populations. A number of investigators have interpreted this curve as reflecting the activity of two different types of regulatory factors during growth. For example, Laird *et al.* (1965) hypothesize that "the interaction between the two opposing, genetically programmed, processes of exponential

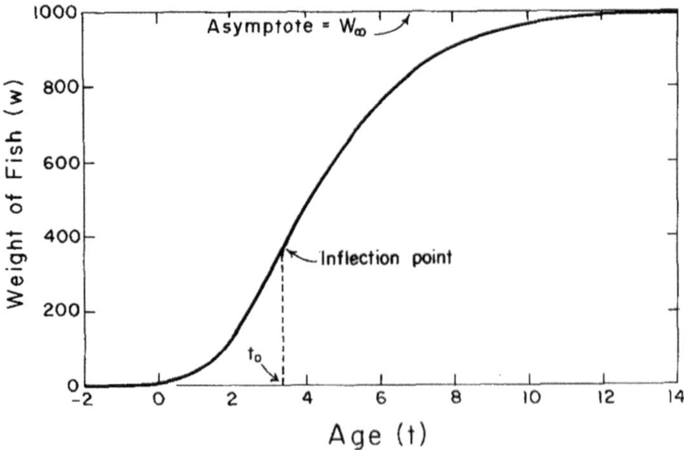

Fig. 10. Example of a Gompertz growth curve.

growth and of exponential decay of the specific growth rate, results in the familiar properties of growth of warm-blooded animals . . . . (1) growth toward a final limiting size, (2) a limiting size that is characteristic of the species or breed . . . , (3) a specific growth rate that decreases constantly as the organism ages, and (4) a decreasing rate of this decrease" (p. 244). In the notation of Eqs. (21)–(23) below, the original instantaneous rate of increase, when $w = w_0$, is represented by the product kg, while g is the instantaneous rate of decrease of that rate.

Weymouth and Thompson (1930) applied the Gompertz curve to the growth of a bivalve, Silliman (1967) used it for fishes, and Zweifel and Lasker (1976) argue that it should be the preferred model for fish growth—including larval growth. On the theoretical side, however, it is very difficult to picture how development could reach a satisfactory conclusion (a weight within the normal range for the species), if it were wholly controlled by two rates fixed in the zygote, in the absence of some kind of feedback mechanism to relate growth to size already achieved. And as a matter of observation, we know that growth rate is influenced by diet and exercise in both warm-blooded and cold-blooded animals. Also, the guinea pig seems to be the only known case where growth in weight from early embryo to adult can be described by a simple Gompertz curve (Laird *et al.*, 1965, Fig. 1). For other mammals and birds Laird *et al.* found that there is a break at birth or hatching; in addition, there is a non-Gompertzian, in fact more or less linear, phase of growth from adolescence to adult size, for which they invented a special, but unconvincing, explanation. Among primates, and man especially, the growth pattern is even more complicated (Laird, 1967). Similarly, Zweifel and Lasker (1976, Fig. 2) found that not one but two Gompertz cycles were needed to describe just the larval growth of anchovies (*Engraulis*). Obviously something considerably more sophisticated than a fixed initial pair of rates must be in control.

To date, then, no satisfactory evidence has been presented that the Gompertz curve is closer to developmental realities than are the other curves in use.* However this does not detract from its usefulness.

---

*Zweifel and Lasker's (1976) enthusiasm for the Gompertz curve has even led them to question the appropriateness of the allometric formula for relating weight to length. They point out that if both length and weight of a given group of fish conform to Gompertz curves, their weight–length relationship cannot be exactly of the type described by Eq. (1), except in two special cases. On the basis that "all experimental evidence indicates that both length and weight can be described by a Gompertz-type curve," they propose a more complex weight–length relationship in their expression (4). No references are given to this experimental evidence; in fact their paper does not cite even one case where Gompertz curves have been fitted to both length and weight of the same group of fish. We may reasonably doubt that existing agreements of lengths and weights to the Gompertz curve are either numerous enough or exact enough to make us adopt a new weight–length relationship on that basis. Zweifel and Lasker's Fig. 1 shows both their new relationship and the ordinary regression of log weight on log length, for anchovies of about 0.02 to 4 mg. The new relationship is slightly the better fit, but a functional regression line would be better than either (Section I,D).

The differential form of the Gompertz curve is of the type shown in Eq. (14), with $f(y) = by(\ln y)$; in weight symbols it is

$$\frac{dw}{dt} = aw - bw(\ln w) \tag{19}$$

Putting $g = b$ and $W_\infty = e^{a/b}$, this takes the form

$$\frac{dw}{dt} = gw(\ln W_\infty - \ln w) \tag{20}$$

Equation (20) has some similarity to the logistic of Eq. (16): Here the instantaneous growth rate is proportional to the difference between the *logarithms* of the asymptotic size and the actual size.

Three commonly used integral forms of Eq. (20) are

$$w = w_0 e^{k(1-e^{-gt})} \tag{21}$$

$$w = W_\infty e^{-ke^{-gt}} \tag{22}$$

$$w = W_\infty e^{-e^{-g(t-t_0)}} \tag{23}$$

where w, biomass at any time t; w_0, biomass at time $t = 0$ (*not* $t = t_0$); W_∞, asymptotic biomass; g, the instantaneous rate of growth when $t = t_0$; k, a dimensionless parameter, such that kg is the instantaneous growth rate when $t = 0$ and $w = w_0$; t_0, the time at which the (absolute) growth rate starts to decrease, that is, the inflection point of the curve. Evidently Eq. (22) can be derived from Eq. (21) by using the transformation $W_\infty = w_0 e^k$. Similarly Eq. (23) can be derived from Eq. (22) using $k = e^{gto}$.

The point of inflection of the Gompertz curve is at $t = t_0$ and $w = W_\infty / e$. Hence it is situated $1/e = 0.3679$ of the distance from the t-axis to the asymptote. At large negative values of t, w asymptotically approaches the t-axis. The instantaneous growth rate at inflection is $dw/wdt = g$.

The Gompertz curve of Fig. 10 has the parameters $W_\infty = 1000$, $k = 5.436$, $g = 0.5$, $w_0 = 4.357$, $t_0 = 3.386$.

In population dynamics studies the Gompertz curve has been used successfully to describe fish growth by Silliman (1967), who says that its form makes it particularly convenient for use with an analog computer.

F. Pütter Growth Curve No. 1

Pütter (1920) was apparently the first to introduce this curve in biology. Brody (1927, 1945) applied it to the growth in weight of domestic animals beyond the inflection point and, applied to length, it has had an energetic proponent in von Bertalanffy (1934, 1938, 1957). In fishery literature it is usually called the Bertalanffy or Brody–Bertalanffy curve, but it seems more appropriate to give it the name of the very original and perspicacious scientist who first proposed it.

Von Bertalanffy distinguished three metabolic types in the animal kingdom: (1) where metabolism is proportional to surface area or to the ⅔ power of weight; (2) where it is proportional to weight; (3) where it is intermediate between these situations. On the basis of oxygen consumption experiments with guppies (*Lebistes*), and a few other species examined by Jost, he assigned fishes to the first category above; although by 1957 data for many other species were available which indicate that most fishes belong to the intermediate type (Winberg, 1956). He then endeavored to establish "a definite and strict connection between metabolic types and growth types, in consequence of a general theory of growth which establishes rational quantitative laws of growth and indicates the physiological mechanisms upon which growth is based" (von Bertalanffy, 1957, p. 223). For this general theory he adopted Pütter's suggestion that rate of increase in biomass (w) is proportional to the difference between a rate of anabolism that is proportional to the ⅔ power of biomass and a rate of catabolism that is proportional to biomass. This implies a relationship of the form of Eq. (15) with $f(y) = by^{2/3}$; or in weight symbols

$$\frac{dw}{dt} = bw^{2/3} - aw \tag{24}$$

Dividing through by $w^{2/3}$, and assuming that weight (w) is proportional to the cube of length (l), we obtain the corresponding expression in terms of length

$$\frac{dl}{dt} = b - al \tag{25}$$

where a and b are positive constants, different from those in Eq. (24). Eq. (25) might be regarded as a bipartite expression of the type of Eq. (15), in which $f(y) = by^0$.

How exactly von Bertalanffy related the concepts of anabolism and catabolism to his metabolic types is never made clear. By "catabolism" he did not mean total metabolism as measured by oxygen consumption; rather he tried to limit it to the breakdown of body tissue. His attempt to measure this catabolism in terms of nitrogen excreted by unfed animals is not convincing: breakdown of fat tissue is not accounted for, and there is no demonstration that values obtained for unfed fish apply to more normal conditions.

"Anabolism," in von Bertalanffy's scheme, would presumably be the sum of catabolism as defined above plus the increase in body size over a given period of time—both expressed in the same units, such as calories. The idea that anabolism, however defined, is proportional to the ⅔ power of weight may have derived from an assumption that the surface area of the gut would limit absorption of nutrients and hence additions to biomass, but there is no basis for such an idea. For example, Szarski *et al.* (1956) have shown that the absorptive area of the gut of the bream (*Abramis*) increases more rapidly than the body's surface area, in fact about as fast as biomass, and other similar findings are cited by Parker and Larkin (1959). Apart from that, area of the gut wall could be a factor limiting food absorption and hence "anabolism" only if food were always available in excess, which is far from being the case in nature.

The fanciful speculations above have been summarized here because they have been rather widely quoted as a solid theoretical basis for the Pütter No. 1 growth curve. In fact, however, neither theory nor data are available to indicate that any one of the asymptotic curves should be preferred to any other, except on purely empirical grounds.

1. *Pütter's Equation*

Putting $K = a$ and $L_\infty = b/a$, Eq. (25) becomes

$$\frac{dl}{dt} = K(L_\infty - l) \tag{26}$$

showing that the (absolute) rate of increase in length is proportional to the difference between the asymptotic length L_∞ and the actual length. This can be integrated to a form used by Pütter (1920) and (in weight symbols) by Brody (1927)

$$l = L_\infty - ce^{-Kt} \tag{27}$$

where l, length of the fish at any time t; L_∞, asymptotic length; K, a parameter that governs the rate at which increase in length decreases, the *Pütter growth coefficient* (also called the Brody coefficient), which has the dimensions of 1/time; c, a parameter equal to the difference between L_∞ and the value of l when $t = 0$. An expression of identical form in chemistry is known as the equation of monomolecular reactions.

An example of Eq. (27) is Curve No. 1 in Fig. 11. By using the transformation

$$t_0 = \frac{\ln(c/L_\infty)}{K} \tag{28}$$

Eq. (27) can be changed into the form used by von Bertalanffy (1934), in which two parameters are in the exponent

$$l = L_\infty\left(1 - e^{-K(t-t_0)}\right) \tag{29}$$

The parameter t_0 is the time at which the fish would have had zero length if it had always grown according to Eq. (29).

Fabens (1965) has an excellent exposition of the meaning of the terms in Eq. (27), and Allen (1950) applied it to some New Zealand brown trout populations, notably those of the Horokiwi Stream (Allen, 1951, p. 123). Subsequently the curve received extensive usage in the Eq. (29) form after it was

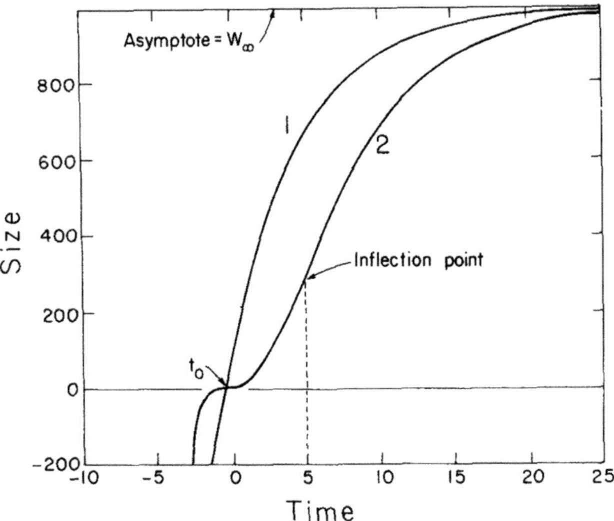

Fig. 11. Examples of Pütter growth curves Nos. 1 and 2 (see Sections IV,F and IV,G).

adopted in Beverton and Holt's (1957) monograph of fish population dynamics.

The parameters of Curve No. 1 of Fig. 11 are $L_\infty = 1000$, $K = 0.2$, $c = 897.0$, and $t_0 = -0.5435$.

2. *Ford's Equation*

Independently of other workers, Ford (1933) developed a curve on the basis of empirical observations on the growth of herring

$$l_{t+1} = L_\infty(1 - k) + kl_t \tag{30}$$

The parameter k, called *Ford's growth coefficient*, can be estimated from the slope of a GM functional line fitted to a graph of one year's length against that of the previous year. Equation (30) can be developed from Eq. (29) by inserting lengths for two successive years, l_t and l_{t+1}, in Eq. (29), subtracting the former from the latter, and putting $k = e^{-K}$. Equation (30) can be used to estimate L_∞, but it does not of itself provide the complete curve (29) because t_0 is lacking. Another useful relationship is

$$l_{t+2} - l_{t+1} = k(l_{t+1} - l_t) \tag{31}$$

3. *Walford Lines*

Walford (1946) was the first to plot a graph of l_{t+1} against l_t, obtaining the straight line indicated by Eq. (30). This line has slope k, and has become known as a *Walford line*. The graph is particularly useful for eliminating from

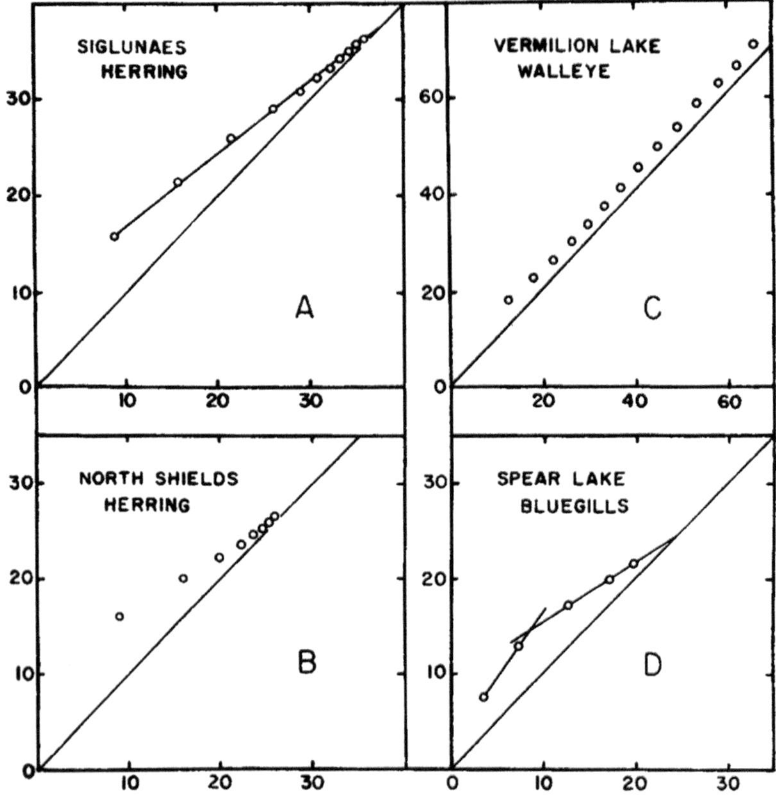

Fig. 12. Examples of Walford lines, l_{t+1} against l_t. (A) An approximately asymptotic line for *Clupea harengus*; (B) a line that becomes parallel to the diagonal at older ages; (C) a line parallel to the diagonal, for *Stizostedion vitreum*; (D) a line that describes the three right-hand points, but is not applicable at the younger ages, for *Lepomis macrochirus*. Both axes are in cm; the first point on each graph represents length at age 2 plotted against length at age 1. (From Ricker, 1975a, *Bulletin of the Fisheries Research Board of Canada*, No. 191; reproduced by permission.)

a series of age–length data the points that do not fit the assumptions required for application of Eqs. (29) and (30). Figure 12 shows several Walford lines.

4. Chapman's Modification

Chapman (1961) suggested using an expression obtained by subtracting l_t from both sides of Eq. (30); after rearrangement this gives

$$l_{t+1} - l_t = L_\infty(1 - k) - l_t(1 - k) \tag{32}$$

Here a regression of $l_{t+1} - l_t$ on l_t has a slope of $-(1 - k)$; its ordinate intercept is $L_\infty(1 - k)$, and its abscissal intercept is L_∞. Again both L_∞ and k can be estimated by fitting a functional straight line.

5. Weight Data

While the Pütter No. 1 curve is commonly fitted to length data, Brody (1927, 1945) and a few other authors have fitted it directly to weight data. It can be shown that the value of K so estimated is often practically identical with that obtained from the lengths of the same group of animals (Ricker, 1958, p. 200), though the value of t_0 is quite different. However, this procedure precludes using weights at ages that lie below the inflection point of the weight–age curve.

G. Pütter Growth Curve No. 2

The cube of Eq. (27) or (29) has sometimes been called the Bertalanffy growth curve, for example, by Ricklefs (1967) and Fletcher (1975), and indeed it comes more directly from the basic Eq. (24). However, it has a very different shape from the No. 1 curve in the region between the time axis and the asymptote, so it is here considered separately. Since biomass is commonly proportional to a power of length that is close to 3, this curve is better written with weight symbols. Putting $W_\infty = (b/a)^3$ and $K = a/3$, Eq. (24 becomes

$$\frac{dw}{dt} = 3Kw\left[(W_\infty/w)^{1/3} - 1\right] \tag{33}$$

Two integral forms of Eq. (33) are

$$w = \left(W_\infty^{1/3} - ce^{-Kt}\right)^3 \tag{34}$$

$$w = W_\infty\left(1 - e^{-K(t-t_0)}\right)^3 \tag{35}$$

where w, mass at any time t, W_∞, asymptotic mass; K, the Pütter coefficient governing the rate at which increase in mass decreases; t_0, the point where the curve becomes tangent to the time axis; c, $W_\infty{}^{1/3}e^{Kt_0}$. Curve No. 2 in Fig. 11 is of this type. The symbols K and t_0 estimate the same quantities as in a corresponding length curve fitted by Eq. (27) or (29), provided w is proportional to l^3. This curve has an inflection point where

$$t = \frac{-\ln\left(W_\infty^{1/3}/3c\right)}{K} = t_0 + \frac{\ln 3}{K} \tag{36}$$

$$w = \frac{8W_\infty}{27} = 0.2963W_\infty \tag{37}$$

The instantaneous rate of growth at the inflection point is $dw/wdt = 3K/2$.

Expressions (34) and (35) become tangent to the t-axis at $t = t_0$, then fall away steeply as t becomes smaller (Fig. 11), but the portion below the time axis is of course not involved in growth modeling.

The parameters of curve No. 2 in Fig. 11 are $W_\infty = 1000$, $K = 0.2$, $c = 8.970$, and $t_0 = -0.5435$.

H. Johnson's Growth Curve

Johnson (1935) and Schumacher (1939) independently proposed this curve, and Krüger (1962, 1964, 1965, 1973) applied it to fish growth using the name "reciprocal function." However he refrained from suggesting any theoretical basis for its applicability, pointing out only that it provides some good fits to data. The general differential form of the Johnson curve is

$$\frac{dy}{dt} = y[a - b(\ln y)]^2 \tag{38}$$

Putting $l = y$, $L_\infty = e^{a/b}$ and $g = b^2$, Eq. (38) becomes

$$\frac{dl}{dt} = gl(\ln L_\infty - \ln l)^2 \tag{39}$$

Equation (39) resembles the Gompertz Eq. (20), but here the instantaneous rate of growth is proportional to the *square* of the difference between the logarithms of asymptotic size and actual size. An integral form of Eq. (39) is

$$l = L_\infty e^{-1/g(t-t_0)} \tag{40}$$

$$\ln l = \ln L_\infty - \frac{1}{g(t-t_0)} \tag{41}$$

where l, length at any time t; L_∞, asymptotic length; g, a parameter with the dimensions of 1/time (1/g = Krüger's a); t_0, the point at which the curve meets the time axis; t_0 usually lies to the left of the vertical axis (i.e., is negative).

The inflection point of a Johnson curve is at

$$t = t_0 + \frac{1}{2g}; \quad l = L_\infty e^{-2} \tag{42}$$

Thus the inflection point is situated $e^{-2} = 0.1353$ of the distance from the time axis to the asymptote. The instantaneous growth rate at the inflection point is $dl/ldt = 4g$. Figure 13 shows a Johnson curve in which $L_\infty = 1000$, $g = 0.4343$, and $t_0 = -5$.

As the Johnson curve approaches the time axis it becomes tangent to it when $t = t_0$. Then it immediately rises vertically to an indefinitely large positive value, from which it descends, as t continues to decrease, until it approaches the asymptote L_∞ from above when $t \rightarrow -\infty$ (Fig. 13). Of course t values less than t_0 are not involved in describing fish growth.

I. The Richards Function

By adding an additional parameter, flexibility can be obtained in respect to the position of the inflection point of a bipartite curve, in combination with different types of curvature. Such a four-parameter function was proposed by

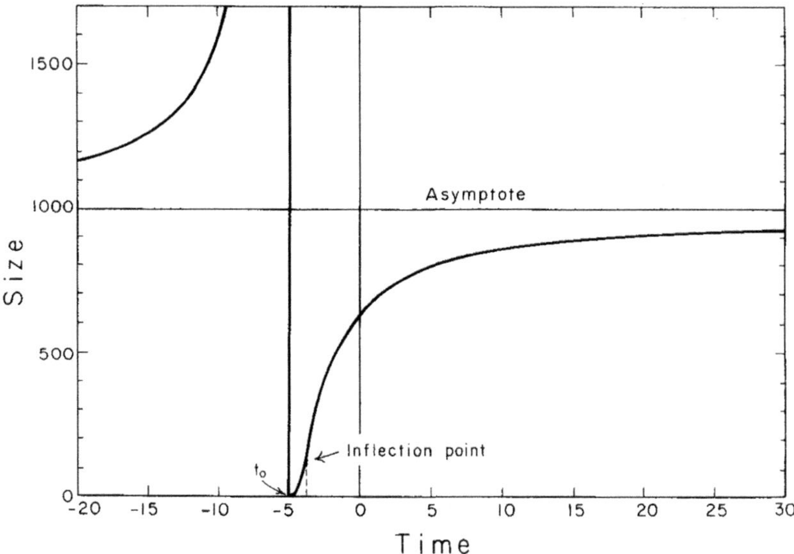

Fig. 13. Example of a Johnson growth curve.

Richards (1959) as a general growth curve, and it has been discussed by various authors since; what follows is adapted from Fletcher (1975). The general differential form of the Richards function is

$$\frac{dw}{dt} = c_1 w + c_2 w^n \tag{43}$$

where n is positive and c_1 and c_2 can be either positive or negative. Given that a and b are positive coefficients, when $c_1 = +a$, $c_2 = -b$ and $n > 1$, Eq. (43) is of the type of Eq. (14), and the integral form is

$$w^{1-n} = \frac{b}{a} + Ke^{a(1-n)t} \quad (n > 1) \tag{44}$$

With $c_1 = -a$, $c_2 = +b$, and $n < 1$, Eq. (43) is of the Eq. (15) type and the integral becomes

$$w^{1-n} = \frac{b}{a} - Ke^{a(n-1)t} \quad (0 < n < 1) \tag{45}$$

For the range that is of interest here ($0 < w < W_\infty$), Eqs. (44) and (45) have been written so that the integration constant K is positive. In both these expressions the upper asymptote is

$$W_\infty = (b/a)^{1/(1-n)} \tag{46}$$

The inflection point is at

$$w = W_\infty n^{1/(1-n)} \tag{47}$$

The absolute growth rate (slope) at inflection is

$$\frac{dw}{dt} = \pm\frac{a(1-n)}{n}\left(\frac{bn}{a}\right)^{1/(1-n)} = m \tag{48}$$

where the negative sign applies to Eq. (44) and the positive to Eq. (45). The instantaneous growth rate at inflection is equal to Eq. (47) divided into Eq. (48), which is $dw/wdt = \pm a(1-n)/n$, the sign being such that the rate is positive.

The Richards function has usually been written with separate equations for $n > 1$ and $n < 1$, as above. However, Fletcher (1975) has derived a single expression which takes care of the whole range of possible values of n:

$$w^{1-n} = W_\infty^{1-n} + K\left[\exp\left(-tmn^{n/(n-1)}W_\infty\right)\right] \tag{49}$$

where W_∞ is the asymptote and m is the slope of the curve at the point of inflection, as defined in Eq. (48).

In theory the inflection point of a Richards curve can be located at any position between the time axis and the asymptote. However Fletcher (1975) points out that inflections close to the upper asymptote require very large values of n and are not very practical.

Chapman (1961) first suggested that Eq. (43) might be used to describe the growth of fishes. By incorporating Eq. (1) he developed its integral form in terms of length. Pella and Tomlinson (1969) used the Richards function to describe the growth in weight of fish *populations*. They applied it to Silliman and Gutsell's aquarium populations of guppies, and to the stock of yellowfin tuna (*Thunnus albacares*) of the eastern Pacific Ocean.

J. Asymptotic Growth: Is It Real?

The logistic, Pütter, Johnson, Gompertz, and Richards formulas all imply that the increase in size of a fish is asymptotic; that is, size will tend toward some fixed limit no matter how long the fish lives. In practice it is an average asymptotic size that is estimated; for individual fish the asymptote may be greater or less than average. Thus asymptotic size, however estimated, should not be called the maximum size of fish in the population concerned; usually a few old individuals will be found that are considerably larger than the asymptotic size computed for the population, particularly in terms of weight. In any event, the magnitude of any computed asymptote depends partly on what growth function is used to estimate it. For example, Krüger (1969, p. 212) fitted both a Pütter No. 1 and a Johnson curve to Ketchen and Forrester's data for female "English" sole (*Eopsetta jordani*). Agreement with observation was slightly better for the Johnson curve, but both fits were satisfactory. However the Pütter asymptote was at 600 mm, whereas the Johnson asymptote was 739 mm—a considerable difference.

Knight (1968) has contended that for fishes asymptotic growth is a mathematical fiction rather than a real phenomenon. Krüger (1969), on the other hand, maintains that it is real, even while pointing out that estimates of the position of the asymptote are subject to wide variation depending on the curve fitted. Knight is right in this sense, that no matter to what age a fish's size is observed to have an asymptotic trend, we can never be sure that it would continue in that fashion if the fish were to survive longer.

A factor that may contribute to a spurious appearance of asymptotic growth is the within-age size-selective mortality discussed in Section III,E. Nearly all the data fitted by asymptotic formulas have been "population" growth curves, the points fitted being the mean size of the survivors at successive ages. Parker and Larkin (1959)* have compared back-calculated lengths of chinook salmon (*Oncorhynchus tshawytscha*) with observed lengths of maturing specimens. The right-hand panel of Fig. 14 shows growth as computed from scales of maturing fish of different ages. These have the pattern, invariable in salmon and very common in other fishes, of slower growth by the fish that survive the longest. The corresponding Walford lines are in the left panel, and only the broken "population" line, formed by joining the observed size of maturing fish in successive years, shows any tendency to approach the diagonal. If that were the only line available, growth would (erroneously) be considered to be asymptotic. A special study would be needed to ascertain how generally this effect contributes to "asymptotism" in other species.

There are also examples of Walford lines that tend to curve parallel to the diagonal at the oldest ages represented (Fig. 12), in spite of being based on surviving fish. Thus it is by no means established that asymptotic growth is characteristic of fishes generally.

Our conclusion must be that the existence of asymptotic growth to an indefinitely great age can be neither proved nor disproved; nor need all kinds of fish be the same in this respect. Also, the question is not one of any great importance. Whether asymptotes really exist or not, asymptotic formulas are a convenient way of modeling many observed growth series, and we may expect them to be used into the indefinite future.

*Parker and Larkin's "concept of growth" has not been considered here because it relates growth to size already achieved rather than to age, and because it has apparently not been used in any practical manner. However their paper contains much useful information and stimulating discussion.

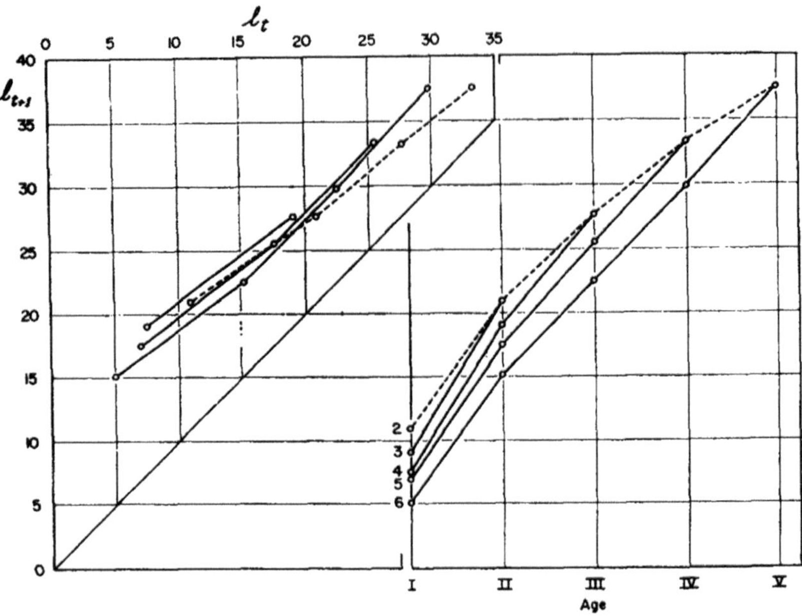

Fig. 14. Right panel: Computed fork lengths of chinook salmon that mature at the successive ages shown by arabic numerals at the left; the final season of growth has been omitted because it is not a full year. Left panel: The solid lines are Walford lines for the same salmon using back-calculated lengths from fish of ages 4–6. The broken line is a Walford line based on the lengths of the fish at their last scale annulus prior to maturity. (From Parker and Larkin, 1959, *Journal of the Fisheries Research Board of Canada;* reproduced by permission.)

K. Fitting Growth Curves

The logistic, Gompertz, Johnson, and the two Pütter curves all have three parameters to be estimated from the data. Many methods of estimation have been proposed, but they all involve successive approximation either by hand or by computer. Here we will describe briefly a simple noncomputer method available for each curve.

For the Pütter No. 1 curve, a simple and adequate method of fitting was described by Beverton (1954) (see also Ricker, 1958, 1975a). A trial estimate of the asymptote (L_∞) is obtained from a Walford line (l_{t+1} against l_t), fitted freehand or by a functional straight line, taking care to eliminate any points at the lower or upper end that do not conform to the line. To estimate t_0, Eq. (29) may be written as

$$\ln(L_\infty - l) = \ln L_\infty + Kt_0 - Kt \tag{50}$$

Thus the observed values of l are subtracted from the trial L_∞ and the natural logarithms of these differences are plotted against age (t). This line will, in general, have some curvature, but this is sensitive to the trial value chosen for L_∞. Additional trials will reveal the best (straightest) line, which can be selected sufficiently well by eye. Recalcitrant values at either the lower or the upper end should be rejected, the former because they indicate that the relationship is not appropriate for the younger ages, the latter perhaps for a similar reason, but also because they will usually be based on few fish. From the line finally chosen, the slope is a new estimate of K, and t_0 can be calculated from the ordinate intercept, which is equal to $\ln L_\infty + Kt_0$. If this line is fitted by least squares, it is of course the ordinary, not the functional, regression that should be used. Another method is described by Rafail (1973), and for computer use a program is given by Allen (1967) and Fabens (1965), while Marquardt's algorithm is recommended by Conway *et al.* (1970).

For the logistic, Gompertz, and Pütter No. 2 curves, Ricklefs (1967) published tables, reproduced in Table I, of "conversion factors" whereby the appropriateness of a given curve for the data at hand may be tested using a trial estimate of the asymptotic value, W_∞ (or L_∞). Observed figures are expressed as a percentage of the trial asymptote, and the corresponding conversion factors are plotted against age. If the line is curved, and if no larger or smaller trial asymptote decreases the curvature, the formula chosen is not suitable for the data. If a substantial part of the curve is straight, the formula is suitable, and additional trial values of the asymptote are used until the maximum number of points fit the line. The slope (S) of this "converted growth curve," fitted by eye or by an ordinary regression line, is equal to $1/W_\infty$ times the differential (absolute growth rate) of the corresponding curve at its inflection point. The value of t at the inflection point is the point at which the converted growth curve meets the t-axis; call this X. The curve parameters are then computed as follows.

Logistic [Eqs. (16)–(18)]: $g=4S$; $t_0=X$; $c=e^{gX}$

Gompertz [Eqs. (20)–(23)]: $g=Se$; $t_0=X$; $k=e^{gX}$

Pütter No. 2 [Eqs. (33)–(35)]: $K=9S/4$*; $t_0=X-(\ln 3)/K$; $c = {W_\infty}^{1/3}/3e^{-KX}$

Johnson [Eq. (40)]: $g=Se^2/4$; $t_0=X-1/(2g)$

It frequently happens that more than one of the three-parameter curves above will fit a given body of data adequately, and only rarely will none of them prove suitable. For such cases the four-parameter Richards function should almost always provide a good fit. However, to fit it requires two-stage iteration that is best done by computer, although Richards (1959) describes a manual trial and error method. Computer programs are given by Nelder (1961) and Causton (1969).

*Ricklefs (1967) inadvertently wrote this as 4S/9 (his Eq. 7).

Table I Ordinates of Straight Lines That Correspond to Ordinates (w/W_∞ or l/L_∞) of Four Growth Curves[a]

	0.00	0.01	0.02	0.03	0.04	0.05	0.06	0.07	0.08	0.09
Logistic curve										
0.0		−1.149	−0.973	−0.869	−0.795	−0.736	−0.688	−0.617	−0.611	−0.578
0.1	−0.549	−0.523	−0.498	−0.475	−0.454	−0.434	−0.415	−0.396	−0.379	−0.363
0.2	−0.347	−0.331	−0.316	−0.302	−0.288	−0.275	−0.261	−0.249	−0.236	−0.224
0.3	−0.212	−0.200	−0.188	−0.177	−0.166	−0.155	−0.144	−0.133	−0.122	−0.112
0.4	−0.101	−0.091	−0.081	−0.070	−0.060	−0.050	−0.040	−0.030	−0.020	−0.010
0.5	0.000	0.010	0.020	0.030	0.040	0.050	0.060	0.070	0.080	0.090
0.6	0.101	0.112	0.122	0.133	0.144	0.155	0.166	0.177	0.188	0.200
0.7	0.212	0.224	0.236	0.249	0.261	0.275	0.288	0.302	0.316	0.331
0.8	0.347	0.363	0.379	0.396	0.415	0.434	0.454	0.475	0.498	0.523
0.9	0.549	0.578	0.611	0.647	0.688	0.736	0.795	0.869	0.973	1.149
Gompertz curve										
0.0		−0.562	−0.502	−0.462	−0.430	−0.403	−0.381	−0.360	−0.341	−0.323
0.1	−0.307	−0.291	−0.276	−0.262	−0.249	−0.236	−0.223	−0.210	−0.198	−0.187
0.2	−0.175	−0.164	−0.153	−0.142	−0.131	−0.120	−0.110	−0.099	−0.089	−0.079
0.3	−0.068	−0.058	−0.048	−0.038	−0.028	−0.018	−0.008	0.002	0.012	0.022
0.4	0.032	0.042	0.052	0.062	0.073	0.083	0.093	0.103	0.114	0.124

0.5	0.135	0.146	0.156	0.167	0.178	0.189	0.201	0.212	0.223	0.235
0.6	0.247	0.259	0.272	0.284	0.297	0.310	0.323	0.337	0.351	0.365
0.7	0.379	0.394	0.410	0.425	0.442	0.458	0.476	0.494	0.512	0.532
0.8	0.552	0.573	0.595	0.618	0.643	0.668	0.696	0.725	0.757	0.791
0.9	0.828	0.869	0.914	0.965	1.024	1.093	1.177	1.284	1.435	1.692
Pütter No. 2 curve										
0.0		−0.380	−0.348	−0.323	−0.302	−0.284	−0.268	−0.252	−0.238	−0.224
0.1	−0.211	−0.198	−0.186	−0.174	−0.163	−0.151	−0.140	−0.129	−0.119	−0.108
0.2	−0.098	−0.087	−0.077	−0.067	−0.057	−0.046	−0.036	−0.026	−0.016	−0.006
0.3	0.004	0.014	0.024	0.034	0.044	0.054	0.064	0.074	0.084	0.095
0.4	0.105	0.115	0.126	0.136	0.147	0.158	0.169	0.180	0.191	0.202
0.5	0.213	0.225	0.236	0.248	0.260	0.272	0.285	0.297	0.310	0.323
0.6	0.336	0.349	0.363	0.377	0.391	0.406	0.421	0.436	0.452	0.468
0.7	0.484	0.501	0.519	0.537	0.556	0.575	0.595	0.616	0.637	0.660
0.8	0.683	0.708	0.733	0.761	0.789	0.820	0.852	0.886	0.924	0.964
0.9	1.008	1.056	1.110	1.171	1.241	1.342	1.425	1.554	1.736	2.045
Johnson curve										
0.0		−0.153	−0.132	−0.116	−0.102	−0.090	−0.078	−0.067	−0.056	−0.046
0.1	−0.036	−0.025	−0.015	−0.005	0.005	0.015	0.025	0.035	0.045	0.055

Continued

Table I Ordinates of Straight Lines That Correspond to Ordinates (w/W_∞ or l/L_∞) of Four Growth Curves—Cont'd

0.2	0.066	0.076	0.087	0.098	0.109	0.120	0.131	0.143	0.155	0.167
0.3	0.179	0.192	0.204	0.218	0.231	0.245	0.259	0.274	0.289	0.304
0.4	0.320	0.336	0.353	0.371	0.389	0.407	0.426	0.446	0.467	0.488
0.5	0.510	0.533	0.557	0.582	0.608	0.635	0.663	0.692	0.723	0.755
0.6	0.789	0.825	0.862	0.901	0.942	0.986	1.032	1.081	1.133	1.188
0.7	1.247	1.310	1.377	1.449	1.527	1.611	1.702	1.801	1.908	2.026
0.8	2.155	2.298	2.457	2.635	2.834	3.060	3.319	3.617	3.964	4.375
0.9	4.867	5.469	6.222	7.189	8.478	10.283	12.990	17.502	26.525	53.592

[a]The ordinates of the observed curves are indicated by the sum of corresponding entries in the top row and the left-hand column. Figures for the logistic, Gompertz, and Pütter No. 2 curves are from page 982 of Ricklefs (1967), and are used by permission of Dr. R. E. Ricklefs and *Ecology*. They are Copyright 1967 by The Ecological Society of America. Figures for the Johnson curve have been computed using a formula derived by Dr. Jon Schnute: $-4e^{-2}[0.5+(\ln l/L_\infty)^{-1}]$.

L. Curves Fitted to the Seasonal Cycle of Growth

A typical seasonal cycle of fish growth in temperate or chilly regions resembles an S-shaped Sachs cycle (Section III,A). Thus the S-shaped curves, particularly the logistic, Gompertz, or Pütter No. 2, are the most appropriate candidates for describing the seasonal course of growth. There are apparently not many examples of such fittings, but Miura *et al.* (1976) used logistic curves to describe seasonal growth in length of four age groups of redspot salmon (*Oncorhynchus rhodurus*).

Lockwood (1974) fitted Pütter No. 1 curves to the seasonal increase in length of trout (*Salmo trutta*) and plaice (*Pleuronectes platessa*). Although the fit was reasonable from the inflection point onward, the early-season phase of accelerating growth could of course not be described by this curve.

Pitcher and MacDonald (1973) suggest two methods of modifying the Pütter No. 1 curve so that it will describe seasonal increase in length over several years. The first uses a cosine function to switch growth off in winter and turn it on again in spring. The second incorporates a sine function that produces a smoother seasonal pattern; however, it includes some shrinkage in winter, which could be appropriate for weight data but less so for length. In either case two parameters are added to the basic Pütter expression, making five in all—which is not excessive considering that several years of growth are described. Either curve gave a reasonable fit to data on growth of a population of minnows (*Phoxinus phoxinus*). However the practical value of these expressions remains obscure.

V. GROWTH IN RELATION TO TEMPERATURE

A. Maximum Size and Age

It has long been known that many fishes tend to live longer and grow larger in the cooler parts of their range, the increase in age being usually more striking than the increase in size. This is true in spite of the fact that for most fishes both maximum age and maximum size must be defined somewhat arbitrarily. On a basis of straightforward sampling, the larger the sample, the greater will be the maximum age or size observed; however, if samples of several thousand large individuals have been examined, the oldest and largest ones obtained can reasonably be considered of the maximum age and size for purpose of discussion. In a similar manner we might use 100 years and 230 cm as limits that are not often exceeded by man.

Another arbitrary but useful upper limit of size is the asymptotic length or weight indicated by one of the growth curves, and the only one that has been used comparatively is the Pütter No. 1 (Taylor, 1958, 1959; Beverton and Holt, 1959). For an upper age limit Taylor uses the age at which, according to the curve, 95% of the asymptotic growth will have been realized (Fig. 15).

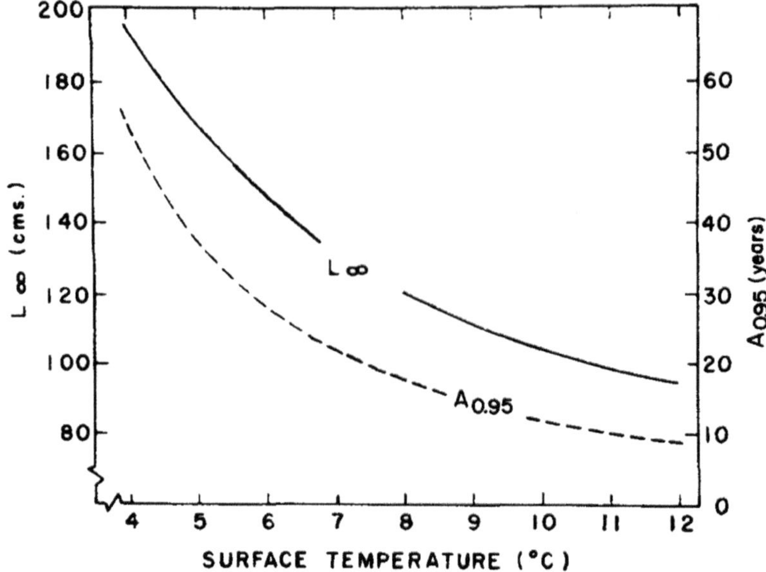

Fig. 15. Relation between ocean temperature and the Pütter asymptotic lengths (L_∞), and an index of maximum age ($A_{0.95}$), for a series of cod populations. (From Taylor, 1958; reproduced by permission.)

The slower growth of fish at more northern latitudes is not unexpected. Like other physiological processes, growth is affected by body temperature, which in most fishes is close to the ambient temperature. Relating growth rate to temperature is complicated by a strong negative relation between growth rate and fish size. The first attempts to circumvent this made use of the inverse relationship; that is, the time required to attain a given size or stage of development was related to temperature.

B. Hyperbolic Relationships

The simplest temperature relationship, and the one most used to date, is a rule that is said to date from Réaumur in 1735 (Hayes, 1949). Development from fertilized egg to hatching requires a fixed number of degree-days (day-degrees, Tagesgrade) and, within limits, is independent of the temperatures that actually prevail. The limits for most trout and salmon are about $3^\circ - 14^\circ$C (Fig. 16). If temperature does not vary, this implies direct proportionality between absolute rate of growth and Celsius temperature, and the equation is the hyperbola

$$t = \frac{K}{T} \tag{51}$$

where t, time in days needed to complete a given stage of growth; T, temperature in degrees Celsius; K, number of degree-days required.

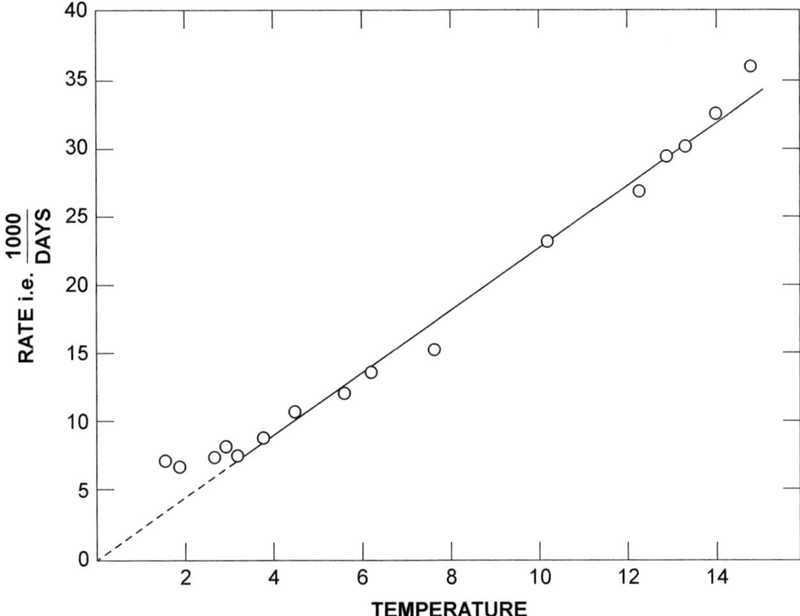

Fig. 16. Reciprocal of number of days required for development of brook trout eggs (fertilization to hatching) plotted against water temperature. (From Hayes, 1949; reproduced by permission.)

If temperature varies, Eq. (51) becomes

$$\sum_{i=1}^{t} T_i = K \tag{52}$$

where T_i is the temperature on day i.

A modification of Eqs. (51) and (52) has been proposed in which the temperature is measured from a point other than 0°C. If T_0 is the new base temperature, the expressions become

$$t = \frac{K}{T - T_0} \tag{53}$$

$$\sum_{i=1}^{t} (T - T_0)_i = K \tag{54}$$

This makes the hyperbolic relationship applicable, approximately at least, to a much larger number of species. For example, Lasker's (1964) observations on incubation times of Pacific sardines (*Sardina caerulea*) between 12° and 20°C follow the relationship fairly well, with $T_0 = 8°C$ and $K = 474$ degree-hours; the extremes of 11° and 21°C are more aberrant.

It is also possible to relate size directly to cumulative degree-days, and when length is used the relationship can be close to linear, even in the wild. For postembryonic growth of mussels (*Mytilus edulis*) near Copenhagen Boëtius (1962) obtained a perfectly straight line over a period of 2 years, using 0°C as the base temperature. Ursin (1963) shows several examples which have some curvature, and he was able to get a straight line for North Sea plaice (*Pleuronectes platessa*) over 4 years' time by making some major adjustments to the temperatures used in calculating the degree-days.

C. Janisch's Catenary Curve

When the hyperbolic degree-day formula came to be applied to experimental data involving a wide range of temperatures, it failed badly; it was found that there was an optimum temperature for growth, beyond which growth slowed down—even when the fish were given all the food they would eat.

Janisch (1927) found that a catenary curve would fit such observations; it has the form

$$t = t_0[\cosh k(T - T_0)] \tag{55}$$

where t, time in days required for growth to a specified size or stage; t_0, time required for that growth when $T = T_0$; T, observed temperature during time t; T_0, the temperature at which growth is fastest; k, a parameter that may be called *Janisch's temperature coefficient*. A catenary is the curve described by a uniform chain or rope suspended between two points at the same elevation. There is no obvious reason why it should describe the rate of growth of fishes or anything else, but this is not an obstacle to employing it if it proves empirically useful.

Curve B in Fig. 17 is a catenary for which most rapid development (50 days) is at 10°C and the temperature coefficient is $k = 0.22$. The limbs of the curve on either side of the minimum are shown of equal length, but in practice the right limb is much shorter, being curtailed by death of the fish at higher temperatures. Curve A in Fig. 17 is the hyperbola for the degree-day rule, when development takes 50 days at 10°C. The two curves have a generally similar shape from about 3° to 9°C. They can be made to provide very similar values within that range by using $k = 0.263$ for Janisch's coefficient in the catenary, while retaining the 10°C optimum point.

Ursin (1963) points out that the hyperbola of Eq. (53), approximating part of the left half of the catenary, can be computed from the tangent of the left inflection point of the bell-shaped reciprocal of the catenary; it has the form

$$(T - T_0 + 2.296/k)t = 2t_0/k \tag{56}$$

where T_0, t_0, and k are all as in Eq. (55). Such an hyperbola passes through the inflection point but of course does not pass exactly through the optimum point.

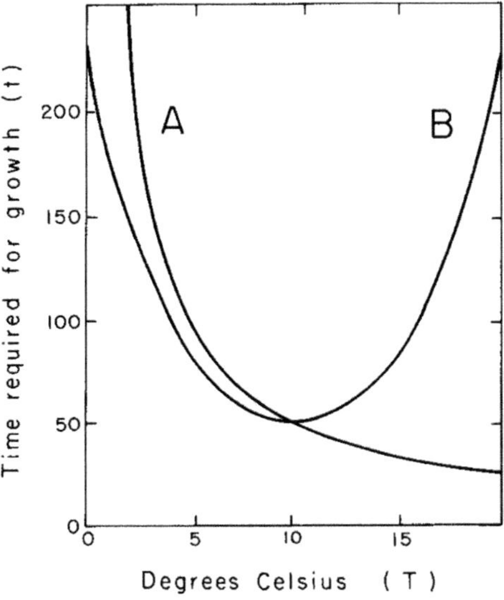

Fig. 17. Example of a catenary curve of duration of growth against temperature, with most rapid growth at 10°C (B); and the hyperbola that passes through the minimum point of the catenary (A).

Ursin (1963) also attempted to combine the catenary temperature relationship with the Pütter No. 1 growth curve. He accepted von Bertalanffy's theoretical background for the Pütter curve and postulated that "anabolic and catabolic processes separately conform to catenary curves" (p. 2). For the resulting abstruse speculations the reader must be referred to Ursin's paper. Like von Bertalanffy, he failed to estimate anabolism and catabolism quantitatively, let alone test their agreement with the catenary relationship, yet he felt able to reach such conclusions as, for example, that the cod is "a fish in which anabolism is more affected by temperature than catabolism" (p. 15).

D. Parabolas

A more familiar curve that provides just as good a fit to most or all of the growth–temperature data available is the quadratic parabola. Here it takes the form

$$t = t_0 + k(T - T_0)^2 \tag{57}$$

The symbols t, t_0, T, and T_0 all have the same meaning as in Eq. (55), but the coefficient k of course has a different value.

Bělehrádek (1930) suggested generalizing the above as the socalled "power parabola":

$$t = t_0 + a(T - T_0)^n \tag{58}$$

This provides a growth–temperature relationship flexible enough to take care of almost any set of observations, but at the expense of using four parameters.

E. Elliott's Growth Curves

Undeterred by the obvious complexity of the relation between growth, temperature and size, Elliott (1975a) has constructed a set of growth curves on the basis of his experiments with brown trout (*Salmotrutta*) of 12 to about 250 g, fed an excess ration at different temperatures for up to 42 days. During this short period of time growth was adequately represented by an exponential curve. For trout of a given size he found that the relation between instantaneous growth rate (G) and Celsius temperature (T) could be approximated by four straight lines, each of the form

$$G = a + bT \tag{59}$$

In the range 3.8°–12.8°C b was positive; in the range 12.8°–13.6°C it was close to zero, although this short interval is ignored; for 13.6°–19.5°C b was negative; and for 19.5°–21.7°C it was negative with a steeper slope.

At any given temperature Elliott found that a plot of the logarithm of instantaneous growth rate (G) against logarithm of mean biomass ($\overline{w}$) produced a straight line (Fig. 18); when antilogged, the resulting equation is

$$G = p\overline{w}^{-q} \tag{60}$$

[Elliott used an arithmetic mean of initial and final weight, which is adequate for the short time spans he used. For a longer period of exponential growth, say n time units in length, the true mean is $w_0(e^{Gn} - 1)/Gn$, where w_0 is the initial weight.]

Equations (59) and (60) were then combined by computing a and b in Eq. (59) for a 1 g trout, and combining it with Eq. (60) when $\overline{w} = 1\,\text{g}$; this gave three expressions of the form

$$G = (a + bT)\overline{w}^{-q} \tag{61}$$

Finally, growth was put into the general form

$$\frac{dw}{dt} = (a + bT)w^{1-q} \tag{62}$$

$$w_t = \left[q(a + bT)t + w_0{}^q\right]^{1/q} \tag{63}$$

where w_0 is initial weight and a, b, and q are the same parameters as in Eq. (61), for the different temperature ranges. Comparison of observed final sizes with those calculated from Eq. (63) shows very close agreement.

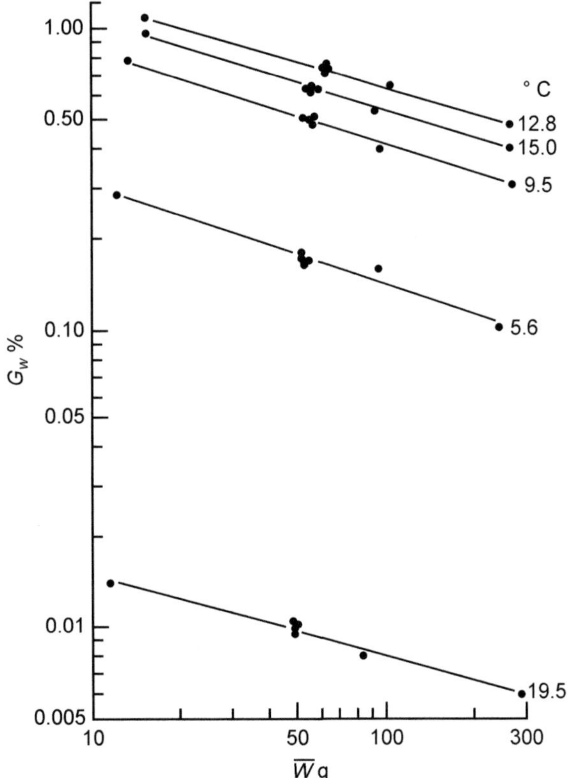

Fig. 18. Double logarithmic relation between relative growth rate, as percent per day (ordinate), and the mean weight of brown trout fed unlimited rations (abscissa), at different temperatures. (From Elliott, 1975a; reproduced by permission of Blackwell Scientific Publications from the *Journal of Animal Ecology*.)

Equation (63) is an elegant summary of Elliott's experimental data but, unreasonably no doubt, we experience a feeling of disappointment. It applies to only a short period of time, it sheds no light on the nature of growth processes, nor is it an expression that can be used routinely to represent fish growth. Elliott is careful to restrict its applicability to brown trout fed excess rations under his own experimental conditions. Even if this form of expression were to prove generally applicable, the experimentation needed to determine the parameters makes it impractical to employ it in any general fashion.

F. Zweifel and Lasker's Temperature Functions

Zweifel and Lasker (1976) combined the Gompertz curve with temperature and developed two functions to describe growth in relation to temperature.

Using length symbols (l instead of w) in Eq. (21) above, the parameter g was itself conceived as a variable that changes with Celsius temperature according to a Gompertz curve of the form of Eq. (21)—see Zweifel and Lasker's expressions (5) and (5a). Finally, they permitted the origin of the temperature scale to vary, and so obtained a six-parameter expression [their Eq. (5b)].

Considered empirically, these functions provided good fits to data for growth of yolk sac larvae of Pacific sardines (*Sardina caerulea*) and the eggs of anchovies (*Engraulis mordax*). However, the large number of parameters makes these expressions cumbersome and presents an onerous problem of curve fitting, while at the same time it is not obvious that they are either theoretically enlightening or practically useful.

G. Brett's Tabular Presentation

Without trying to fit any mathematical curve to his data, Brett (1974, Table 4) tabulated the expected instantaneous growth rates of sockeye salmon (*Oncorhynchus nerka*) fed to satiation three times a day in freshwater. The table is for sockeye from 0.3 to 500 g, and temperatures from 3° to 18°C. In the absence of a meaningful or even a convenient mathematical relationship—whose discovery becomes more and more unlikely as the years go by—this type of presentation emerges as the most useful and informative way to summarize growth–temperature relationships. Figure 19 illustrates the nature of the trends.

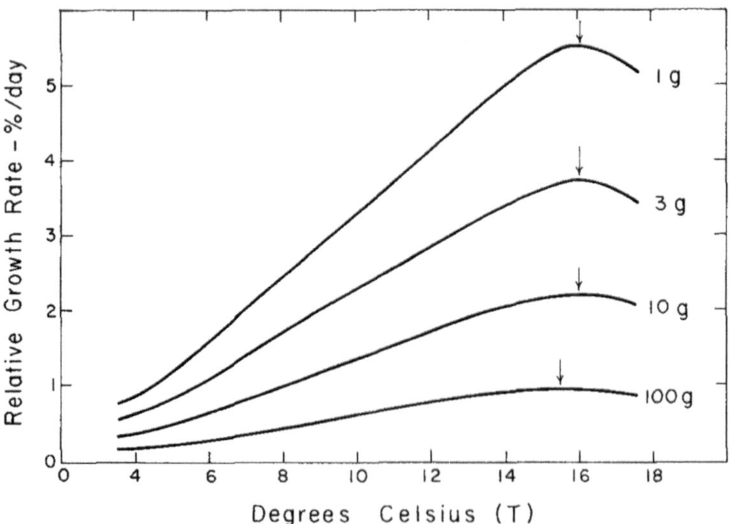

Fig. 19. Instantaneous growth rate plotted against temperature, for sockeye salmon of different sizes fed to satiation three times a day. (Data are from the smoothed values in Table 4 of Brett, 1974.)

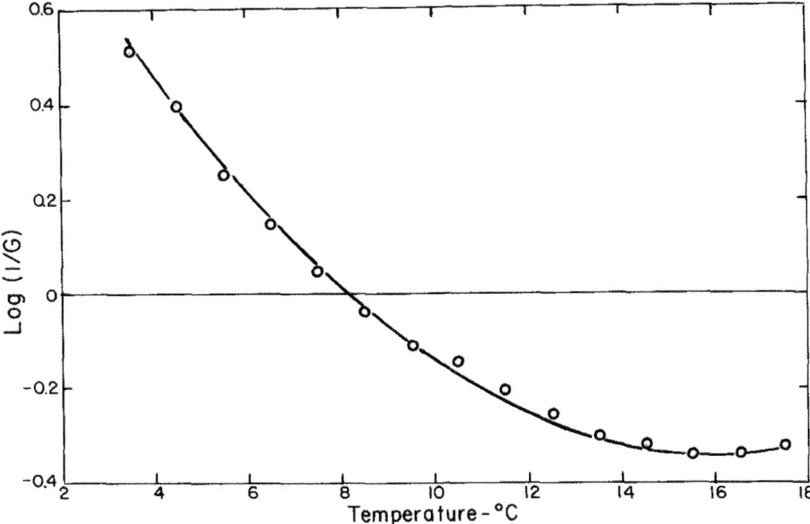

Fig. 20. Relation between the logarithm of the reciprocal of the instantaneous rate of growth of 10-g sockeye salmon and temperature, fitted with a parabola. (Data from Table 4 of Brett, 1974.)

By analogy with Section V,C, the reciprocals of the instantaneous growth rates ($1/G$) can be plotted. The curves obtained are concave upward, but neither a parabola nor a catenary will describe them at all closely. A reasonably good fit can be made of an ordinary parabola to the *logarithm* of the reciprocal of the instaneous growth rate (Fig. 20), which is a relationship of the form

$$\log(1/G) = a + b(T - T_0)^2 \tag{64}$$

T_0 is the temperature for maximum rate of growth, a is the corresponding value of $\log(1/G)$, and b is a constant to be found by trial. Using base-10 logarithms, the curve in Fig. 20 has $a = -0.344$, $b = 0.0056$, $T_0 = 16.0$. But why go to this trouble? Nothing is gained over using Brett's Table 4 directly.

VI. GROWTH IN RELATION TO RATIONS

A. Empirical Representations

The relation between the food consumed by fish and the resulting growth is a matter of great practical importance, so it has been the subject of numerous experiments, among which Ivlev's (1955) work is outstanding. Two recent comprehensive studies are by Brett *et al.* (1969) on sockeye salmon and by Elliott (1975b) on brown trout. Both authors present their information in a series of tables and graphs illustrating various aspects of the phenomenon,

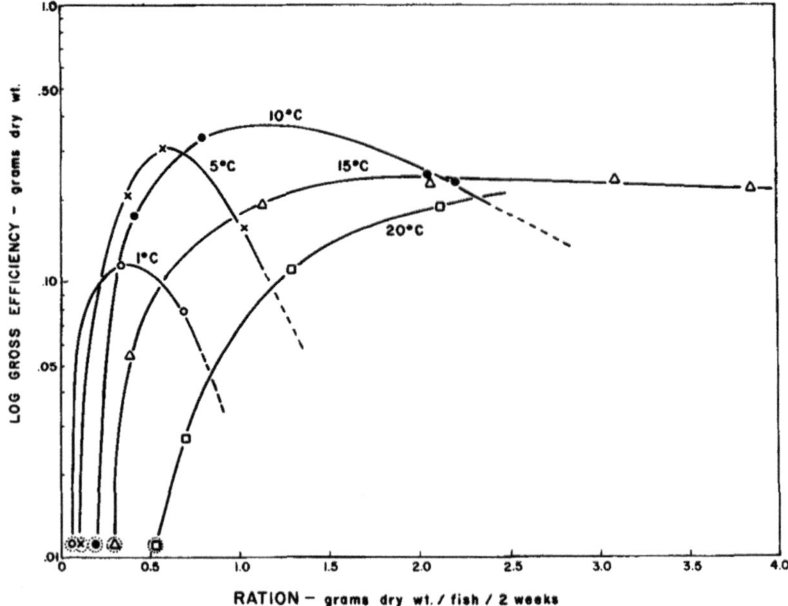

Fig. 21. "*K*-lines" at five temperatures for sockeye salmon forced to swim continuously at a moderate rate. Circled points are interpolated. (From Brett *et al.*, 1969, *Journal of the Fisheries Research Board of Canada;* reproduced by permission.)

particularly its relation to temperature. One of these is shown in Fig. 21. The technique used by Brett *et al.* differed from that of Elliott and most other authors in that they imitated natural conditions to the extent of making their fish swim continuously at a moderate rate. The ordinary procedure has been simply to let the fish perform only voluntary movements in aquaria or ponds.

The most interesting, albeit predictable, result that emerges from all this work is that an increase in temperature will increase growth rate only if it is accompanied by an increase in food consumption more than sufficient to meet the energy requirements of the automatic increase in basal metabolism and of a probable increase in activity. If temperature increases but food consumption does not, growth rate will decrease and may even fall to zero or beyond. For example, in Fig. 21 the ration that permits the maximum rate of conversion of food to growth at 5°C does little more than maintain body size at 20°C. More generally, when food is scarce fastest growth will occur at a lower temperature than what is optimum when food is available in excess.

The sections to follow give brief accounts of a number of attempts to describe the relation between growth rate and ration by mathematical expressions, and even to expand this into general accounts of trophic relationships in nature.

B. Stauffer's Sine Curve

Stauffer's (1973) approach to this problem was to find the best relationship between ration and instantaneous growth rate at a given temperature, in terms of two or more of four critical points in the growth–ration relationship. These points are as follows, described in terms of ration (R), increase in mass (Δw), and instantaneous growth rate (G).

No food:	$R=0$, Δw and G are negative
Maintenance ration:	R small, Δw and $G=0$
Optimum ration:	$\Delta w/R$ (Ivlev's K_1) is a maximum
Maximum ration:	R and G have their maximum value

Stauffer examined three possible mathematical relationships between G and R: a straight line, the Michaelis–Menton curve, and a sine curve. He tested each of them against some data of Brett *et al.* (1969) and found best agreement with a sine curve of the form

$$G = G_{\mathrm{m}} \times \sin\left(\frac{\pi}{2} \times \frac{R - R_n}{R_{\mathrm{m}} - R_n}\right) \tag{65}$$

where G, instantaneous growth rate at ration R; G_{m}, maximum instantaneous growth rate; R_{m}, maximum ration; R_n, maintenance ration. The fit was good at 5° and 10°C, but much less so at 15° and 20°C (Fig. 22).

Although Stauffer (1973, p. 29) states that "this algebraic expression does not explain the growth phenomenon," he also makes the claim that "it does contribute to the understanding of growth," but fails to indicate how.

In practice, there can be considerable difficulty in evaluating G_{m} and R_{m}, for the amount of food that fish will consume when not particularly hungry can vary a good deal with frequency of feeding and with slight changes in environmental conditions, outside stimuli, and so on (Brett, 1974).

C. Elliott's Combined Formula

Elliott (1975b) attempted the more ambitious task of combining growth, ration, and temperature into a single formula. He derived a five-parameter expression that will describe the situation for his experimental brown trout, provided temperature is broken into three separate ranges; thus, fifteen parameters are required in all. The analysis is an impressive *tour de force* but, in contrast to Elliott's excellent empirical data, it is difficult to picture how his formula can be anything more than a mathematical curiosity of no real utility.

D. Paloheimo and Dickie's Analysis

Paloheimo and Dickie (1965, 1966a,b) made a general review and analysis of data on fish growth, apparently in an effort to penetrate more deeply into the

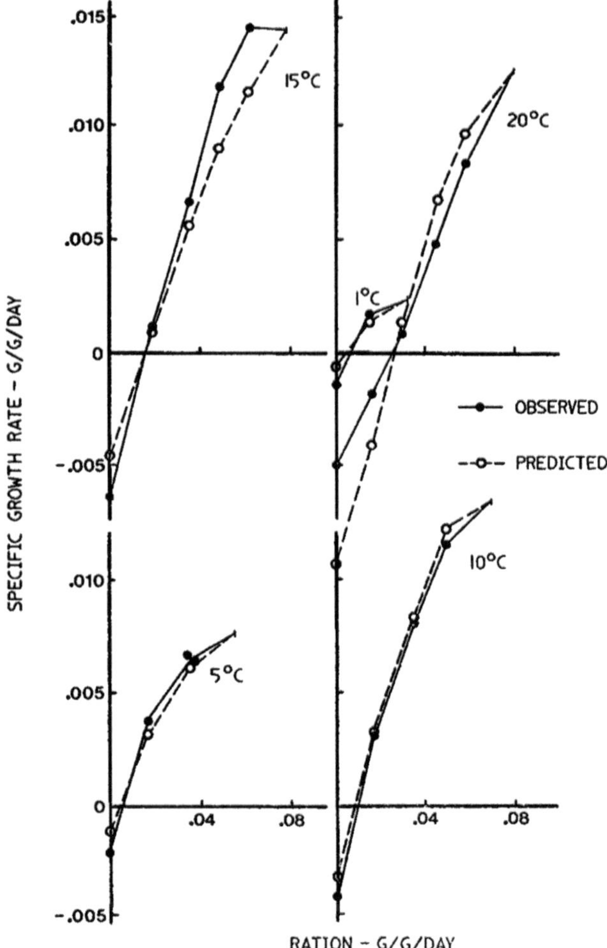

Fig. 22. Observed growth rates of sockeye salmon in relation to ration at different temperatures (solid lines), and growth rates computed from Stauffer's sine formula (broken lines). Both rations and growth are in grams per gram of mass of the fish per day. (From Stauffer, 1973; reproduced by permission.)

physiological processes involved. Their approach is based on two empirically determined relationships. The first of these is a logarithmic relation between metabolism (T) and biomass (w) or, better, body caloric content, at any given temperature:

$$\log T = a + b(\log w) \tag{66}$$

This relationship has been observed in many experiments with many species of fish and other animals, and it has been generally familiar to ichthyologists since the appearance of Winberg's (1956) comprehensive review. The slope b has usually been estimated from the ordinary regression of $\log T$ on $\log w$, but Ricker (1973) showed that it is more appropriate to use a functional regression, which gives metabolism and biomass the same mathematical position in the relationship.

For fish held in respirometers performing voluntary movements only, Winberg's summary indicates considerable variation between experiments and possibly between species, though it is difficult to be sure that different kinds or sizes of fish are being held under equivalent experimental conditions. However, a general average slope for all species of $b=0.80$ at 20°C was obtained, and this value was adopted by Paloheimo and Dickie. More recent experiments summarized by Glass (1969) show that this exponent tends to decrease with increase in temperature over the range 10°–25°C, with very few exceptions*; at 10°C it may average more than 1. In any event a functional regression line should be used to estimate the exponent from Eq. (66), and applying this method to the various experiments summarized by Winberg increased the general average exponent from 0.80 to 0.85 at 20°C (Ricker, 1973). There is also evidence that b becomes considerably larger—approaching 1—when the fish are in active motion (Brett and Glass, 1973).

The other basic relationship used by Paloheimo and Dickie is between the growth efficiency coefficient (K) and food ration (r). Ivlev (1945) defined a coefficient K_1, equal to increase in mass divided by food consumed, and a coefficient K_2 for which the divisor is food absorbed by the alimentary canal*; but since these two are approximately proportional, Paloheimo and Dickie refer to both by the symbol K. A graph of K against r has an ascending left limb starting from a positive value on the r-axis, a maximum, and a descending right limb (Fig. 22). Paloheimo and Dickie discovered that in most published experiments the right limb could be described by a straight line in a semi-logarithmic presentation:

---

*Glass computes exponents in two ways: by the usual logarithmic formula, and also by a nonlinear iterative method that requires successive approximations by computer. He points out that the iterative method is correct if the error term in the function $T = cw^b$ is additive, but that the logarithmic method is correct if the errors are multiplicative. However, in this as in most biological situations the errors are in fact multiplicative, approximately at least, as shown by the fact that variability becomes larger absolutely among larger fish. Thus it would be wrong to adopt the cumbersome iterative procedure. In the examples Glass used, the estimates by iteration were sometimes larger, sometimes smaller, than the logarithmic estimate: the average for all species and temperatures was 0.880 for the former and 0.938 for the latter; for the 20°C experiments only, the corresponding figures were 0.788 and 0.833.

*Ivlev also proposed a K_3 coefficient, but this was shown by Winberg (1956) to be an artificial quantity based on erroneous interpretation of experimental data (see Ricker, 1966, p. 1750, footnote 10).

$$\log K = a - br \tag{67}$$

They refer to this as the "K-line," and it plays a major role in their subsequent theoretical development.

Two aspects of the K-line deserve discussion. One is Paloheimo and Dickie's (1966a) conclusion that for any given absolute ration the value of K is independent of the biomass of the fish. However, data bearing on this generalization are scanty and far from conclusive, consisting essentially of one of Kinne's experiments with a small warmwater cyprinodont. In any event there must be definite limits to its applicability, for a ration that would merely maintain weight ($K=0$) for a fish of 1 kg would permit rapid growth ($K>0.2$) in one of 10 g. Also, Paloheimo and Dickie state that "body weight is the more efficient predictor of growth efficiency," which seems to contradict their generalization.

Kerr (197la) has emphasized that in nature the nutritional–physiological situation for fish that have to obtain food by active foraging is likely to be different from that of aquarium fish, and from the ecological point of view it would seem to make little difference whether K is governed by size of ration or fish size or both, for in nature the two are highly correlated: big fish tend to eat more than small fish. Something that must be considered, of course, is that when a fish matures, a considerable part of its food becomes channeled into the ovaries and testes, so eggs and milt should be included in the new body substance when computing K. If growth is estimated from yearly weighings, or from scale annuli, it will not include the sexual products, and a K so computed may decrease abruptly at maturity.

A puzzling aspect of Eq. (67) is the fact that in some cases the K-line extends linearly back to very small rations. In fact, in none of the experiments cited by Paloheimo and Dickie is there much trace of the left-hand or ascending portion of the K-line, yet it is obvious that the line must have its origin at the maintenance ration—where none of the food ingested is used for growth and $K = 0$. In some experiments the smallest ration used was apparently well above maintenance, but this is not true of those presented in Figs. 6, 7, and 9 and one series in Fig. 10 of Paloheimo and Dickie (1966a). The K-lines that can be computed from Fig. 1 of Elliott (1975b) are similar, although for two of the curves there is a point on the ascending limb. All these experiments were based on the weight rather than the caloric content of the fish, so there may be a plausible explanation: the nutritional conditions experienced by the fish prior to the experiment may have exerted an influence on into the experimental period. For some time prior to an experiment the fish to be used are commonly fed to satiation or nearly so, and so will have the high fat and protein content, and correspondingly low water content, that is characteristic of such a regime. On reduced rations percentage of water increases substantially (Brett *et al.*, 1969; Elliott, 1975b). It seems possible that growth processes such as cell division that are involved in an increase in size have a certain momentum, and do not stop immediately when a fish is suddenly put on low rations. Thus the fish may for a while continue to increase in length and weight by increasing the water content of its body. In the case of LeBrasseur's young chum salmon (*Oncorhynchus keta*) graphed by Kerr (1971a, Fig. 3), another factor may have entered: The smallest fish may still have contained a quantity of egg yolk, which too would make for a spuriously large *K*. None of the experiments from which *K*-lines have been estimated appear to have been started with fish that were previously held on a maintenance ration. It would be instructive to perform such an experiment and find whether, under such circumstances, the *K* values computed from weight would at first increase only quite gradually with size of ration. If so, then long straight *K*-lines would be the exception rather than the rule, and, insofar as they are an essential part of Paloheimo and Dickie's theoretical framework, the latter may be of limited applicability to conditions in the wild.

Notice also that the *K*-lines of Fig. 21 are not of the sort that Paloheimo and Dickie consider typical. Rather, they increase gradually to their maximum, and in fact at 20°C the maximum was not reached at the largest ration the fish would accept. The reason for the absence of any major straight section in these curves must lie in some aspect of the experimental conditions used by Brett *et al.* (1969). Their sockeye were made to swim continuously at a moderate rate, simulating the active life of sockeye in the wild.

It is outside the scope of this chapter to review Paloheimo and Dickie's analysis in detail. However, in spite of the ingenuity and persistence that are exhibited in the development of various aspects of their work, one finishes reading their papers with a feeling of anticlimax. After their bold start from observational data, we had hoped that they would be able to shed greater light on growth processes and, just possibly, produce a new growth equation involving parameters convenient for routine use.

E. Kerr's Analysis

Starting where Paloheimo and Dickie leave off, Kerr (1971a,b,c) has extended their work, taking into consideration the various components of metabolism and the size, density and distribution of food organisms. Much ingenuity is expended in this analysis, and various aspects of it can be subjected to additional experimental testing in future. As with the Paloheimo–Dickie scheme, no outline of the development can be given here. Figures 2 and 3 of Kerr's (1971c) paper compare a computed growth curve with an observed one for the Lake Opeongo lake trout (*Salvelinus namaycush*), both before and after the trout obtained an additional and larger prey species, the cisco (*Coregonus artedii*). However, agreement is obtained by postulating prey densities (different for each of four size categories of trout) that will generate the two growth curves observed, and there is no single mathematical expression that describes either curve.

F. Ursin's Analysis

Another ambitious theoretical analysis of fish growth processes has been made by Ursin (1967). He works from and toward Pütter's generalized growth equation

$$\frac{dw}{dt} = Hw^m - kw^n \tag{68}$$

where w is mass at time t, and H, m, k, and n are coefficients. Whereas Pütter (1920), von Bertalanffy (1934), Beverton and Holt (1957), and others assume that $m = 2/3$ and $n = 1$ in Eq. (68), Ursin estimates these parameters from published experimental data, using a variety of techniques and some rather uncertain assumptions. These all give values of n that are less than 1, with a mean about 0.83; while estimates of m are more variable, with 0.6 as a possible representative figure. However there is no attempt to demonstrate that m and n are constants that represent physiological realities reflecting basic protoplasmic processes. Rather, they are artificial, though not necessarily useless, parameters that emerge automatically once the form of Eq. (68) and various ancillary hypotheses have been assumed.

In any event Ursin's analysis, like those of Elliott, Kerr, and Paloheimo and Dickie, has not produced any relationship for everyday use based on physiologically meaningful concepts. It now seems safe to conclude that no such simple relationship exists.

VII. SUMMARY

Starting with the fertilized egg, the growth of a fish can be divided into a series of stanzas, marked off by developmental or ecological crises such as hatching, transformation from larva to fingerling, migration from salt to fresh water or vice versa, or maturation. Within each stanza growth usually follows an S-shaped "Sachs cycle," although the relative duration of the two branches of the S can vary greatly, and they are very different depending on whether length or weight is under consideration. A distinction must be made between individual growth and the increase in mean length or weight of a year-class, because of the widespread occurrence of mortality that is selective by size. Either the larger or the smaller members of a brood may suffer the greater mortality. Mortality from predation may be more severe among the smaller members of a brood, especially during the first year of life. As the fish grow older and larger, both internal mechanisms and the effects of fishing usually make for greater mortality among faster-growing individuals (Lee's phenomenon), with the result that estimates of growth rates from a comparison of the surviving fish at successive ages are too small, sometimes much too small. However, in many species true growth rates for individual fish can be calculated at yearly intervals from marks on scales or bones, and can be used, for example, to compute the production of the species in its habitat.

The numerical description of growth that is most convenient for combination with other parameters of the stock or of the environment is the instantaneous or specific growth rate; if growth is changing rapidly, it must be measured at short intervals of time. Over somewhat longer periods a double logarithmic relationship may describe growth satisfactorily, particularly during incubation and larval life, but it is difficult to combine this with other parameters. A variety of descriptive asymptotic curves are also available, including the logistic, Gompertz, Pütter Nos. 1 and 2, Johnson, and Richards. These have been used mainly to model growth at yearly intervals, but two have been applied to the seasonal course of growth. Although there have been several suggestions, none of these curves has been demonstrated to have any physiological basis that might qualify it as *the* growth curve for fishes generally, and frequently two or more of them will describe satisfactorily the same series of observations. When such curves are applied to the mean size of the fish in a population at successive ages, rather than to an individual fish, they all have the limitation that they describe only the size achieved by the *survivors* at each age. Thus they cannot be used to measure the true mean growth rate of the fish in the population, in the very common situation where faster-growing individuals tend to die earlier than slower-growing ones.

For fish of a given size, consuming a fixed ration greater than maintenance, the instantaneous growth rate at first increases with temperature, reaches a maximum, then decreases to zero. Continued temperature increase with the same ration causes loss of weight and eventually death. This complex relationship can be simplified by using its reciprocal—the length of time required to achieve a given size—and in that form it has, in some cases, been fitted reasonably well by a parabola or a catenary. Even a simple hyperbola may describe a considerable portion of the range of observations, and this is the basis of the degree-day rule used to predict hatching times of fish eggs.

Several rather complex curves have been used to describe the relation between growth and quantity of food consumed, but none shows real promise of having wide usefulness. To date, the most convenient form for presenting data on growth in relation to temperature or rations is an empirical table of smoothed values, or the corresponding graph.

ACKNOWLEDGMENTS

I am greatly indebted to Dr. J. R. Brett of the Pacific Biological Station, Nanaimo, British Columbia, for encouragement to write this chapter, for copies of and references to pertinent literature, and for penetrating comments. Drs. R. I. Fletcher of the University of Washington and J. Schnute of the Pacific Biological Station have given invaluable assistance with the mathematical aspects of the various growth curves. Dr. G. D. Stauffer of the U.S. National Marine Fishery Service, La Jolla, made available his thesis and additional references.

REFERENCES

Ahlstrom, E. H. (1943). Studies on the Pacific pilchard or sardine (*Sardinops caerulea*). 4. Influence of temperature on the rate of development of pilchard eggs in nature. *U.S. Fish Wildl. Serv., Spec. Sci. Rep.—Fish.* No. 23, 1–26.

Alexander, G. R., and Shetter, D. S. (1961). Seasonal mortality and growth of hatcheryreared brook and rainbow trout in East Fish Lake, Montmorency County, Michigan, 1958–59. *Pap. Mich. Acad. Sci., Arts Lett.* **46,** 317–328.

Allen, K. R. (1950). The computation of production in fish populations. *N.Z. Sci. Rev.* **8,** 89.

Allen, K. R. (1951). The Horokiwi stream: A study of a trout population. *N.Z. Fish. Res. Div. Fish. Res. Bull.* No. 10.

Allen, K. R. (1967). Computer programs available at St. Andrews Biological Station. *Fish. Res. Board Can. Tech. Rep. No.* 20.

Belehrádek, J. (1930). Temperature coefficients in biology. *Biol. Rev. Biol. Proc. Cambridge Philos. Soc.* **5,** 30–58.

Beverton, R. J. H. (1954). Notes on the use of theoretical models in the study of the dynamics of exploited fish populations. *U.S. Fish. Lab., Beaufort, N.C.* Misc. Contrib. No. 2.

Beverton, R. J. H., and Holt, S. J. (1957). On the dynamics of exploited fish populations. *U.K. Min. Agric. Fish., Fish. Invest.* **19,** 1–533.

Beverton, R. J. H., and Holt, S. J. (1959). A review of the lifespans and mortality rates of fish in nature, and their relation to growth and other physiological characteristics. *In* "The Lifespan of Animals." *Ciba Found. Colloq. Ageing* **5,** 142–177.

Boëtius, I. (1962). Temperature and growth in a population of *Mytilus edulis* (L.) from the northern harbour of Copenhagen (the Sound). *Meddelelsen Danmarks Fiskeriog Havunders.* **3,** 339–346.

Brett, J. R. (1974). Tank experiments on the culture of pan-size sockeye (*Oncorhynchus nerka*) and pink salmon (*O. gorbuscha*) using environmental control. *Aquaculture* **4,** 341–352.

Brett, J. R., and Glass, N. R. (1973). Metabolic rates and critical swimming speed of sockeye salmon (*Oncorhynchus nerka*) in relation to size and temperature. *J. Fish. Res. Board Can.* **30,** 379–387.

Brett, J. R., Shelbourn, J. E., and Shoop, C. T. (1969). Growth rate and body composition of fingerling sockeye salmon, *Oncorhynchus nerka*, in relation to temperature and ration size. *J. Fish. Res. Board Can.* **26,** 2363–2394.

Brody, S. (1927). Growth rates. *Mo. Agric. Exp. Stn., Bull.* No. 97.

Brody, S. (1945). "Bioenergetics and Growth." Reinhold, New York.

Cassie, R. M. (1954). Some uses of probability paper in the analysis of size frequency distributions. *Aust. J. Mar. Freshwater* Res. **5,** 513–522.

Causton, D. R. (1969). A computer program for fitting the Richards function. *Biometrics* **25,** 401–409.

Chapman, D. G. (1961). Statistical problems in dynamics of exploited fish populations. *Proc. Berkeley Symp. Math. Stat. Probab., 4th,* pp. 153–168.

Chugunova, N. I. (1959). "Handbook for the Study of Age and Growth of Fishes." Akad. Nauk Press, Moscow. (English transl., "Age and Growth Studies in Fish." Off. Tech. Serv., Washington, D.C., 1963.)

Conway, G. R., Glass, N. R., and Wilcox, J. C. (1970). Fitting nonlinear models to biological data by Marquardt's algorithm. *Ecology* 51, 503–507.

Elliott, J. M. (1975a). The growth rate of brown trout, *Salmo trutta* L., fed on maximum rations. *J. Anim. Ecol.* **44,** 805–821.

Elliott, J. M. (1975b). The growth rate of brown trout, *Salmo trutta* L., fed on reduced rations. *J. Anim. Ecol.* **44,** 823–842.

Fabens, A. J. (1965). Properties and fitting of the von Bertalanffy growth curve. *Growth* **29,** 265–289.

Fletcher, R. I. (1973). A synthesis of deterministic growth laws. Univ. Rhode Island Sch. Oceanogr., Kingston, Rhode Island.

Fletcher, R. I. (1975). A general solution for the complete Richards function. *Math. Biosci.* 27, 349–360.

Ford, E. (1933). An account of the herring investigations conducted at Plymouth during the years from 1924 to 1933. *J. Mar. Biol. Assoc. U.K.* **19,** 305–384.

Gerking, S. D. (1957). Evidence of aging in natural populations of fishes. *Gerontologia* **1,** 287–305.

Gerking, S. D. (1966). Annual growth cycle, growth potential, and growth compensation in the bluegill sunfish in northern Indiana lakes. *J. Fish. Res. Board Can.* **23,** 1923–1956

Glass, N. R. (1969). Discussion of calculation of power function with special reference to respiratory metabolism in fish. *J. Fish. Res. Board Can.* **26,** 2643–2650.

Gompertz, B. (1825). On the nature of the function expressive of the law of human mortality, and on a new mode of determining the value of life contingencies. *Philos. Trans. R. Soc. London* **115,** 515–585.

Graham, M. (1929). Studies of age determination in fish. Part II. A survey of the literature. *U.K. Min. Agric. Fish., Fish. Invest.* **11**(2).

Graham, M. (1935). Modern theory of exploiting a fishery, and application to North Sea trawling. *J. Cons., Cons. Perm. Int. Explor. Mer* **13,** 76–90.

Gray, J. (1928). The growth of fish. II. The growth rate of the embryos of *Salmo fario*. *Br. J. Exp. Biol.* **6,** 110–124.

Hayes, F. R. (1949). The growth, general chemistry, and temperature relations of salmonid eggs. *Q. Rev. Biol.* **24,** 281–308.

Hayes, F. R., and Armstrong, F. H. (1943). Growth of the salmon embryo. *Can. J. Res., Sect. D* **21,** 19–33.

Hogman, W. J. (1968). Annulus formation on the scales of four species of coregonids reared under artificial conditions. *J. Fish. Res. Board Can.* **25,** 2111–2122.

Imbrie, J. (1956). Biometrical methods in the study of invertebrate fossils. *Bull. Am. Mus. Nat. Hist.* **108,** 211–252.

Ivlev, V. S. (1945). The biological productivity of waters. *Úsp. Sovrem. Biol.* **19,** 98–120. [English transl., *J. Fish. Res. Board Can.* **23,** 1727–1759 (1966).]

Ivlev, V. S. (1955). "Experimental Ecology of the Feeding of Fishes." Pishchepromizdat, Moscow. [English transl., Yale University Press, New Haven (1961).]

Janisch, E. (1927). Das Exponentialgesetz als Grundlage einer vergleichenden Biologie. *Abh. Theorie Org. Entwicklung* **2,** 1–371.

Johnson, N. O. (1935). A trend line for growth series. *J. Am. Stat. Assoc.* **30,** 717.

Jolicoeur, P. (1975). Linear regressions in fishery research: Some comments. *J. Fish. Res. Board Can.* **32,** 1491–1494.

Jones, R. (1958). Lee's phenomenon of "apparent change in growth rate," with particular reference to cod and plaice. *Int. Comm. Northwest Atl. Fish. Spec. Publ.* **1,** 229–242.

Kermack, K. A., and Haldane, J. B. S. (1950). Organic correlation and allometry. *Biometrika* **37,** 30–41.

Kerr, S. R. (1971a). Analysis of laboratory experiments on growth efficiency of fishes. *J. Fish. Res. Board Can.* **28,** 801–808.

Kerr, S. R. (1971b). Prediction of fish growth efficiency in nature. *J. Fish. Res, Board Can.* **28,** 809–814.

Kerr, S. R. (1971c). A simulation model of lake trout growth. *J. Fish. Res. Board Can.* **28,** 815–819.

Knight, W. (1968). Asymptotic growth: an example of nonsense disguised as mathematics. *J. Fish. Res. Board Can.* **25,** 1303–1307.

Krüger, F. (1962). Über die mathematische Darstellung des tierischen Wachstums. *Naturwissenschaften* **49,** 454.

Krüger, F. (1964). Neuere mathematische Formulierungen der biologischen Temperaturfunktion und des Wachstums. *Helgol. Wiss. Meeresunters.* **9,** 108–124.

Krüger, F. (1965). Zur Mathematik des tierischen Wachstums. I. Grundlagen einer neuen Wachstumsfunktion. *Helgol. Wiss. Meeresunters.* **12,** 78–136.

Krüger, F. (1969). Das asymptotische Wachstum der Fische—ein Nonsens? *Helgol. Wiss. Meeresunters.* **19,** 205–215.

Krüger, F. (1973). Zur Mathematik des tierischen Wachstums. II. Vergleich einiger Wachstumsfunktionen. *Helgol. Wiss. Meeresunters.* **25,** 509–550.

Laird, A. K. (1967). Evolution of the human growth curve. *Growth* **31,** 345–355.

Laird, A. K., Tyler, S. A., and Barton, A. D. (1965). Dynamics of normal growth. *Growth* **29,** 233–248.

Lasker, R. (1964). An experimental study of the effect of temperature on the incubation time, development and growth of Pacific sardine embryos and larvae. *Copeia* pp. 399–405.

Lee, R. M. (1912). An investigation into the methods of growth determination in fishes. *Publ. Circonstance, Cons. Perm. Explor. Mer* 63.

Lockwood, S. J. (1974). The use of the von Bertalanffy growth equation to describe the seasonal growth of fish. *J. Cons., Cons. Perm. Int. Explor. Mer* **35,** 175–179.

Lumer, H. (1937). The consequences of sigmoid growth for relative growth functions. *Growth* **1,** 140–154.

MacDowell, E. C., Allen, E., and MacDowell, G. G. (1927). The prenatal growth of the mouse. *J. Gen. Physiol.* **11,** 57–70.

Martin, W. R. (1949). The mechanics of environmental control of body form in fishes. *Univ. Toronto Stud., Biol. Ser.* No. 58. (Also called *Publ. Ont. Fish. Res. Lab.* No. 70.)

Minot, C. S. (1891). Senescence and rejuvenation. First paper. On the weight of guinea pigs. *J. Physiol.* (*London*) **5,** 457–464.

Miura, T., Suzuki, N., Nagoshi, M., and Yamamura, K. (1976). The rate of production and food consumption of the biwamasu, *Oncorhynchus rhodurus*, population in Lake Biwa. *Res. Popul. Ecol.* (*Tokyo*) **17,** 135–154.

Nelder, J. A. (1961). The fitting of a generalization of the logistic curve. *Biometrics* **17,** 89–110.

Paloheimo, J., and Dickie, L. M. (1965). Food and growth of fishes. I. A growth curve derived from experimental data. *J. Fish. Res. Board Can.* **22,** 521–542.

Paloheimo, J., and Dickie, L. M. (1966a). Food and growth of fishes. II. Effects of food and temperature on the relation between metabolism and body size. *J. Fish. Res. Board Can.* **23,** 869–908.

Paloheimo, J., and Dickie, L. M. (1966b). Food and growth of fishes. III. Relations among food, body size and growth efficiency. *J. Fish. Res. Board Can.* **23,** 1209–1248.

Parker, R. R., and Larkin, P. A. (1959) A concept of growth in fishes. *J. Fish. Res. Board Can.* **16,** 721–745.

Partlo, J. M. (1955). Distribution, age and growth of eastern Pacific albacore (*Thunnus alalunga* Gmelin). *J. Fish. Res. Board Can.* **12,** 35–60.

Pella, J. J., and Tomlinson, P. K. (1969). A generalized stock production model. *InterAm. Trop. Tuna Comm., Bull.* **13,** 420–496.

Pelluet, D. (1944). Criteria for the recognition of developmental stages in the salmon (*Salmo salar*). *J. Morphol.* **74,** 395–407.

Petersen, C. G. J. (1892). Fiskensbiologiske forhold i Holboek Fjord, 1890–91. *Beret. Dan. Biol. Stn. 1890–1891* 1, 121–183.

Pitcher, T. J., and MacDonald, P. D. M. (1973). Two models for seasonal growth in fishes. *J. Appl. Ecol.* **10,** 599–606.

Pütter, A. (1920). Wachstumsähnlichkeiten. *Pfluegers Arch. Gesamte Physiol. Menschen Tiere* **180,** 298–340.

Rafail, S. Z. (1973). A simple and precise method for fitting a von Bertalanffy growth curve. *Mar. Biol.* **19,** 354–358.

Richards, F. J. (1959). A flexible growth function for empirical use. *J. Exp. Bot.* **10,** 290–300.

Ricker, W. E. (1945). A method of estimating minimum size limits for obtaining maximum yield. *Copeia* No. 2, pp. 84–94.

Ricker, W. E. (1946). Production and utilization of fish populations. *Ecol. Monogr.* **16,** 373–391.

Ricker, W. E. (1958). Handbook of computations for biological statistics of fish populations. *Bull., Fish. Res. Board Can.* No. 119.

Ricker, W. E. (1966). Annotations to Ivlev's paper "The biological productivity of waters" *J. Fish. Res. Board Can.* **23,** 1727–1759.

Ricker, W. E. (1969). Effect of size-selective mortality and sampling bias on estimates of growth, mortality, production and yield. *J. Fish. Res. Board Can.* **26,** 479–541.

Ricker, W. E. (1973). Linear regressions in fishery research. *J. Fish Res. Board Can.* **30,** 409–434.

Ricker, W. E. (1975a). Computation and interpretation of biological statistics of fish populations. *Bull., Fish. Res. Board Can.* No. 191.

Ricker, W. E. (1975b). A note concerning Professor Jolicoeur's comments. *J. Fish. Res. Board Can.* **32,** 1494–1498.

Ricker, W. E., and Foerster, R. E. (1948). Computation of fish production. *Bull. Bingham Oceanogr. Collect.* **11,** 173–211.

Ricklefs, R. E. (1967). A graphical method of fitting equations to growth curves. *Ecology* **48,** 978–83.

Sachs, J. (1874). Über den Einfluss der Lufttemperatur und des Tageslichtes auf die stündlichen und täglichen Änderung des Langenwachstums der Internoden. *Arb. bot. Inst. Wurzburg* **1,** 99–192.

Schaefer, M. B. (1954). Some aspects of the dynamics of populations important to the management of the commercial marine fishes. *Inter-Am. Trop. Tuna Comm., Bull.* 1(2), 27–56.

Schmalhausen, I. (1926). Studien über Wachstum und Differenzierung. III. Die embryonale Wachstumskurve des Hünchens. *Wilhelm Roux' Arch. Entwicklungsmech. Org.* **109,** 322–387.

Schumacher, F. X. (1939). A new growth curve and its application to time-yield studies. *J. For.* **37,** 819 820.

Silliman, R. P. (1967). Analog computer models of fish populations. *U.S. Fish Wildl. Serv., Fish. Bull.* **66,** 31–46.

Stauffer, G. D. (1973). A growth model for salmonids reared in hatchery environments. Ph.D. Thesis Univ. of Washington, Seattle.

Sund, O. (1911). Undersökelser over brislingen i norske farvand vaesentlig paa grundlag av "Michael Sar's" togt 1908. *Aarsberet. Nor. Fisk.* (1910) 3, 357–410.

Szarski, H., Delewka, E., Olechnowiczawa, S., Predygier, Z., and Slankowa, L. (1956). Uklad trawienny leszcza (*Abramis brama* L.). [The digestive system of the bream.] *Stud. Soc. Sci. Torun., Sect. E* (*Zool.*) **3,** 113–146.

Taylor, C. C. (1958). Cod growth and temperature. *J. Cons., Cons. Perm. Int. Explor. Mer* **23,** 366–370.

Taylor, C. C. (1959). Temperature and growth—the Pacific razor clam. *J. Cons., Cons. Perm. Int. Explor. Mer* **25,** 93–101.

Taylor, C. C. (1962). Growth equations with metabolic parameters. *J. Cons., Cons. Perm. Int. Explor. Mer* **27,** 270–286.

Teissier, G. (1948). La relation d'allométrie: sa signification statistique et biologique. *Biometrics* **4,** 14–18.

Tesch, F. W. (1971). Age and growth. *In* "Methods for Assessment of Fish Production in Fresh Waters" (W.E. Ricker, ed.), pp. 98–130, Int. Biol. Program,Handbook No.3,2nd ed. Blackwell Scientific, Oxford and Edinburgh.

Ursin, E. (1963). On the incorporation of temperature in the von Bertalanffy growth equation. *Medd. fra Danmarks Fiskeri- og Havundersøgelser* **4,** 1–16.

Ursin, E. (1967). A mathematical model of some aspects of fish growth, respiration and mortality. *J. Fish. Res. Board Can.* **24,** 2355–2453.

Van Oosten, J. (1929). Life history of the lake herring (*Leucicththys artedi* LeSueur) of Lake Huron, as revealed by its scales with a critique of the scale method. *U.S. Bur. Fish. Bull.* **44,** 265–448.

Vasnetsov, V. V. (1953). Developmental stages of bony fishes. *In* "Ocherki po Obshchim Voprosam Ikhtiologii," pp. 207–217. Akademiya Nauk Press, Moscow (in Russian).

Verhulst, P. F. (1838). Notice sur la loi que la population suit dans son acroissement. *Corres. Math. Phys.* **10,** 113–121.

von Bertalanffy, L. (1934). Untersuchungen über die Gesetzlichkeit des Wachstums. *Wilhelm Roux' Arch. Entwicklungsmech. Org.* **131,** 613.

von Bertalanfly, L. (1938). A quantitative theory of organic growth. *Hum. Biol.* **10,** 181–213.

von Bertalanffy, L. (1957). Quantitative laws in metabolism and growth. *Q. Rev. Biol.* **32,** 217–231.

Walford, L. A. (1946). A new graphic method of describing the growth of animals. *Biol. Bull.* **90,** 141–147.

Weymouth, F. W., and Thompson, S. H. (1930). The age and growth of the Pacific cockle (*Cardium corbis* Martyn). *Bull U. S. Bur. Fish.* **46,** 633–641.

Winberg, G. G. (1956). "Rate of Metabolism and Food Requirements of Fishes. Nauchnye Trudy Belorusskogo Gos. Univ. Minsk, 253 pp. (English version, *Fish. Res. Board Can.* Transl. No. 194, 1960.)

Zar, J. H. (1968). Calculation and miscalculation of the allometric equation as a model in biological data. *Bioscience* **18,** 1118–1120.

Zweifel, J. R., and Lasker, R. (1976). Prehatch and posthatch growth of fishes—A general model. *U.S. Fish Wildl. Serv., Fish. Bull.* **74,** 609–621.

Other volumes in the Fish Physiology series

VOLUME 11A	The Physiology of Developing Fish: Eggs and Larvae *Edited by W. S. Hoar and D. J. Randall*
VOLUME 11B	The Physiology of Developing Fish: Viviparity and Posthatching Juveniles *Edited by W. S. Hoar and D. J. Randall*
VOLUME 12A	The Cardiovascular System, Part A *Edited by W. S. Hoar, D. J. Randall, and A. P. Farrell*
VOLUME 12B	The Cardiovascular System, Part B *Edited by W. S. Hoar, D. J. Randall, and A. P. Farrell*
VOLUME 13	Molecular Endocrinology of Fish *Edited by N. M. Sherwood and C. L. Hew*
VOLUME 14	Cellular and Molecular Approaches to Fish Ionic Regulation *Edited by Chris M. Wood and Trevor J. Shuttleworth*
VOLUME 15	The Fish Immune System: Organism, Pathogen, and Environment *Edited by George Iwama and Teruyuki Nakanishi*
VOLUME 16	Deep Sea Fishes *Edited by D. J. Randall and A. P. Farrell*
VOLUME 17	Fish Respiration *Edited by Steve F. Perry and Bruce L. Tufts*
VOLUME 18	Muscle Development and Growth *Edited by Ian A. Johnston*
VOLUME 19	Tuna: Physiology, Ecology, and Evolution *Edited by Barbara A. Block and E. Donald Stevens*
VOLUME 20	Nitrogen Excretion *Edited by Patricia A. Wright and Paul M. Anderson*
VOLUME 21	The Physiology of Tropical Fishes *Edited by Adalberto L. Val, Vera Maria F. De Almeida-Val, and David J. Randall*
VOLUME 22	The Physiology of Polar Fishes *Edited by Anthony P. Farrell and John F. Steffensen*
VOLUME 23	Fish Biomechanics *Edited by Robert E. Shadwick and George V. Lauder*
VOLUME 24	Behavior and Physiology of Fish *Edited by Katharine A. Sloman, Rod W. Wilson, and Sigal Balshine*

VOLUME 35	Biology of Stress in Fish *Edited by Carl B. Schreck, Lluis Tort, Anthony P. Farrell, and Colin J. Brauner*
VOLUME 36A	The Cardiovascular System: Morphology, Control and Function *Edited by A. Kurt Gamperl, Todd E. Gillis, Anthony P. Farrell, and Colin J. Brauner*
VOLUME 36B	The Cardiovascular System: Development, Plasticity and Physiological Responses *Edited by A. Kurt Gamperl, Todd E. Gillis, Anthony P. Farrell, and Colin J. Brauner*
VOLUME 37	Carbon Dioxide *Edited by Martin Grosell, Philip L. Munday, Anthony P. Farrell, and Colin J. Brauner*
VOLUME 38	Aquaculture *Edited by Tillmann J. Benfey, Anthony P. Farrell, and Colin J. Brauner*
VOLUME 39A	Conservation Physiology for the Anthropocene – A Systems Approach Part A *Edited by Steven J. Cooke, Nann A. Fangue, Anthony P. Farrell, Colin J. Brauner, and Erika J. Eliason*
VOLUME 39B	Conservation Physiology for the Anthropocene – Issues and Applications *Edited by Nann A. Fangue, Steven J. Cooke, Anthony P. Farrell, Colin J. Brauner and Erika J. Eliason*
VOLUME 40A	The 50th Anniversary Issue of Fish Physiology: Physiological Systems and Development *Edited by David J. Randall, Anthony P. Farrell, Colin J. Brauner, and Erika J. Eliason*

Index

Note: Page numbers followed by "*f*" indicate figures and "*t*" indicate tables.

D

E

F

G

H

I

J

K

L

N

O

P

R

T

U

V

W

9780443137310